The Quintessence of Basic and Clinical Research and Scientific Publishing

Gowraganahalli Jagadeesh •
Pitchai Balakumar • Fortunato Senatore
Editors

The Quintessence of Basic and Clinical Research and Scientific Publishing

Editors
Gowraganahalli Jagadeesh
Retired Senior Expert Pharmacologist at the Office of Cardiology, Hematology, Endocrinology, and Nephrology
Center for Drug Evaluation and Research, US Food and Drug Administration
Silver Spring, MD, USA

Distinguished Visiting Professor at the College of Pharmaceutical Sciences
Dayananda Sagar University
Bengaluru, Karnataka, India

Visiting Professor at the College of Pharmacy
Adichunchanagiri University
BG Nagar, Karnataka, India

Visiting Professor at the College of Pharmaceutical Sciences
Manipal Academy of Higher Education (Deemed-to-be University)
Manipal, Karnataka, India

Senior Consultant & Advisor, Auxochromofours Solutions Private Limited
Silver Spring, MD, USA

Pitchai Balakumar
Professor & Director, Research Training and Publications, The Office of Research and Development
Periyar Maniammai Institute of Science & Technology (Deemed to be University)
Vallam, Tamil Nadu, India

Fortunato Senatore
Division Cardiology & Nephrology, Office of Cardiology, Hematology, Endocrinology and Nephrology
Center for Drug Evaluation and Research, US Food and Drug Administration
Silver Spring, MD, USA

ISBN 978-981-99-1283-4 ISBN 978-981-99-1284-1 (eBook)
https://doi.org/10.1007/978-981-99-1284-1

This Springer imprint is published by the registered company Springer Nature Singapore Pte Ltd.
The registered company address is: 152 Beach Road, #21-01/04 Gateway East, Singapore 189721, Singapore

Preface

It is inevitable to create a vibrant environment where active research, innovation and entrepreneurship should become amiably infectious and eventually pandemic for the progression of a nation and the world, bringing new knowledge, productiveness, emulation, equality and opportunity for the substantial benefits of entire creatures.
—*Kumaran Shanmugam and Pitchai Balakumar, 2022*

Biomedical Research is a broad term that comprises both basic and applied research. Basic research focuses on underlying principles for testing the fundamentals of nature, including physiological processes. Its outcome is an understanding of how nature works, including how disease is manifested by pathophysiologic processes. It does not find a cure for a disease, but rather develops our understanding of what causes the disease. Applied research relies on understanding the pathophysiology of a disease, as derived from basic research, and attempts to identify mechanisms to reverse such pathophysiology (i.e., finding a cure for the disease). A specific type of applied research is clinical research, in which a new drug is tested to determine if there can be a cure for a disease, or at least a reduction in its incidence. A clinical trial is a specific type of clinical research that tests a hypothesis about the efficacy of a drug by assigning patients to receive either a drug or a control agent.

All forms of biomedical research involve key principles: (1) asking and refining a question (i.e., generating a hypothesis) such that it can be evaluated in a research plan; (2) designing a research protocol that provides a pathway to testing the hypothesis (i.e., answering the research question); and (3) conducting the research based on the protocol to produce interpretable data that forms the evidence, allowing one to conclude whether the hypothesis is accepted or rejected. The results of the study add further erudition to existing knowledge in a growing field based on contributions from other investigators (i.e., peers).

Drug approval requires significant knowledge, including efficacy and adverse effects in experimental animals, a safe dose range in animals that can be mapped to a safe dose range in humans, and finally, how the drug's mechanism of action can successfully interfere with the pathophysiology of the disease. The fields of research facilitating the acquisition of knowledge in these areas include toxicology, or how poisonous the drug can be at various doses; pharmacodynamics (what the drug does to the body); pharmacokinetics (what the body does to the drug); and pharmacogenomics (how genes affect pharmacokinetics and pharmacodynamics). Conducting research on people is a serious venture that demands open and informed consent. Rules and regulations have been developed across the world to ensure the safety of humans enrolled in clinical trials and that every detail of the protocol—most importantly, what can go wrong—is described very clearly to patients before they are enrolled in a trial.

This book is a crucial resource for people who aspire to engage in biomedical research by fostering an understanding of the determinants of both basic and applied biomedical research. It covers the entire spectrum of research processes starting from idea generation, creative/critical thinking, conceptualization, hypothesis generation, research design including clinical trial design, study/clinical trial implementation, data acquisition and analysis. This book also covers new and evolving topics such as artificial intelligence, drug repurposing approaches, and bioinformatics.

This book elaborately covers publication strategies, including literature search, manuscript preparation, presentation, persuasive discussion, editorial processes, elements of peer review, and bibliometrics. In addition, it focuses on ethics in the conduct of research and publication, including issues of misconduct. Chapters on grantsmanship explore grant application methodologies to enhance the probability of being awarded research grants. Finally, the book addresses key strategies for constructing effective networks, securing effective mentorship, building career advancement opportunities, and understanding the protection of intellectual property and patent filing.

We extend our gratitude to all of the authors and coauthors of this book for their contributions to the wealth of knowledge the book offers. Additionally, we are grateful to Dr. Naren Aggarwal, publication director, Ms. Raman Shukla, editor, and Mr. Kamesh Senthilkumar and his team at Springer Nature for their great support leading to the timely publication of the book.

All opinions expressed in this book are those of the chapter authors and do not necessarily represent the opinions of the editors and their employers.

Silver Spring, MD, USA — Gowraganahalli Jagadeesh
Vallam, Tamil Nadu, India — Pitchai Balakumar
Silver Spring, MD, USA — Fortunato Senatore

Contents

Editors and Contributors

About the Editors

Gowraganahalli Jagadeesh, PhD has worked for more than three decades for the US Food and Drug Administration as a Senior Expert Pharmacologist in the Office of Cardiology, Hematology, Endocrinology, and Nephrology at the Center for Drug Evaluation and Research. He was an Adjunct Scientist at the NIH, Bethesda, while working at the FDA. He had also taught at Banaras Hindu University, Varanasi, India, and was a Canadian Heart Foundation Fellow and NIH staff scientist (at Northeastern University, Boston). Dr. Jagadeesh received B.Pharm., M.Sc., and Ph.D. degrees from RGUS-Bengaluru, All India Institute of Medical Sciences-Delhi, and BHU, respectively. He is currently a Distinguished Visiting Professor at College of Pharmaceutical Sciences, Dayananda Sagar University, Bengaluru; and a visiting professor at College of Pharmacy, Adichunchanagiri University, BG Nagara, Karnataka; and College of Pharmaceutical Sciences, Manipal Academy of Higher Education (Deemed-to-be-University), Manipal, India. More than 60 scientific papers, including book chapters, have been authored by him, and he has edited three books as well.

Pitchai Balakumar, MPharm, PhD is working as Professor & Director, Research Training and Publications, The Office of Research and Development, Periyar Maniammai Institute of Science & Technology (Deemed-to-be-University), Vallam, Tamil Nadu, India. Previously, he was a Professor of Pharmacology at King Khalid University, Saudi Arabia, and Senior Associate Professor and Head of Pharmacology Unit at the Faculty of Pharmacy, AIMST University, Malaysia. He was the Visiting Scientist at the University of Montreal, Canada, in 2009. Prof. Balakumar's name has been featured, for the third consecutive time (since 2020), among the World Ranking Top 2% Scientists in the Field: Pharmacology & Pharmacy, based on a study done by Stanford University scientists. He earned M. Pharm. and Ph.D. degrees in Pharmacology specializing in Cardiovascular Pharmacology from Punjabi University-Patiala, Punjab, India. He has two decades of teaching, research, and editorial experiences in the field of Cardiovascular-Renal Sciences and trained several graduates in India and abroad. He has published around 125 scientific papers (over 5700 citations with h-index 42) and received several awards and honors for excellence in overall academic and research performance. Along with two senior scientists from the US FDA, Prof. Balakumar edited a reference book titled *Pathophysiology and Pharmacotherapy of Cardiovascular Disease* published by Springer in 2015. He also edited a number of focused Special Issues in cardiovascular sciences. Prof. Balakumar is a "Consulting Editor" at Pharmacological Research, Elsevier.

Fortunato Senatore, MD, PhD, FACC received his bachelor's degree in biochemistry and master's degree in biomedical engineering from Columbia University and PhD in chemical engineering from Rutgers University. He served as an assistant professor of chemical engineering at Texas Tech University and focused on artificial organ research and biocompatible prostheses. He attended medical school at the Texas Tech University School of Medicine during his tenure at the Texas Tech University School of Engineering. He trained in Internal Medicine at the Mayo Clinic and Cardiology at Harvard University-Massachusetts General Hospital. He served in the pharmaceutical industry with increasing levels of responsibility for 17 years where he conducted and supervised clinical research in cardiovascular disease, renal disease, and endocrinology. He subsequently joined the FDA in the Division of Cardiology and Nephrology and has been serving as Medical Officer since 2013. For the past 4 years, he has been serving as Lead Physician.

Contributors

Rosalyn Adigun, MD, PharmD, MSc, FASE, FACC Division of Cardiology and Nephrology, OCHEN, Food and Drug Administration, Silver Spring, MD, USA

Fred K. Alavi, PhD, MS Department of Pharmacology and Toxicology, Office of Cardiovascular, Hematology, Endocrinology and Nephrology, CDER, US Food & Drug Administration, Silver Spring, MD, USA

Ali Alqahtani, PhD Department of Pharmacology, College of Pharmacy, King Khalid University, Abha, Kingdom of Saudi Arabia

Russ Altman, AB, PhD, MD Kenneth Fong Professor of Bioengineering, Genetics, Medicine, Biomedical Data Science and (by courtesy, Computer Science), Stanford University, Stanford, CA, USA

Saloni Andhari, PhD MAEER's Maharashtra Institute of Pharmacy, Pune, Maharashtra, India

Pitchai Balakumar, PhD Professor & Director, Research Training and Publications, The Office of Research and Development, Periyar Maniammai Institute of Science & Technology (Deemed to be University), Vallam, Tamil Nadu, India

Aritra Banerjee Department of Pharmacoinformatics, National Institute of Pharmaceutical Education and Research, Mohali, Punjab, India

Wajeeda Bano, PhD Department of Studies and Research in Economics, Mangalore University, Mangalagangothri, Konaje, India

Joanne Berger, MLS FDA Library, Office of Data, Analytics and Research (ODAR), Office of Digital Transformation (ODT), U.S. Food and Drug Administration, Rockville, MD, USA

Yuemin Bian, PhD Center for the Development of Therapeutics, Broad Institute of MIT and Harvard, Cambridge, MA, USA

David W. Boulton Clinical Pharmacology and Quantitative Pharmacology (CPQP), Clinical Pharmacology & Safety Sciences, R&D, AstraZeneca, Gaithersburg, MD, USA

Janessa Bower, PhD Department of Educational Psychology, University of North Texas, Denton, TX, USA

K. Byrappa, PhD Adichunchanagiri University, B.G. Nagara, Mandya, Karnataka, India

Cynthia E. Carr, PhD Research Support Services, School of Medicine, The University of California, Irvine, CA, USA

Ananya Chakraborty, MD Department of Pharmacology, Vydehi Institute of Medical Sciences & Research Centre, Bengaluru, India

Saager Chawla Department of Urology, University of California at San Diego, San Diego, CA, USA

Jenny Cheng Clinical Pharmacology and Quantitative Pharmacology (CPQP), Clinical Pharmacology & Safety Sciences, R&D, AstraZeneca, Gaithersburg, MD, USA

Minjun Chen, PhD Division of Bioinformation and Biostatistics, National Center for Toxicological Research, US Food & Drug Administration, Jefferson, AR, USA

Gemma Cox, MD Hospital for Sick Children, University of Toronto, Toronto, ON, Canada

Michael J. Curtis, PhD, DSP, FHEA, FBPhS Faculty of Life Sciences and Medicine, School of Cardiovascular Medicine & Sciences, King's College, London, UK
Rayne Institute, St Thomas' Hospital, London, UK

Monojit Debnath, PhD Department of Human Genetics, National Institute of Mental Health and Neuro Sciences, Bengaluru, India

Kirti N. Deshmukh Department of Natural Products, National Institute of Pharmaceutical Education and Research, Mohali, India

Akash Dey Department of Natural Products, National Institute of Pharmaceutical Education and Research, Mohali, India

Jin Dong Clinical Pharmacology and Quantitative Pharmacology (CPQP), Clinical Pharmacology & Safety Sciences, R&D, AstraZeneca, Gaithersburg, MD, USA

Efe Eworuke Division of Epidemiology Office of surveillance and epidemiology, US Food & Drug Administration, Silver Spring, MD, USA

Asdrubal Falavigna Department of Neurosurgery, Caxias do Sul University, Caxias do Sul, Brazil

Vincenzo Falavigna Department of Neurosurgery, Caxias do Sul University, Caxias do Sul, Brazil

Michael F. W. Festing, MSc, PhD, DSc, CStat, FIBiol Care of the Medical Research Council, Swindon, UK

Elimika P. Fletcher, PharmD, PhD Guidance and Policy Team, Office of Clinical Pharmacology, Office of Translational Science, Center for Drug Evaluation and Research, U.S. Food and Drug Administration, Silver Spring, MD, USA

Diego A. Forero, MD, PhD School of Health and Sport Sciences, Fundación Universitaria del Área Andina, Bogotá, Colombia

Anuj Gahlawat Department of Pharmacoinformatics, National Institute of Pharmaceutical Education and Research, Mohali, Punjab, India

Charu Gandotra, MD Division of Cardiology and Nephrology, Office of Cardiology, Hematology, Endocrinology, and Nephrology, Center for Drug Evaluation and Research, US Food and Drug Administration, Silver Spring, MD, USA

Prabha Garg, PhD Department of Pharmacoinformatics, National Institute of Pharmaceutical Education and Research, Mohali, Punjab, India

K. Gokulakrishnan, PhD Department of Neurochemistry, National Institute of Mental Health and Neuro Sciences, Bengaluru, India

Alok Goyal Department of Natural Products, National Institute of Pharmaceutical Education and Research, Mohali, India

Ramadevi Gudi, PhD Department of Pharmacology and Toxicology, Office of Cardiology, Hematology, Endocrinology and Nephrology, CDER, US Food & Drug Administration, Silver Spring, MD, USA

Rituja Gupta, MPharm School of Pharmacy, Dr. Vishwanath Karad MIT World Peace University, Pune, Maharashtra, India

Gwendolyn Halford, MLS FDA Library, Office of Data, Analytics and Research (ODAR), Office of Digital Transformation (ODT), U.S. Food and Drug Administration, Rockville, MD, USA

Wendy Halpern, DVM, PhD, DACVP Genetech, San Francisco, CA, USA

Avijit Hazra, MD Institute of Postgraduate Medical Education and Research, Kolkata, India

Praveen Hoogar, PhD Centre for Bio Cultural Studies, Manipal Academy of Higher Education, Manipal, Karnataka, India

Gavin Hou Department of Pharmaceutical Sciences and Computational Chemical Genomics Screening Center, Pharmacometrics and System Pharmacology PharmacoAnalytics, School of Pharmacy, University of Pittsburgh, Pittsburgh, PA, USA

Rachel Huddart, PhD PharmGKB, Stanford, CA, USA

Julie A. Hutt, DVM, DACVP Greenfield Pathology Services, Greenfield, IN, USA

Sanjay M. Jachak, PhD Department of Natural Products, National Institute of Pharmaceutical Education and Research, Mohali, India

Gowraganahalli Jagadeesh, PhD Retired Senior Expert Pharmacologist at the Office of Cardiology, Hematology, Endocrinology, and Nephrology, Center for Drug Evaluation and Research, US Food and Drug Administration, Silver Spring, MD, USA
Distinguished Visiting Professor at the College of Pharmaceutical Sciences, Dayananda Sagar University, Bengaluru, Karnataka, India
Visiting Professor at the College of Pharmacy, Adichunchanagiri University, BG Nagar, Karnataka, India

Visiting Professor at the College of Pharmaceutical Sciences, Manipal Academy of Higher Education (Deemed-to-be University), Manipal, Karnataka, India
Senior Consultant & Advisor, Auxochromofours Solutions Private Limited, Silver Spring, MD, USA

Rekha Kambhampati, MD Division of Cardiology and Nephrology, Office of Cardiology, Hematology, Endocrinology, and Nephrology, Center for Drug Evaluation and Research, US Food and Drug Administration, Silver Spring, MD, USA

Pradnya Kamble Department of Pharmacoinformatics, National Institute of Pharmaceutical Education and Research, Mohali, Punjab, India

Jayant Khandare, PhD School of Pharmacy, Dr. Vishwanath Karad MIT World Peace University, Pune, Maharashtra, India
School of Consciousness, Dr. Vishwanath Karad MIT World Peace University, Pune, Maharashtra, India
Actorius Innovations and Research, Pune, Maharashtra, India

Summon Koul, PhD School of Consciousness, Dr. Vishwanath Karad MIT World Peace University, Pune, Maharashtra, India

Gopala Krishna, PhD, MBA, DABT, Fellow - ATS Pharmaceutical Drug Safety, Krislee Farms LLC, Mt Airy, MD, USA

Shreyanshi Kulshreshtha Department of Natural Products, National Institute of Pharmaceutical Education and Research, Mohali, India

B. R. Prashantha Kumar, MPharm, PhD Department of Pharmaceutical Chemistry, JSS College of Pharmacy, Mysuru, India
Constituent College of JSS Academy of Higher Education & Research, Mysuru, India

B. Sajeev Kumar, MPharm, PhD Department of Pharmaceutics, College of Pharmaceutical Sciences, Dayananda Sagar University, Bengaluru, Karnataka, India

Gunjan Kumar, MD Clinical Studies, Trials and Projection Unit, Indian Council of Medical Research, New Delhi, India

S. C. Lakhotia, PhD, FNA, FASc, FNASc Cytogenetics Laboratory, Department of Zoology, Banaras Hindu University, Varanasi, India

Maria Learoyd Clinical Pharmacology and Quantitative Pharmacology (CPQP), Clinical Pharmacology & Safety Sciences, R&D, AstraZeneca, Cambridge, UK

Lindsay Ellis Lee, PhD Department of Pediatrics, Quillen College of Medicine, East Tennessee State University, Johnson City, TN, USA

Benjamin P. Lewis, PhD Lewis Regulatory Consulting, LLC, Gaithersburg, MD, USA

Rakesh Lodha, MD Department of Pediatrics, All India Institute of Medical Sciences, New Delhi, India

Rajanikanth Madabushi, PhD Guidance and Scientific Policy, Office of Clinical Pharmacology, Office of Translational Science, Center for Drug Evaluation and Research, U.S. Food and Drug Administration, Silver Spring, MD, USA

S. Manikandan, MD Department of Pharmacology, Jawaharlal Institute of Postgraduate Medical Education and Research, Pondicherry, India

Srikumaran Melethil, PhD, JD, FAAPS University of Missouri-Kansas City, Kingwood, TX, USA

Jane E. Miller, PhD Edward J. Bloustein School of Planning and Public Policy, Rutgers University, New Brunswick, NJ, USA

Kirtida Mistry, MD Division of Cardiology and Nephrology, Office of Cardiology, Hematology, Endocrinology, and Nephrology, Center for Drug Evaluation and Research, US Food and Drug Administration, Silver Spring, MD, USA

Sherry J. Morgan, DVM, PhD, DACVP, DABT, DABVT StageBio, Mason, OH, USA

Aparna Mukherjee, MD, PhD Clinical Studies, Trials and Projection Unit, Indian Council of Medical Research, New Delhi, India

Srinivas Mutalik, PhD Department of Pharmaceutics, Manipal College of Pharmaceutical Sciences, Manipal Academy of Higher Education, Manipal, Karnataka, India

Usha Y. Nayak, PhD Department of Pharmaceutics, Manipal College of Pharmaceutical Sciences, Manipal Academy of Higher Education, Manipal, Karnataka, India

Geeta Negi, PhD, DABT Division of Pharmacology and Toxicology, Office of Cardiology, Hematology, Endocrinology and Nephrology, CDER, US Food and Drug Administration, Silver Spring, MD, USA

Rohit Pal, MPharm Integrated Drug Discovery Center, Department of Pharmaceutical Chemistry, Acharya & BM Reddy College of Pharmacy, Bengaluru, Karnataka, India

Vijender Panduga Clinical Pharmacology and Quantitative Pharmacology (CPQP), Clinical Pharmacology & Safety Sciences, R&D, AstraZeneca, Gothenburg, Sweden

Li Pang, MD Division of Systems Biology, National Center for Toxicological Research, US Food & Drug Administration, Jefferson, AR, USA

Ravikumar Peri, PhD, DABT DSRE/PTS, Takeda Pharmaceuticals, Cambridge, MA, USA

N. M. Raghavendra, MPharm, PhD Department of Pharmaceutical Chemistry, College of Pharmaceutical Sciences, Dayananda Sagar University, Bengaluru, Karnataka, India

Mahadevan Raj Rajasekharan, PhD Department of Urology, University of California at San Diego, San Diego, CA, USA

R. Rajkumar Department of Pharmacoinformatics, National Institute of Pharmaceutical Education and Research, Mohali, Punjab, India

Suganthi S. Ramachandran, MD, DM Department of Pharmacology, Jawaharlal Institute of Postgraduate Medical Education and Research, Pondicherry, India

Savithiri Ratnapalan, MBBS, PhD Department of Pediatrics and Dalla Lana School of Public Health, University of Toronto, Toronto, ON, Canada
Divisions of Emergency Medicine, Clinical Pharmacology and Toxicology, The Hospital for Sick Children, University Avenue, Toronto, ON, Canada

Rohan Reddy Department of Urology, University of California at San Diego, San Diego, CA, USA

Venkatesh Pilla Reddy, PhD Clinical Pharmacology and Pharmacometrics, R&D | Clinical Pharmacology & Safety Sciences, AstraZeneca, Cambridge, UK

Mark A. Runco, PhD Social Sciences, Southern Oregon University, Ashland, OR, USA

Martina D. Sahre, PhD Guidance and Policy Team, Office of Clinical Pharmacology, Office of Translational Science, Center for Drug Evaluation and Research, U.S. Food and Drug Administration, Silver Spring, MD, USA

Hardeep Sandhu Department of Pharmacoinformatics, National Institute of Pharmaceutical Education and Research, Mohali, Punjab, India

A. Sankaranarayanan, PhD, FCP Vivo Bio Tech Ltd, Hyderabad, Telangana, India

Pujan Sasmal, M. Pharm Integrated Drug Discovery Center, Department of Pharmaceutical Chemistry, Acharya & BM Reddy College of Pharmacy, Bengaluru, Karnataka, India

Alejandro Schcolnik-Cabrera, MD, PhD Division of Immunology-Oncology, Hôpital Maisonneuve-Rosemont Research Centre, Montréal, QC, Canada

Fortunato Senatore, MD, PhD, FACC Division Cardiology & Nephrology, Office of Cardiology, Hematology, Endocrinology and Nephrology, Center for Drug Evaluation and Research, US Food and Drug Administration, Silver Spring, MD, USA

Kumaran Shanmugam, PhD Periyar Maniammai Institute of Science and Technology (Deemed to be University), Thanjavur, India

Pradeep Sharma, PhD Clinical Pharmacology and Quantitative Pharmacology, Clinical Pharmacology & Safety Sciences, R&D, AstraZeneca, Cambridge, UK

Qiang Shi, PhD Division of Systems Biology, National Center for Toxicological Research, US Food & Drug Administration, Jefferson, AR, USA

Amruthesh C. Shivachar, PhD College of Pharmacy and Health Sciences, Texas Southern University, Houston, TX, USA

Dilip Kumar Singh, PhD Department of Pharmaceutical Sciences, Washington State University, Spokane, WA, USA

Saranjit Singh, PhD Department of Pharmaceutical Analysis, National Institute of Pharmaceutical Education and Research, Mohali, Punjab, India

Lauren E. Smith, PhD Health Services Research, Institute of Health Policy, Management and Evaluation, University of Toronto, Toronto, ON, Canada

Samuel Sorkhi Department of Urology, University of California at San Diego, San Diego, CA, USA

B. N. Srikumar, PhD Department of Neurophysiology, National Institute of Mental Health and Neuro Sciences, Bengaluru, India

Ranga V. Srinivas, PhD National Institute on Alcohol Abuse and Alcoholism, National Institutes of Health, Bethesda, MD, USA

Shamala Srinivas, PhD Scientific Review and Policy, Division of Extramural Activities, National Cancer Institute Shady Grove, National Institutes of Health, Rockville, MD, USA

Anup K. Srivastava, PhD, DABT Division of Pharmacology and Toxicology, Office of Immunology and Inflammation, CDER, US Food and Drug Administration, Silver Spring, MD, USA

P. K. Srividhya, PhD Periyar Maniammai Institute of Science and Technology (Deemed to be University), Thanjavur, India

Felix Stader Modelling and Simulations Group, Simcyp Division, Certara UK Limited, Sheffield, UK

Karupiah Sundram, PhD Faculty of Pharmacy, AIMST University, Bedong, Malaysia

Radhakrishna Sura, BVSc, MS, PhD, DACVP, DABT Nonclinical Safety and Pathobiology, Gilead Sciences, Inc., Foster City, CA, USA

P. M. Gurubasavaraja Swamy, MPharm, PhD Integrated Drug Discovery Center, Department of Pharmaceutical Chemistry, Acharya & BM Reddy College of Pharmacy, Bengaluru, Karnataka, India

Ghanshyam Teli, MPharm Integrated Drug Discovery Center, Department of Pharmaceutical Chemistry, Acharya & BM Reddy College of Pharmacy, Bengaluru, Karnataka, India

Mark Tirmenstein, PhD Toxicology, Nonclinical Development, Sarepta Therapeutics, Cambridge, MA, USA

John R. Turner, PhD University of North Texas, Prosper, TX, USA

N. Udupa, PhD Shri Dharmasthala Manjunatheshwara University, Dharwad, Karnataka, India

Tanmaykumar Varma Department of Pharmacoinformatics, National Institute of Pharmaceutical Education and Research, Mohali, Punjab, India

Ganesan Venkatasubramanian, MD, PhD Department of Psychiatry, National Institute of Mental Health and Neuro Sciences, Bengaluru, India

J. Vijay Venkatraman, MBBS, F Diab, MBA, FPIPA (UK) Oviya MedSafe Pvt Ltd., Coimbatore, India

Maria Eduarda Viapiana Department of Neurosurgery, Caxias do Sul University, Caxias do Sul, Brazil

Beth Williamson Development Science, DMPK, UCB Biopharma, Slough, UK

Xiang-Qun Xie, MD, PhD, EMBA School of Pharmacy, Pittsburg, PA, USA
Drug Discovery Institute, Pittsburg, PA, USA
National Center of Excellence for Computational Drug Abuse Research, Pittsburg, PA, USA
University of Pittsburgh, Pittsburg, PA, USA

Xi Yang, Ph.D. Toxicology Service, Potomac, MD, USA

Jialu Zhang, PhD Office of Biostatistics, Center for Drug Evaluation and Research, US Food and Drug Administration, Silver Spring, MD, USA

Guangyi Zhao, PhD Department of Pharmaceutical Sciences and Computational Chemical Genomics Screening Center, Pharmacometrics and System Pharmacology PharmacoAnalytics, School of Pharmacy, University of Pittsburgh, Pittsburgh, PA, USA

Shuyuan Zhao Department of Pharmaceutical Sciences and Computational Chemical Genomics Screening Center, Pharmacometrics & System Pharmacology PharmacoAnalytics, School of Pharmacy, University of Pittsburgh, Pittsburgh, PA, USA

Part I

Fundamentals of Research

The Roadmap to Research: Fundamentals of a Multifaceted Research Process

1

Gowraganahalli Jagadeesh, Pitchai Balakumar, and Fortunato Senatore

Abstract

Understanding the logic of the research process is essential before beginning to identify a research problem. The conceptualization of a research topic—the starting point for scientific research—is formed after searching and reviewing the literature and identifying a knowledge gap. Brainstorming sessions with a mentor may result in new and expanded knowledge and ideas that facilitate formulating a research question, generating a hypothesis, and specifying an objective, all of which subsequently evolve into a study design. These steps are interconnected to address the study's scientific rationale and relevance. Research proposals or protocols should be evaluated based on whether they are interesting, novel, well-designed, feasible, and impactful. Planning the research will include detailed methods describing how the study will be carried out and how to analyze the data generated from the experiments. After analyzing the data, the researcher should be ready to write a scientific paper. Publication in a peer-reviewed journal is the gateway to building on the collected knowledge, further advancing the field, and helping the researcher obtain funding. The latter is often contingent on high-quality antecedent research with data leading to a new hypothesis as the basis of new research. New research potentially will produce results that can fill knowledge gaps by establishing facts and developing principles

The opinions expressed herein are those of GJ and FS and do not necessarily reflect those of the U.S. Food and Drug Administration.

G. Jagadeesh (✉)
Retired Senior Expert Pharmacologist at the Office of Cardiology, Hematology, Endocrinology, and Nephrology, Center for Drug Evaluation and Research, US Food and Drug Administration, Silver Spring, MD, USA

Distinguished Visiting Professor at the College of Pharmaceutical Sciences, Dayananda Sagar University, Bengaluru, Karnataka, India

Visiting Professor at the College of Pharmacy, Adichunchanagiri University, BG Nagar, Karnataka, India

Visiting Professor at the College of Pharmaceutical Sciences, Manipal Academy of Higher Education (Deemed-to-be University), Manipal, Karnataka, India

Senior Consultant & Advisor, Auxochromofours Solutions Private Limited, Silver Spring, MD, USA
e-mail: GJagadeesh2000@gmail.com

P. Balakumar
Professor & Director, Research Training and Publications, The Office of Research and Development, Periyar Maniammai Institute of Science & Technology (Deemed to be University), Vallam, Tamil Nadu, India
e-mail: directorcr@pmu.edu; pbalakumar2022@gmail.com

F. Senatore
Division Cardiology & Nephrology, Office of Cardiology, Hematology, Endocrinology and Nephrology, Center for Drug Evaluation and Research, US Food and Drug Administration, Silver Spring, MD, USA
e-mail: Fortunato.Senatore@fda.hhs.gov

G. Jagadeesh et al. (eds.), *The Quintessence of Basic and Clinical Research and Scientific Publishing*,
https://doi.org/10.1007/978-981-99-1284-1_1

while discovering unexplored territory in the field.

Keywords

Research process · Novel idea · Research question · Hypothesis · Objective · Scientific rationale

1.1 Introduction

Research is the original intellectual investigation undertaken to gain or establish new knowledge, fill potential knowledge gaps, and understand concepts in any one facet of a major subject specialization. It includes the generation of a multitude of ideas leading to new or substantially improved scientific insights with relevance to societal needs. A goal of research is to validate or reject hypotheses and refine existing knowledge and disseminate the findings for the public's benefit. Whether basic research, applied research, a clinical trial, developmental research, or observational analytical research conducted by academia, industry, or a government regulatory body, the fruits of labor help everyone involved; both in biomedical and clinical research, and the research community in general. A further goal of research is to extend human knowledge by finding answers to questions.

Research is inclusive and based on teamwork; it is inherently a social and cooperative venture, and increasingly, a global one. In this context, senior scientists or managers and senior leadership, including the department head, director, president, and vice chancellor, should facilitate research by instilling a positive research climate, including pathways for collaboration between laboratories, departments, and institutions organizing and making research resources available and accessible. Collaborations should also encourage hands-on workshops, interactive seminars, lectures, and journal clubs, all aimed at mutual respect for individual autonomy while integrating the concatenation of evolving thought. It takes a whole research community to sculpt a novice graduate student into a scientist [1].

Scientific research follows standard processes and norms. Research is both theoretical and empirical. It is theoretical because the starting point for scientific research is the conceptualization of a research topic that stems from ideas crystallized after a search and review of the literature. A successful research project is built on four cornerstones: researchable topic, research question, hypothesis, and objective [2] (Table 1.1). Research is empirical because all the planned studies involve a series of experiments, observations, measurements, analyses of data, manuscript writing, and finally, publication. Understanding the research processes that include all the above stages in sequence (Fig. 1.1) is a key step in the scientific enterprise and is considered essential to success in a scientific career. The time spent researching a problem is just as important as the time spent solving it.

This chapter provides a strategic overview of the logic and organization of sequential steps

Table 1.1 Key elements of the research protocol or research project

Element	Purpose
Research topic[a]	The keystone of the study. It begins, drives, and ends the study
Research question[a]	A relationship between two or more variables is phrased as a question
Hypothesis[a]	Relationships are phrased as declarative statements that the planned research is designed to test
Objective[a]	The purpose of the research question from which the hypothesis is generated
Significance of the study	Why is the research question important? What are the implications of the study?
Materials and methods	Experimental design, methods, subjects, variables, data analyses, and statistics

[a]Considered four cornerstones of a research project or a protocol

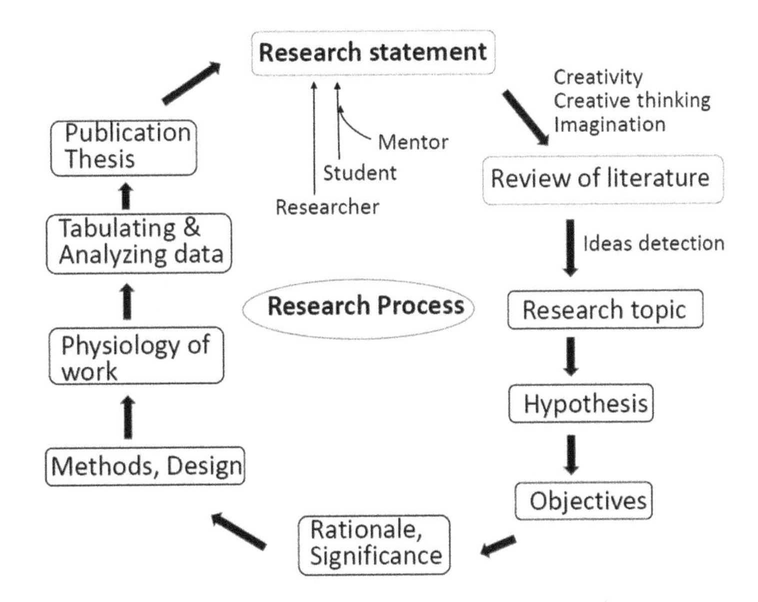

Fig. 1.1 The research process begins with a research statement culminating in the publication of research findings, in the form of original research articles or a thesis

involved in the development of a research project and the dissemination of findings.

1.2 Graduate Students, Novice Researchers, and Mentorship

Education at the graduate level and beyond offers an opportunity to conduct research and pursue a research career, provided one is innovative and has a passion to discover. This can happen by applying logic with perseverance and perspiration, or hard and smart work. It is a long journey; and for this, one needs the ambition and resilience to rise in the competitive world of academia and research. They should act swiftly and be an agile learner. A successful researcher must also be smart, creative, motivated, hard-working, and skillful. Additionally, good communication skills (*see* Chaps. 49 and 50), both oral and written, are required. Together, the abovementioned traits and skills will increase a researcher's probability of a successful scientific career.

As a graduate student, first one must observe others' activities in the laboratory and study a vast amount of literature related to the subject of their research. In addition, one inevitably will attend hands-on training workshops, seminars or journal clubs, scientific meetings, and conferences [3]. All information gleaned from these activities is worthwhile to consider before formulating a research topic and hypothesis or theory.

The research mission can only be accomplished with the help of infrastructure—a fully equipped laboratory, information research center, an environment conducive to study,—and the expert support of a mentor, advisor, guide, or supervisor. The role of the mentor (see Chap. 55) in inspiring, supporting, and imparting research skills to a novice researcher or graduate student is of supreme importance. Thus, a student should choose a mentor whose work they admire and who is well supported by grants and departmental infrastructure [4]. Before meeting with faculty, the student might explore the mentor's current scholarship, research, mentoring experience, and recent publications. That person should be a good match for the student's interests and expectations [5]. In mentor-led student research, the mentor is only a catalyst, the student is the reactant doing the bench work, data analysis, and preparation of the draft manuscript for publication (see chapters in Part VIII). However, the mentor's role in manuscript preparation and promotion may also be as a co-reactant with the student to ensure its successful publication in an indexed journal (Chap. 40). Most graduate students and young researchers have not been exposed to multistep research processes, particularly the development of a study protocol or a project, the first step in scientific research planning. Thus, broad-based training in the scientific methods of research and communication is very necessary for freshmen. The subsequent sections introduce the research processes and discuss in detail their relevance and applications under different scenarios. They are also covered in several chapters of this book.

1.3 Methods of Thinking or Reasoning

How do you develop a research project? Broadly, there are two ways of approaching any problem: deductive or inductive reasoning [3, 6, 7]. Deductive reasoning is a top-down logic. It is based on a series of logical arguments starting with a set of sequential research questions, hypotheses, and objectives, and followed by experimental study, data acquisition, data analysis, and discussion (Fig. 1.2). The hypothesis is generated at the initiation of the research process, whereby the research will lead to either confirmation or rejection of the hypothesis. The researcher considers every aspect of the study and works all the way to a conclusion with a reasonable degree of certainty. As knowledge is expanded and gained during the process, the argument shifts from general to a specific component, or a divergent to convergent structure. For example, a double-blind controlled clinical trial always starts with a hypothesis, e.g., a test drug lowers blood pressure in comparison to the placebo, or a test drug is not inferior to a reference drug. It is further illustrated with an example in Box 1.1.

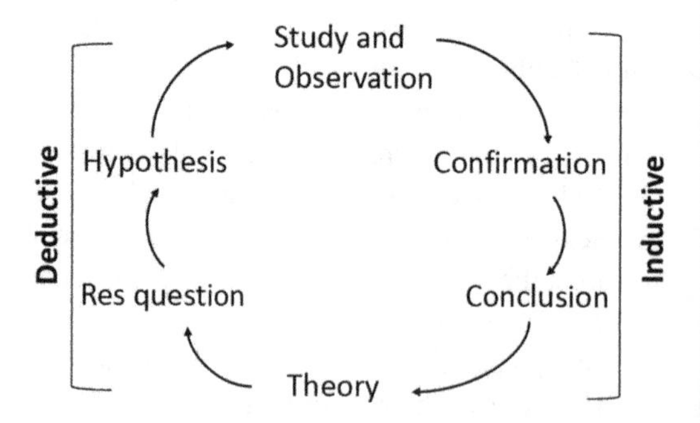

Fig. 1.2 Inductive or deductive research. Research involves both inductive and deductive reasoning and thinking. Inductive reasoning is more open-ended and exploratory at the beginning, while deductive reasoning is narrower and is concerned with testing, confirming or refuting a hypothesis

Inductive or synthetic reasoning or thinking works in the reverse. The conclusion is based on a series of observations or measurements made in the laboratory (Fig. 1.2). No hypothesis or theory is set in the beginning. For example, it is observed that different forms of plant extracts are studied for lowering blood pressure or blood cholesterol. Based on these observations, a hypothesis can be generated. Thus, the emphasis is changed from specific observations to making broader generalizations and theories.

Box 1.1 Deductive Reasoning

Theory [8, 9]

1. Prorenin (PR), renin (R), and (pro)renin receptors [(P)RR] are functionally interrelated in pathophysiological conditions.
2. There is a relationship between elevated levels of circulating (pro)renin and (pro)renin-induced activation of (P)RR and incidences of hypertension, cardiac fibrosis, glomerulosclerosis, diabetes (nephropathy and retinopathy), and end-organ damage.
3. It involves both angiotensin II-dependent and -independent (dual) signal transduction pathways creating additive cardiovascular and renal pathogenesis.
4. (P)RR is essential for cell survival and is an integral part of early organ development.
5. PR, the (P)RR, v-H+-ATPase, and Wnt/β-catenin signaling are all interconnected.

Hypotheses are developed based on the above theory

(continued)

Box 1.1 (continued)

1. Elevated circulating PR plays a pathologic role in the induction of diabetic nephropathy.
2. Nonproteolytic activation of PR (i.e., PR and (P)RR) contributes to the development of cardiac fibrosis.
3. PR and activation of (P)RR cause glomerulosclerosis.
4. Activation of R, PR, and (P)RR is an underlying cause of cardiac and renal damage.
5. (P)PRs do not play a role in pathologic conditions linked to high renin levels. However, (P)RR blockade is beneficial in pathologic conditions associated with high plasma prorenin and low plasma renin levels noted in diabetic subjects.
6. Both the RAAS-dependent and -independent signal pathways are involved in diabetic nephropathy.
7. High and persistent levels of (pro)renin activate not only the angiotensin II-independent signaling cascade but also the canonical Wnt/ß-catenin signaling, triggering renal pathologies.

1.4 The Elements of a Research Protocol or Research Project (See Table 1.1)

The first critical step in a research enterprise is selecting a suitable research topic based on interest, need, opportunity, and motivation. A well-defined topic drives the study and is crucial in moving the project forward. It is articulated by identifying and developing specific research question(s) that stem from a novel scientific idea based on a sound concept. It originates from a review of literature, voracious reading on the subject matter, discussion and brainstorming with the mentor and colleagues, and attending scientific meetings and conferences. After selection of a topic, the deductive reasoning research process begins.

1.4.1 Reviewing the Literature, Getting to Know the Unknown

Research is to see what everybody else has seen, and to think what nobody else has thought. Albert Szent-Györgyi (1937 Nobel Prize winner, Nobel Prize in Physiology or Medicine)

Reviewing the literature is the most important step in pursuing scientific research, providing a critical in-depth understanding of and updating the researcher on a particular topic. The pertinent literature review helps budding researchers identify their research niches, while it helps established researchers to identify the reasons for further research in their fields of expertise [10].

To conduct original research, one should be aware of past and ongoing research in their field. A focused, thoughtful, and critical review of the literature will help in building a cogent problem statement and a clear research question. A strong conceptual framework will solidify the proposal.

A systematic and careful review of literature (a) will provide background material on what has already been done in the area (awareness), (b) helps to avoid duplicating the study, (c) may suggest variables, techniques, or measuring tools that could be useful in planning the study, (d) includes information that could be valuable in interpreting the conclusions of the study, (e) reveals present unique ideas useful to current and future studies, (f) may identify limitations of the previous studies (e.g., unresolved questions), and (g) may identify inconsistencies in findings, controversies, or contradictions among studies. More importantly, the literature search and thorough review facilitate the identification of knowledge gaps in the existing literature and provide an opportunity to extrapolate inventive ideas requiring new research. This process introduces and explains findings that support the new research, synthesizes the main conclusions of literature relevant to the research problem, highlights unresolved issues or questions within the literature, and helps to establish the originality and significance of the new research [10–12].

For a complete review of this topic, including how to search the biomedical literature, see Chap. 37.

1.4.2 The Basis for the Development of Ideas

If I have 300 ideas in a year and just one turns out to work, I am satisfied.
(Alfred Bernhard Nobel 1833–1896, owner of 355 patents)

The best research ideas are those uncovered through a step-by-step developmental process—rather than through a "Eureka!" moment. In other words, a great idea does not arise in an explosive moment but rather forms in a very slow process of aggregation, crystallization, and organization of multiple thoughts into one big idea., e.g., stalactite and stalagmite mineral formation. The idea is the product of creativity and critical thinking. An understanding of creativity (resulting from creative thinking), critical thinking, and logic are important for all researchers, novice or established in generating new concepts and ideas for solving a research problem. There is no force in the universe more powerful than an idea (*quote from Victor Hugo, the French poet, and novelist*). It requires looking at the old and discovering the new. For a more in-depth discussion on the development and formation of ideas, read Chap. 2 by Runco.

Successful research requires several modes of thought: creative thinking, critical thinking, divergent thinking, and convergent thinking. Creative thinking is the force behind problem identification, often as an extrapolating exercise from an antecedent knowledge base. It is one of the bases of choosing a research topic that involves moving between divergent (considering many possibilities) and convergent (closing in on a few possibilities and those most feasible to study) thinking to settle upon an idea from a selection of ideas. Critical thinking is a refinement of the creative process, accomplished by harnessing the creative aspect of problem identification and molding it into a tangible set of concepts from which hypotheses and objectives are generated. A mentor usually assists with this process.

Divergent thinking is the ability to think of and expand on original ideas with fluency and speed, e.g., brainstorming on a broad research topic. Convergent thinking, on the other hand, is the narrowing and refining of those divergent many ideas into one idea for in-depth study, i.e., project selection [13]. Also, see Chap. 3 by Runco on divergent and convergent thinking.

Successful research incorporates all identified thinking paradigms, deployed both sequentially and in parallel. The integration of these modes of thought ultimately forms the basis of judgment, as it uses reason and logical choices before making a final decision [14]. The brilliant, novel idea should be turned into work (research development), innovation, patent, and commercialization (see Chaps. 56 and 57). If not, it is inefficient, and could be considered a waste of time and talent (Fig. 1.3). In research, (e.g., postgraduate and beyond), the idea—either first-tiered (NIVEA) or second-tiered (see Box 1.2)—should generate research questions, hypothesis(es), and objectives. Subsequent laboratory work could eventually produce a thesis; and later, could result in a publication, a degree, and a job. Second-tiered ideas involve the improvement or enhancement of existing ideas, e.g., refining or redoing mechanisms of action, improving or modifying analytical methods, and alternate methods of synthesis.

Originality can be challenging and uncertain, but persistence breeds confidence, which in turn builds strength of character. The ability to produce honest and original work will not only cement the critical thinking abilities of students but also reflect favorably on the institutions that have gone to great lengths to reinforce their overall learning and infrastructure and protect their reputations. There should be no plagiarism of ideas. The project (thesis) should be written in the ink of creativity, intelligence, and hard work. Predicted growth or success in research over time results from the combined forces of intelligence, creativity, thinking, motivation, skill, environment, perseverance, learning strategies, and hard and smart work.

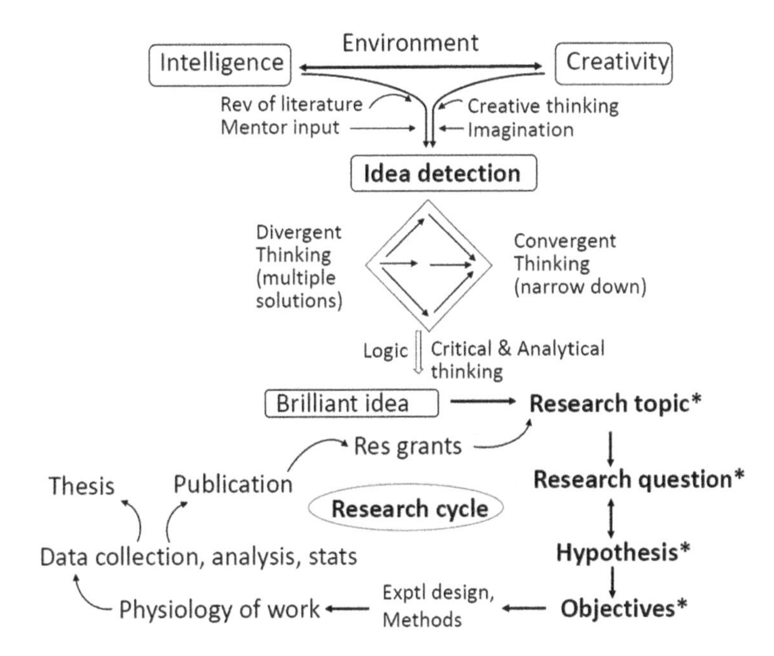

Fig. 1.3 The anatomy and physiology of research

> **Box 1.2 Creativity Is the Idea**
> Creative solutions often emerge amid harsh material constraints, such as unavailability of resources. The generated idea(s) should be
>
> Novel (new and original)
> Intentional (a result of thoughtful effort)
> Valuable/useful/important (advancing the existing knowledge and contributing to society)
> Excellent (noteworthy effort spent to make it the best it can be), and
> Appropriate (achievable under the existing conditions)
> Tractable (ability to answer)

Your idea is your starting point. A good idea will put you in a strong starting position and lead you toward a quality proposal. It should be compelling and novel, aimed at moving your chosen field forward. The basis for generating novel idea (s) arises from in-depth knowledge related to the subject or thesis topic, which may be gained from a strong literature review (see Chaps. 37 and 38).

Extensive reading and text mining may enable researchers to unearth a treasure trove of ideas exploring the potential and significance of their topics. Research experience and voracious reading of scientific articles can be sparks for igniting innovation. This involves staying abreast of recent literature, attending meetings or seminars, participating in discussion sessions, and brainstorming with a mentor. Further, one should possess a curious, sharp mind, and be a good observer in and around the laboratory. The better the idea, the greater the chance of writing a quality protocol and succeeding in finishing the work on time.

Yet, no matter how good your idea is conceptually, you must be capable of pursuing it. For example, it is difficult to do big things in a small laboratory. Certain aspects are beyond your control, such as resources, (e.g., a fully equipped lab, library, available literature, infrastructure), expertise, (e.g., guidance), funding, (e.g., grants), and laboratory personnel. You will have to convince key stakeholders, through carefully thought-out ideas based on antecedent facts leading to a valid hypothesis, that your proposed pathway merits further intellectual and financial consideration.

1.4.3 Creativity and Intelligence

Creativity is the capability or act of conceiving original or unusual ideas. Intelligence is the ability to learn from experience and adapt to changing environments, and may be predictive of scholastic achievement or meaningful cognitive performance. There are "multiple intelligences" [15] such as linguistic, mathematical, visual-spatial, musical, physical skills, (e.g., dancing and sports), and naturalist (e.g., understanding the patterns in nature, studying clouds). Creativity is a key component or aspect of intelligence, or can be a unique trait separate from intelligence.

For the past several thousand years humans have lived on Earth, creativity has played an essential role in the development of civilization, e.g., moving a rock with a lever, the discovery of fire, invention of the wheel, and use of numbers. Creativity involves creative thinkers, a cognitive process applied to knowledge (e.g., a review of literature on a topic), and the ability to be stored in memory. It emanates, in part, through intelligence. Thus, creativity can be construed either as an aspect of intelligence, or as a unique ability separate from intelligence. [16].

It also can be argued that creativity and intelligence are two independent processes or interdependent (interwoven) entities. Creativity can be considered the end result of some intelligent idea that leads to novel outcomes. Outcomes of creativity may depend on the environment, (e.g., an institute, library, lab facilities, and mentors), and the intrinsic logistical capabilities of the individual engaged in the creative process. The honing of a student's research skills is both a social and a cooperative venture. The school needs to establish a positive climate by organizing appropriate resources and making them readily available. Assistance from all corners is needed to accomplish this; it takes an entire school to raise a brilliant student, and a whole research community to bring up a scientist [1].

What about Albert Einstein? He was intelligent and creative; but above all, he was a diligent and hardworking researcher. He said, "It's not that I'm so smart, it's just that I stay with problems longer." Creativity and innovative thinking, (e.g., patenting), are among our most potent tools for thriving in a world that is at once challenging and filled with opportunity. We must explore both the nature and nurture of creativity and hone a capacity for putting these vital skills into everyday practice. For that to happen, researchers, in addition to practicing, must be aware of the core skills of creative thinking. The more you think creatively, the more creative ideas will flow. Runco and team, in the next two chapters, discuss at length the role of ideas in scientific creativity and what can be done to facilitate ideation and the creative process.

Imagination is another trait that contributes to developing a research topic. Albert Einstein is credited with having said, "I believe in intuition and inspiration. Imagination is more important than knowledge. For knowledge is limited to all we now know and understand, while imagination embraces the entire world, and all there ever will be to know and understand." Imagination is the ability to form new images and sensations in the mind that are not perceived through sight, hearing, or other senses. It helps make knowledge applicable to solving problems and is fundamental to integrating experience and the learning process.

Imagination can also be expressed through stories, such as fairy tales or fantasies. Children are good at exercising their imaginations and tend to enjoy creating make-believe situations and stories. Since the mind travels faster than any other known mode of transportation, we can travel to any distant galaxy a billion light-years away in no time. That is the mind's power. Poets, film directors, dance choreographers, and scientists are examples of creative people who harness the power of imagination.

> *"Logic will get you from point A to point B. Imagination will take you everywhere."—Albert Einstein*

"Imagination is the beginning of creation. You imagine what you desire, you will what you imagine and at last you create what you will."—George Bernard Shaw, playwright

1.4.4 Choosing a Research Topic/Problem (Table 1.1)

A Problem Well Stated is Half Solved
—Charles Kettering, head of research at General Motors from 1920 to 1947

The most crucial decision you will make as a graduate or postgraduate student, or as a doctoral candidate is choosing a researchable area or topic that forms the basis for your future research. This activity can be challenging and difficult and calls for careful thought and planning. The topic is the keystone of the study. It begins the project, drives it, moves it forward, and concludes it. Importantly, it dictates the remaining elements of the study; thus, it should not be too narrow, (e.g., yielding trivial results) or too broad, (e.g., not being researchable or taking longer than the allotted time), or unfocused, (e.g., should not cause drift). Because of these potential pitfalls, a good or novel scientific idea must be based on a sound concept built on the knowledge gap discovered in the literature review.

Questions for consideration in choosing a topic include [12, 17–20]:

- Is it original to the field, manageable, and relevant to future areas of research you might wish to pursue?
- Is it similar to work you might have already done, (e.g., for a master's degree), or from your preliminary studies?
- Is it one in which you and your mentor, as well as the research community, are interested and in which you have some expertise?
- Does it fit comfortably with your way of thinking, e.g., analytical? If not, the work will be difficult to finish.
- Is it a topic on which adequate literature is available and accessible to you?
- Is there a strong data-driven scientific rationale for the study?
- Is it feasible to pursue within the limits of laboratory facilities and can it be completed in a timely fashion?
- Does it have a focus, pose questions, and include major variables? Is it distinguishable from the previous study?
- Does it contribute meaningfully to the field being investigated and is it relevant to society?

A research topic is best formulated into a question or a series of questions. The research question is then formulated into a hypothesis; the latter forces us to think carefully about the comparisons needed to answer the research question and establishes the format for applying statistical tests to interpret the results. A question is used to write more detailed objectives, which drive researchers to be particular about their methods and to define key terms.

1.4.5 Research Question

The topic of the research is formulated into a question, which in turn must be clearly defined in terms of specific objectives [21]. It is the starting point of a research project, sets the framework, and will guide the rest of the study design, so it is important to get it right at the beginning [22, 23]. The question should be thought-provoking, and emerge from the title, findings, results, and problems observed in previous studies and from the literature review (Box 1.3). It identifies a knowledge gap in the literature and recently published reports, thus forming a basis for writing the research hypothesis [18]. Its wording should be precise to specify the new knowledge to be gained [21].

A good research question should pass the "FINER" test [24]. It should be:

- **F**easible (within your capabilities, manageable achievable in the allotted time frame and with the available money supply)
- **I**nteresting (piques curiosity, should generate answers)
- **N**ovel (innovative, emanating from a knowledge gap, providing new findings or supplying

an alternate explanation for an existing mechanism of action, advancing a theory)

- **E**thical (receive approval from the institutional board), and
- **R**elevant (to current needs, inspires future research, contributes to existing knowledge, applies to society).

Box 1.3 Research Questions

Examples

1. *Research topic*: Hepatoprotective activities of *Terminalia arjuna* and *Apium graveolens* on paracetamol-induced liver damage in albino rats.
   Research Questions:
   How is paracetamol metabolized in the body? Does it involve P450 enzyme?
   How does paracetamol cause liver injury?
   What are the mechanisms by which drugs can alleviate liver damage?
   Which biochemical parameters are considered part of an index of liver injury?
   Which major endogenous inflammatory molecules are involved in hepatotoxicity?
   What is the extent of hepatoprotective activity offered by these two plant extracts?
   How does the hepatoprotective activity of *Terminalia arjuna* compare with that of *Apium graveolens*?
2. *Research topic*: Evaluation of the antihypertensive effect of test substance X in spontaneously hypertensive rats (SHRs).
   Research Questions:
   Does the test substance lower blood pressure in the SHR?
   What is its magnitude of blood pressure reduction and duration of action?
   How does it differ from that of a reference drug (in terms of 24-h hypertensive area, duration, and dose)?
   Does pharmacokinetic half-life translate into the duration of the antihypertensive effect?
   Will there be an attenuation of the antihypertensive effect over time (tolerance)?
   How do plasma renin activity, test substance plasma concentration, and decline in mean arterial blood pressure correlate with each other for dose and duration?
   What could be the possible cellular mechanism associated with the blood pressure-modulating potential of test substance X?

1.4.6 Hypothesis

The formulated research question is translated or reframed into a workable or testable hypothesis (es) (see Box 1.4). A hypothesis is generated from the results of a previous experiment, from published literature, or just the creative idea(s). It should link a process to an existing biologic pathway, and incorporate measurable results. Additionally, it dictates the type of statistical analysis for the data. Thus, a hypothesis is a specific statement of a predicted relationship between two variables, with the statistics testing this relationship. It contains a proposition to be accepted or rejected by the study and its results.

The study contains at least one dependent variable (predictor or effect, e.g., atherosclerosis, blood pressure) and one independent variable (e.g., cause, such as increased consumption of a high-fat diet, or exposure to a drug or chemical). Examples include:

(a) Increased consumption of a high-fat diet is associated with an increased risk of atherosclerosis.

(b) High dietary sodium and fat are associated with an increased risk of atherosclerosis (and hypertension).

The outcome (prediction) is that the independent variable will be related (positively or negatively) to the dependent variable.

A hypothesis forces us to think carefully about the comparisons needed to answer the research question, and it establishes the format for applying statistical tests to interpret the results [21]. Tests of statistical significance are used to compare findings from two different kinds of treatments (e.g., placebo and drug, or test drug and reference drug). Statistics are used in hypothesis testing to ensure that the results are attributable to the study drug and to gauge with a reasonable degree of confidence that the difference between the drug and the comparator is credible.

A study that utilizes statistics to compare groups of data should have a hypothesis. A good hypothesis helps in selecting an experimental design and drives the study to completion. If the research question or objective uses any of the following terms: greater/more/more likely/less than, causes, leads/similar/related to, compared/associated/correlated with, a hypothesis should be formulated.

Types of hypotheses: The hypothesis is the statement that the study is designed to accept or reject (hypothesis testing). It should be written in the present tense and use declarative sentences, and contain the population and variables. There are two types of hypotheses:

1. Null hypothesis (Ho): a hypothesis stating that there is no difference between the variables or groups being compared, e.g., a group treated with a test antihypertensive drug versus one treated with a placebo (independent variable), to the measured variable, (e.g., blood pressure).
2. Alternative or Research hypothesis (H1): a hypothesis stating that a difference does exist, such as a possible effect of treatment, between the variables or groups being compared to the measured variable, (e.g., blood pressure). It gives a magnitude of difference, (e.g., 20 mmHg) and specifies a direction, (e.g., lowers blood pressure).
Two types of H1:
(a) Two-tailed or non-directional alternative hypothesis: defines the magnitude of the difference without indicating a direction, leaving open whether it is positive or negative relative to the control.
(b) One-tailed or unidirectional: gives a direction in addition to magnitude, and states that the effect of the treatment is better than that of the control, claiming the superiority of the drug being studied. (For more details on statistical comparisons between groups, see Chaps. 29 and 30).

Box 1.4 Writing the Hypothesis

Examples

- Single oral doses of test drug X are sufficiently safe and well-tolerated for dose escalation.
- Plasma concentrations of test substance X increase with increases in the dose, and a linear relationship exists between the dose and PK parameters (Cmax).
- Test article X is neither an inducer nor an inhibitor of cytochrome P450 isoenzymes. Or, test article X does not affect CYP activity.
- Test article X lowers blood cholesterol in animal models.
- There is a difference between olmesartan medoxomil and captopril in lowering blood pressure in hypertensive individuals. Olmesartan is better than captopril in controlling blood pressure.
- The bioavailability and in vivo release characteristics of drug X (or a new modified formulation) are altered by the

(continued)

Box 1.4 (continued)

co-administration of food. Food interferes with the absorption of drug X.

- Supplementation of oral test substance X at 100 mg twice daily for 4 weeks increases endothelial function by increasing nitric oxide bioavailability and improves the ability of blood vessels to dilate in patients with coronary artery disease.
- Title of the article: *Hepatoprotective and antioxidant effects of gallic acid in paracetamol-induced liver damage in mice.* Hypothesis: Gallic acid treatment alleviates the progression of inflammation induced by a high dose of paracetamol.

Is a hypothesis necessary? All biological research, including discovery science, is hypothesis-driven. However, not all studies need a hypothesis These include descriptive studies, (e.g., describing characteristics of a plant or a chemical compound), or studies that examine only one variable. For example, a complete genome sequence can be determined independently of a hypothesis, or a study on the prevalence of high metabolizers in a population does not need a hypothesis except when comparing two different sets of the population, (e.g., European and Indian). Studies in pharmaceutical sciences, including Pharmacognosy and Phytochemistry, Pharmaceutical and Medicinal Chemistry, and Pharmaceutics, are commonly done without hypotheses, as are characterizations of drug candidates for their pharmacokinetic properties, e.g., establishing the absorption rate constant of a new compound using single-pass intestinal perfusion in rats; evaluation of drug activities of plant material, such as ginger or pomegranate; devising (new) methods, such as for measuring plasma concentrations of captopril; and developing drug formulations.

1.4.7 Objective(s)

The purpose of the objective(s) is to describe concisely and realistically what the proposed research is intended to accomplish, defining needed efforts or actions, summarizing what the researchers expect to achieve, and specifying the scientific questions that the study is designed to answer. A statement of objectives also is essential for selecting the factors to be investigated, the response variables to be measured, the data needed to describe the effects of the factors, and the kind(s) of statistical analysis required. Thus, when the study is completed, the results will be compared to the objectives and research questions specified in the protocol. If objectives are not defined, the project will lack clarity.

Objectives are a subset/stepping stone of the aim and goal. Objectives should cover the broad, long-term goals of the study. Researchers should ensure that all the objectives are attainable within the stated time frame, that they are connected and related to each other, and that they are supported by good preliminary data and scientific expertise [25]. The objectives should cover the entire breadth of the project; writers of objectives need to consider contributing factors such as variables and drug treatment.

Objectives should be "SMART": Specific, Measurable, Achievable, Realistic, and Time-Bound [26]. Additionally, they provide accountability, are compatible with the study, and are arranged in a logical sequence. Arranging objectives can also involve combining them or breaking one objective into multiple objectives for the study. They are sometimes organized into hierarchies: primary, secondary, and exploratory, and can be general or specific (Box 1.5). Objectives in the study protocol or project normally use action verbs, such as "describe," "establish," "determine," "compare," "evaluate," or "investigate." Non-action verbs, such as, "understand," "appreciate," or "study," are used less frequently.

Box 1.5 Writing Objectives

Examples

1. To determine the **cause-and-effect** relationship between activation of the extracellular signal-regulated kinase and the effect of protein kinase C on phenylephrine-induced contractions in non-pregnant and pregnant ewe uterine arteries.
2. To determine whether subcutaneous administration of alamandine would influence blood pressure and cardiac hypertrophy in SHRs, and to *preliminarily explore* the molecular signaling pathways underlying its regulatory effects.
3. *Primary*: To evaluate the safety and tolerability of single oral doses of amlodipine in healthy volunteers.
   (a) *Secondary*: To assess the PK profile of amlodipine following single oral doses.
   (b) *Exploratory*: To evaluate the incidence of peripheral edema reported as an adverse event of treatment.
4. To study the blood pressure-lowering effect of losartan in SHR and in sodium-depleted marmosets after single- or multiple-dose oral administration.
5. To evaluate the antioxidant potential of *Curcuma longa* using human blood lymphocytes in an in vitro assay.
6. To assess the antioxidant, cytotoxic, and hepatoprotective activities of the different extracts of *Alhagi maurorum* and its phenolic metabolites.
7. To investigate the effect of ellagic acid on the seizure threshold in two acute seizure tests in male mice, i.e., in the i.v. pentylenetetrazol seizure test and the maximal electroshock seizure threshold test.
8. To characterize the effect of food on the bioavailability and pharmacokinetics of the test substance in healthy human volunteers.
9. To study the effects of a test substance on L- and T-type calcium currents.

1.4.8 Scientific Rationale and Significance of the Proposed Study

Here, we discuss the rationale and the importance of the study based on the background and research question or hypothesis related to the proposed study's objectives. The rationale is the logic behind the research project that leads to the objective and hypothesis. It requires prior knowledge and includes probability and expectation [5]. The statement should include the current state of knowledge relevant to the proposal, gaps that the project is intended to fill, an unmet need that exists and poses a problem if that gap continues to exist, (e.g., development of a vaccine for SARS-CoV-2, malaria, or monkeypox; unmet medical therapy to treat patients with the hepatorenal syndrome). The following two examples illustrate the rationale for conducting a study.

1. Development of an oral suspension as an alternative to the solid oral dosage form. Pharmaceutical products are dispensed in different dosage forms, such as tablets, capsules, suspensions, and parenteral solutions. A laboratory wishes to develop an oral suspension as an alternative to the solid dosage form, with a rationale that it adds convenience for physicians who can dose titrate by volume to achieve the proper dose instead of combining multiple tablet or capsule strengths. Also, an oral suspension helps patients who, due to illness, are not able to safely swallow the solid product or may prefer a liquid formulation. Patients need not manually crush tablets or open capsules and disperse the contents in water or juice, which may lead to inaccuracy and result in suboptimal dosing.
2. Hypertrophic cardiomyopathy (HCM) is an autosomal dominant genetic disease present in approximately one in 500 individuals. Currently, no sarcomere-targeted therapies for

HCM have been developed or tested. Drugs targeting the disease at its source might influence the downstream events to reverse left ventricle remodeling back to normal. In this direction, we expect that the drug, a first-in-class small molecule allosteric modulator of cardiac myosin, will selectively target cardiac myosin and reversibly inhibit its binding to actin. This reduces the aggregate force of systolic contraction and is predicted to facilitate diastolic relaxation and improve dynamic left ventricular outflow tract obstruction in patients with HCM.

The reason/basis/implication for doing the proposed study can be understood from the following questions. [18, 27, 28]

1. Why is the experiment being conducted? (Because benefits of some kind will accrue from the application of the new knowledge.)
2. What makes your research noteworthy?
3. Is this a study to confirm, refute, or extend previous findings or to provide new findings?
4. Is it designed primarily to test research equipment?
5. What are the practical implications and methodological advances of this study?
6. Why is this research topic important at this time, how can the results be applied to further research in this field or in related areas, and what is the potential contribution of this research to the problem(s) addressed?
7. Will it make an original contribution to knowledge in the existing field?
8. What innovations will come about?

Finally, you must demonstrate that the proposed study is designed to provide a clear answer to the question being asked, thus providing a pathway toward expanding knowledge and directing future research.

Examples

1. The proposed studies will provide insight into the mechanisms of the pathogenesis of enhanced alpha 1 adrenergic receptor vasoreactivity in subjects with progressive renal disease. Additional support for a potential link between nitric oxide and sympathetic activity will lay the groundwork for new strategies in the treatment and prevention of vascular disease among the rapidly growing group of individuals with chronic kidney disease.
2. The results of the current study are intended to provide information on the relative bioavailability of compound X under fasted and fed conditions in support of the ongoing clinical study.
3. The methods described here are simple and can be successfully applied in quality control for precise and rapid spectrophotometric determination of compound X and compound Y in combination drug products.
4. These in vivo studies, coupled with extensive in vitro studies on the protective role of Ang (1–7) in various cell types, including cardiomyocytes, cardiofibroblasts, and endothelial cells, are clearly supportive of the protective action of Ang (1–7) in heart disease. Enhancing Ang (1–7) action represents an important targeted therapeutic application in disease processes characterized by an activated renin-angiotensin system. (Reference: *J Mol Med. (2015);93:1003–1013.*)
5. In this study, we show for the first time how multi-layer polymer films that undergo surface or layer-by-layer degradation can provide controlled and sustained drug release. Understanding how drugs are released through this strategy could provide an alternative approach to fine-tuning and achieving a desirable drug release profile from polymeric films. This, in turn, would broaden the uses of these polymers for a myriad of localized drug delivery applications. (Reference: J Pharma Sci. (2010);99,3060.)

1.5 Physiology of Research

Research planning should also include experimental design and detailed methods describing how the research or study will be carried out (Fig. 1.3). This includes

- Plan of study (study outline)
- Experimental design (*see* Chap. 28)
- Statistical analysis (*see* Chaps. 25, 29 and 30)
- Materials
- Methods, instrumentation, and measuring techniques
- Data recording and collection
- Difficulties and limitations
- Ethical and safety issues (*see* Chaps. 58 and 59)
- Timeline (work schedule, experimental targets)

The end result of the research will be the dissemination of findings as detailed articles in peer-reviewed journals (*see* chapters in Part VIII), books, or presentations at academic and scientific meetings (*see* Chaps. 49 and 50).

1.5.1 Dissemination of Research Results

An experiment is a well-conceived plan for data collection, data analysis, and data interpretation. Soon after completing the study and analyzing the data, the researcher should be ready to produce a well-written scientific paper. Postgraduate students should learn how to write a thesis (*see* Chap. 48) and a manuscript that has the potential to advance the field and strengthen their career prospects. Scientists must not only "do" science but also must "write" it. Starting as postgraduate students, scientists are measured primarily by their publications, not by their dexterity in laboratory manipulation, nor by their innate knowledge, wit, or charm [29, 30]. What matters, possibly more than the gathering of data, is how one tells the story of the project in clear, succinct, simple language, weaving in previous work in the field, answering the research question, and addressing the hypothesis set forth at the beginning of the study. Additionally, the paper should detail how the results can be applied to further research in the field and advance our understanding of the study concept. The most tangible result of scientific research is publication. If science is not in a form that can be read, it is dead [31].

Clarity and precision are the most important aspects of scientific writing. Poor scientific writing, implying poor thinking, could impede the career prospects of the writer. For many researchers, it is easier to do experiments than it is to write manuscripts; however, good writing is associated with good science [32]. Given the number of article submissions and the number of researchers vying for space in journals, good-quality writing can help one stand out from the crowd.

That said, many researchers would prefer someone else, such as their mentors or senior faculty, to handle the writing [32, 33]. Writing a research paper can be as challenging as the research itself. It is the key or gateway to successfully promoting their findings, a prospect that can be daunting.

In the past, faculty believed that teaching student was the most important aspect of the job, and teachers devoted all their energy to excelling in this activity. Today, the bar for entry into the teaching profession has been raised; publishing papers is part of the job description at all levels of the academic career. It helps establish an individual as an expert in their field of knowledge.

Peer-reviewed publications provide supporting evidence for evaluating the merits of research funding requests. Prior research experience and preliminary findings supported by publication(s) open the doors for grants (*see* chapters in Part X), which are essential for enabling high-quality research. For a lengthy discussion on scientific writing and publishing, *see* Chaps. 41–47.

Institutions at national and international levels are ranked by organizations offering grants (by $amount) and by publications and their impact on society (in solving problems or providing guidance), and patents, among other factors. To stand out in the crowd, universities and institutes must encourage research, and provide environments in which applicants would like to work. Conducting cutting-edge research offers a win-win situation for both researchers and their institutions.

Following are some of the systematic developmental and structural processes in writing a good scientific paper. It is organized; it has a central idea and a progression of ideas. It involves an organic process of planning, researching, drafting, revising, and updating current knowledge for future study. The language is clear, and the vocabulary is appropriate and varied. Finally, its title is effective in drawing readers [32]. The basic components of a research or original article are (for more details, *see* Chaps. 41–43):

Title: Either an indicative title. stating the purpose of the study. or a declarative one revealing the study conclusion, should make a good impression and encourage readers to at least browse the paper.

Abstract: An abstract may be structured or unstructured and consist of up to 250 words. It contains an extraction of key points from each section.

Introduction: Based on research literature, it gives background details of the work, the research question, the hypothesis-tested objectives, and the rationale and significance of the study, along with a brief description of why you did the research and what it entailed.

Methods: These describe the design and controls, with sufficient detail for replication, and statistical analyses of data.

Results: These describe the findings. Supplemented with tables and figures, they show how the data support the research question (*see* Chaps. 43 and 45).

Discussion: The root of a research paper, it addresses whether the study answered questions outlined in the Introduction and if it supported the hypothesis, as well as states major findings and comments on the strengths of the work (*see* Chap. 46).

Conclusion: This contains a summary of key study findings, states conclusions, and discusses the outlook for future studies.

References: *See* Chaps. 39 and 48.

A well-prepared manuscript conforming to the journal's author guidelines should be marketed through an effective cover letter aimed at creating a positive first impression on journal editors. Once the manuscript fits the journal's standards, the editorial office sends it to a panel of reviewers who are experts in the field. Peer review, which remains the foundation of publishing and an essential element of quality, is a process of subjecting an author's research or ideas to the experts' scrutiny, and is critical to establishing a credible body of knowledge for others to build upon. In the peer review process, these experts act as a filter to keep sloppy, fraudulent, or otherwise bad research out of the journal. Critical assessment of a manuscript is vital to peer review and to the publication process (*see* Chap. 47).

To develop a manuscript that passes successfully through peer review, authors are advised not to write without extensively reading scientific literature in their fields. They need to develop good writing skills. Most importantly, they should follow a plan, without which there can be no experiment and no publication [2]. A good piece of laboratory work combined with good writing improves one's chances of publication. In short, while writing a publishable and citable paper is an arduous job requiring meticulous planning, hard work, and persistence, it can also be extremely rewarding.

1.5.2 Research and Publication Ethics

Research ethics is a complex subject, involving integrity and trust at the bench, in the clinic, and throughout the publication process. It requires a commitment to strong ethical standards, and to avoiding scientific misconduct, namely, behaviors leading to fraudulent data, fabrication, falsification, misinformation, and plagiarism. Scientific misconduct is inimical to the scientific method, leading to knowledge steeped in the truth (*see* Chaps. 58 and 59).

Research ethics are governed by Good Clinical Practice (GCP) in the United States, a guidance that evolved from commissions and reports that exposed and rectified unethical research practices, specifically the Tuskegee Syphilis Experiment conducted from 1932 to 1972 on black Americans without their knowledge of the true intent of the research objectives. The principles of GCP are codified in the U.S. Code of Federal Regulations (CFR). Similar guidance was developed globally and known as the International

Council for Harmonisation (ICH). ICH guidance covers many aspects of clinical research (*see* Chaps. 20–22). The specific guidance similar to GCP is known as ICH-E6 [34]. This guidance was developed in a consensus-driven paradigm including the US, Canada, European Union, non-European Union Nordic countries, Japan, Australia, and the World Health Organization. The basis of ICH-E6 was also rooted in unethical human research, as documented in the Nuremberg Code in 1947 [35]. Both GCP and ICH-E6 are dedicated to ensuring that subjects recruited to participate in a study are not harmed and are fully informed of all potential adverse outcomes (for details, see Chap. 23). Producing honest and original work and avoiding unintentional instances of plagiarism are discussed in Chaps. 58 and 59.

1.6 Conclusion

We have provided an overview of the research process with a focus on undertaking a literature review for idea detection, combined with creativity and critical thinking that progresses to developing the research question and hypothesis for a suitable researchable topic. These areas of focus firmly set up the four elements of a research project: research topic, research question, hypothesis, and objective. The basic procedures of scientific research involve literature search and review, formulating a problem, developing a study protocol or research plan, and describing the reasoning behind the protocol, study designs, statistical issues, results of analysis, and data interpretation. These procedures form a complex chain in any scientific enterprise. Without a well-designed plan, there could be no successful research.

Our vision at the start of a study defines what we will do, how we will do it, how we will analyze it, how we will interpret it, and what its impact will be on society. Later, effective communication of study findings is necessary for the work to be publishable and gather citations (see Chap. 40). The merit or scientific value of a published article stands on its own, not because it is published in a high-impact journal, but because it is interesting, novel, well-designed, feasible, highly cited, and impactful. Each paper is a stepping stone for new and continuing research.

Acknowledgements We would like to thank Joanne Berger, FDA Library, for editing a draft of the manuscript.

Conflict of Interest None.

References

1. Jagadeesh G (2011) Preamble: scientific tools. RGUHS J Pharm Sci 1:96
2. Jenicek M (2010) The four cornerstones of a research project: Health problem in focus, objectives, hypothesis, research question. In: Jagadeesh G, Murthy S, Gupta YK, Prakash A (eds) Biomedical research, from ideation to publication. Wolters Kluwer, New Delhi, pp 27–34
3. Bartz C (2011) Getting started with research: ideas to research process. RGUHS J Pharm Sci 1:176–179
4. Chenevix-Trench G (2006) What makes a good PhD student? Nature 441:252
5. Murthy S (2010) Research: a burning passion to discover. In: Jagadeesh G, Murthy S, Gupta YK, Prakash A (eds) Biomedical research, from ideation to publication. Wolters Kluwer, New Delhi, pp 18–26
6. Meadows KA (2010) An overview of research process. In: Jagadeesh G, Murthy S, Gupta YK, Prakash A (eds) Biomedical research, from ideation to publication. Wolters Kluwer, New Delhi, pp 35–47
7. Rotello CM, Heit E (2010) Relations between inductive reasoning and deductive reasoning. J Exp Psychol Learn Mem Cogn 36(3):805–812
8. Balakumar P, Jagadeesh G (2010) Cardiovascular and renal pathologic implications of prorenin, renin, and the (pro)renin receptor: promising young players from the old renin-angiotensin-aldosterone system. J Cardiovasc Pharmacol 56(5):570–579
9. Balakumar P, Jagadeesh G (2011) Potential cross-talk between (pro)renin receptors and Wnt/frizzled receptors in cardiovascular and renal disorders. Hypertens Res 34(11):1161–1170
10. Kuberappa YV, Kumar AH (2012) Knowing the known to understand the unknown-a systematic approach to reviewing the scientific literature. RGUHS J Pharm Sci. 2:1–7
11. University of Nebraska–Lincoln. The overview of research process. https://researchwriting.unl.edu/overview-research-process. Accessed 30 March 2022
12. Moorhead JE, Rao PV, Anusavice KJ (1994) Guidelines for experimental studies. Dent Mater 10:45–51

13. Reisman FK (2010) Creative and critical thinking in biomedical research. In: Jagadeesh G, Murthy S, Gupta YK, Prakash A (eds) Biomedical research, from ideation to publication. Wolters Kluwer, New Delhi, pp 3–17
14. Reisman FK (2011) Creative, critical thinking and logic in research. RGUHS J Pharm Sci. 1:97–102
15. Sternberg RJ (2012) Intelligence. Dialogues Clin Neurosci 14:19
16. Jung RE et al (2009) Biochemical support for the "Threshold" theory of creativity: a magnetic resonance spectroscopy study. J Neurosci 29:5319–5325
17. Nte AR, Awi DD (2006) Research proposal writing: breaking the myth. Niger J Med 15:373–381
18. Bordage G, Dawson B (2003) Experimental study design and grant writing in eight steps and 28 questions. Med Educ 37:376–385
19. Saunderlin G (1994) Writing a research proposal: the critical first step for successful clinical research. Gastroenterol Nurs 17:48–55
20. Kahn CR (1994) Picking a research problem—the critical decision. N Engl J Med 330:1530–1533
21. Enarson DA, Kennedy SM, Miller DL (2004) Getting started in research: the research protocol. Int J Tuberc Lung Dis 8:1036–1040
22. Durbin CG (2004) How to come up with a good research question: framing the hypothesis. Respir Care 49:1195–1198
23. Boyle EM (2020) Writing a good research grant proposal. Pediatr Child Health 30(2):52–56. https://doi.org/10.1016/j.paed.2019.11.003
24. Hulley SB et al (2007) Designing clinical research, 3rd edn. Lippincott Williams & Wilkins, Philadelphia
25. Preparing Grant Applications. https://deainfo.nci.nih.gov/extra/extdocs/apprep.htm. Accessed 30 Sept 2022
26. Develop SMART Objectives. https://www.cdc.gov/publichealthgateway/phcommunities/resourcekit/evaluate/develop-smart-objectives.html. Accessed 25 Sept 2022
27. Gerin W, Kinkade CK, Page NL (2017) Writing the NIH grant proposal: a step-by-step guide, 3rd edn. SAGE, Los Angeles, pp 61–110
28. Lamanauskas V (2021) Writing a scientific article: focused discussion and rational conclusions. Probl Educ 21st Century 79:4–12. https://doi.org/10.33225/pec/21.79.04
29. SIUE: Southern Illinois University. Why must scientists write? https://www.siue.edu/~deder/rite2.html. Accessed 30 Sept 2022
30. Day RA (1988) How to write and publish a scientific paper. Oryx Press, Phoenix
31. Piel G (1986) Editorial. The social process of science. Science 231(4735):201. https://doi.org/10.1126/science.231.4735.201
32. Van Way CW (2010) How to write a good scientific paper. In: Jagadeesh G, Murthy S, Gupta YK, Prakash A (eds) Biomedical research, from ideation to publication. Wolters Kluwer, New Delhi, pp 411–420
33. University of Nebraska-Lincoln. The basics of scientific writing. https://graduate.unl.edu/connections/scientific-writing. Accessed 30 Sept 2022
34. Guideline for good clinical practice. ICH-E6. https://www.ema.europa.eu/en/ich-e6-r2-good-clinical-practice. Accessed 25 Sept 2022
35. Ethical codes & research standards. http://www.hhs.gov/ohrp/archive/nurcode.html. Accessed 25 Sept 2022

Processes Involved in the Generation of Novel Ideas

2

Mark A. Runco and Janessa Bower

Abstract

This chapter focuses on the creative process. It explains why a focus on process is more practical than the alternatives. It reviews theories and research on the creative process. Some of the earliest process theories are summarized, as are more recent theories. Associative theory and the divergent thinking model are explored in detail and recommendations are offered. These include taking one's time when working, the justification being that this is necessary for the remote associates that tend to be highly original. The value of incubation is also presented. Very recent work using computers to test associations is summarized and leads to the concept of semantic distance. The pros and cons of working in groups are presented. Practical implications are noted throughout the chapter and collected at the end of the chapter.

Objectives of this chapter: Summarizing theories of and research on the creative process, exploring practical implications.

M. A. Runco (✉)
Creativity Research & Programming, Southern Oregon University, Ashland, OR, USA
e-mail: RuncoM@sou.edu

J. Bower
Department of Educational Psychology, University of North Texas, Denton, TX, USA
e-mail: JanessaBower@my.unt.edu

Keywords

Associative theory · Semantic distance · Divergent thinking · Originality · Brainstorming

2.1 Introduction

Scholarship, at its best, is creative. Thus, to understand outstanding scholarship, there is good reason to look to the field of creativity research. That field is quite large. It started small and, up until the 1950s and 1960s, was not recognized as a legitimate scientific endeavor. Creativity was, early on, viewed as inextricable from the arts and as such was assigned to the humanities rather than the sciences [1–3]. That changed in the 1950s and 1960s thanks to the efforts of Guilford [4, 5], Berkeley's Institute for Personality Research and Assessment [6–10] and a handful of other pioneers. They demonstrated that several parts of creativity could be objectively studied. Since then the field of creativity research has boomed. There are now nearly two handfuls of scholarly journals (e.g., Creativity Research Journal, Journal of Creativity), as well as a sizable *Encyclopedia of Creativity* [11]. In addition to the utilization of objective methods, the field has moved well beyond artistic creativity. Creativity is now recognized in a range of domains (e.g., mathematics, technology, design, writing, morality), not just the arts. Quite a few

G. Jagadeesh et al. (eds.), *The Quintessence of Basic and Clinical Research and Scientific Publishing*,
https://doi.org/10.1007/978-981-99-1284-1_2

studies have focused on scientific creativity [12–14]. Much of what we review herein has that focus. Our primary concern here is on the creative process and how it has been and can be used by scientists and scholars. We begin by looking at novelty and originality. These are prerequisites for creativity. What processes lead to novelty and originality? Our approach in this chapter is to survey theory and research on creativity and, when possible, to highlight implications for creative scholarship.

2.2 Novelty and Originality

Novelty is an important part of creative work. Indeed, the standard definition of creativity [15], which is used or assumed in a great deal of the research on the topic, recognizes novelty (or some form of it, including originality) and effectiveness (or some form of it, such as utility). Neither novelty nor originality are synonymous with creativity. They are necessary but not sufficient. That being said, novelty is vital for creative work and can be operationalized such that good empirical work is possible. Indeed, there is quite a bit of good empirical research on novelty and originality.

The most important kind of research on novelty (and, more broadly, creativity) is that which examines the processes involved in the production of original ideas. Process research, more than any other kind of research, offers information about the mechanisms responsible for the production of novel ideas. Process research is often contrasted with research on products, personality, and places [16, 17]. These are the four Ps of Rhodes' [16] model of creativity. The research on *products* focuses on the outcome of creative efforts. These may be patents, publications, works or art, inventions, and the like. The second P focuses on the creative *person*. This includes personality traits, types, or cognitive styles associated with creativity. Personality research has identified autonomy, individualism, open mindedness, flexibility, and wide interests as "core characteristics" common to most creative individuals. The final P in the original framework is *place*. Here the focus is on the settings and external forces or pressures that act upon the creative person or process. Simonton [18] added *persuasion* to the 4P framework, the logic being that creative people influence and persuade others with their creative ideas and products. Runco [17] reorganized the alliterative framework into a hierarchy. This includes all of the categories just summarized but had *Potential* and *Performance* as the two overarching categories and all other approaches subsumed under one of those two. Since this hierarchy added Potential as a category of creativity research, it is known as the 6P framework, with the original 4Ps and then Persuasion and Potential as subsequent additions. Only process research informs us about how novelty and creativity come about. As Jay and Perkins [19] put it, studies of process can explain the mechanism used by the individual to produce novelty and originality. Process research can also lead to clear recommendations.

What has been discovered about the processes? We will next summarize theories and research on the creative process and mention practical implications for scholarship and scientific creativity along the way.

2.3 Early Views of the Creative Process

Process research frequently describes stages (sometimes called phases). Poincaré [20] seems to be the first to propose discrete stages. He described conscious hard work, unconscious dynamics, ideas moving from the unconscious to consciousness, and an expectation to check the results. Wallas [21] extended this and described four stages: preparation, incubation, illumination, and verification. Preparation may involve consciously identifying a task or problem, discerning given information, collecting information, and perhaps putting oneself in the right setting. Incubation involves stepping away, not thinking about the problem, allowing the subconscious to turn over the problem while the individual consciously works on other things [22]. Illumination is the a-ha when ideas click and the solution becomes evident. Often individuals are unaware of the preparation and incubation periods and illumination can feel like an instantaneous moment, but in fact the insight of an illumination is protracted [23]. This is interesting in part because it implies that there is a kind of exchange between the conscious and the unconscious. This view was mentioned long ago by Spearman [24] and, more recently, by Hoppe and Kyle [25]. After the illumination period, individuals go through a verification process of consciously and logically assessing the practicality, effectiveness, and appropriateness of the idea or product. Verification may involve the individual who had the insight, or others, and the idea or product may be accepted or revised. There is often a *recursion* after verification where the individual goes back and revisits one or more of the first three stages [26].

Something more must be said about preparation. That is because (a) good scholarship may depend on it, (b) scholars can improve their preparation, and (c) there is quite a bit of theory and research about it. The research has demonstrated that *problem finding* is an important part of the creative process; and problem finding is an important part of the preparation stage. Problem finding is an umbrella term and may include problem *identification*, problem *definition*, or problem *construction* [19, 26]. Einstein and Inhelder [27, p 83] recognized something akin to problem finding when they wrote: "The formulation of a problem is often more essential than its solution.... To raise new questions, new possibilities, to regard old problems from a new angle, requires imagination and marks real advance in science." Wertheimer [28, p 123] added "often in great discoveries the most important thing is that a certain question is found. Envisaging, putting the productive question is often a more important, often a greater achievement than the solution of a set question." Guilford [4] described "sensitivity to problems" in his structure of intellect theory. Torrance [29, p 16] recognized "the process of sensing gaps or disturbing missing elements and formulating hypotheses" in his theory of creativity. Standing back, it is clear that preparation is important for good creative work, and that this may involve finding a good problem. Getzels [30] actually claimed that a good solution of any sort requires a good problem. Also important is that problems can be redefined such that creative solutions are likely. Scholars need not accept problems or questions as they receive them. They may modify the problem so it is more meaningful.

Research on the incubation stage [22, 31] also has clear practical implications. An obvious implication is that it is useful to take one's time rather than working quickly or rushing. Further, it is useful to think about a topic or problem, but then think about other things. This will allow incubation to occur. Incubation can be useful during the preparation stage and when finding a good problem. Very likely it is wise to invest at least as much time into selecting the topic of one's work as in the exploration of that topic–and perhaps incubate before settling on a problem or focus for one's work.

Importantly, the research on problem finding shows clearly that creative insights often occur when problems are questioned. This fits with the suggestions from the creativity literature about questioning one's assumptions; however, the practical idea, in the context of problem finding, is to ensure that one's investment is on the right topic or problem. It can even help to change how problems are represented. Changing the scale of a problem (zooming in or zooming out) can suggest new and novel ideas as well. Adams [32] went into detail about all of the ways that problems and assumptions can be questioned (also see Runco [33] on tactics for dealing with assumptions).

2.4 More Recent Views of the Creative Process

A more recent conception of the creative process was proposed by Mumford et al. [34]. They started with three premises: (a) creative problem solving requires the production of high quality, original, and elegant solutions to complex, novel, ill-defined problems; (b) problem solving requires knowledge or expertise; and (c) although different performance domains impose different knowledge requirements, and stress processes differently, similar processes would underlie creative thought in most domains. Their model includes eight domain-general, aspects of the creative process. Each step moves linearly but also can move back to revisit a previous step, much like the recursion mentioned earlier. The steps are (a) problem definition, (b) information gathering, (c) concept/case selection, (d) conceptual combination, (e) idea generation, (f) idea evaluation, (g) implementation planning, and (h) adaptive execution.

Another example of a process model of creativity is the two-tier model [35]. It recognizes problem finding and verification, both mentioned above, as well as ideation, which involves the production or generation of new ideas. These three things are all on one primary tier. A secondary tier includes influences on the primary tier. These influences include knowledge and motivation (the latter being one example of an extracognitive influence). Knowledge may be *declarative* or *conceptual* or *procedural* (i.e., "know-how"). Procedural knowledge is often tactical; the individual knows to utilize some tactic. In that sense, he or she knows how to increase the probability of finding original ideas. The questioning of assumptions mentioned earlier is an example of a tactic, as is "change the representation of the problem."

2.5 Divergent Thinking

It is useful at this point to employ a "zoom in" tactic and focus the research specifically on ideation. Ideation is nicely operationalized in models of divergent thinking (DT); and these models of DT are highly testable, include clear descriptions of process, and are highly practical. There are a number of implications for scholarship.

There are two theoretical bases for DT. One is the *structure of intellect* theory that Guilford [4, 5] proposed (and investigated for his entire career). The other is associative theory, which is certainly one of the clearest examples of process. It helps to explain the origins of originality, which ties it to creativity, but also applies quite broadly to all ideation.

2.5.1 Associative Theory

Associative theory describes thinking in terms of associations. The basic supposition is that a person has an idea (or some other mental element), and it somehow leads to another idea, and then another, and then another, and so on. Thinking is, then, a kind of chain. This associative process can lead to creative ideas. It does not always do so, but it can.

Some associative research holds the premise that creative potential and word association ability are correlated such that more creative people can make more word associations when given an initial stimulus [36]. Many studies have supported Mednick's theory [37, 38] although some [39] have questioned the validity of his Remote Associates Test (RAT) [40]. A typical RAT gives a series of three-word stimuli, such as safety/cushion/point, and the examinee must think of a word or concept that relates to all three. In this example, the answer is "pin" (i.e., safety pin, pincushion, pin-point). The more of these associations an individual can make, the higher their RAT score. One problem with the RAT is that it is highly verbal. Indeed, there are questions about it having a *verbal bias*, meaning that individuals with high verbal scores would always do well on the test and individuals with low verbal abilities would always do poorly, regardless of level of creative ability. If this is the case, what is the RAT measuring? Creative potential or verbal ability?

The research on associative processes has clear implications for scholarship. This is apparent in the initial work by Mednick [36] but also in demonstrations by Milgram and Rabkin [41] and Runco [42]. Their investigations showed that ideas tend to become more original as time goes by. Mednick referred to *remote associates* as original ideas that are only found after the initial and more obvious or rote ideas are depleted. The investigations are easy to describe: examinees are asked to generate a list of ideas and given ample time to produce a long list. Next, the halfway point of that list is found and the first half is compared to the second half (the latter representing ideas that were found only late in the associative chains). Originality tends to be higher in the second half of the ideas. Recent research has used more robust methods to examine time and avoided the dichotomization implied by comparisons of halves [43]. Here again, more time leads to more originality. We might say that good scholarship is more likely if there is ample time.

Some fairly recent thinking about associative theory draws from new technologies. Examples of this will be presented below. First, the theoretical bases of DT should be summarized. Associative theory represents one base, the other being the structure of intellect model. The next section summarizes key points from this model and goes into detail about what DT tells us about the creative process.

2.5.2 Structure of Intellect

Guilford's [5] structure of intellect theory eventually described 180 orthogonal cognitive abilities and capacities. Guilford presented this theory as a cube and distinguished between intellectual products, intellectual contents, and intellectual operations. The theory as a whole was not very well received. There were compelling criticisms of the subjective factor analytic methods used. Several groups have continued to use the entire structure of intellect model in their work on creativity and cognition [44, 45] but most work has focused on one particular part of this model, namely the concept of divergent thinking. DT has proven to be distinct from convergent thinking and conventional intelligence and is one of the most influential concepts in the field of creative studies. Divergent thinking tests may be the most commonly used assessment for the potential for creative problem solving (and problem finding), and many training programs focus on divergent thinking.

For the present purposes, what is most important is the idea that original thinking results from a process that diverges. Conventional thinking, in contrast, results from convergent thinking. When thinking convergently the individual takes the available information and looks to correct or conventional ideas. When thinking divergently the individual uses information as a starting point. The individual's thinking literally diverges; it relies on neither convention nor normative logic. Such divergence often leads to original ideas.

Divergent thinking is apparent when the individual produces many ideas ("ideational fluency"), a large proportion of which are novel ("ideational originality") and varied ("ideational flexibility"). Quite a bit of research suggests that one of the best ways to find original ideas is to produce a large number of ideas. This of course is consistent with the idea of remote associates and has implications for scholarship.

2.5.3 Social Influences on the Creative Process

Quite a bit of work looks at the creative process in social contexts. It needs to be recognized, especially in endeavors like scholarship. Runco and Beghetto [46] described a process that is both personal and then social. They described intrapersonal creativity as primary (and first in the process) and interpersonal judgments as secondary. The unique thing about this view is that it focuses on the interpretive bases of creativity. As Runco and Beghetto [46] explained it, human cognition is a constructive process. This is especially obvious in perception. Humans do not merely take in information and process it in its raw form. Instead, we add to the information if we need to do so to make it meaningful, and sometimes we ignore parts for the same reason. Expectations come into play as well. Perception can be viewed as a very basic creative process because meaning is created. Runco and Beghetto used this reasoning to describe intrapersonal creativity. The original aspect of their work was to take it one step further–to a second step in the process. There the result of the intrapersonal interpretation is judged by an audience. The audience may be peers, supervisors, teachers, or whomever. Their secondary creativity involves the construction of meaning by the audience! Just as the individual creator constructs meaning, so too must an audience when it considers the idea produced by the individual.

Other research on the interpersonal stages of the creative process points to the importance of risk taking. This comes as no surprise, given that there is always some risk involved in original thinking. An original idea is by definition novel and new. This means that other people have not thought of it. If they had thought of it, it would not be original. This in turn means that the individual who produces the new idea will not really know how other people will react. Further, if the idea is highly original, it is probably unconventional. This means that the novel idea is not a part of normative thought. If a person works on a problem and has an idea, but it is a conventional idea that fits with current norms, there is little risk. This individual can accurately expect acceptance precisely because the idea is a conventional one. But what of an idea that is novel and thus unconventional? There is no way to predict how others will react. This is why there is always a risk to offer an original idea to an audience.

Interpersonal processes may also be involved as ideas are formed. Any one idea is the result of an individual's thinking (although two or more individuals could think of the same idea at the same time), but procedures like brainstorming attempt to use groups to increase the likelihood that individuals in a group will find original ideas. Admittedly, one of the three guidelines for brainstorming is "quantity over quality." Members of a brainstorming group are told to produce as many ideas as they can (quantity) and to postpone consideration of the quality (including originality) of the ideas. It is postponement, however, because at some point ideas must be evaluated. The third guideline for brainstorming groups is to hitchhike or piggyback. This means that individuals are told to take someone else's ideas and extend that same line of thought. Brainstorming is quite popular, but a great deal of research questions its efficacy [47, 48]. Several things may occur in a group to preclude highly original ideas. There is *social loafing*, for example, which means that when working in a group many people do not put as much effort into the attempt to solve a problem. Risk is also relevant here because a person may take risks with original ideas when alone because there is no one to judge them, but there is a risk if someone else hears the ideas. In fact it is a linear thing: there is no risk when working alone, a bit of risk if there is one other person, a bit more if there are two people, and so on. There is a great deal of risk in a large group. Much of the work on social influences on the creative process suggests that scholars should work alone, at least some of the time. This will facilitate original thinking.

2.6 Discussion and Implications

This chapter focused on the creative process. It distinguished the creative process from the Product, Person, and Place emphases in the creativity research. It also cited research on domain specificity to argue that scholarship, including that leading to scientific breakthroughs, may in some ways be distinct from other kinds of creativity (e.g., musical, technological). Recommendations were offered singly as we moved through this chapter but they can be brought together in this Discussion. We believe that scholars should consider the following: recognize that creativity is (a) not only expressed in the arts and that (b) originality is central. They should find ways to be original, including working alone some of the time and intentionally questioning assumptions. They should not accept problems as given; instead they should consider revising and redefining problems so that they allow originality and are aligned with one's intrinsic interests. It is as important to spend time finding a problem as it is to spend time solving the problem. And it is important to incubate and take time away from one's work and from thinking about the problem at hand. Step away once in a while. Put time into associations to find remote associates.

Several points should be briefly revisited. Consider the parenthetical statement above that original ideas may sometimes be found simultaneously by several individuals. There is quite a bit of work on multiple discoveries [49] which suggests a kind of social influence not yet covered, namely Zeitgeist, or the "spirit of the times."

We should also revisit associative theory. That is because, as noted above, there is fairly recent research and theory made possible by technology, and in particular computers, which is directly relevant to the creative process and supports associative theory. In fact it has implications for creative cognition broadly defined. Simply put, research is now using computers to build models and test theories of creativity, especially associative theory. Less relevant to the present discussion, and yet quite important, is the fact that computers are making the assessment of creativity faster, cheaper, and arguably more reliable [50]. In the next few paragraphs we will summarize some of the research that has something to say about the creative process and associative theory.

The Special Interest Group on the Lexicon of the Association for Computational Linguistics has long examined word sense disambiguation. It has also developed SemEval, which is a program allowing examination of semantic distance [51, 52]. This is the interval between associates. Semantic distance has proven to be predictive of creative potential [53]. Creative individuals may be good at exploring conceptual associates but they may also think in a way that leads to connections that are distant or loosely connected rather than adjacent or contiguous. Some individuals can move from idea A to idea Z but must do so step-by-step (e.g., from idea A to idea B to idea C, and so on). There is not much distance there. Others can go from idea A to idea Z with a kind of conceptual leap covering a great deal of semantic distance. There is an interesting parallel between this line of work and the extensive research which has explained creativity in terms of loose conceptual boundaries [54]. It is also easy to see a parallel with the idea of remote associates [36].

Johnson et al. [55] drew from theories of distributional semantics and proposed the idea of *divergent semantic integration* (DSI). DSI measures the connections of divergent thinking within a narrative. One model of DSI, Bidirectional Encoder Representations from Transformers (BERT), explained 72% of the variance in human ratings of narratives. Furthermore, the results generalized across ethnicity and language proficiency levels. The program is available for open use at osf.io/ath2s. In a similar vein, Runco et al. [56] examined the *idea density* (ID) [57] of narratives and found that ID was correlated with several indicators of originality and creativity. The indicators included citation impact and the hits of TedTalks, both of which are pertinent to scholarship.

There are other investigations showing relationships between creative potential and the creative process with word choice, semantic similarity, and the like [57, 58], but as mentioned above, the point is that technology has provided new tools and in some ways supported and refined earlier theories, including associative theory. The ideas of semantic distance may be the best example of a refinement. A final idea to revisit is that the creative process is not entirely cognitive. Associative theory, DT, and problem-solving and -finding are often treated as if they are entirely cognitive, but this is mostly because of the need to operationalize for research–and because cognitive processes are easier to operationalize and measure compared with emotional and unconscious processes. Nonetheless several extra-cognitive contributions were mentioned above. These no doubt contribute to good scholarship. Risk-taking and risk-tolerance stand out. The two-tier model suggests that motivation is important. In fact there is a view that motivation is central to creative performance. This view points to the intrinsic motivation that was included in the two-tier model. According to this view, an individual who finds something intrinsically interesting will make careful choices about his or her work. Our suggestion is to make the choices that are described above in our recommendations.

2.7 Concluding Remarks

Creativity may be approached from various perspectives, including the person, the place, the product, or the process. This chapter argued that the last of these is the most useful and practical. It comes the closest to explaining the actual mechanisms by which individuals are creative. It also leads to recommendations. This chapter identified a handful of practical recommendations that should be considered when involved in scholarship, including taking risks, investing in originality, taking time to incubate, and sometimes working alone.

Conflict of Interest All authors declare no conflict of interest for this work.

References

1. Cropley AJ (2014) Is there an 'arts bias' in the Creativity Research Journal? Comment on Glăveanu (2014). Creat Res J 26(3):368–371
2. Rocavert C (2016) Democratizing creativity: how arts/philosophy can contribute to the question of arts bias. Creat Res J 28(3):229–237. https://doi.org/10.1080/10400419.2016.1195642
3. Runco MA (2014) Creativity: theories and themes: research, development, and practice, 2nd edn. Elsevier Academic Press, New York, 520p
4. Guilford JP (1950) Fundamental statistics in psychology and education, 2nd edn. McGraw-Hill, New York, 565p
5. Guilford JP (1968) Creativity, intelligence, and their educational implications. Knapp, San Diego, 229p
6. Barron FX (1955) The disposition toward originality. J Abnorm Soc Psychol 51(3):478–485
7. Barron FX (1995) No rootless flower: an ecology of creativity. Hampton Press, Cresskill, 356p
8. Helson R (1999) Institute of personality assessment and research. In: Runco MA, Pritzker SR (eds) Encyclopedia of creativity, 1st edn. Academic Press, San Diego, pp 71–78
9. Taylor CW (1964) Widening horizons in creativity. Wiley, New York, 466p
10. Taylor CW, Barron FX (1963) Scientific creativity: its recognition and development. Wiley, New York, 419p
11. Runco MA, Pritzker SR (2020) Encyclopedia of creativity, 3rd edn. Academic press, New York, 1663p
12. Albert RS, Runco MA (1987) The possible different personality dispositions of scientists and nonscientists. In: Jackson DN, Rushton JP (eds) Scientific excellence: origins and assessment. Sage, Newbury Park, pp 67–97
13. Miller AI (2000) Metaphor and scientific creativity. In: Hallyn F (ed) Metaphor and analogy in the sciences. Kluwer Academic, Amsterdam, pp 147–164
14. Zuckerman H (1977) Scientific elite: nobel laureates in the United States. The Free Press, New York, 335p
15. Runco MA, Jaeger GJ (2012) The standard definition of creativity. Creat Res J 24(1):92–96
16. Rhodes M (1961) An analysis of creativity. Phi Delta Kappan 42(7):305–310
17. Runco MA (2007) A hierarchical framework for the study of creativity. New Horiz Educ 55(3):1–9
18. Simonton DK (1995) Exceptional personal influence: an integrative paradigm. Creat Res J 8(4):371–376
19. Jay E, Perkins DN (1997) Creativity's compass: a review of problem finding. In: Creativity research handbook, vol 1. Hampton Press, Cresskill, pp 257–293
20. Poincaré H (2014) The foundations of science: science and hypothesis, the value of science and method, Reissue edn. Cambridge University Press, Cambridge, 570p
21. Wallas G (1926) The art of thought. Harcourt, New York, 162p
22. Dodds RA, Ward TB, Smith SM (2012) Incubation to problem solving and creativity. In: Runco MA (ed) Creativity Research Handbook. Hampton Press, Cresskill, pp 251–284
23. Gruber HE (1981) On the relation between 'AHA experiences' and the construction of ideas. Hist Sci 19(1):41–59
24. Spearman C (1930) Creative mind. Nisbet & Company, London, 153p
25. Hoppe KD, Kyle NL (1990) Dual brain, creativity, and health. Creat Res J 3(2):150–157
26. Runco MA (1994) Problem finding, problem solving, and creativity. Ablex Publishing, Norwood, 318p
27. Einstein A, Infeld L (1938) The evolution of physics: the growth of ideas from early concepts to relativity and quanta. Cambridge University Press, Cambridge, 226p
28. Wertheimer M (1945) Productive thinking. Harper & Brothers, New York, 224p
29. Torrance EP (1962) Guiding creative talent. Prentice Hall, New Jersey, 278p
30. Getzels JW (1975) Problem-finding and the inventiveness of solutions. J Creat Behav 9(1):12–18
31. Christensen BT, Schunn CD (2005) Spontaneous access and analogical incubation effects. Creat Res J 17(2–3):207–220
32. Adams JL (1980) Conceptual blockbusting: a guide to better ideas, 2nd edn. Norton & Company, New York, 153p
33. Runco MA (2020) Tactics and strategies. In: Runco MA, Pritzker SR (eds) Encyclopedia of creativity, 3rd edn. Elsevier Academic Press, New York, pp 529–532
34. Mumford MD, Mobley MI, Reiter-Palmon R, Uhlman CE, Doares LM (1991) Process analytic models of creative capacities. Creat Res J 4(2):91–122
35. Runco MA, Chand I (1995) Cognition and creativity. Educ Psychol Rev 7(3):243–267
36. Mednick SA (1962) The associative basis of the creative process. Psychol Rev 69(3):220–232
37. Marko M, Michalko D, Riečanský I (2019) Remote associates test: an empirical proof of concept. Behav Res Methods 51(6):2700–2711

38. Wu CL, Huang SY, Chen PZ, Chen HC (2020) A systematic review of creativity-related studies applying the remote associates test from 2000 to 2019. Front Psychol 11:573432
39. Lee CS, Huggins AC, Therriault DJ (2014) A measure of creativity or intelligence? Examining internal and external structure validity evidence of the remote associates test. Psychol Aesthet Creat Arts 8(4): 446–460
40. Mednick SA, Mednick MT (1971) Remote associates test: examiner's manual. Houghton Mifflin, Boston
41. Milgram RM, Rabkin L (1980) Developmental test of Mednick's associative hierarchies of original thinking. Dev Psychol 16(2):157–158
42. Runco MA (1986) Divergent thinking and creative performance in gifted and nongifted children. Educ Psychol Meas 46(2):375–384
43. Paek SH, Alabbasi AM, Acar S, Runco MA (2021) Is more time better for divergent thinking? A meta-analysis of the time-on-task effect on divergent thinking. Think Skills Creat 41:100894
44. Bachelor PA, Michael WB (1997) The structure of intellect model revisited. In: Runco MA (ed) The creativity research handbook. Hampton Press, Cresskill, pp 155–182
45. Meeker MN (1999) Structure of intellect systems. teacher training. Structure of Intellect, Vida
46. Runco MA, Beghetto RA (2019) Primary and secondary creativity. Curr Opin Behav Sci 27:7–10
47. Pauhus PB, Dzindolet MT, Poletes G, Camacho LM (1993) Perception of performance in group brainstorming: the illusion of group productivity. Pers Soc Psychol Bull 19(1):78–89
48. Rickards T, De Cock C (2012) Understanding organizational creativity: toward a multiparadigmatic approach. In: Runco MA (ed) Creativity research handbook, 2nd edn. Hampton Press, New York, pp 1–32
49. Ione A (1999) Multiple discovery. In: Runco MA, Pritzker SR (eds) Encyclopedia of creativity, 1st edn. Academic Press, San Diego, pp 261–271
50. Organisciak P, Dumas D (2020) Open creativity scoring. University of Denver, Denver. https://openscoring.du.edu/
51. Agirre E, Banea C, Cer D, Diab M, Gonzalez-Agirre A, Mihalcea R, Rigau G, Wiebe J (2016) Semeval-2016 task 1: semantic textual similarity, monolingual and cross-lingual evaluation. In: SemEval-2016. 10th international workshop on semantic evaluation, San Diego, 16–17 June 2016. ACL, Stroudsburg, pp 497–511
52. Miura N, Takagi T (2015) WSL: sentence similarity using semantic distance between words. In: Proceedings of the 9th international workshop on semantic evaluation (SemEval 2015), June 2015, pp 128–131
53. Acar S, Runco MA (2014) Assessing associative distance among ideas elicited by tests of divergent thinking. Creat Res J 26(2):229–238
54. Martindale C (1990) The clockwork muse: the predictability of artistic change. Basic Books, New York, 432p
55. Johnson DR, Kaufman JC, Baker BS et al (2022) Divergent semantic integration (DSI): extracting creativity from narratives with distributional semantic modeling. Behav Res. https://doi.org/10.3758/s13428-022-01986-2
56. Runco MA, Turkman B, Acar S, Nural MV (2017) Idea density and the creativity of written works. J Genius Eminence 2(1):26–31
57. Covington MA (2012) CPIDR® 5.1 user manual. Artificial Intelligence Center, The University of Georgia, Athens, 10p
58. Weinstein TJ, Ceh SM, Meinel C, Benedek M (2022) What's creative about sentences? A computational approach to assessing creativity in a sentence generation task. Creat Res J 34(4):419–430

3 Creativity and Critical Thinking Contribute to Scholarly Achievement

Mark A. Runco and Lindsay Ellis Lee

Abstract

This chapter explores theories of creativity as they relate to critical thinking and scholarly achievement. It discusses domain differences in creativity and raises the possibility that scholarly creativity is distinct from other expressions of creativity. It also covers theories and research that show creativity to be distinct from general intelligence and critical thinking. One key idea is that creativity and critical thinking are not entirely distinct but instead sometimes work together. This chapter offers support for the discriminant validity and predictive validity of creativity and for domain specificity. Quite a few implications for scholarship are noted.

Keywords

Threshold theory · Originality · Divergent thinking · Domain specificity · Latent semantic analysis · Citation impact

3.1 Introduction

Scholarship depends on both critical and creative thinking. This chapter examines their contributions. The specific objectives of this chapter are implied by the three issues that are conveyed by its title. One concerns the domain specificity of creativity. Is scholarly creativity independent of other kinds of creativity (e.g., artistic, musical, engineering, mathematical)? A second issue concerns the contribution of creative thinking to achievement in the natural environment, including scholarly achievement. A final issue concerns the collaboration of creative and critical thinking. The discussion of this collaboration includes an exploration of the possibility that creative and critical thinking are separate and distinct instead of both expressions of some more general cognitive capacity. Research and theory are cited and critically examined. Practical implications for scholarship are highlighted. It makes the most sense to begin with the third issue because that requires that creative and critical thinking are each carefully defined. Those definitions can then be used when exploring the other two issues.

3.2 Creative and Critical Thinking

There is more research pointing to a separation of creative and critical thinking than there is recognizing their collaboration. There is a

M. A. Runco (✉)
Southern Oregon University, Ashland, OR, USA
e-mail: RuncoM@sou.edu

L. E. Lee
Department of Pediatrics, Quillen College of Medicine, East Tennessee State University, Johnson City, TN, USA
e-mail: leele1@etsu.edu

G. Jagadeesh et al. (eds.), *The Quintessence of Basic and Clinical Research and Scientific Publishing*,
https://doi.org/10.1007/978-981-99-1284-1_3

historical reason for this: Early on creativity was not separated from general intelligence. It was viewed as a particular expression of intelligence, but dependent on intelligence. Eventually, this view was challenged, and only when data examined the relationships between general intelligence and creativity were collected was the latter appreciated as a distinct talent. In fact, the initial data were unclear. Some of the data supported the theory that creativity was just a particular kind of intelligence [1]. Only when methodologies were improved was the distinctiveness of creativity apparent [2]. Creativity cannot be assessed like intelligence is assessed. If it is, the two seem to overlap. But when creativity assessments allow for original thinking and spontaneity, that overlap mostly disappears. We say "mostly" because even today the predominant view is that there is a threshold of intelligence, with individuals below that threshold unable to be notably creative. This is sometimes called triangular theory [3, 4] because the distribution of data points in a bivariate (creativity and intelligence) scatterplot resembles a triangle. There is dispersion and variability at the moderate and high levels of intelligence (i.e., above the threshold) but nearly no variability below the threshold.

The idea that creativity and general intelligence are distinct has important implications. It means that educators cannot just support general intelligence and expect creative talents to follow along, for example. Creativity must be targeted, and its distinctiveness recognized. Along the same lines, anyone hiring an employee must look specifically at creativity; they can't just hire an intelligent applicant. That intelligent applicant may not be very creative.

The view that creativity is distinct from intelligence was taken too far. When this happened, an interplay was ignored [5]. Simplifying, actual achievement in the natural environment requires more than creative thinking. Creative thinking is usually involved, depending on the nature of the achievement and the domain in which the individual is working, but it is best if the individual can think both creatively and critically. Individuals have an advantage if they can think in an original fashion, but they should be aware of gaps and limits of information as well. The individual has a huge advantage if he or she can evaluate ideas after they are produced.

It is informative to zoom in and considering the components of intelligence and creativity. Let us begin with critical thinking that is mentioned in the title of this chapter. It is usually associated with intelligence. Critical thinking can be viewed as a process that allows the individual to identify gaps and recognize what is missing or wrong with ideas, solutions, or conclusions. Creative thinking, on the other hand, tends to be productive. New ideas are brought into existence. They are original precisely because they are new.

One often-cited model of creative thinking focuses on divergent thinking [4, 6, 7]. Runco and Smith [8] used it in an investigation specifically on what they called the evaluative component of creative thinking. The term evaluative was used to avoid the word "critical." Runco and Smith [8] suggested that "critical" implies that there is something wrong or flawed. They tried to avoid this connotation by discussing evaluative, rather than critical, components. They also described a *valuative* component where the individual is looking at ideas and solutions, trying to find their value instead of their deficiencies and flaws.

Runco and Smith [8] collected data on both intra- and interpersonal evaluations of ideas. They also administered a measure of the preference for ideation. Analyses of their data indicated that intra- and interpersonal evaluative accuracy were moderately correlated ($Rc = 0.63$). Divergent thinking (the production of ideas, discussed in detail below) was correlated with intrapersonal evaluative accuracy ($Rc = 0.45$). Interestingly, participants in this investigation could more accurately identify popular ideas than unique ideas. This is notable because the latter is strongly related to creativity. Popular ideas, on the other hand, are unoriginal. By and large evaluations of ideas were not very accurate. Of several categories examined, the highest level of accuracy was 42%. So, at most 42% of the ideas were correctly categorized as popular or unique. This implies that people are not very good at recognizing their own original ideas, nor accurate

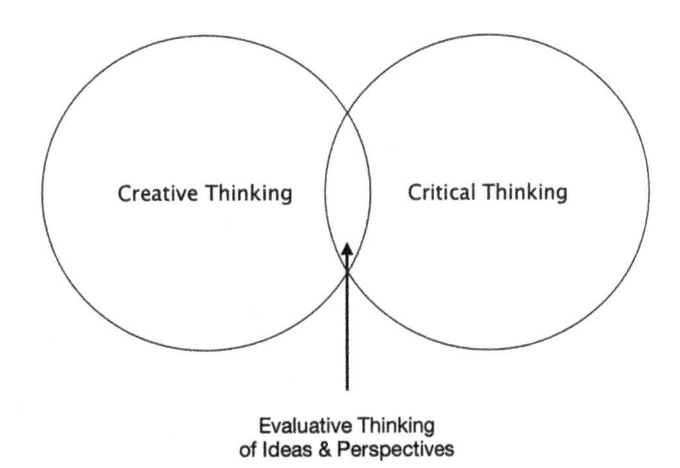

Fig. 3.1 The overlap between creative and critical thinking

at recognizing the original ideas of others. For the present purposes, it is also important that a standardized measure of critical thinking was unrelated to all the idea evaluation scores. There may be a special kind of judgment involved in the evaluation of ideas that is unrelated to the judgments that are required by critical thinking.

This investigation of evaluative, valuative, divergent, and critical thinking is just one example of how the components of creativity and intelligence have been examined. The key conclusions of the research on these things are as follows: (a) creative thinking and its components are largely distinct from intelligence and its components, but (b) in many real-world situations, creative thinking and intelligence work together. See Fig. 3.1.

3.3 Domain Specificity

A very similar research approach has been used when studying one of the other issues mentioned above, namely that involving domains of creative performance. Here again there is an interest in distinctiveness, and in particular an interest in the distinctiveness of areas or domains in which an individual may be creative.

This debate about the domain specificity versus generality of creativity has been debated by scholars for decades [9–13]. The domain-general perspective on creative thinking argues that there are skills, abilities, and aptitudes that can be widely applied regardless of the domain or discipline [14, 15]. One explanation for generality is that there is a transference of abilities across domains [16–21]. A domain-specific perspective, on the other hand, posits that creativity and critical thinking differ from one field and domain (e.g., science, math, economics, music, the arts) to another. The idea of distinct domains goes back a long time. It was often assumed before creativity came under scientific scrutiny, and thus not surprisingly, was one part of some of the earliest studies [22]. Yet it was Gardner [16] who offered a hugely compelling argument for domain specificity. At first, he identified distinct linguistic, musical, logical-mathematical, spatial, body-kinesthetic, intrapersonal, and interpersonal domains, and a bit later added the naturalistic domain. His theory was convincing because he drew from developmental, neuroscientific, psychometric, and experimental research.

Some research has explored the distinctiveness within domains, especially related to creativity [e.g., mathematical creativity; [23, 24]]. For example, mathematical creativity can be defined as the reasoning involved in finding variations of novel solutions to mathematical problems, with possible differences in the various branches of mathematics (e.g., arithmetic, algebra, geometry, calculus, statistics/probability). If an individual has strength in algebraic logic, that does not necessarily mean that his or her exact skills will be useful to create differential equations or applications of Bayesian statistics. A parallel limit may be apparent in the sciences (astronomy, physics, psychology, anthropology, chemistry). Generally speaking, scientific creativity involves "a kind of intellectual trait or ability producing or potentially producing a certain product that is original and has social or personal value, designed with a certain purpose in mind, using given information" [25, p 392]. Similarly, Fiest [26] defined scientific creativity as being able to create novel and useful solutions to natural and social scientific problems (e.g., biology, zoology, chemistry, physics). Simonton [27] argued that scientific creativity requires a "combinatorial process" which allows a cross-pollination across areas.

Domain specificity has been supported by psychometric research on discipline-based creative and critical thinking [13, 28]. For instance, Adey and Hu [25] developed the Scientific Creativity Structure Model (SCSM) and created a 7-item scientific creativity test for secondary students. Domain-specific assessments have been created to focus on *Math* (Creativity Abilities in Mathematics Test [CAMT; [29]], Mathematical Creativity Test [MCT; [24]], Mathematical Creativity Scale [MCS; [30, 31]]), *Visual Arts* (Test of Creative Imagery Abilities [TCIA; [32]], Test for Creative Thinking/Drawing Production [TCT/DP; [32]]), and quite a few groups have adapted the Consensual Assessment Technique [CAT; [33]] for a range of domains. The Consensual Assessment Technique [33, 34] has been used to assess many different types of domain-specific products using a panel of "appropriate" judges [34] For instance, Denson et al. [35] created a web-based adapted version of the CAT to assess engineering design products. There is probably more evidence in favor of domain specificity than there is showing general creative skills and abilities. Still, there is no reason to reject the possibility that there are both domain general and domain specific contributions to creative performances.

There may be individual differences, as well. Consider in this regard the research on *polymathy*. Broadly speaking, polymathy is apparent when an individual is creative in more than one domain. The research of Root-Bernstein and Root-Bernstein [36] suggests that polymathy may best be understood as a particular kind of inter-domain collaboration. They studied winners of the prestigious MacArthur awards and found that, at the highest level, creative individuals often had a creative avocation, such as playing a musical instrument. They also excelled professionally in another creative domain, such as mathematics. What Root-Bernstein and Root-Bernstein demonstrated so clearly was that, in their sample of award winners, the involvement in an avocation supported vocational creativity! Runco and Alabbasi [37] recently reported signs of this same thing–one domain supporting a very different domain–in their work with gifted adolescents.

3.4 Critical and Creative Thinking Leading to Scholarly Achievement

The third and last issue mentioned in the introduction to this chapter concerned the possibility that creative and critical thinking contribute to scholarly achievements. If they do contribute, how? Are there both general creative and critical skills that support scholarship as well as domain specific skills?

One step towards an answer holds that creative thinking is required for good problem solving [5, 38–40] and scholarship often requires the solving of problems. Creative problem-solving is flexible, so the individual can adapt to changes in problems or situations. Creative problem-solving is original, so the individual may think of novel and new solutions and not rely on old ones. This is quite important for advancement in almost every field. Generality is implied by the fact that creative problem-solving helps in a variety of situations, so a person may solve problems in one context (e.g., when collecting data or doing research) but then be able to solve problems that are discovered in different situations (e.g., when writing up results of an investigation or presenting results to one's peers).

Very importantly, creative thinking is not only expressed in problem-solving. It is also expressed in problem *finding* [41]. As a matter of fact, some researchers hold that problem-finding is more important than problem-solving and that a creative solution requires a creative problem [42]. Einstein pointed to the importance of problem-finding when he wrote:

> The formulation of a problem is often more essential than its solution. . .To raise new questions, new possibilities, to regard old problems from a new angle, requires imagination and marks real advances in science [43, p 83].

This quotation refers to problem *formulation*, and as a matter of fact, problem finding is an umbrella concept. It has been called problem construction, problem discovery, problem definition, problem identification, and so on. Some of these are just different labels for problem-finding, but some indicate different problem-finding processes. That is the case for problem identification, which is the label for literally finding a problem, in contrast to problem definition, which may occur well after a problem has been found. Problem definition may include problem re-definition, which is important because sometimes creative insights result from changes made to a problem. The problem is not accepted as it is encountered. Instead, the person makes changes to it, which allow new insights.

Research has demonstrated that some individuals are good at problem-finding but not good at problem-solving, and vice versa. Or course some individuals are good at both. What may be most important is that scholars should invest in problem finding, to identify good research problems. They should also question the assumptions of given problems. They may also question the way problems are represented. Think here about Einstein solving problems with Gedankenexperiments. He used visual thinking to solve problems in physics and mathematics. Root-Bernstein [44] and Adams [45] both suggested changing the way problems are represented as a tactic for original thinking.

These ideas should be related to the domain differences outlined in the section above. That is because both creative and critical thinking are important in most domains, but in some domains, creative achievement may be more dependent on one than the other. In some of the arts, for example, divergent thinking may be more important than convergent [46].

Psychometricians approach the question of how creative and critical thinking contribute to scholarship and achievement with analyses of *predictive validity*. This is one of the more important kinds of validity. (Early in this chapter we discussed the relationship of creativity with intelligence. In psychometric terms, the concern there was *discriminant validity*.) Data indicate that creative thinking does have good predictive validity. Care must be taken, however, because a great deal depends on the criterion used when calculating predictive validity. It may sound tautological but creative thinking tests only have good predictive validity when the criterion depends on creativity. Creativity tests would not have good predictive validity if a measure of general intelligence or conventional thinking was used as a criterion. When the criterion does depend on creative talent, however, predictive validities are good. In fact, one longitudinal study demonstrated that certain creativity tests (measuring *divergent thinking*) had surprisingly good predictive validities after 50 years [47]! Divergent thinking tests were administered in the 1950s and 1960s and used as estimates of creative potential. Criterion measures were administered 50 years later. The correlation between the predictors and the criteria was above 0.30. One study with children also used divergent thinking tests as an estimate of creative potential and reported predictive validities of over 0.50 [13].

We have referred to divergent thinking (DT) several times in this chapter. That is because DT is often used to describe creative thinking. DT tests allow objective research because they are reliable. Thus, they are often used to estimate the potential for creative thinking in the empirical research on creativity. In addition to good reliability, they have good theoretical bases [4, 7]. It would be good to go into more detail about DT here because there are several implications for scholarship.

DT is cognition that moves in different directions. It diverges. This is in contrast to convergent thinking, which is the process used when one particular (correct or conventional) idea is desired. The thinking converges on that particular idea. The individual may need to identify the US State that is surrounded by water, the farthest west of all other states, and the last to enter the US (August 1959). Thinking must converge on Hawaii. What if the individual is asked for a list of all of the unique features of Hawaii? Or asked for modes of transportation that could be used to get to Hawaii? These last two questions are open-ended and allow the individual to generate a large number of answers. That is characteristic of divergent thinking. There is openness and the possibility of finding a large number of ideas. Other DT tasks used in the research ask for alternative uses to some common object (e.g., a pencil), improvements of some product (e.g., a chair), instances of some category (e.g., all of the strong things you can think of), or solutions to some realistic problem (e.g., being out in the rain without an umbrella).

DT tests allow ideation. There are quite a few studies on how to improve ideation. Indeed, there are several meta-analyses of methods to improve ideation [48–50]. These suggest that it is usually best to work alone, avoid hurrying and time limits, and focus on truly divergent (e.g., unusual, weird, surprising) ideas, at least at first. Working alone may come as a surprise, especially if you are familiar with brainstorming. That is essentially group DT, but come to find out, when in groups there is a tendency towards *production loss* (fewer ideas are produced) and *social loafing* (individuals put less effort into the task at hand). It may be that there is sometimes a benefit to working alone and later working with other people.

Just above we suggested looking for weird ideas early in the process. This may be justified by asking, what is good DT? This may be answered by looking at the results of DT. The obvious one is just productivity. This is labeled *ideational fluency*. A person who does this well will produce a large number of options and solutions. Then there is *ideational originality*. This is operationalized in terms of the uniqueness or unusualness of the ideas produced. If a person produces many ideas but they are all quite common and conventional, originality is lacking. But if an individual produces ideas that few others, or perhaps no one else thinks of, originality is high. This is quite important because most scientific definitions of creativity require some sort of originality [51]. Looking to weird ideas sometimes helps when trying for original ideas. The last thing to look for in DT is *ideational flexibility*. This is apparent when the individual produces ideas tapping a diverse set of conceptual categories. In other words, the person's ideas are notably varied. If a person is asked to name strong things and answers Supergirl, Superman, Spiderman, Superboy, the Hulk, and so on, there is not much variation. Thus, this is indicative of low flexibility of thought (a correlate of which is *rigidity*.) But if that person answers Superglue, Superboy, gravity, love, hardwood, and faith, etc., there is more variation. Each of these ideas taps a different conceptual category. Flexibility is high, which is a very good thing. Certainly, the task just used as an example is quite simple (strong things) but there are realistic tasks on tests of DT, and more importantly, problems and tasks encountered when doing scholarly work are often at least partly open-ended and would probably benefit from thinking divergently.

The research on DT has been extended by the fairly recent research on new computer models of creative cognition and the computer scoring of ideas. The hand scoring of divergent thinking tasks is an arduous process. It was used for decades and required the check of inter-rater reliability. The automated computer scoring of DT ideas using text-mining methods, such as *latent semantic analysis* (i.e., semantic-based algorithms; SBA), has gained traction. Semantic analysis involves measuring the semantic distance of words and phrases based on their latent similarity [52, 53]. To assess verbal creativity, Acar et al. [54] compared text-mining systems (the Open Creativity Scoring freeware platform [https://openscoring.du.edu/; [55]], TASA [Touchstone Applied Science Association; [56]], EN 100k [57], and GLoVe [Global Vectors for Word Representation 840B; [58]] to automatically score originality on the activities used in the verbal form of the Torrance Test of Creative

Thinking (i.e., Unusual Uses Test [UUT], Just Suppose Test[JST]). They found that using text-mining methods can quite reliably score originality for UUT and JST. The usage of semantic distance scoring to quickly and automatically score divergent thinking tasks shows promise for making scoring divergent thinking tests more efficient for educational purposes (e.g., gifted identification). Thus, schools could more easily identify students with creative potential in verbal tasks and align them with advanced programs to encourage their creative productivity (and future scholarly achievement in their domains of interest). For the present purposes what may be most important is that the computer research has supported the idea of DT and the conception of originality.

3.5 Discussion and Implications for Scholarship

This chapter points to creativity as required for good scholarship. Critical thinking is also discussed, but mostly we cited the creativity research. In fact, recall here that one issue in the research on creativity, at least early on, concerned the distinctiveness from general intelligence. General intelligence is an umbrella term that includes things like knowledge, logic, convergent and critical thinking. Also keep in mind that the view presented here is that good scholarship requires creative thinking but also things which can be separated from creative thinking–including critical thinking.

Some views of scholarship relegate creativity. Consider in this regard all of the work on citation impact (CI). This research uses algorithms that calculate indices of CI. These algorithms do not include any measure of creative talent. They are based on the degree to which an author or publication is cited by others. It may be that authors and publications which are creative tend to be cited by others, but certainly authors and publications are often cited for reasons unrelated to their creativity [59]. There is also reason to suspect that highly creative works will be misunderstood, disliked, and unrecognized. After all, creative works may not fit with conventional thinking. At the extreme they may suggest paradigm shifts [60] and paradigm shifts will usually threaten individuals who have a large investment in existing paradigms [61].

One new method has been developed to quantify scholarly creativity and replace or supplement CI. Forthmann and Runco [62] started with the idea that impact, like that estimated by CI, is distinct from the quality of scholarship (including originality and creativity). This distinction had been explored earlier [59]. Quoting Forthmann and Runco,

> Following extensive evidence from creativity research and theoretical deliberations, multiple indicators of openness and idea density are operationalized for bibliometric research. Correlations between impact and the various openness, idea density, and originality indicators were negligible in two large bibliometric datasets (creativity research: N = 1643; bibliometrics dataset: N = 2986) [62, p 1].

These negligible correlations supported the discriminant validity of the indicators of creativity that were operationalized by Forthmann and Runco [62]. They were also interested in the convergent validity of those same creativity indicators. These were significant and in line with previous bibliometric research but not large in the sense of effect sizes.

Forthmann and Runco [62] looked deeper with an exploratory graph analysis of "the nomological net of various operationalizations of openness and idea density." They examined several measures of variety because of its relationship to flexibility of thought. These included how many different journals were cited and how many different authors were cited. Disparity measures (also operationalized in terms of journals and authors cited) "were found to be strongly connected nodes in separate clusters. Both idea density indicators (based on abstracts or titles of scientific work) formed a separate cluster."

What is important in this work is that it (a) is exploratory, (b) offers tentative confirmation of the discriminant validity of the originality of published works, (c) offers tentative evidence of the convergent validity of published works, and (d) points to specific things in written works (e.g., the variety of journals and authors cited) that might be indicative of a scholar's original thinking. This last point might be considered a theme of this chapter: originality is vital for creativity, and the creativity of scholarship is not guaranteed by merely being informed, logical, and conventional (i.e., fitting in with what others are doing).

Other concrete suggestions were mentioned above, in addition to the scholar's writing style. One suggestion was that scholars take their time when working. This will increase the likelihood of finding remote *associates*, which are often highly original [63, 64]. Importantly, it is not just time on task that is important. In other words, although it is good to consciously concentrate on a problem, it is also useful to take time away from the task. Quite a bit of research suggests that the incubation that occurs when the individual is not concentrating on a task is very useful [see [40] for a review]. Quite a few suggestions follow from the model of divergent thinking. It is a good idea to have a large number of ideas, for example, and to vary one's perspective.

3.6 Concluding Remarks

Creative thinking may overlap with some kinds of intelligence and critical thinking. It is possible to operationalize each so there is minimal overlap, but in the natural environment, there is overlap, and that overlap is useful. Achievement in the natural environment requires both creative and critical thinking. It probably also requires motivation and drive [65] and, in many domains, expertise. The contributions of each of these things varies from domain to domain. Although research on domain specificity has not examined the distinctiveness of scholarship, that research concludes that creativity differs from domain to domain and, interestingly, that some domains complement other domains (e.g., mathematics and science, music and bodily-kinesthetic abilities). This brief overview of the creativity research, though focused on three issues (discriminant validity, domain specificity, and predictive validity), identified quite a few practical implications for scholars and scholarship.

References

1. Getzels JW, Jackson PW (1962) Creativity and intelligence: explorations with gifted students. Wiley, London
2. Wallach MA, Kogan N (1965) Modes of thinking in young children: a study of the creativity-intelligence distinction. Holt, Reinhart, & Winston
3. Albert RS, Runco MA (1987) The possible different personality dispositions of scientists and nonscientists. In: Jackson DN, Rushton JP (eds) Scientific excellence: origins and assessment. Sage, Beverly Hills, pp 67–97
4. Guilford JP (1968) Intelligence, creativity, and emotional implications. Robert R. Knapp, San Diego
5. Cropley A (2006) In praise of convergent thinking. Creat Res J 18:391–404. https://doi.org/10.1207/s15326934crj1803_13
6. Acar S, Runco MA (2019) Divergent thinking: new methods, recent research, and extended theory. Psychol Aesthet Creat Arts 13:153–158. https://doi.org/10.1037/aca0000231
7. Runco MA (1991) Divergent thinking. Ablex Publishing, New York
8. Runco MA, Smith WR (1992) Interpersonal and intrapersonal evaluations of creative ideas. Personal Individ Differ 13:295–302. https://doi.org/10.1016/0191-8869(92)90105-X
9. Baer J (2010) Is creativity domain specific? In: Kaufman JC, Sternberg RJ (eds) The Cambridge handbook of creativity. Cambridge University Press, Cambridge, pp 321–341. https://doi.org/10.1017/CBO9780511763205.021
10. Gardner H (1998) A multiplicity of intelligences. Sci Am 9:19–23
11. Kim KH, The APA (2011) 2009 Division 10 debate: are the Torrance tests of creative thinking still relevant in the 21st century? Psychol Aesthet Creat Arts 5:302–308. https://doi.org/10.1037/a0021917
12. Plucker JA (1998) Beware of simple conclusions: the case for content generality of creativity. Creat Res J 11:179–182. https://doi.org/10.1207/s15326934crj1102_8
13. Runco MA (1986) Divergent thinking and creative performance in gifted and nongifted children. Educ Psychol Meas 46:375–384. https://doi.org/10.1177/001316448604600211
14. Baer J (2015) Domain specificity of creativity. Elsevier Science, Amsterdam
15. Plucker JA, Beghetto RA (2004) Why creativity is domain general, why it looks domain specific, and why the distinction does not matter. In: Sternberg RJ, Grigorenko EL, Singer JL (eds) Creativity: from

potential to realization. American Psychological Association, Washington, DC, pp 153–167. https://doi.org/10.1037/10692-009
16. Gardner H (1983) Frames of mind: a theory of multiple intelligences. Basic Books, New York
17. Spearman C (1927) The abilities of man. Macmillan, New York
18. Warne RT, Burningham C (2019) Spearman's g found in 31 non-Western nations: strong evidence that g is a universal phenomenon. Psychol Bull 145:237–272. https://doi.org/10.1037/bul0000184
19. Davies M (2013) Critical thinking and the disciplines reconsidered. High Educ Res Dev 32:529–544. https://doi.org/10.1080/07294360.2012.697878
20. Moore T (2004) The critical thinking debate: how general are general thinking skills? High Educ Res Dev 23:3–18. https://doi.org/10.1080/0729436032000168469
21. Smith G (2002) Are there domain specific thinking skills? J Philos Educ 36:207–227. https://doi.org/10.1111/1467-9752.00270
22. Patrick C (1938) Scientific thought. J Psychol 5:55–83
23. Kattou M, Kontoyianni K, Pitta-Pantazi D, Christou C (2013) Connecting mathematical creativity to mathematical ability. ZDM 45:167–181. https://doi.org/10.1007/s11858-012-0467-1
24. Meier MA, Burgstaller JA, Benedek M, Vogel SE, Grabner RH (2021) Mathematical creativity in adults: its measurement and its relation to intelligence, mathematical competence and general creativity. J Intell 9:10. https://doi.org/10.3390/jintelligence9010010
25. Hu W, Adey P (2002) A scientific creativity test for secondary school students. Int J Sci Educ 24:389–403. https://doi.org/10.1080/09500690110098912
26. Feist GJ (1998) A meta-analysis of personality in scientific and artistic creativity. Pers Soc Psychol Rev 2:290–309. https://doi.org/10.1207/s15327957pspr0204_5
27. Simonton DK (2021) Scientific creativity: discovery and invention as combinatorial. Front Psychol 12:1–11. https://doi.org/10.3389/fpsyg.2021.721104
28. Weiping H, Shi QZ, Han Q, Wang X, Adey P (2010) Creative scientific problem finding and its developmental trend. Creat Res J 22(1):46–52. https://doi.org/10.1080/10400410903579551
29. Balka DS (1974) Creative ability in mathematics. Teach Child Math 21:633–636. https://eric.ed.gov/?id=EJ106476
30. Akgul S, Kahveci NG (2016) A study on the development of a mathematics creativity scale. Eurasian J Educ Res 62:57–76. https://doi.org/10.14689/ejer.2016.62.5
31. Jankowska DM, Karwowski M (2015) Measuring creative imagery abilities. Front Psychol 6:1591. https://doi.org/10.3389/fpsyg.2015.01591
32. Urban KK (2004) Assessing creativity: the test for creative thinking-drawing production (TCT-DP)—The concept, application, evaluation, and international studies. Psychol Sci 46:387–397
33. Amabile TM (1982) Social psychology of creativity: a consensual assessment technique. Pers Soc Psychol Rev 43:997–1013. https://doi.org/10.1037/0022-3514.43.5.997
34. Amabile T (1996) Creativity in context: update to the social psychology of creativity. Westview, Boulder
35. Denson CD, Buelin JK, Lammi MD, D'amico S (2015) Developing instrumentation for assessing engineering design. J Technol Educ 27:23–40. https://doi.org/10.21061/jte.v27i1.a.2
36. Root-Bernstein R, Root-Bernstein M (2004) Artistic scientists and scientific artists: the link between polymathy and creativity. In: Sternberg RJ, Grigorenko EL, Singer JL (eds) Creativity: from potential to realization. American Psychological Association, Washington, DC, pp 127–151. https://doi.org/10.1037/10692-008
37. Runco MA, Abdulla Alabbasi AM (2022) Creative activity and accomplishment as indicators of polymathy among gifted and nongifted students. Submitted for publication
38. Basadur M (1994) Managing the creative process in organizations. In: Runco MA (ed) Problem solving, problem finding, and creativity. Ablex, Westport, pp 237–268
39. Brophy DR (1998) Understanding, measuring and enhancing individual creative problem-solving efforts. Creat Res J 11:123–150
40. Dodds RA, Ward TB, Smith SM (2012) Incubation to problem solving and creativity. In: Runco MA (ed) Creativity research handbook, 3 vols. Hampton Press, New York, pp 251–284
41. Runco MA (ed) (1994) Problem finding, problem solving, and creativity. Ablex Publishing, Westport
42. Getzels JW (1975) Problem-finding and the inventiveness of solutions. J Creat Behav 9:12–18. https://doi.org/10.1002/j.2162-6057.1975.tb00552.x
43. Einstein A, Infeld L (1938) The evolution of physics. Simon & Schuster, New York
44. Root-Bernstein RS (1988) Discovering. Cambridge University Press, Cambridge
45. Adams JL (1984) Conceptual blockbusting. Norton, New York
46. Hudson L (1966) Contrary imaginations. Methuen, North Yorkshire
47. Runco MA, Millar G, Acar S, Cramond B (2010) Torrance tests of creative thinking as predictors of personal and public achievement: a fifty-year follow-up. Creat Res J 2010(22):361–368. https://doi.org/10.1080/10400419.2010.523393
48. Hunter ST, Bedell KE, Mumford MD (2007) Climate for creativity: a quantitative review. Creat Res J 19:69–90. https://doi.org/10.1080/10400410709336883
49. Ma H-H (2009) The effect size of variables associated with creativity: a meta-analysis. Creat Res J 21:30–42. https://doi.org/10.1080/10400410802633400
50. Acar S, Runco MA, Park H (2020) What should people be told when they take a divergent thinking test? A meta-analytic review of explicit instructions for

divergent thinking. Psychol Aesthet Creat Arts 14:39–49. https://doi.org/10.1037/aca0000256
51. Runco MA, Jaeger G (2012) The standard definition of creativity. Creat Res J 24:92–96
52. Acar S, Runco MA (2014) Assessing associative distance among ideas elicited by tests of divergent thinking. Creat Res J 26:229–238. https://doi.org/10.1080/10400419.2014.901095
53. Beketayev K, Runco MA (2016) Scoring divergent thinking tests by computer with a semantics-based algorithm. Eur J Psychol 12:210–220. https://doi.org/10.5964/ejop.v12i2.1127
54. Acar S, Berthiaume K, Grajzel K, Dumas D, Flemister CT, Organisciak P (2023) Applying automated originality scoring to the verbal form of Torrance Tests of Creative Thinking. Gift Child Q 67:3–17. https://doi.org/10.1177/00169862211061874
55. Dumas D, Organisciak P, Doherty M (2020) Measuring divergent thinking originality with human raters and textmining models: a psychometric comparison of methods. Psychol Aesthet Creat Arts 15:645–663. https://doi.org/10.1037/aca0000319
56. Landauer TK, Dumais ST (1997) A solution to Plato's problem: the latent semantic analysis theory of acquisition, induction, and representation of knowledge. Psychol Rev 104:211–240. https://doi.org/10.1037/0033-295X.104.2.211
57. Günther F, Dudschig C, Kaup B (2015) LSAfun-An R package for computations based on Latent Semantic Analysis. Behav Res Methods 47:930–944. https://doi.org/10.3758/s13428-014-0529-0
58. Pennington J, Socher R, Manning C (2014) Glove: global vectors for word representation. In: Moscitti A, Pang A, Daelemans B (eds) Proceedings of the 2014 conference on empirical methods in natural language processing. Association for Computational Linguistics, Stroudsburg, pp 1532–1543
59. Runco MA (1995) Insight for creativity, expression for impact. Creat Res J 8:377–390. https://doi.org/10.1207/s15326934crj0804_4
60. Kuhn TS (1962/2012) The structure of scientific revolutions, 4th edn. University of Chicago Press
61. Rubenson DL, Runco MA (1992) The psychoeconomic approach to creativity. New Ideas Psychol 10:131–147. https://doi.org/10.1016/0732-118X(92)90021-Q
62. Forthmann B, Runco MA (2020) An empirical test of the inter-relationships between various bibliometric creative scholarship indicators. MDPI 8:34. https://doi.org/10.3390/publications8020034
63. Mednick S (1962) The associative basis of the creative process. Psychol Rev 69:220–232. https://doi.org/10.1037/h0048850
64. Runco MA (1986) Flexibility and originality in children's divergent thinking. J Psychol 120:345–352
65. Runco MA, Chand I (1995) Cognition and creativity. Educ Psychol Rev 7:243–267. https://doi.org/10.1007/BF02213373

Part II

Non-clinical Research

Writing Research Protocols in Pharmacological Studies

4

Gowraganahalli Jagadeesh and Pitchai Balakumar

Abstract

The effects of drugs or chemicals are studied at every plausible level of complexity involving pure proteins, cells in tissue culture, intact isolated tissue, in situ perfused organs, and intact animals, including normal, surgically modified, and transgenic animals. Each of these models has its own protocols, methodologies, techniques, and data analyses. The research protocol is a roadmap for research that graduate students and early-career researchers can use to build track records and expertise in their fields. It is developed by integrating a multitude of ideas that emanate from the identification of a knowledge gap found while conducting a literature search. Next, it is processed into selection of a research topic of interest, followed by asking intriguing research questions, and the development of hypotheses and objectives. This chain of procedures described in the research protocol follows a well-organized plan of study that forms the basis of a clinical investigation or exploration of cellular mechanisms and molecular targets for potential new chemical entities. The protocol also describes the study background, rationale, and significance; design and methods; and data analysis using the right kinds of statistical tests, among others, within a clinical trial or an experimental study. It is designed to provide a satisfactory answer to the research question. In effect, the protocol is a written reminder of things to do when conducting the study. In this chapter, we describe a systematic and step-by-step approach to planning research protocols, including five examples at three levels of complexity in a hierarchical manner from cells to intact animals.

G. Jagadeesh (✉)
Retired Senior Expert Pharmacologist at the Office of Cardiology, Hematology, Endocrinology, and Nephrology, Center for Drug Evaluation and Research, US Food and Drug Administration, Silver Spring, MD, USA

Distinguished Visiting Professor at the College of Pharmaceutical Sciences, Dayananda Sagar University, Bengaluru, Karnataka, India

Visiting Professor at the College of Pharmacy, Adichunchanagiri University, BG Nagar, Karnataka, India

Visiting Professor at the College of Pharmaceutical Sciences, Manipal Academy of Higher Education (Deemed-to-be University), Manipal, Karnataka, India

Senior Consultant & Advisor, Auxochromofours Solutions Private Limited, Silver Spring, MD, USA
e-mail: GJagadeesh2000@gmail.com

P. Balakumar
Professor & Director, Research Training and Publications, The Office of Research and Development, Periyar Maniammai Institute of Science & Technology (Deemed to be University), Vallam, Tamil Nadu, India
e-mail: directorcr@pmu.edu; pbalakumar2022@gmail.com

G. Jagadeesh et al. (eds.), *The Quintessence of Basic and Clinical Research and Scientific Publishing*,
https://doi.org/10.1007/978-981-99-1284-1_4

Keywords

Research protocol · Research topic · Research question · Hypothesis · Objectives · Data analysis

4.1 Introduction

Before a researcher begins any kind of pharmacological experiment or clinical trial of a drug, a well-organized plan of the study, or the research protocol, is necessary. A protocol is a road map for research that evolves from the outline of the study plan, and eventually forms the basis of competent research and/or a grant proposal. It is the structural support for the research in the form of a thesis, dissertation, or publication [1]. The protocol explains what will be done in the study by detailing each component and how it is to be carried out [2].

The research protocol is procedural and contains a set of instructions and procedures, (i.e., methodology), to be implemented throughout the study. It will serve as a reminder of things to do [3, 4]. On the other hand, a research proposal is a formal and detailed statement of what and why one intends to conduct it. In addition, it presents and justifies a plan of action, shows how the researcher proposes to pursue the study, and sets the scope of the research [5]. The proposal is usually larger than a protocol with a budgetary component. Grant proposals are submitted to a granting agency to request funding for short- or long-term research [3] (*see* Chapters in Part X). A thesis (Chap. 48) and project may have several study protocols. Thus, both protocol and proposal are built on the same platform. Mentorship plays an important part in protocol and proposal development and in the guidance and support of early career researchers [6] (Chap. 55).

Protocols are fundamentally the researcher's plan for how to conduct a research study—be it an experiment or a clinical trial. A clinical trial investigator should deliberate on all the problems that could arise during the trial (*see* Chapters in Part IV). In both kinds of studies, well-defined research questions, hypotheses, objectives, and methods, including experimental design (Chap. 28), data analysis, statistics (*see* Chaps. 29 and 30), and limitations to the study should cover every possible setting or problem that might arise [2]. The law of biological variation makes the study more complicated than it seems on paper; therefore, it is essential for the researcher to extensively review publications on the selected research topic in order to have an in-depth knowledge of research methodology, study design, observations, and measurements; as well as how the data is collected, analyzed, tabulated, plotted or graphed, interpreted, and used in subsequent research (*see* Chaps. 37 and 38). The outcome of such studies, if planned and executed well, should define or establish the mechanism of action of drugs at cellular or molecular levels, or explore the pharmacology and therapeutic properties or benefits of a synthetic or natural compound for additional studies. In this chapter, we review the developmental process of planning for a study, illustrated with a few examples of how a study protocol is structured. This process is initiated immediately after the development of a study outline based on a review of literature and topic selection (*see* Chaps. 1 and 38 for details).

4.2 The Basis for Protocol Development

Research projects generally are comprised of four components: writing a protocol; performing experiments; tabulating, analyzing, graphing the data; interpreting (discussion); and writing a thesis and/or a manuscript for publication. Protocols lack both results and the interpretation of those results that lead to answering research questions. Thus, the protocol is the first step of scientific research planning and an essential planning tool for a study (Box 4.1).

Box 4.1 Important Considerations for a Protocol

It is the critical first step in identifying a process that will lead to answering the research question.

(continued)

Box 4.1 (continued)

It helps the candidate organize their research in a logical, focused, and efficient way.

Writing a research protocol takes a great deal of time, since it requires thoughtful consideration.

A systematic step-by-step approach is essential for planning the details of a study.

It helps the candidate receive feedback from their advisor(s) and senior colleagues.

It helps to ensure that a researcher's work is on course to be completed on time.

It is a written reminder of things to do.

Without a well-designed protocol, there can be no research or experiment.

Early-career researchers who wish to build a track record and expertise in their preferred fields need the supervision of a mentor in writing a protocol, proposal, and thesis. The mentor will provide insight, valuable guidance, and step-by-step instructions in both the anatomy and physiology components of the research. The guidance begins with the study outline, preparation of the draft, and finalizing the protocol. Writing a research protocol takes time, as it requires an extensive review of the literature, followed by brainstorming sessions and discussion with the mentor or advisor, statistician, and colleagues. Once the anatomy aspects are successfully completed, the physiology of the research begins, leading to the successful completion of the study and eventual publication of the work (Fig. 4.1).

4.2.1 Essentials of a Study Protocol

After identifying a research problem or topic, a strong review of literature will help in generating novel ideas. A novel idea will drive the chosen field ahead. Stronger ideas tend to have more robust starting positions. The idea should lead to the enumeration of research questions, followed by generation of hypothesis(es) and objectives (Box 4.2). These elements of a protocol should explain the study by answering the questions: why (question and background information), how (study design and rationale), who (target and study population), what (variables to be measured), and so what (significance of results and contributions to knowledge) [7]. The chances of writing a quality protocol depend on creativity, critical thinking, logic, the current state of

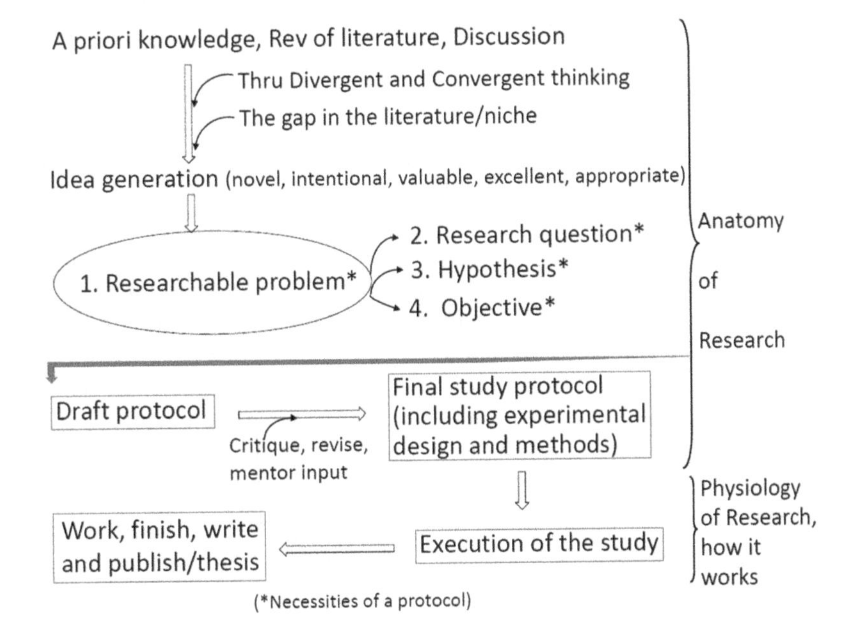

Fig. 4.1 Developmental process in planning for a protocol

knowledge relevant to the proposal, and the gaps that the project is intended to fill.

The literature review, one of the essential parts of research, identifies limitations of previous studies, unresolved or unanswered questions, poor methodology, improper data analysis, and weaknesses in understanding the phenomenon or theory that the authors could not explain. More importantly, it provides background material for the study and suggests variables, helps avoid duplication, and presents ideas, including knowledge gaps useful in designing the study [4]. These aspects should be exploited to establish the originality of the new research.

4.2.2 Research Topic

A clear understanding of the statement of a problem is the first and foremost step in building the protocol [7]. This is an important decision all researchers make before venturing into a study, as it requires careful thinking and planning. Also, a good or novel scientific idea based on a sound concept is essential in formulating a research topic. Choosing an appropriate research topic is a challenging and labor-intensive task and is based on reading published studies and several proposals, brainstorming with the mentor, and attending conferences and scientific meetings.

Creative thinking in choosing a research topic involves moving between divergent (considering many possibilities) and convergent (closes on a few and most feasible to study) thinking [8]. Critical thinking is important in selecting the chosen idea. Thus, critical thinking is convergent thinking, and is part of creative and analytical thinking [8]. It is also judgmental, as it incorporates reasoning and logical choices before making a final decision. A researcher needs to ensure that the topic is original to the field, manageable, and relevant to future areas of research they may wish to pursue. The well-researched topic has great potential to generate research questions and objectives [9]. Above all, a researcher should choose an important subject area that interests and inspires them, and for which they feel they can develop a "passion" [6].

4.2.3 Research Question

The second step in the anatomy of research is the search for a research question. It emerges from the title, observations, results, and problems observed in previous studies, and from the literature. Research problem or statement, theory or a phenomenon observed, and study designs take the form of a series of research questions. This step establishes a relationship between two or more variables and connects with established theory and research by way of thought-provoking questions. The research question sets the framework for additional literature searching and guides the research as the implications of the findings are compared with and draw attention to any limitations. Each section of the research must answer the researcher's chosen questions [10]. Furthermore, research questions should be precisely defined so that their answers will provide new knowledge when the study is completed [6].

4.2.4 Hypothesis

A question should be rendered or reframed to form a hypothesis. It can originate from the completed experiment or an experiment in progress, from published literature, or just the creative idea(s). A hypothesis is a specific testable statement of prediction. It is accepted or rejected by the study and its results. A formalized hypothesis contains at least one dependent, (e.g., measurement of heart rate, blood pressure), and one independent, (e.g., high-fat diet, the effect of a drug), variable and how these might be related. Studies that use statistics to compare groups of data, (e.g., vehicle- or placebo-treated groups vs. drug-treated groups), should have a hypothesis. A good hypothesis helps in the selection of an experimental design and drives the study to timely completion.

4.2.5 Objectives

The formulated research question should be developed into a hypothesis, followed by objectives. Objectives are statements highlighting what can be achieved by conducting the research. These could consist of a single main objective, or several, organized into primary, secondary, and exploratory objectives. They should be clear, precise, concise, and realistic so as to be achieved in a given time frame, and are developed logically from a description of the research problem based on knowledge gaps. The objectives are also linked to research questions, hypotheses, and study designs.

Box 4.2 Elements of a Research Protocol

1. Research topic
2. Research question
3. Hypothesis
4. Objectives
5. Research design and methods
6. Statistical analyses of data
7. Difficulties and limitations
8. Ethical and safety issues
9. Rationale and significance of the proposed research

4.2.6 Summary

The success of research largely depends on a protocol with clear processes or procedures derived from addressing the proposed questions. It relates the defined research questions and objectives to other relevant literature, and dissects the evaluating objectives into a sequence of steps that describe a method for investigating each objective, leading to a reasonable and convincing conclusion. Upon completion of the study, the research work should have made a worthwhile contribution to the field, proposing, supporting, or rejecting a new or improved concept, theory, or model. Also, it should contribute new data or information to existing knowledge, and offer a new or improved solution for analyzing the data, new methods of analysis, or new ways of interpreting the mechanism of action. Finally, the work should be publishable and gather citations.

4.3 Writing Study Protocols: Models and Examples

The expanding number of molecular targets for potential drugs has enlarged the arena for a growing list of pharmacologic studies involving a variety of techniques. Several kinds of experiments with variations within the following three major classes of studies are available in the literature. We describe in brief study protocols for in vitro, ex-vivo, and in vivo studies that are commonly used in pharmacology research to assess the actions of potential drug candidates. In these examples, the background information is restricted to a few paragraphs (although more could be written) on the class of investigative drug and the disease for which it is intended.

4.3.1 *In Vitro* Studies

4.3.1.1 Cell-Based Study

Topic

Activity and potency determination of test drug X in cell-based relaxin family peptide receptor 1 cAMP assay.

Introduction

Human relaxin (relaxin 2) is a peptide endogenous pregnancy hormone secreted predominantly by the corpus luteum. The circulating level of relaxin increases during the first trimester of pregnancy in humans wherein it inhibits uterine contraction and induces growth and softening of the cervix. Also, it induces favorable cardiovascular changes that support the hemodynamic and metabolic demands of gestation [11]. It promotes vasodilation, increases renal blood flow and glomerular filtration rate, and reduces both systemic vascular resistance and renal resistance leading to increased cardiac output and global arterial compliance [12]. Several studies have suggested relaxin is a potential treatment for heart failure as it reverses or attenuates cardiac hypertrophy, fibrosis, and inflammation [13, 14].

Many in vitro studies have demonstrated in-depth insights into the mechanisms and signal transduction pathways associated with relaxin and serelaxin (a recombinant form of human relaxin-2) cardiovascular and renal actions. Relaxin (relaxin2) binds and activates G protein-coupled relaxin family peptide receptor 1 (RXFP1) selectively with greater efficacy than RXFP2. The major signaling and extensively studied signaling mechanism for both RXFP1 and RXFP2 receptors is the ability of relaxin to rapidly increase intracellular levels of cAMP through $G_{\alpha s}$ stimulation of adenylyl cyclase. This has been demonstrated in several cell lines expressing either human, Cynomolgus monkey, rat, or mouse RXFP1 receptors [15]. RXFP1 is expressed in both endothelial and smooth muscle cells in a wide range of arteries and veins [16, 17], and the tubules of the kidney [18]. The binding of relaxin activates a variety of downstream signal transduction pathways besides adenylyl cyclase such as MAP kinases, nitric oxide signaling, and tyrosine kinase phosphorylation [19, 20].

Test substance X is a small molecule agonist for the RXFP1 and is being developed as a potential therapeutic agent for the treatment of a specific cardiovascular disease. It is unlikely to exert any effect on the other relaxin family peptide receptors.

Research Questions

Does the test substance X increase the production of cAMP by activating RXFP1? Is the test substance a full agonist and equipotent to relaxin in the human recombinant RXFP1 cAMP assay? Does the test substance stimulate cAMP production in a concentration-dependent manner? What is the comparative potency of the test substance relative to relaxin in cell lines expressing human, Cynomolgus monkey, or rat RXFP1?

Hypothesis

cAMP elevation is the major signaling pathway for both RXFP1 and RXFP2. Test substance X stimulates RXFP1 resulting in cAMP production in recombinant Chinese hamster ovary (CHO) cells transfected to express the human RXFP1. Like human relaxin, the test substance may exhibit a similar activity in humans, Cynomolgus, and rat RXFP1.

Objectives

Test substance X and reference compound relaxin are tested head-to-head in cAMP accumulation assays in cell lines stably expressing either human, Cynomolgus monkey, rat, or mouse RXFP1 in ovary cells. The primary objective is to assess the species selectivity and potency of test substance X relative to relaxin. The secondary objective is to assess selectivity within the RXFP family, both test substance X and relaxin are tested for their ability to modulate cAMP in cell lines stably expressing the human RXFP2, RXFP3, and RXFP4. Increased production of cAMP in CHO cells is used as the pharmacodynamic endpoint.

Methods

Chinese hamster ovary recombinant cells or cultured human vascular smooth muscle cells expressing either human RXFP1, Cynomolgus monkey RXFP1, rat RXFP1, or mouse RXFP1 receptors are employed. For assay details measuring cAMP production consult the references (luciferase reporter assay). Test substance X and relaxin are tested in duplicate at 12–14 different concentrations. To assess selectivity within the RXFP family, both compounds are tested for their ability to modulate cAMP in CHO cell lines expressing the human RXFP2, RXFP3, and RXFP4.

Data Analysis

Data from individual concentration-response curve replicates are analyzed and best fitted using GraphPad Prism software nonlinear regression sigmoidal dose-response model. Intrinsic activity (defined as the ratio of maximum cAMP production observed for the test substance divided by the observed maximum cAMP production for the reference ligand relaxin), EC50 (the concentration at which the response is one-half-maximal), relative EC50, and potency (defined as the sum of affinity and efficacy relative to reference ligand derived from EC50 values) are calculated from the concentration-response curves. Statistical analyses are performed on raw data using a suitable test, with statistical significance accepted at $p < 0.05$ level.

Projected Results

The results of the protocol should be aligned with the known apriori findings from previous datasets. Similar to relaxin, test article is expected to stimulate cAMP production in CHO-human RXFP1 cells. The native human relaxin is likely a more potent activator of RXFP1 than test substance. It is likely to activate humans, rats, and other species RXFP1 with similar or less but comparable potency to relaxin. Furthermore, test substance is unlikely to elicit any activity on RXFP3 and RXFP4.

Limitations of the Study

As noted in the introduction, in addition to stimulating cAMP accumulation, relaxin also initiates additional intracellular signaling actions such as cGMP accumulation and ERK phosphorylation. Test substance G protein coupling preference, full agonist activity, the kinetics of binding, tachyphylaxis, and saturation in response are not part of this study.

4.3.1.2 Studies in Isolated Tissues

Topic

Effect of test drug X on angiotensin II-induced contractile response in isolated rabbit thoracic aorta.

Introduction

Hypertension is a major risk factor for the development of cardiovascular, cerebral, and renal vascular abnormalities. High blood pressure is also prevalent in patients with diabetes mellitus. Its prevalence increases with advancing age. Various strategies are employed to achieve the target blood pressure. The treatment regimens often involve more than one class of drugs acting through different mechanisms with unique properties that could prove to be superior to conventional therapy. One of the pharmacologic approaches targeting both blood pressure control and related structural and functional improvements of the heart and blood vessels is to interrupt the renin-angiotensin-aldosterone axis [21].

The renin-angiotensin-aldosterone system (RAAS) is an important regulatory component in the maintenance of cardiovascular homeostasis and body fluid composition in normal and hypertensive subjects. Its principal peptide, angiotensin II, elicits several important pharmacologic actions such as vascular contraction and an increase in blood pressure, release of aldosterone from the adrenal cortex, enhancement of sympathetic outflow, and modulation of central effects such as drinking behavior [22]. Molecular biology and the development of angiotensin II receptor blockers (ARBs) have demonstrated the existence of a family of angiotensin II receptor subtypes. Of these, AT-1 and AT-2 receptors are the dominant receptor subtypes activated by angiotensin II. The AT-1 receptor mediates vascular contraction, the release of epinephrine in the adrenal medulla, and renal sodium reabsorption, which together cause blood pressure elevation. Thus, the antagonism of angiotensin II at the AT-1 receptor site constitutes a class of important antihypertensive agents [23].

The discovery of drugs targeting AT-1 receptors dates to saralasin, a peptidergic angiotensin receptor blocker (ARB). It was of limited use because of its short duration of action, injectable form, and partial agonist activity. The breakthrough came with the introduction of losartan, the first orally active non-peptide ARB in 1995. Soon after there was a flood of this class of drugs that subsequently lead to the approval of eight nonpeptidic AT-1 selective angiotensin II receptor antagonists [24].

Current research in a laboratory, for instance, has focused on the development of orally active nonpeptide angiotensin II receptor antagonists as new antihypertensive agents. The investigative substance is purported to antagonize the action of angiotensin II in experimental models.

Research Questions

Does the test substance selectively antagonize the contraction induced by angiotensin II in the rabbit aorta? Is the inhibition of angiotensin II-induced contraction of aortic smooth muscle dose-dependent? What is the nature of antagonism? Is it surmountable competitive antagonist without suppressing the maximum contraction? Does the test substance induce any agonistic effect in this tissue at a maximum concentration? Is the test substance more potent than losartan, a reference drug?

Hypotheses

It is hypothesized that test substance X selectively and competitively antagonizes the contraction induced by angiotensin II. It does not inhibit the contraction induced by other vasoconstrictor substances such as potassium chloride, norepinephrine, and serotonin on aortic tissue. It is not inferior to the reference drug, losartan.

Objective

The primary objective is to evaluate the inhibitory effect of test substance X on the contraction induced by angiotensin II and other vasoconstrictors in isolated rabbit aorta. The secondary objective is to compare the potency of test substance X with losartan.

Methods

Thoracic aorta is excised from male New Zealand White rabbits (2–2.5 kg) (guinea pig or rat aorta can also be used), cut in the size of 3–4 mm, and mount in an organ bath containing Krebs physiological solution. The solution is gassed with 95% oxygen/5% carbon dioxide and maintained at 37 ° C. After an hour of stabilization, angiotensin II, and various vasoconstrictor substances (potassium chloride, norepinephrine, and serotonin) are applied at a single or cumulative dose and after observing a constricting reaction, they are relaxed completely by repeatedly washing with Krebs solution 3–4 times. A maximum constricting reaction is expected by applying angiotensin II or various vasoconstrictor substances cumulatively. After establishing a control response to a vasoconstrictor, tissue is treated with test substance or losartan at two or three concentration levels for 30 min and then, challenged with a vasoconstrictor. The responses are recorded on a polygraph.

Data Analysis and Statistics

A successful concentration-response curve to an agonist reflects the sensitivity of tissue and measures its potency at the EC50 level (the concentration of a drug that induces a response halfway between the baseline and maximum). It should be noted that the curve is also defined by other three parameters, the baseline response, the maximum response, and the slope. A shift in the concentration-response curve to the right in the presence of test substance X or reference compound gives the potency of an antagonist. The data is analyzed by a statistical tool that fits the logEC50 which can be converted to the EC50. Results are expressed as group mean ± SD The relative potency of test substance X is determined based on EC50 derived from the concentration-response curve or pA2 value. The statistical significance of differences between mean data is evaluated using suitable statistical tests. The difference is taken to be significant when $P < 0.05$.

Projected Results

Isolated thoracic aorta has been used to define the selectivity and compare the potency of angiotensin II receptor antagonists at the AT-1 receptor site. The selectivity of ARBs is demonstrated by their inability to relax the contractile effects of KCl, norepinephrine, and serotonin. Additionally, AT-2 receptor antagonists such as PD123177, and CGP42112A do not antagonize the contractile effect of angiotensin II in the aorta. Based on the nature of the antagonism on the concentration-contractile response curve for angiotensin II on this tissue, the antagonists can be competitively surmountable (shifting the curve to the right with no depression of maximal response) or competitively insurmountable (concentration-response curve shifts to the right with the depression of maximal response as a result of a slow off-rate from AT-1 receptor binding sites) [25]. Besides two classes of antagonism, a few antagonists demonstrate inverse agonism, expressing different degrees of negative intrinsic efficacy. This property is studied mostly by biochemical and receptor binding assays (for details, see [25].

Limitations of the Study

The selectivity of a test substance for angiotensin II receptors can only be evaluated by receptor binding techniques using the AT-1 receptor-predominant bovine adrenal cortical or rat lung membrane and AT-2 receptor predominant bovine cerebellar membranes. The rate of association and dissociation of test drugs from AT-1 receptor sites can be demonstrated from radioligand binding studies [25, 26].

4.3.2 Ex Vivo Study

4.3.2.1 Perfusion Preparation

Title

Effect of test substance X on arginine vasopressin (AVP)-induced decrease in renal blood flow in the isolated perfused kidney preparation.

Introduction

In this example, we consider test article X purported to be a vasopressin V1 receptor antagonist. AVP is an endogenous hormone formed in the magnocellular neurons of the hypothalamus and released from the posterior pituitary in response to hypotension, decreased blood volume, decreased intravascular volume, and increased plasma osmolality [27]. Under normal physiological conditions, AVP has a limited role in the regulation of blood pressure, and a drop of >10% stimulates AVP release [28]. During hypotension and hypovolemia, AVP acts as a potent vasoconstrictor with little or no antidiuretic effect as its level surges [29]. The biological effects of AVP are predominantly mediated by three vasopressin receptors (V1, V2, V3) and at a slightly high concentration stimulate oxytocin (OT), and purinergic (P2) receptors with different tissue specificity and intracellular pathways [27, 30]. Pressor V1 (or V1a) receptors are mainly expressed on vascular smooth muscle, cardiomyocytes, platelets, liver, and CNS but are also expressed in high density in the renal medullary interstitial cells, vasa recta, extracellular loop 2, and epithelial cells of the collecting duct and specialized renal cells like glomerular mesangial cells or cells of the macula densa that control the release of renin [27, 31, 32]. The vasoconstriction is the result of increased intracellular calcium via the $G_{q/11}$-phospholipase C-coupled phosphatidyl-inositol-triphosphate and calcium signaling pathway [33].

The V1 receptors in the renal cortical and medullary vasculature facilitate vasoconstriction of renal blood vessels affecting overall renal blood flow [34]. This activity decreases renal medullary blood flow inducing pathological processes such as tissue hypoxia [31] contributing to chronic kidney disease (CKD) progression. It may be noted that CKD is common in people with diabetes and high blood pressure [35]. These findings suggest that drugs that selectively target the action of vasopressin at the V1 receptors of renal blood vessels are well suited for the treatment of CKD. In this context, test article X, presumed to be a vasopressin V1 receptor antagonist, is evaluated by addressing V1 receptor antagonism in an isolated perfused kidney model. The preparation is stable for a long period and has been used for several physiological and biochemical studies including drug disposition studies and effects on renal function given that the functional integrity of the entire nephron is preserved. Thus, it is used to investigate the glomerular filtration rate, the tubular reabsorption of water and sodium and their urine excretion, some aspects of the secretory function of the kidney (e.g., renin secretion), and the renal metabolism [36].

Research Questions

Does test article X selectively antagonize AVP-induced reduction in renal blood flow and venous volume? Does the inhibition of

AVP-induced vasoconstriction within the kidney result in dose-dependent increase in blood flow? Does test article X induce any agonist effect in this tissue at a maximum concentration?

Hypotheses

It is hypothesized that test article X selectively antagonizes V1 receptors in the renal vascular bed increasing overall renal blood flow and venous volume. The high selectivity for the V1 receptor is unlikely to inhibit other vasopressin receptor subtype (V2) in the renal bed and has low potential to cause unwanted off-target related side effects.

Objective

The primary objective is to evaluate the inhibitory effect of test article X on vasopressin-induced vasoconstriction measured by renal blood flow and venous volume.

Methods

The isolated perfused rat/mouse kidney is an appropriate model for the study of the effects of drugs and their modulators on the physiological and biochemical aspects of renal function. Here we describe the effect of test article X on vasopressin-induced decrease in renal blood flow and venous volume as a consequence of vasoconstriction.

Male Sprague Dawley/Wistar rats weighing about 300 g body weight and aged 12–15 weeks are anesthetized. Incisions are made along the midline to open the abdominal cavity. A perfusion cannula is inserted into the abdominal aorta and is advanced to the origin of the left renal artery and fixed in this position. The aorta proximal to the left renal artery is ligated, and perfusion (Krebs-Henseleit buffer) is started in situ. The urethra followed by the renal vein is cannulated, and the left kidney is excised, placed in a thermostatic moistening chamber, and perfused (typically through a peristaltic pump with a pressure sensing transducer placed between the pump and the tissue) at a constant pressure of 100 mmHg [37, 38]. The venous and urethral effluent is drained outside the chamber and collected for additional studies such as analysis of sodium, potassium, glucose, renin secretion/activity, and creatinine clearance [38]. After the stabilization of the system, AVP, and test drug dissolved in perfusate is infused into the arterial limb of the perfusion cycle. After continuous infusion of AVP establishes a reduction of the rate flow and venous volume, a single or cumulative concentration of test article (in the presence of a vasoconstrictor) is added to the dialysate every 4–5 min and the response is continuously recorded.

Data Analysis and Statistics

Data is expressed as group mean ± SD. At least two parameters are calculated, decrease/reduction in flow rate and reduction of venous volume by AVP. Dose-dependent inhibition of vasoconstrictor response on these two parameters by test article is analyzed. A minimal effective concentration (that is statistically significantly different from the control) and concentration-dependent inhibition of agonist-induced reduction in responses are calculated. The statistical significance of differences between mean data is analyzed by a suitable statistical tool. The difference is taken to be significant when $P < 0.05$.

Projected Results

Vasoconstrictors such as vasopressin, angiotensin II, and adenosine (A1 receptor activation) decrease renal venous volume and renal blood flow. A minimal effective concentration and dose-dependent inhibition of agonist-induced reduction in responses are calculated for the test substance.

Limitations to the Study

The perfusion circuit should be air and bubble-free anywhere in the system. Constant perfusion pressure should be established before perfusing with drugs and the drug concentration should be held constant wherever possible. The preparation is likely to develop edema in the animals at some time in the study. It is not a reliable preparation for studying the nature of the antagonism of test substance to vasopressin.

4.3.3 In Vivo Studies

4.3.3.1 Example 1: Evaluating the Antihypertrophic Cardiomyopathic Effect of Test Substance X

Topic

Acute and chronic effect of an orally active test article for anti-hypertrophic cardiomyopathic effect on systemic and left-ventricular hemodynamics, and cardiac performance in conscious chronically instrumented dogs. [It can also be studied in rats or mice.]

Introduction

Hypertrophic cardiomyopathy (HCM) is an obstructive disease with mechanical obstruction to the left ventricular outflow tract characterized by left ventricular hypertrophy, hypercontractility, reduced ventricular chamber size, and diastolic dysfunction or impaired diastolic relaxation [39, 40]. Histopathologically, it includes myocyte hypertrophy and disarray, microvascular remodeling, and fibrosis [39]. The clinical onset of the disease is the result of hyperactivity of the sarcomere, the functional unit of myocytes, which underlies systolic and diastolic dysfunction [41].

Given that HCM is a disease of the sarcomere, approximately 40% of the affected individuals have mutations in one or more genes encoding sarcomeric proteins [42]. Mutations increase myosin heads [43] and that is the basis of hypercontractility [44]. In vitro biochemical analyses suggest that disease-causing mutations in cardiac myosin increase force production relative to wild type by about 50% [45].

One of the targeted pharmacologic approaches to relieve hypercontraction and left ventricular outflow tract obstruction that could improve functional capacity and symptoms is by direct inhibition of cardiac myosin motors during the sarcomere contraction cycle. This is accomplished by inhibiting the transition of myosin from the off-actin state to the on-actin state and thereby promoting diastolic relaxation [42]. Mechanistically, the target drug should decrease the ability of 'myosin' to bind to 'actin' in the strongly bound state (i.e., stabilize the 'weakly-bound' or off-actin state of myosin), which translates to decreased contractile force of the myocardium. The reduction in cardiac contractile force is expected to alleviate LVOT obstruction in HCM patients by counteracting the excessive thickening of the left ventricular wall of the outflow tract since LVOT obstruction is a primary cause of morbidity in the majority of symptomatic HCM patients [46]. A reduction in sarcomere contraction producing a 10–20% reduction in left ventricular ejection fraction (reduction in fractional shortening of cardiac muscle) relative to baseline (control) might be beneficial to HCM patients.

Drugs that selectively target cardiac myosin and reversibly inhibit its binding to actin reduce the aggregate force (and thus power output) of systolic contraction and are predicted to facilitate diastolic relaxation (lusitropy or distensibility) and are predicted to improve dynamic LVOT obstruction in patients with HCM. The recently approved drug mavacamten, the first in the class, has been demonstrated to improve heart function by selectively inhibiting the enzymatic activity of cardiac myosin [47, 48].

The test article is a small molecule allosteric inhibitor of the cardiac sarcomere to oppose the increase in cardiac contractility.

Research Questions

Does decreased contractile force of the myocardium measured as a reduction in ventricular ejection fraction or reduction in fractional shortening of cardiac muscle promote diastolic relaxation? Is a decrease in cardiac contractility exposure dependent? Is there a correlation or dose-response relationship between increased inhibition of fractional shortening and increased relaxation of cardiac muscle leading to heart failure? What is the dose titration to exposure level (Cmax)?

Hypothesis

The primary hypothesis of the study is that cardiac myosin inhibition by investigative compound reduces cardiac hypercontractility, improves left ventricular diastolic filling, and in turn increases stroke volume. Test substance decreases cardiac function, which in turn would decrease cardiac output and arterial blood pressure. This would reduce cardiac contractility (reduction in fractional shortening or ejection fraction) an essential feature for combating hypercontractility prevailing in HCM. The arterial baroreflex is expected to stabilize arterial blood pressure with minimal changes in heart rate. The secondary hypothesis of the study is the functional inhibition (decrease in contractility) of test substance X is exposure (Cmax, AUC) dependent.

Objectives

The primary objective of the study is to evaluate the effect of oral doses (single or repeat doses) of test article X on systemic and left ventricular function correlating to a decrease in cardiac contractility measured as a decrease in ventricular ejection fraction, fractional shortening, and dP/dtmax (LV+dP/dt). The secondary objective is to correlate the blood levels of the drug with hemodynamic responses. The exploratory objective is to learn safety measures as these kinds of drugs are purported to have a narrow therapeutic index. In such a case, the test drug might cause heart failure as a result of excessive diastolic relaxation. What is the hard stop of test drug at LVEF <50? The additional exploratory objective is to find if the administered dose causes any adverse, unexpected effects.

Methods

The studies are performed on male Beagle dogs (approximate age 18–22 months weighing 22–30 kg) placed in individual cages and receiving a measured amount of food at a fixed interval (s) and water *ad libitum*. The animals are chronically instrumented (telemetered) under anesthesia for providing a single lead ECG and systemic (arterial) and left ventricular pressures, coronary blood flow, and LV diameter. Also, catheters are placed in the descending thoracic aorta for arterial blood sampling and blood pressure measurements and in the coronary sinus for myocardial venous blood sampling. Echocardiography is performed for EF and fractional shortening measurements (refer to published papers for methodology details, for example, [49]). Following surgery, all animals are allowed to recover for at least 14 days.

Study Design

Male dogs (n = 5 animals/dose level/treatment) are randomly assigned to receive a vehicle (control group), a single dose of test article, or a single dose of the reference drug. Both drugs are administered in a gelatin capsule or, are dissolved in a 0.9% saline vehicle or suspended in a suitable vehicle and administered orally by gavage for *n* number of days. For a single dose study, a crossover design (*see* Chap. 28) is recommended provided the terminal half-life of the drug is hours and not days.

Based on the results of in vitro studies and having known or established the selectivity and potency of test substance for the target site (inhibition of the enzymatic activity of myosin), it is advisable to begin the study with single ascending oral doses. The doses may be spaced on an arithmetic or linear scale (e.g., 1, 2, 3, 4, 5,... mg/kg) or a narrow increase between doses) rather than on a logarithmic scale (e.g., 1, 3, 10, 30,...mg/kg) given that a small reduction (10–20%) in contractility/ejection fraction is sufficient to achieve the target effect. A substantial reduction (>20%) would result in excessive relaxation of the cardiac muscle resulting in diastolic failure and animals might die from heart failure suggesting a narrow therapeutic index. The chronic oral doses (one or more dose levels on a narrow scale administered daily for *n* number of days) are selected based on the results of a single dose range-finding study. For systemic exposure, blood samples are collected via either the jugular or peripheral vein before dosing and at frequent intervals starting from 30 min to 24 h post-dose.

The following parameters are measured before dosing (at baseline) and at various intervals (e.g., 1, 3, 5, 8, and 24 h following the onset of treatment) and those should be described with the obtained data in the results section. For reference, time-matched cardiovascular data is collected in a

Table 4.1 Acute and chronic hemodynamic, cardiac and electrocardiographic effects of test article and reference drug in conscious healthy dogs at different intervals post-dose

Single or multiple oral dose study	Measurements			
	Hemodynamic	Cardiac performance	Electrocardiographic	Plasma conc, ng/ml
Treatment	MBP, PP, LVEDP, LVESP, dP/dt_{max}, dP/dt_{min}	EF/FS, SV, EDV, ESV, CO, LVd, Vmax	HR, PR, QRS, QT, QTcF	At specified intervals
Baseline (pretreatment)				
Day 1 (single dose), change in %				
Subsequent days (chronic), change in %				

Typical parameter measured pretreatment (baseline) followed by the duration of dosing
CO cardiac output (calculated), L/min), *dP/dt_{max} and dP/dt_{min}* peak rates of left ventricular pressure increase/decrease during systole and diastole (respectively) mmHg/s, *EDP and ESP* left ventricular end-diastolic and end-systolic pressures (respectively), mmHg, *EF* left ventricular ejection fraction, %, *FS* fractional shortening, %, *LVEDV and LVESV* left ventricular end-diastolic and end-systolic volumes (respectively), ml, *LVd* LV diameter, mm, *SV* left ventricular stroke-volume (calculated) ml, *MBP* mean arterial blood pressure mmHg, *PP* pulse pressure mmHg, *Vmax* estimated maximal velocity of contractile element shortening. ECG: *HR* heart rate, bpm, *PR, QRS, and QT* duration of the electrocardiographic intervals reflecting atrioventricular (PR) and ventricular conduction (QRS), as well as repolarization (QT), ms, *QTcF* rate-corrected (Fridericia) QT interval, ms

subset of untreated animals serving as control. Both peak and dose responses are evaluated at a steady state (Table 4.1).

Hemodynamic measurements: mean blood pressure, systolic blood pressure, diastolic blood pressure, left ventricular end-diastolic and end-systolic volumes, dP/dt_{max} and dP/dt_{min} (peak rates of left ventricular pressure increase / decrease during systole and diastole, respectively), aortic pressure, left ventricular diameter.

Electrocardiographic Measurements

Cardiac performance: Ejection fraction, fractional shortening, estimated maximal velocity of contractile element shortening (Vmax), cardiac output (calculated), stroke volume, LV volume, LV diameter.

Fractional shortening is calculated by this equation:

$$FS = 100 \times [(LVEDD - LVESD)/LVEDD]$$

Left ventricular end-diastolic dimension (LVEDD), end-systolic dimension (LVESD) and heart rate are measured by transthoracic echocardiography.

At the end of the chronic study, the heart is removed and analyzed histologically for myocyte hypertrophy.

Statistical Analysis

Descriptive statistics (mean and standard deviation) are used to summarize the measurement results (both absolute values and changes from baseline/pre-dosing values (as % change)). Measured values are expressed as percent change relative to baseline (set as 100%) or relative to reference drug if it is part of the study. For specific cardiac functional parameters, blood pressure, and SVR, comparative statistics are used. Data recorded before and after the administration of test article are compared using a suitable statistical tool. The responses are compared at time points using paired Student's t-test (*see* Chaps. 29 and 30).

Projected Results

Test article and reference drug are expected to decrease LV function characterized by decreased ejection fraction, fractional shortening, and dP/dtmax (LV+dP/dt) to varying extents. There could be a modest increase in heart rate with or without QT interval change. These effects might vary qualitatively and quantitatively with acute and chronic dosing. In addition to LV systolic function measurement, there could be an effect on arterial blood pressure- systolic, diastolic, and MAP. With the oral administration of drugs, the effect on the parameters will be evident after *n* number of days of treatment, peak, and stabilized after *n* number of days. The effect will be more gradual in onset.

Study Limitations

The question remains as to what extent the data are directly related to the target site action (e.g., myosin inhibition), off-target effects, and to what extent are to reflex effects. These questions can be addressed by repeating experiments in the presence of independent variables such as autonomic nervous system blockers and reversing the effect (reduction in contractility) by dobutamine. This helps in defining or narrowing down the mechanism of action of an investigative test article. This can be monitored by the plasma concentration of drugs.

4.3.3.2 Example 2: Evaluating the Antihypertensive Effect of a Test Article

There are several rodent and non-rodent models to study the antihypertensive effect of test drugs with a diverse range of mechanisms of action such as targeting the various stages in the renin-angiotensin-aldosterone axis, SNS, CNS, vasculature, oxidative redox signaling, kidneys, and calcium channels. Recently Jama et al. [50] have described several animal models to study hypertension, including genetic, diet, pharmacologic, and surgical models of hypertension. Here we describe the commonly used spontaneously hypertensive rat (SHR), a genetic model of hypertension that mostly responds to renin inhibitors, ACE inhibitors, and AT-1 receptor antagonists (see [24] for a review on this topic).

Title

Evaluation of the antihypertensive effect of test drug X in the SHR.

Introduction

The text should focus on the development of a new drug and its primary pharmacologic action based on previously conducted *in vitro* and *ex vivo* studies. The following protocol is developed on the assumption that a laboratory focuses on developing drugs that reduce the production of angiotensin II or prevent its binding to the target receptor, AT-1 (for details, see references [24, 51]). It is the assumption that test drug X is a putative AT-1 receptor antagonist. A short introduction to ARBs should precede outlining the objectives and hypothesis of the study. Also, if the investigator wishes to learn the potency and duration of action of test substance X in comparison to established or prototype drugs such as losartan, then comparative parameters should be included in the study design.

Research Questions

Does test article X statistically significantly lower blood pressure relative to baseline or vehicle control in SHR? Is reduction in blood pressure dose-dependent, and is the reduction sustained for *n* hours at ≥ED50? How does the magnitude and duration of the hypotensive effect of test drug X compare with the reference drug? Is there tolerance to blood pressure reduction over time? Is there a good fit between test substance's increased plasma concentration and blood pressure reduction (% change relative to baseline) to dose and time after administration of test article X?

Hypothesis

Test substance lowers blood pressure in SHR. If a comparator or reference drug is used in the study, then the blood pressure-lowering effect of test substance is as effective as equal doses of the reference drug (non-inferiority). For a multiple-dose study, plasma concentrations of test substance increase dose-dependently with a proportionate decrease in blood pressure. For a chronic study, the drug does not accumulate over time and there is no tolerance.

Objectives

The primary objective of the study is to evaluate the potency and duration of action of test substance X. In this study, measuring blood pressure and heart rate are two dependent variables. Additional objectives are specified based on the independent variables chosen. Pharmacologic intervention such as combination with other drugs is considered an independent variable with levels. Consequently, the study design is different from the variables chosen.

(a) To evaluate the relative potency (efficacy may not be the correct choice of word here since it is mainly used with agonists rather than with antagonists) of test substance X and losartan (reference drug), both in single and repeat dose studies.
(b) In a chronic study, the objective of the study is to evaluate the effectiveness and stable maintenance of blood pressure (degree and duration of decrease in blood pressure).
(c) To evaluate the additive effect of the combination of test substance X with diuretic hydrochlorothiazide, or a calcium channel blocker amlodipine.

The secondary objective is to study the peak hypotensive effect and duration of hypotension as a function of the plasma concentration of test article X.

The exploratory objective constitutes studying the adverse effects of test drug X on chronic administration (e.g., 3 weeks or more) on elevations in both plasma creatinine, urea, and potassium. It may be noted that hyperkalemia is one of the listed adverse effects of AT-1 receptor antagonists. It is usually offset when given in combination with a diuretic [24].

Methods

Both acute (single escalating dose of 1, 3, or 10 mg/kg) and chronic (1 or more mg/kg/day a week or more) effects of test drug are studied in conscious male SHR (21–25 weeks of age and 300–470 g body weight). Rats are housed in individual cages and receive food and water *ad libitum*. Rats are chronically instrumented to record blood pressure and heart rate at least a week before the start of the experiment. Animals are randomly assigned to receive the vehicle, test substance, reference drug, or combination (n = 6–7 animals/dose/treatment). Test substance and reference drug are dissolved or suspended in a suitable vehicle and administered orally by gavage. A control group is treated with the vehicle. Blood pressure and heart rate are recorded continuously or during and after cessation of drug treatment in conscious, freely moving rats. For determination of plasma concentration of test substance and reference drug, blood samples are collected either from the same group or from a satellite group at varying time intervals starting from 15/30 min to 24 h post-dose.

Study Design

It can be single escalating doses (e.g., 1, 3, or 10 mg/kg) or chronic/repeat dose study. For the latter study, animals will receive the vehicle, test substance, or reference drug, once daily, by oral gavage at least for 10 days. The additive effects of the drug can be studied by administering it with or without hydrochlorothiazide, or amlodipine (Table 4.2).

Observation and Measurements

For a single dose study, blood pressure and heart rate are continuously monitored for 24 h after administration. For repeat dose study, both parameters are monitored continuously for the first 24 h and at several time intervals daily and after cessation of drug treatment. Such recording is essential to locate the maximal response at X dose at Y h after administration and for ascertaining the duration of blood pressure reduction.

Data Analysis and Statistics

Treatment-induced effects on blood pressure (mmHg) and heart rate (beats per minute, bpm) are expressed as change from baseline (pretreatment) and relative to control at half-

Table 4.2 Study design for a single dose study[a]

Dose (mg/kg)	Mean arterial blood pressure			Heart rate	
	Δ mmHg rel to baseline[b]	% Change rel to control[b]	Hypotensive area, %.h[c]	Δ bpm	% Change rel to control
	0–24 h at 30 min interval	0–24 h at 30 min interval	24-h	0–24 h at 30 min interval	0–24 h at 30 min interval
0 vehicle	x ± SD		–	x ± SD	–
Test article					
1					
3					
10					
30					
Reference drug					
Combination: Test article and HCTZ or amlodipine					

HCTZ hydrochlorothiazide, *rel* relative
[a]A similar design is drawn for a repeat dose study of duration of a few days to a few weeks
[b]Change in MAP relative to its baseline is transformed from mmHg at half-hourly intervals into % change
[c]For an explanation, see under data analysis and statistics

hourly or hourly intervals. Data from each animal are expressed as the group mean ± SD (Table 4.2). Statistical analysis is performed on raw data determining 'minimum detectable difference' (MDD), a statistical indicator of the smallest effect size that can determine in an experiment. Also, the data should determine the dose-dependent effect, time to reach peak response, or maximum change in MAP (if delayed substantially relative to reference drug might indicate the test article is a pro compound and time is needed for the formation of the active metabolite(s)) and the duration of response (varies with the dose that may or may not increase the hypotensive response but prolongs the antihypertensive effect that could last beyond 24 h after dosing which could be reflected in plasma concentrations of test drug). Also, the 'duration' (sustainability of hypotension) and the 'magnitude' of the effect are calculated at each dose for each treatment. The raw data is transformed into % change relative to its baseline and control at each dose for each treatment.

The 24-h hypotensive area (% change of mean blood pressure from pre-dose versus time after administration at hourly intervals, expressed as %.h) is analyzed with the AUC by the trapezoidal rule at each dose group for each treatment. The hypotensive outcome is evaluated by applying an analysis of covariance for change from baseline for 24-h mean blood pressure. Any significant difference between vehicle and treatments (treatment combination or reference drug) is statistically analyzed by ANOVA, followed by Tukey's multiple comparison test. A P value of less than 0.05 is considered statistically significant. This measure determines the relative potency of treatment. If the plasma concentration of test substance X is one of the parameters of the investigation, then the magnitude of blood pressure reduction to plasma concentration and time after dosing is calculated and plotted. Furthermore, a PK/PD relationship is explored. The long PK half-life translates into a prolonged duration of the antihypertensive effect of test article which also can be known from radioligand binding kinetic studies [26].

4.4 Concluding Remarks

The planning of a study must be done step-by-step and in a systematic way. Identifying a research problem, selecting a topic, and writing a protocol are prerequisites in the scientific enterprise and are essential to the success of a study.

The research question and the study design are the two most important components of a study protocol that enables the study to begin. Formulated research questions should be developed into a hypothesis, followed by objectives. A clearly written purpose statement for the study should be included with the protocol. In this chapter, protocols were described at three levels of complexity, with five examples in a tiered manner from cells to intact animals, demonstrating that without a well-designed plan, no research project or experiment can proceed.

Acknowledgments We wish to thank Joanne Berger, FDA Library, for having critically read and edited the draft of the manuscript.

Conflict of Interest None.

Disclaimer This article reflects the views of GJ and should not be construed to represent US FDA's views or policies.

References

1. Jagadeesh G (2010) Writing a research protocol in pharmacology and toxicology. In: Jagadeesh G, Murthy S, Gupta YK, Prakash A (eds) Biomedical research, from ideation to publication. Wolters Kluwer, New Delhi, pp 51–71
2. Al-Jundi A, Sakka S (2016) Protocol writing in clinical research. J Clin Diagn Res 10(11):ZE10–ZE13. https://doi.org/10.7860/JCDR/2016/21426.9316
3. Bordage G, Dawson B (2003) Experimental study design and grant writing in eight steps and 28 questions. Med Educ 37:376–385
4. Moorhead JE, Rao PV, Anusavice KJ (1994) Guidelines for experimental studies. Dent Mater 10: 45–51
5. Nte AR, Awi DD (2006) Research proposal writing: breaking the myth. Niger J Med 15:373–381
6. Boyle EM (2020) Writing a good research grant proposal. Pediatr Child Health 30(2):52–56. https://doi.org/10.1016/j.paed.2019.11.003
7. Enarson DA, Kennedy SM, Miller DL (2004) Getting started in research: the research protocol. Int J Tuberc Lung Dis 8:1036–1040
8. Reisman FK (2010) Creative and critical thinking in biomedical research. In: Jagadeesh G, Murthy S, Gupta YK, Prakash A (eds) Biomedical research, from ideation to publication. Wolters Kluwer, New Delhi, pp 3–17
9. Balakumar P, Srikumar BN, Ramesh B, Jagadeesh G (2022) The critical phases of effective research planning, scientific writing, and co[illegible] Pharmacogn Mag 18:1–3
10. Bodemer N, Ruggeri A (2012) Finding [illegible] research question, in theory. Science 335:1439[illegible]
11. Du XJ, Bathgate RA, Samuel CS et al (2010) Ca[illegible] vascular effects of relaxin: from basic science to cl[illegible]cal therapy. Nat Rev Cardiol 7:48–58
12. Conrad KP (2011) Maternal vasodilation in pregnancy: the emerging role of relaxin. Am J Physiol Regul Integr Comp Physiol 301(2):R267–R275
13. Devarakonda T, Salloum FN (2018) Heart disease and relaxin: new actions for an old hormone. Trends Endocrinol Metab 29(5):338–348
14. Samuel CS, Royce SG, Hewitson TD et al (2017) Anti-fibrotic actions of relaxin. Br J Pharmacol 174: 962–976. https://doi.org/10.1111/bph.13529
15. Halls ML, Bathgate RA, Summers RJ (2006) Relaxin family peptide receptors RXFP1 and RXFP2 modulate cAMP signaling by distinct mechanisms. Mol Pharmacol 70(1):214–226
16. Novak J, Parry LJ, Matthews JE et al (2006) Evidence for local relaxin ligand-receptor expression and function in arteries. FASEB J 20:2352–2362
17. Jelinic M, Leo CH, Uiterweer P et al (2014) Localization of relaxin receptors in arteries and veins, and region-specific increases in compliance and bradykinin-mediated relaxation after in vivo serelaxin treatment. FASEB J 28:275–287. https://doi.org/10.1096/fj.13-233429
18. Bogzil AH, Ashton N (2009) Relaxin-induced changes in renal function and RXFP1 receptor expression in the female rat. Ann N Y Acad Sci 1160:313–316
19. Halls ML, Bathgate RA, Sutton SW et al (2015) International Union of Basic and Clinical Pharmacology. XCV. Recent advances in the understanding of the pharmacology and biological roles of relaxin family peptide receptors 1-4, the receptors for relaxin family peptides. Pharmacol Rev 67(2):389–440
20. Valkovic AL, Bathgate RA, Samuel CS, Kocan M (2019) Understanding relaxin signalling at the cellular level. Mol Cell Endocrinol 487:24–33
21. Balakumar P, Handa S, Alqahtani A et al (2022) Unraveling the differentially articulated axes of the century-old renin-angiotensin-aldosterone system: potential therapeutic implications. Cardiovasc Toxicol 22(3):246–253
22. Balakumar P, Jagadeesh G (2021) The renin-angiotensin-aldosterone system: a century-old diversified system with several therapeutic avenues. Pharmacol Res 174:105929
23. Balakumar P, Jagadeesh G (2014) Structural determinants for binding, activation, and functional selectivity of the angiotensin AT1 receptor. J Mol Endocrinol 53(2):R71–R92
24. Balakumar P, Jagadeesh G (2015) Drugs targeting RAAS in the treatment of hypertension and other cardiovascular diseases. In: Jagadeesh G, Balakumar P, Maung-U K (eds) Pathophysiology and pharmacotherapy of cardiovascular disease. Springer, pp 751–806. https://doi.org/10.1007/978-3-319-15961-4_36

(2021) The concept of ...action in drug discovery ... R (ed) Drug discovery ... 3. Springer, Singapore,

...C (1987) Different affinity states ...rgic receptors defined by agonists ... in bovine aorta plasma membranes. J ...xp Ther 243:430–436

...L, Landry DW, Granton JT (2003) Science ... vasopressin and the cardiovascular system part ...eceptor physiology. Crit Care 7:427–434

28. Bankir L (2001) Antidiuretic action of vasopressin: quantitative aspects and interaction between V1a and V2 receptor-mediated effects. Cardiovasc Res 51:372–390
29. Gordon AC, Mason AJ, Thirunavukkarasu N et al (2016) Effect of early vasopressin vs norepinephrine on kidney failure in patients with septic shock: The VANISH Randomized Clinical Trial. JAMA 316(5): 509–518
30. Koshimizu T, Nakamura K, Egashira N et al (2012) Vasopressin V1a and V1b receptors: from molecules to physiological systems. Physiol Rev 92:1813–1864
31. Bolignano D, Zoccali C (2010) Vasopressin beyond water: implications for renal diseases. Curr Opin Nephrol Hypertens 19:499–504
32. Higashiyama M et al (2001) Arginine vasopressin inhibits apoptosis of rat glomerular mesangial cells via V1a receptors. Life Sci 68:1485–1493
33. Holmes CL, Landry DW, Granton JT (2003) Science review: vasopressin and the cardiovascular system part 2 – clinical physiology. Crit Care 8:15–23
34. Nakanishi K, Mattson DL, Gross V et al (1995) Control of renal medullary blood flow by vasopressin V1 and V2 receptors. Am J Physiol 269(1 pt 2):R193–R200
35. Shu S, Wang Y, Zheng M et al (2019) Hypoxia and hypoxia-inducible factors in kidney injury and repair. Cell 8(3):207. https://doi.org/10.3390/cells8030207
36. Bekersky I (1983) The isolated perfused kidney as a pharmacological tool. TIPS 4:6–7
37. Schurek HJ (1980) Application of the isolated perfused rat kidney in nephrology. Contrib Nephrol 19: 176–190
38. Schweda F, Wagner C, Kraemer BK et al (2003) Preserved macula densa-dependent renin secretion in A1 adenosine receptor knockout mice. Am J Physiol Renal Physiol 284(4):F770–F777
39. Frey N, Luedde M, Katus HA (2012) Mechanisms of disease: hypertrophic cardiomyopathy. Nat Rev Cardiol 9(2):91–100
40. Marian AJ, Braunwald E (2017) Hypertrophic cardiomyopathy: genetics, pathogenesis, clinical manifestations, diagnosis, and therapy. Circ Res 121: 749–770
41. Ho CY (2011) New paradigms in hypertrophic cardiomyopathy: insights from genetics. Prog Pediatr Cardiol 31:93–98
42. Spudich JA (2019) Three perspectives on the molecular basis of hypercontractility caused by hypertrophic cardiomyopathy mutations. Pflügers Archiv 471:701–717. https://doi.org/10.1007/s00424-019-02259-2
43. Sarkar SS, Trivedi DV, Morck MM et al (2020) The hypertrophic cardiomyopathy mutations R403Q and R663H increase the number of myosin heads available to interact with actin. Sci Adv 6:eaax0069
44. Nag S, Trivedi DV, Sarkar SS et al (2017) The myosin mesa and the basis of hypercontractility caused by hypertrophic cardiomyopathy mutations. Nat Struct Mol 24(7):570–577. https://doi.org/10.1038/nsmb.3417
45. Sommese RF, Sung J, Nag S et al (2013) Molecular consequences of the R453C hypertrophic cardiomyopathy mutation on human beta-cardiac myosin motor function. Proc Natl Acad Sci U S A 110:12607–12612. https://doi.org/10.1073/pnas.1309493110
46. Gersh BJ, Maron BJ, Bonow RO et al (2011) 2011 ACCF/AHA guideline for the diagnosis and treatment of hypertrophic cardiomyopathy: a report of the American College of Cardiology Foundation/American Heart Association Task Force on Practice Guidelines. Circulation 124(24):e783–e831
47. FDA approval 2022. https://www.fda.gov/drugs/news-events-human-drugs/fda-approves-new-drug-improve-heart-function-adults-rare-heart-condition. Accessed 25 May 2022
48. Mullard A (2022) FDA approves first cardiac myosin inhibitor. Nat Rev Drug Discov 21:406
49. Hartman JC, del Rio CL, Reardon JE et al (2018) Intravenous infusion of the novel HNO donor BMS-986231 is associated with beneficial inotropic, lusitropic, and vasodilatory properties in 2 canine models of heart failure. JACC Basic Transl Sci 3: 625–638
50. Jama HA, Muralitharan RR, Xu C et al (2022) Rodent models of hypertension. Br J Pharmacol 179:918–937. https://doi.org/10.1111/bph.15650
51. Balakumar P, Jagadeesh G (2017) Editorial. Renin-angiotensin-aldosterone: an inclusive, an invigorative, an interactive and an interminable system. Pharmacol Res 125:1–3. https://doi.org/10.1016/j.phrs.2017.07.003

Basics of Designing General Toxicology Studies

5

Ravikumar Peri

Abstract

Drug development is a complex process and is tailored specifically to a given therapeutic indication, the target patient population, the therapeutic platform/s of choice (small molecules/biotherapeutics/gene therapy) and the route of administration. Non-clinical toxicology studies are critical to establishing the safety of new pharmaceutical candidates to enable first in human (FIH) clinical studies, to support the continued progression through phases of clinical development and finally for the registration with regulatory authorities to support a marketing application. Well designed and executed toxicology studies thus enable a smooth transition of new molecular entities through milestones of clinical development. This chapter aims to explain the general scientific principles and key considerations in the basic design of repeat-dose toxicology studies with a focus on small molecules and the exceptions applicable to study designs for biotherapeutics and gene therapies.

R. Peri (✉)
Pre-Clinical Toxicology, Alexion, AstraZeneca Rare Disease, Wilmington, DE, USA
e-mail: Ravi.peri@alexion.com

Keywords

Repeat-dose toxicity studies · Chronic toxicity studies · Small molecules · Biotherapeutics · Gene therapy · Regulatory guidance

5.1 Introduction

Drug discovery in the pharmaceutical industry is a rigorous, rational iterative process with integrated participation from various scientific disciplines. Safety assessment is an essential component of the marketing application of a new drug and requires the sponsors to conduct a comprehensive spectrum of non-clinical studies to assess the pharmacology, pharmacokinetic and toxicokinetic properties and toxicological effects, in relation to its exposure in humans in clinical use for the proposed therapeutic indications. Non-clinical toxicology assessments to establish human safety are thus an integral part of the regulatory requirements for the conduct of human clinical trials with new molecular entities [1, 2] as well as for the subsequent registrations after the completion of clinical trials. These include a combination of in-vitro and in-vivo assessments to demonstrate the efficacy of the intended therapy, elicit potential toxicities as well as demonstrate human safety. In-vivo studies in pertinent animal models (*see* Chap. 15) conducted in a controlled setting are critical for the assessment of toxicological effects of a drug.

G. Jagadeesh et al. (eds.), *The Quintessence of Basic and Clinical Research and Scientific Publishing*,
https://doi.org/10.1007/978-981-99-1284-1_5

These include single- and repeat-dose toxicity studies, reproduction and development studies (*see* Chap. 6), genotoxicity (mutagenicity) studies (*see* Chap. 7), and carcinogenicity studies (*see* Chap. 8) and are designed to elicit toxicological endpoints and guide human safety in clinical trials (*see* chapters in Part IV) [3–5]. They typically include endpoints such as histopathological assessments, carcinogenicity assessments, embryofetal malformations and developmental milestones that cannot readily be assessed in humans. These studies are conducted in healthy laboratory animals, with rats and mice being the predominant rodent species, and dogs, nonhuman primates, rabbits, and minipigs/pigs as the second large animal species. Dose-response relationships across a wide range of doses and exposures are established in these studies and are necessary to assess toxicokinetics, monitor adverse events/findings, potential translation to humans, and reversibility of those findings during a post-dose recovery period. When target organs of toxicity are identified in repeat-dose studies of up to 1 month duration, it is imperative to investigate if the severity of the finding exacerbates in sub-chronic and chronic toxicity studies of longer duration, and if the findings are recapitulated at lower doses [6].

Several considerations go into the design of various toxicology studies, but in general are governed by the modality/platform of the intended therapy (small molecules/biologics/gene therapies, etc.), the target chemical/product class, the stage of clinical development to be supported by the therapy, the pertinent regulatory guidelines, good laboratory practices, and institutional animal care and use committee (ACUC) guidelines.

A comprehensive description of the considerations that go into the design of all toxicology studies is beyond the scope of a single chapter, hence this article aims at describing the general scientific principles applicable to all studies. Applicable exceptions to study design with biotechnology derived products and gene therapy studies are highlighted in subsequent sections.

5.2 General Scientific Principles

Toxicological assessment starts very early in the discovery process. Initial assessments for new drug targets include the identification of pharmacological receptor engagement and potential consequences of agonism or antagonism utilizing a transgenic rodent model with either target overexpression or knock-out. For drug targets with established or approved therapeutics, potential toxicological liabilities can be obtained from published literature, and or documents submitted to regulatory agencies such as Summary Basis for Approval (SBA) submitted to US FDA or Human medicine European Public Assessment Report (EPAR) submitted to European Medicinal Agency (EMA).

For the results and toxicology study data to be accepted by regulatory authorities over the world, it is imperative that the test facility be accredited by American Association for Accreditation of Laboratory Animal Care (AAALAC) or International Laboratory Accreditation Cooperation (ILAC), and all pivotal toxicity studies are conducted under good laboratory practices (GLP) as specified in the Code of Federal Regulations Title 21, Part 58 (CFR Title 21 Part 58) [7] or in the Organization for Economic Co-operation and Development (OECD) [8], to ensure the quality and integrity of test data. These studies can be conducted either at appropriately qualified sponsor laboratories in-house or at well-established contract research organizations (CROs).

For small molecules, an assessment of genotoxic potential is conducted in accordance with guidelines specified in ICH S2 (R1). Mutagenicity assessments are conducted in-vitro in an AMES study in bacterial reverse mutation assays in the presence and absence of metabolic activation. Clastogenicity assessments investigate chromosomal damage and are conducted in-vitro in mammalian cells or mouse lymphoma thymidylate kinase assays, in the presence and absence of metabolic activation. In-vitro assays are usually followed up with in-vivo micronuclei assessments in rodent bioassays where chromosomal damage/aberration is evaluated in hematopoietic cells (*see* Chap. 7). Safety pharmacology assessments are conducted in compliance with guidelines specified in ICH S7A and ICH S7B and are described later in this chapter. The core battery of studies includes the assessment of effects on cardiovascular, central nervous, and respiratory systems, and should generally be conducted before human exposure. In accordance with guidelines specified in ICH S3A, in-vitro metabolic and plasma protein binding data for animals and humans and toxicokinetic assessments (e.g., absorption, distribution, metabolism and excretion) in the species selected for repeated-dose toxicity studies are evaluated prior to the conduct of repeat-dose toxicological studies. If exposures to metabolites in humans are expected to be greater than 10% of the parent drug-related exposure, or at significantly higher levels than the exposures observed in toxicity studies, additional toxicological assessments of metabolites may also be warranted (*see* Chap. 13).

Pivotal toxicology studies in support of an Investigational New Drug (IND), New Drug Application (NDA) or Biological Licensing Application (BLA) are typically conducted in two mammalian species [3], a rodent (e.g., Sprague Dawley or Han Wistar rats, CD-1 mice) and a non-rodent (Cynomolgus monkeys, Beagle dogs, New Zealand white rabbits, Yorkshire pigs, Yucatan or Gottingen minipigs). The choice of appropriate species is an important consideration in toxicology studies. Interspecies variability in the homology of the receptor target or expression patterns of the receptor target may attribute to differences in binding affinities of drugs, the coupled second messenger signal transduction pathways engaged and may contribute to differences in pharmacology as well as the exaggerated pharmacology and toxicology associated with the manipulation of the target. The differences in tissue distribution of the druggable target in humans and preclinical species may also result in differences in target organ toxicities. Hence, the selection of appropriate animal species for the conduct of toxicological studies is extremely important.

Acute toxicity assessments in two species at appropriate high doses either by the intended route of clinical administration or by parenteral administration are warranted to define the upper bounds of limits of toxicity to establish a maximum tolerated dose (MTD) and to establish potential inter-species and intra-species variabilities. However, one needs to be cognizant of the limitations posed by the high doses, the potential non-linearity of findings to lower doses and exposures, and distortion of physiological and pathological mechanisms as well as drug disposition at high doses due to saturation of the ADME processes. The acute toxicity studies may be followed up with appropriate exploratory dose-range finding toxicity studies designed to illustrate toxicity at the high dose and to establish clear No Observed Effect Levels (NOEL) or No Observed Adverse Effect Levels (NOAEL) to drive exposure multiples for human safety at projected efficacious exposures. These exploratory non-GLP toxicity studies with appropriate toxicokinetic endpoints help delineate the doses selected for pivotal repeat-dose GLP toxicology studies.

The requirements of repeat-dose toxicological studies are dictated by the therapeutic indication, the scope of clinical trial being supported, and the stage of clinical development. The general guidelines for the design of such studies are described by the regulatory authorities https://www.fda.gov/drugs/guidance-compliance-regulatory-information/guidances-drugs, https://www.ema.europa.eu/en/human-regulatory/research-development/scientific-guidelines/non-clinical/non-clinical-toxicology, and are in conformance with the guidance set forth by the international conference on harmonization https://www.ich.org/page/ich-guidelines. Though the specific elements that go into the design of

Table 5.1 ICH Guidance documents for toxicology studies

1	S1A-S1C—Carcinogenicity Studies
2	S2—Genotoxicity Studies
3	S3A-S3B—Toxicokinetics and Pharmacokinetics
4	S4—Toxicity Testing
5	S5—Reproductive Toxicology
6	S6—Biotechnology Product
7	S7A-S7B—Safety Pharmacology Studies
8	S8—Immunotoxicology Studies
9	S9—Nonclinical Evaluation for Anticancer Pharmaceuticals
10	S10—Photosafety Evaluation
11	S11—Non-clinical Biodistribution Considerations for Gene Therapy Products
12	M3—Nonclinical Safety Studies
13	M4—Common Technical Document
14	M6—Gene Therapy
15	M7—Mutagenic Impurities

individual studies are governed by specific guidance documents [9] listed in Table 5.1 below, the general guidelines and principles are described by ICH M3 (R2).

To support the first in human clinical studies, toxicological assessments in two different mammalian species (rodent and non-rodent) of appropriate duration need to be conducted [1]. However, for biotechnology derived products, depending on the lack of expression of the therapeutic target in one of the non-clinical species, studies in a single species may be deemed adequate [10]. The Route of administration of the test drug and the dosing duration and frequency in the non-clinical toxicology studies need to mimic the intended route of clinical administration in humans, and the frequency of administration in humans. Typically, non-clinical toxicology studies of up to 2-week duration are sufficient to support clinical dosing up to 2 weeks, toxicology studies of up to 6-months duration are sufficient to support clinical programs of up to 6 months duration, and chronic toxicity studies (6-months in rodents and 9-months in non-rodent) are required to support clinical studies of greater than 6-months duration. However, to support a marketing application the recommended duration of studies is 1-month in rodents and non-rodents for clinical studies of up to 2-weeks, 3-months in rodents and non-rodents for clinical studies of greater than 2-weeks up to 1-month duration, 6-months in rodents and non-rodents for clinical studies of greater than 1-month to 3-months duration, and chronic toxicity studies (6-months in rodents and 9-months in non-rodent) are recommended to support clinical studies of greater than 6-months duration [11, 12]. In some instances, for some biotechnology derived therapeutics, toxicology studies of up to 6-month duration in non-rodent species may be considered adequate.

5.3 Basic Design Elements of GLP Study Protocols

- **Objectives:** Well defined study title with clearly outlined objectives and scientific justifications for the selection of dose groups and controls chosen in the study. Key endpoints to be assessed in the study and their timing should also be clearly delineated.
- **Key personnel:** Key personnel responsible for various aspects of the study conduct and auditing and their contact details must be clearly outlined in the study protocol [13]. The Study Director has overall responsibility for the technical conduct of the study, as well as for the interpretation, analysis, documentation and reporting of results, and represents the single point of study control [14].

- **Animals:** Clear assignment of animals to each dose group and recovery arms (if applicable) of the study.
  - Typically for rodent studies of 1- to 3-month duration, 10–15 animals/sex/group are assigned to the main study and 5 animals/sex/group to the post-dose recovery period. Animals are assigned at an age of 7–8 weeks unless otherwise justified with a scientific rationale in the study protocol (e.g., for juvenile studies rodents as young as post-natal day 7–8 can be assigned, depending on the pediatric trial being supported). The inclusion of an appropriate post-dose recovery period is critical to monitor the recovery of any adverse findings observed during the in-life part of the study and at the end of the dose necropsy. The assignment of animals to the recovery phase may vary depending on the study and sponsor preferences, with some only assigning controls and high-dose groups to the recovery phase, while others assign post-dose recovery animals to all dose groups. Based on the study design, additional rodents (3–5 animals/sex/group) may be assigned to the study for toxicokinetic assessments at defined time points in the study. Depending on the study design if interim necropsy peel offs are included (e.g., 1-month peel off necropsy on a 3-month study) additional animals would have to be incorporated into each dose group. For chronic toxicity studies in rodents, typically 20 animals/sex/group may be assigned to the main study. Thus, it may be apparent that the requirement of animals in rodent toxicology studies can vary based on the individual study needs and the complexity of the design.
  - For non-rodent studies of 1- to 3-month duration, or for chronic toxicity studies typically 3–4 animals are assigned to the main study and 2 animals to the post-dose recovery period. Dogs are typically 6–9 months of age at study onset, and nonhuman primates (NHP) are ≥2 years old. The assignment of animals to post-dose recovery groups may vary between studies and sponsors. Unlike rodent studies, toxicokinetic assessments in large animals can be obtained from the blood/plasma samples collected from animals assigned to the main study and post-dose recovery phase.
  - 3Rs of animal testing: While it is critical to power the toxicology studies adequately to get meaningful interpretable results from them, scientists need to be cognizant and responsible in the use of animal models. A thorough consideration must be given to the principles of the 3Rs that include replacement, reduction and refinement [15]. While “replacement” may not be feasible in in-vivo toxicology studies emphasis can be given to “reduction” by optimizing the dose groups, using fewer animals per group, combining efficacy, toxicokinetic and toxicological assessments in a single study when feasible and limiting post-dose recovery to controls and high dose groups only, and “refinement” by modifying husbandry and experimental procedures to eliminate animal pain and distress, and improve their welfare. An example of a typical animal allocation table is illustrated in Table 5.2.
- **Test Article:** The test article or the active pharmaceutical ingredient (API) used in the toxicology studies must be representative of the batch used in clinical studies in terms of identity (batch, lot, form, etc.), strength, purity, stability, etc. The impurities present in the tox batch must also be representative of the clinical batch so that they can be qualified as per ICH Q3 and M7 specifications. The excipients or inactive ingredients used in formulations must be adequately qualified based on prior use in toxicology studies, or published literature. The safety profile of any novel excipient/s to be used in the formulation need to be assessed either in a stand-alone study or as a concurrent arm in the proposed toxicology study. The stability of the test

Table 5.2 Animal allocation for repeat dose toxicity studies[a]

Group number	Dose level[b] (mg/kg/day)	Dose conc.[c] (mg/ml)	Number of animals					
			Toxicity[d]		Recovery[e]		Toxicokinetics[f]	
			Male	Female	Male	Female	Male	Female
1. Control	0	0	10/3	10/3	5/2	5/2	3/0	3/0
2. Low	3	1	10/3	10/3	5/2	5/2	3/0	3/0
3. Mid	10	3.33	10/3	10/3	5/2	5/2	3/0	3/0
4. High	30	10	10/3	10/3	5/2	5/2	3/0	3/0

[a]This table represents an example of animal allocation to GLP repeat-dose toxicology studies of 4 weeks or longer duration. If interim necropsies are built into the study design, the dose-groups and number of animals assigned to the peel off phase differ based on the study objectives
[b]Doses in the study typically span 1–2 log units depending on the tolerability and the feasibility of formulate. Dosing frequency may vary and depends on the intended dosing frequency in the clinics. Doses mentioned in this table are just representative examples
[c]The concentration of the dosing solution has to be appropriately adjusted to account for the maximal dose volumes that can be administered by a given route of administration and vary by the species of choice
[d]For the main toxicity assessment, typically 10–15 rodents and 3–4 non-rodent species are assigned to the dose groups
[e]Animal allocation on the recovery arms can vary. Typically, 5 rodents and 2 non-rodents are assigned to the recovery phase. The inclusion of recovery groups also varies among studies. Some studies assign recovery animals to all dose groups, while others may prefer assigning them only to control and high-dose groups, or control, mid-dose and high-dose groups
[f]Allocation of animals for toxicokinetic investigations is optional. For rat studies, if sufficient blood/plasma samples can be collected from the main study animals without interfering with other study endpoints, composite TK samples are collected and separate TK animals are not assigned. Typically, $n \geq 3$/sex/time point are assigned to rodent studies depending on the toxicokinetic design, the number of TK days and samples collected on each day. A composite TK profile is established with any given animal not bled more than 3–4 times over a period of 24 h. Typically, study endpoints other than survival, body weight and food consumption are not collected from the animals assigned to the TK group. For non-rodent species, separate animals are not assigned to the TK group

article in the formulation under the conditions of storage also needs to be established either ahead of the study conduct or as an assessment of an ongoing GLP study. The assessment of homogeneity and concentration of dosing solutions under the conditions of storage and use constitutes an integral part of a toxicity study and needs to be conducted using validated analytical methods. Dose volumes and maximal limits established within the testing organizations are dependent on the test species and route of administration. Hence, test article concentration and the maximal volume that can be administered play a crucial role in deciding the maximal dose of the test article that can be administered.

- **Dose Selection:** Dose selection is given careful consideration in study design and thoroughly whetted out prior to protocol initiation [16–18]. Toxicokinetic and toxicology data generated from shorter duration 5- to 10-day dose range finding exploratory toxicology studies provide valuable information in dose selection for FIH enabling GLP toxicology studies. In the absence of significant toxicities in the exploratory studies, the top dose selected for the GLP studies may be the limit dose of 1000–2000 mg/kg or alternatively the maximal feasible dose. However, if acute toxicities are observed in the range finding studies, or a maximal tolerated dose is established, that is taken into consideration for defining the top dose in the GLP studies to elicit target organ toxicities without significantly compromising the health of the animal or confounding the interpretation of results [16, 17]. The low dose is chosen with the intent to provide exposures equivalent to or greater than the projected human efficacious exposures and is expected to be the NOAEL dose. The mid dose is selected to elicit an intermediate tox profile and is expected to provide 3 to 10-fold higher exposures relative to the NOAEL (*see* Chap. 13).

- **Observations and Parameters Evaluated:** These vary depending on the study design, but commonly evaluated parameters include:
  - Toxicokinetics and metabolite identification/monitoring (mainly for small molecules): It is very important to establish a dose exposure relationship for toxicities, hence toxicokinetics constitutes an integral part of most study designs. Blood samples can be collected at defined intervals post-dose and processed to an appropriate matrix (whole blood/plasma/serum) for bioanalytical verification of drug concentrations using validated methods and subsequent toxicokinetic analysis. Samples are collected early in the study, typically on Day 1 of dosing and towards the end of the study prior to necropsy to ensure the consistency of exposures. In studies of longer duration such as chronic toxicity studies, toxicokinetic analysis can be built in the interim phases as well. For small molecules drug accumulation can happen depending on the saturation of elimination mechanisms or distribution to non-vascular compartments and conversely a decrease in exposures can be observed with repeat-dosing if there is induction of metabolizing enzymes. For biotherapeutics, the formation of neutralizing antidrug antibodies can lead to exposure loss, thereby limiting the scope of long-term toxicity studies. For small molecules, if the metabolic pathways are well characterized and metabolites are likely to be produced in non-clinical species at a concentration >10% it is imperative that the concentrations of metabolites be measured and a toxicokinetic profile of the major metabolites be monitored (*see* Chap. 13).
  - Food Consumption, Body Weights and Body Weight Gains: The food consumption and body weight assessments begin during the pre-study phase and are conducted routinely throughout the dosing phase as well as post-dose recovery period at specified frequency (typically on a weekly interval) and help monitor the effect of the test article on the growth of the animals relative to concurrent control animals assigned to the study and historical controls to provide additional perspective if needed.
  - Clinical Observations: Animals are observed cage side for general behavior and health daily or as per the intervals specified in the study protocol and any unusual observations are brought to the attention of the veterinary staff for follow up. The observations in addition to the general health of the animal must include endpoints such as clinical signs of toxicity, the onset, frequency, duration, and potential reversibility of findings, and moribundity and mortality.
  - Physical exams: These are conducted by trained veterinary staff during the pre-study phase prior to the administration of test articles to ensure the selection of healthy animals for the study and at appropriate intervals during the in-life part of the study to monitor health. The physical examinations include neurological and respiratory evaluations. Ophthalmologic examinations are also routinely conducted in toxicology studies. Additional physical examinations of animals exhibiting overt toxicity are conducted at the discretion of the staff veterinarian.
  - Safety Pharmacology Parameters: Unless the knowledge of the drug target or the chemical class of molecules dictates dedicated stand-alone safety pharmacology studies, these endpoints are built into the repeat-dose toxicology studies [19] typically prior to dose initiation and prior to the terminal necropsies and include respiratory and CNS assessments in rodent studies and respiratory, CNS and cardiovascular assessments in large animals. Additionally, in accordance with the guidance (ICH S7B), especially for small molecules, dedicated cardiovascular safety pharmacology studies are conducted in telemetered animals (dogs or monkeys) to assess the

impact of the test article on hemodynamic parameters and ECG intervals, and to assess any proarrhythmic liability.
  - Clinical Pathology: Clinical pathology encompasses a cumulative assessment of hematology, serum chemistry, urinalysis and coagulation parameters. These are very tightly regulated in the body and any perturbations in these may point to specific target organ toxicity. In rodents, clinical pathology assessments are conducted prior to the onset of dosing and once prior to the terminal necropsy. In large animals the assessments are typically conducted twice prior to dosing and once prior to terminal necropsy. Assessments at additional time points can be built into the study as deemed necessary. Additionally, in some studies, the serum, urine and plasma matrices can be collected for investigating biomarkers of efficacy as well as toxicity.
  - Necropsy and anatomic pathology: Complete necropsy is performed on all animals at the end of dose or post-dose period, and those that are either found dead or euthanized in moribund conditions. A comprehensive list of organs (specified in the protocol) is collected, weighed and stored in an appropriate fixative for microscopic analysis [20]. A comprehensive list of tissues to be collected in repeat-dose studies was published by Bregman et al. [20]. For microscopic analysis some sponsors prefer to analyze the top dose and the control groups and based on the target organs identified specific analysis of those tissues at the mid and low dose groups is conducted. Alternately some sponsors prefer to analyze all tissues at all dose levels. Pathologists can apply special stains to investigate the target organs or perform more detailed electron microscopic analysis as deemed appropriate, in consultation with the study director and appropriate protocol amendments.
- **Analysis and Interpretation of Results:** Analysis of results is the most critical aspect of a toxicology study and must be performed by qualified scientists using specified standard operating procedures and applying appropriate statistical methods where applicable. Pertinent staff including the study director, study monitor (where applicable), toxicologist, analytical and bioanalytical scientists, TK scientist, clinical and anatomic pathologist and other specialists should formulate and review results from their respective sections and document the key findings. Additionally, during an integrated review of the study data, the results must be evaluated in totality and an integrated risk assessment be formulated.
- **Study Report:** All the aspects of a toxicology study in its entirety must be capitulated in an integrated study report. While specialist scientists may interpret results from various parts of the study and author the sub-sections of the report, the Study Director needs to assimilate components of each of the sub-reports into a main toxicology report, write a concise and accurate study summary, and sign the final report prior to official dissemination and archival.

5.4 Key Considerations for Toxicological Assessments of Biotechnology Derived Pharmaceuticals

Biotechnology derived products is a broad term that applies to products derived from a variety of expression systems including bacteria, yeast, insect, plant, and mammalian cells intended for diagnostic, therapeutic, or prophylactic uses. The active pharmaceutical ingredients or diagnostic agents include proteins and peptides, and their derivatives, often produced by cell cultures or by recombinant DNA technology. Some examples include growth factors, enzymes, hormones, recombinant plasma factors (involved in coagulation), cytokines, monoclonal antibodies, etc. Even though the basic tenets of toxicology testing remain similar to those described earlier in this chapter and are governed by ICH M3 (R2) [1], the main guidance driving the nonclinical studies is ICH S6 (R1) [21]. There are several exceptions including the duration of chronic toxicity studies [10, 22] that will be elaborated on in this section. These include:

- Test material specifications: Often the test articles may be produced by an established validated manufacturing process in the initial stages of drug development but may undergo changes (such as the cell lines or reagents used in the manufacturing, processing and purification, formulations etc.) during clinical development. In such instances, it is essential to establish the bioequivalence of the newer product using validated bioanalytical assays and if necessary, further pharmacological and toxicological characterization. Biotechnology derived products are associated with unique impurity profiles that may differ significantly from small molecules. These arise from contaminating host cell proteins (HCP) and can lead to allergic reactions, immunopathological effects and additional risk of infections, and hence need to be tightly controlled during manufacture.
- Biological activity/pharmacodynamics: Biological activity can be evaluated in appropriate in-vitro cell culture assays in cell lines expressing the recombinant species-specific receptors and may be predictive of activity in the preclinical species or humans. Such in-vitro studies help determine receptor occupancy, receptor affinity, and/or pharmacological effects, and in selecting an appropriate animal species for in vivo pharmacology and toxicology studies. For monoclonal antibodies, the immunological properties including antigenic specificity, complement binding, cytokine release, and tissue cross-reactivity should be defined in detail. Tissue cross-reactivity studies should be conducted in a range of animal and human tissues.
- Species selection: Often biotechnology derived products are designed for human use and may not cross-react with a similar specificity and affinity with epitopes of receptors expressed in all animal models. Hence, it is important to conduct tests to identify the relevant animal species for evaluating pharmacological activity as well as for conducting toxicological studies. Though toxicology studies are normally conducted in two animal species, in the absence of two relevant species, it may suffice to conduct the studies in a single species [8]. For example, for many monoclonal antibodies that do not cross react with rodent receptors, it suffices to conduct the studies in a single relevant non-rodent species such as non-human primates. In the absence of a relevant non-clinical species, studies may be conducted using surrogate homologous ligands in appropriate animal models or transgenic animals expressing the relevant human receptors. However, the limitations of such studies should be recognized, and they may not always be predictive of clinical efficacy or safety.
- Dosing (route of administration and frequency): Most biotechnology derived products are administered clinically by a parenteral route and non-clinical studies should mimic this taking into consideration the pharmacokinetics/toxicokinetics and the bioavailability of the product in the species being used. Local tolerance testing is an essential endpoint evaluated in most toxicity studies. If a given formulation and dose volumes pose a challenge, an alternative parenteral route may be acceptable if it provides adequate exposures and bioavailability. Dose selection follows the criterion as described earlier in this chapter and the intent of the toxicology studies is to find a toxic dose, a NOAEL, and an intermediate dose with an appropriate dose-response relationship. There may be instances when the maximum feasible dose may not elicit toxicities. In such instances, an attempt must be made to demonstrate a pharmacodynamic response to demonstrate target engagement and a scientific rationale and an exposure multiple to the proposed clinically efficacious exposure must be provided.

- Immunogenicity, Pharmacokinetic and toxicokinetic effects: Biotechnology derived products such as monoclonal antibodies tend to have a long half-life in plasma. This should be taken into consideration in designing an appropriate dosing regimen both in preclinical toxicology studies as well as human clinical trials. Unlike the small molecules that tend to have a daily dosing regimen, these products may be administered at a reduced frequency of once a week or once every 2 weeks depending on the bioavailability and plasma clearance. Many biotechnology-derived pharmaceuticals intended for humans are immunogenic in animals. Assessment of innate immune responses (cytokine release, hematologic responses, etc.), as well as antidrug antibodies are key elements in these studies. Antibody responses, the onset of response, titer, the number of responding animals, and neutralizing or non-neutralizing, should be well characterized and documented. The effect of ADA on PK/PD correlation, impact on toxicokinetics, adverse events, and the emergence of toxicological events should be taken into consideration while interpreting the results. However, the induction of antibody formation in animals to a foreign human protein may not always be predictive of the potential formation of antibodies in humans or the translatability of animal toxicities to humans. Though humans may develop antibodies against humanized proteins, it may or may not significantly impact the therapeutic responses. However, ADA-mediated loss in exposure in toxicology studies may limit the scope of the toxicology program, as consistent exposures may not be maintained over the duration of the study and may also preclude the conduct of meaningful long-term assessments including the chronic toxicology studies, reproductive toxicology (especially in non-human primates with long gestation period) and juvenile toxicity studies.
- Exceptions to toxicity testing: Biologics, or biotechnology-derived therapeutics, are not typically expected to interact directly with DNA or other chromosomal material, so they do not require genotoxicity assessments which is critical to small molecules. However, genotoxicity assessments may be necessary for certain products, such as antibody-drug conjugates, where the biotherapeutic is linked to a small molecule through an organic linker. Carcinogenicity assessments are also generally not applicable to biologics. Instead, the potential for carcinogenicity should be assessed based on the mechanism of action, biological activity, target patient population, and frequency and duration of dosing. In some instances, carcinogenicity studies may be mandated, but the relevance of the target in the rodent species used in the carcinogenicity studies needs to be given careful consideration.

5.5 Key Considerations for Toxicological Assessments for Gene Therapy

Over the past two decades, significant strides have been made in developing gene therapies for rare genetic disorders including the approval of Luxturna for the treatment of patients with confirmed biallelic RPE65 mutation-associated retinal dystrophy that leads to vision loss and may cause complete blindness in certain patients, and Hemegenix for treatment of the genetic bleeding disorder Hemophilia B. Significant gene-therapy research programs are actively being pursued by pharmaceutical companies for several rare genetic disorders. Even though, these projects present unique opportunities for patients, toxicological assessment of the safety of these therapies poses unique challenges as well [23] This section summarizes some of those challenges [24].

- Unlike small molecules or biotherapeutics such as monoclonal antibodies that are administered for long periods of time and need repeat-dose chronic toxicity studies to support them, gene therapy involves a single administration of the construct of interest and a long term follow up [25].

- Unlike the exposure multiples of 50× or higher that are targeted at the high doses for small molecules, and 10× for monoclonal antibodies, the multiples with viral vector gene therapies are at or slightly above the clinically efficacious exposure. Several factors including the target population, transgene product potency, vector tropism, promoter strength and formulation limits play an important role in determining the high dose. Biodistribution endpoints that are optional for small molecules and biotherapeutics are critical for toxicological assessment of gene therapies.
- Regulatory requirements: Regulatory requirements for gene therapies across the globe are more flexible and are evolving. The demand for rapidly delivering life-transforming gene therapies to patients challenges the extent of non-clinical characterization and toxicity assessments that can be completed expeditiously. Sponsors may typically include assessments in single species or single sex for a given species when appropriate and may reduce the number of animals and or dose groups in studies to condense the timelines, but the true impact of such modifications to study design and to patient safety is gradually evolving.
- Animal species: The animal species chosen for toxicology studies are based on their susceptibility to infection with the viral vector, the efficiency of transduction, the pharmacological response to the transgene, similarity in physiology and anatomy to humans, the ability to tolerate the gene therapy product, and the feasibility of the delivery method. The choice of single species is justified when the pharmacological target organs are transduced, vector safety is known, there is optimal human transgene expression and a predicted response, and the underlying disease state being treated can be capitulated in that model. The use of sexually mature animals in studies to assess reproductive toxicity and germline transmission needs careful consideration. Many of the rare genetic disorders treat pediatric patients, and hence appropriate inclusion of juvenile animals in toxicology studies may be required. Rodent models when pertinent may include the standard strains of rats and mice for toxicological assessments or transgenic or knockout models for efficacy assessments. Non-rodent models are often screened for pre-existing neutralizing and or binding antibodies to viral capsid as exclusion criteria to select animals for the study. The practice of concomitant administration of immunosuppression either intermittently or for the entire duration of dosing to prevent infusion reactions or transient immune responses, and the choice of a such treatment regimen, especially in non-human primate studies vary across the industry.
- Study design: In some instances, the inclusion of appropriate toxicity and biodistribution endpoints in the efficacy studies enables them to be substituted for standalone toxicity studies. The default duration of FIH-enabling study is typically 3–6 months. Since most assessments are performed after the administration of a single dose of gene therapy, it is common for sponsors to build in appropriate interim necropsy intervals to assess biomarkers, biodistribution and toxicity endpoints at the end of 1-, 3- and 6-month intervals post dose, as appropriate. It is typical to include 2–3 dose groups in the study and include 5–20 animals/sex/group for rodent studies and 3–6 animals/sex/group for non-rodent studies. The number of animals enrolled in the study needs to be appropriately modified if interim necropsies are built into the study. Biodistribution assessments are conducted either in exploratory standalone studies, or are included in the pharmacology/toxicology studies, and can be conducted either on a limited subset of tissues or an exhaustive tissue list [26, 27]. Most assessed analytes in these studies measure transgene DNA and or mRNA, using validated quantitative polymerase chain reactions (qPCR) or reverse transcriptase (RT)-PCR methods, and limited immunohistochemical analysis is conducted in targeted tissues to measure the expression of a protein of interest [27].

- Known toxicities with Gene Therapy: Liver is the main tissue target of systemically administered gene therapies and hepatotoxicity is the most common side effect observed non-clinically and in the clinic. In some instances, this may be driven by cytotoxic T lymphocyte responses to the transduced liver cells. The clearance of the transduced cells may also result in the loss of the transgene expression over time. The other known toxicities associated with gene therapy include dorsal root ganglion (DRG) and/or peripheral nerve toxicity, adverse immunogenicity and inflammatory findings, and insertional mutagenesis [26].
- Genomic insertion assessments: Depending on the viral platforms chosen for transduction the construct may either have an episomal expression (adenoviral vectors AAV) or get integrated into the nuclear DNA (lentiviral vectors) [28]. Genomic insertions have the potential to produce frame shift mutations and trigger an oncogenic response, hence assessment of insertional mutagenesis is given important consideration in gene therapies. When necessary, such analysis should be conducted using validated linear amplification-mediated (LAM)-PCR methods.
- Immunogenicity: Immunogenicity assessments are typically included in most gene therapy programs and can include assessment of binding and or neutralizing antibodies to either the capsid or the transgene or a combination of both. Depending on the therapeutic program, additional immunological assessments for hematological endpoints, cytokine release, or other adverse events may also be included in the study design [26, 29, 30].
- Safety Pharmacology: Standalone safety pharmacology studies are not routinely conducted for gene therapy products but appropriate cardiovascular, respiratory and CNS assessments are integrated into toxicology studies at designated peel off intervals in the study. For therapies targeted to the brain, additional CNS endpoints for efficacy and toxicity are built into the study, and extensive histopathological sampling and analysis are routinely incorporated into the studies.
- Regulatory interactions: Since gene therapy represents a gamut of expression platforms to treat inherited disorders for many rare diseases, and the field is still evolving, it is imperative that the sponsors of such studies interact with regulatory agencies where they intend to register the products. Such interactions prior to the conduct of the studies help avoid confusion and streamline the development path for such therapies. These meetings could be either informal meetings such as the initial targeted engagement for regulatory advice (INTERACT) meeting with CBER early in the program, or more formal binding meetings with US FDA or EMA, PMDA, Health Canada, Paul Ehrlich Institute (PEI), etc. Since this is an evolving discipline, one should be wary that there may be potential differences in the opinions and inputs provided by different regulatory agencies, and the sponsor is obliged to design optimal studies to be accepted globally.

5.6 Concluding Remarks

Design of toxicology studies is a multidisciplinary scientific exercise that includes the coordinated interactions among the study director, toxicologist, clinical and anatomic pathologists, formulation specialists, analytical and bioanalytical scientists, statisticians, toxicokinetic scientists, technical staff for the conduct of the study, veterinary staff, and quality assurance department. The specific requirements and scope of toxicology studies vary depending on the stage of development of the therapeutic entity, the target biology and mechanism of action, the modality and platform of therapy (small molecules/biotherapeutics/gene therapy, etc.), the route of administration and the duration of dosing. Well designed and executed toxicology studies facilitate a smooth transition during the regulatory reviews and avoid unnecessary duplication of studies that may become impediments or cause delays in the initiation or execution of clinical studies in humans. As outlined in the chapter, there are multiple regulatory guidelines that detail the essential elements of toxicology studies. All studies must be driven by a well-defined study protocol that clearly enlists the objectives of the study and the key parameters and endpoints to be evaluated. The chapter has tried to highlight the key scientific principles governing toxicology studies with a focus on general principles that apply to small molecules, biotherapeutics, and gene therapy products. However, it does not delve into an exhaustive description of design elements for specialized toxicology studies like phototoxicity, genotoxicity, safety pharmacology studies, reproductive toxicity, juvenile toxicity studies, and carcinogenicity assessments and readers are directed to the appropriate guidance documents and are discussed elsewhere in this book.

Conflict of Interest The author declares no conflict of interest that would influence the preparation of this book chapter.

References

1. Olson H, Betton G, Robinson D, Thomas K, Monro A, Kolaja G et al (2000) Concordance of the toxicity of pharmaceuticals in humans and in animals. Regul Toxicol Pharmacol 32(1):56–67
2. Tamaki C, Nagayama T, Hashiba M, Fujiyoshi M, Hizue M, Kodaira H et al (2013) Potentials and limitations of nonclinical safety assessment for predicting clinical adverse drug reactions: correlation analysis of 142 approved drugs in Japan. J Toxicol Sci 38(4):581–598
3. M3(R2) Nonclinical safety studies for the conduct of human clinical trials and marketing. https://www.fda.gov/regulatory-information/search-fda-guidance-documents/m3r2-nonclinical-safety-studies-conduct-human-clinical-trials-and-marketing-authorization. Accessed 10 Dec 2022
4. Non-clinical: toxicology. https://www.ema.europa.eu/en/human-regulatory/research-development/scientific-guidelines/non-clinical/non-clinical-toxicology. Accessed 10 Dec 2022
5. Rhomberg LR, Baetcke K, Blancato J, Bus J, Cohen S, Conolly R et al (2007) Issues in the design and interpretation of chronic toxicity and carcinogenicity studies in rodents: approaches to dose selection. Crit Rev Toxicol 37(9):729–837
6. Roberts R, Callander R, Duffy P, Jacobsen M, Knight R, Boobis A (2015) Target organ profiles in toxicity studies supporting human dosing: does severity progress with longer duration of exposure? Regul Toxicol Pharmacol 73(3):737–746
7. CFR - Code of Federal Regulations Title 21: PART 58 good laboratory practice for nonclinical laboratory studies. https://www.accessdata.fda.gov/scripts/cdrh/cfdocs/cfcfr/CFRSearch.cfm?CFRPart=58. Accessed 10 Dec 2022
8. Good laboratory practices. https://www.oecd.org/chemicalsafety/testing/good-laboratory-practiceglp.htm. Accessed 10 Dec 2022
9. ICH guidelines. https://www.ich.org/page/ich-guidelines. Accessed 10 Dec 2022
10. Chapman KL, Andrews L, Bajramovic JJ, Baldrick P, Black LE, Bowman CJ et al (2012) The design of chronic toxicology studies of monoclonal antibodies: implications for the reduction in use of non-human primates. Regul Toxicol Pharmacol 62(2):347–354
11. Van Cauteren H, Bentley P, Bode G, Cordier A, Coussement W, Heining P et al (2000) The industry view on long-term toxicology testing in drug development of human pharmaceuticals. Pharmacol Toxicol 86(suppl 1):1–5
12. DeGeorge JJ, Meyers LL, Takahashi M, Contrera JF (1999) The duration of non-rodent toxicity studies for pharmaceuticals. International Conference on Harmonication (ICH). Toxicol Sci 49(2):143–155
13. Organization and Personnel 2015. https://www.fda.gov/inspections-compliance-enforcement-and-criminal-investigations/fda-bioresearch-monitoring-information/organization-personnel. Accessed 10 Dec 2022
14. OECD (1999) The role and responsibilities of the study director in GLP studies, OECD series on principles of good laboratory practice and compliance monitoring, No. 8, OECD Publishing, Paris. https://doi.org/10.1787/9789264078673-en
15. Animal use alternatives (3Rs). https://www.nal.usda.gov/animal-health-and-welfare/animal-use-alternatives. Accessed 10 Dec 2022
16. Buckley LA, Dorato MA (2009) High dose selection in general toxicity studies for drug development: a pharmaceutical industry perspective. Regul Toxicol Pharmacol 54(3):301–307
17. Borgert CJ, Fuentes C, Burgoon LD (2021) Principles of dose-setting in toxicology studies: the importance of kinetics for ensuring human safety. Arch Toxicol 95(12):3651–3664
18. Sewell F, Corvaro M, Andrus A, Burke J, Daston G, Delaney B et al (2022) Recommendations on dose level selection for repeat dose toxicity studies. Arch Toxicol 96(7):1921–1934
19. Luft J, Bode G (2002) Integration of safety pharmacology endpoints into toxicology studies. Fundam Clin Pharmacol 16(2):91–103

20. Bregman CL, Adler RR, Morton DG, Regan KS, Yano BL, SoT P (2003) Recommended tissue list for histopathologic examination in repeat-dose toxicity and carcinogenicity studies: a proposal of the Society of Toxicologic Pathology (STP). Toxicol Pathol 31(2): 252–253
21. Pre-clinical safety evaluation of biotechnology-derived pharmaceuticals (ICH S6[R1]) [cited 2023 05 Jan]. https://www.ema.europa.eu/en/pre-clinical-safety-evaluation-biotechnology-derived-pharmaceuticals-ich-s6-r1. Accessed 10 Dec 2022
22. Clarke J, Hurst C, Martin P, Vahle J, Ponce R, Mounho B et al (2008) Duration of chronic toxicity studies for biotechnology-derived pharmaceuticals: is 6 months still appropriate? Regul Toxicol Pharmacol 50(1):2–22
23. Preclinical assessment of investigational cellular and gene therapy products. https://www.fda.gov/regulatory-information/search-fda-guidance-documents/preclinical-assessment-investigational-cellular-and-gene-therapy-products. Accessed 10 Dec 2022
24. Mullard A (2021) Gene therapy community grapples with toxicity issues, as pipeline matures. Nat Rev Drug Discov 20(11):804–805
25. Bolt MW, Whiteley LO, Lynch JL, Lauritzen B, Fernández de Henestrosa AR, MacLachlan T et al (2020) Nonclinical studies that support viral vector-delivered gene therapies: an EFPIA Gene Therapy Working Group perspective. Mol Ther Methods Clin Dev 19:89–98
26. Baldrick P, McIntosh B, Prasad M (2023) Adeno-associated virus (AAV)-based gene therapy products: what are toxicity studies in non-human primates showing us? Regul Toxicol Pharmacol 138:105332
27. Silva Lima B, Videira MA (2018) Toxicology and biodistribution: the clinical value of animal biodistribution studies. Mol Ther Methods Clin Dev 8:183–197
28. Sherafat R. Toxicity risks of adeno-associated virus (AAV) vectors for gene therapy (GT) 2021. https://www.fda.gov/media/151599/download. Accessed 10 Dec 2022
29. Ertl HCJ (2022) Immunogenicity and toxicity of AAV gene therapy. Front Immunol 13:975803
30. Yang TY, Braun M, Lembke W, McBlane F, Kamerud J, DeWall S et al (2022) Immunogenicity assessment of AAV-based gene therapies: an IQ consortium industry white paper. Mol Ther Methods Clin Dev. 26:471–494

General Design Considerations in Reproductive and Developmental Toxicity Studies

6

Wendy Halpern

Abstract

Developmental and reproductive toxicity studies are conducted with the objective of identifying adverse effects relevant to fertility, pregnancy and development. Many approaches are used to enlighten our understanding of reproductive toxicity. Each segment of the testing paradigm focuses on slightly different objectives, with an overarching goal to understand the mechanistic and exposure relationships of candidate therapeutics that drive toxicity. There are established standard design approaches for these developmental and reproductive toxicity (DART) studies. The DART strategy for each program should carefully consider the intended patient population and existing knowledge regarding the toxicity profile, exposure relationships and expected pharmacology. Overall, mammalian DART studies are large and complex, and require substantial planning and experience for appropriate design and interpretation. DART testing is typically conducted in mammalian rodent species (mouse or rat), with a second non-rodent species used to further evaluate teratogenic potential (often rabbit), but in some cases, studies can be combined, refined, or avoided. In addition, there are non-mammalian tools that can aid in strategy, study design and/or data interpretation. This chapter will explore considerations for the use and design of both 'standard' and alternative study types and tools to inform pharmaceutical and biopharmaceutical drug development.

Keywords

Fertility · Reproduction · Pregnancy · Development · Teratogenicity · ICH guidance

6.1 Introduction

Nonclinical developmental and reproductive toxicity testing (DART) has a special place in drug development because clinical studies rarely assess reproductive function or pregnancy in patients. Therefore, there are detailed international and regional regulatory guidance documents that can help clarify the regulatory expectations for nonclinical DART testing to support both the conduct of clinical trials and eventual labeling. The primary objective of this specialized nonclinical testing is to inform physicians and provide a risk assessment to all patients of reproductive potential, with aspects of development, parturition and lactation further focusing on patients who might become pregnant. Success requires the combination of the knowledge that drug developers have regarding their specific molecule programs, the specialized

W. Halpern (✉)
Genentech, South San Francisco, CA, USA
e-mail: Halpern.wendy@gene.com

G. Jagadeesh et al. (eds.), *The Quintessence of Basic and Clinical Research and Scientific Publishing*,
https://doi.org/10.1007/978-981-99-1284-1_6

functional expertise of reproductive toxicologists and scientists needed for the smooth execution and interpretation of DART studies, and the guidance available to understand and meet regulatory expectations. This chapter will explore, at a high level, strategic and practical approaches to support nonclinical DART testing based on scientific rationale and consistent with the current regulatory guidance. However, this topic is very broad in scope and the reader is encouraged to utilize additional texts for detailed protocols and additional information, e.g. [1–4].

The International Council for Harmonization (ICH) has developed several guidances that each address specific aspects of DART testing for candidate therapeutics. Of these, the recently adopted ICH S5(R3) is a current and comprehensive resource for internationally harmonized approaches to DART testing, mostly focused on *in vivo* DART studies conducted in mammals [5]. However, it also acknowledges the growing role of alternative approaches, including *in vitro*, *ex vivo* and nonmammalian testing that might eventually contribute to an overall reduction in animal use. The ICH S5(R3) also acknowledges general consistency with the previously adopted ICH M3(R2), which specifies the timing for testing relative to the size and population of the clinical trial program [6, 7], with an expansion of options around regional differences for inclusion of patients who could become pregnant in early clinical trials. In addition, it is consistent with the ICH S6(R1) for biopharmaceuticals, acknowledging both the low potential for risk based only on physicochemical characteristics of protein therapeutics, as well as the challenges of DART testing with highly specific biopharmaceuticals. The ICH S6(R1) details an *in vivo* pregnancy study design specific for monoclonal antibodies given to nonhuman primates as a combined embryo-fetal development/ pre-and postnatal development (EFD/PPND) study, as well as other approaches that can be acceptable [8]. Finally, the ICH S9 and the ICH S9 Q/A highlight some exceptions that apply to the development of novel therapeutics intended for use in an advanced cancer setting [9, 10]. These and other guidance documents relevant to drug development can be found through the ICH website (www.ich.org), or through regionally specific health authority sites (e.g., www.fda.gov); the specific references in this chapter are taken from the harmonized guidance publication date as a Federal Register Notice from the US FDA.

Developmental and reproductive toxicity is often described with reference to lifecycle segments, as illustrated in Fig. 6.1. However, as described in the ICH S5(R3), an integrated testing strategy (ITS) for each individual program should be the first step, with accumulated experience informing prioritization and utility of additional testing. Although specific guidance is provided to aid appropriate study and program design, there is no single set of studies appropriate for all programs, and flexibility is intended to enable a program-specific approach. In some cases, combined studies can be used, while other situations may benefit from expanded testing prior to definitive studies. Table 6.1 highlights the different study types described in this chapter with examples of the types of endpoints evaluated. While there are several 'optional' or 'for cause' endpoints that can be included based on specific program concerns, there are others that are routine for many who conduct these studies, especially in routine species for testing.

The timing of when to conduct dedicated DART testing during the drug development process has long been a challenge. These studies are large, complex, and need to be conducted with appropriate dose levels to ultimately support registration and labeling. Because of this, DART data are often not available for early clinical trials, and all regions require the incorporation of stringent measures to prevent pregnancy. Embryo-fetal development studies conducted with pregnant animals of two species are considered a key component for the risk assessment for teratogenic potential of novel therapeutics, and these 'Segment 2' studies are typically the first dedicated DART studies conducted for a drug development program. One exception is for highly specific biopharmaceuticals, such as monoclonal antibodies, which can have a low potential for direct teratogenic effects. Based on

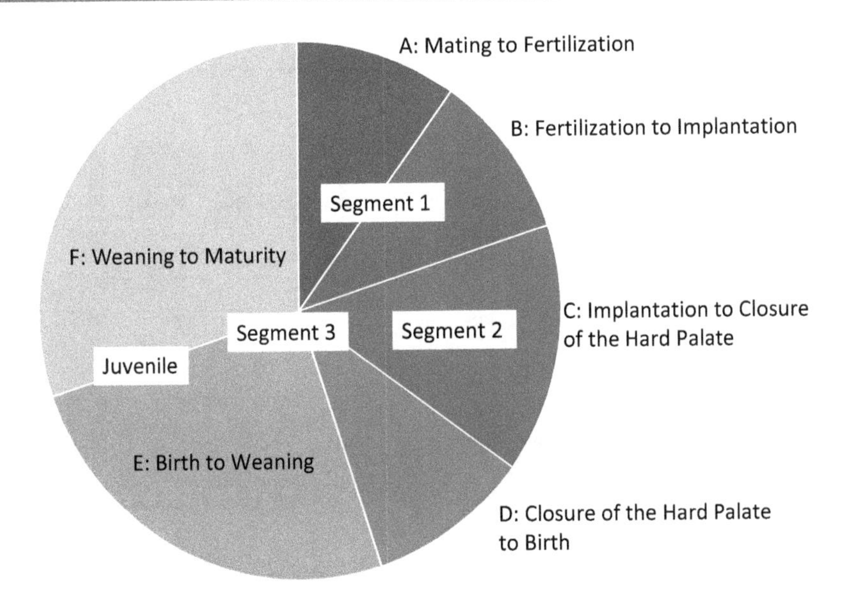

Fig. 6.1 Lifecycle segments for reproduction and development. A = Copulation to Fertilization, B = Fertilization through Implantation, C = Implantation through Closure of the Hard Palate (end of major organogenesis), D = Closure of Hard Palate to Parturition, E = Birth to Weaning, F = Weaning to Maturity. Standard DART testing typically includes Segment 1/FEED studies (A–B), Segment 2/EFD studies (C), and Segment 3/PPND studies (C–F). Juvenile toxicity (E, F) and general toxicity (F) data can also contribute to risk assessment

the ICH S6(R1), nonclinical testing during pregnancy for biopharmaceuticals is typically conducted in parallel with Phase 3 to support registration and labeling [8]. A general example of timing of DART testing relative to clinical development is shown in Fig. 6.2.

The data from the EFD and general toxicity testing have an important role in establishing dose/exposure and dose/toxicity relationships, including potential exposure differences in the dam during pregnancy. These data can then help inform the design of additional reproductive function studies such as fertility and early embryonic development (FEED) studies, and PPND studies, both of which are typically limited to a single rodent species (mouse or rat). In the US, dedicated FEED studies in rodents are typically conducted parallel to Phase 2 clinical efficacy studies for small molecule therapeutics, or parallel to pivotal Phase 3 studies for large molecules with high pharmacologic specificity. It has been shown that large molecules are unlikely to be present at high levels in the semen, and that there is a negligible risk for transmission of intact protein therapeutics to a conceptus of a female partner who is, or may become, pregnant [11, 12].

For drugs intended for use in an advanced cancer setting, and especially those drugs that are known to be genotoxic, the potential reproductive and developmental risk to patients and their offspring can often be assumed. There are also potential risks to maintaining pregnancy with cancer immunotherapies [13], and there can be challenges in achieving clinically relevant exposures in the absence of maternal toxicity in nonclinical studies with anticancer pharmaceuticals. However, for drugs intended to treat indolent or slowly progressive cancers, especially those cancer types that occur in younger patient populations, additional reproductive testing can be warranted. A case-by-case approach to pharmaceuticals intended to treat other (non-oncology) life threatening or severely debilitating conditions can be appropriate with regard to design and timing of nonclinical studies. This also applies to candidate therapies outside the scope of the ICH S5(R3), such as gene, cell and tissue-based therapies, and to some

Table 6.1 DART study endpoints

	General toxicity	Male FEED	Female FEED	pEFD, EFD (dam)	pEFD (fetus)	EFD (fetus)	PPND (dam)	PPND (F1)
InLife Observations	Routine	Routine	Routine	Routine			Routine	Routine
Body Weight (Serial)	Routine	Routine	Routine	Routine	Routine	Routine	Routine	Routine
Organ Weights	Routine	Routine	Routine	Routine				Routine
Macroscopic Pathology	Routine	Routine	Routine	Routine	(exams)	(exams)	Routine	Routine
Microscopic Pathology	Routine	Optional	Optional	Optional			Optional	Optional
Copulation (Vaginal Plug)		Routine	Routine					
Estrous Cycle (Cytology)	Optional	Optional	Routine					
Corpora Lutea		Routine	Routine	Routine				
Uterine Implantation Sites/Scars		Routine	Routine	Routine			Routine	
Viable Fetuses/Offspring (F1)		Routine	Routine		Routine	Routine		Routine
Early Resorption (F1)		Routine	Routine		Routine	Routine		
Late Resorption (F1)		Optional	Optional		Routine	Routine		
Nonviable Fetus/ Offspring (F1)		Routine	Routine		Routine	Routine		Routine
Fetal External Exam (F1)					Routine	Routine		
Fetal Visceral Exam (F1)					Optional	Routine		
Fetal Skeletal Exam (F1)					Optional	Routine		
Pre-Weaning Developmental Milestones (F1)								Routine
Vaginal Opening (Female F1)								Routine
Preputial Separation (Male F1)								Routine
Behavior/Locomotion (F1)								Routine
Learning and Memory (F1)								Routine
Breeding Test (F1), *also includes FEED endpoints (F1/F2)*								Routine (F1/F2)
Exposure/Toxicokinetics	Routine	Optional	Optional	Routine	Optional	Optional	Optional	Optional

F1: offspring of F0 (dosed males or females)
F2: offspring of F1

non-antibody based biopharmaceuticals such as enzyme replacement therapies, where a weight of evidence using existing available information regarding molecule class, mechanism, and disease context will drive the rationale and strategy for additional nonclinical testing [8, 14].

Because DART testing is resource-intense and requires large numbers of laboratory animals for testing, and because laboratory species do not always correctly predict human risk, there is strong interest from both regulators and drug developers to explore alternatives to mammalian *in vivo* testing. The methods that are specifically discussed in the ICH S5(R3) include *in vitro*, *ex vivo* and nonmammalian *in vivo* approaches [5]. At present, none of these alternative approaches are considered equivalent to, or an improvement over, standard and established *in vivo* testing paradigms, and qualification is needed for use as a replacement for *in vivo* studies

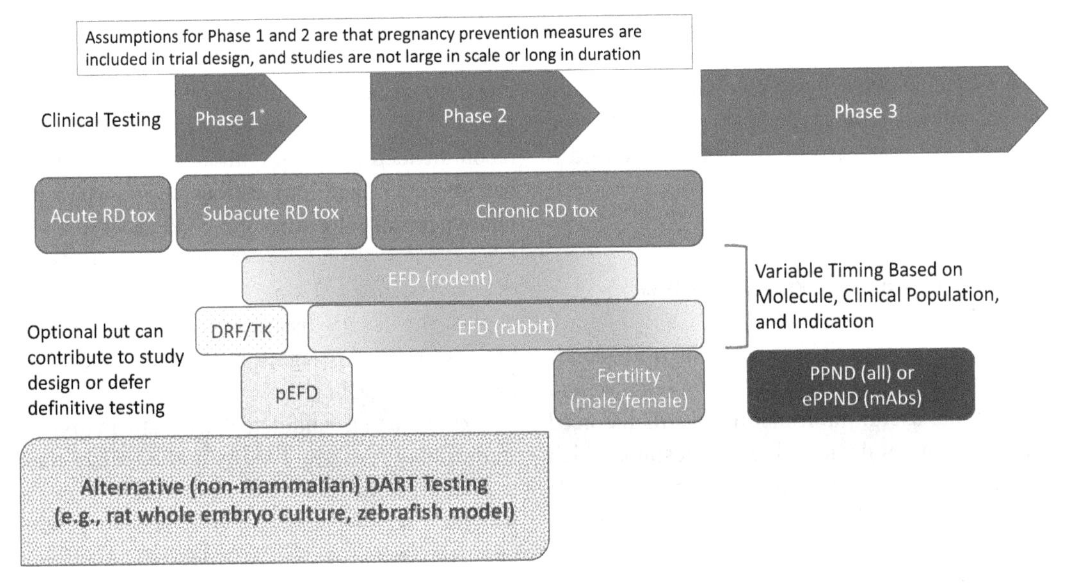

Fig. 6.2 Timing of DART studies relative to clinical development. The figure highlights the general relationship between the availability of DART study data during pre-registrational clinical development. There is variability in the studies conducted and time of completion based on individual program concerns

in a regulatory setting [5, 15]. However, as additional experience is gained, this may change in the future, and there is already progress toward establishing qualified options [16, 17]. In addition to the alternatives discussed in the ICH S5(R3), newer *in vitro* methods such as complex microphysiological systems (MPS), *in silico* methods for prediction of toxicity, as well as the use of -omics data, are also being explored [15, 18]. Already, results from these approaches can contribute to early screening, investigation of mechanisms of toxicity, and the overall weight of evidence for an ITS to support risk assessment. Continued cross-functional communication between physicians, developmental physiologists, drug developers, and environmental and agrochemical toxicologists will be critical in advancing effective alternative approaches to characterize reproductive and developmental toxicity and provide a meaningful risk assessment to patients.

The Key Objective of this Chapter is to provide a general introduction to nonclinical reproductive and developmental toxicity testing by describing current testing paradigms and their regulatory framework, providing additional context for drivers of study design and interpretation, and looking forward to opportunities to advance the science of DART testing.

6.2 Reproductive Toxicity

Reproductive toxicity refers to adverse effects of a drug that result in impaired reproductive function. This can include, but is not limited to, structural toxicity to the reproductive tissues, or to neuroendocrine tissues regulating reproductive function. The value of reproductive toxicity testing also takes into account the intended patient population and expected duration of treatment. A drug expected to be used entirely in an aged or young pediatric patient population might not need extensive reproductive testing, whereas nonclinical data can provide critical information for the chronic or lifetime management of diseases that affect people of reproductive potential.

For the nonclinical assessment, this begins with a knowledge of the molecule class, expected mechanism of activity and associated pharmacology, along with an understanding of reproductive and developmental biology and physiology.

Structural effects on reproductive and endocrine tissues are typically first evaluated in general toxicity studies, and these data can help inform risk communication to patients participating in early clinical trials, especially if a clear hazard is identified.

6.2.1 General Toxicity Studies

Most general toxicity testing in rodents is conducted in young adult and adult animals, and can provide useful information towards the need for and design of dedicated DART testing, which is often conducted with mice, rats and/or rabbits (*see* Chap. 5). The first wave of spermatogenesis begins in male rats at around the time of weaning, with spermarche about a month later [19–22]. Thus, a male rat of 7–8 weeks of age can have the histologic appearance of an adult, although maturation continues for another several weeks as subsequent waves of post-pubertal spermatogenesis are completed. In general, male rats will achieve optimal reproductive performance characteristics when they are at least 12 weeks of age [19].

In contrast to rodent studies, many nonrodent studies are conducted with dogs or macaques that are pre- or peripubertal in age. Macaques, in particular, are poorly suited for routine screening of reproductive pathology as group sizes are relatively small, especially when divided by sex, and age is not reliably correlated with sexual maturity over the age range of animals typically used for toxicity studies (2–4 years) [23]. It is important for the pathologist to be aware of the sexual maturity status of both rodent and nonrodent laboratory animals at the time of necropsy from general toxicity studies, as this is critical for the interpretation of pathology of reproductive tissues and communication of results [19].

In sexually mature males, reproductive pathology endpoints in general toxicity studies typically include a stage-aware microscopic review of spermatogenic progression, as well as the macroscopic and microscopic assessment of epididymides and accessory sex glands [24–26]. Evaluation of sperm or semen is an optional endpoint that can be added for cause [27]. In females, tissue specific findings as well as an integrated assessment of the estrous cycle phase across the reproductive tract tissues is recommended [28, 29]. Although the female dog has infrequent and long estrous cycles which are not synchronized during the conduct of a study, which limits the ability to interpret effects on cycle progression during routine general toxicity testing, mammary tissue development with the first estrus is a straightforward way to confirm maturity [19, 30].

Exposure to a candidate therapeutic in toxicity studies may result in toxicity to the mature male or female reproductive tract, as detected by organ weights, macroscopic and microscopic pathology assessment of tissues. Additional endocrine tissues (especially the pituitary, thyroid, and adrenal glands) can also have pathologic findings that contribute to the overall picture of reproductive health [31]. Finally, there can be general findings indicating stress that can affect the reproductive system and should be considered during study interpretation [32, 33]. Although it is well recognized that hormones are critical in the regulation and progression of both spermatogenesis and estrous or menstrual cycles, it is not practical or reasonably feasible to incorporate sample collections for hormone analysis in studies not specifically designed for that purpose [34, 35].

Depending on the effect identified, additional investigative studies, or expanded testing can be helpful to understand the mechanism of toxicity. For example, a male mating test (to naive females) can be incorporated into a chronic toxicity study in the rat if there are fertility concerns from earlier studies [36]. However, there are also situations where the toxicity is predictable, and no additional testing is warranted. For example, many drugs that kill rapidly dividing cells will have an effect on the spermatogonia, or germ cells, of the testis. Unlike the spermatocytes, these cells are outside of the blood-testis barrier, and are therefore predictably susceptible to toxicity, resulting in progressively reduced spermatogenesis and a typical histologic appearance [24]. The testis is generally considered a toxicologically sensitive tissue [24]. When findings are identified, questions often arise as to the translatability and reversibility of findings. One common pitfall in general toxicity studies is to have a recovery phase of insufficient duration to allow reconstitution of disrupted spermatogenesis. In fact, for relatively short duration repeat dose toxicity studies, it is not uncommon for the microscopic pathology at the end of the recovery phase to be more severe than that at the end of dosing. This does not necessarily indicate an irreversible finding, but rather a finding with inadequate time to observe reversibility. See the previous chapter for a discussion on this topic.

6.2.2 Fertility and Early Embryonic Development (FEED) Studies

The FEED studies are intended to capture potential effects on Segments A and B, fertilization through implantation (see Fig. 6.1). These studies are typically conducted in the rat or mouse, where reproductive and developmental milestones are well characterized, and there is abundant historical control data to provide context for risk assessment [19, 25, 29, 37]. Fertility studies may be conducted earlier in development when there is a mechanistic concern, or may be waived entirely if a specific reproductive hazard has already been identified that precludes the intended value of the study. Studies that include mating during the dosing phase are generally not warranted when the nonhuman primate or dog is the only pharmacologically relevant species. The relatively low 'catch' rate in macaques and single offspring per pregnancy means that breeding studies would be prohibitively large; likewise, the infrequent and brief periods of mating receptivity in female dogs means that breeding studies in dogs are impractical and rarely conducted [27].

While the ICH S5(R3) provides guidance for the design and conduct of fertility studies, there are a number of decision points for the drug developer to consider based on the context of the specific program. The first of these is whether or not a FEED study is warranted based on existing knowledge and patient population. If yes, then the next question is whether to have two FEED studies, a single study where both males and females are dosed prior to mating, or a combination study with either general toxicity testing (males), or the EFD study (females). A full combination or two generation design could also be considered if most appropriate for program concerns. A third consideration is how long to dose the males prior to the breeding phase; this can range from 2 to 10 or more weeks. Finally, a fourth consideration is whether or not to include a recovery phase, especially if reproductive toxicity is expected, or, for example, for neuroactive drugs that may have behavioral effects.

As a basic guide, while combination studies offer the potential to reduce laboratory animal use, the design and approach should consider existing information to ensure that additional *in vivo* studies are unlikely to be requested [5]. The data from these studies are used to inform the risk assessment for reproductive toxicity and recommendations for contraception during drug development, and any findings are included in drug approval and labeling [11]. One common challenge for data interpretation and human risk assessment is that nonclinical species used for toxicity testing generally have a more robust reproductive capacity than humans [38, 39]. In addition, although both mice and humans have hemochorial placentae, the rodent inverted yolk sac can result in differences in placental transfer to the fetus [40].

FEED studies include a mating phase to model aspects of conception, including libido, ability to mate, and fertilization, none of which are captured in routine repeat dose toxicity studies. A general FEED study design is shown in Fig. 6.3. Dosing is initiated prior to cohabitation for mating to capture potential effects on spermatogenesis and epididymal transit in males, and estrous cyclicity in females. In females, a 2 week dosing period prior to cohabitation is adequate, with dosing extended through the end of organogenesis. In males, a longer dosing period prior to mating can be warranted; in some cases a second, recovery phase breeding can further aid interpretation [41]. In studies where the females are dosed, estrous cycle can be monitored via daily

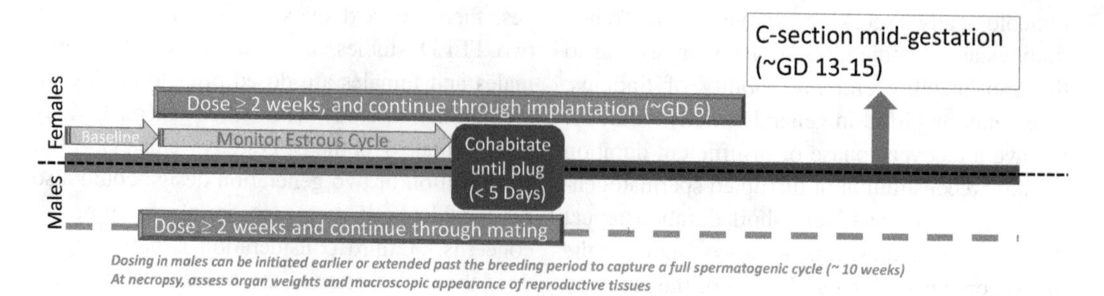
C-section mid-gestation (~GD 13-15)
Females
Males
Dose ≥ 2 weeks, and continue through implantation (~GD 6)
Baseline
Monitor Estrous Cycle
Cohabitate until plug (< 5 Days)
Dose ≥ 2 weeks and continue through mating
Dosing in males can be initiated earlier or extended past the breeding period to capture a full spermatogenic cycle (~ 10 weeks)
At necropsy, assess organ weights and macroscopic appearance of reproductive tissues
*If no reproductive tissue abnormalities in general toxicity studies of at least 2 weeks duration, can combine male and female fertility studies, BUT if findings are expected, should keep them separate, with one naïve rat per breeding pair.

Fig. 6.3 FEED Study. There can be variability in the sex dosed (male, female, or both), the duration of male dosing prior to cohabitation for breeding, and the duration of dosing during pregnancy prior to C-section or parturition

vaginal cytology over 2-week periods both at baseline and during the dosing phase prior to mating [42–44].

There are both male and female-specific endpoints for a FEED study. For **males**, these can include organ weights and macroscopic pathology, with optional microscopic pathology, at the end of the dosing or recovery phase. Sperm evaluation (viability, morphology, motility) was routinely conducted in the past, but is now considered more of an optional endpoint included 'for cause', or when a substantial effect size is expected, as there is often high intra- and interindividual variability in sample quality. In addition, the male mating and fertility indices are recorded to track how many treated males copulate with their paired female, and how many of those females become pregnant after copulation. The **females** also have mating and fertility indices, and are further evaluated for estrous cycle progression, fertility and fecundity by tracking preimplantation losses [fewer implantation sites than corpora lutea (CL)], evidence of early fetal resorptions, viable/nonviable fetuses, and overall litter size. The endpoints included can vary based on the dosing period and on the overall study objectives.

As previously noted, the assessment of functional reproductive endpoints can be conducted with dedicated male and female studies, or as part of a combination study approach. If effects on male fertility, female cyclicity, or embryo-fetal development are expected based on the mechanism of action, or based on existing data from completed general toxicity or embryo-fetal development studies, then the FEED study or studies should be carefully designed to capture new information. For example, if there is evidence of teratogenicity in an EFD study, then a dedicated male fertility study may still inform recommendations for contraceptive use by male patients, if adequate margins for potential exposure of a partner are established. However, a dedicated female fertility study is unlikely to add value for patients who could become pregnant, since a hazard to fetal development has already been identified and pregnancy prevention will be recommended.

6.3 Embryo-Fetal Development Studies

Embryo-fetal development studies are designed to identify developmental toxicants that result in malformations and embryo-fetal lethality (MEFL). Classic examples include **thalidomide**, which resulted in thousands of cases of infant death and phocomelia *in the late 1950s* [45–48], **cyclopamine**, which is a naturally occurring toxin that inhibits the developmentally important Hedgehog (Hh) signaling pathway [49], the retinoid **all-trans retinoic acid**, which resulted in malformations in rats leading to strong warnings against pregnancy [50, 51], and heavy metals such as **mercury**, which can cause neurodevelopmental and musculoskeletal malformations [52, 53]. While not all of these are directly relevant to drug development, they are examples that have shaped the rationale behind, and design of, nonclinical testing strategies to detect teratogenic potential. There are many other potential pathways and mechanisms that could result in malformations during organogenesis in the developing fetus, but deep knowledge of a few examples with clear human relevant correlates has helped us begin to map some common adverse outcome pathways (AOPs). This approach is already being used in the environmental and agrochemical industry setting [54, 55], and may help us eventually predict teratogenic potential for novel drugs based on their mechanism of action and physicochemical characteristics.

Current recommendations are to evaluate embryo-fetal development in two species (a rodent and a nonrodent). Some regions (Europe and Japan) expect preliminary or definitive EFD data from two species to enable inclusion of individuals who could become pregnant in Phase 2 trials, while other regions (US) generally expect both fertility and EFD data prior to the initiation of Phase 3 studies. Under the M3(R2) guidance, the pEFD approach could support inclusion of up to 150 women of childbearing potential in clinical studies of up to 3 months in duration [6, 7]. This guidance was expanded with the ICH S5(R3) to remove patient number and trial duration limits *if* the pEFD is expanded to include (1) additional presumed pregnant dams per group, (2) inclusion of skeletal, in addition to external and visceral, evaluations of the fetuses, and (3) conduct of the study according to Good Laboratory Practices (GLP) [5]. While such studies could potentially expand timely patient access to trials in some regions, an early assessment of the value in the pEFD appears to be predominantly in identifying substantial hazards for MEFL [56].

There are also some situations where definitive studies in a single species can be appropriate: (1) once a positive signal for MEFL has been identified at or near clinically relevant exposures, (2) for vaccine development, (3) in the setting of advanced cancer, life threatening or severely debilitating conditions, or (4) where there is a single pharmacologically relevant species. In addition, there is little value in conducting an EFD (or ePPND) study for biopharmaceuticals where there are no pharmacologically relevant species or species-relevant surrogate molecules. If available, reproductive data from genetically modified animal models can contribute to the knowledge gap, as has been done for some enzyme replacement therapies, cancer immunotherapies, and antibody therapeutics [13, 14, 57–59].

A recent review evaluated the registration information for the 103 new drugs approved by the FDA in 2020–2021 [60]. As noted for other prior, similar, reviews for product approvals since 2014, the rat and rabbit are the most commonly used species for DART testing. Consistent with other systematic reviews [61–63], the embryo-fetal development data from rats and rabbits often show differences in the NOAEL, as well as in susceptibility to maternal toxicity, which, in these reviews, tended to be more severe in the rabbit.

It was also noted that there is a continued decrease in the use of NHP for DART testing of biopharmaceuticals (22% of BLAs approved in 2020–2021 versus 62% of BLAs approved between 2002 and 2015) [60]. It is generally agreed that NHP, often the cynomolgus monkey, should only be used for DART studies when it is the only pharmacologically relevant species. With biopharmaceuticals, and consistent with the ICH S6(R1), alternatives to DART testing in NHP could include (1) the use of a surrogate molecule with comparable pharmacology in a rodent species, (2) a weight of evidence assessment based on existing data relevant to the molecule target, class or mechanism of action, or (3) no testing at all in certain instances such as when there is a bacterial or viral target and no relevant animal model. These options have become of even greater relevance with the reduced availability of cynomolgus monkeys for biomedical research during the SARS-COV-2 pandemic [64].

6.3.1 Design Considerations

A schematic of an EFD study design is shown in Fig. 6.4. The definitive EFD study design, with at least 16 presumed pregnant females at the start of dosing on GD 6 or 7, is powered to reasonably detect treatment related MEFL at the litter level. Ideally, there will be a control group and three distinct dose levels tested, ranging from a clinically relevant low dose to a high multiple (up to 25× MRHD), preferably with minimal or no maternal toxicity [65, 66]. Unfortunately, as acknowledged in the S5(R3), this scenario can be difficult to readily predict or achieve for a variety of reasons, including altered exposures or tolerability by the dam during pregnancy [5]. However, the relationship between dose/exposure and MEFL is critical [65, 66]. Many substances have the potential to result in developmental toxicity if given at high dose levels, and the risk assessment process needs to consider both the identification of MEFL in nonclinical EFD studies, as well as the safety margin between the nonclinical NOAEL for MEFL and the MRHD [67, 68]. In order to best support this risk assessment, it is important to be aware of the iterative process leading to the conduct of the definitive studies, and to collect relevant data from non-pivotal studies, if conducted, to best inform dose selection and design.

For an EFD conducted in the mouse or rat, there may be adequate exposure/toxicity data from the repeat dose general toxicity studies to consider going directly to the pivotal EFD study. However, a pEFD approach can also be appropriate when there are already concerns for developmental or maternal toxicity [56]. In this setting, a definitive study, or second species, might not be needed if a clear hazard at clinically relevant exposures is demonstrated. Also, if limiting maternal toxicity is identified in a pEFD, fewer dams will have been exposed at that toxic dose level, but the study can support the selection of a lower dose range for the definitive study conducted later in development.

In contrast to the rodent species, the first use of the rabbit as a test system may be for the second EFD species, so a PK/tolerability dose range-finding study can be helpful to inform dose selection. This can be conducted in the pregnant or non-pregnant doe, possibly followed by a pEFD for early identification of pregnancy-related hazards. The rabbit has been a useful second species for the detection of MEFL [61]. However, for some therapeutic classes, maternal toxicity can preclude meaningful interpretation of study data [69]. For example, antibiotics can cause marked disruption of the rabbit hindgut flora leading to dysbiosis and reduced food consumption at a time of high metabolic demand, often with whole litter losses as a result. In addition, there can be developmental delays associated with maternal stress [33, 70]

Whether part of a DRF, pEFD or EFD, if there is a separate cohort of pregnant PK animals, it can be helpful to assess fetal drug concentrations at

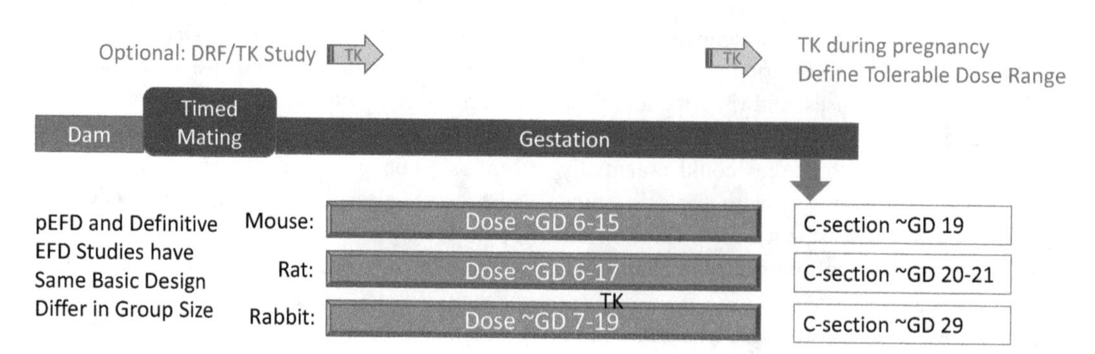

Fig. 6.4 pEFD/EFD Study. The general timing of dosing is noted for studies conducted in the mouse, rat or rabbit. For non-pivotal studies such as DRF, pEFD and pEFD (enhanced) studies, size and endpoints can vary. Not all studies are typically conducted for each species or all programs

the end of the dosing period. This is not a routine or expected endpoint, will not be a full PK curve, and usually can just be compared to systemic exposure of the dam or doe at the last sample timepoint collected, but it can provide some evidence of fetal exposure, or lack thereof, during the period of organogenesis. In general, although feasible, fetal exposures are not routinely collected at the time of the main C-section because maternal systemic exposures may have already cleared between the end of dosing and the time of C-section, and because the blood collection process from the fetus can compromise other critical endpoints such as the fetal visceral and skeletal evaluations.

The endpoints evaluated for an EFD study primarily focus on fetal survival and structure [61]. These are methodically evaluated using protocols to capture the number, sex distribution, viability and size of pups or kits from each litter, as well as a detailed evaluation of the fetal viscera, cranium, and skeleton [4]. In addition, there is a macroscopic evaluation of the viscera of the dam at the time of C-section, including an assessment of implantation sites and the appearance of the placenta at the maternal-fetal interface. A useful lexicon to describe abnormalities in EFD studies has been developed and provides some consistency in data reporting [71].

Maternal toxicity is of substantial concern in the design of EFD studies, as it can lead to adverse reproductive outcomes such as increased resorptions, fetal deaths, and whole litter losses that fall under the category of MEFL. Since malformations can also lead to fetal losses, it is not possible to definitively separate these categories. However, indices of maternal stress or toxicity should be tracked and considered in study interpretation, and the dose selection should also avoid more than minimal maternal toxicity [5].

6.4 Pre- and Postnatal Development Studies

The rodent pre- and postnatal development study builds on existing knowledge from the FEED and EFD studies, and contributes additional information about functional reproductive endpoints such as parturition and lactation, as well as assessing the growth and behavior of offspring. In this study type, the dam is typically dosed from early pregnancy (GD 6–7) through weaning [approximately postnatal day (PND) 21 in the rat] (Fig. 6.5). This is a fairly 'hands off' study design other than dose administration. Both assessments of maternal toxicity and associated toxicokinetic characterization of exposure during pregnancy have been established in the EFD study. The PPND study should not be conducted with dose levels expected to result in more than mild maternal toxicity, but ideally will be designed to achieve exposures 3–10-fold above the anticipated MRHD.

Because dams are allowed to deliver on their own, it is often not possible to evaluate the placenta or to determine accurate numbers of stillborn offspring if there is cannibalism. Rather, the number of viable offspring for each litter are determined on the delivery day, and the dams are necropsied at the time of weaning to assess the number of uterine implantation sites as well as to collect and preserve any macroscopically evident abnormalities from the dam (with paired control tissues). As with other studies in pregnancy, the litter is the unit of measure for the endpoints evaluated.

Pup evaluations begin from parturition, when the number of live pups and their sex are recorded. After this initial assessment, pups are monitored daily for clinical signs and mortality, and are weighed twice weekly during the lactation phase to monitor growth. Early in the postnatal period, usually on about PND 4, the litter size will be culled to 4–5 pups per sex. This should be uniform for all litters in the study, and consistent with available historical data from the study site, as litter size will affect the rate of growth of the pups [72].

The offspring stay with their birth dam, and their survival and growth reflects, in part, her lactational success. There may also be a substantial transfer of the test item into the milk, leading to indirect exposure of the pups. When there is pup mortality, it can be helpful to conduct a

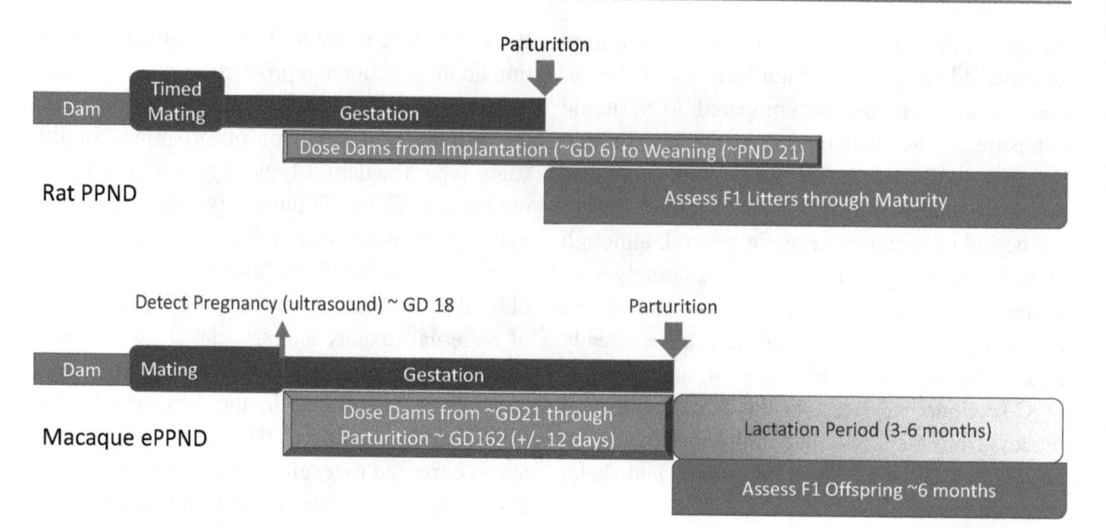

Fig. 6.5 PPND and ePPND Study. In rodent PPND, dosing of the dam occurs from implantation through weaning (end of lactation), and litter evaluations encompass both pregnancy outcomes and subsequent development of offspring. In the nonhuman primate ePPND, dosing occurs from confirmation of pregnancy through parturition. The ePPND also incorporates some EFD endpoints such as post-partum radiographic skeletal assessments and visceral malformation assessment at necropsy of offspring

necropsy to look for abnormalities, but also to record, for example, the presence and character of stomach contents. When there is morbidity and mortality across a whole group, euthanasia of a subset of pups at that dose level can be considered to collect blood for exposure and/or diagnostic purposes, and it can be helpful to consult a relevant histologic atlas [1]. Exposure levels of the pups can be measured, for example from culled pups early in the postnatal period, and can contribute to study interpretation overall, although this is not routinely done. Even when pup exposure is confirmed, it can be complicated to determine the extent of gestational placental versus lactational transfer, the impact of the postnatally immature gastrointestinal system, and the translational relevance to humans [73, 74].

Although body weight increases in the pups are a sensitive measure of growth, it is recommended that some additional preweaning landmarks of development and reflex ontogeny are also assessed. A few options noted in the S5 (R3) are eye opening, pinna unfolding, surface righting, auditory startle, air righting and responses to light; it is not expected that all of these are monitored or that these are the only options available. For example, anogenital distance and nipple retention in male pups might be considered if endocrine effects are expected, for example, on androgen signaling [55, 75, 76]. The selection of appropriate preweaning developmental endpoints can be based on existing product knowledge to best inform the risk assessment for patients.

At the time of weaning, the dams are euthanized for a macroscopic examination of viscera, including an assessment of implantation sites to compare with their litter size. The remaining pups from each litter are allocated to both general and specific postweaning endpoints. For example, all pups will continue to be monitored daily to detect altered behavior, locomotion, morbidity, or mortality, and will be weighed weekly to assess continued growth. Food consumption is an optional endpoint for the postweaning phase, and may be conducted on a 'per cage' basis when pups are socially housed. As the pups mature, the timing of preputial separation (PS) in males and vaginal opening (VO) in females will be recorded. It should be noted that vaginal opening in rats is closely linked to the timing of first estrus and first ovulation, but it is not as closely correlated in the mouse [29]. Individual animals, often 1 male and 1 female from each litter, can also be used for additional detailed assessments. These typically will include (1) an assessment of sensory functions and motor activity, (2) a complex learning and memory task, such as a water maze, (3) a test of breeding (non-littermate pairs from each dose group), and (4) necropsy with macroscopic and microscopic evaluation of tissues. Depending on the litter size and mechanistic concerns, additional testing can be included. Also, necropsy endpoints can be conducted subsequent to other testing (sensory/motor or learning/memory).

6.5 Pre- and Postnatal Development Study, Enhanced Design, in Macaques

When the nonhuman primate is the only pharmacologically relevant species for a biopharmaceutical such as a monoclonal antibody, a pre- and postnatal development study, enhanced design (ePPND) can be appropriate [8, 77]. This single study approach effectively combines segments C through E, based on the expectation that there will be a low placental transfer of antibodies to the fetus during the period of organogenesis [78]. A schematic for the ePPND study in macaques is also included in Fig. 6.5. In this setting, fertility is only assessed indirectly via pathology of the reproductive tract tissues [5].

The ePPND has some species-dependent differences as compared to a rodent EFD or PPND study. Macaques almost always have a single offspring per pregnancy, and adverse pregnancy outcomes are not uncommon, even in control animals. Therefore, it can be difficult to align on optimal group size. The ICH S5 (R3) recommends a minimum group size of 16 pregnant dams, although there is also literature to suggest that 14 can be adequate [5, 79]. As always, interpretability is somewhat dependent on effect size, so a large effect can be apparent with even smaller group sizes, but confidence in setting a NOAEL increases with a larger group design. Dose selection can also be challenging, especially if the desired pharmacology reflects saturation of the target; in this case, establishing a high multiple of the clinical exposure is not always feasible, and additional dose ranging may not add value.

In an ePPND study, pregnancy outcomes are emphasized in reporting, with adverse outcomes that include fetal losses (resorption, abortion or fetal death detected via gestational ultrasound), stillbirths and early postnatal mortality. Morbidity and mortality can be associated with premature birth or dystocia, so gestation length is also tracked for each dam. The postpartum period assessments can include observation of maternal/infant behavior and nursing, neurobehavioral and reflex assessment of the offspring, a radiographic skeletal assessment when the offspring are about a month old, and additional monitoring of growth and mobility milestones during the lactation phase. Offspring are typically followed for at least 1 month, and often several months, with study endpoints driven by specific program concerns. These can include PK (often single time points, usually paired with the dam), immunotoxicity, and terminal phase necropsy of offspring for visceral assessment and pathology.

Overall, ePPND studies in macaques require long lead times and are subject to the availability of mature animals for breeding and appropriate housing. These studies can take several months to enroll the requisite number of pregnant dams, and have both a relatively long (~4–5 months) dosing phase as well as postnatal monitoring; thus the in-life phase of the study often exceeds 1 year. With the currently limited availability of macaques for this type of study, both regulators and drug developers are seeking alternative solutions for DART testing for highly specific large molecules [64]. A weight-of-evidence approach can help prioritize the most impactful studies to conduct [80].

6.6 Concluding Remarks

DART assessments are a critical component of the drug development process, and ultimately contribute to drug labeling and use. There are both detailed protocols and regulatory guidance to aid drug developers in designing an informative nonclinical DART program for risk assessment for patients. This is a broad topic; patients want to know about the potential impact of pharmaceuticals on their current and future fertility, they want to know if they need to observe stringent pregnancy prevention, and they want to know what to do if they become pregnant. There is a wealth of historical knowledge from nonclinical mammalian DART studies which, combined with clinical data from various types of pregnancy registry, can help inform patient risk assessment. At present, the testing paradigms outlined in the recent S5(R3) provide clear guidance to drug developers, with some flexibility for individual programs. As additional alternative approaches continue to be explored and qualified, the paradigms for the most relevant and informative testing may shift. Continued diligence and vigilance in cross-functional communication are critical to continue to advance our ultimate objective of clear and appropriate risk assessment for patients.

Conflict of Interest The author declares no conflict of interest with the content of this chapter. At the time of writing, the author was an employee of Genentech, A Member of the Roche Group.

References

1. Parker GA, Picut CA (eds) (2016) Atlas of histology of the juvenile rat. Academic Press, London, 462p
2. Gupta RC (ed) (2011) Reproductive and developmental toxicology. Academic Press, London, 1220p
3. Baldock R, Bard J, Davidson D, Morriss-Kay G (eds) (2015) Kaufman's atlas of mouse development supplement: with coronal sections. Academic Press, London, 344p
4. Barrow PC (ed) (2013) Teratogenicity testing, vol 947: Methods in molecular biology. Humana Press, 599p. https://link.springer.com/book/10.1007/978-1-62703-131-8. Accessed 14 Dec 2022
5. ICH (2021) S5(R3) detection of reproductive and developmental toxicity for human pharmaceuticals. https://www.fda.gov/regulatory-information/search-fda-guidance-documents/s5r3-detection-reproductive-and-developmental-toxicity-human-pharmaceuticals. Accessed 9 Dec 2022
6. ICH (2010) M3(R2) nonclinical safety studies for the conduct of human clinical trials and marketing authorization for pharmaceuticals. https://www.fda.gov/regulatory-information/search-fda-guidance-documents/m3r2-nonclinical-safety-studies-conduct-human-clinical-trials-and-marketing-authorization. Accessed 13 Dec 2022
7. ICH (2013) M3(R2) nonclinical safety studies for the conduct of human clinical trials and marketing authorization for pharmaceuticals: questions and answers. https://www.fda.gov/regulatory-information/search-fda-guidance-documents/m3r2nonclinical-safety-studies-conduct-human-clinical-trials-and-marketing-authorization. Accessed 13 Dec 2022
8. ICH (2012) S6 (R1) preclinical safety evaluation of biotechnology derived pharmaceuticals. https://www.fda.gov/regulatory-information/search-fda-guidance-documents/s6r1-preclinical-safety-evaluation-biotechnology-derived-pharmaceuticals. Accessed 13 Dec 2022
9. ICH (2018) S9 nonclinical evaluation for anticancer pharmaceuticals questions and answers. https://www.fda.gov/regulatory-information/search-fda-guidance-documents/s9-nonclinical-evaluation-anticancer-pharmaceuticals-questions-and-answers. Accessed 14 Dec 2022
10. ICH (2010) S9 Nonclinical evaluation for anticancer pharmaceuticals. https://www.fda.gov/regulatory-information/search-fda-guidance-documents/s9-nonclinical-evaluation-anticancer-pharmaceuticals. Accessed 14 Dec 2022
11. Bowman CJ, Becourt-Lhote N, Boulifard V, Cordts R, Corriol-Rohou S, Enright B et al (2022) Science-based approach to harmonize contraception recommendations in clinical trials and pharmaceutical labels. Clin Pharmacol Ther. https://doi.org/10.1002/cpt.2602
12. Scialli AR, Bailey G, Beyer BK, Bøgh IB, Breslin WJ, Chen CL et al (2015) Potential seminal transport of pharmaceuticals to the conceptus. Reprod Toxicol 58: 213–221
13. Prell RA, Halpern WG, Rao GK (2016 May) Perspective on a modified developmental and reproductive toxicity testing strategy for cancer immunotherapy. Int J Toxicol 35(3):263–273
14. FDA (2019) Investigational enzyme replacement therapy products: nonclinical assessment. https://www.fda.gov/regulatory-information/search-fda-guidance-documents/investigational-enzyme-replacement-therapy-products-nonclinical-assessment. Accessed 14 Dec 2022
15. Clements JM, Hawkes RG, Jones D, Adjei A, Chambers T, Simon L et al (2020) Predicting the safety of medicines in pregnancy: a workshop report. Reprod Toxicol 93:199–210

16. McNerney M, Potter D, Augustine-Rauch K, Barrow P, Beyer B, Brannen K et al (2021) Concordance of 3 alternative teratogenicity assays with results from corresponding in vivo embryo-fetal development studies: final report from the International Consortium for Innovation and Quality in Pharmaceutical Development (IQ) DruSafe working group 2. Regul Toxicol Pharmacol 124:104984
17. Song YS, Dai MZ, Zhu CX, Huang YF, Liu J, Zhang CD et al (2021) Validation, optimization, and application of the zebrafish developmental toxicity assay for pharmaceuticals under the ICH S5(R3) guideline. Front Cell Dev Biol 9:721130
18. Marx U, Akabane T, Andersson TB, Baker E, Beilmann M, Beken S et al (2020) Biology-inspired microphysiological systems to advance patient benefit and animal welfare in drug development. ALTEX 37(3):365–394
19. Vidal JD, Colman K, Bhaskaran M, de Rijk E, Fegley D, Halpern W et al (2021) Scientific and Regulatory Policy Committee Best Practices: documentation of sexual maturity by microscopic evaluation in nonclinical safety studies. Toxicol Pathol 49(5): 977–989
20. Picut CA, Ziejewski MK, Stanislaus D (2018) Comparative aspects of pre- and postnatal development of the male reproductive system. Birth Defects Res. 110(3):190–227
21. Picut CA, Remick AK (2017) Impact of age on the male reproductive system from the pathologist's perspective. Toxicol Pathol 45(1):195–205
22. Marty MS, Sue Marty M, Chapin RE, Parks LG, Thorsrud BA (2003) Development and maturation of the male reproductive system. Birth Defects Res B Dev Reprod Toxicol 68:125–136
23. Li X, Santos R, Bernal JE, Li DD, Hargaden M, Khan NK (2022) Biology and postnatal development of organ systems of cynomolgus monkeys (*Macaca fascicularis*). J Med Primatol. https://onlinelibrary.wiley.com/doi/10.1111/jmp.12622
24. Vidal JD, Whitney KM (2014) Morphologic manifestations of testicular and epididymal toxicity. Spermatogenesis 4(2):e979099
25. Creasy DM, Chapin RE (2018) Male reproductive system. In: Fundamentals of toxicologic pathology, 3rd edn. Academic Press, London, pp 459–516. https://doi.org/10.1016/b978-0-12-809841-7.00017-4
26. Creasy DM, Chapin RE (2014) Testicular and epididymal toxicity: pathogenesis and potential mechanisms of toxicity. Spermatogenesis 4:e1005511. https://doi.org/10.1080/21565562.2014.1005511
27. Halpern WG, Ameri M, Bowman CJ, Elwell MR, Mirsky ML, Oliver J et al (2016) Scientific and Regulatory Policy Committee points to consider review: inclusion of reproductive and pathology end points for assessment of reproductive and developmental toxicity in pharmaceutical drug development. Toxicol Pathol 44(6):789–809
28. Dixon D, Alison R, Bach U, Colman K, Foley GL, Harleman JH et al (2014) Nonproliferative and proliferative lesions of the rat and mouse female reproductive system. J Toxicol Pathol 27(3–4):1S. https://doi.org/10.1293/tox.27.1S
29. Laffan SB, Posobiec LM, Uhl JE, Vidal JD (2018) Species comparison of postnatal development of the female reproductive system. Birth Defects Res. 110(3):163–189
30. Woicke J, Al-Haddawi MM, Bienvenu JG, Caverly Rae JM, Chanut FJ, Colman K et al (2021) International Harmonization of Nomenclature and Diagnostic Criteria (INHAND): nonproliferative and proliferative lesions of the dog. Toxicol Pathol 49(1):5–109
31. Brändli-Baiocco A, Balme E, Bruder M, Chandra S, Hellmann J, Hoenerhoff MJ et al (2018) Nonproliferative and proliferative lesions of the rat and mouse endocrine system. J Toxicol Pathol 31-(3 suppl):1S–95S
32. Everds NE, Snyder PW, Bailey KL, Bolon B, Creasy DM, Foley GL et al (2013) Interpreting stress responses during routine toxicity studies: a review of the biology, impact, and assessment. Toxicol Pathol 41(4):560–614
33. Uphouse L (2011) Stress and reproduction in mammals. In: Hormones and reproduction of vertebrates, volume 5 (Mammals). Academic Press, London, pp 117–138. https://doi.org/10.1016/b978-0-12-374928-4.10007-0
34. Chapin RE, Creasy DM (2012) Assessment of circulating hormones in regulatory toxicity studies II. Male reproductive hormones. Toxicol Pathol 40: 1063–1078. https://doi.org/10.1177/0192623312443321
35. Andersson H, Rehm S, Stanislaus D, Wood CE (2013) Scientific and regulatory policy committee (SRPC) paper: assessment of circulating hormones in nonclinical toxicity studies III. Female reproductive hormones. Toxicol Pathol 41(6):921–934
36. Mitchard T, Jarvis P, Stewart J (2012) Assessment of male rodent fertility in general toxicology 6-month studies. Birth Defects Res B Dev Reprod Toxicol 95(6):410–420
37. Takakura I, Creasy DM, Yokoi R, Terashima Y, Onozato T, Maruyama Y et al (2014) Effects of male sexual maturity of reproductive endpoints relevant to DART studies in Wistar Hannover rats. J Toxicol Sci 39(2):269–279
38. Working PK (1988) Male reproductive toxicology: comparison of the human to animal models. Environ Health Perspect 77:37–44
39. Scialli AR, Clark RV, Chapin RE (2018) Predictivity of nonclinical male reproductive findings for human effects. Birth Defects Res. 110(1):17–26
40. Schmidt A, Schmidt A, Markert UR (2021) The road (not) taken – Placental transfer and interspecies differences. Placenta 115:70–77
41. Powles-Glover N, Mitchard T, Stewart J (2015) Time course for onset and recovery from effects of a novel

male reproductive toxicant: Implications for apical preclinical study designs. Birth Defects Res B Dev Reprod Toxicol 104(3):91–99
42. Cora MC, Kooistra L, Travlos G (2015) Vaginal cytology of the laboratory rat and mouse: review and criteria for the staging of the estrous cycle using stained vaginal smears. Toxicol Pathol 43(6):776–793
43. Byers SL, Wiles MV, Dunn SL, Taft RA (2012) Mouse estrous cycle identification tool and images. PloS One 7(4):e35538
44. Ajayi AF, Akhigbe RE (2020) Staging of the estrous cycle and induction of estrus in experimental rodents: an update. Fertil Res Pract 6:5
45. Vargesson N (2019) The teratogenic effects of thalidomide on limbs. J Hand Surg Eur 44(1):88–95
46. Ito T, Handa H (2012) Deciphering the mystery of thalidomide teratogenicity. Congenit Anom 52:1–7. https://doi.org/10.1111/j.1741-4520.2011.00351.x
47. Ito T, Handa H (2020) Molecular mechanisms of thalidomide and its derivatives. Proc Jpn Acad Ser B Phys Biol Sci 96(6):189–203
48. Kim JH, Scialli AR (2011) Thalidomide: the tragedy of birth defects and the effective treatment of disease. Toxicol Sci 122(1):1–6
49. Lee ST, Welch KD, Panter KE, Gardner DR, Garrossian M, Chang CWT (2014) Cyclopamine: from cyclops lambs to cancer treatment. J Agric Food Chem 62(30):7355–7362
50. Wang W, Jian Y, Cai B, Wang M, Chen M, Huang H (2017) All-trans retinoic acid-induced craniofacial malformation model: a prenatal and postnatal morphological analysis. Cleft Palate Craniofac J 54(4): 391–399
51. Coluccia A, Belfiore D, Bizzoca A, Borracci P, Trerotoli P, Gennarini G et al (2008) Gestational all-trans retinoic acid treatment in the rat: neurofunctional changes and cerebellar phenotype. Neurotoxicol Teratol 30(5):395–403
52. Yoshitaka H, Kaneki N (1988) How it came about the finding of methyl mercury poisoning in Minamata district: identification of human teratogens. Congenit Anom 28:S59–S69
53. Léonard A, Jacquet P, Lauwerys RR (1983) Mutagenicity and teratogenicity of mercury compounds. Mutat Res 114(1):1–18
54. Beekhuijzen M (2017) The era of 3Rs implementation in developmental and reproductive toxicity (DART) testing: current overview and future perspectives. Reprod Toxicol 72:86–96
55. Palermo CM, Foreman JE, Wikoff DS, Lea I (2021) Development of a putative adverse outcome pathway network for male rat reproductive tract abnormalities with specific considerations for the androgen sensitive window of development. Curr Res Toxicol 2:254–271
56. Barrow P (2023) An assessment of the reliability of 52 enhanced preliminary embryofetal development studies to detect developmental toxicity. Birth Defects Res 115:218. https://doi.org/10.1002/bdr2.2108
57. Scialli AR, Daston G, Chen C, Coder PS, Euling SY, Foreman J et al (2018) Rethinking developmental toxicity testing: Evolution or revolution? Birth Defects Res. 110(10):840–850
58. Yin L, Wang XJ, Chen DX, Liu XN, Wang XJ (2020) Humanized mouse model: a review on preclinical applications for cancer immunotherapy. Am J Cancer Res 10(12):4568–4584
59. Morton JJ, Alzofon N, Jimeno A (2020) The humanized mouse: emerging translational potential. Mol Carcinog 59(7):830–838
60. Barrow P (2022) Review of embryo-fetal developmental toxicity studies performed for pharmaceuticals approved by FDA in 2020 and 2021. Reprod Toxicol 112:100–108
61. Theunissen PT, Beken S, Beyer BK, Breslin WJ, Cappon GD, Chen CL et al (2016) Comparison of rat and rabbit embryo–fetal developmental toxicity data for 379 pharmaceuticals: on the nature and severity of developmental effects. Crit Rev Toxicol 46(10): 900–910
62. Barrow P, Clemann N (2021) Review of embryo-fetal developmental toxicity studies performed for pharmaceuticals approved by FDA in 2018 and 2019. Reprod Toxicol 99:144–151
63. Barrow P (2018) Review of embryo-fetal developmental toxicity studies performed for pharmaceuticals approved by FDA in 2016 and 2017. Reprod Toxicol 80:117–125
64. FDA (2022) Nonclinical considerations for mitigating nonhuman primate supply constraints arising from the COVID-19 pandemic. https://www.fda.gov/regulatory-information/search-fda-guidance-documents/nonclinical-considerations-mitigating-nonhuman-primate-supply-constraints-arising-covid-19-pandemic
65. Andrews PA, Blanset D, Costa PL, Green M, Green ML, Jacobs A et al (2019) Analysis of exposure margins in developmental toxicity studies for detection of human teratogens. Regul Toxicol Pharmacol 105: 62–68
66. Andrews PA, McNerney ME, DeGeorge JJ (2019) Exposure assessments in reproductive and developmental toxicity testing: an IQ-DruSafe industry survey on current practices and experiences in support of exposure-based high dose selection. Regul Toxicol Pharmacol 107:104413
67. Shroukh WA, Steinke DT, Willis SC (2020) Risk management of teratogenic medicines: a systematic review. Birth Defects Res. 112(20):1755–1786
68. Scialli AR (2020) Teratogen? Birth Defects Res. 112(15):1103–1104
69. Janer G, Slob W, Hakkert BC, Vermeire T, Piersma AH (2008) A retrospective analysis of developmental toxicity studies in rat and rabbit: what is the added value of the rabbit as an additional test species? Regul Toxicol Pharmacol 50(2):206–217
70. Scialli AR (1987) Is stress a developmental toxin? Reprod Toxicol 1(3):163–171
71. Makris SL, Solomon HM, Clark R, Shiota K, Barbellion S, Buschmann J et al (2009) Terminology of developmental abnormalities in common laboratory mammals (version 2). Reprod Toxicol Congenit Anom 28(86):371–434

72. Posobiec LM, Laffan SB (2021) Dose range finding approach for rodent preweaning juvenile animal studies. Birth Defects Res. 113(5):409–426
73. Neal-Kluever A, Fisher J, Grylack L, Kakiuchi-Kiyota S, Halpern W (2019) Physiology of the neonatal gastrointestinal system relevant to the disposition of orally administered medications. Drug Metab Dispos 47:296–313. https://doi.org/10.1124/dmd.118.084418
74. Downes NJ (2018) Consideration of the development of the gastrointestinal tract in the choice of species for regulatory juvenile studies. Birth Defects Res 110(1): 56–62
75. Schwartz CL, Christiansen S, Hass U, Ramhøj L, Axelstad M, Löbl NM et al (2021) On the use and interpretation of areola/nipple retention as a biomarker for anti-androgenic effects in rat toxicity studies. Front Toxicol 3:730752
76. Pedersen EB, Christiansen S, Svingen T (2022) AOP key event relationship report: linking androgen receptor antagonism with nipple retention. Curr Res Toxicol 3:100085. https://doi.org/10.1016/j.crtox.2022.100085
77. Weinbauer GF, Fuchs A, Niehaus M, Luetjens CM (2011) The enhanced pre- and postnatal study for nonhuman primates: update and perspectives. Birth Defects Res C Embryo Today 93(4):324–333
78. DeSesso JM, Williams AL, Ahuja A, Bowman CJ, Hurtt ME (2012) The placenta, transfer of immunoglobulins, and safety assessment of biopharmaceuticals in pregnancy. Crit Rev Toxicol 42(3):185–210
79. Luetjens CM, Fuchs A, Baker A, Weinbauer GF (2020) Group size experiences with enhanced pre- and postnatal development studies in the long-tailed macaque (Macaca fascicularis). Primate Biol 7(1):1–4
80. Rocca M, Morford LL, Blanset DL, Halpern WG, Cavagnaro J, Bowman CJ (2018) Applying a weight of evidence approach to the evaluation of developmental toxicity of biopharmaceuticals. Regul Toxicol Pharmacol 98:69–79

7 Genetic Toxicology Studies

Ramadevi Gudi and Gopala Krishna

Abstract

Genetic toxicology is a sub-discipline of toxicology dealing with the potential of compounds to cause damage to the genetic material, deoxyribonucleic acid (DNA), by various endogenous or exogenous mechanisms. DNA damage can range from gene mutations to large chromosomal alterations. Such DNA damage is implicated in many human diseases and disorders including heritable effects and the multistep process of malignancy (cancer). It is vital to screen for compounds with potential genotoxic activity for which there is possible human exposure to control, eliminate, and prevent unwarranted exposure to such compounds. In this chapter, protocols for standard genetic toxicology assays employed to evaluate pharmaceuticals which are accepted by regulatory agencies globally are discussed.

R. Gudi (✉)
Division of Pharmacology and Toxicology, Office of Cardiology, Hematology, Endocrinology and Nephrology, CDER, US Food and Drug Administration, Silver Spring, MD, USA
e-mail: Ramadevi.gudi@fda.hhs.gov

G. Krishna
Pharmaceutical Drug Safety, Krislee Farms LLC, Mt Airy, MD, USA

Keywords

Mutation · Clastogen · Aneugen · Cell lines · Metabolic activation · Cytotoxicity · DNA damage · 3Rs

7.1 Introduction

A physical, chemical or biological agent that can cause DNA damage is known as a genotoxin or a mutagen. Genotoxic agents are ubiquitous. They are present in foods, pharmaceuticals, industrial chemicals, pesticides, environment, tobacco products and cosmic radiation [1, 2]. Humans are exposed to hazardous chemicals frequently by several means including occupational exposure. It has been reported that 4% of all human cancers are caused by occupational exposures and up to 30% among blue-collar workers [3, 4]. Thus human exposure to genotoxic agents is a likely possibility in this day in age. Exposure to genotoxic agents can potentially induce DNA damage in somatic cells which is associated with the multistep process of carcinogenicity and is considered a primary concern. DNA damage in germ cells may lead to heritable damage and/or infertility in offspring [4].

DNA damage or changes in DNA can occur within the sequence of nucleotide bases, or at a larger scale by alteration of chromosome structure (clastogenicity) or number (aneugenicity) induced by direct or indirect mechanisms. The

G. Jagadeesh et al. (eds.), *The Quintessence of Basic and Clinical Research and Scientific Publishing*,
https://doi.org/10.1007/978-981-99-1284-1_7

direct DNA damage is generally caused by chemicals like alkylating agents such as methyl methane sulfonate or active metabolites like benzo (a) pyrene that interact with the DNA helix. Indirect mechanism of DNA damage involves chemical interactions with other cellular macromolecules, e.g., microtubules/spindle apparatus. DNA damage is also produced by reactive oxygen species (ROS) involving environmental toxins that interact with DNA leading to adduct formations. Not all induced changes to the DNA lead to mutations. DNA damage events are common occurrence, and the human body receives tens of thousands of cellular DNA damage per day. When the cell sustains DNA damage, and if the damage is extensive, the cell undergoes apoptosis or programmed cell death, effectively releasing it from becoming a mutant cell. If the damage is less severe in single-strand DNA damage, it can be repaired by cellular DNA repair processes by means of base excision repair, nucleotide excision repair, or mismatch repair returning the cell to its undamaged state. If the damage is incorrectly repaired or incompletely repaired prior to DNA synthesis such as in double-strand breaks, this can lead to a fixed inappropriate base pairing and/or formation of a mutation. Note that not all types of DNA damages will result in a mutation and not all mutations will result in cancer or other health effects. The carcinogenic initiation process of a specific mutation is dependent on the gene and the tissue affected. Most mutations may be largely neutral (e.g., passenger mutations) [5]. However, mutations in genes involved in critical steps cellular and/or biochemical functional pathways (e.g., oncogene, tumor-suppressor gene, or a gene that controls the cell cycle) can result in a proliferative or survival advantage that yields expansion of a clonal cell population [6].

It is conceivable to assume that exposure to chemical mutagens has a high probability of being initiators of carcinogenesis (genotoxic carcinogens). It is important to screen for chemicals with potential mutagenic activity for which there is possible human exposure because carcinogenic effects of genotoxic carcinogens will not be evident for many years, ultimately defeating the purpose of clinical monitoring. Regulatory agencies worldwide have taken measures to detect and /or identify genotoxic compounds to control or eliminate levels prior to human exposure. Genotoxicity testing has become part of safety evaluations of all new chemical entities. The assay systems and the protocol designs used are determined by international consortiums that include participation of regional regulatory agencies for a variety of products. This chapter will focus on the genetic toxicology evaluation of pharmaceuticals for human use.

7.2 Global Approach to Genotoxicity Testing of Pharmaceuticals

In 1990, the International Council for Harmonisation (ICH) of Technical Requirements for Pharmaceuticals for Human Use was established. This organization brought together regulatory authorities and the pharmaceutical industry trade groups to harmonize technical aspects of safety testing for pharmaceuticals through the development of a series of guidelines [7]. The ICH provides globally accepted guidelines on specific aspects of regulatory genotoxicity tests for pharmaceuticals (ICH S2A) [8], A Standard Battery for Genotoxicity Testing of Pharmaceuticals dealing with (ICH S2B) [9]. In 2012, (ICH S2 R1) [for Genotoxicity testing and data interpretation for pharmaceuticals intended for human use, replaced S2A and S2B, [10]. This guideline provides approaches for genotoxicity testing on how and what assays to conduct, as well as data interpretations. In addition, ICH also recommends the latest Organization for Economic Co-operation and Development (OECD) [11] test guidelines and the reports from the International Workshops on Genotoxicity Testing (IWGT) as relevant references for developing genotoxicity assay protocols for pharmaceuticals for human use.

7.3 Genetic Toxicology Assays

Over the past 40 years, more than 200 genetic toxicology assay methods have been developed to measure DNA damage for use in the detection of mutagenic activity and assessment of potential human risks. The objectives of genetic toxicology testing are primarily two-fold, (1) hazard identification with respect to DNA damage at the early stages of drug development. Hazard identification allows for several predictions including a carcinogenic mode of action, heritable germ cell damage, and a study of DNA damage. (2) determine the mechanism of action for carcinogens (genotoxic vs. non-genotoxic agents). Genotoxicity studies are conducted in both in vitro and in vivo test systems to detect DNA damage. The types of DNA damage endpoints detected in the genetic toxicology assays include mutations, changes to chromosome structure or number, single strand (SSBs) and double strand DNA (DSBs) breaks, DNA repair and biomarkers of DNA damage. Additional changes in DNA damage include DNA cross-linking and DNA intercalation [12].

Out of several test methods, a standard 3-test battery has been recommended for genotoxicity assessment of pharmaceuticals for prediction of potential human risks, interpretation of study results, and with a main purpose of improving risk characterization for carcinogenic effects due to alterations in the genetic material (ICH S2 R1). The test battery incudes :

1. Assessment of mutagenicity in a bacterial reverse gene mutation test (commonly known as Ames Test).
2. Assessment of genotoxicity in mammalian cells in vitro.
   or
3. Assessment of genotoxicity in rodents in vivo.

The protocols for the genetic toxicology standard battery are based on the testing guidelines developed by OECD.

OECD 471 Bacterial Reverse Mutation Test
OECD 473 *In Vitro* Mammalian Chromosome Aberration Test
OECD 474 Mammalian Erythrocyte Micronucleus Test
OECD 475 Mammalian Bone Marrow Chromosome Aberration Test
OECD 476 *In Vitro* Mammalian Cell Gene Mutation Assays Using the *Hprt* and *xprt* locus
OECD 487 *In Vitro* Mammalian Cell Micronucleus Test
OECD 488 Transgenic Rodent Somatic and Germ Cell Gene Mutation Assays
OECD 489 *In Vivo* Mammalian Alkaline Comet Assay
OECD 490 *In Vitro* Mammalian Cell Gene Mutation Test using the thymidine Kinase Gene

This chapter will focus on (1) Bacterial Reverse Mutation Test, (2) In Vitro Mammalian Cell Micronucleus Test, and (3) Mammalian Erythrocyte Micronucleus Test, as these are the most commonly used standard test battery for genotoxicity assessment for pharmaceuticals.

7.4 Bacterial Reverse Gene Mutation Test

The bacterial reverse gene mutation test uses specific amino acid(s) requiring bacterial tester mutant strains of *Salmonella typhimurium (S. typhimurium)* and *Escherichia coli (E.coli)* to detect point mutations of one or a few DNA base-pairs [13–15]. In this test, the mutagenic potential of a test article or compound and its metabolites is evaluated by measuring its ability to induce reverse mutations at the histidine locus of several strains of *S. typhimurium* and at the tryptophan locus of *E coli* strain WP2 *uvr*A with or without the use of exogenous metabolic activation system called S9 containing liver microsomal fraction. This test system has been demonstrated to detect a diverse group of chemical mutagens [16]. The principle of this test is to determine whether a test article is mutagenic by testing its capacity to revert mutations present in the parent tester mutant bacteria (auxotroph) and restore the functional capability of the bacteria (prototroph or wild type) to synthesize an essential amino acid histidine/tryptophan, thus the name of the assay is defined as reverse mutation assay. The revertant colonies of the bacteria following the test article treatment are detected by their ability to grow in the absence of the amino acid required by the mutant parent test strain. It has been reported that bacterial mutation assays detect 90% of carcinogens as mutagens suggesting a high correlation between carcinogenicity and mutagenicity. However, this test may not be appropriate for certain classes of chemicals that are cytotoxic to bacteria (e.g., certain antibiotics) and those that interfere with the mammalian cell replication system (e.g., some topoisomerase inhibitors and some nucleoside analogs). In such cases, OECD 471 recommends additional mammalian mutation tests to be conducted.

7.4.1 Test System

The test strains *S. typhimurium* and *E. coli* each have defective genes involved in histidine and tryptophan, respectively, and thus have lost the ability to synthesize histidine or tryptophan. The recommended set of bacterial strains are *S. typhimurium* TA98, TA100, TA1535, TA1537 and *E. coli* WP2 *uvr*A [17]. These strains have been demonstrated to be sensitive to a wide range of mutagenic and carcinogenic chemicals. Each strain only detects a certain type of damage, thus typically five strains are needed to detect the different types of mutational events. These tester strains are primally sensitive to three types of mutational events: (1) basepair substitution mutation involving the substitution of one base to another, (2) frameshift mutation that alters the reading frame of the DNA by insertion or deletion of one or more bases, (3) DNA cross linking agents that covalently bind the two strands of the DNA double helix. Tester strains employed in the assay (TA98, TA100, TA1535, TA 1537, and TA 97) revert from auxotroph (histidine dependence) to prototrophs (histidine independence) when exposed to mutagens that induce a specific type of mutations as described above. Tester strains E.coli WP2 and TA102 are used to detect DNA cross-linking agents, oxidizing mutagens, and hydrazines.

Mutations are rare events and even the simplest organisms have very effective DNA repair systems. To enhance strain sensitivity to detect a wide variety of carcinogens as mutagens, many mutations have been introduced into the strains [18, 19]. The genotype of some of the more commonly used test strains are described in Green and Muriel [17].

7.4.2 Preparation of Tester Strains

The tester strains are obtained from a recognized supplier to ensure that the cells are well characterized and traceable and follow the instructions for cell culture maintenance, and testing [14, 20].

7.4.3 Exogenous Metabolic Activation

Certain promutagens and procarcinogens when tested in vitro require metabolic activation to convert them to their active forms (e.g., polyaromatic hydrocarbons, azo dyes and amines). In vitro test systems generally lack in vivo metabolic properties. Therefore, exogenous activation system (S9) has been developed [21]. Other liver enzyme inducing agents such as a combination of phenobarbital and β-naphthoflavone that are equivalent to Aroclor-induced S9 have also been developed [22, 23]. In vitro S9 activation systems cannot fully compensate mammalian in vivo metabolic systems. Thus, the in vitro test systems do not provide direct correlation on the mutagenic and/or carcinogenic potency of a test article to the in vivo mammalian systems. The S9 is commercially available and is stored frozen at −70 °C for 2 years. The S9 homogenate is characterized by several mutagens to verify for its activity to metabolize (e.g., benzo (a)pyrene, dimethylbenzanthracene, and 2-aminoanthracene) to forms mutagenic to *S. typhimurium* TA100. The S9 is used in combination with a co-factor-supplemented post-mitochondrial fraction (S9 mix) that is prepared immediately before use.

7.4.4 Test Article Formulation

Appropriate solvents to formulate the test articles are determined ahead of time. The commonly used solvents for Ames assay are water, dimethyl sulfoxide (DMSO), ethanol, and acetone to test up to 5 mg/plate. Other solvents may be used to achieve the 5 mg/plate dose including acetonitrile, dimethyl formamide, methanol, polyethylene glycol, propylene glycol, N-methyl-2-pyrrolidone, ethylene glycol methyl ether, and dimethylacetamide.

7.4.5 Positive Controls

The positive controls used concurrently with the assay are listed OECD test guideline 471. Results obtained from positive controls are used to ensure system performance and the validity of the test system but not to provide data for comparison with the test article.

7.4.6 Experimental Design

The bacterial mutation assays are commonly conducted in two phases. A preliminary toxicity assay to establish the concentration range of the test article to use in the definitive mutagenicity assay. The definitive mutagenicity assay is a confirmatory assay to evaluate the mutagenic potential of the test article.

The test article is added to the plates using the options of plate incorporation method or a preincubation method (Fig. 7.1). The plate incorporation method was originally developed by Ames et al. 1975 [13]. In this method, top agar, S9 mix or buffer or blank, tester strain and test article dosing solution are mixed with molten selective top agar, and then plated onto the surface bottom agar plates. After the plated mixture is solidified, the plates are incubated for approximately 48–72 h at 37 °C. This method is commonly used and is sensitive to many mutagens. It is not sensitive to all mutagens (e.g., nitrosamines, divalent metals, aldehydes, azo

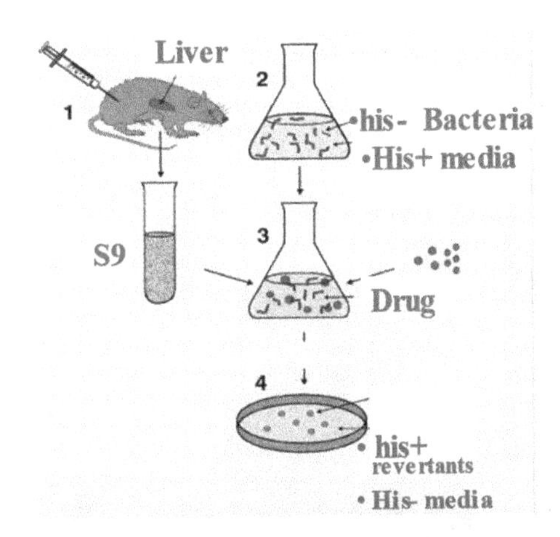

Fig. 7.1 Ames assay standard study design

dyes, pyrrolizidine alkaloids, allyl compounds, and nitro compounds). To overcome the limitation of the plate incorporation method, a preincubation method was developed by Yahagi et al. [24]. In the preincubation method, S9 mix or sham (buffer) is added to preheated culture tubes (37 °C) and to these tubes added tester strain and test article dosing solution. After mixing, the tubes are incubated for 20–60 min at 37 °C. At the end of the incubation, selective top agar is added to each tube and the mixture is overlaid onto the bottom agar plates. After the overlay is solidified, the plates are incubated for approximately 48–72 h at 37 °C. At the end of the incubation period, the plates are removed and the number of revertant colonies per plate are counted or stored in the refrigerator at 2–8 °C until colony counting could be conducted.

7.4.7 Cytotoxicity and Mutagenicity Evaluation

The cytotoxicity is assessed based on the characterization of the bacterial background lawn using a dissecting microscope [25]. Evidence of toxicity is scored in comparison to the vehicle control plate and documented along with the revertant counts for each of the test plate. Toxicity is assessed based on a decrease in the number of revertant colonies per plate and/or a thinning or disappearance of the bacterial background lawn.

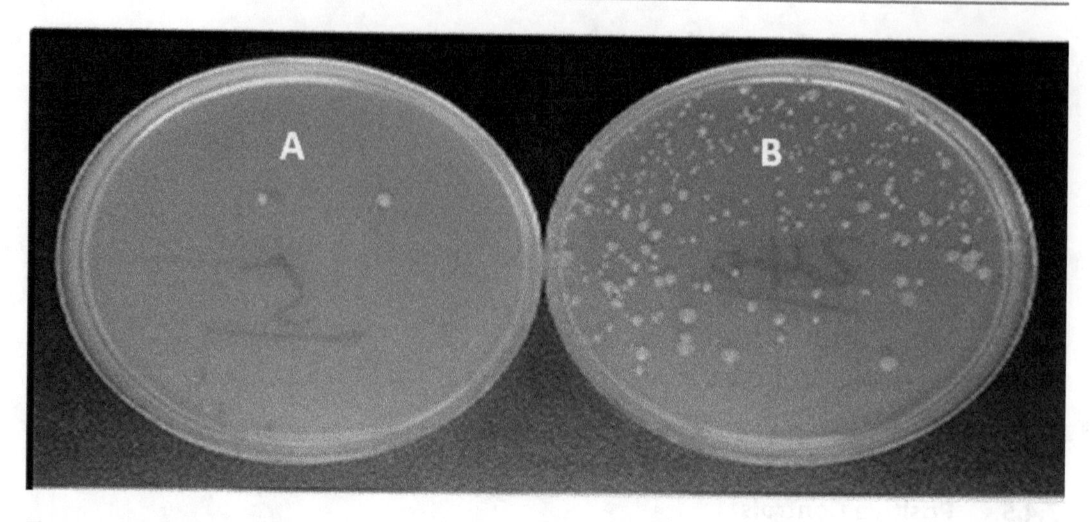

Fig. 7.2 Negative (**a**) and positive (**b**) plates from standard Ames assay showing revertant colonies. (Figure courtesy of Ms. Emily Dakoulas, Senior Scientist, Department of Genetic Toxicology, Inotiv)

The test article precipitation is evaluated after the incubation period by visual examination without magnification. A total five analyzable test article concentrations with a minimum of three concentrations that are non-toxic are required to assess the data. A concentration with a >50% reduction in the mean number of revertants per plate compared to the mean vehicle control value is considered toxic. As appropriate, colonies are enumerated either by hand or by colony counter machine. Revertant colonies generally appear as isolated colonies 1–2 mm size in diameter. Triplicate plating should be used at each test article concentration for adequate estimation of variation. The revertant colony data from individual plates, mean and standard deviation of colonies per concentration with and without S9 should be presented for the test article and positive and negative (untreated and/or solvent) controls.

The criteria to evaluate the study results are based on the evidence of dose-related increase of the mean number of revertant colonies at one or more test article concentrations relative to the vehicle control in presence and/or absence of metabolic activation systems (S9) [26]. For the test article to be considered as positive in an Ames test, it must induce a dose-related increase in the mean number of revertant colonies/plate at least in one tester strain over a minimum of two increasing concentrations of the test article. Biological relevance of the results should be considered first. Statistical methods may be used to support evaluating the test results [27], but not used solely as a determining parameter for a positive response. Generally, the data are concluded positive, if there is a dose-related increase of equal to or greater 2–3-fold in the number of revertant colonies in comparison to vehicle control and clearly above the corresponding historical vehicle control range (Fig. 7.2). Ames assay could be done in a miniaturized version (single plate containing 6 well, generally tested using 2 Ames tester strains TA98 and TA100) prior to standard Ames assay as screening and use small quantity of test article [28]. The results of this assay serve as a guide during the drug discovery process and assist in selecting suitable drug candidates and not intended for regulatory submission.

7.5 In Vitro Mammalian Cell Micronucleus Test

In vitro micronucleus (MN) assay detects damage to chromosomes in mammalian cells (primary cells or established cell lines). Chromosomal damage (aberration) involving structural damage to chromosome (chromosome breakage) or loss of whole chromosome(s) result in the formation of micronucleus or micronuclei (plural). Micronucleus contains chromosome fragment or whole chromosome that is not integrated into the main nucleus. A test article that induces structural chromosome aberrations or chromosome fragment(s) in cells is defined as a clastogen. A test article that interacts with the parts of mitotic and meiotic cell division (spindle apparatus), may lead to loss of whole chromosome(s) resulting in abnormal chromosome number in the cells is defined as an aneugen and the process is called aneuploidy. Thus micronucleus test allows for the detection of both clastogens and aneugens. Micronucleus test can also be performed in vivo utilizing the bone marrow and/or peripheral blood erythrocytes in rodents described later in this chapter. Micronucleus test is an alternative test for in vitro chromosome aberration test. Micronuclei can be detected in interphase cells, and the analysis of the damage is easier, and less time consuming, comparatively less subjective than the chromosome aberration assay. Micronuclei represent chromosome breakage (clastogen) or loss of chromosome (aneuploidy) that has been transmitted to the daughter cell (Fig. 7.3), unlike chromosome aberrations scored in metaphase cells may not be transmitted. In both cases, the cell with MN or aberrant chromosome may not survive depending on the extent of damage.

The in vitro MN test can be conducted with or without the actin polymerization inhibitor cytochalasin B (cyto B). Addition of Cyto B in the treatment medium blocks the separation of daughter cells after mitosis (cytokinesis block), leading to the formation of binucleated cells containing two main nuclei with a smaller nucleus or is otherwise called micronucleus (Fig. 7.4). The cytokinesis blocked method allows for easy identification and analysis of micronuclei. Presence of micronuclei in the binucleated cells represents completion of one mitosis and damage to DNA following the recent treatment with the test article [29, 30].

The in vitro MN test can be conducted in a variety of cell types, and in the presence or absence of cyto B [31, 32]. These include primary human peripheral lymphocytes (HPBL) in the presence of cytochalasin B and TK6, CHO, CHL and L5178Y cell lines, in the presence or absence of cytochalasin B. As an example, in vitro MN test in HPBL with the binucleated cyto B blocked method is discussed to evaluate the potential of a test article to induce micronuclei. The test should be conducted with and without S9 mix, as discussed previously.

7.5.1 Preparation of Cells

A peripheral blood sample (0.5 mL per culture tube) is obtained by venipuncture from a young healthy adult donor (approximately 18–35 years

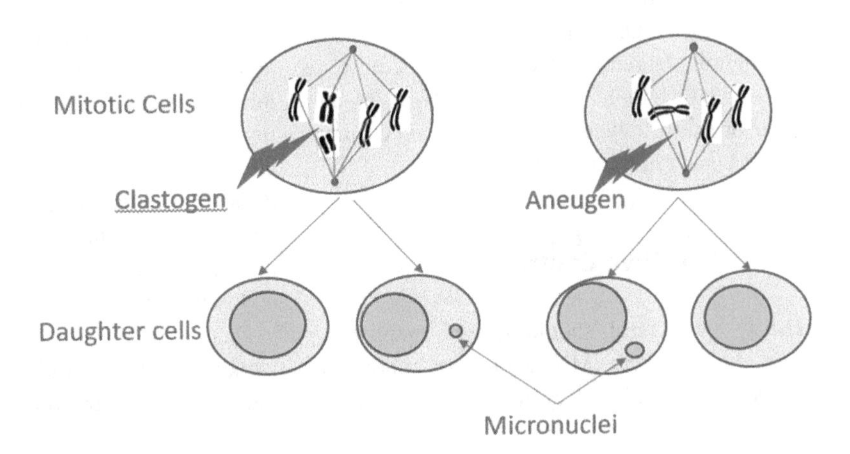

Fig. 7.3 Mechanism of micronucleus formation

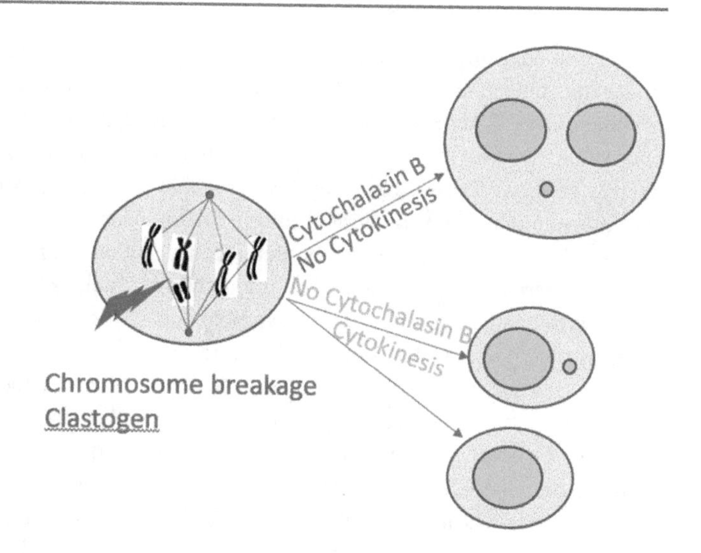

Fig. 7.4 Role of cytochalasin

of age), non-smoking, with no known recent exposures to genotoxic chemicals or radiation. The blood sample is collected directly into tubes containing sodium heparin and then held at room temperature prior to addition to the standard complete culture medium with 2% phytohemagglutinin (PHA) in centrifuge tubes. PHA is added to stimulate lymphocyte division. The cultures are incubated at 37 °C in a humidified atmosphere of 5% CO_2 in the air for 44–48 h prior to treating the cells with the test articles.

7.5.2 Test Article Formulation

Appropriate solvents to formulate the test articles are determined ahead of time. The commonly used solvents for in vitro mammalian cell assays are water, DMSO, ethanol, acetone to test up to 0.5 mg/mL. Organic solvents (DMSO, acetone and ethanol) should not exceed 1% (v/v) and water or saline solvents should not exceed more the 10% (v/v) in the treatment medium.

7.5.3 Positive Controls

The positive controls should be used concurrently with the definitive MN assay are mitomycin C (-S9, clastogen) and cyclophosphamide (+S9, clastogen), and vinblastine (−S9, Aneugen). Additional positive control chemicals can be found in the test guideline OECD 487.

Cyto B obtained commercially is prepared in DMSO to achieve a final concentration of 3–6 μg/mL in the culture medium.

7.5.4 Experimental Design

In vitro MN assay is performed in two phases, a preliminary cytotoxicity assay to determine the test article toxicity to the cells and to select appropriate concentrations to evaluate in the definitive MN assay.

In the preliminary cytotoxicity assay, the highest concentration recommended for pharmaceuticals is 1 mM or 0.5 mg/mL, whichever is lower for freely soluble or non-toxic test articles. For non-toxic articles with limited solubility, the highest concentration, should be the lowest concentration with visible precipitate in cultures, provided the presence of precipitate does not interfere with scoring of the slides. Evaluation of precipitation can be done by naked eye or light microscopy at the end of treatment. In addition, pH and osmolality of the highest test concentration prior to dosing is measured and pH is adjusted to neutral. The osmolality of the highest concentration, lowest precipitating concentration and highest soluble concentration in the treatment medium is measured. Concentrations with excessive osmolality of 20% higher than the vehicle should be avoided, because extreme physiological conditions including pH and osmolality can cause chromosome damage in vitro [33].

Test article cytotoxicity is measured using the cell growth inhibition method compared to vehicle control. The cell growth inhibition when using cyto B is measured using the cytokinesis blocked proliferation index (CBPI) method [34]. The CBPI indicates the average number of nuclei per cell and is used to calculate cell proliferation. For cytotoxic test articles, the test concentrations selected should cover a range cytotoxicity (no cytotoxicity to moderate cytotoxicity) to aid selection of appropriate concentrations for analysis.

Cell culture treatments are performed after 48 h stimulation with PHA. Duplicate cultures are included per concentration in the definitive micronucleus study. All culture tubes are centrifuged and refed with 5 mL complete medium for the tubes assigned to the non-activated treatment for 4 and 24 h or 4 mL culture medium for the tubes assigned to 4 h of S9-activated system + 1 mL of S9 mix. The cell cultures are treated with 50 to 500 μL of vehicle (depending on the type of formulation vehicle selected), or test article dosing solution. After the 4-h treatment in the presence or absence of S9-activated systems, the treatment medium is removed by centrifugation and washing with phosphate buffered saline (CMF-PBS), and re-fed with complete medium containing cyto B at 3–6 μg/mL and returned to the incubator for additional 20 h prior to harvesting. For the 24-h treatment in the absence of S9, cyto B is added at the beginning of the treatment with test article. The treatment is terminated at 72 h from the initiation of cell cultures or at the end of 24 h from the initiation of the treatment (Fig. 7.5).

7.5.5 Cell Harvesting and Staining

Cells are collected by centrifugation, and cell pellets are resuspended in a hypnotic solution (0.75 M potassium chloride) for 1–2 min and fixed with fixative (methanol: glacial acetic acid, 25:1 v/v). A few drops of the evenly suspended cell suspension are applied to the slides and air dried and stained with acridine orange. With acridine orange staining, the main nuclei and the MN stain fluoresce greenish yellow in the background of dull reddish orange cytoplasm (Fig. 7.6).

7.5.6 Slide Analysis

The criteria to score binucleated cells are those with, (1) main nuclei that are separate or that touch and even overlap as long nuclear boundaries can be distinguished and of approximately equal size, (2) main nuclei that are linked by nucleoplasmic bridges. Cells with trinucleated, quadrinucleated, or multinucleated cells or apoptotic should not be included in the scoring. Slides are evaluated by fluorescence microscopy using a blue excitation filter and a yellow barrier filter at ×1000 magnification.

For cytotoxicity, at least 500 cells per culture are evaluated by fluorescence microscopy using a blue excitation filter and a yellow barrier filter for the number of cells with one nucleus (mononucleated cell), two nuclei (binucleated cell) and three or more nuclei (multinucleated cell) for CBPI. The test article toxicity is demonstrated by a decrease in CBPI relative to the concurrent vehicle group.

$$\textbf{CBPI} = \frac{((\text{No. mononucleate cells}) + (2 \times \text{No. binucleate cells}) + (3 \times \text{No. multinucleate cells}))}{(\text{Total number of cells})}$$

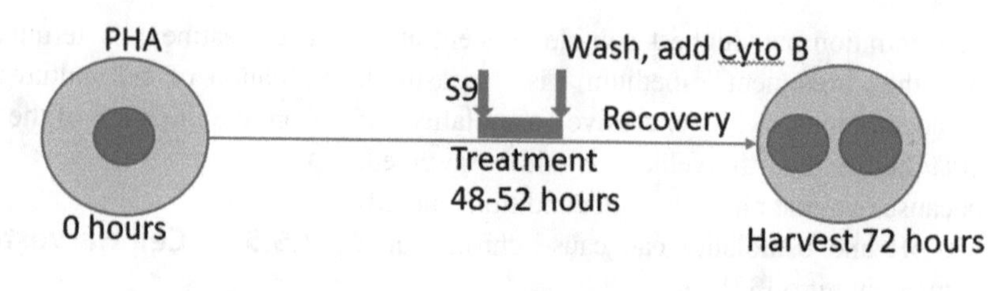

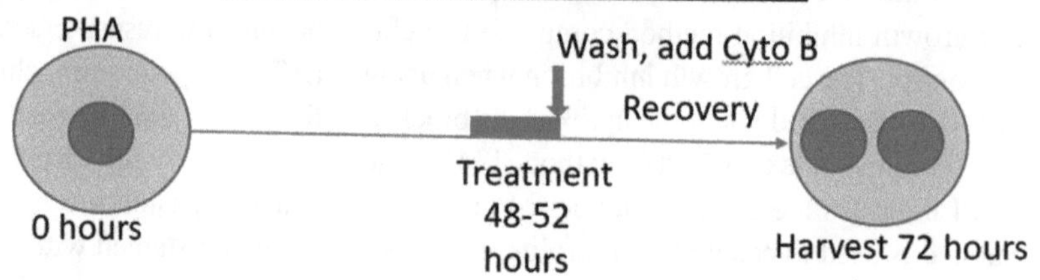

Fig. 7.5 In vitro MN treatment scheme

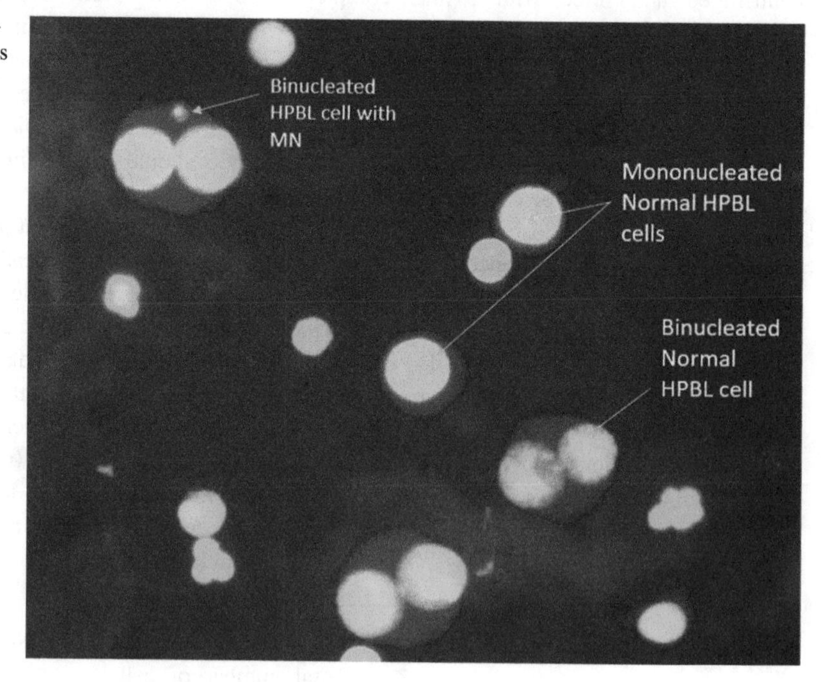

Fig. 7.6 Acridine orange-stained human lymphocytes showing mononucleated and binucleated cells with micronuclei.
(Figure courtesy of Dr. Shambhu Roy, Principal Scientist, Department of Genetic Toxicology, Inotiv)

Thus, a CBPI of 1 (all cells are mononucleate) is equivalent to 100% toxicity.

7.5.7 Selection of Treatment Concentrations

1. For toxic test articles, the highest concentration should be with 55 ± 5% cytotoxicity (reduction in CBPI compared to concurrent vehicle control).
2. For poorly soluble test articles that are not cytotoxic, the highest concentration should be the lowest concentration with visible precipitate at the end of the treatment.
3. For freely soluble and non-toxic test articles, the highest concentration should be 1 mM or 0.5 mg/mL, whichever is lower.

For the definitive MN assay, duplicate cultures should be used per concentration of test, vehicle and positive control articles. Duplicate cultures of one or two concentrations of positive controls are included in each treatment condition. Based on the toxicity profile of the test articles, 4–6 concentrations may be selected to test to cover the non-toxic to the toxic range. The toxicity profile of the test articles may differ in each treatment condition, as new test article concentrations are selected to test in the definitive assay. Concurrent toxicity profile using CBPI method should be determined in the definitive assay. Based on the toxicity profile (CBPI) of the test article, at least three concentrations of the test article, vehicle control and one concentration of positive control should be selected for MN endpoint scoring as described in the selection of treatment concentrations. The microscope slides for MN analysis are prepared as described above fixed and stained with acridine orange.

7.5.8 Micronucleus Scoring

To ensure unbiased data collection, all slides from the study are independently coded before the microscopic analysis. Slides are evaluated by fluorescence microscopy using a blue excitation filter and a yellow barrier filter at ×1000 magnification. The scoring criteria for MN are, (1) the diameter of MN should be less than one-third of the main nucleus, (2) MN should be separate from or marginally overlap with the main nucleus as long as there is clear identification of the nuclear boundary, and (3) MN should have similar staining as the main nucleus. The incidence of MN formation is determined in 1000 definitive binucleated cells (BN) from each culture (2000 binucleated cells from each concentration).

The micronucleated binucleate (MNBN) cells scoring data from the test article and positive controls are analyzed using statistical methods for significance compared to vehicle control groups. In addition, a trend analysis is performed for a dose-response relationship. The test article is considered negative if none of the test article concentrations demonstrate a statistically significant increase in the incidence of MNBN cells compared to vehicle control. The test article is considered positive, if there is a statistically significant increase in the incidence of MNBN cells at least in one concentration with an appropriate dose-response relationship in a trend test, and the increase is above the testing laboratory's historical negative/vehicle/solvent control data [35].

7.6 Mammalian Erythrocyte Micronucleus Test In Vivo

Like the in vitro genotoxicity assays, erythrocyte MN assay is the most widely used in vivo rodent assay to complete the 3-test standard genotoxicity assessment battery [36].

The original Test Guideline OECD 474 for MN assay was adopted in 1983 and has been revised based on the scientific progress made over the years [37]. Latest revisions such as inclusion of genotoxicity assessment into toxicology studies, flow cytometry scoring, and interpretation of the data comprehensively along with drug exposures so 3R (refine, reduce and replace) concept of minimizing animal use is practiced, yet quality data are generated [38, 39]. In vivo mammalian MN test considers metabolism, pharmacokinetics and DNA repair processes that contribute to the totality of whole animal responses to a test article and helps better understand and complements genotoxicity detected by an in vitro system. Like in vitro MN assay, in vivo mammalian MN test detects the test article's ability to cause damage to the chromosomes or spindle apparatus of erythroblasts. This test is generally conducted using erythrocytes sampled either in the bone marrow or peripheral blood cells of mice or rats.

The primary objective of conducting MN test is for hazard identification of test articles that cause cytogenetic damage as identified by using either microscope or automation, resulting in the formation of MN containing either chromosome fragments or lagging chromosome(s) during cell division/maturation process. As bone marrow erythroblasts develop into immature erythrocytes (referred to as a polychromatic erythrocyte (PCE) or reticulocyte), the main nucleus is extruded leaving behind any remaining MN in the cytoplasm. Identification of MN in these cells is easy because they lack the main nucleus. An increase in the frequency of MNPCEs in treated animals over the background levels is an indication of induced structural or numerical chromosomal aberrations. Newly formed MNPCEs are selectively identified and quantitated by special staining followed by either visual scoring using a microscope, or by an automated analysis such as flow cytometry. Enumerating sufficient number of PCEs in the peripheral blood or bone marrow is greatly facilitated by using an automated scoring platform which are acceptable alternative to manual evaluation. Automated systems that can enumerate MNPCEs include, but are not limited to, flow cytometers, image analysis platforms, and laser scanning cytometers. On as needed basis, understanding whether a MN in PCE is due to chromosome fragment(s) or from whole chromosomes, can be verified with specialized fluorescent staining specific to kinetochore or centromeric region of DNA, which are hallmarks of intact chromosomes. Lack of kinetochore or centromeric DNA indicates that the MN contains only fragments of chromosomes (clastogenicity), while the presence suggests chromosome loss (aneugenicity).

In the MN assay, rodent bone marrow is considered as the target tissue since erythrocytes are produced in this tissue. Enumerating MNPCE in peripheral blood is acceptable in other nonrodent species for which adequate sensitivity to detect article that causes structural or numerical chromosomal damage in these cells has been demonstrated and scientific rationale are provided. Frequency of MNPCE is the principal endpoint. However, the frequency of mature erythrocytes (routinely known as normochromatic erythrocytes, NCEs) that contain MN in the peripheral blood also can be used as an endpoint in species without strong splenic selection or interference against MN cells and when animals are treated continuously for a period that exceeds the lifespan of the erythrocyte in the species used (e.g., 4 weeks or more in the mouse). One of the limitations of this assay is: if there is evidence that the test article, or its metabolite(s), will not reach the target tissue, such as bone marrow, MN assay may not be appropriate.

7.6.1 Primary Basis of the Assay

Experimental animals are administered with the test article by an appropriate route, usually the same route as the clinical route of administration. When bone marrow is used for sample collection, the animals are humanely euthanized at an appropriate time(s) after test article treatment, the bone marrow is aspirated from femur, and slide preparations are made and stained. When peripheral blood is used, the blood is collected at an appropriate time(s) after test article treatment, and slide preparations are made and stained. When drug treatment is administered acutely, it is essential to select harvest times at which the treatment-related induction of MNPCE can be detected considering the pharmacokinetic properties of the test article. In the case of peripheral blood sampling, enough time must also have elapsed for these events to appear in circulating blood. Preparations are analyzed for the presence of MN, either by manual visualization using a microscope or by automated methods. Vehicle (solvent) and positive controls are used to verify the assay performance, thus, validity. However, if the laboratory is proficient in conducting MN assay, then, positive control slides/samples from a prior experiment may be incorporated for MN counting, thus reducing unnecessary animal use (3R concept). Overall, the historical control data in each lab and/or published literature facilitate the interpretation of MN assay results, particularly for biological significance and relevance to human conditions.

7.6.2 Assay Methodology

Test system—Generally, the animal selection consists of commonly used strains of healthy young adult mice, rats (6–10 weeks old) or another relevant mammalian species may be used. Conventional laboratory diets may be used with an unlimited supply of drinking water, being housed individually or in small groups (no more than five per cage) of the same sex and treatment group.

7.6.3 Preparation of Test Formulation Doses

Depending on type of test articles, for example: solids should be dissolved or suspended in appropriate vehicles. Liquids may be dosed directly or diluted before dosing. Inhalants can be administered as a gas, vapor, or a solid/liquid aerosol, depending on their physicochemical properties. Wherever possible, it is preferred to use of an aqueous solvent/vehicle, such as: water, physiological saline, methylcellulose solution, carboxymethyl cellulose sodium salt solution, or oil based solvents including olive oil and corn oil. Positive controls (e.g., cyclophosphamide, ethyl methanesulfonate, methyl methanesulfonate, mitomycin C (as clastogens); colchicine or vinblastine (as aneugens), article should reliably produce a detectable increase in MN frequency over the spontaneous level, at a single harvest time.

7.6.4 Assay Conducting Procedure

If there is no toxicological gender difference response then, genotoxicity studies could be conducted with one gender, thus, reducing animal use and practicing 3R principles. Nevertheless, data showing differences between genders (e.g., differing systemic toxicity, metabolism, bioavailability, bone marrow toxicity, etc. including e.g., in a dose-range-finding study), then, it is important to test in both sexes. Keeping maximum tolerated dose in mind, a minimum of five analyzable animals per sex per group with three test article dose groups, concurrent vehicle and positive control groups.

7.6.5 Selection of Doses

Selection of doses is based on a preliminary dose range-finding toxicity study or on existing toxicology information preferably in the same laboratory aimed to identify the maximum tolerated dose (MTD). A minimum of three dose levels separated by a factor of 2, but by no greater than 4 is used. For nontoxic test articles, the highest dose should be 2000 mg/kg/day. For toxic articles, the MTD should be the highest dose administered and the lower dose levels should generally cover a range from the MTD to a dose producing little or no toxicity. Concurrent negative control groups are included for each tissue collection time. When multiple collection times are used, a single positive control group is included at 24-hr sample collection time. Maximum volume that can be administered by gavage or injection at one time should be 5–10 mL/kg (exceptions up to a maximum of 20 mL/kg). Experimental animals should be observed for clinical signs, body weight changes and recorded daily following treatment. During dosing period, at least twice daily, all animals should be observed for morbidity and mortality. Animals with severe toxicities should be humanely euthanized with relevant documentation. Under certain circumstances, animal body temperature may be monitored considering that certain treatment induced hyper- and hypothermia have been implicated in causing spurious results leading to difficult interpretation of data. When possible, a blood sample should be taken at appropriate time (s) to permit investigation of the plasma levels of the test article to demonstrate that exposure of the bone marrow occurred, where warranted and where other exposure data do not exist.

7.6.6 Cell Processing

From the experimental animals, bone marrow cells are obtained using femurs or tibias, by aspirating using a needle attached to syringe containing 0.5–1 mL heat inactivated fetal bovine serum and collected into tubes. Cells are pelleted by centrifugation, and the supernatant is discarded leaving a small amount of serum with the pellet. Cells are re-suspended, and a small drop of the suspension is spread onto a clean glass slide (two slides per animal). Slides are air-dried and fixed in methanol and stained with acridine orange for microscopic evaluation. For the peripheral blood, small volumes of the peripheral blood sample are obtained by bleeding from the tail vein or other appropriate blood vessel according to animal welfare standards. Peripheral blood smear preparations are made and then stained for microscopy with acridine orange (Fig. 7.7) or fixed and stained appropriately for flow cytometric/automated analysis. Additional staining methods including anti-kinetochore antibodies, FISH with pancentromeric DNA probes, or primed *in situ* labelling with pancentromere-specific primers, together with appropriate DNA counterstaining are used to help identify the nature of the MN (chromosome/chromosomal fragment) to determine whether the mechanism of MN induction is due to clastogenic and/or aneugenic activity.

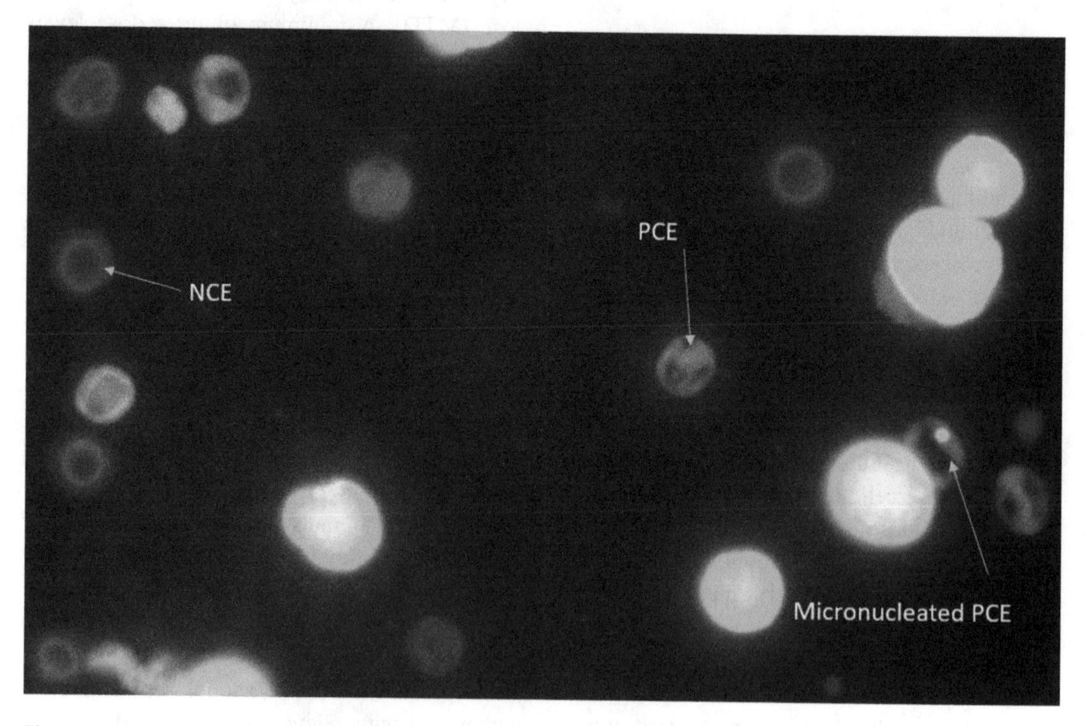

Fig. 7.7 Acridine orange-stained rat bone marrow cells. (Figure courtesy of Dr. Shambhu Roy, Principal Scientist, Department of Genetic Toxicology, Inotiv)

7.6.7 Enumeration of MN

Scoring of slides (including positive and negative controls), should be blind coded before counting to avoid scorer bias of the treatment condition; such coding, however, is not necessary when using automated scoring systems which do not rely on visual inspection and cannot be affected by operator bias. The proportion of PCEs (immature) among total (immature + mature) erythrocytes (TEs) is determined for each animal by counting a total of at least 500 TEs for bone marrow and 2000 TEs for peripheral blood to determine target tissue toxicity and drug exposure. At least 4000 PCEs per animal should be scored for the incidence of MNPCEs. Evidence of exposure of the bone marrow to a test article may include depression of the PCEs to TEs ratio or measurement of the plasma or blood levels of the test article. However, in the case of intravenous administration, evidence of exposure is not needed. Alternatively, absorption, distribution, metabolism and excretion (ADME) data, obtained in an independent study using the same route and the same species can be used to demonstrate bone marrow exposure. It is important to use appropriate statistical tests (Analysis of Variance and Armitage trend test) to aid significance. The criteria for a valid test are (1) Vehicle control MNPCEs must be within 95% control limits of the historical control range, (2) A statistically significant increase in MNPCE is noted in the positive control compared to concurrent negative control, (3) There should be generally, five animals/sex/group or dose. A negative result indicates that there is no significant increase in MNPCE at any dose level compared to the concurrent control. A positive result indicates that there is a significant dose-dependent increase in MNPCE at one or more dose levels compared to concurrent negative control. In cases where the response is not clearly negative or positive, further investigations including analyzing more cells or performing a repeat experiment using modified experimental conditions could be useful. In rare cases, even after further investigations, the data will preclude concluding that the test article produces either positive or negative results, and the study will therefore be concluded as equivocal.

7.7 Follow-Up Studies for Positive Genetic Toxicology Assays

Periodically genetic toxicology studies, particularly in vitro test systems, may produce positive results. Positive genetic toxicology results are expected due to direct interaction with genetic material and are not anticipated to detect non-genotoxic carcinogens (liver enzyme inducers, peroxisome proliferators, hormonal carcinogens). Published literature evaluating a large database of 700 chemicals suggests 75–95% of rodent non-carcinogens are positive in one or more in vitro genotoxicity assays leading to "false" positives or "misleading" positives [40]. It has been reported in a survey that approximately 30% of approved (marketed) drugs are positive in at least one in vitro genetic toxicology test [41]. In vitro mammalian cell assays are positive more often than Ames assay. A clear positive or negative result does not require a verification. Inconclusive results or equivocal results should be repeated with modified doses. There could be several confounding factors for the positive results, e.g., due to the presence of a mutagenic impurity, highly cytotoxic concentration, one outlying culture or animal, the magnitude of the response, reproducibility, etc. A weight-of-evidence approach is taken for regulatory decision making to assess the biological relevance of the positive finding to humans. Appropriate follow-up studies are conducted to verify the reproducibility of the initial positive results [42].

Per (ICHS2(R1)), If the Ames assay is positive, two appropriate in vivo assays should be conducted in two different tissues.

The follow-up assays recommended for Ames positive are

1. In vivo Big Blue transgenic mouse mutation assay with justification on the selection of tissue/organ
2. In vivo *Pig-A* Assay for Ames assay positive without S9

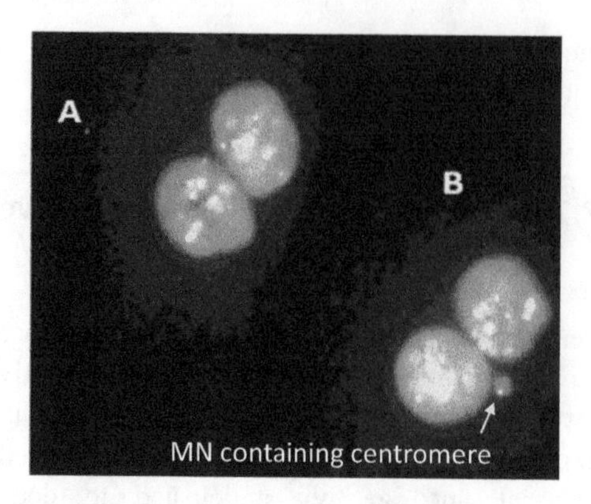

Fig. 7.8 Acridine orange stained binucleated cells (**a**) without MN (**b**) with MN containing centromere

A positive in vitro MN could be the result of either two different modes of action (MOA), a direct acting clastogen or a non-DNA reactive aneugen as discussed previously. It is important to distinguish between these two MOAs. The recommended assay for positive in vitro or in vivo MN is to conduct additional staining methods (anti-kinetochore antibodies, FISH with pancentromeric DNA probes) to identify the contents of MN (Fig. 7.8). A presence of centromere in the MN indicates aneugenic MOA and absence of centromere in the MN indicates clastogenic MOA [43].

If an in vitro chromosome damage assay is positive, then, a second in vivo assay such as comet assay in liver cells is recommended to further confirm the results (OECD TG 489). Under alkaline conditions (>pH 13), the comet assay can detect some forms of DNA damage (e.g., strand breaks, alkali labile sites, adducts). Following the test article treatment, the length of DNA movement from the nucleus is a measure of fragmented DNA (Fig. 7.9).

If the in vivo MN study is positive, while all the in vitro genetic toxicology studies are negative, a nongenotoxic MOA such as disturbed erythropoiesis or physiology (such as hypo/hyperthermia) could be the cause for a contributing factor [44]. Changes in body temperature during the test article treatment can induce MN in vivo. Follow-up studies for positive in vivo MN studies to repeat the in vivo MN study with implantable temperature transponders [45, 46].

7.8 Concluding Remarks

The usefulness of genetic toxicology data in drug development is hazard identification. Further, these studies would assist in the quantitative risk assessment in determining a safe dose considering parameters such as no observed adverse effect level (NOAEL) and/or a low observed adverse effect level (LOAEL). These data serve as a place holder and form the premises for rodent carcinogenicity data.

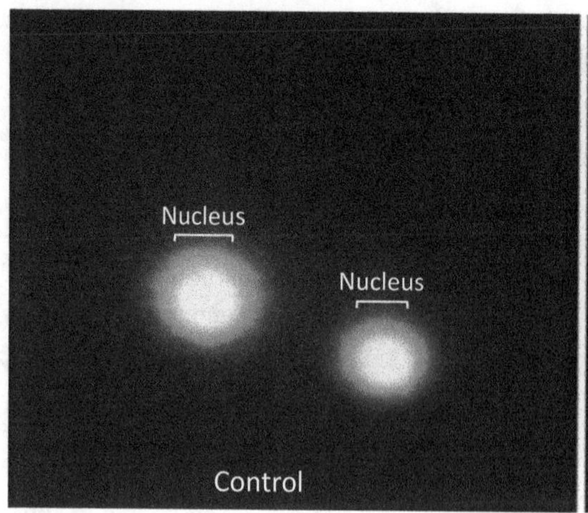

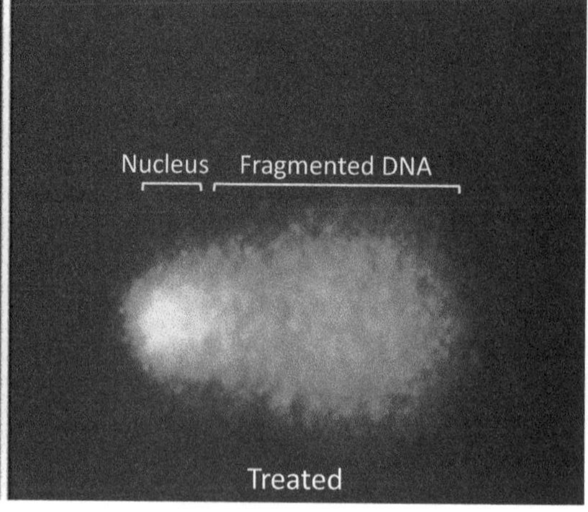

Fig. 7.9 Comet assay (liver cells)

Genotoxicity testing is a key early safety assessment during pharmaceutical development as they cover multiple mechanisms. As a first step, screening is considered with the publicly existing information on the class of drug effects using artificial intelligence techniques, including *in silico* assessment with structure activity relationship (SAR) software such as DEREK, Leadscope. These tools help provide some guidance regarding the structural alerts and help modify the molecule or select a potential drug molecule without the genotoxic safety liability. Then, biological genotoxicity screening tests (non-GLP) such as mini-Ames and mini-in vitro MN assays are used to minimize the need for the use of huge quantity of test article and for a quick read on the genotoxic potential during initial stages of drug discovery. These assays provide further guidance on the potential genotoxicity, if any. As a critical regulatory step, initially 2 in vitro tests that follow good laboratory practices (GLPs) are required to support defining first dose in human as part of Investigational New Drug Applications (IND) for Phase 1. These are: a bacterial mutation test and an in vitro mammalian chromosome damage test such as: chromosome aberration or MN. To support Phase 2 clinical trials which involves repeated exposure to the investigational pharmaceutical, an in vivo test is conducted, typically consisting of an in vivo MN assay in rodents. The in vivo test can be incorporated into a repeat dose general toxicology study. These three tests, referred to as the 'standard battery' are most often completed prior to submission of an IND. These three tests constitute Option 1 for the standard battery (S2R1). Option 2 standard battery that is considered equally suitable includes, generally, a test for bacterial mutagenesis test, prior to Phase 1 clinical trial, and two in vivo tests in two different tissues, (1) mammalian erythrocyte micronucleus test as described above and (2) a Comet assay in liver or DNA strand breakage assay prior to Phase 2 clinical trial. If the results are negative in all three assays (Ames test, In vitro MN test and In vivo MN test in rodents), then, the test article is considered to be non-genotoxic, and supportive of clinical trials in healthy volunteers and the intended patient population. If the results are clearly positive in all three assays, then, the test article is judged as genotoxic and more research to understand carcinogenic potential may be warranted. Whether or not such articles are taken into clinical testing is determined based on overall data considering risk and benefit to human subjects. For example, administration of genotoxic agents may be appropriate for certain populations in the oncology therapeutic area. Conversely, the administration of such agents may not be favored for clinical indications that are not considered severely debilitating or life-threatening. The close association between the genotoxic and carcinogenic properties of chemicals makes genetic toxicology testing an early indicator of carcinogenic potential and therefore constitutes a key prioritization step in the selection of articles in the discovery phase and a critical safety screen prior to initiation of clinical trials.

Conflict of Interest None.

Disclaimer

This article reflects the views of the author (Ramadevi Gudi) and should not be construed to represent US FDA's views or policies.

References

1. Pandey H, Kumar V, Roy BK (2014) Assessment of genotoxicity of some common food preservatives sing *Allium cepa* L. as a test plant. Toxicology 1:300–308
2. Husgafvel-Pursiainen K (2004) Genotoxicity of environmental tobacco smoke: a review. Mutat Res 567: 427–445
3. Keshava N, Ong T (1999) Occupational exposure to genotoxic agent. Mutat Res 437:175–194
4. Hayashi M (2022) Opinion: regulatory genotoxicity: past, present and future. Genes Environ 44:13. https://doi.org/10.1186/s41021-022-00242-5
5. Pon JR, Marra MA (2015) Driver and passenger mutations in cancer. Annu Rev Pathol Mech Dis 10: 25–25
6. Minten EV, Yu DS (2019) DNA repair: translation to the clinic. Clin Oncol 31(5):303–310
7. https://www.ich.org/. Accessed Aug 2022
8. ICH S2A specific aspects of regulatory genotoxicity tests for pharmaceuticals. https://www.fda.gov/media/71959/download. Accessed Aug 2022

9. ICH S2B genotoxicity: a standard battery for genotoxicity testing of pharmaceuticals. https://www.fda.gov/regulatory-information/search-fda-guidance-documents/s2b-genotoxicity-standard-battery-genotoxicity-testing-pharmaceuticals. Accessed Aug 2022
10. ICH S2(R1) genotoxicity testing and data interpretation for pharmaceuticals intended for human use. https://www.fda.gov/regulatory-information/search-fda-guidance-documents/s2r1-genotoxicity-testing-and-data-interpretation-pharmaceuticals-intended-human-use. Accessed Aug 2022
11. OECD guidelines for genetic toxicology (1997) www.oecd.org/dataoecd
12. Khanna KK, Jackson SP (2001) DNA double-strand breaks: signaling, repair and the cancer connection. Nat Genet 27:247–254
13. Ames BN, McCann J, Yamasaki E (1975) Methods for detecting carcinogens and mutagens with the Salmonella/mammalian-microsome mutagenicity test. Mutat Res 31:347–364
14. Maron DM, Ames BN (1983) Revised methods for the Salmonella mutagenicity test. Mutat Res 113:173–215
15. Gatehouse D, Haworth S, Cebula T, Gocke E, Kier L, Matsushima T, Melcion C, Nohmi T, Venitt S, Zeiger E (1994) Recommendations for the performance of bacterial mutation assays. Mutat Res 312:217–233
16. McCann J, Choi E, Yamasaki E, Ames BN (1975) Detection of carcinogens as mutagens in the Salmonella/microsome test: assay of 300 chemicals. Proc Natl Acad Sci 72(12):5135–5139
17. Green MHL, Muriel WJ (1976) Mutagen testing using trp+ reversion in Escherichia coli. Mutat Res 38:3–32
18. Ames BN, Lee FD, Durston WE (1973) An improved bacterial test system for the detection and classification of mutagens and carcinogens. Proc Natl Acad Sci 70(3):782–786
19. McCann J, Spingarn NE, Kobori J, Ames BN (1975) Detection of carcinogens as mutagens: bacterial tester strains with R factor plasmids. Proc Natl Acad Sci 72(3):979–983
20. Wagner VO III, Sly JE, Klug ML, Staton TL, Wyman MK, Xiao S, San RHC (1994) Practical tips for conducting the *Salmonella* and *E. coli* mutagenicity assays under proposed international guidelines. Environ Mol Mutagen suppl 23:70. [EMS Poster 1994]
21. Ames BN, Durston WE, Yamasaki E, Lee FD (1975) Carcinogens and mutagen: a simple test system combining liver homogenate for activation and bacteria detection. Proc Natl Acad Sci 70(8):2281–2285
22. Matsushima T, Sawamura M, Hara K, Sugimura T (1976) A safe substitute for polychlorinated biphenyls as an inducer of metabolic activation system. In: DeSerres FJ, Fouts JR, Bend JR, Philpot RM (eds) In vitro metabolic activation in mutagenesis testing. Elsevier/North-Holland, Amsterdam, pp p85–p88
23. Ong T, Mukhtar M, Wolf CR, Zeiger E (1980) Differential effects of cytochrome P450-inducers on promutagen activation capabilities and enzymatic activities of S-9 from rat liver. J Environ Pathol Toxicol 4(1):55–65
24. Yahagi et al (1977) Mutagenicities of N-nitrosamines on Salmonella. Mutation Res 48:121–129
25. Putman DL, Gudi R, Wagner VO III, San RHC, Jacobson-Kram D (2001) Genetic toxicology. In: Jacobson-Kram D, Keller KA (eds) Toxicology testing handbook, principles, applications, and data interpretation. Marcel Dekker, New York, pp 127–194
26. Claxton LD, Allen J, Auletta A, Mortelmans K, Nestmann E, Zeiger E (1987) Guide for the Salmonella typhimurium/mammalian microsome tests for bacterial mutagenicity. Mutat Res 189:83–91
27. Mahon GAT, Green MHL, Middleton B, Mitchell I, Robinson WD, Tweats DJ (1989) Analysis of data from microbial colony assays. In: Kirkland DJ (ed) UKEMS Sub-Committee on guidelines for mutagenicity testing part II. Statistical evaluation of mutagenicity test data. Cambridge University Press, Cambridge, pp 28–65
28. Diehl MS, Wallaby SL, Snyder RD (2000) Comparison of the results of a modified miniscreen and the standard bacterial reverse mutation assays. Environ Mol Mutagen 36(1):72–77
29. Fenech M, Morley AA (1985) Measurement of micronuclei in lymphocytes. Mutat Res 147:29–36
30. Kirsch-Volders M et al (2000) Report from the In Vitro Micronucleus Assay Working Group. Environ Mol Mutagen 35(3):167–172
31. Lorge E et al (2006) SFTG international collaborative study on in vitro micronucleus test. I General conditions and overall conclusions of the study. Mutat Res 607(1):13–36
32. Clare G et al (2006) SFTG international collaborative study on the in vitro micronucleus test. II. Using human lymphocytes. Mutat Res 607(1):37–60
33. Scott D, Galloway SM, Marshall RR, Ishidate M Jr, Brusick D, Ashby J, Myhr BC (1991) Genotoxicity under extreme culture conditions. A report from ICPEMC Task Group 9. Mutat Res 257:147–204
34. Fenech M (2007) Cytokinesis-block micronucleus cytome assay. Nat Protoc 2(5):1084–1104
35. Honma M (2011) Cytotoxicity measurement in in vitro chromosome aberration test and micronucleus test. Mutat Res 724:86–87
36. Krishna G, Hayashi M (2000) In vivo rodent micronucleus assay: protocol, conduct, and data interpretation. Mutat Res 455:155–166
37. OECD guideline 474 (2016) https://www.oecd.org/env/test-no-474-mammalian-erythrocyte-micronucleus-test-9789264264762-en.htm
38. Krishna G, Krishna PA, Goel S, Krishna KA (2019) Alternative animal toxicity testing and biomarkers. In: Gupta R (ed) Biomarkers in toxicology, 2nd edn. Elsevier, Amsterdam. https://doi.org/10.1016/B978-0-12-814655-2.00008-6
39. Krishna G, Urda G, Theiss J (1998) Principles and practices of integrating genotoxicity evaluation into

routine toxicology studies: a pharmaceutical industry perspective. Environ Mol Mutagen 32:115–120
40. Kirkland D, Aardema M, Henderson L, Muller L (2005) Evaluation of the ability of a battery of three in vitro genotoxicity tests to discriminate rodent carcinogens and non-carcinogens. Sensitivity, specificity and relative predictivity. Mutat Res 584:1–256
41. Snyder RD, Green JW (2001) A review of the genotoxicity of marketed pharmaceuticals. Mutat Res 488:151–169
42. Dearfield KL, Thybaud V, Cimino MC, Custer L, Czich A, Harvey JS, Hester S, Kim JH, Kirkland D, Levy DL, Lorge E, Moore MM, Ouedraogo-Arras G, Schuler M, Suter W, Sweder K, Tarlo K, Benthem J, Goethem F, Witt KL (2011) Follow-up actions from positive results of in vitro genetic toxicity testing. Environ Mol Mutagen 52:177–204
43. Kirsch-Volders M, Sofuni T, Aardema M, Albertini S, Eastmond D, Fenech M, Ishidate M, Kirchner S, Lorge E, Morita T, Norppa H, Surralles J, Vanhauwaert A, Wakata A (2003) Report from the in vitro micronucleus assay working group. Mutat Res 540(2):153–163
44. Tweats DJ, Blakey D, Heflich RH, Jacobs A, Jacobsen SD, Morita JT, Nohmi T, O'Donovan MR, Sasaki YF, Sofuni T, Tice R (2007) Report of the IWGT working group on strategies and interpretation of regulatory in vivo tests. I. Increases in micronucleated bone marrow cells in rodents that do not indicate genotoxic hazards. Mutat Res 627:78–91
45. Shuey DL, Gudi R, Krsmanovic L, Gerson RJ (2007) Evidence that oxymorphone-induced increases in micronuclei occur secondary to hyperthermia. Toxicol Sci 95(2):369–375
46. Asanami S, Shimono K (1997) Hypothermia induces micronuclei in mouse bone marrow cells. Mutat Res 393:91–99

Rodent Carcinogenicity Studies

8

Mark Tirmenstein and Ravikumar Peri

Abstract

Rodent lifetime carcinogenicity studies are used to assess the carcinogenic potential of chemicals and pharmaceuticals to humans and are a regulatory requirement for most pharmaceuticals. In the past, 2-year rat and mouse studies were commonly required for market authorization of pharmaceuticals in development. These studies use a large number of animals, are costly, complex, and require a long time to complete. In recent years, the 2-year mouse carcinogenicity study has been largely replaced by the 6-month Tg.rasH2 transgenic mouse study. The 6-month Tg.rasH2 transgenic mouse study is less expensive, requires fewer animals and provides quicker results relative to the 2-year mouse carcinogenicity study. The current chapter will discuss some factors that need to be considered in determining the experimental design for these rodent carcinogenicity studies. Not all drugs in development require rodent carcinogenicity studies and interactions with health authorities are strongly recommended to determine if these studies are required for market authorization. In addition, it is well known that many rat carcinogens have not been demonstrated to represent a risk to humans. This has led many to call for a re-evaluation of the relevance of 2-year rat carcinogenicity studies. Recent ICH guidelines have shifted away from recommending a 2-year rodent carcinogenicity study as a default requirement. Instead, they have suggested using a weight of evidence approach to determine if a 2-year rat carcinogenicity study would have value in assessing the carcinogenic risk of drugs in development to humans.

Keywords

Carcinogenicity assessments · Drug-development · Rodent · Tg.rasH2 · Transgenic mouse · ICH S1B(R1) · Nongenotoxic · Carcinogens

8.1 Introduction

There is a long history of using lifetime rodent studies to assess the carcinogenic potential of chemicals and pharmaceuticals. The development of standardized 2-year carcinogenicity study

M. Tirmenstein (✉)
Toxicology, Nonclinical Development, Sarepta Therapeutics, Cambridge, MA, USA
e-mail: MTirmenstein@sarepta.com

R. Peri
Pre-Clinical Toxicology, Alexion, AstraZeneca Rare Disease, Wilmington, DE, USA
e-mail: Ravi.peri@alexion.com

G. Jagadeesh et al. (eds.), *The Quintessence of Basic and Clinical Research and Scientific Publishing*,
https://doi.org/10.1007/978-981-99-1284-1_8

protocols in rats and mice dates back to the pioneering work of National Cancer Institute scientists John and Elizabeth Weisburger in the 1960s [1]. The impetus for carcinogenicity testing of pharmaceuticals occurred in 1968 when drug package inserts with a section discussing carcinogenesis were required for newly approved drugs [2].

An important change occurred in carcinogenicity testing in mice in 1997 [2]. In that year, the FDA agreed to allow the use of transgenic animal models in the evaluation of drugs. Once Tg.rasH2 mice became commercially available, this animal model gained widespread acceptance for carcinogenicity testing in mice. Today, the 6-month Tg.rasH2 mouse carcinogenicity study is widely used and has for the most part supplanted the 2-year lifetime mouse carcinogenicity study in drug development [3]. Conducting a 6-month Tg.rasH2 mouse study offers several advantages over the 2-year mouse carcinogenicity study including: lower cost, reduced animal numbers and a shorter duration of testing. However, since no generally accepted alternative for the 2-year rat carcinogenicity study has been accepted to date, the overall time required for carcinogenicity testing in rodents is still approximately 3 years in duration. Despite the obvious advantages of the transgenic mouse carcinogenicity study, a 2-year mouse carcinogenicity study can still be conducted instead of the 6-month transgenic mouse study. In some cases, a scientific rationale may dictate the use of the 2-year mouse carcinogenicity study instead of the 6-month transgenic mouse study. Either study is acceptable for a carcinogenicity assessment in mice.

For now, rodent carcinogenicity studies are required for market authorization for most small molecule drugs in development, and there are several guidance documents that aid in the development of the experimental design of these studies. This chapter will discuss the prerequisites of rodent carcinogenicity testing, the general experimental design of these studies, and the interpretation of results.

8.2 Drugs Requiring Rodent Carcinogenicity Testing Rationale

It should be noted that not all drugs in development require rodent carcinogenicity studies for marketing authorization. Small molecules that are intended for chronic administration (at least 6 months continuous administration) in general and, to a lesser extent, monoclonal antibodies for chronic indications require rodent carcinogenicity studies. Still, drugs intended for shorter duration administration do not. In addition, carcinogenic studies are generally required for pharmaceuticals used frequently in an intermittent manner in the treatment of chronic or recurrent conditions. Regardless of the duration of the intended treatment, carcinogenicity testing may also be required for drugs that elicit a cause for concern. This may include drugs that induce preneoplastic lesions (i.e., hyperplasia) in chronic toxicology studies and classes of drugs that have carcinogenic properties or whose mechanism of action could in theory lead to tumor development. However, there is no point in testing drugs identified as being unequivocally genotoxic. These drugs in the absence of other information are thought to damage DNA and to be carcinogenic in humans as well as animals. A negative finding in a rodent carcinogenicity study will not change the fact that it has been classified as a potential mutagen. Biotechnology products such as antibodies or peptides/proteins generally do not require carcinogenicity testing [4]. A request for a waiver for carcinogenicity studies for biotechnology products is usually accompanied by a weight of evidence rationale for not conducting these studies. This weight of evidence can include data from transgenic, knock-out, or animal disease models, human genetic disease, information on class effects, pharmacology and mechanism of action, in vitro data, data from chronic toxicity studies and clinical data [4].

Many rodent carcinogens have been identified that have not been demonstrated to represent a risk to humans [5], leading some to question the value of rodent carcinogenicity testing. It has been suggested that the histopathology results from subchronic [6] and chronic repeat-dose rat studies can be used in place of the 2-year rodent carcinogenicity study [7]. According to this rationale, noncarcinogens would not be expected to develop preneoplastic lesions in subchronic and chronic repeat-dose toxicity studies and would thereby not require additional testing in a 2-year rodent carcinogenicity study. Only those compounds that develop preneoplastic lesions in the repeat-dose toxicity studies would then be required to be tested in the 2-year rat carcinogenicity study.

Recently, a new ICH guideline (TESTING FOR CARCINOGENICITY OF PHARMACEUTICALS S1B(R1) [8] has backed away from a default requirement for a 2-year rat carcinogenicity study and instead has recommended a weight of evidence approach for deciding whether to conduct this study. This approach is already being used at least to some degree for biotechnology products. It would emphasize a more science-based approach for determining if a 2-year rat carcinogenicity study is required. Eliminating the requirement for a 2-year rat carcinogenicity study would be expected to decrease costs and timelines for drug approvals and reduce the number of rodents used in nonclinical development. According to the guidelines [8], some of the factors to consider in using the weight of evidence approach to decide whether to conduct a 2-year rat carcinogenicity study include:

- Target Biology—how characterized are the biological pathways and class effects
- Secondary Pharmacology—is there off-target activity?
- Histopathology—were there hyperplastic lesions or other lesions of concern in chronic or subchronic studies?
- Hormonal effects—is there evidence of endocrine disruption and/or reproductive organ perturbations?
- Positive Genotoxicity—is there equivocal evidence of positive genotoxicity of uncertain relevance to humans? If so, additional testing may help determine relative risks to humans.
- Immune Modulation—is there evidence of immune effects of uncertain relevance to humans?

Regardless, if the 2-year rat carcinogenicity study is conducted, the guidelines recommend that a mouse carcinogenicity study be conducted in most cases. The mouse carcinogenicity study can be either a 2-year carcinogenicity study or a 6-month transgenic mouse study. As was mentioned previously, there are considerable advantages to conducting the 6-month transgenic mouse study.

Since 2-year carcinogenicity testing can require years to complete, interactions with the FDA are recommended early in drug development to determine if carcinogenicity studies are warranted and prevent unnecessary delays.

8.3 Experimental Design

There are some common elements in the experimental design in the 6-month transgenic mouse and 2-year rodent carcinogenicity studies. There are generally three dose groups of the drug to be tested (low, mid and high) along with a vehicle control group. If the vehicle is not a standard vehicle or the toxicologic properties of the vehicle have not been sufficiently characterized a water control group can also be included for comparison purposes. Importantly, all rodent carcinogenicity studies should be conducted under GLP conditions.

Other standard endpoints that are usually included in all rodent carcinogenicity studies are the following:

- Periodic dose formulation analysis and stability testing (especially important if more than one batch of the drug is used in a 2-year study). Care should be taken not to use drug supplies past their expiration date.
- Daily clinical observations (including palpable mass assessments) and mortality checks.
- Toxicokinetic evaluations on Day 1 and after approximately 6 months of dosing. If major

Table 8.1 Animal allocation to 2-year carcinogenicity studies in rodents

Group[a]	Subgroup	Dose level[b] (mg/kg/day)	Dose[b] concentration (mg/ml)	Number of animals[c]	
				Male	Female
1. Control	1. Main	0	0	50–70	50–70
	2. TK			3–5	3–5
2. Control	1. Main	0	0	50–70	50–70
	2. TK			3–5	3–5
3. Low dose	1. Main	10	1	50–70	50–70
	2. TK			12–15	12–15
4. Mid dose	1. Main	30	3	50–70	50–70
	2. TK			12–15	12–15
5. High dose	1. Main	100	10	50–70	50–70
	2. TK			12–15	12–15

[a]Groups 1 and 2 are control groups. One of the control groups typically is the formulation buffer in which the test article is formulated (vehicle), and the second is typically either water or a normal saline solution

[b]The doses listed are only examples. A detailed strategy to dose selection is described later in the chapter, but the high dose is targeted to give 25× exposure multiple when feasible; the low dose should not elicit toxicity and provide exposure slightly above human efficacious exposure; and the mid dose is selected based on the linearity of toxicokinetics (TK) to elicit some toxicities. Dose volumes and concentrations are typically adjusted based on the route of administration, with 10 mL/kg being the maximal volume for an oral dose

[c]The number of animals allocated to each dose group varies between 50 and 70 for the main study and at least 12–15 animals/time point for toxicokinetic assessment. Only 3–5 animals are assigned to the vehicle TK groups because the analyte is needed just to demonstrate the absence of test article. A large number of animals are allocated to the study to ensure that drug-treated and control groups should have at least 25 rodents per sex per group survive to the end of the study for adequate statistical analysis of tumor incidence as recommended by guidelines

metabolites or metabolites of particular concern have been identified in previous human or animal studies, these should also be characterized in the toxicokinetic evaluations.

- Clinical pathology endpoints are not usually included but particular clinical pathology endpoints may have value to assess pharmacodynamic effects on a case-by-case basis.
- Histopathology of representative organs and tissues (a full tissue list in carcinogenicity studies).
- Rigorous statistical analysis of tumor incidence in treated animals vs. controls.
- An example of animal allocation to rodent carcinogenicity study is described in Table 8.1.

8.3.1 Rat Strains and Number of Animals

In general, the strain of rats used in the 2-year rodent carcinogenicity study is usually the same strain that has been used previously in the rat toxicology program (3- and 6-month repeat-dose toxicity studies). Target organs have been identified and chronic doses determined from previous repeat-dose toxicology studies in this strain of rat, so there is less risk in switching to a new strain of rat for a carcinogenicity study. It should be recognized that there is always the possibility that other strains of rats may be more or less susceptible to the toxicity of the drug in development. The Sprague Dawley (SD) rat is the most commonly used strain of rat in toxicology studies for pharmaceuticals in the United States. However, SD rats are known to have a high incidence of certain background diseases and spontaneous tumor formation [9, 10]. For example, chronic progressive nephropathy is a major cause of mortality in male SD rats. Pituitary tumors are common in male and female SD rats, and female SD rats have a high incidence of mammary gland tumors. There may be advantages in using rat strains that have lower mortality and a lower incidence of spontaneous tumors. Data provided by Envigo (an Inotiv Company) [11], suggests that Wistar Hannover rats, for example, have

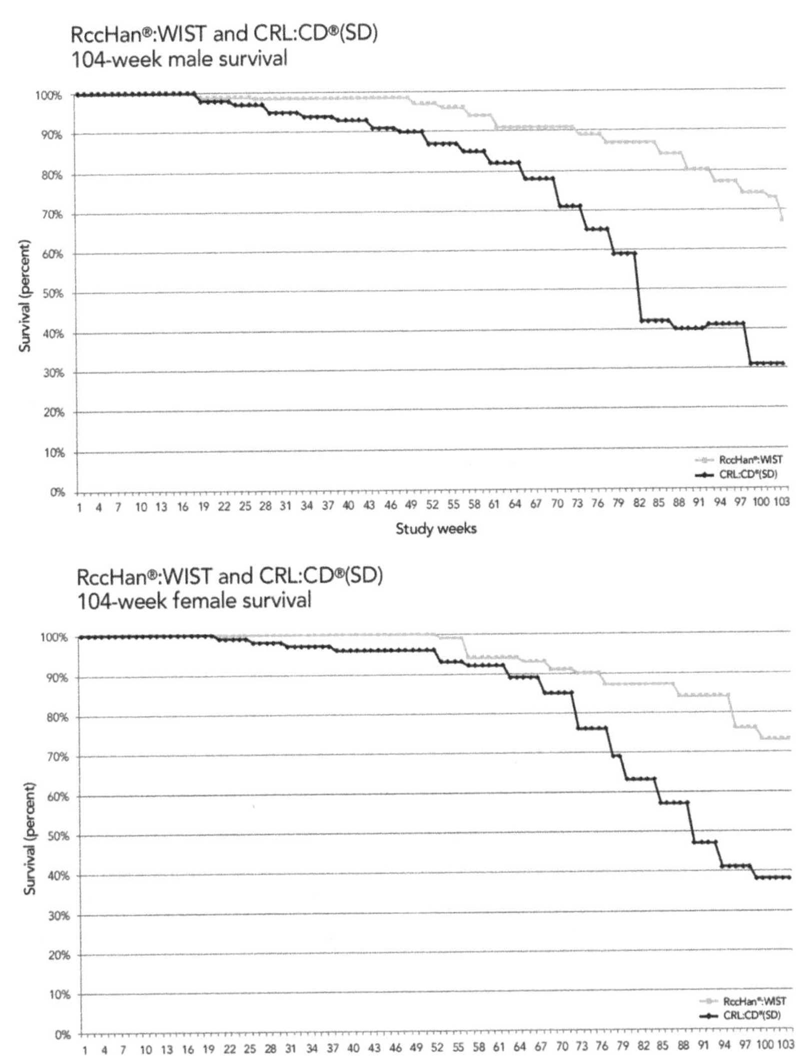

Fig. 8.1 Survival curves for Wistar Hannover and CRL:CD®(SD) rats over 104 weeks. (Top) Survival curves for male Wistar Hannover and CRL: CD®(SD) rats, and (Bottom) survival curves for female Wistar Hannover and CRL:CD®(SD) rats (Envigo White Paper: https://insights.envigo.com/hubfs/resources/white-papers/white-paper-wistar-han-rat-for-carcinogenicity-studies.pdf)

lower mortality in the 2-year rat carcinogenicity study compared to SD rats (Fig. 8.1). However, if a conscious decision is made to choose a strain of rats different from those in which chronic toxicity data is conducted, it might be helpful to conduct a bridging toxicity study with appropriate toxicology and toxicokinetic assessments such as a 3-month dose range-finding study.

Although the guidelines suggest at least 50 male and 50 female rats should be used for each dose group [12], a larger number of SD rats (up to 70 per sex per group) is frequently used to compensate for the high mortality observed in SD rats. It is important to note that FDA guidance indicates that drug-treated and control groups should have at least 25 rodents per sex per group survive to the end of the study for adequate statistical analysis of tumor incidence [12]. Rodent carcinogenicity studies should be designed with this overall goal in mind. If survival becomes an issue, the Executive Carcinogenicity Assessment Committee (ECAC) should be consulted for potential remediation strategies (see Sect. 8.3.5).

8.3.2 Mouse Carcinogenicity Strains and Number of Animals

If a 2-year mouse carcinogenicity study is conducted, the same criteria with regard to the number of groups and the number of male and female mice in each group applies. In most cases, the strain of mice that has been used in dose range-finding studies should be used for the 2-year carcinogenicity study. There are generally three dose groups of the drug to be tested (low, mid and high) along with a vehicle control group. Each group should contain at least 50 male and 50 females. As with the rat, more mice may be included to ensure that at least 25 mice per sex per group survive to the end of the study for adequate statistical analysis of tumor incidence [12].

For the 6-month transgenic mouse study, there are also typically three dose groups for the drug and one or two vehicle control groups depending on the nature of the vehicle, with each group having 25 male and 25 female main study mice with additional mice for toxicokinetics. In addition, a positive control group (a known carcinogen) consisting of 10 males and 10 females is generally included in the 6-month transgenic mouse study [13]. A single intraperitoneal dose (75 mg/kg) of the carcinogen, N-nitrosomethylurea (NMU) is commonly used as the positive control in these studies [3].

8.3.3 Route of Administration

As in all toxicology studies, where feasible, the route of administration should be the same as that used in humans. Rodent carcinogenicity studies where the drug is incorporated into feed can be conducted if necessary. However, there are several factors which make this less desirable than standard oral gavage administration, such as variability in drug administration, difficulties in determining the amount of drug administered, stability of the drug in feed, and palatability issues of the drug incorporated into feed.

8.3.4 Selection of Doses

One of the more difficult tasks in designing rodent carcinogenicity studies is dose selection. Fortunately, extensive guidance is provided in ICH guidelines (DOSE SELECTION FOR CARCINOGENICITY STUDIES OF PHARMACEUTICALS S1C(R2)) [14]. The doses selected for the carcinogenicity study should be tolerated without significant chronic physiological dysfunction or mortality and provide AUC exposure margins that allow an adequate margin of safety over the maximum human therapeutic exposures and permit interpretation of data in the context of clinical use.

8.3.4.1 High Doses

2-Year Rat Carcinogenicity Study: The high dose is usually selected such that anticipated systemic AUC exposures are 25 times the maximum human AUC exposures for the drug [14]. AUC exposures after repeat administration for 6-months in the chronic rat toxicology studies should provide the needed data for this determination. If male and female rats have different AUC exposures, it may be necessary to select different doses for male and female rats in the carcinogenicity study. If the 25 times exposure multiples cannot be achieved, a series of backup strategies in selecting the high dose in the rat carcinogenicity study can be adopted, including the following [14]:

- Exposure Saturation—If there is saturation in exposure to high doses of a drug, this rationale can be used to limit the high dose. If increasing doses do not significantly increase AUC exposures, there is no point in dosing at higher levels. In this case, the high dose would be limited to the lowest dose of the drug that induces saturation in exposure.
- Maximum Tolerated Dose (MTD)—In this case, the high dose would be selected such that it does not produce a pharmacodynamic or toxic response that would be expected to increase mortality in the high dose. The

integrity of the study may be affected if there are not 25 rats remaining after the end of 2-years of drug administration. One of the specific factors that characterizes toxicity in the ICH guidelines [14] is decrease in bodyweight gains. Specifically, doses that induce more than a 10% decrease in body weight gain compared to controls would not be expected to be tolerated in a 2-year carcinogenicity study and may be used to define the MTD.

- Maximum Feasible Dose (MFD)—The high dose can be selected based on the highest dose that can be practically administered to rats. This could include limitations in formulating the drug in the selected vehicle or could be based on local tolerance issues with the drug.
- Limit Dose—If none of the above criteria apply, the high dose can be selected based on the limit dose as defined by ICH guidelines. If the maximum recommended human dose does not exceed 500 mg/kg/day, the limit dose would be 1500 mg/kg/day. If the human dose exceeds 500 mg/day the high dose can be increased up to the MFD. In most cases, other criteria will be used instead of the limit dose for selecting the high dose in rodent carcinogenicity studies.

2-Year Mouse Carcinogenicity Study: The same strategy in selecting the high dose in the 2-year rat carcinogenicity study also applies to the 2-year mouse carcinogenicity study. However, in most cases in drug development, there may be limited experience in conducting long-term repeat-dose toxicity studies in mice, therefore, an appropriate repeat-dose range-finding study (usually 3 months in duration) with mice may be necessary to assist in selecting doses for the subsequent 2-year mouse carcinogenicity study.

6-Month Transgenic Mouse Study: The high dose for the 6-month Tg.rasH2 transgenic mouse study is generally selected based on results of a short term (5–7 days) and 28-day dose range-finding studies [15]. Using these studies, an estimated MTD is established and used as the high dose in the subsequent 6-month transgenic mouse study. The estimated MTD is based on the effects of the drug on body weights, clinical observations, clinical pathology findings and gross and microscopic histopathology findings, and any other parameter that would be expected to increase mortality and/or compromise the well-being of the animals used on the study [15]. However, this traditional way of selecting the high dose has been questioned based on analysis of previous 6-month transgenic mouse studies. An analysis of body weight, mortality, and tumor response in control and treated groups of 29 Tg.rasH2 transgenic mouse studies, suggested that the MTD was exceeded at the high and/or mid-doses in several studies, and the incidence of tumors in high doses was lower when compared to the low and mid-doses of both sexes [15]. The decreased incidence of tumors in the high dose group was due to excessive toxicity resulting in significant decreases in body weight and increased mortality. These effects were related to high exposure toxicity and not carcinogenic effects.

In a recent ICH guideline [8], it was proposed that the high dose in the 6-month transgenic mouse study should be selected based on a 50-fold AUC exposure multiple rather than the estimated MTD. One possible explanation why the MTD is often exceeded in the 6-month studies relates to how the dose range-finding studies are conducted. In most cases, CByB6F1 mice, the wild type littermates of Tg.rasH2 mice, are used instead of the transgenic mice for the dose range-finding studies. The CByB6F1 mouse has the same genetic background as the Tg.rasH2 mouse except for the omission of the transgene (Tg) element but it costs much less than the transgenic mice. Data collected from the 5 or 7-day dose range-finding study are used to set doses for the 28-day dose range-finding study, and the data collected in the 28-day study are used to select doses for the 6-month transgenic mouse study. There are some significant differences between the transgenic mice and their wild type littermates, with the transgenic mice weighing, on average 10% less than their wild type littermates. They also appear to gain weight at a lower rate than their wild type litter mates. This bodyweight difference may make the transgenic mice more susceptible to toxicity than the wild type mice. This seems to be especially true for male transgenic mice. The authors proposed that a possible solution for this problem is to conduct the 28-day dose range-finding study in transgenic mice instead of their wild type littermates [15]. This would, of course, increase costs for conducting the 28-day dose-ranging study but would be expected to provide a more accurate MTD estimation.

8.3.4.2 Low and Mid Doses

The rationale for selecting low and mid doses is similar for the 6-month transgenic mouse and 2-year rodent carcinogenicity studies. Regardless of the method used for selecting the high dose, selecting the mid and low doses for the carcinogenicity study should provide information to aid in assessing the relevance of study findings to humans. The low and mid doses are selected based on rodent and human pharmacokinetic, pharmacodynamic, and toxicity data. The rationale for the selection of these doses should be provided in the protocol. ICH guidelines (dose selection for carcinogenicity studies of pharmaceuticals S1C(R2)) [14] recommend considering the following points in selecting the low and mid doses in rodent carcinogenicity studies: the linearity of pharmacokinetics and saturation of metabolic pathways; human exposure and therapeutic dose; pharmacodynamic response in rodents; alterations in normal rodent physiology; mechanistic information and potential for threshold effects; and the unpredictability of the progression of toxicity observed in short-term studies. In practical terms, in most cases, the low dose should be selected to provide a dose without toxicity and anticipated AUC exposure slightly above human AUC exposures at the maximum human dose. The mid-dose should be selected to explore dose-response relationships and provide AUC exposures between the low (2–3-fold) and high doses (25-fold).

8.3.5 Interactions with the Executive Carcinogenicity Assessment Committee (ECAC) at the US FDA

The ECAC was established to ensure consistency in recommendations and conclusions regarding protocols for and results of carcinogenicity studies across review divisions. The ECAC meets regularly to review all carcinogenicity protocols and final study reports. All rodent carcinogenicity study protocols should be submitted to the ECAC for review and approval prior to initiating these studies. Sponsors usually submit the protocols in requests for special protocol assessments (SPAs), which the Center for Drug Evaluation and Research (CDER) of the FDA must complete within 45 days. The ECAC will review protocols and suggest alternative doses if necessary. Sufficient time should be incorporated into drug development timelines to ensure that feedback from the ECAC is provided.

In addition, the ECAC should be notified as soon as possible when unexpected events occur in rodent carcinogenicity studies. These can include proposed changes in doses due to mortality or morbidity in the high dose group and proposed dosing holidays due to the condition of the animals and even the early sacrifice of entire male or female dose groups so that adequate numbers of animals are present for a robust statistical analysis. For survival issues in rat carcinogenicity studies, the ECAC standard advise usually includes the following:

- Terminate only the drug-treated group when the number of animals in that group declines to 15 before week 100
- Terminate all groups of a sex, including controls, if the number of animals in a drug-treated group declines to 15 at or after week 100
- Terminate all groups of a sex if the number of combined identical control animals declines to 20

Feedback from the ECAC is usually provided in a timely manner and can ensure that the rodent carcinogenicity studies are successfully completed.

Routine communications with the contract research organization (CRO) where rodent carcinogenicity studies are conducted are essential for preventing major issues with these studies.

8.4 Tumor Analysis or Data Evaluation and Statistical Analysis

Data for each sex are analyzed separately and statistically. Survival analyses are performed with a two-sided risk for increasing and decreasing death with dose. Tests are performed for dose response and each drug-treated group against control using Kaplan-Meier product-limit estimates, along with log-rank and Wilcoxon tests. If there is more than one control group, the two identical control groups are combined for analysis.

For each given tumor type, statistical analysis is performed if the incidence in at least one drug-treated group is increased by at least two occurrences over the control group. Tests to compare tumor incidence are performed, with a one-sided risk for increasing incidence with dose. Tests are performed for dose response and each test substance-treated group against the control group. Tumors occurring in animals dying spontaneously or euthanized in extremis during the study are classified as one of the following:

1. Fatal: The tumor was a factor in the demise of the animal.
2. Non-fatal: The tumor was not a factor in the demise of the animal.
3. Uncertain

Fatal and non-fatal tumors are combined for analysis.

Observable or palpable tumors are analyzed using the time to death or time of detection of the tumor. Unadjusted P-values are reported for tumors. An indication of a possible dose-response in the incidence rate of individual tumors is assessed on the basis of whether the tumor is a rare or common type (the common tumor's significance level is <0.005, rare tumor's <0.025). A pairwise comparison of an increase in incidence rates in the treated and control groups are tested at 0.01 or 0.05 significance level for common and rare tumors, respectively [16]. The incidence rate for defining whether a tumor type is rare or common is based on site-specific background historical data. The criteria for combination should be based on Guidelines for Combining Neoplasms for Evaluation of Rodent Carcinogenicity Studies authored by McConnell et al. [17].

8.5 Interpretation of Results

The main value of rodent carcinogenicity studies is to identify nongenotoxic carcinogens. In most cases, genotoxic carcinogens can be identified using genotoxicity assessments and do not require lifetime rodent carcinogenicity studies. It is generally accepted that nongenotoxic carcinogens have thresholds for the development of carcinogenicity. Determination of this threshold and defining exposure multiples that are associated with the development of tumors is important for human risk assessments. Compounds that induce tumor formation at exposures that are much higher than the anticipated human therapeutic exposures elicit less concern than those that induce tumors at exposures that are similar to anticipated human therapeutic exposures. In most toxicology studies, a no-observed-adverse-effect level (NOAEL) is established based on adverse toxicologic findings. In a rodent carcinogenicity study, it is common to establish one NOAEL for target organ toxicity and another NOAEL for carcinogenicity or tumor formation. These NOAELs may be at different dose levels. A different NOAEL for carcinogenicity may also be established for males vs. females. This is especially true if tumors develop in sex organs (i.e., testes for males, mammary glands for females). In some cases, it may be helpful to identify the mechanism of tumor formation. There are established mechanisms for tumor development in rats, for example, that are not relevant to humans [5]. If this is the case, a rational argument can be established why the observed tumor formation is rat or mouse specific and is not a risk to humans. In general, compounds that induce tumors in both mice and rats are more of a concern than those that induce tumors in only one species. However, in the absence of a species-specific scientific rationale, carcinogenic findings in only one species can also raise a concern in human risk assessments.

Tumorigenic findings of marginally increased frequency present a challenge in interpreting the results. In such instances, it helps to compare the data with the historical controls in the laboratories conducting these studies to provide additional perspectives. Historical control data can also be submitted with the study reports as a part of the regulatory submission. It is important to note that some regulatory agencies conduct their own statistical analysis of the data, and preemptive submission of files containing raw data may minimize review questions.

8.6 Concluding Remarks

The requirement for conducting a 2-year rat carcinogenicity is currently undergoing a period of re-evaluation. In the past, 2-year rat carcinogenicity studies were, in most cases, a default requirement. However, based on the knowledge that not all rat carcinogens represent a risk to humans, there is a current emphasis on using a weight of evidence approach to decide whether a 2-year rat carcinogenicity study should be conducted. This weight of evidence shifts the decision to conduct this study to one based on scientific evidence. In recent years, we have seen the 2-year mouse carcinogenicity study largely replaced by the 6-month transgenic mouse study. This has led to reduced costs, shortened timelines and a reduction in the number of mice used in carcinogenicity testing. Overall, deciding rather to conduct a 2-year rat carcinogenicity study on a case-by-case basis guided by scientific evidence would be expected to have similar results. They are shortening timelines, reducing the number of animals used and decreasing costs. As detailed in the current chapter, there are several guidelines detailing how 2-year carcinogenicity in rats and mice and the 6-month transgenic mouse study should be conducted and especially how doses should be selected. The rationale for dose selection should be carefully considered and the justification for the selection of these doses should be documented. It is critical that the study protocol be submitted to the ECAC of the regulatory agency for approval before starting the study. Failure to seek feedback from ECAC could lead to an invalid study and a need to repeat the study. The experimental design should be approved before the study is initiated. The overall goal of rodent carcinogenicity studies is to determine if the drugs being tested represent a carcinogenic risk to humans. To accomplish this goal, a complete list of tissues needs to be evaluated for tumors in sufficient numbers of treated and control animals to provide a robust statistical evaluation of tumor incidence. It is also essential to evaluate systemic exposures following drug treatment so that tumor findings can be evaluated relative to anticipated human exposures in the clinic.

Conflict of Interest The authors declare they have no conflict of interest that would influence the preparation of this book chapter.

References

1. Weisburger J, Weisburger E (1967) Tests for animal carcinogens. Methods Cancer Res 1:307–398
2. Jacobs AC, Hatfield KP (2013) History of chronic toxicity and animal carcinogenicity studies for pharmaceuticals. Vet Pathol 50:324–333
3. Bogdanffy MS, Lesniak J, Mangipudy R, Sistare FD, Colman K, Garcia-Tapia D, Monticello T, Blanset D (2020) Tg.rasH2 mouse model for assessing carcinogenic potential of pharmaceuticals: industry survey of current practices. Int J Toxicol 39:198–206
4. ICH guideline: preclinical safety evaluation of biotechnology-derived pharmaceuticals S6(R1). https://www.ich.org/page/safety-guidelines. Accessed 10 Aug 2022
5. Cohen SM, Klaunig J, Meek ME, Hill RN, Pastoor T, Lehman-McKeeman L, Bucher J, Longfellow DG, Seed J, Dellarco V, Fenner-Crisp P, Patton D (2004) Evaluating the human relevance of chemically induced animal tumors. Toxicol Sci 78:181–186
6. van der Laan JW, Buitenhuis WH, Wagenaar L, Soffers AE, van Someren EP, Krul CA, Woutersen

RA (2016) Prediction of the carcinogenic potential of human pharmaceuticals using repeated dose toxicity data and their pharmacological properties. Front Med (Lausanne) 3:45
7. Sistare FD, Morton D, Alden C, Christensen J, Keller D, Jonghe SD et al (2011) An analysis of pharmaceutical experience with decades of rat carcinogenicity testing: support for a proposal to modify current regulatory guidelines. Toxicol Pathol 39:716–744
8. ICH guideline: testing for carcinogenicity of pharmaceuticals S1B(R1). https://www.ich.org/page/safety-guidelines. Accessed 6 Sept 2022
9. Weber K, Razinger T, Hardisty JF, Mann P, Martel KC, Frische EA et al (2011) Differences in rat models used in routine toxicity studies. Int J Toxicol 30:162–173
10. Weber K (2017) Differences in types and incidence of neoplasms in Wistar Han and Sprague-Dawley rats. Toxicol Pathol 45:64–75. Erratum in: Toxicol Pathol. 2017;45:440
11. Envigo White Paper. The Wistar Hannover rat for carcinogenicity studies. A more effective model, Sept 2018. https://insights.envigo.com/hubfs/resources/white-papers/white-paper-wistar-han-rat-for-carcinogenicity-studies.pdf. Accessed 12 Aug 2022
12. Redbook 2000: IV.C.6. Carcinogenicity studies with rodents, Jan 2006. https://www.fda.gov/regulatory-information/search-fda-guidance-documents/redbook-2000-ivc6-carcinogenicity-studies-rodents. Accessed 10 Aug 2022
13. Paranjpe MG, Denton MD, Vidmar TJ, Elbekai RH (2015) Regulatory forum opinion piece*: retrospective evaluation of doses in the 26-week Tg.rasH2 mice carcinogenicity studies: recommendation to eliminate high doses at maximum tolerated dose (MTD) in future studies. Toxicol Pathol 43:611–620
14. ICH guideline: dose selection for carcinogenicity studies of pharmaceuticals S1C(R2). https://www.ich.org/page/safety-guidelines. Accessed 10 Aug 2022
15. Paranjpe MG, Belich J, Vidmar TJ, Elbekai RH, McKeon M, Brown C (2017) Tg.rasH2 mice and not CByB6F1 mice should be used for 28-day dose range finding studies prior to 26-week Tg.rasH2 carcinogenicity studies. Int J Toxicol 36:287–292
16. Lin KK, Rahman MA (2018) Expanded statistical decision rules for interpretation of results of rodent carcinogenicity studies of pharmaceuticals. In: Peace KE, Chen DG, Menon S (eds) Biopharmaceutical applied statistics symposium, vol 3. Springer, New York, pp 151–183
17. McConnell EE, Solleveld HA, Swenberg JA, Boorman GA (1986) Guidelines for combining neoplasms for evaluation of rodent carcinogenesis studies. J Natl Cancer Inst 76(2):283–289

Designing Studies in Pharmaceutical and Medicinal Chemistry

9

N. M. Raghavendra, B. R. Prashantha Kumar, Pujan Sasmal, Ghanshyam Teli, Rohit Pal, P. M. Gurubasavaraja Swamy, and B. Sajeev Kumar

Abstract

The discovery of medicine started years ago with the use of medicinal plant parts for the benefit or improvement of physiological conditions. The traditional uses of medicinal plants grew tremendously with Indian medicine and the Chinese medicine system. Later on, with time, researchers tried to isolate and synthesize the lead molecule of a natural product to increase its potency. The major source of drugs was plants, microorganisms, animals, and marine. With changes in time and environment, the need for quick recovery from a disease condition was felt for many reasons and thus synthetic-based drugs came to market with many regulations. In the initial days, the cost of drug development was a very high and costly process which may go up to $400-$600 million with the investment of 10-12 years of time. With the development of computer programs, some part of the drug discovery process was made easy with different software and programs. Many in-silico methods were developed by many researchers and scientists which boosted the drug design method and reduced the cost and time. The two main approaches of computer-aided drug design i.e., structure-based drug design and ligand-based drug design changed the process of drug discovery completely. Different visualization techniques of the 3D structure of a protein helped the researchers to understand the nature of the protein and the way to inhibit that protein and receptor. Structure-based drug design tools like homology modeling, molecular docking, and de-novo drug design helped in the discovery of drug molecules based on the detailed structure of protein whereas the structure-based drug design tools like pharmacophore mapping and QSAR techniques contributed to the discovery of drug molecules based on the nature of the previously reported drug and lead molecules. The in-silico ADMET properties prediction of a molecule based on the structural basis of the compound with the application of different rules like Lipinski's rule of five, Veber's rule, Ghosh rule, etc helped to develop a drug molecule with better pharmacokinetic and pharmacodynamic properties, which resists the easy rejection of a developed molecule in the clinical trials. Later with the advancement of

N. M. Raghavendra (✉) · B. Sajeev Kumar
College of Pharmaceutical Sciences, Dayananda Sagar University, Bengaluru, Karnataka, India
e-mail: ppl-pharmacy@dsu.edu.in; sajeevkumar-sps@dsu.edu.in

B. R. P. Kumar
JSS College of Pharmacy, Constituent College of JSS Academy of Higher Education and Research, Mysuru, India
e-mail: brprashanthkumar@jssuni.edu.in

P. Sasmal · G. Teli · R. Pal · P. M. Gurubasavaraja Swamy
Acharya & BM Reddy College of Pharmacy, Bengaluru, Karnataka, India
e-mail: gurubasavaraj@acharya.ac.in

G. Jagadeesh et al. (eds.), *The Quintessence of Basic and Clinical Research and Scientific Publishing*,
https://doi.org/10.1007/978-981-99-1284-1_9

computers, several artificial intelligence approaches were developed such as machine learning and deep learning. These techniques have many applications such as drug-target interactions, drug repurposing, prediction of synthesis and retrosynthesis, etc. Illustration about all these aspects is presented in a systematic way in this chapter.

Keywords

Drug design · Traditional medicine · Computer-aided drug design · Artificial intelligence · Structure based drug design · Ligand based drug design

9.1 Introduction

The discovery of medicine started thousands of years ago with the use of medicinal plants for the benefit or improvement of physiopathological conditions. The traditional uses of medicinal plants grew tremendously. As years passed, researchers tried to isolate and synthesize the lead molecule of a natural product to evaluate its efficacy [1]. The major source of drugs was plants, microorganisms, animals, and marine. With changes in time and environment, the need for quick recovery from a disease condition was realized for several reasons and thus synthetic-based drugs reached the market with many regulations. Many in-silico methods were developed by researchers which boosted the drug design method and reduced the cost and time. The two main approaches of computer-aided drug design *i.e.*, structure-based drug design and ligand-based drug design markedly changed the process of drug discovery [2]. The *in-silico* ADMET properties prediction of a molecule based on the structural basis of the compound with the application of different rules such as Lipinski's rule of five, Veber's rule, Ghosh's rule, paved the way to develop a drug molecule with better pharmacokinetic and pharmacodynamic properties, which resisted the easy rejection of a developed molecule in the clinical trials [3]. With the advancement of computers, several artificial intelligence approaches such as machine learning and deep learning were developed for designing chemical molecules [4]. In this book chapter, we have enlisted the three different domains of drug discovery and development.

9.2 Traditional Methods of Drug Design

9.2.1 Introduction

Drug discovery and development have a long history dating back to the dawn of human civilization. In those primeval times, drugs were not just used for physical medications but were also connected with spiritual and godly curative. Mentors or holy leaders were often the administrators of drugs [5]. The initial drugs or traditional drugs were plagiaristic of plant products and complemented by animal materials and minerals. These drugs were most possibly revealed through an amalgamation of trial-and-error investigation and observation of human and animal reactions as a result of ingesting such compounds [6]. Although these traditional medications likely developed autonomously in various civilizations, there are several resemblances, such as the use of the same basils to treat similar diseases. This was probably done by primeval traders, who may have assisted in the spread of medical acquaintances during their travels [7]. Prior to the 1800s, only medical herbs were accessible to treat and recover from illnesses. In the late 1800s, drug development and discovery evolved to utilize scientific methods. Since then, an increasing number of drugs have been established, scrutinized, and manufactured on a large-scale [8]. The modern pharmaceutical industry developed after World War I has decisively established the use of scientific or systematic principles in drug discovery and development. Despite the abundant use of pharmaceutical drugs today, many holy cultures have conserved their own traditional medicines. In certain cases, pharmaceutical drugs are used in addition to or in place of these folk remedies.

9.2.2 Discovery of Examples of Traditional Medicines

9.2.2.1 Indian Medicine

The traditional system of medicine practiced in India is Ayurveda. The origins of Indian medicine can be traced back between 3000 and 5000 years, and it was once practiced only by the Brahmin elite in the distant past [9]. The term "Vedas" refers to the traditional therapies found in holy texts on medicinal plant concoctions. The materia medica is meant to be descriptive and is primarily focused on the medicinal use of plants [10]. Cardamom, squill, fennel, and cinnamon are just a few of the herbs that have made their way into conventional Western medicine. The use of henbane as antivenin for snakebites was first documented in the fourth century AD by the Indian physician Susruta [11].

9.2.2.2 Chinese Medicine

According to folklore, traditional Chinese medicine (TCM) dates back to 3500 BC, during the era of the mythological Sheng Nong emperor. The TCM scripts of ancient China have been preserved thanks to the dynasty system and scrupulous documentation. Several important medical writings include the Shang Han Lun (Discussion of Fevers), the Huang Di Nei Jing (The Internal Book of Emperor Huang), and the Sheng Nong Ben Cao Jing (The Pharmacopoeia of Sheng Nong, a mythological Emperor) [12, 13]. There are numerous such medications in the Chinese pharmacopeia. Few possible Chinese herb constituents have been adopted in Western medications [14], for instance, the alkaloid ephedrine from Mahuang for the treatment of asthma [15], and reserpine from Rauwolfia for hypertension, emotional and mental management [16].

9.2.2.3 Egyptian Medicine

On ancient papyrus, the written chronicles of ancient Egyptian medical knowledge were preserved. About 877 prescriptions and formulations for eye, skin, and gynaecological internal medicine were discovered in the Ebers papyrus (dating from circa 3000 BC). Another document, the Kahun papyrus from approximately 1800 BC, detailed significant treatments for gynaecological issues [17]. The ancient Egyptian pharmacopeia included a wide range of therapeutic options, including those derived from minerals, metals, animals, and plant sources. Ancient Egyptians utilized a remarkably diverse range of plants, whether it was the entire plant, or only the fruit, leaves, juice, or root [18]. Plants of several kinds were included, such as acacia, anise, barley, cassia, castor bean, etc. Many bioactive secondary metabolites found in plants, including saponins, diterpenes, sesquiterpenes, isochromenes, flavonoids, isoflavonoids, and alkaloids, belong to a wide spectrum of chemical families [19]. The Egyptian pharmacopoeia included antimony, alum, charcoal from charred wood, copper, feldspar, iron, oxide, limestone, red ochre, sodium carbonate, sodium bicarbonate, salt, stibnite, sulphur, and maybe arsenical compounds. In situations of boils, felonies, and burns, these treatments are typically suggested for their localized mild astringent or antiseptic effects. Sulfide of antimony Sulphur is used orally to treat bilharziasis. Calamine soothes, yellow ochre-hydrated iron oxide heals ankylostoma's anaemia, red ochre-natural iron oxide treats haemorrhage, and green copper ore treats eye irritation [20].

9.2.2.4 Greek Medicine

Greek physicians learned medicine from Egyptians, Babylonians, Chinese, and Indians. Linseed was a soothing emollient, laxative, and antitussive, whereas castor oil was a laxative. Other remedies included asafoetida gum resin as an antispasmodic and fennel herb to relieve intestinal colic and gas [21]. It is conceivably the dispelling of the idea that ailments are brought on by supernatural forces or spells that represents the greatest gift of the Greeks to medicine. The Greeks understood that illnesses have natural causes. Hippocrates, the father of medicine, is credited with establishing medical ethics around 400 BC [22].

9.2.2.5 Roman Medicine

As excellent administrators, the Romans established hospitals, but they were primarily utilised for military applications. They ensured that coordinated medical care was available through their labor [23]. Additionally, the Romans expanded the Greeks' use of a pharmacy. Galen and Dioscorides were two eminent physicians of the Roman era. In Dioscorides' Materia Medica, descriptions of treatments based on 80% plant, 10% animal, and 10% mineral items may be found.

9.2.3 Different Sources of Traditional Drug Discovery

There are many physiologically active molecules in national resources. Many modern medications are either produced directly from natural sources or were created using a base component that was primarily obtained from a natural source, such as morphine and quinine. The majority of biologically active natural products are secondary metabolites, which have a variety of chiral centers and quite complex structures. This makes it difficult to synthesize these chemicals. Instead, one must typically extract them from their natural sources, which is a time-consuming, expensive, and ineffective operation. A chemist would never consider manufacturing the fundamentally novel chemical structures seen in many natural compounds. The trioxane ring of the natural antimalarial drug artemisinin, which has a highly unstable look, is one of the most surprising structures [24].

9.2.3.1 Plant

Lead compounds have always been abundant in plants (e.g., morphine, cocaine, digitalis, quinine, nicotine). Many of these lead compounds have been used to create synthetic pharmaceuticals (e.g., local anaesthetics developed from cocaine), while others are helpful medicines themselves such as morphine and quinine. Plants are always a promising source of novel medications. The antimalarial artemisinin from a Chinese plant, the Alzheimer's medication galantamine from daffodils, and the anticancer agent paclitaxel (Taxol) from the yew tree are just a few examples of clinically relevant drugs that have lately been extracted from plants.

9.2.3.2 Microorganisms

Numerous medications and lead compounds have been developed from microorganisms including bacteria and fungi. These organisms produce a variety of antibacterial substances. After the discovery of penicillin (Penicillium molds, principally *P. chrysogenum* and *P. rubens*), the screening of microbes gained enormous popularity. To examine novel bacterial or fungal strains, soil and water samples from all around the globe were collected. This resulted in the creation of an incredible arsenal of antibacterial medicines, such as cephalosporins (*Cephalosporium acremonium*), tetracyclines (*Streptomyces aureofaciens* and *S. rimosus*), aminoglycosides (*Actinomycetes*), rifamycin (*Amycolatopsis rifamycinica*), and vancomycin (*Bacterium Amycolatopsis orientalis*). A few microbial metabolites have served as lead compounds in various medical specialties, although the majority of medications originating from microbes are utilized in antibacterial therapy. For example, the fungus *Aspergillus alliaceus* produces asperlicin, a new antagonist of the peptide hormone cholecystokinin (CCK), which regulates appetite [25, 26]. In the brain, CCK acts as a neurotransmitter and is hypothesized to be associated with anxiety and panic attacks. As a result, asperlicin analogs may be useful in the management of anxiety. Other examples are ciclosporin, a fungal metabolite used to control the immunological response following organ transplants, and lovastatin, the first of the therapeutically relevant statins discovered to lower cholesterol levels [27]. Lipstatin is a natural product that was isolated from *Streptomyces toxytricini* [28]. It blocks pancreatic lipase and contributed as the precursor to the anti-obesity drug called orlistat. Finally, rasfonin, a fungus compound discovered from the fermented mycelium of *Talaromyces sp.*, increases cancer cell apoptosis but not normal cell death. It serves as a promising starting chemical for new anticancer medications [29]. Curacin A from a marine cyanobacterium, *Lyngbya majuscula* shows potent antitumor activity [30].

9.2.3.3 Marine Sources

Discovering lead compounds from marine sources has also received attention. A variety of biologically powerful compounds with significant inflammatory, antiviral, and anticancer activities have been discovered from marine sources such as coral, sponges, and fish. The antitumor medications eleutherobin, bryostatins, dolastatin, cephalostatins, and halichondrin B are from marine sources. An easier version of halichondrin B was authorized for the treatment of breast cancer in 2010.

9.2.3.4 Animal Source

Animals occasionally serve as a source of novel lead compounds. For instance, a group of antimicrobial polypeptides called the magainins were isolated from the skin of the *Xenopus laevis*, an African-clawed frog [31]. These substances shield the frog against illness and could offer guidance for the creation of new antibacterial and antifungal medications for humans. A strong analgesic substance called epibatidine, which is made from the skin extracts of the Ecuadorian poison frog, is another illustration [32].

Animal, plant, snake, spider, scorpion, and insect venoms and poisons have high potency because they frequently interact in very specific ways with a macromolecular target in the body of prey. As a result, they have proven to be valuable resources for research into drugs targeting enzymes, ion channels, and receptors. These poisons frequently take the form of polypeptides, such as cobra bungarotoxin. The tetrodotoxin produced by puffer fish, for example, is a non-peptide toxin that is incredibly potent.

9.2.4 Protocol for the Drug Design from a Natural Source

Extracting and purifying active components from natural cures is part of drug discovery and development. Natural substances have complicated chemical structures that vary in producing species. Several secondary metabolites have pharmacological effects. These metabolites employ their actions on molecular targets that vary from one case to the other. These targets may be enzymes, receptors, transporters, transcription factors, or even nucleic acids. Good knowledge of the chemical composition of plants leads to a better understanding of their possible and specific medicinal value. When existing methods with advanced technologies are applied, it can lead to a modern revelation of drugs, benefitting medicinal purposes. The development of modern technologies has streamlined the screening of natural products in discovering new drugs. Academic and industrial researchers must come forward for extensive research in this field. Research for drug discovery must create robust and prudent lead molecules, which progress from a screening hit to a drug candidate through structural elucidation and structure. Steps associated with the drug discovery process from natural resources are illustrated in Fig. 9.1.

9.3 Computer-Aided Drug Design

9.3.1 Introduction

The journey of a lead molecule from a laboratory to becoming a drug molecule and getting approval from a regulatory body for the betterment of human or animal health is a very complicated and expensive process [33]. In the traditional methods of drug design, we presented information about the old and ancient methods of drug discovery which were based on critical observations and identifications of the effectiveness of mostly plant, animal parts, and marine sources which can be used to treat many diseases and disorders. This needs decades of observations and generation-wise delivery of the information to their newborns for the use and betterment of the knowledge. With time as science and industries evolved, the number of diseases increased, and to fight that the drug discovery process needed to be changed. After the discovery

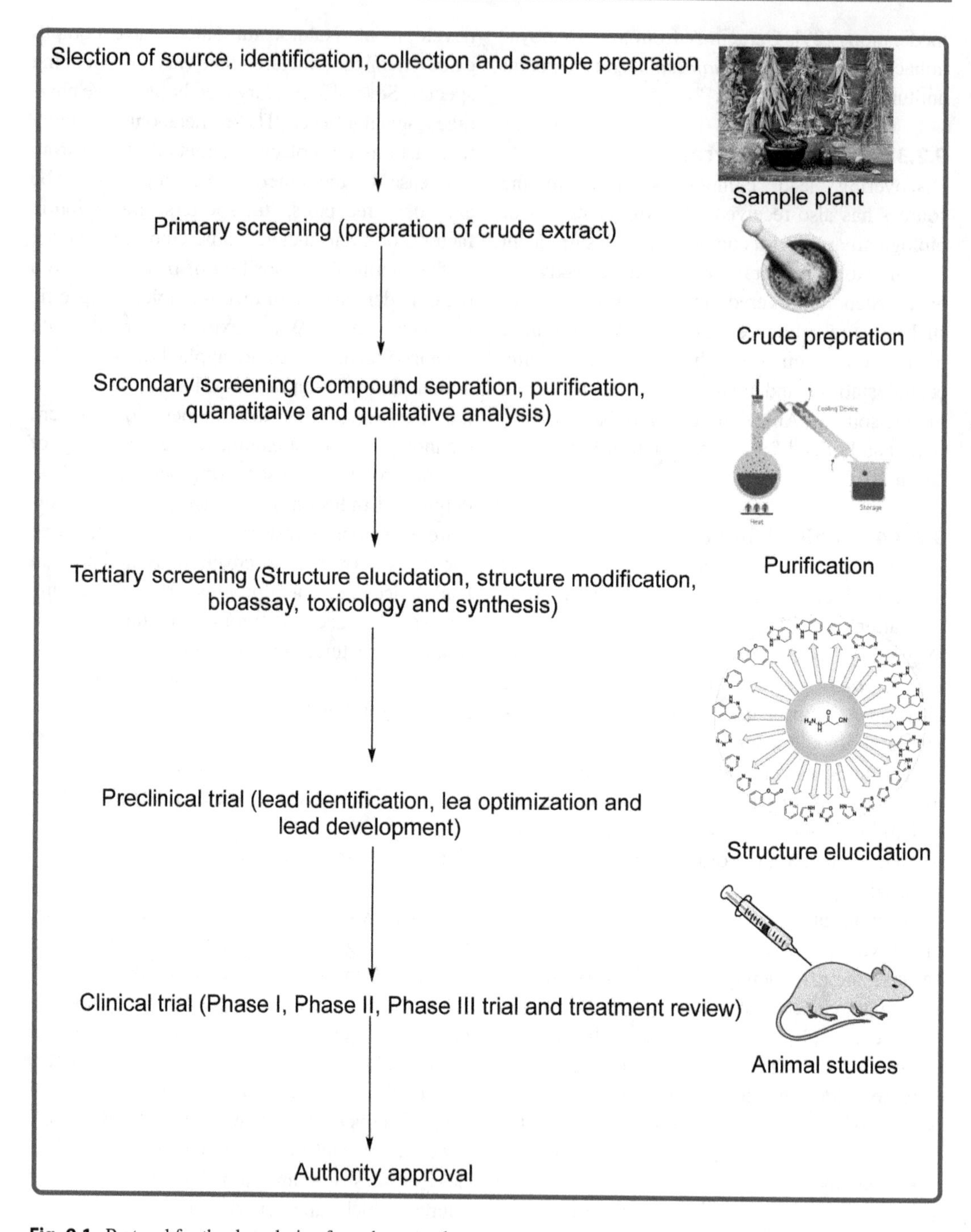

Fig. 9.1 Protocol for the drug design from the natural source

of computers (1855–1945) many things changed including the drug discovery and development procedures. With the implementation of different software and applications, the long process of drug discovery became notably short. Various types of software have evolved parallel at the same time which gave a massive boost to the drug discovery and development process, including the knowledge about the disease at the cellular level, the nature of the protein, the arrangement of

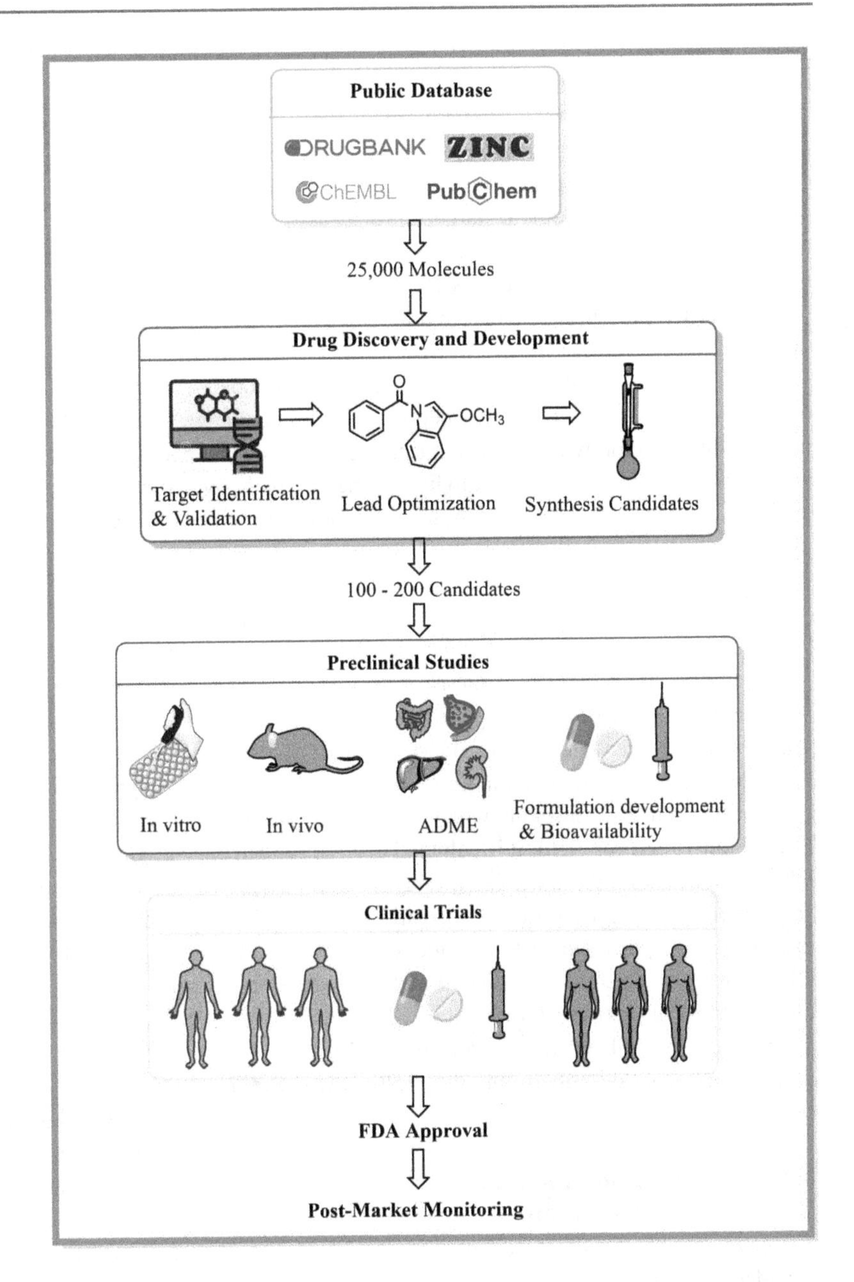

Fig. 9.2 Different stages of drug discovery and development process

amino acids sequences in a protein, analyzing the nature and structure of receptor of a real protein [34]. In Fig. 9.2 the implementation of CADD in the drug development process is illustrated.

There are many software and processes for Computer Aided Drug Design which are mainly grouped into two groups i.e., Structure-Based Drug Design (SB-DD) and Ligand Based Drug Design (LB-DD).

9.3.2 Structure-Based Drug Design

As the name suggests, the structure-based drug design starts with an understanding of the protein or macromolecule structure that causes the disease or disorder at the molecular level [35]. It is a paradigm for drug discovery that integrates the strengths of numerous scientific fields, including X-ray crystallography, nuclear magnetic resonance, medicinal chemistry, molecular modeling, biology, enzymology, and biochemistry [36]. SBDD's current strength relies on enzyme inhibitors as drugs, but recent technological advancements have made it feasible to target more complicated biological targets, such as those that control immune activation and immune suppression. There are many studies where structure-based drug design took a crucial role in the drug discovery process eg., the HIV protease has been the main focus of research on HIV proteins [37]. A study of the non-peptide based inhibitor haloperidol, which was found using the DOCK program [38], has brought to light some of the drawbacks of structure-based design [39]. DOCK searches a chemical database for structures that fit in the rigid form and the chemical property of the binding pocket as a template [40]. When the ketone group in haloperidol (Ki of 100 μM) is replaced by a thioketal ring, a molecule (UCSF8) with a Ki of 15 μM is produced. The HIV-1 protease with its Gln7 to Lys variation has been investigated crystallographically in association with this inhibitor, which also has a Ki of about 15 μM for them. It is believed that the inhibitor binds with the dimerizing enzyme in a single mode. In contrast to the conformation used to generate the prediction, UCSF8 interacts with HIV-1 protease around 4.8 Å distant and rotated by 79° from the expected orientation. Various methods can be utilized in structure-based drug design and are described in the following sections.

9.3.2.1 Target Identification and Validation

The development of medicines (small molecules, peptides, antibodies, or more recent modalities such as short RNAs or cell treatments) that will change the disease state by regulating the activity of a biological target or receptor is the preeminent strategy in drug discovery [41].

Being one of the initial steps of the drug discovery and development process, target identification and validation involves four key pharmacological targets that can be found in organisms which include receptors and enzymes, nucleic acids (DNA and RNA), carbohydrates, and lipids. Though the majority of pharmaceuticals on the market use proteins as a target, the decoded genomes of nucleic acids will lead the researchers' focus toward it in near future [42]. The target identification of B-Raf protein is shown below (Fig. 9.3).

9.3.2.2 Binding Pocket Prediction and Analysis

Researchers have proposed various criteria for choosing a binding site because it is not an easy challenge to choose binding sites [44]. The overall shape, amino acid composition, type of solvation, hydrophobicity, electrostatics, and chemical fragment interactions are the key characteristics that define a protein binding pocket [45]. The majority of currently accessible techniques are based on molecular surface similarity searches for functional site databases like PDB, which offer information on protein structures that have been thoroughly examined and experimentally confirmed. In addition, several additional models, including Support Vector Machines (SVM), Hidden Markov Models (HMM), and Critical Structure Prediction Assessment (CASP9), are used to build some of the approaches [46, 47]. Recently, Sasmal et al. have reported the binding pocket analysis of PD-L1 protein by implementation of SVM technique [48] (Fig. 9.4).

9.3.2.3 Different Visualizing and Modeling Tools for Macromolecules

Using experimental methods to determine three-dimensional structures is expensive and time-consuming. Therefore, employing sequence information in comparative modeling or homology modeling to predict three-dimensional structures accurately results in models that are suitable and applicable in different fields of drug development [49, 50]. When an X-ray crystallographic or NMR-validated structure of a protein is absent, modeling the 3D structure of a protein from a sequence is required for drug development [51, 52]. When a suitable template cannot be found in the PDB database, various modeling

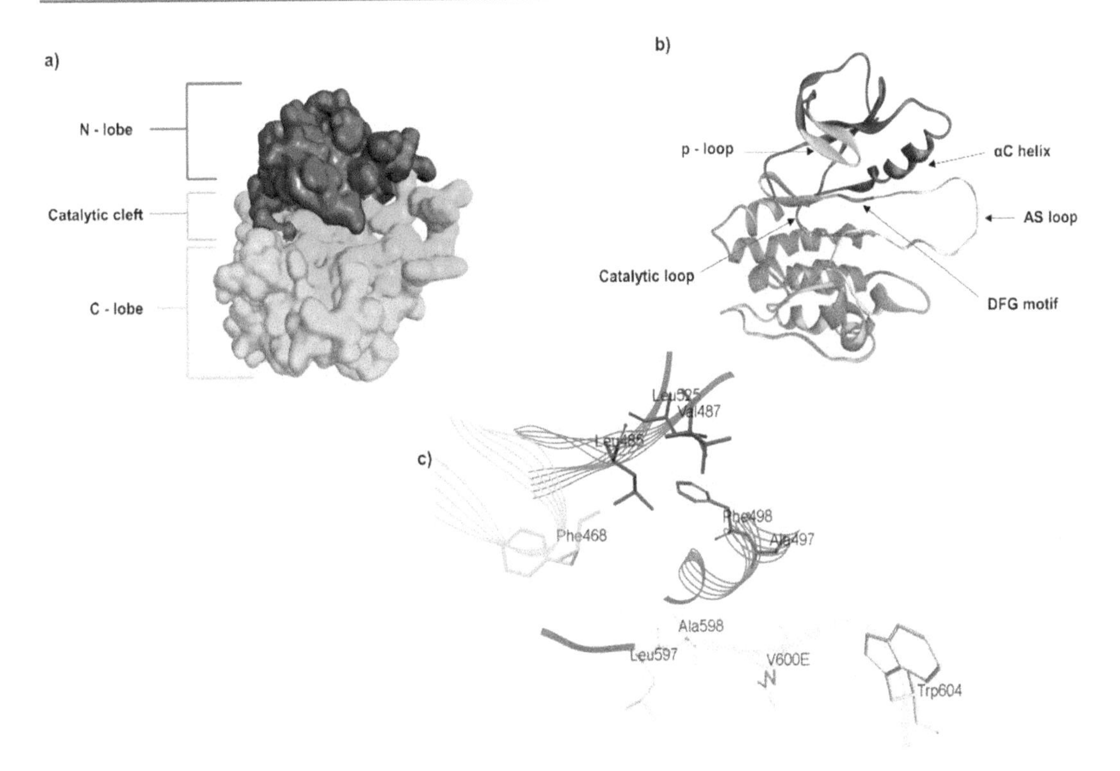

Fig. 9.3 Target identification of B-Raf protein. Reprinted with permission from Ref. [43]. (**a**) B-Raf protomer kinase domain surface representation in different colors. The N-lobe (Red) and the C-lobe (Cyn) of the kinase are connected by a catalytic cleft (Green). (**b**) The key structural regions of B-Raf kinase; P-loop (Cyn), αC-helix (Red), DGF motif (Pink), and AS-loop (Yellow) are highlighted. (**c**) The backbone residues' coloration in this hydrophobic region surrounding amino acid V600E corresponds to where they are in the protein

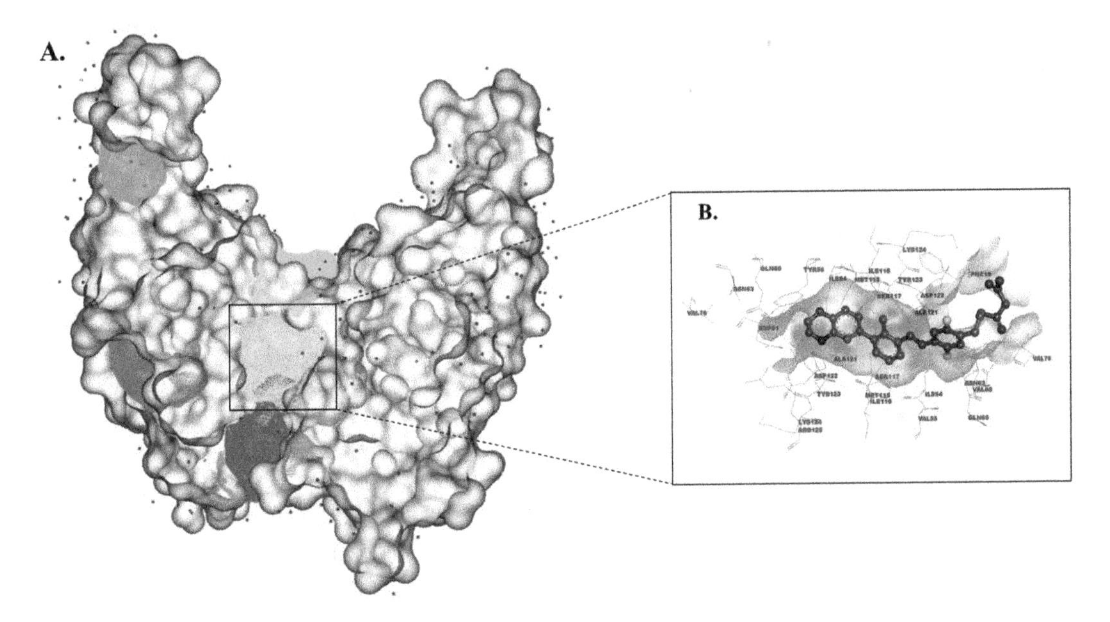

Fig. 9.4 Binding Pocket analysis of PD-L1 protein. Reprinted with permission from Ref. [48]. (**a**) The predicted binding pocket and sub-pockets of PD-L1 protein. (**b**) The main druggable pocket of the protein and crucial amino acids

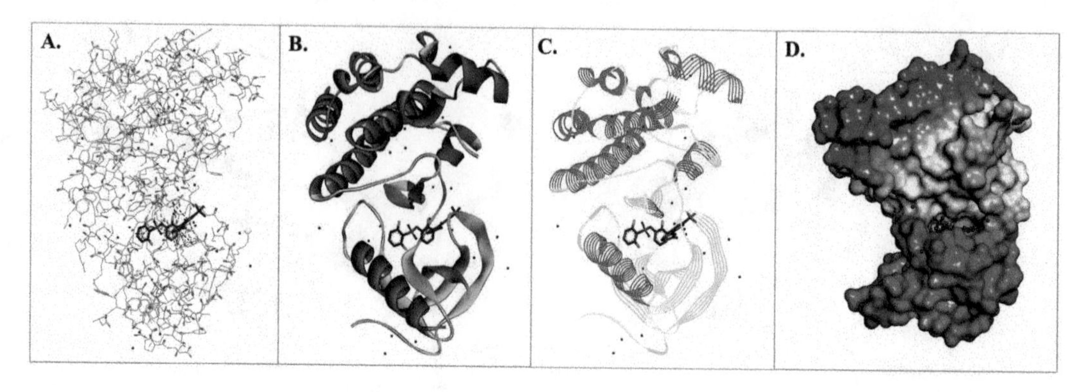

Fig. 9.5 Different visualization style of B-Raf macromolecule (PDB ID: 4XV2) in Biovia Discovery Studio. (**a**) Atoms. (**b**) Solid Ribbon. (**c**) Line Ribbon. (**d**) Surface

techniques like threading, fold recognition along with different ab initio methods are applied [53].

The 3D predicted and experimentally determined atomic structures of macromolecules are seen and analyzed in great detail using visualization tools like RasMol, PyMol, Chimera, and others. Recently, there has been a lot of interest in developing user-friendly simulation platforms based on computer graphics for molecular biologists. It is frequently used by visual editors to create models for a better comprehension of atomic data of 3D coordinates as well as in biology for the display of simulation findings in post-processing or studies. Biovia Discovery Studio Visualizer also provides a great role in visualizing the 3D protein model, ligand structure, and protein-ligand complexes. With the help of a sequence alignment program, it can superimpose different protein structures and also can predict binding sites. The visualization of B-Raf protein in different 3D visualizing models is represented in Fig. 9.5.

9.3.2.4 Molecular Docking

Molecular docking is a computer approach that searches for ligand shapes that can energetically and geometrically fit into a protein's binding site [54]. To forecast the binding behavior of novel hypothetical molecules, docking calculations are necessary. Blind docking is effective when the binding site in the target protein structure is unknown since the entire protein structure is viewed as the binding site region; however, site-specific docking is useful to predict the interactive nature of the ligand molecule if the binding site is known [55]. Because only specific amino acid residues in the binding site cavity are targeted, blind docking often produces less precise findings and requires more time and computer memory than site-specific docking. All energy estimates are based on the assumption that the small molecule will use the binding mode with the lowest free energy within the binding site. The change in free energy that results from the binding is known as the free energy of binding and is expressed as:

$$\Delta G_{binding} = \Delta G_{complex} - \left(\Delta G_{protein} + \Delta G_{ligand}\right)$$

Where, $\Delta G_{binding}$ is free energy after the binding of ligands to the protein. $\Delta G_{complex}$ is the free energy of a protein-ligand complex. $\Delta G_{protein}$ is the energy of protein only and ΔG_{ligand} is the free energy of ligand only.

The force field scoring functions estimate binding affinity by using molecular mechanics force fields. As scoring functions, several docking methods use the nonbonded variables from AMBER and CHARMM [56]. The non-covalent receptor-ligand complex's binding free energy is measured using empirical scoring systems based on chemical interactions [57]. These scoring methods often include separate variables for hydrogen bonds, hydrophobic interactions, ionic interactions, and binding entropy, as is the case with the Böhm scoring functions used in FlexX and SCORE in

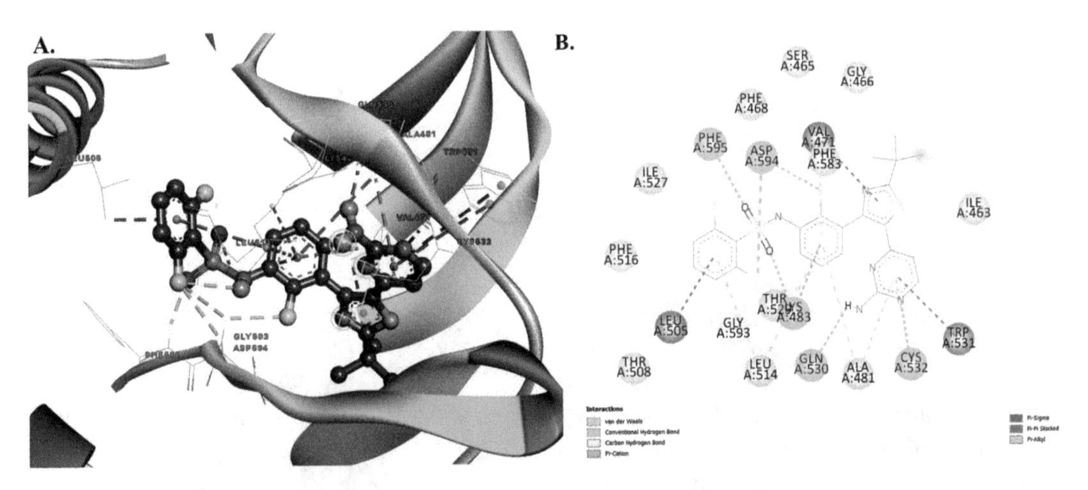

Fig. 9.6 Molecular docking interaction of dabrafenib with B-Raf protein using AMBER force field by AutoDock Vina software. (**a**) 3D interaction of dabrafenib and B-Raf protein. (**b**) 2D in interactions of dabrafenib and B-Raf protein

DOCK4 [58, 59]. Software programs for molecular docking include FlexX [58], FlexiDock [59], DOCK [60], and AUTODOCK [61]. The molecular docking of vemurafenib with B-Raf protein, where the AMBER force field was utilized in AutoDock Vina, is represented in Fig. 9.6.

9.3.2.5 Binding Free Energy Calculations

Numerous attempts have been made to determine the binding free energy of protein-ligand complexes using computational methods since the scoring function utilized has a significant impact on molecular docking's efficacy [62]. Over the past several years, a variety of methods have been suggested as less expensive alternatives to the quick and roughly correct estimate of binding free energy.

Molecular Mechanics Poisson Boltzmann Surface Area (MM-PBSA)

The MM-PBSA method uses a continuum solvent model and a collection of complex, ligand, and receptor conformations obtained from a single molecular dynamics trajectory to calculate binding free energy [63]. It has been tested in a variety of systems, including protein-ligand and RNA-ligand complexes, with surprising effectiveness [64].

XSCORE

XSCORE is an experimental scoring function, which is calibrated using data from experiments [65], considering van der Waals contacts, hydrogen bonds, deformation penalty, and hydrophobic interactions among the receptor and the ligand.

Molecular Mechanics/Generalized Born Surface Area (MM-GBSA)

In MM-GBSA, the free energy of binding is determined by combining the energy of the gas phase (MM), the energy of electrostatic solvation (GB), and the contribution of non-electrostatic forces to the solvation energy (SA) [63, 66]. Typically, a protein and ligand complex's free energy of binding is calculated as the mean of multiple conformations that were derived via MD simulation trajectories.

Molecular Mechanics-Generalized Born with Molecular Volume (MM-GBMV)

The MM/GBMV technique, which is used in CHARMM, breaks down the contributions from various interactions in the free energy of binding of the ligand, the protein, and the complex as follows [67]:

$$\Delta G_{bind} = \Delta H - T\Delta S \approx \Delta E_{MM} + \Delta G_{sol} - T\Delta S \tag{9.1}$$

$$\Delta E_{MM} = \Delta E_{int} + \Delta E_{el-st} + \Delta E_{vdw} \quad (9.2)$$

$$\Delta G_{sol} = \Delta G_{generalized-born} + \Delta G_{molecular\ volume} \quad (9.3)$$

Here, the molecular mechanics energy changes in the gas phase are denoted by ΔE_{MM}, which includes the changes in the internal energy (ΔE_{int}), electrostatic energy ($\Delta E_{el\text{-}st}$), and the van der Waals energy (ΔE_{vdw}). The solute and the implicit solvent interaction energy is denoted as ΔG_{sol}, which is the total of the polar and non-polar contribution to the desolvation-free energy with the polar contribution ($\Delta G_{generalized\text{-}born}$) calculated by the Generalized Born using Molecular Volume (GBMV) model which is implemented in CHARMM and the molecular volume contribution ($\Delta G_{molecular\ volume}$), calculated by implementation of molecular volume within the GB module. As the goal was to determine relative binding free energies, the entropic contribution of the system's vibrational modes to the binding free energy was disregarded [68].

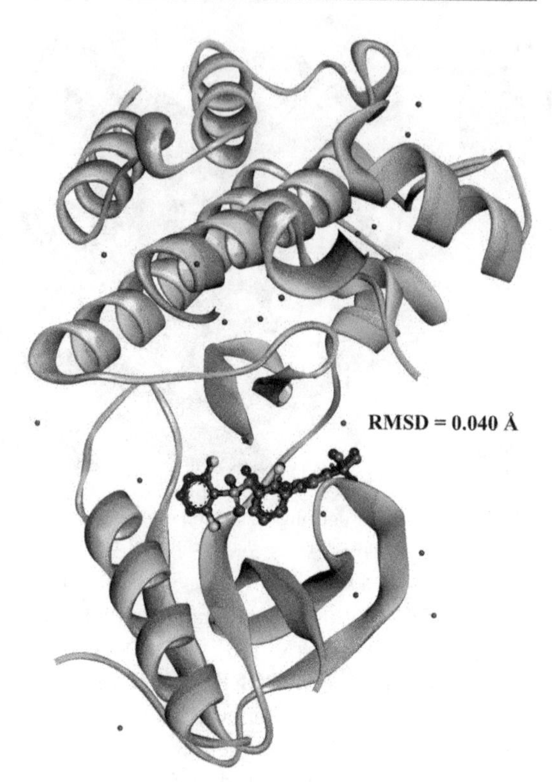

Fig. 9.7 Validation of molecular docking. Reprinted with permission from Ref. [43]. The picture shows the original protein (deep yellow) with the co-crystal ligand (dabrafenib) in deep blue color, superimposed with the image of the protein (light blue) that had been docked again with the re-docked ligand (dabrafenib) in characteristic color

9.3.2.6 Validation of Molecular Docking

The validation of molecular docking is very essential for reporting the docking results. There are several reported techniques for validating molecular docking. Pose selection is a common approach that includes re-docking a substance onto the target's active site with known conformation and orientation, frequently from a co-crystal structure, utilizing docking software [69]. Programs are deemed to be effective if they can return poses that are within a predetermined root mean square deviation (RMSD) value of the known conformation (often 1.0 Å or 1.5 Å based on the size of the ligand). The validation of molecular docking was shown in Fig. 9.7 by superimposing the docked file and the original protein where the RMSD was found to be 0.040 Å.

9.3.2.7 De Novo Drug Design

De novo design is an alternative to molecular docking in that ligands are constructed inside the ligand binding domain as opposed to using ligands that are previously known [70]. In an iterative procedure, the putative ligand is constructed within the receptor groove, fragment by fragment, using the 3D structure of the receptor. In de novo design, two fundamental types of algorithms are frequently employed [71]. The first one is the "outside-in-method," which involves first analyzing the binding site to identify which particular functional groups may bind strongly. The ligands are created by joining these distinct fragments with common linker units. The second approach is known as the "inside-out method," in which molecules are built inside the binding site to enhance occupancy and fit [72]. When the receptor structure is available, the unknown lead compounds can only be developed by de novo design. This technique can be employed in the reverse condition also. Various researchers have created several programs for creating ligands

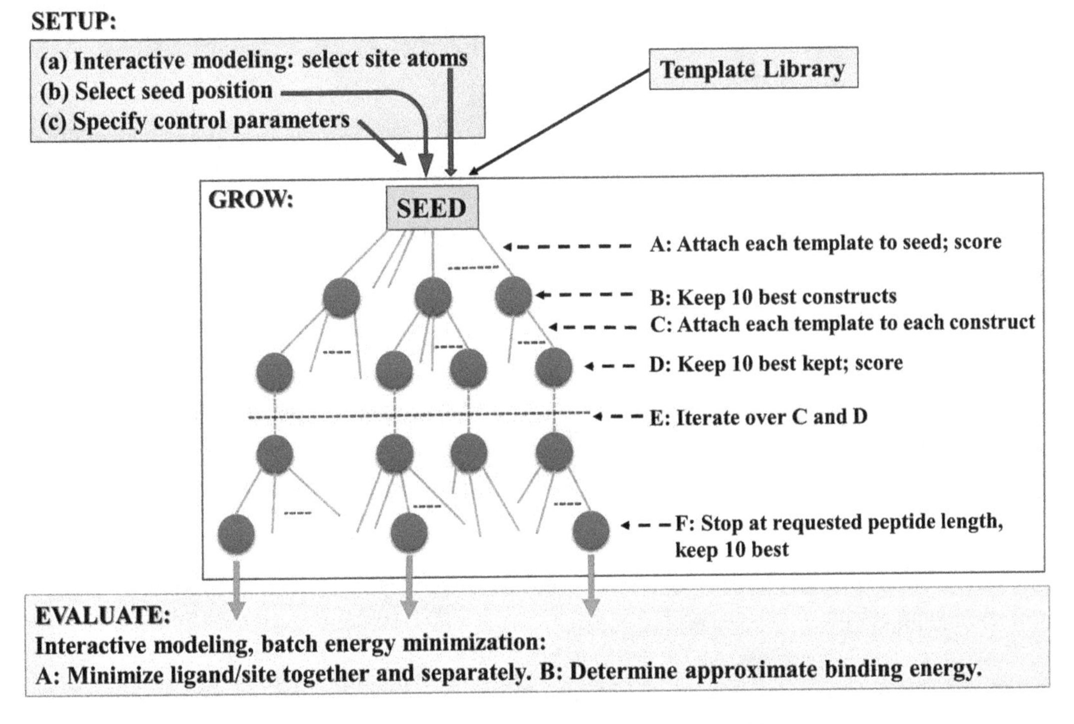
SETUP:
(a) Interactive modeling: select site atoms
(b) Select seed position
(c) Specify control parameters
Template Library
GROW:
SEED
A: Attach each template to seed; score
B: Keep 10 best constructs
C: Attach each template to each construct
D: Keep 10 best kept; score
E: Iterate over C and D
F: Stop at requested peptide length, keep 10 best
EVALUATE:
Interactive modeling, batch energy minimization:
A: Minimize ligand/site together and separately. B: Determine approximate binding energy.

Fig. 9.8 Diagrammatic representation of De Novo Drug Design (GROW) for peptide

from scratch. De novo design tools like GROW (Fig. 9.8) [73], GRID [74], LUDI [75], LEAP-FROG [76], GROUPBUILD [77], and SPROUT [78] have been widely used.

In 2008, Alig et al. used a de novo fragment-based drug design strategy to create a new cannabinoid receptor inhibitor to treat obesity. From known cannabinoid receptor ligands, fragments were created, which were then pharmacophorically separated into several receptor site regions. These were then combined to make novel ligands that inhibit or antagonize the receptor protein or receptor [79].

9.3.2.8 Homology Modeling

Homology modeling or comparative modeling uses the 3D structure of a protein that has been experimentally discovered to forecast the 3D structure of another protein with a related amino acid sequence [80]. The amino acid sequence of a protein determines its structure in a certain way. The structure is more preserved throughout evolution than the sequence, causing related sequences to adopt nearly identical structures and distantly related sequences to exhibit fold similarity [81, 82].

The phases in the multistep homology modeling process are as follows:

Template Recognition and Initial Alignment

After the identification of a homologous protein structure, the target sequence's similarity to a prospective template should be sufficient to be detected by simple sequence alignment techniques like BLAST or FASTA.

Alignment Correction

Aligning the amino acid sequence with the standard template is the next step and for low-similarity sequences, one can use different sequences from related proteins. For instance, the alignment of the sequence LTLTLTLT with the YAYAYAYAY sequence is nearly impossible and only a third sequence TYTYTYTYT, which can align both of them easily, can only solve the issue.

Backbone Generation

The target's backbone may be constructed with the right alignment of the template coordinates. The side-chain coordinates are replicated as well when the residues are similar.

Loop Modeling

After that, the gaps containing areas, caused by insertions and deletions are moldelled by loop modeling. The less precise loop modeling method is used to simulate these gaps. Currently, the issue is approached using two basic methods:

(a) The database searching technique involves locating loops from existing protein structures and then superimposing them with the target protein's major chains. It is possible to employ certain specialist software such as FREAD and CODA.
(b) The ab initio technique creates a large number of random loops and looks for one with a reasonable energy value, an acceptable angle, and a location within the Ramachandran plot.

Side-Chain Modeling

The side chain modeling involves analyzing the protein-ligand interactions at active sites of the protein and protein-protein contact interface interactions. The last interaction energy can be estimated by calculating the side chain's torsion angle's confirmation with nearby atoms and can be utilized to build a side chain and a rotamer library, which contains all the advantageous side-chain torsion angles taken from recognized protein crystal structures, may be utilized.

Model Optimization

An energy reduction strategy for the entire model involves modifying the comparative positions of the atoms to ensure that the entire shape of the molecule possesses the least energy potential. By applying varied stimulation conditions (heating, cooling, putting with water molecules), the atoms gravitate toward a global minimum, which gives Molecular Dynamic Simulation (MDS) a greater chance of identifying the real structure.

Model Validation

Each homology model has flaws. There are two basic causes:

(a) The proportion of target and template sequences that are identical. Crystallographically determined structures can be used to compare the model's precision after getting 90% or more accuracy, and a major inaccuracy arises if it is less than 30%.
(b) The number of template mistakes.

The finished model has to be assessed to confirm the stereochemical characteristics, φ–ψ angles, bond lengths, chirality, and tight contacts by different models such as Modeler, SWISS-MODEL [83], Schrodinger, and 3D-JIGSAW [84]. The choice of template, the method employed, and the model's validation all contribute to its success.

In 2003, Lewis et al. developed the structure of human CYP1A2 by using the CYP2C5 crystallographic structure (PDB ID: 1DT6) as a template [85].

9.3.3 Ligand-Based Drug Design

When existing modeling tools cannot establish the structure of the target protein or it is uncertain, an alternative methodology is employed, ligand-based drug design. To connect ligand activity to structural data, statistical techniques are used. In the sections that follow, various methods for ligand-based drug design are addressed.

9.3.3.1 Pharmacophore Studies

A pharmacophore is the spatial configuration of a group of essential properties found in a chemical entity that interact effectively with the receptor, causing ligand-receptor interaction and producing the therapeutic effect [86, 87]. Pharmacophore models are widely utilized when certain active compounds have been discovered but the 3D target protein structure is unclear [88]. A pharmacophoric map is a three-dimensional representation of a pharmacophore that is created by describing the nature of all the important

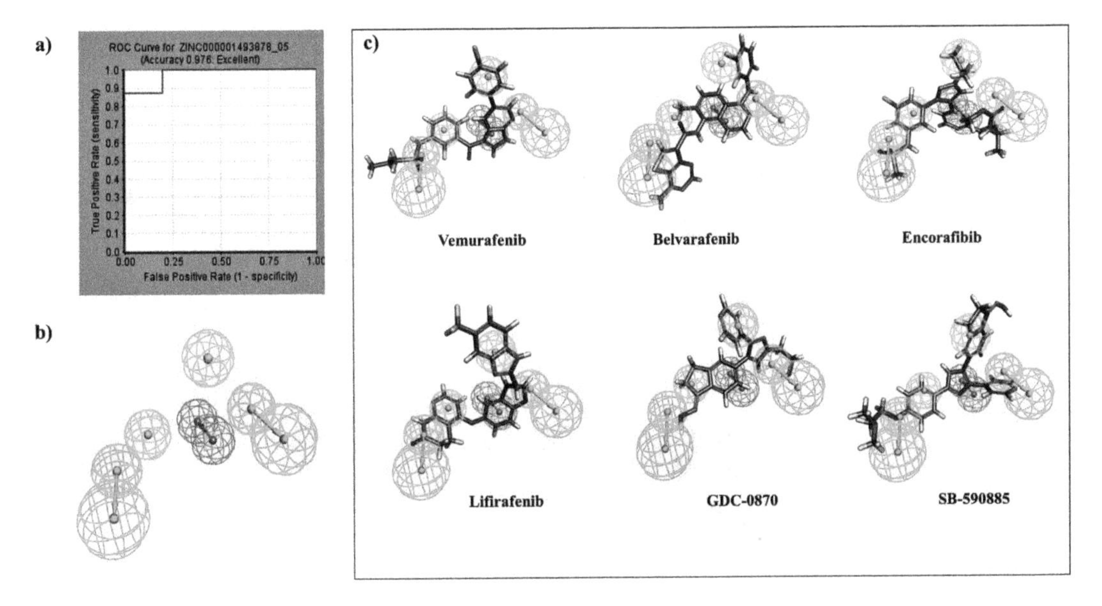

Fig. 9.9 Pharmacophore mapping of natural product-based B-Raf inhibitors. Reprinted with permission from Ref. [43]. (**a**) ROC curve for validation. (**b**) Generated Pharmacophore. (**c**) Matching the compounds with the pharmacophore

pharmacophoric properties and the relationships between those features in three dimensions [89]. The superimposition of the active chemicals can reveal their commonalities by creating a pharmacophore map. In Fig. 9.9 the pharmacophore map of natural B-Raf inhibitors has been shown [43].

A pharmacophore can be mapped using a collection of active compounds in two steps:

1. Examining the compounds to find pharmacophoric characteristics.
2. Aligning the molecules' active conformations to determine the ideal overlay of the relevant characteristics.

The creation and improvement of the molecules, as well as the placement of ligand points and site points (positioning of ligand atoms to macromolecule atoms), are all done during pharmacophore mapping. Hydrophobic areas like the cores of aromatic rings and hydrogen bond acceptors and donors are typical ligand and site points. They can also be utilized to create molecules with exact properties (lead optimization). It also uses pharmacophore fingerprints to assess the variety and comparability of drugs. Additionally, it may be used to align substances based on the 3D arrangement or to develop foresightful 3D QSAR models.

The zinc metal-based enzyme Matrix Metalloprotease-8 (MMP8) breaks down collagen and aids in the development of pannus in rheumatoid arthritis [90]. By using a combination of structure-based and ligand-based methods, especially the pharmacophore hypothesis of AADRH, Kalva et al. have screened potent non-zinc-based inhibitors of MMP8 from the zinc database [90]. By creating a DKA pharmacophore, inhibitors for the target protein HIV-1 integrase were tested and eliminated [91]. Because of the effectiveness of the DKA structural analog, the DKA pharmacophore was transferred to the n-alkyl hydroxypyrimidinone carboxylic acid family's naphthyridine carboxamide core, and the drug raltegravir was the first integrase inhibitor approved by the FDA.

9.3.3.2 Quantitative Structure-Activity Relationship

The quantitative structure-activity relationship (QSAR) method is a tool employed in drug discovery to predict the biological activity of

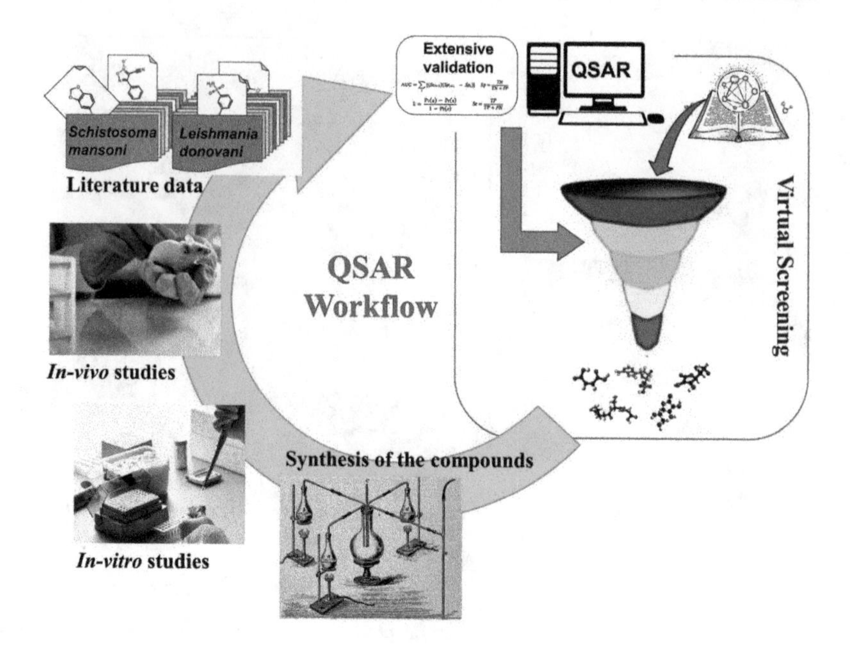

Fig. 9.10 The QSAR Workflow

bioactive molecules [92]. The association between the compound's structural features and the relevant biological behaviors is discovered using mathematical and statistical techniques (Fig. 9.10). As a result, the QSAR model is created utilizing structural factors to forecast a drug's biological characteristics. This may be done by employing descriptors that are simple to calculate, such as molecular weight, the number of rotatable bonds, and log P. The QSAR paradigm was developed by Hansch and Fujita as a result of advancements in physical organic chemistry throughout time and the contributions of Hammett and Taft in linking chemical activity to structure.

There are various approaches to QSAR:

1. 1D QSAR: The affinity linked with pKa, logP, etc.
2. 2D QSAR: The affinity linked with a structural pattern.
3. 3D QSAR: The affinity linked with the three-dimensional structure.
4. 4D QSAR: The affinity linked with multiple representations of ligands.
5. 5D QSAR: The affinity linked with multiple representations of induced-fit scenarios.
6. 6D QSAR: The affinity linked with multiple representations of solvation models.

Case Study

The 5-hydroxytryptamine 1A (5-HT_{1A}) serotonin receptor has long been considered a promising therapeutic target for mood and anxiety disorders such as schizophrenia [93]. Luo et al. in the year 2014, established a QSAR-based virtual screening strategy to uncover novel hit compounds targeting the 5-HT_{1A} receptor to address the significant side effects of marketed medications targeting the 5-HT_{1A} receptor [94, 95]. To begin, binary QSAR models were created utilizing Dragon descriptors and a variety of machine learning algorithms. The created QSAR models were then robustly tested and used for virtual screening in four commercial chemical databases in consensus. Fifteen compounds were chosen for testing, and nine of them were shown to be active at low nanomolar doses. [(8α)-6-methyl-9,10-didehydroergolin-8-yl]methanol, one of the verified hits, showed potent activity with a binding affinity (Ki) of 2.3 nM against the 5-HT_{1A} receptor.

9.3.4 MD Simulation Analysis

MD simulation is a promising biomolecular dynamics area, which involves appropriate models, physical applications, and large-scale calculations [96]. MD simulations have recently been expanded to cellular scales, with simulations of a complete cell done. A molecule in MD simulation may be thought of as a collection of charged points (atoms) connected by springs (bonds) [97]. The force field is utilized to represent the temporal evolution of bond lengths, bond angles, torsions, and non-bonding interactions between atoms [98]. In MD simulation, protein and water molecules create an in-vivo-like environment, and a little time step is used by the atoms of water and protein (in fs time duration). Each atom's forces are calculated and recorded in a file using a force field, which consists of equations, constants, potential energy functions, and potential terms for both bound and unbonded atoms. For MD simulation, several well-known and often used tools are GROMACS [99, 100], AMBER [101], Nanoscale MD (NAMD) [102], and CHARMM-GUI [103]. To execute these MD simulations, more powerful gear and software are required.

9.3.5 ADMET Prediction

One of the key stages in any drug discovery effort is ADMET because it offers information on the pharmacokinetics (ADME, or Absorption, Distribution, Metabolism, and Excretion) and pharmacodynamics (T), or the Toxicity (T), of lead compounds before wet-lab research [104]. As a result, it lessens the chance of experimentation's time, money, and drug failure. Poor pharmacokinetics and pharmacodynamics were found to be the primary reason for the increase in cost and failure in the late-stage of the drug development process. It is now widely accepted that computer-based ADMET prediction must be taken into account in the drug discovery program before moving on to *in vitro* and *in vivo* studies. Effective and safe drugs must have high potency, affinity, and selectivity against the molecular target, as well as acceptable absorption, distribution, metabolism, excretion, and tolerable toxicity. These characteristics are also present in a properly combined manner of pharmacodynamics (PD) and pharmacokinetics (PK). In this regard, a variety of programs, including SwissADME [105], ProTox [106], pkCSM [107], QikProp [108], DataWarrior [109], MetaTox [110], MetaSite [111], and StarDrop [112] to mention a few, are now available for the prediction of ADMET.

The development of computers and the foundations of various QSAR techniques were used for PK research, and one such work was published in 1977 by Timmermans et al. on the relationship between 27 clonidine analogues' lipophilicity and blood-brain barrier (BBB) permeability [113]. In order to associate *in-vivo* primary and secondary PK endpoints to octanol-buffer partition coefficients, Hinderling et al. were pioneers in this field in the 1980s [114]. The study used many β-adrenoceptor antagonists as its foundation. For big datasets, Lipinski et al. were the first to connect several physicochemical characteristics to PK features in the 1990s; this approach later became known as the Rule of Five and is frequently referenced [115]. Other significant medicinal chemistry guidelines, such as those put out by Ghose et al., Veber et al., Egan et al., and Muegge et al., arose in the late 1990s and early 2000s as a result of the drug-likeness concept [86, 116–118]. Lipinski's rule with a few other fundamental properties, such as refractivity, polar surface area (2D), polarizability, and van der Waals surface area, are all that it can currently be limited to. The number of small molecules for which early data on ADMET are available as references have substantially grown because of advances in combinatorial chemistry and high-throughput screening. Some of the rules for better bioavailable drugs are mentioned below.

9.3.5.1 Lipinski's Rule of Five

In the discovery set, 'the rule of 5' predicts that poor absorption or permeation is more likely when:

1. There are more than five H-bond donors.
2. There are more than ten H-bond acceptors.
3. The molecular weight (MWT) is greater than 500.
4. The calculated Log P (CLogP) is greater than 5 (or MlogP > 4.15).

9.3.5.2 Ghosh Rule

In this study, they examined the Comprehensive Medicinal Chemistry (CMC) database and seven various subsets of medicinal compounds from various classes. These include certain drugs that work on the central nervous system as well as conditions that are related to heart, cancer, inflammation, and infection. The qualifying range (including more than 80% of the molecules) for the CMC database is

1. The calculated log P is between −0.4 and 5.6, with an average value of 2.52.
2. For molecular weight, the qualifying range is between 160 and 480, with an average value of 357.
3. For molar refractivity, the qualifying range is between 40 and 130, with an average value of 97.
4. For the total number of atoms, the qualifying range is between 20 and 70, with an average value of 48.

9.3.5.3 Veber's Rule

For more than 1100 drug candidates being researched at SmithKline Beecham Pharmaceuticals (now GlaxoSmithKline), they assessed the relative significance of molecular features considered to impact that therapeutic property for oral bioavailability measurements in rats and suggested the properties of a molecule to be a good bioavailable one. The suggestions are

1. The molecular weight should not exceed more than 500.
2. Rotatable bonds should be 10 or fewer.
3. The polar surface area is equal to or less than 140 $Å^2$.
4. The total number of H-bond donors and acceptors should be 12 or fewer.

Additionally, by incorporating new principle descriptors and using computational tools created by the application of improved algorithms, the accuracy of ADMET tools can be improved. This will speed up the drug discovery process for the identification of novel drug-like compounds at a reasonable cost.

9.4 Artificial Intelligence in Drug Discovery

9.4.1 Introduction

Although CADD has been a fruitful tool in accelerating the drug discovery process, the algorithm and processing are not powerful, which results in low accuracy of results. Screening of large numbers of ligands and receptors was difficult with low precision and accuracy rates. Artificial intelligence has shown the potential to synergize drug discovery with CADD [119]. Many different AI strategies can be employed for drug discovery, to name machine learning models, deep neural networks, transformers, recurrent neural networks, graph neural networks, and convolutional neural networks. Along with this change in learning paradigms, reinforcement learning and self-supervised have also gained a lot of attention nowadays. These AI methods can resolve the issues and barriers in the drug development and design process.

9.4.2 Artificial Intelligence: From Machine Learning to Deep Learning

The neural network, which was created as a computational tool in the 1980s, was used for the first time in 1989 to discover new drugs [120]. Additionally, several neural network approaches have been used in the design and discovery of pharmaceuticals. ANN was the first widely used technique that is helpful in the procedure of variable selection. In modeling research from the 1990s to the 2000s (Fig. 9.11) [121],

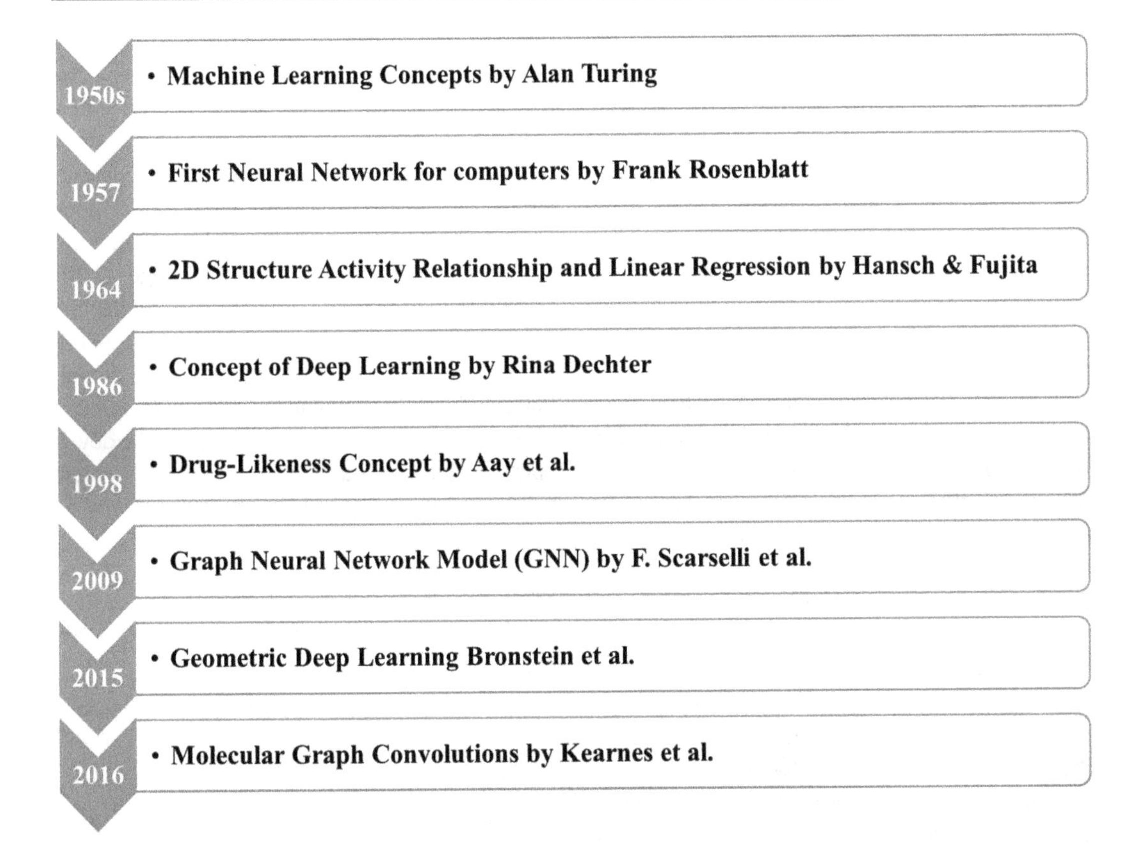

Fig. 9.11 Timeline of Machine Learning events in drug discovery

new machine-learning techniques were created based on nonlinear modeling algorithms including k-nearest neighbors [122], support vector machines [123], and random forest [124] were extensively utilized in place of linear regression. Ajay et al. proposed the idea of Drug-likeness in 1998 to identify whether a molecule can be a drug or not with the help of 1D and 2D molecular descriptors. Based on these models, ML algorithms are widely used in drug discovery. Thus, the field of research of QSAR was established utilizing a prediction algorithm, such as linear regression, and chemical descriptors of the sequences [125].

In the 1980s, the idea of Deep Learning (DL) was first introduced with an artificial neural network. At that time, the data used for model development was insufficient therefore DL did not exhibit a considerable benefit over ML [121]. Deep Learning got a huge push during 2008 due to developments in computer hardware such as GPUs and cloud computing. The Graph Neural Network model (GNN) was proposed in 2009 by F. Scarselli et al.[126] and Molecular Graph Convolutions in 2016 [127].

9.5 Conclusion

In this chapter, we discussed the journey of drug discovery and development from traditional systems to artificial intelligence. In the ancient era, diseases were treated with herbs by several systems such as Indian, Egyptian, Greek, Roman, and Chinese. These systems mostly include herbs as drugs in different forms such as paste, decoctions, extracts, maceration. After some advancements, drugs were isolated from plants, microorganisms, and marine animals such as penicillin, cephalosporins, eleutherobin. In the middle era, drugs were discovered based on hit and trial methods and serendipity. The obtained leads from natural sources were manipulated by the means of synthetic methods and potent drugs were discovered such as captopril, dihydroartemisinin, and artesunate. However, leads from these traditional methods require more than 10 years of development and sometimes they may fail. Towards the end of the eighteenth century and after World War I there were numerous pandemics (plague, cholera, etc.) as a result of low living standards. There were urgent needs for drug discovery through the fast-forward process. To overcome the obligations of the traditional methods, researchers turn their path toward computer-aided drug design (CADD). The time-consuming process of drug discovery was shortened by using various software and applications. With the help of the software, researchers could analyze the protein of microorganisms, amino acid sequencing, and analyze the nature and structure of receptors. In the CADD, there are mainly two techniques such as structure-based drug design and ligand-based drug design. Structure-based drug design is the process in which the structure of a protein is analyzed then the drug is designed using several techniques such as target identification and validation, binding pocket analysis, molecular docking, de novo design, and homology modeling. Many drugs were discovered using structure-based drug designs such as saquinavir and aliskiren. The ligand-based drug design is used when the target protein structure is unknown. Thousands of ligands were downloaded from databases such as ChEMBL, ZINC, and PubChem and calculated the descriptors as per requirements for 2D and 3D QSAR using Dragan, PaDEL, RDkit, etc. After that, the ligands are screened to identify the lead molecules with the help of different types of QSAR models such as 2D and 3D. ADMET and drug-likeness are also predicted using CADD. Pharmacokinetics, pharmacodynamics, and toxicity studies are helping to improve the process of drug discovery using various web tools such as SwissADME, ProTox. ML techniques and current advancements in DL have the potential to boost the efficiency of the drug discovery and development process. In the chapter, we discussed ML and DL algorithms and how they are used in drug discovery. Deep reinforcement learning generates the model through experiments then critics decide whether the model is good or bad. Researchers have been able to use geometric deep learning to take advantage of the symmetries of various unstructured molecular representations, which has increased the flexibility and versatility of the computational models that are currently available for molecular structure generation and property prediction. For the development of ligand-based drugs, these are potent tools that can be highly helpful.

Acknowledgment We sincerely thank funding agencies ICMR (File No. 5/13/79/2020/NCD-III), AICTE (File No. 8-125/FDC/RPS (POLICY-1)/2019-20), and RGUHS (Project code: 19PHA339) for their financial and moral support. We are also grateful to the management and staff of Acharya Institutes and College of Pharmaceutical Sciences of Dayananda Sagar University for their constant support and encouragement.

Conflict of Interest The corresponding author on behalf of all authors declares no conflict of interest.

References

1. Li J, Lu C, Jiang M, Niu X, Guo H, Li L et al (2012) Traditional chinese medicine-based network pharmacology could lead to new multicompound drug discovery. Evid Based Complement Alternat Med 2012: 149762
2. Huang H-J, Yu HW, Chen C-Y, Hsu C-H, Chen H-Y, Lee K-J et al (2010) Current developments of computer-aided drug design. J Taiwan Inst Chem Eng 41(6):623–635
3. Cheng F, Li W, Liu G, Tang Y (2013) In silico ADMET prediction: recent advances, current challenges and future trends. Curr Top Med Chem 13(11):1273–1289
4. Cassidy JW, Taylor B (2020) Artificial intelligence in oncology drug discovery and development. IntechOpen. https://doi.org/10.5772/intechopen.92799
5. Jachak SM, Saklani A (2007) Challenges and opportunities in drug discovery from plants. Curr Sci 92:1251–1257
6. Khan SR, Al Rijjal D, Piro A, Wheeler MB (2021) AI-integration and plant-based traditional medicine for drug discovery. Drug Discov Today 26:982

7. Süntar I (2020) Importance of ethnopharmacological studies in drug discovery: role of medicinal plants. Phytochem Rev 19(5):1199–1209
8. Feng Y, Wu Z, Zhou X, Zhou Z, Fan W (2006) Knowledge discovery in traditional Chinese medicine: state of the art and perspectives. Artif Intell Med 38(3):219–236
9. Vaidya PB, Vaidya BS, Vaidya SK (2010) Response to Ayurvedic therapy in the treatment of migraine without aura. Int J Ayurveda Res 1(1):30
10. Ansari A (2010) Unani system of medicine and development of its materia medica. Iran J Pharm Res 3:21–22
11. Loukas M, Lanteri A, Ferrauiola J, Tubbs RS, Maharaja G, Shoja MM, Yadav A, Rao VC (2010) Anatomy in ancient India: a focus on the Susruta Samhita. J Anat 217(6):646–650
12. Sucher NJ (2013) The application of Chinese medicine to novel drug discovery. Expert Opin Drug Discovery 8(1):21–34
13. Reid D (1996) The Shambhala guide to traditional Chinese medicine. Shambhala Publications, Boulder
14. Zhou J, Xie G, Yan X (2011) Encyclopedia of traditional Chinese medicines-molecular structures, pharmacological activities, natural sources and applications. Springer Berlin, Heidelberg. https://doi.org/10.1007/978-3-642-16744-7
15. Chan E, Tan M, Xin J, Sudarsanam S, Johnson DE (2010) Interactions between traditional Chinese medicines and Western. Curr Opin Drug Discov Devel 13(1):50–65
16. Yamgar RS, Sawant SS (2015) An update on drug discovery and natural products. Asian J Pharma Sci Technol 5(3):137–155
17. Forshaw R (2014) Before Hippocrates. Healing practices in ancient Egypt. In: Medicine, healing and performance. Oxbow Books, Oxford, pp 25–41
18. Jouanna J (2012) Egyptian medicine and Greek medicine. In: Greek medicine from Hippocrates to Galen. Brill, Leiden, pp 1–20
19. Hartmann A (2016) Back to the roots–dermatology in ancient Egyptian medicine. J Dtsch Dermatol Ges 14(4):389–396
20. Metwaly AM, Ghoneim MM, Eissa IH, Elsehemy IA, Mostafa AE, Hegazy MM et al (2021) Traditional ancient Egyptian medicine: a review. Saudi J Biol Sci 28(10):5823–5832
21. Kirsch DB (2011) There and back again: a current history of sleep medicine. Chest 139(4):939–946
22. Watts HE (2014) The plight of the wounded healer: unraveling pain as a precursor to practicing potent psychotherapy. Pacifica Graduate Institute, Carpinteria
23. Prioreschi P (1996) A history of medicine: Roman medicine. Edwin Mellen Press, Lewiston
24. Webster HK, Lehnert EK (1994) Chemistry of artemisinin: an overview. Trans R Soc Trop Med Hyg 88:27–29
25. Singh L, Lewis A, Field M, Hughes J, Woodruff G (1991) Evidence for an involvement of the brain cholecystokinin B receptor in anxiety. Proc Natl Acad Sci 88(4):1130–1133
26. Woodruff G, Hughes J (1991) Cholecystokinin antagonists. Annu Rev Pharmacol Toxicol 31(1): 469–501
27. Amedei A, D'Elios M (2012) New therapeutic approaches by using microorganism-derived compounds. Curr Med Chem 19(22):3822–3840
28. Birari RB, Bhutani KK (2007) Pancreatic lipase inhibitors from natural sources: unexplored potential. Drug Discov Today 12(19-20):879–889
29. Lu Q, Yan S, Sun H, Wang W, Li Y, Yang X et al (2015) Akt inhibition attenuates rasfonin-induced autophagy and apoptosis through the glycolytic pathway in renal cancer cells. Cell Death Dis 6(12): e2005-e
30. Gerwick WH, Proteau PJ, Nagle DG, Hamel E, Blokhin A, Slate DL (1994) Structure of curacin A, a novel antimitotic, antiproliferative and brine shrimp toxic natural product from the marine cyanobacterium Lyngbya majuscula. J Org Chem 59(6): 1243–1245
31. Conlon JM (2004) The therapeutic potential of antimicrobial peptides from frog skin. Rev Med Microbiol 15(1):17–25
32. Angerer K (2011) Frog tales–on poison dart frogs, epibatidine, and the sharing of biodiversity. Innovation 24(3):353–369
33. Singh D, Tripathi A, Kumar G (2012) An overview of computational approaches in structure based drug design. Nepal J Biotechnol 2(1):53–61
34. Nag A, Dey B (2011) Computer-aided drug design and delivery systems. McGraw-Hill Education, New York
35. Ferenczy G (1998) A SZERKEZET-ALAPU GYOGYSZERTERVEZES MODSZEREI. Acta Pharm Hung 68(1):21–31
36. Navia MA, Peattie DA (1993) Structure-based drug design: applications in immunopharmacology and immunosuppression. Immunol Today 14(6):296–302
37. Oakley AJ, Wilce MC (2000) Macromolecular crystallography as a tool for investigating drug, enzyme and receptor interactions. Clin Exp Pharmacol Physiol 27(3):145–151
38. Meng EC, Gschwend DA, Blaney JM, Kuntz ID (1993) Orientational sampling and rigid-body minimization in molecular docking. Proteins 17(3): 266–278
39. Fitzgerald PM (1993) HIV protease-ligand complexes. Curr Opin Struct Biol 3(6):868–874
40. Rutenber E, Fauman EB, Keenan RJ, Fong S, Furth PS, de Montellano PO et al (1993) Structure of a non-peptide inhibitor complexed with HIV-1 protease. Developing a cycle of structure-based drug design. J Biol Chem 268(21):15343–15346

41. Kontoyianni M (2017) Docking and virtual screening in drug discovery. In: Proteomics for drug discovery. Springer, New York, pp 255–266
42. Gashaw I, Ellinghaus P, Sommer A, Asadullah K (2011) What makes a good drug target? Drug Discov Today 16(23-24):1037–1043
43. Chettri S, Sasmal P, Adon T, Kumar BS, Kumar BP, Raghavendra NM (2023) Computational analysis of natural product B-Raf inhibitors. J Mol Graph Model 118:108340
44. Pathak RK, Singh DB, Sagar M, Baunthiyal M, Kumar A (2020) Computational approaches in drug discovery and design. In: Computer-aided drug design. Springer, New York, pp 1–21
45. Stank A, Kokh DB, Fuller JC, Wade RC (2016) Protein binding pocket dynamics. Acc Chem Res 49(5):809–815
46. Schmidt T, Haas J, Cassarino TG, Schwede T (2011) Assessment of ligand-binding residue predictions in CASP9. Proteins 79(S10):126–136
47. Liu B, Liu B, Liu F, Wang X (2014) Protein binding site prediction by combining hidden markov support vector machine and profile-based propensities. ScientificWorldJournal 2014:464093
48. Sasmal P, Babasahib SK, Kumar BP, Raghavendra NM (2022) Biphenyl-based small molecule inhibitors: novel cancer immunotherapeutic agents targeting PD-1/PD-L1 interaction. Biorg Med Chem 73:117001
49. Bodade R, Beedkar S, Manwar A, Khobragade C (2010) Homology modeling and docking study of xanthine oxidase of Arthrobacter sp. XL26. Int J Biol Macromol 47(2):298–303
50. Pathak RK, Taj G, Pandey D, Kasana VK, Baunthiyal M, Kumar A (2016) Molecular modeling and docking studies of phytoalexin(s) with pathogenic protein(s) as molecular targets for designing the derivatives with anti-fungal action on 'Alternaria' spp. of 'Brassica'. Plant Omics 9(3):172–183
51. Hekkelman ML, te Beek TA, Pettifer S, Thorne D, Attwood TK, Vriend G (2010) WIWS: a protein structure bioinformatics Web service collection. Nucleic Acids Res 38(suppl_2):W719–WW23
52. Bagaria A, Jaravine V, Huang YJ, Montelione GT, Güntert P (2012) Protein structure validation by generalized linear model root-mean-square deviation prediction. Protein Sci 21(2):229–238
53. Mutharasappan N, Ravi Rao G, Mariadasse R, Poopandi S, Mathimaran A, Dhamodharan P et al (2020) Experimental and computational methods to determine protein structure and stability. In: Frontiers in protein structure, function, and dynamics. Springer, New York, pp 23–55
54. Ferreira LG, Dos Santos RN, Oliva G, Andricopulo AD (2015) Molecular docking and structure-based drug design strategies. Molecules 20(7): 13384–13421
55. Huang S-Y, Zou X (2010) Advances and challenges in protein-ligand docking. Int J Mol Sci 11(8): 3016–3034
56. Wildman SA (2001) Three-dimensional quantitative structure-activity relationships based on atomic property descriptors. University of Michigan, Ann Arbor
57. Schneider G, Böhm H-J (2002) Virtual screening and fast automated docking methods. Drug Discov Today 7:64–70
58. Rarey M, Kramer B, Lengauer T, Klebe G (1996) A fast flexible docking method using an incremental construction algorithm. J Mol Biol 261(3):470–489
59. Kuntz ID, Blaney JM, Oatley SJ, Langridge R, Ferrin TE (1982) A geometric approach to macromolecule-ligand interactions. J Mol Biol 161(2):269–288
60. Goodsell DS, Morris GM, Olson AJ (1996) Automated docking of flexible ligands: applications of AutoDock. J Mol Recognit 9(1):1–5
61. Morris GM, Goodsell DS, Huey R, Olson AJ (1996) Distributed automated docking of flexible ligands to proteins: parallel applications of AutoDock 2.4. J Comput Aided Mol Des 10(4):293–304
62. Kollman P (1993) Free energy calculations: applications to chemical and biochemical phenomena. Chem Rev 93(7):2395–2417
63. Kollman PA, Massova I, Reyes C, Kuhn B, Huo S, Chong L et al (2000) Calculating structures and free energies of complex molecules: combining molecular mechanics and continuum models. Acc Chem Res 33(12):889–897
64. Gohlke H, Kiel C, Case DA (2003) Insights into protein–protein binding by binding free energy calculation and free energy decomposition for the Ras–Raf and Ras–RalGDS complexes. J Mol Biol 330(4): 891–913
65. Wang R, Lai L, Wang S (2002) Further development and validation of empirical scoring functions for structure-based binding affinity prediction. J Comput Aided Mol Des 16(1):11–26
66. Tsui V, Case DA (2000) Theory and applications of the generalized Born solvation model in macromolecular simulations. Biopolymers 56(4):275–291
67. Lee MS, Salsbury FR Jr, Brooks CL III (2002) Novel generalized Born methods. J Chem Phys 116(24): 10606–10614
68. Kralj S, Hodošček M, Podobnik B, Kunej T, Bren U, Janežič D et al (2021) Molecular dynamics simulations reveal interactions of an IgG1 antibody with selected Fc receptors. Front Chem 9:705931
69. Hevener KE, Zhao W, Ball DM, Babaoglu K, Qi J, White SW et al (2009) Validation of molecular docking programs for virtual screening against dihydropteroate synthase. J Chem Inf Model 49(2): 444–460
70. Liu X, IJzerman AP, van Westen GJ (2021) Computational approaches for de novo drug design: past, present, and future. Artif Neural Netw 190:139–165

71. Fischer T, Gazzola S, Riedl R (2019) Approaching target selectivity by de novo drug design. Expert Opin Drug Discovery 14(8):791–803
72. Schneider G, Fechner U (2005) Computer-based de novo design of drug-like molecules. Nat Rev Drug Discov 4(8):649–663
73. Moon JB, Howe WJ (1991) Computer design of bioactive molecules: a method for receptor-based de novo ligand design. Proteins 11(4):314–328
74. Goodford PJ (1985) A computational procedure for determining energetically favorable binding sites on biologically important macromolecules. J Med Chem 28(7):849–857
75. Böhm H-J (1992) LUDI: rule-based automatic design of new substituents for enzyme inhibitor leads. J Comput Aided Mol Des 6(6):593–606
76. Bharatam PV, Khanna S, Francis SM (2008) Modeling and informatics in drug design. In: Preclinical development handbook: ADME and biopharmaceutical properties, vol 29. Wiley, Hoboken, pp 1–46
77. Rotstein SH, Murcko MA (1993) GroupBuild: a fragment-based method for de novo drug design. J Med Chem 36(12):1700–1710
78. Gillet V, Johnson AP, Mata P, Sike S, Williams P (1993) SPROUT: a program for structure generation. J Comput Aided Mol Des 7(2):127–153
79. Alig L, Alsenz J, Andjelkovic M, Bendels S, Bénardeau A, Bleicher K et al (2008) Benzodioxoles: novel cannabinoid-1 receptor inverse agonists for the treatment of obesity. J Med Chem 51(7):2115–2127
80. Siezen RJ, de Vos WM, Leunissen JA, Dijkstra BW (1991) Homology modelling and protein engineering strategy of subtilases, the family of subtilisin-like serine proteinases. Protein Eng 4(7):719–737
81. Sutcliffe MJ, Haneef I, Carney D, Blundell T (1987) Knowledge based modelling of homologous proteins, Part I: three-dimensional frameworks derived from the simultaneous superposition of multiple structures. Protein Eng 1(5):377–384
82. Khare N, Maheshwari SK, Rizvi SMD, Albadrani HM, Alsagaby SA, Alturaiki W et al (2022) Homology modelling, molecular docking and molecular dynamics simulation studies of CALMH1 against secondary metabolites of Bauhinia variegata to treat Alzheimer's disease. Brain Sci 12(6):770
83. Guex N, Peitsch MC (1997) SWISS-MODEL and the Swiss-Pdb Viewer: an environment for comparative protein modeling. Electrophoresis 18(15):2714–2723
84. Bates PA, Kelley LA, MacCallum RM, Sternberg MJ (2001) Enhancement of protein modeling by human intervention in applying the automatic programs 3D-JIGSAW and 3D-PSSM. Proteins 45(S5):39–46
85. Lewis D, Lake B, Dickins M, Ueng Y-F, Goldfarb P (2003) Homology modelling of human CYP1A2 based on the CYP2C5 crystallographic template structure. Xenobiotica 33(3):239–254
86. Muegge I, Heald SL, Brittelli D (2001) Simple selection criteria for drug-like chemical matter. J Med Chem 44(12):1841–1846
87. Mosberg HI (1999) Complementarity of δ opioid ligand pharmacophore and receptor models. Peptide Sci 51(6):426–439
88. de Groot MJ, Ekins S (2002) Pharmacophore modeling of cytochromes P450. Adv Drug Deliv Rev 54(3):367–383
89. Day BW (2000) Mutants yield a pharmacophore model for the tubulin–paclitaxel binding site. Trends Pharmacol Sci 21(9):321–323
90. Kalva S, Vinod D, Saleena LM (2014) Combined structure-and ligand-based pharmacophore modeling and molecular dynamics simulation studies to identify selective inhibitors of MMP-8. J Mol Model 20(5):1–18
91. Summa V, Petrocchi A, Matassa VG, Gardelli C, Muraglia E, Rowley M et al (2006) 4, 5-Dihydroxypyrimidine carboxamides and N-alkyl-5-hydroxypyrimidinone carboxamides are potent, selective HIV integrase inhibitors with good pharmacokinetic profiles in preclinical species. J Med Chem 49(23):6646–6649
92. Hansch C, Grieco C, Silipo C, Vittoria A (1977) Quantitative structure-activity relationship of chymotrypsin-ligand interactions. J Med Chem 20(11):1420–1435
93. Nichols DE, Nichols CD (2008) Serotonin receptors. Chem Rev 108(5):1614–1641
94. Luo M, Wang XS, Roth BL, Golbraikh A, Tropsha A (2014) Application of quantitative structure–activity relationship models of 5-HT1A receptor binding to virtual screening identifies novel and potent 5-HT1A ligands. J Chem Inf Model 54(2):634–647
95. Neves BJ, Braga RC, Melo-Filho CC, Moreira-Filho JT, Muratov EN, Andrade CH (2018) QSAR-based virtual screening: advances and applications in drug discovery. Front Pharmacol 9:1275
96. Heidari Z, Roe DR, Galindo-Murillo R, Ghasemi JB, Cheatham TE III (2016) Using wavelet analysis to assist in identification of significant events in molecular dynamics simulations. J Chem Inf Model 56(7): 1282–1291
97. Berger A, Linderstrøm-Lang K (1957) Deuterium exchange of poly-DL-alanine in aqueous solution. Arch Biochem Biophys 69:106–118
98. Wolf A, Kirschner KN (2013) Principal component and clustering analysis on molecular dynamics data of the ribosomal L11 23S subdomain. J Mol Model 19(2):539–549
99. Oostenbrink C, Villa A, Mark AE, Van Gunsteren WF (2004) A biomolecular force field based on the free enthalpy of hydration and solvation: the GROMOS force-field parameter sets 53A5 and 53A6. J Comput Chem 25(13):1656–1676
100. Pronk S, Páll S, Schulz R, Larsson P, Bjelkmar P, Apostolov R et al (2013) GROMACS 4.5: a high-throughput and highly parallel open source molecular simulation toolkit. Bioinformatics 29(7):845–854
101. Case DA, Cheatham TE III, Darden T, Gohlke H, Luo R, Merz KM Jr et al (2005) The Amber

biomolecular simulation programs. J Comput Chem 26(16):1668–1688
102. Phillips JC, Braun R, Wang W, Gumbart J, Tajkhorshid E, Villa E et al (2005) Scalable molecular dynamics with NAMD. J Comput Chem 26(16): 1781–1802
103. Brooks BR, Brooks CL III, Mackerell AD Jr, Nilsson L, Petrella RJ, Roux B et al (2009) CHARMM: the biomolecular simulation program. J Comput Chem 30(10):1545–1614
104. Segall M (2014) Advances in multiparameter optimization methods for de novo drug design. Expert Opin Drug Discovery 9(7):803–817
105. Daina A, Michielin O, Zoete V (2017) SwissADME: a free web tool to evaluate pharmacokinetics, drug-likeness and medicinal chemistry friendliness of small molecules. Sci Rep 7(1):1–13
106. Drwal MN, Banerjee P, Dunkel M, Wettig MR, Preissner R (2014) ProTox: a web server for the in silico prediction of rodent oral toxicity. Nucleic Acids Res 42(W1):W53–WW8
107. Pires DE, Blundell TL, Ascher DB (2015) pkCSM: predicting small-molecule pharmacokinetic and toxicity properties using graph-based signatures. J Med Chem 58(9):4066–4072
108. Jorgensen WL, Duffy EM (2002) Prediction of drug solubility from structure. Adv Drug Deliv Rev 54(3): 355–366
109. Sander T, Freyss J, von Korff M, Rufener C (2015) DataWarrior: an open-source program for chemistry aware data visualization and analysis. J Chem Inf Model 55(2):460–473
110. Myshkin E, Brennan R, Khasanova T, Sitnik T, Serebriyskaya T, Litvinova E et al (2012) Prediction of organ toxicity endpoints by QSAR modeling based on precise chemical-histopathology annotations. Chem Biol Drug Des 80(3):406–416
111. Cruciani G, Carosati E, De Boeck B, Ethirajulu K, Mackie C, Howe T et al (2005) MetaSite: understanding metabolism in human cytochromes from the perspective of the chemist. J Med Chem 48(22): 6970–6979
112. T'jollyn H, Boussery K, Mortishire-Smith R, Coe K, De Boeck B, Van Bocxlaer J et al (2011) Evaluation of three state-of-the-art metabolite prediction software packages (Meteor, MetaSite, and StarDrop) through independent and synergistic use. Drug Metab Dispos 39(11):2066–2075
113. Timmermans P, Brands A, Van Zwieten P (1977) Lipophilicity and brain disposition of clonidine and structurally related imidazolidines. Naunyn Schmiedebergs Arch Pharmacol 300(3):217–226
114. Hinderling PH, Schmidlin O, Seydel JK (1984) Quantitative relationships between structure and pharmacokinetics of beta-adrenoceptor blocking agents in man. J Pharmacokinet Biopharm 12(3): 263–287
115. Lipinski CA, Lombardo F, Dominy BW, Feeney PJ (1997) Experimental and computational approaches to estimate solubility and permeability in drug discovery and development settings. Adv Drug Deliv Rev 23(1-3):3–25
116. Ghose AK, Viswanadhan VN, Wendoloski JJ (1999) A knowledge-based approach in designing combinatorial or medicinal chemistry libraries for drug discovery. 1. A qualitative and quantitative characterization of known drug databases. J Comb Chem 1(1):55–68
117. Veber DF, Johnson SR, Cheng H-Y, Smith BR, Ward KW, Kopple KD (2002) Molecular properties that influence the oral bioavailability of drug candidates. J Med Chem 45(12):2615–2623
118. Egan WJ, Merz KM, Baldwin JJ (2000) Prediction of drug absorption using multivariate statistics. J Med Chem 43(21):3867–3877
119. Hopfield JJ (1982) Neural networks and physical systems with emergent collective computational abilities. Proc Natl Acad Sci 79(8):2554–2558
120. Aoyama T, Suzuki Y, Ichikawa H (1989) Neural networks applied to pearmaceutical problems. I. Method and application to decision making. Chem Pharm Bull(Tokyo) 37(9):2558–2560
121. Zhu H (2020) Big data and artificial intelligence modeling for drug discovery. Annu Rev Pharmacol Toxicol 60:573
122. Zheng W, Tropsha A (2000) Novel variable selection quantitative structure– property relationship approach based on the k-nearest-neighbor principle. J Chem Inf Comput Sci 40(1):185–194
123. Burbidge R, Trotter M, Buxton B, Holden S (2001) Drug design by machine learning: support vector machines for pharmaceutical data analysis. Comput Chem 26(1):5–14
124. Sprague B, Shi Q, Kim MT, Zhang L, Sedykh A, Ichiishi E et al (2014) Design, synthesis and experimental validation of novel potential chemopreventive agents using random forest and support vector machine binary classifiers. J Comput Aided Mol Des 28(6):631–646
125. Ajay A, Walters WP, Murcko MA (1998) Can we learn to distinguish between "drug-like" and "nondrug-like" molecules? J Med Chem 41(18): 3314–3324
126. Scarselli F, Gori M, Tsoi AC, Hagenbuchner M, Monfardini G (2008) The graph neural network model. IEEE Trans Neural Netw 20(1):61–80
127. Kearnes S, McCloskey K, Berndl M, Pande V, Riley P (2016) Molecular graph convolutions: moving beyond fingerprints. J Comput Aided Mol Des 30(8):595–608

Experimental Protocols in Phytochemistry and Natural Products: An Ever-Evolving Challenge

10

Sanjay M. Jachak, Alok Goyal, Akash Dey, Shreyanshi Kulshreshtha, and Kirti N. Deshmukh

Abstract

Natural products have been an important source of drugs for the past few centuries. Extraction and isolation of natural products involve extensive efforts, resources and take considerable time. It is therefore of highest importance to create effective and focused extraction and separation procedures for such bioactive natural compounds. To prevent contamination of extracts, decomposition of secondary metabolites, or the formation of artifacts as a result of extraction conditions or solvent impurities, chromatographic separation, it is crucial to minimize interference from compounds that may coextract with the target compounds during the extraction of plant material. The idea, guiding principles, and case studies of extraction and isolation methodology presented in this chapter provide an updated view of the enormous technological effort being made and the wide range of applications being established. To test the effectiveness of extraction, different types of deep eutectic solvents, *viz.*, choline chloride, betaine, and L-proline with various polarity, viscosity, compositions, and solubilization can be employed. Alkaloids, terpenoids, and flavonoids are a class of natural product compounds that are the subject of the case studies in this chapter.

Keywords

Extraction · Isolation · Alkaloids · Terpenoids · Flavonoids · Deep Eutectic solvents

10.1 Introduction

Natural Products (NPs) are molecules with biological properties that derive from naturally occurring sources, such as plants, animals, and microorganisms. Secondary metabolites, also referred to as NPs, are the byproducts of gene expression that are typically not necessary for the reproduction, growth, or development of the plant. They are created as a result of environmental adaptation or perhaps as a defense mechanism against predators; in either case, secondary metabolites are created to aid and improve the survival of the plant [1].

Due to the expanding global demand for natural products for primary healthcare, nutraceuticals, and herbal medicines, manufacturers of medical plant extracts and producers of essential oils are now adopting the best extraction techniques. To generate extracts and essential oils of defined quality

S. M. Jachak (✉) · A. Goyal · A. Dey · S. Kulshreshtha · K. N. Deshmukh
Department of Natural Products, National Institute of Pharmaceutical Education and Research (NIPER), Mohali, Punjab, India
e-mail: sanjayjachak@niper.ac.in

G. Jagadeesh et al. (eds.), *The Quintessence of Basic and Clinical Research and Scientific Publishing*,
https://doi.org/10.1007/978-981-99-1284-1_10

with minimal variations, many techniques are employed. For millennia, people have used herbs and medicinal plants as a source of a wide range of bioactive components. Numerous diseases are treated by generations of indigenous practitioners using the plant raw material or its components [2].

In order to extract active phytochemicals from plants, it is essential to adopt appropriate extraction procedures and techniques that yield extracts and fractions that are high in bioactive components. Thus, the yield, the type of phytochemical yield, etc., depend on the extraction techniques [3].

In this chapter, the experimental protocols in phytochemistry and natural products, challenges faced in extraction and isolation techniques, strategies to overcome the difficulties, extraction protocol optimization and the concept of green extraction technique by using deep eutectic solvents as a modern extraction technique have been discussed.

10.2 Basics of Extraction Procedures and Isolation Techniques from Natural Products

Extraction is the process of utilizing specific solvents to separate the medicinally useful mixture of numerous naturally occurring active chemicals that are often found inside plant components (tissues) [4]. It is also known as the treatment of plant material with a solvent in which the medicinally beneficial components are dissolved but the majority of the inert material is left intact. Therefore, the goal of all extraction processes is to separate the soluble plant metabolites from the residue, an insoluble cellular marc [5].

Three types of extraction processes are typically used, viz.: liquid/solid, liquid/liquid, and acid/base shake out. The active components must be extracted using suitable extraction techniques, which take into account, the plant parts utilized as the starting material, the solvent employed, time required for extraction, size of particles, and stirring/mixing in-between the ongoing extraction [3, 6].

Plant materials can be extracted using a variety of methods that range in price and complexity. Percolation and maceration are two relatively straightforward procedures that work well and are affordable for the majority of applications. However, some specific applications demand more advanced and expensive extraction methods requiring specialized apparatus, such as those used in supercritical-fluid extraction and large-scale steam distillation [7].

10.3 Newer Extraction Methods/Techniques

The development of environmentally friendly extraction methods for natural resources is currently a popular research topic in the multidisciplinary domains of applied chemistry, biology, and technology. There are six guiding principles of green extraction while outlining a comprehensive approach to putting this idea into practice at both the scientific and industry levels. New and innovative technologies in case of extraction and isolation of phytoconstituents from natural sources have been employed. Process intensification or optimization in extraction technique is done now-a-days, for increasing extractive yield. The use of natural deep eutectic solvents (NADES) in extraction and isolation of different classes of natural products is of great significance and also in current trend [8].

10.3.1 General Extraction Protocol

The following section describes extraction of alkaloids, terpenoids, flavonoids and phenolics from selected medicinal plants. The methods described are those extensively carried out in the past from other labs and a few from our lab and are described for the benefit of researchers.

10.4 Alkaloids

A modified alternative of the traditional "acid-base shakeout" approach can be used to preferentially extract alkaloids that contain basic amines. Alkaloids should not be extracted using mineral acids and/or strong bases due to the possibility of formation of artifacts. The pH is changed during partitioning by tartaric acid and Na_2CO_3 in the alkaloid extraction method described [9].

In order for the basic amines to form salts and partition into the aqueous layer, the pH must first be reduced to be at least two units below their pKa values (along with the polar constituents, including quaternary amines and alkaloid N-oxides). The organic layer separates the nonpolar components. The aqueous layer becomes basic after the organic layer has been removed, resulting in the free-base form of the basic alkaloids. With the exception of the N-oxides and quaternary amines, which would stay in the aqueous layer when these alkaloids are partitioned with an organic solvent, they favour the organic layer, resulting in a nearly pure alkaloid fraction [10].

10.4.1 Extraction and Isolation of Phytocompounds from the Dried Powdered Fruits of *Piper cubeba* [11]

Piper cubeba dried powdered fruits were macerated with methanol to extract the phytochemicals from the plant material. On a rotary evaporator, the combined methanol extracts were condensed to dryness to produce the extract. The hexane-ethyl acetate gradient was used to elute the methanolic extract using vacuum liquid chromatography on silica gel (230–400 mesh), beginning with 5% EtOAc and progressing through 10, 20, 40, 80, and 100% EtOAc. All of these fractions underwent TLC analysis before being dried on a rotary evaporator to produce residues.

Compounds **1** (sesamin) and **2** (*trans*-cubebin) were isolated from the fraction that was eluted with 10% EtOAc and subjected to repeated column chromatography (CC) on Silica gel (230–400 mesh) using hexane-ethyl acetate gradient. Repeated column chromatography on Silica gel (230–400 mesh) using gradient of hexane and ethyl acetate from 20% EtOAc, resulted in the isolation of compound **3** (methyl trimethoxycinnamate). Repeated CC on silica gel (230–400 mesh) and Sephadex LH-20 led to the isolation of compounds **4** (pellitorine), **5** (tetrahydro piperine), **6** (piperine), and **7** (piplartine) from a fraction eluted with 40% EtOAc (6 g). Repeated CC using Sephadex LH-20 was used to isolate compound **8** (piperic acid) from the 80% EtOAc fraction (1.5 g) (Fig. 10.1).

10.4.2 Extraction and Isolation of Alkaloids from the Leaves of Atemoya (*Annona squamosa*) [12]

A powder made from dried and ground *A. squamosa* leaves was used. Hexane and MeOH were used to extract the phytoconstituents from the leaves, providing hexane and MeOH extracts after each solvent was removed under reduced pressure.

The MeOH extract, which was initially treated with an acid-base extraction to provide alkaloid and neutral fractions, showed significant concentration of alkaloids according to TLC analysis. With increasing concentrations of hexane, CH_2Cl_2, EtOAc, and MeOH, the alkaloidal fraction was submitted to silica gel CC after being treated with 10% $NaHCO_3$ solution. Fractions were collected. Following TLC analysis, these fractions were pooled .

Asimilobine, mixture of **asimilobine** and **pronuciferine**, and a mixture of three oxoaporphine alkaloids-**lanuginosine, liriodenine** and **lysicamine** were obtained after CF-5 was subjected to preparative TLC eluted with CH_2Cl_2:MeOH (95:5 v/v), three times. A mixture of alkaloids, including **asimilobine** and **anonaine**, as well as the proaporphine alkaloid

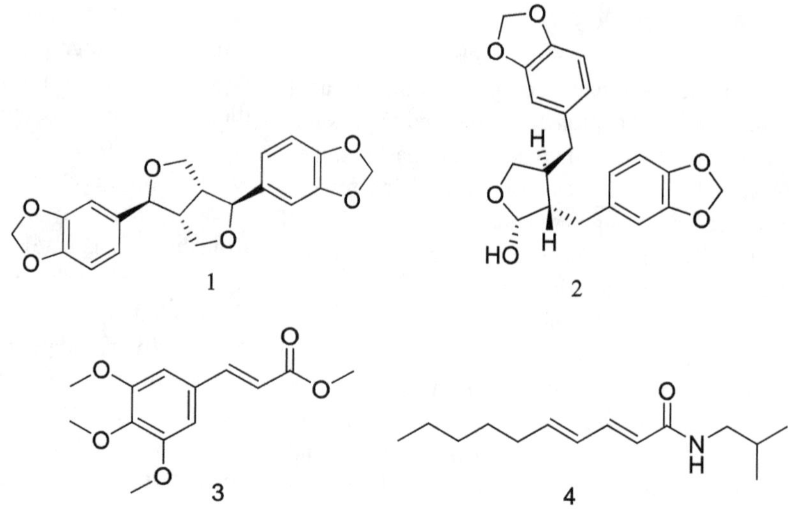

Fig. 10.1 Structures of phytoconstituents isolated from *Piper cubeba* fruits

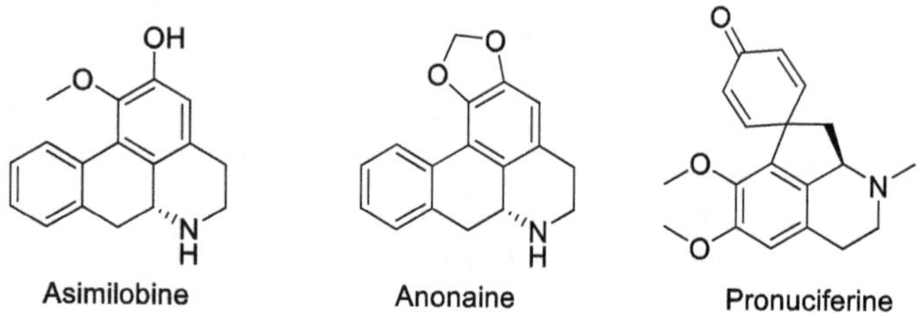

Fig. 10.2 Structures of compounds isolated from *Annona squamosa* leaves

Asimilobine Anonaine Pronuciferine

stepharine, was obtained after pooled fraction 7 was subjected to preparative TLC and eluted with CH_2Cl_2:MeOH (95:5, v/v), three times (Fig. 10.2; Scheme 10.1).

10.4.3 Extraction and Isolation Procedure of Furoquinoline Alkaloids from *Zanthoxylum buesgenii* [13]

Sandjo et al. [14] isolated furoquinoline alkaloids from *Zanthoxylum buesgenii* (Engl.). The solvent system, DCM/MeOH (1:1 v/v) was used to extract phytocompounds from dried aerial plant material for 2 days. The material was first cut into small pieces, then pulverized, and the powder, then underwent extraction. After that, MeOH was used for extraction. Both solutions were combined after 24 h and then evaporated on rotary evaporator under vacuum. The obtained crude extract was subjected to liquid-solid extraction process utilizing methanol (the liquid portion), *n*-hexane and ethyl acetate. Based on TLC profiling, both hexane and ethyk acetate fractions were pooled to obtain fraction A.

The presence of alkaloids was detected in fractions A by TLC using Dragendorff's reagent. Fraction A was subjected to Si gel CC. Elution using 100% *n*-hexane, *n*-hexane/ethyl acetate (gradient) and 100% ethyl acetate produced six pooled fractions. **Maculine**, **Isofagaridine**, **kokusaginine**, and **teclearverdoornine** were isolated and characterized using NMR analysis (Fig. 10.3).

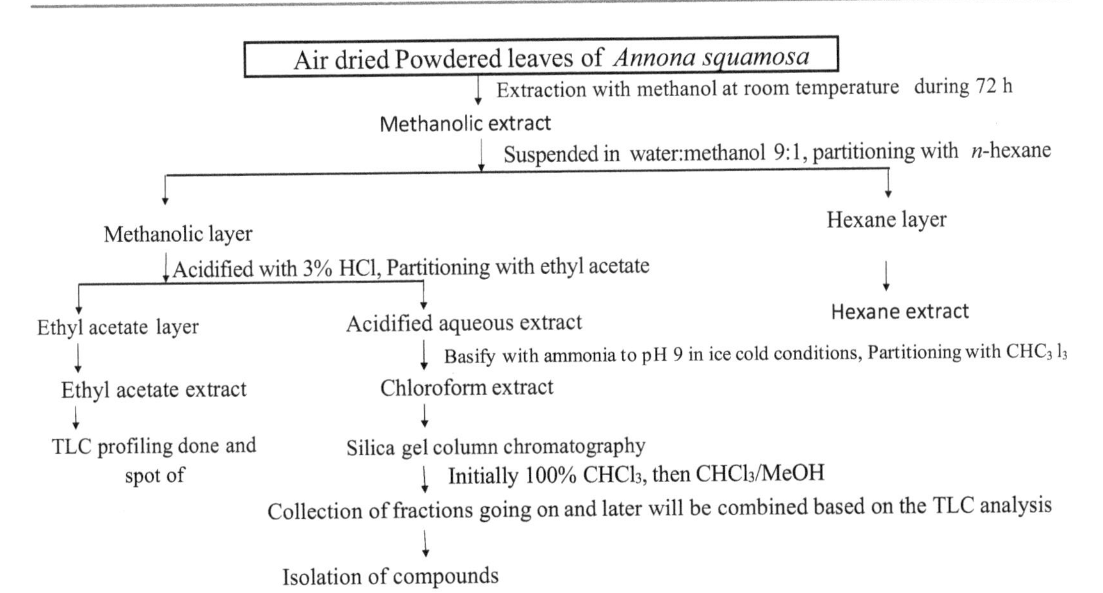

Scheme 10.1 Isolation of alkaloids from Annona squamosa

Maculine

Fig. 10.3 Maculine isolated from *Zanthoxylum buesgenii*

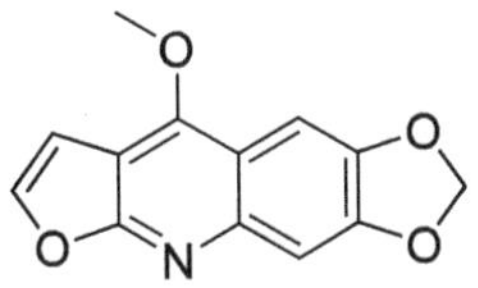

10.4.4 Extraction and Isolation Protocol of Alkaloids from Leaves and Stems of *Murraya koenigii*

Murraya koenigii leaves were dried in an oven set at 40 °C and pulverized into coarse powder. In an Accelerated Solvent Extractor, the powdered plant material (50 g) was successively extracted with *n*-hexane, dichloromethane, ethyl acetate, and methanol. The other extracts were dried on a rotary evaporator and kept in a refrigerator while the *n*-hexane extract was discarded.

From the EtOAc extract *M. koenigii* leaves four alkaloids were isolated. The extract was loaded onto silica gel (60–120 mesh) column and eluted using a gradient of benzene: $CHCl_3$ and then $CHCl_3$: CH_3OH. A total of five fractions were produced using comparable TLC techniques. **Mahanimbine** was obtained from fraction 2 using preparative TLC with benzene: $CHCl_3$ (3: 2) serving as the mobile phase. Re-chromatography of fraction 6 produced **koenimbine**, **koenigicine**, and **clausazoline-K**, respectively [15].

Air dried stems of *Murraya koenigii* were extracted with DCM:MeOH (1:1) over 4 days at room temperature. The solvent evaporated under reduced pressure to obtain extract. Five fractions (MK-I-V) were obtained after pooling the sub-fractions based on their analytical TLC data, from the extract which was primarily subjected to column chromatography using silica gel (100–200 mesh) and eluted by increasing the polarity of the elution solvent system of hexane and EtOAc (100% hexane to 100% EtOAc). Fraction I was further subjected to column chromatography over silica gel, and the pure phytochemicals were obtained when the fraction was eluted.

Four new carbazole alkaloids, **murrayakonine A**, **murrayakonine B**, **murrayakonine C** and **murrayakonine D** were isolated [16] (Fig. 10.4).

10.5 Terpenoids

The largest and the most widespread class of secondary metabolites is terpenoids (also known as isoprenoids). These are mainly found in plants

Fig. 10.4 Alkaloids from *Murraya koenigii*

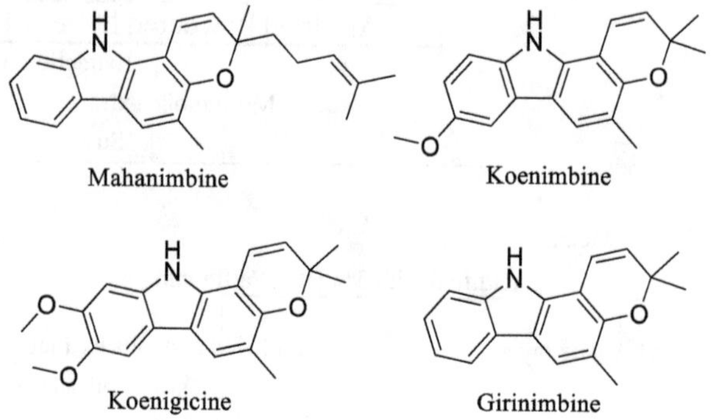

Mahanimbine

Koenimbine

Koenigicine

Girinimbine

Table 10.1 Classification of terpenoids according to the number of isoprene (C_5H_8) units into following classes

Name	No. of isoprene units	No. of carbon atoms	General formula	Example
Hemiterpenoids	1	5	C_5H_8	DMAPP, isovaleric acid, prenol
Monoterpenoids	2	10	$C_{10}H_{16}$	Camphor, geraniol, menthol
Sesquiterpenoids	3	15	$C_{15}H_{24}$	Farnesol, geosmin, humulone
Diterpenoids	4	20	$C_{20}H_{32}$	Ginkgolides, retinol, steviol
Sesterterpenoids	5	25	$C_{25}H_{40}$	Andrastin A, manoalide
Triterpenoids	6	30	$C_{30}H_{48}$	Betulinic acid, amyrin, oleanolic acid, ursolic acid
Tetraterpenoids	8	40	$C_{40}H_{64}$	Carotenoids
Polyterpenoids	>8	>40	$(C_5H_8)_n$	Natural rubber, gutta-percha

and some lower invertebrates. The term 'Terpen' (English 'Terpene') was given by Kekule and was originally used to describe the hydrocarbons found in turpentine oil (German 'Terpentin'). The suffix '*ene*' in Terpenes indicates the presence of olefinic bonds. Hence, chemically, terpenoids are isoprene (2-methyl-1,3-butadiene—C_5H_8) based secondary metabolites [17–19] (Table 10.1; Fig. 10.5).

In natural systems, terpenoids are biosynthesized from acetate activated as acetyl-coenzyme A via acetoacetyl-coenzyme A, 3-hydroxy-3-methylglutarylcoenzyme A, and mevalonate to isopentenyl diphosphate (IPP), which is the first precursor, having branched C_5 isoprene skeleton, in the formation of terpenoids [18]. Terpenoids, mainly monoterpenoids and sesquiterpenoids, are major components of essential oils and are responsible for the characteristic scent, odour and/or smell [19].

Extraction and Isolation

Usually, terpenoids occur as complex mixtures and there are several methods available for their extraction [18, 20].

1. Mixture of lower terpenoids (mono- and sesquiterpenoids) are extracted by subjecting the plant material to steam distillation, in the form of essential oils. Next, the terpenoids can be separated from the essential oil by following methods:
   (a) Chemical method: by treating the essential oil with agents like nitrosyl chloride, phthalic anhydride, etc.
   (b) Physical methods: by fractional distillation method or by using gas chromatography.
2. Higher terpenoids are generally extracted from plant materials by successive solvent extraction or by using liquid/liquid partitioning

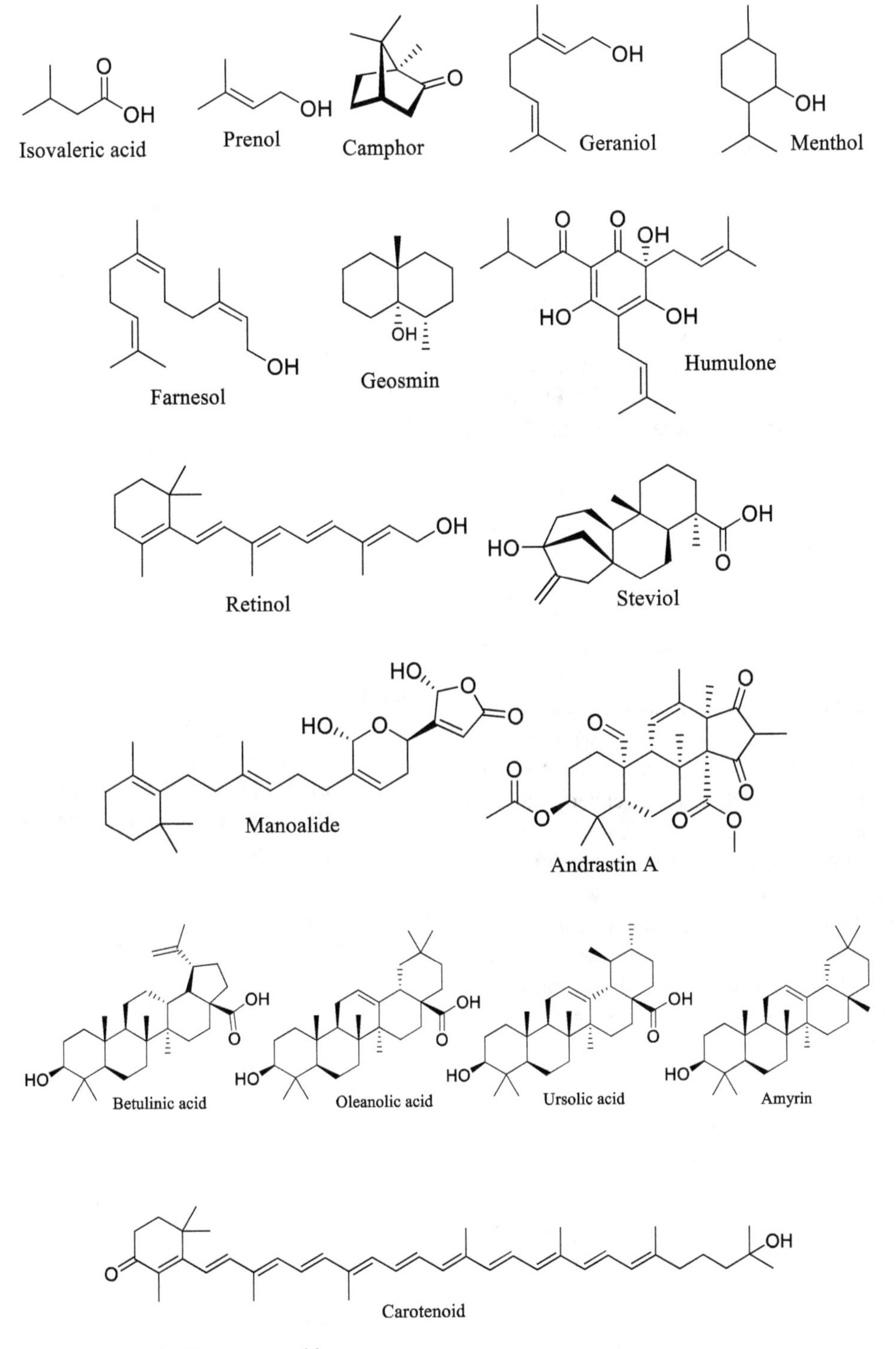

Fig. 10.5 Structures of different terpenoids

(solvents used in order of their increasing polarity) followed by column chromatography to isolate the desired compound.

Apart from the conventional methods to extract and isolate terpenoids, some other newer methods are also used, which are provided in the case studies given below:

10.5.1 Isolation of Sesquiterpenes from Seeds of *Ferula hermonis* Using a Combination of VLC with Silica Gel and Gel Filtration Using Sephadex® LH-20 [21, 22]

Seeds of *Ferula hermonis* were ground and extracted with 1 L of *n*-hexane: ethyl acetate mixture in the ratio of 1:1, using maceration for 24 h at room temperature. Pooled extract was filtered and concentrated under vacuum; oily extract was obtained. The extracted oil was subjected to silica gel VLC using *n*-hexane—ethyl acetate in increasing polarity and fractions were collected. Fraction obtained with 10% thyl acetate in *n*-hexane was subjected to gel filtration on Sephadex® LH-20 and was eluted with chloroform. Fractions were collected. Similar fractions were pooled together after monitoring the TLC. Preparative TLC was performed on fractions selected based on TLC. *n*-hexane—ethyl acetate mobile phase was used as mobile phase which yielded 17 sesquiterpenoids, 3 of which were named **feruhermonins A-C**.

10.5.2 Supercritical Fluid Extraction of Triterpenoids from Dandelion Leaves [23, 24]

Triterpenoids—**β-amyrin** and **β-sitosterol** were extracted from freshly ground leaves of *Taraxacum officinale* (Dandelion) by SFE on a high pressure equipped with a 5 L NATEX extractor vessel using supercritical CO_2. The flow rate of the solvent feed was kept at 7.4 kg CO_2/kg dried leaves per hour. The accumulated product samples were removed at regular intervals and weighed. The extraction procedure was terminated when the increase in yield was less than 0.1% while 10 kg CO_2 passed through the vessel. Extraction at the optimum pressure and temperature condition gave 3.2–4.0% extract (dry weight) of the raw material. It was seen from the TLC analysis that the obtained extract contained **β-amyrin** and **β-sitosterol** (Fig. 10.6).

10.5.3 Isolation of 3-Acetyl-11-keto-β-boswellic Acid (AKBA) and 11-keto-β-boswellic Acid (KBA) from *Boswellia serrata* [25, 26]

B. serrata gum resin was taken, methanolic extract was prepared by Soxhlet extraction method. The extract solution was filtered and concentrated under vacuum to obtain dried extract. Dried extract was taken and processed

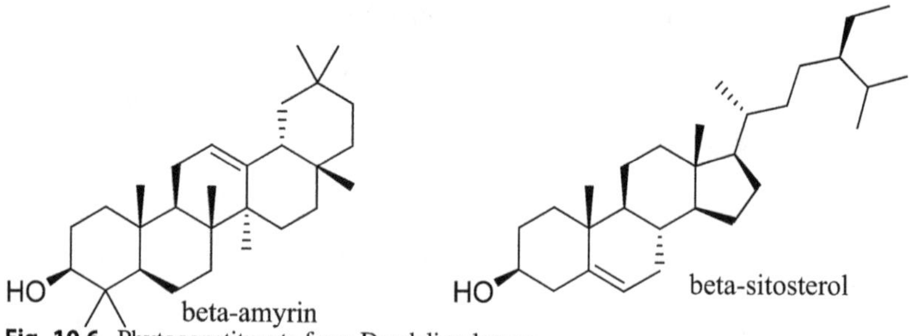

Fig. 10.6 Phytoconstituents from Dandelion leaves

to make KBA enriched fraction. The enriched fraction was subjected to silica gel column chromatography. The column was eluted with hexane-ethyl acetate mobile phase in a gradient pattern and fractions were collected. Fraction 3 (eluted with 15% ethyl acetate in hexane), when analyzed using TLC, confirmed the presence of AKBA. Fraction 4 (eluted with 20% ethyl acetate in hexane) showed the spot for KBA on TLC. Both the fractions were further purified over Sephadex LH-20 column chromatography, followed by semi-preparative HPLC to give 99.9% pure **AKBA** and **KBA** (Fig. 10.7; Scheme 10.2).

10.5.4 Isolation of Arjunolic Acid from *Terminalia arjuna* (Unpublished Work)

Approximately 1 kg dried and powdered heart wood of *T. arjuna* was macerated with methanol thrice for 24 h. The extract obtained above was

Fig. 10.7 AKBA and KBA isolated from *Boswellia serrata*

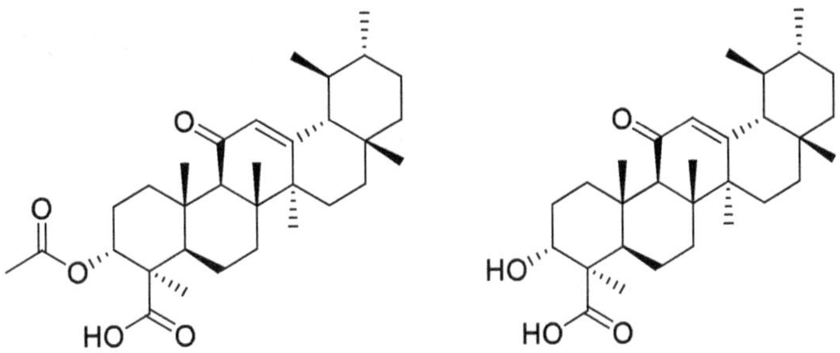

3-Acetyl-11-keto-beta-boswellic acid

11-keto-beta-boswellic acid

Scheme 10.2 Isolation of AKBA and KBA

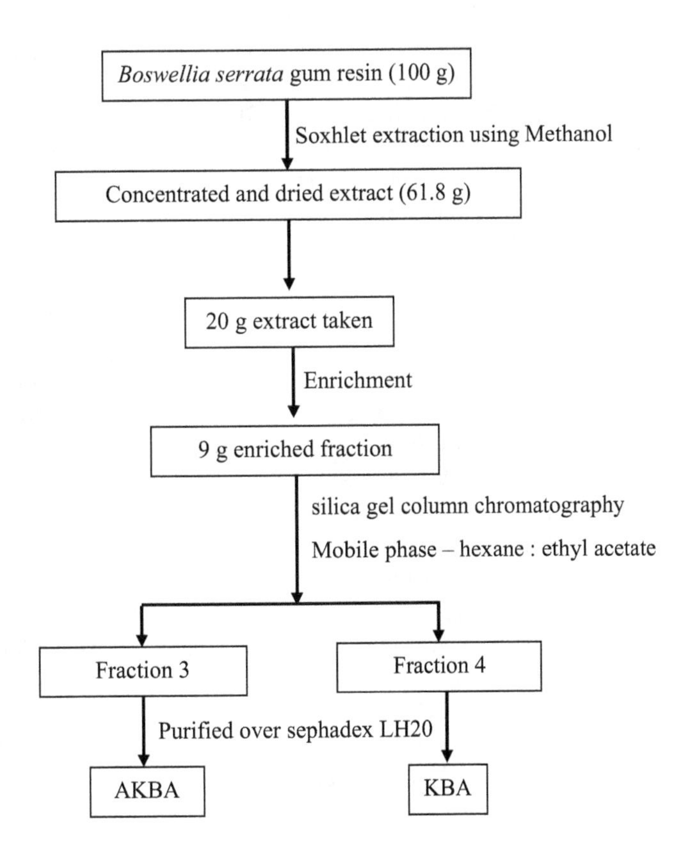

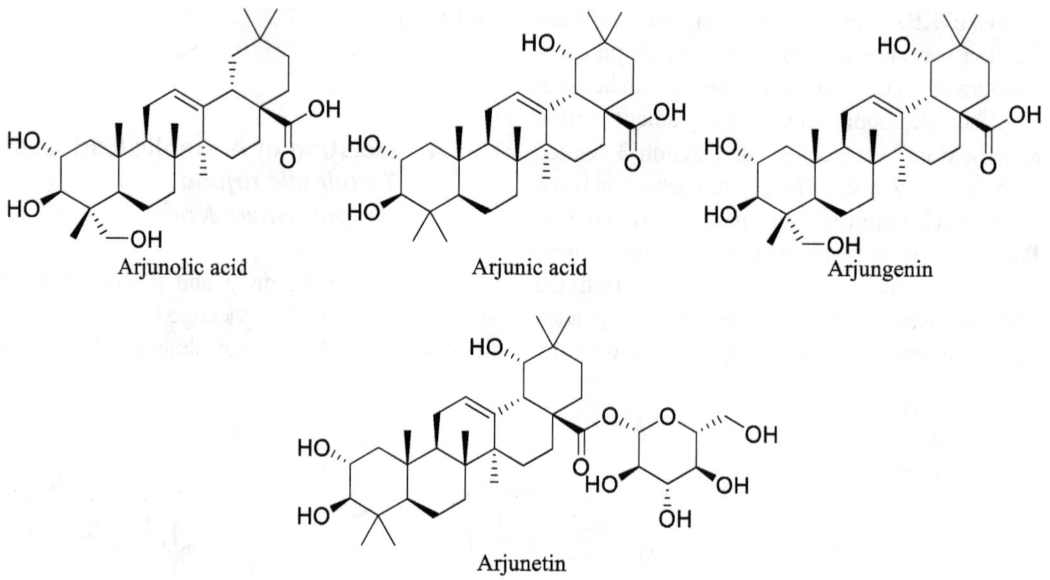

Fig. 10.8 Phytoconstituents from Terminalia arjuna

concentrated upon rotary evaporator. The dried extract obtained was 45.25 g, which was partitioned with ethyl acetate thrice. The ethyl acetate fraction obtained (10.20 g) was subjected to silica gel column chromatography and eluted with hexane-ethyl acetate. Various fractions were collected. The fraction collected with 70% ethyl acetate in hexane had **arjunolic acid**. Other compounds which were isolated from other fractions are **arjunic acid**, **arjungenin** and **arjunetin** (Fig. 10.8).

10.6 Flavonoids and Phenolics

One of the most prevalent families of chemical constituents in plants are polyphenols. An extensive spectrum of biological actions of polyphenolic substances has been reported, because of their chemical constituents [27]. For the treatment of certain modern lifestyle disorders, such as cancer, polyphenols have been promoted as nutraceuticals in human medicine [28]. In addition, these influence the colour and sensory qualities of fruits and vegetables. the polyphenols also play a significant role in plant development and reproduction [29]. According to the number of phenol rings and the structural components that connect these rings to one another, polyphenols are classified into several classes like flavonoids, tannins (hydrolyzable and condensed), stilbenes, lignans, and phenolic acids [30].

Flavonoids are a large class of polyphenolics that have low molecular weight and carbon atoms arranged in C6-C3-C6 configuration with two aromatic rings A and B joined by a 3-carbon bridge, mainly in the form of heterocyclic ring C. Variation in substitution pattern on C ring flavonoids further categorize flavonoids into isoflavones, flavonones, flavonols, anthocyanidines and flavanols (catechin). The structures of the major classes of polyphenols with flavonoids are listed below [31] (Fig. 10.9).

Flavonoids and polyphenols have been reported to have a number of therapeutic and biological effects including anti-inflammatory [32], anti-oxidant [33], anti-cancer [34], antibacterial [35], antifungal [36], antiviral [37], anti-diabetic [38], prevention of cardiovascular diseases [39], anti-ulcer [40], Alzheimer's disease [41], Parkinson's disease [42]. Flavonoids and polyphenols are also used as nutraceuticals now-a-days [43].

Out of all the listed activities, flavonoids and other polyphenols, mainly have key protecting activity for the body against free radicals,

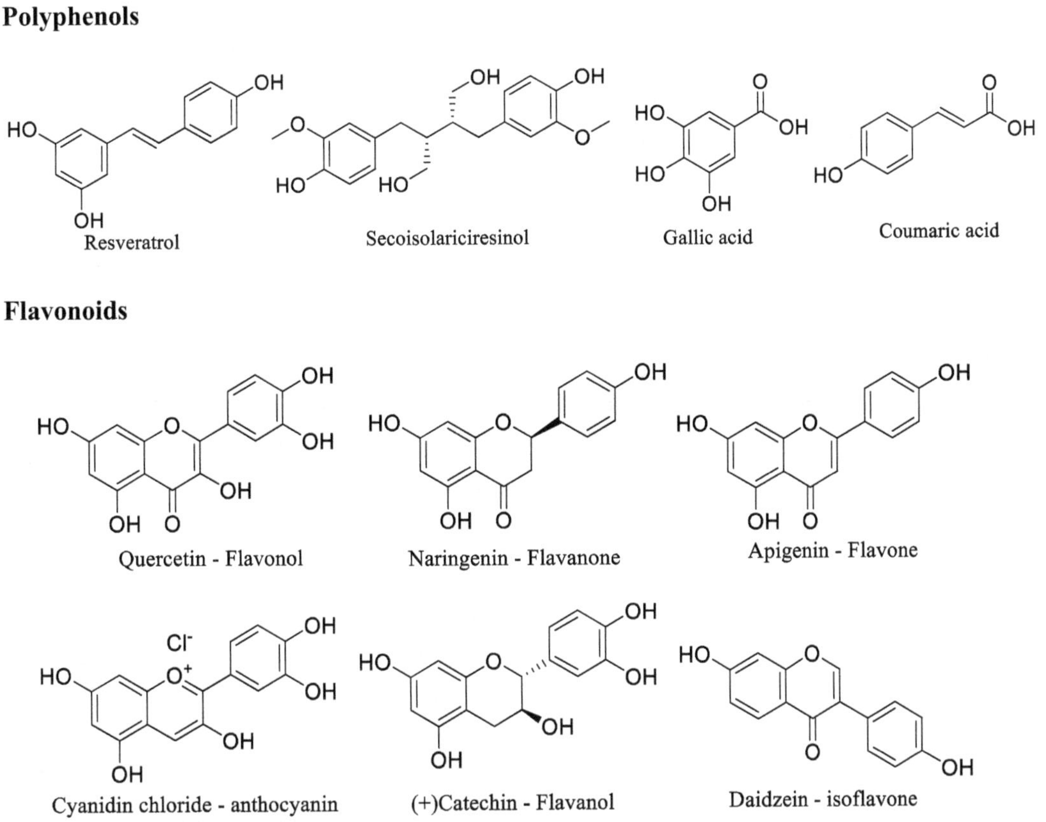

Polyphenols

Resveratrol Secoisolariciresinol Gallic acid Coumaric acid

Flavonoids

Quercetin - Flavonol Naringenin - Flavanone Apigenin - Flavone

Cyanidin chloride - anthocyanin (+)Catechin - Flavanol Daidzein - isoflavone

Fig. 10.9 Some examples of polyphenols

reducing oxidative stress, acting as anti-aging agents and also show anti-cancer activities.

The extraction and isolation of phytochemicals from polyphenols and flavonoids is now an attractive interest area because of all the vital functions of these compounds that have been enumerated above. There is no general standard method for extraction and isolation of polyphenolic compounds [44]. Natural phenolic compounds can be extracted using conventional methods such as heating, boiling, or refluxing; however, these methods have the drawbacks of losing polyphenols due to ionization, hydrolysis, and oxidation during extraction, as well as having a lengthy extraction period [45]. The extraction of polyphenols from plants has recently been made possible by a number of innovative extraction techniques, such as high hydrostatic pressure, supercritical fluid, ultrasound, and microwave assistance these techniques of extraction, are affordable, efficient, and near certainty for standard extraction methods [46].

The extraction efficiency of desired components and their concentration from polyphenols raw material were affected by several factors like temperature, liquid-solid ratio, flow rate, and particle size. For instance, batch liquid-solid extraction at 50 °C compared to room temperature revealed that the phenolic content of almond shell extract was three times higher [47]. Usually aqueous, semi-polar, polar organic solvents (ethyl acetate, methanol and ethanol) or a mixture of solvents are used for extraction of polyphenols which make all polyphenol compounds, soluble/extractable, because of the solubility of free soluble esters and soluble glycosides.

Some phenolic acid and anthocyanin compounds have insoluble bond complexes which are linked together with ester or glycoside linkage these were extracted by using acid or base hydrolysis method. After extraction, they were separated by classical liquid chromatography if they were not able with normal then used counter-current chromatography (CCC) or high-speed counter-current chromatography (HSCCC). Isolated compound structures are elucidated and characterized by using different spectral methods like NMR, IR, and Mass spectrometry [48] (Scheme 10.3).

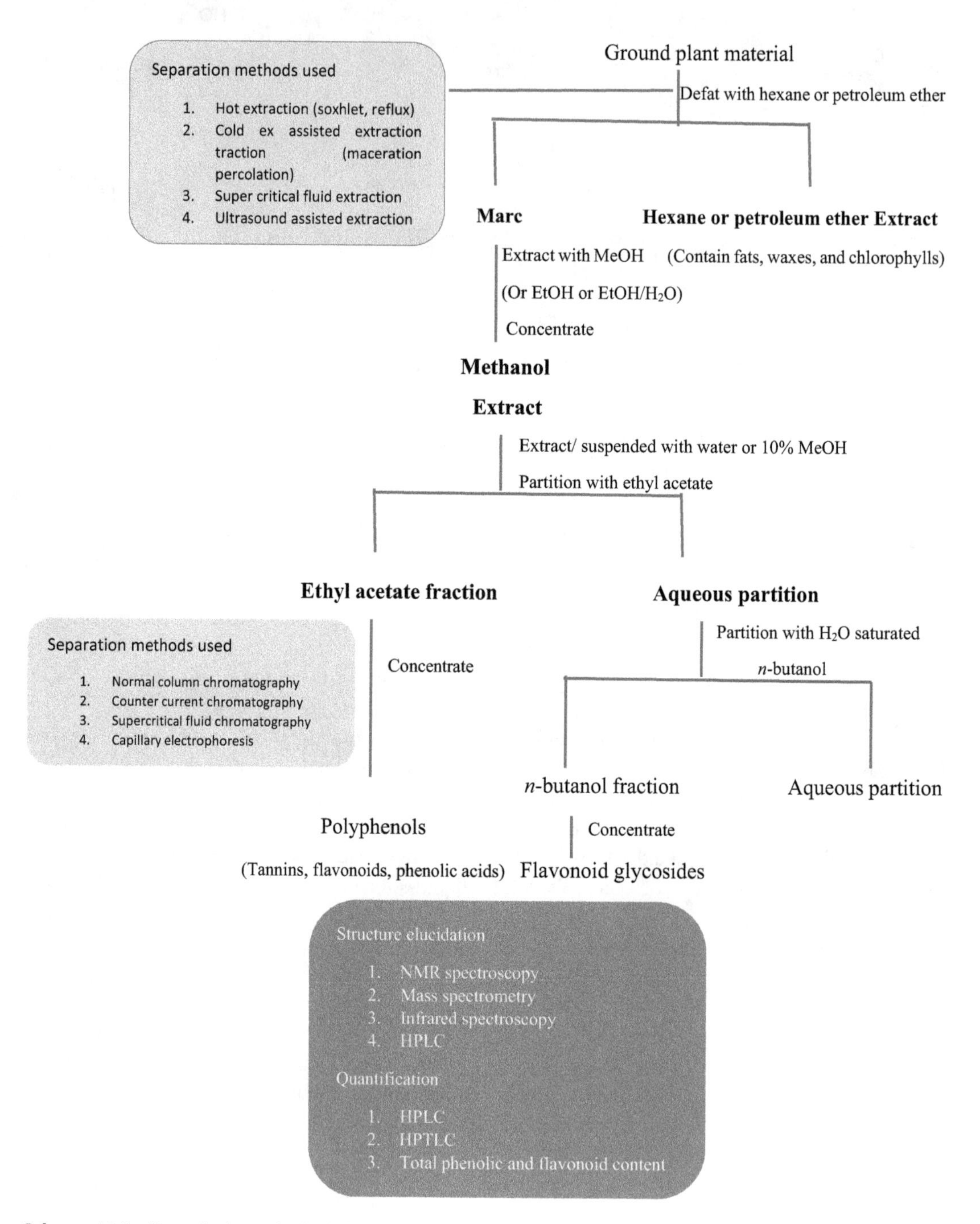

Scheme 10.3 General scheme for isolation of polyphenols and their identification

10.6.1 Isolation of Rotenoids from *Boerhaavia diffusa* [49]

Boerhaavia diffusa L. (Nyctaginaceae) is an ethnomedicinally valuable plant used for the prevention of illnesses such as jaundice, dyspepsia, nephrotic syndrome, convulsions, and enlargement of the spleen, stomach pain, stress, and inflammation, because of the present of rotenoids type of flavonoid.

They were isolated from the dried roots of *Boerhaavia diffusa* L. dried root material was first defatted with hexane and then extracted with methanol using a soxhlet apparatus. Concentrated to dry under reduced pressure using a rotatory evaporator and suspended in water and partitioned with ethyl acetate and dried. Then dried EtOAc fractions were subjected to column chromatography using silica gel (100–200 mesh) and eluted using mobile phase hexane-ethyl acetate. All eluted fractions were collected and combined based on TLC giving six pooled fractions. (BDM-F1-BDM-F6) Further, they were subjected to the Sephadex LH-20 CC and eluted with methanol. Isolated compounds were elucidated by NMR, IR, and HRMS (Fig. 10.10).

10.6.2 Isolation of Iridoids and Flavonoids from *Vitex negundo* [50]

Vitex negundo Linn., is a shrub commonly known as nirgundi in Ayurveda used for antioxidant, analgesisc, antigenotoxic, anti-inflammatory and anti-convulsant activity. It has reported contain for various types of polyphenolic compounds. One of them isolation of iridoids type flavonoids described below.

Powdered leaves of *V. negundo* were extracted using methanol as a solvent for 24 h by soxhlet apparatus. The methanol extract was dried under reduced pressure and subjected to a Diaion resin HP-20 column. Eluted gradient-wise using methanol and water 0%, 20%, 30%, 40-80%, and 100% (F-1 to F-5) sequence. Fractions F-2, F-4 and F-5 were found mixture of nirgundoside and agnuside, subjected to preparative HPLC using methanol and water 30:70, 40:60, 60:40 ratio respectively, at a flow rate of 5 ml/min and fraction F-3 was subjected to CC on sephadex LH-20 and eluted with methanol. From preparative HPLC **negundoside**, **agundoside**, **vitexin**, **isovitexin**, **vitexicarpin**, and from Sephadex isolated **isoorientin**.

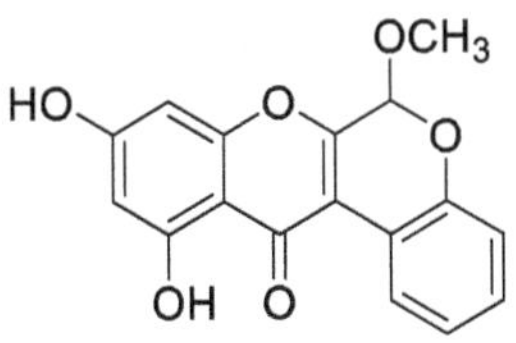

Boeravinone K 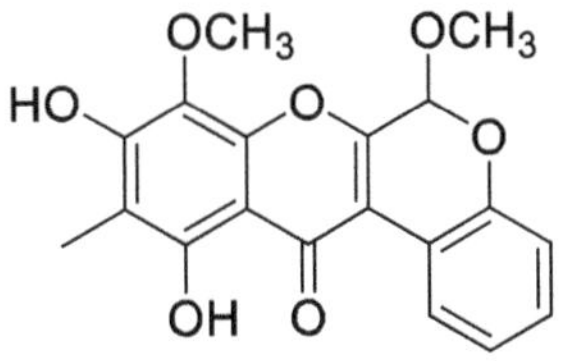   Boeravinone L 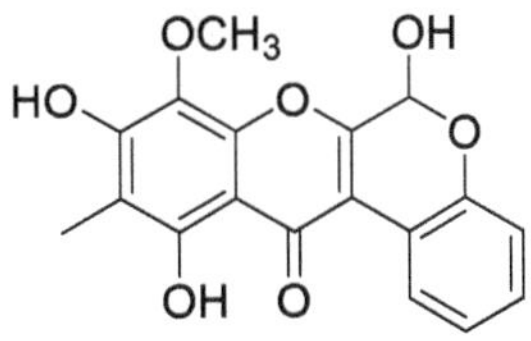   Boeravinone M

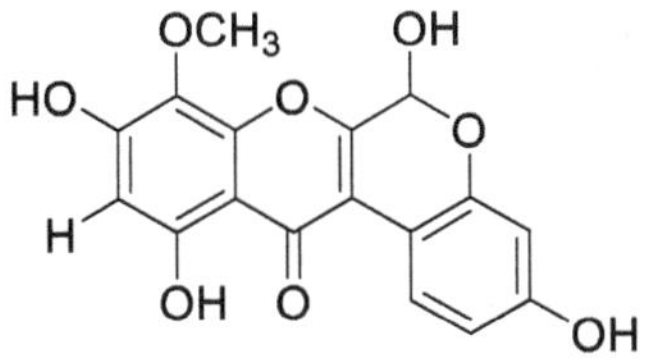

Boeravinone N 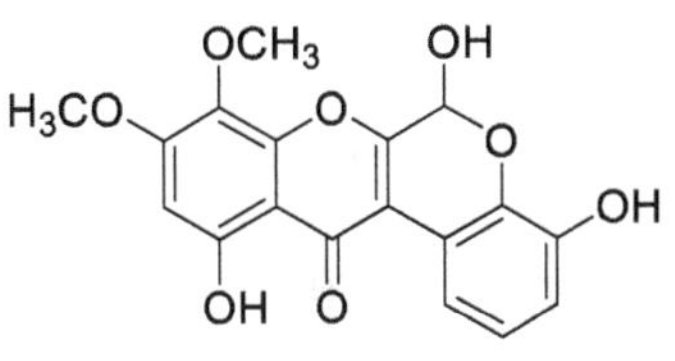   *Boeravinone* O

Fig. 10.10 Rotenoids from *Boerrhavia diffusa*

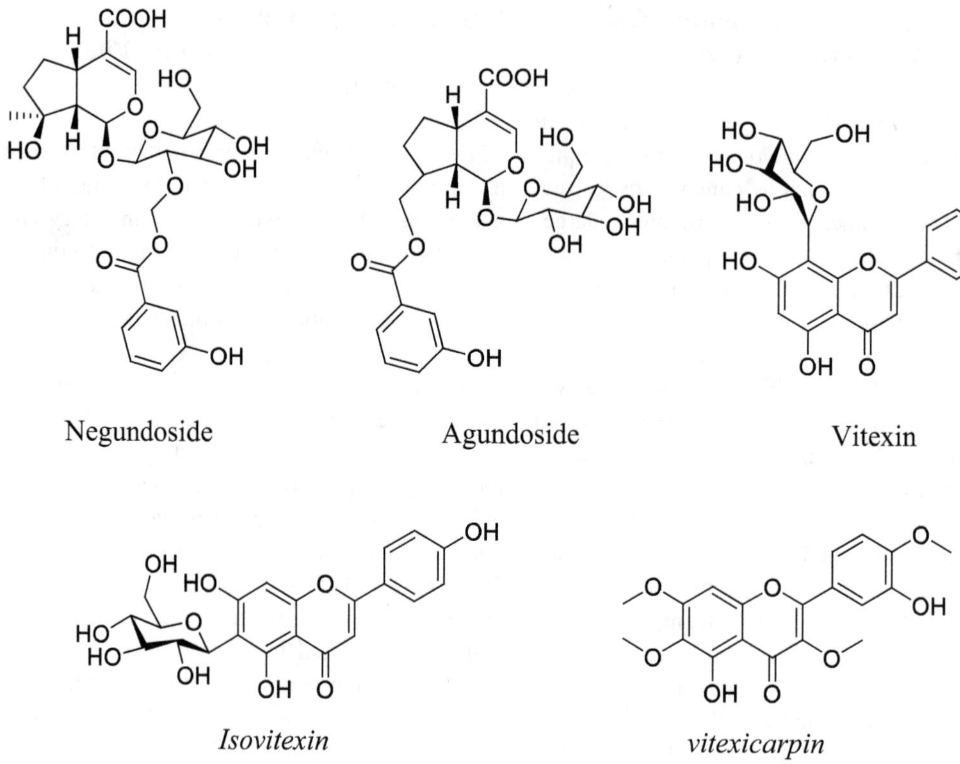

Negundoside Agundoside Vitexin

Isovitexin *vitexicarpin*

Fig. 10.11 Phytoconstituents from *Vitex negundo*

Following is the list of compounds isolated and their structure (Fig. 10.11).

10.6.3 Isolation of Flavonoids from Plant *Tridax procumbens* [51, 52]

Tridax procumbens is also known as coatbuttons. It possesses hepatoprotective, anti-inflammatory, antioxidant and hypotensive properties. Dried powdered aerial part of the plant was sonicated with methanol for 10 min, and extract centrifuged for 5 min (1200 rpm) then the supernatant was filtered and subjected for gel permeation chromatography on sephadex LH-20 column. The sample was eluted with 100% methanol, collected fractions were analysed by TLC (thin layer chromatography), and pooled accordingly. Three fractions showed a single spot on TLC, which was further analysed by HPLC for their purity. These compounds after structure elucidation by NMR, HRMS, and IR found to be **quercetagetin-3,6,4′-trimethoxyl-7-O-β-D-glucopyranoside, 3-O-methyl quercetin-4′-O-β-D-glucopyranoside**, and **luteolin-4′-O-β-D-glucopyranoside**, respectively (Fig. 10.12).

10.6.4 Isolation of Tuberosin from *Pueraria tuberosa* Roots

Pueraria tuberosa tubers contains isoflavonoids which shows antioxidant, anti-inflammatory, and anti-diabetic activity. Dried roots of *P. tuberosa* were powdered and extracted with hexane and then ethanol using soxhlet extractor. Extract was dried and subjected to CC and eluted with organic solvent with increasing polarity. Ethyl acetate fraction was re-subjected to CC using benzene: ethyl acetate (7:3) as an eluent. Isolated tuberosin was recrystallized with benzene, and evaluated for structure elucidation by NMR, IR, and UV [53].

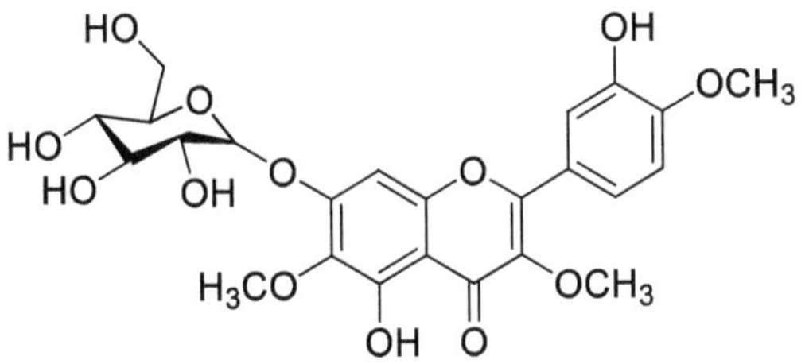

Quercetagetin-3,6,4′-trimethoxyl-7-O-β-D-glucopyranoside

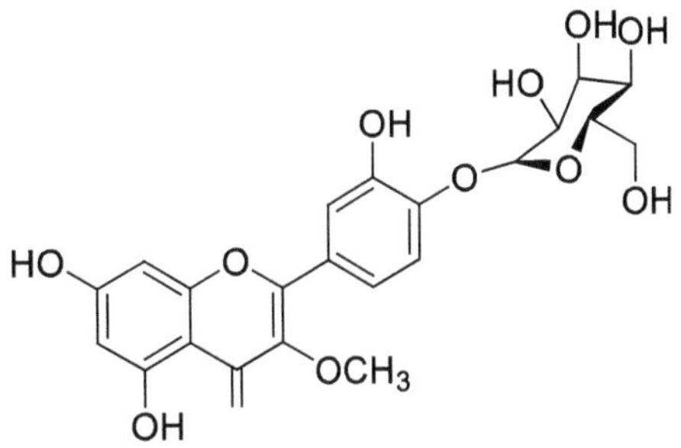

3-O-methylquercetin-4′-O-β-D-glucopyranoside

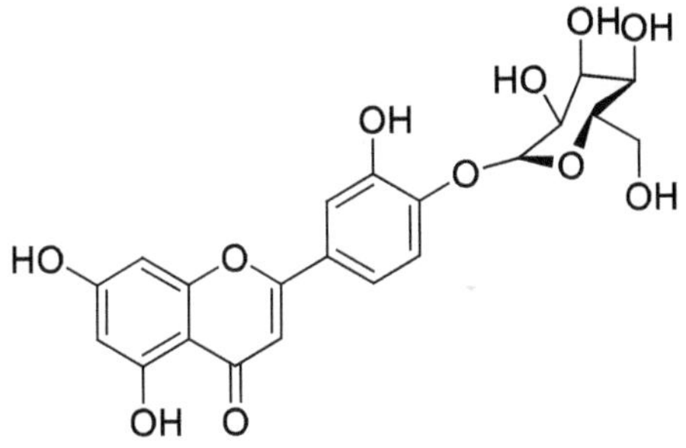

luteolin-4′-O-β-D-glucopyranoside

Fig. 10.12 Isolated compounds from *Tridax procumbens*

10.6.5 Isolation of Flavonoids and other Constituents from *Pueraria lobata* Roots

Puerarin is a marker bioactive isoflavonoid shows antioxidant, anti-inflammatory, and anti-diabetic activity. Compounds from dried roots of plant *P. lobata* were isolated by powdered and refluxed roots with methanol for 3 h then extract concentrate to dryness by reduced pressure on a rotatory evaporator. Then extract was suspended in water and partitioned with *n*-Hexane, chloroform, ethyl acetate and *n*-butanol. *n*-hexane fraction subjected to chromatography on silica gel column using hexane : ethyl acetate as a gradient (100:1 to 0:1) gives compound **lupenone** with other sub-fraction, one of this fraction further loaded on column and eluted hexane: ethyl acetate as a gradient (20:1 to 0:1) to yield **lupeol**, and other sub-fraction which were recrystallized with 100% methanol to produce compound **coumesterol** and **puerarol**. Then the EtOAc fraction was also subjected for CC using CH_2Cl_2-MeOH (1:1 to 0:1, gradient) yield **daidzain, geninstain, puerarine,** and **puerarol B-2-O-glucopyranoside** [54] (Fig. 10.13).

10.7 Challenges Faced in the Experimental Protocols in Phytochemistry and Natural Products

Numerous naturally occurring substances have been identified to be capable of impeding the separation and purification of a desired bioactive plant ingredient. The following broad steps may aid in realising that contamination may have occurred during extraction

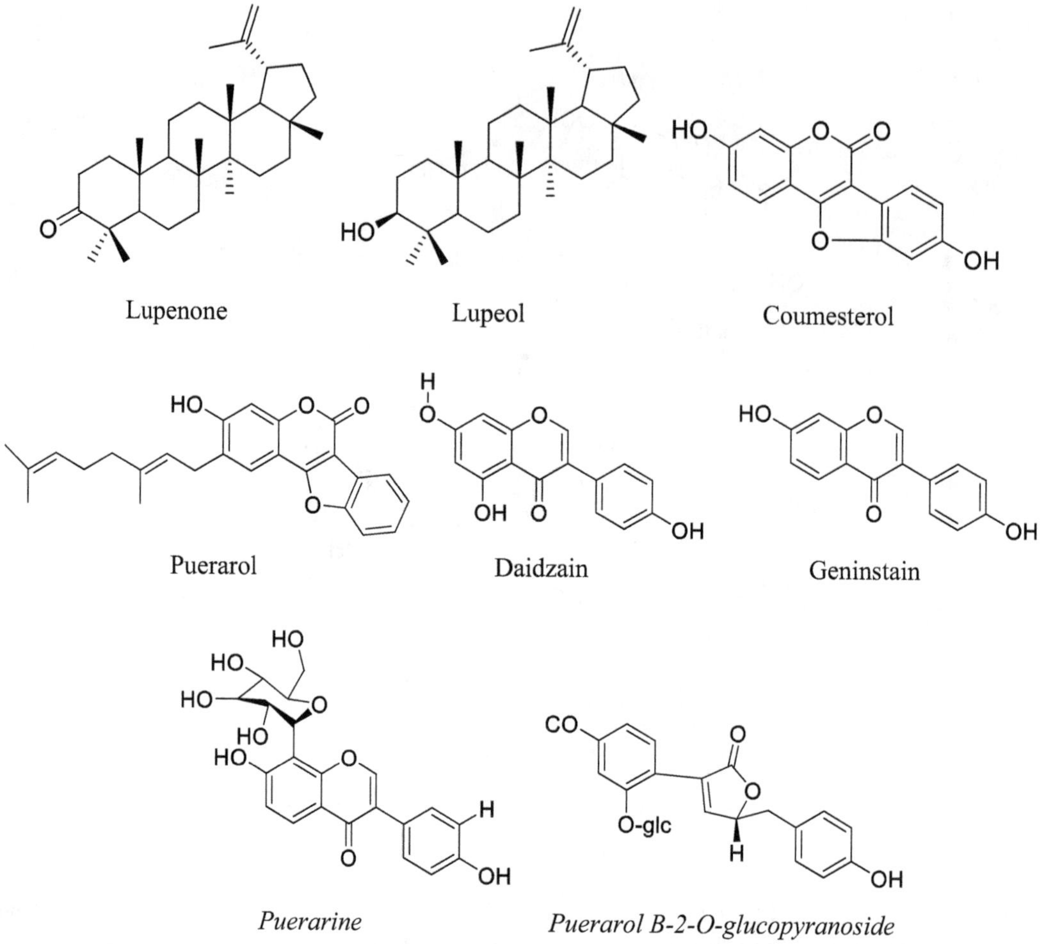

Lupenone Lupeol Coumesterol

Puerarol Daidzain Geninstain

Puerarine *Puerarol B-2-O-glucopyranoside*

Fig. 10.13 Phytoconstituents from *Pueraria* roots

10.7.1 Lipids

Lipids are typically extracted using low polarity solvents, but when polar solvents are utilised, they may coextract. By running a TLC plate and employing iodine vapour in a sealed space to expose brown spots, these compounds can be seen. They can also be seen by doing proton NMR spectral measurement, which shows a high broad peak at roughly chemical shift value range 1.2 to 1.4 [55, 56].

10.7.2 Pigments

Depending on the plant part treated, unwanted pigments like chlorophylls and flavonoids could be in high content. Although they are difficult to get rid of, but some of the following techniques could be used [57].

10.7.3 Tannins

Vegetable tannins are polyphenols that are frequently present in high concentrations in plant extracts, and they frequently provide false-positive results in biological experiments as a result of their propensity to precipitate proteins via multipoint hydrogen bonding. The proanthocyanidins (also known as condensed tannins) and polyesters based on gallic acid and/or hexahydroxydiphenic acid and their derivatives are two structurally distinct groups that can be used to classify vegetable tannins (also known as plant polyphenols) [58].

10.7.4 Plasticizers

Solvents, filter papers, plastic equipment, and chromatographic solid phases kept in plastic containers can all be contaminated by plasticizers. Dioctylphthalate ester has frequently been observed to contaminate plant isolates, and in its purest form, it is a yellow oil that significantly inhibits P-388 murine lymphocytic leukaemia cells from growing. When concentrated sulfuric acid or concentrated sulfuric acid-acetic acid (4:1) is sprayed on silica gel and heated at 110 °C for 5 min with $R_f = 0.4$, TLC reveals a pink-violet spot (TLC solvent system: petroleum ether-ethyl acetate, 19:1) [59].

10.7.5 Grease

Ground-glass joints in extraction equipment, as well as stopcocks in columns and vacuum lines, are lubricated with silicone grease. It can be identified by the following mass spectral fragmentation ions: m/z 429, 355, 281, 207 and 133, and it may contaminate plant samples. The degradation of the aliphatic chains causes losses in the ion pattern at every 14 mass unit intervals when hydrocarbon grease is utilised [60].

10.8 Strategy for Overcoming Difficulties

10.8.1 Strategy to Remove Lipids, Fats and Waxes

The ground plant material can be percolated with petroleum ether or hexanes and left to dry before the entire extraction process in order to remove the fats and waxes from an extract. Alternatively, the material can be directly extracted with the chosen solvent and then defatted. The nonpolar lipid components can also be removed using column chromatography or vacuum-liquid chromatography (VLC), which both employ petroleum ether or hexane as eluents. This enables the extract to be fractionated. By filtering the sample through a reverse-phase chromatographic column, fats and waxes can be removed from a contaminated sample while the lipids are retained. A different strategy is to add enough methanol or methanol-water binary mixture to dissolve the target component [61].

10.8.2 Strategy to Remove Pigments

A selective adsorption phenomenon is a principle by which activated charcoal or activated carbon is known to decolorize solutions. Either the powder is mixed with the liquid to be decolored, let to stand for a while, and then filtered, or the solution can be percolated via a relatively short charcoal column. Heating improves the adsorption's effectiveness. The drawback of charcoal is that several medicinally active substances, including morphine, strychnine, and quinine, can also be adsorbed. Chlorophylls can be retained at the head of a neutral alumina column, however, they are frequently removed from extracts by solvent partitioning in significant quantity [62].

10.8.3 Strategy to Remove Tannins (or Polyphenols)

The generation of a precipitate with ferric chloride can indicate the presence of tannins, and they can be eliminated from aqueous and nonpolar extracts by passing them through polyamide, collagen, sephadex LH-20, or silica gel. A chloroform extract can be washed with 1% aqueous sodium chloride, the top phase can be discarded, and the chloroform phase can then be dried with anhydrous Na_2SO_4 for a few hours to remove tannins. Additionally, proteins, gelatin-sodium chloride solution (5% w/v NaCl and 0.5% w/v gelatin), caffeine, nylon, or other substances may precipitate tannins [62].

10.8.4 Strategy to Remove Plasticizers

By distilling the solvents, passing the sample through a reverse phase chromatographic column, or passing the extract or sample through porous alumina, plasticizers can be removed [60].

10.9 Extraction Protocol Optimization

The strategy for developing environmentally friendly extraction processes for alkaloids from various sources provides a good solution, employing NDES and non-ionic surfactants as a green solvent. The simultaneous interaction effects of independent variables were critically analysed using the suggested mathematical methodology. Instead of using traditional organic solvents like methanol and ethanol, Genapol X-80 and the NDES mixture of choline chloride, fructose, and water can be utilised, with the NDES mixture having a better extraction efficiency [63].

10.10 Use of Eutectic Solvents

For thorough extraction of varous kinds of bioactive natural products, green Deep Eutectic Solvent (DES) extraction method was described Alkaloids, phenolic acids, flavonoids, and saponins were successfully extracted using the custom-made DESs, however, anthraquinones had a reduced extractability. The concentration of water in the DES/water combination and the solid-to-liquid ratio were shown to be the main factors affecting extraction yields after further optimization for alkaloids extraction using Response Surface Methodology (RSM). This offered real-world illustrations demonstrating the ability of DESs that efficiently and selectively extract bioactive chemicals from biomaterials. In addition, DES solvents were the first to be used to extract alkaloids, and Choline Chloride-Lactic Acid (ChCl-La) was found to be far more effective than earlier techniques necessitating the use of organic solvents [64].

10.11 Concluding Remarks

Natural product extraction techniques, as well as their isolation, identification, and uses, are clearly and steadily gaining importance. Modern analytical processes, which are both environmentally benign and economically feasible, place a high priority on research innovation and safe extraction techniques. The extraction and isolation technique for alkaloids, terpenoids, flavonoids and other polyphenols from a few medicinally important plants have been discussed in this chapter. The difficulties faced during extraction and isolation of bioactive components have also been listed along with the strategies employed to resolve those difficulties have been discussed. The concept of the green extraction process and its optimization and use instead of conventional techniques have been discussed. Natural deep eutectic solvents (NDESs), which are environmentally friendly and long-lasting, are now being used to extract bioactive substances or pharmaceuticals effectively. They typically contain neutral, acidic, or basic substances that, when combined in specific molar ratios, result in liquids with a high viscosity. These can be very useful for phytochemistry and natural products in this modern era.

Conflict of Interest The authors declare no conflict of interest.

References

1. Dias DA, Urban S, Roessner U (2012) A historical overview of natural products in drug discovery. Meta 2(2):303–336
2. Dekebo A (2019) Introductory chapter: plant extracts. Plant extracts. IntechOpen, London, p 6
3. Stéphane FFY, Jules BKJ, Batiha GE-S, Ali I, Bruno LN (2021) Extraction of bioactive compounds from medicinal plants and herbs. IntechOpen, London
4. Handa S (2008) An overview of extraction techniques for medicinal and aromatic plants. Extract Technol Med Aromat Plants 1:21–40
5. Azwanida N (2015) A review on the extraction methods use in medicinal plants, principle, strength and limitation. Med Aromat Plants 4(196):2167–0412
6. Harborne A (1998) Phytochemical methods a guide to modern techniques of plant analysis. Springer, Cham

7. Li H-B, Jiang Y, Wong C-C, Cheng K-W, Chen F (2007) Evaluation of two methods for the extraction of antioxidants from medicinal plants. Anal Bioanal Chem 388(2):483–488
8. Chemat F, Vian MA, Cravotto G (2012) Green extraction of natural products: concept and principles. Int J Mol Sci 13(7):8615–8627
9. Sarker SD, Latif Z, Gray AI (2005) Natural product isolation: an overview. Springer, Cham, pp 1–25
10. Jones WP, Kinghorn AD (2012) Extraction of plant secondary metabolites. Nat Prod Isolat 2012:341–366
11. Ahirrao P, Tambat R, Chandal N, Mahey N, Kamboj A, Jain UK et al (2020) MsrA efflux pump inhibitory activity of Piper cubeba lf and its phytoconstituents against Staphylococcus aureus RN4220. Chem Biodivers 17(8):e2000144
12. Rabêlo SV, Costa EV, Barison A, Dutra LM, Nunes XP, Tomaz JC et al (2015) Alkaloids isolated from the leaves of atemoya (Annona cherimola× Annona squamosa). Rev Bras 25:419–421
13. Adamska-Szewczyk A, Glowniak K, Baj T (2016) Furochinoline alkaloids in plants from Rutaceae family-a review. Curr Issues Pharm Med Sci 29(1): 33–38
14. Sandjo LP, Kuete V, Tchangna RS, Efferth T, Ngadjui BT (2014) Cytotoxic benzophenanthridine and furoquinoline alkaloids from Zanthoxylum buesgenii (Rutaceae). Chem Cent J 8(1):1–5
15. Birari R, Roy SK, Singh A, Bhutani KK (2009) Pancreatic lipase inhibitory alkaloids of Murraya koenigii leaves. Nat Prod Commun 4(8): 1934578X0900400814
16. Nalli Y, Khajuria V, Gupta S, Arora P, Riyaz-Ul-Hassan S, Ahmed Z et al (2016) Four new carbazole alkaloids from Murraya koenigii that display anti-inflammatory and anti-microbial activities. Org Biomol Chem 14(12):3322–3332
17. Dev S (1989) Terpenoids. Natural products of woody plants. Springer, Cham, pp 691–807
18. Las Heras B, Rodriguez B, Bosca L, Villar A (2003) Terpenoids: sources, structure elucidation and therapeutic potential in inflammation. Curr Top Med Chem 3(2):171–185
19. Ludwiczuk A, Skalicka-Woźniak K, Georgiev M (2017) Terpenoids. Pharmacognosy. Elsevier, Amsterdam, pp 233–266
20. Jiang Z, Kempinski C, Chappell J (2016) Extraction and analysis of terpenes/terpenoids. Curr Protocols Plant Biol 1(2):345–358
21. Auzi AA, Gray AI, Salem MM, Badwan AA, Sarker SD (2008) Feruhermonins A–C: three daucane esters from the seeds of Ferula hermonis (Apiaceae). J Asian Nat Prod Res 10(8):701–707
22. Reid RG, Sarker SD (2012) Isolation of natural products by low-pressure column chromatography. In: Natural products isolation. Springer, Cham, pp 155–187
23. Nahar L, Sarker SD (2012) Supercritical fluid extraction in natural products analyses. In: Natural products isolation. Springer, Cham, pp 43–74
24. Simandi B, Kristo ST, Kery A, Selmeczi L, Kmecz I, Kemeny S (2002) Supercritical fluid extraction of dandelion leaves. J Supercrit Fluids 23(2):135–142
25. Bairwa K, Jachak SM (2015) Development and optimisation of 3-Acetyl-11-keto-β-boswellic acid loaded poly-lactic-co-glycolic acid-nanoparticles with enhanced oral bioavailability and in-vivo anti-inflammatory activity in rats. J Pharm Pharmacol 67(9): 1188–1197
26. Bairwa K, Jachak SM (2016) Nanoparticle formulation of 11-keto-β-boswellic acid (KBA): anti-inflammatory activity and in vivo pharmacokinetics. Pharm Biol 54(12):2909–2916
27. Tomás-Barberán F, Ferreres F, Gil M (2000) Antioxidant phenolic metabolites from fruit and vegetables and changes during postharvest storage and processing. Stud Nat Prod Chem 23:739–795
28. Garg SK, Shukla A, Choudhury S (2019) Polyphenols and flavonoids. Nutraceut Vet Med 2019:187–204
29. Sangeetha KS, Umamaheswari S, Reddy CUM, Kalkura SN (2016) Flavonoids: therapeutic potential of natural pharmacological agents. Int J Pharm Sci Res 7(10):3924
30. Archivio Filesi D, Di Benedetto R, Gargiulo R, Giovannini C, Masella R (2007) Polyphenols, dietary sources and bioavailability. Ann Ist Super Sanita 43(4):348
31. Ignat I, Volf I, Popa VI (2011) A critical review of methods for characterisation of polyphenolic compounds in fruits and vegetables. Food Chem 126(4):1821–1835
32. González R, Ballester I, López-Posadas R, Suárez M, Zarzuelo A, Martínez-Augustin O et al (2011) Effects of flavonoids and other polyphenols on inflammation. Crit Rev Food Sci Nutr 51(4):331–362
33. Sanbongi C, Osakabe N, Natsume M, Takizawa T, Gomi S, Osawa T (1998) Antioxidative polyphenols isolated from Theobroma cacao. J Agric Food Chem 46(2):454–457
34. Niedzwiecki A, Roomi MW, Kalinovsky T, Rath M (2016) Anticancer efficacy of polyphenols and their combinations. Nutrients 8(9):552
35. Taguri T, Tanaka T, Kouno I (2006) Antibacterial spectrum of plant polyphenols and extracts depending upon hydroxyphenyl structure. Biol Pharm Bull 29(11):2226–2235
36. Sitheeque M, Panagoda G, Yau J, Amarakoon A, Udagama U, Samaranayake L (2009) Antifungal activity of black tea polyphenols (catechins and theaflavins) against Candida species. Chemotherapy 55(3): 189–196
37. Chojnacka K, Skrzypczak D, Izydorczyk G, Mikula K, Szopa D, Witek-Krowiak A (2021) Antiviral properties of polyphenols from plants. Foods 10(10): 2277

38. Domínguez Avila JA, Rodrigo García J, González Aguilar GA, De la Rosa LA (2017) The antidiabetic mechanisms of polyphenols related to increased glucagon-like peptide-1 (GLP1) and insulin signaling. Molecules 22(6):903
39. Tangney CC, Rasmussen HE (2013) Polyphenols, inflammation, and cardiovascular disease. Curr Atheroscler Rep 15:1–10
40. Alsabri SG, Rmeli NB, Zetrini AA, Mohamed SB, Meshri MI, Aburas KM et al (2013) Phytochemical, anti-oxidant, anti-microbial, anti-inflammatory and anti-ulcer properties of Helianthemum lippii. J Pharmacog Phytochem 2(2):86–96
41. Akter R, Chowdhury MA, Rahman MH (2021) Flavonoids and polyphenolic compounds as potential talented agents for the treatment of Alzheimer's disease and their antioxidant activities. Curr Pharm Des 27(3):345–356
42. Magalingam KB, Radhakrishnan AK, Haleagrahara N (2015) Protective mechanisms of flavonoids in Parkinson's disease. Oxidative Med Cell Longev 2015:314560
43. Bahadoran Z, Mirmiran P, Azizi F (2013) Dietary polyphenols as potential nutraceuticals in management of diabetes: a review. J Diabetes Metab Disord 12:1–9
44. Balasundram N, Sundram K, Samman S (2006) Phenolic compounds in plants and agri-industrial by-products: antioxidant activity, occurrence, and potential uses. Food Chem 99(1):191–203
45. Hayouni EA, Abedrabba M, Bouix M, Hamdi M (2007) The effects of solvents and extraction method on the phenolic contents and biological activities in vitro of Tunisian Quercus coccifera L. and Juniperus phoenicea L. fruit extracts. Food Chem 105(3): 1126–1134
46. Huang W, Xue A, Niu H, Jia Z, Wang J (2009) Optimised ultrasonic-assisted extraction of flavonoids from Folium eucommiae and evaluation of antioxidant activity in multi-test systems in vitro. Food Chem 114(3):1147–1154
47. Pinelo M, Rubilar M, Sineiro J, Nunez M (2004) Extraction of antioxidant phenolics from almond hulls (Prunus amygdalus) and pine sawdust (Pinus pinaster). Food Chem 85(2):267–273
48. Mattila P, Kumpulainen J (2002) Determination of free and total phenolic acids in plant-derived foods by HPLC with diode-array detection. J Agric Food Chem 50(13):3660–3667
49. Bairwa K, Singh IN, Roy SK, Grover J, Srivastava A, Jachak SM (2013) Rotenoids from Boerhaavia diffusa as potential anti-inflammatory agents. J Nat Prod 76(8):1393–1398
50. Roy SK, Bairwa K, Grover J, Srivastava A, Jachak SM (2013) Analysis of flavonoids and Iridoids in Vitex negundo by HPLC-PDA and method validation. Nat Prod Commun 8(9):1934578X1300800914
51. Jachak S, Selvam C, Srivastava A, Ahuja V (2007) Evaluation of anti-inflammatory activity and identification of bioactive compounds from Vitex negundo L., Cardiospermum halicacabum L. and Tridax procumbens L. Planta Med 73(9):8
52. Mecina GF, Chia MA, Cordeiro-Araújo MK, do Carmo Bittencourt-Oliveira M, Varela RM, Torres A et al (2019) Effect of flavonoids isolated from Tridax procumbens on the growth and toxin production of Microcystis aeruginos. Aquat Toxicol 211:81–91
53. Pandey N, Tripathi YB (2010) Antioxidant activity of tuberosin isolated from Pueraria tuberose Linn. J Inflamm 7(1):1–8
54. Jin SE, Son YK, Min B-S, Jung HA, Choi JS (2012) Anti-inflammatory and antioxidant activities of constituents isolated from Pueraria lobata roots. Arch Pharm Res 35:823–837
55. Ingkaninan K, Ijzerman A, Taesotikul T, Verpoorte R (1999) Isolation of opioid-active compounds from Tabernaemontana pachysiphon leaves. J Pharm Pharmacol 51(12):1441–1446
56. Ringbom T, Huss U, Stenholm Å, Flock S, Skattebøl L, Perera P et al (2001) Cox-2 inhibitory effects of naturally occurring and modified fatty acids. J Nat Prod 64(6):745–749
57. Lee I-S, Ma X, Chai H-B, Madulid DA, Lamont RB, O'Neill MJ et al (1995) Novel cytotoxic labdane diterpenoids from Neouvaria acuminatissima. Tetrahedron 51(1):21–28
58. Haslam E (1989) Plant polyphenols: vegetable tannins revisited: CUP Archive
59. Banthorpe DV (1991) Classification of terpenoids and general procedures for their characterization. Methods Plant Biochem 7:1–41
60. Coll JC, Bowden BF (1986) The application of vacuum liquid chromatography to the separation of terpene mixtures. J Nat Prod 49(5):934–936
61. Sarker SD, Latif Z, Gray A (2006) An introduction to natural products isolation. Methods Mol Biol 864:1–25
62. Wall ME, Taylor H, Ambrosio L, Davis K (1969) Plant antitumor agents III: a convenient separation of tannins from other plant constituents. J Pharm Sci 58(7):839–841
63. Takla SS, Shawky E, Hammoda HM, Darwish FA (2018) Green techniques in comparison to conventional ones in the extraction of Amaryllidaceae alkaloids: best solvents selection and parameters optimization. J Chromatogr A 1567:99–110
64. Duan L, Dou L-L, Guo L, Li P, Liu E-H (2016) Comprehensive evaluation of deep eutectic solvents in extraction of bioactive natural products. ACS Sustain Chem Eng 4(4):2405–2411

Drug Substance/Product Quality Analysis (Quality Assessment)

11

Dilip Kumar Singh and Saranjit Singh

Abstract

Multiple analytical studies are required on drug substances/products to maintain the quality during their lifecycle. One key regulatory change that has happened over the last few years is a heightened focus on the analysis of micro/trace components, which include not only synthetic impurities and degradation products but also metabolites, drug remnants in environment samples, drugs as adulterants, etc. Strategies have been proposed in the literature for the characterization of each type, and owing to their low concentrations, emphasis is on the use of sophisticated hyphenated instruments. While the pharmaceutical industry is duty bound to carry out the desired analyses, the regulatory directives also offer a good opening for research in academia. The objective of this chapter is to highlight the regulatory requirements for the characterization of the micro/trace components, to discuss the practical steps and protocols involved, and to provide a detailed discussion of the opportunities for academic scientists.

D. K. Singh
Department of Pharmaceutical Sciences, Washington State University, Spokane, WA, USA

S. Singh (✉)
Department of Pharmaceutical Analysis, National Institute of Pharmaceutical Education and Research, S.A.S. Nagar, Mohali, India
e-mail: ssingh@niper.ac.in

Keywords

Pharmaceuticals · Quality assessment · Micro/trace analysis · Nitrosamine impurity · Drug substance · Protocols

11.1 Introduction

The discovery to development to market is a long journey traversed by a new molecule. To make sure that a quality drug product of a new drug reaches the hands of caregivers and patients, there is a requirement for a string of analytical activities (Table 11.1).

Over the years, the demand for drug substance and drug product analysis has grown tremendously, as regulatory expectations have become expanded and stringent. Rather the whole success of regulatory approval, especially in the case of complex generics, is guided by an effective analytical characterization program [1, 2]. The methodology and techniques involved during discovery to market journey depend upon kind of the drug, the step of development, the nature of investigation being undertaken, and the type of product(s) chosen for marketing. Fortunately, there have been advancements in instrumentation, whose range has also expanded to cater to every regulatory directive. Most modern instruments are sophisticated, offering high sensitivity, resolution, and throughput. Irrespective of their cost, all innovative companies rely on the

G. Jagadeesh et al. (eds.), *The Quintessence of Basic and Clinical Research and Scientific Publishing*,
https://doi.org/10.1007/978-981-99-1284-1_11

Table 11.1 Analytical activities during new drug substance and product development

1. Spectral data acquisition and structural characterization of new chemical entity
2. Semi-preparatory and preparatory purification, method development for purity determination
3. Characterization of impurities and degradation products, their synthesis/isolation and quantitation
4. Solid-state analysis
5. Bioanalysis, pharmacokinetics, metabolic profiling
6. Pre-formulation studies
7. Assay method development for phase-appropriate formulations, the applicability of compendial quality evaluation and functionality tests
8. Validation of developed methods and equipment qualification (DQ, IQ, OQ, CQ, PQ, CSV, etc.)
9. Stability testing, followed by an analysis of the samples
10. Defining key performance indicators (KPIs) or key quality indicators (KQIs) and setting specifications
11. Analytical methodology transfer
12. Commercial batch testing (starting materials, intermediates, APIs, and finished formulations)
13. Regulatory compliant documentation
14. Implementation of innovative analytical platforms and technologies for automated/continuous manufacturing, if applicable

Table 11.2 Micro/trace components of interest during the life cycle of drug candidates and their products

Drug life cycle	Micro/trace component(s)
Drug candidate synthesis and manufacturing in later stages	Impurities (organic, residual solvents, elemental, genotoxic, cohorts of concerns (nitrosamines, nitrosamine drug substance related impurities (NDSRIs), azido impurities), etc.
Pharmacological, toxicology and clinical research	Proteomics, metabolomics, biomarkers
DMPK investigations	Drugs (initial and clearance stages of PK profile) and metabolites
Formulation development and stability testing	Drug degradation products; and drug-drug (in case of FDCs), drug-excipient, drug-packaging interaction products
Environmental pollution profiling	Drug traces in the environment

best analytical tools available at a given time, so that the final product quality exceeds normative expectations.

It may be pertinent to highlight here an emphasis by International regulatory, which has brought a sea change in quality testing operations in the industry. It is the requirement of qualitative and quantitative micro/trace analysis (Table 11.2), applicable today not only to new drug substances and drug products but also to generics. Pharmaceutical manufacturers often face recall orders, if any impurity (IMP) or degradation product (DP) exceeds the defined limits. Table 11.3 provides a list of types of IMPs, their applicability, and regulatory/compendial requirements issued for them. Since mid-2018, a spate of recalls has happened owing to nitrosamine IMPs, and there is the latest focus and recalls happening owing to nitrosamine drug substance-related IMPs (NDSRIs) and azido IMPs in marketed drugs. One case example of NDSRI is depicted in Fig. 11.1. These IMPs are considered cohorts of concern, because of their carcinogenic and toxicogenic features. The current regulatory directives require proper control of their levels in the final products [3–7]. This chapter delves into the regulatory requirements for the characterization of the micro/trace components, provides a brief on the practical steps involved, and describes in detail the prospects for academic researchers.

Table 11.3 Types of impurities, their applicability, and regulatory guidelines/compendial general chapters for each (restricted to ICH/USFDA/EMA/USP/EP/JP)

Impurity type	Applicability	Major guidelines/compendial chapters
Organic impurities – Starting materials – Byproducts – Intermediates – Degradation products – Reagents, ligands, and catalysts – Geometric and stereoisomers	Drug substances	– ICH, Q3A – USFDA, ANDAs: impurities in drug substances – USP, <1086> impurities in drug substances and drug products – EMA, control of impurities of pharmacopoeial substances – EP, 5.10 control of impurities in substances for pharmaceutical use – JP, <G0-3-172> concept on impurities in chemically synthesized drug substances and drug products
Degradation products Components arising from – Drug degradation – Drug-impurity interaction – Drug-excipient interaction – Drug-excipient impurity interaction – Drug-residual solvent interaction – Degradation product-residual solvent interaction – Drug-microbe interaction – Drug/degradation product-packaging component interaction – Drug-drug and all other possible interactions in fixed-dose combinations	Drug products	– ICH, Q3B – USFDA, ANDAs: impurities in drug products – USP, <1086> impurities in drug substances and drug products – JP, <G0-3-172> concept on impurities in chemically synthesized drug substances and drug products
Residual solvents	Mainly drug substances but also drug products	– ICH, Q3C – USP, <467> residual solvents – EP, 2.4.24 identification and control of residual solvents – EP, 5.4 residual solvents – JP, 2.46 residual solvents
Elemental impurities – Reagents, ligands, and catalysts – Heavy metals or other residual metals – Inorganic salts – Other materials (e.g., filter aids, charcoal)	Drug substances and products	– ICH, Q3D – USP, <232> elemental impurities-limits – USP, <233> elemental impurities-procedures – EP, 2.4.20 determination of elemental impurities – EP, 5.20 elemental impurities – JP, 2.66 elemental impurities
Mutagenic/genotoxic impurities	Drug substances and products	– ICH, M7
Nitrosamines and other cohorts of concern (nitrosamine drug substance related impurities (NDSRIs) or active substance-derived nitrosamines, azido impurities)	Mainly drug substances but also drug products when the said impurities are the result of degradation	– ICH, M7 – USFDA, control of nitrosamine impurities in human drugs – USP, <1469> nitrosamine impurities – EMA, nitrosamine impurities – EMA, questions and answers for marketing authorisation holders/applicants on the CHMP opinion for the article 5(3) of regulation (EC) No 726/2004 referral on nitrosamine impurities in human medicinal

(continued)

Table 11.3 (continued)

Impurity type	Applicability	Major guidelines/compendial chapters
		products – EP, 2.5.42 N-nitrosamines in active substances

ICH International Council for Harmonisation, *USFDA* United States Food and Drug Administration, *EMA* European Medicines Agency, *USP* United States Pharmacopeia, *EP* European Pharmacopoeia, *JP* Japanese Pharmacopoeia, *ANDA* Abbreviated New Drug Application

11.2 The Requirement and Steps Involved in the Characterization of Micro/Trace Components

When an unidentified IMP/DP peak is encountered in a chromatogram, the first requirement is its characterization. Then only comes organizing availability of its standard (usually prepared through synthesis once the structure is known) and eventually its quantitation. The conventional approach to the characterization of any micro/trace component involves isolation/enrichment to enough quantity, which is suitable for mass, nuclear magnetic resonance (NMR) and infrared (IR) spectral analysis. The modern approach focuses on the use of a variety of sophisticated hyphenated techniques (Table 11.4). The benefit is acquisition of spectral data by direct transfer of the peak of interest (with on-line enrichment, as required) to interfaced mass, NMR and IR spectroscopic systems. The hyphenated mass and NMR tools are the mainstay instruments that have been employed for the characterization of even other minor components, like metabolites, biomarkers, etc. Mass spectrometers, which allow tandem mass analysis, are more popular for the quantitation of micro/trace components. Therefore, one finds their mention in regulatory recommendations for the determination of nitrosamines and NDSRIs [8–10]. Table 11.5 lists various kinds of mass tools and their applicability.

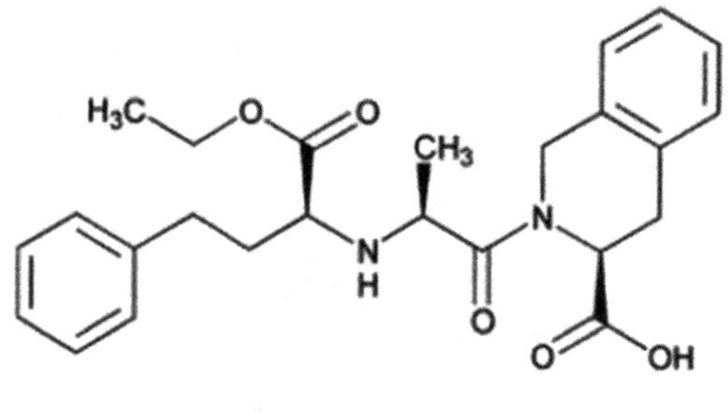

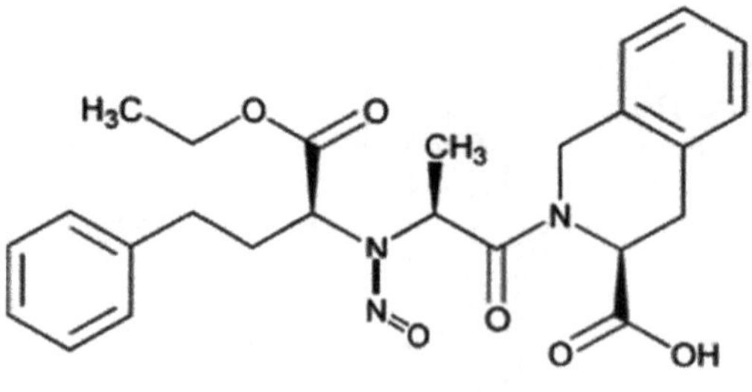

Fig. 11.1 An example of nitrosamine drug substance-related impurity (NDSRI)

While developing a new drug or a generic, the industry ought to perform multiple activities targeted to an analysis of micro/trace components, e.g., (1) their separation from the main constituent(s) on the column or capillary; (2) identification through spiking with pure materials or standards (starting materials, intermediates, reagents, solvents, excipients, etc., as applicable); (3) characterization of unidentified ones through spectral data acquisition; (4) isolation/synthesis of characterized components to best possible purity, or commercial procurement, if the compound of identified structure is pre-known and available; (5) safety evaluation (qualification) through predictive tools, followed by *in vitro* and/or *in vivo* studies; (6) setting of limits and specifications; (7) quantitation and monitoring in laboratory/pilot/ production and routine batch samples, and finally, (8) developing a control strategy. Information

Table 11.4 Variety of sophisticated hyphenated analytical techniques and their role in qualitative and quantitative analysis of the specific category of impurities

Technique	Utility
LC-MS	Separation, mass assessment and quantitation of organic impurities
LC-IR	Separation and IR spectrum recording of organic impurities
LC-NMR	Separation and NMR spectrum recording of organic impurities
CE-MS	Separation, mass assessment and quantitation of organic impurities, including those of chiral nature
CE-NMR	Separation and NMR spectrum recording of organic impurities, including those of chiral nature
HS-GC-MS	Separation, mass assessment and quantitation of residual solvents and volatile impurities
GC-IR	Separation and IR spectrum recording of residual solvents and volatile impurities
ICP-MS	Mass assessment and quantitation of elemental impurities
ICP-OES	Mass assessment and quantitation of elemental impurities

Table 11.5 Various kinds of mass tools and their utility

Variety of LC-MS tools	
LC-MS (single quad) LC-MS (triple quad) LC-MS-TOF LC-MS-Q-TOF LC-MS-TOF-TOF LC-MS-IT (ion trap) LC-MS-Q-IT LC-MS-IT-TOF LC-MS-Orbitrap LC-MS-Q-Orbitrap LC-IT-Orbitrap LC-MS-FTICR LC-MS-Q-FTICR LC-IT-FTICR	
Utility	**MS type used for the purpose**
High resolution mass spectrometry (HR-MS)	TOF, Orbitrap, FTICR
Multiple stage mass spectrometry (MS^n)	Ion trap
Tandem mass spectrometry (MS/MS)	Q-TOF, Q-Orbitrap, Q-IT, IT
Precursor ion, product ion and neutral loss scans	Triple quad
Selected/multiple reaction monitoring (SRM/MRM)	Triple quad
Post-run extracted ion chromatograms	All
Hydrogen/deuterium-exchange mass spectrometry (HDE-MS)	All
Molecular formula generator and RDB calculator	Available with all
Isotopic simulation	Possible with all

and data generated during all these steps are sought by the United States Food and Drug Administration (USFDA) as part of the Chemistry Manufacturing and Control (CMC) dossier, and as relevant, in annual reports. This applies to both New Drug Applications (NDA) and Abbreviated New Drug Applications (ANDA). Based on its experience of missing data in ANDA applications, USFDA has been forced to issue a Refuse to Receive (RTR) mandate for lack of justification of IMP limits in ANDA submissions [11].

11.3 The Opportunity for Academia

If we look into the activity possible in an academic environment, it is mainly the establishment of degradation chemistry of drugs through stress testing approach; or otherwise characterization of metabolites, and establishing the fate of a particular drug from the perspective of environmental pollution. Another kind of study that can be pursued in academia is a survey of multi-source drug substances for relative IMPs originated during synthesis, their extents, and the type and extent of DPs present in multi-source drug products. As the drug degradation profile is intrinsic to drug structure and doesn't vary with the manufacturing route, a well-investigated study in literature, which reports drug degradation behaviour under a variety of extrinsic and intrinsic factors, including temperature, humidity, light, oxidation, pH, etc., along with degradation route and mechanisms in each condition, is straight-way useful to all world-wide generic manufacturers of that drug. Certain regulatory agencies mention that there is no need for stress testing by individual generic manufacturers if a good degradation behaviour study has been reported in the literature [12, 13]. The same is the case with metabolite identification studies, and residue analysis of drugs and their remnants in environmental matrices, which are also intrinsic. Practically, the same set of tools finds application in all these mentioned studies. However, it shall be noted that in an academic environment, one can only take projects on generic drugs, as an innovator involved in new drug development will not easily share the newly discovered molecule with academia, provided confidentiality concerns are well settled in advance.

It may be pertinent to add that projects on simple method development and validation for separation of the above enumerated type of components are no more considered challenging unless the investigation involves the characterization of unidentified components by involving relevant tools.

The strategies/protocols for the characterization of IMPs/DPs/metabolites/drug remnants in environmental samples using sophisticated hyphenated tools have been proposed in the literature, including several from our laboratories. There are many firsts to our credit, like we proposed a guideline for stress testing on drugs [14], published a critical review on the establishment of stability-indicating assay methods [15], outlined the process for the characterization of IMPs and DPs using hyphenated tools [16, 17], and offered a comprehensive strategy for metabolite identification during drug discovery and development based on 'high-quality throughput using minimum resources' approach [18]. Also, we laid down a systematic strategy for the identification and determination of pharmaceuticals in the environment at trace levels [19]. An example of our survey investigation has been the screening of herbal healthcare products for adulteration of PDE-5 inhibitors [20], for which we employed the strategy, reported by us in a separate publication [21]. Another survey study encompassed the evaluation of the presence of 25 steroidal and non-steroidal anti-inflammatory drugs in 58 herbal healthcare products collected from various parts of the country [22].

If one critically evaluates the strategy/protocols given in the referenced texts, it will be found that while the analyses part almost remains the same for both characterization and quantitation, the main difference is in the nature of the sample, and hence the sample preparation. It is simple solubilization of a drug substance to optimal concentration in the mobile phase when the target is IMP/DP analysis. This is even the case of forced degradation studies, where the drug is dissolved in the stressor solution, and the prepared samples are subjected to pre-fixed forced degradation conditions, like high temperature and/or humidity, light, oxidative environment, etc., and then diluted/neutralized before analysis. More experimental details can be found in our guidance paper [14]. For drug products, the best way is to follow the sample preparation method and procedure (based on formulation type) suggested in pharmacopeial monographs under the related substance test. The same procedure can even be employed when a particular formulation of a new drug is being developed and the interest is to check for DPs in samples placed on stability. Before and after the analytical run, help can also be taken from Zeneth, which is asserted to be an expert knowledge-based software that quickly yields accurate forced degradation predictions [23]. The software is even claimed to help determine the chemical structures of DPs detected by a mass spectrometer, or other detection methods, and to deduce the likely degradation pathways. A range of filters can be applied to provide a results tree, which is said to be consistent with the experimental findings. Figure 11.2 briefly outlines the activities related to stress testing experiments. The workflow includes *in cerebro* and/or *in silico* prediction of hypothetical DPs, the conduct of experimental stress testing, development of stability-indicating methods, characterization of potential unidentified DPs, establishment of degradation pathway and mechanism of degradation, as well as *in silico* toxicity prediction of each characterized DP. Strategies to control non-mutagenic and mutagenic DPs are also included.

Nowadays, Quality by Design (QbD) is a well-established systematic approach, which is used by pharmaceutical companies to control the DPs. ICH Q11 emphasizes that a control strategy is required for all drug substances and/or existing drug products to limit the levels of DPs within the given acceptance criteria [24]. Also, risk management of DPs and scientific knowledge, such as understanding their formation, fate, and purge (whether the DP is removed *via* stabilization strategy) is considered to be important. Integration of technology, chemistry, risk management, design space, and control tools is required during the manufacturing of the drug substance, whenever a particular DP, characterized during stress studies and formed during processing, is considered a critical quality attribute (CQA) [25]. At the same time, attention is also paid to the stability of the reaction mass to understand the formation of critical DP due to the presence of a residual reagent. The latter is then also considered as a CQA, along with the DP. The IND/NDA stability studies need to be repeated if the manufacturing process with respect to intermediate or reagent is changed markedly.

In the case of drug products also, apart from conventional formulation stabilization approaches, involving the use of stabilizers, excipients and protective packaging, the roles of technology, chemistry, risk management, manufacturing design space, and control tools have assumed importance to keep a check on the DPs [25, 26]. IMPs in the excipients also have attracted significant attention as these may catalyze the degradation of a drug to a pre-characterized DP [27]. Interestingly, a new focus is on the stress testing of excipients per se [28].

Overall, a big problem for the industry is to predict the exact level of DP, whenever it is considered a CQA, so as to avoid recalls from the market in the future. However, the recently launched software, *viz.*, ASAPprime® and Mirabilis allow companies to make better decisions in such situations early in the development process [29, 30]. These also help in quick reformulation and choosing the optimal combination of packaging, formulation, ingredients, and manufacturing process.

For metabolite identification (metID) studies, the metabolites are generated in either *in vitro* systems (microsomes, S9 fractions, hepatocytes, and recombinant enzymes), or *in vivo* where plasma and excreta are the major sample types. The target of sample preparation herein is to obtain concentrated samples free or almost free of biological matrix. This is usually achieved through protein precipitation and solid-phase extraction (SPE). In one of our studies, we employed a novel additional step of freeze-liquid separation to reduce the loss of polar analytes due to the overloading of SPE cartridges [31]. For metabolite characterization, there is an advantage that LC-MS manufacturers provide *in silico* tools, as part of a software bundle, for both prediction and detection of the metabolites. Usually, the prediction software foretells the biotransformation of any molecule by considering all mammalian phase I and phase II enzyme systems. The predicted structures are listed along with the accurate theoretical mass of their protonated and unprotonated species. The metabolite detection software, when supplied with the LC-HRMS system, facilitates the matching of the accurate mass of metabolites eluted, with those predicted. The software-predicted/detected metabolites and those additionally observed in total ion and UV chromatograms are confirmed using the accurate mass values, ring plus double bonds (RDBs)

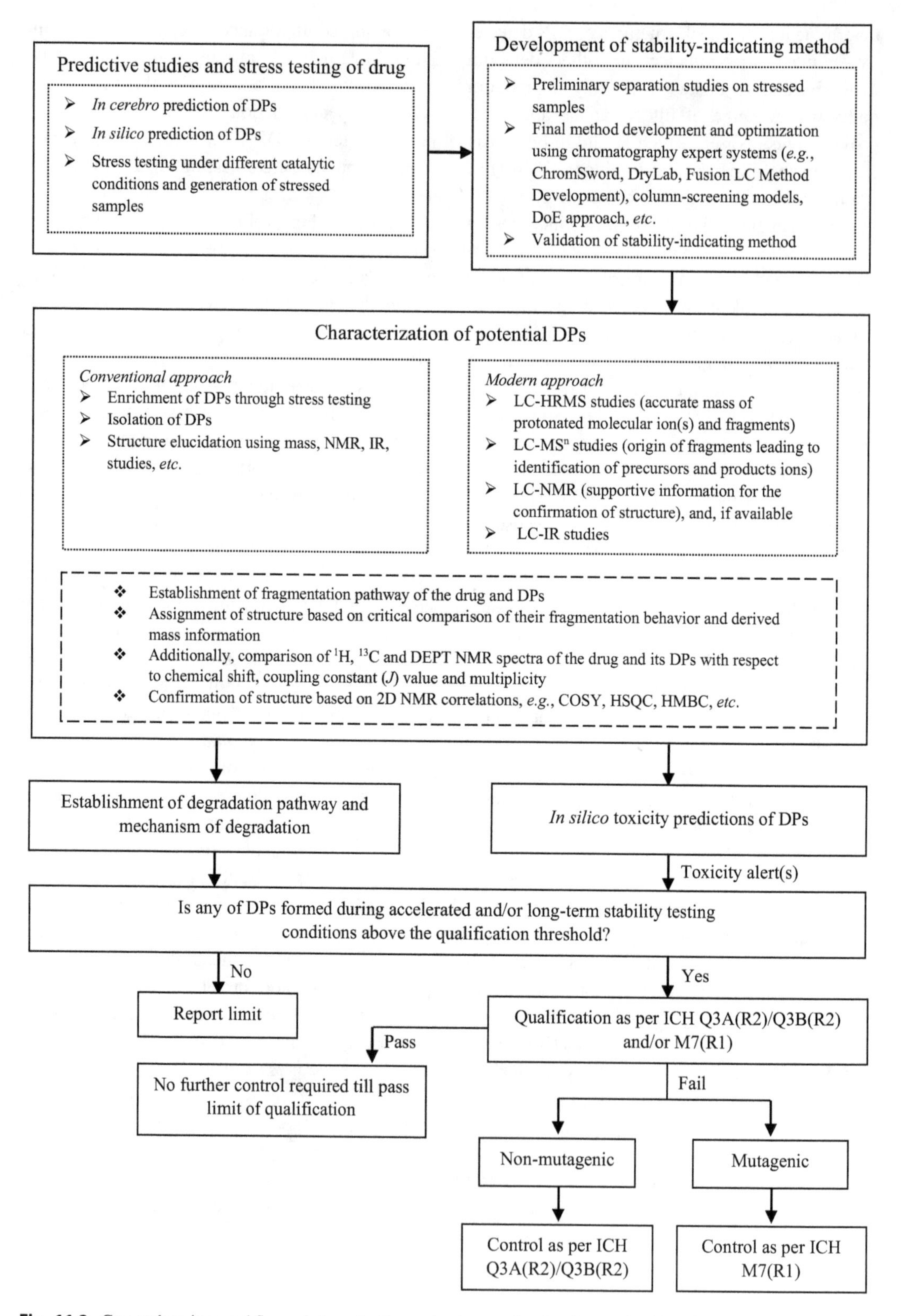

Fig. 11.2 Comprehensive workflow of stress testing and other studies leading to the identification, characterization, toxicity evaluation, and control of degradation products (DPs)

calculations, application of nitrogen rule, and determination of exact mass losses. The site of change in the drug structure because of metabolism is identified through a comparison of MS fragmentation pattern of each metabolite with that of the drug. After structures of metabolites detected in different *in vitro* and *in vivo* samples are elucidated, extracted ion chromatograms (EICs) of the individual metabolites are evaluated to determine their relative amounts.

Pharmaceuticals are produced or consumed in large amounts globally and it is no surprise that environmental pollution with them is rising progressively. Whether consumed or disposed of without use (due to any reason), the potent drug molecules eventually make their way into the environment and may persist as contaminants, either in an intact form, as a DP, or as a metabolite of the parent. Interest in them has got kindled because of tremendous progress in the analytical techniques for trace analysis. Reports exist on the effects of pharmaceutical contaminants on aquatic flora and fauna but long-term eco-toxicological consequences, especially to humans, are still unmapped [32]. There are a few basic questions that one needs to answer before taking up analytical research projects in this area, owing to the vastness of the scope. The first question to answer is - what is the purpose of the study? Is it exploration of major pharmaceutical pollutants (e.g., industrial discharge in lake/water body), or micro/trace level contaminants (e.g., discharge from hospitals/households into sewage). Secondly, whether the interest is in the detection of targeted pollutants only, or to characterize all those present, and so on. The defined purpose hence lays down the extensiveness of subsequent study actions. For example, in our laboratory, we narrowed down the scope to evaluate the presence of residues of forty commonly prescribed drugs in ground drinking water in villages surrounding our institute [19]. The expected concentrations were at micro/trace level, so the first part of sample collection was procuring a minimum required water quantity. The analyte enrichment was the next key step, for which SPE was the chosen method. For identification of the drugs present, the enriched samples were subjected to LC-MS/MS analysis. The identification strategy included matching of retention times against the standards, comparison of MRM transitions, matching of qualifier to quantifier intensity ratio (as suggested by the Environment Protection Agency), comparison of base peak MS/MS profile, and comparison of accurate mass data. Eventually, quantification was done through the use of calibration curves developed using quantifier MRM transitions.

As mentioned earlier, the survey testing done in our laboratory was targeted at observing the adulteration of herbal dietary supplements (HDSs) with synthetic drugs, like phosphodiesterase type-5 (PDE-5)-inhibitors (viz., sildenafil, vardenafil and tadalafil), and multiple steroidal and non-steroidal anti-inflammatory agents. It came to our notice that unapproved analogues of all PDE inhibitor drugs were being found in HDSs, and more seriously, concealed, structurally modified analogues were also being used increasingly. As many of these adulterants are unknown, the likelihood exists of much higher associated risk, because their effects and side effects are not known pre-hand. So, we focused to build a strategy to help identify, not only the approved drugs but also their known and unknown derivatives. This was made possible through a critical study of the reported mass fragmentation behaviour of the drugs and their known derivatives. We could identify one or two common mass fragments, which if observed in the mass spectrum of any peak in the mass chromatogram of the sample, would mean a strong likelihood of an analogue of any one of the PDE inhibitors. In the case of adulteration of AHPs with steroidal and non-steroidal anti-inflammatory agents, the strategy was simpler because of the availability of all involved standard drugs. The study primarily involved comparing ultraviolet and mass spectral data of the standard with similar data for unknown peak(s), comparison of retention time values, and final confirmation through spiking of standard(s) in the sample [22].

We exemplify in Fig. 11.3 a modified strategy/protocol for the characterization of IMPs and DPs over and above the one reported by us earlier [17]. However, it must be understood that, apart

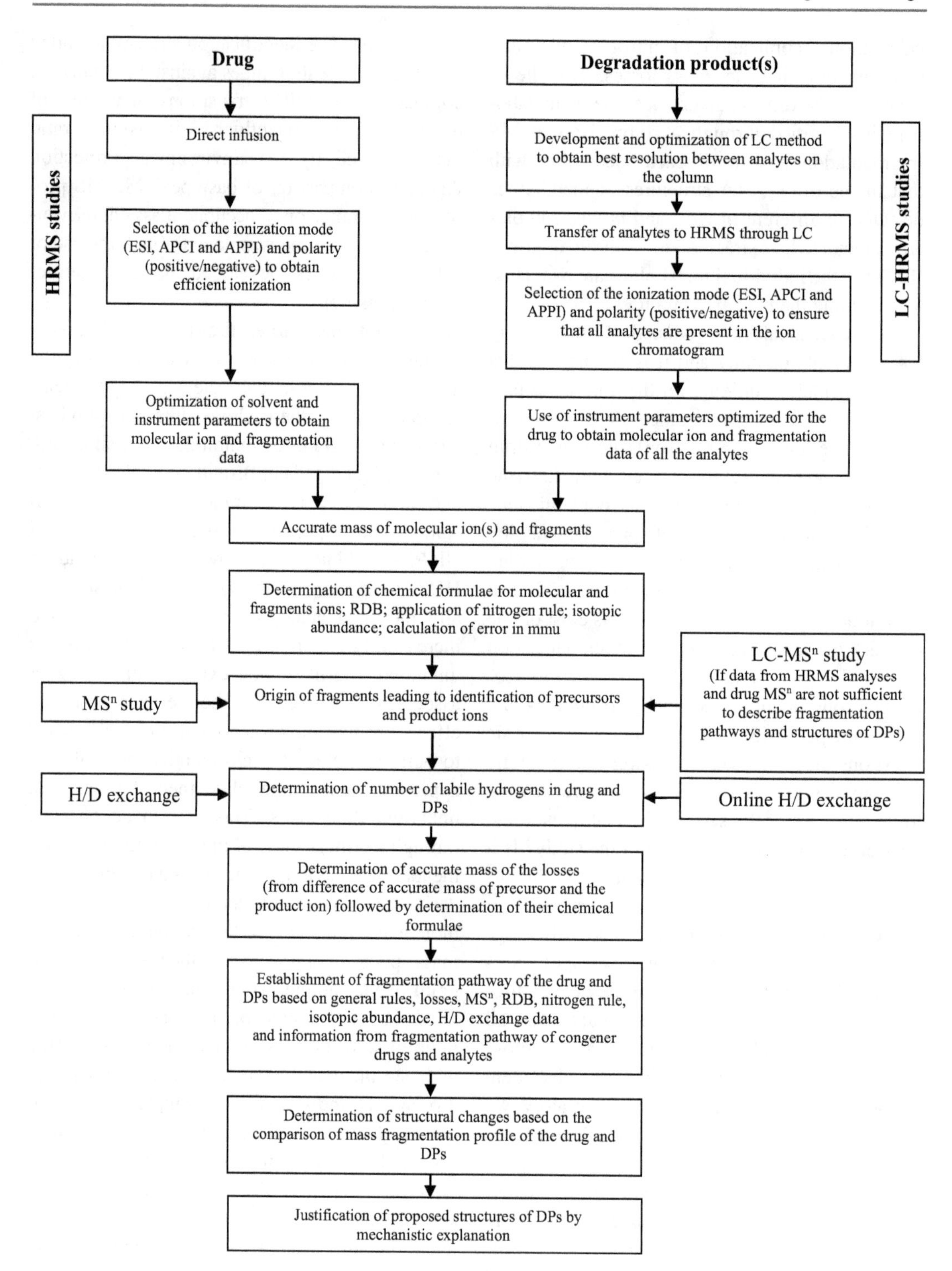

Fig. 11.3 General strategy for the characterization of degradation products (DPs) by LC-MS tools. Duly modified after adaption from [17] with due permission

from a strategy/protocol, there are a lot of practical intricacies involved while carrying out experimental work to generate useful data for the characterization of micro/trace components using sophisticated hyphenated instruments. One can find extensive discussion on these in our published reviews pertaining to data generation using LC-MS [17] and LC-NMR [33]. The nature of sensitivity needed even governs the purchase of instruments for the purpose among various types and models available with the vendors. For example, one can buy simple LC as front-end, or instead UHPLC, or capillary/nano LC systems for much lower analyte concentrations. Similar is the situation with back-end MS and NMR detectors, wherein models are available with ever-improving resolution and sensitivity.

11.4 Concluding Remarks

There is a big advantage of pursuing research in the areas of identification and characterization of micro/trace components, produced either upon transformation (including biotransformation) or when present as contaminants or adulterants. This precursor step is critical to the quantitative assessment of micro/trace components in actual samples, and for exercising controls to comply with stringent regulatory limits. The matter is of deep regulatory interest, as it is the responsibility of regulators worldwide to ensure the availability of high-quality and high-purity products to patients globally.

It is to be acknowledged that the eventual success of the mentioned effort requires a thorough understanding and knowledge of all aspects involved in the steps of planning, execution, and data interpretation. This chapter provides references to resources that can be referred to for the conduct of successful experiments, and to arrive at acceptable inferences. It highlights the nature of investigations possible to be undertaken by scientists in academia.

Author's Declaration This research did not receive any specific grant from funding agencies in the public, commercial, or not-for-profit sectors. The views expressed in this chapter are those of the authors, and do not represent the official positions of their organizations. The list of the authors' organizations merely represents their current (DKS) or past affiliations (SS).

Conflict of Interest None.

References

1. https://grantsforus.io/type-of-eligible-entity/organizations/nonprofits-with-501c3/apply-for-the-development-of-advanced-analytical-methods-for-the-characterization-of-complex-generics-grant-program/. Accessed 29 Nov 2022
2. https://www.fda.gov/media/108937/download. Accessed 29 Nov 2022
3. https://www.fda.gov/regulatory-information/search-fda-guidance-documents/control-nitrosamine-impurities-human-drugs. Accessed 29 Nov 2022
4. https://www.fda.gov/drugs/drug-safety-and-availability/updates-possible-mitigation-strategies-reduce-risk-nitrosamine-drug-substance-related-impurities. Accessed 29 Nov 2022
5. https://www.tga.gov.au/news/safety-alerts/azide-impurity-sartan-blood-pressure-medicines#:~:text=What%20is%20the%20azide%20impurity,individual's%20risk%20of%20developing%20cancer. Accessed 29 Nov 2022
6. https://www.edqm.eu/en/-/risk-of-the-presence-of-mutagenic-azido-impurities-in-losartan-active-substance. Accessed 29 Nov 2022
7. https://rhelaw.com/class-action/irbesartan-losartan-and-valsartan-azido-impurity-class-action/. Accessed 29 Nov 2022
8. https://www.fda.gov/media/130801/download. Accessed 29 Nov 2022
9. https://www.fda.gov/media/138617/download. Accessed 29 Nov 2022
10. https://www.ema.europa.eu/en/documents/referral/nitrosamines-emea-h-a53-1490-questions-answers-marketing-authorisation-holders/applicants-chmp-opinion-article-53-regulation-ec-no-726/2004-referral-nitrosamine-impurities-human-medicinal-products_en.pdf. Accessed 29 Nov 2022
11. https://www.fda.gov/files/drugs/published/ANDA-Submissions-%E2%80%94-Refuse-to-Receive-for-Lack-of-Justification-of-Impurity-Limits.pdf. Accessed 29 Nov 2022
12. https://www.ema.europa.eu/en/documents/scientific-guideline/guideline-stability-testing-stability-testing-existing-active-substances-related-finished-products_en.pdf. Accessed 29 Nov 2022
13. https://www.who.int/publications/m/item/who-guidelines-on-stability-testing-of-active-

pharmaceutical-ingredients-and-finished-pharmaceutical-products. Accessed 29 Nov 2022
14. Singh S, Bakshi M (2000) Guidance on conduct of stress tests to determine inherent stability of drugs. Pharm Tech Online 24:1–14
15. Bakshi M, Singh S (2002) Development of validated stability-indicating assay methods–critical review. J Pharm Biomed Anal 28(6):1011–1040. https://doi.org/10.1016/s0731-7085(02)00047-x
16. Singh S, Handa T, Narayanam M, Sahu A, Junwal SRP (2012) A critical review on the use of modern sophisticated hyphenated tools in characterization of impurities and degradation products. J Pharm Biomed Anal 69:148–173. https://doi.org/10.1016/j.jpba.2012.03.044
17. Narayanam M, Handa T, Sharma P, Jhajra S, Muthe PK, Dappili PK, Shah RP, Singh S (2014) Critical practical aspects in the application of liquid chromatography-mass spectrometric studies for the characterization of impurities and degradation products. J Pharm Biomed Anal 87:191–217. https://doi.org/10.1016/j.jpba.2013.04.027
18. Prasad B, Garg A, Takwani H, Singh S (2011) Metabolite identification by liquid chromatography-mass spectrometry. Trends Anal Chem 30(2):360–387. https://doi.org/10.1016/j.trac.2010.10.014
19. Jindal K, Narayanam M, Singh S (2015) A systematic strategy for the identification and determination of pharmaceuticals in environment using advanced LC–MS tools: application to ground water samples. J Pharm Biomed Anal 108:86–96. https://doi.org/10.1016/j.jpba.2015.02.003
20. Savaliya AA, Shah RP, Prasad B, Singh S (2010) Screening of indian aphrodisiac ayurvedic/herbal healthcare products for adulteration with sildenafil, tadalafil and vardenafil using LC/PDA and extracted ion LC-MS/TOF. J Pharm Biomed Anal 52(3):406–409. https://doi.org/10.1016/j.jpba.2009.05.021
21. Singh S, Prasad B, Savaliya A, Shah RP, Gohil VM, Kaur A (2009) Strategies for the characterization of sildenafil, vardenafil, tadalafil and their analogues in herbal preparations, and detection of counterfeit products containing these drugs. Trends Anal Chem 28(1):13–28. https://doi.org/10.1016/j.trac.2008.09.004
22. Savaliya AA, Prasad B, Raijada DK, Singh S (2009) Detection and characterization of synthetic steroidal and non-steroidal anti-inflammatory drugs in Indian ayurvedic/herbal products using LC-MS/TOF. Drug Test Anal 1(8):372–381. https://doi.org/10.1002/dta.75
23. https://www.lhasalimited.org/products/zeneth.htm
24. Development and manufacture of drug substances (chemical entities and biotechnological/biological entities) Q11. International Conference on Harmonisation, Geneva (Switzerland), 2012
25. Davis B, Lundsberg L, Cook G (2008) PQLI control strategy model and concepts. J Pharm Innov 3:95–104. https://doi.org/10.1007/s12247-008-9035-1
26. Hulbert MH, Feely LC, Inman EL et al (2008) Risk management in the pharmaceutical product development process. J Pharm Innov 3:227–248. https://doi.org/10.1007/s12247-008-9049-8
27. Wasylaschuk WR, Harmon PA, Wagner G et al (2007) Evaluation of hydroperoxides in common pharmaceutical excipients. J Pharm Sci 96:106–116. https://doi.org/10.1002/jps.20726
28. https://ipecamericas.org/sites/default/files/ExcipientQualificationGuide.pdf
29. Burns MJ, Ott MA, Teasdale A et al (2019) New semi-automated computer-based system for assessing the purge of mutagenic impurities. Org Process Res Dev 23:2470–2481. https://doi.org/10.1021/acs.oprd.9b00358
30. Li H, Nadig D, Kuzmission A, Riley CM (2016) Prediction of the changes in drug dissolution from an immediate-release tablet containing two active pharmaceutical ingredients using an accelerated stability assessment program (ASAPprime®). AAPS Open 2:1–9. https://doi.org/10.1186/s41120-016-0010-5
31. Prasad B, Singh S (2009) *In vitro* and *in vivo* investigation of metabolic fate of rifampicin using an optimized sample preparation approach and modern tools of liquid chromatography-mass spectrometry. J Pharm Biomed Anal 50:475–490. https://doi.org/10.1016/j.jpba.2009.05.009
32. Hejna M, Kapuscinska D, Aksmann A (2022) Pharmaceuticals in the aquatic environment: a review on eco-toxicology and the remediation potential of algae. Int J Environ Res Public Health 19:7717. https://doi.org/10.3390/ijerph19137717
33. Sahu A, Balhara A, Singh DK, Kataria Y, Singh S (2019) NMR spectroscopy, techniques, LC-NMR and LC-NMR-MS. In: Worsfold P, Poole C, Townshend A, Miró M (eds) Encyclopedia of analytical science, vol 7, 3rd edn. Elsevier, Amsterdam, pp 220–247. https://doi.org/10.1016/B978-0-12-409547-2.14074-0

Drug Delivery Systems: Lipid Nanoparticles Technology in Clinic

12

Saloni Andhari, Rituja Gupta, and Jayant Khandare

Abstract

Drug delivery systems (DDS) aid the administration of therapeutic cargoes to desired tissues to evoke a pharmacological response with minimal adverse effects. DDS are associated with the delivery of small molecules, proteins, and nucleic acids. The chemical nature and the architecture of the delivery systems facilitate these therapeutic agents to achieve the most efficient pharmacologic responses. In this setting, lipids are used to deliver hydrophobic or hydrophilic drugs in self-assembled vesicles. Along with the utilization of cationic lipids, delivery of nucleic acids on account of charged-based interactions has gained momentum. As the field evolves, ionizable lipids are being synthesized which aid the fabrication of lipid nanoparticles (LNPs) in providing a protective environment for delivery of nucleic acids by avoiding enzymatic degradation. Progression in the development of synthetic lipids has led to the regulatory approval of mRNA vaccines, developed to provide immunity, by utilizing LNPs as delivery systems. Moreover, ionizable biodegradable lipids present immense opportunities for *in vivo* delivery of tools for genome engineering. This chapter highlights the biological importance of lipids, recent developments in the delivery of a variety of therapeutic cargoes using LNPs, types of lipids employed in fabrication and the methodologies for synthesis. Finally, the chapter details the pharmacokinetics and pharmacodynamics achieved using LNPs in advanced therapeutics.

Keywords

Lipid nanoparticles · Drug delivery · Ionizable lipids · mRNA · Liposomes · Nanotherapeutics

12.1 Introduction

The inherent amphiphilic nature of lipids renders them suitable candidates for self-assembly. This trait was first explored for the delivery of drug molecules such as doxorubicin, daunorubicin, amphotericin B and others, using self-assembled

S. Andhari
MAEER's Maharashtra Institute of Pharmacy, Pune, Maharashtra, India
e-mail: saloni.andhari@mippune.edu.in

R. Gupta
School of Pharmacy, Dr. Vishwanath Karad MIT World Peace University, Pune, Maharashtra, India
e-mail: rituja.gupta@mitwpu.edu.in

J. Khandare (✉)
School of Pharmacy, Dr. Vishwanath Karad MIT World Peace University, Pune, Maharashtra, India

School of Consciousness, Dr. Vishwanath Karad MIT World Peace University, Pune, India

Actorius Innovations and Research, Pune, Maharashtra, India and USA
e-mail: jayant.khandare@mitwpu.edu.in; jayant@actorius.co.in

G. Jagadeesh et al. (eds.), *The Quintessence of Basic and Clinical Research and Scientific Publishing*,
https://doi.org/10.1007/978-981-99-1284-1_12

vesicles referred to as the "liposomes" [1]. The approval of the first liposomal drug formulation developed by Gabizon and Barenholz set the stage for the industrial boom in the field of LNPs technology in drug delivery [2]. Liposomes and different types of LNPs have been reported for delivery of small molecules, peptides, nucleic acids as well as gases such as nitric oxides and most recently cannabidiol [3–8].

The first lipid-based formulation was approved by the FDA in 1995 for the administration of a small anti-cancer drug molecule (doxorubicin), while complex nucleic acids delivery for vaccination was provided with emergency approval in 2020 [9]. The need to develop newer and more versatile lipids along with the need to mass produce LNPs gave rise to multiple companies including Lipex Biomembranes Inc., Northern Lipids Inc., Precision NanoSystems, Moderna Inc. and BioNTech SE. Therefore, LNPs currently present an extremely promising class of delivery systems and the sheer number of research articles being published, patents being filed and the sprouting of new companies in the allied fields makes them a noteworthy clinically implicated DDS. The key objectives of the chapter are to (a) delineate the evolution of various types of LNPs over the years, (b) discuss the chemical compositions of LNPs, (c) describe the synthetic approaches and (d) highlight the clinical implications of LNPs.

12.2 Biological Importance of Lipids

Lipids are amphipathic molecules comprising hydrophobic and hydrophilic constituents, whose existence can be verified all around the living world such as in microorganisms, fungi, higher plants and animals [10]. They play a role in the formation of cellular structure, act as energy storage molecules and are involved in many biological processes, including transcription of genes, regulation of metabolic pathways as well as physiological responses [11]. A new definition for lipids based on their structural origin has been put forth by the consortium of lipid metabolites and pathways strategy (Lipid MAPS), according to which, lipids are small molecules with hydrophobic/amphipathic characteristics, which have been generated in entirety or partly, by condensation (carbanion catalyzed) reactions of thioesters and/or by condensation (carbocation catalyzed) reactions of isoprene units [11].

The chemical structure of phospholipids may be divided into the head, core and tail region representing various chemical components (Fig. 12.1). It has been established that lipids play a crucial role by being the backbone of biological membranes as the bi-layered lipid membrane separates cells from the external environment and also compartmentalizes the cells thus providing a special milieu for multiple important biochemical processes [12]. These membranes even serve as a significant matrix to promote the transmembrane protein function. Furthermore, they significantly advance the function of lipid second messengers whilst signal transduction.

12.3 Types of Lipid Nanoparticles

12.3.1 Liposomes

The term 'liposome' was introduced in 1960 and has attracted great interest from the scientific community worldwide [13, 14]. Liposomes are nanosized vesicles formed due to the spontaneous self-assembly of amphipathic molecules in an aqueous milieu. These closed lipid bilayer structures, ranging in the size between 20 and 1000 nm, consist of a hydrophilic core and a hydrophobic corona.

Phospholipids namely phosphatidylcholines, phosphatidylethanolamines, phosphatidylserines, and phosphatidylglycerols, accompanied by stabilizers such as cholesterol, comprise the main structural components of the liposomes [15]. Depending upon the physicochemical properties of the drugs, they can either be encapsulated in the aqueous volume inside or intercalated in the lipophilic fatty acid chains found at the periphery of the liposomes (Fig. 12.2). Since liposomes are majorly

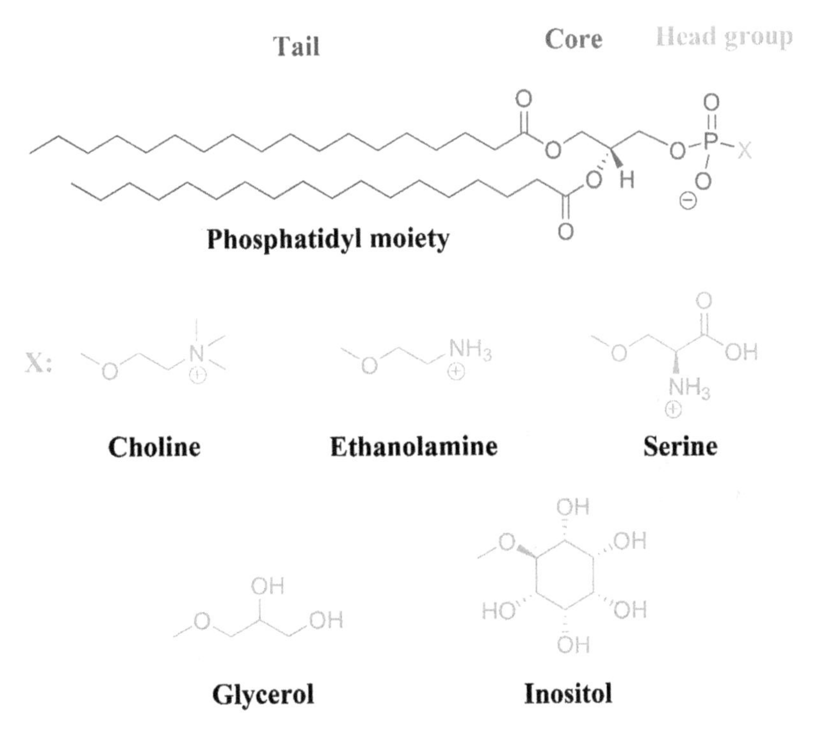

Fig. 12.1 Chemical structures of various naturally occurring phospholipids. The phosphatidyl moiety comprises of the tail and core and head groups which determine the charge of the phospholipid

comprised of lipids and are nanosized in nature, they are duly considered to be the nascent stage of LNPs [17].

DOXIL® and MYOCET® are amongst the first liposomal doxorubicin formulations to be approved by the FDA for commercial use (Fig. 12.2) [2]. DOXIL® and MYOCET® are synthesized using remote/active loading of doxorubicin using ammonium sulphate and citrate gradients to yield respective salt crystals (Fig. 12.3) [18]. The chemical composition of DOXIL®, a PEGylated liposomal preparation, includes hydrogenated soy phophatidylcholine (HSPC) (9.58 mg/mL), *N*-(carbonyl-methoxypolyethylene glycol 2000)-1,2-distearoyl-sn-glycero-3-phosphoethanolamine sodium (DSPE-PEG-2000) (3.19 mg/mL), cholesterol (3.19 mg/mL), doxorubicin (2 mg/mL, 3.45 mM), ammonium sulfate (2 mg/mL), histidine (10 mM, buffer), and sucrose (10%, for isotonicity) with a total lipid to drug weight ratio of 16:2. The particle size being 90 nm. Similarly, the chemical composition of MYOCET®, a non-PEGylated liposomal preparation, includes egg phosphatidylcholine (5.4 mg/mL), cholesterol (2.2 mg/mL), doxorubicin (2 mg/mL, 3.45 mM), citric acid monohydrate (4.4 mg/mL), sodium carbonate (2.2 mg/mL), lactose (10 mg/mL), and sodium chloride injection (7.2 mg/mL) with a total lipid to drug weight ratio of 7.6:2. The particle size being 150 nm.

In 2018, the US-FDA released a guidance document for liposomal drug products in light of the new approvals being sought after [19]. Various liposomal formulations approved by the FDA are listed in Table 12.1.

Being the most versatile drug delivery platform, liposomes have been reported to undergo several clinical trials with an aim to come up with drugs related to the treatment of cancer, inflammation, anaesthesia, bacterial and fungal infections as well as therapies related to gene manipulation [32]. Liposomes can broadly be classified based on their size and method of preparation, namely, small unilamellar vesicles (SUV), large unilamellar vesicles (LUV) or giant unilamellar vesicles (GUV) and multilamellar vesicles (MLV) [33].

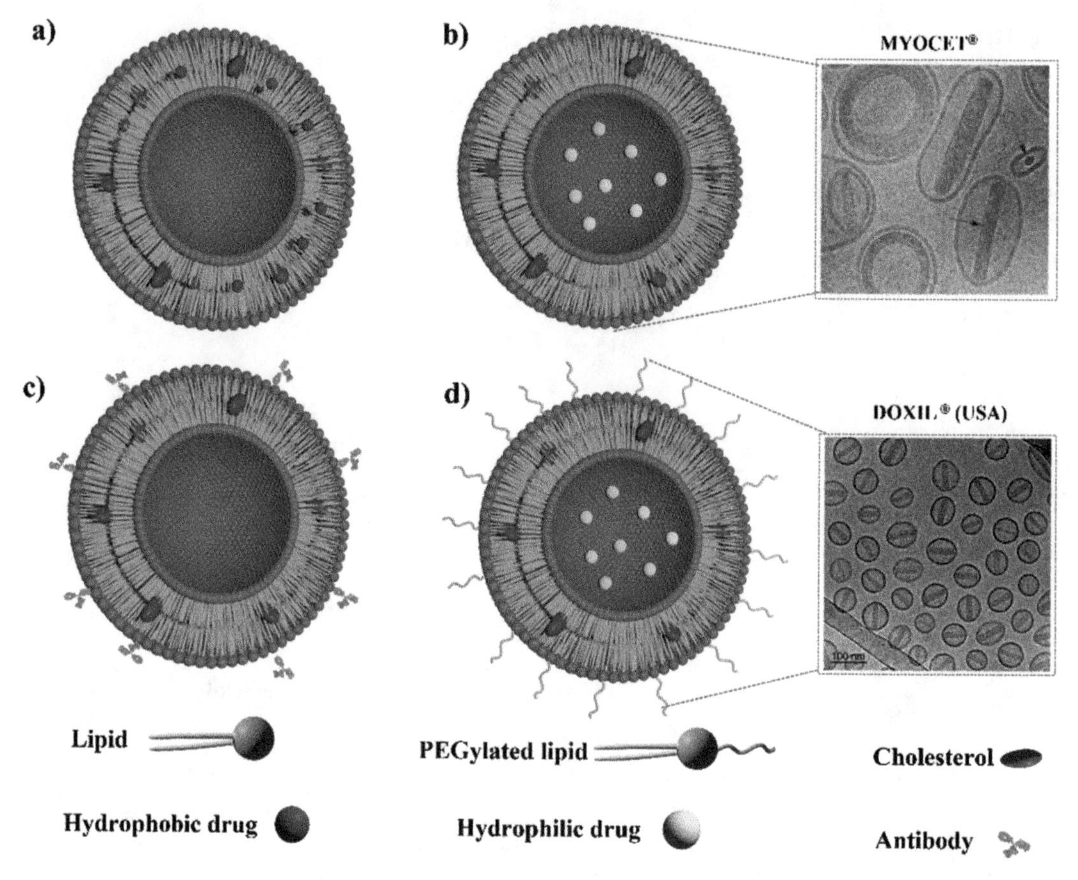

Fig. 12.2 Pictorial representation of various liposomal formulations. (**a**) Hydrophobic drug loaded liposome. (**b**) Hydrophilic drug loaded liposome. The inset represents TEM micrograph of MYOCET®. TEM image reprinted from [16], Copyright © (2001), with permission from Elsevier. (**c**) Immunoliposome. (**d**) Hydrophilic drug loaded PEGylated liposome. The inset represents TEM micrograph of DOXIL®. Scale bar: 100 nm. TEM image reprinted from [2], Copyright © (2012), with permission from Elsevier

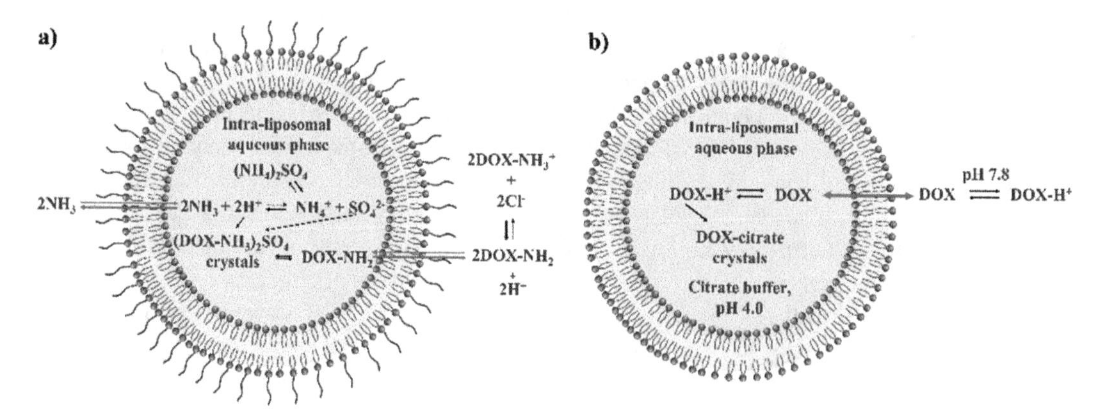

Fig. 12.3 Schematic representation of mechanism of remote loading of doxorubicin in (**a**) DOXIL® and (**b**) MYOCET®. Information adapted from [2, 16], and reproduced here, Copyright © (2012, 2001), with permission from Elsevier

Table 12.1 Clinically approved lipid-based (liposomal) products

S. No.	Products and their APIs	Chemical composition of the lipid bilayer	Approval year	Company	Reference
A.	Liposomes in cancer therapy				
1.	Doxil® (USA)/ Caelyx® (EU) Doxorubicin	HSPC: cholesterol: DSPE-PEG 2000 (56:39:5 % mole ratio)	1995	Sequus Pharmaceuticals Inc., ALZA corporation, Janssen Pharmaceuticals, Johnson & Johnson (USA)	[2, 20]
2.	DaunoXome® Daunorubicin (discontinued)	DSPC: cholesterol (2:1 mole ratio)	1999	NeXstar Pharmaceuticals Inc., (USA)	[21, 22]
3.	Myocet® (TLC D-99) Doxorubicin	EPC: cholesterol (2:1 weight ratio)	2000	Elan Pharmaceuticals (USA), GP Pharm (Spain), Teva Pharmaceuticals (Poland)	[18, 23]
4.	Mepact® Mifamurtide	DOPS: DOPC (3:7 mole ratio)	2004/ 2009	Takeda Pharmaceutical Company Ltd. (Japan)	[24]
5.	Marqibo® (VSLI) Vincristine	SM: cholesterol (60:40 mole ratio)	2012	Talon Therapeutics Inc. (USA)	[25]
6.	Doxorubicin HCl liposome Inj. (Generic of Doxil®) Doxorubicin	HSPC: cholesterol: DSPE-PEG 2000 (56:39:5 % mole ratio)	2013	Sun Pharmaceutical Industries Ltd. (India), Caraco Pharmaceutical Laboratories (USA)	[23]
7.	VYXEOS® Daunorubicin and cytarabine	DSPC: DSPG: cholesterol (7: 2:1 mole ratio)	2017	Jazz Pharmaceuticals Inc. (Ireland, Europe)	[26]
B.	Liposomes for infectious diseases				
8.	Abelcet® (ABLC, lipid complex) Amphotericin B (AmB)	DMPC: DMPG (7: 3 mole ratio)	1995	Sigma-Tau Pharmaceuticals Inc. (USA) (now known as Leadiant Biosciences, Inc., Italy)	[27, 28]
9.	Ambiosome® AmB	HSPC: DSPG: cholesterol (2:0.8:1)	1997	Gilead Sciences Inc. (USA), Astella Pharma Inc. (Japan)	[28, 29]
C.	Liposomes for pain management				
10.	DepoDur® (epidural) Morphine sulphate (discontinued)	DOPC: DPPG: cholesterol: triolein	2004	SkyePharma Plc (France)	[30]
11.	Exparel® Bupivacaine	DEPC: DPPG: cholesterol: tricaprylin	2011	Pacira Pharmaceuticals Inc. (USA)	[31]
D.	Miscellaneous applications				
12.	Visudyne® Verteporfin	DMPC: EPC (1:8 mole ratio)	2000	Novartis (Switzerland and USA)	[28]

API active pharmaceutical ingredient, *PEG* polyethylene glycol, *DSPE* 1,2-distearoyl-*sn*-glycero-3-phosphoethanolamine, *EPC* egg phosphatidylcholine, *DSPC* distearoylphosphatidylcholine, *DOPS* dioleoylphosphatidylserine, *DOPC* dioleoylphosphatidylcholine, *SM* sphingomyelin, *DMPC* dimyristoyl phosphatidylcholine, *DMPG* dimyristoyl phosphatidylglycerol, *DSPG* distearoylphosphatidylglycerol, *DPPG* dipalmitoylphosphatidylglycerol, *DEPC* dierucoylphosphatidylcholine

12.3.2 Solid Lipid Nanoparticles

While liposomes are reported as a versatile drug carrier system, they are accompanied by several shortcomings such as the use of organic solvents, low entrapment efficiency and difficulty in large scale production. A new type of colloidal carrier system known as "solid lipid nanoparticles" (SLNs) has been proposed and found to overcome the mentioned drawbacks with liposomes [15]. Over conventional liposomes, the SLNs are comprised of solid lipids forming a hydrophobic core with a monolayer of the phospholipid coating [34]. The composition of the SLNs comprises lipids and various stabilizing agents such as surface-active agents and their particle size vary between 40 and ~1000 nm.

SLNs offer advantages such as biodegradability, lack of inherent toxicity, stability, reduced drug leakage and do not undergo hydrolysis, and no particle growth. In addition, SLNs are easy to scale up and are stable under common sterilization conditions. Furthermore, SLNs are dispersed in an aqueous medium, have high entrapment efficiency and are capable of releasing drug after a single injection within a few hours or over several days [35].

Various other colloidal drug carrier systems such as liposomes, lipid emulsions, polymeric nanoparticles (NPs) have also been proposed for the controlled delivery of drugs by the intravenous route of administration. Each of these carrier systems offers several benefits as well as limitations. Generally, a low level of systemic toxicity and cytotoxicity is observed in the case of most of the carrier systems including SLNs, polymeric NPs, liposomes, and lipid emulsions. The generation of organic solvent residue by the SLNs and lipid emulsions is almost negligible while it is maximum with the polymeric NPs and in the case of liposomes it varies, depending on its composition. One of the main advantages offered by SLNs, liposomes and lipid emulsions is simple large-scale production. Furthermore, SLNs and lipid emulsions can be easily sterilized as compared to liposomes and polymeric NPs [15].

12.4 Various Lipids Used for Drug Delivery

Cancer therapeutics have been revolutionized by the emergence of mRNA-based delivery systems in the form of different types of immunomodulating agents such as vaccines and chimeric antigen receptor (CAR) T-cell therapies [36–38]. Lipid-based nanocarriers have been proposed to be the most developed tool for mRNA delivery. The use of lipid nanoparticle systems as carriers for mRNA targeting has helped in improving its stability by rendering protection against degradation in the extracellular compartments along with enhancement of cellular uptake and delivery to the intended site of action [39, 40].

The earliest lipid-based systems that have been reported for mRNA delivery are lipoplexes. They consists of positively charged cationic lipids that interact electrostatically with the negatively charged phosphate molecules in the backbone of mRNA. However, in a short period, the lipoplexes displayed several lacunae including *in vivo* instability, decreased efficiency of transfection, and poor chemical tunability [41].

Recently, the hybrid polymeric NPs are being explored as novel mRNA-based DDS owing to their superior properties offered by them over to polymeric NPs and liposomes. The composition of hybrid polymeric LNPs comprises a core of mRNA loaded polymers with a lipid shell. The lipid coating can be arranged as a monolayer or bilayer of a mixture of cationic lipids or ionizable lipids, helper lipids, PEGylated lipids, and/or sterol lipid (cholesterol) (Fig. 12.4). Hybrid NPs, consisting of lipid polymers, offer innumerable benefits such as a small particle size, enhanced efficiency of nucleic acid incorporation, enhanced surface area for further chemical modifications and prolonged circulation time [43]. The structures of various lipids used for mRNA delivery are depicted in Figs. 12.5 and 12.6 while various applications are enlisted in Table 12.2. Currently, multiple LNPs are undergoing clinical trials for utility as vaccines in infectious diseases and cancers which are enlisted in Table 12.3.

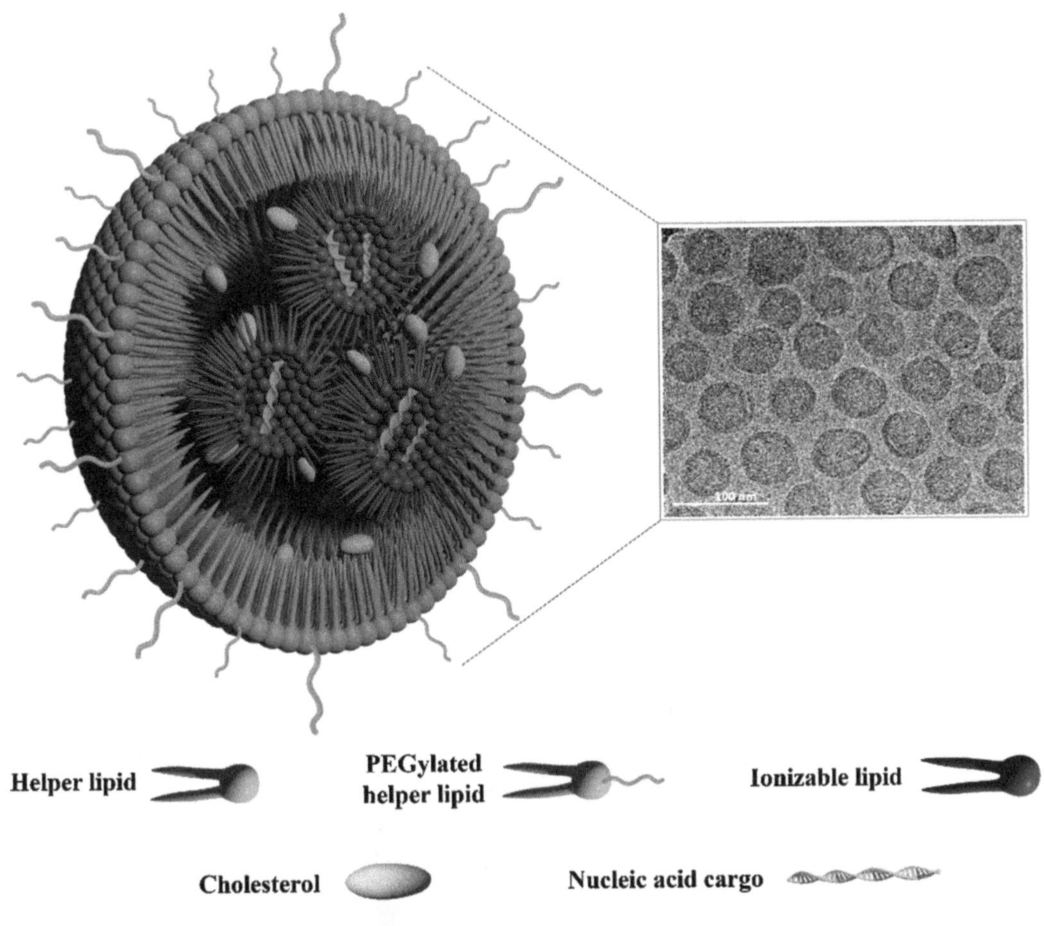

Fig. 12.4 Cartoon representing the different components of LNPs for nucleic acid delivery including the helper lipids, PEGylated lipids, ionizable lipids, cholesterol, and nucleic acid cargo. The inset represents TEM micrograph of LNPs synthesized for the delivery of nucleic acids. Scale bar: 100 nm. TEM image reprinted from [42], Copyright © (2020), with permission from Elsevier

12.5 Approaches for Fabrication of Lipid Nanoparticles

The methods involved in the fabrication of LNPs can be broadly classified as 'conventional' methods such as film hydration, melt homogenization, and micro-emulsification, and 'novel' methods, such as microfluidic hydrodynamic focusing, pulsed jetting and double emulsion droplet formation.

12.5.1 Film Hydration

This method is comparatively easier and relatively simple for fabrication of LNPs [32]. The organic medium comprising of the lipids is well-dried to provide a thin layer of the film at the base of the vessel. Further, this film is hydrated to obtain a liposomal dispersion. Depending upon the hydration conditions, either giant unilamellar vesicles or multilamellar vesicles are formed. Either a probe sonicator or a bath sonicator can be of help in preparation of small unilamellar vesicles [33].

Helper lipids

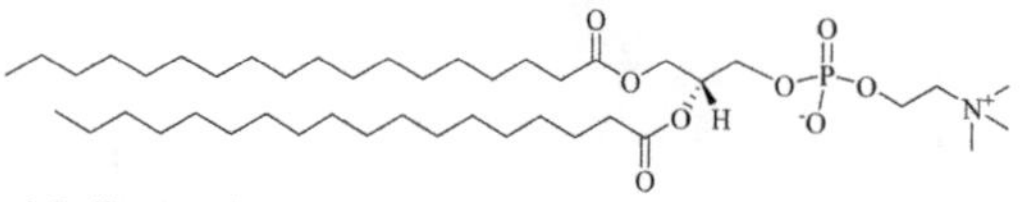

1,2-distearoyl-sn-glycero-3-phosphocholine (DSPC)

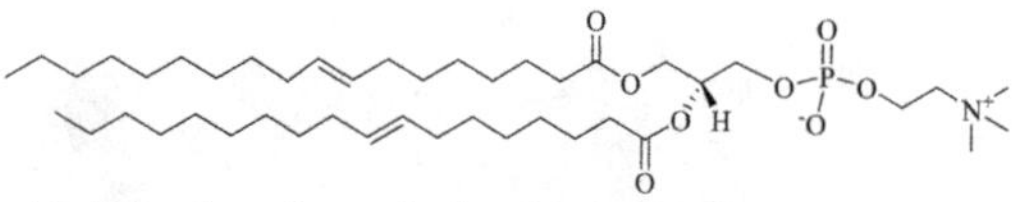

1,2-dioleoyl-sn-glycero-3-phosphoethanolamine (DOPE)

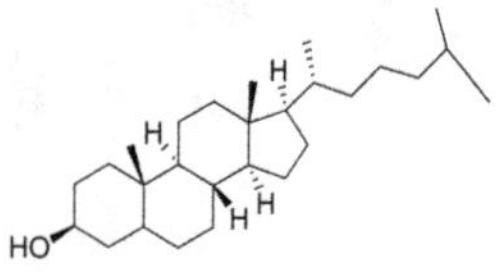

Cholesterol

Stealth lipids

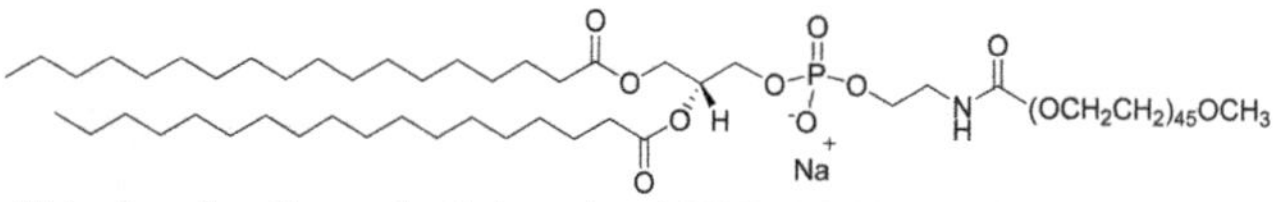

N-(carbonyl-methoxypolyethylene glycol 2000)-1,2-distearoyl-sn-glycero-3-phosphoethanolamine sodium salt (DSPE-PEG 2000)

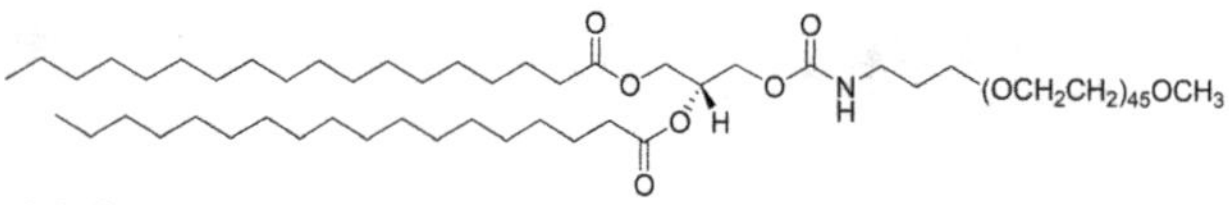

1,2-distearoyl-rac-glycero-3-methylpolyoxyethylene 2000 (DSG-PEG 2000)

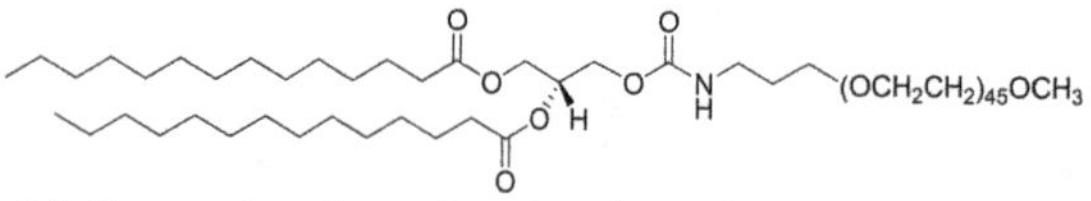

1,2-dimyrstoyl-sn-glycero-3-methoxylpolyethylene glycol 2000 (DMG-PEG 2000)

Fig. 12.5 Chemical structures of various helper and stealth phospholipids utilized in fabrication of lipid NPs. Information adapted from [40]

12.5.2 Melt Homogenization

This method is recommended for the fabrication of lipoidal dispersions that are specifically aimed to be delivered via the intravenous route of administration. It is a two-step process that involves homogenising the lipids in their molten form in an aqueous medium. The process begins by melting the lipids at least 10 °C above their melting point and then dispersing the same in a hot aqueous medium with the help of a mechanical stirrer or ultrasonicator. The resulting pre-mix is further introduced into a high-pressure homogeniser under optimal conditions. The final step includes cooling the hot dispersion at room temperature and thus allowing the solidification of the oil droplets [15].

12.5.3 Micro-emulsification

This method involves emulsification of micro-heterogenous dispersions comprising organic phase, aqueous phase, surfactant, and co-surfactant. The lipids are melted at the required temperature and the resulting

Cationic lipids

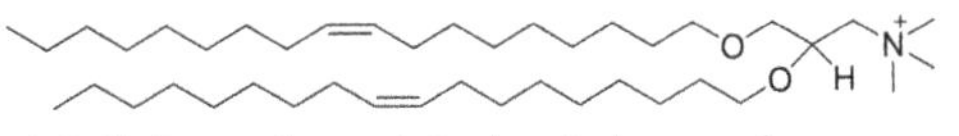

1,2-di-O-octadecenyl-3-trimethylammonium-propane (DOTMA)

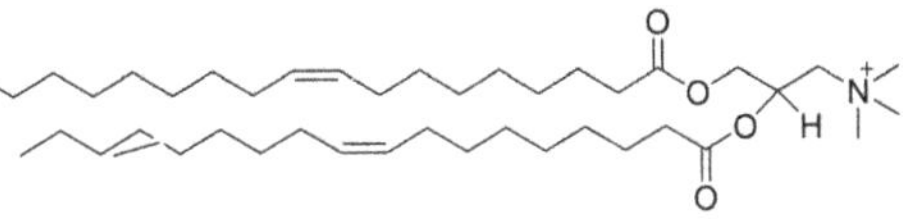

1,2-dioleoyl-3-trimethylammonium-propane (DOTAP)

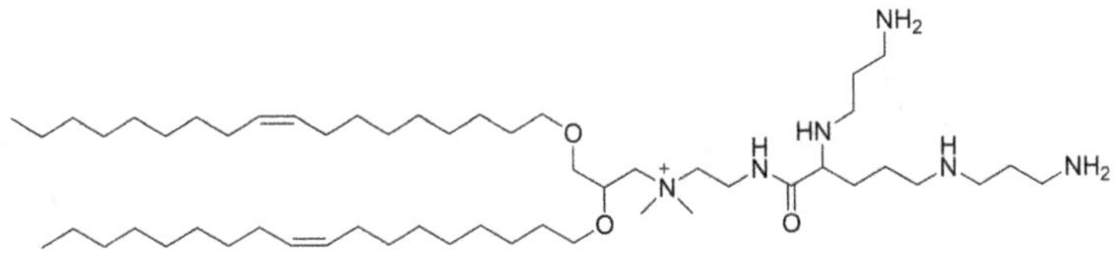

2,3-dioleyloxy-N-[2-(sperminecarboxamido)ethyl]-N,N-dimethyl-1-propanaminium trifluoroacetate (DOSPA)

Ionizable lipids

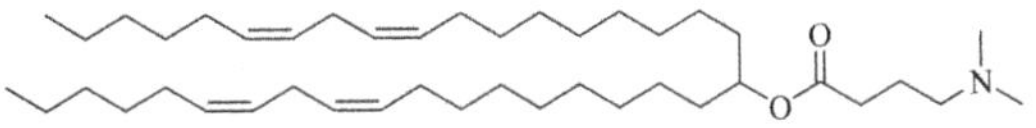

(6Z, 9Z, 28Z, 31Z)-heptatriaconta-6,9,28,31-tetraen-19-yl 4-(dimethylamino) butanoate (DLin-MC3-DMA/MC3)

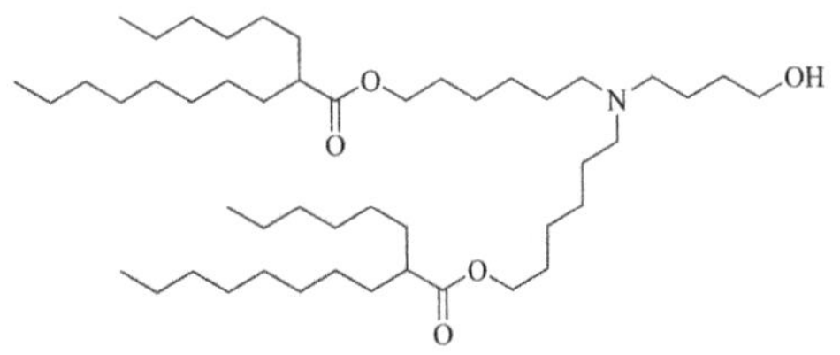

((4-hydroxybutyl)azanediyl)bis(hexane-6,1-diyl)bis(2-heyldecanoate) (ALC-0315)

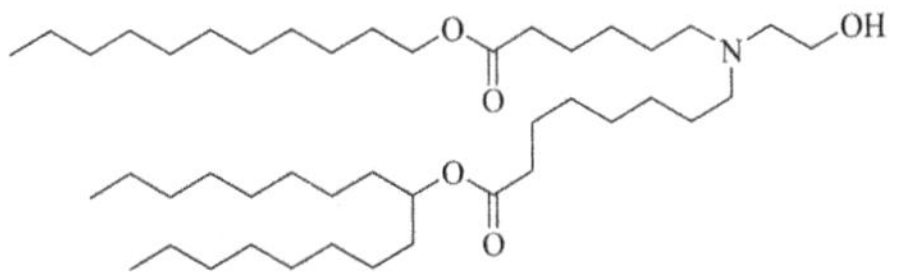

Heptadecan-9-yl 8-((2-hydrocyethyl)(6-oxo-6-(undecyloxy)hexyl)amino)octanoate (Lipid H (SM-102))

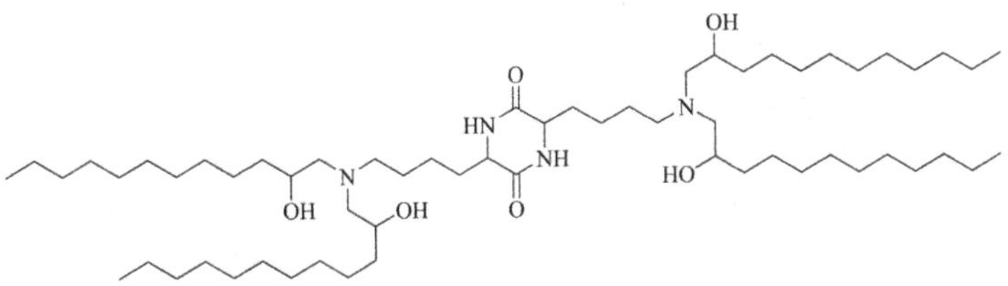

3,6-bis(4-(bis(2-hydroxydodecyl)amino)butyl)piperazine-2,5-dione (cKK-E112)

Fig. 12.6 Chemical structures of various cationic and ionizable phospholipids utilized in fabrication of lipid NPs. Information adapted from [44]

Table 12.2 Lipid excipients intended for delivery of nucleic acids. Information adapted from [45], Copyright © (2020), with permission from Elsevier

S. No.	Formulation		Nucleic acid cargo	Targeted disease site
	Composition	Ratio		
1.	DOTAP: cholesterol: DOPE	1:0.75:0.5	siRNA splicing factor	A549 lung carcinoma
2.	DODAP: DSPC: cholesterol: PEG	40:10:48:2	siRNA tetR	Liver tumor
3.	DODMA: EPC: cholesterol: PEG	45:54:1	siBcl-2, miR-122	Liver tumor
4.	Cationic lipid: EPC: PEG	45:54:1	siRNA LOR-1284	Acute myeloid leukemia
5.	DC-cholesterol: DOPE: PEG	1:1:0.005	siRNA kinesin spindle protein	Ovarian tumor
6.	DOPE: DPPC: cholesterol PEI-nucleic acids	70:15:15	siRNA luciferase pCMV-luc. pEGFP-N1	Ovarian tumor
7.	EPC: cholesterol: DOPE-PEI	90:10:0.5	siMDR-1	Ovarian tumor
8.	DSDAP: DSPC: PEG	40:64:6	miRNA-126	Hindlimb ischemia
9.	DLin-MC3-DMA: DSPC: cholesterol: PEG	50:10:38.5: 1.5	mRNA cmCFTR	Cystic fibrosis
10.	DLin-KC2-DMA: DOPE: cholesterol: PEG	50:10:39:1	pDNA	Luciferase

DOTAP 1,2-dioleoyl-3-trimethylammonium propane, *DOPE* dioleoly-*sn*-glycero-phophoethanolamine, *DODAP* 1,2-dioleoyl-3-dimethylammonium propane, *DODMA* 1,2-dioleyloxy-3-dimethylaminopropane, *DSDAP* 1,2-distearoyl-3-dimethylammonium-propane; *DLin-MC3-DMA* heptatriaconta-6,9,28,31-tetraen-19-yl-4-(dimethylamino) butanoate, *DLin-KC2-DMA* 2,2-dilinoleyl-4-(2-dimethylaminoethyl)-(1,3)-dioxolane

microemulsion is then instantaneously dispersed in an aqueous medium with the help of mechanical stirring. The average diameter of the dispersed phase droplets is mostly observed as less than 100 nm [15].

12.5.4 Microfluidic Hydrodynamic Focusing

This method comprises energetically introducing an alcoholic solution of dissolved lipids to go through the central channel of the microfluidic device. Lipid precipitation and self-assembly of the LNPs are observed due to the phenomenon of reciprocal diffusion occurring across the interface of the device (Fig. 12.7). Crossflow injection and methods using supercritical fluids are amongst a few other recently developed processes for production of LNPs [32, 45, 57].

12.5.5 Pulsed Jetting

This method was first introduced by Funakoshi et al. [58]. It generally resembles the technique of blowing soap bubbles across a loop. Through a micro-nozzle in the assembly, the aqueous solution is sprayed onto a thin lipoidal membrane. Owing to the energy gained from the aqueous solution, the formed bubbles are carried ahead that further compress the lipid membrane, thus, resulting in the formation of a vesicle. Although the method is highly efficient, it faces several limitations including difficulty in encapsulation of large molecules such as proteins, remnants of the residual solvent, bulky assembly, and hypersensitivity towards experimental reagents and conditions [45, 59].

12.5.6 Double-Emulsion Droplet Formation

This method was first described by Pautot et al. [60] and is a single step approach involving the amalgam of the organic phase, such as oleic acid, 2-propanol + oleic acid, 2-butanol, and 1-octanol, with the aqueous phase (Fig. 12.8). It considers the stabilization of a water-in-oil emulsion using phospholipids, followed by the subsequent transfer of the droplets to an aqueous phase. Further,

Table 12.3 Various LNP-mRNA vaccines under clinical trials against infections and cancer

S. No.	Vaccine name and route of administration	Target disease	Encoded antigen	Clinical trial phase and identification number[a]	Reference
A.	Infections				
1.	mRNA-1644 mRNA-1644v2-Core (IM)	HIV	eOD-GT8 60 mer Core-g28v2 60 mer	I *NCT05001373*	[46]
2.	mRNA-1273 (IM)	SARS-CoV-2	Spike	III (EUA and CMA) *NCT04470427*	[47]
3.	BNT 162b2 (IM)	SARS-CoV-2	Spike	III (EUA and CMA) *NCT04368728*	[48]
4.	VAL-506440 (IM)	Influenza H10N8	Haemagglutinin	I *NCT03076385*	[49]
5.	mRNA-1647 (IM)	Cyto-megalovirus	Pentameric complex and B glycoprotein	II *NCT04232280*	[50]
6.	CV7202 (IM)	Rabies virus	G glycoprotein	I *NCT03713086*	[51]
B.	Cancer				
7.	mRNA-4157 (IM)	Skin cancer	Personalized neoantigens	II *NCT03897881*	[52]
8.	Lipo-MERIT (IV)	Skin cancer	NY-ESO-1, tyrosinase, MAGE-A3, TPTE	I *NCT02410733*	[53]
9.	HARE-40 (ID)	HPV positive squamous cell carcinoma	HPV oncoprotein E6 and E7	I/II *NCT03418480*	[54]
10.	W-ova1 (IV)	Ovarian cancer	Ovarian cancer antigens	I *NCT04163094*	[55]
11.	mRNA-5671/V941 (IM)	Colorectal cancer, non-small cell lung carcinoma	KRAS antigens	I *NCT03948763*	[56]

CMA conditional marketing authorization, *EUA* emergency use authorization, *NY-ESO-1* New York oesophageal squamous cell carcinoma 1, *MAGE-A3* melanoma antigen family A, *TPTE* putative tyrosine-protein phosphatase, *HPV* human papillomavirus, *IM* intramuscular, *IV* intravenous, *ID* intradermal

[a] ClinicalTrials.gov identification number

these droplets converge with a second layer of lipids while crossing the oil/water liquid interface, resulting in the formation of unilamellar bi-layered vesicles. This technique has been reported to overcome various liposome manufacturing limitations [62].

12.6 Characterization Techniques

12.6.1 Electron Microscopy

Electron microscopy aids to visualize the inner architecture of single NPs with high resolution capabilities. Negative staining for transmission electron microscopy (TEM) and cryo-transmission electron microscopy (cryo-TEM) are the two most used techniques for imaging LNPs.

Negative staining involves interaction of the LNPs with various heavy metal salts or acids which form a dark contrast surrounding the LNPs and the LNPs themselves appear bright. The commonly used negative stains include phospho-tungstic acid and uranyl acetate. The primary advantage of negative staining is a requirement of less advanced instruments for imaging. However, the disadvantages of negative

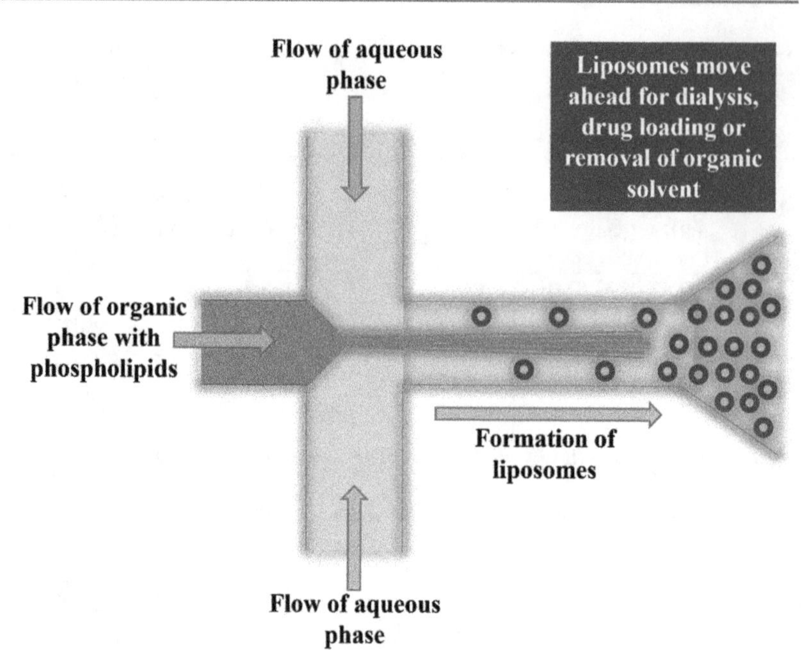

Fig. 12.7 Generation of liposomes using a novel microfluidic hydrodynamic focusing method. The diffusive mixing of the aqueous and organic medium inside the microfluidic channels, with an approximate diameter of 500 μm, results in the formation and collection of varying sizes of self-assembled phospholipids. Information adapted from [33]

staining include the appearance of artifacts due to loss in structural morphology during sample preparation and poor adsorption of LNPs on the grids. The adsorption of LNPs can be enhanced by coating the grids with bacitracin solution, deposition of silica or by glow discharging of grids prior to use [15].

Cryo-TEM has evolved as an essential tool in nano-chemistry for studying the morphological characteristics of biological colloids. Cryo-TEM has been utilized extensively to evaluate the structure of LNPs extensively as it preserves the lipid bilayer without disruptions as in the case of negative staining. Specifically, in the case of active loading of doxorubicin within the liposomes using either ammonium sulfate or citrate gradients, the crystal structure within the liposomal compartments has been established using cryo-TEM. Shamrakov et al. recently utilized cryo-TEM to evaluate the changes in shape of liposomal vesicles upon corresponding changes in drug loading concentrations and demonstrated that the concentration of drug encapsulated in DOXIL® was optimal considering alterations in shape parameters upon increasing drug concentration (Fig. 12.9a–d) [63]. Additional advancements in instrumentation have also been utilized for automated 3D reconstructions to study heterogeneous assemblies using tomography techniques [64].

The latest advances in cryo-EM technology suits even a single particle analysis. It is possible to elucidate the structural changes in macromolecules, including membrane proteins, which may arise due to the influence of environmental conditions [65].

12.7 Pharmacokinetics and Pharmacodynamics of Lipid Nanoparticles

The pharmacokinetics (PK) and pharmacodynamics (PD) of drug molecules play critical role in efficacy of treatment and the physico-chemical properties of the molecules govern the same. On the other hand, for nanotherapeutics, the size, shape and surface charge play a critical role in its PK determining the therapeutic outcomes [66]. The *in vivo* distribution of NPs is facilitated by the circulatory system and upon injection the NPs first interact with the blood components before accumulating into solid tissues [67]. The interaction of NPs with blood proteins results in

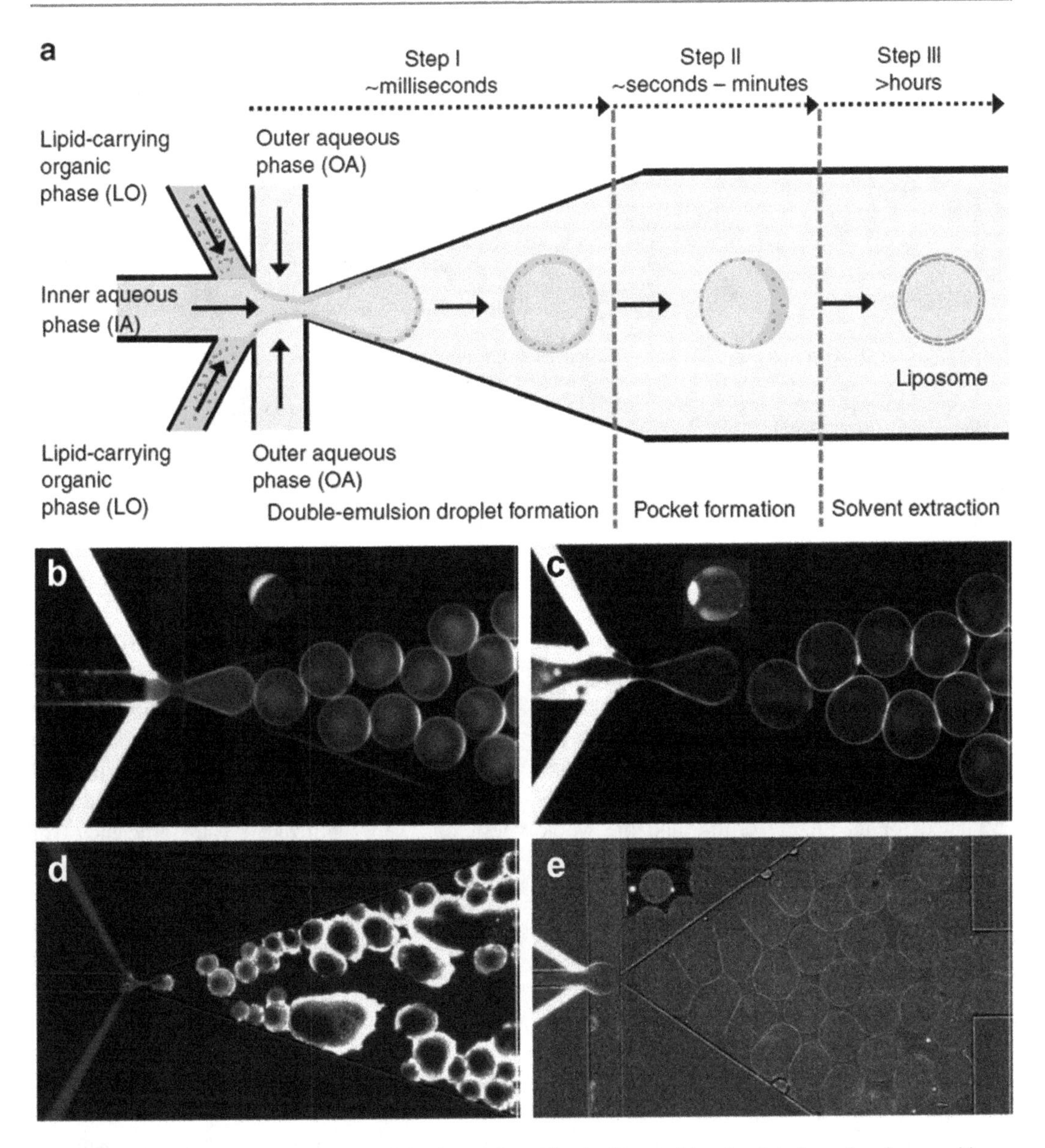

Fig. 12.8 (**a**) A schematic illustration of the 'octanol-assisted liposome assembly', a technique based on the double emulsion droplet formation approach, for the formation of monodispersed liposomes; (**b**–**e**) Fluorescence microscopic images of the liposomes generated using different types of organic phases as a carrier for the lipids. The double emulsion droplets formed undergo rapid conversion into liposomal self-assemblies with an adjacent side pocket of the solvent residue that is subsequently removed during by solvent extraction process. Reprinted from [61] (Open access under the Creative Commons CC BY license)

the formation of a protein coat on the surface of NPs known as the "protein corona" which also governs the accumulation/elimination of NPs in various organs [68].

Certain tissues and organs may act as filters or traps for NPs on a size dependent basis and each tissue/organ processes the NPs using varied mechanisms [69]. In the case of cancer, the leaky tumor vasculature and lack of interstitial

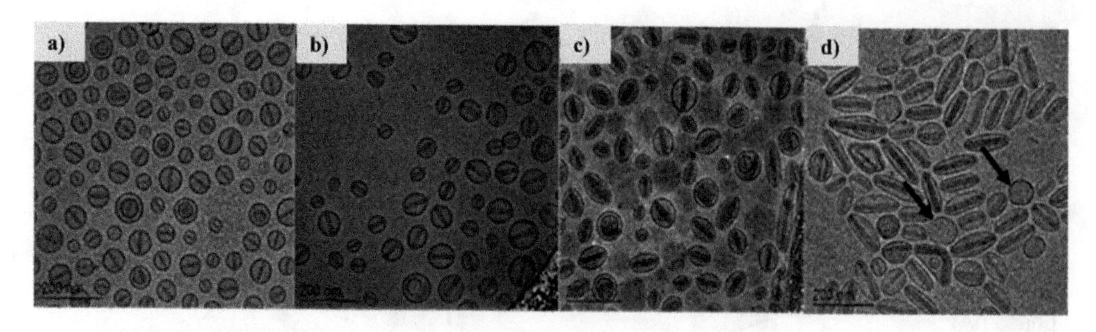

Fig. 12.9 Cryo-TEM micrographs of liposomes loaded with doxorubicin sulfate nanorod crystals with drug loading concentrations of (**a**) 1 mg/mL, (**b**) 2 mg/mL, (**c**) 3 mg/mL and (**d**) 4 mg/mL. Thus, as the initial drug loading concentration was increased the shape of final liposomes was distorted from spherical to elongated shape. Scale bar: 200 nm. Reprinted from [63] (Open access under the Creative Commons Attribution CC-BY license)

convection, well-known as the "enhanced permeation and retention (EPR) effect" is responsible for the accumulation of NPs at the solid tumor site [70]. Thus, considering all these variables it becomes important to study the PK-PD parameters for LNPs and below we describe a few examples of the same.

Liposomal doxorubicin was one of the first lipid-based delivery systems to be thoroughly evaluated for its PK-PD profile and the whole journey from first clinical trials to the approval of DOXIL can be studied as detailed by Barenholz in a review article [2]. The composition of DOXIL, as discussed in the previous section, and the vital advancements offered compared to previous liposomal formulations include the active loading of doxorubicin and PEGylation of lipids. The active loading of drug drastically reduces the leakage of drug from the vesicles in circulation which in turn decrease systemic side effects. Conversely, PEGylation of the lipids hinder interactions with the reticuloendothelial system (RES) and thus prolong the circulation time. Clinically, a dose corresponding to 50 mg/m^2 results in 300 times increase in area under the curve (AUC) compared to free doxorubicin, the clearance reduces 250 times, and the volume of distribution reduces 60 times [71].

Additionally, studies have shown that liposomes can be synthesized for incorporation of multiple anti-cancer agents in a single formulation [72]. Pakunlu et al. reported that liposomal formulations can be utilized for concurrent administration of doxorubicin and antisense oligonucleotides (ASO) targeting the proteins involved in resistance mechanisms [73]. The target proteins were namely, MDR1 gene (encoding for P-glycoprotein pump) and BCL2 (responsible for inhibition of cellular apoptosis). To formulate the liposomes, lipid mixture comprising egg phosphatidylcholine, DSPE-PEG-2000 and cholesterol were utilized. The efficiency of formulations was compared with free doxorubicin in animal models (mice bearing xenografts of multidrug resistant human ovarian cancer cell line A2780/AD). The incorporation of ASO in the liposomal formulation caused a significant down-regulation of MDR1 gene and BCL2 protein. The simultaneous administration of PEGylated liposomal doxorubicin with ASO enhanced the induction of apoptosis in tumor cells, using terminal deoxynucleotidyl transferase mediated d-UTP-fluorescein nick end labelling (TUNEL), as observed in Fig. 12.10. Furthermore, the inhibition of P-glycoprotein and BCL2 enhanced the cytotoxic effects of PEGylated liposomal doxorubicin as well as the growth inhibitory effects on tumors (Fig. 12.11a, b). Thus, liposomes present a unique DDS which can concurrently incorporate tumor inhibiting drug molecules as well as suppressors of resistance mechanisms.

In addition, much like the PK and PD parameters of a drug change upon alterations in

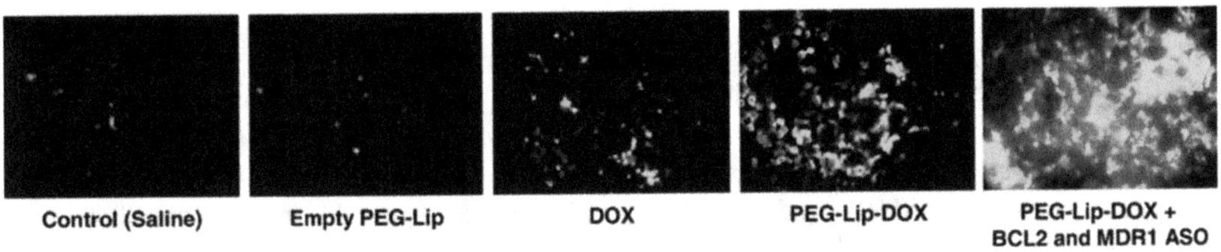

Control (Saline) | Empty PEG-Lip | DOX | PEG-Lip-DOX | PEG-Lip-DOX + BCL2 and MDR1 ASO

Fig. 12.10 Fluorescent micrographs of multidrug resistant tumor tissue labeled by TUNEL. The mice bearing tumors were treated with empty liposomes (Empty PEG-Lip), free doxorubicin (DOX), PEGylated liposomal doxorubicin (PEG-Lip-DOX) and liposomes comprising DOX and ASO coding for MDR1 gene and BCL2 protein (PEG-Lip-DOX + BCL2 and MDR1 ASO). The tissue samples were collected 24 h post treatment. Reprinted from [73], Copyright © (2006), with permission from Elsevier

the structure of a drug molecule, similarly, the PK and PD of LNPs can be altered by chemical modifications in lipid structures. Here, we describe a study by Moderna Therapeutics wherein the clearance and induction of immune response in mice obtained from LNPs comprising of mRNA formulated using novel ionisable lipids were compared with LNPs formulated using MC3 [74].

The novel ionisable lipids used in the study are comprised of ester bonds and are biodegradable. LNPs comprising of mRNA and ionisable lipid MC3 administered intra-muscularly remained at the administration site at a concentration of 50% compared to the maximum concentration after 24 h of injection, indicating a prolonged local exposure. While accumulation in the liver and spleen after 24 h of administration was observed indicating systemic exposure.

On the other hand, mRNA LNPs formulated using novel ionisable lipids degraded rapidly in muscle, spleen, and liver. Though, the lipids were cleared rapidly from the injection site potent immune responses were observed. Owing to the

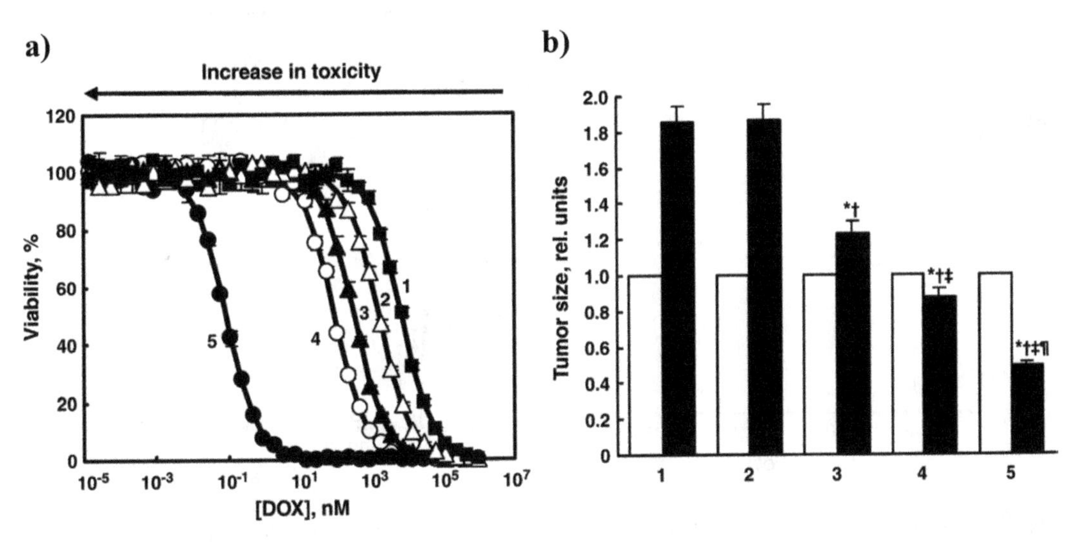

Fig. 12.11 (**a**) Viability of multidrug resistant human ovarian cancer cell lines post treatment with various formulations. (1) free doxorubicin, (2) PEGylated liposomal doxorubicin, (3) PEGylated liposomal doxorubicin with ASO targeting BCL2 protein, (4) PEGylated liposomal doxorubicin with ASO targeting MDR1 gene and (5) PEGylated liposomal doxorubicin with ASO targeting BCL2 protein and MDR1 gene. (**b**) Antitumor efficiency of formulations. The mice bearing tumors were treated with (1) saline, (2) empty liposomes, (3) free doxorubicin, (4) PEGylated liposomal doxorubicin and (5) PEGylated liposomal doxorubicin with ASO targeting BCL2 protein and MDR1 gene. Reprinted from [73], Copyright © (2006), with permission from Elsevier

biodegradability of the novel lipids, reduced inflammation at the injection site was noted and thus improved tolerability towards LNPs was observed.

Clustered regularly interspaced short palindromic repeats (CRISPR) and CRISPR-associated protein 9 (Cas9) were initially identified as the components of anti-phage defence system of prokaryotes [75]. Thereafter, the CRISPR/Cas9 system has been utilized as a technology for genetic editing, delivered primarily using viral vectors, having implications in the treatment of a variety of diseases [76, 77]. However, due to the limitations of viral vectors including immunological responses towards viral components, the focus was shifted toward the utility of LNPs as carriers for CRISPR/Cas9 [78, 79]. Finn et al. reported the development of an LNP-based delivery platform for CRSPR/Cas9 as a *in vivo* genetic editing technology for liver-based diseases [80]. The LNPs comprised of a biodegradable ionizable lipid (LP01), cholesterol, DSPC and PEG2k-DMG as the carriers.

Streptococcus pyogenes Cas9 (Spy Cas9) mRNA and highly modified single guide RNA (sgRNA) were concomitantly incorporated within the LNPs to achieve knockdown of transthyretin (TTR) protein synthesized by the liver. The structural changes in TTR protein, resulting from mutations, are responsible for diseased states such as amyloidosis. Finn et al. demonstrated that the ionisable lipid, LP01 (Fig. 12.12a) was

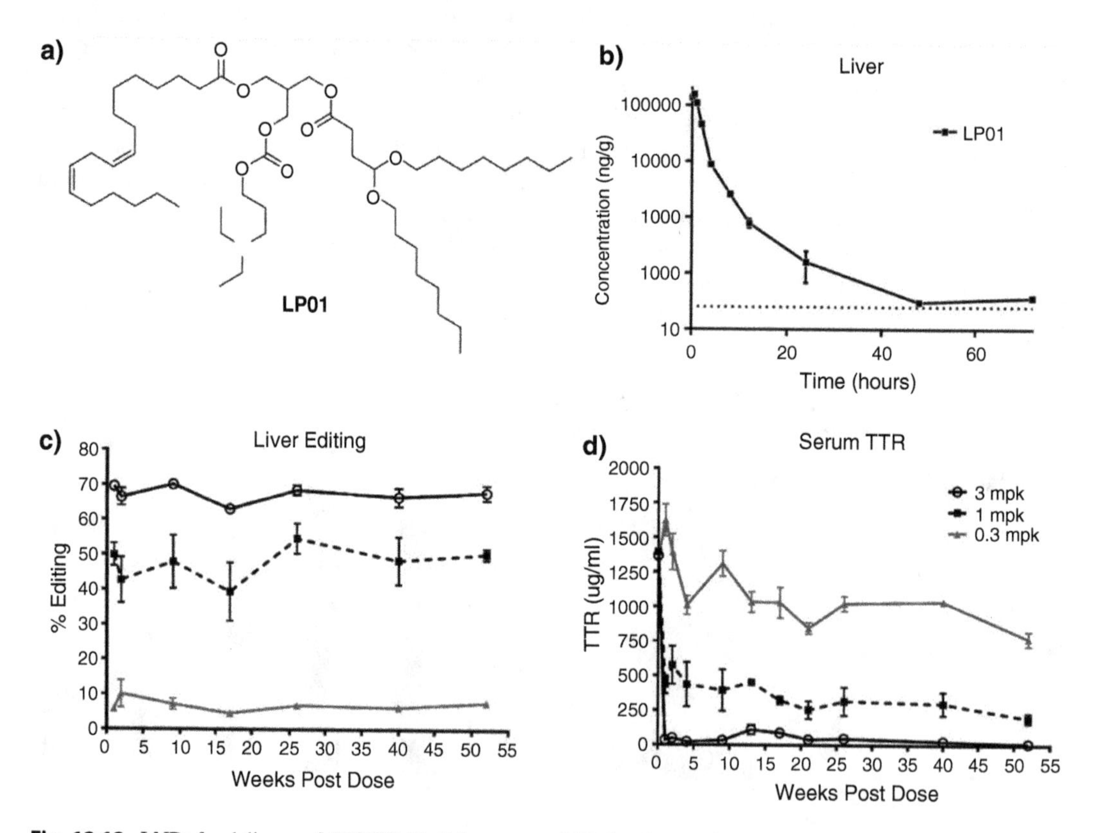

Fig. 12.12 LNPs for delivery of CRISPR/Cas9 for gene editing in the liver. (**a**) Chemical structure of LP01. (**b**) Clearance of LP01 from the murine liver. The limit of quantification is represented by the basal dotted line. (**c**) Editing obtained in the liver of sacrificed mice after systemic injection of Cas9 mRNA and highly modified sgRNA LNP formulation (n = 3). (**d**) Concentration of TTR in plasma after injection of Cas9 mRNA and highly modified sgRNA LNP formulation in mice (n = 5). For (**c, d**) the initial dose of the formulation was either 0.3, 1 or 3 mg/kg (mpk). For all *in vivo* studies CD-1 mice were used. Adapted from [80], Copyright © (2018), with permission from Elsevier

cleared from the liver rapidly (Fig. 12.12b) with a $t_{1/2}$ corresponding to approximately 6 h thus preventing potential toxicity issues arising from carrier components. The therapeutic effect was monitored by determining the genetic editing achieved and the reduction in serum TTR levels post injection of LNP formulations (Fig. 12.12c, d). They determined that the editing achieved in the liver and the corresponding decrease in serum concentrations of TTR were dependent on the initial dose of the formulation and the effect was sustained for 12 months.

Noteworthy PK-PD parameters can be achieved by careful alterations in the design of the delivery system. Changes in the physical form of the cargo can help reduce systemic effects and changes in the surface property of carriers may enhance systemic circulation (both exemplified by DOXIL). Similarly, alterations in chemical structures of the cargo may give a more durable therapeutic effect and the biodegradability of the carrier system may avoid toxic side effects (exemplified by CRISPR/Cas9 LNPs reported by Finn et al. [80].

12.8 Concluding Remarks

The use of LNPs began as early as the 1960s and has sustained the interests of researchers for almost six decades. Advancements are continually being made to scale-up the synthetic approaches for liposomes as well as SLNs. Microfluidic and membrane extrusion techniques are at the forefront of commercialized fabrication approaches. Furthermore, new types of lipids are being designed and fabricated to encapsulate therapeutic molecules. This revival in research and commercial interests in LNPs is led by mRNA vaccines.

As the leading research groups rush to synthesize novel ionisable lipids, the precise mechanisms involved in the entrapment of nucleic acids using ionizable lipids in LNPs need to be explored. In addition, although LNPs provide PK and PD advantages, the fate of lipids and their toxicity require detailed *in vivo* evaluations. In the past, LNPs have assisted in overcoming drawbacks of anticancer drugs. Currently, they are at the frontline for the safe and targeted delivery of nucleic acids and in the future are expected to aid in the *in vivo* delivery of genetic editing technologies such as CRISPR/Cas9 as well as in regenerative medicine.

Conflict of Interest None.

References

1. Elbayoumi TA, Torchilin VP (2010) Current trends in liposome research. In: Weissig V (ed) Methods in molecular biology, vol 605. New York, Humana Press, pp 1–27
2. Barenholz Y (2012) Doxil® - the first FDA-approved nano-drug: lessons learned. J Control Release 160: 117–134
3. Torchilin VP (2005) Recent advances with liposomes as pharmaceutical carriers. Nat Rev Drug Discov 42: 145–160
4. Liu P, Chen G, Zhang J (2022) A review of liposomes as a drug delivery system: current status of approved products, regulatory environments, and future perspectives. Molecules 27:1372
5. Huang SL, Kee PH, Kim H, Moody MR, Chrzanowski SM, MacDonald RC, McPherson DD (2009) Nitric oxide-loaded echogenic liposomes for nitric oxide delivery and inhibition of intimal hyperplasia. J Am Coll Cardiol 54:652–659
6. Shilo-Benjamini Y, Cern A, Zilbersheid D, Hod A, Lavy E, Barasch D, Barenholz Y (2022) A case report of subcutaneously injected liposomal cannabidiol formulation used as a compassion therapy for pain management in a dog. Front Vet Sci 9:892306
7. Duan Y, Dhar A, Patel C, Khimani M, Neogi S, Sharma P, Siva Kumar N, Vekariya RL (2020) A brief review of solid lipid nanoparticles: part and parcel of contemporary drug delivery systems. RSC Adv 10:26777–26791
8. Dhiman N, Awasthi R, Sharma B, Kharkwal H, Kulkarni GT (2021) Lipid nanoparticles as carriers for bioactive delivery. Front Chem 9:580118
9. Anselmo AC, Mitragotri S (2021) Nanoparticles in the clinic: an update post COVID-19 vaccines. Bioeng Transl Med 6:e10246
10. Weil P (2018) Membranes: structure and function. In: Rodwell VW, Bender DA, Botham KM, Kennelly PJ, Weil P (eds) Harpers's illustrated biochemistry, 31th edn. New York, McGraw Hill, pp 467–487
11. Han X (2016) Lipids and lipidomics. In: Desiderio DM, Loo JA (eds) Lipidomics: comprehensive mass spectrometry of lipids. Hoboken, Wiley, pp 1–21
12. Gurr M, Harwood J, Frayn K, Murphy D, Michell R (2016) Lipids: biochemistry, biotechnology and health, 6th edn. Wiley, Chichester, pp 13–43

13. Bangham AD, Standish MM, Watkins JC (1965) Diffusion of univalent ions across the lamellae of swollen phospholipids. J Mol Biol 13:238–252
14. Weissig V (2017) Liposomes came first: the early history of liposomology. Methods Mol Biol 1522:1–15
15. Utreja S, Jain NK (2011) Solid lipid nanoparticles. In: Jain NK (ed) Advances in controlled and novel drug delivery, 6th edn. CBS Publishers and Distributors Pvt. Ltd., New Delhi, pp 408–425
16. Swenson CE, Perkins WR, Roberts P, Janoff AS (2017) Liposome technology and the development of Myocet™ (liposomal doxorubicin citrate). Breast 10: 1–7
17. Gregoriadis G (2016) Liposomes in drug delivery: how it all happened. Pharmaceutics 8:19
18. Lee MK (2019) Clinical usefulness of liposomal formulations in cancer therapy: lessons from the experiences of doxorubicin. J Pharm Investig 49: 203–214
19. Liposome drug products: chemistry, manufacturing, and controls; human pharmacokinetics and bioavailability; and labeling documentation. FDA. https://www.fda.gov/regulatory-information/search-fda-guidance-documents/liposome-drug-products-chemistry-manufacturing-and-controls-human-pharmacokinetics-and. Accessed 11 Nov 2022
20. Tejada-Berges T, Granai CO, Gordinier M, Gajewski W (2002) Caelyx/doxil for the treatment of metastatic ovarian and breast cancer. Expert Rev Anticancer Ther 2:143–150
21. Forssen EA, Ross ME (2008) Daunoxome® treatment of solid tumors: preclinical and clinical investigations. J Liposome Res 4:481–512
22. Forssen EA (1997) The design and development of DaunoXome® for solid tumor targeting in vivo. Adv Drug Deliv Rev 24:133–150
23. Pillai G (2019) Nanotechnology toward treating cancer: a comprehensive review. In: Mohapatra SS, Ranjan S, Dasgupta N, Mishra RK, Thomas S (eds) Applications of targeted nano drugs and delivery systems: nanoscience and nanotechnology in drug delivery. Amsterdam, Elsevier, pp 221–256
24. Liu KK, Sakya SM, O'Donnell CJ, Flick AC, Ding HX (2012) Synthetic approaches to the 2010 new drugs. Bioorg Med Chem 20:1155–1174
25. Silverman JA, Deitcher SR (2013) Marqibo® (vincristine sulfate liposome injection) improves the pharmacokinetics and pharmacodynamics of vincristine. Cancer Chemother Pharmacol 71:555–564
26. FDA approves liposome-encapsulated combination of daunorubicin-cytarabine for adults with some types of poor prognosis AML. FDA. https://www.fda.gov/drugs/resources-information-approved-drugs/fda-approves-liposome-encapsulated-combination-daunorubicin-cytarabine-adults-some-types-poor. Accessed 24 Nov 2022
27. Faustino C, Pinheiro L (2020) Lipid systems for the delivery of amphotericin B in antifungal therapy. Pharmaceutics 12:29
28. Bulbake U, Doppalapudi S, Kommineni N, Khan W (2017) Liposomal formulations in clinical use: an updated review. Pharmaceutics 9:12
29. Cavassin FB, Baú-Carneiro JL, Vilas-Boas RR, Queiroz-Telles F (2021) Sixty years of amphotericin B: an overview of the main antifungal agent used to treat invasive fungal infections. Infect Dis Ther 10:115–147
30. Hartrick CT, Hartrick KA (2008) Extended-release epidural morphine (DepoDur): review and safety analysis. Expert Rev Neurother 8:1641–1648
31. McAlvin JB, Padera RF, Shankarappa SA, Reznor G, Kwon AH, Chiang HH, Yang J, Kohane DS (2014) Multivesicular liposomal bupivacaine at the sciatic nerve. Biomaterials 35:4557–4564
32. Tenchov R, Bird R, Curtze AE, Zhou Q (2021) Lipid nanoparticles - from liposomes to mRNA vaccine delivery, a landscape of research diversity and advancement. ACS Nano 15:16982–17015
33. Pattni BS, Chupin VV, Torchilin VP (2015) New developments in liposomal drug delivery. Chem Rev 115:10938–10966
34. Haider M, Abdin SM, Kamal L, Orive G (2020) Nanostructured lipid carriers for delivery of chemotherapeutics: a review. Pharmaceutics 12:288
35. Mehnert W, Mäder K (2001) Solid lipid nanoparticles: production, characterization and applications. Adv Drug Deliv Rev 47:165–196
36. Kranz LM, Diken M, Haas H, Kreiter S, Loquai C et al (2016) Systemic RNA delivery to dendritic cells exploits antiviral defence for cancer immunotherapy. Nature 534:396–401
37. Pardi N, Hogan MJ, Porter FW, Weissman D (2018) mRNA vaccines - a new era in vaccinology. Nat Rev Drug Discov 17:261–279
38. Foster JB, Barrett DM, Karikó K (2019) The emerging role of in vitro-transcribed mRNA in adoptive T cell immunotherapy. Mol Ther 27:747–756
39. Wadhwa A, Aljabbari A, Lokras A, Foged C, Thakur A (2020) Opportunities and challenges in the delivery of mRNA-based vaccines. Pharmaceutics 12:102
40. Guevara ML, Persano F, Persano S (2020) Advances in lipid nanoparticles for mRNA-based cancer immunotherapy. Front Chem 8:589959
41. Xue HY, Guo P, Wen WC, Wong HL (2015) Lipid-based nanocarriers for RNA delivery. Curr Pharm Des 21:3140–3147
42. Samaridou E, Heyes J, Lutwyche P (2020) Lipid nanoparticles for nucleic acid delivery: current perspectives. Adv Drug Deliv Rev 154-155:37–63
43. Guevara ML, Persano S, Persano F (2019) Lipid-based vectors for therapeutic mRNA-based anti-cancer vaccines. Curr Pharm Des 25:1443–1454
44. Hou X, Zaks T, Langer R, Dong Y (2021) Lipid nanoparticles for mRNA delivery. Nat Rev Mater 6: 1078–1094

45. Filipczak N, Pan J, Yalamarty SSK, Torchilin VP (2020) Recent advancements in liposome technology. Adv Drug Deliv Rev 156:4–22
46. A Phase 1 study to evaluate the safety and immunogenicity of eOD-GT8 60mer mRNA vaccine (mRNA-1644) and Core-g28v2 60mer mRNA vaccine (mRNA-1644v2-Core). ClinicalTrials.gov. https://clinicaltrials.gov/ct2/show/NCT05001373. Accessed 11 Nov 2022
47. A study to evaluate efficacy, safety, and immunogenicity of mRNA-1273 vaccine in adults aged 18 years and older to prevent COVID-19. ClinicalTrials.gov. https://clinicaltrials.gov/ct2/show/NCT04470427. Accessed 24 Nov 2022
48. Study to describe the safety, tolerability, immunogenicity, and efficacy of RNA vaccine candidates against COVID-19 in healthy individuals. ClinicalTrials.gov. https://clinicaltrials.gov/ct2/show/study/NCT04368728. Accessed 25 Nov 2022
49. Safety, tolerability, and immunogenicity of VAL-506440 in healthy adult subjects. ClinicalTrials.gov. https://clinicaltrials.gov/ct2/show/NCT03076385. Accessed 25 Nov 2022
50. Dose-finding trial to evaluate the safety and immunogenicity of cytomegalovirus (CMV) vaccine mRNA-1647 in healthy adults. ClinicalTrials.gov. https://clinicaltrials.gov/ct2/show/NCT04232280. Accessed 26 Nov 2022
51. A study to assess the safety, reactogenicity and immune response of CureVac's candidate rabies mRNA vaccine in healthy adults. ClinicalTrials.gov. https://clinicaltrials.gov/ct2/show/NCT03713086. Accessed 27 Nov 2022
52. An efficacy study of adjuvant treatment with the personalized cancer vaccine mRNA-4157 and pembrolizumab in participants with high-risk melanoma (KEYNOTE-942). ClinicalTrials.gov. https://clinicaltrials.gov/ct2/show/NCT03897881. Accessed 27 Nov 2022
53. Evaluation of the safety and tolerability of i.v. administration of a cancer vaccine in patients with advanced melanoma. ClinicalTrials.gov. https://clinicaltrials.gov/ct2/show/NCT02410733. Accessed 25 Nov 2022
54. HPV anti-CD40 RNA vaccine. ClinicalTrials.gov. https://clinicaltrials.gov/ct2/show/NCT03418480. Accessed 11 Nov 2022
55. Ovarian cancer treatment with a liposome formulated mRNA vaccine in combination with (neo-)adjuvant chemotherapy. ClinicalTrials.gov. https://www.clinicaltrials.gov/ct2/show/NCT04163094. Accessed 26 Nov 2022
56. A study of mRNA-5671/V941 as monotherapy and in combination with pembrolizumab (V941-001). ClinicalTrials.gov. https://clinicaltrials.gov/ct2/show/NCT03948763. Accessed 24 Nov 2022
57. Koynova R, Tenchov B (2015) Recent progress in liposome production, relevance to drug delivery and nanomedicine. Recent Pat Nanotechnol 9:86–93
58. Funakoshi K, Suzuki H, Takeuchi S (2007) Formation of giant lipid vesiclelike compartments from a planar lipid membrane by a pulsed jet flow. J Am Chem Soc 129:12608–12609
59. Funakoshi K, Suzuki H, Takeuchi S (2006) Lipid bilayer formation by contacting monolayers in a microfluidic device for membrane protein analysis. Anal Chem 78:8169–8174
60. Pautot S, Frisken BJ, Weitz DA (2003) Production of unilamellar vesicles using an inverted emulsion. Langmuir 19:2870–2879
61. Deshpande S, Caspi Y, Meijering AE, Dekker C (2016) Octanol-assisted liposome assembly on chip. Nat Commun 7:10447
62. Silva BF, Rodríguez-Abreu C, Vilanova N (2016) Recent advances in multiple emulsions and their application as templates. Curr Opin Colloid Interface Sci 25:98–108
63. Nordström R, Zhu L, Härmark J, Levi-Kalisman Y, Koren E, Barenholz Y, Levinton G, Shamrakov D (2021) Quantitative cryo-TEM reveals new structural details of Doxil-like PEGylated liposomal doxorubicin formulation. Pharmaceutics 13:123
64. Lengyel JS, Milne JLS, Subramaniam S (2008) Electron tomogrpahy in nanoparticle imaging analysis. Nanomedicine 3:125–131
65. Yao X, Fan X, Yan N (2020) Cryo-EM analysis of a membrane protein embedded in the liposome. Proc Natl Acad Sci U S A 117:18497–18503
66. Khandare J, Calderón M, Dagia NM, Haag R (2012) Multifunctional dendritic polymers in nanomedicine: opportunities and challenges. Chem Soc Rev 41:2824–2848
67. Liu E, Zhang M, Huang Y (2016) Pharmacokinetics and pharmacodynamics (PK/PD) of bionanomaterials. In: Zhao Y, Shen Y (eds) Biomedical nanomaterials, vol 8, 1st edn. Wiley-VCH, Weinheim, pp 1–60
68. Walkey CD, Chan WCW (2012) Understanding and controlling the interaction of nanomaterials with proteins in a physiological environment. Chem Soc Rev 41:2780–2799
69. Bertrand N, Leroux JC (2012) The journey of a drug-carrier in the body: an anatomo-physiological perspective. J Control Release 161:152–163
70. Matsumura Y, Maeda H (1986) A new concept for macromolecular therapeutics in cancer chemotherapy: mechanism of tumoritropic accumulation of proteins and the antitumor agent smancs. Cancer Res 46:6387–6392
71. Gabizon A, Shmeeda H, Barenholz Y (2003) Pharmacokinetics of pegylated liposomal Doxorubicin: review of animal and human studies. Clin Pharmacokinet 42:419–436
72. Minko T, Pakunlu RI, Wang Y, Khandare JJ, Saad M (2006) New generation of liposomal drugs for cancer. Anti Cancer Agents Med Chem 6:537–552
73. Pakunlu RI, Wang Y, Saad M, Khandare JJ, Starovoytov V, Minko T (2006) In vitro and in vivo

intracellular liposomal delivery of antisense oligonucleotides and anticancer drug. J Control Release 114:153–162
74. Hassett KJ, Benenato KE, Jacquinet E, Lee A, Woods A et al (2019) Optimization of lipid nanoparticles for intramuscular administration of mRNA vaccines. Mol Ther Nucl Acids 15:1–11
75. Jinek M, Chylinski K, Fonfara I, Hauer M, Doudna JA, Charpentier E (2012) A programmable dual-RNA-guided DNA endonuclease in adaptive bacterial immunity. Science 337:816–821
76. Platt RJ, Chen S, Zhou Y, Yim MJ, Swiech L et al (2014) CRISPR-Cas9 knockin mice for genome editing and cancer modeling. Cell 159:440–455
77. Ehrke-Schulz E, Schiwon M, Leitner T, Dávid S, Bergmann T, Liu J, Ehrhardt A (2017) CRISPR/Cas9 delivery with one single adenoviral vector devoid of all viral genes. Sci Rep 7:17113
78. Liu C, Zhang L, Liu H, Cheng K (2017) Delivery strategies of the CRISPR-Cas9 gene-editing system for therapeutic applications. J Control Release 266: 17–26
79. Zhen S, Li X (2020) Liposomal delivery of CRISPR/Cas9. Cancer Gene Ther 27:515–527
80. Finn JD, Smith AR, Patel MC, Shaw L, Youniss MR et al (2018) A single administration of CRISPR/Cas9 lipid nanoparticles achieves robust and persistent in vivo genome editing. Cell Rep 22:2227–2235

Pharmacokinetic Studies for Drug Development

13

Fred K. Alavi

Abstract

This chapter introduces investigators to the role of pharmacokinetics (PK) and toxicokinetics (TK) play in drug development. The chapter describes the type of studies, the timing of such studies, general methods and reasons for conducting them in animals. Pharmacokinetic studies can provide a firsthand understanding of how an animal's body absorbs, distributes, metabolizes and eliminates a new molecular entity (NME) before it's ever assessed in humans. Despite some differences and variations among species, nonclinical test species share considerable physiological, biochemical and cellular structures with humans. As such test species serve as invaluable models for safety testing of NMEs, with TK serving as the pivotal link for establishment of a safety margin for the therapeutic dose. Although this chapter does not cover modeling, the PK data from animals has been effectively used to estimate PK parameters in humans. Toxicokinetic studies, which are generally recommended for most pivotal toxicology studies (ICH M3 (R2): Guidance for industry: nonclinical safety studies for the conduct of human clinical trials and marketing authorization for pharmaceutics, 2010). M3(R2) Nonclinical safety studies for the conduct of human clinical trials and marketing authorization for pharmaceuticals. FDA), provide a link between target organ toxicity signals and drug exposure. Furthermore, exposure data (i.e., AUC, Cmax) is essential for the interpretation of toxicology findings and clinical relevance.

Keywords

ADME · Pharmacokinetics · Autoradiography · CYP enzymes · Protein binding · Drug transporters · Toxicokinetics · NOAEL · Safety margin

13.1 Introduction

Pharmacokinetics (PK) is simply the quantitative analysis of processes involved in drug absorption (A), distribution (D), metabolism (M) and excretion (E), collectively known as ADME (Fig. 13.1). Each process is affected by multiple factors that may vary between individual animals and across species as a whole. Despite variations among species, nonclinical ADME studies can be instrumental in our understanding of human PK. These nonclinical studies are generally separated into PK and TK sections. The nonclinical PK studies aim to gain as much knowledge as

F. K. Alavi (✉)
Division of Non-Malignant Hematology, Center for Drug Evaluation and Research, US Food and Drug Administration, Silver Spring, MD, USA
e-mail: fred.alavi@fda.hhs.gov

G. Jagadeesh et al. (eds.), *The Quintessence of Basic and Clinical Research and Scientific Publishing*,
https://doi.org/10.1007/978-981-99-1284-1_13

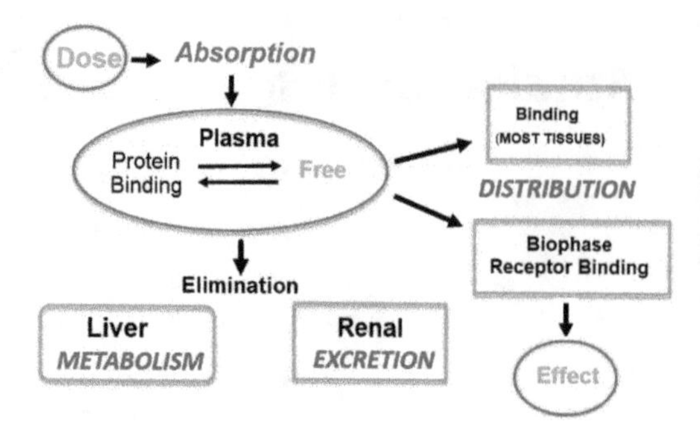

Fig. 13.1 Graphical overview of drug kinetics (absorption, distribution, metabolism and excretion) in the body

possible about the kinetics of an NME while the primary goal of the TK studies is to link the toxicity signal identified in the toxicology studies to drug exposure in animals and clinical relevance and safety measures to avoid adverse effect in humans.

The objective of this chapter is to introduce the reader to general principles of nonclinical PK, why and how they are done and how the information gained from these studies applies to the clinical development of a new compound.

13.2 Exploratory PK

An exploratory PK study in rodents conducted early on in drug development, when there is limited data for selecting appropriate doses for the pivotal nonclinical study, can be very productive. Pharmacokinetic data collected from an exploratory PK study can guide future studies with regard to: (a) selecting more relevant nontoxic doses for future well refined PK studies in rodents and non-rodent models, (b) vehicle composition, (c) route of administration, (d) blood sample collection intervals to cover the elimination phase, (e) validation and establishment of the analytical methods for parent and potential metabolites, (f) initial assessment of area under the concertation-time curve to determine a rough estimate of bioavailability. The information gained from a single dose PK study in rodents can guide more refined PK studies in rodents and other species to predict PK parameters in humans using interspecies scaling (Box 13.1).

Box 13.1 Why to Do Nonclinical ADME Studies

- To get an initial assessment of absorption and distribution of a new drug to different tissues
- To understand the limits of absorption after oral dosing (bioavailability, F)
- To collect data not feasible in humans
- To support the pharmacodynamic properties of a new drug
- To evaluate the metabolic profile of a new drug, what to expect and how to assess their safety
- To understand the potential hepatic first-pass effect
- To predict human PK (i.e., interspecies scaling)

It should be noted that even before an exploratory PK study is conducted, several factors need to be considered which can change the direction of the study. A small molecule should be approached differently than a large molecule/protein. A PK study with a small molecule may start with an oral route of administration while a large molecule has to be administered parenterally. The investigator should establish validated analytical methods beforehand. Knowing the physiochemical properties of a small molecule may guide the selection of a vehicle for oral delivery while large molecules may need to be buffered to reduce injection site reaction and improve absorption [1].

13.3 PK Study Design

1. Animal model: The rat is well suited for initial exploratory PK studies.
2. Dose selection: Use a dose that produces a concertation range detectable by the analytical method validated prior to the study. The dose may also be based on pharmacology studies.

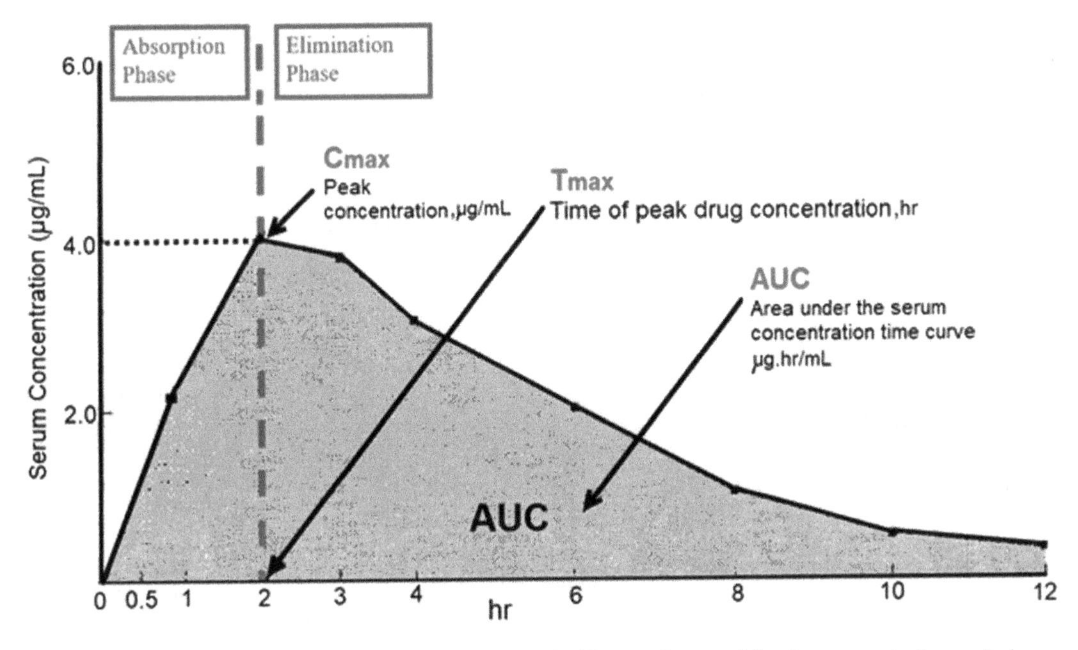

Fig. 13.2 A typical concentration-time curve for a PK study. The matrices used for drug concentration analysis can range from serum, plasma, blood, RBC, CSF to saliva

3. Route of administration: For small molecules, the oral route is likely to be most successful. For large molecules (proteins and peptides), SC is favored over IV or IM.
4. Blood collection: 0.5 mL over 24 h (i.e., 0, 0.5, 1, 2, 4, 8, 12, and 24 h) via a tail vein under a heat lamp is well suited. The total should be limited to 10 mL/kg/day/rat. Blood collection timing is critical. Every effort should be made to collect the sample at the pre-determined time. Drug concentration at 60 min post dose is different from blood collected at 65 min post dose. Any change in the time blood is collected should be noted as it represents a different time interval.
5. Analysis: LC/LCMS for small molecules and RIA or ELISA for proteins/peptides
6. Data evaluation:
   a. Plot of AUC: AUC defined as the area under the serum/plasma concentration by time curve. It can be plotted entering plasma drug concentrations collected from test animal at different time intervals post dose using a software application (Fig. 13.2). Software application can determine the AUC and other PK parameters. (i.e., Cmax, AUC, $t_{1/2}$, Vd, Cl, F, MRT).[1] AUC is a derived parameter and presented as $AUC_{0\text{-}t}$ (to the last detectable concertation), $AUC_{0\text{-}\infty}$ (zero to infinity) or AUC_{tau} (between dosing intervals at steady state).
   b. An estimate of absolute bioavailability can be predicted from the AUC determined after oral and intravenous administration. The AUC for the IV route of administration (ROA) will need to be dose normalized if the IV dose is different than the oral dose which is normally the case.

[1] Definitions: Cmax is the peak drug concertation (μg/mL, ng/mL); Tmax, the time of peak drug concertation (h, min). AUC: the area under the curve is the integral of drug blood levels over time (μg h/mL or ng h/mL); Cl: systemic clearance, a measure of the body's ability to eliminate drug (volume/time, i.e., mL/min), t1/2: half-life, the time it takes for the drug concentration to decrease by 50%. Vd: apparent volume of drug distribution in the body presented in liters (i.e., L), MRT: mean residence time, a measure of overall persistence of compound in the body, h.

7. Once all of the exploratory PK samples are collected, analyzed and the PK parameters are evaluated, one can proceed with formal PK studies with greater efficiency. Standard PK studies conducted in clinically relevant species are formally used to determine PK parameters for submission to regulatory agencies. Using PK data from drug-sensitive animal models can be used to predict clinical PK using interspecies scaling methods.

13.4 Absorption

Absorption is the process of uptake of a compound from the site of administration into the systemic circulation (i.e., blood). The plasma concentration of a drug is the best measurable correlate of biological activity. Following oral administration, a drug can face a gauntlet of biological barriers before reaching the blood (central compartment).

Some of the most common factors that can affect drug absorption are:

- Physiochemical characteristics such as ionization, solubility, particle size, molecular weight and drug formulation can play a significant role in drug absorption and are usually considered when a new drug candidate is selected for clinical development.
- Gastrointestinal states such as the presence of food, anatomy, and physiology can impact drug absorption. A decrease in gastric emptying, larger meal size, higher viscosity and/or acidity may allow for greater drug absorption.
- Hepatic and intestinal first pass effect: The intestine and liver are both effective metabolizing organs and as such, they are capable of removing absorbed drugs in their first pass through organs. Recent studies have shown that changes in the liver microflora may also alter the metabolic profile of drugs and result in the formation of genotoxic metabolites (see metabolism).
- Drug transport: gastrointestinal absorption
  - Passive diffusion is the most common process, requiring no energy, is unsaturable, and follows a concentration gradient (from high concentration to low concentration).
  - Pore Transport is a relatively minor transport mechanism, limited by pore size thus allowing small molecules less than 100 MW.
  - Active transport also known as carrier-mediated transport often employs cell membrane-bound proteins that require energy and are thus saturable. Active transporters can transport polar/ionic drugs against a concentration gradient.
  - Pinocytosis mode of transport engulfs drugs/particles from the luminal membrane into a vesicle that then transverses the membrane to release the contents inside of the cell. This mode of transport is ideal for polar and large molecules (peptides, proteins).

$$\text{Bioavailability}, F = \frac{\text{AUC}_{\text{Oral}}}{\text{AUC}_{\text{IV}}} \times \frac{\text{Dose}_{\text{IV}}}{\text{Dose}_{\text{Oral}}} \times 100$$

An initial indication of the extent of drug absorption can be derived from the AUC collected from a single dose study without intravenous dose with some basic assumption [2]. This will provide a rough initial estimate of bioavailability, but a more accurate measure of absolute bioavailability is determined following administration of the same drug by both the oral and intravenous routes. If the oral and IV dosages are not the same, dose normalization is required (Fig. 13.3). Absolute bioavailability is defined as the fraction (F) of the drug dose that reaches the systemic circulation unchanged. It is determined as a percentage of intravenous AUC with the assumption that intravenous dose absorption is 100% [3]. When a drug cannot be administered intravenously, the relative bioavailability can be determined through comparison to a reference drug [4] administered via the same route (i.e., two oral formulations).

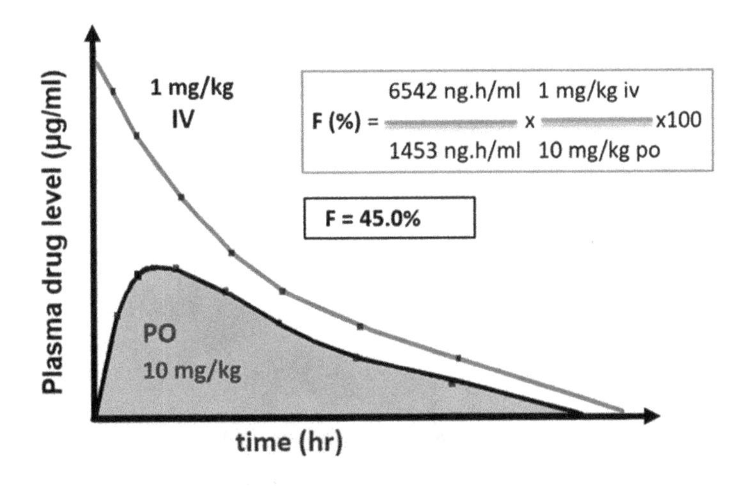

Fig. 13.3 Example of bioavailability (F) determination following oral (10 mg/kg) and intravenous (1 mg/kg) administration of drug X to rats

13.5 Distribution

Drug distribution is defined as translocation of a drug throughout the body following absorption. It is impacted by many factors including protein binding, carrier-mediated transporters, and endocytosis. In this section, I will be discussing the methods used to address drug distribution such as protein binding, transporters and their significance on drug reaching the site of action. Distribution as part of ADME, generally refers to the apparent volume of distribution (Vd), a calculated mathematical parameter, governed by the protein binding, physiochemical properties of the test molecule and the various forms of transporters distributed throughout the body.

$$V_D = \frac{\text{total amount of drug in the body}}{\text{drug blood plasma concentration}}$$

I will discuss the methods used to address drug distribution i.e., protein binding, transporters, and their significance on drug reaching the site of action later in the chapter. Nonclinical distribution studies can serve multiple purposes: (a) reveal the drug concertation/radioactivity in major tissues at a given time, (b) provide data on the relative affinity for a specific organ/tissue, (c) role of transporters, (d) evidence of fetal access and (e) presence in the milk, (f) provide an estimate of distribution in humans. In vitro studies can address protein binding and transporters while in vivo studies are generally carried out in rodents with a radiolabeled drug to determine the overall extent of the parent drug and its metabolites distribution throughout the body. Since the pharmacological activity and the toxicity of a drug are dependent on the drug concertation at the site of action, it is paramount to address all of the factors that can affect drug distribution. As a drug is distributed across the body to larger and larger volumes, drug concentrations can drop precipitously to a fraction of the initial dose [5] and if the distribution is impacted by properties of the molecule (biological vs. small molecules), protein binding and active or passive transporters, the amount of the drug at the site of action can be significantly diminished (Fig. 13.4) while toxicity in off-target organs can be significantly elevated.

It is paramount to determine drug levels in different tissues in order to address the potential sites (organs/tissues) of drug toxicity. The most common approach used to conduct a tissue distribution study is to employ a radiolabeled form of the NME. In such a study, a small amount of radiolabeled drug is administered to rats and various tissues are collected and the total radioactivity is analyzed. Tissues that demonstrate a disproportionate accumulation of drugs may represent potential targets of toxicity.

Drug tissue distribution studies are generally performed early in drug development using mass balance studies that employ a radiolabeled form of the investigational drug [6, 7]. However, before a mass balance study is initiated, a

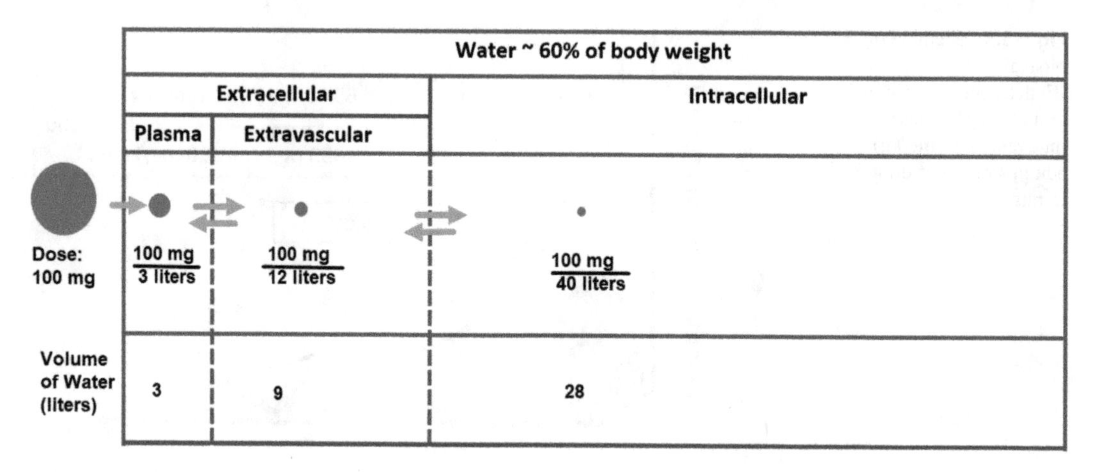

Fig. 13.4 Hypothetical dilution of a drug as it is distributed from the delivery site to blood to extracellular volume to intracellular volume. As one might expect, for a well distributed drug, only a fraction of the dose will reach the cellular target. Data source: Levine 2nd edition, 1973 [5]

radiolabeled form of the compound has to be synthesized where ^{14}C or ^{3}H is chemically attached to the main structure of the molecule without leading to a loss of pharmacological activity. The radiolabeled drug should be free of labeled fragments as such impurities may distort the distribution data. Once the radiolabeled compound is produced a small amount of the radiolabeled drug is administered orally or intravenously to rats or mice.

There are alternative methods such as whole animal autoradiograms where the radioactivity in the whole animal is determined. By measuring the amount of radioactivity in different tissues, an investigator can determine if their new investigational drug reaches the anticipated target tissue (s) or where tissues with the highest concentrations could pose a safety issue.

13.5.1 Wet Tissue Radioactivity

For the wet tissue radioactivity method, animals are sacrificed at different time intervals and blood and tissue samples are collected from every tissue possible [8]. Tissue samples are weighed, digested and homogenized. The collection time may follow existing experience gained from a successful PK study and its collection time interval. To a small aliquot of a tissue sample, a specific amount of liquid scintillator is added, and the radioactivity is measured in a liquid scintillation counter. Wet tissue radioactivity is labor intensive but represents a highly quantitative method of measuring distribution of parent drug and metabolite radioactivity in different tissues. Collection and analysis of bile, urine and feces can further contribute to characterization of intact parent drug and metabolites. One of the limitations of the wet tissue study is precision of tissue distribution and difficulty measuring exposure in subsections of tissues collected, such as measuring radioactivity in choroid plexus or renal cortex.

13.5.2 Autoradiography

Another common and relatively quick method is autoradiography where the whole-body radioactivity is measured following administration of a radiolabeled investigational compound. Animals are sacrificed and frozen and sections are scanned for radioactivity (Fig. 13.5).

Although autoradiography in the past has been semiquantitative and generally exploratory. Newer methods utilizing thinner cryosections (20–50 μm) with precise analytical tools have

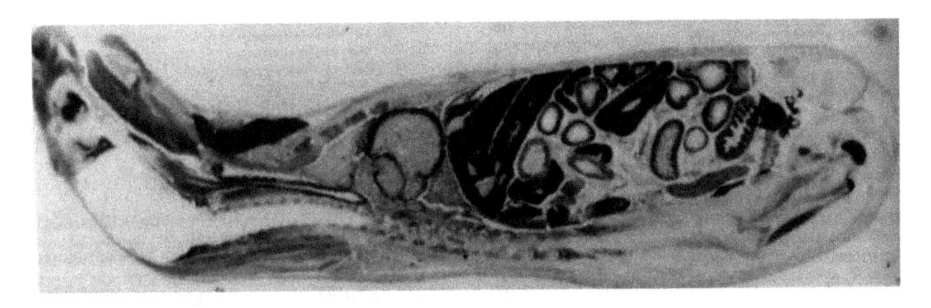

Fig. 13.5 Autoradiographic image of a radiolabeled drug in a rat [9]. Minimal presence of radioactivity in the brain is due to blood-brain barrier, separating brain interstitial space from blood

allowed for the collection of highly qualitative and quantitative drug distribution data for small animals [10]. These new methods permit greater localization of radioactivity if the investigator is interested in determining radioactivity in specific regions of the tissue/organs.

13.5.3 Protein Binding

The objective of the protein binding studies is to determine the extent of drug binding to plasma proteins (albumin, α1-acid glycoproteins, globulins, lipoproteins) and occasionally to tissue proteins. Albumin is a large molecular weight protein (66,500 kDa) that binds to acidic and neutral compounds and constitutes about 59% (3.5–5 g/dl) of total plasma proteins. Albumin is a high capacity (6 binding sites), low affinity protein with binding that is rarely saturable while α-acid glycoprotein, AAG (40,000 kDa, 55–140 mg/dl) is a low capacity but high affinity protein with 2 bindings sites that has a greater affinity to basic drugs. The remaining proteins in the plasma that makeup about 2–2.5 g/dl of plasma proteins are α, β, Y globulins. They generally bind to cortisone, T4 (α1globulin), Vitamin A, D and Vitamin K (α2-globulin). Lipoproteins generally bind to basic, lipophilic, steroids and heparin.

Order : Albumin > AAG > Lipoproteins > Globulin

Protein binding studies are relatively straightforward. There are multiple in vitro methods for measuring protein binding (equilibrium dialysis bags, semipermeable chambers, ultrafiltration, ultracentrifugation, rapid equilibrium dialysis, chromatography) [11–13]. The general principle is the same. For studies utilizing an equilibrium dialysis bag, a small known volume of plasma is placed in a bag separated by a semipermeable membrane from a buffer solution containing a known amount of the test compound (Fig. 13.6). Other techniques may utilize a slightly different approach. Methods for Tests are usually done at

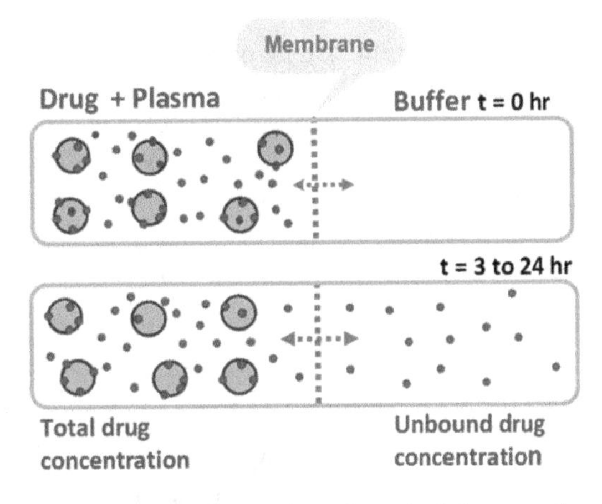

Fig. 13.6 Depending on the study method, the test molecule will reach equilibrium. The semipermeable membrane allows only small molecules to move freely across the membrane. Newer methods can enhance the time required for a test compound to reach equilibrium. When comparing protein binding data from tests done in triplicate, one should consider the variations around the mean protein binding. For example, the difference between mean human (94%) and rat (96%) protein binding may not be realistically different if there is an overlap in % binding (Rat: 96%, 95%, 97% versus humans: 96%, 94%, 92%). Small differences in mean protein binding are particularly irrelevant when the calculated values in humans (98%) and rats (99%) are very high

several drug concentration levels in triplicate. After equilibrium is reached, drug concentrations are measured by a validated analytical method, and the percent bound, and hence free fraction (fu) is determined for each species plasma.

An interaction between a drug molecule and plasma/tissue protein (macromolecules) leading to formation of a drug-macromolecule complex is governed by protein binding affinity. Protein binding affinity (Ka) is determined by the ratio of k1(binding rate constant) and k2 (disassociation rate constant). Depending on the type of bond (Vander Wall forces, ionic bond, hydrophobic bond of α1-AG, hydrogen bond and irreversible covalent bond), drug-protein binding can be reversible or irreversible. Higher binding affinity ($k1 > k2$) will favor the drug-protein complex formation. Generally speaking, protein binding is considered negligible (0–50%), moderate (50–90%), high (90–99%) or very high (>99%).

$$[\mathbf{Protein}] + [\mathbf{Drug}] \underset{k2}{\overset{k1}{\rightleftarrows}} [\mathbf{Protein-Drug\ Complex}]$$

A drug with protein binding of 96% ± 3% in rats is similar to 98% ± 3% in dogs when there is overlap even though the mean values would suggest twice as much free fraction in rats. Note that under steady state conditions, plasma protein binding displacement does not affect the free drug concentration (C_{free}) for most drugs [14].

Box 13.2 Why Do Protein Binding Studies

- Only free unbound drug is pharmacologically active and available for distribution and metabolism and elimination
- Protein binding can be a limiting factor for low extraction ratio drugs in the liver (hepatic clearance)
- Highly bound drugs will have a limited reach and lower volume of distribution in the body
- Expect changes in potential toxicity when % bound varies greatly between species
- Drug displacement interaction-One drug can displace another when both bind the same site on plasma proteins.
- High drug concentrations can saturate protein binding sites, leading to a non-linear increase in unbound drugs in the plasma
- Disease affecting the levels of α-acid glycoprotein and albumin in the plasma may change the free unbound fraction (fu) in the plasma (but not necessarily the drug concentration (C_{free}) in plasma)

13.5.4 Drug Transporters

Drug transporters are proteins that can move drugs across biological membranes by active transport via ion mobilization or by passive process across a concentration gradient. The passive (concertation gradient dependent) or active (ATP dependent) translocation of drugs across biological barriers into (influx) and out (efflux) of organs/tissues (i.e., intestine, brain, kidney, liver) can affect the process of drug absorption, distribution and excretion.

Currently, there are more than 70 transporters that are divided into three superfamilies: ATP-Binding Cassette (ABC), Solute-Linked Carrier (SLC) and the Solute Carrier Organic anion (SLCO) superfamily [15]. The location and role of several well recognized human drug transporters are depicted in Fig. 13.7.

The ABC transporter superfamily has at least 7 members and these transporters hydrolyze ATP to translocate compounds unidirectionally across the cell membrane. The ABC superfamily members include P-gp, MRP1, and BCRP.

- Multidrug Resistance-associated Protein 1 is known to facilitate drug resistance and can efflux anions, the antioxidant glutathione (GSH) and the proinflammatory leukotriene C4.
- P-glycoprotein is an efflux transporter expressed on the apical side of enterocytes in the small intestine, the blood-brain barrier in

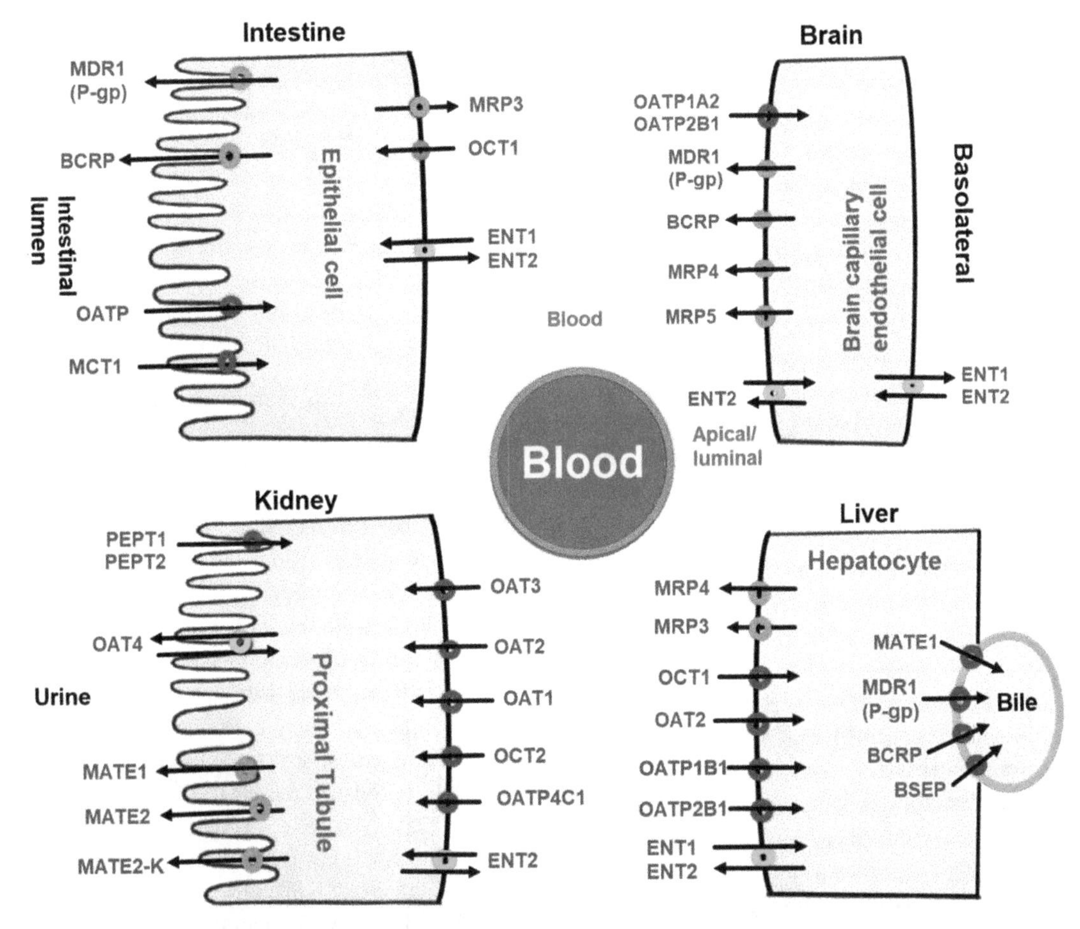

Fig. 13.7 The location and function of transporters generally investigated for an experimental drug are presented below. Transporters recommended by the FDA in the investigation of drug-interactions are depicted as red (i.e., MDR1, BCRP, OATP2B1, OATP1B3, OATP1B1, OAT1 and OAT2, OAT3. Abbreviations: *MATE* multidrug and toxin extrusion, *MRP* multidrug-resistance protein, *BSEP* bile salt export pump, *OAT* organic anionic transporters, *OCT* organic cationic transporters, *ENT* equilibrative nucleoside transporters, *OATP* organic anion transporting polypeptide, *PEPT1* proton coupled oligopeptide transporter, *MDR* multidrug resistance; *P-gp* P-glycoprotein also referred to as MDR, *BCRP* breast cancer resistance protein

the brain and the blood-retinal barrier in the eye. P-glycoprotein can remove (pump out) hydrophobic drugs from cells (i.e., intestinal lumen) and although this activity can be beneficial for blocking toxins, it can prevent drugs from reaching their site of action.

- Breast Cancer Resistance Protein is expressed in many tissues and has broad substrate recognition.

The SLC superfamily has more than 65 families and includes OCT and OAT that are capable of transporting a diverse range of compounds and ions. These transporters have an important role in the distribution and elimination of many potentially toxic endogenous and exogenous organic cations and anions.

The SLCO (formerly SLC21) superfamily, also known as the OATP superfamily, has at least 6 families that are predominantly involved in the transport of large molecules such as bile salts, prostaglandin, xenobiotics.

In vitro nonclinical studies are frequently conducted to address the potential role of BCRP, P-gp, OCT2, MATE1 and MATE2-K in

Table 13.1 Examples of in vitro systems to investigate transporter-mediated drugs [16]

Transporter	In vitro systems
ABC transporters	
BCRP, P-gp	Caco-2 cells, commercial or in-house membrane vesicles, knockout/down cells, transfected cells (MDCK, LLC-PK1, etc.)
Solute carrier (SLC) transporters	
OATP1B1/3	Hepatocytes, transfected cells (CHO, HEK293, MDCK, etc.)
OAT 1/3, OCT2	Transfected cells (CHO, HEK293, MDCK, etc.)
MATEs[a]	Commercial or in-house membrane vesicles, transfected cells (CHO, HEK293, MDCK)

CHO Chinese hamster ovary cell, *HEK293* human embryonic kidney 293 cell, *LLC-PK1* Lilly Laboratory cancer porcine kidney 1 cell, *MDCK* Madin-Darby canine kidney cell

[a] The function of MATEs depends on the driving force from oppositely directed proton gradient; therefore, the appropriate pH of a MATE assay system should be employed

the transport of an NME for regulatory submission (Table 13.1).

- In vitro Caco-2 cell assay to assess efflux transporter BCRP. The Breast Cancer Resistance Protein is expressed by the intestinal epithelial cells, bile canaliculus, kidney, blood-brain barrier and placenta. The breast cancer resistance protein can actively pump small molecules from enterocytes into the intestinal lumen, from hepatocytes into the bile duct and from the brain into the bloodstream, effectively reducing drug efficacy and toxicity.
- In vitro Caco-2 cell assay to assess efflux transporter P-gp. P-gp is expressed in a wide variety of tissues including the intestine, liver, kidney and brain. It actively pumps small molecules from enterocytes into the intestinal lumen and bile duct, from proximal tubular cells to tubules, and from the brain into the blood.
- In vitro assay to assess drug effect on human OCT2, Multidrug and Toxin Extrusion 1 and Multidrug and Toxin Extrusion 1 (MATE1) and MATE2-K transporters. The OCT2 transporter is located on the basolateral membrane of the proximal tubular cells of the kidney. The MATE1 and MATE2-K transporters are located on the lumen side of tubular cells of the kidney and actively excrete drugs from the tubular cells into the urine. Drugs affecting these transporters may result in drug-drug interactions that can impact drug toxicity and efficacy.

The role of P-gp and BCRP in transport of the NME can be determined in vitro using Caco-2 cells. A general flow chart described by Giacomini and Huang [15] may be considered when evaluating the role of transporters involved in the absorption, distribution and clearance of an NME. Briefly, if haptic or biliary secretion of an NME is ≥25%, you may consider determining if the NME is substrate for OATP1B1 and/or OATP1B3 in vitro. If the NME clearance by the kidney ≥25%, you may consider determining whether the NME is substrate for OAT1, OAT3, OCT2 and/or MAT1 in vitro. OATP1B1 and/or OATP1B3 in vitro. Frequently used Knockout animal models [17] recommended by International Transporter Consortium (ITC) when investigating role of different transporters are listed in Table 13.2 [17]

13.6 Metabolism

Biotransformation is a process by which a compound is converted to another compound by a biological system. This takes place as soon as a compound enters the body and involves nearly every organ system with some capacity to contribute to the overall breakdown of a compound. The liver is the principal organ involved in metabolism followed by the intestine, kidney, lung, skin and blood. The primary focus of the discussion here is small molecules. Biological products are high molecular weight entities that are typically composed of peptides, proteins and globulins

Table 13.2 Knockout animal models to address the role of specific transporters

Transporter	Knockout models (KO)
P-glycoprotein (P-gp) (ABCB1, commonly MDR1)	*Mdr1a*$^{-/-}$ mice *Mdrla*$^{-/-}$ rats *Mdr1a/1b*$^{-/-}$ mice
BCRP (ABCG2)	*Bcrp*$^{-/-}$ *mice* *Bcrp*$^{-/-}$ *rats*
P-gp+ BCRP (MDR1 + ABCG2)	*Mdr1a/1b/Bcrp*$^{-/-}$ *mice*
MRP2 (ABCC2)	*Mrp2*$^{-/-}$ *mice* *Mrp2*$^{-/-}$ *rats*
OATP1B1, OATP1B3 (SLC01B, 1B3)	*Oat1a/1b*$^{-/-}$ *mice* *OATP1B1 or 1B3 humanised* *Oatp1a/1b*$^{-/-}$ *mice*
OCT2 (SLC22A2)	*Oct1/2*$^{-/-}$ *mice*
MATE1, MATE2K, MATE2 (SLC47A1-2)	*Mate1*$^{-/-}$ mice
OAT1, OAT3 (SLC22A6, 8)	*Oat1*$^{-/-}$ *and Oat3*$^{-/-}$ *mice*

and/or their combination (peptides, proteins and antibodies). Biological products are generally administered via a parenteral route (IV, SC or IM) with distribution largely limited to the plasma and extracellular fluid. Since biological compounds are degraded by proteases and peptidases to smaller peptides and/or amino acid fragments, their degradation and metabolism are minimally impacted by hepatic CYP enzymes. However, they can affect liver enzymes responsible for the metabolism of small molecules.

The liver enzymes that catalyze the conversion of small molecules into inactive metabolites are the predominant determinant for the duration of drug action in the body. These complex liver metabolic reactions are divided into phase 1 and phase 2 metabolism pathways. Dividing the reactions into two phases does not mean one phase takes precedence over the other. A drug can go through both in any sequence, once or repetitively, or only take one route.

13.6.1 Phase 1 Metabolism

Phase 1 metabolism, also known as non-synthetic metabolism, is a process whereby one of the functional groups below are introduced into the compound. Phase 1 involves microsomal enzymes to catalyze lipid soluble compounds and non-microsomal enzymes to catalyze less lipid soluble substances. The addition of a functional group can produce a short lived reactive/toxic metabolite.

- Oxidation (oxygen added)
- Reduction (oxygen removed)
- Hydrolysis
- Introduction of polar functional group

Some of the enzymes involved in phase 1 reactions are: Cytochrome P450, Flavin monooxygenase, alcohol dehydrogenase, Aldehyde dehydrogenase, Aldo-keto reductase, Xanthine oxidase, esterase, amidase, peptidase and epoxide hydrolase.

13.6.2 Phase 2 Metabolism

Phase 2 metabolism, also known as synthetic metabolism, generally renders a compound more water soluble and biologically inactive with some exceptions (i.e., acyl glucuronides). The five well recognized reactions are glucuronidation, sulfation, acetylation, methylation and amino acid conjugation. Glucuronidation is a low-affinity/high-capacity reaction and represents the most common Phase 2 metabolic pathway.

Conjugation occurs with compounds at the site of the functional groups. The functional groups are amide (–NH_2), carboxyl (–COOH), sulfhydryl (–SH) and hydroxyl (–OH). By conjugating with the functional group on a drug, glucuronidation increases drug solubility and facilitates renal filtration. Enzymes responsible for phase 2 reactions are: UDP-glucuronosyl transferase, sulfotransferase, glutathione S-transferase, B-acetyl transferase, glutamine, B-acetyl-transferase, glycine B-acyl-transferase and methyl transferase. UDP-glucuronosyl transferase (UGT) is the most prominent phase 2 enzyme. UDP-glucuronosyl transferase is a membrane bound enzyme that requires UDP-glucuronic acid as co-factor and is expressed predominantly in the liver but is also found in the intestine and kidney.

13.6.3 Enzyme Induction and Inhibition

Hepatic metabolism of drugs is governed by the concentration of the metabolizing enzymes [19]. Since high liver enzyme concentrations can metabolize more drugs, enzymatic induction can result in lower levels of therapeutic agents. The reverse is also true. Lower levels of liver enzymes will metabolize less drug and taken together enzyme induction and enzyme inhibition represents significant factors when developing an NME.

Cytochrome P450 (CYP450) is an intracellular-membrane bound-heme-containing protein that catalyzes oxidative reactions and is by far the most prominent enzyme in the liver. Although the liver has the highest concentration of CYP450, other organs such as the intestine, lung and nasal mucosa also express moderate levels. There is substantial variation in the expression of CYP450 isozymes among individuals and species and there are at least 12 known families and 22 subfamilies of CYP450 (see the example for CYP3A4, Fig. 13.8).

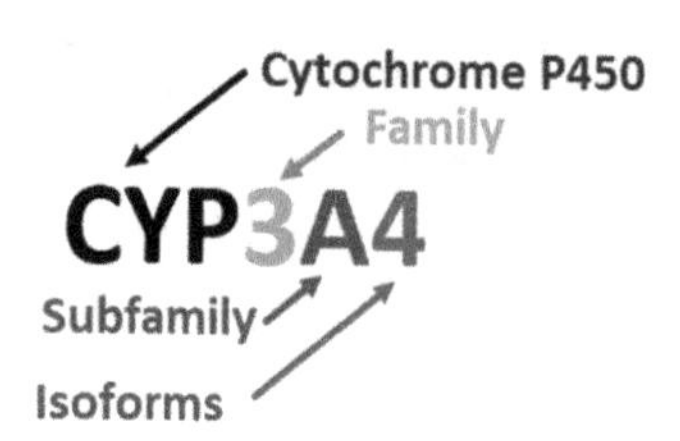

Fig. 13.8 General name classification of CYP enzymes

Some of the most common liver enzyme inducers are barbiturates, rifampin, phenytoin and alcohol. In contrast, ketoconazole, verapamil and grapefruit juice are well known inhibitors. Although inducers (i.e., rifampin inducer of CYP2A19) can significantly reduce the concentration and efficacy of therapeutic agents, inhibitors may pose an even greater concern. Inhibitors that elicit a significant increase in plasma drug levels have the potential to increase the risk of drug-related toxicity. It is on this basis that the development of any new investigational compound must include an assessment of the potential for liver enzyme induction and inhibition. Tables 13.3 and 13.4 list some of the most potent CYP450 isozyme inhibitors and inducers.

13.6.4 Factors Affecting Drug Metabolism

- Age (from neonatal to mature to geriatric)
- Individual genetic variation (normal vs. slow vs. fast vs. ultra metabolizers)
- Interspecies differences in liver enzymes
- Body condition (obese vs. lean, liver or kidney function)
- Pregnancy
- Sex differences (androgenic regulated P450 enzymes)
- Nutrition (high fat, fiber, grapefruit juice)
- Route of drug admin. (First pass effect: liver and GI)
- Dose (high dose → shift to less common pathways)
- Intestine microflora

Effect of Age There is growing evidence that age plays a critical role in drug metabolism and excretion. The fetal and neonatal liver has immature CYP3A Sub-family, 2D6 and lower glucuronidation capacity (i.e., chloramphenicol toxicity) but normal ability to sulfate. Limited neonatal renal function can also slow renal clearance (i.e., antibiotics) while larger total body water and lower protein binding can affect free drug concentrations.

Table 13.3 Prominent cytochrome P450 enzyme inhibitors (derived from FDA guidances on metabolites) [18]

CYP * enzymes	Strong inhibitors >fivefold increase in AUC or >80% decrease in CL	Moderate inhibitors >2 but <fivefold increase in AUC or 50-80% decrease in CL	Weak inhibitors >1.25 but <twofold increase in AUC or 20–50% decrease in CL
CYP1A2	Ciprofloxacin, enoxacin, fluvoxamine	Methoxsalen, mexiletine, oral contraceptives, phenylpropanolamine, thiabendazole, zileuton	Acyclovir, allopurinol, caffeine, cimetidine, daidzein. disulfiram. echinacea. famotidine, norfloxacin, propafenone, propranolol, terbinafine, ticlopidine, verapamil
CYP2B6			Clopidogrel, ticlopidine, prasugrel
CYP2C8	Gemfibrozil		Fluvoxamine, ketoconazole, trimethoprim
CYP2C9		Amiodarone, fluconazole, miconazole, oxandrolone	Capecitabine, cotrimoxazole, etravirine, fluvastatin, fluvoxamine, metronidazole, sulfinpyrazone, tigecycline, voriconazole, zafirlukast
CYP2C19	Fluconazole, fluvoxamine, ticlopidine	Esomeprazole, fluoxetine, moclobemide, omeprazole, voriconazole	Allicin (garlic derivative), armodafinil, carbamazepine, cimetidine, etravirine, human growth hormone (Rhgh), felbamate, ketoconazole, oral contraceptives
CYP3A	Boceprevir, clarithromycin, conivaptan, grapefruit juice, indinavir, itraconazole, ketoconazole, lopinavir/ritonavir, mibefradil, nefazodone, nelfinavir, posaconazole, ritonavir, saquinavir. telaprevir, telithromycin. voriconazole	Amprenavir. aprepitant, atazanavir. ciprofloxacin. darunavir/ritonavir, diltiazem. erythromycin, fluconazole, fosamprenavir, grapefruit juice, imatinib, verapamil	Alprazolam, amiodarone, amlodipine, atorvastatin, bicalutamide, cilostazol, cimetidine, cyclosporine, fluoxetine, fluvoxamine, ginkgo, goldenseal, isoniazid, nilotinib, oral contraceptives, anitidine, ranolazine, tipranavir/ritonavir, zileuton
CYP2D6	Bupropion, fluoxetine, paroxetine, quinidine	Cinacalcet, duloxetine, terbinafine	Amiodarone, celecoxib, cimetidine, desvenlafaxine, diltiazem, diphenhydramine, echinacea, escitalopram, febuxostat, gefitinib, hydralazine, hydroxychloroquine, imatinib, methadone, oral contraceptives, propafenone, ranitidine, ritonavir, sertraline, telithromycin, verapamil

* Common CYP enzymes involved in metabolism of variety of drugs

Similar changes also occur in the elderly marked by diminished CYP3A4 levels and induction capacity and reduced renal function. It is estimated that renal function decline by 1ml/min/year after the age of 40. Theoretically, normal renal function (120 mL/min) could decrease to 80 mL/min by the age of 80.

Effect of Pregnancy Placenta has high levels of CYP1A family. Profound induction may increase exposure to toxic metabolites not expected in a non-pregnant person. Furthermore, hormonal changes can affect CYP enzymes (↑ CYP2D6, ↑ CYP3A4, ↓ CYP1A2) thus affecting drug metabolism.

Table 13.4 Prominent cytochrome P450 enzyme inducers (derived from FDA guidances) [18]

CYP enzymes	Strong inducers >80% decrease in AUC	Moderate inducers 50–80% decrease in AUC	Weak inducers 20–50% decrease in AUC
CYP1A2		Montelukast, phenytoin, smokers Versus non-smokers	Moricizine, omeprazole, phenobarbital
CYP2B6		Efavirenz, rifampin	Nevirapine
CYP2C8		Rifampin	
CYP2C9		Carbamazepine, rifampin	Aprepitant, bosentan, phenobarbital, St. John's Wort
CYP2C19	Rifampin	Artemisinin	
CYP3A	Avasimibe, carbamazepine, phenytoin, rifampin, St. John's Wort	Bosentan, efavirenz, etravirine, modafinil, nafcillin	Amprenavir, aprepitant, armodafinil, echinacea, pioglitazone, prednisone, rufinamide
CYP2D6	None known	None known	None known

Route of Administration After oral administration, the drug travels from the gut to the portal vein to the liver. Microsomal enzymes in the intestinal wall and gut microflora may result in a notable first pass effect. Rapid metabolism in the liver may leave little drug left to get into the systemic circulation. In contrast, SC and IV routes will reduce the initial liver and intestinal first pass effect.

Effect of Sex Although male and female rodents generally have similar CYP profiles, concentrations of some CYP enzymes are affected by sex hormones. Examples of some differences in some CYP enzymes in mice, rats and humans are listed below.

Species	CYP enzymes
Mice	CYP2D9 is specific to male mice
Rats	CYP2C11 and CYP2A2 are specific to male rats
	CYP2C12 is specific to female rats
Humans	CYP3A more abundant in men than women
	CYP enzyme may account for up to 40% differences in PK
	CYP1A2 is more in men than women (i.e., caffeine)
	CYP3A4 is more in women than men (i.e., erythromycin)
	CYP2B6 is more in women than men

Intestinal Microflora The role of the intestine, one of the largest organs involved in metabolism, is further complicated by the microflora associated with it. Studies have indicated that intestinal microorganisms are capable of metabolizing drugs [31] and under certain disease conditions, a change in the microbiota can result in the production of genotoxic metabolites [19]. In other instances, a bacterial enzyme can break down the conjugated metabolite. For example, a glucuronidated drug is susceptible to glucuronidase, a bacterial enzyme, and the removal of the glucuronide (by glucuronidase) can result in the release of the parent drug.

Inter-Species Variations Although most CYP enzymes are well conserved across nonclinical test species (i.e., CYP2E1), there are enough differences in the concentrations of some (CYP1A, -2C, -2D and CYP3A) species to have an impact on drug metabolism profiles [20–23]. For example, humans have one CYP2D isoform (CYP2D6) while rats have at least 5 isoforms (Table 13.5). CYP2C11, an androgen regulated enzyme, plays an important role in male rats but is nearly inactive in other species and female rats. Humans metabolize amphetamine by deamination while rats and dogs metabolize it by aromatic hydroxylation. Guinea pigs (occasionally used in toxicology studies) have little sulfotransferase activity while humans possess substantial sulfotransferase activity and variations like these may result in disproportional or unique metabolites. Smaller animals have higher relative

Table 13.5 Cytochrome P450 Isoforms[a] in different species

P450	Human	Monkey	Rabbit	Rat	Mouse
CYP1A	1A1, 1A2	1A1, 1A2	1A1, 1A2	1A1, 1A2	1A1, 1A2
CYP2A	2A6, 2A7	–	2A10, 2A11	2A1, 2A2, 2A3	2A4, 2A5
CYP2B	2B6	2B17	2B4, 2B5	2B1, 2B2, 2B3	2B9, 2B10, 2B13
CYP2C	2C8, 2C9, 2C18, 2C19	2C20, 2C37	2C1, 2C2, 2C3, 2C4 2C5, 2C14, 2C15, 2C16	2C6, 2C7, 2C11, 2C12, 2C13, 2C22, 2C23, 2C24	2C29
CYP2D	2D6	2D17		2D1, 2D2, 2D3, 2D4, 2D5	2D9, 2D10, 2D11, 2D12, 2D13
CYP2E	2E1	2E1	2E1, 2E2	2E1	2E1
CYP3A	3A4, 3A5	3A8	3A6	3A1, 3A2, 3A9	3A11, 3A13, 3A16
CYP4A	4A9, 4A11		4A4, 4A5, 4A6, 4A7	4A1, 4A2, 4A3, 4A8	4A10, 4A12, 4A14

CYP cytochrome P450 subfamily, – no information available to date

[a] These are found predominantly, but not exclusively, in the liver of the relevant species [22]

amounts of CYP enzymes/kg body weight compared to larger animals and humans and thus one would expect greater metabolism in smaller animals.

13.6.5 Nonclinical Metabolism Studies

Metabolism studies generally follow two paths: (a) in vitro tests conducted generally with hepatic microsomes, hepatocytes, or liver slices, and (b) in vivo metabolism studies incorporated into mass balance and toxicology studies conducted in nonclinical animal models (Box 13.3).

Box 13.3 Why Do Metabolism Studies

- To determine the extent of metabolism (changes in drug levels can impact PD)
- To determine the metabolic pathways and which hepatic enzymes are involved in metabolism
- To identify metabolites and toxicity concerns whether they are pharmacologically active
- To determine if the test drug will induce or inhibit any metabolic enzymes and evaluate the risk of potential drug-drug interactions
- To identify if the metabolism profile in animals is similar to humans (any evidence of disproportional metabolite)?
- To determine if there is a potential sex difference in metabolism and associated toxicity

In vitro microsomal studies are widely used to characterize the test drug metabolism by phase 1 enzymes (i.e., CYP enzymes). Since liver microsomal preparations are commercially available for multiple species (rats, mice, rabbits, dogs, monkeys and humans), it is possible to characterize the metabolic profile for animal models of interest. Since in vitro microsomal studies are standardized and well established, most are conducted by contracting commercial labs similar to standard genotoxicity tests. Metabolism studies for small molecules may include:

- In vitro microsomal assay using liver microsomes from a wide range of test species (i.e., mouse, rat, rabbit, dog, monkey) and humans to profile and identify metabolites formed by microsomes from each species.
- In vitro assay using hepatocytes from a wide range of nonclinical animal models (i.e., mouse, rat, rabbit, dog, monkey) and humans

to profile and identify metabolites formed by hepatocytes from each species.

- In vitro microsomal assay to identify the major cytochrome P450 enzymes involved in drug metabolism. The assay is usually directed to common CYP enzymes involved in metabolism of small molecules (CYP1A2, CYP2B6, CYP2C8, CYP2C9, CYP2C19, CYP2D6 and CYP3A). Selective CYP enzyme inducers and inhibitors are used to determine the potential for drug-drug interactions.
- In vitro microsomal assay using human hepatocytes to determine the induction activity of a drug on common CYP enzymes (CYP1A2, CYP2B6, and CYP3A4). Increased concentration of these CYP enzymes can enhance the clearance of co-administered drugs. Selective CYP enzyme inducers are used as a positive control (i.e., Omeprazole, Phenobarbital, Rifampicin, Phenacetin, Midazolam, Acetaminophen, Hydroxy bupropion).
- In vitro microsomal assay using human liver microsomes to determine the inhibitory activity of a drug on common CYP enzymes (CYP1A2, CYP2B6, CYP2C8, CYP2C9, CYP2C19, CYP2D6 and CYP3A). Selective CYP enzyme inhibitors are used as a positive control (i.e., α-Naphthoflavone, Ticlopidine Hydrochloride, Montelukast Sodium Salt, Sulfaphenazole, (S)-(+)-N-3-Benzylnirvanol, Quinidine hydrochloride, Ketoconazole).

Information gained from the in vitro metabolism studies can be used to predict in vivo biotransformation, metabolic pathways, and identity of metabolites. That said, in vitro studies by their nature, may not represent what happens in vivo. In vivo data collected from mass balance and toxicology studies are likely more reflective of the actual drug metabolism and metabolite formation. In vivo studies can determine whether exposure to metabolites provides safety coverage to metabolites formed in humans. Metabolites formed by phase 1 are more likely to be chemically reactive and pharmacologically active while the opposite is true for phase 2 which usually renders a drug viable for renal elimination.

13.6.6 Disproportional and Unique Human Metabolites

Although there is substantial shared liver enzyme activity across species, there are instances where metabolism studies identify a unique or disproportional human metabolite that may require nonclinical safety assessment [32]. Per FDA human metabolite guidance [24], when the exposure for a human metabolite is less than 10% of total drug exposure in humans, with no evidence of reactivity or genotoxicity, no nonclinical safety assessment is necessary. It should be noted that the 10% rule does not apply when the disproportional human metabolite is reactive or genotoxic.

A toxicological assessment of a particular metabolite may not be necessary if the exposure levels achieved in one of the relevant nonclinical test species is equal to or greater than the metabolite exposure levels observed in humans. When a disproportional metabolite is not expressed in any of the relevant test species, a nonclinical safety assessment should be considered (Fig. 13.9). Demonstrating that a metabolite is not pharmacologically active does not eliminate the need for a toxicological assessment of a disproportional metabolite (off-target activity).

Box 13.4 Points to Consider When a Human Metabolite Is Suspected

- Any known or new toxicity signal
- Potential for genotoxicity (QSAR test)
- In vitro versus in vivo metabolite profile
- Is there biliary exposure in animal models
- The extent of human exposure compared to test species is unique or disproportional
- Does the metabolite accumulate with repeated dosing?
- Reliability and validation of the analytical assay
- Synthesis of the metabolite
- Alternate animal model
- Background knowledge of the drug class

(continued)

Box 13.4 (continued)

- Metabolite is pharmacologically active or inactive. Pharmacologically inactive are not devoid of toxicity
- For drug doses <10 mg/day, a greater fraction of the drug related material might be a more appropriate trigger for testing
- Is metabolite a phase I or phase II product? Phase II products-generally less toxic, and more water soluble (glucuronide and glutathione, metabolites with hydroxyl group) vs. more toxic reactive acyl glucuronide metabolites
- Resolve any human metabolite issues early in the drug development

13.7 Excretion

Drug excretion (elimination) is defined as irreversible removal of a drug from the body by all possible routes. Drug clearance involves many organs including the liver (CL_H), kidney (CL_R), bile (CL_B), lung (CL_P) and many minor routes (Fig. 13.10). It is by far one of the most important processes. Collectively, they clear the body of the drug and represent the total drug clearance (CL). As a calculated value, CL is determined using mathematical formulas. CL is one of the most consequential PK parameters.

Renal clearance (CL_R) of a drug is the combined process of glomerular filtration, tubular secretion and tubular reabsorption. Renal excretion to urine is a major route of elimination for many drugs. The kidneys can readily filter water soluble and low molecular weight compounds (<300 kDa). Large molecules cannot be passively filtered readily due to their size except under certain disease conditions or age-related deterioration in renal function. Large molecules, such as albumin, can leach into the renal tubules in older male rats with glomerulosclerosis. In comparison, tubular secretion is an energy dependent (active) transport process that is capable of excreting large molecules with high molecular weights and protein-bound drugs. Tubular reabsorption, that operates in the opposite direction, can return highly lipophilic compounds from the renal tubules back to the blood.

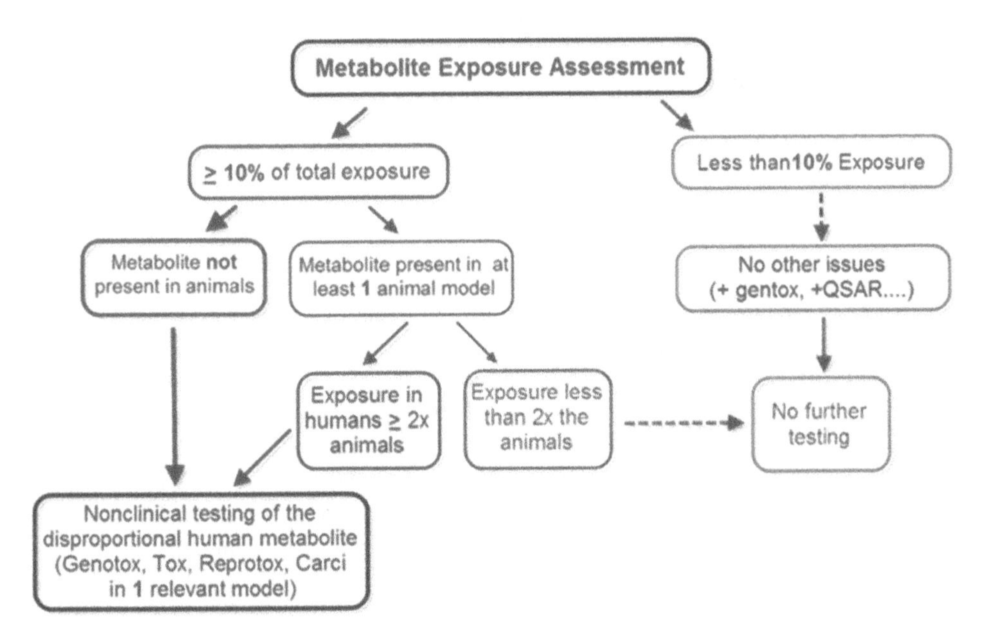

Fig. 13.9 Human metabolite decision chart (FDA metabolite safety assessment guidance)

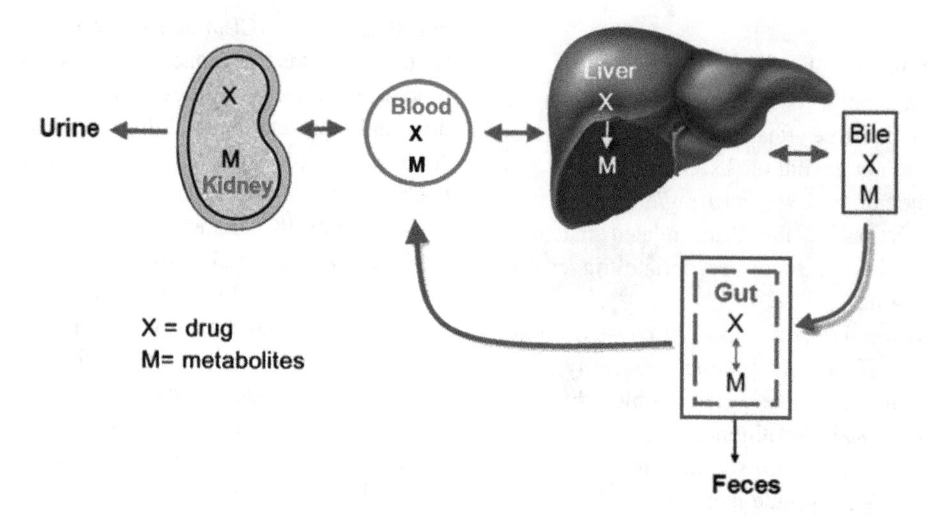

Fig. 13.10 Routes of elimination and metabolism: The three most common pools of excretion pathways are bile, urine and feces. All are accessible for sample collection in rodents. Enterohepatic recirculation occurs when a compound is absorbed from the gut, metabolized by the liver, excreted to bile then to the gut and reabsorbed back into blood. This recycling can extend the half-life and expose the drug to more intestinal degradation

Renal Clearance = Filtration + Secretion − Reabsorption

The rate at which filtration, tubular secretion and reabsorption operate is governed by the amount of drug filtered through glomeruli (A), unbound drug in plasma (C_u) and glomerular filtration rate. Changes in any of these parameters (disease, decline in renal function with age, high BP) can significantly reduce drug clearance with a resultant increase in drug exposure. Renal function (GFR ~120 ml/min) declines in humans by approximately 1 mL/min/year over age 40, indicating that renal function should be considered if a drug is cleared primarily via renal filtration. Similarly, active tubular secretion (acid and base pumps) can be inhibited resulting in drug accumulation in the plasma and represents a notable safety concern. Knowing the role of tubular section in drug excretion has been exploited in some instances when higher plasma drug levels are desirable. Probenecid, a potent renal acid pump inhibitor has been used to decrease renal tubular excretion in the clinical setting.

$$A = C_u \cdot GFR$$

13.8 Toxicokinetics

Toxicokinetic studies are essentially PK studies at the toxicological doses used in toxicity studies. The objectives of TK studies are to determine the relationship between systemic exposure and drug toxicity and how it may relate to humans. Toxicokinetic studies are generally conducted during the course of drug development and range from acute to chronic toxicity studies in rodents (i.e., rat, mouse) and nonrodents (i.e., dogs, monkeys) to reproductive toxicity studies in rats and rabbits. Toxicokinetic data collection from carcinogenicity studies may not be necessary if there are TK data available from chronic toxicity studies.

Toxicokinetic data can demonstrate the systemic drug burden (API and metabolites). Plotting the AUC data can demonstrate the rate, extent and duration of exposure. The toxicity may be related to the rate of exposure (threshold plasma concentration, mode of administration (gavage vs diet mixture) and/or dosage form (solution, capsule or micronization). As line 1 in Fig. 13.11 demonstrates, the AUC exposure may increase proportional to dose or increase non-linearly and reach absorption saturation as in line 2. Line 3 is

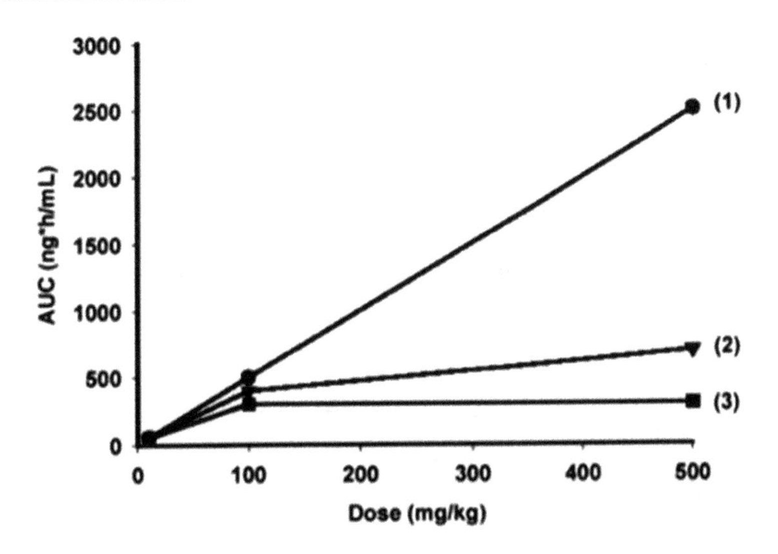

Fig. 13.11 Theoretical presentation of AUC exposure data representing: (1) dose proportional increase in AUC with an increase in dose; (2) non-linear increase in AUC representing absorption saturation, (3) exposure plateau due to absorption saturation, poor delivery and/or rapid degradation

indicative of absorption saturation, intolerance (i.e., emesis in dogs) and/or rapid drug degradation.

The three primary TK parameters of interest in toxicological studies are Cmax, AUC and Tmax. The Cmax is expressed as the concentration of the drug in a volume of plasma (i.e., μg/mL or ng/mL) while AUC is generally expressed as drug concentration time/volume of plasma (i.e., μg h/mL or ng h/mL). Since the primary objective of TK data is to relate drug exposure to toxicity, the determination of the AUC and Cmax are often sufficient to achieve this goal (Box 13.5). Pharmacokinetic parameters such as $t_{1/2}$, drug clearance (Cl), and volume of distribution (Vd) have limited value. Drug exposures at toxicological doses may exceed the gastrointestinal (GI) absorption capacity and be excreted in the feces which is indicative of poor bioavailability (F). High doses can also exceed the clearance capacity in animals and result in drug accumulation and potential toxicity.

Box 13.5 Why to Do TK Studies

- To interpret toxicology findings in relation to drug concentrations in plasma (see ICH S3A, 1995)
- To bridge toxicology findings to the clinical exposure
- To determine dose linearity/proportionality (increase in AUC and Cmax proportional to the dose). Dose linearity would suggest that absorption and metabolism are similar over the dose range
- To assess the potential for drug accumulation that may occur with repeated dosing (first vs. the last dose)
- To explore drug receptor or enzymatic activity at Cmax
- To monitor parent alone or parent + metabolite(s) and their similarities/dissimilarities among species
- To gain an understanding of "effective" drug exposure for unanticipated findings
- To determine gender differences in disposition
- To establish exposure multiples (safety margins) between animals and clinical doses based on plasma drug exposure (AUC, Cmax)

Toxicokinetic studies in small animals (Box 13.6) with small blood volume (i.e., mouse and rat) are often carried out in satellite animals treated identically to the main toxicology study

Box 13.6 General Design of TK Satellite Animals in a 14-day Oral Toxicity Study in Rats

- Dose groups: 0, 20, 100 and 500 mg/kg/day for 14 days by oral gavage
- Bleeding schedule (3 rats/gender/time-point/dose-group)
- Sampling days: Day 1 and Day 14 (end of the study)
- Blood sampling consideration (10 mL/kg BW/24 h period, rats: 2–3 blood samples per rat; mouse: 1 sample per mouse per time period).
- Blood sample collection from control can demonstrate absence of drug in the blood.

Sampling time, hr	Rat #	Sampling Day
0.5	1, 2, 3	Day 1 and Day 14
2	4, 5, 6	Day 1 and Day 14
4	7, 8, 9	Day 1 and Day 14
8	10, 11, 12	Day 1 and Day 14
24	13, 14, 15	Day 1 and Day 14

animals but used for blood sample collection only [24, 25]. In larger animals, blood samples for TK analysis are collected from the main study animals. In an effort to use fewer rodents in toxicity studies, in the spirit of the 3Rs (Replace, Reduce, Refine), microsampling is being employed more often. As investigators gain experience and precision analytical instruments become more widely available, the use of microsampling techniques to draw small quantities of blood (<50 μL) from main study animals has become more commonplace [26–28]. By collecting small blood volumes, repetitive sampling can be conducted in the main toxicology study without any significant impact on blood hematocrit and hemodynamics. Investigators can reduce or eliminate satellite animals, thus reducing the number of animals needed by half. An additional benefit of the microsampling technique is that it allows for the analysis of drug levels for each animal in the main study to be used in the toxicity signal evaluations for the same individuals, which is far superior to an indirect comparison of exposure between satellite animals and main study animals.

13.8.1 Safety Margin Calculations

The objective of toxicology studies is to address the safety of an NME in humans (Fig. 13.12). This is most often achieved by conducting toxicology studies in two species (rodents and nonrodents) at doses that will address the toxicity profile of an NME [32]. The dose level that produces no adverse effect (NOAEL) [2] in a nonclinical test species is generally used to determine the maximum recommended starting dose in humans (MRSD) and to determine the safety margin (more accurately, exposure multiples) for the maximum recommended human dose (MRHD). As the name implies, the MRSD is the first dose used in human clinical trials that are most often conducted in healthy human subjects. In comparison, the MRHD is the maximum dose that can be supported by the nonclinical data.

[2] NOAEL is the highest dose level that does not produce a significant adverse effect in the nonclinical test species.

Fig. 13.12 Toxicology and TK data gained from animals are used to ensure safety in humans

13.8.2 First Dose in humans

In the absence of clinical experience and availability of exposure data (AUC or Cmax), which is normally the case with an NME, the NOAEL (mg/kg) established in animal toxicology studies is converted to the human equivalent dose (mg/m^2) based on body surface area (BSA). Key elements you may consider for the estimating the MRSD are listed in Box 13.7.

Box 13.7 Major Elements Needed for MRSD in Humans

- Review and evaluate animal data (rodent and non-rodent)
- Determine the no observed adverse effect levels (NOAEL)
- Convert NOAEL to human equivalent dose (HED)
- Select the most sensitive species or most relevant toxicity signal for assessing human risk
- Apply a safety factor (i.e., [9]) to increase assurance of safety for the first dose in humans

Converting the NOAEL in mg/kg to mg/m^2 BSA is a reasonably reliable method for the normalization of NOAEL and is, in general, consistent with the NOAEL based on AUC for small molecules. It is therefore considered acceptable to estimate the MRSD based on a NOAEL that has been normalized for BSA. Table 13.6 derived from FDA guidance is used for estimating a safe starting dose. The table lists the conversion factors for most common lab animal species used toxicological assessment [29].

After NOAEL is converted to HED, a tenfold safety factor is generally applied to increase the assurance that the first dose in humans will not elicit an adverse effect. Humans may be more sensitive, or a drug may have greater bioavailability in humans than in animals. However, you may increase or decrease the standard tenfold safety factor based on your knowledge on the pharmacological class and/or known toxicity profile of the test compound (Table 13.7).

Two examples of MRSD calculations using normalized NOAEL based on body surface area provided are provided in Box 13.8. The MRSD estimates in the two examples were similar suggesting greater concordance between the two species and more confidence in the estimates. When the two test species (rodent versus non-rodent) result in two different MRSD estimates, it is prudent to go with most sensitive species that yields lower MRSD.

Once there is clinical AUC data, the best approach for determining a safety margin (exposure multiples) is to divide the AUC for the NOAEL by the AUC for the MRHD. The AUC representing drug exposure in the blood is the most relevant measure of drug exposure in animals and humans and is therefore preferred over safety margins based on BSA. However, in rare conditions when blood drug levels and AUC

Table 13.6 Conversion of animal doses to human equivalent doses based on body surface area (derived from FDA Guidance for Industry [29])

Species	To convert animal dose in mg/kg to dose in mg/m^2, Multiply by k_m	To convert animal dose in mg/kg to HED[a] in mg/kg, either	
		Divide Animal dose by	Multiply Animal dose by
Human	37	–	–
Child (20 kg)[b]	25	–	–
Mouse	3	12.3	0.08
Hamster	5	7.4	0.13
Rat	6	6.2	0.16
Ferret	7	5.3	0.19
Guinea pig	8	4.6	0.22
Rabbit	12	3.1	0.32
Dog	20	1.8	0.54
Primates			
Monkeys[c]	12	3.1	0.32
Marmoset	6	6.2	0.16
Squirrel monkey	7	5.3	0.19
Baboon	20	1.8	0.54
Micro-pig	27	1.4	0.73
Mini-pig	35	1.1	0.95

[a] Assumes 60 kg human. For species not listed or for weights outside the standard ranges, HED can be calculated from the following formula: HED = animal dose in mg/kg × (animal weight in kg/human weight in kg)$^{0.33}$
[b] This km value is provided for reference only since healthy children will rarely be volunteers for phase 1 trials
[c] For example, cynomolgus, rhesus, and stumptail

Table 13.7 Factors to consider when deciding what safety factor to use in the calculation of MRSD

Increase the safety factor to greater than 10	Decrease the safety factor to less than 10
• Steep dose response curve Severe toxicities or deaths in animals • Non-monitorable toxicity • Toxicities without prodromal indicators • Variable bioavailability • Irreversible toxicity • Unexplained mortality • Large variability in dose or AUC levels eliciting effect • Questionable study design or conduct • Novel therapeutic agent and target • Animal model with limited utility	• Pre-clinical studies are well designed and high caliber • NOAEL was based on chronic toxicity studies • Similar metabolic profile and bioavailability • Toxicities that are easily monitorable • Toxicity finding is minimal and reversible • Minimal toxicity experience with similar compounds • Shallow dose-response curve • Toxicities represent exaggerated pharmacology in normal animals

cannot be determined, safety margins based on BSA may represent a reasonable alternative.

$$\text{Safety Margin} = \left[\frac{\text{NOAEL dose in animals}_{(\text{AUCormg/m}^2)}}{\text{Maximum Recommended Human Dose}_{(\text{AUCormg/m}^2)}}\right]$$

13.9 Concluding Remarks

Pharmacokinetic studies in animal models are a critical component of the nonclinical safety assessment for any investigational drug. Nonclinical PK studies can guide and impact every aspect

Box 13.8 MRSD Calculations Based on BSA

Rat NOAEL: 10 mg/kg/day

Get Human Equivalent dose: 10 mg/kg x 0.162 = 1.62 mg/kg
For a 60 kg person HED = 60 kg x 1.62 mg/kg = 97.2 mg
Divide HED by the **Safety Factor of 10** (97.2 mg / 10 = 9.7 mg)
MRSD in humans ≈ **10 mg**

Dog NOAEL: 5 mg/kg/day

Get Human Equivalent dose: 5 mg/kg x 0.541 = 2.7 mg/kg
For a 60 kg person HED = 60 kg x 2.7 mg/kg = 162 mg
Divide HED by the **Safety Factor of 10** (162 mg / 10 = 16.2 mg)
MRSD in humans ≈ **16 mg**

of a nonclinical development program. Pharmacokinetic and TK data gained from nonclinical study can guide and predict clinical PK parameters with reasonable accuracy before a drug is ever assessed in humans. Finally, TK data can serve as the critical link between toxicity signals and drug exposure and aid in the interpretation of the toxicology findings and how they relate to humans.

Acknowledgment The author is grateful for diligent edits and recommendations of Dr. Jeff Quinn, CDER, FDA.

Conflict of Interest None.

Disclaimer The content of this chapter are the views of the author and should not be considered as FDA's views or policies.

References

1. Jibaldi M, Perrier D (1982) Absorption kinetics and bioavailability. In: Swarbrick J (ed) Pharmacokinetics. Dekker, New York, pp 156–198
2. Atkinson A, Daniels C, Dedrick R, Grudzinskas C, Markey S (2001) Principals of clinical pharmacology. Academic Press, New York, pp 35–41
3. Levine RR, Walsh CT, Schwartz-Bloom RD (1978) Levine's pharmacology: drug actions and reactions, 2nd edn. Parthenon Publication Group, New York
4. Chen A et al (1995) Metabolism and metabolite pharmacokinetics of BRB-I-28, a class Ib antiarrhythmic agent. Eur J Drug Metab Pharmacokinet 20:151–161
5. Alavi F et al (1991) Disposition of BRB-I-28 (7-benzyl-7-aza-3-thiabicyclo[3.3.1]-nonane hydroperchlorate), a novel antiarrhythmic agent. Drug Invest 3(5):317–323
6. Rifai N (2018) Tietz textbook of clinical chemistry and molecular diagnostics. Elsevier, St. Louis
7. Pardridge WM, Oldendorf WH, Cancilla P, Frank HJL (1896) Blood-brain barrier: interface between internal medicine and the Brain. Ann Intern Med 105:92–98
8. Solon E, Kraus L (2002) Quantitative whole-body autoradiography in the pharmaceutical industry. J Pharmacol Toxicol Methods 46:73–81
9. Waters N et al (2008) Validation of a rapid equilibrium dialysis approach for the measurement of plasma protein binding. J Pharm Sci 97:10
10. Alavi F, Rolf L, Clarke C (1993) The pharmacokinetics of sulfachlorpyridazine in channel catfish, Ictalurus punctatus. J Vet Pharmacol Therp 16:232–236
11. Lee K-J et al (2003) Modulation of nonspecific binding in ultrafiltration protein binding studies. Pharm Res 20:1015–1021
12. Toutain A, Bousquet-Melou A (2002) Free drug fraction vs. free drug concentration: a matter of frequent confusion. J Vet Pharmacol Ther 25:460–463
13. Giacomini, Huang (2013) Transporters in drug development and clinical pharmacology. Clin Pharmacol Ther 94(1):3–9. https://doi.org/10.1038/clpt.2013.86

14. FDA. In vitro drug interaction studies - cytochrome P450 enzyme- and transporter-mediated drug interactions guidance for industry. In vitro drug interaction studies — cytochrome P450 enzyme- and transporter-mediated drug interactions guidance for industry
15. Drug development and drug interactions. Table of substrates, inhibitors and inducers. FDA.
16. Cao A et al (2022) Commensal microbiota from patients with inflammatory bowel disease produce genotoxic metabolites. Science 378:369
17. Martignoni M, Geny M, Groothuis M, de Kanter R (2006) Species differences between mouse, rat, dog, monkey and human CYP-mediated drug metabolism, inhibition and induction. Expert Opin Drug Metab Toxicol 2(6):875–894
18. Guengerich F (1997) Role of cytochrome P450 enzymes in drug-drug interactions. Adv Pharmacol 43:7–35
19. Nelson DR et al (1996) P450 superfamily: update on new sequences gene mapping accession numbers and nomenclature. Pharmacogenetics 6:1–42
20. Lewis D et al (1998) Cytochromes P450 and species differences in xenobiotic metabolism and activation of carcinogen. Environ Health Perspect 106(10):633–641
21. FDA (2020) FDA metabolite guidance for industry: safety testing of drug metabolites. https://www.fda.gov/regulatory-information/search-fda-guidance-documents/safety-testing-drug-metabolites
22. ICHS3A: guideline for industry toxicokinetics: the assessment of systemic exposure in toxicity studies
23. ICH S3A Q&As: note for guidance on toxicokinetics: the assessment of systemic exposure in toxicity studies
24. Jonsson O, Palma Villar R, Nilsson LB, Norsten-Höög C, Brogren J, Eriksson M, Königsson K, Samuelsson A (2012) Capillary microsampling of 25 µl blood for the determination of toxicokinetic parameters in regulatory studies in animals. Bioanalysis 4(6):661–674
25. White S, Hawthorne G, Dillen L, Spooner N, Woods K, Sangster T, Cobb Z, Timmerman P (2014) EBF: reflection on bioanalytical assay requirements used to support liquid microsampling. Bioanalysis 6(19):2581–2586
26. US Food and Drug Administration (2020) Guidance for industry—clinical drug interaction studies-cytochrome P450 enzyme- and transporter-mediated drug interactions
27. FDA (2005) FDA guidance for industry: estimating the maximum safe starting dose in initial clinical trials for therapeutics in adult healthy volunteers. https://www.fda.gov/regulatory-information/search-fda-guidance-documents/estimating-maximum-safe-starting-dose-initial-clinical-trials-therapeutics-adult-healthy-volunteers
28. Cannady E, Katyayan K, Patel N (2022) ADME principles in small molecule drug discovery and development: an industrial perspective. In: Haschek and Rousseaux's handbook of toxicologic pathology, 4th edn. Academic Press, London, pp 51–76. https://doi.org/10.1016/B978-0-12-821044-4.00003-0
29. Zamek-Gliszczynski A et al (2013) ITC recommendations for transporter kinetic parameter estimation and translational modeling of transport-mediated PK and DDIs in humans. Am Soc Clin Pharmacol 94:64–79
30. Barzegar-Jalali M (1990) Determination of absolute bioavailability without intravenous administration. Int J Pharm 65:R1–R3
31. Shamat M (1993) The role of the gastrointestinal microflora in the metabolism of drugs. Int J Pharm 97:1–13
32. FDA (2010) ICH M3 (R2): guidance for industry: nonclinical safety studies for the conduct of human clinical trials and marketing authorization for pharmaceutics. M3(R2): nonclinical safety studies for the conduct of human clinical trials and marketing authorization for pharmaceuticals

14 New Alternative Methods in Drug Safety Assessment

Xi Yang, Qiang Shi, Minjun Chen, and Li Pang

Abstract

In the past few decades, despite advancements in biomedical research, the success rate of bringing a new drug candidate to the market has been very low. Additionally, over the same time period, some approved drugs have been removed from the market due to toxicity identified after years of in-human use. Using new alternative methods (NAMs), such as in vitro models with human cells or tissues, specific organ models, in silico tests, and humanized animals may improve the accuracy of toxicity prediction and reduce adverse effects in humans. As the drug-induced liver and heart injury are the leading causes of drug attrition and drug withdrawal, this chapter summarizes the current development of human cell-based emerging technologies for hepatic and cardiovascular toxicity prediction and quantitative structure-activity relationship-driven strategies in drug safety assessment. The limitations and challenges in routinely adopting the new alternative methods in drug development are also discussed.

Keywords

Drug development · Safety assessment · New alternative methods (NAMs) · Microphysiological systems (MPS) · Human induced pluripotent stem cells (hiPSC) · Quantitative structure-activity relationship (QSAR)

14.1 Introduction

Historically, nonclinical studies have relied heavily on rodent and nonrodent animals to determine the safety profile of a drug candidate before testing it in a first-in-human clinical trial. Primary objectives of nonclinical animal studies are to identify harmful effects, identify dosages that cause those effects, and determine safe exposure limits [1]. Nevertheless, while nonclinical studies have shown great accuracy in predicting safe starting doses and have guided appropriate safety monitoring decisions for clinical trials, animal studies do not always detect or predict all adverse effects in human patients. For example, some drugs have been withdrawn because of unexpected post-market hepatotoxicity and cardiotoxicity [2, 3]. This could have been due

X. Yang
Toxicology Service, Potomac, MD, USA

Q. Shi · L. Pang (✉)
Division of Systems Biology, National Center for Toxicological Research, US FDA, Jefferson, AR, USA
e-mail: qiang.shi@fda.hhs.gov; li.pang@fda.hhs.gov

M. Chen
Division of Bioinformation and Biostatistics, National Center for Toxicological Research, US FDA, Jefferson, AR, USA
e-mail: minjun.chen@fda.hhs.gov

G. Jagadeesh et al. (eds.), *The Quintessence of Basic and Clinical Research and Scientific Publishing*,
https://doi.org/10.1007/978-981-99-1284-1_14

to factors such as translational deficits between species, age group differences, the use of healthy animals without the inclusion of disease models, genetic diversity within the human population, and/or the relatively low number of human patients involved in the clinical studies.

Over the years, significant efforts have been made to promote fundamental changes in drug safety assessment, including complementing the traditional two species animal testing approach, to consider the use of mechanism-based in vitro models, combined with high-throughput assays, and computational bioinformatic tools for data analysis and risk assessment. Studies are being done to develop a better understanding of non-animal methods, such that data from alternative approaches could be used to guide decision making about safety and to make recommendations for human clinical trial monitoring.

The U.S. Food and Drug Administration (FDA) has a long history of fostering innovation and advancing new technologies to meet twenty-first century pharmaceutical safety challenges [4]. The recent advances in New Alternative Methods (NAMs), including in vitro platforms, ex vivo tissue models, and in silico modeling, hold the promise to improve the prediction of safety signals in conjunction with traditional non-clinical studies.

The development of induced pluripotent stem cell (iPSC) technology has created much excitement to support drug safety assessment, including disease- and patient-specific mechanistic studies. Unlike the simplified immortalized cell lines, which only retain some basic functions of their origin, iPSC-derived cells are closer in phenotype to that of primary cells. In addition, primary cells are limited by cell source variability and de-differentiation in prolonged culture; two limitations that may be circumvented with iPSCs. In fact, iPSC-derived cells may have potential to be the preferred cell type for use with microphysiological systems (MPS) [5], if the challenges in cell differentiation and maturation are overcome in the near future.

Currently, there is no consensus on what constitutes an MPS; however, definitions generally encompass technologies such as 3D spheroids, multiple cell type co-cultures, and organ-on-chips with the microfluidic flow of cell culture media (Table 14.1). The promise and excitement of MPS rise from the ability to generate human-relevant data to reduce the reliance on animal models in research and regulatory decision making. However, challenges still exist that

Table 14.1 Overview of cell-based NAMs

Category	Platform properties	Ready to use on-platform assay	Throughput	Complexity	Variability	Contribution in regulatory decisions
iPSC in 2D cell culture	Grown on multi-wells, homogeneous monolayer	Relatively easy, many standardized assays	Medium to high (96- or 384-well)	Low complexity	Relative low variability	A few case studies [6]
Spheroid/ organoid	Multiple organ-specific cell types, recapture specific function of the organ	Specific functional assays are possible, challenges in distributing test article	Medium to high, require skilled end-user	Requires time (days to weeks) for the organoid formation	Medium to high variability	Rare
Organ-on-a-chip	Seeded into a chip with fluid transport (pump or passive flow), and/or with the incorporation of mechanical stimulation	Specific functional assays are possible, costly, low reproducibility	Low, require skilled end-user	Variable complexity depends on platform design, requires time	Medium to high variability	Rare

need to be addressed before these systems can be widely used, including extrapolation of results across species, exposure extrapolations from in vitro to in vivo, difficulties with replicating the whole organ, and maintenance in long-term culture [5].

As hepatotoxicity and cardiotoxicity are the leading causes of attrition of a candidate drug in drug development and withdrawal post-approval, NAMs for cardiovascular (CV) and liver are being more widely used in the routine drug development process, as compared to other organs [7]. In this chapter, we provide an overview of the NAMs used in drug safety assessment including liver MPS and human iPSC-derivatives that may improve drug-induced hepato- and cardiotoxicity detection, and general considerations of quantitative structure-activity relationship (QSAR) modeling for toxicity prediction.

14.2 Liver MPS to Aid the Assessment of Drug Hepatotoxicity

Drug-induced hepatotoxicity has been a leading cause of acute liver failure and post-marketing drug withdrawals for the past several decades [8]. Conventional animal experiments and clinical trials may fail to accurately identify drugs with low incidences of clinically significant hepatotoxicity that later become more apparent as having drug-related hepatoxic effects when prescribed to larger patient populations in post-market settings [8]. Primary liver cells are considered the gold standard in vitro model for drug hepatotoxicity and drug metabolism, but their predictive value is far from perfect, because key hepatic functions, particularly drug metabolizing enzymes, are decreased over time after cell isolation. To address this issue, various MPS have been developed. Primary liver cells are cultured in physiologically-relevant environments so that hepatic functions can be better preserved. These liver MPS have been reported to be more effective in predicting hepatotoxicity than conventional culture methods and are being investigated as an alternative approach in safety testing to supplement and reduce animal tests in the future.

There are over 20 types of cells in human livers, but hepatocytes along with Kupffer cells, liver sinusoidal endothelial cells (LSECs), and stellate cells, which are the four primary types of non-parenchymal cells (NPCs), account for approximately 65%, 15%, 10% and 8%, respectively, of all cells in the liver. The percentage of other cell types is very small and their role in predicting drug hepatotoxicity has not been extensively studied. Immortalized hepatic cell lines and iPSC-derived hepatocyte-like cells have been used in liver MPS for hepatotoxicity studies [9], but these cells lack many key characteristics of mature primary live cells.

14.2.1 Liver Spheroids

Isolated liver cell suspension can aggregate spontaneously, forming hepatic spheroids. Typically, spheroid culture is achieved by maintaining cells as hanging drops [10] or using ultra-low attachment plates [11], both of which are commercially available. Hepatocytes cultured as spheroids can be maintained for up to two weeks and are more sensitive in detecting drug hepatotoxicity as compared to conventional approaches, such as hepatocytes cultured as two-dimensional (2D) monolayers, mainly because essential liver functions are better preserved in spheroids [10, 11]. Hepatocytes can be co-cultured with NPCs, forming two-cell-type or multiple-cell-type spheroids, which are useful in detecting immune-response mediated hepatotoxicity as the presence of the NPCs confers an immune response [12, 13]. However, due to unknown reasons, not all hepatocytes will form spheroids. A recent report showed that cells from 30% of human donors did not grow as spheroids [13]. Fortunately, spheroid-qualified hepatocytes are now commercially available.

The performance of hepatic spheroids for the prediction of hepatotoxicity has been directly compared to conventional 2D culture. Using liver spheroids produced by the hanging-drop method, 110 drugs, among which 49 were classified as non-hepatotoxic and 61 as hepatotoxic, were tested using the same cells cultured in 2D format. The accuracy achieved in predicting hepatotoxicity was 70% with spheroids and 53% with 2D cultured hepatocytes [10], suggesting that spheroids may improve the evaluation of liver safety. Similarly, using the ultra-low attachment plates to test 100 drugs, the liver spheroids had an accuracy of 68%, as compared to 53% with 2D cultured hepatocytes, in predicting drug hepatotoxicity [12]. In contrast, 2D cultured hepatocytes have also been reported to have an accuracy of 65% in predicting the clinical hepatotoxicity of 34 FDA-approved protein kinase inhibitors [14]. Furthermore, an early study involving 344 compounds reported that primary human hepatocytes cultured in conventional 2D predicted hepatotoxicity with an accuracy of 71% [15]. Thus, further studies involving additional cell sources, drugs, and drug classes are needed to further explore the potential for increased predictivity of spheroids over 2D culture.

14.2.2 Liver-on-a-Chip

Primary cells can attach to certain scaffolds that mimic in vivo growth environments, such as cell-to-cell communications and polarity. The attached cells can then be perfused using a culture medium, emulating the blood flow and shear pressure that cells are exposed to under in vivo conditions. This culturing approach is commonly called Organ-on-a-Chip. Many Liver-on-a-Chip platforms (Liver-Chips) are commercially available. Some of them have been investigated by third parties and published data suggest that these platforms perform better in maintaining cell viability, protein synthesis functions of the liver, and key drug metabolizing enzyme activities when compared with conventional 2D culture methods. Furthermore, the data indicated that the hepatotoxicity of some compounds can be reproduced in these systems [2, 16–18]. However, because Liver-Chips are generally low-throughput, only a small number of compounds have been examined so far and, therefore, the general applicability of Liver-Chips to the prediction of drug hepatotoxicity remains unclear. Of note, in a recent preprint publication that will be peer reviewed, it was found that the Emulate Liver-Chips predicted drug hepatotoxicity with an accuracy of 78%, but only 27 compounds were examined [19].

14.2.3 Synthetic Hemoglobin in Liver MPS

The in vitro primary cell cultures that include hepatocytes usually have lower levels of mitochondrial oxidative phosphorylation, because the cells are forced to switch to glycolysis for energy production. To restore the oxygen levels and associated mitochondrial functions, a synthetic hemoglobin has been developed and evaluated in a liver MPS platform; wherein, it was shown that the primary hepatocytes maintained higher levels of mitochondrial oxidative phosphorylation, albumin production, and drug metabolizing enzyme activities, as compared to 2D culture [20]. Since mitochondrial liability is an established predictor of drug hepatotoxicity [21, 22], these data are consistent with synthetic hemoglobin in MPS cultures potentially aiding the prediction of drug hepatotoxicity. However, there is currently only one commercially available liver MPS with synthetic hemoglobin, and its usefulness in drug hepatotoxicity predictivity has not been examined to date. Another liver MPS platform has been designed to mimic the physiological oxygen gradients in different zones of the liver, but this system has not been tested for drug hepatotoxicity or commercialized and its performance in maintaining in vivo liver functions remains to be confirmed by independent groups [23].

14.2.4 Multi-Organ MPS Involving the Liver

One group has demonstrated the ability to integrate 13 cell types from 13 different organs in an

MPS platform [24]; however, this is only a proof of concept study and the system has not yet been standardized or used for safety evaluation. A triple-organ MPS involving the heart, liver and skin has been reported to examine drug hepatotoxicity and cardiotoxicity simultaneously [25]. This system showed that acetaminophen, the most implicated drug in clinical hepatotoxicity, also caused significant toxicity to cardiomyocytes [25]. Acetaminophen is one of only four drugs that have been tested in this triple-organ MPS [25] and further investigations are needed to confirm its value in the assessment of liver and cardiac risks. In addition, key hepatic and cardiac functions have not been well characterized in this system. Other multi-organ MPS involving the liver have also been reported, but only immortalized cell lines were tested and therefore their clinical relevance might be low [26].

14.2.5 Issues to be Addressed

Though liver MPS has shown some promise in improving the prediction of drug hepatotoxicity, many issues remain to be addressed.

First, almost all liver MPS are designed or optimized for cells of human origin. Literature data on liver MPS using animal cells are scarce, even though the greatest experience with preclinical risk assessment and prediction of clinical safety is based on nonclinical work in animal species. Thus, it remains unclear how MPS approaches correlate with traditional methods and, therefore, correlate with current risk assessment practices. As one major purpose of developing MPS is to reduce animal use, MPS data collected using animal cells are needed to observe the correlation between in vivo observations and MPS findings. Such correlation may help build confidence in using liver MPS data to help predict human clinical hepatotoxicity. In fact, in vivo to in vitro correlation is easier to investigate in animals than in humans, as human hepatotoxicity data are not always straightforward. Indeed, establishing a causative relationship between a human hepatoxicity response and the drug is the most challenging issue in this area [27].

Second, the liver MPS platforms need to be characterized more comprehensively. Almost all platforms focus on the preservation of hepatocyte functions, while the viability and function of NPCs are usually unexamined or under-investigated. A notable example is stellate cells that reside in a non-proliferative, quiescent state under healthy conditions, but are activated during injury, inflammation, infection, and the process of cell isolation and purification. Maintaining cultured stellate cells in the quiescent state is a challenging issue, and this has not been addressed in any Liver MPS platforms [28]. Comprehensive characterization of primary liver cells is important because standardized approaches to prepare these cells have not been established and vendors use different ways to isolate, purify and store the cells, which can impact the integrity and usability of the cell stocks for MPS.

Thirdly, although liver MPS makes it possible to co-culture the major liver cell types in an integrated system, the adaptive or regenerative responses observed in vivo have not been investigated. The liver is unique in that it has a remarkable capacity to regenerate, that is, to produce new cells when a liver injury occurs. The extent of regeneration is a determinant of the patient outcome when drug hepatotoxicity occurs. Conventional culture methods involve only hepatocytes and thus do not allow the study of regeneration, which requires the complicated interaction between multiple liver cell types. Thus, liver MPS should be characterized for its ability to mimic liver regeneration, as this may help improve the accuracy in predicting drug hepatotoxicity.

14.3 Preclinical Cardiac Safety Assessment with Human iPSC-Derivatives

Drug-induced cardiotoxicity represents another leading cause of discontinuation of clinical trials and post-approval drug withdrawals [3]. Species differences in the CV system, may fail to highlight drugs that are toxic to humans which can sometimes go unidentified in nonclinical toxicology and safety pharmacology studies, and vice versa [29]. A second possible explanation for the failure to detect drug-induced cardiotoxicity during drug development is that, at present, cardiac safety assessment is normally only conducted in healthy hearts/animals, which are potentially less sensitive to adverse side effects, as compared to diseased subjects. The presence of comorbidities and/or cotreatments may make a subtler toxicity from less potent toxicants more apparent, which could involve a variety of mechanisms, such as affecting ion channel gene expression/function and myocyte contractility, modifying electromechanical coupling and mitochondrial function, and inducing structure remodeling of the myocardium [3]. The development of various disease models of NAMs may improve cardiotoxicity detection and prediction, which could be beneficial for drug development and patients.

Human iPSC technology has significantly enhanced drug discovery and disease modeling. Herein, we summarize the utility of human iPSC derivatives for cardiac safety evaluation at the early preclinical stage. The potential of human cardiac MPS to model human disease is also discussed concerning how it may enhance postmarket surveillance via detection and prediction of risk for drug-related cardiotoxicity in specific patient populations that were not fully evaluated in clinical trials.

14.3.1 The Usefulness of Human iPSC Technology on Drug-Induced Cardiotoxicity Detection and Prediction

Drug testing with isolated adult human primary cardiomyocytes and ex vivo human heart slices resembles the potential pharmacological and toxicological actions of test compounds as seen in humans; however, due to the limited availability of human heart samples, applying such methods in routine toxicology studies and drug development is very difficult. Although long-term culture of human primary cardiomyocytes and heart slices is feasible, it is highly debated whether or not the long-term cultivation systems sufficiently preserve the cell/tissue viability and functionality [30, 31]. Moreover, it is impossible to study the development of cardiac toxicity and pathological characteristics of CV disease with human heart tissue due to inherent limitations and compromised tissue integrity, as such samples are obtained from patients who underwent cardiac transplantation for end-stage CV disease or postmortem.

Advancements in iPSC technology offer the capability to reprogram adult somatic cells into pluripotent stem cells, which can be further differentiated into many different cell types, such as cardiomyocytes, cardiac fibroblasts, endothelial cells, vascular smooth muscle cells, etc. [32]. The theoretically unlimited supply of iPSC-derived cardiomyocytes (iPSC-CMs) from donors with diverse genetic backgrounds may allow us to address the challenges that cannot be answered with the existing methods and tools currently used in preclinical studies; for example, to identify and understand factors that contribute to the variable incidence of drug-induced cardiotoxicity in different patient subgroups. In fact, since the launch of the first commercial line of human iPSC-CMs in 2009, iPSC-CMs are increasingly used in drug-induced cardiotoxicity detection and prediction, particularly for drug proarrhythmic potential assessment and cancer therapy-related cardiotoxicity screening [6, 30, 33]. A multi-site validation study on the predictability and reproducibility of electrophysiological assays with commercially available human iPSC-CMs [34] has led to the inclusion of in vitro cardiomyocyte assays as appropriate nonclinical evaluations to assess the potential of drugs to induce delayed ventricular repolarization in the International Council for Harmonization (ICH) guidance for industry ICH E14/S7B Questions and Answers (Q & As), which provides guidance on best practices for the design, conduct, analysis, interpretation and reporting of in vitro, in silico, and in vivo nonclinical assays to support nonclinical and clinical evaluation of drug-induced delayed ventricular repolarization [6, 35].

Numerous high- and medium-throughput assay platforms have been developed to assess the function of human iPSC-CMs in monolayer cultures; however, human iPSC-CMs in conventional 2D culture exhibit a fetal phenotype and may not reflect the responses of adult human cardiomyocytes and cardiac tissue [36]. Seeded human iPSC-CMs in 3D scaffold-based hydrogels or other materials more accurately mimic the alignment of cardiomyocytes in extracellular matrix as seen in native heart tissue. When electrical stimulation is applied, 3D cardiac microtissues displayed adult-like gene expression profiles, sarcomere structure, and electromechanical properties [36], which suggests that the 3D culture approach may have improved predictive potential and further supports their use for drug screening and disease modeling. A blinded multicenter study evaluated the responses to 36 compounds with known positive, negative, or neutral inotropic effects of human iPSC-CMs in 2D monolayers compared to 3D engineered heart tissues. Compared to the 78% accuracy reported for animal models in predicting the contractility changes of the human heart, the reported accuracy of contractility-related assays with human iPSC-CMs was 85% for 2D monolayers and 93% for 3D engineered heart tissues [37]. Nevertheless, contaminations of cytokines or growth factors in natural hydrogel materials may limit the use of microtissue models for drug development [31].

As is the case with the liver, the heart is composed of multiple cell types. Co-culture of pre-differentiated cardiomyocytes, endothelial cells, and cardiac fibroblasts in more complex 3D environments (e.g., spheroids) may be more physiologically relevant to adult hearts and may better recapitulate the effects of therapeutics. For example, human iPSC-CMs co-cultured with cardiac endothelial cells and cardiac fibroblasts in spheroids have been reported to show a similar sensitivity (73%) but improved specificity, for evaluation and detection of structural cardiotoxins (84%), as compared to the monolayer culture of human embryonic stem cell-derived cardiomyocytes (74%) [38]. By integrating other types of cells into the culture, cardiac spheroids hold the potential to build highly integrated MPS that may enable the vascularization, innervation, and crosstalk between immune cells and cardiomyocytes.

Organoid culture is another widely used scaffold-free 3D culture technique that mimics the cellular heterogeneity of cardiac tissue. In contrast to 3D spheroid MPS models, the multiple distinguishable cell types of cardiac organoids are differentiated from iPSC and self-organized into 3D structures. Several recently developed cardiac organoid models have built an endothelial network into the organoids that resemble some chamber-like features during early heart development [32]. Due to the embryonic and fetal-like characteristics, the cardiac organoid models are currently mainly used for the research of heart development and regeneration. The lack of reproducibility is another major obstacle limiting the use of cardiac organoid models [31].

In recent years, heart-on-a-chip MPS models (Heart-Chips) have been developed to reproduce the mechanisms and reactions of heart cells to test drugs. The constant perfusion and mechanical stimulation in Heart-Chips mimic the molecular transport and mechanics of the heart under in vivo conditions. Nevertheless, most Heart-chips only resemble the ventricular-like tissue, given that it may be desirable to mimic the four-chamber structure and the electrophysiology of the heart to reliably predict the potential risk for in vivo cardiotoxicity [31]. The multiorgan-on-a-chip is another interesting development, which connects several organs in one MPS platform and enables the investigation of the systemic effects of drugs and their metabolites. For example, the integration of tumor, liver, and heart models can evaluate the efficacy of anticancer drugs and off-target cardiotoxicity of the parent drugs and their metabolites [32]. The installation of endothelial barriers into multiple organ chips via vascular perfusion has been shown to mimic the systemic transportation of small molecules between organs and enabled the prediction of the distribution of molecules in multiple organs [32]. Although these complex MPS models hold promise to provide more physiological, human-relevant data for in vitro to in vivo extrapolation (IVIVE), it will likely be necessary to reduce the cost and improve the throughput of the heart- or multiorgan-on-a-chip technologies to enable efficient cardiac safety evaluation of drugs and drug candidates in pre-clinical studies. In addition, the absorption of small molecules by polydimethylsiloxane (PDMS), a widely used material to build microfluidic organ-on-chip devices, should be considered in IVIVE. Developing MPS with animal cells and meaningful comparisons of the results obtained from traditional animal toxicology models after exposure to well-characterized positive and negative reference compounds will be essential for the wide adoption of the MPS approaches in drug development and regulatory use in risk assessments.

14.3.2 Potential to Enhance Post-Marketing Surveillance with Patient- and Disease-Specific Human iPSC-Derivatives

Although drug safety is evaluated in patient populations prior to market approval decisions, there may be patient subgroups with increased sensitivity to some adverse effects that don't become apparent until the post-market phase. These cardiotoxic effects could include inhibition of cardioprotective survival signaling pathways or enhanced deleterious effects in patient subgroups with CV comorbidities [3]. The establishment of NAMs for various CV disease models could facilitate the evaluation of the potential risk in certain patient subgroups that were outside the scope of clinical trials in the initial market application. Thus, CV disease model NAMs could be used in post-marketing surveillance to further evaluate patient subgroups, such as those carrying genetic variants/mutations or those with underlying CV diseases.

Human iPSC-derivatives carrying donor-specific genetic information can be valuable models for disease modeling. Specifically, human iPSC-CMs have been used to study the pathogenesis of several major inherited cardiomyopathies, including Long QT syndromes, Brugada syndrome, catecholaminergic polymorphic ventricular tachycardia, arrhythmogenic right ventricular dysplasia, hypertrophic cardiomyopathy, dilated cardiomyopathy, arrhythmogenic cardiomyopathy, Duchenne muscular dystrophy, Barth syndrome and Pompe disease (metabolic cardiomyopathy) etc. [39, 40]. Patient-specific iPSC-derived endothelial cells and smooth muscle cells have effectively modeled disease phenotypes and/or mechanisms of idiopathic and family pulmonary arterial hypertension, calcific aortic valve disease, Marfan syndrome, supravalvular aortic stenosis and Williams-Beuren syndrome [40]. Moreover, human iPSC-CMs have been reported to recapitulate the inter-individual variation of clinical susceptibility to oncology drug-induced cardiotoxicity[40]. Therefore, growing evidence suggests that genetic screening and drug toxicity testing with donor-specific iPSC derivatives may enable patient stratification and identification of individuals who may be genetically predisposed to drug-induced CV toxicity.

Several drugs have been removed from the market due to cardiotoxicity in patient subgroups, such as those with diseased hearts, that weren't identified in pre-market clinical trials [3]. Thus, drug testing in human iPSC-based disease models may aid in revealing "hidden" cardiotoxicity, which could avoid unnecessary harm to patients. Human iPSC-CMs have also been employed to model the pathophysiological changes of common non-hereditary cardiomyopathy. For example, chronic exposure of human iPSC-CMs to norepinephrine generated an in vitro model of heart failure, which mimicked features of clinical heart failure such as cardiomyocyte hypertrophy, contractile dysfunction, cell death, and release of N-terminal pro B-type natriuretic peptide, a commonly used clinical diagnostic biomarker of heart failure [39]. In addition, human iPSC-CMs exposed to hydrogen peroxide have been shown to mimic ischemic injury. A potential cardioprotective agent was identified in the human iPSC-CM-based ischemic injury model, which was then shown to reduce ischemia-reperfusion injury in mice [40].

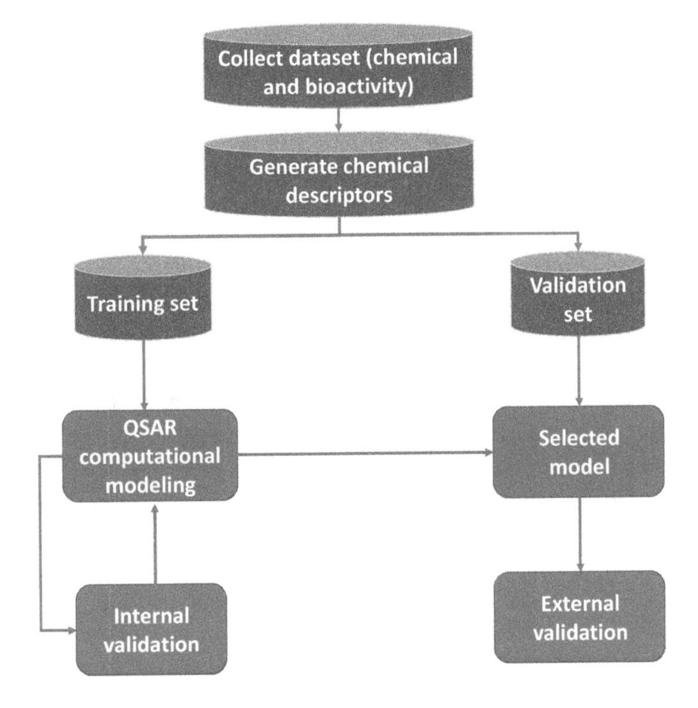

Fig. 14.1 The basic flowchart for developing a QSAR model

Currently, patient- or disease-specific iPSC-CMs and disease models of CV MPS are used primarily to investigate the pathophysiology of CV disease or mechanisms underlying certain disease phenotypes and to screen potential therapeutics. Although such disease models hold potential to increase our understanding of drug mechanisms of action, it is important to verify if the models can be unitized to reveal "hidden" cardiotoxicity, as seen in clinical practice. Careful validation of the in vitro disease models and accurate data explanation is needed to enhance post-marketing surveillance and avoid inducing severe CV adverse reactions in those vulnerable patients.

14.4 QSAR Modeling

QSAR and other emerging in silico models are attractive NAM approach for drug safety assessment. QSAR is an approach using computational modeling to build a relationship between biological activities and the structural properties of chemicals. Currently, the QSAR model has been widely used in drug development due to its rapid attainment of results, and the fact that physical drug substances are unnecessary. The basic flowchart for developing a QSAR model is illustrated in Fig. 14.1. To develop a QSAR model, three basic components generally need to be considered: (1) toxicity dataset and molecular descriptors of compounds, (2) computational algorithm, and (3) model evaluation.

14.4.1 Toxicity Dataset and Molecular Descriptors of Compounds for QSAR Modeling

The first component of QSAR modeling is the toxicity dataset, which is composed of a set of chemicals with known chemical structures and measured bioactivity endpoints for prediction. The bioactivity endpoints are usually study-specific but can include toxicity endpoints that assess carcinogenicity, cytotoxicity, or other measurements such as lipophilicity. For developing a QSAR model, the dataset is usually divided into a training set for building the model and a validation set to evaluate the model performance.

The basic assumption for QSAR modeling is that similar compounds share similar biological activities. Based on this assumption, the chemical structures of chemicals can be compared to infer their potential bioactivity. To construct a QSAR model, chemical structures need to be transformed into numerical descriptors, which represent the structural characteristics of the compounds. Specially designed algorithms can be used to generate mathematical representations of chemical structures and physicochemical properties, referred to as molecular descriptors.

Molecular descriptors can be qualitative or quantitative. As a numeric representation of chemical structures, they can range from simple counts of a particular atom type, such as sulfur (S) or nitrogen (N), to the distribution of properties, such as charge across a molecule surface. Additionally, a molecular descriptor could also be used to describe the value of the molecule's physiochemical properties. For instance, logP is a widely used descriptor representing the lipophilicity of chemicals, which can be calculated using a commercial or public chemometric tool or can be experimentally measured through the partitioning of the molecule between lipophilic and aqueous phases.

The most used are topological descriptors, which consider connectivity along with atom and bond labels of chemical structure. Topological descriptors can stem from dimensionality or the origin of structure. Based on dimensionality, it can be categorized as 0D-descriptors such as count of specific atoms, 1D-descriptors such as structural fragments, 2D-descriptors such as topological indices, 3D-descriptors such as quantum-chemical descriptors (e.g., size, steric, surface and volume descriptors), and 4D-descriptors, such as those derived from GRID or CoMFA methods. Meanwhile, the origin-based descriptors take into account graph theory-based topological descriptors, geometrical descriptors such as distances, valence angles, and surfaces, constitutional descriptors such as functional group count, and thermodynamic descriptors such as entropy, and quantum-chemical descriptors. A 3D-QSAR descriptor approach, 3D-SDAR, reported reliable and reproducible modeling of human Ether-à-go-go Related Gene (hERG) potassium channel affinity [41], resulting in the identification of a three-center toxicophore composed of two aromatic rings and an amino group, which is similar to an earlier reported phospholipidosis toxicophore.

Many commercial or open-source software package algorithms have been published or privately developed for calculating molecular descriptors from chemical structure [42], such as Mold2 (https://www.fda.gov/science-research/bioinformatics-tools/mold2), and several commercial software packages [43]. The Mold2 package was developed by the FDA's National Center for Toxicological Research (NCTR). Mold2 is free to the public and can generate 777 chemical descriptors, most of which are calculated from the 2D chemical structure and are derived from those well-documented molecular descriptors reported in the literature.

14.4.2 Computational Algorithms for Developing QASR Models

After toxicity data is collected and molecular descriptors are generated, a computational algorithm is used to develop QSAR models. Machine learning has been widely applied to develop QSAR models for predicting toxicity. As a branch of artificial intelligence, machine learning imitates the way that humans learn to analyze data. Using statistical methods and software, it can extract and discover hidden or non-obvious patterns encoded in molecular descriptors or bioactivity data, upon which it bases predictions of the toxicity profile of new compounds.

Many machine learning algorithms have been used in QSAR modeling for identifying chemical patterns in molecular descriptors and their relationship with toxicity. Some notable machine learning algorithms include Neural Networks, Deep learning, Logistic Regression, Naïve Bayes, K-Nearest Neighbors, Decision Forest/Trees, and Support Vector Machines [44]. This discussion will focus on neural networks and deep learning methods.

Neural networks are a branch of machine learning and the foundation of deep learning algorithms. Inspired by information processing in the human brain, the neural network algorithm mimics the way that biological neurons signal to one another and is constructed by connected units called artificial neurons, which are formed as layers. In a neural network, there is an intermediate set of hidden layers that receive the data modified by the previous layer in the sequence, and the consecutive layers are then connected using weights. This parallel neural network gets molecular descriptors input data from the first layer, which is then translated to the toxicity endpoints in the last layer.

Deep neural networks or deep learning refers to a family of neural networks with multiple hidden layers, which has demonstrated success in the last few years. The "deep" in deep learning is refers to the depth of layers in a neural network. Compared with classical, or "non-deep", machine learning, deep learning is less dependent on human intervention for learning. For example, deep learning is better at processing unstructured data, such as text and images, and it could automatically extract important features from data, minimizing the dependency on human experts. Convolutional neural networks (CNN) and Recurrent Neural Networks (RNNs) are the two most common deep learning algorithms. CNNs work better for data presented in multiple arrays including computer vision and image classification, while RNNs are more successful for tasks such as speech and language recognition, as it leverages sequential or time series data. Notably, deep learning also inherits the same problems associated with neural networks: overfitting and long computation time [45].

Kang et al. [46] developed a deep neural network-based model for predicting the risk of drug-induced liver injury. They used extended connectivity fingerprints of diameter 4 (ECFP4) to generate molecular descriptors and used the curated data from DILIrank [47], LiverTox [48], and other literature as the toxicity endpoints. A stratified tenfold cross-validation was used to select the best model, and the applicability domain was defined. Evaluation from the external validation data set showed their model's performance with an accuracy of 0.731, a sensitivity of 0.714, and a specificity of 0.750. The model was further evaluated with four external data sets and 15 drugs with DILI cases reported after the year 2019. Li et al. [49] developed a deepDILI model with a Matthews correlation coefficient value of 0.331, which performed better than several "non-deep" machine learning algorithms.

14.4.3 Statistical Metrics for Evaluating Model Performance

Another critical step in QSAR modeling is to evaluate the model performance. The basic method to evaluate a model is to compare the predictions from the candidate model with the known outcomes using a statistical metric. The commonly used statistical metrics for QSAR models include area under the receiver operating characteristic (ROC) curve, accuracy, sensitivity, specificity, balanced accuracy, and Matthews correlation coefficient (MCC).

The area under the ROC curve is a quantitative metric for QSAR modeling which does not need the predefined cut-off threshold for prediction. A ROC curve is a graphical plot created by plotting the True Prediction Rate (TPR) against the False Prediction Rate (FPR) when given a variety of cut-off thresholds.

$$\mathrm{TPR} = \mathrm{TP}/(\mathrm{TP} + \mathrm{FN})$$

$$\mathrm{FPR} = \mathrm{FP}/(\mathrm{FP} + \mathrm{TN})$$

Here, True Positive (TP)—the correct prediction of the positive class, True Negative (TN)—the correct prediction of the negative class, False Positive (FP)—the incorrect prediction of the positive class, False Negative (FN)—the incorrect prediction of the negative class.

Accuracy is the most used statistical metric and is defined as the percentage of correct predictions versus the total number of samples. Its formula for accuracy: Accuracy = (TP + TN)/(TP + TN + FP + FN). Sensitivity and specificity represent the predictive accuracy in the positive and negative samples, respectively. These formulas are: Sensitivity = TP/(TP + FN) and Specificity = TN/(TN + FP). Accuracy can be biased when the dataset is unbalanced with regard to the ratio of positives and negatives. *MCC* is an alternative measure of accuracy for an unbalanced dataset, which is calculated through the Pearson product-moment correlation coefficient between actual and predicted values. The formula for MCC is as follows:

$$\mathrm{MCC} = (TP * TN - FP * FN) / \sqrt{(TP+FP)*(TP+FN)*(TN+FP)*(TN+FN)}$$

MCC falls within the range of [−1, 1], with −1 and +1 representing perfect misclassification and perfect classification, respectively; when MCC = 0, the prediction is random.

Balanced accuracy is another statistical metric to evaluate model performance in the unbalanced dataset with the formula below

$$\text{Balanced accuracy} = (\text{Sensitivity} + \text{Specificity})/2$$

14.4.4 A Case Study of QSAR Modeling for Drug-Induced Liver Injury

QSAR models have been broadly developed to identify the potential liability of drug-induced liver injury at the early stage of drug discovery [50]. Here, a study to develop a QSAR model for predicting DILI will be discussed, which has applied external validation to ensure the model performance [51, 52].

A QSAR model was developed based on Mold2 molecular descriptors and a decision forest algorithm [53] using a training set of 197 drugs, which were annotated for the risk of drug-induced liver injury in humans based on FDA-approved drug labels [47, 54]. The developed QSAR model was first evaluated by internal validation that employed a 2000-run of tenfold cross-validation based on the training set. The mean accuracy from cross-validation was 69.7%, with a relative standard deviation (RSD) of 2.9%. The developed QSAR model was further externally validated by 3 independent datasets, including an NCTR validation dataset comprised of 190 drugs [51], the Greene et al. dataset with 328 drugs [55], and the Xu et al. dataset with 241 drugs [15]. The accuracy from the validation results for these three datasets were 69.7%, 61.6% and 63.1%, respectively. The performances evaluated by internal and external validation by different datasets are consistent, suggesting that the developed QSAR model is robust and could be useful in assessing a drug's risk for causing drug-induced liver injury in humans.

14.5 Concluding Remarks

The development of NAMs is promising. Currently, some of the NAMs are beginning to be utilized in regulatory applications. For example, human cardiomyocyte-based in vitro assays and in silico modeling have been accepted into the international regulatory guidance ICH S7B/E14 Q & A to supplement traditional hERG assay and animal studies for cardiac safety assessment [6]. Also, in vitro models of hepatocyte spheroids and in silico modeling had been utilized to support the hepatotoxicity assessment of Ubrogepant [56]. Furthermore, an in vitro-in silico quantitative systems pharmacology approach has been recommended by the FDA for the efficacy evaluation of Naloxone [56]. Despite this notable progress, most of the emerging technologies, particularly various MPS, are not yet normally submitted in support of regulatory decision making and have not been rigorously tested or qualified for regulatory use. It is imperative that NAM approaches continue to be critically evaluated to see if their use can support drug development by increasing the understanding of drug mechanisms of action and drug safety, including how the predictivity of these approaches compares to existing risk assessment methods and tools currently used to support regulatory applications. There have been active discussions regarding the potential regulatory use of NAMs among industry, academia, and regulatory authorities. Continued discussions among stakeholders on the utility, benefits, limitations, and challenges of NAMs are essential for applying the novel technologies to strengthen industry's internal review process and to inform regulatory decision making. The eventual regulatory acceptance of NAMs will require general consensus of stakeholders on performance criteria of the methods and strategies to determine which and how the NAMs can be moved forward for regulatory use.

Acknowledgements The authors would like to thank Drs. Laura Schnackenberg, Jessica Hawes, and Nakissa Sadrieh for reviewing and editing the manuscript.

Conflict of Interest The authors declare that there are no conflicts of interest regarding the publication of this article.

Disclaimer The views presented in this article do not necessarily reflect those of the Food and Drug Administration.

References

1. Dorato MA, Buckley LA (2006) Toxicology in the drug discovery and development process. Curr Protoc Pharmacol. https://doi.org/10.1002/0471141755.ph1003s32.
2. Shi Q, Arefin A, Ren L, Papineau KS, Barnette DA, Schnackenberg LK et al (2022) Co-culture of human primary hepatocytes and nonparenchymal liver cells in the emulate(R) liver-chip for the study of drug-induced liver injury. Curr Protoc 2(7):e478. https://doi.org/10.1002/cpz1.478
3. Ferdinandy P, Baczko I, Bencsik P, Giricz Z, Gorbe A, Pacher P et al (2019) Definition of hidden drug cardiotoxicity: paradigm change in cardiac safety testing and its clinical implications. Eur Heart J 40(22):1771–1777. https://doi.org/10.1093/eurheartj/ehy365
4. Advancing Alternative Methods at FDA. https://www.fda.gov/science-research/about-science-research-fda/advancing-alternative-methods-fda. Accessed 17 Aug 2022
5. Rusyn I, Sakolish C, Kato Y, Stephan C, Vergara L, Hewitt P et al (2022) Microphysiological systems evaluation: experience of TEX-VAL tissue chip testing consortium. Toxicol Sci 188(2):143–152. https://doi.org/10.1093/toxsci/kfac061
6. Yang X, Ribeiro AJS, Pang L, Strauss DG (2022) Use of human iPSC-CMs in nonclinical regulatory studies for cardiac safety assessment. Toxicol Sci. https://doi.org/10.1093/toxsci/kfac095
7. Weaver RJ, Valentin JP (2019) Today's challenges to de-risk and predict drug safety in human "mind-the-gap". Toxicol Sci 167(2):307–321. https://doi.org/10.1093/toxsci/kfy270
8. Woo SM, Alhaqqan DM, Gildea DT, Patel PA, Cundra LB, Lewis JH (2022) Highlights of the drug-induced liver injury literature for 2021. Expert Rev Gastroenterol Hepatol 2022:1–19. https://doi.org/10.1080/17474124.2022.2101996
9. Bircsak KM, DeBiasio R, Miedel M, Alsebahi A, Reddinger R, Saleh A et al (2021) A 3D microfluidic liver model for high throughput compound toxicity screening in the OrganoPlate(R). Toxicology 450:152667. https://doi.org/10.1016/j.tox.2020.152667
10. Proctor WR, Foster AJ, Vogt J, Summers C, Middleton B, Pilling MA et al (2017) Utility of spherical human liver microtissues for prediction of clinical drug-induced liver injury. Arch Toxicol 91(8):2849–2863. https://doi.org/10.1007/s00204-017-2002-1
11. Bell CC, Dankers ACA, Lauschke VM, Sison-Young-R, Jenkins R, Rowe C et al (2018) Comparison of hepatic 2D sandwich cultures and 3D spheroids for long-term toxicity applications: a multicenter study. Toxicol Sci 162(2):655–666. https://doi.org/10.1093/toxsci/kfx289
12. Li F, Cao L, Parikh S, Zuo R (2020) Three-dimensional spheroids with primary human liver cells and differential roles of kupffer cells in drug-induced liver injury. J Pharm Sci 109(6):1912–1923. https://doi.org/10.1016/j.xphs.2020.02.021
13. Baze A, Parmentier C, Hendriks DFG, Hurrell T, Heyd B, Bachellier P et al (2018) Three-dimensional spheroid primary human hepatocytes in monoculture and coculture with nonparenchymal cells. Tissue Eng Part C Methods 24(9):534–545. https://doi.org/10.1089/ten.TEC.2018.0134
14. Zhang J, Ren L, Yang X, White M, Greenhaw J, Harris T et al (2018) Cytotoxicity of 34 FDA approved small-molecule kinase inhibitors in primary rat and human hepatocytes. Toxicol Lett 291:138–148. https://doi.org/10.1016/j.toxlet.2018.04.010
15. Xu JJ, Henstock PV, Dunn MC, Smith AR, Chabot JR, de Graaf D (2008) Cellular imaging predictions of clinical drug-induced liver injury. Toxicol Sci 105(1):97–105. https://doi.org/10.1093/toxsci/kfn109
16. Rubiano A, Indapurkar A, Yokosawa R, Miedzik A, Rosenzweig B, Arefin A et al (2021) Characterizing the reproducibility in using a liver microphysiological system for assaying drug toxicity, metabolism, and accumulation. Clin Transl Sci 14(3):1049–1061. https://doi.org/10.1111/cts.12969

17. Eckstrum K, Striz A, Ferguson M, Zhao Y, Sprando R (2022) Evaluation of the utility of the beta human liver emulation system (BHLES) for CFSAN's regulatory toxicology program. Food Chem Toxicol 161:112828. https://doi.org/10.1016/j.fct.2022.112828
18. Eckstrum K, Striz A, Ferguson M, Zhao Y, Welch B, Solomotis N et al (2020) Utilization of a model hepatotoxic compound, diglycolic acid, to evaluate liver organ-chip performance and in vitro to in vivo concordance. Food Chem Toxicol 146:111850. https://doi.org/10.1016/j.fct.2020.111850
19. Ewart L, Apostolou A, Briggs SA, Carman CV, Chaff JT, Heng AR, et al Qualifying a human liver-chip for predictive toxicology: performance assessment and economic implications. bioRxiv. 2022: 2021.12.14.472674. https://doi.org/10.1101/2021.12.14.472674.
20. Shoemaker JT, Zhang W, Atlas SI, Bryan RA, Inman SW, Vukasinovic J (2020) A 3D cell culture organ-on-a-chip platform with a breathable hemoglobin analogue augments and extends primary human hepatocyte functions in vitro. Front Mol Biosci 7:568777. https://doi.org/10.3389/fmolb.2020.568777
21. Pessayre D, Fromenty B, Berson A, Robin MA, Letteron P, Moreau R et al (2012) Central role of mitochondria in drug-induced liver injury. Drug Metab Rev 44(1):34–87. https://doi.org/10.3109/03602532.2011.604086
22. Zhang J, Salminen A, Yang X, Luo Y, Wu Q, White M et al (2017) Effects of 31 FDA approved small-molecule kinase inhibitors on isolated rat liver mitochondria. Arch Toxicol 91(8):2921–2938. https://doi.org/10.1007/s00204-016-1918-1
23. Kang YBA, Eo J, Bulutoglu B, Yarmush ML, Usta OB (2020) Progressive hypoxia-on-a-chip: an in vitro oxygen gradient model for capturing the effects of hypoxia on primary hepatocytes in health and disease. Biotechnol Bioeng 117(3):763–775. https://doi.org/10.1002/bit.27225
24. Miller PG, Shuler ML (2016) Design and demonstration of a pumpless 14 compartment microphysiological system. Biotechnol Bioeng 113(10):2213–2227. https://doi.org/10.1002/bit.25989
25. Pires de Mello CP, Carmona-Moran C, McAleer CW, Perez J, Coln EA, Long CJ et al (2020) Microphysiological heart-liver body-on-a-chip system with a skin mimic for evaluating topical drug delivery. Lab Chip 20(4):749–759. https://doi.org/10.1039/c9lc00861f
26. Messelmani T, Morisseau L, Sakai Y, Legallais C, Le Goff A, Leclerc E et al (2022) Liver organ-on-chip models for toxicity studies and risk assessment. Lab Chip 22(13):2423–2450. https://doi.org/10.1039/d2lc00307d
27. Weber S, Gerbes AL (2022) Challenges and future of drug-induced liver injury research-laboratory tests. Int J Mol Sci 23(11):6049. https://doi.org/10.3390/ijms23116049
28. Kamm DR, McCommis KS (2022) Hepatic stellate cells in physiology and pathology. J Physiol 600(8): 1825–1837. https://doi.org/10.1113/JP281061
29. Hamlin RL (2020) Species differences in cardiovascular physiology that affect pharmacology and toxicology. Curr Opin Toxicol 23-24:106–113. https://doi.org/10.1016/j.cotox.2020.07.004
30. Pang L, Sager P, Yang X, Shi H, Sannajust F, Brock M et al (2019) Workshop report: FDA workshop on improving cardiotoxicity assessment with human-relevant platforms. Circ Res 125(9):855–867. https://doi.org/10.1161/CIRCRESAHA.119.315378
31. Kreutzer FP, Meinecke A, Schmidt K, Fiedler J, Thum T (2022) Alternative strategies in cardiac preclinical research and new clinical trial formats. Cardiovasc Res 118(3):746–762. https://doi.org/10.1093/cvr/cvab075
32. Thomas D, Choi S, Alamana C, Parker KK, Wu JC (2022) Cellular and engineered organoids for cardiovascular models. Circ Res 130(12):1780–1802. https://doi.org/10.1161/CIRCRESAHA.122.320305
33. Gintant G, Burridge P, Gepstein L, Harding S, Herron T, Hong C et al (2019) Use of human induced pluripotent stem cell-derived cardiomyocytes in preclinical cancer drug cardiotoxicity testing: a scientific statement from the American Heart Association. Circ Res 125(10):e75–e92. https://doi.org/10.1161/RES.0000000000000291
34. Blinova K, Dang Q, Millard D, Smith G, Pierson J, Guo L et al (2018) International multisite study of human-induced pluripotent stem cell-derived cardiomyocytes for drug proarrhythmic potential assessment. Cell Rep 24(13):3582–3592. https://doi.org/10.1016/j.celrep.2018.08.079
35. ICH E14/S7B Implementation Working Group (2022) E14/S7B Q & As. https://database.ich.org/sites/default/files/E14-S7B_QAs_Step4_2022_0221.pdf. Accessed 17 Aug 2022
36. Denning C, Borgdorff V, Crutchley J, Firth KS, George V, Kalra S et al (2016) Cardiomyocytes from human pluripotent stem cells: from laboratory curiosity to industrial biomedical platform. Biochim Biophys Acta 1863(7):1728–1748. https://doi.org/10.1016/j.bbamcr.2015.10.014
37. Saleem U, van Meer BJ, Katili PA, Mohd Yusof NAN, Mannhardt I, Garcia AK et al (2020) Blinded, multicenter evaluation of drug-induced changes in contractility using human-induced pluripotent stem cell-derived cardiomyocytes. Toxicol Sci 176(1): 103–123. https://doi.org/10.1093/toxsci/kfaa058
38. Archer CR, Sargeant R, Basak J, Pilling J, Barnes JR, Pointon A (2018) Characterization and validation of a human 3D cardiac microtissue for the assessment of changes in cardiac pathology. Sci Rep 8(1):10160. https://doi.org/10.1038/s41598-018-28393-y
39. Li J, Hua Y, Miyagawa S, Zhang J, Li L, Liu L et al (2020) hiPSC-derived cardiac tissue for disease modeling and drug discovery. Int J Mol Sci 21(23): 8893. https://doi.org/10.3390/ijms21238893

40. Paik DT, Chandy M, Wu JC (2020) Patient and disease-specific induced pluripotent stem cells for discovery of personalized cardiovascular drugs and therapeutics. Pharmacol Rev 72(1):320–342. https://doi.org/10.1124/pr.116.013003
41. Stoyanova-Slavova IB, Slavov SH, Buzatu DA, Beger RD, Wilkes JG (2017) 3D-SDAR modeling of hERG potassium channel affinity: a case study in model design and toxicophore identification. J Mol Graph Model 72:246–255. https://doi.org/10.1016/j.jmgm.2017.01.012
42. Guha R, Willighagen E (2012) A survey of quantitative descriptions of molecular structure. Curr Top Med Chem 12(18):1946–1956. https://doi.org/10.2174/156802612804910278
43. Yap CW (2011) PaDEL-descriptor: an open source software to calculate molecular descriptors and fingerprints. J Comput Chem 32(7):1466–1474
44. Wu Z, Zhu M, Kang Y, Leung EL, Lei T, Shen C et al (2021) Do we need different machine learning algorithms for QSAR modeling? A comprehensive assessment of 16 machine learning algorithms on 14 QSAR data sets. Brief Bioinform 22(4):321. https://doi.org/10.1093/bib/bbaa321
45. LeCun Y, Bengio Y, Hinton G (2015) Deep learning. Nature 521(7553):436–444. https://doi.org/10.1038/nature14539
46. Kang MG, Kang NS (2021) Predictive model for drug-induced liver injury using deep neural networks based on substructure space. Molecules 26(24):7548. https://doi.org/10.3390/molecules26247548
47. Chen M, Suzuki A, Thakkar S, Yu K, Hu C, Tong W (2016) DILIrank: the largest reference drug list ranked by the risk for developing drug-induced liver injury in humans. Drug Discov Today 21(4):648–653
48. Health NIo (2017) LiverTox: clinical and research information on drug-induced liver injury. Nih gov. https://livertox.nih.gov. Accessed 17 Aug 2022
49. Li T, Tong W, Roberts R, Liu Z, Thakkar S (2021) DeepDILI: deep learning-powered drug-induced liver injury prediction using model-level representation. Chem Res Toxicol 34(2):550–565. https://doi.org/10.1021/acs.chemrestox.0c00374
50. Przybylak KR, Cronin MT (2012) In silico models for drug-induced liver injury - current status. Expert Opin Drug Metab Toxicol 8(2):201–217. https://doi.org/10.1517/17425255.2012.648613
51. Chen M, Hong H, Fang H, Zhou G, Kelly R, Borlak J et al (2013) Quantitative structure-activity relationship models for predicting drug-induced liver injury based on FDA-approved drug labeling annotation and using a large collection of drugs. Toxicol Sci 136(1):242–249
52. Hong H, Zhu J, Chen M, Gong P, Zhang C, Tong W (2018) Quantitative structure–activity relationship models for predicting risk of drug-induced liver injury in humans. In: Drug-induced liver toxicity. Springer, Cham, pp 77–100
53. Decision Forest. https://www.fda.gov/science-research/bioinformatics-tools/decision-forest. Accessed 17 Aug 2022
54. Chen M, Vijay V, Shi Q, Liu Z, Fang H, Tong W (2011) FDA-approved drug labeling for the study of drug-induced liver injury. Drug Discov Today 16(15-16):697–703
55. Greene N, Fisk L, Naven RT, Note RR, Patel ML, Pelletier DJ (2010) Developing structure-activity relationships for the prediction of hepatotoxicity. Chem Res Toxicol 23(7):1215–1222. https://doi.org/10.1021/tx1000865
56. Strauss DG (2022) Advancing alternative methods for regulatory use. Presentation to the FDA science board. https://www.fda.gov/media/159235/download. Accessed 3 Oct 2022

15 Animal Models for the Study of Human Disease

Sherry J. Morgan, Julie A. Hutt, and Radhakrishna Sura

Abstract

The use of conventional healthy animals (CHAs) has long been a mainstay of nonclinical safety testing as well as evaluation of efficacy of new chemical entities (NCEs) in the discovery setting. However, the potential value of directed animal models of human disease (AMDs) has increasingly been of value in evaluation of both disciplines. While more frequently employed in the evaluation of efficacy (discovery phase), there has been increasing interest in utilizing AMDs in investigation of the safety (development phase), particularly in the development of targeted NCEs that may not be appropriate for testing in CHAs. While it is accepted that no animal model (neither CHA nor AMD) will accurately predict all risks (or potential benefits) of an NCE, careful selection of the model with an understanding of the pros and cons of each model will optimize the results of investigations. This chapter will outline some of the major organ systems, human diseases, and associated AMDs along with considerations for their use.

Keywords

Animal models · Nonclinical efficacy evaluation · Pharmaceutical development/ products · Nonclinical research and development · Risk management · Safety assessment

15.1 Introduction

Historically, studies utilizing conventional healthy animals (CHAs) have provided all or the preponderance of nonclinical safety information to support the safety of new chemical entities (NCEs) prior to and during clinical development. Conventional healthy animals have been defined as animals that have not been genetically, surgically, or chemically altered to produce disease [1]. Animal models of human disease (AMDs) have long been an accepted and integral component in determination of potential efficacy of many NCEs. More recently, the potential value of AMDs in the prospective evaluation of safety and/or retrospective elucidation of mechanisms of toxicity of NCEs has been a topic of considerable interest and selected use of AMDs has provided considerable value in support of clinical development. While AMDs have great potential value, their use, model selection and study design

S. J. Morgan (✉)
StageBio, Mason, OH, USA
e-mail: smorgan@stagebio.com

J. A. Hutt
Greenfield Pathology Services, Greenfield, IN, USA
e-mail: jhutt@gfpath.com

R. Sura
Gilead Sciences, Inc., Foster City, CA, USA
e-mail: radha.sura@gilead.com

G. Jagadeesh et al. (eds.), *The Quintessence of Basic and Clinical Research and Scientific Publishing*,
https://doi.org/10.1007/978-981-99-1284-1_15

Table 15.1 When are animal models of disease useful or appropriate?

Study intent	Comments
Proof of concept efficacy studies	• Most common use of animal models in drug discovery
Early discovery safety information	• Targeted set of safety endpoints and biomarkers • Aid in the identification of target organ toxicities or consequences of exaggerated pharmacology
Mechanism of toxicity/ hypothesis testing	• Provide insight into the pathogenesis and/or relevance of toxicities noted in CHAs or understand the pathogenesis of unexpected toxicity in a clinical setting • Of value in cases where the pharmacodynamic effect of drugs under health physiologic conditions is different from a drug's effect in disease states
Target engagement when CHAs lack the target	• Evaluate on-target toxicity of on-target adverse

Table 15.2 What are some challenges or limitations in utilizing an animal model of disease?

Consideration	Comments
Analogy to human disease	• AMD should reliably exhibit all critical manifestations of the disease relevant to the investigation
Homogeneity with respect to disease	• Consistency with respect to disease manifestation is critical
Background changes	• Evaluation of background changes prior to model selection/study start
Unintended manifestation	• Long term testing may not be feasible (especially for control and low dose groups such as with hyperglycemic or hypertensive model)

should be carefully considered to optimize their utility in providing clarity for path forward in nonclinical and clinical development.

A previous publication has provided general guidance or recommendations for when animal models are useful or appropriate and considerations for challenges and limitations [2]. The salient recommendations and considerations are outlined in Tables 15.1 and 15.2.

This chapter will outline some of the available models of human disease by an organ/system approach. While adverse effects are often identified in CHAs, some may be more easily predicted in an AMD. From a clinical perspective, adverse responses involving the hepatic, cardiovascular, neurologic, and gastrointestinal system are the most frequent adverse events resulting in termination of clinical development [3, 4]. Examples of some of the primary AMDs for these four organ systems as well as the renal and respiratory systems are outlined in this chapter. While the listing is not intended to provide all examples of human diseases and associated AMDs, sufficient examples and detail are included to provide the reader with an overall understanding that there is no one model that is optimal or even acceptable for evaluation of a human disease state. Instead, multiple AMDs are typically available, and selection of the model must be made based on the question or hypothesis to be addressed.

15.2 Hepatic Disease Models

15.2.1 Hepatic Toxicity: Relevance to Efficacy Evaluation (Discovery) and Drug Attrition (Drug Development)

Given the worldwide prevalence of liver diseases (LD), there is a need to develop animal models to help researchers identify therapeutic targets and develop effective treatments [5]. Understanding of liver injury pathogenesis and progression is imperative to developing new therapeutics. However, the current rodent models are unable to mimic human liver diseases completely, especially all the clinical aspects seen in humans [6]. A good animal model should be able to replicate the etiology and progression observed in human disease through the various stages starting from hepatomegaly to steatosis, fibrosis/cirrhosis and lastly the end stage of hepatocellular carcinoma [7]. Most models recapitulate only certain aspects of the human disease and hence can be used to answer only specific research questions [8].

15.2.2 Hepatic Toxicity: Reason for Insufficient Translation from Animal to Human

Alcohol-related [AFLD] and non-alcoholic fatty liver disease (NAFLD) are the most prevalent metabolic liver diseases [9]. Both diseases share similar pathogenesis and histopathological findings spanning from fatty liver, steatosis, steatohepatitis leading to advanced chronic liver disease manifesting as fibrosis, cirrhosis and finally primary liver cancer [5]. Due to the complex underlying pathogenesis, which has not been elucidated yet, there are no animal models that can mimic human disease progression. Most models replicate one or more specific pathways and have played a pivotal role in elucidating pathophysiological mechanisms underlying these diseases. Taking the example of phosphodiesterase-4 inhibitor ASP9831, which selectively inhibited activated macrophages and Kupffer cells and lowered alanine transaminase (ALT) levels, necroinflammation and fibrosis in two ALD animal models (acute hepatitis model and a methionine and choline deficient (MCD) diet NASH model), it did not have any results in human clinical trials [10]. Thus, model selection is important in order to understand the process being investigated. Some of the commonly utilized models are outlined in Table 15.3.

15.3 Cardiovascular Disease Models

15.3.1 Cardiac Toxicity: Relevance to Efficacy Evaluation (Discovery) and Drug Attrition (Drug Development)

Cardiac toxicity is a major cause of drug attrition during clinical development and post-approval drug withdrawal of new drugs [15]. Drug attrition due to cardiovascular toxicity is relatively rarely encountered early in Phase I development, being approximately 9% of total drug attrition [16] but is a major cause of attrition of drugs withdrawn post-approval ranging from approximately 16% [17] to approximately 45% of total drug attrition [4]. The reason for the higher attrition rate later in clinical development and post-approval may be related to longer periods of administration of the compound to subjects/patients with higher risk factors and other factors. In addition to being a major cause of drug attrition during clinical development, spontaneous cardiovascular disorders are common in humans, thus necessitating models to evaluate the potential therapeutic benefit of NCEs in their amelioration.

15.3.2 Cardiac Toxicity: Reason for Insufficient Translation from Animal to Human

While there are general similarities in anatomy of the cardiovascular system between animals and humans, there are a multitude of physiologic differences, particularly when considering differences between humans and rodents that may preclude accurate prediction of the potential for an effect on the cardiovascular system.

Small animal models (mice and rats) are commonly utilized, with general advantages being a short reproductive cycle, large litters, a well-known genome, and relative low cost. While large animals (primarily pigs and dogs but also rabbits, sheep, and nonhuman primates) may be

Table 15.3 Examples of available models to predict hepatic toxicity: advantages and limitations

Species	Animal model	Relevant pathology findings	Pros and cons	References
Alcoholic liver disease (ALD) model				
Rat/ mouse	Alcohol in drinking water; ad libitum	• Steatosis, increased inflammatory cell infiltrates	• Pros: easy to perform, steatosis • Inflammatory infiltrates • Cons: no progression beyond steatosis	[5, 7]
Rat/ mouse	Meadows-Cook (MC)	• Steatosis and inflammatory cell infiltrates	• Pros: easy model, animal fed regular diet, and alcohol is mixed in drinking water, model of early alcoholic liver injury closer representation of human alcoholic liver injury than Lieber-DeCarli (LD) model • Cons: less steatosis, no change in neutrophils (inflammatory infiltrates)	[11]
Rat/ mouse	Lieber-DeCarli diet	• Steatosis, inflammatory cell infiltrates	• Pros: early stages of ALD—early hepatic lesions of ALD: steatosis, activation of Kupffer cells, ROS generation, and hepatocyte cell death. Easy to handle, accurate, reliable, inexpensive model • Cons: no liver fibrosis, more steatosis than MC model, low elevation of liver enzymes; no change in neutrophils	[7, 12]
Rat/ mouse	NIAA (National Institute on Alcohol Abuse and Alcoholism-NIAAA) model	• Severe steatosis, hepatocellular damage, and hepatic neutrophil infiltration	• Pros: less costly, more time efficient, easy to perform, closely represents progression of human alcoholic hepatitis, marked elevation of fatty liver and enzymes indicating liver injury; mild fat accumulation in liver cells, slightly elevated liver enzymes indicating damage, and little or no inflammation • Cons: high blood alcohol levels	[13]
Rat/ mouse	Tsukamoto-French model	• Severe hepatic steatosis and necrosis	• Pros: progressive changes, fibrosis with Kupffer cell activation and inflammatory infiltrates • Cons: limited use owing to its technical difficulty, and its requirement for intensive medical care and expensive equipment, long-term feeding with a high mortality rate	[5, 13]
Non-alcoholic liver disease (NALD) model				
Rat/ mouse	Methionine and choline deficient (MCD) diet	• Steatohepatitis, fibrosis, and necro-inflammatory foci containing lymphocytes and neutrophils	• Pros: oxidative stress and changes in cytokines and adipokine; elevated AST, ALT • Cons: only minor inflammation and fibrosis, significant weight loss (~40%)	[5, 10]
Rat/ mouse			• Pros: a significant amount of fibrosis, increased ALT	[5, 10]

(continued)

Table 15.3 (continued)

Species	Animal model	Relevant pathology findings	Pros and cons	References
	Choline-deficient l-amino acid-defined diet	• Steatohepatitis, mild ballooning degeneration, fibrosis	• Cons: metabolic features of human NAFLD still fail to appear when used in the same time frame as the MCD diet	
Rat/mouse	High-fat diet (HFD)	• Steatosis, fibrosis and hepatic necrosis, inflammation	• Pros: phenotype similar to the human disease, characterized by obesity (after 10 weeks), insulin resistance (hyperinsulinemia after 10 weeks and glucose intolerance after 12 weeks) and hyperlipidemia, increased ALT, AST • Cons: steatosis and inflammation are substantially less pronounced, minimal fibrosis, requires a large sample size due to variability in steatosis, inflammation and fibrosis	[5, 10]
Rat/mouse	Modified HFD	• Severe steatosis, ballooned hepatocytes, inflammatory infiltrates, satellitosis, Mallory-Denk bodies	• Pros: significantly higher plasma triglycerides, higher ALT levels and more steatosis • Cons: significantly higher plasma triglycerides, higher ALT levels and more steatosis	[5, 10]
Liver fibrosis				
Rat/mouse	Chemical induced: CCl4	• Inflammation, fibrosis, early cirrhosis	• Pros: low cost to develop, easy implementation • Cons: animal welfare concerns, variable tolerability, differences in administration route, dose, etc., can cause discrepancies between studies	[5, 14]
Rat/mouse	Thioacetamide	• Fibrosis, cirrhosis	• Pros: suitable for the study of connective tissue metabolism in fibrotic and cirrhotic models; not hepatotoxic; more periportal inflammatory cell infiltration and more pronounced ductal hyperplasia • Cons: liver injury and fibrosis are dependent on CYP2E1	[5]

better models for predicting cardiovascular toxicity in a clinical setting, their use may be complicated by other factors (cost, higher test article requirements, and maintenance challenges).

15.3.3 Atherothrombotic Disease

Atherothrombotic disease in humans is responsible for considerable morbidity and mortality in humans, particularly in developed countries [18]. The atherothrombotic disease is a complex, multifactorial disease with different etiologies that synergistically promote lesion development. Mouse models have been of particular use to study the development and progression of these complex lesions. Some of the commonly utilized models are outlined in Table 15.4.

Table 15.4 Examples of available models to predict cardiac toxicity (atherothrombotic disease): advantages and limitations

Species	Animal model	Relevant pathology findings	Pros and cons	References
Mouse	LDLR−/−	• Deficient in LDLR → delayed clearance of VLDL and LDL from plasma → moderate increase of plasma cholesterol with associated atherosclerosis lesions (foam-cell fatty streak)	• Pro: accelerate severity of hypercholesterolemia and lesions by feeding a high-fat, high cholesterol diet • Con: response to treatment varies from lowering plasma cholesterol without effect on atherosclerosis to weak lesion reduction with or without lower plasma cholesterol	[19, 20]
	ApoE−/−	• Marked increase in plasma levels of LDL and VLDL due to a failure in clearance through the LDLR and LDLR-related proteins	• Pro: under normal dietary conditions, dramatically elevated levels of cholesterol → extensive widely distributed lesions in aorta; exacerbate by high-fat diet • Con: infrequency of plaque rupture and thrombosis (common complications of human disease)	[19, 21]
	Diabetes-accelerated atherosclerosis	• Type I diabetes induced by streptozotocin or viral injection (models include LDLR−/− and apoE −/− mice)	• Pro: diabetes induction does not markedly change plasma lipid levels and thus mimics accelerated atherosclerosis in the human disease • Pro: useful in demonstrating the importance of inflammatory and immunological mechanisms in the formation and progression of atheroma plaque; monitor with non-invasive imaging	[19, 22, 23]
Rabbit	Atherosclerosis with an inflammatory component	• High-cholesterol diet + repeated or continuous intimal injury (e.g., indwelling aortic catheter, balloon angioplasty) → resemble human plaques (inflammatory component)	• Pro: evaluate influence of inflammation on atherosclerotic plaques (hyperlipidemic rabbits + arthritis) • Pro: model for plaque rupture—aggressive vascular injury + hyperlipidemic diet—resembles human familial hypercholesterolemia; monitor with non-invasive imaging	[19, 24, 25]
Pig	Diabetes-induced accelerated atherosclerosis	• Diabetes + hypercholesterolemia → model for "vulnerable" plaque	• Pro: Probably the best way to recreate human plaque instability—evaluate plaque-stabilizing therapies • Cons: High cost, difficult handling, few genomic tools	[19, 26]

15.3.4 Abdominal Aortic Aneurysms (AAA)

Animal models of atherothrombotic abdominal aortic aneurysms are essential in the nonclinical evaluation of potential NCEs for the suppression of aneurysmal degeneration. Recent insights into the development of human AAA have come from studies in mouse models, with elastase-induced AAA appearing to recapitulate many of the features of human AAA. Some of the primary models to investigate development and/or potential for benefit of NCEs for AAA are outlined in Table 15.5.

Table 15.5 Examples of available models to predict cardiac toxicity (abdominal aortic aneurysms): advantages and limitations

Species	Animal model	Relevant pathology findings	Pros and cons	References
Mouse	Calcium chloride induced	• $CaCl_2$ applied between renal arteries and iliac bifurcation	• Pro: significant dilatation of aorta within 14 days; no need for mechanical intervention • Pro: developed for other species (e.g., rabbit)	[19]
	Elastase induced	• Infuse a segment of the aorta with elastase → degradation of elastic fibers → aortic wall inflammation and dilatation	• Pro: significant dilatation of aorta within 14 days (100% increase in diameters) • Pro: recapitulates many features of human disease (increased expression of MMPs, cathepsin, and other proteases) • Pro: developed for other species (e.g., rabbit)	[19, 27, 28]
	Angiotensin II-induced	• Inflammation of aortic wall → vessel dissection and rupture; severity is higher in apoE−/− and LDLR−/− males	• Con: location different from humans (suprarenal vs. infrarenal) • Con: No clear association between angiotensin II and aneurysms in human	[19]
	Spontaneous mutants	• Blotchy mouse—spontaneous mutation on the X chromosome → abnormal copper absorption → weak elastic tissue due to failed crosslinking of elastin and collagen → aortic aneurysms	• Con: due to other systemic effects, cardiovascular effects may be difficult to interpret	[19, 29]
Rabbit	Various methods (e.g., $CaCl_2$ or elastase-induced)	• See $CaCl_2$ and elastase-induced above	• Pro: rabbit aneurysms more closely resemble human aneurysms hemodynamically and histologically (vs. mouse)	[19]
Pig	Balloon angioplasty + collagenase/ elastase solution	• Gradual expansion with degradation of aortic wall elastic fibers, inflammatory cell infiltrate and persistent smooth muscle cell loss	• Pros: numerous similarities with human abdominal aortic aneurysms • Cons: complex animal handling, special housing, surgical room facilities, animal cost, reduced sample sizes	[19, 30]

15.3.5 Heart Failure

As is the case with other cardiovascular disorders in humans, heart failure is increasingly associated with morbidity and mortality, particularly in developed countries with an aging population. Both small and large animal models are available for evaluation of the pathogenesis or potential efficacy of NCEs. Some of the primary models are outlined in Table 15.6.

Table 15.6 Examples of available models to predict cardiac toxicity (heart failure): advantages and limitations

Species	Animal model	Relevant pathology findings	Pros and cons	References
Rat/ mouse	Surgical method—left coronary artery ligated or cauterized	• Myocardial infarction and potential for effect on ventricular function	• Pros: efficient and reproducible; temporary or permanent occlusion	[31, 32]
	Pharmacological method	• β-1 adrenergic receptor agonist (e.g., isoproterenol)—administration before ischemia → cardioprotective action but higher doses induce cardiomyocyte necrosis, left ventricular dilatation/ hypertrophy	• Pros: efficient and reproducible	[33]
Pig	Balloon catheterization of descending coronary artery	• Myocardial infarction and subsequent sequelae	• Pros: collateral coronary circulation and arterial anatomy of pigs like human • Cons: requires specialized equipment, dedicated surgical facilities and skilled personnel	[34]
Rabbit	Surgical method—left coronary artery ligated	• Myocardial infarction and subsequent sequelae	• Pros: less expensive than pig; sarcomeric proteins in rabbits like human	[35]
	WHHL rabbit model (does not require surgery)	• Myocardial infarction and subsequent sequelae	• Pros: as with conventional rabbit • Cons: does not have plaque rupture (conventional rabbit does not either)	[36]

15.4 Nervous System Disease Models

15.4.1 Nervous System Toxicity: Relevance to Efficacy Evaluation (Discovery) and Drug Attrition (Drug Development)

Animal models are essential to understand the underlying pathology and molecular mechanisms of nervous system toxicity (neurological disorders/ neurodegenerative disease); and serve as valuable tools in nonclinical settings to evaluate new drug candidates. However, these models have limitations, most of which can be attributed to the complexity of the nervous system. In nervous system models, the complexity is further highlighted as there is a gap in understanding of the underlying mechanisms for the disorders or diseases, whether they have a genetic origin and how some of these conditions are heterogenous in nature with a lot of subjective variability which is not easily discernible in animals due to their inability to describe symptoms [37].

15.4.2 Nervous System Toxicity: Reason for Inadequate Translation from Animal to Human

Neurological and neurodegenerative conditions are believed to arise either due to developmental deficits or functional impairment. Animal models are central in developing a basic understanding of neurological disease mechanisms such as initial cell death and repair in stroke, pathologies underlying Parkinson's disease (motor and non-motor pathologies), etc. However, these models often fail when they are used for nonclinical testing of drug candidates due to lack of translatability [38] as many observations such as anxiety, dizziness, headache are not discernible in the animal models [39]. Also, some neurological effects go undetected in animal models, especially due to a pre-existing condition.

Most neurodegenerative disease models for Alzheimer's disease (AD), Parkinson's disease (PD), frontotemporal dementia (FTD) and amyotrophic lateral sclerosis (ALS) only recapitulate the initial proteinopathy or some feature of the human disorder without fully recapitulating the entire human disease [40]. This limitation is mainly attributed to the complexity of the intact human nervous system and the lack of corresponding glial complexity, complex neuronal circuits, absence of vascular and immunologic components in the animal models for these disease conditions [40]. Additionally, the shorter life span of rodents leads to incomplete development of neurodegeneration, thus falling short of modeling the aging-related changes observed in these diseases. Genomic differences between rodents and humans such as lack of well conserved RNA protein binding in rodents, RNA processing alterations all contribute towards less than perfect animal models for modeling neurodegenerative diseases [40].

Despite the limitations, these models provide valuable information on the mechanism and treatment when they are used appropriately to answer specific questions [37]. Most models have good predictivity if they are used for compounds which act through known or established mechanisms. The majority of the models were developed for a particular class of medicines, for example, the forced swim test (FST) was first developed for tricyclic antidepressants, however, it needed to be refined when used for selective serotonin reuptake inhibitors (SSRIs] [37]. Another example is pre-pulse inhibition, which is deficient in Schizophrenic patients, but can be recapitulated in rodents when treated with amphetamine. Likewise, certain transgenic mouse models such as R1/2 for Huntington's disease are able to produce aberrant huntingtin protein and demonstrate characteristics of Huntington's disease such as motor, mood and cognition deficits. Moreover, it has been demonstrated through transcriptional profiling that there is good concordance in the gene expression changes between animal models and humans. Recently developed transgenic models for Alzheimer's disease (AD) and Amyotrophic lateral sclerosis (ALS) are also well validated and have robust behavioral assays [37]. Some of the commonly utilized models are outlined in Table 15.7.

Besides the rodent models, other organisms such as *Drosophila*, *C. Elegans*, *Danio rerio*, yeast and other models have been tapped into to gain insight into the protein pathology underlying these neurodegenerative diseases which manifest as cellular and organism pathology. However, we are yet to develop more robust models that are translatable [40]. The current animal models have to be used in the context of a specific question as the majority of the models are capable of answering one or more questions but do not recapitulate the entire disease pathology.

15.5 Gastrointestinal Disease Models

15.5.1 Gastrointestinal (GI) Toxicity: Relevance to Efficacy Evaluation (Discovery) and Drug Attrition (Drug Development)

In general, there is a high degree of concordance between GI disorders identified during nonclinical testing and those identified in humans during clinical stages of drug development, particularly for studies in dogs [45, 46]. GI toxicity is estimated to contribute to the attrition of approximately 3.4% of drugs in the late stages of development [47] and is responsible for approximately 2% of post-approval drug withdrawals [4, 17]. Animal models of human GI diseases are most frequently used in drug discovery to elucidate disease pathogenesis and test the efficacy of NCEs. Additionally, drug safety liabilities identified in the GI tract of CHAs may evoke

Table 15.7 Examples of available models to predict CNS toxicity: advantages and limitations

Species	Animal model	Relevant pathology findings	Pros and cons	References
Parkinson's disease models				
Pharmacologic based models of PD				
Rat	Reserpine model	• No morphologic changes	• Pros: produces symptoms similar to those observed in the early stages of PD, good for early preclinical stages, detects motor deficits, short course • Cons: no Lewy bodies or morphologic changes, lack of selectivity for dopamine	[39, 41]
Rat	Pesticide induced model: rotenone and paraquat	• Lewy bodies, and selective loss of neurons (dopaminergic)	• Pros: detects motor deficits, short course-progressive • Cons: replicability is a challenge due to high mortality	[39, 41]
Rat/primate	1-Methyl 4-phenyl-1,2,3,6-tetrahydro-pyridine [MPTP] model	• Selective loss of dopaminergic neurons	• Pros: detects motor deficits, short course • Cons: metabolite [MPP +] is selectively taken up by dopaminergic neurons, where it inhibits complex I of the respiratory chain	[39, 41]
Rat	6-Hydroxy-dopamine [6-OHDA] model	• Selective loss of dopaminergic neurons	• Pros: detects motor deficits, short course-acute • Cons: no Lewy bodies	[39, 41]
Genetic-based models of α-synuclein pathology				
Transgenic mouse	Mutations in α-synuclein cause autosomal dominant PD	• Characteristic Lewy pathology • Neurodegeneration	• Pros: exhibit robust non-dopaminergic deficits including anxiety, gastrointestinal dysfunction, non-DA related motoric dysfunction • Cons: substantial neurodegeneration in absence of loss of DA neurons	[40]
Huntington's disease models				
Transgenic mouse	Transgenic models [R6 line of transgenic mice R6/1; R6/2, C57BL/6J BACHD]	• No histologic change	• Pros: very aggressive, rapidly progressing form, overt behavioral symptoms, progressive motor deficit • Cons: classical Huntington's-like motor impairments seen in this transgenic model, do not occur due to degeneration of the striatum	[39, 42, 43]
Alzheimer's disease [AD] models				
Genetic-based models of amyloid pathology				
Rat/mouse			• Pros: administer defined amounts of a specific Aβ	[44]

(continued)

Table 15.7 (continued)

Species	Animal model	Relevant pathology findings	Pros and cons	References
	Amyloid precursor protein [APP]/PSEN1 and PSEN2 mutant mouse models	• Inflammation, microglial activation, and limited cell loss	species of known sequence and length to develop pathological changes, deliver experimental results, including plaque pathology, within a timeframe of a few weeks • Cons: brain alterations surpass the effect of aging on AD progression	
Transgenic mouse	Amyloid precursor protein [APP]- genetically modified mice overexpressing APP or Aβ42	• Cerebral amyloid angiopathy, astrocytosis, microgliosis, mild hippocampal atrophy	• Pros: model amyloid deposition in senile plaques and some cases cerebrovascular amyloid • Cons: lack of prominent tau accumulation	[37, 40]
Genetic based models of tau pathology				
Transgenic mouse	Transgenic overexpression of mutations that cause FTD with Parkinsonism linked to chromosome 17 [FTD-MAPT]	• Overt neurodegenerative changes • Pick bodies (intracytoplasmic spherical inclusions) than the classic AD neurofibrillary tangle [NFT] pathology	• Pros: cognitive deficits, progression is exhibited • Cons: exhibit only a subset of pathology	[40]
Amyotrophic lateral sclerosis (ALS) and frontotemporal dementia (FTD) models				
Transgenic rat/mouse	Models of SOD1-related ALS	• Cortical and spinal motor neuron loss • Glial activation • Accumulation of misfolded protein	• Pros: severity of disease determined by the level of a transgene, while SOD1 gene deletion does not lead to motor neuron disease • Cons: TDP-43 pathology not recapitulated	[40]
Transgenic rat/mouse	Models of TDP-43 pathology	• Motor neuron degeneration	• Pros: TDP-43 is a major component of sporadic ALS and FTD • Cons: neuronal degeneration and neuromuscular denervation occur without paralysis or reduced lifespan	[40]
Stroke				
Rat/mouse	Occlusion model	• Focal ischemia/ thromboembolism	• Pros: ischemic damage observed in discrete regions • Cons: irreversible and difficult to control	[37]
Rat/mouse	Emboli model	• Major infarct/ischemic damage	• Pros: significant Ischemic damage • Cons: not temporary and difficult to control	[37]
Pain				
Rat/mouse	Chronic constriction injury [CCI] model	• Nerve injury/damage (neuropathy, degeneration of nerve fibers)	• Pros: reproducible • Cons: technically challenging, time consuming	[37]

(continued)

Table 15.7 (continued)

Species	Animal model	Relevant pathology findings	Pros and cons	References
Rat/mouse	Spinal nerve ligation [SNL] model	• Nerve injury/damage (neuropathy, degeneration of nerve fibers)	• Pros: reproducible • Cons: technically challenging, time consuming	[37]

more severe GI toxicity in humans with GI disease. In such cases, additional safety evaluations in an animal model of human GI disease may be helpful when the target human population has a high prevalence of the disease.

15.5.2 Gastrointestinal Toxicity: Reasons for Insufficient Translation from Animal to Human

Species variations in GI tract anatomy and physiology, the distribution and abundance of drug metabolizing enzymes, and the absence of the requisite drug target in the appropriate cell type as well as variations in diet and microbiome all may contribute to insufficient translation from animals to humans during nonclinical development with CHAs [48]. The rodent inability to vomit, in particular, is a hindrance to their use in some circumstances. Differences in disease pathogenesis between an AMD and a human disease may also contribute to inadequate translation from animals to humans.

15.5.3 Inflammatory Bowel Disease (IBD) Models

IBD is the term for two chronic inflammatory diseases affecting the GI tract: Crohn's disease and ulcerative colitis. The exact cause of IBD is unknown, but a complex pathogenesis involving a dysregulated immune response to environmental triggers in genetically susceptible individuals has been proposed. Murine models of intestinal inflammation have provided major insights into the pathogenesis of IBD in recent years. Bowel inflammation in these models can be classified into four broad categories according to the pathogenic mechanisms involved in the induction of inflammation: (1) mucosal barrier disruption; (2) impaired innate immunity; (3) exaggerated effector cell responses; and (4) defects in regulatory cells [49]. Chemically induced and genetically engineered models are the most commonly used models in drug development; they are primarily used for efficacy testing and pharmacokinetic studies. Over 74 genetically engineered mouse strains have been created for studying IBD, 20 of which constitute susceptibility genes identified in human IBD. A discussion of these transgenic models is beyond the scope of this chapter; however, an excellent review of transgenic mouse models of IBD is provided by Mizoguchi et al. [50]. Some of the more salient models of IBD are outlined in Table 15.8.

15.5.4 Gastrointestinal Ulcer Disease

Gastroduodenal ulcers were traditionally considered to have resulted from the excess gastric acid secretion in association with diet and stress. Infection with *Helicobacter pylori* and the use of NSAID anti-inflammatory drugs are now considered the major risk factors for development of most gastroduodenal ulcers. *H. pylori* infection is associated with pangastritis or antral predominant gastritis leading to hypochlorhydria or hyperchlorhydria, respectively, and ulcer formation in the stomach or duodenum [56]. NSAID anti-inflammatory drugs induce ulcers by inhibiting the release of cyclooxygenase 1 (COX-1)-derived prostaglandins, resulting in decreased gastric mucus and bicarbonate secretion and reduced mucosal blood flow

Table 15.8 Mouse models of inflammatory bowel disease

Model	Method of induction	Features of disease/model	Pros and cons	References
Dextran sodium sulfate (DSS)	Addition to drinking water	• Epithelial cell injury/ colonic barrier disruption • Activation of innate/ adaptive immune response • Immune dysregulation/ loss of tolerance to commensal bacteria • Changes in mucin content, microbiome, and metabolome	• Pros: rapid, simple, reproducible; useful for testing NCEs and investigating immunologic aspects of Crohn's disease, including cytokine response/ oral tolerance mechanisms • Cons: massive epithelial damage/microbial invasion may not be relevant for human IBD	[51–53]
Trinitrobenzene sulfonic acid (TNBS)	Intrarectal TNBS (a hapten)	• Haptenation of proteins • Inflammation via Th1-mediated delayed hypersensitivity, with features of Crohn's disease • Chronic TNBS colitis in BALB/c mice was used to investigate mechanisms of lamina propria fibrosis.	• Pros: rapid, simple reproducible; can induce acute, chronic, or relapsing colitis; colon neoplasia via initiator-promoter mechanism if animals pre-dosed with carcinogen; used to test the efficacy of Crohn's disease immunotherapies • Cons: high mortality under some conditions	[52, 53]
Oxazolone	Intrarectal oxazolone (a hapten)	• Haptenation of proteins • Distal colon inflammation resembles ulcerative colitis • Prior sensitization (subcutaneous injection) induces longer duration colitis after intrarectal administration	• Pros: rapid, simple, reproducible; IL-13, NKT cells involved in pathogenesis; useful for testing NCEs • Cons: NKT cell subset in oxazolone colitis are different than in ulcerative colitis	[52, 53]
Adoptive Transfer	Wild type naïve T cells (CD4+ CD45RB^{high}) transferred to SCID or Rag1/2 knockout mice	• Colitis does not occur in germ-free mice, implicating altered barrier function/ invasion by commensal bacteria in the pathogenesis • Established the role of Treg cells in inflammation/ maintenance of mucosal homeostasis	• Pros: reproducible; useful for testing NCEs • Cons: expensive and labor intensive; the precise role of Tregs in human IBD not established	[52–54]
SAMP1/Yit inbred mouse	Selective breeding	• Barrier defect/ dysfunctional Tregs • Spontaneous inflammation of the terminal ileum/cecum (primary site of Crohn's disease)	• Pros: reproducible; useful for investigating pathways that precede disease onset	[49, 52, 55]

[56]. Antibiotic treatment has reduced the prevalence of *H. pylori*-associated gastroduodenal ulcers. [57] Indeed, gastroduodenal ulcers are now most frequently associated with the use of NSAID drugs [58]. However, the development of antibiotic resistant strains of *H. pylori* highlights the need for continued research into new antibiotics and alternative strategies for preventing and treating *H. pylori*-associated gastritis and gastroduodenal ulcers.

Animal models have contributed substantially to our understanding of gastric physiology and mechanisms of control of acid secretion. Animal models of *Helicobacter*-related gastric ulceration are limited to gnotobiotic pigs and Mongolian gerbils, with inconsistent reports of *Helicobacter*-related gastric ulceration in mice [59, 60]. Animal models of gastroduodenal ulceration in rodents induced by a variety of methods have been used to investigate the pathogenic mechanisms contributing to gastroduodenal ulcer formation and the efficacy of preventative and therapeutic NCEs. These were recently reviewed, and the features of these models are summarized in Table 15.9 [61, 62].

15.6 Respiratory Disease Models

15.6.1 Respiratory Toxicity: Relevance to Efficacy Evaluation (Discovery) and Drug Attrition (Drug Development)

Quantitative evaluation of respiratory function in single-dose safety pharmacology studies combined with microscopic evaluation of the respiratory tract in repeat dose studies is generally sufficient for detecting respiratory toxicity in nonclinical studies. As such, respiratory toxicity is generally not a major cause of drug attrition during the late stages of drug development [47] and is responsible for only 2–3% of post-approval drug withdrawals [4, 17]. Animal models of human respiratory disease are frequently used in drug discovery to elucidate disease pathogenesis and test the efficacy of NCEs. However, if respiratory tract changes are identified in nonclinical safety studies in CHAs, it may be useful to perform additional safety evaluations in an AMD, especially if there is a high prevalence of the disease in the target patient population.

15.6.2 Respiratory Toxicity: Reasons for Insufficient Translation from Animal to Human

Species variations in respiratory tract anatomy and physiology, the distribution and abundance of drug metabolizing enzymes, and the absence of the requisite drug target in the appropriate cell type all may contribute to insufficient translation from animals to humans during nonclinical development using CHAs. When using AMDs, differences in disease pathogenesis between the AMD and the human disease may also be a contributing factor for inadequate translation.

15.6.3 Chronic Obstructive Pulmonary Disease (COPD)

COPD is an inflammatory airway disorder with features of both chronic bronchitis and emphysema, characterized by bronchial obstruction and decreased airflow during expiration. The capture of both chronic bronchitis-related or emphysematous changes in a single model has proven a challenge. Two excellent recent reviews of available animal models of COPD and their advantages and disadvantages are summarized in Table 15.10 [63, 64].

15.6.4 Asthma

Asthma is a heterogeneous disease characterized by chronic airway inflammation, increased mucus production, intermittent airflow obstruction, airway hyperresponsiveness (AHR) and airway remodeling. Animal models of asthma are developed in two phases: sensitization via subcutaneous or intraperitoneal allergen injection or intranasal instillation, followed by later allergen challenge via intranasal/intratracheal instillation or inhalation. Commonly used allergens include ovalbumin, house dust mites and cockroach extracts [69]. Immunologic, pathologic, and physiologic parameters are frequently evaluated in AMDs.

A variety of species have been used to develop models of asthma [69–71]. The features of the models and the pros and cons related to some of these species are summarized in Table 15.11.

Table 15.9 Animal models of gastrointestinal injury and ulceration

Model	Mechanism	Ulcer location/common use of model	Reference
Water immersion or cold water/cold restraint stress	• Endogenous histamine release • Increased gastric motility	• Gastric; widely used model • Evaluating mucosal/cytoprotective NCEs • Testing effects of NCEs on ulcer healing	[61, 62]
NSAIDs (oral)	• Suppression of prostaglandin	• Gastric; most commonly used model • Testing antisecretory/cytoprotective NCE efficacy	[61, 62]
Ethanol (oral)	• Membrane damage • Increased gastric acid secretion • Reduced mucosal blood flow • Oxidative stress	• Gastric • Testing cytoprotective NCE efficacy • Not useful for testing antisecretory NCEs due to the mechanism of ulcer induction	[61, 62]
Acetic acid (submucosal injection/oral)	• Chronic ulcers	• Gastric • Testing the efficacy of NCEs for healing chronic ulcers • Testing antisecretory/cytoprotective NCE efficacy	[61, 62]
Histamine (subcutaneous)	• Increased gastric acid secretion • Decreased mucus production • Vasodilation	• Gastric • Testing antisecretory NCE efficacy • Testing H2- receptor antagonist efficacy	[61, 62]
Reserpine (oral)	• Mast cell degranulation • Increased acid secretion/gastric motility	• Gastric	[61, 62]
Serotonin (oral)	• Decreased mucosal blood flow	• Gastric	[61]
Pylorus ligation	• Accumulation of gastric acid	• Gastric • Testing antisecretory/cytoprotective NCE efficacy	[61, 62]
Diethyldithiocarbamate (subcutaneous)	• Oxidative stress	• Gastric • Testing antioxidant/cytoprotective NCE efficacy	[61]
Methylene blue (oral)	• Ischemia • Oxidative stress	• Gastric and duodenal • Screening anti-ulcer NCEs	[61]
Ischemia-reperfusion	• Ischemia • Oxidative stress	• Gastric • Evaluating anti-ulcer NCEs	[60]
Cysteamine	• Increased gastric acid secretion • Oxidative stress • Decreased mucus secretion (Brunner's glands)	• Duodenal • Acute and chronic	[61, 62]
Indomethacin/histamine (subcutaneous)	• Increased gastric acid secretion • Impaired bicarbonate secretion	• Duodenal • Investigating the pathogenesis of duodenal ulcers • Screening anti-duodenal ulcerogenic NCEs	[61, 62]
Ferrous iron/ascorbic acid (gastric wall injection)	• Oxidative stress	• Gastric	[61]
Acetic acid (subserosal)/ *H. pylori* (oral)		• Gastric	[61]

Table 15.10 Animal models of chronic obstructive pulmonary disease

Species	Method of induction	Relevant pathology findings	Pros and cons	References
Mouse/rat/hamster	Protease instillation (e.g., elastase, papain)	• Air space enlargement	• Pros: rapid disease induction; procedurally simple, low cost; used in combination with cigarette smoke exposure • Cons: precise dosing required; lack of immune component in pathogenesis	[63–65]
Mouse/rat	Lipo-polysaccharide instillation/inhalation	• Airway inflammation/remodeling • Air space enlargement	• Pros: rapid disease induction; procedurally simple, low cost; useful in understanding infection-induced exacerbations of COPD	[63, 64]
Mouse/rat/guinea pig/dog/primate	Cigarette smoke inhalation	• Airway/alveolar inflammation • Airway remodeling • Air space enlargement	• Pros: Pathogenesis similar to human COPD • Cons: Procedurally complicated; expensive	[63, 64]
Mouse	Gene manipulation/natural mutants	• Various, including air space enlargement	• Pros: used in combination with cigarette smoke/protease exposures • Cons: constitutive genetic changes may cause lung developmental anomalies; limited similarities to human disease pathogenesis	[63, 64, 66, 67]
Mouse	Anti-elastin autoimmunity	• Air space enlargement	• Pros: useful for studying immunologic features of COPD	[64, 68]

15.6.5 Acute Lung Injury/Acute Respiratory Distress Syndrome (ALI/ARDS)

ALI/ARDS refers to the clinical syndrome in critically ill human patients characterized by acute onset of progressive hypoxemia with dyspnea/tachypnea, decreased lung compliance and diffuse alveolar infiltration in the lung. ALI/ARDS is caused by direct or indirect lung injury resulting in disordered inflammation with disruption of alveolar endothelial and epithelial barriers [74]. The acute syndrome is exudative and characterized by diffuse alveolar damage, with necrosis of alveolar pneumocytes and endothelial cells, edema, hyaline membranes, hemorrhage, neutrophilic/macrophage infiltrates, and atelectasis. Repair of ALI/ARDS may result in pulmonary fibrosis.

Four main features of experimental ALI/ARDS in AMDs and the most relevant methods to assess these features have been identified [75, 76]. The main features include histological evidence of injury, alteration of the alveolar capillary barrier, presence of an inflammatory response, and evidence of physiological dysfunction. Various AMDs have been used to investigate the pathogenesis of ALI/ARDS and test the efficacy of NCEs. Most attempts to reproduce in animals the known risk factors for ALI/ARDS in humans. Species differences in the innate immune system are an important consideration in the selection of species for the model. The features of various methods for inducing ALI/ARDS in AMDs are summarized in Table 15.12.

15.6.6 Pulmonary Fibrosis (PF)

PF is a chronic interstitial lung disease that results from lung injury secondary to numerous inciting causes [86]. The cause is often unknown (idiopathic pulmonary fibrosis [IPF]). Fibrosis may be

Table 15.11 Animal models of asthma

Species	Features of disease/model	Pros and cons	References
Mouse	• Strain variation in susceptibility • Short term challenge: eosinophil/ Th2 cell infiltration; increased mucus; AHR • Prolonged challenge: eosinophil (including intraepithelial)/ lymphocyte (Th2 and B cells) infiltration; AHR; airway remodeling	• Pros: low cost; species-specific reagents; ease of handling, sensitization, challenge; IgE is a major antibody; well-known genetics; ease of genetic manipulation • Cons: nonphysiological late-phase bronchoconstriction; limited airway smooth muscle; inflammation distribution different from human asthma; tolerance development	[69–71]
Rat	• Strain variation in susceptibility • Th2 dominated response with eosinophilia; antigen-specific IgE; airway remodeling; non-specific AHR	• Pros: low costs; ease of handling, sensitization, challenge; larger size than mice; IgE is the major antibody; long-lasting airway hyperreactivity; immediate- and late-phase airway responses • Cons: paucity of species-specific reagents; tolerance development	[69–71]
Guinea Pig	• Eosinophilic/neutrophilic airway inflammation; AHR	• Pros: ease of sensitization and challenge; early and late phase asthmatic responses; natural AHR; useful for testing bronchodilator efficacy • Cons: high cost; inbred strain shortage; paucity of species-specific reagents; IgG1 is a major antibody; tolerance development	[69–73]
Dog	• Eosinophilia; acute physiological constriction in response to allergen inhalation; AHR; increased IgE	• Pros: natural allergen susceptibility; development of atopy; eosinophils naturally found in airways; long-term changes in pulmonary function; selective breeding for animals with high IgE levels • Cons: high cost; labor intensive; larger airways; the paucity of species-specific reagents	[69, 71]

reversible with the removal of known inciting agents and early therapeutic intervention. In contrast, IPF is chronic and progressive, often lethal, and lacks effective therapeutics [87].

Alveolar epithelial injury is a common factor in the pathogenesis of PF. Progressive fibrosis in IPF is thought to involve recurrent subclinical injuries to an aging and/or genetically vulnerable alveolar epithelium resulting in type II cell dysfunction, aberrant reparative responses, and deposition of increased collagen [87]. A variety of familial and acquired gene mutations contribute to the susceptibility to IPF in humans [87] and mice [88]. Genetically modified mouse models have been used to evaluate the role of specific genes and aging in the pathogenesis of IPF [89, 90]. A role for lung dysbiosis has also been demonstrated [91].

Mice and rats administered a single dose of bleomycin by intratracheal instillation or oropharyngeal aspiration are the most clinically relevant models of PF, owing to the histological and radiological similarities to IPF in humans [92]. Mice and rats are considered the first- and second-line choice of animal species, respectively, for preclinical testing of NCEs for PF [93]. Initial lung injury with acute inflammation (days 1–7) is followed by fibroproliferation (days 7–14), and fibrosis/airway remodeling (days 14–28). The spontaneous resolution begins after day 28 [89]. Other drugs and chemicals have also been used in rodents to induce PF [90, 92, 94].

Transgenic mice overexpressing profibrotic genes (e.g., TGF-β, TNF-α, IL-13, and IL-1β) have been used to explore the relevance specific signaling pathways in the development of PF

Table 15.12 Animal models of acute lung injury/acute respiratory distress syndrome

Primary site of injury	Agent	Mechanism/relevant pathology findings	Pros and cons	References
Capillary endothelium	Oleic acid (intravascular/ intracardiac)	• Models lipid embolism after bone fracture • Endothelial necrosis; swelling/necrosis of Type I cells; oleic acid deposition in airspaces; microvascular thrombosis; neutrophilic alveolar inflammation, edema, hemorrhage, fibrin exudation	• Pros: reproducible; useful for exploring the impact of ventilation strategies • Cons: questionable relevance for humans; few cases of ALI/ARDS are due to long bone fracture	[77, 78]
	LPS (intravenous)	• Endothelial apoptosis; neutrophil recruitment/entrapment in capillaries; interstitial edema	• Pros: reproducible; easy administration • Cons: variations in LPS purity; no pneumocyte injury; species variations in response depending on presence/absence of pulmonary intravascular macrophages	[77]
Alveolar pneumocyte	Acid (HCL) aspiration	• Injury to airway epithelium/ alveolar pneumocytes; capillary endothelium secondarily • Acute neutrophilic inflammation, hemorrhage, edema/impaired alveolar fluid clearance; systemic leukopenia/ thrombocytopenia	• Pro: reproducible; useful for evaluating hemodynamic changes, mechanisms of neutrophil recruitment, and impact of ventilation strategies • Cons: instillate pH needed is lower than gastric juice; absence of other gastric content; narrow range of useful acid concentrations	[44, 77, 79]
	Surfactant depletion (warm saline lavage)	• Alveolar collapse/mechanical injury; impaired host defense • Neutrophilic inflammation, alveolar fluid accumulation, hyaline membranes when combined with mechanical ventilation/LPS	• Pros: useful for evaluating the impact of ventilation strategies • Cons: requires anesthesia/ intubation	[77]
	Mechanical ventilation	• Direct damage to alveolar pneumocyte through mechanical stretch	• Pros: useful for evaluating the impact of ventilation strategies • Cons: requires anesthesia/ intubation	[44, 77, 80, 81]
	Bleomycin (intratracheal)	• Alveolar pneumocyte death; neutrophilic infiltration without hyaline membranes	• Cons: questionable physiologic relevance	[77]
	Hyperoxia	• Free radical damage (speculated) • Type I pneumocyte and endothelial cell death; neutrophilic inflammation; fluid exudation into alveoli	• Cons: rodent strain variation in susceptibility; questionable relevance to humans; special equipment required	[77, 82, 83]
Capillary endothelium/ alveolar pneumocytes	Ischemia/ reperfusion	• Clamp pulmonary or pulmonary/ bronchial arteries for a specific period of time • Alternative involves clamping non-pulmonary vascular beds • Alveolar edema; neutrophilic infiltration; hemorrhage	• Pros: reproduces known clinical conditions in humans • Cons: requires anesthesia/ surgery; lung injury is mild if clamping non-pulmonary vascular beds	[77, 84]

(continued)

Table 15.12 (continued)

Primary site of injury	Agent	Mechanism/relevant pathology findings	Pros and cons	References
	Sepsis (intravenous bacteria)	• Microvascular injury • Increased pulmonary vascular permeability with intravascular neutrophil sequestration, thrombosis, and interstitial edema • No intra-alveolar neutrophils or hyaline membranes	• Cons: unclear relevance to human ALI/ARDS	[77]
	Sepsis (cecal ligation/ puncture)	• Microvascular injury • Variable neutrophilic alveolar infiltration, interstitial/alveolar edema; mild lung injury; minimal hyaline membrane formation	• Cons: requires surgery	[77, 85]

[90]. The use of humanized mice intravenously injected with cells from human IPF lung biopsy/ explants into immunosuppressed mice have also been used as models of PF and interstitial lung remodeling [95].

15.7 Renal Disease Models

15.7.1 Renal Toxicity: Relevance to Efficacy Evaluation (Discovery) and Drug Attrition (Drug Development)

Renal toxicity is not a commonly encountered reason for late-stage withdrawal of drugs from the market [17], but drug-induced renal toxicity is a frequent cause of drug attrition in earlier stages of development. Similar to the situation with the liver, drug-induced toxicity may be related to the metabolism of specific NCEs. However, as compared to the liver, the potential for renal regeneration is more limited [96]. Renal models of disease may be of value in understanding which NCEs may post a particular risk in populations which have compromised renal function and also play a critical role in elucidation of the potential therapeutic benefit of NCEs.

15.7.2 Renal Toxicity: Reason for Inadequate Translation from Animal to Human

While animal models have enhanced the understanding of renal injury in humans because of the many similarities with humans, there are considerable differences in pathophysiology between humans and animal models [39]. Examples include enhanced urine concentrating ability in rodents, differences in medullary thickness, sex-related susceptibility of certain drugs/ chemicals to kidney injury/neoplasia (e.g., α-2 microglobulin related), differences in renal metabolism between species and strains (e.g., glutamine synthetase in the rat but not other species, differences in CYP-450 isozymes between rat strains).

15.7.3 Acute Kidney Injury (AKI) Disease Models

Small animal models (mice and rats) are most commonly utilized based on the availability of genetically induced models and also due to the potential variability in expression between animals and potential for significant toxicity/

Table 15.13 Examples of available models of acute kidney injury (AKI) disease models

Species	Animal model	Relevant pathology findings	Pros and cons	References
Rat/ mouse	Ischemia/ reperfusion	• Large areas of renal cortex not perfused due to clamping of renal artery → diffuse cortical necrosis • Considerable differences in susceptibility between strains and ages of rat/mouse	• Pros: reproducible lesion, surgery relatively simple • Cons: difference from human AKI—multifocal damage in human vs. diffuse damage with ischemia model	[97–99]
	Toxin/drug (e.g., cisplatin and folic acid)	• Cisplatin → impaired glomerular filtration and proximal tubular injury → AKI • Folic acid → folic acid crystals in tubules → tubular necrosis, epithelial regeneration, and fibrosis	• Cons: difficulty in interpretation due to complex mechanisms of AKI; not commonly utilized.	[39]
	Infection (sepsis)	• Induced by cecal ligation and puncture	• Cons: difficulty in interpretation due to complex mechanisms of AKI; not commonly utilized	[39]

shortened lifespan and the need to ensure sufficient numbers of animals for evaluation. Some of the more commonly utilized models are outlined in Table 15.13.

15.7.4 Chronic Kidney Disease Models

Chronic kidney diseases are commonly implicated in morbidity and mortality, particularly with respect to complications of diabetes and hypertension. As is the case with acute kidney disease models, small animal models are most commonly utilized for evaluation of efficacy. Some of the more commonly utilized models are outlined in Table 15.14.

15.8 Concluding Remarks

AMDs have made significant contributions to our understanding of the pathophysiology of disease and have been instrumental in translating biomedical discoveries into treatments for human and animal diseases. While rodents have traditionally been used to model human disease, large animal species have been used increasingly to complement rodent models and in circumstances where their larger size and longer lifespan are advantageous [103]. In drug development, AMDs are primarily used during drug discovery in proof-of-concept efficacy studies and early discovery safety assessments.

In contrast, the clinical development of NCEs is most commonly supported by nonclinical safety testing in young CHAs. While the use of CHAs is and will remain the primary focus of safety evaluation. Under some circumstances, AMDs have advantages over CHAs that are relevant to nonclinical safety testing. Two recent surveys were conducted to understand how AMDs are currently used across the pharmaceutical and biotechnology industries [104, 105]. The surveys showed that AMDs were used primarily in proof-of-concept efficacy studies and early discovery safety assessment, but also to address adverse events identified in clinical trials, to better understand toxicities associated with exaggerated pharmacology in CHAs or when the target is only expressed in the disease state, and/or in response to requests from global regulatory authorities. The use of AMDs for safety evaluation as an alternative to CHAs was the least common industry practice [105].

In summary, while imperfect, AMDs have contributed significantly to our understanding of disease pathophysiology, and have been used in early testing of NCE efficacy and safety, and for addressing specific concerns in the later stages of drug development.

Table 15.14 Examples of available chronic kidney disease models

Species	Animal model	Relevant pathology findings	Pros and cons	References
Rat	Type 1 diabetes mellitus: chemically induced—streptozotocin (STZ) or alloxan	• Toxic to pancreatic β cells → insulin deficiency • Mesangial expansion and glomerular scarring—also characteristic of human	• Pros: utilize for testing new formulations of insulin, transplantation, treatments that may prevent beta cell death • Cons: toxicity to other organ systems; incidence of alloxan-induced ketosis and mortality higher than that with STZ	[100, 101]
Rat/mouse	Type 1 Diabetes Mellitus: Spontaneous autoimmune—NOD mice, BB rats	• β cell destruction (autoimmune process)	• Pros: utilize for investigating treatments that may prevent beta cell death, manipulate autoimmune process • Cons: NOD mice must be kept in specific pathogen-free conditions, also significant gender conditions and unpredictability of onset	[100]
Mouse	Type 1 diabetes mellitus: genetically induced—AKITA mice	• Severe insulin-dependent diabetes begins at 3–4 weeks of age	• Pros: rapid onset • Cons: untreated homozygotes rarely survive longer than 12 weeks	[100]
Rat/mouse/gerbil	Type 2 diabetes mellitus: obese models—(mono- or polygenetic)	• Obesity-induced hyperglycemia	• Pros: relevant model based on similarity to human pathogenesis and relationship to obesity • Cons: variability between models with respect to severity; need to carefully select model to be suitable for the proposed duration of evaluation	[100]
Rat	Hypertension: dahl salt sensitive, stroke-prone hypertensive, ren2 transgenic	• Recurrent or progressive injuries in glomeruli, tubules, interstitium, and/or vasculature	• Cons: models tend to be strain, gender, or age dependent—can be considerable variability	[102]

References

1. Sura R, Hutt J, Morgan S (2021) Opinion on the use of animal models in nonclinical safety assessment: pros and cons. Toxicol Pathol 49(5):990–995
2. Morgan SJ, Elangbam CS, Berens S, Janovitz E, Vitsky A et al (2013) Use of animal models of human disease for nonclinical safety assessment of novel pharmaceuticals. Toxicol Pathol 41:508–518
3. Olson H, Betton G, Robinson D, Thomas K, Monro A et al (2000) Condordance of the toxicity or pharmaceuticals in humans and animals. Regul Toxicol Pharmacol 32:56–67
4. Stevens JL, Baker TK (2009) The future of drug safety testing: expanding the view and narrowing the focus. Drug Discov Today 14:162–167
5. Nevzorova YA, Boyer-Diaz Z, Cubero FJ, Gracia-Sancho J (2020) Animal models for liver disease - a practical approach for translational research. J Hepatol 73(2):423–440
6. Malečková A, Tonar Z, Mik P, Michalová K, Liška V et al (2019) Animal models of liver diseases and their application in experimental surgery. Rozhl Chir 98(3):100–109
7. Brandon-Warner EW, Schrum L, Schmidt CM, McKillop IH (2012) Rodent models of alcoholic liver disease: of mice and men. Alcohol 46(8): 715–724
8. Liu Y, Meyer C, Xu C, Weng H, Hellerbrand C et al (2013) Animal models of chronic liver diseases. Am J Physiol Gastrointest 304(5):G449–G468
9. Zhang P, Wang W, Mao M, Gao R, Shi W et al (2021) Similarities and differences: a comparative review of the molecular mechanisms and effectors of NAFLD and AFLD. Front Physiol 12:710285
10. Van Herck MA, Vonghia L, Francque SM (2017) Animal models of nonalcoholic fatty liver disease-a starter's guide. Nutrients 9(10):1072
11. Alharshawi K, Aloman C (2021) Murine models of alcohol consumption: imperfect but still potential

source of novel biomarkers and therapeutic drug discovery for alcoholic liver disease. J Cell Immunol 3(3):177–181
12. Delire B, Stärkel P, Leclercq I (2015) Animal models for fibrotic liver diseases: what we have, what we need, and what is under development. J Clin Transl Hepatol 3(1):53–66
13. Lamas-Paz A, Hao F, Nelson LJ, Vázquez MT, Canals S et al (2018) Alcoholic liver disease: utility of animal models. World J Gastroenterol 24(45): 5063–5075
14. Bao YL, Wang L, Pan HT, Zhang TR, Chen YH et al (2021) Animal and organoid models of liver fibrosis. Front Physiol 12:666138
15. Laverty HG, Benson EJ, Cartwright EJ, Cross MJ, Garland C et al (2011) How can we improve our understanding of cardiovascular safety liabilities to develop safer medicines? Br J Pharmacol 163:675–693
16. Sibille M, Dergar N, Janin A, Kirkesseli S, Durand DV (1998) Adverse events in phase-I studies: a report in 1015 healthy volunteers. Eur J Clin Pharmacol 54: 13–20
17. Siramshetty VB, Nickel J, Omieczynski C, Gohlke BO, Drwal MN et al (2016) Withdrawn–a resource for withdrawn and discontinued drugs. Nucleic Acids Res 44:D1080–D1086
18. Mushenkova NV, Summerhill VI, Silaeva YY, Deykin AV, Orekhob AN (2019) Modeling of atherosclerosis in genetically modified animals. Am J Transl Res 11(8):4614–4633
19. Zaragoza C, Gomez-Guerrero C, Martin-Ventura JL, Blanco-Colio L, Lavin B et al (2011) Animal models of cardiovascular diseases. J Biomed Biotechnol 2011:497841
20. Bentzon JF, Falk E (2010) Atherosclerotic lesions in mouse and man: is it the same disease? Curr Opin Lipidol 21(5):434–440
21. Plump AS, Smith JD, Hayek T, Aalto-Setala K, Walsh A et al (1994) Severe hypercholesterolemia and atherosclerosis in apolipoprotein E deficient mice created by homologous recombindation in ES cells. Cell 71:343–353
22. Shen X, Bornfeldt KE (2007) Mouse models for studies of cardiovascular complications of type 1 diabetes. Ann N Y Acad Sci 1003:202–217
23. Weinreb DB, Aguinaldo JGS, Feig JE, Fisher EA et al (2007) Non-invasive MRI of mouse models of atherosclerosis. NMR Biomed 20(3):256–264
24. Largo R, Sanchez-Pernaute O, Marcos ME, Moreno-Rubio J (2008) Chronic arthritis aggravates vascular lesions in rabbits with atherosclerosis: a novel model of atherosclerosis associated with chronic inflammation. J Am Coll Cardiol 32(7):2057–2064
25. Shimizu T, Nakai K, Morimoto Y, Ishihara M (2009) Simple rabbit model of vulnerable atherosclerotic plaque. Neurol Med Chir 49(8):327–332
26. Gerrity RG, Natarajan R, Nadler JL, Kimsey T (2009) Diabetes-induced accelerated atherosclerosis in swine. Diabetes 50(7):1654–1665
27. Anidjar S, Salzmann JL, Gentric P, Lagneau P, Camilleri JP et al (1990) Elastase-induced experimental aneurysms in rats. Circulation 82(3):973–981
28. Lizarbe TR, Tarin C, Gomez M, Lavin B, Aracil E et al (2009) Nitric oxide induces the progression of abdominal aortic aneurysms through the matrix metalloproteinase induced EMMPRIN. Am J Pathol 175(4):1421–1430
29. Brophy CM, Tilson JE, Braverman IM, Tilson MD (1988) Age of onset, pattern of distribution and histology of aneurysm development in a genetically predisposed mouse model. J Vasc Surg 8(1):45–48
30. Molacek J, Treska V, Kober J, Certik B, Skalicky T et al (2008) Optimization of the model of abdominal aortic aneurysm – experiment in an animal model. J Vasc Res 46(1):1–5
31. Pfeffer MA, Pfeffer M, Fisbein MC (1979) Myocardial infarct size and ventricular function in rats. Circ Res 44(4):503–512
32. Michael LH, Entman ML, Harley CJ, Younker KA, Zhu J et al (1995) Myocardial ischemia and reperfusion: a murine model. Am J Phys 269(6):H2147–H2154
33. Zbinden G, Bagdon RE (1963) Isoproterenol-induced heart necrosis, an experimental model for the study of angina pectoris and myocardial infarct. Rev Can Biol 22:257–263
34. Suzuki Y, Lyons K, Yeung AC, Ikeno F (2008) In vivo porcine model of reperfused myocardial infarction: in situ double staining to measure precise infarct/area at risk. Catheter Cardiovasc Interv 71(1):100–107
35. Gonzalez GE, Seropian M, Krieger PJ, Verrilli L et al (2009) Effect of early versus late AT receptor blockade with losartan on postmyocardial infarction ventricular remodeling in rabbits. Am J Phys 297(1): H375–H386
36. Shiomi M, Ito T, Yamada S, Kawashima S, Fan J (2003) Development of an animal model for spontaneous myocardial infarction (WHHLM1 rabbit). Arterioscler Thromb Vasc Biol 23(7):1239–1244
37. McGonigle P (2014) Animal models of CNS disorders. Biochem Pharmacol 1:140–149
38. Chesselet MF, Carmichael ST (2012) Animal models of neurological disorders. Neurotherapeutics 9(2): 241–244
39. Morgan SJ, Elangbam CS (2016) Animal models of disease for future toxicity predictions. In: Olaharsi AJ, Jeffy BD (eds) Drug discovery toxicology: from target assessment to translational biomarkers. Wiley, Hoboken
40. Dawson TM, Golde TE, Lagier-Tourenne C (2018) Animal models of neurodegenerative diseases. Nat Neurosci 21(10):1370–1379

41. Leal PC, Lins LC, de Gois AM, Marchioro M, Santos JR (2016) Commentary: evaluation of models of Parkinson's disease. Front Neurosci 10:283
42. Cepeda C, Cummings DM, André VM, Holley SM, Levine MS (2010) Genetic mouse models of Huntington's disease: focus on electrophysiological mechanisms. ASN Neuro 2(2):103–114
43. Ribeiro FM, Camargos ERD, DeSouza LCD, Teixeira AL (2013) Animal models of neurodegenerative diseases. Rev Bras Psiquiatr 35(Suppl 2):S82
44. Reiss LK, Uhlig U, Uhlig S (2012) Models and mechanisms of acute lung injury caused by direct insults. Eur J Cell Biol 91(6-7):590–601
45. Clark M, Steger-Hartmann T (2018) A big data approach to the concordance of the toxicity of pharmaceuticals in animals and humans. Regul Toxicol Pharmacol 96:94–105
46. Monticello TM (2015) Drug development and nonclinical to clinical translational databases: past and current efforts. Toxicol Pathol 43(1):57–61
47. Guengerich FP (2011) Mechanisms of drug toxicity and relevance to pharmaceutical development. Drug Metab Pharmacokinet 26(1):3–14
48. Sanger GJ, Holbrook JD, Andrews PL (2011) The translational value of rodent gastrointestinal functions: a cautionary tale. Trends Pharmacol Sci 32(7):402–409
49. Kolios G (2016) Animal models of inflammatory bowel disease: how useful are they really? Curr Opin Gastroenterol 32(4):251–257
50. Mizoguchi A, Takeuchi T, Himuro H, Okada T, Mizoguchi E (2016 Jan) Genetically engineered mouse models for studying inflammatory bowel disease. J Pathol 238(2):205–219. https://doi.org/10.1002/path.4640
51. Eichele DD, Kharbanda KK (2017) Dextran sodium sulfate colitis murine model: an indispensable tool for advancing our understanding of inflammatory bowel diseases pathogenesis. World J Gastroenterol 23(33): 6016–6029
52. Baydi Z, Limami Y, Khalki L, Zaid N, Naya A et al (2021) An update of research animal models of inflammatory bowel disease. Sci World J 2021: 7479540
53. Kiesler P, Fuss IJ, Strober W (2015) Experimental models of inflammatory bowel diseases. Cell Mol Gastroenterol Hepatol 1(2):154–170
54. Liao CM, Zimmer MI, Wang CR (2013) The functions of type I and type II natural killer T cells in inflammatory bowel diseases. Inflamm Bowel Dis 19(6):1330–1338
55. Ishikawa D, Okazawa A, Corridoni D, Jia LG, Wang XM et al (2013) Tregs are dysfunctional in vivo in a spontaneous murine model of Crohn's disease. Mucosal Immunol 6(2):267–275
56. Lanas A, Chan FKL (2017) Peptic ulcer disease. Lancet 390(10094):613–624
57. Groenen MJ, Kuipers EJ, Hansen BE, Ouwendijk RJ (2009) Incidence of duodenal ulcers and gastric ulcers in a Western population: back to where it started. Can J Gastroenterol 23(9):604–608
58. Musumba C, Jorgensen A, Sutton L, Van Eker D, Moorcroft J et al (2012) The relative contribution of NSAIDs and Helicobacter pylori to the aetiology of endoscopically-diagnosed peptic ulcer disease: observations from a tertiary referral hospital in the UK between 2005 and 2010. Aliment Pharmacol Ther 36(1):48–56
59. Burkitt MD, Duckworth CA, Williams JM, Pritchard DM (2017) Helicobacter pylori-induced gastric pathology: insights from in vivo and ex vivo models. Dis Model Mech 10(2):89–104
60. Ansari S, Yamaoka Y (2022) Animal models and *Helicobacter pylori* infection. J Clin Med 11(11): 3141
61. Adinortey MB, Ansah C, Galyuon I, Nyarko A (2013) *In vivo* models used for evaluation of potential antigastroduodenal ulcer agents. Ulcers 2013: 796405. https://doi.org/10.1155/2013/796405
62. Mishra AP, Bajpai A, Chandra S (2019) A comprehensive review on the screening models for the pharmacological assessment of antiulcer drugs. Curr Clin Pharmacol 14(3):175–196
63. Ghorani V, Boskabady MH, Khazdair MR, Kianmeher M (2017) Experimental animal models for COPD: a methodological review. Tob Induc Dis 15:25
64. Tanner L, Single AB (2020) Animal models reflecting chronic obstructive pulmonary disease and related respiratory disorders: translating pre-clinical data into clinical relevance. J Innate Immun 12(3):203–225
65. Serban KA, Petrache I (2018) Mouse Models of COPD. Methods Mol Biol 1809:379–394
66. Brusselle GG, Bracke KR, Maes T, D'hulst AI, Moerloose KB et al (2006) Murine models of COPD. Pulm Pharmacol Ther 19(3):155–165
67. Fujita M, Nakanishi Y (2007) The pathogenesis of COPD: lessons learned from in vivo animal models. Med Sci Monit 13(2):19–24
68. Gu BH, Sprouse ML, Madison MC, Hong MJ, Yuan X et al (2019) A novel animal model of emphysema induced by anti-Elastin autoimmunity. J Immunol 203(2):349–359
69. Aun MV, Bonamichi-Santos R, Arantes-Costa FM, Kalil J, Giavina-Bianchi P (2017) Animal models of asthma: utility and limitations. J Asthma Allergy 10: 293–301
70. Shin YS, Takeda K, Gelfand EW (2009) Understanding asthma using animal models. Allergy, Asthma Immunol Res 1(1):10–18
71. Zosky GR, Sly PD (2007) Animal models of asthma. Clin Exp Allergy 37(7):973–988
72. Ricciardolo FL, Nijkamp F, De Rose V, Folkerts G (2008) The guinea pig as an animal model for asthma. Curr Drug Targets 9(6):452–465
73. Sagar S, Akbarshahi H, Uller L (2015) Translational value of animal models of asthma: challenges and promises. Eur J Phamacol 15(759):272–277
74. Mokrá D (2020) Acute lung injury - from pathophysiology to treatment. Physiol Res 69(Suppl 3):S353–S366

75. Matute-Bello G, Downey G, Moore BB, Groshong SD, Matthay MA et al (2011) An official American Thoracic Society workshop report: features and measurements of experimental acute lung injury in animals. Am J Respir Cell Mol Biol 44(5):725–738
76. Kulkarni HS, Lee JS, Bastarache JA, Kuebler WM, Downey GP et al (2022) Update on the features and measurements of experimental acute lung injury in animals: an official American Thoracic Society workshop report. Am J Respir Cell Mol Biol 66(2):e1–e14
77. Matute-Bello G, Frevert CW, Martin TR (2008) Animal models of acute lung injury. Am J Physiol Lung Cell Mol Physiol 295(3):L379–L399
78. Gonçalves-de-Albuquerque CF, Silva AR, Burth P, Castro-Faria MV, Castro-Faria-Neto HC (2015) Acute respiratory distress syndrome: role of oleic acid-triggered lung injury and inflammation. Mediat Inflamm 2015:260465
79. Gramatté J, Pietzsch J, Bergmann R, Richter T (2018) Causative treatment of acid aspiration induced acute lung injury - recent trends from animal experiments and critical perspective. Clin Hemorheol Microcirc 69(1-2):187–195
80. Yehya N (2019) Lessons learned in acute respiratory distress syndrome from the animal laboratory. Ann Transl Med 7(19):503
81. Joelsson JP, Ingthorsson S, Kricker J, Gudjonsson T, Karason S (2021) Ventilator-induced lung-injury in mouse models: is there a trap? Lab Anim Res 37(1): 30
82. Amarelle L, Quintela L, Hurtado J, Malacrida L (2021) Hyperoxia and lungs: what we have learned from animal models. Front Med 8:606–678
83. Lv R, Zheng J, Ye Z, Sun X, Tao H et al (2014) Advances in the therapy of hyperoxia-induced lung injury: findings from animal models. Undersea Hyperb Med 41(3):183–202
84. Fard N, Saffari A, Emami G, Hofer S, Kauczor HU et al (2014) Acute respiratory distress syndrome induction by pulmonary ischemia-reperfusion injury in large animal models. J Surg Res 189(2):274–284
85. Mishra SK, Choudhury S (2018) Experimental protocol for cecal ligation and puncture model of polymicrobial sepsis and assessment of vascular functions in mice. Methods Mol Biol 1717:161–187
86. Noble PW, Barkauskas CE, Jiang D (2012) Pulmonary fibrosis: patterns and perpetrators. J Clin Invest 122(8):2756–2762
87. Lederer DJ, Martinez FJ (2018) Idiopathic pulmonary fibrosis. N Engl J Med 378(19):1811–1823
88. Nureki SI, Tomer Y, Venosa A, Katzen J, Russo SJ et al (2018) Expression of mutant Sftpc in murine alveolar epithelia drives spontaneous lung fibrosis. J Clin Invest 128(9):4008–4024
89. Yasutomo K (2021) Genetics and animal models of familial pulmonary fibrosis. Int Immunol 33(12): 653–657
90. Tashiro J, Rubio GA, Limper AH, Williams K, Elliot SJ et al (2017) Exploring animal models that resemble idiopathic pulmonary fibrosis. Front Med 28(4): 118
91. O'Dwyer DN, Ashley SL, Gurczynski SJ, Xia M, Wilke C et al (2019) Lung microbiota contribute to pulmonary inflammation and disease progression in pulmonary fibrosis. Am J Respir Crit Care Med 199(9):1127–1138
92. Li S, Shi J, Tang H (2022) Animal models of drug-induced pulmonary fibrosis: an overview of molecular mechanisms and characteristics. Cell Biol Toxicol 38(5):699–723
93. Jenkins RG, Moore BB, Chambers RC, Eickelberg O, Königshoff M et al (2017) ATS assembly on respiratory cell and molecular biology. An official American Thoracic Society Workshop Report: use of animal models for the preclinical assessment of potential therapies for pulmonary fibrosis. Am J Respir Cell Mol Biol 56(5):667–679
94. Miles T, Hoyne GF, Knight DA, Fear MW, Mutsaers SE et al (2020) The contribution of animal models to understanding the role of the immune system in human idiopathic pulmonary fibrosis. Clin Transl Immunol 9(7):e1153
95. Habiel DM, Espindola MS, Coelho AL, Hogaboam CM (2018) Modeling idiopathic pulmonary fibrosis in humanized severe combined immunodeficient mice. Am J Pathol 188(4):891–903
96. Pereira CV, Nadanaciva S, Oliveira PJ, Will Y (2012) The contribution of oxidative stress to drug-induced organ toxicity and its detection in vitro and in vivo. Expert Opin Drug Metab Toxicol 8:219–237
97. Gobe G, Willgoss D, Hogg N, Schoch E, Endre Z (1999) Cell survival or death in renal tubular epithelium after ischemia-reperfusion injury. Kidney Int 56: 1299–1304
98. Lu X, Li N, Shushakova N, Schmitt R, Menne J et al (2012) C57BL/y and 129sv mice: genetic differences to renal ischemia-reperfusion. J Nephrol 5:738–743
99. Kusaka J, Koga H, Hagiwara S, Hasegawa A, Kudo K et al (2012) Age-dependent responses to renal ischemia-reperfusion injury. J Surg Res 172:153–158
100. King AJF (2012) The use of animal models in diabetes research. Br J Pharmacol 166:877–894
101. Szkudelski T (2001) The mechanism of alloxan and streptozotocin action in β cells of the rat pancreas. Physiol Res 50:537–546
102. Yang HC, Zuo Y, Fogo AB (2010) Models of chronic kidney disease. Drug Discov Today Dis Model 7:13–19
103. Beck AP, Meyerholz DK (2020) Evolving challenges to model human diseases for translational research. Cell Tissue Res 380(2):305–311
104. Tomohiro M, Okabe T, Kimura Y, Kinoshita K, Maeda M et al (2019) Toxicologic pathology forum: current status on the use of animal models of human disease in the pharmaceutical industry in japan in nonclinical safety assessment-opinion paper. Toxicol Pathol 47(2):108–120
105. Butler LD, Guzzie-Peck P, Hartke J, Bogdanffy MS, Will Y (2017) Current nonclinical testing paradigms in support of safe clinical trials: an IQ consortium DruSafe perspective. Regul Toxicol Pharmacol 87 (Suppl 3):S1–S15

Part III

ADME: Clinical Pharmacology

16 Physiologically Based Pharmacokinetic Modelling in Drug Discovery and Clinical Development: A Treatise on Concepts, Model Workflow, Credibility, Application and Regulatory Landscape

Pradeep Sharma, Felix Stader, Vijender Panduga, Jin Dong, and David W. Boulton

Abstract

Physiologically based pharmacokinetic modelling (PBPK) approach considers the body as a multi-compartment system where physiological organs and kinetics of drug transfer between them are modelled using mathematical expressions. It is a mechanistic approach capable of handling complex clinical scenarios, fast gaining regulatory acceptance for the potential to predict PK during drug discovery and development in a target population of interest. Recently PBPK has been used to answer various clinical pharmacology questions related to drug-drug interactions (DDIs), food effects, formulation effects, PK in organ impaired populations and specific populations like paediatrics and pregnancy. This chapter describes the basics of PBPK modelling for small and large molecule drugs, its role/engagement during different stages of discovery, preclinical and clinical development, approaches to construct and assess the credibility of the model, applications in clinical development pharmacology, regulatory guidance, potential challenges, and future developments in the area of PBPK modelling.

Keywords

PBPK modelling · Clinical development · Model workflow · Regulatory landscape · Biopharmaceutic modelling · Drug-drug interaction

16.1 Introduction

Pharmacokinetics (PK) is the science that describes the time-course of a drug concentration in the body resulting from the administration of a particular drug dose. PK in its simplest form is how the body processes the drug after administration, i.e., absorption, distribution, metabolism, and excretion (ADME) of a drug. Understanding these processes play a very significant role in the selection and development of a new chemical

P. Sharma (✉)
Clinical Pharmacology and Quantitative Pharmacology (CPQP), Clinical Pharmacology & Safety Sciences, R & D, AstraZeneca, Cambridge, UK
e-mail: Pradeep.Sharma@astrazeneca.com

F. Stader
Modelling and Simulations Group, Simcyp Division, Certara UK Limited, Sheffield, UK
e-mail: felix.stader@certara.com

V. Panduga
Clinical Pharmacology and Quantitative Pharmacology (CPQP), Clinical Pharmacology & Safety Sciences, R & D, AstraZeneca, Gothenburg, Sweden
e-mail: Vijender.Panduga1@astrazeneca.com

J. Dong · D. W. Boulton
Clinical Pharmacology and Quantitative Pharmacology (CPQP), Clinical Pharmacology & Safety Sciences, R & D, AstraZeneca, Gaithersburg, MD, USA
e-mail: Jin.Dong1@astrazeneca.com; David.Boulton2@astrazeneca.com

G. Jagadeesh et al. (eds.), *The Quintessence of Basic and Clinical Research and Scientific Publishing*,
https://doi.org/10.1007/978-981-99-1284-1_16

entity (NCE) as a prospective drug. The fundamental aspects of PK had been reviewed extensively in the past [1] and are out of the scope of this chapter.

There are two major approaches for quantitative study of various kinetic processes of drug PK, the model-based approach (pharmacokinetic modelling) and the model-independent approach (also called non-compartment analysis, NCA). In the model-based approach, mathematical models are used to describe the changes in drug concentration in the body with time. This is a traditional and frequently used approach for pharmacokinetic characterization of drugs. There are two most common model-based approaches, namely, compartment modelling and physiologically based PK (PBPK) modelling. All model-based approaches use differential and algebraic mathematical equations to describe spatial and temporal distribution of a drug in the body. Table 16.1 shows a comparison of compartment and PBPK modelling approaches.

In the compartmental modelling approach, the body is represented as a series of interconnected compartments arranged either in series or parallel to each other. These compartments are not a real

Table 16.1 Comparison of mamillary compartment modelling and PBPK modelling approaches

Attribute	Classical compartment modelling	PBPK modelling
Definition	Evaluating the kinetics of a drug in the body by lumping similar organs/tissues as one compartment, and creating simplified mathematical models, which can describe plasma concentration-time profile well	Evaluating the kinetics of drug in the body by considering major organ systems as distinct physiological compartments and use of mathematical differential equations together with systems parameters e.g., organ blood flow rate and drug parameters e.g., pKa, to describe the absorption, distribution, and elimination
Model structure	Empirical in nature, mostly 1, 2 or 3 compartmental models	Mechanistic in nature with a multi-compartment model structure
Model complexity	Simple, non-intensive computations, less advanced software platforms needed	Complex structure organisation, several differential equations involved necessitating the need for complicated advanced computing platforms
Data requirements	Less input data needed. Generally, concentration time data in plasma/urine or target body matrix	Extensive input data is needed. This includes drug data (e.g., physiochemical properties, metabolism, plasma protein binding, active / passive permeability) and systems data (e.g., organ volumes, blood flow rates, proteomics)
Model workflow	Use of generally well-established predefined empirical compartment models often in the templated form	Follows a specific workflow depending upon drug characteristics and desired objective. Generally, involved model construction, verification, validation and application
High throughput	Compartmental models can be used as a standard templates to deliver analysis quickly in a high throughput preclinical environment	Time consuming approach where the model specific to each drug is developed depending upon its ADME properties
Population approach	Used in conjunction with statistical methods to evaluate PK in a large PK group (ideally more than 6 individuals) called Population based PK (or 'PopPK' analysis)	Used with demographic characteristics and variation of phenotypic/genotypic properties in population, it is called Population based PBPK (or PopPBPK analysis)
Model application	Useful PK data analysis methods mostly applied to analyse data generated from clinical studies to describe PK quantitatively, simulate untested dosage regimens, check bioequivalence of formulations, compare PK in different populations, assessment of food effects, etc.	Useful not only to describe clinical PK but to prospectively predict untested clinical scenarios. This involves the prediction of drug-drug interactions, extrapolation of observed DDIs to different dosage regimens or interacting drugs, prediction of PK in specific populations e.g., paediatrics, prediction of PK in patients with co-morbidities e.g., renal and/or hepatic organ impairment

Table 16.2 Advantages and disadvantages of PBPK modelling

Advantages	Disadvantages
Models are mechanistic and therefore more physiologically relevant than empirical models	Need rich data set to enable modelling application
Can be used *a priori*, i.e., for prospective predictions to forecast likely PK of the candidate drug in discovery given preclinical properties	Time consuming to build and validate models
Capable of handling complex scenarios, for example, DDI prediction in hepatic impaired population	Lack of systems data e.g., proteomics data may be impeding for application of PBPK modelling
Extrapolation of PK to various clinical scenarios may be possible, e.g., from a small number of individuals to a larger and/or different patient population/ethnic group (pharmacogenetics)	Assumptions are required to patch gaps and unknown parameters in the model
Specific population groups may be modelled	Need special skill set and costly modelling infrastructure
• Paediatrics to take account of ontogeny in physiology	
• Pathophysiological impact on drug PK in diseased populations like renal/hepatic impairment	
Gaining regulatory acceptance so can avoid clinical studies	
Can be used to inform efficient adaptive clinical trial designs for efficient read out and /or reduce undue exposures	

physiological or anatomical regions but hypothetical mathematical compartments. Tissues of similar drug distribution characteristics are lumped together as one compartment. Drug transfer between these compartments is generally described by first order kinetics. The central compartment includes blood and highly perfused organs and tissues such as the heart, lungs, liver, and kidneys. In these organs, the administered drug usually equilibrates rapidly. Peripheral compartment(s) include(s) those organs that are less well-perfused such as adipose and skeletal muscle, and therefore the administered drug will equilibrate more slowly in these organs. The number of compartments needed to describe the PK of a drug is decided by fitting observed plasma concentration data to different compartment models utilizing non-linear regression methods.

In contrast to compartmental modelling, the PBPK approach considers the body as a multi-compartment system where physiological organs and kinetics of drug transfer between them are modelled using mathematical expressions. Thus, in PBPK approach modelling takes into account various physiological parameters of organs like blood flow, organ volume, expression of proteins, etc. PBPK modelling is a fast evolving science of systems pharmacology due to the advent of high computational power in combination with an an abundance of extensive biological and clinical data. Table 16.2 illustrates various advantages and disadvantages of the PBPK modelling approach.

In non-compartmental analysis, drug kinetics or plasma concentration data is analysed by model independent approaches. This type of analysis relies upon algebraic equations to estimate PK parameters, making the analysis less complex than compartmental methods.

This chapter describes the basics of PBPK modelling for small and large molecule drugs, its role/engagement during different stages of discovery, preclinical and clinical development, approaches to construct and assess the credibility of the model, applications in clinical development pharmacology, regulatory guidance, potential challenges, and future developments in the area of PBPK modelling.

16.2 Basic Framework of PBPK Modelling

PBPK modelling utilizes drug properties (Drug model) to predict PK given specific physiology (Systems parameters) by the use of mathematical equations (Model structure) [2]. Therefore, three fundamental aspects of PBPK modelling framework are drug model, systems parameters, and structural models. These are discussed in the following sections.

16.2.1 Drug Model

The PK of a drug is dependent on its specific physiochemical and biological properties which determine its capacity to be absorbed, distributed, and eliminated from the body. Figure 16.1 shows the different kinds of data generated successively during the discovery, preclinical and clinical development of the drug project, and the evolution of PBPK drug models over the life cycle of the drug project contributing to various aspects of pre-clinical and clinical development [3].

Small molecule drugs and large molecule/biologics drugs generally have different pharmacokinetic processes. Therefore, drug model building for small molecule drugs and biologics require different parameters. Tables 16.3 and 16.4 list key drug properties needed for the PBPK model building of small and therapeutic monoclonal antibodies (mAb). PBPK models for advanced biologic modalities e.g., antibody drug conjugate (ADC), oligonucleotides, etc., have also been reported in the recent literature reports [4]. Section 16.3 further describes different approaches used for building the drug PBPK model and Sect. 16.4 illustrates the typical workflow for the development and validation of the drug PBPK model.

16.2.2 Systems Parameters

Systems parameters constitute quantitative estimates of biological parameters and their distribution in human population. These include demographic (e.g., age, sex, height, weight), genomic (allelic forms of enzymes/transporters), phenotypic (extensive, intermediate, ultra-rapid and poor metabolisers), proteomic (expression of CYPs and transporters) and physiological parameters (e.g. organ volumes, blood flow rates, expression of CYPs, transporters) [5]. Large-scale gene expression data from publicly available sources can be downloaded, processed, stored, and customized such that they can be used directly in PBPK model building [6].

Due to exponential growth in fundamental sciences, there is emergence of comprehensive data sets that can be utilized to create a virtual human being, and by randomisation of distribution of variability in these systems parameters, a 'virtual population' can be created. The inclusion of disease specific pathophysiological changes allows to create 'virtual patient population', e.g., oncology patient population [7]. Thus, it is possible to simulate virtual clinical trials, where drug PK can be simulated in a target population set that takes into account the influence of disease, genotypes, and co-morbidities [8, 9].

Development of population models involves adaptation of normal healthy population models to include target population specific attributes, followed by validation to check if it can simulate PK well for multiple sets of model drugs. Different ethnic populations (Caucasians, Chinese, Japanese, etc.) may be developed either as completely new populations or by adapting a previously built population for relevant demographics, physiological parameters, and incorporation of genetic polymorphisms of metabolizing enzymes and transporters. Comparison of observed interethnic PK differences of the model drugs with that of PK predicted by PBPK modelling may be considered an essential component of validation of an ethnic population [9]. In the case of specific populations, changes in system parameters of the healthy volunteer population that describe the population of interest such as paediatric, pregnancy, renal/hepatic impairment, geriatrics, and obesity are made based upon either mechanistic evidence or fitting of the model parameters to observed clinical data [10]. For example, to establish a paediatric population, ontogeny changes in enzymes and transporters need to be explored by both quantifications of the enzyme amount and evaluation of observed in vivo clearance of probe substrates [11]. Recently many population models are reported that are for disease populations (oncology [7], non-alcoholic steatohepatitis (NASH) [12] and specific

Drug data | PBPK model evolution | PBPK milestones and work packages

Discovery Phase

High throughput preliminary 'fit for purpose' model

Physicochemical properties
MW, LogP, pKa, HBD, PSA

In vitro DMPK properties (high throughput screening assays)
CL_{int}, P_{app}, solubility, B/P, fu

Preclinical PK (Cassette dosing)
IV & PO plasma PK

Initial drug PBPK model building
Clearance IVIVE, Vss prediction, absorption simulation

Inform

Selection of best in class candidate drug

Prospective prediction of PK profile of candidate drugs using 'bottom up' approach using in vitro parameters and from PBPK modelling in preclinical species

* IVIVE with microsomes & hepatocytes
* V_{ss} predicted by passive diffusion
* Absorption predicted using in silico models
* Rank ordering of candidate drugs with competitive PK profiles

Preclinical Development

Human PK prediction with associated uncertainty

Comprehensive DMPK parameters
V_{max} & K_m, K_i/IC_{50}, K_p, K_{app}/K_{inact}, fu_{mic} and fu_{inc}

Pharmaceutical parameters
Biorelevant media solubilities, salt /solid forms, initial formulation properties, dog PK

In vivo preclinical
Renal & biliary clearance
Tissue concentrations

Prospective PBPK model
Cross-functional input to model

Apply

Predict Human PK at target dosage regimen

Prospective prediction of PK profile of candidate drug at the target dosage regimen with more mechanistic modelling to include:

* Tissue: plasma partition scaler (K_p) from animal PK
* Prediction of f_a
* Impact of complex DDIs (auto-induction/MBI)
* Inclusion of urinary/biliary elimination

DDI risk assessments: Perpetrator and victim DDI liability due to CYP and/or transporter inhibition

Simulation of target tissue concentrations and link to translation PKPD

Prediction of food effect and formulation attributes

Early Clinical Phase (Phase I/II)

Refinement of Human PBPK model and validation

Early clinical PK Data
Single and multiple ascending dose PK, microdosing, IV dose, mass balance/hADME, DDI, food effect, DDI biomarkers

Pharmaceutical data
Absolute and relative bioavailability

Emerging DMPK Data
Additional findings

Validate drug PBPK model
Refine key model parameters

Apply

Assessment of intrinsic and extrinsic parameters on PK of drug to inform drug label and support clinical development

PBPK model to inform first in man clinical study protocol e.g. exclusion of interacting comedications in recruitment of patients/volunteers

Model validation by comparison of observed versus predicted PK:

* Gap analysis to identify uncertain parameters
* Parameter optimization using middle out approach: combine bottom-up model with top-down optimization

PK prediction in target patient population by incorporating diseased related impact on PK

Inclusion of biomarker data for transporter and CYP DDIs

Confirm in vivo fm, refine /extrapolate DDIs, impact of co-morbidities (e.g. organ impairment) on drug PK

Late Clinical Phase (Phase II-IV)

Late Stage Application of PBPK Model

Late stage clinical PK and real world data
DDI studies, population PK, PK from genotyped subjects & in specific population

Final PBPK Model
Simulate unstudied DDIs, additional populations and DDIs in specific populations, life cycle management and extensions

Support late clinical development and regulatory discussions

* Exploring complex DDI scenarios
* Inform clinical trial designs
* Impact of CMC / formulation changes on PK

PK in ethnic groups, pediatrics/geriatrics/pregnancy

Fig. 16.1 Evolution of drug PBPK model along the value chain of drug project depicting various multi-disciplinary cross-functional model inputs and successive applications in different phases of research and development.

population groups (geriatrics [13], paediatrics [14], pregnancy [15]).

16.2.3 Model Structure

PBPK modelling is basically a mechanistic modelling approach that is based on the combination of various mathematical techniques employing algebraic, differential, and statistical concepts. The fundamental 'structural model' is developed by taking physiological parameters into account such that drug kinetics between different body tissues/organs mimic the real life scenarios, and to this is added 'statistical models' to capture the distribution of variability, and 'error models' to capture 'uncertainties' in predictive power of model and parameter estimates. Monte Carlo simulation-based approaches are used to incorporate variability in pharmacokinetic factors in mechanistic, physiologically based models, together with demographic data and specific information on the genetic variability of enzymes, to predict the drug exposure and time-course of drug concentrations in various body tissues in healthy and patient populations.

Various modelling platforms are available which may be used for conducting PBPK modelling either by using coding language or by using a graphic user interface (GUI). Table 16.5 enlists a few common software platforms used in PBPK modelling. PBPK modelling may be considered a modular approach, where each PK process, namely absorption, distribution, metabolism, and elimination may be modelled by selecting one among various choices of mechanistic models available in the literature to capture drug kinetics by these processes. By appropriate choice of mechanistic models for ADME and linking them together, one can construct a full structural PBPK model fit for purpose for the desired objective. Therefore, it is possible to develop different PBPK drug models by choice of various modular mechanistic ADME modules to answer different specific questions raised during drug discovery and development. The choice of modular mechanistic models depends upon available drug data, rationale assumptions, desired outcomes, or questions to be answered. Generally, models are simpler and less mechanistic at the early stages of drug discovery and become more complex and mechanistic at later stages of clinical development when there is more understanding of drug PK (e.g. evolving the model from minimal PBPK model to full PBPK model. Development of the good structural full PBPK is active area of research for future development of PBPK modelling approach [16]. Figure 16.2 is a holistic view that schematically describes the choice of some commonly known mechanistic models that can be conjoined to develop a structural drug PBPK model for an orally administered small molecule drug.

16.3 PBPK Modelling Strategies

Drug PBPK models may be developed by employing any of following three main strategies.

16.3.1 Bottom-Up Approach

When in vitro and in silico data from various preclinical studies are used to construct a model and predict PK of a drug prospectively without using any clinical data, then it is classed as 'bottom-up' approach. This is also called 'a priori' approach, which is generally employed at pre-clinical stage and heavily dependent on availability of high-quality data. This approach is useful in comparing PK of various candidate drugs and rank order them against a targeted PK profile [32]. Although, not compliant for submission to regulatory agencies, they are a useful internal decision-making tool to help guide fist in man

Fig. 16.1 (continued) Abbreviations are defined in Table 16.3 and CL_{int} intrinsic clearance, K_p tissue: plasma partition coefficient, *hADME* human absorption, distribution, metabolism and excretion, *DDI* drug-drug interactions, *IVIVE* in vitro–in vivo extrapolation, *PKPD* pharmacokinetic-pharmacodynamic, *CMC* chemistry, manufacturing, and controls

Table 16.3 Key drug properties used in building PBPK model for small molecule drugs

Drug property	Symbol	Definition and use in PBPK model
Molecular weight	MW	Used for calculation of drug concentration, prediction of passive permeability across the membranes
Partition coefficient	Log P	Lipophilicity of drug, i.e., the extent of drug partitioning into lipidic phase
Ionization potential	pKa	Ionization potential of drug that determines % of drug unionized at different pH in various tissue compartments
Compound type		A compound may be neutral, monoprotic base or acid, diprotic base, or acid or an ampholyte. This terminology refers to the type (acidic or basic or none) and the number of ionisable centres present in a molecule. This information along with pKa data helps in modelling the tissue distribution of drug
Fraction unbound in plasma	fu	Fraction of drug freely available in plasma that transport, metabolise or distribute into tissues
Blood-to-plasma ratio	B/P ratio	The blood-to-plasma ratio (often referred to as $K_{b/p}$ or B:P ratio) is the ratio of the concentration of drug in whole blood (i.e., red blood cells and plasma) to the concentration of drug in plasma, namely C_B/C_P
ADMET predictors	HBD, PSA	Hydrogen bond donor (HBD) and polar surface area (PSA) are molecular descriptors used for in silico predictions of membrane permeabilities
Caco-2 permeability	P_{app}	The permeability of a compound is the amount of compound that has moved through a membrane in a given time per unit surface area of the membrane
Solubility	S	Intrinsic solubility is the solubility of the compound in its free acid or free base form. pH dependent solubility, surface, and bulk solubility, bile micelle mediated solubility, supersaturation ratio, etc., are needed to model drug release from oral solid dosage forms
Intrinsic metabolic clearance	$CL_{int,vitro,M}$	Drug depletion rate due to metabolism is referred to as intrinsic metabolic clearance. This can be mechanistically characterised by maximal velocity of rate of metabolic degradation (V_{max}) and affinity to metabolising enzyme (K_m)
Transporter activity	$CL_{int,vitro,T}$	Drug transport rate across membranes due to transporters. This can be mechanistically characterised by maximal velocity of rate of transport (V_{max}) and affinity to transporter (K_m)
Inhibition constants	K_i or IC_{50}	Ki (inhibition constant) and IC_{50} (inhibitory concentration 50%) are biochemical properties of the drug to characterise its in vitro inhibition potential for metabolising enzyme or transporting protein. IC_{50} is the drug concentration to inhibit 50% of biological activity, and for reversible inhibition, $K_i = IC_{50}/2$, when the substrate concentration [S] in the medium is such that [S] = Km and Ki ≈ IC_{50}, when [S] ≤10 × Km
Mechanism based inhibition (MBI)	k_{inact} and K_{app}	This occurs when a compound undergoes a catalytic transformation by an enzyme to a species that, prior to release from the active site, inactivates the enzyme by either covalent or non-covalent binding. Mechanism based inhibition is more complex than reversible inhibition as it is time- and concentration-dependent and results in a net loss of active enzyme. The main concern about MBI is that the inhibitory effect may persist in vivo even after the elimination of the inactivating species, and that active enzymes can only be recovered by de novo synthesis. The two major in vitro kinetic parameters that characterise MBI are k_{inact} and K_{app}—the maximal inactivation rate constant and the inhibitor concentration leading to 50% of k_{inact}, respectively
Induction or suppression	E_{max} and EC_{50}	Induction or suppression occurs when a compound up-regulates or downregulates the synthesis of an enzyme or transporter protein. Nonlinear regression analysis to a four-parameter sigmoidal equation is implemented to produce E_{max} (maximum fold induction or suppression), EC_{50} values (concentration of drug which produces a fold induction or suppression of 50% of the calculated E_{max})
Fraction of drug unbound in vitro	fu_{mic} and fu_{inc}	fu_{mic} refers to the fraction of drug unbound to microsomes and fu_{inc} refers to the fraction of drug unbound to cells during incubation experiments. These parameters help to correct for non-specific binding of a drug to in vitro test systems for estimation of various drug parameters

(continued)

Table 16.3 (continued)

Drug property	Symbol	Definition and use in PBPK model
Fraction of drug absorbed	f_a	Fraction of drug absorbed after oral or transdermal administration
Absorption rate constant	k_a	First order rate constant to describe drug absorption
Volume of distribution	V_d and V_{ss}	The volume of distribution (V_d) is defined as a hypothetical volume in which the total amount of drug in the body is required to be dissolved to reflect the drug concentration in plasma. The volume of distribution at steady state (V_{ss}) represents the volume in which a drug would appear to be distributed during a steady state
Fraction metabolised	f_m	Fraction of drug metabolised by a specific metabolic enzyme. These are normally determined from human mass balance and distribution studies along with in vitro biotransformation studies
Clearance	CL	Drug clearance is defined as the volume of plasma in the vascular compartment cleared of drugs per unit of time by the processes of metabolism and excretion. CL_{iv} is clearance of drug after intravenous administration and CL_{po} is oral clearance such that $CL_{oral} = CL_{iv}/F$, where F is oral bioavailability (fraction of an administered dose of unchanged drug that reaches the systemic circulation)

Table 16.4 Key drug properties used in building PBPK model for therapeutic monoclonal antibodies

Drug property	Symbol	Definition and use in PBPK model
Molecular weight	MW	Used to calculate distribution through pores in the endothelial cell layer
Isoelectric point	pI	Isoelectric point is the pH at which the antibody has no net electrical charge. Net charge on antibodies influence their renal filtration
Fraction unbound in plasma	f_u	Fraction of drug freely available in plasma to transport or distribute into tissues
Blood to plasma ratio TMDD with Michaelis-Menten	B/P ratio	The blood-to-plasma ratio (often referred to as $K_{b/p}$ or B:P ratio) is the ratio of the concentration of drug in whole blood (i.e., red blood cells and plasma) to the concentration of drug in plasma, namely C_B/C_P
Fraction of drug absorbed	f_a	Fraction of drug absorbed after transdermal administration
Absorption rate constant	k_a	First order rate constant to describe drug absorption
Binding affinity to Neonatal Fc receptor (FcRn)	$K_{d,FcRn}$	Equilibrium dissociation constant for FcRn is used to model the distribution and FcRn recycling of monoclonal antibodies
Target mediated drug disposition	TMDD	Target-mediated drug disposition (TMDD) is the phenomenon in which a drug binds to its pharmacological target site receptor resulting in an impact on its own pharmacokinetic characteristics. There are various types of TMDD modelling approaches that utilise different drug input parameters:
		• Full TMDD model: k_{on} (rate constant for binding free receptors), k_{off} (rate constant for dissociation), k_{int} (rate constant for drug-complex internalisation)
		• TMDD with Quasi-equilibrium: k_d (equilibrium dissociation constant) and k_{int}
		• TMDD with Quasi-steady state: k_{int}; K_{ss} (steady state constant)
		• TMDD with Michaelis-Menten kinetics: V_{max} (maximum elimination rate) and K_m (Michaelis-Menten constant of elimination kinetics)
Catabolic clearance	CL_{cat}	Non-specific clearance of monoclonal antibodies not bound to FcRn in the endosomal space. Unlike TMDD this is non-saturable and linear over a large dose range

Table 16.5 PBPK modelling platforms widely used in drug discovery and development

Software/platforms	References
Modular GUI platform	
• Simcyp (https://www.certara.com/software/simcyp-pbpk/)	[10]
• Gastroplus (http://www.simulations-plus.com)	[17]
• PK-Sim (http://www.systems-biology.com)	[6]
• Cloe PK (http://www.cyprotex.com/cloepredict/)	[18]
• PKQuest (http://www.pkquest.com)	[19]
• MEDICI-PK, Computing in Technology, http://www.cit-wulkow.de/	[20]
• PhysPK, https://www.physpk.com/	[21]
• Monolix http://lixoft.com/	[22]
• MATLAB-simulink and Simbiology, The Mathworks Inc	[23]
Mathematical programming language/Coding / scripting-based platforms	
• acslXtreme, Aegis Technologies (http://www.acslx.com)	[24]
• MATLAB, The Mathworks Inc. (http://www.mathworks.com)	[25]
• Berkeley Madonna (http://www.berkeleymadonna.com/)	[26]
• ADAPT 5, University of Southern California, http://bmsr.usc.edu/	[27]
• SAAM II, University of Washington, http://depts.washington.edu/saam2/	[28]
• MCSIM, http://www.gnu.org/software/mcsim/	[29]
• Phoenix Winnonlin, Certara, Princeton, NJ (https://www.certara.com/software/phoenix-winnonlin/)	[30]
• R (https://www.r-project.org/)	[31]

dosing and clinical study protocols in absence of any clinical PK.

16.3.2 Top-Down Approach

When clinical data (concentration-time profiles from single and/or multiple ascending doses with summary of PK parameters) is used for parameter estimations and development of drug PBPK model, then it is called as 'top down' approaches. This is also called as 'a posteriori' approach and Population based pharmacokinetic analysis (PopPk) using empirical classical models is top-down approach that is used extensively to quantify PK variability and pivotal covariates.

16.3.3 Middle-Out Approach

When PBPK model is developed by using in vitro /preclinical data as well as clinical data by employing both bottom up and top-down approaches, then is called middle out approach. This approach is by far the most commonly used approach in pharmaceutical industry where one starts PBPK model building using bottom up approach during drug discovery and switch to middle-out approach during during clinical development to apply PBPK modelling to assess impact of intrinsic and /or extrinsic factors on drug PK.

16.4 Workflow of PBPK Model Development

PBPK model development is an iterative process that may involve multiple cycles of "predict, learn, confirm, apply paradigm." The exact workflow to accomplish PBPK modelling is dependent upon the key question to be answered and various objectives might need different approaches specific to accomplish that task under question. However, PBPK model development still broadly follows generic steps and systematic progression through milestones, which is shown in Fig. 16.3 and illustrated below.

16.4.1 Problem Statement and Structural Model Identification

The first step in PBPK modelling is to clearly defining the objective of PBPK modelling. This

Modular mechanistic model and key features

Absorption

First order absorption, Gut as a single compartment associated with a single first order absorption rate constant (ka) and fraction of dose absorbed (fa).

CAT (Compartmental Absorption and Transit) and ACAT (Advanced CAT) models, transit model describing the GIT as a series of 7-9 compartments. Rate of change of drug is calculated as a function of the transit of dissolved drug , dissolution of solid particles, precipitation of dissolved drug, degradation of drug in solution, absorption of drug and exit of drug from the compartment.

ADAM (Advanced dissolution, Absorption and Metabolism) and M-ADAM (Multi-layered ADAM) model, GI Tract is divided into 9 anatomical segments drug absorption in each segment is described as function of drug release from the formulation, dissolution, precipitation, luminal degradation, permeability, metabolism, transport, and transit from one segment to another.

Distribution

Minimal PBPK model: Model that describes distribution of drug only in specific organ / tissues of interest and lumps all other into one compartment.
Full PBPK model: Model describe distribution into all major organ systems.

Perfusion limited distribution: If permeability is very high, then the amount of dug that partitions into the tissue is limited by blood flow rate. A partition coefficient (Kp) is used to calculate the concentration of drug in tissue by following 2 common mechanistic models:

- Poulin and Theil method: Predicts Kp for adipose and non-adipose tissues using drug lipophilicity (Log P).
- Rodger and Rowland method: Predicts Kp by considering extent of ionisation of compound at local pH of compartment.

Permeability-limited Distribution: If permeability if low, the amount of drug that partitions into the tissue will be limited by the permeability and the membrane surface area. Mechanistically, it includes differential equations to model both passive and active transport processes.

Metabolism

In Vitro to In vivo Extrapolation (IVIVE) : Mathematically scale in vitro metabolic rate of drug degradation from sub-cellular/cellular systems to full organ using physiological scalars, then to whole body and finally model distribution in population.

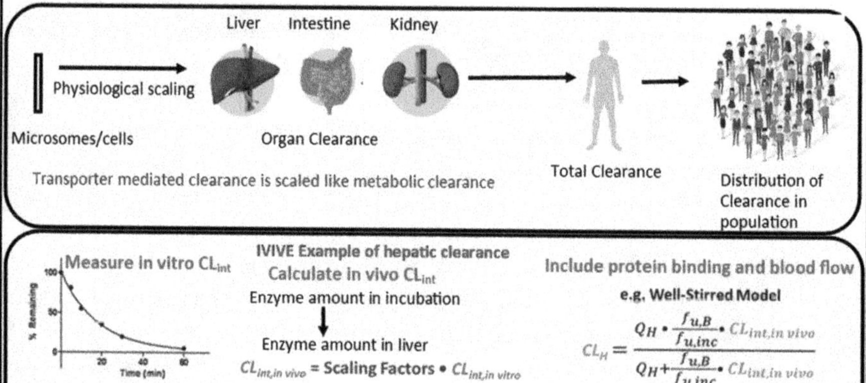

Complex metabolic and transporter interactions: Auto-induction, mechanism based inhibition, reversible inhibition.

Elimination

PopPK clearance: Simple PBPK models with non-mechanistic clearance e.g. oral clearance from population PK model or IV clearance from human mass balance study.
Mechanistic clearance: Total clearance as sum of biliary and /or renal clearance using transporter kinetics along with IVIVE metabolic clearances from eliminating organs.

Fig. 16.2 Schematic diagram showing choices of modular mechanistic ADME models for building structural PBPK models. Figures in metabolism section designed from brgfx / Freepik

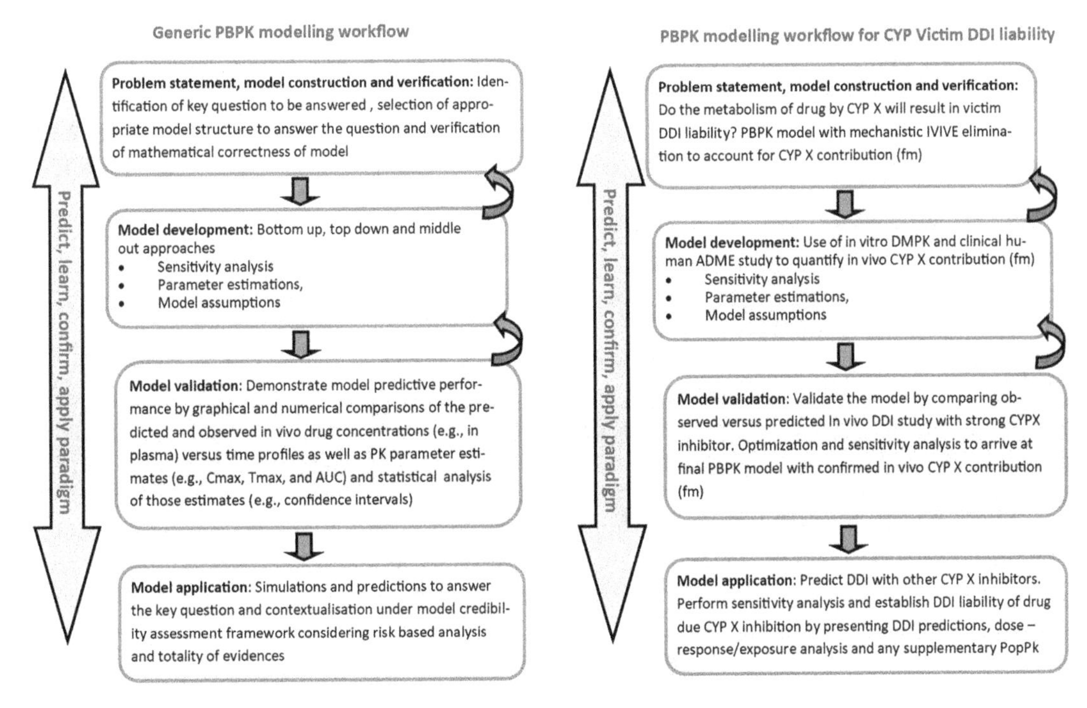

Generic PBPK modelling workflow
Problem statement, model construction and verification: Identification of key question to be answered, selection of appropriate model structure to answer the question and verification of mathematical correctness of model
Model development: Bottom up, top down and middle out approaches
Sensitivity analysis
Parameter estimations,
Model assumptions
Model validation: Demonstrate model predictive performance by graphical and numerical comparisons of the predicted and observed in vivo drug concentrations (e.g., in plasma) versus time profiles as well as PK parameter estimates (e.g., Cmax, Tmax, and AUC) and statistical analysis of those estimates (e.g., confidence intervals)
Model application: Simulations and predictions to answer the key question and contextualisation under model credibility assessment framework considering risk based analysis and totality of evidences
Predict, learn, confirm, apply paradigm
PBPK modelling workflow for CYP Victim DDI liability
Problem statement, model construction and verification: Do the metabolism of drug by CYP X will result in victim DDI liability? PBPK model with mechanistic IVIVE elimination to account for CYP X contribution (fm)
Model development: Use of in vitro DMPK and clinical human ADME study to quantify in vivo CYP X contribution (fm)
Sensitivity analysis
Parameter estimations,
Model assumptions
Model validation: Validate the model by comparing observed versus predicted In vivo DDI study with strong CYPX inhibitor. Optimization and sensitivity analysis to arrive at final PBPK model with confirmed in vivo CYP X contribution (fm)
Model application: Predict DDI with other CYP X inhibitors. Perform sensitivity analysis and establish DDI liability of drug due CYP X inhibition by presenting DDI predictions, dose – response/exposure analysis and any supplementary PopPk
Predict, learn, confirm, apply paradigm

Fig. 16.3 Diagrammatic representation of (**a**) generic PBPK modelling workflow and (**b**) example of workflow with specific application to answer question on DDI liability of drug due inhibition of specific elimination pathway mediated by CYP X

is followed by the identification of the model structure. The model may be completely built by the user from basics or by the use of modular PBPK platforms listed in Table 16.5.

16.4.2 Model Verification

The assessment of the correctness of the mathematical model structure including details of the differential equations used, computer codes and the parameterisations of the model is defined as model verification. If a commercial platform is used such model verification may be performed by the vendor as part of design qualification (DQ) and then by the user as installation qualification (IQ) [2].

16.4.3 Model Development

Building a PBPK model for a drug by integrating its physicochemical properties, in vitro data that are relevant to the key question to be addressed, and estimated sensitive or critical parameters from clinical pharmacokinetic (PK) data.

16.4.3.1 Parameter Sensitivity Analysis

Global sensitivity analysis (GSA) identifies sensitive model parameters among the in vitro-generated input parameters. Alternatively, there can be uncertainty in the true value of some of the parameters, for example, the absence of a specific parameter or unreliability of the in vitro data. In these cases, it is useful to assess the impact of uncertainty in those specific parameters or specific modelling assumptions may have on the simulation outcome. Local sensitivity analysis may be used where the selected parameters are changed within a given reasonable range and a selected set of endpoints are investigated. Identifying whether an input parameter has a significant impact on the outcome of a simulation is highly valuable, as it assists with making decisions.

16.4.3.2 Model Assumptions

The modelling exercise may involve assumptions that are scientifically justified with supportive information and data, when available. The effect of these assumptions on model structure and/or parameter(s) is assessed.

16.4.3.3 Parameter Estimations

Parameter estimation is the process of computing a model parameter value from measured data. There are two approaches for estimating parameters: individual-mode and population-mode. In individual-mode estimation, the data for each subject is fitted independently of the data of all other subjects. In population-mode estimation, individual subject values are assumed to be distributed according to a lognormal (population) distribution with a given population mean and population (also called inter-subject) variance. In addition, the individuals' parameters also condition the individuals' data, as in the individual fitting mode. The difference here is that population means, and variances are estimated, in addition to the individual parameter values. This is different from estimating a population mean and variance from a set of individual-mode fitting parameter values (the so-called naïve approach), because, in the population-mode approach population parameters, individual parameters, and measurement errors condition each other. Inter-subject variability in PK and PD parameters can be caused by differences in age, sex, ethnicity, drug-drug interaction, or random physiological traits.

16.4.4 Model Validation

Model validation refers to an assessment of the model's predictive performance in comparison with observed in vivo data. This is generally done qualitatively graphically by overlaying the simulated plasma concentration profiles on observed clinical profiles from multiple clinical scenarios, and also quantitatively by comparing observed versus predicted summary PK parameters (area under the curve (AUC), Cmax (maximum plasma concentration) and Tmax (time to reach Cmax). Furthermore, the predictive power of models may also be tested employing predictive metrics including geometric mean fold error (GMFE), absolute average fold error (AAFE) and percent prediction error (PPE%) [33, 34].

16.4.5 Model Application

The use of model to simulate untested clinical scenarios to help answer a question under consideration is defined as model application. Section 16.5 explains how model application should be viewed before starting with any PBPK modelling work under the model credibility assessment network by carefully examining risk-based analysis and totality of evidence from overall clinical studies [35]. Section 16.6 explains various applications of PBPK modelling to support clinical pharmacology of drugs.

16.5 Credibility Assessment Framework of PBPK Modelling

Due to the widespread application of PBPK modelling and consequent emergence of regulatory guidance, there has been an immense focus on establishing clearly defined approaches to gain trustworthiness of modelling and contextualisation of outcomes in making decisions that may impacts on patients [35, 36]. The term credibility refers to trust in the predictive capability of a model for a particular context of use (CoU).

Recently, a framework for the credibility assessment of PBPK models was proposed by US FDA [35], which involves examination of model by considering CoU and grade the credibility of model by assessing model influence and decision consequence as follows:

16.5.1 Model Influence

Model influence is described as the role of the model considering all available evidence in addressing the question of interest. Model influence may be graded as:

16.5.1.1 Low Impact

Model provides minor evidence; substantial nonclinical and clinical data are available to inform the decision. Low impact analyses are considered descriptive analyses with very limited impact on decision-making for the overall development program (e.g., a PBPK analysis to gain a more mechanistic understanding of observations).

16.5.1.2 Medium Impact

Model provides supportive evidence; some clinical trial data are available to inform the decision. Medium impact analyses provide supportive evidence and contribute to decision-making along with clinical data.

16.5.1.3 High Impact

Model provides substantial evidence; no clinical trial data relevant to the context of use or limited clinical trial data from similar scenarios are available to inform the decision. High impact analyses provide new evidence in the absence of respective clinical data and their results contribute exclusively to decision-making (e.g., PBPK studies in lieu of clinical studies to inform prescribing information).

16.5.2 Decision Consequence

Decision consequence is the impact of an incorrect decision based on all available evidence. Adverse outcomes resulting from wrong decisions could include (but may not be limited to) the risk of therapeutic failure or risk to patient safety. The significance can be driven by the number of patients likely to be impacted by the wrong decision, the severity of the potential harm, and/or the likelihood of occurrence. Decision consequences may be graded as:

16.5.2.1 Low

Incorrect decisions would not result in adverse outcomes in patient safety or efficacy.

16.5.2.2 Medium

Incorrect decisions could result in minor to moderate adverse outcomes in patient safety or efficacy.

16.5.2.3 High

Incorrect decisions could result in severe adverse outcomes in patient safety or efficacy.

Using the model impact and decision consequence definitions described above, model risk may be assessed (Fig. 16.4). The model risk levels can then be used to select the extent of

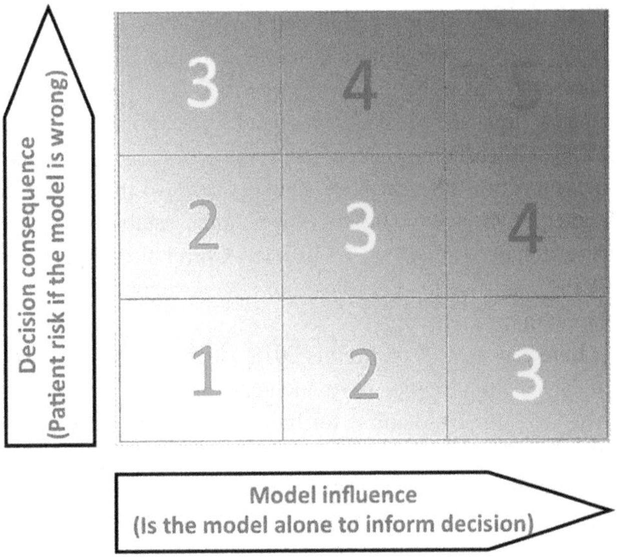

Fig. 16.4 Hypothetical model risk assessment matrix where model risk moves from low (levels 1–2) to medium (level 3) to high (levels 4–5) as model influence or decision consequence increases. The grading for model influence and decision consequence should be determined independently

verification and validation activities and define outcomes that will provide evidence to demonstrate credibility for a COU. More rigorous activities may be selected for models that have greater risk and thus require more evidence to demonstrate credibility. EMA guidelines on PBPK modelling illustrate some examples/case studies which may be considered as low/moderate or high impact PBPK analysis and qualification requirements to generate model credibility for those scenarios [37].

16.6 PBPK Modelling Applications

16.6.1 Small Molecule Drugs

PBPK modelling may be used to inform the impact of intrinsic (genetic, physiological, and pathological characteristics of an individual, e.g., age, renal/hepatic impairment) and extrinsic factors (environmental, regional, or related to lifestyle e.g., diet, concomitant medication use, and smoking habits) on PK of a drug [38].

Briefly, for small molecule drugs, PBPK models can be applied in the following scenarios:

16.6.1.1 Biopharmaceutic Modelling Applications

PBPK is increasingly used as a tool for various biopharmaceutic applications, namely, prediction of the impact of formulation and CMC changes on PK during clinical development and /or post-marketing, assessing pH-mediated DDIs in patients treated with proton pump inhibitors (PPIs)/acid-reducing agents (ARAs), effect of food-on drug PK, support development of a biopredictive in vitro dissolution method, establish mechanistic In Vitro to In Vivo Correlation (IVIVC) and predict the impact of beverage consumption on exposure.

Between year 2008 and 2018, a total of 24 NDA submissions have included of PBPK modelling and simulations for biopharmaceutics-related assessment. In these submissions, PBPK absorption modelling and simulation served as an impactful tool in establishing the relationship of critical quality attributes (CQAs) including formulation variables, specifically in vitro dissolution, to the in vivo performance [39]. These cases encompass both immediate release (IR) (18 out of 24, i.e., 75%) and extended-release(ER) (6 out of 24, i.e., 25%) solid oral dosage formulations involving all categories of Biopharmaceutical Classification System (BCS) classification drugs (71% of these drugs are BCS II or IV drugs). The applications can be categorized as (1) setting clinically relevant dissolution specifications that can ideally reject batches with undesired in vivo performance, which involves both the selection of bio-predictive dissolution methods (i.e., a set of testing conditions for which in vitro dissolution profiles are capable of predicting PK profiles) and the establishment of clinically relevant dissolution acceptance criteria; (2) setting clinically relevant specifications for Critical Material Attributes (CMAs) and Critical Process Parameters (CPPs) (e.g., in support of particle size distribution specification based on the effect of particle size on in vivo absorption); and (3) supporting quality risk assessment and possible risk-based biowaiver request (e.g., via Physiologically Based In Vitro and In Vivo correlation/ relationship (PB-IVIVC/R) or virtual Bioequivalence (BE) trial simulation).

Prediction of food effect on drug PK is less developed but is a fast-growing area, with multiple publications proposing flowchart-based strategies to use PBPK modelling in this area [40]. Currently, applications of PBPK modelling to address questions related to absorption related DDIs due to PPIs at the stage of regulatory submissions. In the years 2013–2017, PBPK modelling has been used in two New Drug Application submissions to predict the liability of pH-dependent DDIs to support the labelling recommendations [41].

16.6.1.2 PK Modelling Applications

PBPK modelling has been used to support the following key clinical pharmacology areas:

- Prediction of drug metabolizing enzyme(s)- or drug transporter(s)-mediated drug-drug interactions (DDIs) to inform inclusion/

exclusion criteria, support dose selection, and waive clinical DDI studies or studies that have difficulty in enrolling patients and inform drug label

- Prediction of dosing regimens, enable sampling timepoint selection, and propose dose in paediatric patients from newborns to adolescents,
- Prediction of PK and DDIs in patients with renal and/or hepatic organ impairment to inform study design or the decision to waive studies
- Estimate maternal–foetal drug disposition during pregnancy to achieve the best therapeutic benefit/risk ratio
- Translation of PK among different ethnic groups (Caucasians, Chinese, Japanese, etc.), including PK and DDI in genetically polymorphic sub-populations

Figure 16.5 is plotted based on data described by Grimstein et al. [42] and shows the increasing numbers of PBPK modelling in regulatory submissions. Between 2008 and 2017, the Office of Clinical Pharmacology (OCP), Centre for Drug Evaluation and Research (CDER), United States Food and Drug Administration (US FDA) received 130 investigational new drug (IND) applications and 94 new drug applications (NDAs) containing PBPK analyses [42]. The intended purpose of the PBPK analyses in these regulatory submissions is primarily to assess enzyme-based drug-drug interactions (60%), followed by applications in paediatrics (15%), DDI with transporter (7%), hepatic impairment (6%), renal impairment (4%), absorption including food effect (4%), and pharmacogenetics (2%). In addition, reviewers conducted de novo (i.e., OCP reviewer initiated) PBPK analyses to inform regulatory decisions for 30 submissions. More recently, from the years 2018 to 2019, OCP received PBPK-related submissions in 56 investigational new drugs (INDs), 57 new drug applications (NDAs), and 3 biologics license applications (BLAs) [43]. It is noteworthy that PBPK modelling is not equally mature in all areas of applications. For example, there is more confidence in DDI predictions due to CYP3A4 modulation when compared to DDI predictions due to transporters [44]. Figure 16.6 from the US FDA clinical pharmacology review document of the ibrutinib is an example of a successful PBPK application to influence the label of ibrutinib in informing DDIs and circumventing the need to conduct clinical trials. Ibrutinib is susceptible to interactions with strong inhibitors and inducers of CYP3A4. The ibrutinib PBPK models were built

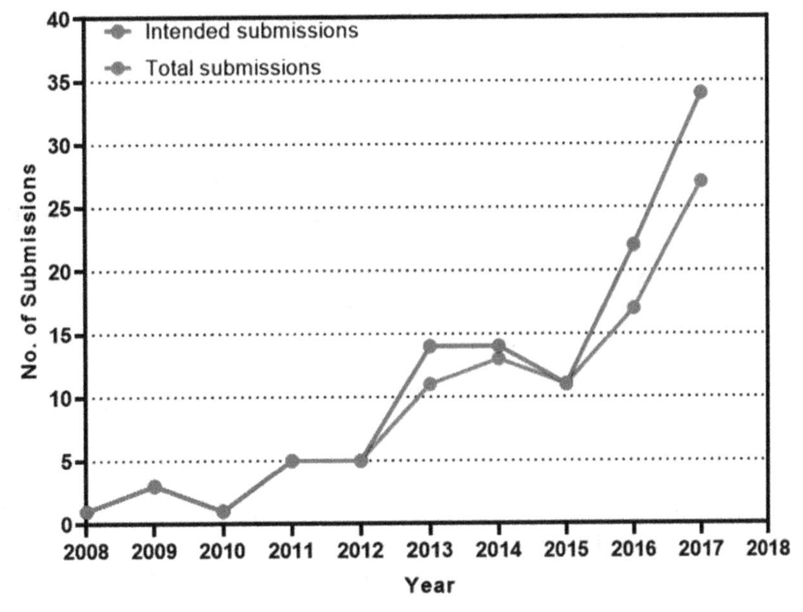

Fig. 16.5 Graph showing the increasing number of PBPK modelling related submissions to OCP US FDA. One NDA submission of PBPK model may have multiple intended submissions, the total number of intended PBPK applications exceeds the number of NDA submissions containing PBPK analyses

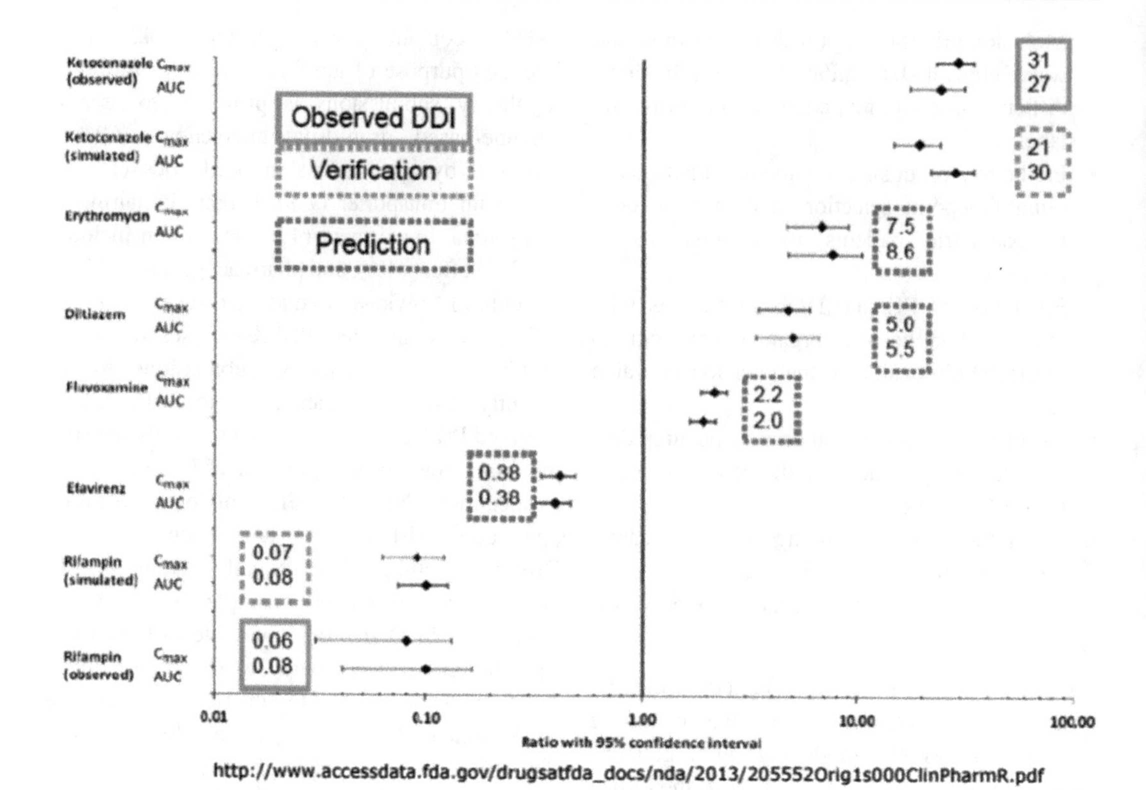

Fig. 16.6 Simulated and observed ibrutinib Cmax ratios and AUC ratios with 95% confidence intervals in presence of weak, moderate and strong inhibitors and moderate and strong inducers of CYP3A4

in the Simcyp Simulator using in vitro data and were validated using clinical data on the observed effects of both a strong CYP3A4 inhibitor and a strong inducer on ibrutinib exposure. Simulations then predicted the effects of a moderate CYP3A4 inducer and other CYP3A4 inhibitors (strong, moderate and weak) on ibrutinib exposure, as well as investigating the impact of dose staggering and dose adjustment. The final drug label included 24 individual claims for untested DDI scenarios (without the need for clinical trials) and provided a dose optimization strategy aligned to individuals with different metabolic profiles.

16.6.2 Large Molecule Drugs

PBPK modelling of biologics has a different remit by virtue of distinct PK processes from small molecule drugs. Compared to small molecule drugs, PBPK modelling applications for large molecules is much less mature but it is an area of great interest due to the high number of medicines approved or in development that are large molecules.

16.6.2.1 DDI Assessment of Therapeutic Proteins (TP) and Antibody Drug Conjugates (ADCs)

TPs that are proinflammatory cytokines (e.g., peginterferon) or TPs that cause increases in proinflammatory cytokine levels can down-regulate the expression of cytochrome P450 (CYP) enzymes, thereby decreasing the metabolism of drugs that are CYP substrates and increasing their exposure levels [45]. Conversely, TPs that reduce cytokine levels (e.g., TNF inhibitors) can relieve the CYP down-regulation from an inflammatory environment (e.g., rheumatoid arthritis), thereby increasing CYP expression and activity and reducing exposure for CYP substrates. TP affects human physiological processes that can in turn alter the pharmacokinetic profiles of co-administered medications (e.g., GLP-1 receptor agonists such as dulaglutide and albiglutide result in delayed gastric emptying). In this case, the sponsor should evaluate the TP as a perpetrator. The application of PBPK modelling in the evaluation of the DDI potential of a TP is an emerging area. PBPK modelling has a potential role in understanding the underlying mechanism of a DDI for TPs. The US FDA in recent guidance adopted an open approach and encouraged pharmaceutical companies to contact them when proposing to use PBPK modelling to evaluate the DDI potential of TPs [45]. For antibody-drug conjugates (ADCs), the small molecule drug component (the 'war head') conjugated to the antibody component can be released into an unconjugated form. Therefore, for ADCs DDI assessment should include small molecule DDIs for released war head /payload and antibody components separately [46]. PBPK modelling of the ADC polatuzumab vedotin was used to successfully inform the label for DDI liability without dedicated clinical DDI trials [47].

16.6.2.2 PBPK-PD Modelling

Large molecule PBPK models are built using (1) non-specific uptake via fluid-phase pinocytosis into vascular endothelial cells, (2) pH-dependent binding to FcRn in the acidic environment of the endosome, (3) proteolytic degradation of unbound mAb in the lysosome, (4) pH-dependent release of bound mAb at the cell surface into the plasma or interstitial fluid via exocytosis, and (5) exit of interstitial mAb into the lymph via convective flow. The complex nature of mAb disposition with a variable human PK, together with the recent evolution of in vitro assays provides an opportunity for predicting in vivo behaviour using in vitro data in a PBPK framework analogous to small molecule. PBPK models of therapeutic antibodies that include subsystems representing the fundamental mechanisms responsible for antibody transport such as paracellular exchange, nonspecific binding, FcRn interaction and transcytosis, can be informed by in vitro assays designed to characterize these interactions and processes. The resulting models can be used to predict the plasma and tissue disposition of antibodies. An in silico-based metric representing the positive charge in antibody complementarity-determining region was incorporated into a PBPK model for predicting antibody PK in humans [48]. Recently, a PBPK modelling framework was used to explore the predictive potential of 14 in vitro assays designed to measure various antibody physiochemical properties, including nonspecific cell-surface interactions, FcRn binding, thermal stability, hydrophobicity, and self-association. It demonstrated the utility of the proposed PBPK-PD model-based framework that integrates physiochemical characteristics of antibodies to predict PK profiles in humans, with the goal of facilitating antibody screening and engineering in early development stages [49].

For therapeutic mAbs and mAbs adducts, binding to receptors not only triggers downstream effects/pharmacological response but also its disposition (TMDD), and thus PK and pharmacodynamics (PD) are intricately linked. PBPK-PD approaches may be a useful application in characterising the dose exposure relations if there is sufficient information on target receptor expression. The target receptor expression may be directly experimentally determined or estimated with modelling approaches [50]. In some circumstances, target expression is not important (e.g., infection) making TMDD irrelevant in mAb PK. A Bamlanivimab PBPK-PD model was recently used to propose the first in man dose that was expected to result in maximum therapeutic effect before the first in human clinical trial [51]. Production of anti-drug antibodies (ADA) and binding to Fc-γ receptors on immune cells lead to increased mAb clearance and these complex immune responses are currently challenging to be modelled mechanistically in PBPK models.

16.6.2.3 PBPK Model of mAbs in Specific Populations

The anatomy and physiology of children change rapidly in the first few months of their life. Since mAbs have long half-lives spanning few weeks to months, continuous maturation of physiology over the course of single dosing may need to be considered within PBPK models aimed at predicting mAb exposure in very young children. Continual maturation of physiology has been incorporated in a few of recently reported paediatric mAb PBPK models where the mAb PK parameters and/or plasma concentrations were predicted within two-fold of observed values [52]. The published paediatric mAb PBPK models incorporate ontogeny of organ volumes, blood flows, lymph flows, interstitial volumes, and haematocrit. Ontogeny of lymph flow differs between the models, one used a uniform scalar in young children (0–1 year) [53] and others allometrically scale adult lymph flows [54], both resulting in increased lymph flow in paediatrics compared to adults, in line with animal data. PBPK modelling of biologics in pregnancy is not reported but is a potential area for future research. Similarly, there is less knowledge on ethnic differences or the impact of organ impairment on the PK of biologics.

16.7 Regulatory Guidance on PBPK Modelling

PBPK modelling had been applied in past in allied sciences for example in assessing the impact of environmental pollutants on human health quite extensively and much before its use was recognised in the pharmaceutical industry. Therefore, considerable regulatory/policy recommendations existed already (Table 16.6) historically which are interesting to read as may be viewed as a good foundation for subsequent tremendous growth and application of PBPK modelling in the pharmaceutical industry. In recent times, there had been a surge of white papers, policy documents, and regulatory guidelines. Many cross-industry working groups have recently published consensus white papers on specific topics on PBPK modelling including regulatory submissions [2], induction [55], pharmaceutical product development [56], transporters [3], and organ impairment [57].

Specific guidances from regulatory agencies have been released to advise the pharmaceutical industry on reporting PBPK analysis (Table 16.6). These guidances provide an outline of key contents in the PBPK analysis report to help enable regulatory agencies to assess PBPK regulatory submissions in a rational way [37, 58]. In addition, PBPK guidance is issued on specific topics as well, for example on biopharmaceutical modelling to advise on scope, methodology and reporting for modelling impact of CMC changes on the PK of drugs [59]. Since PBPK had a widespread application in every aspect of clinical pharmacology (Sect. 16.6), it is noteworthy that now most guidance documents being issued on several different aspects now include a brief description of PBPK analysis in them as an important tool to aid understanding of PK by extrinsic and intrinsic factors.

Fundamental to all this regulatory guidance is positive support for model-based analysis to aid drug discovery and development. The basic expectation is a transparent and clearly explained approach with a scientific basis on methodology and outcomes to help reduce the burden of excessive clinical trials on patients and speed up the availability of novel medicines without compromising efficacy and safety. It is expected that PBPK analysis will continue to have an increasing role in regulatory submissions and further guidance documents will emerge as our understanding matures with historical data being generated.

16.8 Challenges and Future Directions

PBPK modelling is a fast emerging approach that has many challenges that are potential areas of ongoing and future research. Some of these high-level challenges are highlighted below:

- Lack of systems and drug data: key data on the absolute expression of drug metabolising enzymes/transporters of small molecule drugs and target receptors for biologics is unavailable. Also generating extensive data to enable

Table 16.6 Regulatory guidance/policy documents on PBPK modelling of xenobiotics (drugs, environmental pollutants, and food components)

Guidance /policy document title	Issuing body, year[a]	Brief description
1. Core regulatory guidance for the pharmaceutical industry		
Physiologically based pharmacokinetic analyses—format and content: guidance for industry [58]	US FDA, 2018	Recommend format and content for a sponsor or applicant to submit PBPK analyses to the FDA
Guideline on the reporting of physiologically based pharmacokinetic (PBPK) modelling and simulation [37]	EMA, 2018	What to include in a PBPK modelling report including, details of the predictive performance of the drug model and supportive data are expected to qualify a PBPK platform
Guidelines for analysis reports involving physiologically based pharmacokinetic models [60]	PMDA, 2020	Recommend format and content for a sponsor or applicant to submit PBPK analyses to the PMDA
The use of physiologically based pharmacokinetic analyses-biopharmaceutics applications for oral drug product development, manufacturing changes, and controls: guidance for industry [59]	US FDA, 2020	Recommendations regarding the development, evaluation, and use of (PBPK) analyses for biopharmaceutics applications with a focus only on orally administered, systemically active drug products. For other areas, it is considered on a case-by-case basis
In vitro drug interaction studies—Cytochrome P450 enzyme- and transporter-mediated drug interactions: guidance for Industry [61]	US FDA, 2020	Provides a detailed framework of static, semi-mechanistic, and dynamic PBPK modelling for DDI predictions
Clinical drug interaction studies—Cytochrome P450 enzyme- and transporter-mediated drug interactions guidance for industry [62]]	US FDA, 2020	Provides guidance on the use PBPK analysis in supporting the conduct of clinical drug interaction studies
Guideline on the investigation of drug interactions [63]	EMA, 2012	Recommends use of PBPK at different stages during drug development to inform study design, estimate the potential for drug-interactions qualitatively as well as estimate an interaction effect quantitatively
Clinical pharmacology considerations for Antibody-Drug Conjugates (ADC): guidance for Industry [46]	US FDA, 2022	Refers to PBPK modelling of ADCs for DDI predictions
Drug-Drug Interaction assessment for Therapeutic Proteins (TPs): guidance for industry [45]	US FDA, 2020	Refers to PBPK modelling of TPs for DDI predictions
Evaluation of gastric pH-dependent drug interactions with acid-reducing agents: study design, data analysis, and clinical implications: guidance for industry [64]	US FDA, 2020	Refers to exploring the use of PBPK simulations to assess the potential for pH dependent DDIs and inform clinical study designs
Pharmacokinetics in patients with impaired renal function—Study design, data analysis, and impact on dosing and labelling [65]	US FDA, 2020	Recommends use of PBPK modelling to inform clinical trial designs for renal impaired populations
2. Guidance/policy documents from allied areas e.g., environment protection		
Guidance document on the characterisation, validation and reporting of physiologically based kinetic (PBK) models for regulatory purposes [66]	OECD, 2021	Contextual information on the scientific process of PBPK model characterisation and validation. Guidance on how to use PBPK models for specific regulatory purposes was out of scope
ICH Guideline M12 on drug interaction studies [67]	ICH, 2022	Best practice considerations for use of PBPK modelling for the evaluation of DDIs a
Principles of characterising and applying PBK models in risk assessment [68]	WHO, 2010	Best practices for characterizing and applying PBPK models in risk assessment

(continued)

Table 16.6 (continued)

Guidance /policy document title	Issuing body, year[a]	Brief description
Approaches for the application of physiologically based pharmacokinetic models and supporting data in risk assessment [69]	US EPA, 2006	Evaluation and use of PBPK models for predicting internal dose at target organs in risk assessment applications
Use of physiologically based pharmacokinetic models to quantify the impact of human age and interindividual differences in physiology and biochemistry pertinent to risk [70]	US EPA, 2006	Communicate a framework developed for the extrapolation and integration of in vitro-derived measures of chemical metabolism, including those that define human interindividual variability
Scientific Opinion on good modelling practice in the context of mechanistic effect models for risk assessment of plant protection products [71]	EFSA, 2014	Prepare a scientific opinion on good modelling practice in the context of mechanistic effect models for risk assessment of plant protection products

[a]Full forms of abbreviations used: *EFSA* The European Food Safety Authority, *EMA* European Medicines Agency, *ICH* The International Council for Harmonisation of Technical Requirements for Pharmaceuticals for Human Use, *OECD* The Organisation for Economic Co-operation and Development, *PMDA* Pharmaceuticals and Medical Devices Agency (Japanese regulatory agency), *US EPA* United States Environmental Protection Agency, *US FDA* United States Food and Drug Administration, *WHO* World Health Organisation

PBPK modelling approaches may need additional costly and time taking studies, leading to delay in medicines approval or making assumptions in models

- Mechanistic understanding of PK processes: Mechanisms of transporter inhibition, lysosomal trapping, etc., for small molecules and the effect of pH on endosomal entrapment of mAbs during FcRn recycling, etc., are not quantitatively characterised or well understood
- Computation tools: More and better tools for computation for faster and reliable complex computations in practical industry work environments are needed
- Lack of clinical data: For validation of PBPK models in specific populations e.g., pregnancy, lack of clinical data is a deterrent in regulatory acceptance
- Specific case-by-case challenges: each drug project has specific modelling challenges which are driven by typical drug properties. Unavailability of PK data after i.v. administration of drugs to validate the models in the oncology therapy area where i.v. PK is not necessarily generated

Notwithstanding the aforementioned bottlenecks, PBPK modelling had emerged successful tool in model informed drug development and is expected to grow further in the future.

16.9 Concluding Remarks

PBPK modelling is an important tool that provides a qualitative and quantitative understanding of drug PK. It has now been widely used in academia, the pharmaceutical industry, and regulators for effective decision making. Well characterised PBPK models for small molecule and large molecule drugs are available which can be developed through flowchart workflows and complied with a credibility assessment framework for high quality regulatory submissions.

Disclosures and Conflict of Interest Pradeep Sharma, Jin Dong, Vijender Panduga and David W Boulton are employees and/or shareholders of AstraZeneca, and Felix Stader is an employee of Simcyp Division (Certara UK Limited). Simcyp® is a proprietary PBPK modelling software platform of the Simcyp division of Certara UK Ltd.

References

1. Sharma P, Patel N, Prasad B, Varma MVS (2021) Pharmacokinetics: theory and application in drug discovery and development. In: Poduri R (ed) Drug discovery and development: from targets and molecules to medicines. Springer Singapore, Singapore, pp 297–355
2. Shebley M, Sandhu P, Emami Riedmaier A, Jamei M, Narayanan R, Patel A et al (2018) Physiologically based pharmacokinetic model qualification and

reporting procedures for regulatory submissions: a consortium perspective. Clin Pharmacol Ther 104(1): 88–110

3. Guo Y, Chu X, Parrott NJ, Brouwer KLR, Hsu V, Nagar S et al (2018) Advancing predictions of tissue and intracellular drug concentrations using in vitro, imaging and physiologically based pharmacokinetic modeling approaches. Clin Pharmacol Ther 104(5): 865–889
4. Sharma S, Li Z, Bussing D, Shah DK (2020) Evaluation of quantitative relationship between target expression and antibody-drug conjugate exposure inside cancer cells. Drug Metab Dispos 48(5):368–377
5. Brown RP, Delp MD, Lindstedt SL, Rhomberg LR, Beliles RP (1997) Physiological parameter values for physiologically based pharmacokinetic models. Toxicol Ind Health 13(4):407–484
6. Meyer M, Schneckener S, Ludewig B, Kuepfer L, Lippert J (2012) Using expression data for quantification of active processes in physiologically based pharmacokinetic modeling. Drug Metab Dispos 40(5): 892–901
7. Schwenger E, Reddy VP, Moorthy G, Sharma P, Tomkinson H, Masson E et al (2018) Harnessing meta-analysis to refine an oncology patient population for physiology-based pharmacokinetic modeling of drugs. Clin Pharmacol Ther 103(2):271–280
8. Zhou L, Tong X, Sharma P, Xu H, Al-Huniti N, Zhou D (2019) Physiologically based pharmacokinetic modelling to predict exposure differences in healthy volunteers and subjects with renal impairment: ceftazidime case study. Basic Clin Pharmacol Toxicol 125(2):100–107
9. Zhou L, Sharma P, Yeo KR, Higashimori M, Xu H, Al-Huniti N et al (2019) Assessing pharmacokinetic differences in Caucasian and East Asian (Japanese, Chinese and Korean) populations driven by CYP2C19 polymorphism using physiologically-based pharmacokinetic modelling. Eur J Pharm Sci 139:105061
10. Butrovich MA, Tang W, Boulton DW, Nolin TD, Sharma P (2022) Use of physiologically based pharmacokinetic modeling to evaluate the impact of chronic kidney disease on CYP3A4-mediated metabolism of saxagliptin. J Clin Pharmacol 62(8):1018–1029
11. Jo H, Pilla Reddy V, Parkinson J, Boulton DW, Tang W (2021) Model-informed pediatric dose selection for dapagliflozin by incorporating developmental changes. CPT Pharmacometrics Syst Pharmacol 10(2):108–118
12. Sjöstedt N, Neuhoff S, Brouwer KLR (2021) Physiologically-based pharmacokinetic model of morphine and morphine-3-glucuronide in nonalcoholic steatohepatitis. Clin Pharmacol Ther 109(3):676–687
13. Chetty M, Johnson TN, Polak S, Salem F, Doki K, Rostami-Hodjegan A (2018) Physiologically based pharmacokinetic modelling to guide drug delivery in older people. Adv Drug Deliv Rev 135:85–96
14. Pan X, Stader F, Abduljalil K, Gill KL, Johnson TN, Gardner I et al (2020) Development and application of a physiologically-based pharmacokinetic model to predict the pharmacokinetics of therapeutic proteins from full-term neonates to adolescents. AAPS J 22(4):76
15. Chaphekar N, Dodeja P, Shaik IH, Caritis S, Venkataramanan R (2021) Maternal-fetal pharmacology of drugs: a review of current status of the application of physiologically based pharmacokinetic models. Front Pediatr 9:733823
16. Dong J, Park MS (2022) A myth of the well-stirred model: Is the well-stirred model good for high clearance drugs? Eur J Pharm Sci 172:106134
17. Hens B, Seegobin N, Bermejo M, Tsume Y, Clear N, McAllister M et al (2022) Dissolution challenges associated with the surface pH of drug particles: integration into mechanistic oral absorption modeling. AAPS J 24(1):17
18. Brightman FA, Leahy DE, Searle GE, Thomas S (2006) Application of a generic physiologically based pharmacokinetic model to the estimation of xenobiotic levels in human plasma. Drug Metab Dispos 34(1): 94–101
19. Levitt DG, Levitt MD (2016) Human serum albumin homeostasis: a new look at the roles of synthesis, catabolism, renal and gastrointestinal excretion, and the clinical value of serum albumin measurements. Int J Gen Med 9:229–255
20. Telgmann R, von Kleist M, Huisinga W (eds) 2006 Software supported modelling in pharmacokinetics. In: Computational life sciences II. Springer, Berlin
21. Prado-Velasco M, Borobia A, Carcas-Sansuan A (2020) Predictive engines based on pharmacokinetics modelling for tacrolimus personalized dosage in paediatric renal transplant patients. Sci Rep 10(1):7542
22. Lixoft (2022) PBPK glucose-insulin model exploration: Lixoft.com. https://mlxplore.lixoft.com/case-studies/pbpk-glucose-insulin-model-exploration/
23. Peters SA (2008) Evaluation of a generic physiologically based pharmacokinetic model for lineshape analysis. Clin Pharmacokinet 47(4):261–275
24. Chenel M, Bouzom F, Aarons L, Ogungbenro K (2008) Drug-drug interaction predictions with PBPK models and optimal multiresponse sampling time designs: application to midazolam and a phase I compound. Part 1: comparison of uniresponse and multiresponse designs using PopDes. J Pharmacokinet Pharmacodyn 35(6):635–659
25. Stader F, Penny MA, Siccardi M, Marzolini C (2019) A comprehensive framework for physiologically-based pharmacokinetic modeling in Matlab. CPT Pharmacometrics Syst Pharmacol 8(7):444–459
26. Gufford BT, Barr JT, González-Pérez V, Layton ME, White JR Jr, Oberlies NH et al (2015) Quantitative prediction and clinical evaluation of an unexplored herb-drug interaction mechanism in healthy volunteers. CPT Pharmacometrics Syst Pharmacol 4(12):701–710

27. Li X, DuBois DC, Almon RR, Jusko WJ (2020) Physiologically based pharmacokinetic modeling involving nonlinear plasma and tissue binding: application to prednisolone and prednisone in rats. J Pharmacol Exp Ther 375(2):385–396
28. Watanabe T, Kusuhara H, Maeda K, Shitara Y, Sugiyama Y (2009) Physiologically based pharmacokinetic modeling to predict transporter-mediated clearance and distribution of pravastatin in humans. J Pharmacol Exp Ther 328(2):652–662
29. Tsiros P, Bois FY, Dokoumetzidis A, Tsiliki G, Sarimveis H (2019) Population pharmacokinetic reanalysis of a Diazepam PBPK model: a comparison of Stan and GNU MCSim. J Pharmacokinet Pharmacodyn 46(2):173–192
30. Fu Q, Sun X, Lustburg MB, Sparreboom A, Hu S (2019) Predicting paclitaxel disposition in humans with whole-body physiologically-based pharmacokinetic modeling. CPT Pharmacometrics Syst Pharmacol 8(12):931–939
31. Carter SJ, Chouhan B, Sharma P, Chappell MJ (2020) Prediction of clinical transporter-mediated drug–drug interactions via comeasurement of pitavastatin and eltrombopag in human hepatocyte models. CPT Pharmacometrics Syst Pharmacol 9(4):211–221
32. Naga D, Parrott N, Ecker GF, Olivares-Morales A (2022) Evaluation of the success of high-throughput physiologically based pharmacokinetic (HT-PBPK) modeling predictions to inform early drug discovery. Mol Pharm 19(7):2203–2216
33. Pepin XJH, Moir AJ, Mann JC, Sanderson NJ, Barker R, Meehan E et al (2019) Bridging in vitro dissolution and in vivo exposure for acalabrutinib. Part II. A mechanistic PBPK model for IR formulation comparison, proton pump inhibitor drug interactions, and administration with acidic juices. Eur J Pharm Biopharm 142:435–448
34. Chen Y, Cabalu TD, Callegari E, Einolf H, Liu L, Parrott N et al (2019) Recommendations for the design of clinical drug-drug interaction studies with itraconazole using a mechanistic physiologically-based pharmacokinetic model. CPT Pharmacometrics Syst Pharmacol 8(9):685–695
35. Kuemmel C, Yang Y, Zhang X, Florian J, Zhu H, Tegenge M et al (2020) Consideration of a credibility assessment framework in model-informed drug development: potential application to physiologically-based pharmacokinetic modeling and simulation. CPT Pharmacometrics Syst Pharmacol 9(1):21–28
36. Viceconti M, Pappalardo F, Rodriguez B, Horner M, Bischoff J, Musuamba TF (2021) In silico trials: verification, validation and uncertainty quantification of predictive models used in the regulatory evaluation of biomedical products. Methods 185:120–127
37. European Medicines Agency (EMA) (2019) Guideline on the reporting of physiologically based pharmacokinetic (PBPK) modelling and simulation. https://www.ema.europa.eu/en/reporting-physiologically-based-pharmacokinetic-pbpk-modelling-simulation-scientific-guideline. Accessed 26 Apr 2023
38. Lin W, Chen Y, Unadkat JD, Zhang X, Wu D, Heimbach T (2022) Applications, challenges, and outlook for PBPK modeling and simulation: a regulatory, industrial and academic perspective. Pharm Res 39(8): 1701–1731
39. Wu F, Shah H, Li M, Duan P, Zhao P, Suarez S et al (2021) Biopharmaceutics applications of physiologically based pharmacokinetic absorption modeling and simulation in regulatory submissions to the U.S. Food and Drug Administration for new drugs. AAPS J 23(2):31
40. Riedmaier AE, DeMent K, Huckle J, Bransford P, Stillhart C, Lloyd R et al (2020) Use of physiologically based pharmacokinetic (PBPK) modeling for predicting drug-food interactions: an industry perspective. AAPS J 22(6):123
41. Dong Z, Li J, Wu F, Zhao P, Lee SC, Zhang L et al (2020) Application of physiologically-based pharmacokinetic modeling to predict gastric pH-dependent drug-drug interactions for weak base drugs. CPT Pharmacometrics Syst Pharmacol 9(8):456–465
42. Grimstein M, Yang Y, Zhang X, Grillo J, Huang S-M, Zineh I et al (2019) Physiologically based pharmacokinetic modeling in regulatory science: an update from the U.S. Food and Drug Administration's Office of Clinical Pharmacology. J Pharm Sci 108(1):21–25
43. Zhang X, Yang Y, Grimstein M, Fan J, Grillo JA, Huang SM et al (2020) Application of PBPK modeling and simulation for regulatory decision making and its impact on US prescribing information: an update on the 2018-2019 submissions to the US FDA's Office of Clinical Pharmacology. J Clin Pharmacol 60:S1
44. Wagner C, Zhao P, Pan Y, Hsu V, Grillo J, Huang S et al (2015) Application of physiologically based pharmacokinetic (PBPK) modeling to support dose selection: report of an FDA public workshop on PBPK. CPT Pharmacometrics Syst Pharmacol 4(4):226–230
45. US Food and Drug Administration (US FDA) (2020) Drug-drug interaction assessment for therapeutic proteins guidance for industry. https://www.fda.gov/regulatory-information/search-fda-guidance-documents/drug-drug-interaction-assessment-therapeutic-proteins-guidance-industry. Accessed 26 Apr 2023
46. US Food and Drug Administration (USFDA) (2022) Clinical pharmacology considerations for antibody-drug conjugates guidance for industry: guidance for industry. https://www.fda.gov/regulatory-information/search-fda-guidance-documents/clinical-pharmacology-considerations-antibody-drug-conjugates-guidance-industry. Accessed 26 Apr 2023
47. Samineni D, Ding H, Ma F, Shi R, Lu D, Miles D et al (2020) Physiologically based pharmacokinetic model-informed drug development for polatuzumab vedotin: label for drug-drug interactions without dedicated clinical trials. J Clin Pharmacol 60(suppl 1):S120–SS31
48. Jones HM, Zhang Z, Jasper P, Luo H, Avery LB, King LE et al (2019) A physiologically-based pharmacokinetic model for the prediction of monoclonal antibody pharmacokinetics from in vitro data. CPT: Pharmacometrics Syst Pharmacol 8(10):738–747

49. Hu S, Datta-Mannan A, D'Argenio DZ (2022) Physiologically based modeling to predict monoclonal antibody pharmacokinetics in humans from in vitro physiochemical properties. MAbs 14(1):2056944
50. Kalra P, Brandl J, Gaub T, Niederalt C, Lippert J, Sahle S et al (2019) Quantitative systems pharmacology of interferon alpha administration: a multi-scale approach. PLoS One 14(2):e0209587
51. Chigutsa E, Jordie E, Riggs M, Nirula A, Elmokadem A, Knab T et al (2022) A quantitative modeling and simulation framework to support candidate and dose selection of anti-SARS-CoV-2 monoclonal antibodies to advance bamlanivimab into a first-in-human clinical trial. Clin Pharmacol Ther 111(3):595–604
52. Gill KL, Jones HM (2022) Opportunities and challenges for PBPK model of mAbs in paediatrics and pregnancy. AAPS J 24(4):72
53. Malik PRV, Edginton AN (2020) Integration of ontogeny into a physiologically based pharmacokinetic model for monoclonal antibodies in premature infants. J Clin Pharmacol 60(4):466–476
54. Hardiansyah D, Ng CM (2018) Effects of the FcRn developmental pharmacology on the pharmacokinetics of therapeutic monoclonal IgG antibody in pediatric subjects using minimal physiologically-based pharmacokinetic modelling. MAbs 10(7):1144–1156
55. Hariparsad N, Ramsden D, Taskar K, Badée J, Venkatakrishnan K, Reddy Micaela B et al (2022) Current practices, gap analysis, and proposed workflows for PBPK modeling of cytochrome P450 induction: an industry. Perspective 112(4):770–781
56. Mitra A, Suarez-Sharp S, Pepin XJH, Flanagan T, Zhao Y, Kotzagiorgis E et al (2021) Applications of physiologically based biopharmaceutics modeling (PBBM) to support drug product quality: a workshop summary report. J Pharm Sci 110(2):594–609
57. Heimbach T, Chen Y, Chen J, Dixit V, Parrott N, Peters SA et al (2021) Physiologically-based pharmacokinetic modeling in renal and hepatic impairment populations: a pharmaceutical industry perspective. Clin Pharmacol Ther 110(2):297–310
58. US Food and Drug Administration (US FDA) (2018) Physiologically-based-pharmacokinetic-analyses—format-and-content: guidance-for-industry. https://www.fda.gov/regulatory-information/search-fda-guidance-documents/physiologically-based-pharmacokinetic-analyses-format-and-content-guidance-industry. Accessed 26 Apr 2023
59. US Food and Drug Administration (US FDA) (2020) The use of physiologically based pharmacokinetic analyses—biopharmaceutics applications for oral drug product development, manufacturing changes, and controls guidance for industry. https://www.fda.gov/regulatory-information/search-fda-guidance-documents/use-physiologically-based-pharmacokinetic-analyses-biopharmaceutics-applications-oral-drug-product. Accessed 26 Apr 2023
60. Pharmaceuticals and Medical Devices Agency (PMDA) (2020) Guidelines for analysis reports involving physiologically based pharmacokinetic models. https://www.pmda.go.jp/files/000239317.pdf. Accessed 26 Apr 2023
61. US Food and Drug Administration (US FDA) (2020) In vitro drug interaction studies — cytochrome P450 enzyme- and transporter-mediated drug interactions: guidance for industry. https://www.fda.gov/regulatory-information/search-fda-guidance-documents/in-vitro-drug-interaction-studies-cytochrome-p450-enzyme-and-transporter-mediated-drug-interactions. Accessed 26 Apr 2023
62. US Food and Drug Administration (US FDA) (2020) Clinical drug interaction studies — cytochrome P450 enzyme- and transporter-mediated drug interactions: guidance for industry. https://www.fda.gov/regulatory-information/search-fda-guidance-documents/clinical-drug-interaction-studies-cytochrome-p450-enzyme-and-transporter-mediated-drug-interactions. Accessed 26 Apr 2023
63. European Medicines Agency (2012) Guideline on the investigation of drug interactions. https://www.ema.europa.eu/en/documents/scientific-guideline/guideline-investigation-drug-interactions-revision-1_en.pdf. Accessed 26 Apr 2023
64. US Food and Drug Administration (US FDA) (2020) Evaluation of gastric pH-dependent drug interactions with acid-reducing agents: study design, data analysis, and clinical implications: guidance for industry. https://www.fda.gov/regulatory-information/search-fda-guidance-documents/evaluation-gastric-ph-dependent-drug-interactions-acid-reducing-agents-study-design-data-analysis. Accessed 26 Apr 2023
65. US Food and Drug Administration (US FDA) (2020) Pharmacokinetics in patients with impaired renal function — study design, data analysis, and impact on dosing and labeling: guidance for industry. https://www.fda.gov/regulatory-information/search-fda-guidance-documents/pharmacokinetics-patients-impaired-renal-function-study-design-data-analysis-and-impact-dosing-and. Accessed 26 Apr 2023
66. Organisation for Economic Co-operation and Development (OECD) (2021), Guidance document on the characterisation, validation and reporting of Physiologically Based Kinetic (PBK) models for regulatory purposes, OECD Series on Testing and Assessment, No. 331, Environment, Health and Safety, Environment Directorate, OECD. https://www.oecd.org/chemicalsafety/risk-assessment/guidance-document-on-the-characterisation-validation-and-reporting-of-physiologically-based-kinetic-models-for-regulatory-purposes.pdf. Accessed 26 Apr 2023
67. International Council for Harmonisation of Technical Requirements for Pharmaceuticals for Human Use

(ICH) (2022) ICH guideline M12 on drug interaction studies. chrome-extension://efaidnbmnnnibpcajpcglclefindmkaj/https://www.ema.europa.eu/en/documents/scientific-guideline/draft-ich-guideline-m12-drug-interaction-studies-step-2b_en.pdf. Accessed 26 Apr 2023
68. International Programme on Chemical Safety & Inter-Organization Programme for the Sound Management of Chemicals (2010) Characterization and application of physiologically based phamacokinetic models in risk assessment. World Health Organization. https://apps.who.int/iris/handle/10665/44495. Accessed 26 Apr 2023
69. US Environmental Protection Agency (US EPA) (2006) Approaches for the application of physiologically based pharmacokinetic (PBPK) models and supporting data in risk assessment, U.S. Environmental Protection Agency, Washington, D.C., EPA/600/R-05/043F, 2006. https://cfpub.epa.gov/si/si_public_record_Report.cfm?Lab=NCEA&dirEntryID=157668. Online report accessed on 26 Apr 2023
70. U.S. Environmental Protection Agency (US EPA). (2006) Use of physiologically based pharmacokinetic models to quantify the impact of human age and inter-individual differences in physiology and biochemistry pertinent to risk. U.S. Environmental Protection Agency, Washington, D.C., EPA/600/R-06/014A. https://cfpub.epa.gov/ncea/risk/recordisplay.cfm?deid=151384. Online report accessed on 26 Apr 2023
71. EFSA (2014) Scientific Opinion on good modelling practice in the context of mechanistic effect models for risk assessment of plant protection products. EFSA J 12(3):3589

Design and Conduct of Pharmacokinetics Studies Influenced by Extrinsic Factors

17

Maria Learoyd, Beth Williamson, Jenny Cheng, and Venkatesh Pilla Reddy

Abstract

The objective of this chapter is to provide an overview of the design and conduct of clinical pharmacokinetic (PK) studies when investigating extrinsic factors that may influence the PK and pharmacodynamic relationship of small molecule drug candidates in drug development. The clinical pharmacology package has to reflect and evaluate the impact of extrinsic (food effect, co-medications, smoking, etc) and intrinsic factors (age, race, organ dysfunction, etc.) on PK in humans. Information from preclinical absorption, distribution, metabolism and excretion studies provides initial guidance on the potential impact of extrinsic factors on a drug and subsequently, clinical studies are required to characterise the PK of the drug in various clinical settings and form an important part of the marketing application to health authorities and for the drug label. Drug-Drug Interactions (DDI) represent one of the major extrinsic factors to be evaluated during drug discovery and development. Knowledge of the major metabolic pathway is also key information for the investigational drug as a drug may be liable to be a victim of DDI when co-dosed with perpetrator drug that inhibits or induces that metabolic pathway and hence alters the exposure levels in patients. Drugs may inhibit drug metabolizing enzymes and/or transporters competitively, time dependently, and may induce their activity or expression. In theory, all situations can be tested clinically. However, ethical and practical issues may limit the numbers of studies one can conduct. Advances in modelling and simulation approach allow some situations to be predicted in lieu of, or before clinical studies.

Keywords

Clinical pharmacology · Extrinsic factors · Drug-drug interactions · Food effect · ADME · Pharmacodynamic

M. Learoyd · V. Pilla Reddy (✉)
Clinical Pharmacology and Quantitative Pharmacology (CPQP), Clinical Pharmacology & Safety Sciences, R & D, AstraZeneca, Cambridge, UK
e-mail: venkatesh.reddy@astrazeneca.com

B. Williamson
Development Science, DMPK, UCB Biopharma, Slough, UK

J. Cheng
Clinical Pharmacology and Quantitative Pharmacology (CPQP), Clinical Pharmacology & Safety Sciences, R & D, AstraZeneca, Gaithersburg, MD, USA

Abbreviations

ADME	Absorption, distribution, metabolism and excretion
CL	Clearance
DDI	Drug-drug interactions

G. Jagadeesh et al. (eds.), *The Quintessence of Basic and Clinical Research and Scientific Publishing*,
https://doi.org/10.1007/978-981-99-1284-1_17

DME	Drug metabolizing enzyme
DMTA	Design make test analyse
EMA	European Medicine Agency
Fabs	Fraction absorbed
FDA	Food and Drug Administration
Fg	Gut bioavailability
IV	Intravenous
IVIVE	In vitro in vivo extrapolation
Ka	Absorption constant
MID3	Model-informed drug discovery and development
M-M	Michaelis-Menten kinetics
MRCT	Multiregional clinical trials
PBPK	Physiological based pharmacokinetic
PK	Pharmacokinetics
PKPD	Pharmacokinetic-pharmacodynamic
PKPD/E	Pharmacokinetic-pharmacodynamic efficacy
PopPK	Population based pharmacokinetics
SAR	Structural activity relationship
t1/2	Half life
TDI	Time dependent inhibition
Vss	Volume of distribution at steady state

17.1 Introduction

17.1.1 Pre-clinical ADME

Absorption, distribution, metabolism and excretion (ADME) studies completed primarily in drug discovery provide an estimate and guidance of what the human body will do to the candidate drug. Ensuring the most optimal compound enters drug development is crucial due to the significant cost and time required in the clinic and one of the key factors attributed to drug failure is drug exposure either in terms of safety and/or efficacy.

To predict compound exposure and thus overcome potential drug failure, high throughput ADME studies to understand drug concentrations and kinetics across tissues are pivotal in early discovery and design/make/test/analyse (DMTA) cycles provide an iterative strategy to optimise compounds toward candidates. These data are subsequently used to inform and guide the drug development path in the clinic (*see* Chap. 13).

17.1.2 Clinical Pharmacology, Pharmacokinetics, Pharmacodynamics and the Therapeutic Window

Clinical pharmacology is the study of drugs in humans; it is a multidisciplinary field and contributes to the understanding of therapeutic efficacy and safety. Historically, up to half of a drug's prescribing information is provided by Clinical Pharmacology. Clinical pharmacology comprises the following concepts:

- Pharmacokinetics (PK): describes the concentration of the drug over time (in other words, what the body does to the drug's elimination (ADME).
- Pharmacodynamics (PD): describes the drug's effect on the body at a given concentration (or, what the drug does to the body if used at a given concentration). The effect can be positive e.g., receptor occupancy, or negative e.g., drug side effects.
- Pharmacokinetic/Pharmacodynamic relationships (PK/PD): Understanding how PK and PD are related and defining the relationship between the effect observed and the drug concentration. It is most informative for clinical use as it describes the effect of the drug over time, assuming a given concentration. PK/PD together can be thought of as a drug exposure/response relationship.

Exposure-response information is the key data to define the safety and effectiveness of a drug. That is, a drug can be determined to be safe and effective only when the relationship between the beneficial and adverse effects of a defined exposure are known. The therapeutic window (Fig. 17.1) of a drug is determined by the minimal exposure necessary to reach the desired drug response and the exposure above which adverse effects are deemed unacceptable. These therapeutic limits are initially predicted from preclinical ADME data and animal PK studies before a compound enters the clinic. However, it is important to consider intrinsic and extrinsic factors and their impact on a drug's PK.

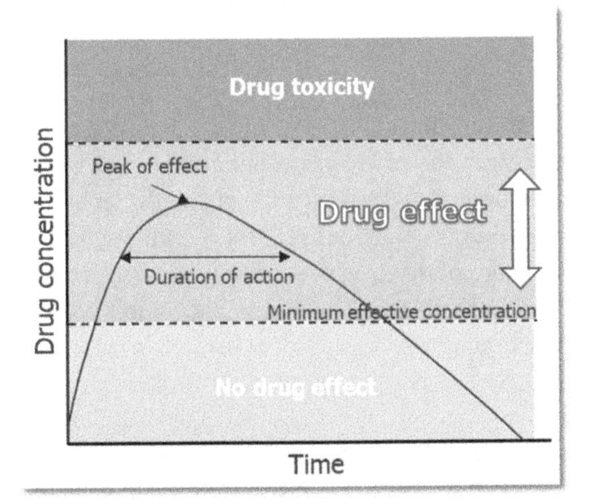

Fig. 17.1 Schematic of drug concentration over time with safety and efficacy target concentrations from which the therapeutic index is defined

Exposure-response data may provide reassurance that even patients with increased plasma concentrations do not have increased adverse effects. Alternatively, if exposure in specific patient subgroups, e.g., renally or hepatically impaired patients, is expected to exceed the therapeutic window then the dose may need to be adjusted. For victim drugs with a wide therapeutic window/index, the impact of extrinsic factors is not always meaningful and/or requires dose adjustments.

The clinical pharmacology package evaluates the impact of extrinsic (food effect, smoking, co-medications) and intrinsic factors (age, race, organ dysfunction, disease state, gender, genetics, pregnancy/lactation) on PK in humans. Information from all these studies is important to characterise the PK of a drug in various routine clinical settings and forms an important part of the drug dossier to regulatory authorities as well as part of the drug label. As drug development programs become more global, careful consideration must be given to variability in drug exposure and response resulting from intrinsic/extrinsic factors, for example, variations in the use of herbal supplements. This is true whether an industry sponsor is planning multiregional clinical trials or whether they wish to submit for marketing approval in multiple countries or in a single country where there is significant ethnic heterogeneity (e.g., USA). Figure 17.2 shows a selection of extrinsic and intrinsic factors that need to be

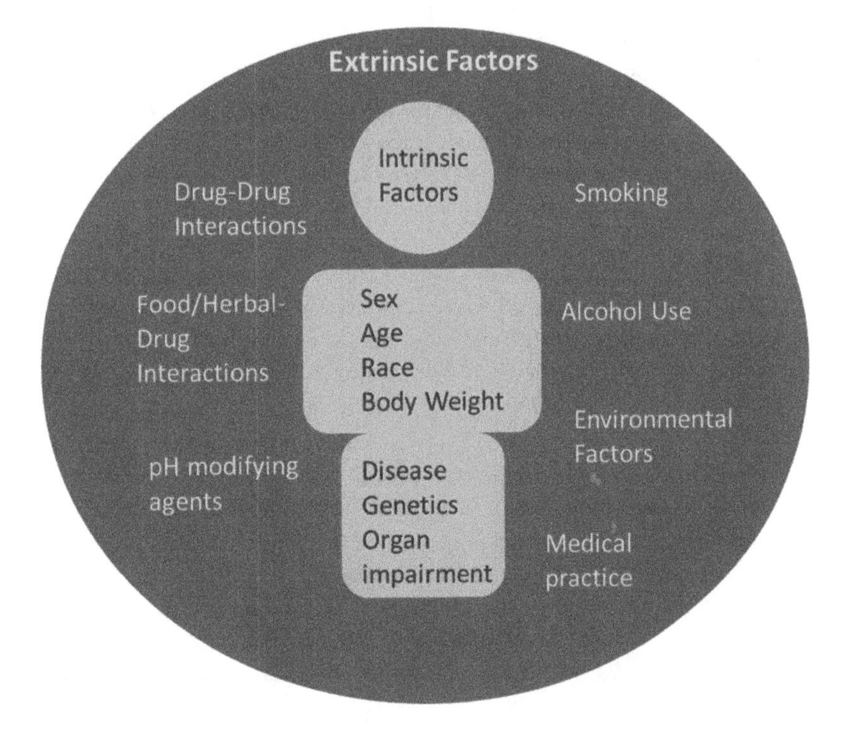

Fig. 17.2 Overview of intrinsic and extrinsic factors that may lead to clinically relevant differences in treatment response via altered pharmacokinetics and pharmacodynamics response of a drug

considered. In this chapter, we focus on the design and conduct of PK studies influenced by extrinsic factors.

17.2 What Pre-clinical ADME Data Could Be Helpful to Predict the Effect of Extrinsic Factors Before Human Clinical Trials?

17.2.1 Absorption

For an orally administered drug, the dissolution and subsequent fraction absorbed from the gut into the systemic circulation (Fa), the fraction escaping first pass metabolism in the gut (Fg) and the fraction escaping first pass metabolism (and transport) in the liver (Fh) define the oral bioavailability (F) [1]. However, these processes are themselves dependent on a multitude of parameters including PK and physicochemical properties which can result in diverse oral exposure. In the clinic, low F (<25%) can result in exposure variability insufficient to achieve drug concentrations at the target site for the necessary duration. Hence, understanding and optimising the PK and physicochemical properties is a key goal during drug discovery for orally administered drugs.

Utilising *in vitro* permeability assays e.g., Caco2 cells or Ussing Chamber, and transporter cell assays e.g., MDCK-MDR1 transfected cells, an assessment of intestinal intrinsic passive permeability [2] and intestinal efflux can be made [3]. These data, alongside measurements of physicochemical properties e.g., aqueous solubility or solubility in simulated gastric and intestinal fluid (and thus pH and food effect), LogD, hydrogen bond donor count and animal PK provides a holistic and multi-factorial estimation of human Fabs which can be further improved through the use of physiologically based pharmacokinetic (PBPK) models e.g., Gastroplus [4].

Compounds with pH-dependent solubility may demonstrate differential absorption and subsequent systemic concentrations when co-administered with gastric acid–reducing agents (ARA) [5–7]. An elevated gastric pH may also affect drug release from a drug product that may be sensitive to the surrounding pH, such as a pH-sensitive delayed-release formulation. Therefore, it is important to understand early in discovery the key physicochemical characteristics of a drug that may be subject to pH-dependent drug-drug interactions (DDIs), i.e., the chemical structure of the drug substance (e.g., weak acid or weak base), drug solubility and stability as a function of pH, drug product formulation features (e.g., whether the formulation contains enteric coating, acidic ingredients, or solubilizing agents to mitigate the pH-dependent DDI potential).

Food changes the gastric emptying rate and thus the time the compound is exposed in the stomach and gastrointestinal tract. In ADME studies, solubility and stability are often assessed in simulated gastric and intestinal fluids from which the PK sampling schedule in the clinic can be defined. An example of a drug regimen whose PK is dependent on a diet is the triple therapy Trikafta (Elexacaftor/tezacaftor/ivacaftor) used to treat cystic fibrosis. To achieve required systemic exposures, the dose is taken with a high fat containing meal such as peanut butter, eggs, nuts. The fat in the diet increases the extent and rate of absorption of Trikafta [8].

17.2.2 Distribution

Volume of distribution at steady state (Vss) represents a hypothetical volume into which a given dose of drug is apparently distributed from its sample site throughout the body in a reversible manner. The ion class, compound charge status, plasma protein binding, lipophilicity and polarity can be used to predict Vss. Previous work has provided guidance on general rules used to predict Vss [9] e.g., bases demonstrate Vss >3 L/kg *versus* acids which often demonstrate Vss <1 L/kg.

Accurate assessment of fraction unbound (fu) is critical due to the impact on DDI prediction, PKPD relationship understanding and therapeutic index. However, optimisation of fu is not advised due to the limited impact the parameter has on *in vivo* efficacy in isolation

[10]. Measurement of plasma protein binding, tissue binding and for CNS targeting drugs, brain binding, using equilibrium dialysis are important values to define and correctly understand the PKPD relationship. Evaluation of concentration dependent binding is also completed as a compound progress towards the clinic to allow appropriate modelling to be completed prior to the clinic.

Animal PK and fraction unbound across species are particularly beneficial when trying to predict human Vss as drug tissue affinities are often conserved across species hence free Vss (Vss,u) is generally consistent across species. Whilst one-compartmental modelling can be limited particularly for compounds with differential rates of distribution, the use of PBPK modelling e.g., SimCYP software, with tissue specific blood flow provides additional clarity on distribution rates into and out of tissues allowing the net effect across multiple compartments to modelled.

17.2.3 Metabolism and Elimination

Chemical modification of drugs by enzymes produces molecules that are often easier to eliminate by the body. The liver has the highest expression of these enzymes, but they are also present in the gut, lungs, kidney and skin. Drug metabolism is mediated in two phases: "functionalization" by cytochrome P450 enzymes (CYPs) and "conjugation" by uridine 5′-diphospho-glucuronosyltransferase (UGT). Phase 1 enzymes, the oxygenases, are responsible for the reactions such as oxidation, dealkylation, and hydrolysis. Phase 2 enzymes, the transferases, are responsible for the addition of different functional elements e.g., glucuronides.

Figure 17.3 provides an overview of the key enzymes and elimination routes responsible for the clearance of commonly prescribed drugs (in the US) from the body. Similarly, 89 of the 100 most prescribed drugs in five European countries and Australia are metabolized and/or known to be transported and only 11 drugs are not subject to either drug metabolism or drug transport [12].

As human *in vivo* clearance cannot be measured in drug discovery the use of human *in vitro* matrices e.g., hepatocytes, and animal PK provides guidance as to the route and rate of elimination (total clearance defined as: the sum of hepatic metabolic, renal, biliary and extra-hepatic metabolism) which can be extrapolated to human.

Project teams aim to synthesise compounds with low total clearance to ensure sufficient concentrations are achieved and maintained over time to engage the target. Hepatic metabolic clearance remains the predominant pathway for the elimination of drugs (Fig. 17.3) thus the use of hepatocytes and plated hepatic co-cultured systems offers an efficient tool to predict *in vivo* clearance. Algorithms such as *in vitro in vivo* extrapolation using the well-stirred model [13, 14] provide confidence that *in vitro* assays assessing (predominantly) CYP and UGT enzymes can account for the clearance observed *in vivo* or if additional pathways or extra-hepatic enzymes may be contributing to the drug clearance. Successful extrapolation of animal data provides confidence the same technique can be applied to humans.

17.2.4 Transporters

Drug transporters are proteins present in many tissues that can alter systemic drug levels by regulating the amount of drug that enters and/or exits the tissue. Many drugs rely on transporters for uptake into and efflux out of cellular compartments (Fig. 17.4). In the gut lumen, transporters contribute to the extent of drug absorption (either due to uptake transporters actively transporting the drug from the gut lumen into the blood or by efflux transporters actively removing the drug from the blood and back into the gut lumen for elimination) and in both the gut and liver they influence how much drug escapes first pass metabolism. Additionally, transporters can influence elimination from the kidney. Finally, transporters at the blood brain barrier provide a crucial barrier to ensure exogenous and endogenous molecules do not enter the brain (Fig. 17.4).

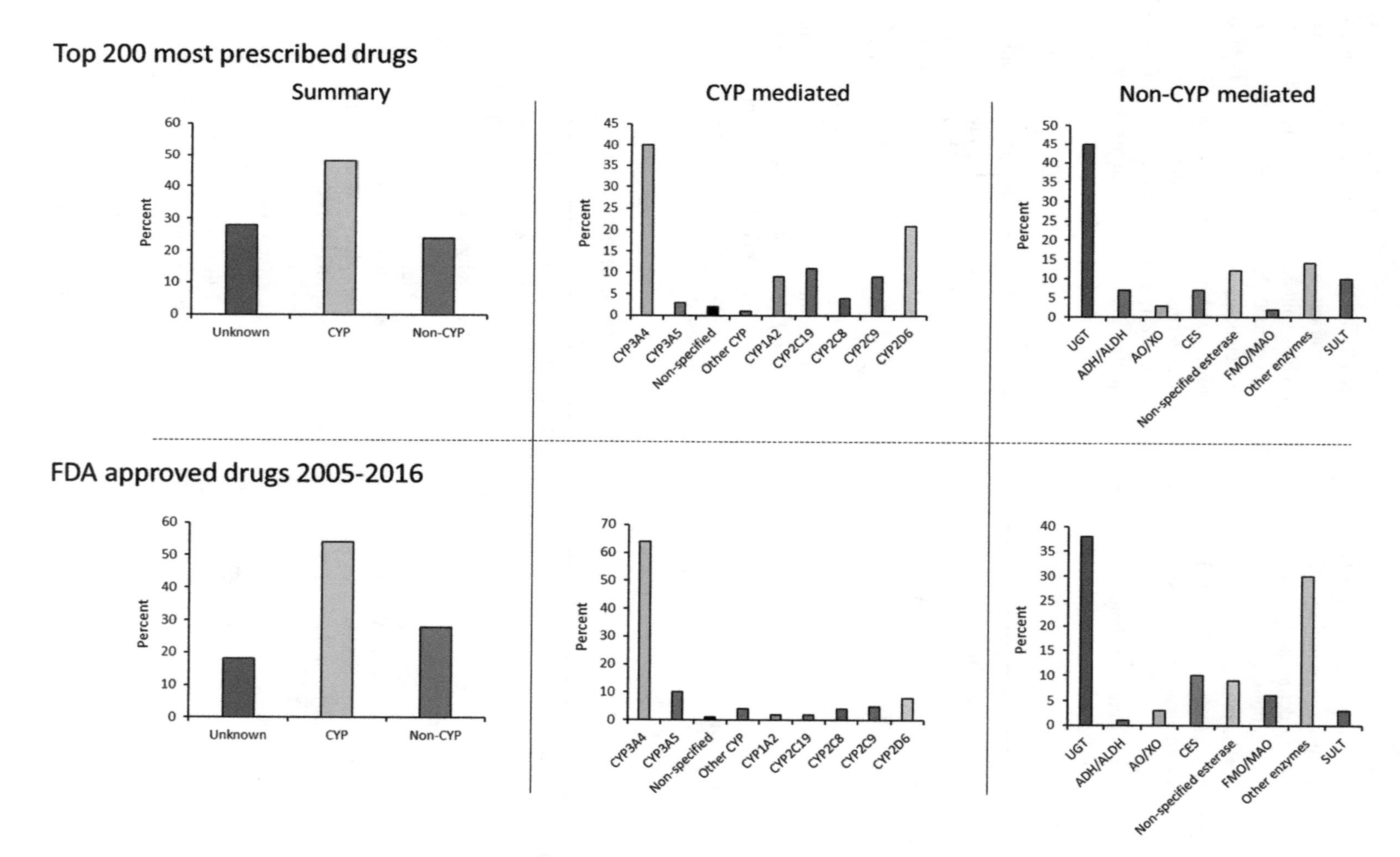

Fig. 17.3 An overview of the contribution of major enzymes to the elimination of drugs. (Adapted from Saravanakumar et al. [11])

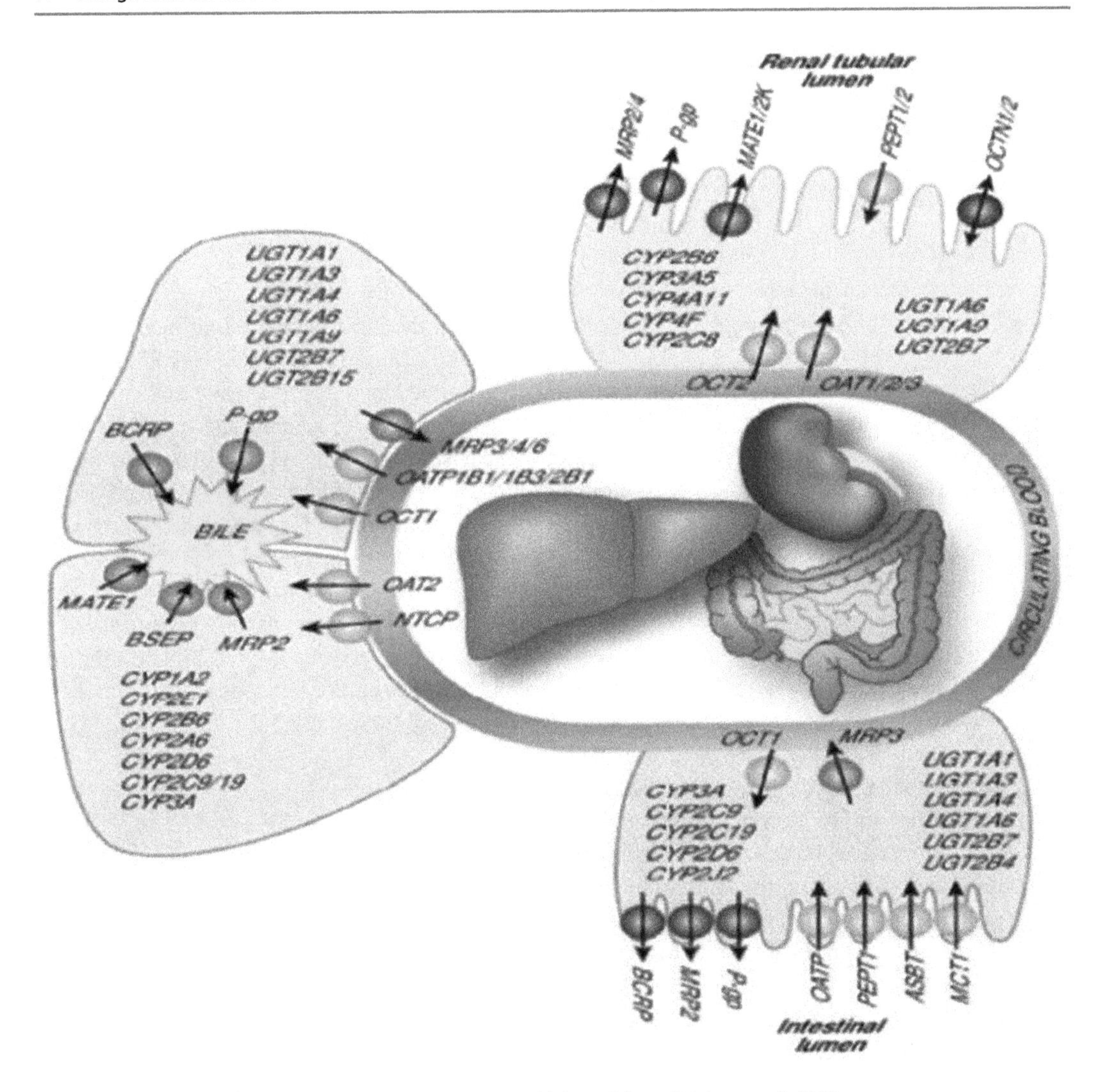

Fig. 17.4 Drug metabolizing enzymes and transporters (Adapted from Srinivas et al. [15])

If a compound is a target for the brain, MDCK cells transfected with the efflux transporters, P-glycoprotein or breast cancer resistance protein are routinely assessed to ensure the compound is not a substrate of the transporter and thus, brain exposure will not be limited.

17.2.5 Pre-clinical Evaluation of Enzyme and Transporter Drug-Drug Interaction Potential

DDIs resulting in altered drug exposure can occur when two drugs are co-administered, and one affects the PK of the other. One drug is the *victim* of the DDI, and the other is the *perpetrator* of the DDI. The *perpetrator* drug changes the blood/tissue level of the *victim* drug. There are two possible outcomes:

- V*ictim* exposure increases because the perpetrator is an inhibitor of the metabolism or efflux transporter.
- V*ictim* exposure decreases because the perpetrator is an inducer of metabolism or an inhibitor of an uptake transporter.

The most common DDIs are associated with changes in the activity of CYPs. Approximately 54% of approved drugs by the FDA (2005–2016) undergo CYP metabolism and of those approximately 64% are metabolised by CYP3A4 (Fig. 17.3) [11]. Risk assessment of a potential

DDI as early as possible can help to identify risks, risk mitigation strategies for the drug development process and design of clinical trials. Metabolism via a number of pathways may be desirable to overcome strong DDI risks.

In drug discovery, *in vitro* assays to evaluate CYP and transporter competitive inhibition, enzyme time dependent inhibition (TDI) and enzyme induction are routinely completed throughout the DMTA cycles.

Competitive inhibition assays often utilise human liver microsomes incubated with a specific enzyme isoform substrate to assess if the test compound (over a range of concentrations) decreases the formation of the known metabolite for the enzyme of interest. The generation of an IC_{50} (the test compound concentration which results in 50% inhibition) can be used early in the static models to assess risk (See Sect. 17.3.2.4). The potential for a test compound to be an inhibitor of the transporters is also regularly evaluated, often with the use of transfected Hek293 cells containing the transporter of interest, or with the use of transporter expressing vesicles. In contrast to metabolite formation in the enzyme inhibition assay, the transporter inhibition assay assesses the impact of the test compound on the movement of a known transporter substrate across the membrane and an IC_{50} is defined (see Drug Metabolizing Enzyme and Transporter DDIs for known transporter mediated DDIs).

TDI is also an important parameter to be evaluated. It is routinely completed using human liver microsomes from which the maximum inactivation rate (Kinact) and the concentration of the compound resulting in 50% inhibition (KI) are characterised. This comprehensive assay is often completed as the compound enters the clinic and can be used to decide if a clinical interaction study is required. However, in discovery, an IC_{50} shift assay is often utilised. The competitive inhibition enzyme assay is completed with and without a pre-incubation step to discriminate between the inhibition mechanism. An IC_{50} shift >1.5-fold is deemed significant and denotes the test compound as a TDI that warrants further evaluation.

The test compound is also characterised for its potential to induce CYP1A2, CYP2B6 or CYP3A4 in cultured human hepatocytes or HepaRG cells. The test compound is incubated over a range of concentrations (if known, over the predicted human efficacious dose range) in the system for 48 h (if evaluating mRNA levels) and/or 72 h (if evaluating catalytic activity). At the end of the incubation the fold induction relative to the control is calculated. A fold induction >2 is defined as a positive result and can be fed back into design to remove the liability or in a clinical study design. As recommended by the regulators (see DDI Regulatory Guidance) a positive *in vitro* result for CYP3A4 indicates the need for additional evaluation of CYP2C8, CYP2C9 and CYP2C19 as these isoforms are also induced via activation of the pregnane X receptor (PXR).

Clinical impact of inhibition, TDI or induction of the enzymes and/or transporters is discussed in further detail in Sect. 17.3.2.

17.2.6 Non-linear PK

Often, the desired observation is that compound exposure increases proportionally to the amount dosed, e.g., a three-fold increase in dose would result in a three-fold increase in drug exposure. However, some drugs display non-linear PK which means drug exposure is not proportional to the amount administered. Observations of non-linear PK in the animal can provide additional guidance on the possible clinical observations. There are many reasons for non-linear exposure, including (but not limited to):

- Solubility limited absorption, e.g., saquinavir [16]
- Saturation of transport processes, e.g., methotrexate [17]
- Saturation of protein binding, e.g., vismodegib [18]
- Increased clearance, e.g., induction or autoinduction, as observed with carbamazepine [19]

- Time dependent auto inhibition, e.g., Ritonavir
- Toxicity, e.g., nephrotoxicity induced by aminoglycosides, reduces organ function [20]
- Enterohepatic recycling, e.g., isotretinoin [21]

Understanding the Michaelis-Menten kinetics with measured Vmax and Km parameters guides the potential for non-linear PK in humans and aids in guiding dose and dose regimen. The use of PBPK and population PK modelling with first order and first order conditional estimation algorithms provides simulations of possible clinical outcomes. Additional details of the DDI related effects are described in more detail in the Sect. 17.3.2.

17.3 Clinical Impact and Evaluation of Extrinsic Factors Effect on PK

17.3.1 Food-Drug Pharmacokinetic Interactions

Food can alter the PK of drugs to the extent that may affect patient safety and/or drug effectiveness. Grapefruit juice is a well-known example of a food that can affect the PK of drugs such as cyclosporine and felodipine primarily by inhibition of CYP3A4 [22]. Food-drug interactions can be complicated if food causes functional changes in the gastrointestinal tract, e.g., alterations in gastric emptying kinetics, increased luminal bile salt concentrations or increased hepatic perfusion [23, 24]. For patient convenience and to improve compliance drugs should be given with or without food; of 40 small molecule oncology drugs approved between 1999 and 2017, food was identified as an extrinsic factor affecting PK in a small number of cases, 12.5% (5/40) [25].

17.3.1.1 Clinical Food Effect Study Planning and Design

A typical food effect study is a single dose, 2-treatment period, 2 sequence, cross-over design; whereby a single dose of the test drug is given fasted or fed. A washout period of at least five half-lives between the two treatment periods is required. The plasma concentration-time profiles after fasted and fed administration of the test drug are compared. Analogous to the evaluation of bioequivalence, the presence of a food effect is assessed primarily in terms of changes in the area under the plasma-concentration-time curve (AUC), the maximum concentration in the plasma (Cmax) and some cases, the time at which this concentration is observed (tmax).

Key Considerations for the Design of a Clinical Food Effect Study

Study Population and Number of Subjects

Food effect and DDI studies are preferably conducted in healthy subjects and the data is extrapolated to patients. Interindividual variability is reduced in healthy subjects who have normal kidney and liver function, are not using a concomitant medication, and are devoid of other interfering factors (e.g., smoking). For some drugs, the safety profile can mean the study must be conducted in patients, and these typically require a larger sample size due to more significant PK variability.

The number of subjects included in a food effect or DDI study should be sufficient to provide a reliable estimate of the magnitude and variability of the interaction. The equivalence approach is recommended when the purpose is to demonstrate PK equivalence e.g., between fed and fasted conditions. In practice, this is accomplished by calculating a 90% confidence interval (CI) for the geometric mean ratio (test/reference) of AUC and, in most cases Cmax. The CI should lie within predefined boundaries, which are set to ensure comparable clinical performance, i.e., similarity in terms of safety and efficacy. For bioequivalence, 0.8–1.25 are the default boundaries, which fulfill most regulatory requirements. Other clinically relevant limitations may be used if scientifically justified, i.e., if they can be supported by dose/exposure-response relationships. If the 90% CI of the geometric mean ratio is within the predefined boundaries, then PK equivalence has been determined. The sample size calculation is based on the likelihood (power) of being able to demonstrate equivalence, given the anticipated true AUC-(or Cmax) ratio (test/reference) and the anticipated true intra- or inter-subject variability, depending on the design. Details on the statistical analysis are given in the FDA Guidance 'Statistical Approaches to Establishing Bioequivalence'.

To determine the number of subjects when an interaction is expected, from food effect or in a DDI study, the Estimation Approach is recommended. This approach estimates an effect/difference with adequate precision. The 90% CI for the geometric mean ratio of AUC (test/reference) is calculated in addition to the point estimate of the geometric mean ratio. This sample size calculation is based on the anticipated intra- or inter-subject variability (for cross-over or parallel group study designs, respectively) and the desired level of precision. Precision is expressed as the ratio between the upper 90% confidence limit and the lower 90% confidence limit and is independent of the true and observed ratio.

Choice of Meal

The current FDA and EMA guidelines specify the meal should be high-caloric (800–1000 kcal) and high-fat (500–600 kcal of total calories derived from fat) for the investigation of food effect on oral drug bioavailability [26, 27]. This meal is intended to cause a maximum physiological response and represent a worst-case scenario. Both FDA and EMA guidelines contain an example of the composition of such a test meal.

Administration

Fasted treatment: Subjects fast overnight for ≥10 h and are given the drug with 240 mL of water. No food should be allowed for at least 4 h post-dose. Fed treatment: Following an overnight fast of at least 10 h, the meal should be taken within 30 min, and the drug is given 30 min after the beginning of meal consumption with 240 mL water.

PK Sample Collection

For both fasted and fed treatment periods, timed samples in the biological fluid, usually plasma, should be collected from the subjects to permit characterization of the complete plasma concentration-time profile for the drug and, if relevant, metabolites.

Data Analysis and Labelling

The PK parameters which are calculated for the fed and fasted state may include: AUC0-inf, AUC0-t, Cmax, tmax and half life ($t_{1/2}$). The 90% CI for the ratio of population geometric means between fed and fasted products should be calculated for AUC0-inf, AUC0-t, and Cmax and compared to the pre-defined boundaries, e.g., 0.8–1.25, to determine the effect of food on PK. The effect of food on the absorption of a drug should be described in the label as well as instructions for taking the drug in relation to food based on clinical relevance.

17.3.2 Drug-Drug Pharmacokinetic Interactions

It is critical that DDIs are investigated to ensure patient safety. In the past, extreme safety concerns caused by DDIs have led to multiple market withdrawals, such as those of mibefradil, terfenadine, cisapride, and cerivastatin in the late 1990s and early 2000s [28–31]. Unintentional and mismanaged DDIs have been reported as a common reason for preventable adverse events and 20–40% of adverse drug reactions [23, 32]. During hospitalisation drug combinations or concomitant therapy, are frequently required to treat certain diseases (for example, cardiovascular disease, cancer and infections) or to treat patients with ≥2 different conditions. Of 40 small molecule oncology drugs approved between 1999 and 2017, CYP inhibition or induction was identified as an extrinsic factor affecting PK in 62.5% (25/40) of cases [25]. Drug transporters and acid reducing agents were also identified as factors affecting PK for a limited number of drugs, 7.5% (3/40) and 10% (4/40), respectively [25].

17.3.2.1 Drug Metabolizing Enzyme and Transporter DDIs

Phase I and Phase II enzymes and transporters can be inhibited or induced, resulting in changes in the rate of drug clearance and systemic exposure. Competitive inhibition results in increased exposure to the victim drug and the DDI effect is alleviated upon elimination (or clearance) of the

Table 17.1 Examples of transporter mediated drug-drug interactions and their relationship to efficacy and safety attributes ([12, 39–41], Fig. 17.4)

Transporter	Example DDI	Clinical risk
P-gp	Substrates: digoxin, atorvastatin, omeprazole, losartan, fexofenadine	Dependent upon combination
	Inducers: rifampicin (multiple dosing), verapamil	
	Inhibitors: verapamil	
BCRP	Substrate: rosuvastatin and inhibitor: febuxostat	Two-fold increased exposure to rosuvastatin, increased cholesterol lowering and muscle-toxicity
OATP1B1/3	Statins	Increased Rhabdomyolysis
	HCV agents	Reduced efficacy
OCT1/2/ MATEs	Metformin	Lactic acidosis, GI tolerability, CV/Metabolic AEs
	Fenoterol	Reduced efficacy
OAT1/3	Furosemide	Hypotension
	Rivaroxaban	Hypokalaemia
		Increased bleeding risk

Where, *HCV* hepatitis C virus, *OATP* organic anion transporting polypeptide, *OCT* organic cation transporter, *OAT* organic anion transporter, *SLC* solute carrier (uptake), *ABC* ATP-binding cassette (efflux)

inhibitor. Whereas, for TDI the increased exposure to the victim drug is sustained after the removal of the inhibitor. Therefore, DDIs resulting from TDI can be potentially more harmful because any toxicity resulting from TDI can be prolonged. Additionally, the DDI magnitude increases with multiple doses of the inhibitor (greater than would be expected from reversible inhibition alone). As TDI destroys enzymes, de-novo synthesis of the enzyme is necessary to restore activity.

Induction of a clearance enzyme by a perpetrator drug causes decreased exposure to the victim drug, typically resulting in loss of efficacy. Less commonly, increased toxicity via induction of metabolism may occur; the strong CYP3A4 inducer and PXR agonist rifampicin given with lorlatinib has been associated with liver toxicity [33]. The DDI effect is sustained upon removal of the inducer and enzyme degradation is necessary to restore activity to baseline (~14 days) [34, 35].

Complex DDI scenarios can arise; for example, the PARP inhibitor Lynparza [36] is a reversible inhibitor, TDI and inducer of CYP3A4, CYP3A4 is also the predominant enzyme responsible for the clearance of Lynparza, and overall the net effect is evidenced as time-dependent PK and weak CYP3A inhibition [37].

DDIs have been applied positively for therapeutic effect; the use of the HIV protease inhibitor ritonavir, a potent CYP3A inhibitor, has become a standard boosting agent for co-administered HIV protease inhibitors. Given the potent inhibition of CYP3A4 by ritonavir, subtherapeutic doses of ritonavir are used to increase plasma concentrations of other HIV drugs oxidized by CYP3A4, thereby extending their clinical efficacy [38] (Table 17.1).

17.3.2.2 Classic Examples of Clinical DDI

Intravenous midazolam is used globally for sedation during minor operations and intensive care treatment, and itraconazole is a widely used antifungal agent. The interaction between the CYP3A4 substrate midazolam and CYP3A4 inhibitor itraconazole was first reported by Olkkola et al. [42]. Itraconazole increased the area under the midazolam concentration-time curve 10.8-fold ($p < 0.001$) and mean peak concentrations 3.4-fold ($p < 0.001$) compared with placebo. Also, itraconazole increased the $t_{1/2}$ of midazolam from 2.8 ± 0.6 h to 7.9 ± 0.5 h. The higher concentrations of midazolam during treatment with antimycotics were associated with profound sedative effects. Subsequently, this DDI has been further characterised by several investigators [42–46]. As a result of this interaction, a number of drugs which are CYP3A4 substrates are contraindicated with itraconazole capsules. Increased plasma concentrations of these CYP3A4 substrates, caused by coadministration with itraconazole, may increase or prolong both therapeutic and adverse effects to such an extent that a potentially serious situation may occur. For example, increased plasma concentrations of some of these drugs can lead to QT interval prolongation and ventricular tachyarrhythmias, including occurrences of torsade de pointes, a potentially fatal arrhythmia (Itraconazole SPC).

DDIs between herbal supplements and drugs may also occur. For example, a review of published data and case reports showed that St John's Wort (SJW) had a clinically significant effect on several commonly prescribed drugs, including warfarin, oral contraceptives and selective serotonin re-uptake inhibitors [47]. The mechanism for these DDI is likely the induction of CYP enzymes and P-gp following multiple dosing of St John's Wort. Subsequently, information about the interactions was provided to health care professionals and patients in Sweden and the UK. The product information of the licensed medicines involved were amended to reflect these identified interactions and SJW preparations were voluntarily labelled with appropriate warnings.

17.3.2.3 DDI Regulatory Guidance

To ensure patient safety, regulatory guidance from health authorities (US FDA, EU EMA, and Japanese PMDA) are available, which provides detailed information on how to assess DDI liabilities of drugs, and this often involves static, semi-mechanistic or dynamic modelling approaches (see Sect. 17.3.2.4). Figure 17.5 shows the evolution of DDI interactions to date, with final guidance to be replaced with ICH M12 DDI guidance. Regulatory guidances are also available for food effects and smoking.

17.3.2.4 DDI Risk Assessment Approaches

Cell and vector based *in vitro* assays can be used to evaluate interactions with transporters and the enzymes responsible for the metabolism [48]. Knowledge of the expected concomitant drugs is required to assess DDI risk. A

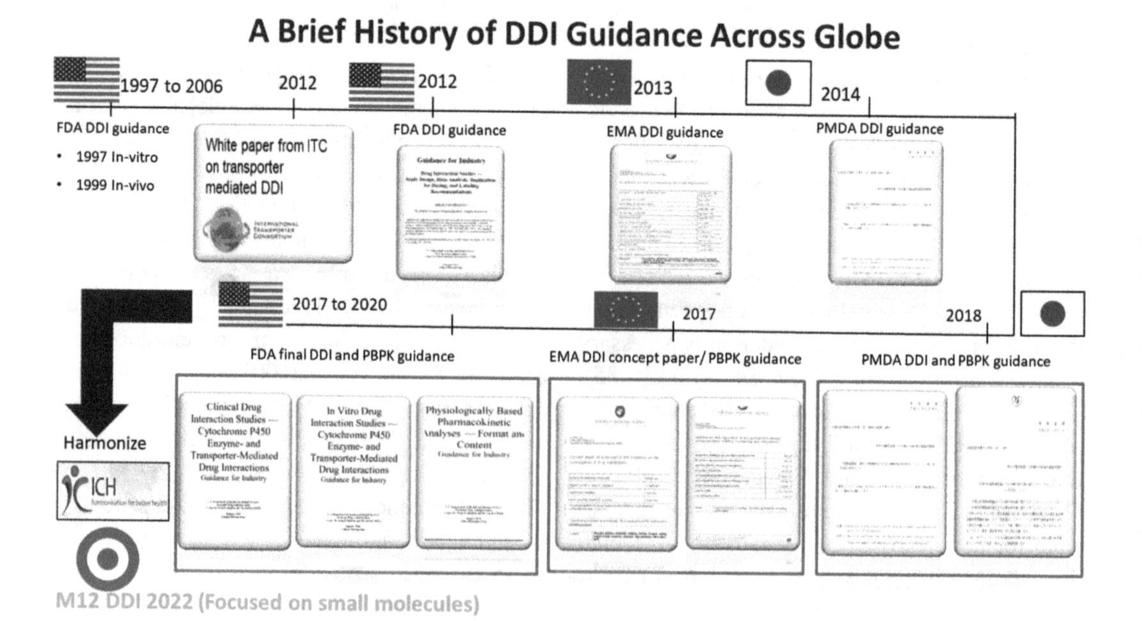

Fig. 17.5 Pharmacogenomics and consequences of polymorphism on DDIs

combination of *in vitro* data, including fraction metabolised (fmCYP) and predicted human PK in static and dynamic PBPK modelling provides an estimation of risk early in discovery, allowing project teams to assess alternative chemical designs to move away from the liability.

Subsequently, when a drug is in clinical development, a decision tree is used to interpret whether *in vitro* DDI finding(s) are likely to be clinically significant and if a clinical DDI study is warranted [26, 27]. In its most simplistic form, a static assessment relates an expected maximal or physiologically relevant plasma concentration to an inhibition parameter and a desired outcome (Eq. 17.1).

Equation 17.1 Static assessment of risk at unbound Cmax, e.g., CYP inhibition

$$K_{i,u} > 50 \times C_{max,u} \left(\text{i.e.,} \frac{C_{max,u}}{K_{i,u}} < 0.02 \right)$$

Where, Ki,u Unbound inhibition constant; Cmax u, unbound maximum concentration.

More complex mechanistic static equations integrate an assessment of DDI components to predict an outcome in a PK parameter, e.g., a change in AUC. The most complex, dynamic and PBPK models make a temporal assessment of risk at physiologically relevant concentrations of inhibitor/substrate interactions (see Sect. 17.4.5). Furthermore, with PBPK modelling food/drug interactions or the impact of genetics, e.g., CYP2D6 polymorphisms, can be estimated. With the increase in complexity, more comprehensive data and a greater understanding of a drug's PK are required. Table 17.2 provides an overview of DDI workflow that could potentially follow from lead optimization to life cycle management stages.

17.3.2.5 Clinical Drug-Drug Interaction Study Planning and Design

The primary objective of a DDI study is to determine the ratio of a measure of victim drug exposure in the presence and absence of a perpetrator drug (e.g., AUC ratio). The design of every study needs to be based on careful evaluation of the available data to ensure safety and maximise the usefulness of the study [49].

The timing of a DDI study is dependent on its purpose. A DDI study may be conducted before undertaking large clinical studies in patients (Phase 2 or Phase 3) to evaluate whether patients given certain drugs can be included in the trial. For drugs that are given as combinations, a DDI study can investigate if there is an interaction between the drugs; for example ceftazidime-avibactam is an antibiotic given with a non-β-lactam β-lactamase inhibitor to overcome antibiotic resistance [50]. Later in development, parallel to Phase 3, a DDI study may be performed to support the label dosing recommendations, for instance, to assess the impact of a CYP3A4 inducer and inhibitor on exposure to the CYP3A substrate oncology drug Tagrisso [51]. Lastly, a DDI study may be completed post-approval to enable expansion to new patient populations [52].

17.3.2.6 Types of DDI Clinical Investigation and Studies

Biomarkers for Metabolism and Transporter Related DDIs

Endogenous compounds can be used to assess the perpetrator's impact on a specific metabolic or transporter as a measure of the activity of the pathway. The main advantage of this approach is that it does not require the additional intervention of the index substrate(s). In early clinical drug development, these data can indicate potential DDI. The endogenous biomarker needs to be selective for the enzyme or transporter under investigation and unaltered by disease or diet. Well-validated endogenous biomarkers that could be used as an alternative to DDI studies are lacking. For example, 4-β-hydroxycholesterol, 6β-hydroxycortisol, and 6β-hydroxycortisone could be used as sensitive biomarkers of CYP3A activity, unconjugated bilirubin as a nonselective marker of UGT1A1, and coproporphyrin as a marker of OATP1B1/

Table 17.2 DMPK and clinical pharmacology framework and workflow for mitigating and designing DDIs in the drug discovery and development process

	Lead Optimisation	CDID	Phase I/Dose Escalation	Phase II III/NDA/paediatrics	LCM
Activity	◆ Determine the optimal in-vitro assays ◆ SAR ◆ Competitor analyses	◆ Project the DDI risk using Dynamic PBPK modelling ◆ Impact of patient variability across LoS	◆ Support clinical dose and schedule selection ◆ DDI risk analysis	◆ Optimal dosing for patients with individual extrinsic and intrinsic factors ◆ Drug label	◆ Prioritise dose / schedule options for other line of sight ◆ Combinations
Key experiments / analyses/principles	• Static Equations for DDI risk flag • PBPK M&S for competitors for aiding in drug design and SAR building in early research • DDI/ preclinical PBPK work for compound nomination	• Static Eq. for DDI risk flag • Characterisation of disposition pathways in-vitro fmCYP/fmT • Sensitivity analyses • Preclinical insights of exposure accumulation auto-inhibition/ induction • Impact of disease or PGX • Understand human metabolites vs preclinical • Contextualize the human DDI risk to preclinical ER relationship	• Refine PBPK model with clinical PK • Generate additional in-vitro, non-clinical data required to qualify the PBPK model to test: i.e CYP contribution (fmCYP or transporter) to drug clearance • Develop PBPK models for combination partner drugs • Use of PBPK for DDI study design, special population etc	• Further PBPK model verification with new data/ethnicity/PGX • Compare human DDI against that predicted by PBPK M&S if data available before MAA • Calibrate DDI simulations and optimize if PBPK model fails to recover DDI data • E-R understating and PBPK for extrapolate to other modulators and inform the drug label	• The appropriate management strategies for clinically significant DDIs • Bridging formulations
	In-Vitro DDI Guidance			Clinical DDI Guidance	

Table 17.3 Endogenous biomarkers for CYP and transporter mediated DDI

DDI mechanism	Biomarker	Recommendation
CYP 3A4 induction	4b-OH-Cholestrol [53]	It is comparable to midazolam clearance as a marker of CYP3A4 induction, and each may be used to evaluate CYP3A4 induction in clinical trials evaluating drug-drug interactions for new drugs
		Measure for all FTIH studies
		Narrow dynamic range. Mostly qualitative assessment and/or disconnects with *in vitro*
OATP1B	CP-I [54]	Measure for all (*in vitro*) OATP1B inhibitors, starting from SAD/MAD studies and in dedicated DDI study (if conducted); Design sample collection based on the objective of the study, i.e., qualitative vs. quantitative assessment of OATP1B activity
		Apply the PBPK approach to refine/confirm the *in vitro* OATP1B Ki based on the changes in plasma CP-I exposure
	Conjugated Bilirubin	A potential OATP1B biomarker, but less sensitive than CP-I
		It might be better to measure CP-1 than bilirubin. It could be considered an add-on to CP-1 to further understand its utility
OAT1/3	4-pyridoxic acid [55]	Current evidence suggests it is a reliable biomarker, but data is limited; Opportunity to understand further by including it in our clinical programs for *in vitro* OAT inhibitors
OCT/MATE	N-Methyl nicotinamide [56]	Current evidence suggests it is a reliable biomarker, but limited data
		Opportunity to understand further by including in our clinical programs for *in vitro* OCT/MATE inhibitors
	Creatinine	Not a reliable marker of renal transporter inhibition. However, typically collected in clinical studies

1B3, respectively, but they are only suitable in the initial characterization of DDI risks (Table 17.3).

17.3.2.7 DDI Studies with Index Perpetrators and Index Substrates

The golden standard for clinical DDI studies is a prospective crossover study, usually performed in healthy subjects. Perpetrators (inhibitors or inducers) and substrates (victims) with well-understood and predictable PK and DDI properties with regard to the level of inhibition, induction, or metabolic pathway are known as "index drugs". The results and findings from DDI studies with index perpetrators or substrates can be extrapolated to concomitant medications sharing the same DDI properties.

17.3.2.8 DDI Studies with Expected Concomitant Drugs

It can be informative to conduct studies investigating DDIs between the investigated drug and drugs likely to be administered to the target population. These studies can also be considered when a drug is used as an add-on to other therapies or as part of a fixed dose combination [50].

17.3.2.9 Cocktail Approach

A cocktail study includes the simultaneous administration of substrates of multiple enzymes and/or transporters to study subjects. This allows a drug to be studied as a perpetrator for several pathways in the same clinical study.

17.3.2.10 In Silico DDI Studies

PBPK models can be used in lieu of some prospective DDI studies. For example, PBPK models have predicted the impact of weak and moderate inhibitors on the substrates of some CYP isoforms (e.g., CYP2D6, CYP3A) and the impact of weak and moderate inducers on CYP3A substrates [37].

17.3.2.11 Key Considerations for the Design of a Clinical DDI Study

Study Population and Number of Subjects
Please refer to Section "Study Population and Number of Subjects".

Formulation
It is preferable to use the final commercial clinical formulation for the DDI study. However, if that isn't feasible, then the data from a DDI study using a development formulation may be extrapolated to the commercial formulation if the formulations are bioequivalent. If the formulations are not bioequivalent, then the data may be bridged using modelling approaches [37]. However, a robust model is needed that can describe the PK of both formulations.

17.3.2.12 Perpetrator and Victim Selection

The Investigational Drug as a Perpetrator of a DDI
If the investigational drug is potentially a DDI *perpetrator* then its effect is measured using an index substrate for a specific pathway, for example, midazolam for CYP3A4. The potential therapeutic dose level of the investigational drug needs to be considered to ensure a relevant dose is selected in the DDI study. The dose of a perpetrator drug may be selected to show the maximal inhibition or induction effect expected clinically.

An index victim drug needs to be selective (the fraction metabolised by the enzyme of interest >80%) and sensitive (>five-fold increase in AUC when given with an index inhibitor) to the pathway being investigated. Monitoring a specific metabolic-parent ratio can be helpful. To minimise the impact on study subjects, the substrate should have a short half-life (to minimise the duration of exposure, effect on enzymes and PK sampling) and lack of toxicity. An index substrate should also have linear PK. FDA and EMA regulatory guidance provides recommendations for clinical index substrates and inhibitors. There are several sensitive and selective index substrates for some CYP enzymes (e.g., CYP1A2, CYP2D6, and CYP3A), whereas sensitive index substrates for some CYPs (CYP2A6, CYP2B6, CYP2C9, and CYP2J2) are currently not available [49]. Moderately sensitive index substrates (whose AUC values increase 2- to <5-fold when co-administered with a known strong index inhibitor) can be used if a sensitive index substrate is not available for an enzyme (e.g., CYP2C9) (https://www.fda.gov/drugs/drug-interactions-labeling/drug-development-and-drug-interactions-table-substrates-inhibitors-and-inducers).

If exposure to the victim drug is expected to increase to the extent that it may be a safety concern, then the dose should be reduced to ensure that exposure is within the therapeutic window. On the other hand, when the effect of a strong inducer is investigated, a relatively high victim drug dose may be necessary to allow quantification of its concentrations.

17.3.2.13 The Investigational Drug as a Victim of DDI

To test whether an investigational drug is a victim of DDIs, an index perpetrator should be used, for example itraconazole as a strong CYP3A inhibitor. A *strong* index inhibitor or inducer should be used if one is available for the specific metabolic pathway of interest, otherwise a moderate can be used. Some perpetrators can also affect other metabolism and/or transporter pathways such as rifampicin (inducer and inhibitor of P-gp, OATP1B1) of the investigational drug and this needs to be considered when selecting an appropriate perpetrator. If a DDI study with a strong index perpetrator indicates that no DDI is present, no further DDI studies of this pathway are required. If a DDI study with strong index perpetrator indicates that there is a clinically significant interaction, then the impact of moderate index perpetrators needs to be understood to guide dosing recommendations. The effect can be evaluated in a clinical DDI study or through modelling and simulation approaches, such as PBPK modelling with verified perpetrator (inhibitor or inducer) and substrate models.

17.3.2.14 Dose Level and Single vs. Multiple Doses

When the investigational drug is the victim, probe or index inhibitors can be utilized to reveal the contributions of specific pathways to the drug's PK. In this situation, a single dose of the victim drug is usually adequate if the drug does not have time-dependent PK.

The mechanism of the perpetrator DDI also informs whether the interaction should be studied following a single dose or multiple dosing. If a perpetrator drug is predicted to be a reversible inhibitor of a CYP enzyme then a single dose study may be adequate. For TDI or induction multiple-dosing is required to ensure the maximal effect is measured.

17.3.2.15 Parallel Versus Crossover Studies

A cross-over design involves given the victim drug alone and then also in combination with a perpetrator to the *same* subject, this may either be randomised by the sequence they receive these or all subjects may follow the same sequence. A parallel group design involves giving one group of subjects the victim drug and one group the victim and perpetrator drugs combined.

The crossover design reduces the effect of interindividual variability because individuals act as their own controls. In a crossover study, the washout period between study phases should be long enough to allow the drugs, metabolites, and their effects to be completely eliminated before the next phase, even when their elimination is impaired by strong inhibitors. If a washout period is not feasible, for instance if a metabolite is slowly eliminated metabolite or if the study needs to be conducted in patients and it is unethical to allow exposure to be lower than is efficacious then a parallel group design may be used. A parallel group design requires a greater sample size.

17.3.2.16 Timing of Drug Administration

The perpetrator and substrate drugs can be administered at the same time for the majority of DDI. The victim drug needs to be dosed at a time to allow measurement of maximal DDI. For example dose staggering, rather than simultaneous administration, may allow maximum inhibitor concentrations at the site of inhibition (e.g., an interval of 1 h between the inhibitor and victim drug administration).

17.3.2.17 Co-Medications and Other Extrinsic Factors Affecting DDIs

Other extrinsic factors (e.g. herbal supplements) which may affect the expression or function of enzymes and transporters should be restricted for a sufficient time before subject enrolment.

17.3.2.18 PK, Pharmacodynamic and Genetic Sampling

Appropriate PK blood sample collection, storage and analytical methods should be developed to cover >80–90% of the AUC of the victim drug and metabolites. Perpetrator PK should also be measured.

Monitoring of PD effects may be required for safety and can be useful when evaluating the clinical significance of the DDI. PD endpoints are particularly useful in cases where the victim drug has active metabolites or where organ specific disposition of the drug is anticipated to be altered (e.g., entry to the central nervous system via blood–brain barrier).

17.3.3 Data Analysis and Labelling

The PK parameters which are key for interpreting the DDI are AUC0-inf and Cmax. The 90% CI for the ratio of population geometric means with and without perpetrator should be calculated. The results of a DDI study are interpreted based on the no-effect boundaries for the substrate drug. A no-effect boundary is the interval within which a change in a systemic exposure measure is considered not significant enough to warrant clinical action (e.g., dose or schedule adjustment, or additional therapeutic monitoring). The no effect boundary should preferably be based on the therapeutic window derived from exposure-response analyses alternatively a pre-defined boundary may be used, e.g. 0.8–1.25.

If an investigational drug is a CYP inhibitor or inducer, it can be classified as a strong, moderate, or weak based on its effect on an index CYP substrate. DDI clinical management and prevention strategies should result in drug concentrations of the victim drug that are within the no-effect boundaries and may include:

- Contraindicating or avoiding concomitant use
- Temporarily discontinuing one of the interacting drugs
- The dose of the substrate drug may be adjusted
- Staggering drug administration (e.g., administer the new drug at a different time than an acid-reducing agent, eg. Acalabrutinib and Ibrutinib)
- Implementing specific monitoring strategies (e.g., therapeutic drug monitoring as is recommended for digoxin, laboratory testing)

The label should include essential DDI information that is needed for the safe and effective use of the drug. For more specific recommendations on content for labelling sections please refer to Health Authority guidance.

17.4 Other Extrinsic Factors Which May Affect PK

17.4.1 Alcohol Consumption

Alcohol can be metabolized by aldehyde dehydrogenase (ALDH) to acetate. It can also be metabolized by CYP enzymes, mainly CYP2E1. It can also be either an inducer or inhibitor of CYP2E1. Chronic, heavy alcohol consumption [57] (5 drinks/occasion) induces the activity of CYP2E1, while short-term heavy consumption inhibits CYP2E1's activity by competing with other substrates [58]. Alcohol intake can alter the PK of medications, including their absorption and metabolism. For example, acetaminophen is metabolized primarily through glucuronidation or sulfation (90–96%) and also through CYP2E1 (4–10%). The FDA recommends that patients who drink more than three alcoholic drinks per day should consult their physician prior to any OTC pain reliever use [59], and patients should always be warned about the increased risk of liver injury when acetaminophen-containing products are taken with alcohol. Conversely, alcohol pharmacokinetics can also be altered by medications. Furthermore, alcohol intake can play a negative role in certain disease states such as diabetes mellitus [60].

17.4.2 Smoking

Cigarette smoking remains highly prevalent in most countries. It can affect drug therapy by both PK and PD mechanisms. Enzymes induced by tobacco smoking may also increase the risk of cancer by enhancing the metabolic activation of carcinogens. Polycyclic aromatic hydrocarbons (PAHs) are some of the major lung carcinogens found in tobacco smoke. PAHs are potent inducers of the hepatic CYP isoforms 1A1, 1A2, and, possibly, 2E1 [61]. After a person quits smoking, an important consideration is how quickly the induction of CYP1A2 dissipates. The primary PK interactions with smoking occur with drugs that are CYP1A2 substrates, such as caffeine, clozapine, fluvoxamine, olanzapine, tacrine, and theophylline [62]. Inhaled insulin's PK profile is significantly affected, peaking faster and reaching higher concentrations in smokers compared with non-smokers, achieving significantly faster onset and higher insulin levels [63]. The primary PD drug interactions with smoking are hormonal contraceptives and inhaled corticosteroids. The most clinically significant interaction occurs with combined hormonal contraceptives [64]. The use of hormonal contraceptives of any kind in women who are 35 years or older and smoke 15 or more cigarettes daily is considered contraindicated because of the increased risk of serious cardiovascular adverse effects. The efficacy of inhaled corticosteroids may be reduced in patients with asthma who smoke. On the other hand, the existing literature indicates that menstrual phase and/or sex hormones (e.g., progesterone and/or estradiol) influence smoking-related symptoms and, possibly cessation outcomes, in women [65]. The use of oral contraceptives is related to increased nicotine metabolism and physiological stress response [66].

17.4.3 Environmental Factors

Environmental factors may play a significant role in the rates of drug metabolism between different individuals. Intentional or unintentional exposure to environmental chemicals can induce or inhibit the activity of hepatic enzymes e.g. CYPS, that metabolise drugs and other foreign chemicals, as well as endogenous substrates such as steroid hormones [67]. A major source of such exposure may be occupational. Exposure to the heavy metal, lead, has been shown to inhibit drug metabolism enzymes, whereas intensive exposure to chlorinated insecticides, and other halogenated hydrocarbons such as poly chlorinated biphenyls, has been shown to enhance the metabolism of test drugs such as antipyrine and phenylbutazone. Altitude has also been negatively correlated with the activity of a herbal antioxidant [68]. Temperature, humidity and clothing types have also been investigated to assess the dermal absorption of chemical vapours. Whilst full body clothing had minimal impact on the dermal absorption of 2-butoxyethanol vapour, increased temperature and humidity was significantly correlated with increased dermal absorption of the vapour [69].

17.4.4 Plasma Protein Binding and Importance in DDIs

Compounds demonstrating a narrow therapeutic index, high plasma protein binding (PPB) (>99% bound) and a low hepatic clearance (or IV drugs with a high hepatic clearance) have the potential to be subject to interactions resulting from PPB displacement [70, 71]. Examples of DDIs mediated by PPB displacement include the interactions between the highly bound and enzyme inhibitor aspirin with the highly bound substrate valproate. Valproate is displaced from its plasma proteins in addition to an inhibition of the valproate oxidation by aspirin resulting in a clinically significant increase of free valproate concentrations. Similarly, omeprazole inhibits CYP2C19 as well as displacing phenytoin from its plasma proteins [72]. However, clinically relevant interactions mediated through PPB displacement are rare. Patient disease state can result in variable PPB, for example, an increase in plasma water for renally impaired individuals or decreased protein synthesis in hepatically impaired individuals can decrease the level of PPB which must be considered in efficacious dose prediction and clinical study design [73].

17.4.5 Physiologically Based Pharmacokinetic modelling

PBPK models attempt to mathematically portray the mammalian body as a series of compartments that represent tissues and organs, connected by the arterial and venous pathways, which are arranged to reflect anatomical layout. These models use actual physiological data such as tissue volumes and blood flow rates, along with compound specific data, to describe the kinetics of the predominant processes that govern the ADME of the compound being investigated. Since PBPK models are mechanistically structured, meaningful predictions can be made, such as extrapolations from high dose to low dose, adult PK to pediatrics, route of exposure to route of exposure and species to species. Chapter 16 by Sharma et al. covers PBPK aspects in more detail.

17.4.6 Population-Based PK modelling

Population PK (popPK) method has been increasingly used to detect and estimate the effect of various comedications. So called an indicator approach popPK DDI. Which relies on identifying significant differences in estimates of PK parameters between subjects using concomitant medication and those who do not. A dichotomous indicator with values of 1 or 0 corresponding to the presence or absence of the concomitant medication is used in the modelling. This 'indicator approach' is used to calculate the 90% CI (by method such as asymptotic SE or bootstrap) of the ratios of the parameters between the presence and absence of interacting drug [74].

17.5 Concluding Remarks

This chapter summarised the key parameters to be considered during the design and conduct of PK studies influenced by extrinsic factors that may influence the PK and PD relationship and therapeutic index of small molecule drug candidates in drug development. The free concentration of drug reaching the target tissue is influenced by many factors; for example, DDI, one drug may interfere with the exposure of other medicines, for example by inhibiting the proteins responsible for their metabolism. As such understanding DDI as an extrinsic factor is pivotal for ensuring accurate dosing when two or more drugs are co-administered. Currently, documents published by regional regulatory agencies are used to inform studies for evaluating DDI potential during drug development. Harmonising study design requirements, design considerations and data interpretation will help reduce uncertainty for the industry and bring drugs to patients faster.

Conflict of Interest V.P.R., M.L., and J.C., are full-time employees and shareholders of AstraZeneca. B.W. is a full-time employee and shareholder of UCB. All other authors declared no competing interests for this work.

Appendix: Example of DDI Study Design (Midazolam with Itraconazole)

This is an open-label, fixed sequence cross-over study conducted at a single study center to assess the PK of midazolam in healthy volunteers when administered alone and in combination with multiple doses of Itraconazole. This study will consist of three treatment periods.

Midazolam (1 mg) will be administered as a single dose on two occasions ~5 days apart: prior to dosing with Itraconazole and after multiple administration of Itraconazole.

Study Plan (example)	Visit 1	In house		In house		Visit 4
Period	Screening	Period 1		Period 2	Period 3	Follow up
Study day		Day -1	Days 1 to 2	Day 2 to X-1	Days X to Y	3–7 days after last PK sample
Inclusion/exclusion criteria	X	X				
Demographic data	X					
Weight and height	X					
Medical history	X					
Concomitant medication	X	X	X	X	X	X
Urinary drug/alcohol screen	X	X				
Serology	X					
Informed consent	X					
Study residency						
Check-in		X				
Check-out					X	
Non-residential visit	X					X
Midazolam oral dosing			Day 1 (0 h)		Day X (0 h)	
Test Drug administration:				Day 2-X-1	Day X (0 h) until Day Y	
Safety/tolerability						
Adverse event questioning	Only					
SAEs	Only SAEs	X	X	X	X	
Blood pressure and pulse rate (supine):	X	X	X[a]		X[a]	
12-lead ECG	X					X
Pulse oximetry			X (0–4 h)		X (0–4 h)	
Clinical laboratory evaluations	X		X	X-1	Y	X
Physical examination	X	X (brief	X (brief)	X (brief)	X (brief)	X
Pharmacokinetics						
Plasma for midazolam[b]			X		X	
Plasma for Itraconazole				Day 2-X-1 (trough)	Day X (trough and at Tmax)	
Blood sample for genotyping		X				
4 β-OH Cholesterol		X			X (pre dose)	
CYP3A4 biomarker (Spot urine)		X			X (pre dose)	

a Vitals on midazolam dosing days: pre-dose, 0.5, 1, 2,3, 4, 8 and 24 h post-dose

[b]Midazolam sampling: pre-dose, 0.25,0.5, 0.75,1, 1.5,2,3,4,6,8,12,16, and 24 h post = dose

References

1. Rowland M, Tozer T (1989) Clinical pharmacokinetics concepts and applications, 2nd edn. Lea and Febiger, Philadelphia
2. Peters SA, Jones CR, Ungell AL, Hatley OJ (2016) Predicting drug extraction in the human gut wall: assessing contributions from drug metabolizing enzymes and transporter proteins using preclinical models. Clin Pharmacokinet 55(6):673–696
3. Shugarts S, Benet LZ (2009) The role of transporters in the pharmacokinetics of orally administered drugs. Pharm Res 26(9):2039–2054
4. Williamson B, Colclough N, Fretland AJ, Jones BC, Jones RDO, McGinnity DF (2020) Further considerations towards an effective and efficient oncology drug discovery DMPK strategy. Curr Drug Metab 21(2):145–162
5. Budha NR, Frymoyer A, Smelick GS, Jin JY, Yago MR, Dresser MJ et al (2012) Drug absorption interactions between oral targeted anticancer agents and PPIs: is pH-dependent solubility the Achilles heel of targeted therapy? Clin Pharmacol Ther 92(2): 203–213
6. Dolton MJ, Chiang PC, Ma F, Jin JY, Chen Y (2020) A physiologically based pharmacokinetic model of vismodegib: deconvoluting the impact of saturable plasma protein binding, pH-dependent solubility and nonsink permeation. AAPS J 22(5):117
7. Sieger P, Cui Y, Scheuerer S (2017) pH-dependent solubility and permeability profiles: a useful tool for prediction of oral bioavailability. Eur J Pharm Sci 105: 82–90
8. Kuek SL, Ranganathan SC, Harrison J, Robinson PJ, Shanthikumar S (2021) Optimism with caution: elexacaftor-tezacaftor-ivacaftor in patients with advanced pulmonary disease. Am J Respir Crit Care Med 204(3):371–372
9. Smith DA, Beaumont K, Maurer TS, Di L (2015) Volume of distribution in drug design. J Med Chem 58(15):5691–5698
10. Riccardi K, Cawley S, Yates PD, Chang C, Funk C, Niosi M et al (2015) Plasma protein binding of challenging compounds. J Pharm Sci 104(8):2627–2636
11. Saravanakumar A, Sadighi A, Ryu R, Akhlaghi F (2019) Physicochemical properties, biotransformation, and transport pathways of established and newly approved medications: a systematic review of the top 200 most prescribed drugs vs. the FDA-approved drugs between 2005 and 2016. Clin Pharmacokinet 58(10):1281–1294
12. Iversen DB, Andersen NE, Dalgard Dunvald AC, Pottegard A, Stage TB (2022) Drug metabolism and drug transport of the 100 most prescribed oral drugs. Basic Clin Pharmacol Toxicol 131(5):311–324
13. Yang J, Jamei M, Yeo KR, Rostami-Hodjegan A, Tucker GT (2007) Misuse of the well-stirred model of hepatic drug clearance. Drug Metab Dispos 35(3): 501–502
14. Gillette JR (1971) Factors affecting drug metabolism. Ann N Y Acad Sci 179:43–66
15. Srinivas MTG, Pratima S (2017) Cytochrome p450 enzymes, drug transporters and their role in pharmacokinetic drug-drug interactions of xenobiotics: a comprehensive review. Peertechz J Med Chem Res 3(1): 001–011. https://doi.org/10.17352/ojc.000006
16. Kigen G, Edwards G (2018) Enhancement of saquinavir absorption and accumulation through the formation of solid drug nanoparticles. BMC Pharmacol Toxicol 19(1):79
17. Inoue K, Yuasa H (2014) Molecular basis for pharmacokinetics and pharmacodynamics of methotrexate in rheumatoid arthritis therapy. Drug Metab Pharmacokinet 29(1):12–19
18. Smith SA, Waters NJ (2018) Pharmacokinetic and pharmacodynamic considerations for drugs binding to alpha-1-acid glycoprotein. Pharm Res 36(2):30
19. Kudriakova TB, Sirota LA, Rozova GI, Gorkov VA (1992) Autoinduction and steady-state pharmacokinetics of carbamazepine and its major metabolites. Br J Clin Pharmacol 33(6):611–615
20. Tulkens PM (1989) Nephrotoxicity of aminoglycoside antibiotics. Toxicol Lett 46(1–3):107–123
21. Colburn WA, Vane FM, Bugge CJ, Carter DE, Bressler R, Ehmann CW (1985) Pharmacokinetics of 14C-isotretinoin in healthy volunteers and volunteers with biliary T-tube drainage. Drug Metab Dispos 13(3):327–332
22. Dahan A, Altman H (2004) Food-drug interaction: grapefruit juice augments drug bioavailability--mechanism, extent and relevance. Eur J Clin Nutr 58(1):1–9
23. Fleisher D, Li C, Zhou Y, Pao LH, Karim A (1999) Drug, meal and formulation interactions influencing drug absorption after oral administration. Clinical implications. Clin Pharmacokinet 36(3):233–254
24. O'Shea JP, Holm R, O'Driscoll CM, Griffin BT (2019) Food for thought: formulating away the food effect - a PEARRL review. J Pharm Pharmacol 71(4):510–535
25. Reyner E, Lum B, Jing J, Kagedal M, Ware JA, Dickmann LJ (2020) Intrinsic and extrinsic pharmacokinetic variability of small molecule targeted cancer therapy. Clin Transl Sci 13(2):410–418
26. European Medicines Agency (EMA) GotIoDI, Committee for Human Medicinal Products (CHMP) (2012) https://www.ema.europa.eu/en/documents/scientific-guideline/guideline-investigation-drug-interactions-revision-1_en.pdf. Accessed 15 Oct 2022
27. Food and drug Administration (FDA) GfIF-EBaFBS (2002) https://www.fda.gov/files/drugs/published/Food-Effect-Bioavailability-and-Fed-Bioequivalence-Studies.pdf. Accessed 15 Oct 2022
28. Po AL, Zhang WY (1998) What lessons can be learnt from withdrawal of mibefradil from the market? Lancet 351(9119):1829–1830
29. Monahan BP, Ferguson CL, Killeavy ES, Lloyd BK, Troy J, Cantilena LR Jr (1990) Torsades de pointes occurring in association with terfenadine use. JAMA 264(21):2788–2790

30. van Haarst AD, van't Klooster GA, van Gerven JM, Schoemaker RC, van Oene JC, Burggraaf J et al (1998) The influence of cisapride and clarithromycin on QT intervals in healthy volunteers. Clin Pharmacol Ther 64(5):542–546
31. Furberg CD, Pitt B (2001) Withdrawal of cerivastatin from the world market. Curr Control Trials Cardiovasc Med 2(5):205–207
32. Dechanont S, Maphanta S, Butthum B, Kongkaew C (2014) Hospital admissions/visits associated with drug-drug interactions: a systematic review and meta-analysis. Pharmacoepidemiol Drug Saf 23(5):489–497
33. Hu W, Lettiere D, Tse S, Johnson TR, Biddle KE, Thibault S et al (2021) Liver toxicity observed with lorlatinib when combined with strong CYP3A inducers: evaluation of cynomolgus monkey as a non-clinical model for assessing the mechanism of combinational toxicity. Toxicol Sci 182(2):183–194
34. Pilla Reddy V, Walker M, Sharma P, Ballard P, Vishwanathan K (2018) Development, verification, and prediction of osimertinib drug-drug interactions using PBPK modeling approach to inform drug label. CPT Pharmacometrics Syst Pharmacol 7(5):321–330
35. Hariparsad N, Ramsden D, Taskar K, Badee J, Venkatakrishnan K, Reddy MB et al (2022) Current practices, gap analysis, and proposed workflows for PBPK modeling of cytochrome P450 induction: an industry perspective. Clin Pharmacol Ther 112(4): 770–781
36. McCormick A, Swaisland H, Reddy VP, Learoyd M, Scarfe G (2018) In vitro evaluation of the inhibition and induction potential of olaparib, a potent poly (ADP-ribose) polymerase inhibitor, on cytochrome P450. Xenobiotica 48(6):555–564
37. Pilla Reddy V, Bui K, Scarfe G, Zhou D, Learoyd M (2019) Physiologically based pharmacokinetic modeling for olaparib dosing recommendations: bridging formulations, drug interactions, and patient populations. Clin Pharmacol Ther 105(1):229–241
38. Rock BM, Hengel SM, Rock DA, Wienkers LC, Kunze KL (2014) Characterization of ritonavir-mediated inactivation of cytochrome P450 3A4. Mol Pharmacol 86(6):665–674
39. Hyafil F, Vergely C, Du Vignaud P, Grand-Perret T (1993) In vitro and in vivo reversal of multidrug resistance by GF120918, an acridonecarboxamide derivative. Cancer Res 53(19):4595–4602
40. Lund M, Petersen TS, Dalhoff KP (2017) Clinical implications of P-glycoprotein modulation in drug-drug interactions. Drugs 77(8):859–883
41. Lehtisalo M, Keskitalo JE, Tornio A, Lapatto-Reiniluoto O, Deng F, Jaatinen T et al (2020) Febuxostat, but not allopurinol, markedly raises the plasma concentrations of the breast cancer resistance protein substrate rosuvastatin. Clin Transl Sci 13(6): 1236–1243
42. Olkkola KT, Backman JT, Neuvonen PJ (1994) Midazolam should be avoided in patients receiving the systemic antimycotics ketoconazole or itraconazole. Clin Pharmacol Ther 55(5):481–485
43. Ahonen J, Olkkola KT, Neuvonen PJ (1995) Effect of itraconazole and terbinafine on the pharmacokinetics and pharmacodynamics of midazolam in healthy volunteers. Br J Clin Pharmacol 40(3):270–272
44. Backman JT, Kivisto KT, Olkkola KT, Neuvonen PJ (1998) The area under the plasma concentration-time curve for oral midazolam is 400-fold larger during treatment with itraconazole than with rifampicin. Eur J Clin Pharmacol 54(1):53–58
45. Templeton I, Peng CC, Thummel KE, Davis C, Kunze KL, Isoherranen N (2010) Accurate prediction of dose-dependent CYP3A4 inhibition by itraconazole and its metabolites from in vitro inhibition data. Clin Pharmacol Ther 88(4):499–505
46. Prueksaritanont T, Tatosian DA, Chu X, Railkar R, Evers R, Chavez-Eng C et al (2017) Validation of a microdose probe drug cocktail for clinical drug interaction assessments for drug transporters and CYP3A. Clin Pharmacol Ther 101(4):519–530
47. Henderson L, Yue QY, Bergquist C, Gerden B, Arlett P (2002) St John's wort (Hypericum perforatum): drug interactions and clinical outcomes. Br J Clin Pharmacol 54(4):349–356
48. Iwatsubo T (2020) Evaluation of drug-drug interactions in drug metabolism: Differences and harmonization in guidance/guidelines. Drug Metab Pharmacokinet 35(1):71–75
49. Tornio A, Filppula AM, Niemi M, Backman JT (2019) Clinical studies on drug-drug interactions involving metabolism and transport: methodology, pitfalls, and interpretation. Clin Pharmacol Ther 105(6): 1345–1361
50. Das S, Li J, Armstrong J, Learoyd M, Edeki T (2015) Randomized pharmacokinetic and drug-drug interaction studies of ceftazidime, avibactam, and metronidazole in healthy subjects. Pharmacol Res Perspect 3(5): e00172
51. Vishwanathan K, Dickinson PA, So K, Thomas K, Chen YM, De Castro CJ et al (2018) The effect of itraconazole and rifampicin on the pharmacokinetics of osimertinib. Br J Clin Pharmacol 84(6):1156–1169
52. FDA. Postmarketing studies and clinical trials—implementation of Section 505(o)(3) of the Federal Food, Drug, and Cosmetic Act Guidance for Industry
53. Jones BC, Rollison H, Johansson S, Kanebratt KP, Lambert C, Vishwanathan K et al (2017) Managing the risk of CYP3A induction in drug development: a strategic approach. Drug Metab Dispos 45(1):35–41
54. Shen H, Christopher L, Lai Y, Gong J, Kandoussi H, Garonzik S et al (2018) Further studies to support the use of coproporphyrin I and III as novel clinical biomarkers for evaluating the potential for organic anion transporting polypeptide 1B1 and OATP1B3 inhibition. Drug Metab Dispos 46(8):1075–1082
55. Shen H, Holenarsipur VK, Mariappan TT, Drexler DM, Cantone JL, Rajanna P et al (2019) Evidence for the validity of pyridoxic acid (PDA) as a plasma-

based endogenous probe for OAT1 and OAT3 function in healthy subjects. J Pharmacol Exp Ther 368(1): 136–145
56. Miyake T, Kimoto E, Luo L, Mathialagan S, Horlbogen LM, Ramanathan R et al (2021) Identification of appropriate endogenous biomarker for risk assessment of multidrug and toxin extrusion protein-mediated drug-drug interactions in healthy volunteers. Clin Pharmacol Ther 109(2):507–516
57. Onder G, Pedone C, Landi F, Cesari M, Della Vedova C, Bernabei R et al (2002) Adverse drug reactions as cause of hospital admissions: results from the Italian Group of Pharmacoepidemiology in the Elderly (GIFA). J Am Geriatr Soc 50(12): 1962–1968
58. Moore AA, Whiteman EJ, Ward KT (2007) Risks of combined alcohol/medication use in older adults. Am J Geriatr Pharmacother 5(1):64–74
59. FDA proposes alcohol warning for all OTC pain relievers. U.S. Department of Health and Human Services. https://www.fda.gov/drugs/bioterrorism-and-drugpreparedness/use-caution-pain-relievers. Accessed on Novemeber 14 2023
60. Fraser AG (1997) Pharmacokinetic interactions between alcohol and other drugs. Clin Pharmacokinet 33(2):79–90
61. Zevin S, Benowitz NL (1999) Drug interactions with tobacco smoking. An update Clin Pharmacokinet 36(6):425–438
62. Kroon LA (2007) Drug interactions with smoking. Am J Health Syst Pharm 64(18):1917–1921
63. Malhotra P, Akku R, Jayaprakash TP, Ogbue OD, Khan S (2020) A review of the impact of smoking on inhaled insulin: would you stop smoking if insulin can be inhaled? Cureus 12(7):e9364
64. Allen AM, Weinberger AH, Wetherill RR, Howe CL, McKee SA (2019) Oral contraceptives and cigarette smoking: a review of the literature and future directions. Nicotine Tob Res 21(5):592–601
65. Lynch WJ, Sofuoglu M (2010) Role of progesterone in nicotine addiction: evidence from initiation to relapse. Exp Clin Psychopharmacol 18(6):451–461
66. Carroll ME, Anker JJ (2010) Sex differences and ovarian hormones in animal models of drug dependence. Horm Behav 58(1):44–56
67. Banerjee BD, Kumar R, Thamineni KL, Shah H, Thakur GK, Sharma T (2019) Effect of environmental exposure and pharmacogenomics on drug metabolism. Curr Drug Metab 20(14):1103–1113
68. Liu W, Yin D, Li N, Hou X, Wang D, Li D et al (2016) Influence of environmental factors on the active substance production and antioxidant activity in Potentilla fruticosa L. and its quality assessment. Sci Rep 6: 28591
69. Jones K, Cocker J, Dodd LJ, Fraser I (2003) Factors affecting the extent of dermal absorption of solvent vapours: a human volunteer study. Ann Occup Hyg 47(2):145–150
70. Rolan PE (1994) Plasma protein binding displacement interactions--why are they still regarded as clinically important? Br J Clin Pharmacol 37(2):125–128
71. Di L (2021) An update on the importance of plasma protein binding in drug discovery and development. Expert Opin Drug Discov 16(12):1453–1465
72. Sandson NB, Marcucci C, Bourke DL, Smith-Lamacchia R (2006) An interaction between aspirin and valproate: the relevance of plasma protein displacement drug-drug interactions. Am J Psychiatry 163(11):1891–1896
73. Benet LZ, Hoener BA (2002) Changes in plasma protein binding have little clinical relevance. Clin Pharmacol Ther 71(3):115–121
74. Duan JZ (2007) Applications of population pharmacokinetics in current drug labelling. J Clin Pharm Ther 32(1):57–79

Specific Populations: Clinical Pharmacology Considerations

18

Rajanikanth Madabushi, Martina D. Sahre, and Elimika P. Fletcher

Abstract

The ultimate goal of new drug development is to not only show the safety and effectiveness of a drug, but also to ensure that the right dosage is determined for the entire spectrum of the target patient population. Late phase clinical trials are generally designed to confirm the safety and efficacy of the new molecular entity in patients with the condition to be treated, but often critically important subpopulations are left out for a variety of reasons, leading to a gap in providing adequate prescribing recommendations. These subpopulations include patients generally referred to as "Special or Specific Populations", for example, those with kidney disease, liver disease, pregnant individuals, lactating individuals, pediatric patients, older adults, etc. Intrinsic to the specific populations are alterations in underlying physiological processes that may affect the drug pharmacokinetics and pharmacodynamics and, ultimately, response to drugs. Understanding the various clinical pharmacology considerations is critical for deriving the appropriate dosage instructions for specific populations. To generate the necessary information to derive dosing adjustments, stand-alone specific population studies are generally considered. The US FDA has issued several guidance documents pertaining to various specific populations. These guidances provide detailed recommendations intended to assist with study design and translation of the information to develop dosing recommendations.

Keywords

Specific populations · Clinical pharmacology · Intrinsic factors · Organ dysfunction · Pregnancy and lactation · Pediatrics

18.1 Introduction

Early clinical development of new molecular entities (NMEs) generally involves stand-alone studies in healthy volunteers (often referred to as Phase 1 studies), intended to provide an understanding of tolerability of the drug in humans and to characterize the clinical pharmacology characteristics of the drug. Generally, these studies have relatively small sample sizes. The knowledge gained from these studies is used to inform the subsequent phases of the clinical development. These include clinical trials that aim to establish proof-of-concept, inform dose selection, and eventually generate evidence in support of

R. Madabushi (✉) · M. D. Sahre · E. P. Fletcher
Guidance and Policy Team, Office of Clinical Pharmacology, Office of Translational Science, Center for Drug Evaluation and Research, US Food and Drug Administration, Silver Spring, MD, USA
e-mail: Rajanikanth.Madabushi@fda.hhs.gov; Martina.Sahre@fda.hhs.gov; Elimika.Fletcher@fda.hhs.gov

G. Jagadeesh et al. (eds.), *The Quintessence of Basic and Clinical Research and Scientific Publishing*,
https://doi.org/10.1007/978-981-99-1284-1_18

marketing registration of the drug in patients with the disease that is intended to be treated. These later phases of clinical development often involve progressively increasing sample size (*see* chapters in Part IV).

Despite the large sample size of the late phase trials, they are not always fully representative of the overall patient population. Moreover, in an attempt to maximize the chances of demonstrating treatment effect, late phase clinical trial designs often aim to limit the anticipated heterogeneity in treatment effects. This is generally achieved by narrowing the study population, often by excluding patients, such as those with kidney disease, liver disease, pregnant individuals, lactating individuals, pediatric patients, older adults, obese patients, patients with genetic polymorphisms, etc. (Fig. 18.1). This strategy can lead to either limited patient experience or lack of clinical clinical experience in some of these subsets compared to those who may receive therapy in the real world. These patient subsets also exhibit alteration in underlying physiological processes that may affect their response to drugs necessitating specific dosing considerations. Under-representation or exclusions of these specific populations in the late phase trials deprives that much needed clinical experience and can lead to a gap in providing adequate prescribing recommendations.

Regulatory agencies have advocated against the exclusion of specific patient subsets from clinical trials without scientific basis and promoted the development of information to support the use in specific populations. For example, the Code of Federal Regulations Title 21, Section 201.57 provides the requirements for the content and format of labeling for human prescription drug and biological products including specific expectations for "Use in specific populations".

To bridge the aforementioned labeling gaps, standalone clinical pharmacology evaluations are often relied upon to determine specific

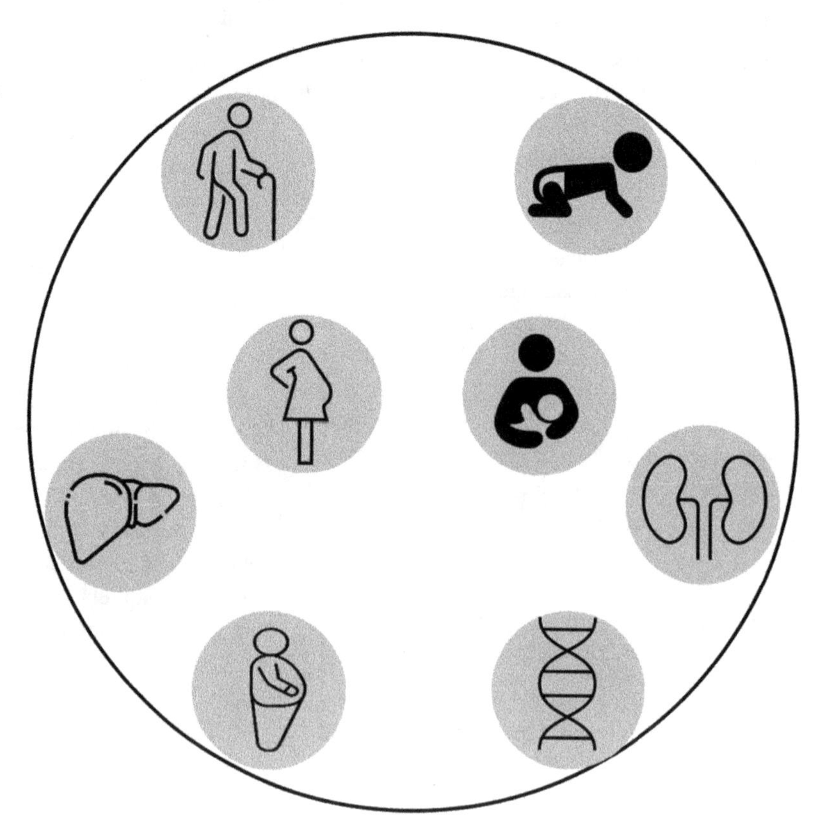

Fig. 18.1 Generally under-represented or excluded patient subsets in late phase clinical trials. These patient subsets are also referred to as specific or special populations

Table 18.1 Various US FDA regulatory guidance documents (Draft/Final) to inform the study design consideration for conducting clinical pharmacology evaluations in specific populations

Specific population/ intrinsic factor	Regulatory guidance
Organ dysfunction	PK in patients with impaired hepatic function: Study design, data analysis and impact on dosing and labeling (Final; 2003) [1]
	PK in patients with impaired renal function: Study design, data analysis and impact on dosing and labeling (Revised Draft; 2022) [2]
Pediatrics	ICH E11 (R1) Clinical investigation of medicinal products in the pediatric population (Final; 2018) [3]
	ICH E11A Pediatric extrapolation (Draft; 2022) [4]
	General clinical pharmacology considerations for pediatric studies of drugs, including biological products (Draft; 2022) [5]
	Pediatric Study Plans: Content of and Process for Submitting Initial Pediatric Study Plans and Amended Initial Pediatric Study Plans (Final; 2022) [6]
	General clinical pharmacology considerations for neonatal studies for drugs and biological products (Final; 2022) [7]
Pregnancy	PK in pregnancy—Study design, data analysis, and impact on dosing and labeling (Final; 2004) [8]
Lactation	Clinical lactation studies: Considerations for study design (Draft; 2019) [9]
Pharmacogenomics	Clinical pharmacogenomics: Pre-market evaluation in early-phase clinical studies and recommendations for labeling (Final; 2013) [10]
Older adults	ICH E7 Studies in support of special populations: Geriatrics (Final; 1994) [11]
	ICH E7 Questions & Answers (2012) [12]

considerations for drug dosing. Intrinsic to specific populations are alterations in the physiology often resulting in changes to drug pharmacokinetics (PK) and/or pharmacodynamics (PD). The standalone clinical pharmacology studies also known as "Specific/Special Population Studies" or "Intrinsic Factor Studies" are aimed to characterize the magnitude of the changes in the drug PK/PD due to the intrinsic factor. Dosage adjustment for specific population subsets can then be derived to account for the magnitude of change in the drug PK/PD based on matching exposure/response to controls or clinical experience in the late phase trials. This approach relies on a fundamental assumption that the drug's exposure-response relationships for efficacy and safety are similar between the studied population and the specific population subset. When such studies are conducted early during drug development, they can also inform the enrollment of specific population subsets with *a priori* dosage adjustment in the late phase trials, which would make it possible to obtain valuable clinical experience with the adjusted dosing in the specific patient subset. It should be noted that clinical pharmacology evaluations in a specific population comes with unique challenges and requires careful consideration of the study design elements. The US FDA has issued several guidance documents pertaining to various specific populations that provide detailed recommendations to assist with study design, conduct and translation of the information to develop dosing recommendations (see Table 18.1).

In the following sections of this chapter, we shall get into further details about the clinical pharmacology considerations and unique aspects for some of the specific populations.

18.2 Organ Dysfunction

Among the intrinsic factors, the dysfunction of organs involved in the elimination (metabolism and excretion) of drugs is a factor of specific interest. The two organs most involved in the elimination of drugs are the liver and the kidneys. There is less known about the impact on PK in patients with dysfunction of the lungs or whether skin diseases could impact the PK of topically applied drugs, therefore this section will focus on dysfunction of the liver and kidneys leading to alteration in the drug PK, herein referred to as hepatic and renal impairment. In the following subsections we will discuss the circumstances when standalone assessment is important, how to classify participants based on organ dysfunction and some key study design considerations.

18.2.1 Hepatic Impairment (HI)

For drugs that are metabolized and/or show biliary transport via the liver accounting for a significant portion of total clearance, hepatic impairment can lead to changes in PK. Often these changes result in increased systemic exposure to the drug and may increase the risk of treatment related adverse events. In these cases, an evaluation of the impact on PK is recommended. Characterization of the impact of hepatic impairment on PK should also be considered when the metabolic route of the drug is unknown or not sufficiently elucidated and when a drug has a narrow therapeutic range [1].

For the standalone HI study, the Child-Pugh score (CP score) is most frequently considered to assess the degree of hepatic impairment. The CP score uses the functional markers bilirubin, albumin, and prothrombin time, or more contemporaneously, international normalized ratio (INR), as well as two clinical markers, encephalopathy grade and presence and degree of ascites [13]. As shown in Table 18.2, the study participants are classified as having mild HI, moderate HI, or severe HI.

Given the complexity of processes determining the metabolism and transport of drugs in the liver, and potential differences in the degree of changes in histology, blood flow and available hepatocytes makes it difficult to utilize a single marker of liver function. A marker commonly used in the oncology setting is the National Cancer Institute Organ Dysfunction Working Group's score (NCI-score), where total bilirubin and AST are used to assess the degree of hepatic impairment [14].

18.2.2 Renal Impairment (RI)

An RI study should be considered when the fraction of systemically available parent drug or active metabolite is significantly eliminated via renal route [2]. Given that increasing molecular weight limits the filtration of proteins, it is unlikely that proteins with a molecular weight greater than albumin (69 kDa) are filtered to a significant degree, unless there is damage to the filtration barriers. Therefore, for peptides and proteins with a molecular weight below 69 kDa, and where more than 30% of the systemically active moiety are found excreted in the urine, the impact of RI on the PK of these drugs becomes important [2].

Drugs that are not significantly excreted through the kidneys, but undergo significant metabolism may still show an impact of RI on their PK. Therefore, characterization of the impact of RI on PK should also be considered for these molecules [2].

In case a drug is used in a population that is likely to undergo renal replacement therapy, either by intermittent dialysis (hemodialysis or peritoneal dialysis), or via continuous renal replacement therapy (CRRT), a standalone evaluation to assess the degree of drug removal via the dialysis or hemofiltration membrane should be considered.

Table 18.2 Classification of study participants in standalone HI studies

Hepatic impairment classification	CP score
Mild Impairment (CP-A)	5–7
Moderate Impairment (CP-B)	8–9
Severe Impairment (CP-C)	10–15

CP Child-Pugh

Table 18.3 Classification of study participants in standalone RI studies

RI classification	GFR or CLcr in mL/min
Mild renal impairment	60 to <90
Moderate renal impairment	30 to <60
Severe renal impairment	15 to <30
Kidney failure	<15 or dialysis patients on off-dialysis days

CLcr creatinine clearance, *GFR* glomerular filtration rate

The glomerular filtration rate (GFR) is generally considered a good marker of renal function. The GFR can be measured using exogenous substances such as inulin, iothalamate, iohexol, etc., or measured by approximation using a creatinine clearance in urine. It should be noted that creatinine clearance (CLcr) tends to overestimate GFR, as creatinine is both filtered and secreted in the tubules.

As measurement of GFR or CLcr is usually time-consuming, equations to estimate GFR or CLcr have been developed. One of the first equations was the Cockcroft-Gault equation [15], utilizing the serum concentration of creatinine as well as a patient's age, sex, and weight to estimate CLcr. In overweight individuals, the use of alternative body weight metrics such as ideal body weight or adjusted body weight when calculating the CLcr is likely to provide a more accurate estimate [16]. A common criticism of this equation is that it was derived using non-standardized creatinine measurements and that it represents a CLcr which usually overestimates GFR.

Common current equations for GFR estimations are the Modification of Diet in the Renal Disease Study Equation (MDRD) [17] and the Chronic Kidney Disease Epidemiology Collaboration (CKD-EPI) [18] equation. In clinical practice, eGFR values are standardized to a body surface area (BSA) value of 1.73 m^2 and expressed and reported in units of mL/min/1.73 m^2. For the purposes of characterization of the impact of renal function on drug pharmacokinetics, it is recommended to individualize GFR. To individualize GFR for drug dosing, multiply the standardized GFR by the individual's body surface area calculated using an appropriate formula [19, 20] and divide by 1.73.

When enrolling participants in the renal impairment study, the classification stages in Table 18.3 are intended to ensure enrollment across a wide range of renal function. For the renal impairment categories, it is important to ensure that the study participants have stable renal function, i.e., to avoid participants where renal function may be acutely changing and thus unstable.

18.2.3 Study Design Considerations for Organ Dysfunction Studies

An enrollment plan that includes participants with mild, moderate, and severe organ dysfunction groups in addition to the normal control group is often referred to as a "Full Study Design". The full study design allows to develop dosing recommendations across the entire spectrum of organ dysfunction. A full study design is generally considered for drugs that are predominantly cleared by the kidneys or the liver. To ensure that factors affecting the PK of the drug are not related to organ dysfunction, participants in the normal control group are typically matched to the participants in the organ dysfunction groups by age, sex, or other factors.

In some situations, a "Reduced Study Design" can be considered. This is an abbreviated study design involving the comparison of one category of impaired organ dysfunction with the normal control group.

- A reduced HI study design can be considered in situations such as in patients with metastatic cancer, hypoalbuminemia, encephalopathy, and ascites that may be related to cancer cachexia. The reduced study design consists of a comparison between the moderate HI group and the normal control. The findings of the study can then be applied to the mild HI group, while the dosing in the severe category would generally be contraindicated or may require further characterization in the severe category.
- In reduced RI study design is generally considered for drugs that are not significantly cleared by the kidneys. This study design consists of a comparison between the severe RI group and the normal renal function control group and is intended to evaluate a "worst-case" scenario [21]. If no effect is seen in the severe RI group, it is not expected for the mild and moderate groups, either. However, if an effect is observed, the moderate and mild groups may still need to be studied, subsequently.

Lastly, in some situations, additional organ dysfunction studies may be considered. For some drugs that are predominantly renally cleared and used in patients with kidney failure treated with dialysis, it may be important to study the effect of these dialytic therapies on the PK of the drug. The most common dialysis method is intermittent hemodialysis (IHD). An IHD study involves each participating study subject receiving a single dose of the drug on two different occasions, once with the dose administered prior to a dialysis session, with dialysis typically commencing just before the anticipated time to maximum concentration is reached (Tmax) following administration of the drug. On the second occasion, the drug should be administered in such a way that it reflects the exposures expected during an off-dialysis day (e.g., not around the end of a dialysis session). For critical care medications likely to be used in patients on CRRT, the findings from IHD studies might not be sufficient to derive dosing recommendations for patients using this modality and may necessitate a separate evaluation. Such an evaluation requires careful and unique considerations owing to practical challenges encountered in a critical care setting [22].

A summary of key study design features for standalone organ dysfunction studies is presented in Table 18.4.

18.2.4 Data Analysis Considerations for Organ Dysfunction Studies

The PK parameters of interest are generally the area under the concentration-time curve (AUC), peak concentration (Cmax), apparent clearance (CL/F) as well as renal and non-renal clearances (CLr, CLnr), where available for RI studies, as well as the apparent volume of distribution (V/F) and the half-life (t1/2). The pharmacokinetic parameters of active metabolites can include the AUC, Cmax, CLr, and t1/2. Non-compartmental and/or compartmental modeling approaches to parameter estimation can be employed. The PK parameters should be summarized by the organ dysfunction categories and compared to the corresponding normal function controls to estimate the impact of organ dysfunction on the PK. Alternatively, a regression approach can also be considered to characterize the relationship where both organ dysfunction and corresponding PK measure are treated as continuous variables. This approach allows for better identification of renal function thresholds that require dose adjustment.

In case that organ impairment alters the PK of a drug to such an extent that a different dose is needed for patients with hepatic or renal impairment, the most common approach to determine an appropriate dose is to match the exposure to the group with normal renal function. Alternatively, the group to match exposure to can be a renal impairment group with acceptable benefit risk findings.

Traditionally, various approaches for exposure-matching have been used, for example, establishing an exposure interval within which a change is not considered clinically significant, the so-called "no-effect boundary". Another approach

Table 18.4 Key study design features of organ dysfunction studies

	Hepatic impairment	Renal impairment
Goal	To evaluate the impact of hepatic function on drug PK/PD	To evaluate the impact of renal function on drug PK/PD
Design	Parallel group design	
Study arms	Control group Mild HI Moderate HI Severe HI	Control group Mild RI Moderate RI Severe RI Kidney failure on dialysis: on a dialysis day and on an off-dialysis day
Typical sample size	N = 6–8 per group or depending on the variability of the drug	
Matching	Based on factors that can influence PK, such as age, sex, race/ethnicity, etc.	
Determination of the level of organ function	Child-Pugh score In some oncology studies, NCI score is considered	eGFR by, e.g., CKD-EPI or MDRD equation CLcr by Cockcroft-Gault equation Measured GFR or CLcr
Number of doses	Single-dose, unless there is time-dependent or non-proportional PK, where a multiple dose study should be considered	
Types of designs		
Full design	Mild, moderate, and severe HI and normal hepatic function control group	Mild, moderate, severe RI, and normal control group
Reduced design	Moderate HI and control group	Severe RI and control group
Sample Collection and Analysis	Plasma or whole blood samples should be collected and analyzed for the parent and metabolites of interest; Plasma protein binding	Plasma or whole blood and urine samples should be collected and analyzed for the parent and metabolites of interest; Plasma protein binding
Duration of sample collection	The frequency and duration of sample collection should be sufficient to accurately estimate relevant PK parameters of parent and relevant metabolites	

could be to project a dose that achieves those exposures that fall between the fifth and 95th percentile of the reference group's exposures.

Matching the exposures assumes that the exposure-response relationships for safety and efficacy are the same, or similar, between normal function and patient subsets with hepatic or renal impairment [23]. This may be true for many drugs, but systematic assessments of this question are lacking. There are, however, examples where the drug response needs to be considered. In the case of the direct acting oral anticoagulants, patients with chronic kidney disease show different bleeding diathesis as well as response to the anticoagulant [24]. Therefore, when deriving dosing recommendations in patients with hepatic or renal impairment, the knowledge about the drug's exposure and renal function and the broader understanding of the exposure-response relationships of the drug should be taken into consideration particularly when the disease itself confounds the treatment effect in addition to the altered PK/PD relationship due to the organ dysfunction.

18.3 Pediatrics

Historically, medications have been used in pediatric patients off-label with limited studies and dosing information. However, US legislative efforts and authorities have led to some progress. Marketing exclusivity incentives of 6 months for pediatric assessments were introduced in the 1997 FDA Modernization Act and reauthorized in the 2002 Best Pharmaceuticals for Children Act (BPCA). In addition, the FDA was given authority by congress to require pediatric studies under the 2003 Pediatric Research Equity Act (PREA) for applications for a new active ingredient, indication, dosage form, dosing regimen, or route of administration. Both BPCA and PREA were made permanent in 2012 and a requirement for early submission of an initial Pediatric Study Plan (iPSP) was added. In addition, the FDA has published numerous guidance documents relevant to pediatric drug development to share the agency's current thinking with sponsors, investigators, IRBs and other relevant stakeholders [5, 6].

Safety, efficacy and dosing in pediatric patients cannot be assumed to be the same as in adults due to differences in size and maturation of physiological processes including those that affect the processing of a drug in the body. The PK, PD, safety and efficacy can differ among pediatrics (0 to <17 years of age) and between pediatrics and adults. The absorption, distribution, metabolism and excretion (ADME) processes mature at different rates across the pediatric age range and approach adult levels at different rates. Related to absorption, pediatric patients show differences in gastric pH, emptying, permeability, and transporter, and metabolism enzyme expression/activity. In addition, skin, muscle, and fat characteristics differ in children affecting the absorption. The drug distribution in pediatric patients differs resulting from aspects such as total body water, adipose tissue, transporters, and plasma protein binding. The elimination of drugs from the body differs in pediatric patients due to differences in aspects such as the rates of metabolism, pathways of metabolism and renal excretion [25]. Therefore, evaluation in pediatric patients can inform dosing, safety and effectiveness and account for differences in ADME and in some cases differences in the disease or PD response.

18.3.1 General Approach to Pediatric Development

It is important to note that human subject protections apply to pediatric studies. Before conducting studies in pediatrics, one needs to consider the benefit-risk. The goal is to maximize the "prospect of direct benefit" (PDB), minimize unnecessary risk and facilitate timely access to safe and effective medications for children [26]. A multidisciplinary approach including clinical pharmacology is needed to inform whether pediatric patients in a clinical trial would have a PDB. PDB is supported by evidence of proof of concept (biological plausibility, nonclinical data, clinical data from adults or other pediatrics) and sufficient or adequately justified dosing regimen and treatment duration.

Pediatric evaluations generally can include PK, PK/PD, safety and effectiveness studies. The studies can range from small dose ranging studies to large adequate and well controlled trials with a clinically meaningful endpoint. The extent of the pediatric program depends on the confidence in the ability to extrapolate effectiveness from adults or other pediatric populations with available data. Pediatric extrapolation is defined as "an approach to providing evidence in support of effective and safe use of drugs in the pediatric population when it can be assumed that the course of the disease and the expected response to a medicinal product would be sufficiently similar in the pediatric [target] and reference (adult or other pediatric) population." The extent of extrapolation will depend upon the existing data and the gaps that still exist in knowledge to support the effectiveness and safety of the use of a product in pediatrics [4].

18.3.2 Clinical Pharmacology Considerations in Pediatric Development

Clinical pharmacology studies in the pediatric population should be conducted in individuals with the disease which the drug is intended to treat, or in rare instances, in those who are at risk of this disease. The clinical pharmacology evaluation generally spans the entire pediatric age range in which the drug will be used. The following table provides a general guide of various pediatric population age groups to consider in clinical pharmacology studies to ensure development of information over the entire age range.

Clinical pharmacology has a central role in planning and informing the design and conduct of pediatric clinical trials. Clinical pharmacology can inform the types of pediatric evaluations/studies needed, dose selection, pediatric formulation, sample size, sample collection and analysis and data analysis. To meet the requirements of PREA, a sponsor is required to submit an initial Pediatric Study Plan (iPSP) early in drug development. The PSP should contain an overview of the disease condition in pediatrics, the drug and planned extrapolation and a plan for waiver or deferral, if applicable. The PSP also contains a summary of nonclinical and clinical studies, plans for an age-appropriate formulation, planned pediatric studies, a timeline of the pediatric development and a summary of agreements with other regulatory agencies. Clinical pharmacology informs many of these sections of an iPSP [6]. In addition, modeling and simulation play a central role in pediatric drug development. Modeling can be used to inform evidence of effectiveness, trail design and dose selection [27]. Some of the clinical pharmacology aspects that play a central role are further discussed below:

18.3.2.1 Dose Selection

Doses are generally labeled by age and/or body size to account for ontogeny. The doses can be flat dosing, body size-based dosing (mg/kg, mg/m^2) or body size tiered dosing (e.g. X mg/kg for 1–2 YO, Y mg/kg for 2–6 YO) or flat dose tiered dosing. Approaches such as allometric scaling and PBPK modeling are used to select starting doses to evaluate in studies generally with a goal of exposure matching with systemic exposures seen in adults [28]. Some important considerations in selecting a dose(s) to study in a pediatric study include: the relative bioavailability of the formulation, the acceptability of using an exposure matching approach, the age groups being studied, the known PK/PD and safety data, ontogeny impacting ADME for the specific drug and variability in PK [5]. More than one dose can be evaluated as part of the pediatric development program.

18.3.2.2 Formulation

An age-appropriate formulation should be addressed as part of an iPSP. For older pediatric patients (e.g., adolescents) the adult formulation may be adequate. However, in cases of formulations such as tablets, a formulation that is more appropriate for younger pediatric patients (e.g., <6 years of age) would be needed if the development program includes this age. The formulation should accommodate the dosing needed in the relevant pediatric age groups. In addition, the types and levels of excipients need to be considered especially for neonates. In general, to support a new formulation, the bioavailability compared to the available adult formulation is assessed in adults as these studies can generally not be conducted in pediatrics as they lack a PDB. This age-appropriate formulation can be used within the pediatric study. Otherwise, the study formulation will need to be adequately bridged to the planned marketed formulation [29].

18.3.2.3 Sample Size

The sample size within a pediatric study should consider having sufficient patients in various age groups. Some age groups are traditionally used (see Table 18.5), however, categorizations should take into account maturational changes of transporters and metabolizing enzymes. In addition, when neonates are studied, further grouping and classification may be necessary [7].

18.3.2.4 Sample Collection

Careful consideration of appropriate blood volume and frequency is needed based on the pediatric age groups included. This is especially challenging and relevant for neonates where

Table 18.5 A general age-based grouping to consider in pediatric studies

Pediatric group	Age
Neonates	Birth–1 month
Infants	>1 month to 2 years
Children	>2 years to 12 years
Adolescents	>12 years to <17 years

blood volume is limited. Therefore, the use of alternative approaches such as sparse PK sampling and micro sampling can be considered. Modeling simulations can be performed to optimize a sampling schedule [5].

18.3.2.5 Covariates

Several factors that have the potential to affect PK and dosing. The factors include age, body size, race/ethnicity, gender, concomitant medications, and pharmacogenomic markers known to affect the drug and laboratory measures of renal and hepatic function. The factors should be captured and the effects of these covariates on the pharmacokinetics (e.g., clearance) can be assessed as part of the PK analysis using population PK approaches [5].

18.3.2.6 Pharmacodynamics

Whenever possible, PD data should be collected along with PK. PD data can include the effect of the drug on biomarkers or clinical endpoints for both effectiveness and safety. The availability of robust PK-PD relationships can be useful to assess the similarity of exposure-response relationships between pediatrics and adults and determine the extent of extrapolation and inform dosing strategies in pediatric patients.

18.3.2.7 Data Analysis

In general, the development of relevant PK/PD models should occur throughout the pediatric development program. This includes leveraging prior knowledge from the adult program to design the pediatric studies, performing comparative analyses to characterize the PK/PD differences in pediatrics, and informing pediatric dosing strategies. Population PK approaches are most amenable to leveraging the data from adults and pediatrics. When relevant information is available, exposure-response analyses are also conducted. In some instances when it is possible to obtain intensive PK sampling, generally in standalone evaluations, noncompartmental analyses (NCA) can also be considered to summarize the PK data in pediatrics. The PK parameters from an NCA generally include AUC, Cmax, CL/F, V/F, and t1/2.

18.4 Pregnancy and Lactation

There are limited clinical data on pregnancy and lactation. A survey of 290 new molecular entities FDA approved between 2010 and 2019 reported that 90% of the labels had pregnancy data based on animal studies and only 11% from human studies. In addition, 49% had lactation data from animal studies and 3% from human studies. The Pregnancy and Lactation Labeling Rule (PLLR) was proposed in 2008 and finalized in 2014. The PLLR was proposed to update the Specific Populations section defined in the Physician's Labeling Rule (PLR). Although PLLR has improved the format and presentation of labeling, there is still a need for pregnancy and lactation data to inform labeling [30].

18.4.1 Pregnancy

Studies in drug development are generally performed in nonpregnant individuals. Studies in pregnancy are important to understand the safety, effectiveness, and dosing of a drug in mother and the safety of the fetus. In general, pregnant individuals are understudied and excluded from clinical trials, while 90% take at least one drug during pregnancy [30]. Physiological changes in pregnancy can result in changes in PK and/or PD of some drugs (Table 18.6). Therefore, doses used in non-pregnant individuals may not be appropriate for pregnant individuals [31]. For many drugs, these physiological changes can result in decreases in PK exposure in pregnancy, although increases have been observed for some drugs. This physiological adaptation in pregnancy makes sense as the body tries to excrete toxic substances to protect the fetus [32].

Data in pregnant individuals is generally collected post-marketing and most of the available data are safety data and not PK. The data are generally collected from people already prescribed the drug by healthcare providers. However, there are circumstances where enrollment of pregnant individuals premarket has been

Table 18.6 Summary of changes in the physiology during pregnancy that can impact the pharmacokinetics

PK parameters	Impact of pregnancy
Absorption	Changes in blood flow, GI transit time and emptying, gastric pH, transporters, and increased frequency of nausea and vomiting
Distribution	Increased cardiac output, plasma volume, total body water, body fat and decreased protein binding and levels
Metabolism	Increased activity of CYP3A, CYP2D6, CYP2C9 and UGTs and decreases in CYP1A2 and 2C19 activity
Elimination	Increased GFR and changes in transporters

considered. Adequate nonclinical data are needed to support studies in pregnant individuals. FDA has published guidance to address the scientific and ethical considerations for studies in pregnancy [33].

Data from PK in pregnancy studies can inform avoiding the drug, a specific dose adjustment, or limitations on the use of a drug. One example is cobicistat, a PK booster, that had exposures in pregnancy decreased by >80% for cobicistat and drugs that it boosted (darunavir and elvitegravir). This change resulted in the FDA updating labeling for cobicistat containing regimens to recommend that they not be initiated in pregnancy [34].

18.4.1.1 Study Design

Ideally, PK/PD studies in pregnancy would include a pre-pregnancy phase (for baseline comparison) and targeted evaluation during all three trimesters. However, given the challenges of enrolling participants before pregnancy, the evaluations are often limited to the second and third trimesters with baseline assessments obtained in the postpartum period. Such a design is referred to as Longitudinal Study Design [8]. This design is particularly amenable for drugs that are administered chronically or for several treatment cycles during pregnancy. A longitudinal study design minimizes interindividual variability across gestational ages.

The other approach to characterize PK in pregnancy is a population PK approach [35]. This approach involves collecting sparse PK and/or PD data in enough non-pregnant individuals and pregnant individuals with representation across second and third trimesters. Population PK approach with a nonlinear mixed effects modeling technique is then applied to characterize the PK differences between pregnant and non-pregnant women and across different trimesters. It should be noted that this approach is likely to detect large differences in PK.

18.4.1.2 Data Analysis

Modeling and simulation can play an important role in complementing studies for PK in pregnancy. Population PK can be used to analyze data when sparse sampling is available from pregnant individuals within a clinical trial. In addition, physiologically based pharmacokinetic (PBPK) modeling and quantitative systems pharmacology (QSP) are emerging areas that could be used to simulate PK and/or PD data in pregnancy. There is an increasing number of publications on the application of PBPK in pregnancy and it is used to predict both maternal and infant PK [36, 37] (*see* Chap. 16).

18.4.2 Lactation

Global and Public Health organizations encourage breastfeeding due to its benefits in infant health and bond development. It is reported that up to four prescription drugs are used by individuals during the lactation period, yet most drug labeling has no human data on use in breastfeeding [30]. Understanding drug exposure in milk is important to inform healthcare providers and patients about the use of medications while breastfeeding.

PLLR specified that the Lactation subsection of drug labeling includes a risk summary [38]. The risk summary should summarize information on the presence of a drug in human milk, the effects on a breastfed child, and the effects on milk production (when available). If the drug is present in human milk, the label must include concentrations in milk, the actual/estimated infant daily dose, and a comparison to the approved pediatric dose or maternal dose [38]. A clinical lactation study should be considered when a drug is anticipated to be used or is commonly used by women of reproductive age.

18.4.2.1 Study Design

In general, there are three types of lactation study designs based on whether the infant is enrolled and whether plasma samples are collected

(a) Lactating woman (milk-only) study: This type of study can be used to detect the presence of a drug in breast milk. When the drug levels are above the limit of quantification of the bioanalysis method, the amount of the drug in breast milk can also be estimated. If the levels are deemed clinically significant, this may lead to further evaluation. Lastly, a milk-only study can also be used to evaluate the impact of a new drug on milk production.
(b) Lactating woman (milk and plasma) study: This study design provides PK characterization of the drug in lactating women in addition to characterizing the amount of drug in breast milk and evaluating the impact of a new drug on milk production.
(c) Mother-infant pair study: These studies include an additional assessment of drug concentration in breastfed infants. In addition to characterizing the PK of a drug in lactating women and the amount in the breast milk, mother-infant pair studies also provide information pertaining to the absorption of the drug in breastfed infants. Such a study is considered when there is prior information indicating that clinically significant amounts of the drug are likely to be found in milk and the drug is likely to be absorbed.

The lactation study could be a single dose or multiple dose study. Multiple dose studies are generally conducted because they tend to enroll individuals already taking the drug as part of standard care. However, there may be situations where the drug is given as part of the study and breastfeeding is interrupted during the study.

18.4.2.2 Sampling

Milk sampling is a unique consideration for lactation studies. Specific aspects include milk type (e.g., foremilk versus hindmilk) and when the sampling is performed postpartum (e.g., colostrum versus mature milk). It is important to routinely collect specific timing of milk sample collection both relative to the timing of the dose as well as days postpartum. Ideally, entire milk from both breasts over 24 h should be collected. Sampling should occur when drug exposure is at a steady state during chronic maternal dosing. For drugs with dosing intervals of more than 24 h, consideration should be made to collect milk over the entire dosing interval or to collect 24-h samples during the expected time to peak plasma concentration. Care should be taken to balance the need for adequate data collection with feasibility.

Measurement of infant milk intake will be critical for estimating infant exposures. A milk intake of 150 mL/kg/day is considered reasonable. Measurement of milk volume and weighing the infant before and after feeding can also be useful in estimating the milk volume.

18.4.2.3 Data Analysis

In addition to computing conventional plasma PK parameters, clinical lactation studies should also include milk PK. The area under the milk concentration curve (AUC) should be calculated. The AUC should be based on data from multiple time points and the average concentration can be based on the AUC. It is also important to capture peak and trough milk concentrations as well as time to reach the maximum peak concentration in breast milk.

There are several equations for calculating the dosage the infant receives from breastfeeding including daily infant dose, estimated daily infant dose, and relative infant dose (RID). A commonly used paradigm is that drugs with a RID of <10% are safe, 10–25% should be used with caution, and >25% are likely unsafe. This was established in the 1980s and is still well accepted. Although it is a great starting point, a single RID number may not reflect the risk assessment across the relevant breastfeeding age range (considering ontogeny in ADME) or account for drugs having different therapeutic windows. A literature review on serious acute adverse drug reactions in infants from drugs in breastmilk reported that 79% of reported reactions occurred in the first 2 months of life. This suggests that careful consideration of ontogeny and development in the breastfeeding infant should be considered as part of the risk assessment [39, 40].

The results of a lactation study can be used to inform whether the drug can be safely used during breastfeeding, whether exposures might warrant considering the benefit to the mother versus the risk to the breastfeeding infant, or the possibility of mitigation strategies (e.g., avoiding breastfeeding for a period of time). Therefore, these studies can be used to guide patients and healthcare providers in making treatment decisions.

18.5 Concluding Remarks

Though late phase randomized clinical trials are considered as the gold standard for determining the safety and efficacy of new drug (see chapters in the next section), they are limited in their ability to be fully representative of the patient population in the real world. Special populations, such as patients with organ dysfunction, pregnant individuals, lactating individuals, pediatrics, etc. often fall into this category of population subsets excluded from late phase clinical trials. In such settings, stand-alone clinical pharmacology studies provide the critical understanding of the factors that alter the drug PK/PD and inform the dosing recommendations. The general clinical pharmacology principles can also guide study design considerations in other special population settings such as older adults, patients with obesity, genetic polymorphisms, etc. As we continue to move towards better dose individualization for many complex situations, it is vital to consider complementary clinical pharmacology evaluations to fill data gaps and enable access of new drugs across various subsets of the patient population.

Conflict of Interest The authors have no conflicts of interest to declare.

References

1. US Food and Drug Administration (2003) Guidance for industry - pharmacokinetics in patients with impaired hepatic function: study design, data analysis, and impact on dosing and labeling. https://www.fda.gov/media/71311/download. Accessed 10 Dec 2022
2. US Food and Drug Administration (2020) Guidance for industry (draft) - pharmacokinetics in patients with impaired renal function — study design, data analysis, and impact on dosing and labeling. https://www.fda.gov/media/78573/download. Accessed 10 Dec 2022
3. ICH E11(R1) - Clinical investigation of medicinal products in the pediatric population (2018). https://database.ich.org/sites/default/files/E11_R1_Addendum.pdf. Accessed 10 Dec 2022
4. ICH E11A - Pediatric extrapolation (2022). https://database.ich.org/sites/default/files/ICH_E11A_Document_Step2_Guideline_2022_0404_0.pdf. Accessed 10 Dec 2022
5. US Food and Drug Administration (2022) Guidance for industry - general clinical pharmacology considerations for pediatric studies of drugs, including biological products. https://www.fda.gov/media/90358/download. Accessed 10 Dec 2022
6. US Food and Drug Administration. Guidance for industry - pediatric study plans: content of and process for submitting initial pediatric study plans and amended initial pediatric study plans. https://www.fda.gov/media/86340/download. Accessed 10 Dec 2022
7. US Food and Drug Administration (2022) Guidance for industry - general clinical pharmacology considerations for neonatal studies for drugs and biological products. https://www.fda.gov/media/129532/download. Accessed 10 Dec 2022
8. US Food and Drug Administration (2004) Guidance for industry - pharmacokinetics in pregnancy – study design, data analysis, and impact on dosing and labeling. https://www.fda.gov/media/71353/download. Accessed 10 Dec 2022
9. US Food and Drug Administration (2019) Guidance for industry - clinical lactation studies: considerations

for study design. https://www.fda.gov/media/124749/download. Accessed 10 Dec 2022
10. US Food and Drug Administration (2013) Guidance for industry - clinical pharmacogenomics: pre-market evaluation in early-phase clinical studies and recommendations for labeling. https://www.fda.gov/media/84923/download. Accessed 10 Dec 2022
11. ICH E7 - studies in support of special populations: geriatrics (1994). https://database.ich.org/sites/default/files/E7_Guideline.pdf. Accessed 10 Dec 2022
12. ICH E7 - studies in support of special populations: geriatrics. Questions & Answers (2012) https://database.ich.org/sites/default/files/E7_Q%26As_Q%26As.pdf. Accessed 10 Dec 2022
13. Child CG, Turcotte JG (1964) Surgery and portal hypertension. Major Probl Clin Surg 1:1–85
14. NCI (2019) Protocol template for organ dysfunction studies. https://ctep.cancer.gov/protocolDevelopment/docs/CTEP_Organ_Dysfunction_Protocol_Template.docx. Accessed 10 Dec 2022
15. Cockcroft DW, Gault MH (1976) Prediction of creatinine clearance from serum creatinine. Nephron 16(1):31–41
16. Pai MP (2010) Estimating the glomerular filtration rate in obese adult patients for drug dosing. Adv Chronic Kidney Dis 17(5):e53–e62
17. Levey AS, Bosch JP, Lewis JB, Greene T, Rogers N, Roth D (1999) A more accurate method to estimate glomerular filtration rate from serum creatinine: a new prediction equation. Modification of Diet in Renal Disease Study Group. Ann Intern Med 130(6):461–470
18. Levey AS, Stevens LA, Schmid CH et al (2009) A new equation to estimate glomerular filtration rate. Ann Intern Med 150(9):604–612
19. Dubois D, Dubois EF (1916) A formula to estimate the approximate surface area if height and weight be known. Arch Intern Med 17:863–871
20. Mosteller RD (1987) Simplified calculation of body surface area. N Engl J Med 317:1098
21. Zhang L, Xu N, Xiao S, Arya V, Zhao P, Lesko LJ, Huang SM (2012) Regulatory perspectives on designing pharmacokinetic studies and optimizing labeling recommendations for patients with chronic kidney disease. J Clin Pharmacol 52(1 Suppl):70S–90S
22. Nolin TD, Aronoff GR, Fissell WH, Jain L, Madabushi R, Reynolds K, Zhang L, Huang SM, Mehrotra R, Flessner MF, Leypoldt JK, Witcher JW, Zineh I, Archdeacon P, Roy-Chaudhury P, Goldstein SL (2015) Pharmacokinetics assessment in patients receiving continuous RRT: perspectives from Kidney Health Initiative. Clin J Am Soc Nephrol 10(1):159–164
23. Sahre MD, Milligan L, Madabushi R et al (2021) Evaluating patients with impaired renal function during drug development: highlights from the 2019 US FDA Pharmaceutical Science and Clinical Pharmacology Advisory Committee Meeting. Clin Pharmacol Ther 110(2):285–288
24. Lutz J, Menke J, Sollinger D, Schinzel H, Thürmel K (2014) Haemostasis in chronic kidney disease. Nephrol Dial Transplant 29(1):29–40
25. Kearns GL, Abdel-Rahman SM, Alander SW, Blowey DL, Leeder JS, Kauffman RE (2003) Developmental pharmacology--drug disposition, action, and therapy in infants and children. N Engl J Med 349(12):1157–1167
26. Bhatnagar M et al (2021) Prospect of direct benefit in pediatric trials: practical challenges and potential solutions. Pediatrics 147(5):e2020049602
27. Bi Y, Liu J, Li L, Yu J, Bhattaram A, Bewernitz M, Li RJ, Liu C, Earp J, Ma L, Zhuang L, Yang Y, Zhang X, Zhu H, Wang Y (2019) Role of model-informed drug development in pediatric drug development, regulatory evaluation, and labeling. J Clin Pharmacol 59 (suppl 1):S104–S111
28. Green FG, Park K, Burckart GJ (2021) Methods used for pediatric dose selection in drug development programs submitted to the US FDA 2012-2020. J Clin Pharmacol 61(suppl 1):S28–S35
29. Khong YM, Liu J, Cook J, Purohit V, Thompson K, Mehrotra S, Cheung SYA, Hay JL, Fletcher EP, Wang J, Sachs HC, Zhu H, Siddiqui A, Cunningham L, Selen A (2021) Harnessing formulation and clinical pharmacology knowledge for efficient pediatric drug development: overview and discussions from M-CERSI pediatric formulation workshop 2019. Eur J Pharm Biopharm 164:66–74
30. Byrne JJ, Saucedo AM, Spong CY (2020) Evaluation of drug labels following the 2015 pregnancy and lactation labeling rule. JAMA Netw Open 3(8):e2015094
31. Dinatale M, Roca C, Sahin L, Johnson T, Mulugeta LY, Fletcher EP, Yao L (2020) The importance of clinical research in pregnant women to inform prescription drug labeling. J Clin Pharmacol 60(suppl 2):S18–S25
32. Feghali M, Venkataramanan R, Caritis S (2015) Pharmacokinetics of drugs in pregnancy. Semin Perinatol 39(7):512–519
33. US Food and Drug Administration (2018) Guidance for industry - pregnant women: scientific and ethical considerations for inclusion in clinical trials. https://www.fda.gov/media/112195/download. Accessed 10 Dec 2022
34. Boyd SD, Sampson MR, Viswanathan P, Struble KA, Arya V, Sherwat AI (2019) Cobicistat-containing antiretroviral regimens are not recommended during pregnancy: viewpoint. AIDS 33(6):1089–1093
35. US Food and Drug Administration (2022) Guidance for industry – population pharmacokinetics. https://www.fda.gov/media/128793/download. Accessed 10 Dec 2022
36. Coppola P, Kerwash E, Cole S (2021) Physiologically based pharmacokinetics model in pregnancy: a regulatory perspective on model evaluation. Front Pediatr 9:687978
37. Quinney SK, Gullapelli R, Haas DM (2018) Translational systems pharmacology studies in pregnant women. CPT Pharmacometrics Syst Pharmacol 7(2):69–81

38. US Food and Drug Administration (2015) Guidance for industry – pregnancy, lactation, and reproductive potential: labeling for human prescription drug and biological products — content and format lactation. https://www.fda.gov/media/92565/download. Accessed 10 Dec 2022
39. Anderson PO, Manoguerra AS, Valdés V (2016) A review of adverse reactions in infants from medications in breastmilk. Clin Pediatr (Phila) 55(3): 236–244
40. Ito S (2018) Opioids in breast milk: pharmacokinetic principles and clinical implications. J Clin Pharmacol 58(suppl 10):S151–S163

Impact of Genetic Variation on Drug Response

19

Rachel Huddart and Russ Altman

Abstract

Pharmacogenomics investigates the influence of genetics on drug response, with the ultimate aim of giving patients the right drug at the right dose and at the right time. Almost all patients carry one or more genetic variants which can significantly alter their response to a drug, causing anything from a lack of response to serious, even fatal, drug toxicity. This chapter introduces the reader to this field of precision medicine, using well-characterized examples which are already used in the clinic to optimize drug prescribing. It also introduces the star allele nomenclature system and describes the process of building a solid evidence base for clinical pharmacogenomic guidelines. The importance of working with diverse cohorts is critical when designing pharmacogenomics studies. Finally, it presents several barriers to the implementation of pharmacogenomics in the clinic and discusses methods for overcoming these.

R. Huddart
Department of Biomedical Data Science, Stanford University, Stanford, CA, USA
e-mail: feedback@pharmgkb.org

R. Altman (✉)
Kenneth Fong Professor of Bioengineering, Genetics, Medicine, Biomedical Data Science and (by courtesy, Computer Science), Stanford University, Stanford, CA, USA
e-mail: russ.altman@stanford.edu

Keywords

Pharmacogenomics · Drug response · Star allele · Clinical implementation · Evidence base · Clopidogrel

19.1 Introduction

Pharmacogenomics is the study of how drug response is affected by genetic variation and has the potential to impact anyone who takes a drug in their lifetime. The response to drugs is mediated by gene products that are involved in either the absorption, distribution, metabolism or excretion of the drug (pharmacokinetics, PK) or the mechanism of action of the drug (pharmacodynamics, PD). Pharmacokinetics can be described as 'what the body does to the drug' and can encompass processes including metabolism. Pharmacodynamics, or 'what the drug does the body' tends to focus on the activity of the drug at its target sites and downstream effects. Variation in the gene products involved in PK and PD can lead to variation in drug response. Indeed, at least 95% of patients carry a genetic variant that can influence their response to at least one drug [1]. Likely, most patients contain variants that will affect many drugs. The vision for pharmacogenomics is that with routine access to individual genome information, drugs can be prescribed with knowledge of how the patient genetics may affect response; drugs with expected decreased efficacy

G. Jagadeesh et al. (eds.), *The Quintessence of Basic and Clinical Research and Scientific Publishing*,
https://doi.org/10.1007/978-981-99-1284-1_19

or increased side effects will be avoided and drugs expected to be efficacious will be favored.

19.1.1 Aminoglycoside Pharmacogenomics

A notable example of the effect of pharmacogenomics on drug pharmacodynamics is the interaction between aminoglycoside antibiotics, used to treat bacterial infections and variants in the mitochondrial gene *MT-RNR1*. Aminoglycosides act by binding the 16S ribosomal RNA (rRNA) subunit of bacterial ribosomes. This binding blocks protein synthesis and ultimately kills the bacterial cell.

MT-RNR1 encodes the 12S rRNA subunit in humans, which is structurally similar to the bacterial 16S subunit. The presence of certain variants within *MT-RNR1* can increase this structural similarity to the point where aminoglycosides will bind to 12S rRNA, leading to inhibition of translation and cell death. This is particularly devastating in cell types that cannot regenerate, such as inner ear hair cells (Fig. 19.1). Indeed, there are many reports of patients

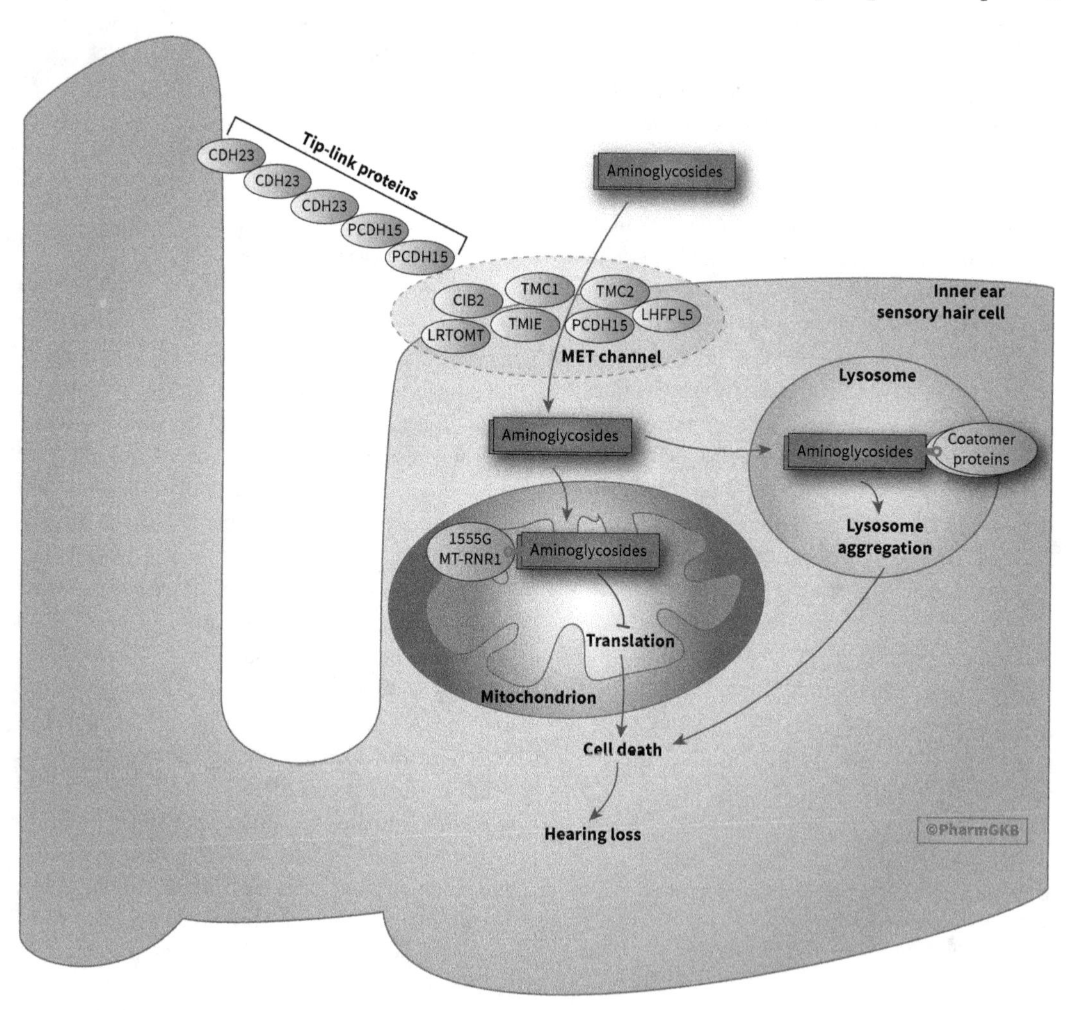

Fig. 19.1 PharmGKB aminoglycoside ototoxicity pathway, adverse drug reaction. The 1555G variant is used here as an example variant in *MT-RNR1* which causes aminoglycoside-induced hearing loss. An interactive version of this pathway can be found at https://www.pharmgkb.org/pathway/PA166254101. (Image made available under a CC BY-SA 4.0 license [2])

experiencing sensorineural hearing loss following the administration of aminoglycoside antibiotics [3–6] and there is now a clinical guideline that recommends that aminoglycosides be avoided in patients carrying certain *MT-RNR1* variants [7]. Annotation of this guideline, in addition to annotations on other published pharmacogenomic prescribing guidelines, can be found on the Pharmacogenomics Knowledgebase (PharmGKB; https://www.pharmgkb.org) [2, 8].

19.1.2 Clopidogrel Pharmacogenomics

A widely used example of the effect of genetic variation on a drug's pharmacokinetics is the effect of variation in the gene *CYP2C19* on the antiplatelet drug clopidogrel, which is used to prevent the formation of blood clots which could cause ischemic events such as strokes or heart attacks. *CYP2C19* encodes the enzyme CYP2C19, part of the cytochrome P450 superfamily of enzymes. The CYP2C19 enzyme metabolizes clopidogrel, a prodrug, to its active metabolites which inhibit platelet aggregation (Fig. 19.2).

Initial studies identified an association between the *CYP2C19*2* allele, which causes a loss of CYP2C19 function, and a reduced clinical effect of clopidogrel [10–13]. This was exemplified in a genome-wide association study (GWAS) showing that the risk of cardiovascular ischemic event or death in patients carrying the *CYP2C19*2* allele was over twice as high as in patients without the allele [14].

19.2 Star Allele Nomenclature

Pharmacogenomics specialists use 'star allele' nomenclature to refer to haplotypes of key genes involved in pharmacogenomics (e.g. *1, *2, *3). Haplotypes in pharmacogenomics can consist of a single variant or multiple variants on the same chromosome. The star allele nomenclature offers enough flexibility to be applied equally to all alleles, regardless of how many individual variants each allele contains. Precise curation and correct use of this nomenclature are vital to allow results from different publications to be compared against each other and for prescribing guidance to be issued at a per-allele level.

The Pharmacogene Variation Consortium (PharmVar; www.pharmvar.org) is a centralized repository of star allele nomenclature for many key genes in pharmacogenomics [15–17]. Other genes, such as *TPMT* and the *UGT*s, have their own nomenclature committees which oversee allele naming [18, 19]. All welcome submission of novel alleles from those in the research community.

For many genes which play a central role in pharmacogenomics, a patient's diplotype (e.g. CYP2C19*2/*3) is translated into a phenotype. This translation is dependent on each allele being assigned to a term which describes the allele's function level (e.g. increased function, normal function, decreased function, no function). Combinations of allele function terms are then used to assign a metabolizer phenotype, which can be linked to specific prescribing recommendations. Table 19.1 shows a simplified version of this process for another cytochrome P450 enzyme, CYP2C9.

Determining the function of rare or newly discovered pharmacogenomic variants can be a challenging process. In vivo metabolism data for these variants is seldom available and in vitro data, while slightly more common, may not truly reflect the in vivo function of the protein. Computational algorithms which are routinely used to predict the effects of a variant on protein function have poor predictive power with pharmacogenomic variants as these algorithms tend to use evolutionary conservation as a predictive feature. This does not apply to pharmacogenomic variants as they are only subject to low selective pressures compared to the higher evolutionary constraints seen with disease-causing variants [20, 21]. As such, there is a need to develop pharmacogenomic-specific prediction algorithms which can account for the unique conditions which apply to variants in pharmacogenes [21, 22].

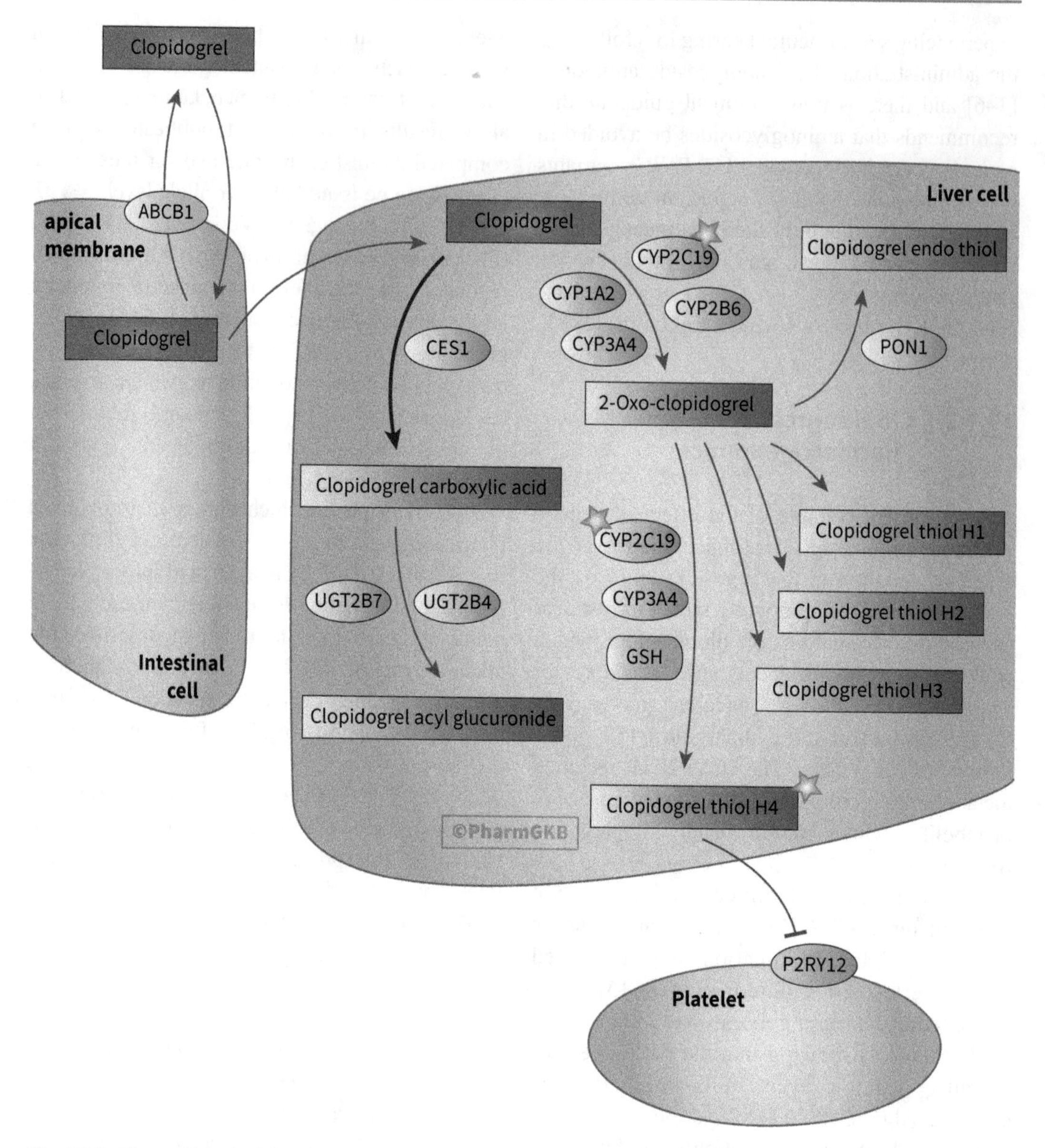

Fig. 19.2 PharmGKB clopidogrel pathway, pharmacokinetics. An interactive version of this pathway can be found at https://www.pharmgkb.org/pathway/PA154424674. (Image made available under a CC BY-SA 4.0 license [9])

Additionally, many pharmacogenomic influences on drug response are likely to result from the combined small effects of variants in multiple genes rather than the larger effect of a single variant or haplotype in a single gene. These effects need to be further characterized before they can be fully implemented in the clinic and will require the development of polygenic scores for pharmacogenomics [23, 24].

Technological advances in biosciences, including induced pluripotent stem cells (iPS cells) and genome editors such as CRISPR/Cas9 [25, 26], can help to characterize the function of pharmacogenomic alleles or identify new variants. The use of iPS cells and/or CRISPR-Cas9 can allow researchers to conduct high-throughput screens to find new genes or alleles with pharmacogenomic implications or assess

Table 19.1 Assignment of CYP2C9 metabolizer phenotype status based on allele function, based on materials available at https://cpicpgx.org/

Allele 1 function	Allele 2 function	Metabolizer phenotype
Normal function	Normal function	Normal metabolizer
Normal function	Decreased function	Intermediate metabolizer
Normal function	No function	Intermediate metabolizer
Decreased function	Decreased function	Intermediate metabolizer
Decreased function	No function	Poor metabolizer
No function	No function	Poor metabolizer

Readers should note that this translation process varies between genes and is currently not standardized between the various organizations issuing pharmacogenomic-based prescribing guidance

pharmacogenomic associations on the background of a specific genetic ancestry [27–30]. However, care must be taken to ensure that these systems are as accurate a representation of their in vivo counterparts as possible. For example, hepatocytes derived from iPS cells are more similar to cells derived from fetal liver samples than those from adults [31]. This means that the expression of key enzymes, including many of the CYP enzymes, will not reflect adult expression patterns [32] and that further treatment of the cells may be required to induce adult CYP expression [31].

19.3 Building the Pharmacogenomics Evidence Base

Returning to the earlier discussion of the effects of *CYP2C19* variation on clopidogrel, subsequent studies have replicated the association between decreased CYP2C19 function and decreased clinical efficacy of clopidogrel. Many of these studies have been curated by PharmGKB. PharmGKB collects, curates, and disseminates knowledge about gene-drug associations and genotype-phenotype relationships. As pharmacogenetic information is added into PharmGKB, it is used to create Clinical Annotations, which are summaries of the curated evidence base for a particular gene-drug or variant-drug relationship. Clinical Annotations are assigned a Level of Evidence (Fig. 19.3) to indicate the strength of the curated evidence base underlying the association.

Clinical Annotations can move up the Levels of Evidence as evidence is curated and added to the annotation. As evidence accumulates, it can be assessed by regulatory agencies (e.g. the U.S. Food and Drug Administration; FDA) or professional organizations such as the Clinical Pharmacogenetics Implementation Consortium (CPIC; www.cpicpgx.org) [33] or the Dutch Pharmacogenetics Working Group (DPWG) [34] and form the basis for prescribing guidelines in clinical practice, or directions in drug labels as determined by regulatory authorities. In the case of clopidogrel, both CPIC and the DPWG have issued recommendations stating that, for some

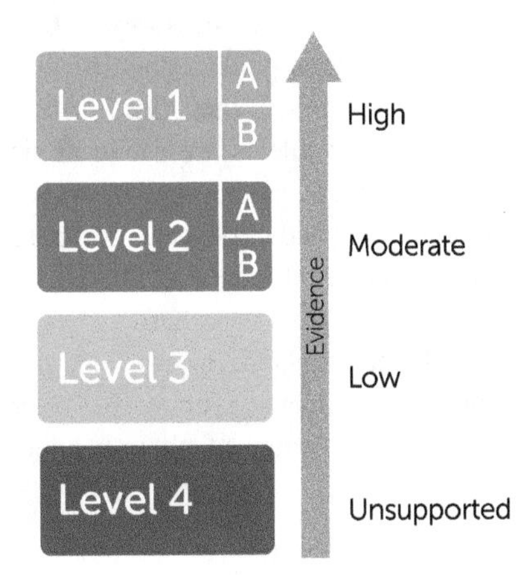

Fig. 19.3 Level of Evidence scale for PharmGKB Clinical Annotations [2]. Levels run from Level 4, for pharmacogenetic associations unsupported by the majority of the curated literature, to Level 1 for associations backed by high levels of evidence, including those implemented clinically. More information is available at https://www.pharmgkb.org/page/clinAnnLevels. (Image made available under a CC BY-SA 4.0 license [9])

indications, patients carrying two no function alleles (e.g. *2) should avoid clopidogrel and be prescribed an alternative therapy [35, 36]. The FDA-approved drug label for clopidogrel also highlights the potential risk of clopidogrel in patients with reduced CYP2C19 function. The boxed warning on the label states that an alternative therapy should be considered for patients who are CYP2C19 poor metabolizers.

This boxed warning was added to the clopidogrel label in 2010 following studies, like those referenced at the beginning of this chapter, which showed a reduced efficacy of clopidogrel in patients with loss-of-function *CYP2C19* alleles. The fact that this pharmacogenomic relationship was not detected during the initial studies on clopidogrel raises questions about study design and highlights the dangers of using a non-representative study cohort.

19.4 Important Factors in Pharmacogenomic Study Design

The frequency of different pharmacogenomic alleles can vary considerably between populations of different genetic ancestries. In the case of clopidogrel, the loss-of-function CYP2C19*2 and *3 alleles are found at significantly higher frequencies in populations of East Asian or Pacific Islander descent compared to those of European or African descent. This variation makes it imperative for pharmacogenomics studies or clinical trials of new drugs to have study cohorts that reflect a diverse patient population carrying a wide range of pharmacogenomic alleles.

19.4.1 Hawaiian Clopidogrel Lawsuit

In 2014, the State of Hawaii filed a lawsuit against the makers of Plavix, the originally licensed form of clopidogrel [37]. Data from the Hawaiian Department of Health found that, between 2007 and 2009, patients of native Hawaiian ancestry who have recently suffered an acute myocardial infarction (AMI) were almost twice as likely to die than those of European ancestry. Many of these patients had been treated with clopidogrel, which was the only antiplatelet drug available at the time. Following the introduction of alternative antiplatelet drugs, such as ticagrelor, the percentages of death following AMI were similar between the two ancestries.

The FDA-approved drug label for clopidogrel includes details on several clinical trials which showed the efficacy of clopidogrel in preventing ischemic events. One of these is the CAPRIE study, which compared the efficacy of clopidogrel against aspirin in preventing stroke. 95% of participants in the CAPRIE trial were of European ancestry, where the *2 and *3 alleles are typically found at frequencies of 14% and 0.16% respectively. This is in stark contrast to Oceanian populations, with *2 and *3 frequencies of 60% and 14%, respectively. As a result, patients with the *2 or *3 alleles were significantly underrepresented in the study, and, accordingly, the risks to patients who could not appropriately metabolize clopidogrel were not detected.

19.4.2 Cohort Diversity in Pharmacogenomics

In addition to ensuring that study cohorts are appropriately diverse, care should also be taken to try to accurately identify as many pharmacogenomic alleles as possible within study participants. This not only allows a fuller understanding of a gene-drug relationship, but it is also possible that new pharmacogenomic alleles could be identified and characterized. This is especially valuable in populations which are underrepresented in pharmacogenomics research.

Like other areas of genomics, pharmacogenomics research is overwhelmingly conducted in participants of European ancestry, to the detriment of individuals from other genetic backgrounds [38, 39]. Research in understudied populations is therefore more likely to result in the discovery of clinically significant variants and improve the applicability of pharmacogenomic guidance in these patient groups [40].

This is particularly true in the context of genome-wide association studies (GWAS) and the development of polygenic scores, as discussed earlier in the chapter. A 2016 analysis found that only 20% of published GWAS were carried out in non-European populations, with the vast majority of those coming from Asian populations [39]. This has a knock-on effect on polygenic scores, as many scores are based on GWAS results. Polygenic scores have already been shown to perform better in patients of European descent compared to patients from other ancestries [40] and the use of European-optimized polygenic scores in the clinic can greatly exacerbate existing health disparities.

The continuing challenge of racial and ethnic disparities in both access to and use of healthcare resources has impeded the development of comprehensive pharmacogenomics knowledge, based on careful consideration of all the genetic variations that are seen in a diversity of populations. In some cases, underrepresented populations have been reluctant to take part in clinical studies due to a combination of past ethical abuses by researchers and continuing systemic racism within the healthcare system [41]. Too often, understudied populations are not approached or even considered by researchers designing pharmacogenomic studies, who instead recruit from population groups seen as more amenable to participation (i.e., those of European descent). Nonetheless, understudied populations are willing to take part in research if cultural and community priorities are respected by the researchers [42, 43]. Indeed, research on the correlation between a patient's ancestry and their attitude toward pharmacogenomics found that Black patients were significantly more likely than White patients to agree that their personal genetic information should be used to guide treatment decisions [44].

The increasing use of biobanks in pharmacogenomic research has already provided a glimpse into the benefits of looking for pharmacogenomic alleles in large, diverse cohorts. Analysis of the UK Biobank found that of the 14 pharmacogenes investigated, 99.5% of biobank participants carried at least one non-typical drug response diplotype [1]. Based on their pharmacogenomic alleles, the average participant was predicted to require a change from standard dosing for 10.3 drugs (Fig. 19.4) [1].

However, despite their large cohort size, biobanks can still fail to accurately represent the general population [45]. Efforts are now being made to increase participant diversity in biobanks, with the All of Us and Gen V projects being notable examples [46, 47]. Practical steps to improve biobank diversity can include a deliberate focus on recruiting from underrepresented populations and removing barriers to participation, for example, by allowing recruitment and participation to occur in multiple languages.

19.4.3 Assignment of Star Alleles

Many pharmacogenomic studies genotype specific SNPs to identify star alleles in a population. In cases such as these, it is important to be clear on what each SNP does and does not identify. A common example is a CYP2D6*10 allele, which is routinely identified by the presence of the T allele of rs1065852 (commonly referred to as the 100C > T SNP). However, at the time of writing, the T allele is also included in the definitions of 23 other CYP2D6 alleles. Some of these are decreased function alleles, like the *10 allele, while others are no function alleles or have not yet had a function term assigned to them.

Another consideration is the assignment of the *1 allele of pharmacogenes. *1 is typically considered to be the reference allele of the gene (a notable exception to this is CYP2C19, where the *38 allele is the reference). However, in almost all pharmacogenomic studies, *1 is assigned if none of the variants genotyped for in the study were found. It is therefore important to realize that an assignment of *1 is not a guarantee that the individual carries the reference allele. There is a significant possibility that the individual has variants which affect protein function but were not included in the genotyping assay.

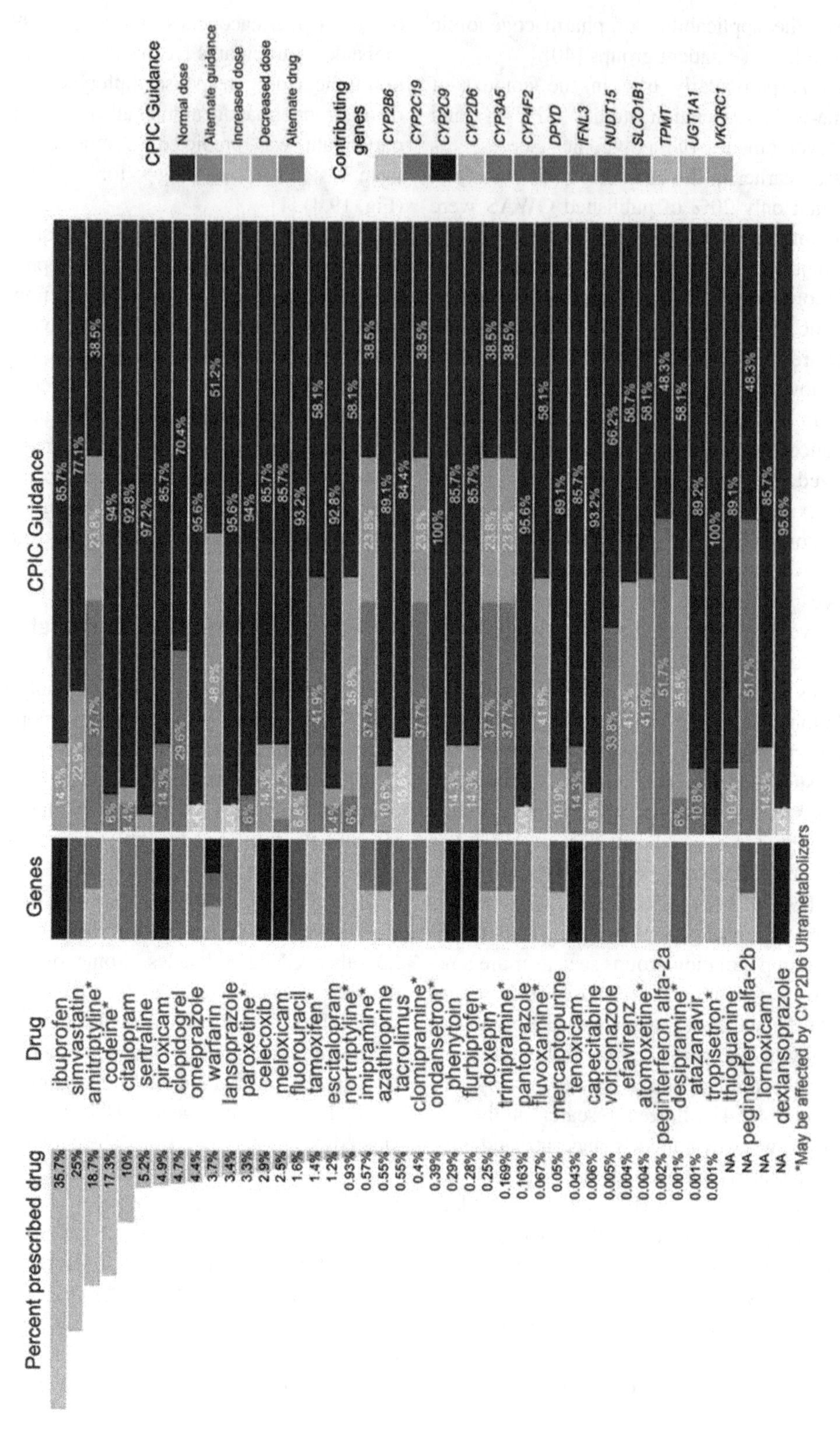

Fig. 19.4 CPIC pharmacogenomic guidance applied to genotypes from the UK Biobank. From left to right, the columns show the percentage of the cohort who have been prescribed each drug; the drug name; the genes analyzed in the study which support the CPIC guidance; and the distribution of CPIC recommendations in the cohort. (Figure reproduced with permission from McInnes et al. [1])

19.5 Bringing Pharmacogenomics into the Clinic

Pharmacogenomics is frequently portrayed as the 'low-hanging fruit' of precision medicine which can be widely implemented in the clinic within the next few years and has been shown to improve patients' perceptions of the quality of their care [48]. However, there remain several barriers and ongoing discussions which must be resolved to allow full implementation of pharmacogenomics. This chapter will give a brief overview of some of the major barriers, but these issues have been discussed in detail elsewhere [49–51].

Despite mounting evidence that genetic variation can cause clinically relevant alterations in drug response, it can be difficult to know when to order a pharmacogenomic test for a patient or how to change a prescribing decision based on the patient's test results. Both CPIC and the DPWG aim to bridge this gap by critically reviewing the available literature and issuing genotype- or phenotype-specific prescribing guidelines to assist clinicians in the decision-making process [33, 34]. CPIC publishes guidelines under the assumption that the patient's test results are already in hand as the expectation is that pharmacogenomic testing of all patients will become a routine part of clinical care in the future. However, the DPWG and other guideline organizations have started to include advice about when to order a pharmacogenomic test for patients.

Implementation of pharmacogenomics typically requires the formation of a pharmacogenomics consult service. Early implementers have described a mixture of pharmacist-led and physician-led services [52–54], but there is a general consensus that multidisciplinary teams are vital to successful implementation [54, 55]. However, members of these teams commonly have a specialized background in pharmacogenomics, something which may not be shared by their colleagues who are relied upon to refer patients to the pharmacogenomics service and use test results in their clinical decision making. Indeed, knowledge gaps and a lack of confidence in interpreting pharmacogenomic test results are frequently cited as a barrier to the implementation of pharmacogenomics [56–59].

The unclear regulatory status of pharmacogenomic tests has also created some barriers. In 2018, the FDA issued a warning to advise patients and healthcare providers against making any changes to their medications based on the results of "unproven" pharmacogenomic tests [60, 61]. This was followed in 2019 by direct communication from the FDA to several pharmacogenomic test providers about removing references to drug-gene relationships which were "not established", according to the FDA [62, 63]. This included drug-gene pairs which were found to have sufficient supporting evidence to be included in CPIC guidelines. However, there was no indication from the FDA about which evidence sources, other than FDA drug labels, it had assessed before coming to these conclusions. In response to criticism from the wider pharmacogenomics community about this lack of transparency, the FDA published its Table of Pharmacogenomic Associations in 2020 [64, 65].

The education barrier can be partially circumvented using well-designed clinical decision support (CDS) tools [66, 67]. CDS primarily consists of an alert system which integrates into a clinician's workflow and can notify the clinician of potentially relevant pharmacogenomic interactions. This can lead to the clinician either ordering a pharmacogenomic test for the patient or, if the patient already has test results available, assessing their prescribing decision based on this information. Care must be taken when designing CDS tools to ensure maximum user engagement and that clinicians do not encounter 'alert fatigue' [66]. Research into CDS alerts for drug-drug interactions (DDIs) found that such alerts were ignored or overridden more than 90% of the time [68]. As a result, it is generally necessary to design multiple types of CDS to fully integrate pharmacogenomics into clinical care [66].

In addition to designing user-friendly CDS tools, consideration must also be given to how to get pharmacogenomic test results into a patient's electronic health record (EHR) so that they can be used by CDS tools. Test results are typically released as unstructured data (e.g., PDF files) which can be difficult to add to an EHR in a way that the results can be easily found by other healthcare providers in the future. Recently developed tools such as PharmCAT [69, 70] can assist implementers in this process by automatically annotating variant call format (VCF) files with diplotypes, phenotypes, and prescribing recommendations.

The cost and, in some healthcare systems, insurance coverage of pharmacogenomic tests represent another significant barrier to implementation [59], although there are early indications that this could be changing. In the United States, payer coverage of pharmacogenomic tests began to slowly expand in the 2010s and this is continuing into the 2020s [71]. In the United Kingdom, the National Health Service (NHS) now routinely screens for specific variants in the gene *DPYD* in all patients beginning fluoropyrimidine chemotherapy [72] and there are growing calls for wider implementation of pharmacogenomics within the service [73].

19.6 Concluding Remarks

Pharmacogenomics is already being used in the clinic to improve prescribing outcomes for patients. However, there are many complexities in the field which need to be considered by researchers, clinicians, and implementers. Resources such as PharmGKB, PharmVar, CPIC, and the DPWG provide valuable materials to help maintain standards in pharmacogenomics and expand its use in the clinic. The next steps for pharmacogenomics are clear. First, we must learn how to interpret the rare variants found in all patients, but not in sufficient numbers to analyze them statistically. Either computational methods or high-throughput experimental methods must be developed to assess the significance of rare variants seen in individuals, families, or small groups. Second, we must learn how to predict or measure the impact of several variants on a single drug. Sometimes both the PK and the PD might be altered, but it may not be clear how to combine these effects to arrive at a summary recommendation about the likely utility of a particular drug for a particular patient. Third, pharmacogenomics must be integrated into the practice of clinicians and pharmacists so that it is not disruptive, but allow them to perform their job tasks seamlessly while using pharmacogenetic information to optimize prescribing. This will certainly require the creation of easy to use, informative (and not irritating or fatiguing) user interfaces. Fourth, healthcare systems must be committed to paying for pharmacogenetic testing. This may require cost-effectiveness studies, randomized trials, or other evidence of clear clinical utility. It is likely that pharmacogenetics will be most effective when all genetic variation is measured through inexpensive genome sequencing, is stored securely, and then can be inspected each time a new prescribing decision is made. Finally, patients must be educated about the benefits and risks of using their genetic information for drug prescribing. They must trust that their genetic information will be stored securely, used appropriately, and not represent any threat of discrimination, inequity, or bias.

Conflict of Interest Statement RBA has been an advisor and stockholder to Personalis, Myome, Nebula Genomics and 23andme. He is an advisor to the All of Us Project, the United Kingdom Biobank, the Danish National Genome Center, and the Swiss Personalized Health Network. RBA and RH have been funded by the US National Institutes of Health to help build the Pharmacogenomics Knowledgebase.

References

1. McInnes G, Lavertu A, Sangkuhl K, Klein TE, Whirl-Carrillo M, Altman RB (2021) Pharmacogenetics at scale: an analysis of the UK Biobank. Clin Pharmacol Ther 109(6):1528–1537
2. Whirl-Carrillo M, Huddart R, Gong L, Sangkuhl K, Thorn CF, Whaley R et al (2021) An evidence-based framework for evaluating pharmacogenomics knowledge for personalized medicine. Clin Pharmacol Ther 110(3):563–572
3. Prezant TR, Agapian JV, Bohlman MC, Bu X, Oztas S, Qiu WQ et al (1993) Mitochondrial ribosomal RNA mutation associated with both antibiotic-induced and non-syndromic deafness. Nat Genet 4(3):289–294

4. Thyagarajan D, Bressman S, Bruno C, Przedborski S, Shanske S, Lynch T et al (2000) A novel mitochondrial 12SrRNA point mutation in parkinsonism, deafness, and neuropathy. Ann Neurol 48(5):730–736
5. Wu L, Li R, Chen J, Chen Y, Yang M, Wu Q (2018) Analysis of mitochondrial A1555G mutation in infants with hearing impairment. Exp Ther Med 15(6): 5307–5313
6. Zhao H, Li R, Wang Q, Yan Q, Deng JH, Han D et al (2004) Maternally inherited aminoglycoside-induced and nonsyndromic deafness is associated with the novel C1494T mutation in the mitochondrial 12S rRNA gene in a large Chinese family. Am J Hum Genet 74(1):139–152
7. McDermott JH, Wolf J, Hoshitsuki K, Huddart R, Caudle KE, Whirl-Carrillo M et al (2022) Clinical Pharmacogenetics Implementation Consortium guideline for the use of aminoglycosides based on MT-RNR1 genotype. Clin Pharmacol Ther 111(2): 366–372
8. Whirl-Carrillo M, McDonagh EM, Hebert JM, Gong L, Sangkuhl K, Thorn CF et al (2012) Pharmacogenomics knowledge for personalized medicine. Clin Pharmacol Ther 92(4):414–417
9. Sangkuhl K, Klein TE, Altman RB (2010) Clopidogrel pathway. Pharmacogenet Genomics 20(7):463–465
10. Kim KA, Park PW, Hong SJ, Park JY (2008) The effect of CYP2C19 polymorphism on the pharmacokinetics and pharmacodynamics of clopidogrel: a possible mechanism for clopidogrel resistance. Clin Pharmacol Ther 84(2):236–242
11. Trenk D, Hochholzer W, Fromm MF, Chialda LE, Pahl A, Valina CM et al (2008) Cytochrome P450 2C19 681G>A polymorphism and high on-clopidogrel platelet reactivity associated with adverse 1-year clinical outcome of elective percutaneous coronary intervention with drug-eluting or bare-metal stents. J Am Coll Cardiol 51(20):1925–1934
12. Mega JL, Close SL, Wiviott SD, Shen L, Hockett RD, Brandt JT et al (2009) Cytochrome p-450 polymorphisms and response to clopidogrel. N Engl J Med 360(4):354–362
13. Collet JP, Hulot JS, Pena A, Villard E, Esteve JB, Silvain J et al (2009) Cytochrome P450 2C19 polymorphism in young patients treated with clopidogrel after myocardial infarction: a cohort study. Lancet 373(9660):309–317
14. Shuldiner AR, O'Connell JR, Bliden KP, Gandhi A, Ryan K, Horenstein RB et al (2009) Association of cytochrome P450 2C19 genotype with the antiplatelet effect and clinical efficacy of clopidogrel therapy. JAMA 302(8):849–857
15. Gaedigk A, Casey ST, Whirl-Carrillo M, Miller NA, Klein TE (2021) Pharmacogene Variation Consortium: a global resource and repository for pharmacogene variation. Clin Pharmacol Ther 110(3):542–545
16. Gaedigk A, Ingelman-Sundberg M, Miller NA, Leeder JS, Whirl-Carrillo M, Klein TE (2018) The Pharmacogene Variation (PharmVar) Consortium: incorporation of the Human Cytochrome P450 (CYP) Allele Nomenclature Database. Clin Pharmacol Ther 103(3):399–401
17. Gaedigk A, Whirl-Carrillo M, Pratt VM, Miller NA, Klein TE (2020) PharmVar and the Landscape of Pharmacogenetic Resources. Clin Pharmacol Ther 107(1):43–46
18. UGT Alleles Nomenclature Home Page: UGT Nomenclature Committee (2005). https://www.pharmacogenomics.pha.ulaval.ca/ugt-alleles-nomenclature/
19. TPMT nomenclature committee (TPMT Alleles): TPMT Nomenclature Committee. https://liu.se/en/research/tpmt-nomenclature-committee
20. McInnes G, Sharo AG, Koleske ML, Brown JEH, Norstad M, Adhikari AN et al (2021) Opportunities and challenges for the computational interpretation of rare variation in clinically important genes. Am J Hum Genet 108(4):535–548
21. Zhou Y, Mkrtchian S, Kumondai M, Hiratsuka M, Lauschke VM (2019) An optimized prediction framework to assess the functional impact of pharmacogenetic variants. Pharmacogenomics J 19(2):115–126
22. McInnes G, Dalton R, Sangkuhl K, Whirl-Carrillo M, Lee SB, Tsao PS et al (2020) Transfer learning enables prediction of CYP2D6 haplotype function. PLoS Comput Biol 16(11):e1008399
23. Lanfear DE, Luzum JA, She R, Gui H, Donahue MP, O'Connor CM et al (2020) Polygenic score for β-blocker survival benefit in European Ancestry patients with reduced ejection fraction heart failure. Circ Heart Fail 13(12):e007012
24. Li JH, Szczerbinski L, Dawed AY, Kaur V, Todd JN, Pearson ER et al (2021) A polygenic score for type 2 diabetes risk is associated with both the acute and sustained response to sulfonylureas. Diabetes 70(1): 293–300
25. Takahashi K, Yamanaka S (2006) Induction of pluripotent stem cells from mouse embryonic and adult fibroblast cultures by defined factors. Cell 126(4): 663–676
26. Jinek M, Chylinski K, Fonfara I, Hauer M, Doudna JA, Charpentier E (2012) A programmable dual-RNA-guided DNA endonuclease in adaptive bacterial immunity. Science 337(6096):816–821
27. Deguchi S, Yamashita T, Igai K, Harada K, Toba Y, Hirata K et al (2019) Modeling of hepatic drug metabolism and responses in CYP2C19 poor metabolizer using genetically manipulated human iPS cells. Drug Metab Dispos 47(6):632–638
28. De Masi C, Spitalieri P, Murdocca M, Novelli G, Sangiuolo F (2020) Application of CRISPR/Cas9 to human-induced pluripotent stem cells: from gene editing to drug discovery. Hum Genomics 14(1):25
29. McDermott U (2019) Large-scale compound screens and pharmacogenomic interactions in cancer. Curr Opin Genet Dev 54:12–16

30. Ouyang Q, Liu Y, Tan J, Li J, Yang D, Zeng F et al (2019) Loss of ZNF587B and SULF1 contributed to cisplatin resistance in ovarian cancer cell lines based on Genome-scale CRISPR/Cas9 screening. Am J Cancer Res 9(5):988–998
31. Takayama K, Hagihara Y, Toba Y, Sekiguchi K, Sakurai F, Mizuguchi H (2018) Enrichment of high-functioning human iPS cell-derived hepatocyte-like cells for pharmaceutical research. Biomaterials 161: 24–32
32. de Wildt SN, Tibboel D, Leeder JS (2014) Drug metabolism for the paediatrician. Arch Dis Child 99(12):1137–1142
33. Relling MV, Klein TE, Gammal RS, Whirl-Carrillo M, Hoffman JM, Caudle KE (2020) The Clinical Pharmacogenetics Implementation Consortium: 10 years later. Clin Pharmacol Ther 107(1):171–175
34. Swen JJ, Wilting I, de Goede AL, Grandia L, Mulder H, Touw DJ et al (2008) Pharmacogenetics: from bench to byte. Clin Pharmacol Ther 83(5): 781–787
35. Lee CR, Luzum JA, Sangkuhl K, Gammal RS, Sabatine MS, Stein CM et al (2022) Clinical Pharmacogenetics Implementation Consortium guideline for CYP2C19 genotype and clopidogrel therapy: 2022 update. Clin Pharmacol Ther
36. Dutch Pharmacogenetics Working Group guidelines: KNMP. https://www.knmp.nl/dossiers/farmacogenetica
37. Wu AH, White MJ, Oh S, Burchard E (2015) The Hawaii clopidogrel lawsuit: the possible effect on clinical laboratory testing. Per Med 12(3):179–181
38. Popejoy AB (2019) Diversity in precision medicine and pharmacogenetics: methodological and conceptual considerations for broadening participation. Pharmgenomics Pers Med 12:257–271
39. Popejoy AB, Fullerton SM (2016) Genomics is failing on diversity. Nature 538(7624):161–164
40. Martin AR, Kanai M, Kamatani Y, Okada Y, Neale BM, Daly MJ (2019) Clinical use of current polygenic risk scores may exacerbate health disparities. Nat Genet 51(4):584–591
41. Corbie-Smith G, Thomas SB, St George DM (2002) Distrust, race, and research. Arch Intern Med 162(21): 2458–2463
42. Erves JC, Mayo-Gamble TL, Malin-Fair A, Boyer A, Joosten Y, Vaughn YC et al (2017) Needs, priorities, and recommendations for engaging underrepresented populations in clinical research: a community perspective. J Community Health 42(3):472–480
43. George S, Duran N, Norris K (2014) A systematic review of barriers and facilitators to minority research participation among African Americans, Latinos, Asian Americans, and Pacific Islanders. Am J Public Health 104(2):e16–e31
44. Saulsberry L, Danahey K, Borden BA, Lipschultz E, Traore M, Ratain MJ et al (2021) Underrepresented patient views and perceptions of personalized medication treatment through pharmacogenomics. NPJ Genom Med 6(1):90
45. Fry A, Littlejohns TJ, Sudlow C, Doherty N, Adamska L, Sprosen T et al (2017) Comparison of sociodemographic and health-related characteristics of UK Biobank participants with those of the general population. Am J Epidemiol 186(9):1026–1034
46. Mapes BM, Foster CS, Kusnoor SV, Epelbaum MI, AuYoung M, Jenkins G et al (2020) Diversity and inclusion for the all of Us research program: a scoping review. PLoS One 15(7):e0234962
47. (MCRI) MCsRI. Gen V Project. https://www.genv.org.au/
48. McKillip RP, Borden BA, Galecki P, Ham SA, Patrick-Miller L, Hall JP et al (2017) Patient perceptions of care as influenced by a Large Institutional Pharmacogenomic Implementation Program. Clin Pharmacol Ther 102(1):106–114
49. Klein ME, Parvez MM, Shin JG (2017) Clinical implementation of pharmacogenomics for personalized precision medicine: barriers and solutions. J Pharm Sci 106(9):2368–2379
50. Duarte JD, Dalton R, Elchynski AL, Smith DM, Cicali EJ, Lee JC et al (2021) Multisite investigation of strategies for the clinical implementation of pre-emptive pharmacogenetic testing. Genet Med 23(12):2335–2341
51. Nicholson WT, Formea CM, Matey ET, Wright JA, Giri J, Moyer AM (2021) Considerations when applying pharmacogenomics to your practice. Mayo Clin Proc 96(1):218–230
52. Eadon MT, Desta Z, Levy KD, Decker BS, Pierson RC, Pratt VM et al (2016) Implementation of a pharmacogenomics consult service to support the INGENIOUS trial. Clin Pharmacol Ther 100(1):63–66
53. Liko I, Corbin L, Tobin E, Aquilante CL, Lee YM (2021) Implementation of a pharmacist-provided pharmacogenomics service in an executive health program. Am J Health Syst Pharm 78(12):1094–1103
54. Dunnenberger HM, Biszewski M, Bell GC, Sereika A, May H, Johnson SG et al (2016) Implementation of a multidisciplinary pharmacogenomics clinic in a community health system. Am J Health Syst Pharm 73(23): 1956–1966
55. Zierhut HA, Campbell CA, Mitchell AG, Lemke AA, Mills R, Bishop JR. Collaborative counseling considerations for pharmacogenomic tests. Pharmacotherapy 2017;37(9):990–9
56. Abdela OA, Bhagavathula AS, Gebreyohannes EA, Tegegn HG (2017) Ethiopian health care professionals' knowledge, attitude, and interests toward pharmacogenomics. Pharmgenomics Pers Med. 10:279–285
57. Hundertmark ME, Waring SC, Stenehjem DD, Macdonald DA, Sperl DJ, Yapel A et al (2020) Pharmacist's attitudes and knowledge of pharmacogenomics and the factors that may predict future engagement. Pharm Pract (Granada) 18(3):2008

58. Bank PC, Swen JJ, Guchelaar HJ (2017) A nationwide survey of pharmacists' perception of pharmacogenetics in the context of a clinical decision support system containing pharmacogenetics dosing recommendations. Pharmacogenomics 18(3):215–225
59. Smith DM, Namvar T, Brown RP, Springfield TB, Peshkin BN, Walsh RJ et al (2020) Assessment of primary care practitioners' attitudes and interest in pharmacogenomic testing. Pharmacogenomics 21(15):1085–1094
60. FDA tells patients, docs to take caution when using unapproved PGx tests to make treatment decisions: GenomeWeb. 2018. https://www.genomeweb.com/regulatory-news/fda-tells-patients-docs-take-caution-when-using-unapproved-pgx-tests-make-treatment#.YxEC2ezMKw1
61. FDA Statement Jeffrey Shuren, M.D., J.D., director of the FDA's Center for Devices and Radiological Health and Janet Woodcock, M.D., director of the FDA's Center for Drug Evaluation and Research on agency's warning to consumers about genetic tests that claim to predict patients' responses to specific medications: U.S. Food and Drug Administration (2018) https://www.fda.gov/news-events/press-announcements/jeffrey-shuren-md-jd-director-fdas-center-devices-and-radiological-health-and-janet-woodcock-md
62. Ray T (2019) FDA stepping up actions against PGx testing, forcing some labs to stop reporting drug information: GenomeWeb. https://www.genomeweb.com/regulatory-news-fda-approvals/fda-stepping-actions-against-pgx-testing-forcing-some-labs-stop#.Yw_DYOzMIkg
63. WARNING LETTER, Inova Genomics Laboratory, MARCS-CMS 577422 — APRIL 04, 2019: U.S. Food and Drug Administration. https://www.fda.gov/inspections-compliance-enforcement-and-criminal-investigations/warning-letters/inova-genomics-laboratory-577422-04042019?utm_campaign=040419_PR_FDA%20issues%20warning%20letter%20to%20Inova%20Genomics%20Laboratory&utm_medium=email&utm_source=Eloqua
64. Ray T (2020) FDA defends actions against PGx testing labs as necessary to protect public health: GenomeWeb. https://www.genomeweb.com/regulatory-news-fda-approvals/fda-defends-actions-against-pgx-testing-labs-necessary-protect-public#.Yw_DO-zMIkg
65. Table of Pharmacogenetic Associations: U.S. Food and Drug Administration. https://www.fda.gov/medical-devices/precision-medicine/table-pharmacogenetic-associations
66. Wake DT, Smith DM, Kazi S, Dunnenberger HM (2022) Pharmacogenomic clinical decision support: a review, how-to guide, and future vision. Clin Pharmacol Ther 112(1):44–57
67. Smith DM, Wake DT, Dunnenberger HM (2023) Pharmacogenomic clinical decision support: a scoping review. Clin Pharmacol Ther 113:803
68. Yeh ML, Chang YJ, Wang PY, Li YC, Hsu CY (2013) Physicians' responses to computerized drug-drug interaction alerts for outpatients. Comput Methods Prog Biomed 111(1):17–25
69. Klein TE, Ritchie MD (2018) PharmCAT: a Pharmacogenomics Clinical Annotation Tool. Clin Pharmacol Ther 104(1):19–22
70. Sangkuhl K, Whirl-Carrillo M, Whaley RM, Woon M, Lavertu A, Altman RB et al (2020) Pharmacogenomics Clinical Annotation Tool (PharmCAT). Clin Pharmacol Ther 107(1):203–210
71. Empey PE, Pratt VM, Hoffman JM, Caudle KE, Klein TE (2021) Expanding evidence leads to new pharmacogenomics payer coverage. Genet Med 23(5):830–832
72. Clinical Commissioning Urgent Policy Statement: Pharmacogenomic testing for DPYD polymorphisms with fluoropyrimidine therapies (2020) NHS England
73. Royal College of Physicians and British Pharmacological Society (2022) Personalised prescribing: using pharmacogenomics to improve patient outcomes. Royal College of Physicians and British Pharmacological Society, London

Part IV

Clinical Research

Phases of Clinical Trials

20

Charu Gandotra

Abstract

Clinical trials are trials conducted on human subjects. Clinical trials of an investigational drug product (IDP) that are intended to support drug approval typically evaluate safety, tolerability, pharmacokinetics (PK), pharmacodynamics (PD) and efficacy. These clinical trials are conducted in four sequential phases. Key aspects of clinical trial design such as study endpoints, population, size, duration and monitoring vary across the phases of the clinical trials and are informed by the results of preceding trials. This chapter provides an overview of the various phases of clinical trials, and includes a snapshot of phases of clinical trials and key regulatory interaction timepoints with the United States (US) Food and Drug Administration.

Keywords

Clinical trials · Trial phases · Human trials · Drug safety · Drug efficacy · Drug development

20.1 Introduction

Clinical trials are trials conducted in humans to understand safety, tolerability, pharmacokinetics (PK), pharmacodynamics (PD) and efficacy of an investigational drug product (IDP), device or other types of intervention. Clinical trials are required to provide substantial evidence of effectiveness and safety of an IDP to support drug approval to treat patients with the disease state of interest. Clinical trials are initiated after an adequate characterization of the IDP in nonclinical studies (animal studies and in vitro studies using human or other mammalian cells). Generally, nonclinical studies should have elucidated the mechanism of action, primary and secondary pharmacodynamics, safety pharmacology, pharmacokinetics, and toxicology of the IDP prior to dosing in humans.

There are four successive phases of clinical trials. Each phase has distinct objectives. As clinical trials progress from phase 1 to 4, the trial size and duration increase, the study population advances from healthy volunteers to the intended patient population, there is increasing focus on efficacy and expansion of safety database. For a given phase, the planned treatment duration must be adequately supported by the duration of exposure in nonclinical studies and any preceding clinical trial(s).

Clinical trial design depends on various factors such as the objectives of the trial, phase of the clinical trial, intended population, properties of the IDP, anticipated treatment effect, etc. Clinical

C. Gandotra (✉)
Division of Cardiology and Nephrology, Center for Drug Evaluation and Research, US Food and Drug Administration, Silver Spring, MD, USA
e-mail: charu.gandotra@fda.hhs.gov

G. Jagadeesh et al. (eds.), *The Quintessence of Basic and Clinical Research and Scientific Publishing*,
https://doi.org/10.1007/978-981-99-1284-1_20

trials are conducted according to a pre-specified study plan described in a clinical trial protocol. Details of elements of clinical trial protocols for each of these phases are discussed in Chap. 21 of Part IV. Prior to initiation of a clinical trial involving an investigational drug, the trial protocol should be approved by the relevant Institutional Review Board(s) (IRB), trial participants should provide Informed Consent (IC), and the trial should be conducted in accordance with the International Council for Harmonization (ICH) of Technical Requirements for Pharmaceuticals for Human Use (ICH) Good Clinical Practice (GCP) Guidelines [1]. The ICH is a cooperative effort between the drug regulatory authorities and the pharmaceutical company professional organizations in the European Union, Japan, and the United States to harmonize drug development across the participating geographic regions and reduce duplicative testing conducted during the research and development of new drugs [2].

Safety monitoring in early phase clinical trials is generally conducted by a medical monitor or a safety review committee. Safety monitoring in late phase clinical trials is generally conducted by an independent Data Safety and Monitoring Board (DSMB). The DSMB periodically reviews unblinded safety data and recommends changes to dosing and/or trial design, as needed, to ensure safety of the study participants. A detailed discussion of the format and functioning of a DSMB is beyond the scope of this chapter.

The key objective of this chapter is to provide an overview of the phases of clinical trials, including a discussion of trial design, objectives, population and treatment duration.

20.2 Phase 1 Clinical Trials

Phase 1 clinical trials are conducted to evaluate safety, tolerability, PK and PD of an IDP generally in healthy volunteers, and sometimes in a specific patient population. Phase 1 trials include first-in-human, single- and multiple-ascending dose, and clinical pharmacology trials.

20.2.1 First-in-Human Trial

A first-in-human (FIH) clinical trial is generally a randomized, double-blinded, placebo-controlled, single-ascending dose (SAD) trial that enrolls healthy volunteers in sequential cohorts with the primary objective of evaluating safety and tolerability of a range of doses of the IDP. In some cases, a FIH trial can be a two-part trial that includes single- and multiple-ascending dose (MAD) parts. The starting dose is based on the nonclinical data that defines the no-observed-adverse-effect level (NOAEL) and safety margin, potential target organs of toxicity and means for monitoring these toxicities. Other objectives of a FIH trial include evaluation of PK and PD of the IDP. In cases where the PD biomarker cannot be evaluated in healthy volunteers (HV), a FIH trial may be conducted in a patient population provided there is adequate safety monitoring, for example, oncology trials.

The study size and duration depend on factors such as number of doses to be evaluated, half-life of the IDP and toxicity profile. The study size usually ranges from 20 to 100 study participants with 6–8 participants in each dose cohort randomized in a 2:1 ratio to IDP versus placebo. Alternatively, the trial can have various treatment dose cohorts and one placebo cohort. The study duration, defined as the period between first patient first visit and last patient last visit, is usually several months [3].

Review of safety data from a low dose cohort informs dosing and timing of initiation of subsequent single- and multiple-dose cohorts. Other key design elements for a SAD/MAD trial include defining dose-escalation rules and maximum tolerated dose, choice of dosing interval, evaluation of exposure-response, subject- and study-stopping rules, safety monitoring and management of adverse events.

Depending on the potential toxicity profile, a FIH trial may be conducted entirely or partially in an inpatient or an outpatient setting with frequent follow-up visits. To further identify and/or mitigate risk to healthy volunteers in a FIH trial, a strategy of sentinel dosing can be employed.

Sentinel dosing is where the first dose of an IDP is administered to one study participant, in advance of the rest of the cohort. This sentinel subject is observed for any adverse events for a pre-defined duration and all available data are reviewed prior to dosing the rest of the cohort. In the case of a placebo-controlled trial, two participants can be dosed simultaneously, one with the IDP and the other with placebo. Depending on the PK, PD and level of uncertainty with the IDP, sentinel dosing may also be employed when initiating subsequent dose cohorts.

A placebo-controlled design and an adequate size SAD/MAD trial are usually preferred to single-arm small trials to better characterize early safety, tolerability and PK profile of the IDP.

20.2.2 Clinical Pharmacology Trials

Clinical pharmacology trials are designed to understand the PK / PD properties of an IDP. The spectrum of clinical pharmacology trials conducted depends on the characteristics of the IDP such as the route of administration, mechanism of action, and the pathway by which the product is metabolized. Some examples of clinical pharmacology trials include bioavailability, mass-balance, bioequivalence, drug-drug interaction (DDI), food-effect, PK studies in patients with hepatic- or renal-impairment, and cardiac electrophysiology trials. An in-depth review of clinical pharmacology studies is provided in Part III.

20.2.3 Phase 1 Studies in Patient Population

After initial phase 1 studies in HVs, subsequent phase 1 trials are conducted in patients to further characterize safety, tolerability, PK and PD.

If an IDP is not found to be safe and/or well-tolerated in early phase studies, then the clinical development for that IDP is unlikely to reach phase 3.

20.3 Phase 2 Clinical Trials

Phase 2 clinical trials include dose-finding, proof-of-concept trials. The key objective of the phase 2 trials is to understand if the IDP is likely to work in the intended population. Other objectives include evaluation of exposure-response for efficacy and safety, and to identify appropriate dose (s) of the IDP to be evaluated in phase 3 trials.

Other drug characteristics that may need to be evaluated in phase 2 include dosing interval, route of administration and immunogenicity. A dose-finding study that evaluates 2–3 doses of the IDP can be used to meet the stated objectives. The efficacy endpoint(s) in phase 2 trial(s) can be PD biomarkers that are expected to predict the anticipated clinical benefit in phase 3 trials. Generally, phase 2 trials are designed as randomized controlled trials (RCTs), enroll several hundred patients, and are conducted over several months to 2 years [3].

Phase 2 trials, while not mandatory, are advisable to guide the decision to conduct and the design of phase 3 trials.

20.4 Phase 3 Clinical Trials

Phase 3 clinical trials are trials designed to provide proof of efficacy of the IDP, while continuing to expand the safety database in the intended population and are planned to be confirmatory (pivotal) trials to support drug approval.

Phase 3 trials are generally designed as RCTs that evaluate the IDP versus placebo or an active comparator, administered on top of standard of care. These trials typically enroll several hundred to thousands of patients and are conducted over 1–4 years [3]. Depending on the therapeutic area, proposed indication and trial endpoint(s), one or two pivotal trials may need to be conducted to support drug approval. Key elements of phase 3 pivotal trial(s) are as follows:

(a) Study design: A randomized controlled study design is preferred as it allows determination of treatment effects while controlling for measured and unmeasured sources of bias.

(b) Study endpoint(s): Study endpoint(s) should measure treatment effect on how a patient feels, functions, or survives. It is advisable to use established endpoints that have previously supported drug approval within a therapeutic area. For novel endpoints, appropriate discussion(s) and agreement with the relevant regulatory agency should be considered.

For phase 3 trial designed to evaluate multiple endpoints with the intention to support multiple claims, the hierarchy of statistical testing should be carefully planned. Some considerations for this include the aspects of the disease condition that the IDP is likely to impact, expected number of endpoint events and their clinical relevance.

(c) Study population: The study population should be selected to represent the intended population. To optimize trial size, the study population is commonly enriched with patients who are at high risk for experiencing the study endpoint(s). Enrichment factors should be carefully selected such that they do not negatively impact generalizability of the trial results to a broader population with the disease of interest.

(d) Statistical analysis plan: The statistical analysis plan for a phase 3 trial should be finalized prior to data lock to ensure trial integrity.

20.5 Phase 4 Clinical Trials

Phase 4 clinical trials are conducted after approval of the IDP, generally, to address residual uncertainty about safety in the intended population or in subpopulations that were not studied in phase 1–3 studies of the approved IDP. Phase 4 trials may also be conducted to confirm efficacy of the IDP when used in a real-world setting compared to a controlled environment of phase 3 trial(s) [4]. Depending on the objective(s), a phase 4 trial can be an RCT, post-marketing registry, or pharmacovigilance study.

Figure 20.1 summarizes the new drug development process, including phases of clinical trials, and key regulatory interaction timepoints with the United States (US) Food and Drug Administration (FDA) [2].

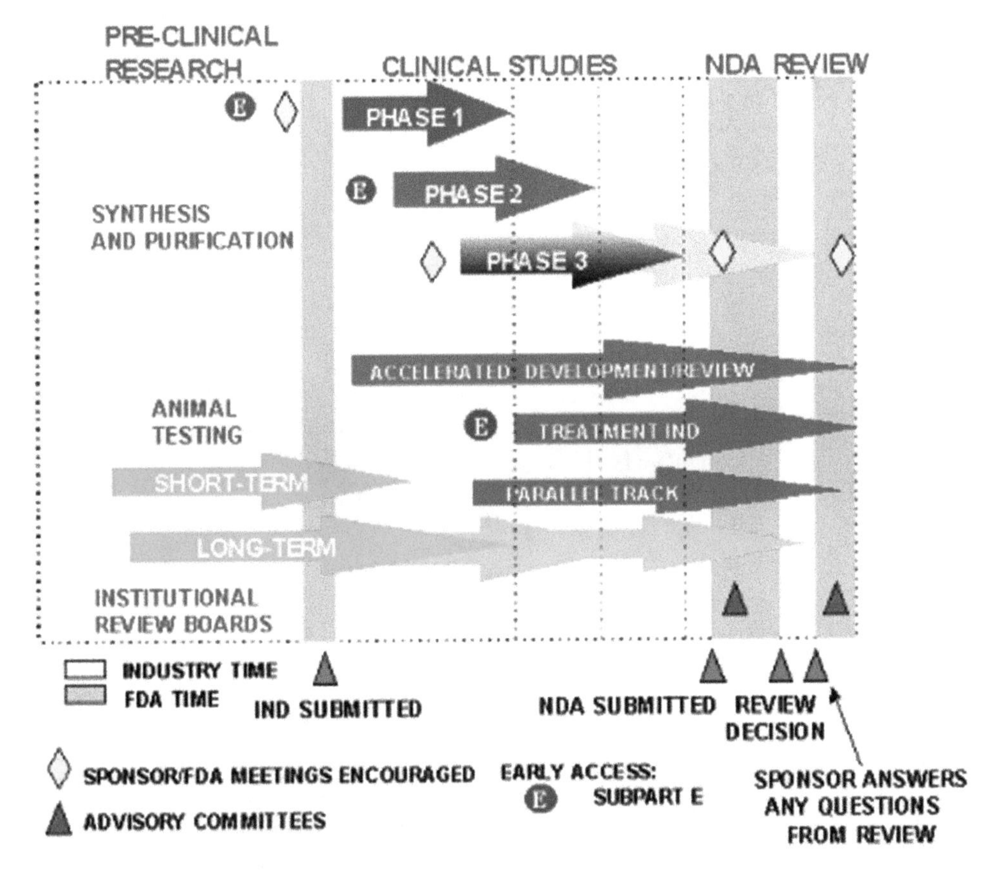

Fig. 20.1 New Drug Development Process and Review. (Source: FDA) [2]. *FDA* Food and Drug Administration, *IND* Investigational New Drug, *NDA* New Drug Application, *BLA* Biologics Licensing Application

▲ IND Submitted: Prior to initiation of clinical trial(s) in the United States, an IND needs to be submitted to FDA. NDA Submitted: After completion of pivotal trials an NDA /BLA is submitted to the FDA. Review Decision: FDA reviews the NDA to decide on approvability of the investigational drug product

▲ Advisory Committee: FDA decides on the need for an Advisory Committee to provide input on key review issues. Not all NDAs need an advisory committee

◇ Sponsor/FDA Meetings: Key points of Sponsor meetings with FDA include prior to IND submission, at end of phase 2 trials, and during NDA review. Additional meetings are encouraged, as needed by the sponsors to seek guidance from the FDA on various aspects of the development program

20.6 Conclusion

Typically, there are four phases of clinical trials, conducted in a sequential fashion with results of early phase trials informing the design of subsequent trials. An adequate nonclinical program is needed before the initiation of clinical trials. Depending on the phase, clinical trials have distinct key design elements.

Disclaimer This chapter reflects the views of the author and should not be construed to represent Food and Drug Administration's views or policies.

Conflict of Interest

None.

References

1. https://database.ich.org/sites/default/files/E6_R2_Addendum.pdf. Accessed 27 Nov 2022

2. https://www.fda.gov/drugs/cder-small-business-industry-assistance-sbia/new-drug-development-and-review-process. Accessed 27 Nov 2022
3. https://www.fda.gov/patients/drug-development-process/step-3-clinical-research. Accessed 27 Nov 2022
4. Smith PG, Morrow RH, Ross DA (eds) (2015) Chapter 22: Phase IV studies. In: Field trials of health interventions: a toolbox, 3rd edn. OUP Oxford, Oxford

Common Clinical Trial Designs

21

Rosalyn Adigun

Abstract

Clinical Trials are studies conducted for the purpose of assessing the effect of an intervention in a defined population, or to evaluate one or more interventions compared to placebo or an active control in human subjects. The objective of a clinical trial is to answer a scientific or medical question related to the effect of an intervention in healthy volunteers or a well-defined patient population with an established condition. As such, one of the most important aspects of clinical trial design is to define the objectives of a study and select the most appropriate trial design to answer the question. To draw meaningful conclusions from clinical trial results, it is important to understand the strengths of each trial design, recognize their limitations, and identify the optimal study design to generate interpretable data of the effect of an intervention without compromising the interpretability of the results. This chapter will provide an overview of common trial designs in clinical medicine.

Keywords

Randomized Controlled Trials · Investigational drug product · Standard of care · Nonrandomized Trials

21.1 Introduction

Clinical Trials are comparative studies designed to evaluate the safety and efficacy of an intervention (drug, device, procedure) independently or relative to another intervention or placebo. Attributes of a clinical trial are inherent in its design and important to consider before the trial commences. The choice of a trial design influences treatment allocation strategies, structure of control groups, and the ability to modify certain aspects of the trial after the study begins.

Historical records of early clinical trials date back to the biblical era, where medical observations were made after subjects were systematically exposed to an intervention (dietary, medical procedure, drugs) and anecdotal reports of these observations recorded. The first modern era physician credited with conducting a controlled clinical trial in medicine was James Lind whose anecdotal recordings of systematically allocating twelve sailors with scurvy to six treatment groups to test the hypothesis of citrus fruits as a treatment for scurvy was positive. Although his findings were published in 1753 in a book titled "A Treatise of the Scurvy", most of this

R. Adigun (✉)
Division of Cardiology and Nephrology, Office of Cardiology, Hematology, Endocrinology, and Nephrology, Center for Drug Evaluation and Research, U.S. Food and Drug Administration, Silver Spring, MD, USA
e-mail: Rosalyn.Adigun@fda.hhs.gov

G. Jagadeesh et al. (eds.), *The Quintessence of Basic and Clinical Research and Scientific Publishing*,
https://doi.org/10.1007/978-981-99-1284-1_21

innovative work was not appreciated until years later [1, 2].

To appropriately assess new therapies, there is a need for critical appraisal of current trial designs, understanding the estimands framework, and selection of the optimal design to demonstrate substantial evidence of effectiveness and ensure approval of safe and effective therapies for public use [2, 3].

Landmark moments in the regulatory oversight over food and drug products have informed the standards of clinical trials and drug development in the United States. In 1906, the United States Congress passed the Pure Food and Drug Act to prevent the misbranding and adulteration of food and drugs. Although the scope of this act was limited in the enforcement authority provided to regulate the approval of new therapeutics, in 1938 the U.S. Food, Drug, and Cosmetic Act subjected new drugs to pre-market safety evaluation. Shortly after the thalidomide tragedy of 1961, the Kefauver–Harris Amendment to the Food, Drug, and Cosmetic Act empowered the FDA to require premarket scientific testing to demonstrate substantial evidence of a drug's effectiveness, and approval of drugs would be based on the submission of adequate and well-controlled studies (for details, *see* Chap. 24).

Important concepts to consider in trial design include strategies for allocating subjects to treatment groups, control groups, study duration, and potential for modifications to design. Randomization in clinical trials minimize allocation bias. Study subjects can be randomized individually or in clusters. These decisions are typically influenced by the research question, study size, and setting of a trial. Control groups are typically enrolled concurrently; however, historical controls can be employed based on the intervention studied. Conventional trial designs are typically more rigid in allowing modifications to the clinical protocol when the trial is ongoing, while adaptive designs allow for prospectively planned modifications based on accumulating trial data.

Ultimately, well-designed clinical trials should be capable of answering the scientific question being asked, and consideration should be given to the acuity of illness, number of interventions compared, estimated effect size, duration of study, dropout rates, and the setting within which the study is to be conducted. Insights into these elements of a protocol will guide the decision of the optimal design of a clinical trial [4].

The key objective of this chapter is to provide a brief overview of various randomized clinical trial designs, as well as trials that are not based on randomization.

21.2 Randomized Control Trials

Considered the gold standard for the valid assessment of therapeutic efficacy, randomized controlled trials (RCT) are prospective comparative studies that measure the effect of a molecular entity, device, or other intervention in a treatment group compared to a control group that has not been exposed to the intervention. It is worth noting that the term "control" is not necessarily synonymous with the absence of treatment. It could reflect no treatment, a different treatment or standard of care, similar treatment but at a different dose or administered on a different schedule. All participants in a RCT have a random chance of being assigned to either the intervention or the control arm. The advantages of RCTs are that they minimize bias in treatment allocation and balance all known and unknown confounders present in the treatment groups.

Subject level data from RCTs are frequently analyzed in two ways, according to the group a subject was randomized into (intention-to-treat) or based only on subjects who completed treatment in the group they were allocated to at randomization (per protocol). There are other variations not commonly used. The intention-to-treat analysis is preferred as it preserves randomization. It is important that the details of the planned analysis are pre-specified in the protocol prior to unblinding of the data.

Limitations of RCTs, as well as the other trial designs discussed, are associated with understanding the meaningfulness of smaller treatment effects over control within the evolving standard of care. As more and more drugs enter the market, improvements in outcomes render any incremental benefit from a trial drug to be smaller and smaller, unless the study drug is purported to produce a breakthrough treatment benefit. The smaller improvements over control require a larger sample size to detect the therapeutic benefit with at least 95% confidence, thus greatly increasing the scope and cost of running a clinical trial. Prohibitive costs may attenuate further drug development. This problem is also compounded by some disease states classified as rare. In these settings, there may not be a sufficient number of patients to satisfy the statistical requirements to ensure adequate data interpretability, thus impeding the ability to draw an adequate conclusion about the drug's benefit. This phenomenon has prompted evolving perspectives on how to obtain evidence to satisfy the statutory requirements for a regulatory decision [5]. Randomized trials have several design features that are often used in drug development, such as the parallel group trial design, factorial design, cross-over design, and placebo-controlled withdrawal design. Comparison of the test drug to a control is governed by a prespecified hypothesis, usually a superiority hypothesis where the test drug is purported to be superior to the control in reducing the incidence of a primary endpoint by a specific amount, commonly known as the treatment effect. Although superiority hypothesis trials are common, drug development strategies also employ alternative hypotheses such as equivalence or non-inferiority testing. Equivalence trials are designed to demonstrate that the mean treatment effect of a test drug is equal to that of a control drug. Since the hypothesis specifies no difference in treatment effect, the sample size requirement may become unwieldy. Consequently, an alternative strategy is the non-inferiority hypothesis. The various types of RCTs, as well as specific features of the non-inferiority hypothesis, will be described below.

21.3 Parallel Group Trial Design

The parallel group design is probably the most commonly used randomized trial design (Fig. 21.1). Subjects are randomized to a treatment group and exposed to one intervention for the duration of the study. The subjects are followed simultaneously from the time of treatment allocation to the end of study timepoint. The main objective of this study design is to compare one treatment to another (e.g., treatment A vs. treatment B, and treatment A vs. treatment C, and/or treatment B .vs. treatment C). The complexity of the comparisons should be pre-specified in the analysis plan along with strategies to minimize and handle missing data. Plans to prevent and control subjects from crossing over to another treatment arm should also be outlined before the trial begins. The database should be complete and clean (correction of mis-entries), to ensure interpretability of the data so as to be able to draw conclusions.

21.4 Factorial Design

Factorial design studies are designed to evaluate two or more interventions in comparison to control in one study simultaneously (Fig. 21.2). While a 2 × 2 factorial design is commonly seen, more complex comparisons can be designed using this approach. This type of study can be appealing because of the ability to run multiple comparisons at once, answer more than one

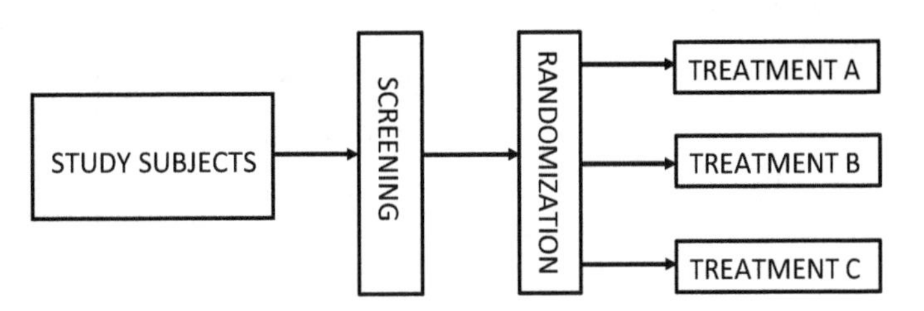

Fig. 21.1 Parallel Group Trial Design

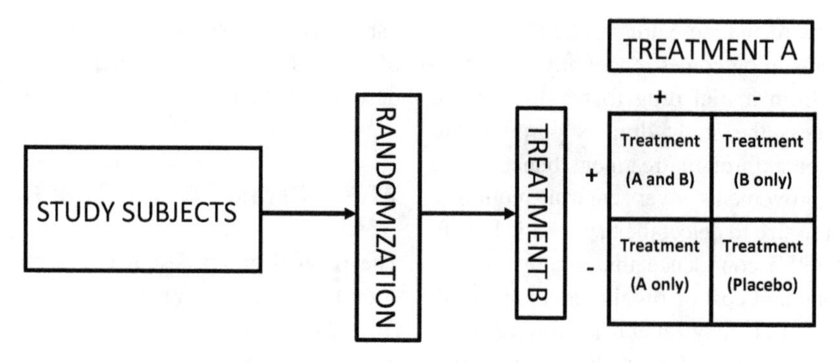

Fig. 21.2 Factorial (2 × 2) Trial Design

research question, and maintain statistical power with a limited sample size. However, the factorial design also has disadvantages. If an interaction exists between the treatments being tested, there usually is a loss of statistical power; and data analyses and interpretation become quite complex. It can also be challenging to screen for and identify patients who meet criteria for multiple research questions simultaneously [4].

21.5 Crossover Trial Design

A crossover trial design is implemented in the setting when each patient receives two or more treatments in a predetermined sequence and allows a subject to serve as his/her own control. In this type of trial design subjects are randomized to a group and each group receives different sequences of the investigation drug product in different orders. The most common scenario is a two group, two-period cross over design (Fig. 21.3). One group is assigned to the intervention while the other group is assigned to placebo in the first period of the study and the group alternates their assignments in the next period after a pre-determined washout period [4, 6].

The advantage of this type of study design is that it allows for within-patient comparisons between treatments, as each patient serves as a control and thus reduces variability, leading to smaller samples sizes required for this type of study [6].

Some factors to consider with this type of clinical trial are the potential for carry-over effects from therapies that subjects are exposed to in the earlier sequence of the study. This can be addressed by extending the washout period before the next sequence. For instance, if an intervention offered in the first sequence of a crossover study design resolves the symptoms of a study participant, this will affect the assessment of treatment effect of the intervention in the second sequence of the crossover trial and can affect interpretability of the study. Other considerations include effects of the order of treatment on the disease (order effect) which may lead to fluctuations in disease severity (e.g., as with allergies) and complicate the ability to analyze the data. Missing data from subject dropout can also be an issue in a crossover study because each patient

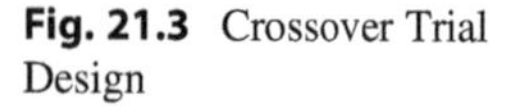
Fig. 21.3 Crossover Trial Design

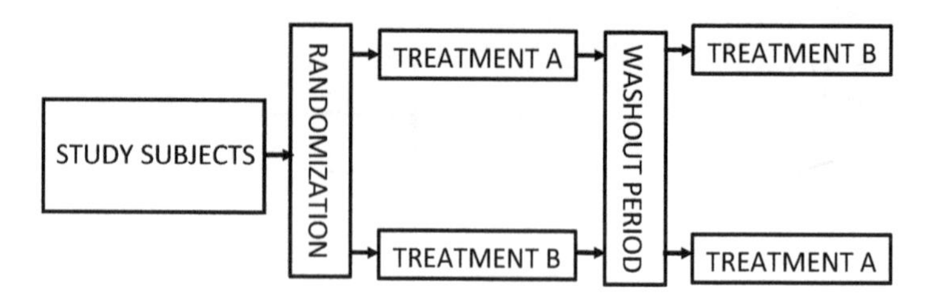

contributes significantly to the trial data and the loss of a single patient is more impactful in comparison to a parallel study design. Ultimately, crossover trials are appropriate for the study of chronic diseases, investigational drugs with relatively short half-lives and reasonable washout periods, and for studies with short, planned study duration.

21.6 Placebo-Controlled Withdrawal study design

The objective of this type of study is to assess a subject's response to dose reduction or discontinuation of an investigational drug product after the subject has received the drug for some time. After an initial open label period during which every subject receives the investigational drug product for a pre-specified period, responders are then randomized to either continue the investigational drug or placebo (Fig. 21.4). The outcome of interest is usually relapse of symptoms treated by the investigational drug, and data from the withdrawal period are usually what is analyzed in this type of trial. One goal of this type of trial is to demonstrate long-term efficacy of an investigational drug (e.g., an antidepressant) by showing that subjects who have been on the drug for some time and then stop the drug lose efficacy compared to those who remain on the drug. The run-in phase also increases the power of the trial by enriching the study with a study population of interest (e.g., subjects with a condition optimally controlled by the intervention) ahead of the withdrawal phase [4]. This study design is beneficial in situations where the effect of treatment discontinuation is likely to be observed within the timeframe of a short randomized withdraw period. An advantage of this design is that subjects don't need to be randomized longer-term to placebo as occurs in a typical randomized, placebo-controlled trial.

21.7 Noninferiority Trials

Noninferiority (NI) trials test whether a new intervention is not too much worse than, or at least as good as, an established intervention. There are a number of reasons why these trials are important (e.g., unethical to administer placebo; for comparative effectiveness assessment; because of non-feasibility of an equivalence trial). However, there are several important aspects that must be considered for this study design to be appropriate. The standard of care or active control must have established superiority over placebo. The metrics (patient population, concomitant therapy, dose) upon which the active control previously demonstrated superiority should be similar in the NI trial. The active control group event rate in a NI trial should be estimated based on the evidence used to demonstrate the superiority of the active control over placebo. An acceptable NI margin should be pre-specified in the trial design, and there should be assay sensitivity (the active comparator should perform as expected) to allow for a valid comparison between the investigational drug and comparator.

Selecting a NI margin is a critical aspect of this study design and should be agreed upon with the regulatory authority. The objective of the NI study is to show that the effectiveness of the investigational drug is not unacceptably smaller than that of the active control using a prespecified NI margin. The identified margin cannot be larger than the presumed entire effect of the active

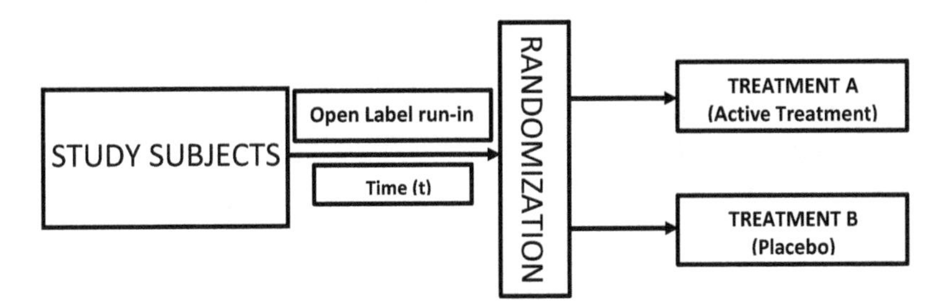

Fig. 21.4 Withdrawal Trial design

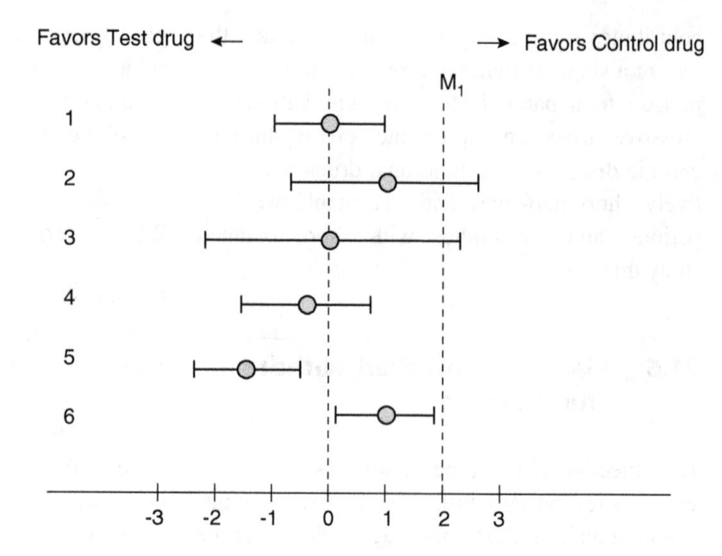

Fig. 21.5 Non-inferiority Trial (Source: FDA)

control in the NI study. It is worth mentioning that the effect of the active control is not measured in the NI trial (there is no placebo arm), but rather is estimated based on past performance of the active control [7, 8].

For a drug to be declared non-inferior to a control drug, the upper bound of the 95% confidence interval for the treatment difference (control minus test drug) must not cross the non-inferiority margin (see Fig. 21.5). In the example shown in Fig. 21.5, note that the non-inferiority margin (M1) is different from the null line set at zero. Results to the left of the null line represents a better mean treatment effect for the test drug over the control drug. Results to the right of the null line represents a worse treatment effect for the test drug over the control drug. In the 5 illustrated trials in the figure, the first trial result showed no difference between test drug and control, but since the NI margin was not crossed, it is considered non-inferior to the control drug. Results from trials # 2 and # 3 demonstrate failure to meet the non-inferiority hypothesis. The null line is crossed in trial # 4, showing no statistically significant treatment effect compared to the control drug. However, non-inferiority was demonstrated. Trial # 5 showed superiority of the test drug compared to the control drug. Trial # 6 showed inferiority of the test drug to the control drug despite meeting the non-inferiority hypothesis.

The NI margin must make clinical sense and ensure that there is residual efficacy of the investigational drug, and that the treatment effect of the investigational drug is not unacceptably less effective than that of the control. Declaration of non-inferiority in the setting where the entire 95% confidence interval is on the right side of the null line, as illustrated in trial # 6 in the figure, will give rise to skepticism about the results and would become a serious review issue if these results were submitted in an application to approve a drug based on non-inferiority.

Analyzing noninferiority by comparing the confidence interval (CI) of the relative risk to a predefined margin (M1).

1. Point estimate of C-T is 0, suggesting equal effect of C and T; upper bound of the 95% CI for C-T is 1, well below M1; NI is demonstrated.
2. Point estimate of C-T favors C; the upper bound of the 95% CI for C-T is >2, above M1; NI is not demonstrated.
3. Point estimate of C-T is zero, which suggests an equal effect; but the upper bound of the 95% CI for C-T is >2 (i.e., above M1), so that NI is not demonstrated.
4. Point estimate favors T; NI is demonstrated, but superiority is not demonstrated.
5. Point estimate favors T; superiority and NI are demonstrated.
6. Point estimate of C-T is 1, favoring the control. The upper bound of the 95% CI for C-T is <M1, demonstrating NI (the entire effect of C

has not been lost) but at the same time the 95% CI for C-T is above zero, indicating that T is actually inferior to C, even while meeting the NI standard.

21.8 Nonrandomized Trials

Nonrandomized controlled trials are studies in which participants in a trial are not assigned by chance to different treatment groups. This type of trial design could consist of a single study arm without a comparator group, but other schemes can be used. Under certain scenarios specified in a protocol, participants may be allowed to choose a treatment group, they may also be assigned to a treatment group by the investigator, or an algorithm could be used to assign participants to treatment groups in order to optimize baseline characteristics. Nonrandomized trials are utilized when the scientific/medical question cannot be adequately addressed by a RCT (e.g., when it is unethical to conduct an RCT or the outcome is rare) and available external controls (comparator arms from prior studies or historical cohorts) can be informative. The studies can be case series, cohort, case-control, or cross-sectional studies to name a few. These studies are inherently limited in their ability to estimate a causal effect.

Nonrandomized trials are easier to design and implement and can also be performed at a reduced cost. However, the potential for allocation bias, challenges establishing comparability between the treatment arms in the study (or with an external control arm) and controlling for confounding factors limit the interpretability of these studies [4].

21.9 Concluding Remarks

Well-designed clinical trials are important for the evaluation and development of new therapies to influence clinical practice patterns, modify the natural history of disease, and to improve public health. As the treatment standards for management of diseases expand, assessment of the incremental benefit of newer therapies becomes complex. Particular attention devoted to the selection an optimal clinical trial designed to evaluate the effect size of an intervention, in a well-defined patient population, over a specified duration with systematic assessment of the benefit-risk profile, and critical appraisal of the results are paramount to success.

The review provided in this chapter is by no means a comprehensive assessment, but an overview of the common designs encountered in drug development. I covered the strengths and limitations of each trial design and scenarios that favor a particular approach.

Ultimately, a clear understanding of the study objectives, the condition to be evaluated, patient population of interest, appropriate trial controls, and statistical design are some of the critical elements that influence successful clinical trial design and execution.

Disclaimer This article reflects the views of the author and should not be construed to represent US Food and Drug Administration.

Conflict of Interest
None.

References

1. Bhatt A (2010) Evolution of clinical research: a history before and beyond James Lind. Perspect Clin Res 1:6–10
2. Chalmers I (2001) Comparing like with like: some historical milestones in the evolution of methods to create unbiased comparison groups in therapeutic experiments. Int J Epidemiol 30:1156–1164
3. Junod S. FDA and clinical drug trials: a short history, U.S. Food and Drug Administration, Feb. 1, 2018.
4. Friedman LM, Furberg C, DeMets DL, Reboussin D, Granger CB (2015) Fundamentals of clinical trials, 5th edn. Springer, Cham, pp 89–118
5. Frieden TR (2017) Evidence for health decision making - beyond randomized, controlled trials. N Engl J Med 377(5):465–475
6. Louis TA, Lavori PW, Bailar JC 3rd, Polansky M (1984) Crossover and self-controlled designs in clinical research. N Engl J Med 310(1):24–31
7. Mauri L, D'Agostino RB (2017) Challenges in the design and interpretation of noninferiority trials. N Engl J Med 377:1357–1367
8. Liu CS-C, J-pei. (2014) Design and analysis of clinical trials: concepts and methodologies, 3rd edn. Wiley, Hoboken, NJ, pp 165–210

Elements of Clinical Trial Protocol Design

22

Rekha Kambhampati and Kirtida Mistry

Abstract

Clinical trials are conducted for many reasons, including to determine whether a new drug is safe and effective for a therapeutic use. A clinical study protocol acts as a how-to guide for the conduct of a study and is essential for successful trial execution. When drugs or biologics are being developed for regulatory approval, the Food and Drug Administration (FDA) requires that a clinical protocol be submitted for each planned clinical study or trial enrolling participants in the United States. According to the US Code of Federal Regulations (21 CFR Part 312), the specific elements and level of detail for a protocol can vary depending on the phase of the study. The clinical study protocol for a phase 1 study should be designed with a focus on safety and dosing; protocols for phase 2 and 3 studies should include detailed descriptions for all aspects of the study, including safety, efficacy, and statistical considerations. The US Code of Federal Regulations (21 CFR Part 312) describes specific elements that should be included in all clinical protocols, regardless of the clinical phase. The International Conference on Harmonization Good Clinical Practice Guidelines (ICH GCP E6, Section 6) specify additional sections that should generally be included in a protocol. The objective of this chapter is to describe the elements that should be included in a clinical trial protocol.

Keywords

Clinical trial phases · Code of Federal Regulations · International Conference on Harmonization · Good Clinical Practice · Protocol elements · Protocol design

22.1 Introduction

Clinical trials are conducted for many reasons, including to determine whether a new drug is safe and effective for use, to study different ways to use currently approved treatments to increase their efficacy or safety, and to learn how to use a treatment safely in a population for which it was not previously tested [1]. A clinical study protocol acts as a how-to guide for the conduct of a study; it describes the background, rationale, objectives, design, methodology, statistical considerations, and organization of a clinical

The opinions expressed herein are those of the authors and do not necessarily reflect those of the US Food and Drug Administration.

R. Kambhampati (✉) · K. Mistry
Division of Cardiology and Nephrology, Center for Drug Evaluation and Research, US Food and Drug Administration, Silver Spring, MD, USA
e-mail: Rekha.Kambhampati@fda.hhs.gov; Kirtida.Mistry@fda.hhs.gov

G. Jagadeesh et al. (eds.), *The Quintessence of Basic and Clinical Research and Scientific Publishing*,
https://doi.org/10.1007/978-981-99-1284-1_22

study [2]. A well-designed protocol is essential for successful trial execution. When drugs or biologics are being developed for regulatory approval, the Food and Drug Administration (FDA) requires that a clinical protocol be submitted for each planned clinical study or trial enrolling participants in the United States [3]. The objective of this chapter is to describe the elements that should be included in a clinical trial protocol.

22.2 Clinical Trial Phases

Per the US Code of Federal Regulations (21 CFR Part 312), the specific elements and level of detail for a protocol can vary depending on the phase of the study [4].

22.2.1 Phase 1

The clinical study protocol for a phase 1 study should be designed with a focus on safety and dosing [5]. Prior to designing a phase 1 clinical protocol, investigators should have an understanding of the expected and/or potential risks of the investigational product. Preclinical studies and other phase 1 studies can provide information on such expected risks, including clinical safety margins and the monitorability of each risk.

According to the US Code of Federal Regulations (21 CFR 312.23(a)(6)(i)), protocols for phase 1 studies can be less detailed and more flexible compared to protocols for phase 2 and 3 studies [4]. The phase 1 protocol should include details on aspects of the study that are critical to participant safety (e.g., safety monitoring of vital signs, laboratory, and other assessments), and should include an outline of the study, comprising elements such as an estimate of the number of participants to be enrolled, a description of safety exclusions, and information on dosing, including the duration and rationale [4]. The justification for the dose should be based on the totality of the existing preclinical and clinical data, and should include information on the pharmacokinetics of the drug and clinical safety margins for predicted human exposures based on preclinical studies.

Phase 1 clinical trials are usually conducted in a small number of healthy volunteers. Sometimes, the first study submitted to an Investigational New Drug Application (IND) is the first time humans will be exposed to an investigational product. Study participants are usually randomized to a single dose of investigational product or placebo in each dose group (e.g., approximately 8 participants per dose level randomized 6:2 to the investigational product or placebo) and observed for a period of time; this is referred to as a single ascending dose (SAD) study. For some first-in-human studies, dosing can begin with one or two participants in a dose cohort (sentinel participants) and if there are no safety concerns, the remainder of the cohort is dosed. Safety, tolerability, pharmacokinetic, and pharmacodynamic data are collected. The protocol for an SAD study should clearly outline the criteria that should be met to begin dosing for the next higher dose level, as well as stopping criteria for halting further dosing in a dose group and halting further dose escalation. The SAD study may also include a cohort that evaluates the effect of food on pharmacokinetics (see Chap. 17).

In general, the next step after an SAD study is to evaluate the safety, tolerability, pharmacokinetics, and pharmacodynamics of multiple doses of the investigational product in a multiple ascending dose (MAD) study. The study conduct is similar to the SAD study, except that multiple doses (e.g., for 1 week) are administered to participants in a particular dose cohort. The MAD study can be conducted in healthy volunteers or patients with the disease of interest. In addition to the aforementioned criteria for protecting the safety of participants in an SAD study, the protocol for an MAD study should also include criteria for halting further dosing of the investigational product in a participant in the dose cohort.

22.2.2 Phases 2 and 3

Phase 2 and 3 studies are typically conducted in patients with the disease or condition of interest [4]. Compared to phase 1 protocols, phase 2 and 3 clinical study protocols should include detailed descriptions for all aspects of the study [4]. In addition to a description of safety-related aspects, the protocol should also include details regarding statistical considerations, such as a description of the sample size, handling of missing data, and handling of protocol deviations (see Chap. 25).

According to the US Code of Federal Regulations (21 CFR 312.23(a)(6)(i)), phase 2 and 3 protocols should be designed to include built-in contingencies for expected protocol deviations [4]. For example, a protocol for a controlled short-term study could include an option for non-responders to cross-over to an alternative therapy [4].

22.3 Anatomy of a Clinical Study Protocol

22.3.1 Required Information for All Protocols

The US Code of Federal Regulations (21 CFR 312.23(a)(6)(iii)) describes specific elements that should be included in all clinical protocols, regardless of the clinical phase.

22.3.1.1 Statement of Objectives and Study Purpose

The protocol should clearly state the objectives and the purpose of the study within the framework of the phase of the clinical trial. The study purpose should explicitly state the goal of the trial.

The study objectives are concise statements that reflect how the study purpose will be achieved. The study objectives should be succinct yet contain sufficient detail to understand the intent of the trial (e.g., to determine the effect of Drug A on loss of kidney function in patients with chronic kidney disease) [6]. We recommend that this section also include the study endpoints, and that the objectives align with these endpoints. A clinical trial may have primary, secondary, and exploratory objectives, and it is good practice to separate these objectives in the protocol. In general, a protocol should not include more than one to two primary and key secondary objectives [6].

22.3.1.2 Investigator Qualifications and Research Facilities

The protocol should include the name and address of each primary investigator and the name of each sub-investigator working under the supervision of the primary investigator(s). A statement of qualifications (e.g., curriculum vitae) for each primary investigator should also be included with the protocol submission. The protocol should also include the name and address of the research facilities to be used and the name and address of each reviewing Institutional Review Board [4].

22.3.1.3 Enrollment Criteria and Sample Size

The protocol should clearly state the inclusion and exclusion criteria, as well as an estimate of the number of patients expected to be enrolled [4]. Eligibility criteria should identify the population of interest and ensure the safety of study participants (e.g., by excluding patients known to be at high risk of severe adverse events). The eligibility criteria may also be used to improve the interpretability of safety signals (i.e., reduce the noise in the safety signals).

In general, enrollment criteria typically include the following components:

- The participant is willing and able to provide signed informed consent.
- The sex and age range for eligible participants (e.g., males and females ages 18 to 65 years).
- For studies enrolling patients with a disease of interest, the protocol should include criteria that will be used for the diagnosis of the disease, including thresholds for biomarkers of interest.

- Contraception requirements for women of childbearing potential. Depending on the risk profile of the investigational product, the enrollment criteria may also include contraception requirements for male study participants and/or female sexual partners of male study participants.
- Key concomitant medications that patients will be allowed to continue or should discontinue while in the study. For medications that will be continued, the criteria should specify what constitutes a "stable" dose of the concomitant medication, when applicable. For medications that will be discontinued, the protocol should specify how long the patient must be off of the medication to be eligible for study enrollment.
- Specific criteria for relevant comorbid conditions. For example, a protocol may exclude patients with a documented history of New York Heart Association Class III or IV heart failure or clinically significant coronary artery disease within six months before screening.
- Thresholds for relevant vital signs and laboratory criteria.
- Exclusion of patients who, in the opinion of the investigator, are unable to adhere to the requirements of the study.

The enrollment criteria can also be used to enrich the study population to select a subset of patients in which the potential effect of an investigational product can be more readily demonstrated [7]. Enrichment can also be used to improve the ability to assess the safety of an investigational product [7]. The enrollment criteria can be tailored to address various enrichment strategies, including:

- A strategy to decrease variability: Including eligibility criteria that will enroll patients with baseline disease characteristics or biomarkers of the disease within a narrow range and will exclude patients whose disease or symptoms improve spontaneously or whose measurements are highly variable. Such a strategy will increase the study power [7].
- Prognostic enrichment: Including patients with a greater likelihood of having a disease-related endpoint event or a substantial worsening of their disease or condition. Such a strategy aims to identify high-risk patients, thus increasing the absolute effect difference between groups (without altering the relative effect) [7].
- Predictive enrichment: Including patients who are more likely to respond to the drug treatment than other patients with the disease or condition based on a specific disease characteristic or biomarker that is related to the drug's mechanism of action. Such a strategy aims to identify more-responsive patients and can lead to larger absolute and relative effect sizes [7].

22.3.1.4 Study Design

The protocol should include a detailed description of the study design. The study design should include the following components [2]:

- A statement of the primary, secondary, and exploratory endpoints to be evaluated during the trial. For information on clinical trial endpoints, see Chap. 20.
- A description of the type of trial that will be conducted (e.g., double-blind, randomized, multicenter, placebo-controlled, parallel design trial). For information on trial designs, see Chap. 21. The description should include a definition of when the patient is considered as formally enrolled in the study. Details on the number of study visits (including follow-up), procedures at each visit, visit windows, and length of study visits should also be included. If any sub-studies are planned, these should also be added to the description. To aid interpretation, it is helpful to also include a diagram of the study design which includes the various phases of the study and treatment arms. Figure 22.1 provides an example of a study design diagram.
- Measures taken to minimize bias, such as randomization and/or blinding. When applicable, the protocol should also include a description

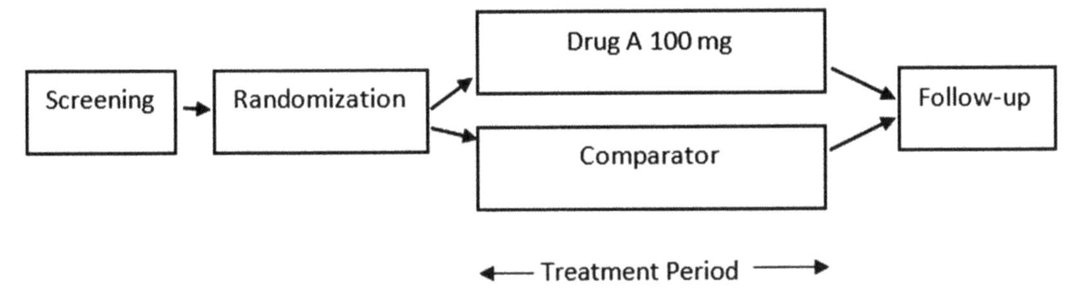

Fig. 22.1 Example Study Design Diagram

of how the randomization codes will be maintained and the procedures for breaking those codes [2].

- The expected duration of each patient's participation in the study.
- Accountability procedures for the investigational product(s) and the comparator(s) [7].

22.3.1.5 Information on Dosing

The protocol should explicitly name the treatments to be administered and the method for determining the dose(s) to be administered [2, 4]. The description for each treatment should include the route of administration, dosing schedule, planned maximum dosage, and the duration of individual patient exposure to the investigational product [2, 4].

22.3.1.6 Observations and Measurements to Fulfill the Study Objectives

The observations and measurements that will be obtained during the study can be broadly divided based on assessments of efficacy and safety. The protocol should clearly specify the efficacy and safety parameters that will be evaluated. For each parameter, the protocol should include the methods and timing for assessing, recording, and analyzing these parameters. For each efficacy parameter, the baseline should be clearly defined, including which measurement will be used as the baseline measurement (e.g., the last serum creatinine measured prior to initiation of investigational product). The protocol should indicate what measurements will be used for missing data (e.g., substitution of spot first morning urine protein measurements for missing 24-hour urine protein measurements).

The preclinical toxicology profile and safety data from early/previous clinical studies should be used to determine the expected risk profile of the investigational product, the specific safety parameters to be evaluated, and the frequency of safety monitoring. The protocol should include procedures for follow-up of abnormal values.

The protocol should include a section that describes the procedures for reporting adverse events. This section should include the definition of an adverse and serious adverse event, time period and frequency for collecting adverse event information, a description of where adverse events will be recorded (e.g., case report form), methods for detecting adverse events, methods for follow-up of adverse events, and the procedure(s) for regulatory reporting requirements for serious adverse events.

22.3.1.7 Procedures to Minimize Risk to Participants

The protocol should include measures to mitigate and adequately monitor for the potential risks. Examples of such measures include enrollment criteria that exclude patients at high-risk for severe adverse events, frequent laboratory monitoring of relevant risks, and inclusion of drug discontinuation and/or dose modification criteria based on expected risks. Table 22.1 includes examples of potential risks and clinical protocol measures that can be implemented to monitor for and mitigate those risks.

The protocol should also include a summary of the Safety Review Committee and/or Data

Table 22.1 Examples of potential risks and clinical protocol measures to mitigate those risks

Potential risk	Protocol measures to mitigate risk
Hepatotoxicity	• Exclusion of participants with hepatic disorders at screening • Inclusion of frequent monitoring of liver laboratory assessments in the Schedule of Assessments • Drug-discontinuation criteria based on elevations in liver laboratory assessments
Fluid overload	• Exclusion of patients with NYHA Class III or IV heart failure or patients with the inability to tolerate an increase in fluid • Inclusion of frequent monitoring of weight, blood pressure, and targeted physical exams in the Schedule of Assessments • Inclusion of a management plan for participants who meet certain thresholds for increase in weight (e.g., initiation of diuretic medications, dose modification, investigational product discontinuation)

Abbreviations: *NYHA* New York Heart Association

Monitoring Committee roles and responsibilities. It is acceptable for details of the roles and responsibilities to be included in a separate charter.

22.3.2 Additional Protocol Information

Clinical study protocols should include sufficient information for investigators to conduct the study systematically in a safe and efficient manner. In addition to the information required by the US Code of Federal Regulations, the ICH Good Clinical Practice Guidelines (ICH GCP E6, Section 6) specify additional sections that should generally be included in a protocol [2]. While these elements are not specific to a phase of clinical development, they generally apply to phase 2 and 3 clinical protocols.

22.3.2.1 ICH E6 Protocol Checklist

As a reference, a Clinical Protocol Checklist based on the ICH Good Clinical Practice Guidelines is included below (Table 22.2) [2].

22.3.2.2 Protocol Synopsis

A brief synopsis of the protocol provides an easy-to-access and quick reference for the key elements of the protocol. In general, elements of the protocol synopsis include the title, protocol number, study phase, number of study sites and participating countries, number of patients, as well as a brief description of each of the following: objectives, methodology and study design, key eligibility criteria, dosing information, endpoints, and statistical methods.

22.3.2.3 Background Information

The background section can provide important context for the disease of interest and the clinical development program, and its inclusion is strongly encouraged. Relevant information on the disease of interest, a description of the affected patient population, rationale for the treatment of the disease with the investigational product (including a description of the mechanism of action), a summary of findings from preclinical studies that have clinical significance, and findings from previous clinical trials that are relevant to this trial should be included. A description of the clinical development program to date, a summary of potential benefits and risks, and a rationale for conducting the study are also recommended. Often, protocols will include a statement that the trial will be conducted in compliance with Good Clinical Practice and applicable regulatory requirements. This statement can be included as part of the background information.

22.3.2.4 Participant Withdrawal and Loss of Follow-up

The protocol should specify the procedures for participant withdrawal. Pre-specified criteria for participant withdrawal, including voluntary withdrawal of consent by the patient or withdrawal due to investigator discretion, should be included. The protocol should include a statement indicating that participants will be free to

Table 22.2 Clinical Protocol Checklist [2]

ICH E6 section	Content/comments
General information	1. Protocol title, protocol identifying number, and date. Any amendment(s) should also bear the amendment number(s) and date(s) 2. Name and address of the sponsor and monitor (if other than the sponsor) 3. Name and title of the person(s) authorized to sign the protocol and the protocol amendment(s) for the sponsor 4. Name, title, address, and telephone number(s) of the sponsor's medical expert (or dentist when appropriate) for the trial 5. Name and title of the investigator(s) who is (are) responsible for conducting the trial, and the address and telephone number(s) of the trial site(s) 6. Name, title, address, and telephone number(s) of the qualified physician (or dentist, if applicable), who is responsible for all trial-site related medical (or dental) decisions (if other than investigator) 7. Name(s) and address(es) of the clinical laboratory(ies) and other medical and/or technical department(s) and/or institutions involved in the trial
Background Information	1. Name and description of the investigational product(s) 2. A summary of findings from nonclinical studies that potentially have clinical significance and from clinical trials that are relevant to the trial 3. Summary of the known and potential risks and benefits, if any, to human participants 4. Description of and justification for the route of administration, dosage, dosage regimen, and treatment period(s) 5. A statement that the trial will be conducted in compliance with the protocol, GCP, and the applicable regulatory requirement(s) 6. Description of the population to be studied 7. References to literature and data that are relevant to the trial and that provide background for the trial
Trial Objectives and Purpose	A detailed description of the objectives and the purpose of the trial
Trial Design	1. A specific statement of the primary endpoints and the secondary endpoints, if any, to be measured during the trial 2. A description of the type/design of trial to be conducted (e.g., double-blind, placebo-controlled, parallel design) and a schematic diagram of trial design, procedures, and stages 3. A description of the measures taken to minimize/avoid bias, including: (a) Randomization, (b) Blinding 4. A description of the trial treatment(s) and the dosage and dosage regimen of the investigational product(s). Also include a description of the dosage form, packaging, and labelling of the investigational product(s) 5. The expected duration of participant participation, and a description of the sequence and duration of all trial periods, including follow-up, if any 6. A description of the "stopping rules" or "discontinuation criteria" for individual participants, parts of trial, and/or entire trial 7. Accountability procedures for the investigational product(s), including the placebo (s) and comparator(s), if any 8. Maintenance of trial treatment randomization codes and procedures for breaking codes 9. The identification of any data to be recorded directly on the CRFs (i.e., no prior written or electronic record of data), and to be considered to be source data
Selection and Withdrawal of Participants	1. Participant inclusion criteria 2. Participant exclusion criteria 3. Participant withdrawal criteria (i.e., terminating investigational product treatment/trial treatment) and procedures specifying: (a) When and how to withdraw participants from the trial/investigational product treatment (b) The type and timing of the data to be collected for withdrawn participants (c) Whether and how participants are to be replaced (d) The follow-up for participants withdrawn from investigational product treatment/trial treatment

(continued)

Table 22.2 (continued)

ICH E6 section	Content/comments
Treatment of Participants	1. The treatment(s) to be administered, including the name(s) of all the product(s), the dose(s), the dosing schedule(s), the route/mode(s) of administration, and the treatment period(s), including the follow-up period(s) for participants for each investigational product treatment/trial treatment group/arm of the trial 2. Medication(s)/treatment(s) permitted (including rescue medication) and not permitted before and/or during the trial 3. Procedures for monitoring participant compliance
Assessment of Efficacy	1. Specification of the efficacy parameters 2. Methods and timing for assessing, recording, and analyzing efficacy parameters
Assessment of Safety	1. Specification of safety parameters 2. The methods and timing for assessing, recording, and analyzing safety parameters 3. Procedures for eliciting reports of and for recording and reporting adverse event and intercurrent illnesses 4. The type and duration of the follow-up of participants after adverse events
Statistics	1. A description of the statistical methods to be employed, including timing of any planned interim analysis(ses) 2. The number of participants planned to be enrolled. In multicenter trials, the numbers of enrolled participants projected for each trial site should be specified. Reason for choice of sample size, including reflections on (or calculations of) the power of the trial and clinical justification 3. The level of significance to be used 4. Criteria for the termination of the trial 5. Procedure for accounting for missing, unused, and spurious data 6. Procedures for reporting any deviation(s) from the original statistical plan (any deviation(s) from the original statistical plan should be described and justified in the protocol and/or in the final report, as appropriate) 7. The selection of participants to be included in the analyses (e.g., all randomized participants, all dosed participants, all eligible participants, evaluable participants)
Direct Access to Source Data / Documents	Ensure that it is specified in the protocol or other written agreement that the investigator(s)/institution(s) will permit trial-related monitoring, audits, IRB/IEC review, and regulatory inspection(s), providing direct access to source data/ documents
Quality Control and Quality Assurance	
Ethics	Description of ethical considerations relating to the trial
Data Handling and Recordkeeping	
Financing and Insurance	Financing and insurance, if not addressed in a separate agreement
Publication Policy	Publication policy, if not addressed in a separate agreement
Supplements	(NOTE: Since the protocol and the clinical trial/study report are closely related, further relevant information can be found in the ICH Guidance for Structure and Content of Clinical Study Reports)

Abbreviation: *GCP* Good Clinical Practice

withdraw consent and/or discontinue participation in the study at any time without prejudice. The protocol should include a detailed description of the withdrawal procedures, including the type and timing of follow-up visits and data to be collected. The protocol should also clearly state whether withdrawn patients will be replaced, and the procedure for doing so.

To minimize missing data, patients should be encouraged to remain in the study. Premature discontinuation of treatment should be distinguished from withdrawal of consent for follow-up visits and from withdrawal of consent for non-patient contact follow-up (e.g., medical records checks). Patients who withdraw should be explicitly asked about the contribution of possible adverse events to their decision, and any adverse event information elicited should be documented. Preferably, the patient should withdraw consent in writing, and if the patient or the patient's representative refuses or is physically unavailable, the site should document and sign the reason for the patient's failure to withdraw consent in writing.

Every effort should be made to minimize the loss of follow-up. The protocol should clearly outline efforts that investigators will be expected to make to contact patients who fail to return for scheduled visits. Patients should be considered as lost to follow-up only after reasonable and documented attempts to reach the patient are unsuccessful. Examples of such attempts include (with patient consent): contacting all telephone numbers for the patient and listed contacts, contacting the patient's primary care physician or other healthcare professional, sending certified letters to all the patient's addresses and contacts, reviewing available medical records for details that may indicate the status of the patient (e.g., hospitalization), conducting an internet search for additional contact information, and checking local and national public records as allowed by the law.

22.3.2.5 Concomitant Medications

The protocol should include a description of key concomitant medications that are permitted and prohibited during the study. Relevant concomitant medications to include in this section are standard of care treatments for comorbid conditions (e.g., medications permitted and/or prohibited for the treatment of hypertension), other treatments for the disease of interest, medications that are strong inhibitors or activators of the investigational product (i.e., notable drug-drug interactions), and medications that would put the patient at higher risk for potential toxicities with the investigational product. Where applicable, the protocol should also include a description of permitted rescue therapies for patients who meet pre-specified criteria. For each prohibited medication, the protocol should indicate how long the medication will be prohibited for (e.g., during the entirety of the treatment period).

22.3.2.6 Statistics

The description of the statistical methods included in the protocols for phase 1 and exploratory and proof-of-concept phase 2 studies can be brief. Phase 1 studies often focus on descriptive statistics. For phase 3 studies, the protocol should include a detailed description of the statistical methods that will be used to assess the investigational product's efficacy and safety, including the timing of pre-specified interim analyses. The protocol should clearly indicate the patient population to be included in the analyses of efficacy and safety (e.g., intent-to-treat population, per protocol population). The efficacy component of the statistical section should include the number of patients planned to be enrolled, the reason for the choice of sample size, and power calculations. Other aspects to include in the statistical section include the level of significance to be used, criteria for termination of the trial (if applicable), procedures to account for and handle missing data, methods to account for intercurrent events, and any pre-specified sensitivity analyses. The protocol should indicate the procedures for reporting any deviations from the original statistical plan [2]. The clinical protocol should contain sufficient details on the planned statistical analyses, and it is acceptable for the full details to be described in a separate Statistical Analysis Plan (for details, see Chap. 25).

22.4 Considerations for Protocols for Pediatric Studies

The enactment of the Pediatric Research and Equity Act (PREA) in 2003 led to the requirement of the conduct of pediatric studies for drugs and biological products if certain criteria are met [8]. The goal of the studies is to obtain evidence-based labeling for the product for the proposed indication in children. Studies must use appropriate formulations for each age group. Sponsors are encouraged to discuss the pediatric development program with relevant regulatory authorities before submitting the pediatric study protocol.

Children are a vulnerable population who cannot provide consent for themselves and are, therefore, afforded additional safeguards under Subpart D (21 CFR 50) when participating in clinical trials [9]. The ethical considerations under Subpart D must be addressed when enrolling pediatric patients in clinical studies.

Protocols for pediatric studies should include additional information above that provided for adult study protocols to address pediatric-specific efficacy and safety considerations (e.g., plans for extrapolation of efficacy from adult or other pediatric populations, appropriateness of endpoints, the safety of excipients and inactive ingredients at the proposed maximum exposures for pediatric participants, details on the dosing regimen, rationale for the proposed dose selection). The clinical protocol should use age-appropriate normal ranges for vital signs, electrocardiograms, and laboratory assessments for monitoring the safety of pediatric patients. Additional safety monitoring may be required (e.g., based on information from juvenile animal toxicity studies or because of developmental immaturity; see Chap. 5) than was obtained in adults [10]. The ICH E11 guidance titled, "Clinical Investigation of Medicinal Products in the Pediatric Population" provides additional information [10].

22.5 Protocol Amendments

According to the US Code of Federal Regulations (21 CFR Part 312) a protocol amendment should be submitted when there are any significant changes [4]. For phase 1 studies, a protocol amendment should be submitted for any meaningful change that affects the safety of patients [4]. For phase 2 and 3 studies, a protocol amendment should be submitted for any change that significantly affects patient safety, the scope of the investigation, or the scientific quality of the study [4]. In general, key efficacy assessments and endpoints should be agreed upon with regulatory agencies early in the development process, ideally, before initiation of the trial. Changes to efficacy assessments and key efficacy endpoints during the conduct of a trial could undermine the credibility of the findings.

Examples of changes that require a protocol amendment include the following [4]:

- An increase in dosage or duration of exposure to the investigational product that is higher or longer than what is currently specified in the protocol
- A significant increase in the number of patients under study
- A significant change to the study design (e.g., addition of a treatment arm)
- The addition of a new test or procedure that is intended to improve monitoring for, or reduce the risk of, a side effect or adverse event; or the dropping of a test intended to monitor safety

Once the significant change has been made, prior to the amended protocol going into effect, the sponsor must submit the amended protocol to FDA for review and the change must be approved by the appropriate Institutional Review Board [4]. An exception to this is for a change made in response to elimination of an apparent immediate hazard. In this situation, the amended protocol may be implemented immediately as long as the FDA and Institutional Review Boards have been notified [4].

22.6 Concluding Remarks

A clinical trial protocol acts as a step-by-step guide for the conduct of a study and is required for all original IND applications in the United States. The US Code of Federal Regulations (21 CFR Part 312) and the ICH Good Clinical Practice Guidelines (ICH GCP E6, Section 6) provide guidance on the information to include in a clinical trial protocol. This chapter provides details on the elements that should be included in a clinical study protocol to ensure that it contains sufficient and clear information for optimal study conduct.

Conflict of Interest None.

References

1. Basics about clinical trials. https://www.fda.gov/patients/clinical-trials-what-patients-need-know/basics-about-clinical-trials#:~:text=to%20determine%20whether%20a%20new,or%20decrease%20certain%20side%20effects. Accessed 16 Jan 2023
2. E6(R2) Good clinical practice: integrated addendum to ICH E6(R1) https://www.fda.gov/regulatory-information/search-fda-guidance-documents/e6r2-good-clinical-practice-integrated-addendum-ich-e6r1. Accessed 16 Jan 2023
3. FDA. IND applications for clinical investigations: clinical protocols. https://www.fda.gov/drugs/investigational-new-drug-ind-application/ind-applications-clinical-investigations-clinical-protocols. Accessed 16 Jan 2023
4. 21 CFR 312. https://www.accessdata.fda.gov/scripts/cdrh/cfdocs/cfcfr/CFRsearch.cfm?CFRPart=312. Accessed 16 Jan 2023
5. FDA. The drug development process. https://www.fda.gov/patients/drug-development-process/step-3-clinical-research. Accessed 16 Jan 2023
6. Al-Jundi A, Sakka S (2016) Protocol writing in clinical research. J Clin Diagn Res 10(11):ZE10–ZE13
7. FDA (2019) Enrichment strategies for clinical trials to support approval of human drugs and biological products: guidance for industry. https://www.fda.gov/regulatory-information/search-fda-guidance-documents/enrichment-strategies-clinical-trials-support-approval-human-drugs-and-biological-products. Accessed 16 Jan 2023
8. How to comply with the Pediatric Research Equity Act: draft guidance for industry. 2005. https://www.fda.gov/regulatory-information/search-fda-guidance-documents/how-comply-pediatric-research-equity-act. Accessed 17 Jan 2023
9. Ethical considerations for clinical investigations of medical products involving children: draft guidance for industry, sponsors, and IRBs. 2022. https://www.fda.gov/regulatory-information/search-fda-guidance-documents/ethical-considerations-clinical-investigations-medical-products-involving-children. Accessed 17 Jan 2023
10. ICH E11 clinical investigation of medicinal products in the pediatric population 2000. https://www.fda.gov/regulatory-information/search-fda-guidance-documents/e11-clinical-investigation-medicinal-products-pediatric-population. Accessed 17 Jan 2023

23 Good Clinical Practice in Clinical Trials, Substantial Evidence of Efficacy, and Interpretation of the Evidence

Fortunato Senatore

Abstract

Sponsors seeking marketing authorization for a drug in the United States of America are required to demonstrate substantial evidence of effectiveness with benefits that outweigh the risks of the drug. The data from adequate and well controlled clinical trial(s) or sources of confirmatory evidence supporting an application for drug approval must be clear and interpretable. Every precaution must be taken to ensure subject safety. The Investigator is required to disclose all risks to the enrolling subject, thereby facilitating the ability of the enrolling subject to provide informed consent with the knowledge that the subject can withdraw from the trial at any time without penalty. These attributes are derived from the principles of Good Clinical Practice that was born from a history of human rights violations and now form the core of how clinical research is performed. The key objectives of this chapter are to describe: (1) Good Clinical Practice; (2) the design of clinical trials based on Good Clinical Practice; (3) the scientific and logistical under-pinning of data interpretation; (4) the operational aspects of filing and reviewing applications submitted by drug companies for approval of a drug; and (5) evaluation of the benefit/risk profile of a drug under consideration for approval.

Keywords

Good Clinical Practice · Substantial Evidence of Effectiveness · Informed Consent · Randomized Clinical Trial · Data Interpretation · Benefit/Risk Evaluation

23.1 Introduction

In the latter half of the twentieth century, there was an evolution in the field of clinical research based on core principles of Good Clinical Practice (GCP) and the International Conference on Harmonization (ICH Guidance). These principles are steeped in protection of subject safety, full disclosure of risk by way of informed consent, and trial conduct designed to produce high quality and interpretable data leading to evidence of a clinically meaningful treatment effect.

In this chapter, the history and principles of GCP and the International Conference on Harmonization-E6 (ICH-E6) will be reviewed. This will segue into the design of clinical trials based on these principles. This will lead to an explanation of how to interpret data and how to evaluate whether the statutory requirements for

F. Senatore (✉)
Division Cardiology & Nephrology, Office of Cardiology, Hematology, Endocrinology and Nephrology, Center for Drug Evaluation and Research, US Food and Drug Administration, Silver Spring, MD, USA
e-mail: Fortunato.Senatore@fda.hhs.gov

G. Jagadeesh et al. (eds.), *The Quintessence of Basic and Clinical Research and Scientific Publishing*,
https://doi.org/10.1007/978-981-99-1284-1_23

substantial evidence of effectiveness have been established. Finally, the methodology of establishing a benefit/risk profile will be reviewed.

23.2 Good Clinical Practice and International Conference on Harmonization-E6

GCP is a set of standards designed to ensure that the rights, safety and well-being of trial subjects are protected. GCP also ensures that the data and reported trial results are interpretable. The research being conducted should be designed to evaluate a clinically significant outcome, as opposed to evaluating a research question that serves to satisfy a whimsical curiosity and perhaps at the expense of patient safety. Two key principles underly the design of clinical trials with the objective to produce high quality data toward answering a clinically relevant question for the benefit of humanity: respect for persons, and beneficence. Research involving humans should be scientifically justified and described in a clear, detailed protocol.

The ICH is a worldwide organization tasked with providing a unified standard for the European Union, Japan, Canada, Australia, the World Health Organization, and the USA to facilitate mutual acceptance of clinical data by the regulatory authorities in those jurisdictions. The specific document ICH-E6 was designed to minimize the need for redundant research [1], thus facilitating protection of the public by not performing research that may not be necessary. The document ICH-E6(R1) is a guidance that involves audit trails, system validations, standard operation procedures, and backups. The document ICH-E6(R2) adds a requirement that both the drug company (i.e., Sponsor) and Investigator or Institution where the Investigator is located, maintain their respective essential documents in a system that provides processes for locating and identifying the document, and other revisions designed to provide practical standardizations for the conduct of clinical trials. Both ICH-E6 (R1) and ICH-E6(R2) reflect a modernizing and evolving research landscape while maintaining the core of GCP-ICH. These features are designed to provide effective monitoring ensuring both subject protection and high-quality trial data in clinical trials that are intended to be submitted to regulatory authorities. Operationally, investigators/institutions that design clinical trials should ensure that before the trial is initiated, foreseeable risks and inconveniences should be weighed against the anticipated benefit for the individual adult subject participating in the trial and the target patient population. A trial should be initiated and continued only if the anticipated benefits justify the risk. The rights, safety, and well-being of trial subjects are the most important considerations and should prevail over the interests of science and society. Clinical trials should be scientifically sound, and described in a clear, detailed protocol. Clinical trials are required to receive approval from Institutional Review Boards (IRBs) or Independent Ethics Committees (IECs) before initiation. The IRB is an independent committee at a site, or group of sites, comprised of medical, scientific, and non-scientific members. The IRB/IEC is tasked with ensuring the protection of the rights, safety, and well-being of human subjects. To do this, the IRB/IEC reviews and ultimately approves key trial documents and assesses submitted reports of serious and unexpected adverse events. The IRB should pay special attention to trials that may include vulnerable subjects. Vulnerable subjects are defined as those who have a compromised ability to provide informed and un-coerced consent, and include prisoners, persons with diminished mental capacity, and persons who are educationally or economically disadvantaged. IRBs or IECs should be provided with all documents related to the clinical trial, including the protocol and any protocol amendments, Informed Consent Form (ICF), Investigator's Brochure (IB), subject recruitment procedures, available safety information, and information about payment to subjects. Further, individuals conducting a clinical trial should be qualified to do so by education, training, and experience to perform his/her respective tasks.

The evolution of GCP and ICH had its roots in dramatic violations of human rights both in Europe and the USA dating back to World War II in Europe and earlier in the USA.

The Nuremberg code was developed in 1947 [2] and the Declaration of Helsinki was written in 1964 [3]. The Nuremberg Code was written in direct response to atrocities committed by Nazi Germany during World War II, where prisoners were experimented on without their consent. The key contribution of the Nuremberg Code was to merge Hippocratic ethics (i.e., do no harm) and the protection of human rights into a single code. The Declaration of Helsinki was developed by the World Medical Association as a statement of ethical principles for medical research involving human subjects, including research on identifiable human material and data. The document was created to set a balance between the interests of humanity and individual patients who are part of clinical trials.

Neither the Nuremberg Code nor the Declaration of Helsinki had the force of law in the USA. Between 1932 and 1972, the United States Public Health Service (USPHS) and the Tuskegee Institute conducted research on Black Americans to understand the natural course of syphilis to support increased treatment efforts for this population. A total of 600 black men (399 with syphilis and 201 without it) were enrolled without informed consent. Subjects were told they were being treated for "bad blood". Penicillin treatment for syphilis was initiated by the USPHS in 1947, but the Tuskegee subjects were not informed about this new treatment, nor were they treated. A newspaper story regarding this experiment was published in 1972 that led to a review of this program by a Federal Advisory Panel. The panel found that the subjects entered the trial freely but were not given the facts required to provide informed consent. The panel concluded the trial was "ethically unjustified". Consequently, the Department of Health and Human Services (HHS) assistant secretary for Health and Scientific Affairs ended the trial in 1972. A class-action lawsuit was brought on behalf of the trial subjects in 1973. In 1974, a settlement was reached. An act of Congress provided lifetime and burial benefits for participants. In 1975, widows and offspring were added to the program [4]. Legislation known as the National Research Act (1974) created the National Commission for the Protection of Human Subjects of Biomedical and Behavioral Research. Work by this Commission lead to the Belmont Report in 1979, which summarized basic ethical principles and guidelines that address ethical issues arising from the conduct of research with human subjects [5]. The Belmont Report led to the FDA rule on protection of human subjects in 1980 (21 Code of Federal Regulations (CFR) part 50) and similar regulations for federally sponsored research in 45 CFR. GCP was finalized in 1996 and became effective in 1997 but was not enforced by law at that time. The Medicines for Human Use Regulations (2004) and the European Union Directive on GCP changed the world's perspective. GCP compliance is now a legal obligation in the UK/Europe for all trials involving the investigation of medicinal products [6].

In a parallel time frame (1962), the world was shocked by the severe limb deformities linked to the use of maternal thalidomide designed as a sedative for the treatment of morning sickness in pregnant women. This drug reaction was discovered after 10,000 infants were born in over 20 countries worldwide. The USA did not yet approve thalidomide for marketing. The manufacturer of thalidomide assessed these adverse events to be unrelated to the drug. To prevent this devastation from happening in the USA, the Kefauver-Harris Amendments, sponsored by Senator Estes Kefauver (Democrat from Tennessee) and Representative Oren Harris (Democrat from Arkansas) were signed into law by President Kennedy on October 10, 1962. These amendments established a framework that required drug manufacturers to prove scientifically that a medication was not only safe, but effective. Prior to these amendments, the original Food, Drug and Cosmetic Act of 1938 had serious shortcomings, such as a manufacturer being able to sell a drug if the FDA did not act within 60 days to prevent its marketing. Following the passage of these amendments, the FDA had the authority to prevent unproven medicines from

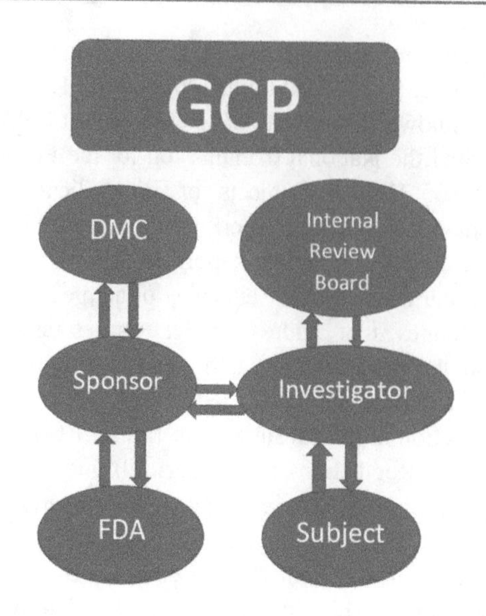

Fig. 23.1 Network of Good Clinical Practice (GCP) Compliance

pouring into pharmacies and subsequently to patients by enforcing demonstration of both safety and efficacy.

Today, GCP compliance is maintained and ensured by the research network as illustrated in Fig. 23.1. The network includes all relevant stakeholders: the Sponsor, the Investigator, the subjects of the clinical research, the IRB, the data monitoring committee (DMC), and the US-FDA. The Sponsor mainly interacts with the DMC and the FDA and vice-versa. The Sponsor and Investigators interact with each other.

The Investigator is responsible for the conduct of the clinical trial at the trial site. The Investigator interacts with the site-specific IRB. The Investigator is also responsible for recruiting subjects and providing details and risks of the trial to the subjects in order to ensure informed consent pursuant to GCP principles. The Investigator is tasked with maintaining all trial documentation, performance of data verification, matching case report forms with source data, and making data available for external monitors. The Sponsor, tasked with site and Investigator selection, is responsible for the initiation, management, and financing of the clinical trial. The Sponsor is tasked with implementing and maintaining quality systems to ensure that trials are conducted, and data are generated, documented and reported in compliance with the protocol and applicable regulatory requirements. A Sponsor may transfer any or all of their trial-related duties and functions to a contract research organization (CRO) in a written document. The ultimate responsibility for the quality and integrity of the trial data always resides with the Sponsor, but if functions are transferred to a CRO, the CRO shares responsibility with the Sponsor for GCP obligations related to those functions.

23.3 Design of Clinical Trials Based on Principles of GCP

The clinical trial is procedurally described by the protocol, a document that delineates the hypothesis/objective, primary endpoint which is the variable the trial is designed to evaluate via the intervention being tested, as well as detailed procedures and organization of the trial. The protocol should include contact information for investigators, trial sites, investigator qualifications, clinical laboratories and other medical and/or technical departments involved in the trial. A rationale for the trial should be provided, as well as a summary of relevant findings from previous studies. A summary of the known and potential risks must be provided. The protocol must also contain a description of and justification for the route of administration, dosage, and dosage regimen. The protocol must contain a statement that the trial will be conducted in compliance with the protocol, GCP, and other regulatory requirements. The protocol should provide an expected duration of subject participation, a description of the sequence and duration of all trial periods, and follow-up if any. There must be a description of discontinuation criteria for individual subjects in the trial as well as stopping rules for the trial, in whole or in part.

The protocol, designed for a regulatory approval, is often a randomized clinical trial (RCT) design, where consenting subjects could be randomized to either a drug arm or a placebo (i.e., comparator) arm. Other trial designs could include multiple doses of the drug, comparing each dose with each other and with placebo. This strategy facilitates evaluation of a dose-response for the primary efficacy endpoint. The comparator arm does not need to be placebo, but rather another drug that has a treatment effect on the same primary efficacy endpoint. The hypothesis may be superiority of the drug versus comparator, or it may be equivalence, or non-inferiority. Statistical techniques form the basis of the analytical plan to evaluate the treatment effect of the drug with respect to these various designs.

The design of the trial should ensure that the results are due to the intervention being tested (i.e., drug is the intervention) compared to a comparator, rather than other variables that may confound the outcome. These confounders may easily occur in a non-randomized trial.

The protocol should contain procedures for maintaining randomization codes and procedures for breaking codes in the event it may be necessary to safeguard subject safety. The protocol should specify inclusion and exclusion criteria and how data will be collected. A description of statistical methodology should be provided in the protocol to explain the details of the data analysis to ensure interpretable results. The RCT should be a blinded trial, where the Sponsor and the investigators do not have knowledge of what arm of the trial the subject was randomized to (i.e., drug arm or placebo arm). This approach will minimize potential bias that could raise a question about data integrity.

The protocol should be accompanied by an IB. The IB is a compilation of data relevant to the trial of products in humans. The purpose is to disclose information regarding the rationale for key protocol features in terms of safeguarding subject welfare.

IRBs must approve the protocol at the site where it is to be conducted before the trial is initiated.

The ICF must clearly delineate the objectives of the trial in understandable language, the risks of participation, all procedures, trial duration; the subject must be informed that participation is voluntary and he/she can withdraw at any time. The information contained in the ICF must be sufficient to facilitate the patient making an informed decision to participate. The Investigator at the site where the research will be conducted is responsible for obtaining informed consent. Neither the Investigator nor the trial staff should coerce or unduly influence a subject to participate or to continue participating in a trial. None of the oral and written information concerning the trial should contain language that: (1) causes the subject to waive or to appear to wave any legal rights; and (2) releases or appears to release the Investigator, the institution, the Sponsor, or their agents, from liability for negligence. In those cases where obtaining informed consent is not possible due to mental status changes of the subject(s) (clinical trial in patients in a comatose state), and there is no legally authorized representative to provide consent, a pathway for informed consent is available as codified in 21 CFR 50.24 known as Exemption of Informed Consent (EFIC). The conditions under which EFIC is applicable include the following:

- The subject's life is threatened, available therapy is not satisfactory or proven, and collection of evidence is necessary to determine the safety and efficacy of the test therapy;
- ***AND*** obtaining consent is not feasible because of the patient's medical condition, and immediate intervention is required, and a legal representative is not available in a timely manner;
- ***AND*** the trial could not practicably be conducted without a waiver;
- ***AND*** the trial is monitored by an independent DMC.

When the trial is finished, the data is rendered "clean" while the Sponsor and investigators are still blinded. This is usually done by assuring alignment of data between the case report form (i.e., forms on which data are collected) and source documents (i.e., hospital records that have been censored to avoid subject personal identification). Once the datasets are declared clean, the Sponsor and the investigators are unblinded. The data are then analyzed in accordance with the pre-specified statistical analysis plan to ascertain the effectiveness of the drug, as well as its safety. Data interpretability is contingent on data cleanliness and the four elements described in the next section.

23.4 Data Interpretation

The approach to data interpretation is quite simple: what do the data look like and what are they telling you? The first overall question when looking at a dataset is: can you distinguish a difference between drug and control? Is the result interpretable and conclusive? Is it statistically significant (does the difference between drug and control meet pre-specified criteria for success, i.e., beyond a reasonable doubt)? Are the results clinically meaningful (i.e., does the drug confer a clinical benefit to patients.)?

To adequately interpret data, four data elements should be ascertained:

1. ***Data Collection***: Were the data properly collected from the trial sites to the database to conduct an analysis?
2. ***Data Analysis***: Was the analysis plan acceptable? Was the analysis performed according to a pre-specified statistical analysis plan?
3. ***Data Stability/Fragility***: Could the results of the trial withstand a slight perturbation of the dataset (e.g., just a few patients experiencing an endpoint altering the outcome of the results)?
4. ***Data Integrity***: Are the data objective or subjective, such that in the latter case, data are obtained by a standard that could have wide variability? Are there too many missing data points?

The clinical trial should have been designed to ensure the adequacy of these four data elements. Clinical trial review metrics for data interpretation are shown in Table 23.1. The trial should have adequate trial duration to detect the desired effect. The patient population in the trial should reflect the patients in the real world who would be prescribed the drug. The primary efficacy endpoint of the clinical trial should be agreed to by the regulatory authorities. The analysis of the data should be performed according to a pre-specified statistical analysis plan where the trial was appropriately powered to detect the hypothesized treatment effect. Monitoring visits should occur in accordance with a pre-specified monitoring plan. Medication dispensing should ensure compliance with the protocol regarding the schedule of dosing administration. If the trial had a randomized design, the type of concomitant mediations and baseline characteristics should be balanced between the arms of the trial, so as to avoid confounding the results by confounding

Table 23.1 Review Metrics for Data Interpretation

Metrics	Key questions
Trial Duration	Adequate for the desired effect and proposed claim?
Patient Population	Appropriate inclusion/ exclusion criteria and reflective of the population likely to use the drug if approved?
Efficacy Measure	Was the primary efficacy endpoint agreed upon? If statistically significant, is it clinically meaningful?
Randomization Scheme	Did the trial design avoid potential bias?
Statistical	Was the analysis performed according to the pre-specified statistical analysis plan; was the trial appropriately powered?
Trial Visits	Was the frequency appropriate to ensure adequate monitoring; was there a monitoring plan?
Medication Dispensing	Were the subjects in the trial compliant in taking the drug?
Concomitant Medications	Was the use of concomitant medications balanced between treatment arms to avoid confounding?
Trial Structure	Were the data safety monitoring board, adjudication committee, steering committee, operational committee adequately organized with corresponding charters to perform the job?

covariates. The trial structure and trial governance: steering committee, adjudication committee, data safety monitoring board, monitoring committee, and other operational committees should be adequately organized with clear tasks from organizational charters such that the four data elements and the rights of subjects in the trial are ensured.

In order for the drug to be approved in the USA, substantial evidence of effectiveness needs to be demonstrated.

23.5 NDA Filing and Review

When a drug company submits a new drug application (NDA) seeking approval of their drug for a proposed indication, the immediate question for a regulatory reviewer to answer is whether the NDA is complete so as to be able to conduct a substantive review. The answer to this question will determine whether the NDA can be filed. The action of filing an NDA is not related to whether the data submitted in the NDA is adequate for approval. The determination of whether there is substantial evidence of effectiveness occurs during the review process.

A substantive review of an NDA is defined as a process of evaluation after it has been determined that all required elements of the NDA package, necessary for an adequate evaluation of the safety and efficacy of the drug, are contained in the submission package. A substantive review will lead to the determination of whether the drug company (i.e., Applicant) presented substantial evidence of efficacy and whether the drug is safe enough for the benefit to outweigh the risk.

The NDA framework to assure that all required data elements have been submitted for a substantive review is illustrated in Table 23.2. Key information is placed in each of the five modules.

Module 1 contains the proposed labeling (the official description of the drug, the claims about the drug's effectiveness based on the evidence provided, a description of side effects, warnings and precautions, a description of the trials and key results leading to the claim, and a description of the pharmacokinetics -what the body does to the drug- and pharmacodynamics -what the drug does to the body-). Module # 1 also contains financial disclosures, thus providing a list of who was paid and how much was paid to individuals participating in the clinical trial. This is very important to establish whether bias may have occurred. Module 1 also contains a reviewer guide that facilitates the medical officer's ability to understand the meaning of the variables and the location of all key elements in the submission package. Finally, Module 1 contains a history of interactions between the Applicant and the FDA. Often, the medical officer assigned to review the NDA may not have been present during the initial communications. All FDA-Applicant meeting minutes describing requirements and agreements should be provided to help guide the review process and decision making in the setting of antecedent agreements.

Module 2 contains key documents, such as the Clinical Overview, Integrated Summary of

Table 23.2 NDA Framework

Module 1	Labelling, Reviewer Guides, Financial Disclosures, FDA-Drug Company Interactions
Module 2	Clinical Overview, Integrated Summary of Safety, Integrated Summary of Efficacy, Statistical Analysis Plan, Benefit/Risk Evaluation
Module 3	Drug Product, Regulatory Information
Module 4	Pharmacology, Toxicology
Module 5	Table of Studies, Case Trial Reports, Literature References, Datasets

Safety, Integrated Summary of Efficacy, Statistical Analysis Plan, and Benefit/Risk Evaluation. The contents of Module 2 informs the medical officer of the Applicant's rationale for developing the drug, the overall results, how the data was analyzed, and how the benefits outweigh the risk from the Applicant's perspective.

Module 3 contains information on how the drug was manufactured; Module 4 contains information on the pharmacology and toxicology of the drug. Module 5 is large as it contains the case study report for each trial conducted to serve as the basis of substantial evidence. Module 5 also contains the datasets from which the medical officer, accompanied by the review team's statistician, may reproduce the data analysis and ascertain whether the Applicant's results are reproducible.

The specific submission requirements for a substantive review are listed in Table 23.3. The Applicant should submit the results of the requisite number of adequate and well-controlled studies that were agreed upon prior to the NDA submission. The sample size (i.e., the number of subjects from all the clinical trials submitted) should be large enough to establish a safety profile for the drug (i.e., the incidence of adverse events, including serious adverse events; and distribution of adverse events among the various organ classes potentially leading to a safety signal). Narratives are usually required on specific subjects who died, experienced a serious adverse event, or discontinued the trial due to an adverse event.

All datasets should be navigable and complete. If there are missing data points due to loss of patient/subject follow-up, imputation methods that were previously agreed upon should have been implemented, along with sensitivity analyses to determine whether other imputation methods produced different results.

Subject-specific information is usually placed on forms called Case Report Forms (CRFs). These forms standardize information for each subject (i.e., subject identifier code-usually a numerical sequence identifying the site and the specific patient without disclosing personal information; subject demographics and characteristics-age, race, concomitant illness, concomitant medications; laboratory values and other measurements such as ECG/echocardiography results at baseline and specific timepoints per protocol; efficacy results for the primary and secondary endpoints; and adverse event/serious adverse events data). The data could be represented as "check-boxes" associated with descriptors that allow for data analysis using standard programming (e.g., SAS, JUMP). Electronic CRFs allow for direct data entry into analytical tools for expeditious output regarding efficacy and safety signal evaluation. Each subject may require several hundred pages of CRFs, which, when multiplied by the sample size, yield a huge database. Consequently, when the phase 3 program is discussed between the Applicant and the FDA, the requisite data and the consequent number of CRFs should be negotiated so as to attenuate the submission of an intractably large dataset.

If key areas of the requisite dataset or elements in each module are missing during submission, it may be determined that a substantive review would not be possible. If the Applicant cannot

Table 23.3 Specific Submission Requirements for a Substantive Review

Efficacy	The requisite number of adequate and well-controlled studies consistent with previous agreements with the reviewing Division
Safety	Safety data consistent with guidelines, coding dictionary, safety narratives
Datasets	Available, complete, navigable, capable of performing independent analyses
Case Report Forms	Death, serious adverse events, drop-outs due to adverse events
Statistical Analysis Plan	A clear description of how data would be analyzed
Financial Disclosures	Payments to investigators, a potential conflict of interest
Statement of GCP	A declarative statement that all principles of GCP were adhered to

remediate identified liabilities impeding a substantive review, the FDA may refuse to file the NDA, which can be re-submitted at a later point in time when the liabilities have been resolved.

If it has been determined that all the requisite information is present to perform a substantive review, then the NDA is filed and a review commences. For the drug to be approved for the Applicant's proposed indication, the review of the NDA must yield substantial evidence of effectiveness.

The legal pathway to the current definition of substantial evidence began in 1962 under the Food, Drug and Cosmetic Act (FDCA). Under this statute, approval of a drug for a proposed indication required results from two adequate and well-controlled clinical trials, usually designed as randomized, double-blind, placebo-controlled trials [7]. The performance of two adequate and well-controlled trials may be considered problematic with respect to a long duration of time, as well as expense, thus putting into question the feasibility of satisfying the statutory requirement.

In 1997, under the Food and Drug Administration Modernization Act (FDAMA), legislation was implemented that amended the FDCA to add the discretion for substantial evidence to come from one adequate and well-controlled trial, plus confirmatory evidence as described in a draft guidance published in December 2019 [8]. In general, FDA guidances are not binding on either the FDA or the public but serve to assist drug companies in planning to submit an NDA. Alternative approaches may be used if they satisfy the requirements of the applicable statutes and regulations. Based on the guidance, the FDA will consider a number of factors when determining whether reliance on a single adequate and well-controlled clinical investigation plus confirmatory evidence is appropriate. These factors may include the persuasiveness of the single trial; the robustness of the confirmatory evidence; the seriousness of the disease, particularly where there is an unmet medical need; the size of the patient population; and whether it is ethical and practicable to conduct more than one adequate and well-controlled clinical investigation. The amended statute did not, however, define "confirmatory evidence". Over many applications, "confirmatory evidence" was interpreted in several ways. The guidance on substantial evidence [8] stated that confirmatory evidence may be derived prior to or after one adequate and well-controlled clinical trial submitted for a regulatory decision. Confirmatory evidence may be based on: (1) an adequate and well-controlled trial in a related disease; (2) real world evidence derived from data on outcomes that also show a lack of effect in a control group; (3) compelling mechanistic evidence in the setting of well-understood disease pathophysiology or well-documented natural history of the disease; and (4) scientific knowledge about the effectiveness of other drugs in the same pharmacological class.

From a practical perspective regarding data interpretation and ascertainment of substantial evidence, the review team considers both statistical significance and clinical relevance.

Statistics is a tool that sets boundaries for determining drug efficacy. The boundary of statistical significance is to rule out with a certain level of confidence that a type 1 error will have been committed by concluding that a drug is efficacious. A type 1 error occurs when the null hypothesis is rejected when it is true. If the hypothesis is superiority of the drug versus comparator in reducing the incidence of a primary efficacy endpoint by a certain percentage, the null hypothesis is the drug is not superior to placebo in reducing the incidence of a clinical endpoint by that percentage. The statistical analysis plan designed to evaluate a difference between the drug and comparator may be "one-sided" or "two-sided". A one-sided test evaluates a treatment-effect in one direction (e.g., drug is superior to comparator in reducing the incidence of a primary efficacy endpoint). In this case, the term "alpha" (α), defined as the probability of committing a type 1 error, is set as 0.05 in the direction of superiority. A two-sided test evaluates a "change" (increase or decrease) in the parameter of interest (i.e., primary efficacy endpoint). In this case, α is split between both directions (superiority and inferiority). Hence, the α is set at 0.025 in each direction. It is easier to reject the null hypothesis with a one-sided test than with a two-sided test, as long as the effect is in the specified direction. Therefore, one-sided tests have a lower type 2 error rate and more power than do two-sided tests (see following paragraphs on type 2 errors and power). A two-sided alternative hypothesis can be used when one wishes to set the type 1 error against a very precise null hypothesis (the effect is exactly zero-no effect without any variability).

As an example, consider testing a drug against a comparator where your hypothesis is simply superiority of the drug compared to comparator in reducing the incidence of a clinical endpoint by a certain percentage (i.e., one-sided test). The trial is designed to test this hypothesis. If the null hypothesis is rejected when it is true, a type 1 error will have been committed, thereby approving a drug for a proposed indication when the drug is actually ineffective (false positive). In setting α at 0.05 (the accepted boundary), there is at least a 95% chance that if an adequate and well-controlled clinical trial shows a statistically significant effect, the result is correct. However, one accepts a 5% chance that a positive result may be a false positive. If one desires high or near 100% certainty that an observed treatment effect is correct, then there is a risk that effective drugs may be declared to be ineffective (i.e., type II error, accepting the null hypothesis when it is false: false negative). A useful analogy is serving as a juror in a criminal case. To be reasonably certain that an innocent suspect is not convicted (type 1 error by rejecting the null hypothesis-not guilty until proven otherwise, when it is true), then the chance of finding a guilty suspect not-guilty increases. The term "beta" (β) is the probability of committing a type 2 error. In clinical trials designed for a regulatory decision with Health Authorities, acceptable boundaries include α set at 0.05 and β set at 0.2 (i.e., accepting a probability of 20% that a type 2 error may be committed). This is analogous to the "beyond a reasonable doubt" paradigm in American jurisprudence, which is distinct from "beyond all doubt".

The required sample size to test the hypothesis set forth in the clinical trial is dependent on a number of factors: the predicted magnitude of the observed difference between drug and placebo, the variance in the measures, and the pre-set values of α and β. If the clinical trial is designed to have a 99% probability (α 0.99) of ruling out a type 1 error and a 95% probability (β 0.95) of ruling out a type 2 error, the sample size required to satisfy these boundaries will be very high so as to render the clinical trial unfeasible to perform. Hence, the regulated industry tolerates a certain level of uncertainty. The term 1- β represents the power of a trial. Thus, with β set at 0.2, the trial has 80% power to detect the purported difference in treatment effect between drug and placebo.

The observed treatment effect must be clinically meaningful, i.e. statistical significance alone may not be sufficient for an approval. A small, statistically significant but clinically meaningless treatment effect in an over-powered trial may not result in an approval. For more discussion on these statistical issues in clinical trials, readers are directed to read the next chapter by Dr. Zhang.

Drug approval involves demonstration of safety in addition to efficacy. The ultimate decision for drug approval will be based on the benefit/risk profile whereby the benefit outweighs the risk, as discussed in the next section.

23.6 Benefit/Risk Evaluation

The most important evaluation stemming from an NDA review is the benefit/risk profile. The benefit/risk evaluation is the culmination of a review process stemming from adequate and well-controlled trials meeting GCP-ICH guidance.

Virtually all therapeutic drugs carry some risks, even if minor. Therefore, the basic question is whether the tradeoff between the benefits and risks of the drug is acceptable. The benefit/risk evaluation usually involves a qualitative balancing of benefits and risks, although sometimes a quantitative benefit risk assessment may be conducted.

An example of a benefit/risk analysis is shown in Fig. 23.2. This figure demonstrates a benefit/risk tree that delineates the most important

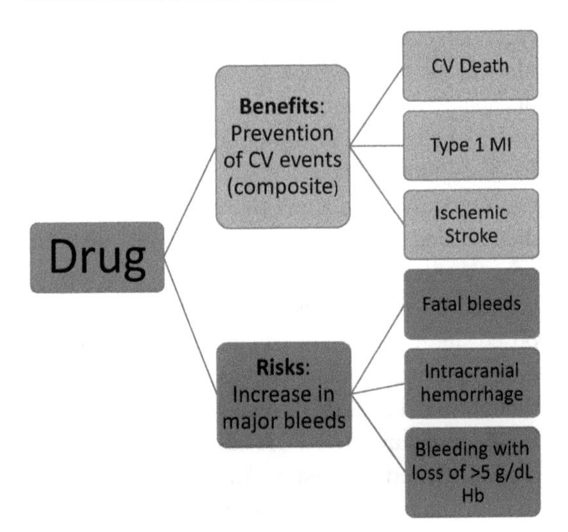

Fig. 23.2 Benefit/Risk Tree

benefits and the most serious risks. It is essential that the development of a benefit/risk tree remain free of bias by listing risks that are not as inherently dangerous compared to the accrued benefit, or by listing benefits that are not as clinically meaningful as the risks.

The example provided refers to an anti-platelet drug designed to prevent platelet aggregation and consequently attenuate blood coagulation in patients who experienced a myocardial infarction (i.e., heart attack). The development strategy for this anti-platelet drug was to prevent recurrent heart attacks, strokes, or death due to cardiovascular disease after the initial heart attack. These events comprised the primary efficacy endpoint in the pivotal clinical trial. This benefit was weighed against the risk of bleeds usually associated with any agent that prevents blood coagulation. The bleeds can be mild or serious. The delineated bleeds are all considered serious: fatal bleed, intracranial hemorrhage, or major bleed causing the loss of greater than 5 g/dL hemoglobin, collectively called key safety events.

In this case, the clinical significance of the benefit appears equally weighed by the seriousness of the risk. The benefit of preventing cardiovascular death is weighed against the risk of a fatal bleed. The benefit of preventing an ischemic stroke is weighed against the risk of an intracranial hemorrhage that can and usually leads to a hemorrhagic stroke. Finally, the benefit of preventing a recurrent heart attack is weighed against the risk of losing a sufficient amount of blood to possibly provoke an ischemic event, potentially leading to a recurrent heart attack. In this scenario, the initial benefit/risk evaluation requires quantification of the number of primary endpoint events and the number of key safety events.

Knowing the number of primary efficacy endpoint events in each arm of the trial, the absolute risk reduction can be calculated. For example, if the primary efficacy endpoint event rate is 50% in the placebo arm and 40% in the drug arm, the absolute risk reduction is 50% - 40% = 10%. From this, the number of patients needed to be treated to prevent one episode of the primary efficacy endpoint (number needed to treat = NNT) is the inverse of the absolute risk reduction (1/(absolute risk reduction) = 1/0.1 = 10). Therefore, it is necessary to treat 10 patients for the prevention of one primary efficacy endpoint event. The same exercise can be performed for each component of the primary efficacy endpoint. These numbers do not tell you that the drug is effective, but rather 1 patient will respond for every 10 treated, either by the drug or by some other mechanism.

Similarly, knowledge of the key safety event rates in each arm of the trial would allow for the calculation of the number of patients required to be treated to cause one patient to experience one of the key safety events. For example, if the key safety event rate (i.e., fatal bleed, intracranial hemorrhage, bleeding causing loss of hemoglobin exceeding 5 g/dL) is 8% in the placebo arm and 12% in the drug arm, the absolute risk increase is 4% (i.e., 12% - 8%). The number of patients needed to be treated to observe one of these harmful events (NNH) is 1/(absolute risk increase) = 1/[(incidence of event rate in drug arm)-(incident of event rate in the placebo arm)] = 1/0.04 = 25. Therefore, it is necessary to treat 25 patients to experience one of the key safety events. The same calculation can be performed for each component of the key safety event.

The NNT and NNH calculations assume that the benefit and the risk are equally important as in the illustration depicted in Fig. 23.2. However, the setting of equally important benefit and risk is unusual. Also, these calculations may or may not have an implicit time factor. That is, if all events happen in short order around the time of the intervention, then there is no useful time component. If they are scattered through time, then the NNT and NNH must be evaluated over some period of time.

A favorable benefit/risk profile would have a low value of NNT and a high value of NNH assuming equal importance of benefit and risk over the same time period. If every patient treated with the drug experienced the purported benefit, and if every 1000th patient treated with the drug was harmed, the benefit would clearly outweigh the risk. However, this may not be the case if the benefit and risk were not of equal importance. If the benefit was attenuation of minor itching after 1 year of treatment and the risk was mortality after the first dose, then the low NNT and high NNH values would clearly indicate that the risk outweighs the benefit.

In the example cited in Fig. 23.2, the benefit/risk evaluation could be impacted if the benefit was driven only by heart attack (no effect on death or stroke), and the risk was driven only by fatal bleeding. Therefore, the harm experienced by one patient after 25 patients take the drug is more consequential than the benefit of preventing a survivable adverse event after 10 patients take the drug. There could be additional benefit/risk evaluation impacts based on the severity of the outcome harboring similar pathophysiology underlying the benefit/risk evaluation. As an example, if the ischemic stroke being prevented is considered mild, but the intracranial hemorrhage that can harm the patient produces a severely debilitating stroke, then the benefit/risk evaluation will be negatively impacted. Hence, the calculated numbers, NNT and NNH, by themselves do not constitute an adequate benefit/risk evaluation. Weights should be associated with each outcome on the benefit/risk tree that could more accurately describe the impact of preventing or causing one of those events. The assignment of such weights should not be arbitrary, but rather based on factors such as patient preferences, clinical outcomes as defined by the healthcare community and epidemiological research, and the price society is willing to pay for the benefit accrued by the drug being developed for a proposed indication. The regulatory decision for drug approval should be based on a carefully thought-out benefit/risk evaluation by considering these factors. The FDA has recently deployed a Multi-Criteria Decision Analysis (MCDA) [9] to inform a regulatory decision. The MCDA is an analytic tool for benefit-risk assessment that involves specifying numeric trade-offs between outcomes. These trade-offs are used to derive relative weights specific to the decision context and the best and worst possible performance values for each outcome (defined by using a 95% confidence interval of the incidence). Weights are then combined with the performance of the drug and the comparator control on each outcome and summed to produce the overall total value of each alternative.

23.7 Concluding Remarks

The history of GCP and ICH is based on ensuring that past unethical behavior in clinical trials is prevented by implementing the principles of doing no harm and designing/conducting clinical trials that lead to clinically meaningful data that can be clearly interpreted. This has been codified by statutory requirements for substantial evidence of effectiveness. The clinical trial design aspects stemming from a carefully thought-out hypothesis involve a well constructed analysis plan and maintenance of subject safety, in the setting of ensuring subject autonomy via a clearly delineated informed consent process.

The data retrieved under the principles of GCP-ICH are intended to facilitate an adequate benefit/risk evaluation, leading to the regulatory decision whether to approve or not-approve a drug applcation for marketing.

Disclaimer This article reflects the views of the author and should not be construed to represent US FDA's views or policies.

Conflict of Interest

None declared.

References

1. https://www.fda.gov/regulatory-information/search-fda-guidance-documents/e6r2-good-clinical-practice-integrated-addendum-ich-e6r1. Accessed 20 Nov 2022
2. https://www.hhs.gov/ohrp/archive/nurcode.htm. Accessed 20 Nov 2022
3. http://www.wma.net/policies-post/wma-declaration-of-helsinki-ethical-principles-for-medical-research-involving-human-subjects/#:~:text=The%20World%20Medical%20Association%20(WMA,identifiable%20human%20material%20and%20data. Accessed 20 Nov 2022
4. http://www.cdc.gov/tuskegee/. Accessed 20 Nov 2022
5. http://www.hhs.gov/ohrp/regulations-and-policy/belmont-report/index.html. Accessed 20 Nov 2022
6. http://www.imperial.ac.uk/clinicalresearchoffice. Accessed 20 Nov 2022
7. http://www.fda.gov/regulatory-information/search-fda-guidance-documents/demonstrating-substantial-evidence-effectiveness-human-drug-and-biological-products. Accessed 20 Nov 2022
8. https://www.fda.gov/media/131881/download. Accessed 20 Nov 2022
9. Lackey L, Garnett C, Senatore F (2021) Applying decision analysis to inform the US Food and Drug Administration's benefit-risk assessment of Ticagrelor for primary prevention of myocardial infarction or stroke based on THEMIS. Circulation 144:655–658. https://doi.org/10.1161/CIRCULATIONAHA.120.053294

Grandfathered Drugs of 1938 in the United States

24

Benjamin P. Lewis

Abstract

The chapter discusses how a drug became known as a "grandfathered drug of 1938" also called a "pre-1938" drug and includes a comprehensive listing of all drugs and therapeutic chemicals that were on the market in 1938. With the passage of the U.S. Federal Food, Drug, and Cosmetic Act (FFDCA) of 1938, it repealed the original 1906 Pure Food and Drug Act and required that all new drugs to undergo a premarket review of safety by the FDA and make a threshold determination whether or not a drug was a "new drug." The Act required all new drugs to be reviewed for safety prior to approval and before entering the marketplace. For those drug products already on the market in 1938, they were never evaluated for safety as per the statutory requirements of the 1938 Act but were assumed to be safe. The 1938 Act contained a grandfather clause that exempted all pre-1938 drugs from the safety requirement of submitting a new drug application to the FDA. The premise being that all pre-1938 drugs were considered safe but must contain the same chemical composition, indications, and other conditions for use to be considered a grandfathered drug. The products already on the market in 1938 were known as "grandfathered drugs of 1938" or "pre-1938 drugs."

Keywords

Grandfathered Drugs of 1938 · Pre-1938 Drugs · The U.S. Food & Drug Act of 1938 · Premarket · Safety · NDA

24.1 Introduction

In 1906 the original Pure Food and Drug Act was passed, also known as the "Wiley Act." The 1906 Act was the first law to provide for regulation of the food and pharmaceutical industries. Whereas, in 1938 the Federal Food Drug and Cosmetic Act (FFDCA) was passed by Congress and became law. It established regulations for the quality standards for food, drugs, medical devices, and cosmetics manufactured and sold in the U.S. The 1938 Act required that all new drugs entering the marketplace be reviewed and approved for safety. However, those drug products already on the market in 1938 were never evaluated for safety as per the statutory requirements of the Act but were assumed to be safe and were allowed to stay on the market. These pre-1938 drugs were known as "grandfathered drugs of 1938." In addition, this chapter explains the reason for creating and

B. P. Lewis (✉)
Lewis Regulatory Consulting, LLC (formerly with the U.S. Food and Drug Administration), Gaithersburg, MD, USA
e-mail: benlewisphd@gmail.com

G. Jagadeesh et al. (eds.), *The Quintessence of Basic and Clinical Research and Scientific Publishing*,
https://doi.org/10.1007/978-981-99-1284-1_24

establishing the first drug law in the U.S. known as the Pure Food and Drug Act of 1906. Later in 1938, this law was repealed and the FFDCA of 1938 was created that strengthened the 1906 law that required FDA to conduct a premarket review of all new drugs prior to entering the marketplace. With passage of the 1938 Act, it created a "grandfather clause" that assumed all drugs on the market prior to 1938 drugs were safe and that they must contain the same chemical composition, indications, and other conditions for use to be considered as a grandfathered drug of 1938. At the end of the chapter are two tables that lists all pre-1938 drugs and pre-1938 therapeutic chemicals on the market at that time.[1]

24.2 Historical Background

24.2.1 An Unsuspecting Cause for a New Drug Law

It was 1906 when famed author Upton Sinclair published a classic novel, *"The Jungle."* It was a fictionalized account of Chicago's meat-packing industry in the early twentieth century. The novel exposed the appalling working and unsanitary conditions of companies, primarily Armour & Co. the leading Chicago meat packer. Sinclair saw men in the pickling room with skin diseases, men who used knives on a sped-up assembly lines with lost fingers, workers with tuberculosis who coughed constantly and spit blood on the floor. No toilets existed and workers urinated in a corner. Sinclair described unsanitary conditions that endangered health, how diseased, rotten, and contaminated meat products were processed, doctored by chemicals, and mislabeled for sale to the public [1, 2]. He wrote that workers would process dead, injured, and diseased animals after regular hours when no meat inspectors were around, and canned pork fat and beef scraps labeled as "potted chicken."

After reading *The Jungle*, the public reacted with outrage about the filthy and falsely labeled meat. The White House was bombarded with mail to reform the meat-packing industry. President Roosevelt appointed a special commission to investigate Chicago's slaughterhouses. The commission recommended that inspections take place at every stage of the processing of meat. They also called for the Secretary of Agriculture to make rules requiring the cleanliness and wholesomeness of animal products [1]. From this, President Roosevelt was able to enact the Meat Inspection Act of 1906. How does the Meat Inspection Act relate to food and drugs?

24.2.2 New Federal Law: A Reason for Food and Drug Reform

The exposé in *The Jungle*, by Upton Sinclair, depicting unscrupulous practices and unsanitary conditions rampant in the meat-packing industry opened the way for Congress to approve a long-blocked law to regulate the sale of most foods and drugs [1]. During that same period, Dr. Harvey W. Wiley, Chief Chemist, Bureau of Chemistry, Department of Agriculture, had been leading a "pure food crusade" for over 20 years with his men called the "Poison Squad." They had tested chemicals added to preserve foods and found many to be dangerous to human health. The public uproar over *The Jungle* revived Dr. Wiley's lobbying efforts in Congress for federal food and drug regulation [1, 3].

24.2.3 The U.S. Food and Drug Act: 1906

The Food and Drug Act of 1906, also called The Pure Food and Drug Act, or the "Wiley Act" was signed into law on June 30, 1906, by President Roosevelt [3–5]. Coincidently it was signed on the same date as the Meat Inspection Act [6]. The Food and Drug Act became effective six months later on January 1, 1907 [3]. It was originally a policing statute that prohibited placing misbranded or adulterated drugs or food on the market. This primarily required drugs to be properly labeled and not be contaminated [5]. The

[1] Marketed today, meaning as of 1978 when the FDA conducted a thorough review of those products.

statute did not require that drugs be pre-approved by FDA in order to be marketed [7]. This Act is the first law to protect the public's health (i.e., consumer protection) from ineffective or dangerous drugs [8]. There were no other requirements in the Act [9]. At the time of signing the Act the FDA did not exist, provisions of this Act were enforced by the Bureau of Chemistry, Department of Agriculture. The Bureau eventually became the U.S. Food and Drug Administration in 1930 with Dr. Wiley as its first Commissioner [8]. The Bureau of Chemistry was responsible for enforcement of the Pure Food and Drug Act prior to becoming the FDA; and the Bureau of Animal Industry, a division of the U.S. Department of Agriculture, was responsible for meat inspections under the Meat Inspection Act of 1906 [10].

24.2.4 The U.S. Federal Food, Drug, and Cosmetic Act of 1938

The Food and Drug Act of 1906 was repealed, and the Federal Food, Drug, and Cosmetic Act (FFDCA) of 1938 was passed by Congress to strengthen the FDA's regulatory authority [9, 11]. The 1938 Act included provisions that enabled the U.S. government to control the introduction of unsafe new drugs into interstate commerce [9]. In addition, the 1938 FFDCA provided for "new drugs" to be subject to a premarket review for "safety" whereas, prior to the 1938 Act (i.e., 1906 Act) no provision for safety existed and drugs only had to be labeled properly to show contents and not be contaminated (i.e., must be pure). Proponents (i.e., Sponsors) of new drugs would be required to file a new drug application (NDA) with the FDA to show evidence of compliance with the safety requirements under the 1938 Act. The FFDCA of 1938 completely overhauled the public health system and was the first to demand premarket evidence of safety for new drugs. This essentially shifted the burden of proof of safety from the FDA to the manufacturer [12]. If FDA did not disapprove of the NDA, the drug could be produced and marketed [9].

24.2.5 Exemption for Safety to Qualify: Grandfather Clause of 1938

In order to be exempted from premarketing procedures of the 1938 statute, i.e., a new drug review for "safety," the FDA was required to make a threshold determination whether or not a drug is a "new drug." If it is classified as a new drug, the Sponsor must apply for approval as an NDA before the drug can be shipped in interstate commerce. However, to be exempted from premarket procedures, the drug must fit the statutory exemption which is known as the "grandfather clause" [9]. The active ingredients in many currently marketed drugs were first introduced at least in some form before June 25, 1938 (the date on which the FFDCA was enacted) [7]. Drugs on the market prior to that date were exempted or "grandfathered" from new drug status under the grandfather clause and, therefore, are exempt from the requirement of submitting an NDA to the FDA, provided it contains the same chemical composition, indications, and other conditions for use as the grandfathered drug [7]. The premise was that all pre-1938 drugs were considered safe, and if the manufacturer did not change the product formulation or indication then an NDA was not required. All drugs in the marketplace today that were marketed before 1938 have been called "pre-1938" or "grandfathered drugs."

If a drug obtained FDA approval between 1938 and 1962, the FDA generally permitted identical, related, or similar (IRS) drugs to an already approved drug to be marketed without independent approval [7]. Many manufacturers also introduced drugs onto the market between 1938 and 1962 based upon their own conclusion that the products were generally recognized as safe (GRAS), and thus exempt from the statutory new drug definition or based upon a formal opinion from FDA that the products were not new drugs [7].

24.2.6 A Grandfather Clause for Efficacy in the 1962 Amendment

Due to a growing concern for pharmaceutical practices, Congress passed the 1962 Drug Efficacy Amendment (also known as the Kefauver—Harris Amendment) to the Federal Food, Drug, and Cosmetic Act of 1938 [13]. The Act required that a new drug should demonstrate that it is "effective," as well as "safe," to obtain FDA approval [7, 9]. Under a "grandfather clause" included in the 1962 Drug Amendments, a drug is exempt from the "effectiveness" requirement, if its composition and labeling have not changed since October 10, 1962 (the date on which the 1962 Drug Amendments were enacted), and if, on the day before the 1962 Drug Amendments became effective, the drug was: used or sold commercially in the United States; not a new drug as defined by the FFD & C Act at the time; and not covered by an effective application [7]. In order to be exempted from premarket procedures, the drug must fit into one of the statutory exemptions which is known as the "grandfather clauses" [9, 14].

24.3 List of Grandfathered Drugs of 1938

A listing of proprietary products that were manufactured in 1938 and also on the market 40 years later in 1978 in which the trade name of the product had not changed in the intervening 40 years, see Table 24.1 Pre-1938 Drugs, and Table 24.2 Pre 1938 Therapeutic Chemicals [15, 16]. It was from 1977 to 1978 that FDA undertook a comprehensive review of all drugs that were still manufactured in 1938 to determine which drugs were still being produced and classified/listed them as "pre-1938" drugs. Some of the names listed are not considered trade names by today's standards but are generic names and were marketed under that name.

Some of the products have had additional dosage forms approved since 1938; there has been a change in the formulation; there has been a change in an indication; a firm had decided to submit an NDA on its product for some reason. The products listed may or may not require a prescription, owing to changes instituted in the original drug law, such as the Durham-Humphrey Amendment passed in 1951, which required a prescription for drugs that cannot be used safely without adequate medical supervision. Under this amendment, drugs were divided into two classes: those that could be used safely in self-medication and sold without a prescription (known as over-the-counter or OTC drugs) and those that required medical supervision and required a prescription (Rx Only). This amendment required drug manufacturers of prescription-restricted drugs to label with the Rx legend and made it illegal to place this legend on drugs not so restricted [17, 18].

24.4 Concluding Remarks

The U.S. Congress repealed the Food and Drug Act of 1906 and passed the Federal Food Drug and Cosmetic Act of 1938 to strengthen FDA's regulatory authority. The 1938 Act required that all new drugs undergo a premarket review procedure for safety prior to being approved and being shipped into interstate commerce. To be exempted from this premarket requirement the drug must fit the statutory exemption known as the "grandfather clause." Drugs on the market prior to the passage of the 1938 Act (June 25, 1938) were exempted from the new drug requirements under the grandfather clause and therefore exempted from submitting an NDA to

Table 24.1 PRE 1938 DRUGS

Name of Drug	Manufacturer 1938	Manufacturer 1977—1978	AHFS[a] category	Current NDA status[b]/comments
Adrenalin	Parke-Davis	Parke-Davis	21:12, 52:24	
Algic	Prague	Unimed		
Alurate	Roche	Roche	28:24	
Aminophylline	Searle	Searle	40:28; 86:00	Numerous NDAs
Amyl nitrate	Calco	Burroughs-Wellcome, Lilly	24:12	Merged to become GlaxoSmithKline
Amytal	Lilly	Lilly	28:24	
Anabolin	Harrower	Alto	08:08	
Anthralin	Abbott	Dermik	84:28	
Aphrodex	Rex	Rex		
A.P.L.	Ayerst	Ayerst	68:18	N17055
Aspasmon	Norgine	Kretschmar		
Atabrine	Winthrop	Winthrop	08:08, 10:00	N15052 injectable only
Atrobarb # 1, 2, 3	Jenkins	Jenkins		
Auralgan	Doho	Ayerst		Disc US 2015
Barbidonna	Van Pelt & Brown	Mallinckrodt		
Belbarb	Haskell	Arnar-Stone		
Belbarb #2	Haskell	Arnar-Stone		
Belladenal	Sandoz	Sandoz		
Bellafoline	Sandoz	Sandoz	12:08	
Bellergal	Sandoz	Dorsey		
Ben-caine B-B	Ulmer	Ulmer	52:16, 84:08	
Benzedrine	Smith Kline & French	Smith Kline & French	28:20	N17017—spansule N83900—tablet
Berocca	Roche	Roche	88:28	N00240—elixir
Betalin S	Lilly	Lilly	88:08	N80853—injectable only
Butasol sodium	McNeil	McNeil	28:24	N00793—tablet
Butyn sulfate	Abbott	Abbott	52:16	
Caffeine sodium benzoate	Various	Various	28:20	
Calcidrine	Abbott	Abbott	48:00	
Calciphen	Rorer	O'Neal, Jones & Feldman		
Calcium chloride solution ampules	Various	Various	68:24	
Calcium gluconate 10% ampules	Various	Various	68:24	
Calcium levulinate ampules	Crookes	Corvit	68:24	
Calpholac	Nestles	Century		
Carbarsone	Lilly	Lilly	8:04	
Carbromal	Calco; Upjohn; Bluline	Bluline; Freeman	28:12	N01592—tablets N01177—tablets
Cenolate	Abbott	Abbott		
Cerose	Wyeth	Ives	48:00	

(continued)

Table 24.1 (continued)

Name of Drug	Manufacturer 1938	Manufacturer 1977—1978	AHFS[a] category	Current NDA status[b]/comments
Cheracol	Upjohn	Upjohn	4:00; 48:00	
Choline chloride	Merck	Baker; City Chemical		
Colchicine	Abbott	Abbott	92:00	
Congo Red	Kirk	Harvey; Consolidated Midland	36:08	
Copavin	Lilly	Lilly	28:08, 86:00, 48:00	
Coramine	Ciba	Ciba	28:20	
Corex	Roth Labs	Econo-Rx		
Cosanyl	Parke-Davis	Parke-Davis	48:00	
Decholin sodium	Riedel de Haen	Dome	56:16	
Diacin	Diacin Labs	Danal Labs		
Digifortis	Parke-Davis	Parke-Davis	24:04	
Digiglusin	Lilly	Lilly	24:04	
Digitalis	Various	Various	24:04	
Digoxin	Burroughs-Wellcome	Burroughs-Wellcome	24:04	N83391—injectable only
Dihydrotachysterol	Winthrop	Philips Roxane	68:24	
Dilantin	Parke-Davis	Parke-Davis	28:12	N84349—capsules N08762—suspension
Dilaudid	Knoll, Merck	Knoll	28:08	
Dionin	Merck	Merck		
Donnatal	Robins	Robins	12:08	
Emeracol	Upjohn	Upjohn	48:00	
Emetine hydrochloride	Various	Various	8:04	
Empirin Compound #2 and #3	Burroughs-Wellcome	Burroughs-Wellcome	52:32	
Ephedrine hydrochloride alkaloid sulfate	Various	Various	12:12	
Ephedrine and Amytal	Lilly	Lilly		
Ephedrine and Nembutal	Abbott	Abbott		
Ephedrine and phenobarbital	Massengill	Various		
Epinephrine	Various	Various	52:24	
Epragen	Lilly	Lilly		
Ergotrate H	Lilly	Lilly	76:00	
Eschatin	Parke-Davis	Parke-Davis	52:08, 68:04	
Estrolin	Lakeside	Laser		
Estrone	Ayerst; Lilly	Various	68:16	N04823—injectable
Ethyl Chloride	Amsco	Various	84:08	
Euresol	Knoll	Knoll	84:28	
Extralin	Lilly	Lilly	20:04.08	
Follutein	Squibb	Squibb	68:18	N17056—injectable

(continued)

Table 24.1 (continued)

Name of Drug	Manufacturer 1938	Manufacturer 1977—1978	AHFS[a] category	Current NDA status[b]/comments
Gluco-calcium	Lilly	Lilly		
Glypectol	Schieffelin	International Chemical		
Gomenol	Bard	Various		
Gynergen	Sandoz	Sandoz	12:16	
Hexalet	Riedel	Webcon	8:36	
Histamine phosphate	Parke-Davis; Breon	Various	36:64	N03862—injectable N02854—injectable N80697—injectable
Indigo carmine	Hynson, Wescott & Dunning	Various	36:40	
Iron carodylate	Various	Various	20:04.04	
Ithyphen	Strauss	Universal		
Ivyol	Burroughs-Wellcome	Burroughs-Wellcome	84:36	
Ketochol	Searle	Searle	20:04.04	
Lextron	Lilly	Lilly		
Lipo-lutin	Parke-Davis	Parke-Davis	68–32	
Liver extract	Lederle; Lilly; Parke-Davis; Valentine	Lederle; Lilly; Parke-Davis; Valentine	20:04.08	
Luminal	Winthrop	Winthrop	28:12, 28:24	
Luminal sodium	Winthrop	Winthrop	28:12, 28:24	
Magnesium sulfate	Various	Various	28:12, 56:12	
Mebaral	Winthrop	Winthrop	28:12, 28:24	
Mecholyl	Merck	Baker	12:04	
Methenamine	Abbott	Various	8:36	
Methylene blue	Abbott; Breon, Calco; Endo	Various	8:36	
Metrazol	Knoll	Knoll	28:20	
Metycaine	Lilly	Lilly	56:12, 72:00	
Myochrysine	Merck Sharpe & Dhome		60:00	
Naftalan	Stiewe; Donner	Hark		
Nembutal	Abbott	Abbott	28:24	N83244—elixir
Neo-Synephrine	Sterns	Winthrop	12:12, 52;24, 52;32	
Nitroglycerin	Abbott	Various	24:12	
Novocain	Winthrop	Winthrop	72:00	
Nupercaine	Ciba	Ciba	56:12, 84;08, 72:00	
Octin	Knoll; Merck	Knoll	86:00	N06420—injectable
Oreton	Schering	Schering	68:08	

(continued)

Table 24.1 (continued)

Name of Drug	Manufacturer 1938	Manufacturer 1977—1978	AHFS[a] category	Current NDA status[b]/comments
Organidan	Wampole	Wallace	48:00	
Otolgesic	Bluline	Bluline	84:00	
Ouabain	Merck	Lilly	24:04	
Oxoids	Professional	Lemmon		
Pancreatic hormone	Grant	Amfre-Grant		
Pantopon	Roche	Roche	28:08	
Papaverine	Roche	Various	86:00	
Parathyroid	Squibb	Lilly	68:24	
Pentobarbital sodium	Lilly	Lilly	28:24	N02368—capsules
Pentothal Sodium	Abbott	Abbott	28:24	
Phenacaine hydrochloride	Werner	Muro		
Phenacetin	Winthrop	Various	28:08	
Phenobarbital	Squibb	Various	28:12, 28:24	
Phenolsulphonphthalein	Hynson, Westcott & Dunning	Hynson, Westcott & Dunning	36:40	Also spelled, Phenolsulfonphthalein
Piperazine	Winthrop	Various	8:08	N09437—syrup; N80671; N80774; N80963; N83756; N80681—tablet
Pitocin	Parke-Davis	Parke-Davis	76:00	N13508—tablet
Pitressin	Parke-Davis	Parke-Davis	68:28	
Podophyllin	Abbott; Lilly; Lloyds; Merrell	Robinson	84:28	
Pontocaine hydrochloride	Winthrop	Winthrop	52:16, 72:00	
Progesterone	Breon	Various	68:32	N17338—injectable
Proluton	Schering	Schering	68:32	
Prostigmin	Roche	Roche	12:04	N00654—solution N02449—injectable
Prunicodeine	Lilly	Lilly	36:56	
Pyridium	Merck	Warner-Chilcott	8:36	
Rectocaine	Kirk	Moore Kirk		
Reton	Roche-Renand	Tri-State		
Salyrgan-theophylline	Winthrop	Winthrop		
Sarapin	High	High		
Seconal	Lilly	Lilly	28:24	N07392—injectable
Serenium	Squibb	Squibb	8:36	
Siderol	Wyeth	Doral		
Skiodan	Winthrop	Winthrop		
Sodium cacodylate	Various	Various		
Sodium thiosulphate	Various	Various		
Somnos	Merck Sharpe & Dohme	Merck Sharpe & Dohme	28:24	
Strontium bromide	Breon; Endo; Kirk	Various	28:12	
Strychnine	Houde Wallace	Baker; Humco		N00443—HT; N00444—HT; N00445—HT
Sulfanilamide	Baker	Baker		N01734; powder

(continued)

Table 24.1 (continued)

Name of Drug	Manufacturer 1938	Manufacturer 1977—1978	AHFS[a] category	Current NDA status[b]/comments
Testosterone propionate	ProMedico	Various	68:08	N0188—injectable; N80254; N80262; N80276; N80676; N80741; N80742; N80743; N83595
Theamin and Amytal	Lilly	Lilly	86:00	
Theelin	Parke-Davis	Parke-Davis	86:16	N03977—injectable
Theobromin with phenobarbital	Premo; Upjohn	Paramount	40:28	
Thiamine hydrochloride injectable	Abbott; Lederle; National; Squibb	Invergy; Jenkins; Robinson; West-ward	88:08	Numerous NDAs
Thyroid	Various	Various	68:36	
Thyroxin	Squibb	Various	68:36	
Thyroxine	Roche	Various	68:36	
Trilax	Barksdale	Drug Industries		
Tuberculin	Winthrop	Lederle; Lilly	36:84	
Urolax	High	Century		

[a]Since 1938, manufacturers of drugs/therapeutic chemicals have submitted to the FDA NDAs for the same or different dosage forms

[b]Since 1938, manufacturers of drugs/therapeutic chemicals have submitted to the FDA New Drug Applications (NDAs) for the same or different dosage forms

Table 24.2 PRE 1938 THERAPEUTIC CHEMICALS

Name of chemical	AHFS category[a]	Current NDA status[b]
Acetanilid		
Acetophenetidin	28:08	Now phenacetin
Alcohol, absolute	28:08, 36:36	
Apomorphine hydrochloride	56:20	
Atropine	12:08, 52:24	
Atropine sulfate	12:08, 52:24	NDA—tablets, injections, ophthalmics
Barbital		
Barbital sodium		
Belladonna leaves, powder	12:08	
Benzocaine	56:16, 84:08	
Bismuth subsalicylate solution	8:28	
Calomel		
Cantharides		
Chloral hydrate	28:24	
Chloroform		
Cocaine hydrochloride	52:16	
Codeine phosphate	28:08, 48:00	
Codeine sulfate	28:08, 48:00	NDA—tablets, injections
Dextrose	40:20	I.V. plastic bags
Eserine sulfate	12:04, 52:20	
Fluorescein	52:36	
Guaiacol		
Hexamethylenamine (methenamine)	8:36	
Homatropine hydrobromide	12:08, 52:24	
Hyoscine (scopolamine)	12:08, 52:24	
Hyoscyamine hydrobromide	12:08	
Hyoscyamine sulfate	12:08	
Iodoform Iron, peptonized		
Lithium carbonate	28:16.12	NDA—tablets, capsules
Magnesium carbonate (heavy)	56:04	
Magnesium hydroxide		NDA—liquids
Magnesium oxide (light)	56:04	NDA—tablets
Mercury, ammonated	84:04.16	
Methylene blue	8:36	
Morphine sulfate	28:04	Injectable, tablets
Nux vomica		
Opium	28:08, 56:08	
Pancreatin	56:16	
Papain	44:00	NDA—tablets
Paraldehyde	28:24	
Pilocarpine hydrochloride	52;20	
Pilocarpine nitrate	52:20	
Pilocarpine salicylate	52;20	
Potassium acetate injection	40:12	
Potassium bromide	28:12	
Potassium chloride	28:12	
Potassium chloride solution	40:12	Numerous NDAs
Potassium iodide solution	48:00	
Potassium permanganate	84:04.16	

(continued)

Table 24.2 (continued)

Name of chemical	AHFS category[a]	Current NDA status[b]
Potassium phosphate	40:12	
Potassium thiocyanate		NDA—tablets
Procaine hydrochloride	72:00	Numerous NDAs
Quinidine sulfate	24:04	Numerous NDAs
Resorcinol	84:28	
Silver nitrate	52:04.12, 84:12, 84:28	
Sodium acetate		
Sodium acid phosphate	40:04	
Sodium bicarbonate solution	40:08, 56:04	
Sodium chloride solution	40:12, 40:36	Numerous NDAs
Sodium fluoride	92:00	
Sodium hydroxide		
Sodium iodide	36:68	
Sodium nucleate		
Sodium phosphate	56:12	
Sodium salicylate	28:08	
Sodium succinate		NDA—injection
Theobromine		
Theophylline	86:00	NDA—elixir
Thymol		
Tin metal		
Tin oxide		
Urea	40:28, 84:16	NDA—injection
Yohimbine hydrochloride		
Zinc chloride	52:28	
Zinc peroxide	84:04.16	
Zinc sulfate	52:04.12	
Zinc sulfocarbolate (Zinc phenolsulfonate)		

[a]*AHFS* American Hospital Formulary Service published by the American Society of Health-System Pharmacists
[b]Since 1938, manufacturers of drugs/therapeutic chemicals have submitted to the FDA New Drug Applications (NDAs) for the same or different dosage forms

FDA. The premise was that all "pre-1938" drugs were considered safe and must contain the same chemical composition, indications, and other conditions for use to be considered as a grandfathered drug of 1938.

Conflict of Interest None.

References

1. Constitutional Rights Foundation. https://www.crf-usa.org/bill-of-rights-in-action/bria-24-1-b-upton-sinclairs-the-jungle-muckraking-the-meat-packing-industry.html. Accessed 28 Sep 2022
2. Wikipedia. https://en.wikipedia.org/wiki/Upton_Sinclair#Early_life_and_education. Accessed 28 Sep 2022
3. Wikipedia. https://en.wikipedia.org/wiki/Pure_Food_and_Drug_Act/. Accessed 29 Sep 2022
4. Food and Drug Administration. https://www.fda.gov/about-fda/changes-science-law-and-regulatory-authorities/part-i-1906-food-and-drugs-act-and-its-enforcement. Accessed 28 Sep 2022
5. University of Dayton. Law review. https://ecommons.udayton.edu/udlr/vol5/iss1/8/. Accessed 28 Sep 2022
6. Britannica. https://www.britannica.com/topic/Meat-Inspection-Act. Accessed 29 Sep 2022
7. Karst KR (2007) Marketed unapproved drugs—past. Present and future? RA Focus. pp 37–42
8. Kille JW (2017) Regulatory toxicology. In: A comprehensive guide to toxicology in nonclinical drug development. Diethylene glycol (Elixir sulfanilamide). https://www.sciencedirect.com/topics/agricultural-

and-biological-sciences/pure-food-and-drug-act. Accessed 30 Sep 2022
9. Patton-Hulce VR (1980) Regulating Laetrile: constitutional and statutory implications. Univ Dayton Law Rev 5(1):155–176. https://ecommons.udayton.edu/udlr/vol5/iss1/8/. Accessed 30 Sep 2022
10. Ballotpedia. https://ballotpedia.org/Federal_Food,_Drug,_and_Cosmetic_Act_of_1938#:~:text=The%20Federal%20Food%2C%20Drug%2C%20and,and%20enforcement%20of%20these%20standards. Accessed 30 Sep 2022
11. https://library.weill.cornell.edu/about-us/snake%C2%A0oil%C2%A0-social%C2%A0media-drug-advertising-your-health/food-and-drug-administration-continued. Accessed 30 Sep 2022
12. Kille JW (2017) A comprehensive guide to toxicology in nonclinical drug development, 2nd edn. https://www.sciencedirect.com/topics/agricultural-and-biological-sciences/pure-food-and-drug-act. Accessed 30 Sep 2022
13. Wikipedia. https://en.wikipedia.org/wiki/Kefauver%E2%80%93Harris_Amendment. Accessed 28 Sep 2022
14. There are two "grandfather clauses" in the current FFD&C Act (1938, 1962). Each clause exempts certain drugs from compliance (i.e., enforcement action) with the Act
15. Lewis BP, Castle RV (1978) Grandfathered drugs of 1938. Am Pharm 18(1):36–39. (Note. The list was compiled and sanctioned by the FDA; The procedure used in retrieving the pre-1938 drugs and chemicals was as follows: (1) A comparison was made between the 1977–1978 and the 1938 Red Book (Publ. Drug Topics, Medical Economics Company, Oradell, NJ). (2) Druggists Circular, Red Book, Price List Section, 88th Revision, Vol. LXXXII, No. 5, New York, NY (May) 1938)
16. No search of pre-1938 or grandfathered drugs have been conducted or necessary since 1978 as reported by Lewis, B.P. and Castle, R.V., 1978. See above reference
17. J Public Law 13:1. Emory University Law School, Atlanta, GA, 1964, p 214
18. A Brief Legislative History of the Food, Drug and Cosmetic Act, Committee on Interstate and Foreign Commerce, U.S. House of Representatives, Subcommittee on Public Health and Environment, 93rd Congress, 2nd Session, Committee Print No. 14, U.S. Government Printing Office, Washington, DC, 1974

General Overview of the Statistical Issues in Clinical Study Designs

25

Jialu Zhang

Abstract

This chapter introduces statistical issues often encountered in confirmatory clinical studies in a typical setting - randomized trials with parallel group design. Starting with some basic statistical concepts such as hypothesis testing and type I error, the chapter covers fundamental statistical issues that one should understand when designing a clinical trial, for example, how to plan the sample size, why we need multiplicity adjustment in a trial, what is an estimand, how to prevent missing data. Adaptive design is also briefly discussed in the last section. Although the concepts are discussed under the confirmatory clinical study setting, many principles in this chapter also apply to exploratory studies.

Keywords

Type I error rate · Sample size · Multiplicity adjustment · Estimand · Missing data · Adaptive design

25.1 Introduction

A clinical development program of a medical intervention usually starts with animal studies. If the intervention is proven to be safe enough to test in human, early phase studies in human, e.g., phase I and phase II studies, will be conducted. Phase I trials are mostly conducted in a small number of health volunteers to test dose limiting toxicity and dose response. Oncology drugs can be exceptions that phase I trials may directly use patients rather than health volunteers. Phase II trials are exploratory and hypothesis generating in nature. Information on efficacy and safety from the phase II trials are used to design the confirmatory clinical trials. A confirmatory trial is designed to address the clinical question of interest and confirm clinical hypotheses. Statistical issues vary in clinical studies with different objectives. This chapter will focus on the statistical issues in the confirmatory clinical trials in a typical setting – randomized trials with a parallel group design. And since majority of the confirmatory clinical trials use frequentist approach, only frequentist statistical methods are discussed in this chapter. This does not mean that Bayesian methodology should not be used in the confirmatory clinical trials, but such discussion will need a separate chapter or even a book.

In this chapter, Sect. 25.2 introduces the framework that's commonly used in clinical trials to determine the effectiveness of a treatment and the concept of type I error rate. Section 25.3

J. Zhang (✉)
Office of Biostatistics, Center for Drug Evaluation and Research, US Food and Drug Administration, Silver Spring, MD, USA
e-mail: Jialu.zhang@fda.hhs.gov

G. Jagadeesh et al. (eds.), *The Quintessence of Basic and Clinical Research and Scientific Publishing*,
https://doi.org/10.1007/978-981-99-1284-1_25

discusses the sample size planning in a clinical trial, which is an important part of a study design to ensure precision. Section 25.4 explains multiplicity adjustment. In a clinical trial with multiple endpoints or doses or patient populations, multiple hypotheses will be tested. Multiplicity adjustment will keep the false positive conclusion at a low level. Section 25.5 reviews a relatively new concept that was brought into clinical trials in 2017, estimands. By examining what treatment effect we really want to estimate, the estimand framework intends to bring clarity in describing the benefit and risk of a treatment. Section 25.6 touches upon the concept of adaptive design. Adaptive design offers flexibility in conducting confirmatory clinical trials under specific statistical methodology and in a pre-planned setting. By introducing a few most simple and well-studied adaptive design, the section offers readers a flavor of adaptive design.

25.2 Type I Error Rate

Hypothesis testing is commonly used framework to determine whether a treatment is effective in a clinical trial. A statistical test is conducted to determine whether the data are such that the null hypothesis, e.g., the experimental drug has no treatment effect, can be rejected, leading to the conclusion that the treatment is effective.

Let H_0: $\mu = 0$ be the null hypothesis and H_1: $\mu \neq 0$ be the alternative hypothesis. μ is the true treatment effect and u is the observed or sample treatment effect. The null hypothesis that the experimental treatment has no effect can be expressed as H_0: $\mu = 0$. Rejecting the null hypothesis leads to the conclusion that the alternative hypothesis of an effective treatment is true. In fact, when the null hypothesis is rejected, it only means that the results are unlikely to occur if the null hypothesis is true. $P(u > 0| \mu = 0)$ is the probability that study data show a treatment effect when the true treatment effect is zero. This probability is called the type I error rate.

Controlling the overall type I error rate in a clinical study is important. A higher type I error rate means a higher chance of false positive, where an ineffective treatment is concluded to be effective. Such a false conclusion can lead to ineffective experimental treatments receiving marketing approval for patients to use. Confirmatory trials used for regulatory approval often have a controlled overall type I error rate. A conventional level of two-sided type I error rate in a clinical trial is $\alpha = 0.05$, which means that there is a 2.5% chance of drawing a false conclusion that the treatment is effective when the treatment has zero effect. Note that the two-sided type I error stems from two-sided hypothesis testing where alternative hypothesis is H_1: $\mu \neq 0$. A two-sided type I error rate of $\alpha = 0.05$ means there is a 5% chance of drawing a false conclusion that the treatment is different from the placebo, which is equivalent to a 2.5% chance of concluding the treatment is better than the placebo and 2.5% chance of concluding the treatment is worse than the placebo.

Let α be the level that the overall type I error rate should be controlled for a study. To test a hypothesis, a p-value based on the observed study data needs to be computed. P-value is the probability of observing a result at least as extreme as the observed study result if the null hypothesis is true. A p-value of 0.01 means that if the treatment indeed has no effect, there is only 1% chance to observe a study result that's more extreme than the current result. In other words, the chance that the null hypothesis is true is small. The P-value calculated from the observed study data is compared with a pre-specified α level, e.g., 0.05. If the p-value is smaller than α, the null hypothesis is rejected.

It is possible that a hypothesis test fails to reject the null hypothesis, therefore one cannot conclude that the treatment is effective, when the experimental treatment is indeed effective. Such an erroneous conclusion is called a type II error, or false negative. Often this occurs when the sample size is not sufficiently large to overcome the uncertainty around the estimate of the treatment effect. In a clinical trial, one also wants to keep the type II error rate reasonably low so that a study can demonstrate the effectiveness of an experimental treatment if the treatment is indeed effective. Table 25.1 illustrates type I and type II

Table 25.1 State of nature and decision

		Decision	
		Null hypothesis is rejected and hypothesis testing concludes there is a treatment effect	Null hypothesis is NOT rejected and hypothesis testing concludes there is no treatment effect
State of Nature	An experimental drug has NO treatment effect	Type I error (false positive)	Conclusion is correct
	An experimental drug has a treatment effect	Conclusion is correct	Type II error (false negative)

error rates. Power is the probability of drawing the correct conclusion that a treatment is effective when the treatment is effective. Computationally, power equals 1 minus the probability of a type II error. A study with good power would have a high probability to reject the null hypothesis for a meaningful treatment effect size and demonstrate the efficacy of the treatment when the treatment is truly effective. This is referred to as a sufficiently powered study. In practice, one may want to have at least 80% power for a study at a meaningful treatment effect size.

A clinical trial should be designed with the chance of false positive and false negative kept at a reasonably low level.

25.3 Sample Size

Sample size calculation in a clinical trial is a critical step. The computation of sample size depends on the objective of the study as well as the metrics we choose to use. A properly designed trial with good sample size can provide a reliable estimate of the treatment effect size with good precision. If one were to demonstrate that an experimental treatment is more effective than the control treatment in a randomized clinical trial, several factors should be considered for the sample size planning: (1) what is the expected treatment effect for the experimental treatment on the endpoint of interest? (2) what's the uncertainty around that expected treatment effect? (3) what is the overall type I error rate for the study? (4) what is the power at a selected meaningful effect size? The sample size calculation usually is based on such information. The larger the expected treatment effect size, the smaller the sample size to have the desired power. But if the variability surrounding the estimated treatment effect is also large, then one needs to consider a larger sample size to overcome the uncertainty and demonstrate the efficacy of the treatment.

An important consideration for the sample size calculation is the expected treatment effect. The selected value is often based on prior studies and existing knowledge. This value should be a clinically meaningful effect size. A tiny estimated treatment effect can lead to rejecting the null hypothesis when the sample size is very large. For example, one can conduct a hypertension study using thousands of subjects and show an estimated treatment effect of a decrease of 1 mmHg or less in systolic blood pressure that leads to rejecting the null hypothesis. But this treatment effect size is fairly trivial and not clinically meaningful.

In some pediatric or rare disease studies, the sample size is limited by feasibility. While treatment may be effective, there are not enough patients to recruit for routine hypothesis testing on a traditional clinical endpoint. Statistical methodologies and endpoints beyond the conventional way, e.g., Bayesian borrowing, real-world data, and bridging biomarkers, can be utilized to demonstrate the treatment efficacy.

If the primary objective of a study is to evaluate the safety profile of a treatment, the sample size consideration can be different from what is discussed above. In a safety study, the sample size should be large enough to rule out an unacceptable increase of critical adverse events. The ICH

E1 guidance provided some guidelines and principles for sample size and drug exposure for safety studies [1]. To characterize the adverse drug events occurring within the first few months of the treatment, ICH E1 suggested that 300 to 600 exposed subjects should be adequate to "observe whether more frequently occurring events increase or decrease over time as well as to observe delayed events of reasonable frequency (e.g., in the general range of 0.5% to 5%)". ICH E1 also recommended that "the total number of individuals treated with the investigational drug, including short-term exposure, will be about 1500" and 100 patients exposed for a minimum of one year. Such a general standard is not applicable in some circumstances and exceptions are discussed in ICH E1.

25.4 Multiplicity Adjustment

A clinical trial usually has multiple endpoints and may have multiple doses or include multiple patient populations. Multiple hypotheses testing each at a type I error rate of 5% can lead to an increased chance of at least one false positive conclusion. For example, assume that a single hypothesis testing in a study has a 5% of chance of drawing a false conclusion that the treatment works when the treatment is not effective. If one tests two different endpoints with the treatment, there are two null hypotheses for hypotheses testing, each with 5% chance of false positive. What is the probability of making a false conclusion about the treatment? Under the condition that the two hypotheses are independent of each other, the probability can be written as follows.

P (at least one null hypothesis is rejected when the treatment is not effective) $=1-0.95 \times 0.95 = 0.0975 > 0.05$.

The chance of drawing a false conclusion that the treatment works almost doubled. This inflation of the overall type I error rate in a trial can be substantial as the number of hypotheses increases without any adjustment to the type I error rate for the individual tests. To control the overall type I error rate, proper multiplicity adjustment needs to be implemented to ensure that the chance of making at least one false positive conclusion is capped at a given level. Confirmatory phase 3 studies should include a pre-specified plan for multiplicity testing and control of the overall type I error rate.

There are many ways to adjust for multiplicity. Some well-known multiple step methods are listed below.

25.4.1 Bonferroni Method

Bonferroni adjustment simply divides the total α level by the total number of hypotheses to be tested. If we set the overall type I error rate to be 5% for a study with three endpoints to be tested, the Bonferroni method divides the 5% among three endpoints $0.05/3 = 0.0167$ so that each hypothesis will be tested at $\alpha = 0.0167$. Bonferroni method is a simultaneous testing method. The rejection of a hypothesis does not depend on other hypotheses and the order of testing does not matter. Bonferroni method can be very conservative for individual tests when the number of hypotheses tested becomes large. If there are 10 hypotheses tested, each test needs to show a p-value $<0.05/10 = 0.005$ to reject the null hypothesis. The power for an individual test will decrease quickly as the number of hypotheses increases.

25.4.2 Holms Procedure

Holms procedure is a step up procedure [2]. One needs to first order the p-values from all hypothesis tests from smallest to largest. Assuming there are m hypotheses, the Holm procedure first compares the smallest p-value to α/m. If the smallest p-value is less than α/m, then the p-value is considered significant. One can then move on to the second smallest p-value and compare it with $\alpha/(m - 1)$. If this second smallest p-value passes the test, one can move on to the third smallest p-value. The procedure continues until a p-value cannot pass the test. All p-values beyond that are considered non-significant. This multiple testing procedure is less conservative than the Bonferroni method.

25.4.3 Hochberg Procedure

Hochberg procedure is a step down procedure [3]. In this procedure, p-values are ordered and compared to critical values α, α/2, α/3, . . ., α/m. But instead of starting with the smallest p-value, the Hochberg procedure starts with the largest p-value and moves down to smaller p-values. Unlike the Bonferroni method and Holms procedure, the Hochberg procedure is not assumption free and may not control the overall type I error rate in some situations when the results of the statistical tests are negatively correlated.

25.4.4 Sequential Test Procedure

In this procedure, hypotheses are tested according to the prespecified order. The order is generally based on the study objectives and clinical question of interest and should be pre-specified before any of the hypothesis testing. Full α can be passed down to the next hypothesis as long as the prior hypothesis wins. However, once a test fails to reject the null hypothesis, all hypotheses after that can no longer be tested. For example, one plans to test the time to hospitalization for heart failure endpoint after the 6-minute walk distance (6MWD) endpoint. The clinical trial has two-sided $\alpha = 0.05$. If the 6MWD endpoint fails to show statistical significance, the testing will stop at the 6MWD endpoint even if the hospitalization for heart failure endpoint may have a p-value<0.05, which is nominally significant. The sequential testing procedure is often used in clinical trials due to its simplicity. One needs to carefully think of the order of the hypotheses at the planning stage, e.g., which hypothesis should be tested first.

25.4.5 Gatekeeping Test Procedure

This procedure groups the primary endpoints and secondary endpoints into families [4]. Each family may contain multiple hypotheses. The procedure dictates how to pass α from one family to the other. It can be further classified into serial gatekeeping and parallel gatekeeping. Parallel gatekeeping can branch out and pass partial α to the next family while serial gatekeeping cannot.

25.4.6 Graphical Approach

Bretz et al. proposed an approach that uses graphic illustration to display how the alpha can be passed among multiple hypotheses [5, 6]. The approach offers flexible α allocation and intuitive illustration. The overall type I error is controlled as long as the following rules are followed. Each hypothesis is assigned an initial α level and the sum of the initial α levels needs to be equal to the total α. Once a hypothesis is rejected, the α assigned to that hypothesis is allowed to be passed to other hypotheses. Fractions of that initial α (between 0 to 1) can be passed to other hypotheses, but the total weight should sum up to 1.

Many multiple procedures mentioned above can be illustrated using the graphical approach. Figure 25.1 is a graphic illustration of a multiplicity testing procedure. H1 is the hypothesis for the primary endpoint and the full $\alpha = 0.05$ is initially allocated to H1. H2, H3 and H4 are the three hypotheses for the three secondary endpoints. The initial α allocated to H2, H3 and H4 are zero so only the primary endpoint (H1) is tested first. Of the three secondary endpoints, H2 and H3 are key secondary endpoints and considered

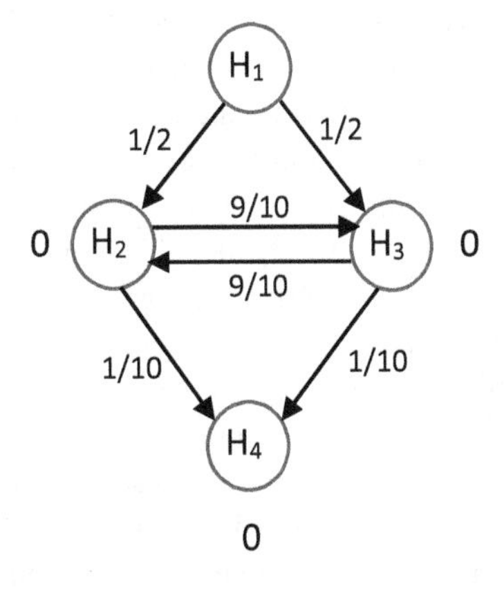

Fig. 25.1 Graphic illustration

more important. If H1 is not rejected, then the total α is used up and nothing can be passed down to any other hypotheses. The testing will stop. If H1 is rejected, the α allocated to H1 will be divided equally and passed down to H2 and H3. H2 and H3 each will get $\alpha/2 = 0.025$ and be tested after the primary endpoint. If only one of these two hypotheses (e.g., H2) is rejected, the majority of its alpha (9/10 of $\alpha/2$) will be recycled to the other key secondary endpoint (in this case, H3). H3 can be tested again using $\alpha/2 + \alpha/2 \times (9/10) = 0.025 + 0.025 \times 0.9 = 0.0475$. This will increase the chance for H3 to be rejected. If both H2 and H3 are rejected, the α allocated to both H2 and H3 will be passed down to test H4. If H2 is rejected but H3 is not rejected despite the alpha recycling, H4 still can be tested since a small portion of α will be passed from H2 to H4. But H4 will be only tested at a very low alpha level $\alpha/2 \times (1/10) = 0.005$. The overall type I error rate of making at least one type I error is controlled at α. Alpha allocation can be flexible depending on the objective of a trial. In this case, the procedure recycles alpha to increase the chance of rejecting the hypotheses for the two key secondary endpoints. H4 is put to low priority and only receives a very small fraction of alpha for testing unless both key secondary endpoints win.

Multiplicity problems are common in clinical trials. Multiple doses, treatment arms, visits, endpoints and patient populations all can lead to multiple hypothesis testing. Depending on the objectives of a study, not all scenarios require a multiplicity adjustment but one should be cautious in interpreting results when there are multiple hypotheses tested and multiplicity adjustment is not used. For example, subgroup analyses are regularly performed in clinical trials to further investigate the treatment effect in various subgroups. The number of subgroups examined in a trial can range from a few to over 30. Because these subgroup analyses usually are to examine the consistency of treatment effect among subgroups for exploratory purpose, no multiplicity adjustment is done. If a specific subgroup showed a much larger or smaller treatment effect than the rest of the population, such results should be considered hypothesis generating and need further confirmation.

25.5 Estimand

A clinical trial is to understand and answer specific clinical questions of interest. In a 6-month trial to study an experimental treatment on a 6-min walk distance in patients with pulmonary arterial hypertension (PAH), a patient may encounter various events before the end of the study. The patient may: (1) die; (2) discontinue the treatment due to some adverse reaction; (3) discontinue the treatment due to lack of efficacy; (4) start to use some prohibited medicine or a rescue therapy; (5) discontinue the study and withdraw consent. If we estimate the treatment effect in the study, e.g., the change from baseline in 6-min walk distance at Month 6, exactly what treatment effect is it that we estimate? Is it the treatment effect regardless of whether patients took rescue therapy? Is it the treatment effect observed until patients went off their treatment? Is it the treatment effect regardless of whether patients died? Is it the treatment effect where patients who died are considered failures? The answers to these questions are closely related to the clinical question of interest that the study is designed to address.

ICH E9 R1 (Statistical Principles for Clinical Trials: Addendum: Estimands and Sensitivity Analysis in Clinical Trials) in 2017 introduced the estimand framework [7]. The intention is to provide clarity in describing the benefit and risk of the treatment by aligning the clinical questions of interest with the study design, conduct and analysis. In ICH E9 R1, estimand "defines the target of estimation for a particular trial objective". An estimand defines what is to be estimated.

The one important concept brought by estimand framework is the intercurrent events. ICH E9 R1 defines intercurrent events as "the events occurring after treatment initiation that affect either the interpretation or the existence of the measurements associated with the clinical question of interest." In addition to the study population and the endpoint, the patient journey through the trial period is also taken into consideration in the clinical question of interest in the estimand framework. Using the same example above, in a 6-month PAH trial, a patient may: (1) die; (2) discontinue the treatment due to some adverse reaction; (3) discontinue the treatment due to lack of efficacy; (4) start to use some prohibited medicine or a rescue therapy; (5) discontinue the study and withdraw consent. In the above scenarios, (1)–(4) are intercurrent events. Measurements affected by the intercurrent events should be differentiated from missing data. For example, the measurements after discontinuation of treatment or initiation of rescue therapy may not be relevant to the treatment effect of interest, but the measurements after the intercurrent events may still be collected and therefore are not considered missing data. Intercurrent events will occur in clinical practice so there is no need to avoid intercurrent events. Rather one should consider all possible intercurrent events and how they should be handled within the chosen estimand framework.

It is of note that in the estimand framework, discontinuation from study or administrative censoring is not considered intercurrent events. The Estimand framework redefines missing data as "data that would be meaningful for the analysis of a given estimand but were not collected".

Missing data should be minimized and avoided in the clinical trials. In the study design stage, steps can be implemented to retain patients and prevent missing data. Efforts should be made to continue following patients after they discontinued treatment. Patients can still be followed if they discontinued treatment in the study but not after they discontinued the study. Therefore, a study protocol should clearly differentiate treatment discontinuation from study withdrawal. Reasons for treatment discontinuation should be sought from the patients or practitioners during the trial. The reasons for treatment discontinuation provide useful information and can imply very different outcomes. For example, is the discontinuation of treatment due to an adverse event or ineffective treatment? In addition, site investigators can be trained about the importance and steps to prevent missing data. Patient participation can be encouraged by including a statement in the consent form to educate patients about the scientific importance of their data even if they discontinue study treatment early.

When designing a clinical trial, it is best to start with an estimand of interest and design the trial based on this estimand. Rather than diving directly into the analysis methods, the estimand framework requires one to first think of what to estimate in the context of a clinical question of interest before getting to the analysis methods. To define an estimand in a trial, five attributes are needed.

1. Treatment
2. Population
3. Endpoint
4. Intercurrent events
5. Population-level summary

Identifying any potential intercurrent events and planning the approach to address these intercurrent events are crucial at the study design stage. There are five different strategies to address how the intercurrent events should be handled.

Treatment policy strategy: This strategy is most frequently used and well accepted. If we are interested in estimating the treatment effect regardless of the intercurrent events, e.g., patients discontinued the treatment or used rescue therapy, then a treatment policy strategy should be used. Patients in many cardiovascular outcome trials were followed until they had a primary event or reached the end of the study regardless of concomitant medication or treatment discontinuation. This is a treatment policy strategy.

Composite variable strategy: The intercurrent event becomes part of the endpoint in this strategy. For example, in a trial comparing the treatment failure rate in the two treatment groups, specific intercurrent events, such as death or taking rescue medication, can be considered as a failure event, and included in the analysis.

Hypothetical strategy: In a trial where patients are allowed to take rescue medication, but one wants to estimate the treatment effect as if the rescue therapy is not available, a hypothetical strategy should be used in such a case. This strategy is assumption-based. Imputation for the measurements after the rescue medication would be based on certain hypothetical assumptions as if the rescue medication were not available for these patients.

While-on-treatment strategy: This strategy is often used in safety studies where one is interested in assessing safety signals when patients are exposed to the treatment.

Principal stratum strategy: This strategy is associated with a subpopulation due to certain intercurrent events that become the clinical question of interest. For example, one may be interested in understanding the treatment effect in the patient population who were able to tolerate the treatment in a study.

If a substantial number of patients had intercurrent events in a trial, e.g., a large number of patients took rescue medication or discontinued treatment, the choice of different strategies can lead to different conclusions. Generally speaking, treatment policy type estimand provides the best reliable estimate. However, sometimes hypothetical estimand or other types of estimand may better address the clinical question of interest. If a different estimand was proposed for the primary analysis in the study, it can be helpful to include supplementary analyses that are based on all observed data (i.e., a treatment policy approach). Hypothetical type estimand is usually based on one specific potentially plausible assumption, for example, in some renal trials, one may assume that the kidney function in patients who took rescue medication would deteriorate at a similar rate as the placebo patients over the trial period if the rescue medication was not available. Nevertheless, the approach still relies on the assumption that the patients' eGFR trajectory would not progress too rapidly over the trial period. So additional sensitivity analyses that systematically and comprehensively explore the space of plausible assumptions should be conducted.

25.6 Adaptive Design

Over the years, statistical methodologies have been developed to allow more flexible study designs. If properly done, adaptive design allows a study to make adjustments based on the new information collected from the trial without inflating the type I error rate or affecting the study integrity. While adaptive designs allow certain flexibility to make changes in an ongoing trial, it does not mean changes can be made post-hoc and without proper adjustment. Careful planning is more important than ever in adaptive design.

The most well-studied adaptive design is group sequential design. By pre-specifying stopping criteria, a trial can be stopped early for futility or efficacy. Futility stopping allows trials to be terminated early and avoid wasting trial resources if early data indicate an ineffective treatment. Futility stopping boundary is determined such that the treatment has very low chance to succeed based on the interim look. It's better to make the futility boundary non-binding. Non-binding futility boundary means that the boundary is set in the way that non-compliance to the futility stopping rule would not affect the overall type I error rate in the trial. Sometimes one may not be ready to terminate a trial yet even if the pre-specified futility boundary is met. In such a case, if the futility boundary is non-binding, the overall type I error rate would not be affected.

Efficacy stopping boundary allows a trial showing overwhelming efficacy results to be stopped earlier, thus reducing the number of trial subjects needed and speeding up the process to make the effective treatment available. There are a number of approaches in early efficacy stopping [8–12]. Most efficacy stopping criteria allocate a very small α for early efficacy stopping to incorporate the greater uncertainty in the early estimate of the treatment effect. A trial should only be stopped early for efficacy when the treatment shows overwhelming treatment effect at the interim look. Unlike a non-binding futility stopping, it is important to adhere to the pre-specified efficacy stopping rule. Adjusting the number of interim looks or the timing of the interim looks based on the ongoing study information can inflate the type I error rate and raise concerns on the study integrity.

Another commonly used adaptive design is sample size re-estimation. While group sequential design is intended to save sample size, sample size re-estimation is often used to maintain trial power in case the true treatment effect is clinically meaningful but smaller than expected in the initial sample size planning. Sample size re-estimation can be either blinded or unblinded.

Blinded sample size re-estimation is a well-characterized adaptive design. The sample size re-estimation is based on the treatment-blind information in the study, e.g., pooled variance or overall event rate. The sample size can be updated with the new estimated pooled variance. Although in theory, blinded sample size re-estimation can be done at any time during the trial and would not inflate the type I error, it is a good practice to pre-plan the timing of the blinded sample size re-estimation in a confirmatory trial and keep good documentation so that the sample size change can be verified by independent parties.

Many cardiovascular outcome trials are event-driven trials. Such studies are designed with a fixed number of primary endpoint events. If the total number of events is fixed, the total number of subjects to recruit can be increased or decreased if the overall event rate is lower or higher than expected. This change in sample size does not affect the overall type I error rate.

Sample size re-estimation can also be unblinded. In the unblinded sample size re-estimation, by-arm information, e.g., estimated treatment effect at an interim look, will be used to re-estimate the sample size. Specific analysis methods and adaptation rules are needed in the unblinded sample size re-estimation to control the overall type I error rate. Statistical methodologies have been developed, e.g., combining test statistics or p-values from data prior and post sample size re-estimation [13–15]. A methodology using conventional unweighted test statistic was also developed with the condition that sample size is only increased when interim results are promising [16, 17]. Such methodology requires even higher standards of compliance to the pre-specified analysis and adaptation rule.

Other adaptive designs include adaption to patient population or treatment arms or endpoints [18]. In all adaptive designs, emphasis should be made to pre-planning and study integrity. To maintain study integrity, analysis methods and adaptation rules should be clearly pre-specified and be fully complied with. Failure to do so can substantially compromise the type I error probability and the credibility of study conclusion.

Also, a detailed operating procedure should be created and followed to minimize operational bias. For example, maintaining confidentiality to interim results is important to prevent operational bias. In general, by-arm results should not be released to the public based on an interim analysis while the study is ongoing. Such knowledge of interim results can potentially impact the ongoing conduct of the trial in ways that are difficult for one to predict [19].

25.7 Concluding Remarks

It's important to mentioned that for any clinical trials intended to support a market approval of an experimental treatment, details on study design should be clearly written before the trial starts. Statistical analyses on key endpoints should be prespecified and statistical analysis plan should be finalized before a considerable amount of data is accumulated in the trial.

This chapter introduces some key statistical concepts in designing a clinical trial and mainly focuses on the statistical issues in confirmatory clinical studies. Many other designs, such as a non-inferiority trial or a cross-over study, are not discussed here. Nevertheless, for anyone who is not a statistician, hopefully this chapter provides a glimpse of the statistical world in the clinical trial.

Disclaimer The views and opinions expressed in this chapter are those of the author and do not necessarily reflect the views or positions of the Agency.

Conflict of Interest
The author does not have any conflict of interest.

References

1. ICH E1 The extent of population exposure to assess clinical safety for drugs intended for long-term treatment of non-life threatening conditions (1994)

2. Hochberg Y, Tamhane AC (1987) Multiple comparison procedures. Wiley, New York, NY
3. Holm SA (1979) A simple sequentially rejective multiple test procedure. Scand J Stat 6(2):65–70
4. Dmitrienko A, Tamhane AC, Wiens BL (2008) General multistage gatekeeping procedures. Biom J 50: 667–677
5. Bretz F, Maurer M, Brannath W, Posch M (2009) A graphical approach to sequentially rejective multiple test procedures. Stat Med 28:586–604
6. Bretz F, Posch M, Glimm E, Klinglmueller F, Maurer W, Rohmeyer K (2011) Graphical approaches for multiple comparison procedures using weighted Bonferroni, Simes, or parametric tests. Biom J 53(6): 894–913
7. ICH E9(R1) Statistical principles for clinical trials: addendum: estimands 73 and sensitivity analysis in clinical trials (2017)
8. DeMets DL, Lan KK (1994) Interim analysis: the alpha spending function approach. Stat Med 13: 1341–1352
9. Jennison C, Turnbull BW (1999) Group sequential methods with applications to clinical trials. CRC Press
10. Lan KG, DeMets DL (1983) Discrete sequential boundaries for clinical trials. Biometrika 70(3): 659–663
11. O'Brien PC, Fleming TR (1979) A multiple testing procedure for clinical trials. Biometrics 35(3):549–556
12. Pocock SJ (1977) Group sequential methods in the design and analysis of clinical trials. Biometrika 64(2):191–199
13. Bauer P, Kohne K (1994) Evaluation of experiments with adaptive interim analyses. Biometrics 50(4): 1029–1041
14. Cui L, Hung HM, Wang SJ (1999) Modification of sample size in group sequential clinical trials. Biometrics 55(3):853–857
15. Müller HH, Schäfer H (2001) Adaptive group sequential designs for clinical trials: combining the advantages of adaptive and of classical group sequential approaches. Biometrics 57(3):886–891
16. Chen YHJ, DeMets DL, Lan KKG (2004) Increasing the sample size when the unblinded interim result is promising. Stat Med 23:1023–1038
17. Mehta CR, Pocock SJ (2011) Adaptive increase in sample size when interim results are promising: A practical guide with examples. Stat Med 30:3267–3284
18. Adaptive design clinical trials for drugs and biologics guidance for industry (2019)
19. Guidance for clinical trial sponsors: establishment and operation of clinical trial data monitoring committees (2006)

Introduction to Pharmacoepidemiology and Its Application in Clinical Research 26

Efe Eworuke

Abstract

Pharmacoepidemiology, a sub-specialty of epidemiology involves the study of the use and effects (beneficial and safety) of drugs on large numbers of people using principles and methods of epidemiology and biostatistics. Originally, pharmacoepidemiology focused on the study of the adverse events associated with medication exposure. However, the field has evolved, and includes drug utilization studies such as studies to assess medication adherence, adherence to treatment guidelines, identifying diseased undertreated population as well as effectiveness assessments such as understanding the effectiveness in real-world clinical settings, dose-response relationships, and drug-drug interactions. Pharmacoepidemiology involves the application of study design methods and analytical techniques on large comprehensive databases to obtain measures of risk or effectiveness associated with drug therapies. Interpretation of these measures should be conducted in the context of the study design methodology and the limitations of the databases. In this chapter, I present a broad overview of pharmacoepidemiology methods, and the measures of risk or effectiveness associated with each study design method, highlighting the types of bias that can be introduced by conducting these analyses.

Keywords

Pharmacoepidemiology · Epidemiology · Observational study · Bias · Non-randomized · Database

26.1 Introduction

Epidemiology is the study of the distribution and determinants of health-related events or states in specified populations [1]. Epidemiology is referred to as the basic science of public health because it is applied to control or address health problems [2]. To examine the distribution of disease, we assess the frequency (number of events, rate, or risk) of the disease and patterns of disease occurrence. Epidemiology is a quantitative discipline building upon the knowledge of probability, statistics, and research methods. Statistics or statistical inference is the process of drawing conclusions about an entire population based on the data from a sample.

Pharmacoepidemiology is a sub-specialty of epidemiology. It involves the study of the use and effects of drugs on large numbers of people. It applies principles and methods of epidemiology

E. Eworuke (✉)
Division of Epidemiology, Office of Surveillance and Epidemiology, Center for Drug Evaluation and Research, US Food and Drug Administration, Silver Spring, MD, USA
e-mail: efeeworuke@gmail.com

G. Jagadeesh et al. (eds.), *The Quintessence of Basic and Clinical Research and Scientific Publishing*,
https://doi.org/10.1007/978-981-99-1284-1_26

to study the effects (effectiveness and safety) of drugs on a large number of people [3]. Initially, pharmacoepidemiology stemmed from studying adverse drug effects *i.e.*, safety, but has now expanded to other topic areas such as drug utilization studies (medication adherence, adherence to treatment guidelines, identifying diseased undertreated populations) and effectiveness assessments including understanding the effectiveness in real-world clinical settings, dose-response relationships, and drug-drug interactions. Pharmacoepidemiology applies study design methods and analytical techniques on large, comprehensive databases. Findings from these studies, *i.e.*, measures of risk or effectiveness must be interpreted in the context of limitations. In this section, I cover topics in pharmacoepidemiology: hypothesis generating and hypothesis testing designs, sources of systematic errors (bias) and data sources used in clinical research.

26.2 Hypothesis Generating Study Designs

Hypothesis generating studies are initially used to identify patterns of association between exposure and outcome. Often, further studies are necessary to formally test the hypothesis developed to prove causation or association. These studies describe the occurrence of disease by place, time or population, or present a hypothetical association between exposure and outcome. Hypothesis generating studies include case series and cross-sectional studies conducted at the individual patient level and ecologic studies conducted at a group (of patients) level respectively.

26.2.1 Case Series

A case report describes a single patient with a particular outcome (e.g., adverse drug event), following drug exposure or disease occurrence. A case series is a series of related case reports. It presents reports of multiple patients with the same adverse drug event looking at prior drug exposures to investigate a potential association or reports of patients exposed to the same drug. The case series provides detailed information about the exposure-outcome relationship under investigation including patient characteristics such as gender, race or age and other clinical details on concomitant exposure or comorbidities. These data have some limitations. Case series cannot provide the incidence or frequency of the disease or adverse event because the population at risk for the outcome is unknown. There is also no control or comparison group and often outcomes that occur long after exposure can be challenging to evaluate in a case series. An example of a case series is the FDA Adverse Event Reporting System (FAERS) database (FAERS Public Dashboard). This database contains valuable information on adverse events and medication error reports. Since its inception, FDA has received millions of cases submitted voluntarily by consumers or healthcare professionals or mandatorily submitted by drug manufacturers [4].

To evaluate causation between exposure and outcome in a case series, we rely on select Bradford Hill criteria [5, 6]. These criteria include the expected temporality between exposure and outcome, evidence of positive de-challenge and/or rechallenge, consistency of the association with other products in the same pharmacological class, supporting evidence from other studies and the absence of alternative explanations. Data mining methods [7, 8] such as the proportional reporting ratio (PRR), the reporting odds ratio (ROR), the information component (IC), and the Empirical Bayes Geometric Mean (EBGM) have been developed to quantify the association between a drug and an adverse event or a drug-associated adverse event.

26.2.2 Cross-Sectional Studies

Cross-sectional studies are another type of study design used for hypothesis generation. In these studies, information on exposures and outcomes are collected at the same time. The data are collected from a sample, or a cross-section of a population to get a snapshot of that population at a certain point in time. A cross-sectional study might be repeated over time to look at trends, but the same individuals are not being followed over time. Cross-sectional studies can give measures of frequency called prevalence. Prevalence measures the proportion of individuals with a certain characteristic, risk factor or disease at one point in time or during a specific period. National surveys such as the National Survey of Children's Health (NSCH) or the National Survey on Drug Use and Health (NSDUH) are examples of cross-sectional studies.

The measures of association between exposure and outcome in cross-sectional studies include the prevalence ratio, prevalence odds ratio and prevalence difference. The prevalence ratio (PR) is the ratio of the prevalence of adverse events in the exposed group divided by the prevalence of the adverse events in the unexposed or comparator group. The prevalence odds ratio (POR) is the odds of an adverse event in the exposed group divided by the odds of an adverse event in the unexposed or comparator group. The prevalence difference (PD) is the prevalence of the adverse event in the exposed group minus the prevalence of the adverse event in the unexposed or compactor group. When the PR or POR is equal to 1, the prevalence of disease in the exposed and in the unexposed or comparator are the same; when PR or POR is less than 1, the prevalence of disease in the exposed group is less than the prevalence in the unexposed group and when PR or POR is greater than 1, the prevalence of disease in the exposed group is greater than the prevalence in the unexposed group. For PD, the prevalence in both exposed and unexposed are the same when $PD = 0$ and prevalence is greater in the exposed when $PD > 0$ and lower in the exposed group when $PD < 0$ compared to the comparator group.

A major limitation of the cross-sectional study design is the inability to confirm the temporal sequence of exposure and outcome because these data are collected at the same time. As noted previously, temporality is one of the key features in establishing causation between exposure and outcome. Another potential bias is the incidence-prevalence bias also known as selective survivor bias. Because the measure of association between drug and exposure is conducted in a sample of the population that reflects the risk of getting the disease, duration of disease and likelihood of surviving the disease up to the point that the data is inherently collected. Therefore, for outcomes such as mortality, the measures of association from a cross-sectional study may be underestimated in an exposed group (if the exposure of interest reduces the risk of mortality) compared to an unexposed group. Cross-sectional survey studies are also subject to non-response or volunteer bias if the participants in the survey are systematically different from the nonparticipants.

26.3 Hypothesis Testing Study Designs

In this section, I will discuss the study designs that can be used to test the exposure-outcome association inferred from case series or cross-sectional study designs. Randomized clinical studies remain the gold standard study design to test the hypothesis between exposure and outcome (also, *see* Chaps. 21–23). Random allocation and blinding of exposure assignment are desirable features of a study to determine the causal association between exposure and outcome. Randomization ensures that the exposed and unexposed groups are exchangeable, therefore only the exposure of interest could have led to the outcome. Blinding the treatment group to both investigator and participants removes subjective assessment of certain outcomes. In observational studies, there is no random allocation of treatment assignment, rather the receipt of one treatment over the other is influenced by patient characteristics and physician prescribing behavior. Due to the absence of randomization, exposure groups in observational studies are not exchangeable and causal inference can only be derived by implementing study design and analytical approaches to address potential bias. Potential biases and proposed remedies are discussed below. Observational studies include cohort, case-control, and self-controlled study designs.

26.3.1 Cohort Study Design

A cohort study is a longitudinal study design, proceeding from exposure to outcome to determine the association between the exposure and the outcome. In a cohort study, an exposed and unexposed group, free of the disease or outcome are identified and followed forward in time to determine the presence of the outcome (Fig. 26.1). If the exposed group has a higher incidence of the outcome relative to the unexposed group, the exposure is said to be associated with an increased risk of the outcome.

The measure of disease frequency derived from a cohort study is the incidence. There are two measures of incidence. The incidence proportion (risk) is the number of patients with the new outcome divided by the total population-at-risk at the start of the follow-up. The incidence rate is the number of patients with the new outcome divided by the total person-time at risk (total time all individuals contributed at the time the outcome is assessed). There are therefore two measures of association that relate the exposure and outcome, the risk ratio and the rate ratio. The risk ratio, also known as relative risk, is calculated by dividing the incidence proportion in the exposed group by the incidence proportion in the unexposed group (Table 26.1).

The rate ratio is calculated by dividing the incidence rate in the exposed group by the incidence rate in the unexposed group (Table 26.2).

The relative risk and risk ratio (RR) can be interpreted similarly. When the RR is equal to 1, there is no association between the exposure and the outcome, or the risk of the outcome in the exposed and unexposed groups are the same. When the RR is less than 1, the risk of the outcome in the exposed is less than the risk of the outcome in the unexposed group. When the RR is greater than 1, the risk of outcome in the exposed is greater than the risk of outcome in the unexposed group.

There are several strengths of cohort study designs. The temporal sequence between exposure and the outcome ensures the temporality criteria in assessing the causal association between exposure and outcome. Cohort studies can also be used to study relatively rare exposures and can be used to examine multiple outcomes at the same time. This study design also allows for direct measures of disease frequency such as incidence rates, relative risk, or attributable risk. There are also disadvantages. Cohort studies cannot be used to study rare outcomes or outcomes that take a long time to occur because a long time is needed to accrue before the outcome occurs.

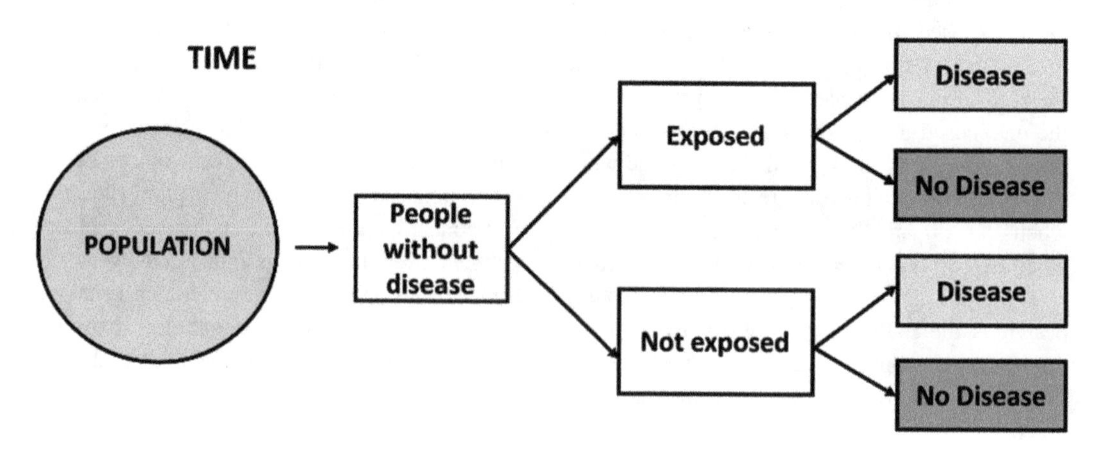

Fig. 26.1 Illustration of a cohort study. In a population, among individuals free of the disease, the presence or absence of exposure is identified and followed over time for the presence or absence of disease

Table 26.1 Calculation of Risk Ratio in a Cohort Study

	Disease	No disease	Total
Exposed	50	150	200
Unexposed	30	170	200
Total	80	320	400

Risk in the exposed: 50/200 (0.25 cases per person) and risk in the unexposed: 30/200 (0.15 per person). Risk ratio: 0.25/0.15 (1.7). In this study, the exposed group was 1.7 times more likely to develop the disease than those not exposed

Table 26.2 Calculation of Rate Ratio in a Cohort Study

	Disease	No disease	Person years at risk
Exposed	50	150	175
Unexposed	30	170	190
Total	80	320	365

Incidence rate in the exposed: 50/175 (0.29 cases per person-year) and the incidence rate in the unexposed: 30/190 (0.16 cases person-year). Rate ratio 0.29/0.16 (1.8). In this study, the exposed group was 1.8 times more likely to develop the disease than those not exposed

26.3.1.1 Study Design Considerations for Cohort Studies

There are certain study design considerations necessary to conduct cohort studies. Appropriate inclusion and exclusion criteria should be applied in defining the study cohorts. These criteria depend on the study question and the outcome of interest. For example, patients with the necessary patient characteristics such as age or alcohol use in the population of interest would need to be selected for study entry in a study that examines the effect of a drug on the elderly or a study that examines risk with alcohol use. Patients with the outcome of interest before the start of this study would have to be excluded from the cohort to obtain the frequency of new outcomes (risk). Other study design considerations include the selection of an appropriate comparator group to minimize differences between the exposure and the unexposed group and accurate measures of exposure and outcome ascertainment to ensure the validity of the observed exposure-outcome association. Systematic error can be introduced into cohort study designs when there is bias in the selection of study participants or measure of exposure and outcome.

26.3.2 Case-Control Study Design

While the cohort study begins with disease free patients categorized into exposed and unexposed who are followed for the presence or absence of the outcome, the case control study design selects patients based on their disease status and collects information on their prior exposure history. Specifically, patients with the disease (cases) and those without the disease (controls) are selected from the source population and the prevalence of the exposure among cases and controls are determined and compared (Fig. 26.2). If the prevalence of the exposure among the cases is higher than the prevalence of exposure among the controls, then the exposure is associated with an increased risk of the outcome.

The measure of association for the control study is the odds ratio. The odds ratio is the ratio of the odds of the outcome in the exposed compared to the odds of the outcome in the unexposed group. Table 26.3 displays a two by two (contingency) table showing how to calculate the odds ratio.

When the odds ratio is equal to 1 the cases and the controls have equal odds of having the expo-

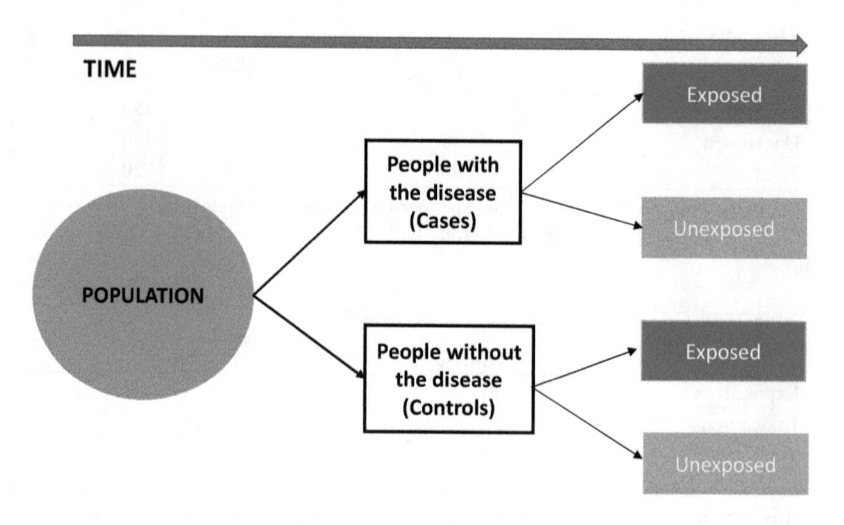

Fig. 26.2 Illustration of case control study design. In a source population individuals with disease (cases), and individuals without disease (controls) are identified. Cases and controls are followed in time for exposed and unexposed status

Table 26.3 Calculation of Odds Ratio in a Case-control Study

	Disease	No disease	Total
Exposed	A	B	A + B
Unexposed	C	D	C + D
Total	A + C	B + D	

Odds in the exposed: A/B; Odds in the unexposed: C/D; Odds ratio: $\frac{AD}{BC}$

sure, indicating no effect. If the cases have higher odds of exposure than the controls, then the odds ratio is greater than 1 suggesting that there is an increased risk of the outcome associated with the exposure. Similarly, if the odds ratio is less than 1 it indicates that the odds of having exposure among the cases are less than the odds of having exposure in the controls, suggesting a protective effect.

Despite having the parameters to calculate a risk ratio from the contingency table for the case-control study, we do not calculate the risk ratio because it is not a reliable measure of association in case-control studies. This is because, in a case-control study, the true denominators (A + B and C + D) are not known since both measures are directly influenced by the study investigators. However, the odds ratio approximates the risk ratio when the disease is rare. Additionally, when A and C become smaller (decreased disease prevalence), the odds ratio approximates the risk ratio, because B and D will contribute to the calculation of both ratios.

Case-control studies are useful for studying rare outcomes or outcomes that take a long time to develop since cases and controls are sampled rather than following patients based on exposure status. It is also possible to evaluate multiple exposures and their risk of a single outcome. A major challenge of case-control studies is the selection of an appropriate control group. The control group must be selected from the same or similar populations as the cases such that both groups are similar in all aspects except for not having the outcome. Case and control selection should also be independent of exposure status, to assure an unbiased estimate of the exposure prevalence in both groups. Lastly, case-control studies are subject to recall bias, specifically when the data are collected after the exposure, or the outcome has occurred. Cases may be more likely to accurately report exposure compared to the controls.

26.3.2.1 Study Design Considerations for Case-Control Studies

Case definitions should be precise to allow for accurate control definitions. It is also preferred to select incident cases to ensure that the etiology of the disease is related to the exposure. A nested case-control design is a special type of case-control study where the controls for a given case is selected at the same time the incident case is selected. The selection of cases and controls is nested in the source population, hence the term "nested case-control". This study design is desirable because temporality between the exposure and outcome can be inferred, especially when the case-control study is nested in the source population of new initiators of a drug. Generally, in a case-control study, the exposure should precede the outcome and the window between outcome and exposure should be the same for both cases and controls. The exposure window must also be etiologically relevant to the outcome of interest. Both the cohort and case-control study designs are referred to as "between-person" designs because we compare groups of patients. However, there are study designs where a single person contributes both the exposed and unexposed experience in the study. We refer to these as self-controlled study designs. In Sect. 26.3.3, I introduce these study designs and challenges in the conduct of these studies.

26.3.3 Self-Controlled Study Designs

Self-controlled study designs or within-person study designs are another types of observational study where the risk of the outcome during an exposed segment of time is compared to the risk of the outcome during an unexposed segment of time in the same patient. Self-controlled study designs should be considered when: (1) appropriate comparators are not available for a cohort study; (2) we cannot measure important differences between the exposed and unexposed group and these differences are not expected to change over time; (3) the risk window between the exposure and outcome is short; (4) the outcome is acute i.e., [develops suddenly and lasts a short time] and (5) the outcome can be detected. There are two broad categories of self-controlled designs, outcome-anchored (case) sampling and exposure-anchored (cohort) sampling study designs [9].

26.3.3.1 Outcome-Anchored (Case) Sampling Designs

Case sampling designs are a modification of the case-control study design. Generally, a sample of cases (incident cases preferably) are selected, and exposure windows hypothesized to be causally related to the outcome are assessed. The following are variations of the case-control sampling designs:

Case-crossover (CCO) design

In the CCO design [10, 11], after the cases are identified, the investigator must first identify a time window to assess the prevalence of exposure. Often this is the time just before the index event and is hypothesized to be the window of time when the exposure is likely to cause the outcome. This time window is referred to as the case time window (Fig. 26.3). One or more windows distant from the index event date are selected to serve as control time windows. During the control time windows, it is hypothesized that the exposure that caused the outcome is unlikely to occur.

Exposure prevalence in the case time window compared to the prevalence in the control time windows is assessed in matched analyses since the same individual acts as both case and control. The interpretation of the relative risk estimate between the case and control time window is different from the relative risk estimate from a between patient study design (cohort or case-control study). The estimate quantifies the risk of the outcome associated with the exposure in the period just before the outcome compared to distant control time windows. There are limitations to the CCO design. First, time trends in drug use such as market uptake of drug

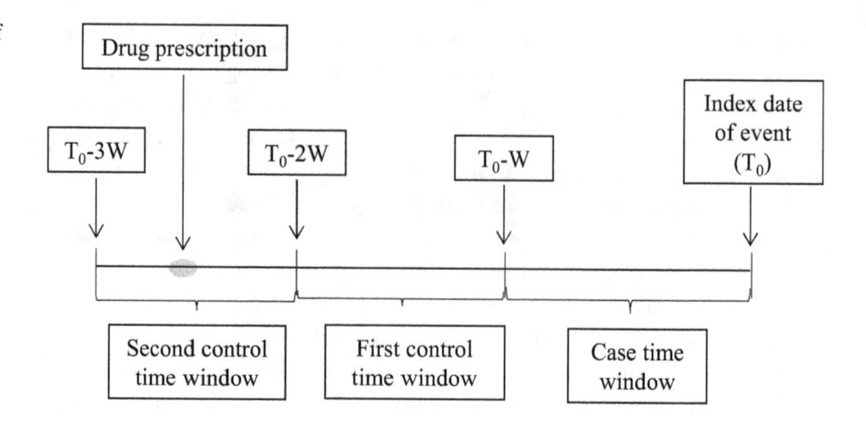

Fig. 26.3 Illustration of the case-crossover study design in pharmacoepidemiology

products or changes in prescribing patterns can alter the probability of identifying exposures in windows that are ordered by calendar time. Second, if exposure in one window affects the exposure in a subsequent window, then "exposure exchangeability" is violated, i.e., the case or control time windows are no longer exchangeable. Lastly, changes in patient health status that can impact the outcome can create bias.

Case-Time-Control (CTC) Design

The CTC design was developed to address the problem of exposure time trends in the CCO design [12]. This design may simply be referred to as a case-control crossover design [13], as it simply selects controls (patients with another event other than the case event) at the same time cases are identified in the CCO and examines the same length of case time and control windows in the cases and controls. By examining the time periods in the controls, the exposure trends in the source population can be estimated and accounted for by comparing the risk estimates from the CCO among the cases to the estimates from the CCO among the time-matched non-cases. Although the exposure trends have been adequately adjusted for in the CTC design, there is no adjustment between the cases and controls in the CTC design, thus bias can still be introduced in the comparative risk estimate if the cases and controls differ.

Case-Case-Time-Control (CCTC) Design

The CCTC design was proposed to circumvent the challenge of using a control CCO as proposed in the CTC design [14]. In the CCTC design, only future cases can serve as controls. This helps to preserve the within patient assessment which is desirable for control of confounding. This design may be affected by limited power since only a subset of the cases will have sufficient history to serve as controls.

26.3.3.2 Exposure-Anchored (Cohort) Sampling Design

When exposure is intermittent during follow-up in a cohort and the outcomes are transient and occur recurrently and independently, i.e., the occurrence of an outcome does not affect subsequent outcomes, cohort sampling designs are efficient within patient designs to assess exposure-outcome relationships. Exposure-anchored designs are initiated on the date of exposure, while outcome-anchored designs are initiated on the date of the outcome. However, in both designs only exposed cases (or controls) contribute to the analyses. There are two variations of the exposure-anchored sampling designs: self-controlled case series and self-controlled risk interval study designs.

Self-Controlled Case Series (SCCS) Design

The SCCS is set up similarly to a cohort study; individuals are followed over time with fixed exposure history and the events are occurring randomly [15, 16]. The total events that occur in the observation period are fixed and the individuals are not censored for an event. Therefore, all exposure periods that occur before and after the event are included in the analysis

(Fig. 26.4). This is different from the CCO designs where individuals are anchored on the event date and exposures relative to the event are assessed.

Exposed and unexposed periods during the observation period are defined based on treatment exposure. In Fig. 26.4, individuals were considered fully exposed while on treatment followed by a sequence of five 35-day periods (for a maximum of 175 days) after which the patient was considered unexposed (baseline period). The relative risk of the outcome during exposed and unexposed periods was compared.

Self-Controlled Risk Interval (SCRI) design

The SCRI is a simplified version of SCCS [17], initially developed for vaccine safety. In the SCCS design for vaccine studies, the exposure period is defined as the time period immediately following vaccination and the control period is defined as all the remaining periods in the observation period. The SCRI modifies the SCCS design by shortening the control period, by selecting a period of similar length to the exposed period, either before or after the exposure period. In practice, the shortened control window is selected close to the exposure period to avoid time-varying confounding.

26.4 Systematic Errors: Bias in Pharmacoepidemiology Studies

The validity of an epidemiologic study is dependent on internal and external validity. Internal validity of a study refers to the extent to which the observed results represent the truth in the studied population and are not due to errors. Systematic error and random error are threats to the internal validity of a study. If the study is internally valid, we can assess external validity. External validity refers to the extent to which the results of a study can be applied to the target population.

The systematic error also known as bias is the error inherent to the study design or conduct. Random error, on the other hand, is the error remaining after the elimination of systematic error. This is directly related to the sample size of the study population. As the study sample increases, so does the random error decrease due to a decrease in variability. We group systematic error (bias) into three broad categories – selection bias, information bias and confounding.

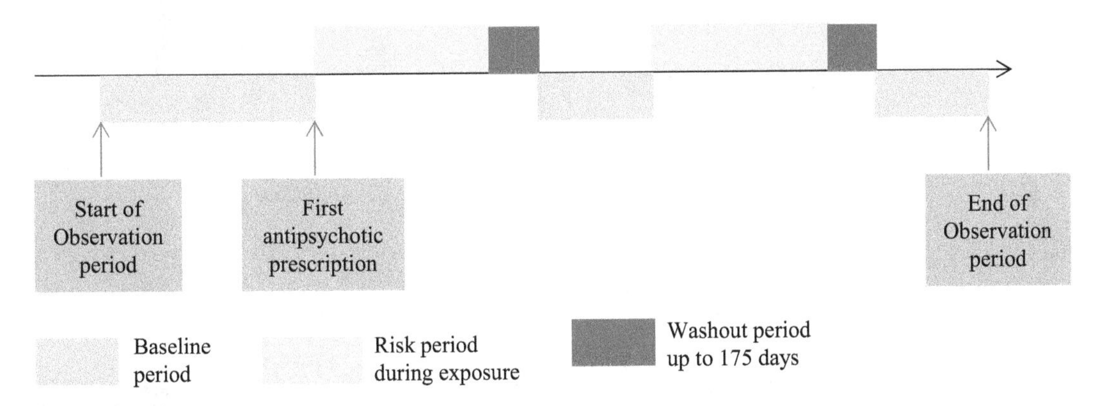

Fig. 26.4 Pictorial representation of a SCCS study design. The study examined antipsychotic drug use during an observation period. All participants had at least one prescription for an antipsychotic and the outcome (single incident stroke)

26.4.1 Selection Bias

Selection bias occurs when individuals have different probabilities of being selected or remaining in the study due to exposure and outcome status [18] (Fig. 26.5). When selection bias is present when the association between exposure and outcome among those selected for analysis is different from the association in the source population. Three common scenarios that can result in selection bias are inappropriate control selection, differential loss of follow-up, and prevalence user bias.

26.4.1.1 Inappropriate Selection of Controls in a Case-Control Study

This can occur when the selection of controls is derived from a population that is not representative of the source population. Consider a case-control study that aims to examine the association between smoking and acute myocardial infarction (AMI) in a hospital (Fig. 26.6). The investigator selects cases as patients admitted for AMI and controls admitted for other conditions. The selection of hospitalized controls is inappropriate because the non-cases could also have a higher prevalence of smoking-related to their hospitalization, for example, hospitalized controls with stroke, lung cancer, or COPD likely have a high prevalence of smoking compared to non-hospitalized controls in the source population. Due to the high prevalence of smoking in the hospitalized controls, the association between smoking and AMI will be underestimated.

26.4.1.2 Differential Loss of Follow-up

Selection bias can also occur if there are differences in the rate of study completion (or dropouts) by exposure and outcome status. Consider a randomized controlled trial that aimed to compare a new drug A to an old drug

With Disease
Without Disease
Source Population
Study Population
Exposed
Unexposed

Ideal situation: the study population comprise of individuals with equal probability of selection, regardless of exposure and outcome status

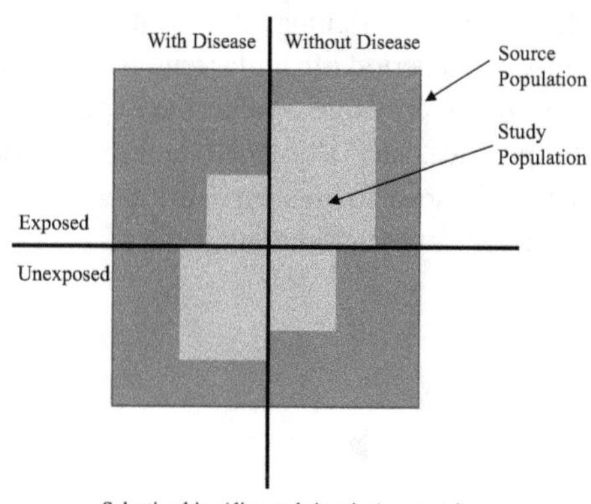

Fig. 26.5 Schematic depicting selection bias

Unbiased Control selection (population controls)

	AMI	No AMI
Smoker	60	30000
Non-smoker	40	60000

Odds ratio: $\frac{60/30000}{40/60000} = 3.0$

Biased Control selection (hospitalized controls)

	AMI	No AMI
Smoker	60	240
Non-smoker	40	160

Odds ratio: $\frac{60/240}{40/160} = 1.0$

Fig. 26.6 Inappropriate selection of controls in a case-control study leading to selection bias

Complete follow-up

	MACE	No MACE
Drug A	100	900
Drug B	100	900

Odds ratio: $\frac{100/900}{100/900} = 1.0$

With differential loss of follow-up

	MACE	No MACE
Drug A	100*0.8=80	900*0.9=810
Drug B	100*0.5=50	900*0.8=720

Odds ratio: $\frac{80/810}{50/720} = 1.42$

Fig. 26.7 Differential loss of follow-up by exposure and outcome status, study leading to selection bias

B on their ability to reduce myocardial adverse cardiovascular events (MACE) (Fig. 26.7). If patients receiving the old drug B were more likely to drop out of the study due to ineffectiveness and thus more likely to develop MACE, then selection bias can occur when we assess the association between exposure and outcome among patients who complete the study.

26.4.1.3 Prevalent User Bias

The inclusion of patients who are prevalent users can lead to selection bias. Patients who develop outcomes likely stop taking the drug of interest, whereas prevalent users tend to be patients who benefit from the treatment. Therefore, the inclusion of prevalent users will distort the study population by selecting patients who are at low risk for the outcome. Selection bias is best addressed by study design and cannot be addressed using analytical techniques. Generally, the study sample of exposed and unexposed patients, cases, and controls should be representative of the source population. Efforts should also be taken to minimize loss to follow-up and monitor the reasons for drop-out from the study.

26.4.2 Information Bias

Information bias occurs due to poor measurement or incorrect classification of study variables (exposure, outcome). When information bias occurs with a continuous variable, we refer to the bias as measurement bias and when it occurs with a binary variable, we refer to the bias as misclassification bias. Most epidemiology analyses, employ category or binary variables, therefore the focus of information bias will be misclassification bias. Two basic types of information bias are non-differential and differential misclassification. For non-differential misclassification bias, the degree of misclassification is the same between the study comparison groups and for differential misclassification, the degree of misclassification between the study group is different for the study comparison groups. Differential and non-differential biases can occur with exposure or outcome.

26.4.2.1 Misclassification of Exposure

Binary, Non-differential Misclassification of Exposure

The extent of misclassification of exposure is the same regardless of outcome status. In the example below (Fig. 26.8), 20% of exposed patients in both the outcome (AE+) and non-outcome (AE-) are classified as unexposed. In other words, 20% of 20 (n = 4) and 20% of 10 (n = 2) of the exposed patients are misclassified as unexposed in the study (observation) and are added to the unexposed group for AE+ and AE-, respectively. Similarly, 10% of the unexposed patients are misclassified as exposed (10% of 80 [n = 8] and 10% of 90 [n = 9]) and are added to the exposed group for AE+ and AE-, respectively. Because of the misclassification, there is a different distribution of exposed and unexposed patients in the study even though, the total number of patients with and without the outcome is the same.

In the study, we added 4 and 7 more patients to the exposed group, AE+ and AE-, and removed the same number of patients from the unexposed group. The estimated OR in our study is lower, i.e., the estimate has moved toward the null (OR = 1) due to non-differential misclassification. Misclassification of exposure

Truth

	AE+	AE-
Exp+	20	10
Exp-	80	90

True OR: 2.25
(20x90)/(80x10)

Non-differential misclassification of exposure

	AE+	AE-
Exp+	20↓4	10↓2
Exp-	80↑8	90↑9

Observed

	AE+	AE-
Exp+	24	17
Exp-	76	83

Estimated OR:1.54
(24x83)/(76x17)

Fig. 26.8 Schematic depicting binary, non-differential misclassification of exposure

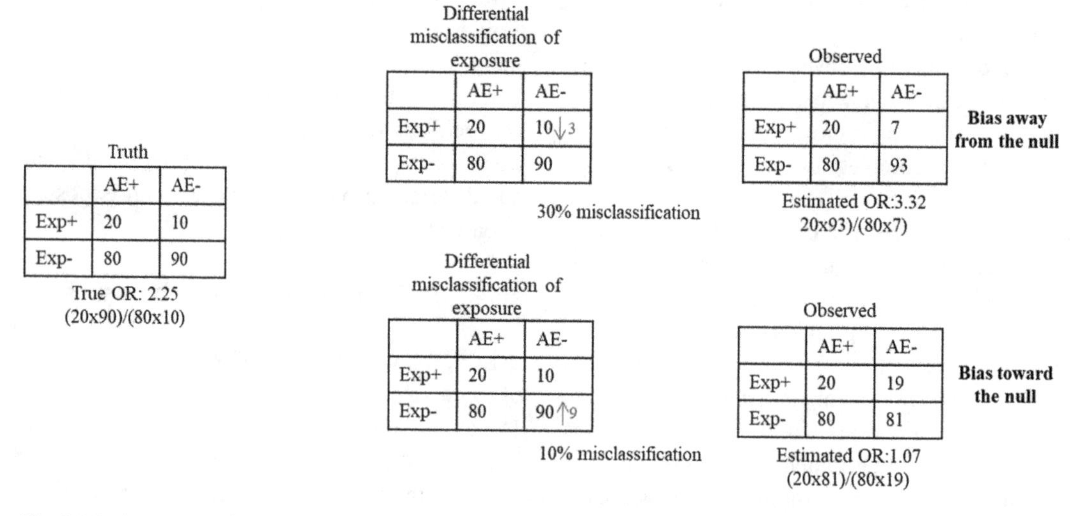

Fig. 26.9 Schematic depicting binary, differential misclassification of exposure

occurs when there is an inaccurate categorization of exposure status. For example, in an observational database relying on pharmacy dispensing insurance claims to ascertain exposure, there could be misclassification of unexposed patients as exposed, if the researcher classified them as exposed when they are not active on their medication, for example, non-compliers to the drug. In another scenario, a researcher may misclassify exposed patients as unexposed, because the unexposed patients procured the treatment from other sources, for example, over-the-counter version of a drug.

Binary, Differential Misclassification of Exposure

The extent of misclassification of exposure is different by outcome status, i.e., exposure misclassification may be higher or lower among patients with the outcome compared to those without the outcome. The example in Fig. 26.9 shows two different scenarios that can lead to two opposing outcomes. In scenario 1, there is 0% misclassification among those who had the outcome, whereas there is a 30% misclassification of exposed patients, i.e., 3 (30%) patients who were truly exposed reported unexposed status. This can occur with recall bias, where patients who do not have the outcome recall their exposure history differently from those who experienced the outcome. As shown in Fig. 26.6, this results in a higher OR than the truth (3.32 vs. 2.25). In scenario II, there is also no misclassification of exposure among patients with the outcome; however, misclassification of exposure is introduced among patients who did not experience the outcome. In this example, 10% (n = 9) of the unexposed patients were classified as exposed in the study. This misclassification resulted in a bias towards the null, in other words, the estimated OR was lower than the true OR. The impact of non-differential misclassification bias cannot be predicted. It can bias the ORs in any direction.

26.4.2.2 Misclassification of Outcome

Misclassification of outcome is analogous to the misclassification of exposure and the effects are similar. In misclassification of the outcome, the patients with the outcome are misclassified as not having the outcome and vice versa. Though the impact of misclassifying outcome status for differential and non-differential misclassification as previously described for exposure misclassification, there is one special case. When there is non-differential misclassification of outcomes because of incomplete capture of patients with the outcome only (i.e., all patients without the outcome are correctly classified), the relative risk will not be biased (Fig. 26.10). In this example, only patients who had the outcome were affected by misclassification, as shown by the red arrows moving patients from AE+ to AE-, but no patients are moving from AE- to AE+. Specificity, the measure of accuracy for determining non-cases is 100%; however, for sensitivity, the measure of classifying cases is 50%.

To address misclassification bias, prospective studies should ensure accurate measurement of exposure and outcome by using appropriate tools or instruments and relying on defined protocols for measurements. Studies that rely on secondary data would need to ensure that algorithms used to establish exposure and outcome for the study have been validated against gold standard definitions. Researchers can also understand how data are generated to be able to identify how measurement errors can be introduced into the study.

26.4.3 Confounding

Confounding occurs when the association between exposure and outcome is distorted by a third factor. A confounder must meet the following characteristics: (1) it is a risk factor for the outcome; (2) it is associated with the exposure (a predictor of the exposure); (3) it is not an intermediate variable on the causal pathway between exposure and the outcome. Intermediate variables can also be confounders, the timing of the variable is important for determining whether a variable is a confounder or not. An example is a study that examines the effect of an anti-HIV drug on HIV complications (Fig. 26.11). Pre-treatment CD4+ count is a pre-treatment variable that is a predictor of exposure and is also a risk factor for the outcome. However, post-treatment CD4+ count is an intermediate variable because the anti-HIV drug can only impact the risk of HIV complications only through a reduction of the CD4+ count.

26.4.3.1 Confounding by Indication and Disease Severity

Indication for treatment or disease severity will often predict the selection of treatment(s) and the indication or disease severity often is associated

Truth

	AE+	AE-	Total
Exp+	80	920	1000
Exp-	40	960	1000

True OR: 2.25
(80/1000)/(40/1000)

	AE+	AE-	Total
Exp+	80	920	1000
	40 → (50%)		
Exp-	40	960	1000
	20 → (50%)		

Non-differential misclassification of outcome. Specificity=100%

Observation

	AE+	AE-	Total
Exp+	80	920	1000
Exp-	40	960	1000

Estimated OR: 2.25
(40/1000)/(20/1000)

Fig. 26.10 Schematic depicting misclassification of exposure

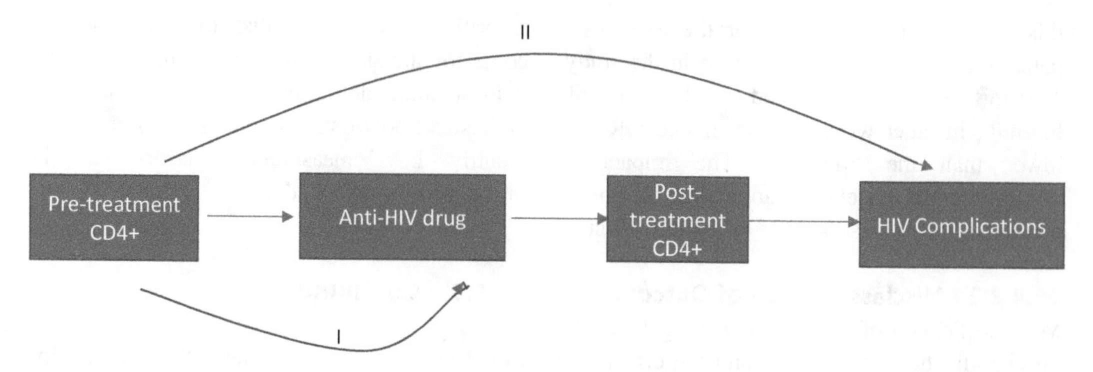

Fig. 26.11 Pre-treatment CD4+ is a confounder because it is a predictor of exposure (I), a risk factor for the outcome (II) and is not an intermediate between exposure and outcome. On the other hand, post-treatment CD4+ is an intermediate variable and not a confounder

with the outcome. The impact of confounding by indication and disease severity will depend on the choice of the comparator used in the study. When the choice between exposure and comparator is strongly driven by the underlying disease or severity and is a risk factor for the outcome, the stronger the confounding bias. This often happens when comparing exposed versus unexposed patients because the unexposed patients either do not have the indication or have lower disease severity. To minimize confounding, the researcher can select an active comparator with a similar indication or disease severity which would result in a similar probability of receiving the exposure of interest or the comparator and thus create similar groups. A good understanding of the treatment guidelines and clinical understanding of treatment selection should guide the appropriate choice of a comparator. Confounding can also be minimized by restriction i.e., excluding patients with factors that are strong confounders to create a homogenous population or employing analytical techniques such as matching patients on key confounders or use of multivariate analyses.

26.4.3.2 Effect-Measure Modification

Effect-measure modification occurs when the effect of an association between exposure and outcome differs by levels of a third variable i.e., the variable modifies the effect of the association. Effect modifiers help to identify susceptible groups to ensure comprehensive communication of the effects or safety of drugs. In an example of a study of the effect of oral contraceptives on deep vein thrombosis (DVT), an overall RR of 3 is observed. Examining the risk by smoking status revealed a risk ratio of 7 among smokers and 1.29 among non-smokers (Fig. 26.12), suggesting that smoking status modifies the effect of the association between oral contraceptives and DVT. The effect from the overall population represents the average effect across all levels of the effect modifier. It is important to note that confounding and effect-modification can be present at the same time in a study.

26.5 Data Sources for Clinical Research

In conducting pharmacoepidemiology research, we apply methods that incorporate study design and analytical techniques on databases which need to be large, comprehensive, validated, and include the appropriate data elements for analyses. Finally, we examine the measures of risk or measures of effectiveness and then interpret the findings in the context of the study limitations. In this section, we introduce data sources and discuss their pros and cons.

An ideal database requires that records that are contained are verifiable, reliable, and updated regularly. The database should also allow for

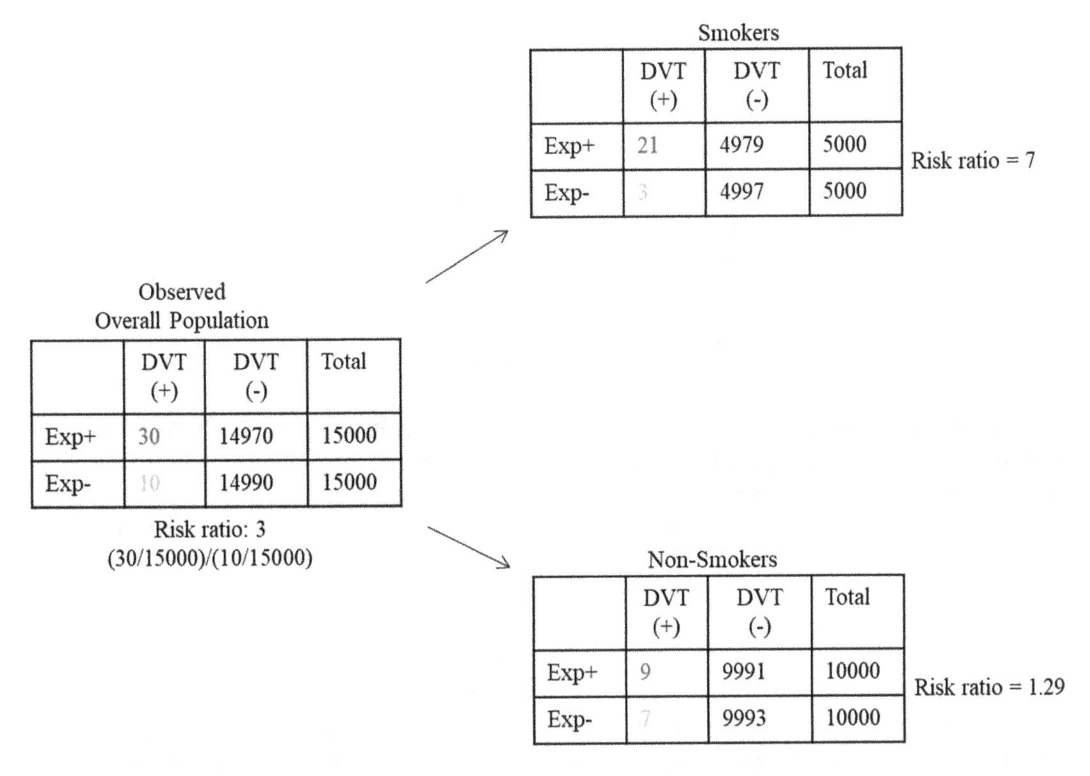

Smokers

	DVT (+)	DVT (-)	Total
Exp+	21	4979	5000
Exp-	3	4997	5000

Risk ratio = 7

Observed Overall Population

	DVT (+)	DVT (-)	Total
Exp+	30	14970	15000
Exp-	10	14990	15000

Risk ratio: 3
(30/15000)/(10/15000)

Non-Smokers

	DVT (+)	DVT (-)	Total
Exp+	9	9991	10000
Exp-	7	9993	10000

Risk ratio = 1.29

Fig. 26.12 Effect modification of the risk of deep vein thrombosis by smoking status

linkage to other databases to pull in additional information using unique identifiers. No database is ideal, however, there are published guidelines [19] on how we may select a database for research. The database should (1) include the appropriate population of interest and the study variables should be updated regularly, (2) allow for linking to other data sources to expand patient information and potentially increase the study sample, (3) include information on how to extract the study population, (4) variables should also be readily available and (5) include information to ensure confidentiality and privacy.

Broadly, data sources can be characterized as primary and secondary data sources. For primary data sources we have prospective registries and surveys, and for secondary data, we have surveys, registries, claims or administrative data sources, and electronic health record data sources.

26.5.1 Primary Data Sources

Primary data are collected prospectively to answer a study question. Data can be collected through questionnaires, interviews, or chart reviews. The advantage of these data sources includes improved capture of study-specific details, such as exposure and outcome information and the ability to appropriately define the population of interest. The disadvantage is that expenditure and time to collect the data can be resource intensive. It is also limited by small sample size often to accommodate intense use of resources.

26.5.1.1 Registries

Registries are described as a collection of patient records on a particular subject. The purpose of registries includes describing the natural history of the disease, determining clinical or cost effectiveness in real world data, measuring and monitoring the safety and harm of products in terms of active surveillance, and measuring the quality of care. Product registries are a collection of patients exposed to a particular drug. A common example are the biologics registries which can also be repurposed to answer other questions. In that situation, registries serve as secondary data sources. Pregnancy registries are a type of product registry; however, they focus on exposure during pregnancy. Disease registries are a collection of patients with a particular disease or condition, for example, the cystic fibrosis registry is a registry of patients with cystic fibrosis. And lastly, we have health services registries which are a collection of patients exposed to a specific healthcare service that is often under evaluation.

26.5.1.2 Surveys

Surveys are a collection of data or knowledge, attitudes, behavior, practices, opinions, or beliefs from selected individuals normally from a sampling frame. Questions will be answered in person, via post, or over the phone. Surveys are generally cross-sectional, but a repeated collection of the same data in the same population can form longitudinal assessments. Usually, surveys are designed to answer a specific research question, via primary data collection. They may also be used for other research questions. In this case, they are regarded as secondary data. Surveys can be used to assess prescriptions, health, nutritional status, or other lifestyle factors, like smoking. They can be used to examine the distribution of unmeasured confounding to improve data quality for other observational studies.

26.5.2 Secondary Data Sources

Secondary data are pre-existing data already collected from a previous research question or as part of administrative records. Examples include electronic health records (EHR) or claims data. These data can be used in outcome research, including pharmacoepidemiology, pharmacovigilance (see the next chapter), assessing prescribing practices, drug utilization, and medication adherence studies. The pros of these data sources are that they are already collected and thus readily available, and therefore less resource intensive. However, a limitation of this data source is that the data were collected for other research questions and have been repurposed. Therefore, the study design and analyses must incorporate the differences in study design to assure valid findings.

26.5.2.1 Administrative or Claims Data

Administrative or claims data arise from a medical or pharmacy claim or a bill that healthcare providers submit to a patient's health insurance provider. The claim contains unique codes for the service or care administered during a patient visit. These codes are generated from the claims data used to define the presence of drug exposure or diagnosis in pharmacoepidemiology studies. For example, when a patient fills a dispensed drug, a specific drug code is generated, and the pharmacy bills the insurance company. When the patient goes to a physician or is hospitalized, a code is generated for the service (e.g., a diagnostic, or procedure code) and a bill is sent to the patient's insurance company. With a common patient identifier, the pharmacy and medical claims can be linked to form a longitudinal medical record. For claims data, pharmacy data are usually more accurate than medical data because filing incorrect drug claims is considered fraud. On the other hand, there is less accuracy in the diagnosis because there is often no incentive to provide a precise diagnosis. These inconsistencies have to be considered when using secondary data sources. To improve the richness of data, you can link claims data to the primary data source. An example of an administrative data source is the Medicare claims database which covers patients 65 years and older.

26.5.2.2 Electronic Medical/Health Record

With the advancement of automated medical care, data generated during health encounters are now easily accessible and increasingly being used for outcomes research. There is a difference between EHR and electronic medical records (EMR.) While the EHR is a longitudinal digital record of the patient's information generated during health encounters, the EMR is a digital version of the paper charts in medical practice. These sources provide information about the healthcare encounter and provide complete clinical detail compared to administrative claims data. However, the lack of standardization across these EMR and EHR systems can often be problematic when considering data from several data sources.

26.5.2.3 Multi-database

Multi databases involve pooling data from different databases to improve generalizability and precision. Precision is often improved because of the very large sample size. Examples include the FDA Sentinel system, Vaccine Safety Datalink, and the Canadian Network of Observational Drug Effects. Often these databases require a common data model to transform the local data into a predefined common data structure which allows for implementing the same analytical scripts on the contributing data sources.

FDA's Sentinel system is a multi-database that the FDA uses to answer questions about the safety of approved drug products. Currently, the data has over 71 million members contributing data with more than 600 million person-years contained in the database. There are over 11 billion pharmacy dispensing and 15 billion unique medical encounters. FDA's Sentinel system comprises several data partners including administrative data sources, EMR/EHR clinical data, data from registries, mother-infant linked data, and other auxiliary data sources. By using a common data model, individual partners can convert their local data, into a common language allowing for the use of the same analytical script across several data partner databases.

26.6 Concluding Remarks

Pharmacoepidemiology is an applied public health science with widespread application in understanding the use, effectiveness, and safety of medications in real-world settings. Pharmacoepidemiology requires careful consideration of study design and analytical techniques to address a study question. The interpretation of study findings should be discussed in the context of the limitations of the study design.

Acknowledgments I would like to thank the staff at the Division of Epidemiology, Office of Surveillance and Epidemiology, Center for Drug Evaluation and Research, FDA, for their contribution to the content of this chapter.

Disclaimer This chapter reflects the views of the author and do not necessarily represent FDA's views or policies.

Conflict of Interest

I have no conflict of interest to declare.

References

1. Last JM (1988) Dictionary of epidemiology, vol 42. Oxford University Press
2. Lesson 1: introduction to epidemiology. https://www.cdc.gov/csels/dsepd/ss1978/lesson1/section1.html#print. Accessed 30 Nov 2022
3. International Society of Pharmacoepidemiology: about pharmacoepidemiology. https://www.pharmacoepi.org/about-ispe/about-pharmacoepidemiology/. Accessed 30 Nov 2022
4. Fang H, Su Z, Wang Y et al (2014) Exploring the FDA adverse event reporting system to generate hypotheses for monitoring of disease characteristics. Clin Pharmacol Ther 95(5):496–498. https://doi.org/10.1038/clpt.2014.17
5. Fedak KM, Bernal A, Capshaw ZA, Gross S (2015) Applying the Bradford Hill criteria in the 21st century: how data integration has changed causal inference in molecular epidemiology. Emerg Themes Epidemiol 12(1):14. https://doi.org/10.1186/s12982-015-0037-4
6. Hill AB (1965) The environment and disease: association or causation? Proc R Soc Med 58(5):295–300
7. Sakaeda T, Tamon A, Kadoyama K, Okuno Y (2013) Data mining of the public version of the FDA adverse event reporting system. Int J Med Sci 10(7):796–803. https://doi.org/10.7150/ijms.6048
8. Patel NM, Stottlemyer BA, Gray MP, Boyce RD, Kane-Gill SL (2022) A pharmacovigilance study of adverse drug reactions reported for cardiovascular disease medications approved between 2012 and 2017 in the United States Food and Drug Administration Adverse Event Reporting System (FAERS) Database.

Cardiovasc Drugs Ther 36(2):309–322. https://doi.org/10.1007/s10557-021-07157-3
9. Cadarette SM, Maclure M, Delaney JAC et al (2021) Control yourself: ISPE-endorsed guidance in the application of self-controlled study designs in pharmacoepidemiology. Pharmacoepidemiol Drug Saf 30(6):671–684. https://doi.org/10.1002/pds.5227
10. 'Chris' Delaney JA, Suissa S (2009) The case-crossover study design in pharmacoepidemiology. Stat Methods Med Res 18(1):53–65. https://doi.org/10.1177/0962280208092346
11. Bird ST, Etminan M, Brophy JM, Hartzema AG, Delaney JAC (2013) Risk of acute kidney injury associated with the use of fluoroquinolones. Can Med Assoc J 185(10):E475–E482. https://doi.org/10.1503/cmaj.121730
12. Schneeweiss S, Stürmer T, Maclure M (1997) Case–crossover and case–time–control designs as alternatives in pharmacoepidemiologic research. Pharmacoepidemiol Drug Saf 6(S3):S51–S59. https://doi.org/10.1002/(SICI)1099-1557(199710)6:3+<S51::AID-PDS301>3.0.CO;2-S
13. Suissa S (1995) The case-time-control design. Epidemiology 6:3
14. Wang S, Linkletter C, Maclure M et al (2011) Future cases as present controls to adjust for exposure trend bias in case-only studies. Epidemiology 22(4):568–574. https://doi.org/10.1097/ede.0b013e31821d09cd
15. Petersen I, Douglas I, Whitaker H (2016) Self controlled case series methods: an alternative to standard epidemiological study designs. BMJ 354:i4515. https://doi.org/10.1136/bmj.i4515
16. Douglas IJ, Smeeth L (2008) Exposure to antipsychotics and risk of stroke: self controlled case series study. BMJ 337(aug28 2):a1227. https://doi.org/10.1136/bmj.a1227
17. Li R, Stewart B, Weintraub E (2016) Evaluating efficiency and statistical power of self-controlled case series and self-controlled risk interval designs in vaccine safety. J Biopharm Stat 26(4):686–693. https://doi.org/10.1080/10543406.2015.1052819
18. Gerhard T (2013) Bias and Confounding presented at the 28th International Conference on Pharmacoepidemiology and Therapeutic Risk Management, Barcelona, Spain
19. Hall GC, Sauer B, Bourke A, Brown JS, Reynolds MW, Casale RL (2012) Guidelines for good database selection and use in pharmacoepidemiology research. Pharmacoepidemiol Drug Saf 21(1):1–10. https://doi.org/10.1002/pds.2229

27 Pharmacovigilance Through Phased Clinical Trials, Post-Marketing Surveillance and Ongoing Life Cycle Safety

Ananya Chakraborty and J. Vijay Venkatraman

Abstract

The evolution of science has gifted humans with safe and effective treatment modalities. Drugs are monitored from clinical trials to real world usage. Pharmacovigilance started in a reactive way but today it is a proactive process. Under the umbrella of regulatory bodies, drugs are monitored during the development and marketing phases. The stakeholders such as healthcare set-ups, the pharma industry, government agencies, consumers, non-profit organizations, and legal bodies collectively contribute to the development of guidelines, safety process, and risk minimization measures for the reduction and prevention of harm caused by drugs. Recently, the effective use of automated tools and integration of artificial intelligence has started helping in integrating data throughout the globe. Also, it is predicted to further simplify the pharmacovigilance process, and reduce the overall cost. This chapter elaborates on the pharmacovigilance process during phased clinical trials, and post-marketing surveillance.

Keywords

Pharmacovigilance · Adverse event · Safety signal · Post-marketing surveillance · MedWatch · MedDRA

27.1 Introduction

27.1.1 What is Pharmacovigilance?

The World Health Organization (WHO) defines pharmacovigilance as the science and activities relating to the detection, assessment, understanding and prevention of adverse effects or any other medicine-related problem. The goal of pharmacovigilance is to improve overall medication safety [1].

27.1.2 What Led to the Evolution of Pharmacovigilance?

Pharmacovigilance has evolved after a series of disasters in the past. In 1848, a little girl from the north of England named Hannah Greener died after receiving chloroform anaesthesia. She was undergoing a toenail surgery. The cause was attributed to lethal arrhythmia, pulmonary aspiration, or an overdose of the drug. She had successfully received ether several months earlier during the removal of another toenail. The case raised an alert regarding the safe use of anaesthetics

A. Chakraborty (✉)
Department of Pharmacology, Vydehi Institute of Medical Sciences and Research Centre, Bengaluru, India
e-mail: dr_ananya@yahoo.com

J. V. Venkatraman
Oviya MedSafe Pvt Ltd, Coimbatore, India
e-mail: vijay.j@oviyamedsafe.com

G. Jagadeesh et al. (eds.), *The Quintessence of Basic and Clinical Research and Scientific Publishing*,
https://doi.org/10.1007/978-981-99-1284-1_27

[2]. However, the first milestone in drug regulation was much later. In June 1906, the US Federal Food and Drugs Act was passed. The main purpose of the act was to prohibit adulterated or spurious food and drugs [3].

Much later, in 1937, there were 107 deaths in the USA due to the use of sulfanilamide elixir. During World War II, soldiers were trained to sprinkle wounds with "SULFA POWDER" to prevent infection. The anti-infective properties of the drug were promising. Hence, a manufacturer decided to market the drug in a liquid form for sore throat. But there was a problem. The powder would not dissolve in water or alcohol. Later, it was found to dissolve in a solvent, "diethylene glycol". The drug was consequently marketed in this solvent. However, no clinical tests were performed before marketing. And it caused the deaths. Consequently, the Federal Food, Drug and Cosmetic Act was established in 1938. It became a priority to prove that medicine was safe before marketing it. This act saved the USA from the thalidomide tragedy many years later when the drug was discovered and marketed in Europe. In the USA, it was declined approval without proof of experiments on pregnant animals [3, 4].

In Europe, thalidomide was a widely used drug in the late 1950s and early 1960s. It was marketed for the treatment of nausea during the early trimester of pregnancy. There was no evidence to prove that the drug was safe before marketing. Slowly, thalidomide treatment resulted in severe birth defects in thousands of children. Children were born with seal limbs, a condition called Phocomelia. This disaster led to a big change in European Pharmacovigilance. In 1964, the "Yellow card" (YC) was structured in the UK. YC is a specific form to compile a spontaneous report of drug toxicity. Spontaneous reporting is a passive mode of reporting adverse event (AE) by a healthcare professional or consumer to a pharmaceutical industry or the regulatory authorities [5–9] (also *see* Chap. 23).

In the USA (1962), the Kefauver-Harris amendment, which mandated the safety and efficacy data of drugs before premarketing submission, was approved. The safety data also required teratogenicity tests in three different animals. In 1968, the WHO Programme for International Drug Monitoring was instituted. In 1992, the International Society of Pharmacovigilance (ISoP) was formed. Later, there was the introduction of pharmacovigilance databases equipped with computerized automated statistical approaches. They needed to be regulatory compliant. They provided key information for risk detection and the ongoing evaluation of the risk-benefit profile of drugs [3, 6–9].

Pharmacovigilance started reactively, but today it is a structured proactive process. Drugs are monitored for AE during clinical trials, and post-marketing surveillance (PMS) of the real-world population. This monitoring helps take pre-emptive measures to reduce the risks associated with drugs as 17% of the failed phase 3 trials are due to safety issues. PMS also leads to drug withdrawals, restrictions in use and labelling changes. Proactive monitoring of drugs helps design drug utilisation practices, essential drugs programmes, standard treatment guidelines and national and institutional formularies [10, 11].

27.1.3 What is the Scope of Pharmacovigilance?

Drug development starts with the evaluation of efficacy and safety in pre-clinical models (see Chaps. 4–15). This is followed by an evaluation of the benefit-risk profile during various phases of clinical trials (*see* Chaps. 20–22). Pre-marketing clinical trials are conducted on a limited number of eligible participants. The frailest and special populations like women, ethnic minorities, and vulnerable populations are usually under-represented. This leads to intrinsic limitations and a chance to miss a population-specific AE, or AEs due to drug-drug interactions that may be unique to specific populations.

It is important to monitor a drug throughout the life cycle for AE. Post-marketing PMS and phase 4 clinical trials are conducted towards the evaluation of safety based on real world evidence (RWE) [12, 13].

The scopes of pharmacovigilance are as follows:

- Collect and record AEs/adverse drug reaction (ADR)s
- Conduct causality assessment and analysis of ADRs
- Collate and code in a database
- Compute benefit-risk and suggest regulatory action
- Communicate the significance of the safe use of drugs among stakeholders

27.1.4 What is the Minimum Dataset Required for a Reportable AE, and What Are Their Sources?

The minimum dataset required to report an AE is (1) an identifiable patient (2) an identifiable reporter (3) product exposure, and (4) an event. This is shown in Fig. 27.1.

AEs are reported by solicited and unsolicited sources.

Solicited reports are obtained from organized data collection systems. They include clinical trials, registries, post-approval patient use programs, other patient support and disease management programs, surveys of patients or healthcare providers, or information gathering on efficacy or patient compliance. Solicited reports are termed as study reports. The site study team and the marketing authorization holder (MAH) are responsible to provide the dataset and to report these cases.

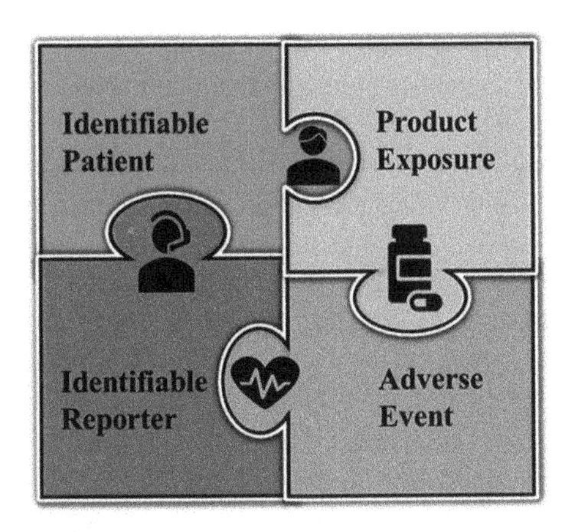

Fig. 27.1 Data set required to report an AE

Unsolicited sources refer to spontaneous reporting done by healthcare workers, patients and their caregivers, government agencies, the pharma industry, etc. Safety data identified by MAH's review of scientific and medical literature as well as reports from verified news in lay press or media are also categorised under spontaneous reports [12–14].

27.1.5 What are the Common Terms Used in Pharmacovigilance?

The following terms are commonly used in pharmacovigilance [10, 15–17].

- **Adverse event (AE)**: An AE is any unwanted and noxious effect of a drug when used in recommended doses. AE does not have a causal association with the study drug.
- **Adverse drug reaction (ADR)**: AE with an established causal association with the study drug is called an adverse drug reaction (ADR).
- **Serious adverse event (SAE)**: AEs are categorized for reporting purposes according to the seriousness and expectedness of the event. An AE that is associated with death, in-patient hospitalization, prolongation of hospitalization, persistent or significant disability or incapacity, a congenital anomaly, or is otherwise life-threatening is termed a serious adverse event (SAE).
- **Adverse events of special interest (AESI)**: AESI is one of the scientific and medical concerns specific to the sponsor's product or programme, for which ongoing monitoring and rapid communication by the investigator to the sponsor could be appropriate.
- **Adverse Event Following Immunization (AEFI):** AEFI is any untoward medical occurrence that follows immunization, and which does not necessarily have a causal relationship with the usage of the vaccine.

- **Expected and unexpected ADR**: Based on expectedness, ADRs are graded as expected and unexpected. For expected reactions, the nature and severity are listed in the reference safety information (RSI) document e.g., core safety information in the investigator's brochure (IB), protocol, informed consent form for phase 2 and 3 trials, summary of product characteristics (SmPC, Europe), product labelling (package inserts, USA) in case of approved drugs. Unexpected reactions are those that are not listed in the same specificity or severity in any of the above documents.
- **Relatedness of AE**: Relatedness is based on various factors such as biological plausibility, prior experience with the product, and temporal relationship between the exposure to the drug and onset of the event, as well as de-challenge (discontinuation of the drug to determine if the AE resolves) and rechallenge (reintroduction of the drug to determine if the AE recurs). Various standard scales are used to describe the degree of causality. Relatedness is graded based on terms like certainly, definitely, probably, possibly, likely related, not related.
- **Labelled and unlabelled ADR:** For a drug with an approved marketing application, any reaction which is not mentioned in the official product information is "unlabelled." If it is included, it is termed "labelled."
- **Spontaneous reporting**: Spontaneous reporting is a passive surveillance method. Spontaneous reporting is usually an unsolicited communication by a healthcare professional, consumer to a company, regulatory authority, or other organization (e.g., WHO, Regional Centre, Poison Control Centre) about an AE.
- **Individual case safety reports (ICSRs)**: Individual case safety report is a document in a specific format for the reporting of one or several suspected AEs in a single patient at a specific point in time.
- **Serious and unexpected suspected adverse reactions (SUSAR)**: SUSAR is a serious unexpected adverse reaction. This term is used in the context of clinical trials. The investigator needs to notify EC within a day. Once an SAE report is received by the sponsor, the team decides based on RSI whether the event should be considered a SUSAR. In most countries, the sponsor gets 7 calendar days to report a SUSAR to the regulatory authorities if the event is fatal or life-threatening. If not, they get 15 days.
- **Development safety update report (DSUR)**: DSUR is a periodic pre-marketing safety report. DSUR contains safety information from all ongoing clinical trials. DSUR has usually submitted annually to regulatory agencies and sometimes also to EC.
- **Periodic safety update report (PSUR)**: PSUR is a post-marketing aggregate safety report. It contains the worldwide safety experience with a drug at a defined time after its marketing authorisation. It is usually submitted every 6 months for the first two years [-European Medicines Agency (EMA)] and every 6 months for the first three years [Japanese Pharmaceuticals and Medical Devices Agency (PMDA)], and then annually.
- **Periodic adverse drug experience report (PADER)**: PADER is a post-marketing aggregate safety report. It is required to be submitted by a sponsor or MAH to the US Food and Drug Administration (FDA). It is submitted every 3 months for the first three years post-marketing authorization, and then annually.
- **Periodic benefit-risk evaluation report (PBRER)**: PBRER is a post-marketing aggregate safety report. It is submitted to the European Union every 6 months for the first two years post-marketing authorization, and then annually.
- **Data lock point (DLP)**: DLP is the date designated as the cut-off date for data to be incorporated into a particular safety update (PSUR, PBRER, etc.) This is usually based on the international birth date (IBD) of the new drug.
- **Safety signal**: A signal is new and important safety information or a new aspect of an already known association that warrants further investigation to accept or refute. It may necessitate remedial actions. Usually, more

than a single case report is required to generate a signal, depending on the seriousness of the event and the quality of the information. If the signal is verified and evidence of causality between a drug and an AE is established, the regulators may issue a recall, change a drug's labelling, or withdraw a medication from the market.

- **Signal detection or data mining:** Data mining is an analytical process. In this, large datasets (ISCRs) from suitable databases are analysed or "mined" in search of meaningful patterns, relationships, or insights, called signals.
- **Quantitative signal detection:** Signal detection where computational or statistical methods are used to identify drug-event pairs (or higher-order combinations of drugs and events) that occur with disproportionately high frequency in large spontaneous report databases.
- **Line listing:** It is a report about the ICSR that can be generated regarding a drug from pharmacovigilance databases.
- **ICH E2E guidelines**: The International Council for Harmonisation of technical requirements for pharmaceuticals for human use (ICH) is a non-profit organization. The goal of ICH is to bring regulatory authorities and the pharmaceutical industry throughout the globe together to harmonize scientific and technical aspects of drug registration. ICH E2E guidelines aid in planning pharmacovigilance activities, especially in preparation for the early post-marketing period of a new drug.
- **Council for International Organizations of Medical Sciences (CIOMS)**: CIOMS was formed in 1949 by WHO and UNESCO. CIOMS form is the standard reporting format of serious adverse events while conducting clinical trials worldwide to respective regulatory authorities.
- **Marketing Authorization Holder (MAH)**: A MAH is a person or company that is licensed to distribute, sell and commercialize a medical product.
- **Medical Dictionary for Regulatory Activities (MedDRA)**: MedDRA is an international medical terminology. It is used for coding cases of AEs in clinical study reports and pharmacovigilance databases, and to facilitate searches in these databases. It is helpful in all phases of drug development.
- **MedDRA hierarchy**: The MedDRA hierarchy consists of five levels. These are System Organ Class (SOC), High Level Group Term (HLGT), High Level Term (HLT), Preferred Term (PT), and Lowest Level Term (LLT). E. g., for SOC term gastrointestinal disorder, HLGT is gastrointestinal signs and symptoms, HLT is nausea and vomiting symptoms; PT is nausea, LLT can be feeling queasy.

27.1.6 What are the Methods in Pharmacovigilance?

Passive Surveillance Passive surveillance is the backbone of the safety monitoring of drugs. Such surveillance is conducted through spontaneous reports by healthcare workers and consumers, and a review of ADR case reports from published sources by sponsors.

Spontaneous reports play a major role in the identification of safety signals once a drug is marketed. Such a report is an unsolicited, voluntary communication by a healthcare professional, or a consumer to a pharmaceutical firm, regulatory authority, or other organisation (e.g., WHO, Regional Centres, Poison Control Centre). The system was developed after the thalidomide disaster. The data is documented in an ADR reporting form, and forwarded to a central or regional database. Although the data capture process may vary a little across the world, the identity of the reporter and patient usually remain anonymous. Other patient-related details like country, age, gender, pre-existing comorbidities, concomitant medications, seriousness, and drug related details like batch number, company details, etc. can be recovered from the reporting forms. However, the most important shortcoming of such surveillance is underreporting, variations in the quality of the information provided, and missing data.

Examples of spontaneous reporting systems include the "Yellow card scheme" (UK), the MedWatch (USA), the suspected ADR reporting scheme (India), etc. These are respectively linked to the databases of the Medicines and Healthcare products Regulatory Agency (MHRA), the FDA Adverse Event Reporting System (FAERS), the WHO Global Individual Case Safety Report (ICSR) database (VigiBase™), etc. [18–20].

At present, the focus is on systematic methods of data mining for the detection of safety signals from spontaneous reports. Techniques have been developed to evaluate the reports in terms of quality and quantity. Quantitative signal detection is most commonly evaluated by disproportionality statistics. These methods include the calculation of the reporting odds ratio (ROR), and proportional reporting ratio (PRR), as well as the use of Bayesian methods and the information component (IC). Data mining techniques have also been used to examine drug-drug interactions and their role in causing ADR [21].

Active Surveillance In this type of reporting, case safety is monitored entirely through a pre-planned process. The process may involve systematic screening of medical records or questioning patients to find out AEs. Such surveillance can be drug-based where adverse events in patients taking certain products are identified, setting based where AE in patients at sentinel healthcare sites (single or small number of health facilities) are monitored, or event-based when an event (e.g., liver failure) is traced back to a drug. Active surveillance is also conducted through observational studies, pharmacoepidemiologic studies, cohort event monitoring programs, and registry studies. The FDA launched its largest multisite distributed database in the world in the year 2016. The initiative named, Sentinel, uses distributed analytic tools and curated real-world health insurance claims data to generate insights on medication safety. Recently, a lot of focus is on active surveillance of drugs based on social media platforms, electronic medical record (EMR) data, and genome-wide association studies. A current publication tried to map ADR mentioned at a social site to the USA national database. They found out that three ADRs of amoxicillin, hypersensitivity, nausea, and rash, shared similar profiles on FAERS and Twitter. They concluded that Twitter and other social media platforms can be a unique and complementary source of early detection of AE [18–22].

Stimulated Reporting It is practiced during the early post-marketing phase when pharmaceutical firms actively provide health professionals with safety information and at the same time encourage the cautious use of new products and the submission of spontaneous reports when an adverse event is identified. In Japan, early post-marketing phase vigilance (EPPV) is conducted for new drugs or for existing drugs for which there is a new usage. This system is a type of stimulated reporting. The recent pandemic saw a lot of stimulated reporting [21, 23, 24].

Targeted Clinical Investigations Such studies are conducted to elucidate the benefit-risk profile of a drug outside of the traditional clinical trial setting. It may be conducted to fully quantify the risk of a critical but relatively rare AE observed during clinical trials. It may also include vulnerable populations; the elderly, children, or patients with renal or hepatic disorders [21].

Literature Surveillance A key component of regulations concerning pharmacovigilance is literature surveillance and reporting. Hence, it is a regular task within the pharmacovigilance workflow. Oftentimes, the task is challenging considering the huge volume of publications across the globe. Of late, there has been the development of artificial intelligence (AI) based techniques, and software driven approaches to ease the process [22, 23].

27.1.7 Who Are the Stakeholders of Pharmacovigilance?

The key stakeholders (Fig. 27.2) directly involved in pharmacovigilance activities are healthcare professionals and establishments, regulatory authorities, consumers, and the pharmaceutical industry (MAH). Other stakeholders include media, advocacy groups, and legal representatives. These stakeholders facilitate the development of new and robust drug policies and decisions. They also highlight the deficiencies and weaknesses in existing drug safety policies and practices [10, 24].

Pharmaceutical firms discover and develop drugs, share risk management plans, submit safety reports, respond to risk minimization measures, and educate and update other stakeholders about their products. Academic centres of pharmacology and pharmacy associated with healthcare centres play an important role through education, training, policy development, clinical research, ethics committees (institutional review boards), and clinical services. In many medical institutions, monitoring of adverse drug reactions (ADR) and medication errors is recognized as an essential quality assurance activity. They report the cases through regional, and national centres to regulatory authorities of various countries. The reports are then analysed, assessed, and compiled for signal generation. They are routed to the regulators. The regulators and global pharmacovigilance teams facilitate international collaboration [6, 11]. Recently, a lot of focus is on patient organizations', education and patient initiated spontaneous reporting of AE [25].

27.2 Pharmacovigilance During Phase 3 Trials

During clinical trials, the safety of a participant is the responsibility of all stakeholders. In case of any undesirable experience, the trial investigator is mainly responsible for the medical care of the participant. Such care needs to be provided irrespective of the association of the injury to the trial product. The study sites capture AE through electronic data capture (EDC) systems. EDC is monitored remotely by sponsor personnel and checked by monitors during field visits. SAE and AESI are flagged in the EDC systems. The flag in EDC provides a good first alert to the sponsor. All AEs are documented and signed by qualified team members in the subject's medical record. The AEs are further coded, and analysed for needful actions by the safety monitoring team.

In most countries, the principal investigator is required to promptly report (within 24 h) a SAE

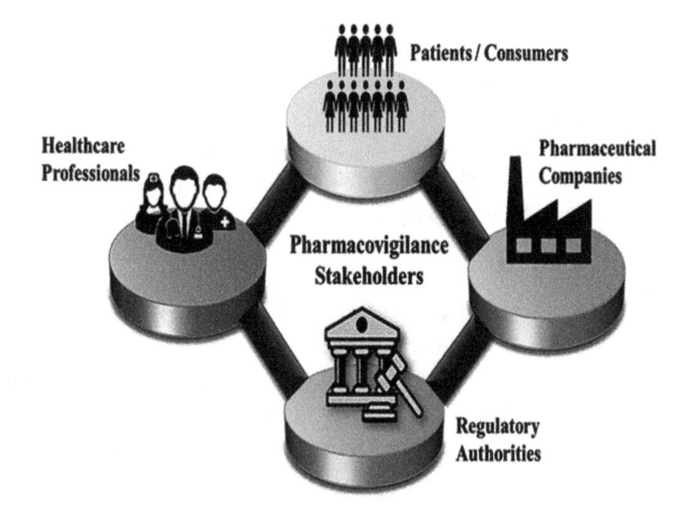

Fig. 27.2 Pharmacovigilance stakeholders

to the ethics committee (EC), and sponsor. The investigator also needs to further send a detailed report after due analysis to the EC, regulator, and the head of the institution where the trial is being conducted, usually within 15 calendar days. The sponsor team reviews the reactions and provides directives to the sites, and also reports to the regulatory authority. The trial sponsor takes up the responsibility of ongoing safety evaluation, reporting to the regulatory authority, and providing compensation if required as per the regulatory guidelines.

The ethics committee reviews the AEs on an interim basis, and SAEs on an emergency basis. They review global data, and if there are too many AESI e.g., cardiac events or gastrointestinal haemorrhage at the study site or other sites, they may seek an explanation from the investigator. If deemed risky, they may direct early closure, temporary stoppage of the trial, or a re-consent process with an update of the risk in the informed consent form.

In the case of an SAE, EC reviews the case, may seek the opinion and queries from the investigator, and determine the causality by taking into consideration the case narrative, and other source documents. They adhere to the ethical principles of beneficence, justice, and non-maleficence. And provide a report to the regulator with causality analysis, and financial compensation requirements within the specified timeline (usually 30 days after notification of SAE). The regulator gives a final decision on the amount of compensation to be given by the sponsor or the sponsor's representative to the grieving party [4, 26].

A higher-than-expected rate of expected AEs (AESI) also requires reporting. Another point to note is that the IB is updated from time to time on an ongoing basis with new important safety information during the trial period [4, 27]. The safety information is compiled into the DSUR periodically. After the successful completion of phase 3, sponsors perform an integrated safety analysis of aggregate data from all study sites and submit the updated DSUR for filing new drug applications (NDA) and marketing authorization [27, 28].

27.3 Post-Marketing Surveillance (PMS)

27.3.1 What is PMS?

After the phase 3 trial, the sponsor requests marketing authorization. Once approved for marketing, the drug is monitored for safety in real world settings. PMS is conducted as an observational, non-interventional phase 4 trial and registry study in a naturalistic setting. ADRs are reported through passive surveillance by health professionals and consumers, and actively by MAH. Until recently, spontaneous reporting was considered the cornerstone of PMS. Lately, there's much focus on electronic health record (EHR), and AI driven data extraction tools to detect ADR. There is hope that this approach will complement and strengthen existing PMS.

27.3.2 What is the Scope of PMS?

During PMS, the true safety profile of a drug is characterized by a spontaneous adverse event monitoring system and aggregate reporting by the sponsor or MAH. PMS attempts at capturing the rare ADRs that occur in less than 1 in 3000–5000 patients. The rare ADRs are unlikely to be detected in Phase I – III investigational clinical trials due to limited sample sizes, short duration, and strict inclusion and exclusion criteria. These ADRs may be unknown at the time of approval of a drug. It has been found that 20% of drugs acquire new black box warnings during this phase. Also, 4% of the drugs are ultimately withdrawn for safety reasons [12, 18].

27.3.3 What is the Process of Spontaneous Reporting Around the Globe?

For spontaneous reporting of AE, every country has its own pharmacovigilance system based on WHO guidelines. There are ADR monitoring units in healthcare organizations to detect, evaluate, and report ADR. Also, there are direct reporting structures through mobiles, toll-free phone numbers, web-based systems, and apps. This allows the regulators to collect information directly into their own database.

In most countries, the pharmacovigilance program is affiliated with the WHO program for International drug monitoring (PIDM). PIDM was initiated in the year 1968. With representation from more than 170 countries, the program covers about 99% of the world's population. It has the largest ADR database in the world (VigiBase™). The spontaneous reports from member countries are routed to the Uppsala Monitoring Centre (UMC) through the national pharmacovigilance centres. UMC is an independent, non-profit foundation that works with the WHO, engages the stakeholders of pharmacovigilance, and collaborates with global drug monitoring. It is based in Uppsala, Sweden. UMC provides international service and scientific research in the field of pharmacovigilance.

Most countries follow the ICH E2E guidelines for pharmacovigilance activities. The European Union (EU), Japan and the USA are the founding members of ICH. ICH has established pharmaceutical regulatory systems of the highest level worldwide. The PV systems are based on long-standing experiences, and empirical knowledge gained from intensive international collaboration. ICH guidelines encourage harmonisation and consistency, prevent duplication of effort, and provide benefits to public health programs that include new drugs throughout the world [21, 29].

In the USA, PMS and spontaneous reporting are through MedWatch. One can voluntarily report AR, SAE, product quality problems, product use errors, or therapeutic equivalence or failure issues. MedWatch interacts with the FDA information database FAERS. The reports in FAERS are evaluated by multidisciplinary staff safety evaluators, epidemiologists, and other scientists in the Centre for Drug Evaluation and Research's (CDER) office of surveillance and epidemiology. The process helps detect safety signals, assess risk mitigation and evaluation strategies (REMS), and recommend regulatory actions [29, 30].

The FDA equivalent in Japan that regulates the safety of drugs is the pharmaceuticals and medical devices agency (PMDA). PMDA works together with the Ministry of Health, Labour, and Welfare. PMDA collects AE information from healthcare professionals and MAHs into their database. They analyse the data, routinely disseminate the safety information, and proactively participate in international harmonization activities. Japan also requires MAH to conduct early post-marketing phase vigilance (EPPV) for the first 6 months after the launch of a new drug [31, 32].

PMS in other Asian countries are also conducted as hospital based and industry based approaches. Each country has its drug regulatory agency. Spontaneous reports are collected by voluntary submission of suspected ADRs reports. The forms are based on CIOMS template. The ADRs are routed to the regulatory database, and then to the UMC. In India, pharmacovigilance was initiated in the year 1986 with a formal ADR monitoring system under the supervision of the drug controller of the country. The program evolved over years and is currently known as the pharmacovigilance program of India (PvPI). Under the umbrella of PvPI, ADR monitoring centres are established at medical schools and other healthcare organizations throughout the country. The national coordinating centre is located at the Indian Pharmacopoeia Commission (IPC) in Ghaziabad. PvPI routes ICSRs to UMC through Vigiflow [32–34].

In the United Kingdom, PMS is through the YC ADR reporting system. The UK also has a process of labelling drugs with an inverted black triangle when they are first marketed. The idea is to bring the AE to the notice of the authorities. This intensive monitoring of the drug continues for a minimum of 2 years from the launch of the product. It is extended further if necessary. In the European economic area, the European Medicines Agency (EMA) provides extensive guidance to enable all stakeholders to meet their legal pharmacovigilance obligations. EMA operates a large database of ADR, EudraVigilance. From this database, ISCRs are directly routed to the UMC thus facilitating global pharmacovigilance activities [35, 36].

In Canada, the PMS program is through Health Canada. It collects and assesses the reports of ADRs. And then conducts a complete analysis of probable safety concerns. Health Canada further recommends pharmaceutical manufacturers change product labels. It also communicates new safety information to healthcare professionals and the public through Dear Healthcare Professionals (DHP) letters [37].

27.3.4 What is the Process of Aggregate Safety Reporting During PMS?

For regulatory reporting, SUSARs are reported continually and non-serious ones are compiled and reported periodically. The periodic aggregate safety reports are periodic safety update reports (PSUR), periodic adverse drug experience reports (PADER), periodic benefit-risk evaluation reports (PBRER), or addendum to clinical overview (ACO). MAH is required to submit the country specific report to the respective regulatory agency. These documents are usually prepared by the safety writers of the sponsor or MAH [30–38].

PSUR PSUR contains the title page, worldwide marketing authorization status, steps taken for safety purposes during the reporting interval, changes in reference safety information, assessed patient exposure, individual case history, a summary of significant findings, signal assessment, and risk evaluation, and overall safety. It is accepted throughout the world. In the USA, the counterpart of PSUR is PADER. PSUR can also be submitted to the US FDA but with a waiver from the need to submit US-style reports.

PADER contains individual case narratives for SAE with fatal outcomes and events of special interest or serious unlisted events. It is a relatively less complex document with five sections. It is prepared as per the US 21 code of federal regulation (CRF) 314.80. It is submitted every 3 months (quarterly) for the first three years post-marketing authorization, and then annually to FDA. Usually, different PADERs are prepared for the investigational drug based on formulations, dosage forms, or indications. PADER is prepared based on the line listings. It includes a summary of all the 15-day alerts regarding serious unlisted domestic and foreign cases and non-15-day alerts of serious listed, non-serious unlisted, and non-serious listed domestic cases reported during the reporting interval. It also includes regulatory actions like 'dear doctor letters', labelling changes, etc. After obtaining the IBD, MAH is required to submit the first quarterly PADER within 30 days.

PBRER contains a detailed risk-benefit evaluation of a drug after the approval of a new drug application. It is prepared as per ICH E2C R2 and European Medicines Agency module VII. It is submitted to the European Union and the rest of the world. It is a relatively complex document with 20 sections. The sections include regulatory updates, cumulative and interval exposure, interventional and non-interventional clinical trials, an overview of signals, and benefit-risk assessment. Usually, one PBRER is prepared for the investigational drug irrespective of formulations, dosage forms, or indications. It is submitted every 6 months for the first two years post-marketing authorization, and then annually.

ACO ACO is an aggregate safety report that is submitted for renewal of the marketing authorization license of a drug. The structure and specifications of the ACO and PBRER are similar. But the information is presented with greater emphasis on the data received since the last renewal/approval rather than discussing cumulatively.

27.4 Patient Narrative

A patient narrative is a brief summary of adverse events experienced by patients, during a clinical trial. They are prepared based on standardized guidelines (ICH). They are submitted along with the clinical study report (CSR). They are prepared from primary source data, line listings in databases, and secondary data like regulatory notifications, etc. They are produced on an interim basis, or after study completion. Based on when they are prepared, they are called either a pre-database lock (DBL) narrative or a post-database lock narrative. Pre-DBL data narratives are updated or modified based on post-DBL data. These narratives are considered clean and final. Final narratives are submitted with the CSR as an appendix. Recently with the advancement of digitalization, narratives can be automated. This tool is very helpful for studies with large volumes of narratives (usually >150) [39, 40]. The contents of a patient narrative are as follows:

- Subject ID (most serious adverse event/events)
- Age, gender, ethnicity, had a medical history including past history and history at the time of the start of the trial
- Upon study admission, the presentation of the subject and the reason for receiving the study drug
- Randomization and dosing regimen
- Description of the most serious adverse event (e.g., prolonged hospitalization or death) including the day it started, its effect, and the action taken (e.g., withdrawal of treatment or change in dosing regimen) if any
- Description of the nature (intensity) and relationship of this event to the study medication (causality information)
- The treatment used for this event and the outcome (resolved, persisted, etc.)
- Other adverse events whether listed and related to these adverse events and their outcomes
- List and description of laboratory results/vital signs which are related to the SAE and the outcome and/or a statement stating that no significant lab values/vital signs were reported
- The last sentence states, "The subject completed the study on which day or the study was terminated on which day and the reasons

27.5 Quantification of ADR Reports

ADR reports are quantified after assessment with standard scales. The process helps evaluate the incidence, type, causality, preventability, and severity of the ADR. To determine the predictability or incidence, the published literature is screened for regional, national and global data. To find out the causality, the WHO scale or Naranjo's algorithms are used. WHO scale is the most widely used causality analysis tool (Fig. 27.3). WHO criteria categorise events into certain, probable, possible, unlikely, unclassified,

	Time relationship between drug use and adverse event	Other drugs/ disease ruled out	Response to drug withdrawal/ dose reduction (Dechallenge)	Response to drug re-administration (Rechallenge)
Certain	✓	✓	✓	✓
Probable/Likely	✓	✓	✓	✗
Possible	✓	✗	✗	✗
Unlikely	✗	✗	–	–
Conditional/ Unclassified	• Report suggesting an adverse event • More data for proper assessment needed or additional data under examination			
Unassessable/ Unclassifiable	• Report suggesting an adverse reaction • Cannot be judged because information is insufficient or contradictory • Data cannot be supplemented or verified			

Fig. 27.3 WHO-UMC causality assessment scale

and unclassifiable. Naranjo scale categorizes them into unlikely, possibly, probably, or definite associations. To evaluate the type, an assessment is done based on Willis and Brown classification. The types are augmented, bizarre, chronic, delayed, withdrawal, familial, genetic, hypersensitivity, and unclassified. To demonstrate the severity, tools such as the Modified Hartwig and Siegel scale are used. It categorizes AE into mild, moderate, and severe categories. To assess the preventability, a modified Schumock and Thornton scale is used. It classifies events into categories like definitely preventable, probably preventable, and non-preventable. The impact of ADR on quality of life (QOL) is measured by the WHO Quality of Life BREF scale. Recently, the focus is on AI based tools to quantify ADR [41–46].

27.6 Risk Management Plan

A risk management plan (RMP) is a document that describes the current knowledge about the safety and efficacy of a drug. It is submitted to regulatory authorities at the time of application of a marketing authorization. RMPs include information on safety profiles, plan for prevention and minimization of risks, plans for studies and other activities to gain more knowledge about the safety and efficacy, and measurement of the effectiveness of risk-minimisation measures. RMPs are continually modified and updated throughout the lifetime of a drug. RMP and good pharmacovigilance practice (GVP), medication error minimization guidelines are available on regulatory websites [47–49].

27.7 Pharmacovigilance Process Flow

The process of pharmacovigilance starts with an event collection. Once a case is received, the case processor looks for the minimum information. The case is then triaged for date, seriousness, causality, and expectedness. It is then processed as per the priority. This ensures that expedited reporting is processed and submitted to the regulatory authorities within timelines. After triage, the processor looks for duplicate data. If there is duplicate data, the processor is required to add a follow-up to the existing case and process the information. If there is no duplicate data; the processor creates a new case to process the information. The case processor then performs the complete data entry with all available information. The AE is then coded in the MedDRA. The case processor can auto code the event term when the term has an exact match in MedDRA or code it manually if the exact match is not available. This ensures the use of the same language across countries, companies, and regulatory bodies. The case is then subjected to causality analysis, narrative writing, a self-quality check, and drafting of queries for the missing information. The cases are submitted through the available safety database. The cases are subjected to signal surveillance, detection, evaluation, governance, communication, and documentation. They are also exposed to benefit-risk management [50, 51]. The process is shown in Fig. 27.4.

27.8 PV During Health Emergencies

During any health emergency, monitoring drug safety is a paramount task. And the recent coronavirus disease (COVID-19) pandemic has shown that the efforts of the WHO, ISoP, several national pharmacovigilance programmes, and other pharmacovigilance stakeholders are vital to the success of public health initiatives across the globe. During the first waves of the Covid-19 pandemic, the absence of vaccines and drugs led to a rush to repurposed drugs, and the use of off-label. Also, expedited discovery and approvals led to a hastened development of vaccines. Pharmacovigilance was thus crucial to be able to adhere to the ethical principle of "do no harm". WHO released the vaccine safety surveillance manual. It facilitated the early detection, investigation, and analysis of adverse events following immunization (AEFIs) and adverse events of special interest (AESIs). The global collaboration and coordination between the stakeholders

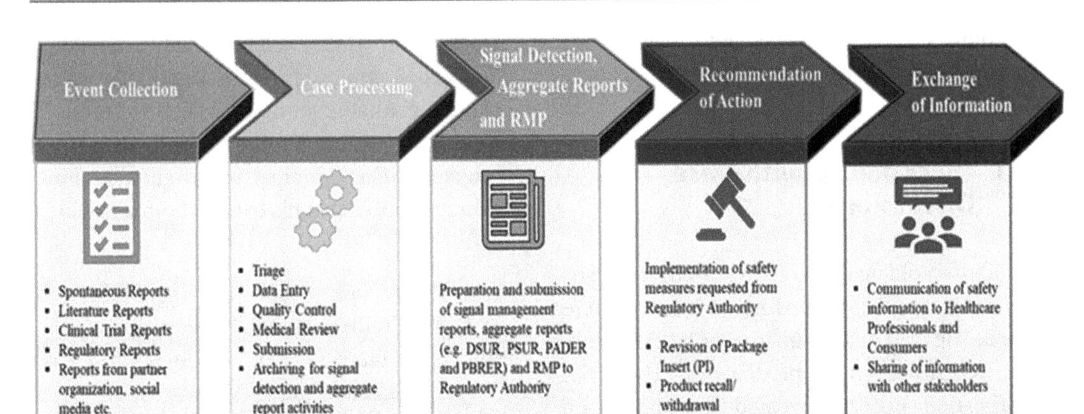

Fig. 27.4 Pharmacovigilance process

ensured an appropriate and rapid response. Drugs received emergency use approvals, expedited reviews, and repurposing. But because of the already existing collaboration across the globe and sharing of data to the UMC VigiBase™, it was possible to manage the functioning of pharmacovigilance. ICSRs were managed, signals were detected, and benefits and risks were assessed and communicated to the healthcare workers and the general public. Information was available in the public domain, and care was taken to minimize confusion during the infodemic [52–57].

27.9 AI Based Approaches in Pharmacovigilance

Safety monitoring starts with an incident of AE. The report is then documented, processed, and analysed. It requires a lot of motivation and willingness on the part of healthcare workers, and consumers to detect, diagnose, and report the events. Hence, a lot of data is missed. Also, significant information present in electronic records, and health claims are missed. There is a prediction that the use of AI can help reduce such errors, and save manpower, and cost. AI can analyse a large amount of data simultaneously to identify patterns and trends, improve the accuracy of the information, and thus reduce the burden and time of case processing and interpretation [58, 59].

AI and machine learning may also be useful in pharmacovigilance for the following:

- The automatic case report entry and processing
- The identification of clusters of adverse events
- The extraction of data from pharmacoepidemiologic, and genetic studies
- The conduction of probabilistic matching within datasets and data linkage
- The prediction and prevention of AE related to drug/food interaction
- The automated literature screening and data mining

27.10 Concluding Remarks

Pharmacovigilance is an essential element of quality healthcare delivery. The process starts during clinical trials and continues throughout the life cycle of a drug. The methods include the detection of an ADR, reporting it based on the regulatory requirements, listing, then generation of signals, and risk management plans. Once manual, the system today relies on sophisticated, proprietary databases for its operation. The system has evolved in terms of tracking activities at the evidentiary level of standards. And addressing the security issues against global hackers, ensuring compliance with local and global standards, evolving with a focus on ecopharmacovigilance, and safety of advanced therapy medicinal products. With the introduction of digital therapeutics, the field promises to further evolve and streamline the medication safety process throughout the globe.

27.11 Quick Recap with Case Discussions

1. A 50-year-old male with hypertension visits his primary care clinic for a blood pressure check. He also gives a history of left ventricular hypertrophy and prediabetes. His brief medication history revealed that he is on tablets, amlodipine 10 mg daily, lisinopril 40 mg daily, and hydrochlorothiazide 25 mg daily. He was on this regimen for the past 5 years. He did not use any supplements or over-the-counter medication regularly. His blood pressure was found to be 160/100 mm Hg. The other vital signs were within normal limits. On physical examination, his gum tissue was firm, pink, moist, and overgrown, displacing most teeth. The patient mentions that his gums were healthy before he started his medications. The physician diagnoses the case as amlodipine-induced gingival overgrowth (AIGO). The physician changes the drug to a different calcium channel blocker and reports the case as an ADR.
   (a) What is the process of spontaneous reporting of ADR?
   (b) What are the tools for causality analysis?
   (c) What's the role of the literature review in the above case?
2. A 60-year old male patient suffering from chronic lymphocytic leukaemia enrolls in a phase 3 clinical trial. After receiving the study drug, he complains of vomiting and generalised weakness. After a couple of hours, he develops breathlessness. His blood pressure was unrecordable, blood gases revealed severe metabolic acidosis, high lactates, and hyperkalaemia. It was decided to shift him to the intensive care unit.
   (a) What are the responsibilities of stakeholders in the above case?
   (b) What are the tools for causality analysis?
   (c) Can such a case have an impact on the global trial of the product?
3. During the global safety review, an ethics committee comes across a couple of cases of serious cardiac events for a diabetic product. The events were not listed in the investigator's brochure. The principal investigator is, however, confident that his trial participants will be safe.
   (a) What are the responsibilities of stakeholders in the above case?
   (b) Is there room for a risk management plan?
   (c) Does this require updating of any trial document?

Acknowledgments The authors thank Ms. Akshya Balaji, Pharmacovigilance Associate, Oviya MedSafe Pvt. Ltd., Coimbatore, India, for drawing the figures used in this chapter.

Conflict of Interest None declared.

References

1. World Health Organization. Pharmacovigilance. Regulation and prequalification (who.int). Accessed 23 Nov 2022
2. Knight PR, Bacon DR (2002) An Unexplained death: hannah greener and chloroform. Anesthesiology 96: 1250–1253. https://doi.org/10.1097/00000542-200205000-00030
3. Fornasier G, Francescon S, Leone R et al (2018) An historical overview over pharmacovigilance. Int J Clin Pharm 40(4):744–747. https://doi.org/10.1007/s11096-018-0657-1
4. FDA consumer magazine (1981) Sulfanilamide disaster. Accessed 23 Dec 2022
5. Gliklich RE, Dreyer NA, Leavy MB (2014) Registries for evaluating patient outcomes: a user's guide [Internet], 3rd edn. Agency for Healthcare Research and Quality (US), Rockville, MD. Adverse Event Detection, Processing, and Reporting. Available from: https://www.ncbi.nlm.nih.gov/books/NBK208615/
6. Vargesson N (2015) Thalidomide-induced teratogenesis: history and mechanisms. Birth Defects Res C Embryo Today 105(2):140–156. https://doi.org/10.1002/bdrc.21096
7. YellowCard. Yellow Card—Making medicines and medical devices safer (mhra.gov.uk). Accessed 24 Nov 2022
8. ISoP—ESOP/ISoP History. ESOP/ISoP History—ISOP (isoponline.org). Accessed on 22 Dec 2022

9. Bihan K, Lebrun-Vignes B, Funck-Brentano C et al (2020) Uses of pharmacovigilance databases: an overview. Therapie 75(6):591–598. https://doi.org/10.1016/j.therap.2020.02.022
10. World Health Organization (2002) The importance of pharmacovigilance Safety monitoring of medicinal products. Accessed 23 Nov 2022
11. Fogel DB (2018) Factors associated with clinical trials that fail and opportunities for improving the likelihood of success: a review. Contemp Clin Trials Commun 7(11):156–164. https://doi.org/10.1016/j.conctc.2018.08.001
12. Alomar M, Tawfiq AM, Hassan N et al (2020) Post marketing surveillance of suspected adverse drug reactions through spontaneous reporting: current status, challenges and the future. Ther Adv Drug Saf 10(11):2042098620938595. https://doi.org/10.1177/2042098620938595
13. Duijnhoven RG, Straus SMJM, Raine JM et al (2013) Number of patients studied prior to approval of new medicines: a database analysis. PLoS Med 10(3): e1001407. https://doi.org/10.1371/journal.pmed.1001407
14. ICH (2003) Post-approval safety data management: definitions and standards for expedited reporting. ICH E10. Accessed on 24 Nov 2022
15. FDA guidance for industry and investigators. Safety Reporting Requirements for INDs and BA/BE Studies. Accessed 25 Nov 2022
16. Chakraborty BS (2015) Pharmacovigilance: a data mining approach to signal detection. Indian J Pharmacol 47(3):241–242. https://doi.org/10.4103/0253-7613
17. WHO. Adverse effects following immunization. Adverse events following immunization (AEFI) (who.int). Accessed 24 Nov 2022
18. Zhou Z, Hultgren KE (2020) Complementing the US food and drug administration adverse event reporting system with adverse drug reaction reporting from social media: comparative analysis. JMIR Public Health Surveill 6(3):e19266. https://doi.org/10.2196/19266
19. Noren GN, Edwards IR (2009) Modern methods of pharmacovigilance: detecting adverse effects of drugs. Clin Med (Lond) 9(5):486–489. https://doi.org/10.7861/clinmedicine.9-5-486
20. Choi YH, Han CY, Kim KS, Kim SG (2019) Future directions of pharmacovigilance studies using electronic medical recording and human genetic databases. Toxicol Res 35(4):319–330. https://doi.org/10.5487/TR.2019.35.4.319
21. ICH topic E2E. E 2 E Pharmacovigilance planning (Pvp) (europa.eu). Accessed 27 Nov 2022
22. Shetty KD, Dalal SR (2011) Using information mining of the medical literature to improve drug safety. J Am Med Inform Assoc 18(5):668–674. https://doi.org/10.1136/amiajnl-2011-000096
23. Sorbello A, Ripple A, Tonning J et al (2017) Harnessing scientific literature reports for pharmacovigilance. Prototype software analytical tool development and usability testing. Appl Clin Inform 8(1):291–305. https://www.ncbi.nlm.nih.gov/pmc/articles/PMC5373771
24. World Health Organization. Stakeholders in COVID-19 vaccines safety surveillance. Accessed on 24 Nov 2022
25. Chinchilla K, Matos C, Hall V et al (2021) Patient organizations' barriers in pharmacovigilance and strategies to stimulate their participation. Drug Saf 44:181–191. https://doi.org/10.1007/s40264-020-00999-0
26. Coates S, Täubel J, Lorch U (2019) Practical risk management in early phase clinical trials. Eur J Clin Pharmacol 2019(75):483–496. https://doi.org/10.1007/s00228-018-02607-8
27. Yao B, Zhu L, Jiang Q, Xia HA (2013) Safety monitoring in clinical trials. Pharmaceutics 5(1):94–106. https://doi.org/10.3390/pharmaceutics5010094
28. Sil A, Das NK (2017) Ethics of safety reporting of a clinical trial. Indian J Dermatol 62(4):387–391. https://doi.org/10.4103/ijd.IJD_273_17
29. WHO. VigiBase, WHO's global database—UMC (who-umc.org). Accessed 28 Nov 2022
30. FDA. Office of surveillance and epidemiology (OSE)—Divisions—FDA. Accessed 28 Nov 2022
31. PMDA. Post-marketing Safety Measures—Pharmaceuticals and Medical Devices Agency (pmda.go.jp)
32. Biswas P (2013) Pharmacovigilance in Asia. J Pharmacol Pharmacother 4(1):S7–S19. https://doi.org/10.4103/0976-500X.120941
33. Kalaiselvan V, Thota P, Singh GN (2016) Pharmacovigilance programme of India: recent developments and future perspectives. Indian J Pharmacol 48(6):624–628. https://doi.org/10.4103/0253-7613
34. Huang YL, Moon J, Segal JB (2014) A comparison of active adverse event surveillance systems worldwide. Drug Saf 37(8):581–596. https://doi.org/10.1007/s40264-014-0194-3
35. The Yellow Card Scheme: guidance for healthcare professionals, patients and the public. www.gov.uk/guidance/the-yellow-card-scheme-guidance-for-healthcare-professionals Accessed 8 Dec 2022
36. European Medicines Agency. EudraVigilance—European Medicines Agency (europa.eus). Accessed 30 Nov 2022
37. Raj N, Fernandes S, Charyulu NR et al (2019) Postmarket surveillance: a review on key aspects and measures on the effective functioning in the context of the United Kingdom and Canada. Ther Adv Drug Saf 26(10):2042098619865413. https://doi.org/10.1177/2042098619865413
38. Kulkarni TN, Kulkarni NG (2019) Authoring a periodic adverse drug experience report. . .here's what you need to know! Perspect Clin Res 10(2):95–99. https://doi.org/10.4103/picr.PICR_126_18

39. Ledade SD, Jain SN, Darji AA, Gupta VH (2017) Narrative writing: effective ways and best practices. Perspect Clin Res 8(2):58–62. https://doi.org/10.4103/2229-3485.203044
40. ICH topic E3. Structure and content of clinical study reports. Accessed 2 Dec 2022
41. UMC. The use of the WHO-UMC system for standardised case causality assessment. Accessed 2 Dec 2022
42. Naranjo CA, Busto U, Sellers EM, Sandor P, Ruiz I, Roberts EA, Janecek E, Domecq C, Greenblatt DJ (1981) A method for estimating the probability of adverse drug reactions. Clin Pharmacol Ther 30(2): 239–245. https://doi.org/10.1038/clpt.1981
43. Wills S, Brown D (1999) A proposed new means of classifying adverse drug reactions to medicines. Pharm J 262:163–165
44. Hartwig SC, Siegel J, Schneider PJ (1992) Preventability and severity assessment in reporting adverse drug reactions. Am J Hosp Pharm 49:2229–2232
45. Schumock GT, Thornton JP (1992) Focusing on the preventability of adverse drug reactions. Hosp Pharm 27:538
46. Stricker BH, Psaty BM (2004) Detection, verification, and quantification of adverse drug reactions. BMJ 329(7456):44–47. https://doi.org/10.1136/bmj.329.7456.44
47. Lavertu A, Hamamsy T, Altman RB (2021) Quantifying the severity of adverse drug reactions using social media: network analysis. J Med Internet Res 23(10):e27714. https://doi.org/10.2196/27714
48. EUA. Risk management plan. Accessed 5 Dec 2022
49. EMA. Guideline on good pharmacovigilance practices (GVP). Accessed 8 Dec 2022
50. FDA. Risk minimization action plans (RiskMAPs) for Approved Products. Accessed 8 Dec 2022
51. Bhangale R, Vaity S, Kulkarni N (2017) A day in the life of a pharmacovigilance case processor. Perspect Clin Res 8(4):192–195. https://doi.org/10.4103/picr.PICR_120_17
52. Tam HL, Chung SF, Lou CK (2018) A review of triage accuracy and future direction. BMC Emerg Med 18(58):1–7. https://bmcemergmed.biomedcentral.com/articles/10.1186/s12873-018-0215-0#citeas
53. Chandler RE, McCarthy D, Delumeau JC et al (2020) The role of pharmacovigilance and ISoP during the global COVID-19 pandemic. Drug Saf 43(6): 511–512. https://doi.org/10.1007/s40264-020-00941-4
54. WHO. COVID 19 vaccines: safety surveillance manual. covid19vaccines_manual_aesi.pdf (who.int). Accessed 28 Nov 2022
55. Gold MS, Lincoln G, Cashman P et al (2021) Efficacy of m-Health for the detection of adverse events following immunization - The stimulated telephone assisted rapid safety surveillance (STARSS) randomised control trial. Vaccine 39(2):332–342. https://doi.org/10.1016/j.vaccine.2020.11.056
56. Pal SN, Duncombe C, Falzon D, Olsson S (2013) WHO strategy for collecting safety data in public health programmes: complementing spontaneous reporting systems. Drug Saf 36(2):75–81. https://doi.org/10.1007/s40264-012-0014-6
57. Ball R, Dal Pan G (2022) "Artificial intelligence" for pharmacovigilance: ready for prime time? Drug Saf 45(5):429–438. https://doi.org/10.1007/s40264-022-01157-4
58. Liang L, Hu J, Sun G et al (2022) Artificial intelligence-based pharmacovigilance in the setting of limited resources. Drug Saf 45:511–519. https://doi.org/10.1007/s40264-022-01170-7
59. Bate A, Stegmann JU (2021) Safety of medicines and vaccines - building next generation capability. Trends Pharmacol Sci 42(12):1051–1063. https://doi.org/10.1016/j.tips.2021.09.007

Part V

Experimental Design and Statistical Analyses

The Design and Statistical Analysis of Randomized Pre-clinical Experiments 28

Michael F. W. Festing

Abstract

This chapter provides an introduction to the design of pre-clinical experiments using the "Completely randomized (CR)" and the "Randomized block (RB)". These are the only randomized experimental designs suitable for widespread use in pre-clinical research. Both designs can have any number of treatments as well as additional "factors" such as both sexes or more than one strain of animals. In both designs, one of the treatments, chosen at random, is assigned to each caged subject so that those receiving different treatments are randomly intermingled in the research environment.

The CR design can have unequal treatment group sizes, but the whole experiment needs to be done at the same time. In contrast, the more powerful RB design is split into several "blocks" each of which has a single subject receiving each of the treatments. These blocks can be spread over time and/or location to provide some assurance of reproducibility. The "Matched pairs" design is a special case of the RB design, with only two treatments.

The chapter includes the raw data and the statistical analysis of four examples of well-designed experiments, analyzed using "MINITAB" statistical software. It also includes a brief account of measures of association using "Correlation" and "Regression".

Scientists should always state the name of the experimental designs which they have used as this provides some assurance that they have used a statistically valid design.

Keywords

Experimental design · Randomized block · Completely randomized · Analysis of variance · Regression · Correlation

28.1 Introduction

Pre-clinical research involving laboratory animals is characterized by high levels of irreproducibility. This is probably due to bias caused by the failure of scientists to design their experiments correctly, using either the "Completely randomized" or the "Randomized block" designs.

This chapter describes these two designs and explains how they should be statistically analyzed, if they are to give reproducible results. It includes a brief introduction to measures of association using "Regression" and "Correlation", although these are not "randomized" experiments.

M. F. W. Festing (✉)
Care of the Medical Research Council, Swindon, UK
e-mail: michaelfesting@aol.com

G. Jagadeesh et al. (eds.), *The Quintessence of Basic and Clinical Research and Scientific Publishing*,
https://doi.org/10.1007/978-981-99-1284-1_28

28.2 Randomized, Controlled Experiments

Randomized, controlled experiments were invented by the mathematician R.A. Fisher in the 1920s [1]. His aim was to develop methods for comparing different varieties of agricultural crops, and/or the effects of different fertilizer treatments on yield, in the presence of environmental variation.

Fisher invented two main designs: the "Completely randomized" (CR) and the "Randomized block" (RB). Both designs involve "replication" and "randomization" with each subject receiving one of the treatments, assigned at random. So, the caged subjects (or plots in a field) receiving different treatments are randomly intermingled in the research environment. This avoids bias due to variation in the research environment.

The advantage of the CR design is that it can have unequal treatment group sizes. In pre-clinical studies the whole experiment should be done at the same time. This is the design which, with considerable elaboration, is used in clinical trials. In contrast, the RB design is split into several equal sized "blocks" each of which has a single subject receiving each of the treatments, assigned at random. These blocks can be spread over time and/or location. The whole experiment consists of "N" blocks, where N is the sample size. The results across all blocks are combined in the statistical analysis. Although the RB design is normally designed with equal treatment group sizes, modern statistical software can cope with unequal sized treatment groups where necessary. The RB is the most widely used design in agricultural and industrial research.

Both the CR and RB designs are analyzed using an "analysis of variance", which Fisher also invented.

The textbook "Statistical methods" by G.W. Snedecor, was first published in 1937 with later editions being co-authored by William Cochran. The seventh edition was published in 1980 [2]. This book is highly influential in making the use of the CR and RB designs widely available in agricultural and industrial research. The CR design also forms the basis of clinical trials (*see* Chap. 25).

The high level of irreproducibility found in pre-clinical research [3] is probably due to the failure of scientists to use these designs. To give just one example, Scott et al. [4] found that more than 50 publications had described therapeutic agents which appeared to extend the lifespan of a mouse model of amyotrophic lateral sclerosis (ALS). But none of them were effective in humans. So, with advice from a statistician, they re-screened all fifty, plus another twenty compounds, using the "Matched pairs" experimental design. It took them five years and used 18,000 mice across 221 studies. They found that *when the experiments were correctly designed, none of the drugs extended the lifespan of these mice.* All fifty previous studies had given false positive results, presumably due to environmental effects being mistaken for the effects of the treatment.

The widespread failure by scientists in academia to use the "Randomized block/"Matched pairs" (RB)/MP or the "Completely randomized" (CR) design probably accounts for most of the excessive levels of irreproducibility currently found in pre-clinical research.

28.3 Treatments and Factors

A "Factor" is an alphanumeric variable with values specified by the investigator. For example, an experiment usually has a factor called "Treatment". This will have two or more "levels" which may, for example, be designated "Control", "Low dose", and "High" dose. These alphanumeric names, starting with a letter, are specified by the investigator. In some experiments "Sex" or "Strain" may be the main, or additional factors, with levels "Male" and "Female" or "CD-1" and "C57BL/6" strain mice.

An experiment can include more than one factor, such as both "Treatment" and "Sex". In this case, it is called a "Factorial design".

Factorial designs often provide extra information, such as whether the two sexes respond in the same way to the treatments, at little or no extra cost, apart from a slightly more complex statistical analysis. They are discussed below, with examples.

A CR design with a single factor such as "treatment" is called a "one-way" design. It is analyzed using a one-way analysis of variance (ANOVA). If it has an additional factor such as two or more strains of animals it is a "Two-way" design, requiring a two-way ANOVA. An RB design with both treatments and blocks is also a "two-way" design. But if it has another factor such as sex, it becomes a "three-way" design. This is discussed in the examples below.

28.4 Planning a Pre-clinical Experiment

When planning a pre-clinical experiment, the investigator should:

1. State the purpose of the experiment
2. Specify the species, sex, age and strain(s) of animals to be used.
3. State the number, name, and nature of the treatment groups.
4. Decide whether any additional factors (such as sex or strain) are to be included.
5. State the name of the experimental design which is to be used (CR, RB/MP or other).
6. Specify the sample sizes, as discussed below.

28.4.1 The Determination of Sample Size

When using the CR design, the experimental animals should be as uniform as possible otherwise the experiment will lack "power" (the ability to detect a treatment effect).

However, if an RB design is to be used, the subjects can be more variable provided those within each block are closely matched. Any differences between the blocks are removed in the statistical analysis a "two-way analysis of variance without interaction" (see below).

An appropriate sample size (the number of subjects in each treatment group) can be estimated using either the "Resource equation" or a "Power analysis" [5]. The resource equation method is simple. It assumes that the available animals are relatively uniform and suggests that:

The total number of subjects, minus the number of treatments should be between ten and twenty.

For example, with three treatments (e.g., Control, Low, and High dose) a sample size of between five and seven subjects per treatment group is suggested. If the experiment is to include more than one factor, a slightly larger sample size may be appropriate.

When there are only two treatments, and a reliable estimate of the standard deviation of the character to be measured is available from previous studies, a "Power analysis" can be used [5]. This can provide an estimate of the magnitude of response which is likely to be detectable for a given sample size and standard deviation.

28.4.2 Randomization and Blinding

Randomization seems to be widely misunderstood. The animals are not randomized. It is the assignment of one of the treatments, *chosen at random*, to each subject. Correct randomization means that the cages of animals receiving different treatments are randomly intermingled in the animal house environment.

Such randomization is essential because the research environment is not homogeneous. For example, the top shelf in a rack of cages has a different environment from the second shelf. Edge cages have a different environment from those in the middle of a rack. The first few animals to be caught and treated may not be as expertly handled as later animals. The last few animals in a gang cage may be exhausted from having been run around the cage to avoid being caught. Circadian rhythms may result in the first few animals being treated or having their outcomes measured being physiologically different from later ones. And when measuring outcomes, the staff may become more skilled at catching and handling the animals as the experiment progresses. Randomization ensures that these environmental factors are distributed reasonably equally among the treatment groups.

Once the subjects have been treated, they should only be identified by their i.d. number, so that investigators are "blinded" to the treatments when assessing the outcomes.

28.4.3 Randomization in Practice

Randomization *should be done in the office before going to the animal house*, using EXCEL. Suppose, for example, that a CR experiment is to have three treatments "Control", "Low" and "High" dose, with a sample size of six. The row numbers in EXCEL can represent the eighteen animal's ID. Column one should have "Control", "Low" and "High" each written six times. Column two should have eighteen random numbers generated by typing "=rand()" in the first cell and pulling down on the small box in the lower right of the cell, to generate a column of random numbers. Columns one and two should then be marked and sorted on column two (the random numbers). This will assign one of the treatments, chosen at random, to each numbered subject, and maintain the sample sizes. The random numbers, which change when accessed, can then be deleted.

A similar randomization procedure should be used if the "Treatment" is an attribute such as a genotype. For example, if the investigator wishes to compare two or more strains of mice for some measurement character, the different strains need to be randomly intermingled within the animal house environment.

If an RB design is to be used with, for example, three treatments "L", "C" and "H" and six blocks (i.e., a sample size of six), then two sorts are required. The row numbers, 1–18, will represent the individual IDs. Column one should have the treatments such as "C", "Low" and "Hi" each repeated six times. Column two should have the block numbers 1–6 repeated three times. Column three should have eighteen random numbers using "=rand()". The three columns should then be marked and sorted first on the random numbers and then on the block numbers. This will assign one of the treatments chosen at random to each of the three animals within each block.

The EXCEL sheet containing the i.d. numbers and treatment assignments should be saved, printed, and taken to the animal house when setting up the experiment.

28.4.3.1 Blinding

Scientists usually want a particular outcome, such as there being "statistically significant" differences between the treatment groups. If they know which treatments the animals have received when measuring the outcomes, they may "adjust" the measurements to obtain the desired result. *Bias can be avoided by ensuring that the experimental subjects are only identified by their ID number when the outcomes are being measured and recorded.*

28.4.4 Assumptions About the Data and Data Transformations

The statistical analysis is based on the properties of the "normal" or "Gaussian" bell-shaped distribution. The experimental subjects or "units" must be *independent* so that any two of them can receive different treatments. Here it is assumed that the experimental unit is a single animal in a cage. If there are two animals in a cage and the treatment is given in food or water, then the two animals cannot receive different treatments. In that case, the pair of caged animals would represent a *single* experimental unit and the statistical analysis should be based on their mean value.

The statistical analysis is based on the assumption that the "residuals" (the deviation of each observation from the group mean) should have a "normal", Gaussian, distribution. This will usually be the case with characters such as body weight and hematology. But it may not be the case with "bounded" data such as percentages or proportions, which cannot be less than zero or more than one hundred.

Minitab can produce a number of residuals diagnostic plots of the residuals (deviation of each observation from its expected value) as part of the statistical analysis (see Example 28.2, below). A normal probability plot of the residuals should give a reasonably good straight line. If this is not the case, a "scale transformation" may be needed, followed by a re-analysis of the transformed data.

The two most common scale transformations are:

1. If the variation increases as the means of the groups increase, a logarithmic transformation may be appropriate; each observation should be replaced by it's logarithm.
2. Counts, with a low mean, often have a "Poisson" distribution with mostly low numbers but an occasional high one. In this case, each observation should be replaced by its square root.

28.5 Animal Welfare

The animal division of the institute should ensure that Investigators have received training in handling the animals and in the use of simple techniques such as giving intraperitoneal injections.

Special care should be taken to ensure that the animals experience the least possible pain and distress. The proposed experiments need to be approved by the local IACUC (Institutional Animal Care and Use Committee) and the Director of the animal division.

Animal welfare organizations sometimes suggest that, as rats and mice are social animals, they should not be housed singly. However, housing two or more male mice or rats in the same cage may lead to fighting and unpredictable results. Also, if the animals were to receive different pharmaceutical treatments there could be cross contamination because mice and rats are coprophagic. So, animals should usually be housed singly while they are the subjects of an experiment.

28.6 Choice of the Experimental Design

The "Randomized block"/"Matched pairs" (RB/MP) is the most widely used design in agricultural and industrial research because it is powerful and convenient to use. The individual blocks or pairs can be separated in time and/or location, to suit the investigator. This also provides some assurance of repeatability because treatment effects will only be observed if the individual blocks are in reasonable agreement, as assessed by the statistical analysis.

Studies involving animal models of development can use a version of the RB design with unequal group sizes. In that case, a litter represents a "block" and subjects within each litter can receive different treatments [6].

However, in a recent survey of one hundred published papers involving mice or rats [7], three of them had used an MP design when comparing genetically modified with wildtype litter mates, but the remaining ninety-seven papers had not named the design which they had used.

Journal editors and referees should reject any paper involving a pre-clinical experiment that fails to state the name of the experimental design which has been used.

When unequal treatment group sizes cannot be avoided, such as when there is only enough of a test reagent to treat a small number of subjects, or when only a few GM animals are available to be compared with wild-type ones, then the CR design with unequal sample sizes can be used.

28.7 Software

Designed experiments, with quantitative outcomes, are usually analyzed using a t-test if there are just two groups or an "Analysis of variance" (ANOVA) when there are three or more treatment groups. Several software packages are available. "MINITAB" (https://www.minitab.com), used below, has been used by this author for many years. It is relatively easy to use, with well formatted output. It or some other high quality software package should be made available (within the Animal division?) to all scientists who use laboratory animals.

As an alternative, "R-Commander" (Rcmdr) is free. It can be downloaded from the "CRAN" website. First, the "R" statistical language needs to be installed and then Rcmdr can be downloaded from it. Further details are given by Festing et al. [8]. However, Rcmdr is not as easy to understand and use as MINITAB.

The MINITAB interface is like a spreadsheet. If the experimental treatments and results (still in random order) have been recorded in EXCEL, they can be copied and pasted directly into MINITAB. This has a menu along the top: "File", "Edit", "Data", "Calc", "Stat", "Graph" etc.

With designed experiments, one of the "Stat", or "ANOVA" options should be chosen. The columns in EXCEL which contain the treatment designations, factors, blocks (if an RB design is being analyzed) and data need to be indicated. A menu offers various outputs, such as the ANOVA table and comparisons of treatments as well as various plots, including those of fits versus residuals. This should be a reasonably straight line. If that is not the case the raw data may need to be transformed, as noted above. Numerical examples are given below.

28.8 Conclusions

Pre-clinical research in Academia is characterized by high levels of irreproducibility which may be costing as much as twenty eight billion dollars per year in the USA alone [3]. In contrast, it is highly unlikely that such high levels of irreproducibility would be tolerated in the pharmaceutical industry.

An obvious difference between the two is that the pharmaceutical industry employs many statisticians, but few applied statisticians are available to assist scientists in Academia.

Maybe the organizations funding pre-clinical research should ensure that statistical advice is available to all their scientists. The animal division of most academic institutions is headed by well qualified veterinarians. An applied statistician could be employed within the animal divisions of the larger research institutes with smaller institutions sharing their services.

28.9 Numerical Examples

The rest of this chapter consists of worked examples of CR and RB designs including some with a factorial arrangement of treatments. Investigators can use these to become familiar with MINITAB, or other statistical software.

1. A "Matched pairs (MP)" design. A comparison of the body weights of rats fed on roast or raw peanuts. A *one-sample t-test* of the signed differences between treated and control subjects. (Data from reference [2] with one slight modification).
2. A "Randomized block" design with three treatments: The effect of two drugs on apoptosis in rat thymocytes. This was a pilot experiment to test whether the response to these drugs was reproducible. The three blocks were spread over a three-week period.
3. A "Completely randomized" (CR) design with three treatments: "Do BALB/c strain mice treated with either Methylcholanthrene (M) or Urethane (U), show an increase in the number of micronuclei in the peripheral blood lymphocytes, compared with the vehicle control mice?
4. A "two strains by two treatments" completely randomized factorial experimental design. Do mouse strains C3H and CD-1 differ in their red blood cell response to chloramphenicol, administered i.p at a dose of zero or 2000 mg/kg?

Example 1: A "Matched Pairs" Design
This experiment aimed to compare the biological value of roast versus raw peanuts as judged by weight gain in rats. Nine pairs of rats, matched for age and weight, were used. One of each pair was fed raw and the other on roast peanuts. At the end of the study, the body weights were recorded in EXCEL as shown in Table 28.1. (Data from reference [2], with one alteration.) The results are

Table 28.1 The raw data. Body weight

Pair	Raw	Roast	Raw-roast
1	61	55	6
2	60	54	6
3	56	47	9
4	63	59	4
5	56	51	5
6	63	61	2
7	59	57	2
8	56	54	2
9	61	58	3

clear-cut and hardly need statistical analysis, but a p-value is usually required when publishing the results.

Table 28.1 The column headings and all the numerical data, stored in EXCEL, were pasted into MINITAB. A "one-sample t-test" of the hypothesis that the mean difference in weight (Col. 4) between the two groups (column 4) does not differ statistically from zero, as shown in Table 28.2. The analysis was done in MINITAB using "Stat, "basic statistics", "one-sample t" of "Raw-roast". The results are shown below.

Conclusion

The 95% Confidence interval (CI, Table 28.2) for the mean difference does not span zero and the p-value is 0.001. So, it can be concluded that the rats on the Roast diet had significantly ($p < 0.01$) lower body weights than the rats given the raw diet.

No graphs are shown. They are not helpful with such a small set of data.

Example 2: A "Randomized Block" (RB) Design with Three Blocks and Three Treatments

This was an *in-vitro* experiment using fresh rat thymocytes. The scientists wanted to confirm that "apoptosis" (programmed cell death) could be reliably induced in rat thymocytes using the drugs "CPG" and "STAU", compared with the vehicle control treatment "C".

Once a week, for three weeks, a rat was sacrificed and a suspension of thymocytes was prepared. A measured aliquot of these cells was added to three tissue culture tubes. One of the three drug treatments: "CPG", STAU" or "C" (medium control), chosen at random, was added to each tube and, after a suitable incubation period, the number of cells undergoing apoptosis was scored. A "two-way analysis of variance without interaction" was used to analyse the results, shown below.

Note that the sample sizes were less than those suggested by the "resource equation" method. This can be risky as the experiment may lack power. However, RB designs are intrinsically more powerful than CR designs and this was an *in-vitro* study with better control of variation than a full animal study. Also, the investigators only wanted to check that they could get repeatable results and did not plan to analyze or publish the results.

Statistical analysis using MINITAB.

1. The raw data in Box 28.1 including the headings, was pasted into MINITAB.
2. A two-way ANOVA *without interaction* (as required for an RB design) was specified. The raw data and results are shown in Boxes 28.1 and 28.2. It is assumed that the

Table 28.2 Minitab output for a one-sample t-test for Raw-Roast

One-sample T: raw-roast

Test of mu = 0 vs not = 0

Variable	N	Mean	StDev	SE Mean	95% CI	T	P
Raw-Roast	9	4.33	2.39	0.80	(2.49, 6.18)	5.42	0.001

investigator has had some training in statistics and is familiar with the Analysis of Variance. Briefly, this breaks the total variation into parts associated with the differences between blocks, treatments ("Drug" in this case), and "Error", or what is left over. It gives a "P-value" for the blocks, drugs, error and total, but in this case, only the "Drug" p-value of $p = 0.023$ is of interest. This is less than the usual 0.05 significance level, so the results are judged to be "Statistically significant".

Below the ANOVA table are some 95% confidence intervals for the three treatments. If these do not overlap the three treatment group means, as in this case, then CPG and STAU differ both from each other and the control C.

Box 28.1 The Raw Data (Note Random Order Within Each Block)

Block	Drug	Score
1	STAU	421
1	CPG	398
1	C	365
2	C	423
2	CPG	432
2	STAU	459
3	STAU	329
3	CPG	320
3	C	308

Conclusions

The normal probability plot of the residuals is a good straight line and the plots of fitted versus residuals show no pattern, which is good (Fig. 28.1). The ANOVA in Box 28.2 shows a statistically significant effect of the drugs ($p = 0.023$) which also differ from each other. The 95% CIs do not span the control means for either drug, so both are significantly different from the controls and each other at $p < 0.05$. There were large differences between blocks but the effects were removed in the statistical analysis. The conclusion is that the results were reproducible, with statistically significant differences between the three treatments.

Example 3: A Completely Randomised (CR) Design with Three Treatments

The purpose of this experiment was to see whether BALB/c strain mice treated with either Methylcholanthrene (M) or Urethane (U), two known carcinogens, show an increase in the number of micronuclei in the peripheral blood lymphocytes of mice, compared with the vehicle treated control mice.

The mice were placed individually in numbered cages and one of the treatments, *chosen at random,* using the "=rand()" function in EXCEL as described in Sect. 28.4.1, above, was assigned to each mouse. Following treatment and a short interval, the micronuclei in a sample of one thousand cells were counted using a laser scanning cytometer. Box 28.3 shows the raw data (subject number, the treatment, and the micronucleus count for each mouse.)

The raw data was pasted into Minitab (without the row numbers). The "Stat", "ANOVA", and "Oneway" options were chosen.

The results are shown in Box 28.4. Animals numbered four and five were outliers but deleting them or transforming the data to the log10 of each observation makes no difference to the conclusions. Overall, the treatment effect was statistically significant ($P = 0.001$) but only the mice treated with Urethane differed significantly at the 5% level from the control mice (the 5% confidence intervals in Box 28.4 cover the control for MCA). The adjusted R-sq shows that the treatments accounted for 42.6% of the total variation.

The normal probability plot of the residuals is a reasonably straight line, as required (see Fig. 28.2). If this were not the case the raw data might have needed to be transformed to a different scale.

The plot of fitted versus residuals is of some slight concern as it suggests that there is a bit more variation in the higher dose group than in the other two groups. This is not a large effect, but had it been larger, a logarithmic transformation of the raw data might have been appropriate.

In conclusion, the number of micronuclei was significantly increased when the mice were

Box 28.2 Output from MINITAB

Two-Way ANOVA: Score Versus Block, Drug

Source	DF	SS	MS	F	P
Block	2	21764.2	10882.1	114.82	0.000
Drug	2	2129.6	1064.8	11.23	0.023
Error	4	379.1	94.8		
Total	8	24272.9			

S = 9.735 R-Sq = 98.44% R-Sq(adj) = 96.88%

```
                  Individual 95% cis for mean based on pooled StDev
Drug    Mean      -----+---------+---------+---------+----
C       365.333   (-------*------)
CPG     383.333         (-------*------)
STAU    403.000              (-------*------)
                  -----+---------+---------+---------+----
                      360       380       400       420
```

Note that R-squared is the proportion of the total variation which is accounted for by the statistical analysis. In this case it was 97%.

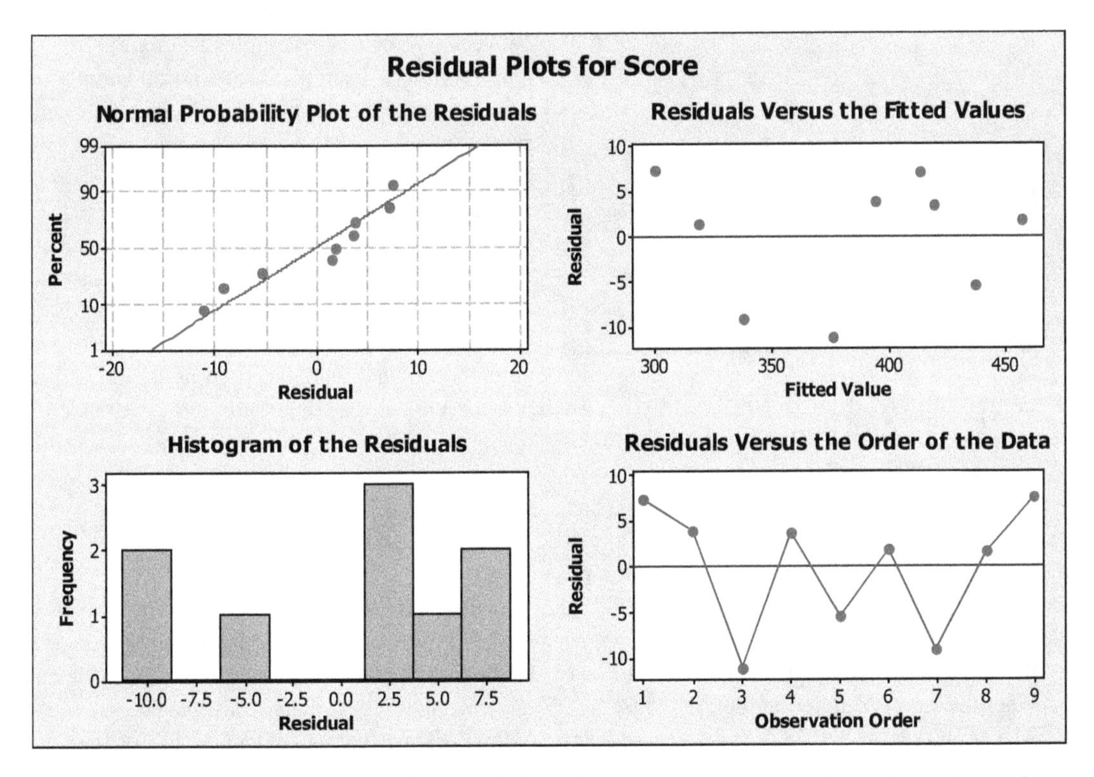

Fig. 28.1 The normal probability plot of the residuals shows a good straight line, as required for a satisfactory analysis. And the plot of residuals versus fitted values shows the required random scattering of points. The histogram is too sparse to be useful, and there is no indication that the orde in which the data was collected had any particular pattern

treated with Urethane, but not when they were treated with Methylcholanthrene, at the concentrations which were used.

Box 28.3 Raw data

Subject	Treatment	Count
1	M	2.00
2	C	1.87
3	C	2.06
4	U	5.39
5	U	6.1
6	M	0.75
7	C	2.15
8	C	1.90
9	C	0.66
10	M	1.87
11	C	1.59
12	M	1.98
13	M	2.27
14	U	3.48
15	U	3.56
16	U	3.85
17	C	1.22
18	U	2.15
19	U	1.57
20	M	2.13
21	U	2.34
22	C	1.23
23	M	0.83
24	M	1.76

Example 4: A "Two Strains by Two Treatments Completely Randomized Factorial Experimental Design"

The aim of this experiment was to see whether there were strain differences in the red blood cell counts of mice following treatment with chloramphenicol. The raw data is shown in Table 28.3. It was analysed using MINITAB using a "two-way analysis of variance *with* interaction" to see whether the strains differed in their response.

The ANOVA Table 28.4 shows significant strain differences ($p = 0.003$), an overall treatment effect that is not quite significant at the 5% level ($p = 0.057$) and a strain by treatment interaction which is significant at $p = 0.016$ with a large decline in RBC counts in response to the chloramphenicol in C3H but not in CD-1, as shown in the boxplot (Fig. 28.3 left panel). The normal probability plot of fits versus residuals is a good straight line (Fig. 28.3 right panel), so there is no problem with the distribution of the data.

In conclusion, strains C3H and CD-1 differ significantly ($p = 0.016$) in their response to chloramphenicol given at a dose of 2000 mg/kg, with C3H being susceptible, but CD-1 being relatively resistant.

28.10 Correlation and Regression

These two statistical methods are used to measure the "association" between two variables. They do not normally involve randomization.

"Correlation" quantifies an association between two or more variables, *without implying causation*. For example, there is, apparently, a positive correlation between ice cream sales and shark attacks on beaches (presumably in Australia?).

The "coefficient of correlation", designated "r" can range from −1.0, a strong negative association to +1.0, a strong positive association between the two variables. In contrast, "Regression" quantifies a presumed causal association between an "independent" variable such as the dose of a pharmaceutical agent, and a "dependent" response such as red blood cell counts.

28.10.1 Correlation

The Pearson "Coefficient of correlation" is a quantification of the *linear* association between two variables. It is quantified by the "Coefficient of correlation" usually designated by "r". MINITAB was used to calculate "r" for the data in Table 28.5.

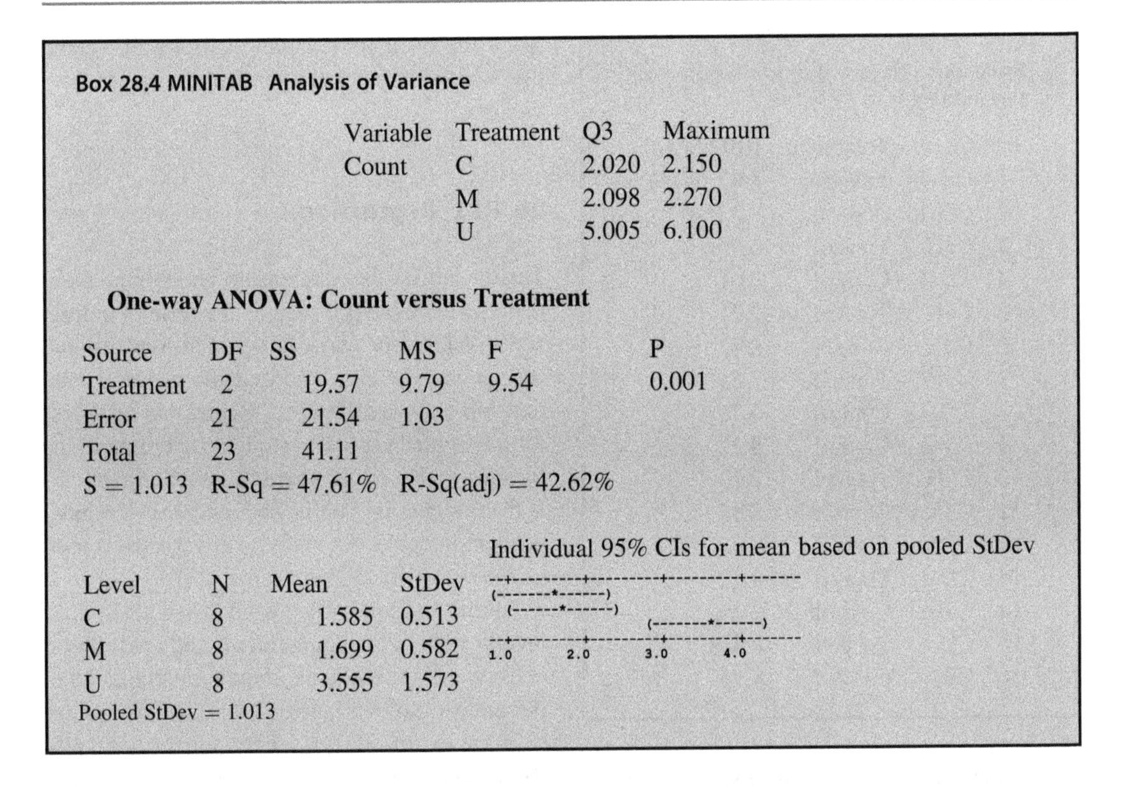

Box 28.4 MINITAB Analysis of Variance

Variable	Treatment	Q3	Maximum
Count	C	2.020	2.150
	M	2.098	2.270
	U	5.005	6.100

One-way ANOVA: Count versus Treatment

Source	DF	SS	MS	F	P
Treatment	2	19.57	9.79	9.54	0.001
Error	21	21.54	1.03		
Total	23	41.11			

S = 1.013 R-Sq = 47.61% R-Sq(adj) = 42.62%

Individual 95% CIs for mean based on pooled StDev

Level	N	Mean	StDev
C	8	1.585	0.513
M	8	1.699	0.582
U	8	3.555	1.573

```
--+---------+---------+---------+-------
(-------*------)
 (-------*-------)
                  (-------*------)
--+---------+---------+---------+-------
1.0       2.0       3.0       4.0
```

Pooled StDev = 1.013

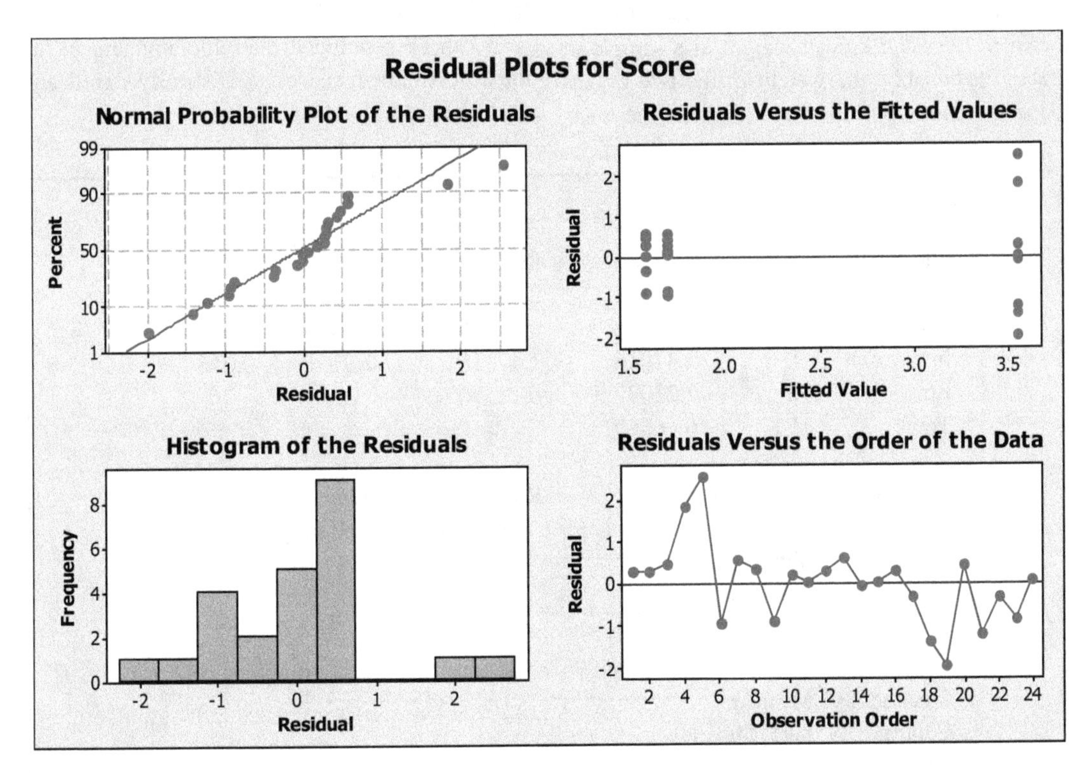

Fig. 28.2 Residual diagnostic plots. Note that the residuals versus fitted value plot suggests greater variation in the urethane treated group. However, transforming the data to a Log10 scale (not shown) makes no difference to the over-all conclusion that there was a response to urethane but not to methylcholanthrene

Table 28.3 The raw data was collected in random order

ID	Strain	Treatment	RBC Score
1	CD-1	Treated	8.47
2	CD-1	Control	8.42
3	C3H	Treated	6.96
4	CD-1	Control	9.01
5	C3H	Treated	7.21
6	C3H	Control	8.77
7	C3H	Treated	7.81
8	CD-1	Control	7.76
9	C3H	Control	8.22
10	C3H	Treated	7.10
11	C3H	Control	7.85
12	CD-1	Treated	8.67
13	CD-1	Treated	9.18
14	C3H	Control	8.48
15	CD-1	Treated	8.31
16	CD-1	Control	8.83

The data and plot (Fig. 28.4) show a positive correlation of r = 0.62 with a p-value of 0.006 between liver and kidney weight in a sample of mice. Presumably, the weight of the two organs depends mainly on the body weight of the mice. So, if the mice are relatively uniform in weight, the correlation will be lower than if they are more variable.

28.10.2 Regression

This quantifies the relationship between an *independent variable (x)*, such as the dose of a drug, and a *dependent variable (y)* such as red blood cell count. The aim is to quantify a causal relationship between the two. "Regression analysis" estimates the best fitting straight according to the formula y = a + bX, where "a" is a constant, "b" is the "regression coefficient" and X is the independent variable. An analysis of variance is used to assess statistical significance.

In this example, mice were dosed at six dose levels with chloramphenicol and red blood cells were counted, as shown in Table 28.6. An analysis of variance (Table 28.7) provides a p-value of 0.001. The fitted line plot (Fig. 28.5) gives the regression formula: RBC = 9.81– 0.058 × dose.

It can be concluded that chloramphenicol, at the doses studied, causes a statistically significant decline in red blood cell counts.

Table 28.4 Two-way ANOVA: score versus strain, treatment

Source	DF	SS	MS	F	P
Strain	1	2.44141	2.44141	13.13	0.003
Treatment	1	0.82356	0.82356	4.43	0.057
Interaction	1	1.47016	1.47016	7.91	0.016
Error	12	2.23077	0.18590		
Total	15	6.96589			

S = 0.4312 R-Sq = 67.98% R-Sq(adj) = 59.97%

```
                    Individual 95% CIs for mean based on pooled StDev
Strain     Mean     ---+---------+---------+---------+------
                    (-------*-------)
C3H        7.80000                 (--------*-------)
CD-1       8.58125  ---+---------+---------+---------+------
                      7.60      8.00      8.40      8.80

                    Individual 95% CIs for mean based on pooled StDev
Treatment  Mean     ------+---------+---------+---------+---
                                  (----------*----------)
Control    8.41750  (----------*-----------)
Treated    7.96375  ------+---------+---------+---------+---
                         7.80      8.10      8.40      8.70
```

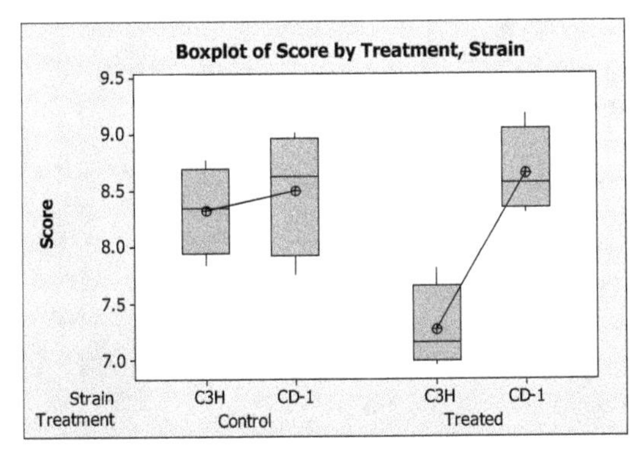

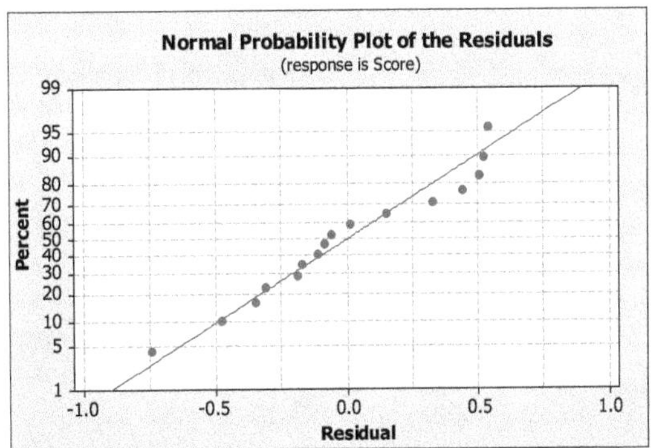

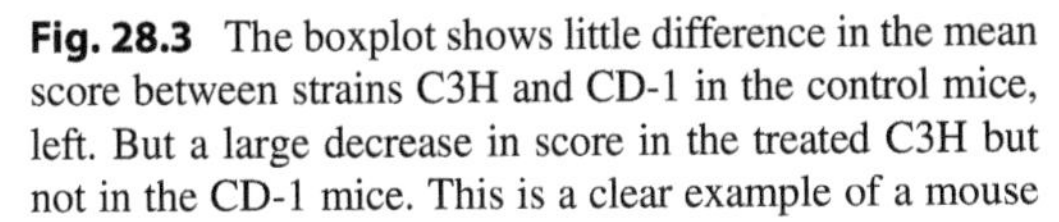

Fig. 28.3 The boxplot shows little difference in the mean score between strains C3H and CD-1 in the control mice, left. But a large decrease in score in the treated C3H but not in the CD-1 mice. This is a clear example of a mouse strain by treatment interaction. The normal probability plot in the right-hand panel shows that the residuals in both strains have a good normal distribution, as required for a valid statistical test

28.11 Conclusions and Recommendations

Randomized, controlled experiments are an essential tool for comparing the effect of different treatments on the physiology and behaviour of laboratory animals. But they must be designed and analyzed correctly if they are to give valid, reproducible, results.

There are only two experimental designs suitable for widespread use: The "completely randomized" ("CR") design needs a uniform group of subjects with the whole experiment being done at the same time. Its main advantage is that it can have unequal treatment group sizes.

The more powerful "Randomized block" (RB) design is split into several "blocks", each of which has a single subject receiving each of the treatments. The blocks can be separated in time and/or location. This can provide some assurance of reproducibility. The results across all the blocks are combined in the statistical analysis.

Table 28.5 Liver and kidney weight in individual mice

Liver wt.	Kidney wt.
0.99	0.29
1.19	0.34
1.28	0.28
1.19	0.32
1.06	0.29
1.04	0.28
1.09	0.29
0.83	0.25
1.03	0.29
1.1	0.24
0.9	0.26
0.99	0.25
0.98	0.24
0.93	0.24
0.8	0.26
0.99	0.29
1.03	0.26
0.86	0.24

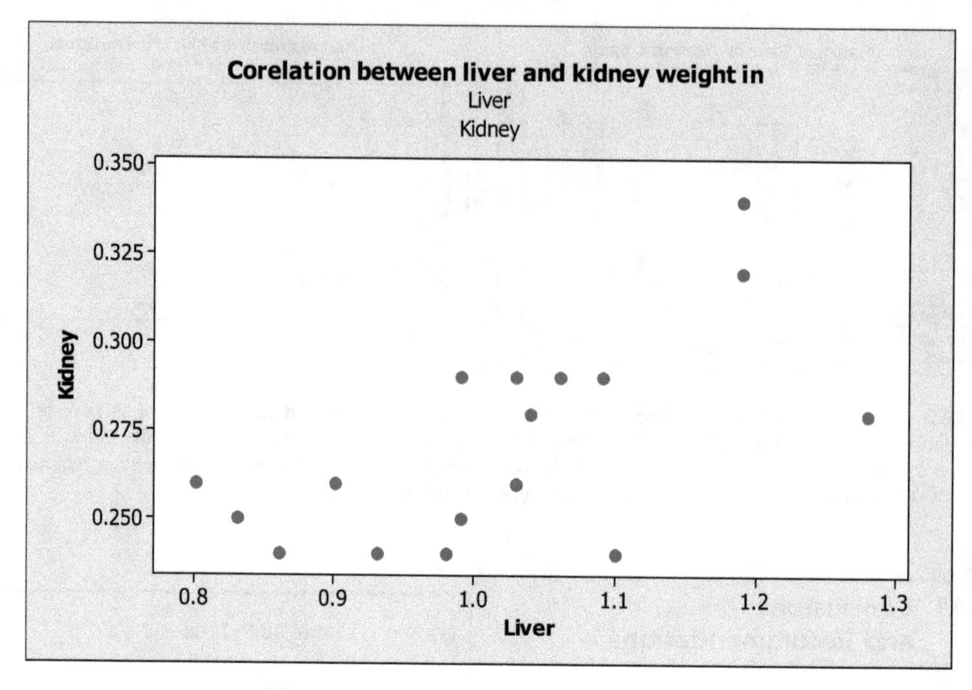

Fig. 28.4 Plot of the correlation between liver and kidney weight in a sample of mice. The Pearson correlation between liver and kidneys in these mice, calculated using MINITB, was r = 0.618 with a statistically significant P-Value of 0.006

Both designs are quite versatile. They can have any number of treatment groups as well as additional factors such as both sexes or more than one strain of animals. These can provide additional information at little or no extra cost.

There are several other designs such as the "Latin square", the "Split-plot", the "Lattice square" and "Sequential" designs. These are only used in exceptional circumstances. Where necessary, details can be found in textbooks on experimental design.

When publishing the results of an experiment the authors should always state the name of the design which they have used: "Completely randomized" (CR) or "Randomized block"/ "matched pairs" (RB/MP). This will provide some assurance that they have designed their experiments correctly.

Funding organizations should find a way of ensuring that an applied statistician is available, possibly within the animal division, of each institute to assist the scientists in designing and analyze the experiments correctly.

Table 28.6 Red blood cell response to varying doses of Chloramphenicol

Dose	RBC
0	9.60
0	9.56
0	9.14
5	9.27
5	9.16
5	9.53
10	9.61
10	9.82
10	9.44
15	9.81
15	9.83
15	9.83
20	8.18
20	8.82
20	8.24
25	7.83
25	8.07
25	7.83

Table 28.7 Analysis of variance table

Source	DF	SS	MS	F	P
Regression	1	4.4037	4.4037	15.03	0.001
Residual Error	16	4.6891	0.2931		
Total	17	9.0928			

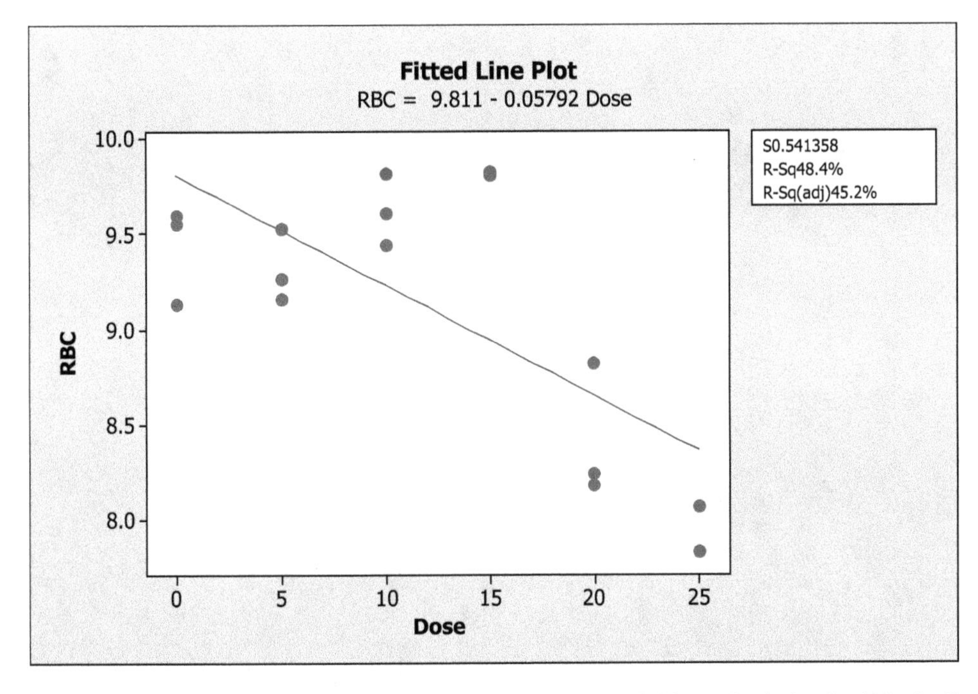

Fig. 28.5 The fitted line plot for the linear association between the dose of chloramphenicol and red blood cell count in mice

Badly designed experiments that lead to incorrect conclusions are scientific, welfare and economic issue. The local IACUC "Institutional Animal Care and Use Committee" should exert pressure to ensure that the experiments are being done correctly and should insist that statistical advice is made available, where necessary, to all scientists doing pre-clinical research.

Conflict of Interest The author has no conflicts of interest. This contribution has been done without funding.

References

1. Fisher RA (1960) The design of experiments. Hafner Publishing Company, Inc, New York
2. Snedecor GW, Cochran WG (1957) Statistical methods, 5th edn. Iowa State University Press, Ames, Iowa
3. Freedman LP, Cockburn IM, Simcoe TS (2015) The economics of reproducibility in preclinical research. PLoS Biol 13(6):e1002165
4. Scott S, Kranz JE, Cole C, Lincecum JM, Thompson K, Kelly N, Bostrom A, Theodoss J, Bashar M, Al-nakhala FG, Ramasubbu J, Heywood JA (2008) Design, power, and interpretation of studies in the standard murine model of ALS. Amyotroph Lateral Scler 9:4–15
5. Festing MF (2018) On determining sample size in experiments involving laboratory animals. Lab Anim 52(4):341–350

6. Festing MF (2006) Design and statistical methods in studies using animal models of development. ILAR J 47:5–14
7. Festing MFW (2020) The "completely randomised" and the "randomised block" are the only experimental designs suitable for widespread use in pre-clinical research. Sci Rep 10(1):17577
8. Festing MFW, Overent P, Cortina Borjs M, Berdoy M (2016) The design of animal experiments. Sabge Publications, London

Descriptive and Inferential Statistics in Biomedical Sciences: An Overview

29

Avijit Hazra

Abstract

Statistical literacy is essential for understanding medical literature and conducting biomedical research. Data constitute the raw material for statistical work. Descriptive statistics summarizes data from a sample or population. Categorical data are described in terms of percentages or proportions. With numerical data, individual observations tend to cluster about a central location, with more extreme observations being less frequent. Measures of central tendency summarize the extent to which observations cluster while the spread is described by measures of dispersion. There is no way of assessing true population parameters by observing samples alone. We can, however, obtain a standard error and use it to define a range in which the true population value is likely to lie with a given level of uncertainty. This is the confidence interval (CI), Conventionally, the 95% CI is used. The commonly encountered pattern in data sets is the normal distribution which appears as a symmetrical bell-shaped curve. Much of medical research begins with a research question that can be framed as a hypothesis. Inferential statistics starts with a null hypothesis that reflects the conservative position of no change or no difference in comparison to the baseline or between groups. Usually, the researcher believes that there is some effect which is the alternative hypothesis. Thinking of the research hypothesis as addressing one of five generic research questions helps in selection of the right hypothesis test. This chapter aims to introduce the basic tenets of descriptive and inferential statistics without delving into the mathematical depths.

Keywords

Descriptive statistics · Inferential statistics · Measures of central tendency · Measures of dispersion · Confidence interval · Normal distribution

29.1 Introduction

Biostatistics encompasses the application of statistical principles to the understanding and exploration of living systems. Biomedical research, in its myriad applications, will simply not be possible without the use of statistics to design and conduct biomedical experiments (see Chap. 28) and to interpret the data captured through observational studies. The major reason for this is the enormous variations exhibited by living systems coupled with our inability, to a large degree, to understand the sources of such variation and to control for them in the course of our observations

A. Hazra (✉)
Department of Pharmacology, Institute of Postgraduate Medical Education and Research, Kolkata, India
e-mail: blowfans@yahoo.co.in

G. Jagadeesh et al. (eds.), *The Quintessence of Basic and Clinical Research and Scientific Publishing*,
https://doi.org/10.1007/978-981-99-1284-1_29

and experiments. Variations may stem from peculiarities of individual subjects, the background exposures or the interventions offered, measurement errors or simply unknown 'chance' factors. Further, most of the time researchers in the life sciences must look out for small changes or differences rather than large effects. The application of statistics allows the researcher to make sense of these small changes or differences while adjusting for variations, usually working with representative samples drawn from larger populations. Simply put, in biomedical research, statistics is a tool for drawing meaningful conclusions in the face of inevitable uncertainty. Outside the ambit of research and audit, statistical description is also used in everyday data presentation.

Biostatistical principles and methods can broadly be envisaged as covering three domains—descriptions of patterns in numbers through various descriptive measures (**descriptive statistics**), drawing conclusions regarding populations by interpreting data captured from representative samples through various statistical tests (**inferential statistics**), and exploring the relationship between variables, sometimes with the goal of prediction (**statistical modeling**). This chapter looks at the various uses of statistics without delving into mathematical depths. The mathematical underpinnings of statistics are undeniable but the aim here is to present the concepts and applications from the point of view of the applied user of biostatistics.

29.2 Data and Variables [1]

Data refers to records of measurement or observations or simply counts and constitutes the raw material for statistical work. A **variable** refers to a particular character on which a set of data are recorded. Data are thus the values of a variable. The nature of the variables of interest will influence the way in which observations are undertaken, the manner in which they will be summarized and the choice of statistical tests that will be used. Figure 29.1 presents the basic classification of data or variable types.

Note that we may assign numbers (scores) to nominal and ordinal categories, although the differences among those numbers do not have numerical meaning. However, category counts do have numerical significance. A special case may exist for either categorical or numerical variables, when only two values are possible for the variable in question; this is **binary** or **dichotomous** data, as opposed to **non-binary** or **polychotomous** data that can take more than two values.

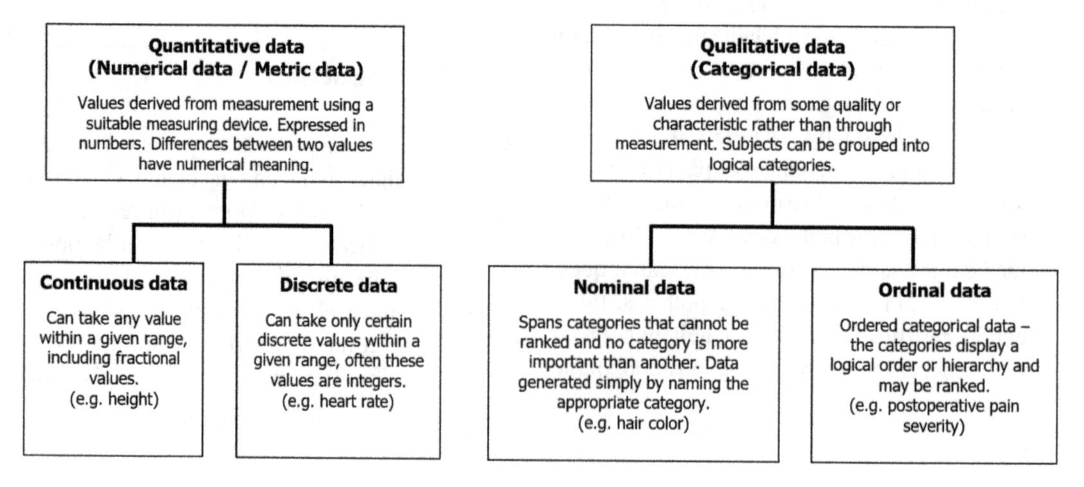

Fig. 29.1 Basic data types

Often certain numerical variables, like age or blood pressure, are treated as district variables although strictly speaking, they are continuous. Numerical data can be recorded on an interval scale or a ratio scale. On an **interval scale**, the differences between two consecutive numbers carry equal significance in any part of the scale, unlike the scoring of an ordinal variable ('ordinal scale'). For example, when measuring height, the difference between 50 and 52 cm is the same as the difference between 150 and 152 cm. With interval data the zero value can be arbitrary, such as the position of zero on Celsius and Fahrenheit temperature scales—the Fahrenheit zero is at a different position to that of the Celsius scale. The **ratio scale** is a special case of recording interval data that has a true zero, such as the Kelvin (absolute) scale of temperature, where 0 °K denotes no heat. Only on a ratio scale, can differences be judged in the form of ratios. 0 °C is not zero heat, nor is 26 °C twice as hot as 13 °C, whereas such value judgments hold with the Kelvin scale.

Changing data scales is possible so that numerical data may become ordinal, and ordinal data may become categorical (even dichotomous). This may be needed because the researcher has concerns about the accuracy of the measuring instrument, is unconcerned about the loss of fine detail, or where group numbers are not large enough to adequately represent a variable of interest. It may also make clinical interpretation easier. For example, in ECG monitoring, the extent of ST-segment depression indicates the degree of myocardial ischemia. Although, theoretically a continuous variable, it is generally accepted that ST-segment depression greater than 1.0 mm indicates significant ischemia, so that ST-segment depression less than this value is categorized as 'no ischemia'. This results in some loss of detail, but clinically is more convenient to deal with and is therefore widely accepted.

Numerical or categorical variables may sometimes need to be **ranked**, that is arranged in ascending order and new values assigned to them in serial order. Values that tie are each assigned the average value of the ranks they encompass. Thus, a data series 20, 30, 30, 30, 30, 50, 70, 90, 150 can be ranked as 1, 3.5, 3.5, 3.5, 3.5, 6, 7, 8, 9 (since the four 30 s encompass ranks 2, 3, 4, 5 giving an average rank value of 3.5). Note that when a numerical variable is ranked, it gets converted to an ordinal variable. The ranking does not apply to nominal variables because their values do not follow any order.

While exploring the relationship between variables, values of some can be considered as a dependent (**dependent variable**) on another (**independent variable**). For example, when exploring the relationship between height and age, it is obvious that height depends on age, at least till a certain age. Thus, age is the independent variable, which influences the value of the dependent variable height. When exploring the relationship between multiple variables, usually in a modeling situation, the value of the **outcome (response) variable** depends on the value of **predictor (explanatory) variables**, while some variables, called **confounding variables (confounders)**, cannot be accurately measured, or controlled, but may confound the results. Thus, in a study of bronchodilator effectiveness, atmospheric pollution may affect bronchial responsiveness and may confound the results. Smoking can be another confounder.

29.3 Descriptive Statistics [1, 2]

Descriptive statistics aims at summarizing data obtained from a series of measurements or observations (a data series or raw data). Categorical data are commonly summarized in terms of frequencies or proportions. With numerical data, individual observations within a sample or population tend to cluster about a central location, with more extreme observations being less frequent. The extent to which observations cluster is summarized by **measures of central tendency** while the spread can be described by **measures of dispersion**. The shape can be described by the property of **kurtosis**, but we will ignore this at the moment as it is seldom required. Box 29.1 summarizes the measures of central tendency while Box 29.2 summarizes the measures of dispersion.

Box 29.1 Measures of Central Tendency

Mean (Arithmetic Mean) or Average

- Calculated as the sum of the individual values in a data series, divided by the number of observations.
- Need not be an actual value in the data series being summarized.
- Most commonly used measure of central tendency to summarize numerical data.
- Usually reliable, unless there are extreme values (outliers) that can distort the mean.
- Generally, not to be used in describing categorical variables because of the arbitrary nature of category scoring. May, however, be used to summarize category counts.

Geometric Mean

- Geometric mean of a series of n observations is the nth root of the product of all the observations.
- Always equal to or less than the arithmetic mean.
- Not often used but can be a more appropriate measure of central tendency when data in a series varies over several orders of magnitude or have been recorded on a logarithmic scale.
- Logarithm of the geometric mean is the arithmetic mean of the logarithms of the observations. Calculated by taking the antilog of the arithmetic mean of the log values of the observations.

Other Varieties of the Mean

- **Harmonic mean** for a set of non-zero positive numbers is obtained as the reciprocal of the arithmetic mean of the reciprocals of these numbers. Used for calculating average rates. Seldom used in biostatistics.
- **Quadratic mean** is the root mean square value. Values are squared, the average of squared values is derived and the square root of this is taken as the quadratic mean.
- **Trimmed mean** or **truncated mean** is the arithmetic mean calculated after removing a small, specified percentage of the largest and smallest values (e.g. 5% trimmed mean would imply 2.5% of the observations are eliminated from each end of the data series). Helps eliminate the influence of outliers that may unduly affect the traditional arithmetic mean.
- **Pooled mean** or **grand mean** is the average of the means of multiple samples from the same population, so long as the samples have the same number of data points. For example, if we sample 5 persons each from 10 groups, and then measure the weight of each person, we can calculate 10 sample means and finally calculate the average of these 10 sample means as the pooled mean.
- **Weighted mean** is used to combine average values from different sized samples of the same population. Each mean is weighted (multiplied) by the sample size, the products are summed, and the average is taken to get the weighted mean.

Median

- If values in a data series are ranked (either in ascending or descending order), this is the actual middle value (for an odd number of observations) or the average of the two middle values (for an even number of observations).
- Denotes the point in a data series at which half the observations are larger and half are smaller.
- Same as the 50th percentile value.

(continued)

Box 29.1 (continued)

- A useful summary measure, particularly if the distribution of the data is not symmetrical, since it is less sensitive to extreme values than the mean.
- For a positively skewed distribution (long tail to the right) median is less than the mean. The reverse holds true for a negatively skewed distribution (long tail to the left). In a normal distribution mean and median coincide.
- Divides the distribution curve of the data into two equal-area portions. The portions will be symmetrical if the distribution curve is normal.

Mode

- The most frequently occurring value in a data series.
- Not often used, for the simple reason that it is difficult to pinpoint a mode if no value occurs with a frequency markedly greater than the rest.
- Also, two or more values may occur with equal frequency, making the data series bimodal or multimodal.

Box 29.2 Measures of Dispersion

Range

- The interval between minimum and maximum values.
- Affected by the two extreme observations.
- Does not provide much information about the overall distribution of observations.

Percentile Related Measures

- **Centiles** or **percentiles** are obtained by arranging the data in order of their values and then grouping them into 100 equal parts (in terms of the number of values in the series).
- It is then possible to state the range covered by any two percentiles values such as 5th to 95th, 10th to 90th, and so on. The median represents the 50th percentile.
- The range of values covered by the middle 50% of the observations about the median (i.e. 25th to 75th percentile values) denotes the **interquartile range** (IQR), which is a particularly useful measure of spread if the data series is skewed.

Variance and Standard Deviation

- A better method of measuring variability about the central location is to estimate how closely the individual observations cluster about it. The **variance** represents the mean square deviation and is calculated as the sum of the squares of individual deviations from the mean, divided by the number of observations. The squaring removes the effect of negative values.
- The **standard deviation** (SD) of a data series is simply the square root of the variance. Unlike the variance, which must be denoted in squared units that are difficult to comprehend, SD retains the basic unit of observation.
- SD is particularly useful for normally distributed data because the proportion of values in this distribution (i.e. the proportion of the area under the curve) is a constant for a given number of standard deviations above or below the mean.
- The formulae for the variance (and standard deviation) for a population has the value 'n' as the denominator. However, the expression (n – 1) is used when calculating the variance (and standard deviation) of a sample. The quantity (n – 1) denotes the **degrees of freedom**, which is the number of independent

(continued)

Box 29.2 (continued)

observations or choices available. For instance, if a series of ten numbers is to add up to 100, we can assign different values to the first nine, but the value of the last is fixed by the first nine choices and the condition imposed that the total must be 100. Thus, this data series will show 9 degrees of freedom. The degree of freedom concept is used when calculating the variance (and standard deviation) of a sample because the sample mean is a predetermined estimate of the population mean, and, in the sample, each observation is free to vary except the last one which takes on a defined value.

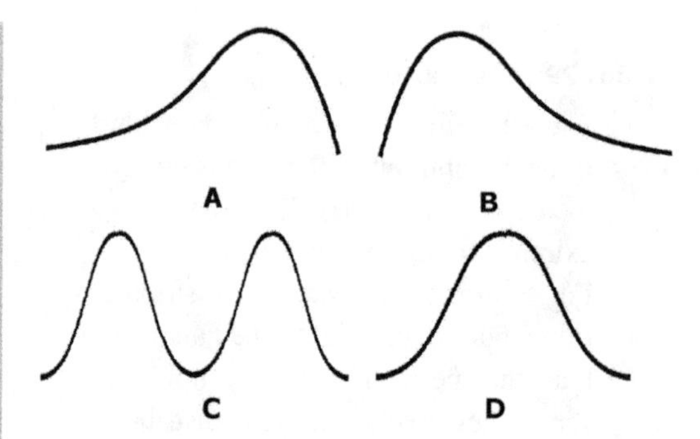

Fig. 29.2 Examples of frequency distributions—(**a**) Asymmetric, negatively skewed (**b**) Asymmetric, positively skewed (**c**) Symmetric but not normal (**d**) Symmetric and normal

29.4 Frequency Distributions and Tables [3, 4]

It is useful to summarize a set of observations with a frequency distribution. The summary may be in the form of a table or a graph (plot). Many frequency distributions are encountered in medical literature, and it is important to have a clear idea of the more commonly encountered ones.

The majority of distributions that quantitative clinical data follow are **unimodal**, that is the data have a single peak (mode) with a tail on either side. The more common of these unimodal distributions are symmetrical; however, some are **skewed** with a substantially longer tail on one side (Fig. 29.2). A **positively skewed** distribution has a longer tail on the right; with most values being relatively low with a smaller number of extremely high values. A **negatively skewed** distribution has a longer tail to the left; with the extreme values being markedly low in comparison to the rest of the dataset. In this instance, the mean, being unduly influenced by the extremely low values on the left, will be smaller than the median. On the other hand, in a positively skewed distribution, the mean will be greater than the median. **Bimodal** or multimodal distributions are uncommon in biology and may even be artifacts. A distribution with two peaks (bimodal) may actually be reflecting a combination of two unimodal distributions, for instance, one for each gender or different age groups. In such cases, appropriate subdivision, categorization, or even recollection of the data may be required to eliminate multiple peaks.

Oftentimes data is presented as a **frequency table**. A frequency table of numerical data may report the frequencies for class intervals (the entire range covered being broken up into a convenient number of classes) rather than for individual data values. In such cases, we can calculate the weighted average by using the mid-point of the class intervals. However, in this instance, the weighted mean may vary slightly from the arithmetic mean of all the raw observations.

29.5 Measures of Precision

The **coefficient of variation** (CV) of a data series denotes the SD expressed as a percentage of the mean. Thus, it denotes the magnitude of the SD relative to the mean. An important source of variability in biological observations is measurement imprecision and CV is often used to quantify this. Thus, CV is commonly used to describe the variability of measuring instruments - CV up to 5% is generally considered acceptable reproducibility. CV can be conveniently used to compare variability between studies, as, unlike SD, its magnitude is independent of the units employed.

Another measure of precision for a data series is the **standard error of the mean** (SEM), which is simply calculated as the SD divided by the square root of the number of observations. Since SEM is a much smaller numerical value than SD, it is often presented in place of SD. However, this is erroneous since SD is meant to summarize the spread of data, while SEM is a measure of precision and is meant to provide an estimate of a population parameter from a sample statistic in terms of the confidence interval.

It is self-evident that when we make observations on a sample, and calculate the sample mean, this will not be identical to the population ('true') mean. However, if our sample is sufficiently large and representative of the population, and we have made our observations or measurements carefully, then the sample mean would be close to the true mean. If we keep taking repeated samples and calculate a sample mean in each case, the different sample means would have their own distribution, and this would be expected to have less dispersion than that of all the individual observations in the samples. In fact, it can be shown that the different random sample means would have a symmetrical distribution, with the true population mean at its central location, and the standard deviation of this distribution would be nearly identical to the SEM calculated from individual samples. This is the essence of the **central limit theorem** in probability theory.

In general, however, we are not interested in drawing multiple samples, but rather in how reliable our one sample is in describing the population. We use the standard error to define a range in which the true population value is likely to lie, and this range is the **confidence interval** [5], with its two terminal values being called **confidence limits**. The width of the confidence interval depends on the standard error and the degree of confidence required. Conventionally, the 95% confidence interval (95% CI) is used in biomedical research. From the properties of a normal distribution curve (see below) it can be shown that the 95% CI of the mean would cover a range of 1.96 standard errors on either side of the sample mean and will have a 95% probability of including the population mean; while 99% CI will span 2.58 standard errors either side of the sample mean and will have 99% probability of including the population mean. Thus, a fundamental relation that needs to be remembered is

$$95\%\text{CI of mean} = \text{Sample mean} \pm 1.96 \times \text{SEM}$$

It is evident that the confidence interval would be narrower if SEM is smaller. Thus, if a sample is larger, SEM would be smaller, and the CI would be correspondingly narrower and thus more 'focused' on the true mean. Large samples, therefore, increase precision. However, although increasing sample size improves precision, it is a somewhat costly approach to increasing precision, since halving SEM entails a fourfold increase in sample size.

Confidence intervals can be used to estimate most population parameters from sample statistics (means, medians, proportions, correlation coefficients, regression coefficients, odds ratios, relative risks, etc.). In all cases, the general pattern of estimating the confidence interval remains the same, that is

$$95\%\text{CI of a parameter} = \text{Sample statistic} \pm 1.96 \times \text{Standard error for that statistic}$$

The formula for estimating standard error however varies for different statistics, and in some instances is quite elaborate. The situation, therefore, is usually managed by relying on computer software to do the calculations.

29.6 Data Transformation [6]

Manipulation of a dataset to alter its distribution is called **data transformation**. There are many

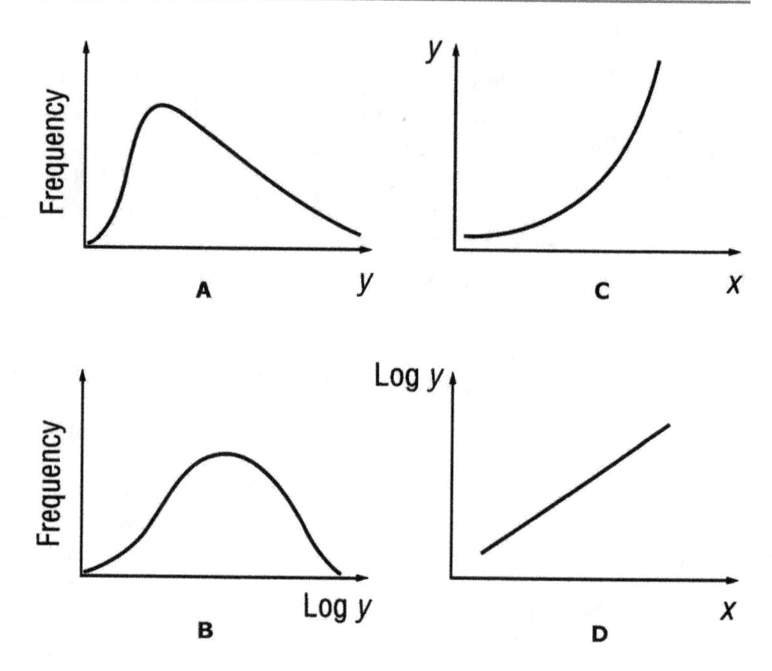

Fig. 29.3 Two uses of logarithmic transformation of data. (**a**, **b**) Making positively skewed data normally distributed, and (**c**, **d**) Linearizing in case of exponential relationship between two variables

different transformations, such as logarithmic, square root, reciprocal, logit transformation, and so on. There are certain advantages to working with symmetrical rather than asymmetrical data sets, and the most used transformation to make positively skewed data symmetrical is the **logarithmic transformation**. In this, every value in the dataset is replaced by its logarithm. Logarithms are defined with respect to a base, the most common being base e (natural logarithms) or base 10 (common logarithm). The end result is independent of the base chosen, provided the same base is used throughout. Notice that in log transformation, the differences in the transformed values are larger at the lower end of the scale. The logarithmic transformation stretches out the lower end and compresses the upper end of a distribution, with the result that positively skewed data will tend to become more symmetrical in shape. Calculations and statistical tests can be carried out on the transformed data before converting the results back to the original scale. A linear relationship between variables is desirable in regression analysis, and the logarithmic transformation is also useful in linearizing data, if an exponential relationship exists between two variables (Fig. 29.3).

29.7 The Normal Distribution

Many biological variables tend to cluster around a central value, with a symmetrical positive and negative dispersion about this point. The values become less frequent the further they lie from the central point. These features characterize a normal distribution (Fig. 29.4); the term ‘normal’ probably relates to the wide prevalence of this distribution. It is also referred to as a Gaussian distribution after the German mathematician, Karl Friedrich Gauss (1777–1855), although Gauss was not the first person to describe such a distribution.

The properties of a normal distribution are:

- Unimodal, bell-shaped distribution.
- The mean, median and mode coincide.
- These values all represent the peak of the distribution.
- The distribution then falls symmetrically around the mean, or, in other words, the curve flattens symmetrically as the variance is increased.
- The skewness (a measure of asymmetry) of the distribution is 0.
- The area delimited by one SD on either side of the mean covers 68.2%, two SD 95.4%, and

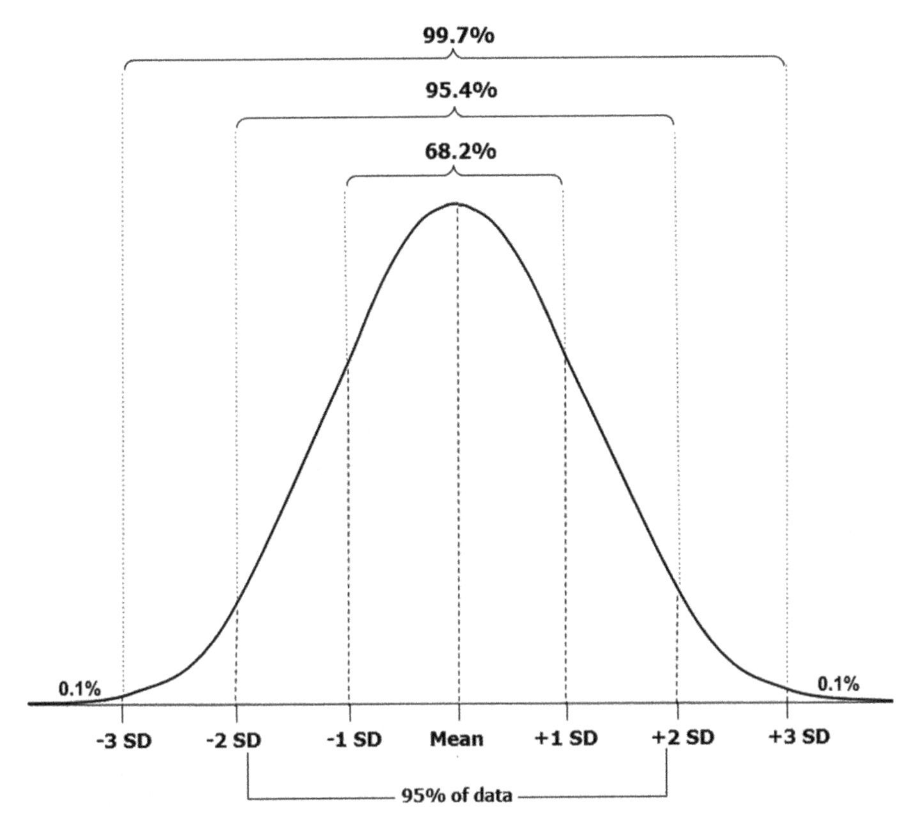

Fig. 29.4 The normal distribution. Note that the boundaries depicted by 1, 2 and 3 standard deviations (SD) on either side of the mean cover 68.2%, 95.4% and 99.7% of the area under the curve. The range covered by 1.96 SD on either side of the mean encompasses 95% of the data

three SD 99.7% of the total area under the curve. This is sometimes referred to as the empirical rule.

- 95% of the values lie within 1.96 standard deviations on either side of the mean. Hence this range is often taken as the normal range or **reference range** for many physiological variables.
- The kurtosis (shape or thickness of the tails) equals 3. The normal distribution is said to be mesokurtic.

If we look at the formula for the normal distribution, it is evident that two parameters define the curve, namely μ (the mean) and σ (the standard deviation):

$$f(x) = \frac{1}{\sigma\sqrt{2\pi}} e^{-\frac{1}{2}(x-\mu)^2/\sigma^2}$$

The standard normal distribution curve is a particular normal distribution for which probabilities have been calculated. It is a symmetrical bell-shaped curve with a mean of 0 and a variance (or standard deviation) of 1. This is also known as the **z distribution**, since **standardized normal deviates** or **z scores** of a random variable x can be calculated using this distribution as follows:

$$z = \frac{x-\mu}{\sigma}$$

The z value thus tells us how many standard deviations the corresponding value of x lies above or below the mean of the normal distribution. Tables of z scores (in statistics books) can be used to find out what proportion of any normal distribution lies above any given z score, and not just z scores of 1, 2, or 3. We can also do the converse, that is use z scores to find the score that divides the distribution into specified proportions. Z scores also allow us to determine the probability of a randomly picked element being above or below a particular score.

As the number of observations increases (say, n > 100), the shape of a sampling distribution will approximate a normal distribution curve even if the distribution of the variable in question is not normal. This is implied by the central limit theorem and is one reason why normal distribution is so important in biomedical research.

Many statistical techniques require an assumption of the normality of the dataset. The sample data do not need to be normally distributed, but they should represent a population that is normally distributed.

29.8 Presenting Data

Once summary measures of data have been calculated, they need to be presented in tables and graphs. Regarding data presentation in tables, it is helpful to remember the following:

- The mean is to be used for numerical data and symmetric (non-skewed) distributions.
- The median should be used for ordinal data or numerical data if the distribution is skewed.
- The mode is generally used only for examining bimodal or multimodal distributions.
- The range may be used for numerical data to emphasize extreme values.
- The standard deviation is to be used along with the mean.
- Interquartile range or percentiles should be used along with the median.
- Standard deviations and percentiles may also be used when the objective is to depict a set of norms ('normative data').
- The coefficient of variation may be used if the intent is to compare variability between datasets measured on different numerical scales.
- 95% confidence intervals should be used whenever the intent is to draw inferences about populations from samples.

For presenting data graphically, it is usually necessary to obtain the frequency distribution or relative frequency distribution (e.g., percentages) of the data. This can then be utilized to draw different types of graphs (or charts or plots or diagrams).

Pie chart depicts the frequency distribution of categorical data in a circle (the 'pie'), with the sectors of the circle proportional in size to the frequencies in the respective categories. A particular category can be emphasized by pulling out that sector. All sectors are pulled out in an 'exploded' pie chart. Pie charts can be made highly attractive, by using color and three-dimensional design enhancements, but become cumbersome if there are too many categories.

Bar chart (also called **column chart**) depicts categorical or discrete numerical data as a series of vertical or horizontal bars, with the bar heights being proportional to the frequencies. The separation between bars is of little significance other than to indicate that the bars denote discrete values or categories. The separation distance is usually kept equal. Bars depicting subcategories can be stacked one on top of another (compound, segmented or stacked bar chart). Two or more data series can be depicted on the same bar chart by placing corresponding bars side by side—different patterns or colors are used to distinguish the different series (clustered or multiple bar chart).

Histogram resembles a bar chart but is used for summarizing continuous numerical data and hence there are no gaps between the bars. The bar widths correspond to the class intervals. The alignment of the bars can be vertical or horizontal. A histogram is popularly used to depict the frequency distribution in a large data series. Accordingly, the class intervals should be chosen so that the bars are narrow enough to illustrate patterns in the data but not so narrow as to become too large in number. A histogram should be labeled carefully to clearly depict where the boundaries lie.

Dot plot (Fig. 29.5) depicts frequency distribution like histograms and can also be used for summarizing discrete numerical data. Instead of bars, it has a series of dots for each value or class

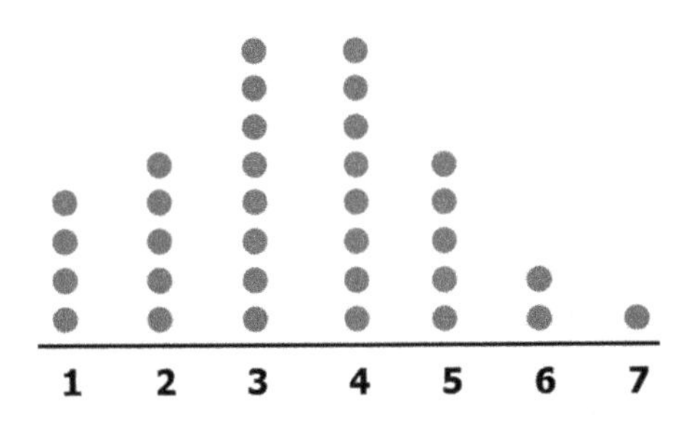

Fig. 29.5 Example of a dot plot. The plot is depicting the days on which subject recruitment occurred in a clinical study. Obviously, there is minimum recruitment over the weekend

interval—each dot representing one observation. The alignment can be vertical or horizontal. Dot plots are conceptually simple but become cumbersome for large data sets.

Stem-and-leaf plot (Fig. 29.6) is like a combination of a diagram and a table. It can depict frequency distribution, as well as individual values for numerical data. The data values are examined to determine their last significant digit (the 'leaf' item) and this is attached to the previous digits (the 'stem' item). The stem items are usually arranged in ascending or descending order vertically and a vertical line may be drawn to separate the stem from the leaf. The number of leaf items should total up to the number of observations. However, it becomes cumbersome with large data set.

STEM	LEAF
01	9
02	4 5
03	2 3 6 8
04	1 2 5 7 7
05	1 1 2 2 4 5 7 7 8
06	1 3 3 3 5 5 6 8 8 9 9
07	2 2 4 5 5 6 6 6 7 8 8
08	4 7 8 8
09	2 5
10	0

Fig. 29.6 Example of a stem-and-leaf plot. The plot is depicting examination scores (out of 100) for 50 students. In this case, the stem value is multiplied by 10 and added to the unitary leaf value to get an individual score. Note that both individual scores and the shape of data distribution can be discerned from the plot

Box-and-whiskers plot (or **box plot**) is a graphical representation of numerical data based on a five-figure summary—minimum value, 25th percentile, median (50th percentile) value, 75th percentile and maximum value (Fig. 29.7). A rectangle is drawn extending from the lower quartile to the upper quartile, with the median dividing this 'box' but not necessarily equally. Lines ('whiskers') are drawn from the ends of the box to the extreme values. Outliers may be indicated beyond the extreme values by dots or asterisks—in such 'refined' box plots, the whiskers have lengths not exceeding 1.5 times the interquartile range. The plot may be presented horizontally or vertically. Box plots are ideal for summarizing large samples. Multiple box plots arranged side by side, allow ready comparison of data sets.

In addition to these commonly used plots used for summarizing data and depicting underlying patterns, many other plots are used in biostatistics for depicting data distributions, time trends in observations, relationships between two or more variables, exploring goodness-of-fit to hypothesized data distributions and drawing inferences by comparing data sets.

29.9 The Need for Inferential Statistics

Descriptive statistics, such as the central location and the dispersion, mostly provide information about a sample data set that we have studied. However, we are often interested in understanding entire populations rather than just a sample drawn from them. This is the arena of **inferential statistics**, the need for which arises out of the fact that a sample, even though representative of a population, is not expected to represent the population perfectly and that any sampling strategy naturally incurs some sampling error. Although measures used in descriptive statistics, such as the 95% confidence interval (CI), also allow us to draw inferences regarding populations, for the most part, statistical inference means making and testing propositions about populations (called

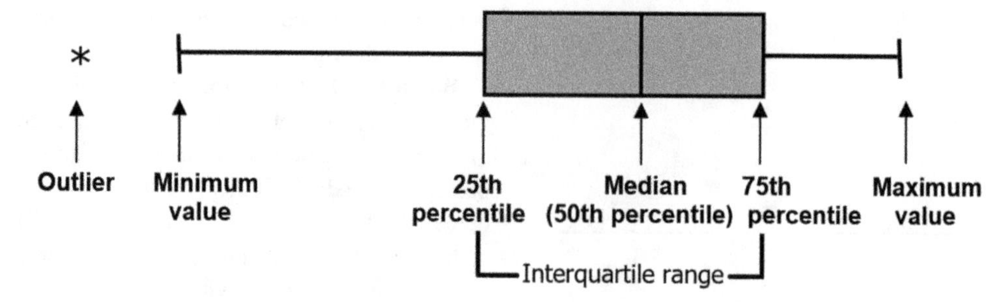

Fig. 29.7 Example of a box-and-whiskers plot. The plot provides a ready 5-figure summary of numerical data and can also depict outliers

hypotheses) using data obtained from a section of the population of interest via some form of sampling. Appropriate sample size and sampling strategy are necessary to ensure that the sample is truly representative of the population under consideration.

29.10 Hypothesis Testing

Much of medical research begins with a research question that can be framed as a hypothesis (see Chap. 1). The statistical convention is, to begin with, a **null hypothesis** that reflects the conservative position of no change or no difference in comparison to the baseline or between groups. In contrast to this, the researcher believes that there is some effect—indeed this is usually the reason for the study in the first place! The researcher, therefore, proceeds to study samples and measure outcomes in the hope of finding evidence strong enough to be able to reject the null hypothesis.

While putting the null hypothesis to the test, two types of error may occur (Fig. 29.8):

- **Type I error** is that made in incorrectly rejecting the null hypothesis. The real situation is that the null hypothesis is true and there is no difference between the two data sets that are being compared; however, the conclusion reached is that there is a difference. This is akin to a false positive error, and its probability is denoted as α.
- **Type II error** is made in incorrectly accepting the null hypothesis. The real situation is that there is a difference that has unfortunately not been found. This is a false negative error, and its probability is denoted as β.

Rather than speak in terms of Type II error, it may be more intuitive to speak in terms of the probability of correctly rejecting the null hypothesis when it is false or, in other words, the probability of detecting a real difference where it does exist. This is denoted by the quantity $(1 - \beta)$ and is called the **power** of the study.

Type I and Type II error probabilities bear a reciprocal relationship and, for a given sample size, both cannot be minimized at the same time. Hence, the strategy is to strike an acceptable balance between the two a priori. Conventionally, this is done by setting the acceptable value of α at 0.05 (i.e. 5%) or less and the value of β at 0.2 (i.e. 20%) or less. The latter is more usually expressed as a power of at least 80%. The chosen values of α and β affect the sample size that needs to be studied—the smaller the values, the larger the size. In addition, the sample size is affected by the standard deviation of the outcome measure in the population in question and the smallest difference that is clinically important and is therefore sought to be detected.

The concept of the ***p* value** is almost universally used in hypothesis testing. It denotes the maximum probability of getting the observed result assuming the null hypothesis to be true. Expressed alternatively, it is the maximum

Fig. 29.8 The difference between Type I and Type II errors. This is illustrated in the context of what is applicable regarding the null hypothesis (in columns) and the decision reached (in rows)

	Null Hypothesis is TRUE	Null Hypothesis is FALSE
Reject null hypothesis	**Type I error (False positive)**	**CORRECT**
Fail to reject null hypothesis	**CORRECT**	**Type II error (False negative)**

probability of getting the observed outcome by chance. Statisticians have agreed that a 1 in 20 chance of an accidental result is acceptable in the real world—thus the cut-off for the *p* value is taken at 0.05 (i.e. 5%). One can be stricter than this and settle for a cut-off of say 0.01 (i.e. 1%) but a more lenient p value is unacceptable. Ideally, all tests should be done in a **2-tailed** situation ('tails' refers to the ends of a distribution curve) implying that no presumption is made regarding the direction of change or difference if it does exist. A one-tailed test is more powerful in detecting a difference, but it should not be applied unless one is sure that change is possible only in one direction.

Hypothesis testing, as it stands now, proceeds through the following steps:

- Select a research design and sample size (see the previous chapter) appropriate to the hypothesis to be tested.
- Decide the test of statistical significance that is to be applied.
- Apply the test and determine the p value from the results one has observed.
- Compare it with the critical value of *p*, say 0.05 or 0.01.
- If the *p* value is less than the critical value, reject the null hypothesis (and rejoice) or otherwise accept the null hypothesis (and reflect on the reasons why no difference was detected).

With the increasing availability of computers and access to statistical software the drudgery involved in statistical calculations is now a thing of the past. There is also no need to look up mysterious statistical tables to get an approximate *p* value from the test statistic that is laboriously calculated. The biomedical researcher is therefore free to devote one's energy to optimally designing the study, selecting the tests to be applied based on sound statistical principles and taking care in conducting the study well. Once this is done, the computer will work on the data that is fed into it and take care of the rest. The argument that statistics is tedious and time consuming is absolutely no excuse now for not doing the appropriate analysis.

29.11 Selecting the Hypothesis Test to Be Applied [7, 8]

While computers and statistical software make the actual calculation part easy, the daunting part is to select the right statistical test of inference to be applied. Over 100 different tests have been described in biomedical literature. Fortunately, most hypotheses can be tackled through a much more limited basket of tests. To help in selecting the appropriate test, it is useful to think of the study in terms of some generic research questions rather than the study specific research questions. Most situations would be covered by just five such generic questions:

Question 1. Is There a Difference Between Groups: Unpaired (Independent Groups) Situation?

Numerical data		Categorical data	
Parametric	Otherwise	2 groups	> 2 groups
2 groups **Unpaired Student's *t* test** >2 groups **Analysis of variance (ANOVA)** or ***F* test**	2 groups **Mann-Whitney *U* test** **Wilcoxon's rank sum test** >2 groups **Kruskal-Wallis *H* test (Kruskal-Wallis ANOVA)**	**Chi-square [χ^2] test** **Fisher's exact test**	**Chi-square [χ^2] test**

Post-hoc (multiple group comparison) tests are to be applied, if ANOVA or its non-parametric counterpart shows a significant difference, to detect between which two groups the significant difference lies. Examples of such tests are:

- Parametric data: **Tukey's honestly significant difference test (Tukey-Kramer test), Student-Newman-Keuls test, Dunnett's test, Scheffe's test, Bonferroni's test**, etc.
- Non-parametric data: **Dunn's test**.

Data sets are considered to be independent if there is no possibility of values in one set being related to or influencing the other. This is not always obvious. For instance, will intraocular pressure in one eye influence that in the other when only one eye gets treated and the other serves as control—if it does then measurements from the two eyes do not constitute independent data sets.

Question 2. Is There a Difference Between Groups: Paired Situation?

Numerical data		Categorical data	
Parametric	Otherwise	2 groups	>2 groups
2 groups **Paired Student's *t* test** >2 groups **Repeated measures ANOVA**	2 groups **Wilcoxon's matched pairs signed rank test** >2 groups **Friedman's ANOVA**	**McNemar's χ^2 test**	**Cochran's *Q* test**

Once again, post-hoc (multiple group comparison) tests are to be applied, if repeated measures ANOVA or its non-parametric counterpart shows a significant difference, to deduce between which two data sets the significant difference lies. These include:

- Parametric data: Tukey's test.
- Non-parametric data: Dunn's test

There is often confusion over which data sets to treat as paired. The following iteration provides a guide:

- Subjects are recruited in pairs, deliberately matched for key potentially confounding variables such as age, sex, disease duration, disease severity, etc. One subject gets one treatment, while his paired counterpart gets the other.
- Before-after or time series data: A variable is measured before an intervention, and the measurement is repeated at the end of the intervention. There may be periodic interim measurements as well. The investigator is interested in knowing if there is a significant change from the baseline value with time.
- A crossover study is done, and both arms receive both treatments, though at different times. The comparison needs to be done within a group in addition to between groups.
- Variables are measured in other types of pairs e.g., right-left, twins, parent-offspring, etc.

Question 3. Is There an Association Between Variables?

Numerical data		Categorical data	
Both variables parametric	Otherwise	2 × 2 data	Otherwise
Pearson's (product moment) correlation coefficient *r*	**Spearman's (rank) correlation coefficient ρ** **Kendall's (rank) correlation coefficient τ**	**Relative risk [Risk ratio]** **Odds ratio [Cross-products ratio]**	**Chi square for trend** **Log linear analysis** **Logistic regression**

If two numerical variables are linearly related to one another, a **simple linear regression** analysis (by least squares method) enables the generation of a mathematical equation to allow the prediction of one variable, given the value of the other. **Multiple linear regression** generalizes this to more than one numerical variable linearly related to one numerical outcome variable.

Question 4. Is There Agreement Between Assessment Techniques?

Numerical data	Categorical data
Intraclass correlation coefficient (quantitative method) **Bland-Altman plot** (graphical method)	**Cohen's Kappa statistic**

Note that assessment techniques may refer to laboratory screening or diagnostic techniques, clinical diagnosis and various types of rating.

Question 5. Is There a Difference Between Survival (Time to Event) Trends?

2 groups	> 2 groups
Cox-mantel test **Gehan's (generalized Wilcoxon) test** **Log-rank test**	**Peto and Peto's test** **Log-rank test**

Note that survival data are always treated as non-parametric. There is also likely to be censoring with such data, meaning that complete information over time may not be available for all subjects. Survival (or time to event) trends are commonly depicted by **Kaplan-Meier plots** (see Chap. 8) and two or more such plots can be compared by the versatile **log-rank test.**

It is evident from the above schemes that it is important to distinguish between parametric tests (applied to data that are normally distributed) and non-parametric tests (applied to data that are not normally distributed or whose distribution is unknown). Parametric tests assume that:

- Data are numerical
- The distribution in the underlying population is normal
- Observations within a group are independent of one another
- The samples have been drawn randomly from the population
- The samples have the same variance ('homogeneity of variance')

If it is uncertain whether the data are normally distributed or not, they can be plotted as a **histogram** or the **quantile versus quantile (QQ) plot** and visually inspected, or tested for normality, using one of several **goodness-of-fit tests** such as the **Shapiro-Wilk, Kolmogorov-Smirnov, Lilliefors, Anderson-Darling** or **D'Agostino-Pearson tests** [9, 10]. These tests compare the sample data with a normal distribution and derive a p value; if $p < 0.05$ then the null hypothesis that there is no difference between a normal distribution and the sample distribution stands rejected and the data cannot be normally distributed. The Shapiro-Wilk test is considered the most powerful among all the tests listed above in detecting departures from normality.

Non-normal or skewed data can be transformed so that they approximate a normal distribution. The commonest method is a **log transformation**, whereby the natural logarithms of the raw data are analyzed. If the transformed data are shown to approximate a normal distribution, they can then be analyzed with parametric tests. Large samples (say $n > 100$) approximate a normal distribution and can nearly always be analyzed with parametric tests. This assumption often holds even when the sample is not so large but say is not less than 30. However, with the increasing availability of non-parametric analysis in computer software, data transformation is now seldom required, and it is more apt to apply the appropriate non-parametric test to skewed data.

The requirement for observations within a group to be independent means that multiple measurements from the same set of subjects cannot be treated as unrelated sets of observations. Such a situation requires specific **repeated measures analysis**. The requirement for samples to be drawn randomly from a population and to have the same variance are not always satisfied but most commonly used hypothesis tests are robust enough to give reliable results even if these assumptions are not fully met.

29.12 Merits and Demerits of the *p* value [11]

The traditional practice is to treat results as statistically 'significant' or 'non-significant', based on the *p* value being smaller than some prespecified critical value, commonly 0.05. This practice is now becoming increasingly obsolete, and the use of exact *p* values (say to 3 decimal places) is now preferred. The use of statistical software allows the ready derivation of such exact *p* values, unlike in the past when statistical tables had to be used. Further, the use of a cut-off for statistical significance can foster the misleading notion that a 'statistically significant' result is the real thing. However, recall that a *p* value of 0.05 means that one out of 20 results would show a difference at least as big as that observed just by chance. Thus, a researcher who accepts a 'significant' result as real will be wrong 5% of the time (committing a type I error). Similarly, dismissing an apparently 'non-significant' finding as a null result may also be incorrect (committing a type II error), particularly in a small study, in which the lack of statistical significance may simply be due to the small sample size rather than the real lack of effect. Both scenarios have serious implications in the context of accepting or rejecting a new treatment. The presentation of exact *p* values allows one to make an informed judgment as to whether the observed effect is likely to be due to chance and this, taken in the context of other available evidence, will result in a better conclusion being reached.

The *p* value does not clearly indicate the clinical importance of an observed effect. A small *p* value simply indicates that the observed result is unlikely to be due to chance. However, just as a small study may fail to detect a genuine effect, a large study may yield a small *p* value (and a very large study a very small *p* value) based on a small difference that is unlikely to be of clinical importance. For instance, a large study of fasting blood glucose lowering treatment may show a lowering of 5 mg/dL as statistically 'highly significant' i.e., $p < 0.01$. This is obviously unlikely to help much in clinical practice. Decisions on whether to accept or reject a new treatment necessarily must depend on the observed extent of change, which is not reflected in the *p* value.

The *p* value provides a measure of the likelihood of a chance result, but it is worthwhile to express the result with appropriate confidence intervals. It is becoming the norm that an estimate of the size of any effect, expressed by its 95% CI, be presented for meaningful interpretation of results. A large study is likely to have a small (and therefore 'statistically significant') *p* value, but a 'real' estimate of the effect would be provided by the 95% CI. If the intervals overlap between two treatments, then the difference between them is not so clear-cut even if $p < 0.05$. Conversely, even with a non-significant *p*, CI values that are apart suggest a real difference between interventions. Thus the 95% CI does what the *p* value is telling us, and in addition, gives us further information on the likely range for the difference that is detected in the underlying population. Increasingly, statistical packages are getting equipped with the routines to provide 95% CI values for a whole range of statistics.

29.13 Concluding Remarks

Descriptive statistics remains local to the sample, describing its central tendency and variability, while inferential statistics focuses on making statements about the population. The central maxim of inferential statistics is a generalization from the sample to the population. To do this it resorts largely to the hypothesis testing approach whose goal is to reject the null hypothesis. Many hypothesis tests are available and the one to be applied depends upon the research question, the nature of the variables of interest and the number of data sets that are to be compared. This chapter has introduced several commonly used tests. The next chapter discusses the applications of most of the parametric and non-parametric tests introduced here. However, the assimilation of this knowledge can only occur through number crunching with one's own data sets. With the increasing availability and user friendliness of statistical software, this task is well within every biomedical researcher's reach.

Conflicts of Interest None.

References

1. Everitt BS (2006) Medical statistics from A to Z: a guide for clinicians and medical students. Cambridge University Press, Cambridge
2. Mishra P, Pandey CM, Singh U, Gupta A, Sahu C, Keshri A (2019) Descriptive statistics and normality tests for statistical data. Ann Card Anaesth 22:67–72
3. Chan YH (2003) Data presentation. Singap Med J 44: 280–285
4. In J, Lee S (2017) Statistical data presentation. Korean J Anesthesiol 70:267–276
5. Hazra A (2017) Using the confidence interval confidently. J Thorac Dis 9:4125–4130
6. Lee DK (2020) Data transformation: a focus on the interpretation. Korean J Anesthesiol 73:503–508
7. du Prel JB, Röhrig B, Hommel G, Blettner M (2010) Choosing statistical tests: part 12 of a series on evaluation of scientific publications. Dtsch Arztebl Int 107: 343–348
8. Nahm FS (2016) Nonparametric statistical tests for the continuous data: the basic concept and the practical use. Korean J Anesthesiol 69:8–14
9. Lee S, Lee DK (2018) What is the proper way to apply the multiple comparison test? Korean J Anesthesiol 71:353–360
10. Razali N, Wah Y (2011) Power comparisons of Shapiro-Wilk, Kolmogorov-Smirnov, Lilliefors and Anderson-Darling tests. J Stat Model Anal 2:21–33
11. Lee DK (2016) Alternatives to P value: confidence interval and effect size. Korean J Anesthesiol 69: 555–562

Principles and Applications of Statistics in Biomedical Research: Parametric and Nonparametric Tests Including Tests Employed for Posthoc Analysis

30

S. Manikandan and Suganthi S. Ramachandran

Abstract

The statistical analysis for comparing two or more groups is divided into parametric and nonparametric tests. Parametric tests use probability distribution to estimate the P value. Parametric tests are in general more robust than nonparametric tests. Every parametric test has a corresponding nonparametric test. Parametric tests have three assumptions viz.—random sampling, the groups have equal variance and data follows the normal distribution. If assumptions are not satisfied, nonparametric tests can be used. t test (independent, paired) and analysis of variance (ANOVA)—one-way and repeated measures are parametric tests. Wilcoxon test, Kruskal Wallis test and Friedman's test are the respective nonparametric counterparts. The Chi square test is used to evaluate the association between categorical variables. The construction of contingency table is an important step in chi square test. The strength of association in a chi square test is provided by the odds ratio (retrospective study) or relative risk (prospective study). ANOVA is used when three or more groups/datasets have to be compared. ANOVA provides a global assessment, and the individual comparisons are given by the post-hoc tests. Correlation can be done to study the association between two quantitative variables and regression analysis is performed when you want to predict one variable from the other. It is advisable to report the confidence interval so that the clinician can know the magnitude of the expected effect in the population. The key objective of this chapter is to explain the principles of various statistical tests and the interpretation of their results.

Keywords

Wilcoxon test · t test · ANOVA · Chi square test · normal distribution · contingency table · expected frequency

30.1 Introduction

Current medical research is dependent on statistics from the early stages (designing the study type, estimation of sample size) to the final analysis of data and publication of results. Statistical methods can be broadly divided into descriptive statistics and inferential statistics. Descriptive statistics refers to describing, organizing and summarizing data. Inferential statistics involves making inferences about a population by analysing the observations of a sample. Hypothesis testing is a form of inferential statistics.

S. Manikandan (✉) · S. S. Ramachandran
Department of Pharmacology, Jawaharlal Institute of Postgraduate Medical Education and Research (JIPMER), Pondicherry, India
e-mail: manikandan001@yahoo.com; drsugiram@gmail.com

G. Jagadeesh et al. (eds.), *The Quintessence of Basic and Clinical Research and Scientific Publishing*,
https://doi.org/10.1007/978-981-99-1284-1_30

Table 30.1 How to choose a statistical test?

			EXPOSURE VARIABLE					
			C1				C2	C3
			Comparison				Association of two variables	Regression analysis
			of 2 datasets		of > 2 datasets			
			Paired	Unpaired	Paired	Unpaired		
			C1a	C1b	C1c	C1d		
OUTCOME VARIABLE	R1	Normally distributed continuous data (Summarized as means)	Paired t test	Unpaired t test	Repeated Measures ANOVA[a]	One-way ANOVA[a]	Pearson correlation	Linear regression
	R2	Scores, ranks and non-normally distributed continuous data (Summarized as medians)	Wilcoxon signed rank test	Wilcoxon Rank Sum test / Mann Whitney U test	Friedman test[a]	Kruskal – Wallis test[a]	Spearman's Rank correlation	Non-parametric regression
	R3	Dichotomous data (Summarized as proportions)	McNemar's test	Chi square test or Fisher Exact test	Cochran's Q test[a]	Chi square test or Fisher Exact test[a]	Contingency coefficient	Logistic regression

Reproduced with modifications from Raveendran R, Gitanjali B, Manikandan S. A Practical Approach to PG dissertation. 2nd ed. Hyderabad: PharmaMed Press. 2012

[a]If the result turns significant, a suitable post-hoc test should be performed (*see* Sect. 30.6).

Statistical tests involving two groups are the most commonly used univariate analysis. A clear understanding of the principles and concepts underlying these statistical tests is essential for the conduct and interpretation of results. Table 30.1 summarizes the parametric and non-parametric tests enumerated in this chapter. It also provides a simplified way of choosing an appropriate statistical test for a given set of variables, study design and type of analysis [1]. Parametric tests are listed in the first row (R1). Non-parametric tests are listed in a row (R2). Tests for analysis of categorical variables are listed in Row (R3). If one knows the type of analysis and exposure and outcome variables, then the statistical test can be identified by looking at a specific row and cell to arrive at the statistical test. Readers are advised to read related Chaps. 28 and 29.

At the end of this chapter, readers should be able to

1. List the various parametric and nonparametric statistical tests.
2. Explain the principle of various statistical tests and their assumptions.
3. Enumerate the methods to check the assumptions and interpret the results of the analysis.

30.2 Parametric Tests for Comparing Two Groups/Datasets

30.2.1 Etymology

A numerical value that describes the population is known as a parameter and that which describes the sample is called a statistic [2]. A mnemonic to remember this is provided in Box 30.1. For example, the mean of the population is a parameter whereas the mean of the sample is a statistic.

Box 30.1

Mnemonic
P for P and S for S
Parameter—**P**opulation; **S**tatistic—**S**ample

Statistical tests which estimate the underlying population's parameter and then use it to test the null hypothesis are known as parametric tests [3]. These tests are based on the probability distribution for making inferences. Statistical tests which are not based on probability distribution and do not test hypotheses concerning parameters are called nonparametric tests. These are also known as distribution free tests.

30.2.2 Parametric Tests

There are various parametric tests, and they are categorized according to the number of groups/dataset and the design of the study. Figure 30.1 provides a flowchart for selecting various parametric tests.

30.2.3 Student's t Test

30.2.3.1 History

Student's t test was developed by William S Gossett who was employed in a brewery (Arthur Guiness Son & Co) in Dublin, Ireland. He was entrusted with quality control in the brewery and he wanted to perform this with a small number of samples. As Guiness Brewery prohibited its employees from publishing the work done in the brewery, he published it under the pseudonym 'Student'. There is a speculation that he conducted these works during the afternoon breaks and hence he named the distribution 't' (tea) distribution [4].

30.2.3.2 Principle

t test is carried out when the mean of two groups/datasets are to be compared (see below Sect. 30.2.3.3). The difference in the mean between the two groups/datasets with relation to the

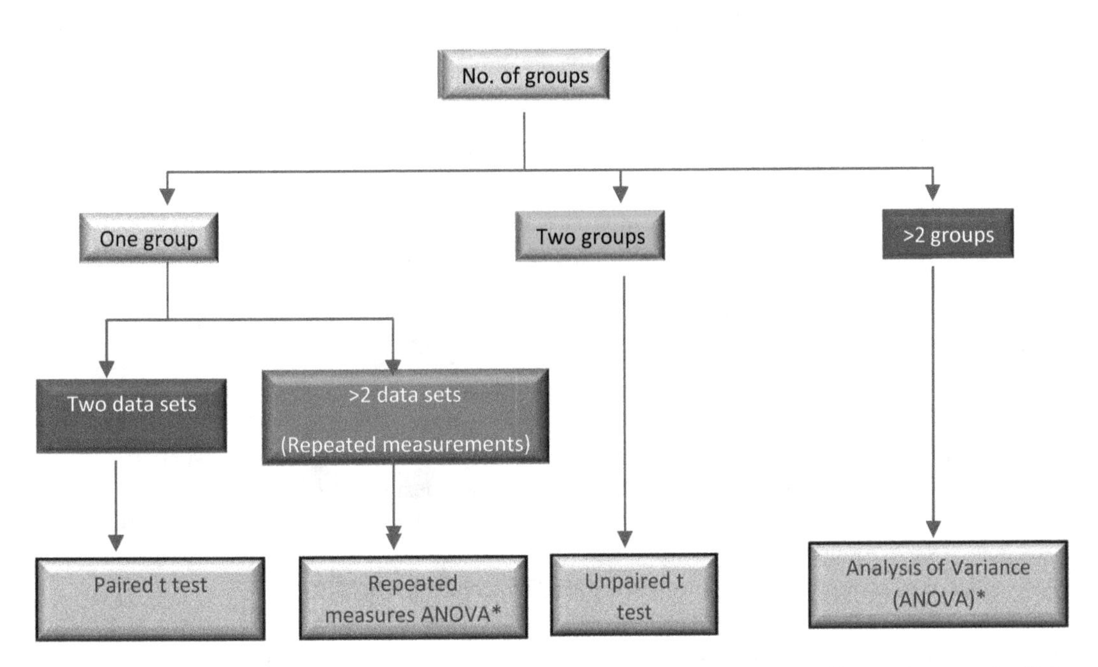

Fig. 30.1 Parametric tests for analysis of data. * if the result turns significant, a suitable post-hoc test should be performed (see Sect. 30.6)

standard error of the difference is calculated. If this ratio is more than or equal to 1.96, then it is less likely to occur by chance ($P<0.05$) [5].

30.2.3.3 Criteria for Performing t Test

Student (independent) t test can be done under the following circumstances.

Outcome variable: Numerical continuous (summarized as mean)

Analysis type: Comparison of mean

Number of groups: two groups/datasets

Study design: Unpaired (two independent groups)

Distribution of data: Normal distribution

30.2.3.4 Assumptions of t Test

All parametric tests have some assumptions. There are three assumptions of t test. They are

1. The samples are selected from the respective population randomly.
2. The variances (square of standard deviation) of both groups are equal.
3. The variable follows a normal distribution.

30.2.3.5 Methods for Checking the Normal Distribution

A frequency histogram of the observed data may be created for each variable. Then the normal distribution curve may be superimposed on it. Now by comparing these two visually, we can easily find out whether the data follows normal distribution [6]. There are statistical tests like the Kolmogorov Smirnov test, Shapiro Wilk test for checking the normality of data. These tests have their own assumptions. So, using one significance test conditional on another significance test is not recommended [5].

Departure from the symmetry of the distribution is called skewness. Skewed distribution can be detected from the summary statistics. If the mean is smaller than twice the standard deviation, then the data are likely to be skewed.

30.2.3.6 The Way Out for Assumptions

Most often convenience sampling is followed (and not random sampling) in medical research. If the intervention is randomly allocated (randomization), then the difference in mean between the two groups behaves like the difference between two random samples [7]. Hence if randomization is followed, the assumption of random sampling will be satisfied.

The same argument placed for checking the normality of data holds good for checking equality of variance. As a rule of thumb, if the ratio of the two standard deviations (larger standard deviation ÷ smaller standard deviation) is greater than 2, then it may be considered that the variance is not equal. Most of the software for statistical analysis provides output for equal and unequal variance. So, if the variances of the two groups are not equal, the output for an unequal variance may be considered.

If the data is not extremely skewed and if the number of samples (sample size) is the same in both groups, then t test will be valid. If the data is extremely skewed and if we wish to perform t test, then transformation of the data should be attempted. Log transformation is preferred among all transformations as back transformation is possible and meaningful [8]. If the data does not follow normal distribution even after transformation, then a nonparametric test should be done.

30.2.4 Paired t Test

30.2.4.1 Conditions for Conducting a Paired t Test

The paired t test is used when there is one group and two datasets (before and after intervention).

Paired t test is carried out under the following circumstances.

Outcome variable: Numerical continuous (summarized as mean)

Analysis type: Comparison of mean

Number of groups: One group and two datasets (baseline and after intervention)

Study design: Paired (matched)

Distribution of data: Normal distribution

30.2.4.2 Principle

In this type of analysis, every participant serves as his/her own control. The observations are paired as both baseline and after intervention measurements are made on the same subject. Thus, interindividual variation is eliminated in this type of study. The difference between the baseline and after intervention measurement is estimated. Then the standard deviation of the difference is calculated (the standard deviation of the difference is not equal to the difference between the standard deviations of the baseline and after intervention data). For the mean difference the t statistic, significance level, and 95% confidence interval are calculated.

30.2.4.3 Applications of Paired t Test

Apart from its application for analysing two datasets from one group of individuals (baseline and after intervention), paired t test may be used to analyse data from studies that have eliminated the interindividual variability. For example, in patients having a bilateral fungal infection of the hand, the intervention is administered on one side (say left side) and the standard therapy (comparator) on the other side. These two datasets are analysed by paired t test.

30.2.5 Example for t Test (Unpaired and Paired)

A pilot study was conducted in patients with type 2 diabetes mellitus to evaluate the efficacy of a new hypothetical drug jipizide as compared to glipizide. The drug was administered for one month. The postprandial blood glucose levels are provided in Table 30.2. The baseline postprandial glucose and the postprandial glucose after one month of therapy are measured in the study. For analyzing this data, we need to create a new variable viz. reduction in postprandial glucose after one month of therapy. Student t test has to be performed for comparing the reduction in postprandial blood glucose between the two groups. The values of postprandial glucose obtained after one month of therapy should not be used as such for analysis as this depends upon the baseline value and the baseline value is different in each patient. The baseline postprandial glucose level and the level after one month of therapy within a group may be compared by a paired t test to assess if the drug is effective. If the efficacy has to be compared between the two groups, then the student t test has to be used.

30.2.6 Interpretation of Results

Before interpreting the results, we need to make sure that the assumptions are satisfied. The P value indicates the probability of the results occurring by chance alone. If the P value is less than 5% (0.05), it is considered that there is a significant difference between the two groups/ interventions (But still there is a 5% probability

Table 30.2 The antidiabetic effect of jipizide in patients with type 2 diabetes mellitus

Glipizide Therapy—Postprandial glucose (mg/dL)			Jipizide Therapy—Postprandial glucose (mg/dL)		
Baseline	After 1 month of therapy	Reduction in postprandial glucose after 1 month of therapy	Baseline	After 1 month of therapy	Reduction in postprandial glucose after 1 month of therapy
220	200	20	230	193	37
180	156	24	190	154	36
174	146	28	164	122	42
160	131	29	158	119	39
190	160	30	210	163	47
240	208	32	224	183	41
200	176	24	189	150	39
185	163	22	188	148	40
158	122	36	166	122	44
210	184	26	220	180	40

N = 10 in each group

that this can happen purely by chance). If we are concluding that there is no significant difference ($P > 0.05$), we should calculate the power and confirm that the study was adequately powered (>0.8 or 80%) to detect the difference.

A significant P value (statistical significance) does not mean clinical or biological significance. When data from groups with large sample sizes are tested, even a small difference will be picked up as statistically significant which need not be clinically significant. For example, in a clinical trial involving 10,000 participants, the new drug produced a mean decrease in systolic blood pressure by 4 mm Hg compared to the reference drug ($P < 0.05$). Now the reduction in systolic blood pressure is statistically significant but not clinically significant as the mean reduction in systolic blood pressure is just 4 mm Hg.

P value does not provide any information regarding the effect size. But 95% confidence interval indicates that there is a 95% chance of including the population parameter in the given interval. For example, if the 95% confidence interval of the mean difference in systolic blood pressure is −12 to −5 mm Hg. It means that if the new drug is used in the real-world population (hypertensives), then we can expect a reduction in systolic blood pressure by 5 to 12 mm Hg.

30.3 Nonparametric Tests for Comparing Two Groups/Datasets

30.3.1 Nonparametric Tests

Nonparametric tests are to be used when the population distributions do not follow a normal distribution. These tests are simple to perform and do not have any assumptions about the population distribution. This does not mean that we can do away with the parametric tests and use the nonparametric tests for all data. As parametric tests are more powerful, they are preferred over nonparametric tests [9]. The various nonparametric tests are mentioned in Fig. 30.2. These are the corresponding nonparametric tests for the parametric tests provided in Fig. 30.1.

30.3.2 Etymology

If the variable in the population does not follow a normal distribution, then we need to use nonparametric tests. These tests are not based on probability distribution. For any meaningful inference about the population, we need to compare the parameters. Hence nonparametric test is a misnomer.

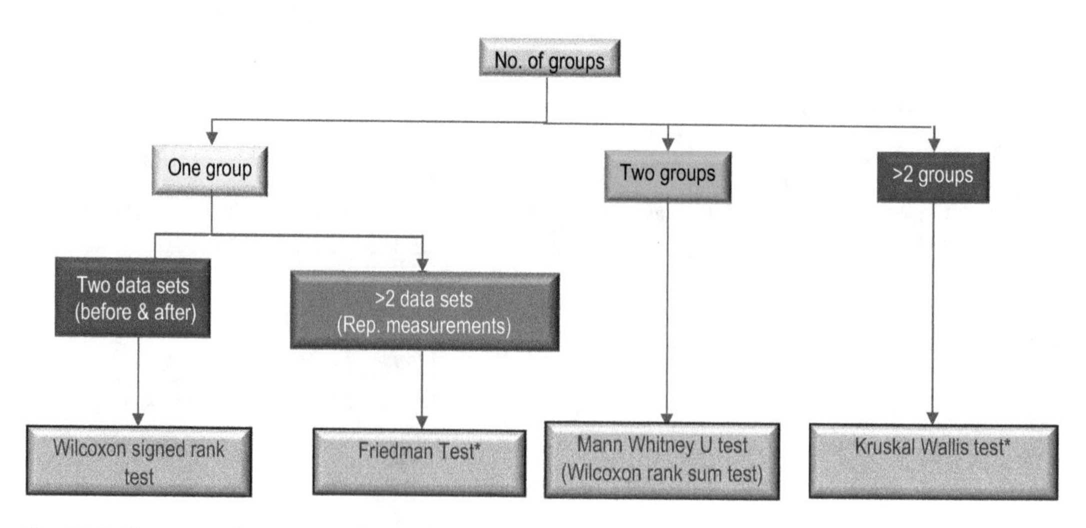

Fig. 30.2 Nonparametric tests according to the number of groups/datasets.* if the result turns significant, a suitable post-hoc test should be performed (see Sect. 30.6)

30.3.3 Wilcoxon Test

30.3.3.1 Types

Based on the number of groups, Wilcoxon test can be of two types viz.—Wilcoxon rank sum test and the Wilcoxon signed rank test. A simple mnemonic for this is provided in Box 30.2. Mann and Whitney also described the rank sum test independently. By convention, the Wilcoxon test is now ascribed to paired data and the Mann Whitney U test to unpaired data.

Box 30.2

Mnemonic
U for U
Unpaired (two independent groups)—Wilcoxon rank sum test
or
Mann Whitney **U** test

30.3.3.2 Principle

The Mann Whitney U test checks whether the medians (as opposed to mean by t test) of the two independent groups are different. The first step is to arrange all the observations (both groups put together) in ascending or descending order. Then rank all these observations ignoring their group. When the observations/scores are equal, the average of the ranks is assigned to all the tied observations. After all the observations are ranked, they are groupwise arranged and the sum of the ranks of each group is calculated (hence the name rank sum test). These ranks are analysed as though they are the original observations to find the level of significance (P value) [10].

30.3.3.3 Example for Assigning Ranks

The adverse effects of a new inhaled beta-2-adrenergic receptor agonist for bronchial asthma are compared with a reference drug, salbutamol, in a pilot study of a sample size of 20 (10 in each group). The tremors are scored 0 to 5 (0, – no tremor; 5, – intense tremor). The observed data are provided in Table 30.3.

Arrange all observations in ascending order

0, 0, 1, 1, 1, 1, 1, 2, 2, 2, 2, 2, 2, 2, 3, 3, 3, 3, 4, 4

Assign ranks

1, 2, 3, 4, 5, 6, 7, 8, 9, 10, 11, 12, 13, 14, 15, 16, 17, 18, 19, 20

Find the average of the ranks for the tied observations and assign the average of the ranks of the tied observations.

1.5, 1.5, 5, 5, 5, 5, 5, 11, 11, 11, 11, 11, 11, 11, 16.5, 16.5, 16.5, 16.5, 18.5, 18.5

The rank assigned for each observation is mentioned in Table 30.4.

T_1—the sum of ranks of group 1

T_2—the sum of ranks of group 2

T—Smaller of T_1 and T_2

Mean sum of ranks (m) $= \{n_1(n_1 + n_2 + 1)\}/2$

$Z = \{\text{Modulus } (m - T) - 0.5\}/\text{S.D}$

If $Z < 1.96$, the null hypothesis is accepted ($P > 0.05$)

Table 30.3 The tremor score of the new drug as compared to salbutamol in patients with bronchial asthma

Patient. No	Tremor score	
	Salbutamol (n = 10)	New drug (n = 10)
1	2	1
2	2	1
3	3	2
4	4	3
5	0	2
6	1	2
7	2	1
8	1	0
9	3	2
10	4	3

Table 30.4 Rank for the data given in the example in Sect. 30.3.3.3

Patient No	Assigned Ranks	
	Salbutamol (n = 10)	New drug (n = 10)
1	11	5
2	11	5
3	16.5	11
4	18.5	16.5
5	1.5	11
6	5	11
7	11	5
8	5	1.5
9	16.5	11
10	18.5	16.5
Sum of ranks	114.5	93.5

If $Z > 1.96$, the null hypothesis is rejected at a 5% significance level.

If $Z > 2.58$, $P < 0.01$ and if $Z > 3.29$, $P < 0.001$

30.3.3.4 Wilcoxon Signed Rank Test

The Wilcoxon signed rank test is done to assess the difference in median between two datasets from a group (baseline scores and scores after intervention). The individual difference in score is calculated for each pair of observations. These are arranged in ascending order (ignoring their sign) and ranked accordingly. The ranking is done as explained previously in Sect. 30.3.3.2. The zero difference values are ignored [11]. Then the ranks of the positive signed difference values and the negative signed ones are separated into two groups (hence the name signed rank test). These are then analysed to find out the significance level.

30.3.3.5 Example for Wilcoxon Signed Rank Test

A pilot study was conducted to assess the reduction in salivary secretion by a new anticholinergic drug. Salivation was scored 1 to 10 (1, – completely dry; 5, – normal secretion; 10, – profuse secretion). The data and the steps in assigning ranks are provided in Sect. 30.3.3.3 (Table 30.5).

T+—sum of ranks of positive value = 40

T−—sum of ranks of negative values = 15

T—Smaller of T+ and T−

T = 15

Sum of positive and negative ranks = n (n + 1)/2

Table 30.5 Effect of a new anticholinergic on salivary secretion

S. No	Salivary secretion score			Rank (ignoring sign)	The rank of positive values (difference)	The rank of negative values (difference)
	Baseline	After drug	Difference			
1	5	3	2	5.5	5.5	
2	5	6	−1	2		2
3	6	2	4	9	9	
4	4	6	−2	5.5		5.5
5	4	3	1	2	2	
6	6	2	4	9	9	
7	5	1	4	9	9	
8	5	7	−2	5.5		5.5
9	7	8	−1	2		2
10	6	4	2	5.5	5.5	

Mean sum of ranks (m) = n(n + 1)/4

Z = {Modulus (T-m) − 0.5}/S.D

If Z < 1.96, the null hypothesis is accepted (P > 0.05)

If Z > 1.96, the null hypothesis is rejected (P < 0.05)

30.4 Parametric Tests for Comparing Three or More Groups/Datasets

30.4.1 Analysis of Variance (ANOVA)

Analysis of variance (ANOVA) is a statistical tool to be used when more than two independent group means are to be compared. In statistical terms, the exposure variable is categorical with more than two levels, and the outcome variable is numerical continuous one. It is a parametric test like t test and was developed by Fisher [12].

30.4.2 Principle

The principle behind ANOVA is to test the mean differences among the groups by estimating variance across the groups and variance within the group and hence the name analysis of variances. The ratio of these variances is represented by the F ratio. When the variability across the groups is greater than that of the variability within the group, we conclude that at least one group mean is significantly different from other group means. Thus, ANOVA is an omnibus test and tells only whether one group mean is different from the rest of the mean. If ANOVA returns significant result, a further post-hoc test should be performed to identify which specific pair/s of means are significantly different (See 6).

30.4.3 Assumptions of ANOVA

1. Each data point in the group is independent of the others and is randomly selected from the population. It should be planned at the design stage itself.
2. The data in each group follows a normal distribution.
3. There is a homogeneity of variance among groups i.e., variance is similar in all the groups. It is commonly assessed by doing Bartlett's test or Levene's test in statistical packages.

   These assumptions must be checked before performing the actual analysis.

30.4.4 Types of ANOVA

Depending on the number of factors in exposure variables, the ANOVA model can be a one-way, two-way, or multifactorial ANOVA (three-way ANOVA). If there is a relatedness of the datasets, then repeated measures ANOVA is used. In this section, one-way, two-way, and one-way repeated measures of ANOVA are discussed. Interested readers may refer to Doncaster and Davey [13] for learning other complex models [nested, split plot, mixed model ANOVA and analysis of covariance (ANCOVA)].

30.4.5 One-Way ANOVA

One-way ANOVA is performed when a continuous outcome variable needs to be compared across one exposure variable with more than two levels. It is called one-way because the exposure groups are classified by just one variable. It is also referred to as factor and hence one factor ANOVA. The experimental design corresponding to one-way ANOVA is termed a completely randomized design.

30.4.5.1 Example for One-Way ANOVA

A researcher wishes to evaluate the effect of three different treatments (drugs A, B, and C) on the reduction of blood cholesterol levels in mg/dL at the end of one month in cafeteria-diet induced obesity rat model against positive control drug statin and vehicle control. Table 30.6 provides the data set.

Table 30.6 Effect of treatment on change in blood cholesterol in experimental rats

Vehicle control (n = 6)	Drug A (n = 6)	Drug B (n = 6)	Drug C (n = 6)	Statin (n = 6)
2	11	30	40	42
5	5	35	41	50
10	10	32	50	42
6	8	31	42	42
5	9	35	45	45
5	11	30	46	49

- Outcome variable: Blood cholesterol (continuous)
- Number of exposures with their levels: One (Drug treatment, five levels)

Table 30.7 provides the output of ANOVA results. One-way ANOVA divides the total variation which is represented by the total sum of squares into two distinct components: the sum of squares (SS) due to differences between the group means and the sum of squares due to differences between the observations within each group (known as the residual sum of squares or residuals/error). The degree of freedom for between-group SS is k – 1, where k is the number of groups and the residual SS has a degree of freedom (n – k), where n is the total observations. The F ratio is estimated by dividing between group mean SS by within group mean SS under the abovementioned pairs of a degree of freedom. The F value test statistic is compared with the F distribution table to locate the critical threshold. If the test statistic exceeds the critical value, we reject the null hypothesis ($P < 0.05$).

30.4.5.2 Interpretation of Results

In the example scenario, since the P value is less than 0.05, we can infer that there are statistically significant differences in the mean decrease in cholesterol among the five groups as determined by one-way ANOVA ($F(4,25) = 238.35$, $P = 0.000$). It only points out that at least one group mean is significantly different from others. To identify which specific pair/pairs of means are different, post -hoc analysis needs to be done (Refer Sect. 30.6).

After performing a post hoc test, Tukey's honestly significant difference (HSD) (refer Sect. 30.6), further comparisons can be made. When compared to vehicle, the highest mean reduction in cholesterol was observed in the statin group [39.5 (CI: 34.0,44.5); $P < 0.001$] followed by drug C [38.5 (CI: 33.4, 43.5); $P < 0.001$] and drug B [26.7 (CI:21.6,31.7); $P < 0.001$]. Drug A did not show a mean reduction in cholesterol as compared to vehicle. The mean reduction in cholesterol observed with drug B is lower as compared to both drug C and statin.

30.4.6 Two-Way ANOVA

Two-way ANOVA is done to compare continuous variables across two exposure variables. In the above example of one-way ANOVA (Sect. 30.4.5.1), if the researcher wants to evaluate the gender effect (male/female) in addition to the treatment effect (five treatment groups), the data sets (5) would be stratified into ($5 \times 2 = 10$) data sets. The total variation would be partitioned for both the exposure factors and their interaction [12].

Table 30.7 Results of one-way Analysis of variance (ANOVA)

ANOVA table	Sum of Squares (SS)	Degree of freedom (Df)	Mean Sum of squares (MS) = (SS/Df)	$F = \frac{\text{Between group MS}}{\text{Within group MS}}$	P value
Between groups (Treatment)	8555.1	4	2138.8	238.35 (4,25)	0.0000
Within groups (Residuals)	224.33	25	8.973		

Table 30.8 Effect of drug treatment by antagonist on depletion of catecholamines in rabbit heart

	Placebo	Drug A	Drug B
Control	420, 500, 460, 520	2, 12, 9, 3	80, 130, 60, 160
Antagonist	460, 530, 440, 560	230, 280, 330, 200	360, 380, 260, 390

30.4.6.1 Example of Two-Way ANOVA

In an experiment, the effect of an antagonist on agonist-induced depletion of catecholamines measured in a nanomolar unit from the rabbit heart was studied. There were three levels of treatment namely, control (placebo), drug A and drug B, and each of these treatments was tested in the presence and absence of an antagonist. Thus, there were six (3 × 2) factor level combinations. Rabbits (4 per treatment) were randomly assigned to each of the six factor level combinations. We are interested to know any possible interaction between treatments and the antagonist in addition to the individual effects of each factor. The data are given in Table 30.8.

- Outcome variable: catecholamine levels in nanomolar units (continuous variable)
- Number of exposure variables with their levels (Factors): Two
  Factor 1: Treatment- three levels
  Factor 2: Presence of Antagonist—two levels

30.4.6.2 Interpretation of Results

As shown in Table 30.9, the F ratio is determined for treatment, antagonist and the drug and antagonist interaction. A statistically significant effect has been noted for both treatment and antagonist status, i.e., at least one of the two drugs had a significant effect on depleting catecholamines as compared to the untreated control. Also, the antagonist had a significant effect on the depletion of catecholamines induced by agonists relative to the placebo (control). It can be deciphered that drugs responded differently under the influence of antagonists since the interaction between agonist and antagonist status shows a statistically significant interaction.

30.4.7 Repeated Measures ANOVA (RM ANOVA)

Repeated measures ANOVA is considered as an extension to paired t test when there is one group and three or more related datasets. It is applied to situations where repeated measurements of the same variable are taken at different time points or under different conditions. The assumption that is specific to RM ANOVA is that the variances of the differences between all combinations of related groups must be equal. It is assessed by Mauchly's test of sphericity. If this assumption is violated, some corrections like Greenhouse-Geisser or Hunyh-Feldt needs to be done to avoid inflating type II error.

30.4.7.1 Example of One-Way Repeated Measures ANOVA

A physician started prescribing a new antihypertensive available on the market. He wants to assess the efficacy of the drug in a group of 10 newly diagnosed hypertensive individuals. He measures

Table 30.9 Results of two-way analysis of variance

ANOVA table	Sum of squares (SS)	Degree of freedom (Df)	Mean sum of squares (MS) = (SS/Df)	$F = \frac{\text{Between group MS}}{\text{Within group MS}}$	P value
Between group (Treatment)	2	534516.33	267258.17	113.47	<0.0001
Between group (Antagonist)	1	177504	177504	75.37	<0.0001
Treatment*Antagonist interaction	2	67233	33616.5	14.27	0.0002
Error		42394	18	2355.2222	

Table 30.10 Effect of the new drug on blood pressure recordings at three time-points

Mean blood pressure (mm of Hg) (n = 10)			
Baseline	1 week	1 month	3 months
160	140	133	120
156	142	138	122
159	146	137	128
162	140	139	118
154	144	130	117
150	139	129	125
152	137	130	127
160	147	129	115
158	142	135	120
154	140	140	121

the blood pressure at baseline before prescribing the drug and collected the mean systolic blood pressure readings of the same individuals at 1 week, 1 month and 3 months after prescribing the drug. The dataset is given in Table 30.10.

30.4.7.2 Interpretation of Results

The results are depicted similarly to two-way ANOVA results with time and participants as between group variations (Table 30.11). A one-way repeated measures ANOVA was run on a sample of 10 participants to determine if there were a reduction in BP after taking a new drug after three months of therapy. The results showed that the new drug elicited statistically significant differences in mean BP reduction over its time course, $F(3, 27) = 139.48$, $P < 0.005$.

30.5 Non-parametric Tests for More Than 3 Datasets

As shown in Fig. 30.2, Kruskal- Wallis and Friedman are two non-parametric tests used to analyze more than two datasets.

30.5.1 Kruskal-Wallis Test

It is a non-parametric test equivalent to one-way ANOVA for comparison of three or more datasets. It is used in the setting of non-normally distributed continuous variables or data sets corresponding to small, unbalanced sample sizes with no homogeneity of variances or in the case of ordinal/discrete variables. The calculation of test statistic requires rank ordering of data like the Mann-Whitney U test. It also performs only global assessments like ANOVA and a suitable post-hoc test needs to be performed to identify which group is significantly different from others.

30.5.1.1 Example of Kruskal-Wallis ANOVA

A physician wants to compare the visual analogue (VAS) score for the two new formulations of an analgesic compared to the standard drug in the treatment of osteoarthritis (n = 10). VAS ranged from 1 to 10. One indicates less pain and ten indicates severe pain (Table 30.12).

Table 30.11 Results of repeated measures analysis of variance

ANOVA table	Sum of squares (SS)	Degree of freedom (Df)	Mean sum of squares (MS) = (SS/Df)	$F = \frac{\text{Between group MS}}{\text{Within group MS}}$	P value
Between groups (time)	6502.675	3	2167.55	139.48	0.0000
Between groups (participants)	143.125	9	15.903	1.02	0.4469
Within group (residuals)	419.575	27	15.539		

Table 30.12 Effect of three formulations on change in VAS (Visual analog scale) score

Change in Visual analog scale (VAS) score from baseline		
Standard drug	Formulation 1	Formulation 2
4	0	4
−1	−1	3
3	2	-1
2	2	3
3	3	4
5	−3	3
4	−2	0
3	1	1
0	1	3
4	−1	2

30.5.1.2 Interpretation of Results

Table 30.13 depicts the results of Kruskal Wallis test showing the rank sum for each group and the chi-squared statistic with P value. Results can be reported as Kruskal-Wallis test showed that there was a statistically significant difference in the VAS Score among the three groups, $\chi^2(2) = 8.473$, $p = 0.0145$. Subsequently, Dunn test should be carried out to identify the individual median differences.

30.5.2 Friedman Test

Friedman test is a non-parametric counterpart to repeated measures ANOVA for outcome variables being ordinal or non-normal continuous data. It is developed by Nobel Prize-winning economist Milton Friedman in 1937 [4]. An example scenario for using Friedman test would be to analyse the effect of two interventions on a numerical pain rating scale measured at three different time points. Paired Wilcoxon test with Bonferroni correction should be performed as post hoc if the result of Friedman test is significant.

30.6 Post-hoc Tests

ANOVA performs global assessment. It does not convey which pair of group means are different from each other. To compare the two groups, we know that t-test needs to be performed. The fallacy of doing multiple t tests is explained by an example. We wish to compare the mean reduction in HbA1C at the end of three months of therapy with three different anti-diabetic drugs A, B and C. For each comparison, we need to do a t test. Hence, three t tests need to be done. When multiple analyses are done with the same experimental dataset, the likelihood of finding a false positive test (type-1 error) increases. This is known as the familywise error rate. For the above example, if we do three t tests, the inflated alpha error can be calculated from the simple formula $1 - (1 - \alpha)^N$, where N is the number of groups, which is 14.6% which is unacceptable as it exceeds the conventional error rate 5%.

To account for such multiple comparisons testing, post hoc tests should be employed in order to do pairwise comparisons. There are numerous post-hoc tests available. Some commonly used

Table 30.13 Results of Kruskal-Wallis equality-of-populations rank test

Group	Observations	Rank sum
1	10	199.50
2	10	91.50
3	10	174.00

chi-squared with ties = 8.473 with 2 d.f
probability = 0.0145

Table 30.14 Characteristics of some common post-hoc tests

Post-hoc test	Characteristics
Fisher's Least Significant difference	All possible pairwise comparisons No safeguard against type I error Not used now-a-days
Tukey's Honestly significant difference	All possible pairwise comparisons Adequate safeguard against type I error Specifically for equal size groups. Tukey Kramer can be used for unequal variances.
Scheffe test	For complex comparisons (a large number of groups) Less likely to return a significant difference Adequate safeguard against type I error
Holm-Sidak	All possible pairwise comparisons and specific comparisons Does not give a confidence interval
Bonferroni test	All possible pairwise comparisons & specific comparisons Sensitive for small groups.
Games Howell test	Preferred in the setting of unequal variances. Mainly used in unequal group sizes or very small sample size
Newman-Keul's test	Only for all possible comparisons No adequate safeguard against type I error
Dunnett's test	Used to compare one control group with all other treatment groups
Dunn's test	Used for non-parametric tests (Kruskal Wallis)

tests and their characteristics are given in Table 30.14. These tests basically differ in how they safeguard against α errors [14, 15]. For example, Bonferroni correction adjusts the α value by dividing it by the number of comparisons to be made. In the above example of three comparisons, the adjusted P value becomes 0.016 (0.05/3) instead of 0.05. The tests that are stringent in adjusting the α error are termed as conservative and those that are not are labelled as liberal. Scheffe test is the most conservative of all and is used in complex comparisons including a large number of groups. Tukey HSD is the commonly used post-hoc test in biomedical research settings. Games Howell is preferred for samples with unequal variances and Dunn's test is used for non-parametric tests. Each has its own pros and cons, and no single test is universally applied in all settings. A researcher has to choose a suitable test based on their research objective.

30.7 Chi-square Test

Chi-square test (χ^2) (pronounced as ki as in kite [10]) is a statistical tool to assess the relationship between two categorical variables. The categorical variables can be either in nominal or ordinal scale. It is a versatile test that can assess both the association of variables as well as test the difference in proportions of two variables [12]. The test in principle compares the observed numbers in each of the four categories in the 2 × 2 contingency table with the numbers to be expected if there were no difference between the two groups.

30.7.1 Criteria to Do Chi-Square Test

Outcome variable: Categorical (nominal/ordinal) summarized as a proportion

Analysis type: Comparison of proportion/association

Number of groups: two/more than two groups

Study design: unpaired

Distribution of data: dichotomous distribution

30.7.2 Contingency Table

Constructing a contingency table is the first step in finding the relationship between two categorical variables. Both exposure and outcome

Table 30.15 Contingency table (2 × 2) showing an association between smoking and Lung cancer

	Lung cancer		
	Yes	No	
Smokers	(a)	(c)	(a + c)
Non-smokers	(b)	(d)	(b + d)
	(a + b)	(c + d)	(a + b + c + d)

variables with two levels are cross-tabulated to form a frequency distribution matrix, conventionally known as 2 × 2 contingency table [12]. In this table, the two levels of variables are arranged as rows and columns. It is a convention that exposure variables are depicted in rows and the outcome variables in columns. Individuals are assigned to one of a cell of the contingency table with respect to their values for the two variables. The data in the cell should be entered as frequencies/counts not any other derived parameter like percentages. Table 30.15 shows the 2 × 2 contingency table for testing the association between smoking and lung cancer. If there are more than two levels in the variable, it is written as r × c tables, where r denotes the number of rows and c denotes the number of columns. Even numerical discrete or continuous variables can be grouped as categorical variables and presented as large contingency table.

30.7.3 Assumptions for Chi-Square Test

Assumptions to be satisfied for performing the Chi-square test are listed below [5]:

1. The data should be random observations from the sample.
2. The expected frequency (not observed) should be 5 or more in 80% of cells. And no cell should have an expected frequency of less than 1.
3. The sample size should be at least the number of cells multiplied by 5. For a 2 × 2 table, the minimum sample should be at least 20.

30.7.4 Example for Chi-Square Test

A new antibiotic is compared with a reference antibiotic in curing patients with urinary tract infections. A total of 380 patients were randomly assigned to receive either reference or new antibiotic and 160 achieved microbiological cures in the new antibiotic group and 120 in the reference antibiotic group. The microbiological cure is recorded by the absence of organisms in urine culture after seven days of treatment.

The null hypothesis for this scenario is that there is no difference in proportions of cure rate between new and reference antibiotic groups and vice versa is the alternate hypothesis. Assuming the null hypothesis, the expected frequency of each cell can be calculated by multiplying the marginal row and marginal column total divided by the grand total. Thus, in the given example, for the first cell (a), the expected frequency is (200 × 280)/380 = 147.4 and similarly, the values for other cells are estimated (Refer Table 30.16).

Table 30.16 Contingency table (2 × 2) depicting two antibiotic groups and their cure rates

	Cured	Not cured	Row total
New antibiotic group	160 (147.4)	40 (52.6)	200
Reference antibiotic group	120 (132.6)	60 (47.4)	180
Column total	280	100	380 (Grand total)

The number in each cell denotes the observed frequency. The values in parentheses represent the expected frequency in each cell assuming the null hypothesis

The difference between observed (O) and expected (E) is then derived (O − E). The Chi-square (X^2) test statistic is obtained from the formula (Box 30.3):

Box 30.3 Chi-square Statistic Formula

$$x^2 = \frac{\sum (O-E)^2}{E}$$

where, O, the observed number; E, expected number; X^2 is the test statistic.

Box 30.4 Worked Out Example

$$x^2 = \frac{(12.6)^2}{147.4} + \frac{(-12.6)^2}{132.6} + \frac{(-12.6)^2}{52.6} + \frac{(12.6)^2}{47.4} = 8.641$$

The degree of freedom for Chi-square statistic is (row-1) × (column −1) [(2 − 1) (2 − 1) = 1]. From the Chi-square distribution table, under the specified degree of freedom (1) and level of significance at 0.05, the χ^2 value is found to be 3.84. Since the test statistic X^2 (8.861) is higher than Chi-square critical value (3.84), we reject the null hypothesis. We can conclude that the new antibiotic has a significantly higher cure rate than the reference drug ($P < 0.05$).

30.7.5 Strength of Association and Its Interpretation

The Chi-square test only informs whether the proportion of one group is significantly different from the other group; it does not tell the magnitude of the relationship between two variables. The strength of association should be determined by absolute measure (risk difference) or relative measures like Relative risk (RR) or Odds ratio (OR).

30.7.5.1 Risk Difference

Risk difference measures the difference of outcome variable between two groups. In the given example, the cure rate with the new antibiotic is (160/200) 80% and the reference drug is (120/180) 66.6%. The risk difference is 13.4%. The interpretation is straightforward, and the precision of the estimate can be given by calculating the confidence interval for this estimate.

30.7.5.2 Relative Risk (RR)

Relative risk is obtained by computing the ratio of the risk of occurrence of an event in the test (exposed) group and the risk of occurrence of an event in the control (unexposed) group. RR of one indicates that there is no difference in the occurrence of events between the exposed and unexposed groups. RR greater than one implies more events occur in the exposed group. Conversely, RR less than one indicates the occurrence of events is less in the exposed group as compared to the non-exposed group. The risk for each group is calculated by the number of events in that group divided by the total population at risk in that group. Thus, it can be calculated from a 2 × 2 contingency table by the formula [a/(a + c)/b/(b + d)] (see Table 30.4). For the example scenario, the relative risk is estimated to be 160/200/120/180 = 1.2. It means there is a 20% higher probability of cure in the new antibiotic group as compared to the reference drug group. Relative risk is the preferred measure in the setting of prospective study designs like cohort and randomized controlled trials as the population at risk is defined in these settings.

30.7.5.3 Odds Ratio

It is the ratio of the odds of occurrence of events in the exposed group and the odds of occurrence of an event in the unexposed group. Odds for each group are calculated by the number of individuals who had events divided by the number of individuals who did not have events. It is computed from 2 × 2 contingency by the formula (ad/bc). The odds ratio of one indicates there is no difference in the odds of the occurrence of events in the exposed and unexposed groups. An odds ratio greater than one denotes higher odds of an event in the exposed group and the reverse is true for odds less than one. For the example scenario, the odds ratio is 2 [160/40/120/60]. It means there are two-fold increased odds of cure in the new antibiotic group compared to reference antibiotic group. For retrospective studies, the odds ratio should be reported. It can also be used in prospective study designs especially when it has to be adjusted for confounding factors [4].

30.7.6 Closely Related Tests to the Chi-Square Test

- If the sample size of the study is too small (n less than 20) or if any of the expected cells takes a value less than 5, then the Chi-square test assumption would be violated, and it can lead to increased type II error. In these situations, Chi-square with Yate's continuity correction or Fisher's exact test should be performed. Yate's continuity correction is done only for the 2 × 2 contingency table.
- If the observations are paired data/related samples, McNemar's Chi-square test can be employed if there are only two datasets and Cochran's Q in case of more than two sets of observations. It is used in the setting of paired design as in paired t test and matched case-control studies [12]. However, for matched case-control studies, conditional logistic regression would be more appropriate.
- When the levels of groups are more than two, the significance derived from Chi-square test provides only global assessment like ANOVA (Refer to Sect. 30.4.2). To find the significance between the specific pair/pairs of groups of proportions, a partitioned Chi- square should be used.
- Cochrane Armitage trend test can be used to test the dose-response relationship of categorical variables to assess whether there is an increasing (or decreasing) trend in the proportions over the exposure categories. This is useful in pharmacogenetic studies to identify the association between different allele frequencies and a clinical phenotype. It is also useful in toxicity studies where teratogenicity is evaluated at various dose levels [4].

30.8 Correlation and Regression

The correlation and regression analysis are performed when the association between two quantitative variables is studied.

30.8.1 Correlation

Correlation is the statistical tool to assess the association between two quantitative variables. Both variables are measured in the same individuals. The first step is to create a scatter plot graphically (Fig. 30.3) and visualize the association between two variables. In a scatter plot, one variable is plotted on the X-axis and the other variable is plotted on the Y-axis. As a convention, the exposure (independent) variable is plotted on the X-axis and the outcome (dependent) variable is plotted on the Y-axis. Assumptions for performing correlation are random sample selection and the two variables, X and Y follow bivariate normal distribution [10]. The strength of association is given by Pearson's correlation coefficient (r) which ranges from −1 to +1, which is a parametric test. The plus and minus symbols indicate the direction of the relationship and the value 1 indicates there is a strong correlation and '0' indicates no correlation. If the assumption of a normal distribution is violated or if either of the variables is ordinal, a non-parametric equivalent, Spearman rank correlation (rho) can be employed. Correlation only expresses the strength of association and does not tell the magnitude of change that happens from one variable to another variable. Also, statistical correlation does not imply causal association.

30.8.1.1 Example of Correlation

A researcher is interested to study the relationship between mid-arm circumference and body weight in children. He collected the body weight and

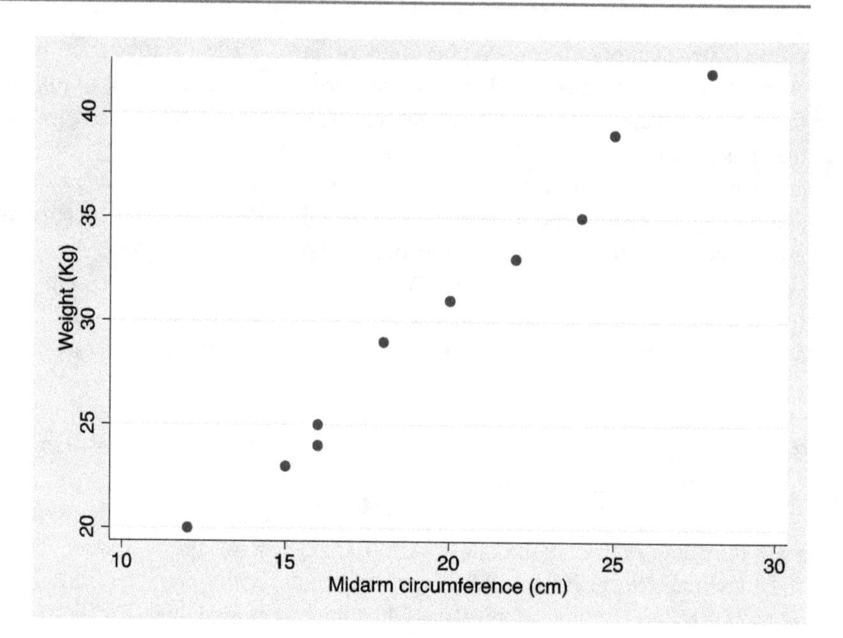

Fig. 30.3 Scatter plot of data mentioned in Table 30.17

mid-arm circumference of 10 school-going children. Table 30.17 provides a data set and Table 30.18 gives the results.

30.8.1.2 Interpretation of Results

In this scenario, the r value is close to 1 which implies a strong correlation. Thus, results can be written as the mid-arm circumference is strongly associated with the body weight of school-going children with a Pearson correlation coefficient of 0.997 ($P<0.0001$). The r^2 in Table 30.18 depicts the coefficient of determination which indicates the proportion of variance in one variable that is explained by the other variable.

30.8.2 Regression

Regression analysis is done in medical research for two main purposes. First, is when the research question focuses on the prediction of one variable from the other variable. Second, is when we need to infer an association between two variables by estimating the effect size after adjusting for the potential confounders and effect modifiers. Suppose in the above example (Sect. 30.8.1.1), if the researcher wants to predict the body weight of individuals if he measures the mid-arm circumference itself, regression analysis can be performed, provided all the variations in body

Table 30.17 The midarm circumference and body weight of children

Mid-arm circumference (cm)	Weight (Kg)
12	20
15	23
16	24
16	25
18	29
20	31
24	35
28	42
25	39
22	33

Table 30.18 Results of Pearson correlation test

Pearson	
Correlation coefficient (r)	0.9927
R squared	0.984
P value	
P value (two tailed)	<0.0001
No. of observations (XY pairs)	10

weight are captured by mid-arm circumference. In this analysis, mid-arm circumference is the independent variable (predictor or explanatory variable) and body weight is the dependent variable (outcome variable). Depending on the type of outcome variable, regression analysis may be logistic, linear, cox or Poisson regression as described in Table 30.20 [16]. When multiple independent variables are assessed, it is called multiple regression. When we use a regression model to predict, the predictor values must be within the range of values that have been used to develop the model (refer to Sect. 30.8.2.3).

30.8.2.1 Simple Linear Regression

In simple linear regression, one independent variable is explored to predict an outcome. The different types of regression based on the outcome variable are provided in Table 30.19. A scatter plot is made as in correlation and the data points are joined by a line of best fit. The differences between observed data points and the predicted line are called residuals. The sum of squares of residuals should be minimum to get the best fit line by the least squares method. To perform linear regression, the following assumptions are checked. For each value of X, the Y values should follow a normal distribution, the mean of Y values lie on the predicted regression line (linearity assumption) and the other important assumption is that variances of Y are similar for each value of X (homoscedasticity). The simple linear regression model is given by $Y = \beta_0 + \beta_1 X + €$, which is quite similar to the straight-line equation studied in elementary geometry ($y = a + bx$). β_1 (regression coefficient) is the slope of the line that estimates the extent of change in Y when X changes by 1 unit. β_0 is the Y-intercept when X is 0. € is the error term included in statistical model to account for random variations.

30.8.2.2 Example of Simple Linear Regression

A researcher wants to investigate the association of body weight and serum creatinine and collected both variables in ten individuals. He also wants to know whether body weight can predict the serum creatinine level. Table 30.20 provides a dataset for both variables.

30.8.2.3 Interpretation of Results

From the results Table 30.21, the regression equation can be written as serum creatinine = −0.402 + 0.019 body weight in Kg, where −0.402 is the value of y-intercept (β_0) and 0.019 is the slope of the regression line (β_1). For a one kg increase in body weight, serum creatinine increases by 0.019 (CI:0.016,0.02). From the regression equation, we can easily predict the value of serum creatinine value of an individual weighing 80 kg to be 1.118. This model should not be used for prediction of those who weigh less than 70 or more than 100, since the regression

Table 30.19 Types of regression based on the outcome variable

	Outcome	Exposure
Linear regression	Numerical	Continuous and/or categorical
Logistic regression	binary	
Cox regression	Time to event	
Poisson regression	Counts of Rare events	

Table 30.20 Dataset showing body weight and serum creatinine levels

Body weight (Kg)	Serum creatinine (mg/dL)
70	0.9
73	0.95
76	0.99
78	1
85	1.2
88	1.25
90	1.3
94	1.32
99	1.4
100	1.45

Table 30.21 Results of simple linear regression analysis

Serum creatinine	Co-efficient	Standard error	t test	P value	Confidence interval
Body weight	0.0185108	0.000845	21.91	0.000	0.0165623, 0.0204593
const	−0.4029697	0.0725866	−5.55	0.001	−0.5703546, −0.2355848

Model $R^2 = 98.16$

model is constructed from individuals weighing 70 to 100 kg.

R^2 can be interpreted as follows: according to the model, body weight in kilogram accounts for a 98.16% variation in serum creatinine in mg/dL. We should keep in mind that such near perfect explanation of a physiological variable by another variable is very rare and the variation is often explained by the combination of variables. In that case, R^2 implies the explanatory power of all the variables in the model acting together.

30.9 Concluding Remarks

Statistical methods are used for drawing inferences about the population from the observations made in the sample. The statistical test for analysis of data is selected based on the type of variable, distribution of data, number of groups/datasets and study design. Most statistical tests come with a set of assumptions that the data must fulfill. These assumptions have to be checked before interpreting the results. These statistical tests calculate the probability of the results occurring by chance alone. The parametric tests are more powerful than the nonparametric tests and should be preferred. P value indicates the probability of the results occurring by chance alone. A significant P value does not mean clinical significance. In addition to the statistical significance, the point estimate of the effect and its precision (95% confidence interval) should be reported.

Conflict of Interest Nil.

References

1. Raveendran R, Gitanjali B, Manikandan S (2012) A practical approach to PG dissertation, 2nd edn. PharmaMed Press, Hyderabad
2. Glaser AN (2014) High-yield biostatistics, epidemiology and public health, 4th edn. Wolters Kluwer, Philadelphia
3. Driscoll P, Lecky F (2001) Article 6. An introduction to hypothesis testing: parametric comparison of two groups – I. Emerg Med J 18:124–130
4. Norman GR, Streiner DL (2000) Biostatistics the bare essentials, 2nd edn. B.C.Decker Inc., Hamilton
5. Campbell MJ, Swinscow TDV (2009) Statistics at square one, 11th edn. Wiley-Blackwell, Chichester
6. Altman DG, Bland JM (1999) Statistical notes. Variables and parameter. BMJ 318:1667
7. Altman DG, Bland JM (1999) Statistical notes. Treatment allocation in controlled trials; Why randomize? BMJ 318:1209
8. Bland JM, Altman DG (1996) Statistical notes. The use of transformation when comparing two means. BMJ 312:1153

9. Applegate KE, Tello R, Ying J (2003) Hypothesis testing III: counts and medians. Radiology 228:603–608
10. Dawson B, Trapp RG (2004) Basic & clinical biostatistics, 4th edn. McGraw-Hill, New York
11. Petrie A, Sabin C (2009) Medical statistics at a glance, 3rd edn. Wiley-Blackwell, Chichester
12. Kirkwood RB, Sterne ACJ (2003) Essential medical statistics, 2nd edn. Blackwell, Massachusetts
13. Doncaster CP, Davey AJH (2007) Analysis of variance and Covariance. Cambridge University Press, New York
14. Field A (2009) Discovering Statistics Using SPSS, 3rd edn. SAGE Publications Ltd., London
15. Kim H-Y (2015) Statistical notes for clinical researchers: post-hoc multiple comparisons. Restor Dent Endod 40(2):172–176
16. Harrell FE (2015) Regression Modeling Strategies, 2nd edn. Springer, New York

Part VI

Drug Discovery Research

31 Artificial Intelligence Generative Chemistry Design of Target-Specific Scaffold-Focused Small Molecule Drug Libraries

Yuemin Bian, Gavin Hou, and Xiang-Qun Xie

Abstract

The *de novo* design of scaffold-focused and target-specific molecular structures using deep learning generative modeling introduces a promising solution to the discovery of novel and potent bioactive drug compounds. Deep learning generative modeling exhibits the creativity that machine intelligence can offer in composing, painting, and even the scratching of novel molecular structures. This chapter mainly covers that how generative chemistry can be effectively applied to the design and generation of scaffold-focused and target-specific small molecules. To this emerging paradigm, the chapter starts with a brief history of artificial intelligence (AI) in drug discovery. Chemical databases, molecular representations, and cheminformatics related tools are covered as the infrastructure. Two example applications of using generative adversarial networks (GAN) and recurrent neural networks (RNN) to realize the *de novo* compound generation towards the cannabinoid receptor 2 (CB2) are discussed in the chapter. Summary, challenges, and future perspectives follow.

Keywords

Artificial intelligent · Drug discovery · Machine learning · Deep learning · PharmacoAnalytics · Generative chemistry · Hit identification · Cannabinoid CB2 receptor

Y. Bian
Center for the Development of Therapeutics, Broad Institute of MIT and Harvard, Cambridge, MA, USA

Department of Pharmaceutical Sciences and Computational Chemical Genomics Screening Center, Pharmacometrics and System Pharmacology (PSP) PharmacoAnalytics Program, School of Pharmacy, Pittsburgh, PA, USA
e-mail: ybian@broadinstitute.org

G. Hou
Department of Pharmaceutical Sciences and Computational Chemical Genomics Screening Center, Pharmacometrics and System Pharmacology (PSP) PharmacoAnalytics Program, School of Pharmacy, Pittsburgh, PA, USA

National Center of Excellence for Computational Drug Abuse Research, Pittsburgh, PA, USA

X.-Q. Xie (✉)
Department of Pharmaceutical Sciences and Computational Chemical Genomics Screening Center, Pharmacometrics and System Pharmacology (PSP) PharmacoAnalytics Program, School of Pharmacy, Pittsburgh, PA, USA

National Center of Excellence for Computational Drug Abuse Research, Pittsburgh, PA, USA

Drug Discovery Institute, Pittsburgh, PA, USA

Departments of Computational Biology and Structural Biology, School of Medicine, University of Pittsburgh, Pittsburgh, PA, USA
e-mail: xix15@pitt.edu

G. Jagadeesh et al. (eds.), *The Quintessence of Basic and Clinical Research and Scientific Publishing*,
https://doi.org/10.1007/978-981-99-1284-1_31

31.1 Introduction

Drug discovery is expensive, and the cost keeps increasing. Now it can take 2.8 billion USD and over 12 years to finish the discovery process for the development of a novel drug [1, 2]. To escalate the discovery process and confront the growing cost, investigating new but efficient strategies is critical and in demand. For drug hit identification, high-throughput screening (HTS) dramatically speeds up the task by evaluating candidate compounds in large volume [3, 4]. Virtual screening (VS) in parallel enriches potential active molecules and filters out undesired structures. VS underwent quick advancement along with the improvement in computational power. Commonly adopted VS strategies are structure-based and ligand-based. With the protein structure of the target of interest available, structure-based approaches including molecular docking [5–8], molecular dynamic simulations [5, 9–11], fragment-based approaches [8, 12], etc., can be conducted to virtually evaluate receptor-ligand interactions for a large compound set for finding plausible hits. With confirmed active compounds or probes of the given target available, ligand-based approaches such as pharmacophore modeling [13, 14], scaffolding hopping [15, 16], structural similarity search [17], etc., can be performed to further optimize known hits. The rapid advancement in Artificial Intelligence (AI) presents a machine learning (ML)-based decision-making model [18, 19] as an alternative solution to contribute to VS campaigns. Drug discovery is a data-enriched area of research. ML methods are favored as they demonstrate the ability to handle big data to excavate hidden patterns, and to use detected patterns to facilitate the future data prediction in an efficient and effective manner.

Drug discovery campaigns with applications of the above-mentioned HTS and VS approaches have been fruitful in the past decades. In confronting the increasing cost of drug discovery, challenges still remain in exploring pioneering techniques and strategies that can better comprehend complex systems. In recent years, the flourishing of deep learning generative modeling provides putative novel solutions to the field. From generated human faces [20], to literature production tools [21], deep learning generative models inspire our perception of the machine intelligence to a new level. It is noted that a generative pre-trained transformer (GPT) for chemistry is a type of generative chemistry that uses deep learning models to generate novel molecules with desired properties. Recently, the immersion of drug discovery and generative modeling investigates the feasibility of automated generation of scaffold-focused and target-specific molecular structures. The discovery of DDR1 kinase inhibitors within 21 days exemplified the capability of using deep learning generative models to offer promising and compelling outcomes [22] This chapter provides examples and discussions based on the authors' applications of using generative chemistry to realize the design of scaffold-focused and target-specific libraries. The chapter starts with a brief evolution of AI in drug discovery, and the infrastructures in deep learning generative chemistry. Two examples using generative adversarial networks (GANs) and recurrent neural networks (RNNs) are detailed to discuss their fundamental architectures as well as their applications in the *de novo* drug design.

31.2 Artificial Intelligence in Drug Discovery

Artificial intelligence (AI) is the study of developing and implementing techniques that enable the machine to behave with intelligence [23]. The concept of AI can be traced back to the 1950s. Researchers were trying to answer whether intelligent tasks that usually are accomplished by mankind can be handled with computers [24]. Initially, researchers assume that a human-level AI system can be realized by specifying explicit rules to navigate knowledge (Fig. 31.1a). This strategy is also known as symbolic AI [25]. Symbolic AI functions as a solution to a set of logical problems including chess playing. But when it comes to image recognition,

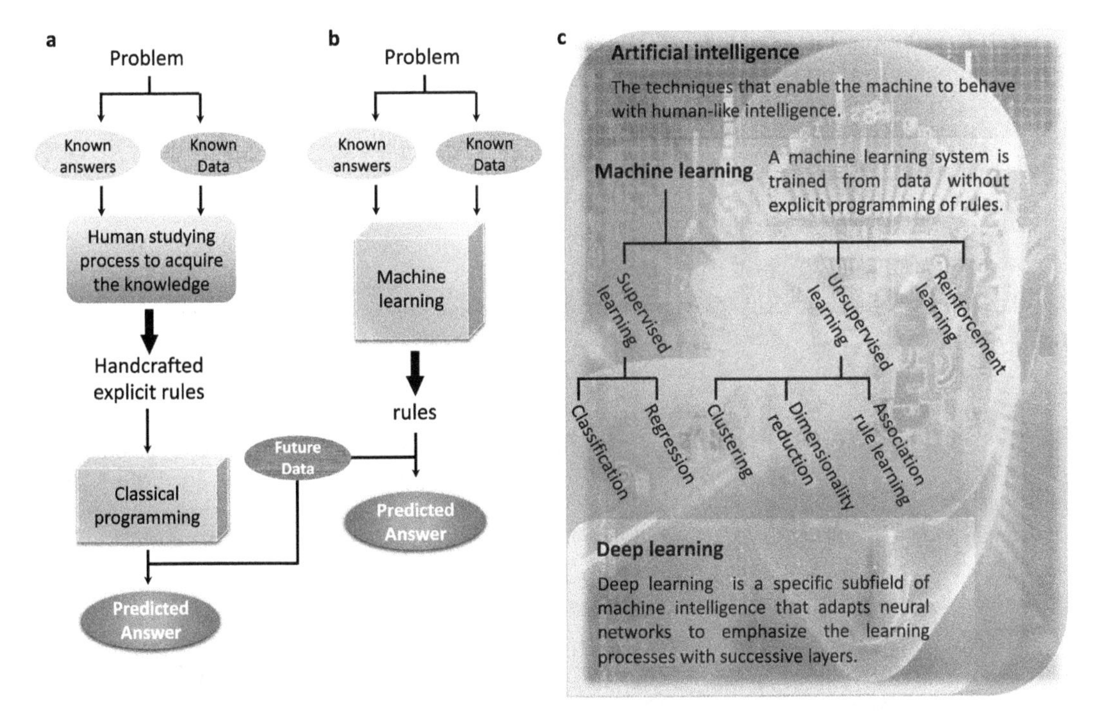

Fig. 31.1 Artificial intelligence to deep learning. (**a**) The programming paradigm for symbolic AI. (**b**) The programming paradigm for machine learning (ML). (**c**) The relationship among artificial intelligence, machine learning, and deep learning

distinguishing active and inactive molecules for a given target, the performance of symbolic AI is greatly compromised. These types of problems usually come with blurry, unclear, and distorted knowledge, which can hardly be transferred into specific rules. Criteria can be defined to select general drug-like compounds that follow Lipinski's rule of five [26]; but exhaustively specifying rules for selecting agonists of a given target is almost impossible [27].

Different from the symbolic AI, ML allows computers to solve problems by learning on their own [28, 29]. Instead of waiting for human specified rules, ML algorithms can summarize patterns by directly learning from the raw data (Fig. 31.1b). Both known data and known answers are keyed for ML algorithms to generate rules to make associations. With future data, generated rules will be used to make predictions. ML methods became a dominant player in the field of AI since 1990s [30]. In drug discovery, supervised learning, unsupervised learning, and reinforcement learning are commonly encountered ML categories (Fig. 31.1c). Both raw data and data labels are fed into the algorithm in supervised learning [31]. Discrete labels make supervised classifications, while continuous numeric labels make supervised regressions. One typical example of supervised classification is the prediction of the protein subtype selection for query molecules [18, 27, 32, 33], while a well-trained regressor is supposed to be able to predict quantitative structure-activity relationships for molecules sharing a similar scaffold [8, 34]. In unsupervised learning, algorithms are trained with unlabeled data. For instance, molecules can be categorized into groups based on structural diversity using unsupervised clustering methods [35, 36]. In reinforcement learning, the learning system can choose actions according to its observation of the environment, and get a penalty (or reward) in return [37]. To achieve the lowest penalty (or highest reward), the system must learn and choose the best strategy by itself.

Deep learning (DL) is a subfield of ML that adapts neural networks to emphasize the learning process with successive layers (Fig. 31.1c). Representations can be learned from raw data and transferred to abstract levels with DL methods [38]. The word "deep" in DL reflects the incorporation of successive layers of representations to process and learn from the raw data [39]. The advancement of DL generative architectures promotes the prevalence of generative chemistry. For example, as a specific type of Recurrent Neural Network (RNN), Long Short-Term Memory (LSTM) models [40], have been used in text generation, which further inspired the compound generation based on the simplified molecular-input line-entry system (SMILES). Another example is the usage of the Generative Adversarial Network (GAN) models [41] for image generation, which motivates graph-based and fingerprint-based molecular scratching. With a simplified and widely-adaptable representation learning process [24, 38], it is foreseen that DL can bring the process of molecular design to the next level.

31.3 Data Sources and Machine Intelligence Infrastructures

The quantity and the quality of data are the foundation of a ML campaign. Molecular and biological data functions as the prerequisite for the development of generative chemistry models. Table 31.1 lists commonly used databases that document chemical and biological information.

The vectorization of collected molecules is necessary to prepare chemical structures into machine-readable representations. Table 31.2 lists commonly used molecular representations. Besides string-based and fingerprint-based descriptions, graph-based and tensorial representations are popular alternative options. Most molecules can be represented in 2D graphs. A graph can be defined as the connectivity relations between a set of atoms (nodes) and a set of bonds (edges) [60]. Tensorial representation is another approach that stores molecular information into atom types, bond types, and connectivity information in tensors.

Table 31.1 Illustrated cheminformatics databases available for drug discovery [42]

Database	Description	Web linkage	Examples of usage
UniProt [43]	Protein sequence and functional information	https://www.uniprot.org	Sequence alignment, protein clustering, protein sets for species, etc.
RCSB PDB [44]	Experimental and computational 3D structures for biological systems	https://www.rcsb.org	Retrieving protein structures for structure-based drug discovery
PDBbind [45]	Experimentally measured binding affinity data for PDB	http://www.pdbbind.org.cn	Benchmarking receptor-ligand interactions
PubChem [46]	A collection of chemical information	https://pubchem.ncbi.nlm.nih.gov	Finding chemical structures, physical properties, biological activities, etc.
ChEMBL [47]	A library of manually curated bioactive compounds	https://www.ebi.ac.uk/chembl/	Collecting target-specific molecules
SureChEMBL [48]	A search engine for patented compounds	https://www.surechembl.org/search/	Searching patented compounds
DrugBank [49]	Information about drugs, drug targets, indications, etc.	https://www.drugbank.ca	Off-target analysis, drug repurposing, etc.
ZINC [50]	A collection of commercially available compounds	https://zinc.docking.org	Providing molecules for virtual screening
Enamine	A vendor to provide chemicals	https://enamine.net	Exploring large chemical space
ASD [51]	A resource documents allosteric modulators	http://mdl.shsmu.edu.cn/ASD/	Studying allosteric modulation
GDB [52]	A library with enumerated compounds	http://gdb.unibe.ch/downloads/	Expanding synthetic feasible chemical space

Table 31.2 Examples of commonly used molecular representations [42]

Representation	Description
SMILES [53]	A line notation for describing the structure of chemical species using short ASCII strings
InChI [54]	The International Chemical Identifier (InChI) is a textual identifier for chemical substances to encode molecular information in a standard way
Physical-chemical properties	Describing molecules using observed or measured characteristics, for example, molecular weight, logP, etc.
Molecular fingerprints	Using a series of binary digits to represent the presence or absence of substructures in a molecule. Examples including MACCS Keys [55], Circular [56, 57], Path [58], Tree [59], and Atom Pair [58] fingerprints
Molecular graphs	Describing a molecule using the connectivity relations between a set of atoms (nodes) and a set of bonds (edges)

Table 31.3 Commonly used cheminformatics and machine learning packages [42]

Package	Description	Web linkage
RDKit [61]	An open-source toolkit for operating cheminformatics tasks	https://www.rdkit.org
Open Babel [62]	A chemical toolbox for manipulating chemical data	http://openbabel.org/wiki/Main_Page
CDK [63]	Open-source modular Java libraries for Cheminformatics	https://cdk.github.io
KNIME [64]	An open-source workflow environment	https://www.knime.com
TensorFlow [65]	An open-source ML platform	https://www.tensorflow.org
CNTK [66]	A unified DL toolkit that describes neural networks as a series of computational steps via a directed graph	https://github.com/microsoft/CNTK
Theano [67]	A Python library and optimizing compiler for manipulating and evaluating mathematical expressions	https://github.com/Theano/Theano
PyTorch [68]	An open-source ML library based on the Torch library.	https://pytorch.org
Keras [69]	A high-level neural networks API on top of TensorFlow, CNTK, or Theano	https://keras.io
Scikit-Learn [70]	A free software ML library for the Python programming language	https://scikit-learn.org/stable/

After the data collection and vectorization, it is time to conduct the chemical data analysis and start a ML-based application. Table 31.3 illustrates examples of frequently considered cheminformatics toolkits and machine learning packages.

31.4 Application of GAN on Designing Small Molecule Library

31.4.1 A Brief Overview

A deep convolutional generative adversarial network (dcGAN) model was developed in this study to design target-specific novel compounds for cannabinoid (CB) receptors [71]. In the adversarial process of training, two models, the Discriminator D and the Generator G, were iteratively trained. D was trained to discover the hidden patterns among the input data to have the accurate discrimination of the authentic compounds and the "fake" compounds generated by G; and G is trained to generate "fake" compounds to fool the well-trained D by optimizing the weights for matrix multiplication of data sampling. To determine the appropriate architecture and the input data structure for the involved convolutional neural networks (CNNs), the combinations of various network architectures and molecular fingerprints were explored. Well-

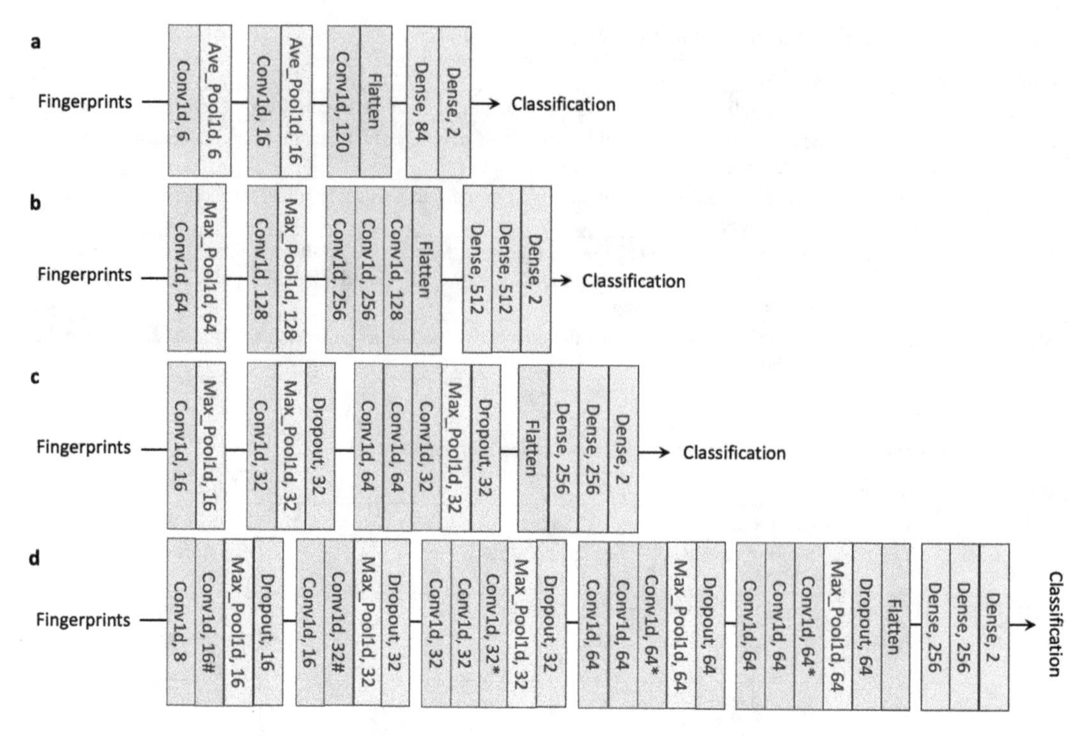

Fig. 31.2 The modified and applied CNN architectures in this study. (**a**) LeNet-5. (**b**) AlexNet. (**c**) ZFNet. (**d**) VGGNet.

developed CNN models including LeNet-5, AlexNet, ZFNet, and VGGNet were investigated. Four types of fingerprints including MACCS, ECFP6, AtomPair, and AtomPair Count were calculated to describe the small molecules with diverse structural characteristics. Generative models with convolutional networks provide promising opportunities for small molecules screening and design.

31.4.2 Modified CNN Architectures and Predictive Analytics

The study started with the determination of the architecture for the Discriminator. Compounds with reported *Ki* values for CB receptors (CB1 and CB2) were retrieved as the training data [27]. CNN is a subtype of Deep Neural Networks (DNNs) with well-developed architectures during recent decades. Preliminary studies were conducted to compare and determine which architectures can better fit the molecular fingerprint-based input data structure and make accurate classifications of CB ligands (Fig. 31.2).

LeNet-5 is a classic seven-layer architecture pioneered in 1998 [72] (Fig. 31.2a). In the adapted model, the first convolutional layer uses six kernels to filter input bit vectors from fingerprints. It is a one-dimensional convolution that each kernel has the size of five with a stride of four. An average pooling layer is followed by six kernels that have the size of two and the stride of one. The stacking of a convolutional and a pooling operation is repeated to give the layers three and four. The fifth layer uses 120 kernels to perform the convolutional operation. After the output matrix is flattened, densely connected layers are followed.

AlexNet was the champion in the 2012 ImageNet competition [73] (Fig. 31.2b). The adapted AlexNet-based architecture in this study stacks convolutional and pooling layers to offer the first four layers. Different from the LeNet-5, the number of kernels for each layer is dramatically increased here. Three more convolutional

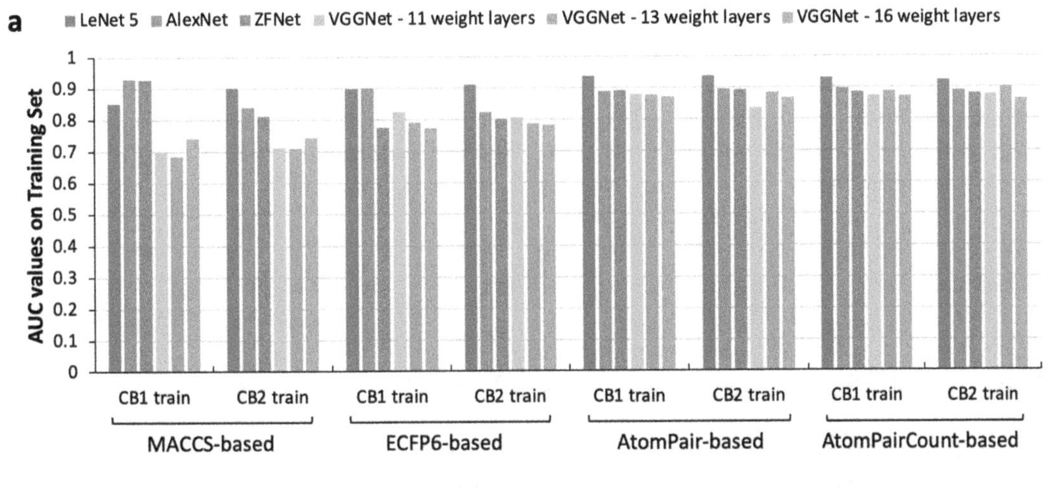

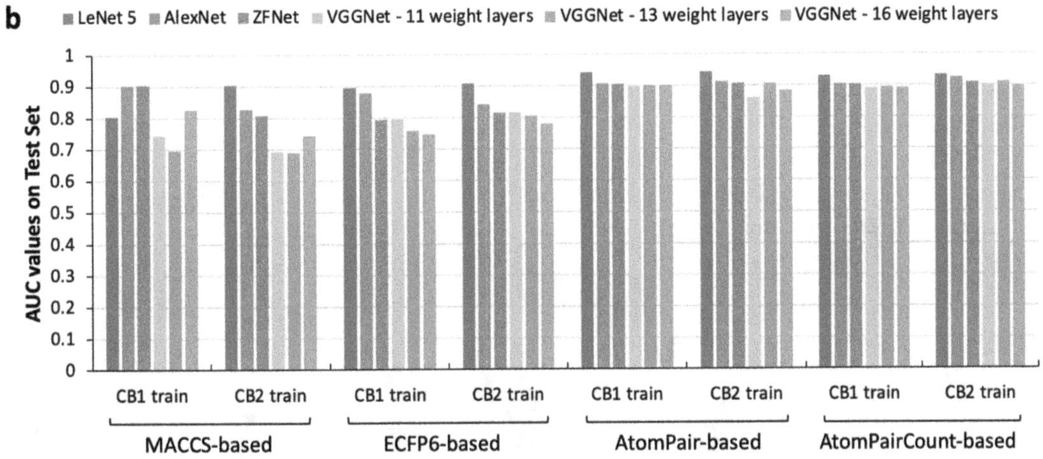

Fig. 31.3 AUC values of all CNN models with each dataset on Training Set (**a**) and Test Set (**b**)

operations are followed before flattening the output matrix. Two densely connected layers are deployed before the output classification.

ZFNet was the winner for the ILSVRC2013 [74] (Fig. 31.2c). The ZFNet maintains the general architecture and the arrangement of the AlexNet but has the hyperparameters adjusted. The adapted ZFNet-based architecture possesses less kernels, different kernel sizes, and varied strides. Notably, two additional dropout layers are inserted after pooling layers. The dropout rate is 0.5.

VGGNet [75] was published in 2014 with uniformed building blocks in its architecture. The adapted VGGNet16-based architecture (Fig. 31.2d) is equipped with 13 convolutional layers, five pooling layers, five dropout layers, and two densely connected layers. Equipped convolutional and pooling layers have uniformed kernel sizes and strides. Dropout layers maintain a consistent dropout rate of 0.5. Densely connected layers have 256 hidden neurons. VGGNet13 and VGGNet11-based architectures share the framework but with less convolutional layers.

Figure 31.3 combines the AUC values from training and testing of CNN models across six architectures (LeNet-5, AlexNet, ZFNet, VGGNet11, VGGNet13, and VGGNet16) on 8 datasets. Consistent model performance on both the Training Set and the Test set can be observed. In most cases, the architecture adapted from the LeNet-5 can give the highest AUC. The architecture adapted from the AlexNet produced the best scores for the CB1 training set when molecules were described in MACCS and ECFP6. The architecture adapted from the ZFNet gave the top score for the CB1 test set when molecules were described in MACCS. Molecular fingerprints in bit-vectors represent sparse matrixes with low data density. Information loss can be expected when the architecture goes deeper. Dropout layers and L1/L2 penalties may relieve the problem of overfitting. But deep architectures can suffer from having more parameters than can be justified by the training data.

Distinctive data structures were retained in training datasets as molecules were vectorized into four different types of fingerprints. MACCS fingerprints are sub-structural keys to denote whether a certain sub-structure exists in a compound. ECFP6 fingerprints, which were hashed into 1024 bits, describe circular topological information of atom neighborhoods. AtomPair fingerprints encode each atom and enumerate all distances between atom pairs. The length was hashed to 1024 bits as well. The combination of AtomPair fingerprints and the LeNet-5-based architecture turns out to be the optimal option in the classification of active and inactive/random compounds for cannabinoid receptors.

31.4.3 The Architecture of the Generated dcGAN Model

Based on the preliminary architecture comparison described above, the model adapted from the LeNet-5 network was therefore selected for the generation of the Discriminator. AtomPair fingerprints were calculated to describe CB molecules as the input data.

The deep convolutional generative adversarial network (dcGAN) is constituted with a Generator G and the Discriminator D (Fig. 31.4). Generally, G is constructed with the reverse convolutional process (Fig. 31.4a). Activation function *relu* is

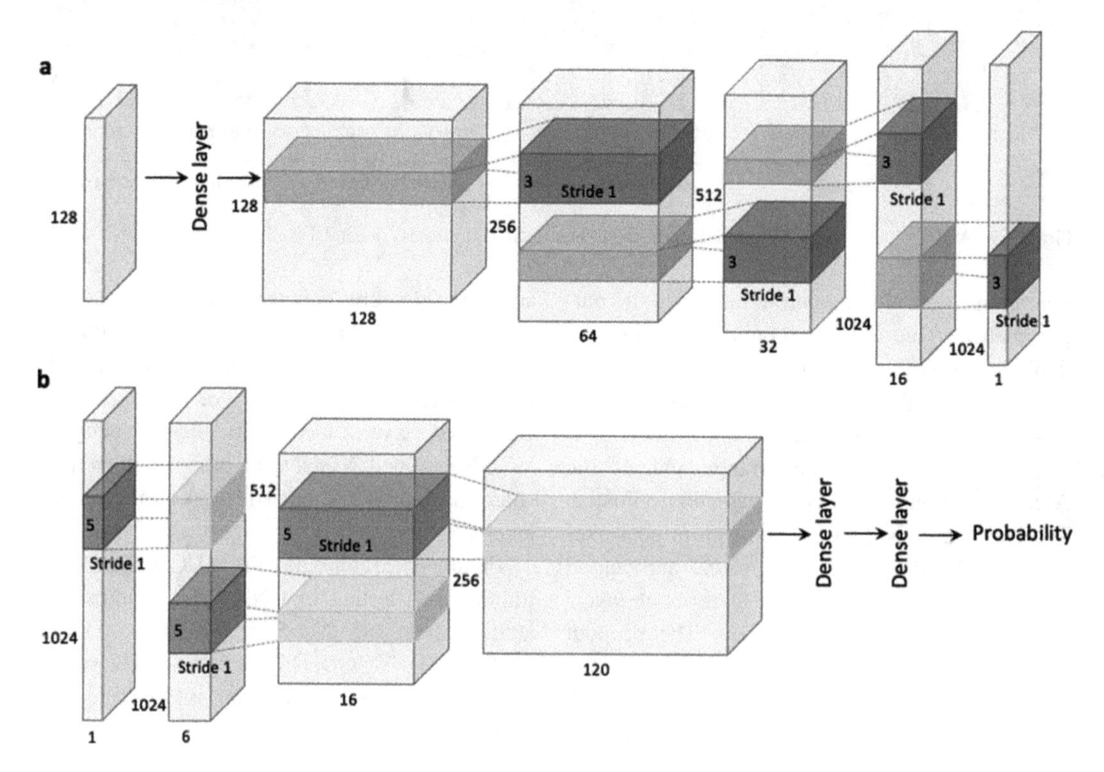

Fig. 31.4 The schematic illustration of GAN Generators and Discriminators. (**a**) The architecture of the Generator. (**b**) The architecture of the Discriminator. (Figure reproduced with permission from [70])

Table 31.4 Layer information for the Generator

Layer	Description
Dense layer	A dense layer transforms the noise input to a matrix of 128*128
Dropout layer	A dropout layer is applied to the matrix with a dropout rate of 0.25. The matrix is up sampled to have the shape of 256*128
Convolutional layer	A reverse convolutional layer with 64 kernels of size 3 and a stride of 1. The matrix is up sampled to have the shape of 512*64
Convolutional layer	A reverse convolutional layer with 32 kernels of size 3 and a stride of 1. The matrix is up sampled to have the shape of 1024*32
Convolutional layer	A reverse convolutional layer with 16 kernels of size 3 and a stride of 1. The matrix is transformed to have the shape of 1024*16
Convolutional layer	A reverse convolutional layer with 1 kernel of size 3 and a stride of 1. To report generated outcome molecular fingerprints

selected for both the dense and reverse convolutional layers. An L2 penalty is assigned to the reverse convolutional layers. Detailed information for each layer of the G is specified in Table 31.4. D generally follows the LeNet-5-based architecture (Fig. 31.4b). Dropout layers with a rate of 0.25 are inserted after each convolutional layer.

31.4.4 The Training and Performance of the Generated dcGAN Model

As a zero-sum non-cooperative game, the GAN model can only be converged if the Nash equilibrium has been reached by the D and G [76]. There are two objectives for the training process to accomplish simultaneously. First, D should be trained to make accurate classifications of fingerprints from known active CB ligands and G generated compounds. Second, G should be trained to generate fingerprints that cannot be distinguished from known active molecules. In other words, both the discriminative loss and the generative loss should be minimized simultaneously to achieve these two objectives (Fig. 31.5). Minimizing the discriminative loss optimizes the weights of the D to make correct distinctions, while minimizing the generative loss optimizes the weights of the G to mislead the D.

Fingerprints of generated compounds from G are labeled as 0 while fingerprints of real active CB compounds are labeled as 1 to train the D and calculate the discriminative loss (Fig. 31.5a). On the other hand, to calculate the generative loss, fingerprints from both generated compounds and known active CB ligands are labeled with 1 to train the G-D stack and report the generative loss (Fig. 31.5b). The first step of reporting discriminative loss is to train the D. The convergence of the discriminative loss reflects D's improvement on making correct classifications. The second step of reporting generative loss is to train the G-D stack. The convergence of the generative loss suggests G's capability of creating compelling fingerprints to mislead a well-trained

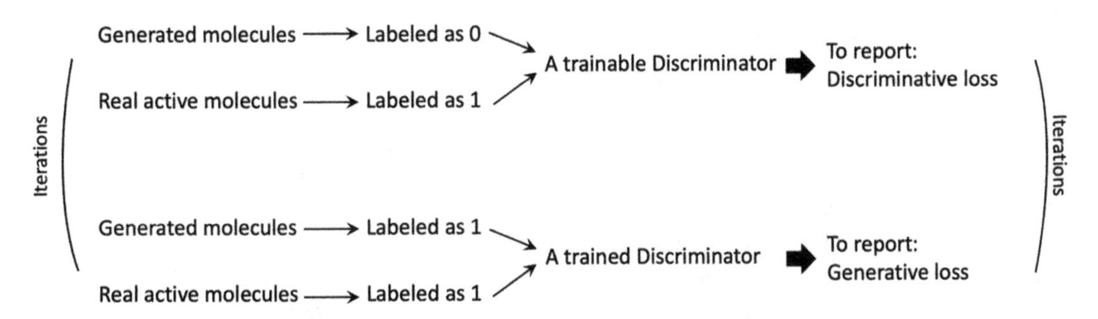

Fig. 31.5 The training iterations of the GAN model

D. Generating fingerprints has the limitation that it is almost impossible to draw the chemical structure out of a molecular fingerprint. As a compromised solution, a molecular similarity search using a generated fingerprint towards a large chemical library can suggest a proximate structure.

31.5 Application of RNN LSTM on Designing Scaffold-Focused and Target-Specific Molecules

31.5.1 A Brief Overview

Enumerating scaffold-focused and target-specific compounds functions as a common strategy in drug discovery. To realize the automated virtual enumeration by generating *de novo* structures, the Deep-learning Molecule Generation Model (DeepMGM) was designed and developed [77]. Using SMILES of drug-like molecules as the input, a long short-term memory (LSTM)-based recurrent neural network was trained to produce a general model (g-DeepMGM). A case study of preparing an indole-scaffold focused library was illustrated to exhibit the feasibility. Then, after a transfer learning process using known CB2 ligands, a Target-specific version (t-DeepMGM) was constructed. A sampling practice showed that generated compounds can maintain similar physical-chemical properties to reported active CB2 molecules. A supervised classifier for CB2 ligands was incorporated at the end to close a virtual design-test circle.

31.5.2 General and Target-Specific Molecule Generation Models (g-DeepMGM and t-DeepMGM)

The scheme of the workflow implemented in DeepMGM is exhibited in Fig. 31.6. A half million diversified ZINC compounds were collected to function as training data for preparing the g-DeepMGM (Fig. 31.6a). The majority (87.2%) of collected molecules don't have violations to the RO5. Molecules were described in SMILES strings. The iterative training process allowed the model to comprehend the grammar of composing valid SMILES strings to represent mostly drug-like molecules. However, the chemical space was not only limited to be drug-like, as there were still 12.8% of training molecules that possess violations to RO5. The One-hot encoding method was used to construct binary vectors to encode SMILES strings into a format that is machine-readable. The g-DeepMGM is a RNN with LSTM layers. The g-DeepMGM was trained to predict the probability distribution of the $n + 1$th SMILES character given the input of a string with n characters. The categorical cross-entropy was selected as the loss function to evaluate the model training process. The molecular generation can be initiated by providing SMILES of a starting scaffold to sample the remaining part of the compound.

The transfer learning of the prepared g-DeepMGM on collected bioactive CB2 ligands was conducted to result in the CB2 specific t-DeepMGM (Fig. 31.6b). A same set of CB2 compounds with reported *Ki* values was used here [27, 71]. These compounds were one-hot encoded into binary vectors as well. These CB2 compounds give a limited chemical space coverage with different physical-chemical properties in comparison to a half million diversified ZINC compounds. In the process of transfer learning, the model combined learned patterns behind a half-million ZINC compounds to further extract features from CB2 ligands. Five case studies using known CB2 seed scaffolds were carried out for enumeration using the trained t-DeepMGM. An MLP-based supervised ML classifier was prepared to virtually assess enumerated compounds from the t-DeepMGM (Fig. 31.6c).

31.5.3 Recurrent Neural Networks (RNN) Model Training and Sampling

The architecture of the g-DeepMGM includes LSTM and dense layers for processing SMILES characters (Fig. 31.7). An 80:20 ratio was adapted

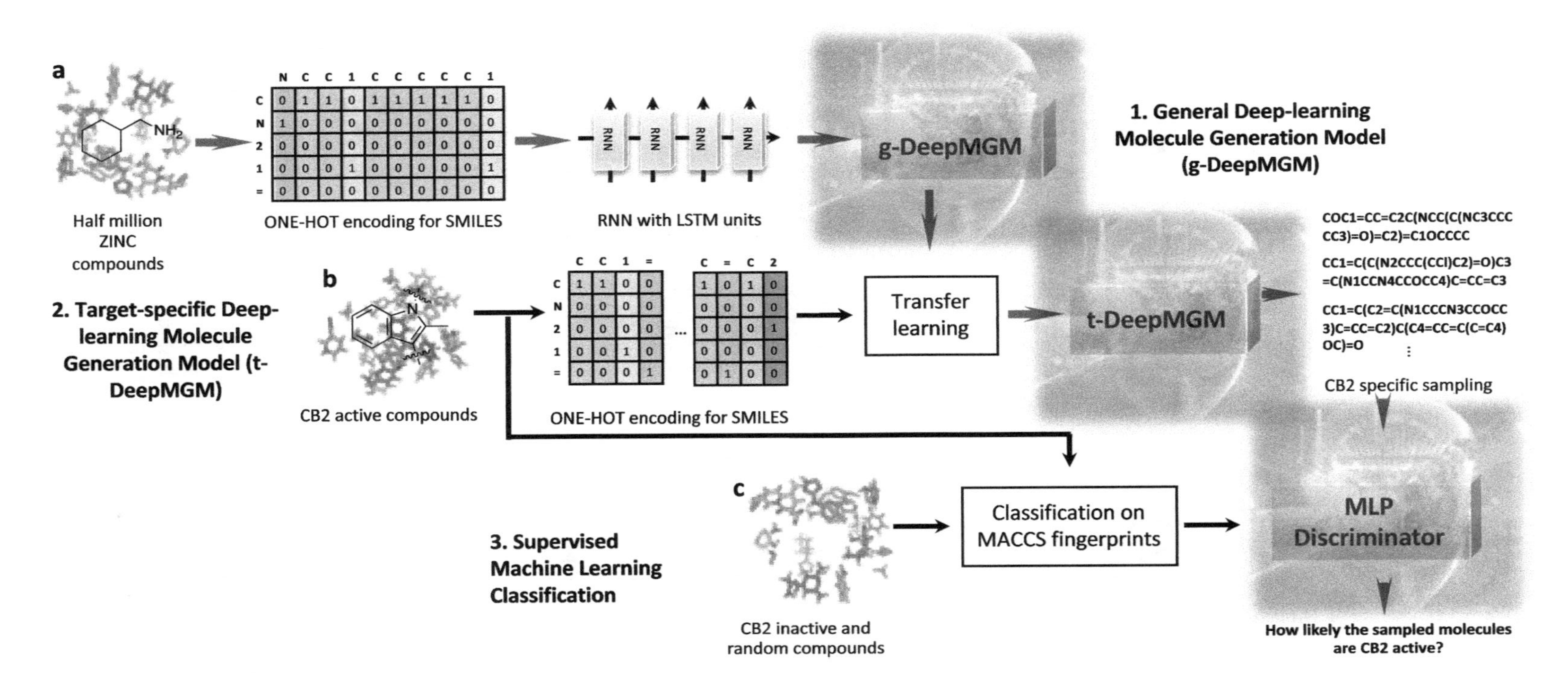

Fig. 31.6 The schematic overview of development pipeline for the General Deep-learning Molecule Generation Model (g-DeepMGM) and the Target-specific Deep-learning Molecule Generation Model (t-DeepMGM) in combination with Supervised Machine-learning Discriminator of multilayer perceptron (MLP)

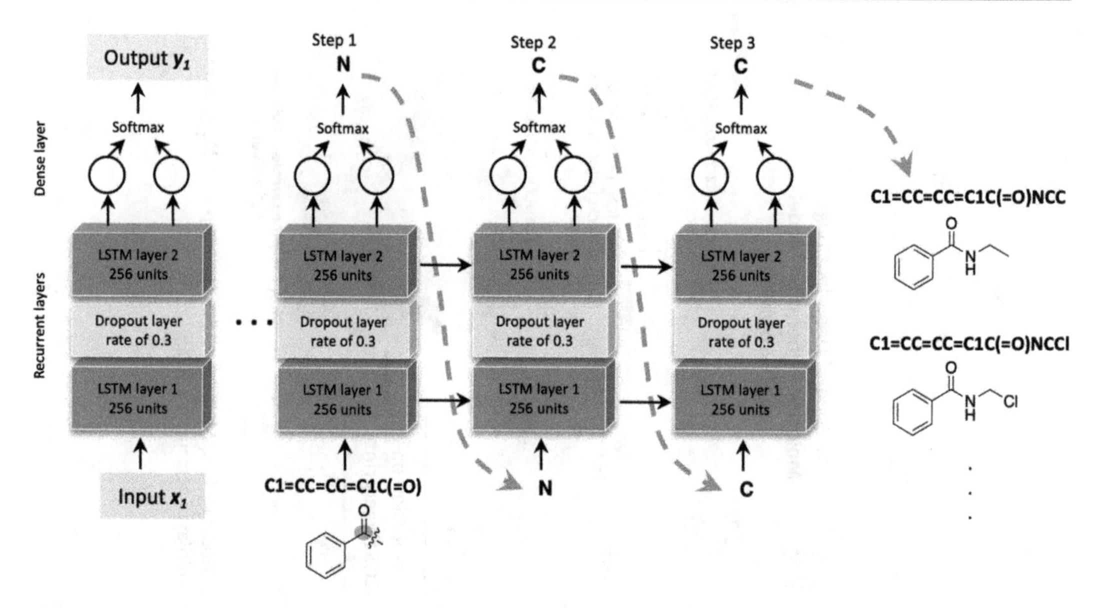

Fig. 31.7 The training and sampling of the recurrent neural networks (RNN) model with long-short-term memory (LSTM) units

to split ZINC compounds into training and test sets. The training process involved using mini-batches of 512 molecules for 100 epochs. The loss and accuracy were calculated for saved models at each epoch.

The training process results in a probability distribution of the next SMILES character given the input string. The resulted distribution is then used as the reference to suggest the next character in the process of sampling. Suggested characters are appended to the initial input afterwards in sequence. The sampling is an iterative process to compose SMILES strings for outcome molecules. The SMILES string from a lead fragment can be provided as the initial input to sample a scaffold-focused chemical library. Figure 31.7 gives an example of benzaldehyde. To enumerate compounds with moieties attached to the carbonyl group, the SMILES string "C1=CC=CC=C1C(=O)" is used as the input to initiate the g-DeepMGM in step 1. The SMILES character "N" is reported as the output. In step 2, SMILES character "N" is appended to the initial string to function as the new input for predicting the next character. The SMILES character "C" is reported as the output. The process is iterated until the generation of a complete molecule is finished. Additional modification points are available throughout the benzaldehyde scaffold. The atom for modification can be selected specifically by adjusting the direction and the starting atom to compose the SMILES strings [78]. In the current work, the assessment of sampling temperatures [79] was conducted as well. The sampling temperature affects the conditional probability distribution of the characters in the SMILES vocabulary. In other words, the sampling of the next character under a lower temperature tends to give more conservative predictions, while the sampling process under a higher temperature favors the selection of permutating characters.

31.5.4 Indole Scaffold-Specific Compounds Generation

The sampling practice on the indole ring is exemplified here to show g-DeepMGM's capability of generating a scaffold-focused compound library. The saved out model at six different training epochs, 10, 20, 40, 60, 80 and 100 were deployed to sample molecules independently under four sampling temperatures, 0.5, 1.0, 1.2,

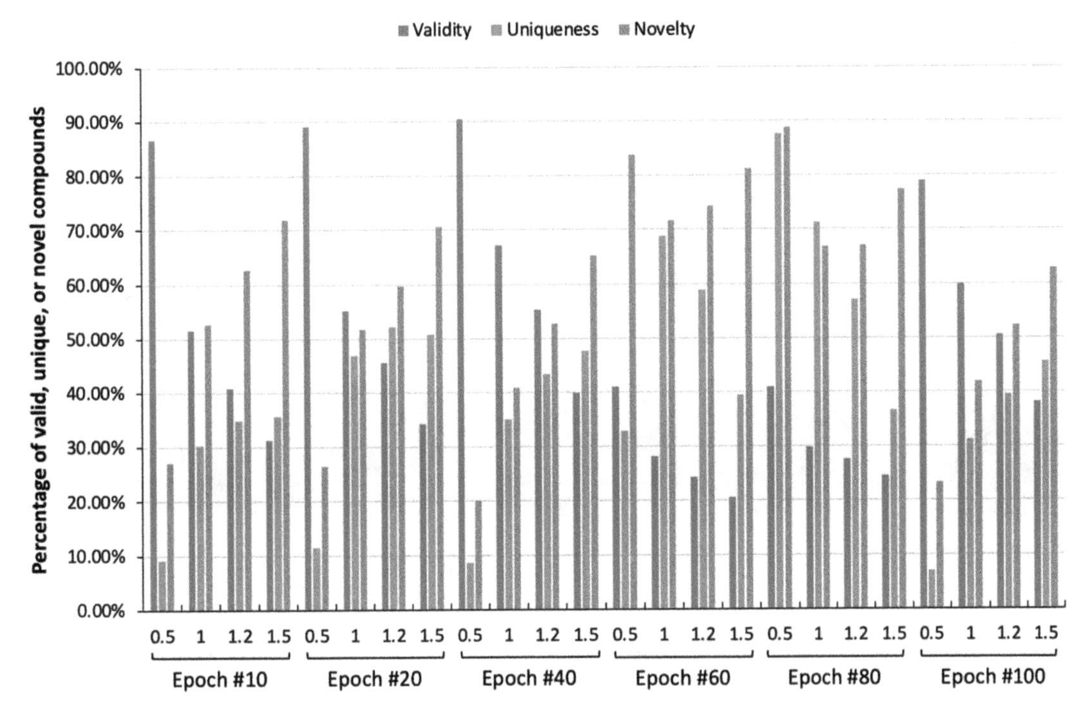

Fig. 31.8 The validity, uniqueness, and novelty of sampled indole molecules under different epochs and sampling temperatures of the g-DeepMGM

and 1.5. Three criteria were adopted here to assess sampled compounds: (1) validity that assesses whether molecular structures can be transformed from sampled SMILES strings; (2) uniqueness that assesses whether valid SMILES strings are unique from each other; and, (3) novelty that assesses whether valid and unique SMILES strings are different from molecules in the training set.

Different scores of validity, uniqueness and novelty can be perceived as molecules were generated under diverse epochs and various sampling temperatures (Fig. 31.8). At epoch 40, g-DeepMGM exhibited the highest capability of creating valid SMILES strings. The improvement sophistication of generating drug-like molecules was expected after training epochs as the purpose of the training process is to allow the model to learn and grasp proper ways of writing valid SMILES strings. With the increase of the temperature, it was noticed that both uniqueness and novelty were improved. Conservative predictions can favor valid but sometimes redundant SMILES strings under low temperatures. Aggressive predictions can produce unique and novel but sometimes invalid SMILES strings in contrast.

A graphical illustration of generated indole-focused compounds may provide intuitional understanding (Fig. 31.9a). At the lowest temperature (T = 0.5), aliphatic chains and ring systems (such as cyclohexane, phenyl, as well as piperazine rings) can both be sampled. Interestingly, amines (primary, secondary, and tertiary) and methoxy moieties can also be produced at a relatively low temperature. Notably, there are limited atom types (carbon, nitrogen, and oxygen) among sampled compounds. When the temperature is increased to T = 1.0, diversified atom types appear. The appearance of a trifluoromethyl group, a bromine substitution, and even a sulfonyl amine moiety appreciably expanded the coverage of potential chemical space by sampled structures. It is also noticed that basic carbon

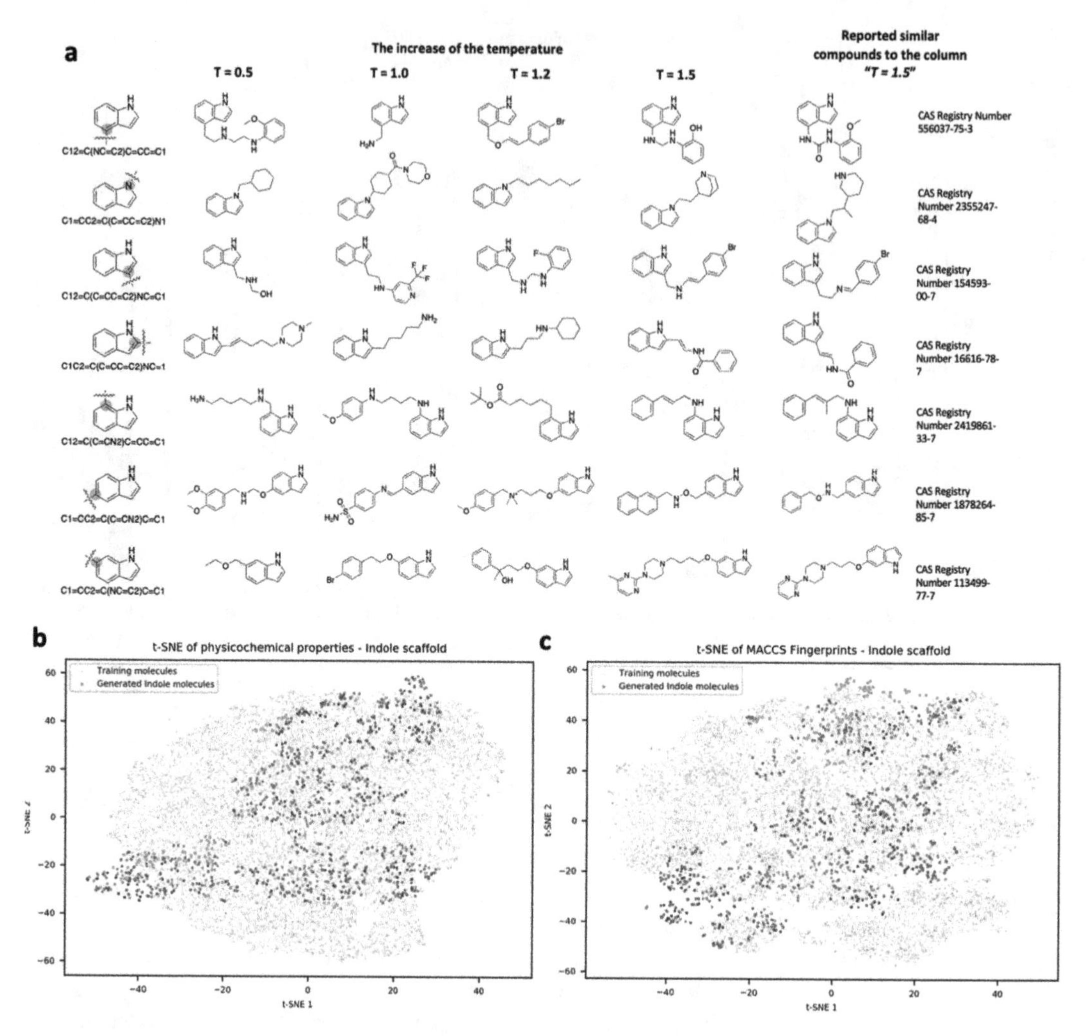

Fig. 31.9 Sampling examples on indole scaffold and t-SNE analysis. (**a**) Randomly selected sampling outcome for the indole scaffold under four temperatures with the g-DeepMGM at epoch 40. Seven SMILES strings representing seven addition positions on the indole were fed as the initial input. Reported similar compounds to the generated molecules in the column "T = 1.5" are listed for comparison. Using both physical-chemical properties-based (**b**) and MACCS fingerprints-based (**c**) t-SNE analysis to compare generated indole molecules and training compounds

chains can be produced at a relatively high temperature. For example, when the T = 1.2, an aliphatic heptyl chain is sampled. Meanwhile, a quaternary ammonium cation is generated to possess a positive charge. The generation of charged molecules besides the neutral compounds may further expand the application scope of the g-DeepMGM [77]. At the highest temperature (T = 1.5), various atom types and mixed moieties (pyrimidine, quinuclidine, naphthalene, etc.) are frequently sampled among generated structures. Although a high degree of structural diversity is observed for sampled molecules under T = 1.5, structurally similar compounds can still be identified from the SciFinder, which suggests the generation of realistic structures.

To investigate the chemical space coverage of sampled indole-focused structures, t-SNE analysis was conducted on both physical-chemical properties (Fig. 31.9b) and MACCS fingerprints

(Fig. 31.9c). A half-million training molecules (blue dots) generally defined the overall boundary of the drug-like chemical space. Sampled indole-focused molecules (colorful dots) were sparsely spread within this space. Colorful dots not only co-localize with blue dots, but also form clusters. Such an observation supports the hypothesis of generating scaffold-focused libraries that emphasizes a certain chemical space.

31.5.5 Transfer Learning for Cannabinoid Receptor Subtype 2 (CB2)

After the transfer learning using collected CB2 ligands, t-DeepMGM was applied to sample molecules on four well-known CB2 scaffolds. Figure 31.10a exemplifies sampling outcomes in comparison with known CB2 active ligands. The amide bond was cut on the **JTE-907** [80, 81] to result in a scaffold for enumerating structures. A sampling outcome (**1**) exhibited the reconstruction of the amide bond. The original benzodioxole ring can be replaced by cyclohexane (**1**), morpholine (**2**), and pyrrolidine (**3**) during the generation. The process also explored the length of the linker in between.

The diethylamine moiety of the biamide CB2 inverse agonist [82] was chopped for structural enumeration. Interestingly, moieties, isopentane (**4**), cyclopropylamine (**5**), and methoxypropane (**6**) that mimic the size of the diethylamine group were generated.

Structural modifications on **AM630**, a classic CB2 antagonist [83, 84] usually occurred on the indole scaffold. The compound enumeration was conducted to explore the replacement of moieties connected to the indole scaffold. With the benzaldehyde group chopped, t-DeepMGM generated both an aliphatic chain (ethyl bromide **9**) and rings (pyrrolidine **7** and morpholine **8**). With the ethylmorpholine moiety chopped, compound **10** was sampled with one more carbon in the linker that connected the indole and the morpholine. Compound **12** was sampled to have a thiomorpholine replace the morpholine. The sulfur in the thiomorpholine and the oxygen in the morpholine are both capable of forming hydrophilic interactions.

The fourth scaffold comes from triaryl sulfonamide derivatives [85]. The sulfonamide was chopped for enumerating compounds. The recurrence of the sulfonamide was observed in sampled compound **13**, which also has a pyridine ring attached.

To investigate the chemical space coverage of t-DeepMGM sampled molecules, t-SNE analysis was conducted on MACCS fingerprints (Fig. 31.10b). Again, a half-million training molecules (blue dots) generally defined the overall boundary of the drug-like chemical space. Known CB2 ligands (orange dots) had a concentrated distribution at focused regions. Molecules sampled by the t-DeepMGM distributed within the space covered by orange dots. The transfer learning process forged the t-DeepMGM to produce scaffold-focused chemical libraries with a preference for the CB2 target [77].

31.6 Concluding Remarks

At this point, the following four areas present challenges and opportunities for AI generative chemistry: (1) the synthetic viability of the generated molecular structures; (2) alternate molecular representations that more accurately depict a structure; (3) the generation of macromolecules; and (4) close-loop automation in conjunction with experimental validations. Some of the current methods for assessing synthesizability are based on synthetic pathways and molecular structural data, which calls for a sophisticated and thorough heuristic definition. The inability to integrate generative models with medicinal chemistry synthesis is revealed to be a major barrier. Small compounds can currently be described well using molecular representations like SMILES strings and molecular fingerprints. However, it will be interesting if the new representations can be made to additionally take into account geometry data in three dimensions. Chiral substances

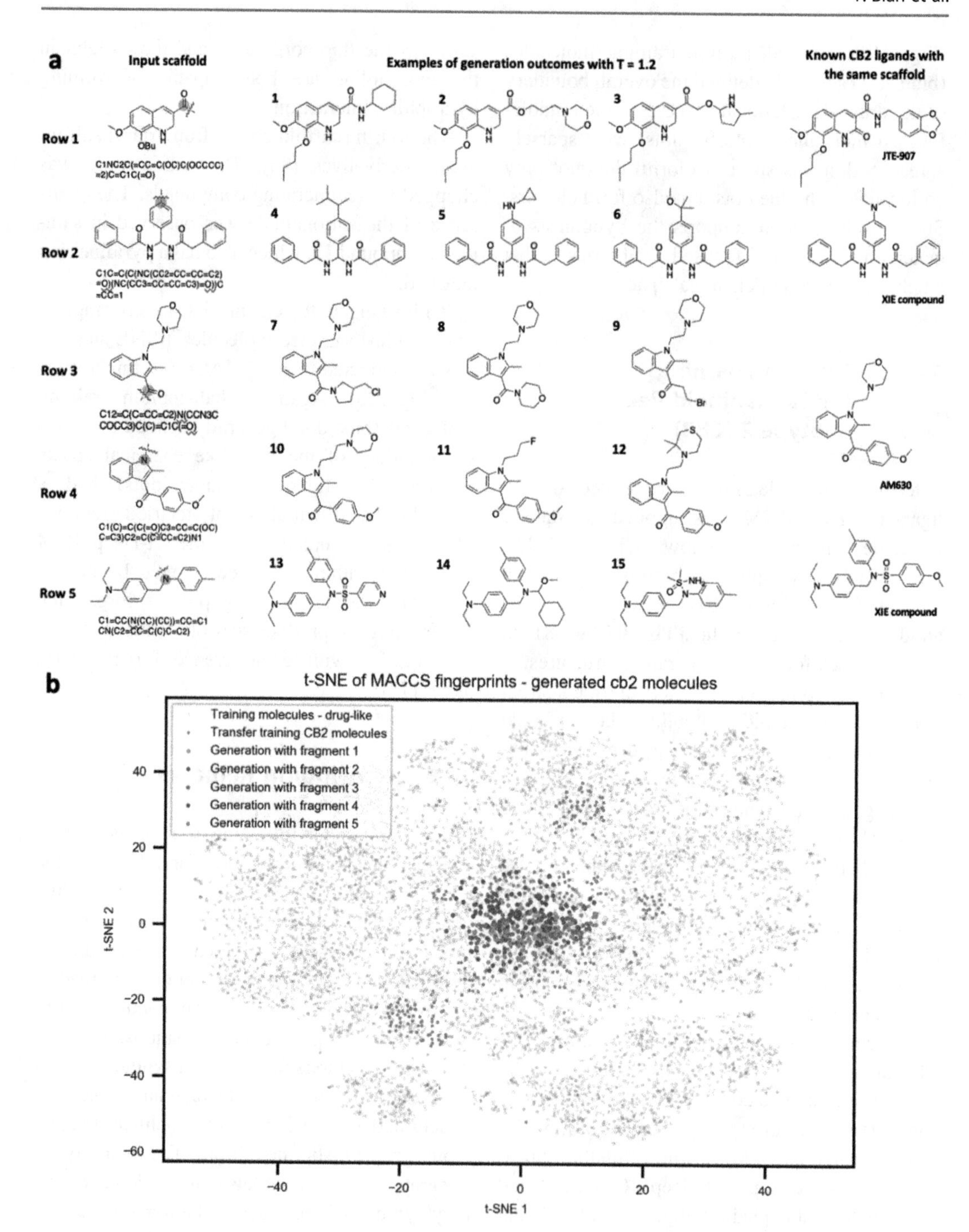

Fig. 31.10 Using t-DeepMGM to sample CB2 target-specific molecules. (**a**) Molecular sampling examples from the t-DeepMGM for four known CB2 scaffolds. (**b**) Using MACCS fingerprints-based t-SNE analysis to compare generated molecules, known CB2 ligands, and initial half-million training compounds

may have many effects on a biological system, and even a minor molecule's change in conformation might affect how a receptor interacts with a ligand. The efficacy of feature extraction on spatial patterns is enhanced by the additional consideration of bond type, length, and angles. Research in deep learning is data-driven. Due to the easier access to chemical data, most current applications of generative chemistry are geared toward designing tiny molecules. De novo protein creation attempts are anticipated as protein-related database construction increases. Folding and its conformation are even more important in determining function, so better representations are certainly needed to describe proteins. Finally, it is worth noting how generative chemistry can be integrated into drug design frameworks to close the loop on this automated process.

Acknowledgments The authors acknowledge the funding support to the Xie Laboratory and CDAR Center from the NIH (R01DA052329, P30PDA035778A and R56AG074951).

Conflict of Interest All authors have no conflict of interest to declare.

References

1. DiMasi JA, Grabowski HG, Hansen RW (2016) Innovation in the pharmaceutical industry: new estimates of R & D costs. J Health Econ 47:20–33
2. Wouters OJ, McKee M, Luyten J (2020) Estimated research and development investment needed to bring a new medicine to market, 2009-2018. JAMA 323(9): 844–853
3. Yasi EA, Kruyer NS, Peralta-Yahya P (2020) Advances in G protein-coupled receptor high-throughput screening. Curr Opin Biotechnol 64:210–217
4. Blay V et al (2020) High-Throughput Screening: today's biochemical and cell-based approaches. Drug Discov Today 25(10):1807–1821
5. Ge H et al (2019) Significantly different effects of tetrahydroberberrubine enantiomers on dopamine D1/D2 receptors revealed by experimental study and integrated in silico simulation. J Comput Aided Mol Des 33(4):447–459
6. Pagadala NS, Syed K, Tuszynski J (2017) Software for molecular docking: a review. Biophys Rev 9(2): 91–102
7. Bian Y-M et al (2019) Computational systems pharmacology analysis of cannabidiol: a combination of chemogenomics-knowledgebase network analysis and integrated in silico modeling and simulation. Acta Pharmacol Sin 40(3):374
8. Bian Y et al (2017) Integrated in silico fragment-based drug design: case study with allosteric modulators on metabotropic glutamate receptor 5. AAPS J 19(4): 1235–1248
9. Kwon JJ et al (2022) Structure–function analysis of the SHOC2–MRAS–PP1C holophosphatase complex. Nature 609(7926):408–415
10. Wang J et al (2004) Development and testing of a general amber force field. J Comput Chem 25(9): 1157–1174
11. Vanommeslaeghe K et al (2010) CHARMM general force field: a force field for drug-like molecules compatible with the CHARMM all-atom additive biological force fields. J Comput Chem 31(4):671–690
12. Hajduk PJ, Greer J (2007) A decade of fragment-based drug design: strategic advances and lessons learned. Nat Rev Drug Discov 6(3):211–219
13. Yang S-Y (2010) Pharmacophore modeling and applications in drug discovery: challenges and recent advances. Drug Discov Today 15(11–12):444–450
14. Wieder M et al (2017) Common hits approach: combining pharmacophore modeling and molecular dynamics simulations. J Chem Inf Model 57(2): 365–385
15. Liu Z et al (2020) Discovery of orally bioavailable chromone derivatives as potent and selective BRD4 inhibitors: scaffolding hopping, optimization and pharmacological evaluation. J Med Chem 63(10): 5242–5256
16. Hu Y, Stumpfe D, Bajorath J (2017) Recent advances in scaffold hopping: miniperspective. J Med Chem 60(4):1238–1246
17. Muegge I, Mukherjee P (2016) An overview of molecular fingerprint similarity search in virtual screening. Expert Opin Drug Discovery 11(2):137–148
18. Fan Y et al (2019) Investigation of machine intelligence in compound cell activity classification. Mol Pharm 16(11):4472–4484
19. Minerali E et al (2020) Comparing machine learning algorithms for predicting drug-induced liver injury (DILI). Mol Pharm 17(7):2628–2637
20. Karras T, et al (2019) Analyzing and improving the image quality of stylegan. arXiv preprint arXiv:1912.04958
21. Wen T-H, et al (2015) Semantically conditioned lstm-based natural language generation for spoken dialogue systems. arXiv preprint arXiv:1508.01745
22. Zhavoronkov A et al (2019) Deep learning enables rapid identification of potent DDR1 kinase inhibitors. Nat Biotechnol 37(9):1038–1040
23. Turing AM (2009) Computing machinery and intelligence. In: Parsing the turing test. Springer, pp 23–65

24. Chollet F (2018) Deep learning mit Python und Keras: das Praxis-Handbuch vom Entwickler der Keras-Bibliothek. MITP-Verlags GmbH & Co. KG
25. Segler MH, Preuss M, Waller MP (2018) Planning chemical syntheses with deep neural networks and symbolic AI. Nature 555(7698):604–610
26. Lipinski CA (2016) Rule of five in 2015 and beyond: target and ligand structural limitations, ligand chemistry structure and drug discovery project decisions. Adv Drug Deliv Rev 101:34–41
27. Bian Y et al (2019) Prediction of orthosteric and allosteric regulations on cannabinoid receptors using supervised machine learning classifiers. Mol Pharm 16(6):2605–2615
28. Lo Y-C et al (2018) Machine learning in chemoinformatics and drug discovery. Drug Discov Today 23(8):1538–1546
29. Jing Y et al (2018) Deep learning for drug design: an artificial intelligence paradigm for drug discovery in the big data era. AAPS J 20(3):58
30. Vamathevan J et al (2019) Applications of machine learning in drug discovery and development. Nat Rev Drug Discov 18(6):463–477
31. Bian Y et al (2023) Target-driven machine learning-enabled virtual screening (TAME-VS) platform for early-stage hit identification. Front Mol Biosci 10:1163536
32. Korotcov A et al (2017) Comparison of deep learning with multiple machine learning methods and metrics using diverse drug discovery data sets. Mol Pharm 14(12):4462–4475
33. Ma XH et al (2009) Comparative analysis of machine learning methods in ligand-based virtual screening of large compound libraries. Comb Chem High Throughput Screen 12(4):344–357
34. Verma J, Khedkar VM, Coutinho EC (2010) 3D-QSAR in drug design-a review. Curr Top Med Chem 10(1):95–115
35. Fan F et al (2019) The integration of pharmacophore-based 3D QSAR modeling and virtual screening in safety profiling: a case study to identify antagonistic activities against adenosine receptor, A2A, using 1,897 known drugs. PLoS One 14(1):e0204378
36. Gladysz R et al (2018) Spectrophores as one-dimensional descriptors calculated from three-dimensional atomic properties: applications ranging from scaffold hopping to multi-target virtual screening. J Chem 10(1):9
37. Nguyen TT, Nguyen ND, Nahavandi S (2020) Deep reinforcement learning for multiagent systems: a review of challenges, solutions, and applications. IEEE Trans Cybernet 50:3826–3839
38. LeCun Y, Bengio Y, Hinton G (2015) Deep learning. Nature 521(7553):436–444
39. Goodfellow I, Bengio Y, Courville A (2016) Deep learning. MIT Press
40. Hochreiter S, Schmidhuber J (1997) Long short-term memory. Neural Comput 9(8):1735–1780
41. Goodfellow I et al (2014) Generative adversarial nets. In: Advances in neural information processing systems
42. Bian Y, Xie X-Q (2021) Generative chemistry: drug discovery with deep learning generative models. J Mol Model 27:1–18
43. The UniProt Consortium (2017) UniProt: the universal protein knowledgebase. Nucleic Acids Res 45(D1): D158–D169
44. Berman HM et al (2000) The protein data bank. Nucleic Acids Res 28(1):235–242
45. Wang R et al (2005) The PDBbind database: methodologies and updates. J Med Chem 48(12): 4111–4119
46. Kim S et al (2019) PubChem 2019 update: improved access to chemical data. Nucleic Acids Res 47(D1): D1102–D1109
47. Gaulton A et al (2017) The ChEMBL database in 2017. Nucleic Acids Res 45(D1):D945–D954
48. Papadatos G et al (2016) SureChEMBL: a large-scale, chemically annotated patent document database. Nucleic Acids Res 44(D1):D1220–D1228
49. Wishart DS et al (2018) DrugBank 5.0: a major update to the DrugBank database for 2018. Nucleic Acids Res 46(D1):D1074–D1082
50. Sterling T, Irwin JJ (2015) ZINC 15–ligand discovery for everyone. J Chem Inf Model 55(11):2324–2337
51. Huang Z et al (2014) ASD v2. 0: updated content and novel features focusing on allosteric regulation. Nucleic Acids Res 42(D1):D510–D516
52. Ruddigkeit L et al (2012) Enumeration of 166 billion organic small molecules in the chemical universe database GDB-17. J Chem Inf Model 52(11):2864–2875
53. Weininger D (1988) SMILES, a chemical language and information system. 1. Introduction to methodology and encoding rules. J Chem Inf Comput Sci 28(1): 31–36
54. Heller SR et al (2015) InChI, the IUPAC international chemical identifier. J Chem 7(1):23
55. Durant JL et al (2002) Reoptimization of MDL keys for use in drug discovery. J Chem Inf Comput Sci 42(6):1273–1280
56. Glen RC et al (2006) Circular fingerprints: flexible molecular descriptors with applications from physical chemistry to ADME. IDrugs 9(3):199
57. Rogers D, Hahn M (2010) Extended-connectivity fingerprints. J Chem Inf Model 50(5):742–754
58. Hert J et al (2004) Comparison of fingerprint-based methods for virtual screening using multiple bioactive reference structures. J Chem Inf Comput Sci 44(3): 1177–1185
59. Pérez-Nueno VI et al (2009) APIF: a new interaction fingerprint based on atom pairs and its application to virtual screening. J Chem Inf Model 49(5):1245–1260
60. Jiang D et al (2021) Could graph neural networks learn better molecular representation for drug discovery? A comparison study of descriptor-based and graph-based models. J Chem 13(1):1–23

61. Landrum G (2016) Rdkit: open-source cheminformatics software. GitHub and SourceForge 10:3592822
62. O'Boyle NM et al (2011) Open Babel: an open chemical toolbox. J Chem 3(1):33
63. Willighagen EL et al (2017) The Chemistry Development Kit (CDK) v2. 0: atom typing, depiction, molecular formulas, and substructure searching. J Chem 9(1):33
64. Arabie P, et al (2006) Studies in classification, data analysis, and knowledge organization. https://doi.org/10.1007/3-540-35978-8_34
65. Abadi M et al (2016) Tensorflow: a system for large-scale machine learning. In: 12th {USENIX} symposium on operating systems design and implementation ({OSDI} 16)
66. Etaati L (2019) Deep learning tools with cognitive toolkit (CNTK). In: Machine learning with microsoft technologies. Springer, pp 287–302
67. Team T, et al (2016) Theano: a Python framework for fast computation of mathematical expressions. https://doi.org/10.48550/arXiv.1605.02688
68. Paszke A et al (2019) PyTorch: an imperative style, high-performance deep learning library. In: Advances in neural information processing systems
69. Chollet F (2015) keras is an open-source neural-network library written in Python. GitHub. https://github.com/fchollet/keras
70. Pedregosa F et al (2011) Scikit-learn: machine learning in Python. J Mach Learn Res 12:2825–2830
71. Bian Y et al (2019) Deep convolutional generative adversarial network (dcGAN) models for screening and design of small molecules targeting cannabinoid receptors. Mol Pharm 16(11):4451–4460
72. LeCun Y et al (1995) Comparison of learning algorithms for handwritten digit recognition. In: International conference on artificial neural networks, Perth, WA
73. Krizhevsky A, Sutskever I, Hinton GE (2012) Imagenet classification with deep convolutional neural networks. In: Advances in neural information processing systems
74. Zeiler MD, Fergus R (2014) Visualizing and understanding convolutional networks. In: European conference on computer vision. Springer
75. Simonyan K, Zisserman A (2014) Very deep convolutional networks for large-scale image recognition. arXiv preprint arXiv:1409.1556
76. Heusel M, et al (2017) Gans trained by a two time-scale update rule converge to a nash equilibrium. 12(1):arXiv preprint arXiv:1706.08500
77. Bian Y, Xie X-Q (2022) Artificial intelligent deep learning molecular generative modeling of scaffold-focused and cannabinoid CB2 target-specific small-molecule sublibraries. Cells 11(5):915
78. Prykhodko O et al (2019) A de novo molecular generation method using latent vector based generative adversarial network. J Chem 11(1):1–13
79. Moret M et al (2020) Generative molecular design in low data regimes. Nat Mach Intellig 2(3):171–180
80. Iwamura H et al (2001) In vitro and in vivo pharmacological characterization of JTE-907, a novel selective ligand for cannabinoid CB2 receptor. J Pharmacol Exp Ther 296(2):420–425
81. Ueda Y et al (2005) Involvement of cannabinoid CB2 receptor-mediated response and efficacy of cannabinoid CB2 receptor inverse agonist, JTE-907, in cutaneous inflammation in mice. Eur J Pharmacol 520(1–3):164–171
82. Yang P et al (2012) Lead discovery, chemistry optimization, and biological evaluation studies of novel biamide derivatives as CB2 receptor inverse agonists and osteoclast inhibitors. J Med Chem 55(22): 9973–9987
83. Pertwee R et al (1995) AM630, a competitive cannabinoid receptor antagonist. Life Sci 56(23–24): 1949–1955
84. Ross RA et al (1999) Agonist-inverse agonist characterization at CB1 and CB2 cannabinoid receptors of L759633, L759656 and AM630. Br J Pharmacol 126(3):665
85. Yang P et al (2013) Novel triaryl sulfonamide derivatives as selective cannabinoid receptor 2 inverse agonists and osteoclast inhibitors: discovery, optimization, and biological evaluation. J Med Chem 56 (5):2045–2058

32 Artificial Intelligence Technologies for Clinical Data PharmacoAnalytics Case Studies on Alzheimer's Disease

Guangyi Zhao, Shuyuan Zhao, and Xiang-Qun Xie

Abstract

Artificial intelligence (AI) technologies are used extensively in health science-related fields. In this book chapter, we will briefly introduce AI applications in clinical data studies focusing on the recent works with AI techniques on Alzheimer's disease-related topics. This chapter will use three case studies as examples of AI applications covering different categories (statistical modeling, machine learning/deep learning, causal analysis/discovery) in Alzheimer's disease clinical pharmacoanalytics. We will dissect the study design, discuss the significance of the findings, and finally analyze the advantages and drawbacks of each model used in the studies. Further readings and resources are also provided for readers interested in carrying out studies independently. Lastly, we will touch upon the future opportunities and challenges of AI technology application in clinical research and the efforts made to overcome those challenges and difficulties. After reading this chapter, the readers should have a brief idea of the available pharmacoanalytics predicting models, their strengths, and their limitations, thus helping readers decide the appropriate analytics models suitable for further studies.

Keywords

Artificial intelligence · Clinical data pharmacoanalytics · Alzheimer's disease · Machine learning · Deep learning · Causal inference

32.1 Introduction

In general definition, artificial intelligence (AI) is the simulation of human intelligence with programmed machines. Artificial intelligent computer systems are extensively used in health

G. Zhao · S. Zhao
Department of Pharmaceutical Sciences and Computational Chemical Genomics Screening Center, Pharmacometrics and System Pharmacology (PSP) PharmacoAnalytics, School of Pharmacy, Pittsburgh, PA, USA

National Center of Excellence for Computational Drug Abuse Research, Pittsburgh, PA, USA
e-mail: guz22@pitt.edu; SHZ115@pitt.edu

X.-Q. Xie (✉)
Department of Pharmaceutical Sciences and Computational Chemical Genomics Screening Center, Pharmacometrics and System Pharmacology (PSP) PharmacoAnalytics, School of Pharmacy, Pittsburgh, PA, USA

National Center of Excellence for Computational Drug Abuse Research, Pittsburgh, PA, USA

Drug Discovery Institute, Pittsburgh, PA, USA

Departments of Computational Biology and Structural Biology, School of Medicine, University of Pittsburgh, Pittsburgh, PA, USA
e-mail: xix15@pitt.edu

G. Jagadeesh et al. (eds.), *The Quintessence of Basic and Clinical Research and Scientific Publishing*,
https://doi.org/10.1007/978-981-99-1284-1_32

science, and the applications consist of information sharing, human assistance, clinical decision support, drug discovery, among others [1]. Among them, clinical decision support, including disease diagnosis and prediction, and imaging analysis, are the most popular fields of research and application, especially with the machine learning/deep learning models, which have significantly advanced in recent years [2]. Other hot topics are using AI techniques for structure-based drug discovery [3–8], antigen/antibody/ligand discovery, and design [9, 10]. The AI computer system can search and model pharmaceutical/medical/clinical data and uncover insights that are very hard for physicians and researchers to discover on their own. AI techniques have been employed to fight against major human diseases, including cancer, cardiovascular disease, hypertension, diabetes, Alzheimer's disease, and others [11].

Artificial Intelligence and machine learning are increasingly used in basic research and clinical neuroscience, including Alzheimer's disease [12], Parkinson's disease [13], and substance use disorder (SUD) [14–16]. This book chapter focuses on a few representative AI techniques applied to Alzheimer's disease clinical research. We will also dissect a few research studies as examples to show the data source, study design, data processing, evaluation metrics, and outcomes. Although this chapter uses Alzheimer's Disease (AD) as an example, most models and methodologies such as statistical analyses and ML/DL predicting models are transferable to other disease models.

AD is a progressive neurodegenerative disease caused by neural brain damage characterized by the formation of amyloid plaques between neurons and the accumulation of phosphorylated tau within neurons [17]. Unfortunately, AD is irreversible and gets worse with time, which is the primary cause of dementia. Patients will first experience memory, language, and thinking problems, and then as more neurons are damaged, patients cannot carry out daily living activities. Eventually, as the lesion extends to the brain parts that control essential body functions such as walking and swallowing, patients will have to stay in bed and require continuous care. AD is fatal, either by itself or by other comorbidities, due to loss of body function.

AD has long been a global health problem. Over 6.5 million Americans are living with Alzheimer's or other types of dementia. One out of nine people aged 65 or older has dementia due to AD [18]. It is estimated that the healthcare cost for AD patients is to be more than $300 billion, and it ranks as the seventh leading cause of death in 2020 and 2021 in the United States [19].

AI techniques have been a trending field in AD research and have drawn attention [20]. Eventually, the number of publications related to AD research using AI techniques has exponentially increased over the last decade (Fig. 32.1).

Through long-term clinical and research efforts, AI tools and expertise have been set up to identify early-stage brain change in AD [21]. Those tools must be fine-tuned before physicians can confidently use them in disease diagnosis and decision-making [22]. There has been a continuous effort to improve the algorithms and data accessibility [23]. In the following section, we will discuss the AI techniques available in AD clinical data research and conduct case studies on three representative research papers.

The objective of this chapter is to provide the readers with the latest AI related technique research examples on AD and help them to understand the advantages and disadvantages of each technique, as well as to introduce related

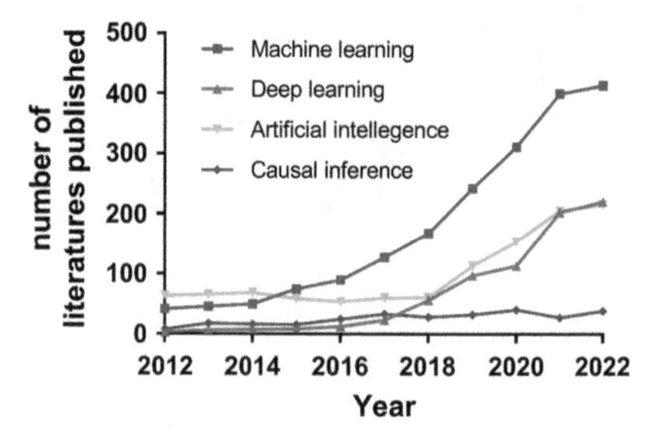

Fig. 32.1 Number of publications on PubMed on Alzheimer's disease research with AI techniques in recent ten years

resources, so they can be utilized in conducting their own research.

32.2 AI Techniques/Models and Case Study with AD Research

Before diving into the case study, we will mention data preprocessing briefly. Even though many research papers did not emphasize it in the text, it is a critical step before any model can be faithfully established. Data preprocessing is about enhancing the quality and unifying the raw data, such as outlier removal, missing value imputation, data conversion, and variables encoding [24].

In the field of data science, it is said that roughly 80% of the time is devoted to data cleaning and processing. Given the nature of the clinical data and medical records, many variables are recorded in different scales, in text, or are even missing [25]. All the variables should be converted into a relatively consistent form and should be normalized/standardized and checked for correlation; while this is not within the scope of this article, we would like our readers to be aware of the scale and importance of the data pre-procession step which should be taken care of very carefully.

Next, we will use three peer-reviewed research articles to review how the statistical, deep learning, and causal analysis models can be utilized in exploring the clinical data of AD.

32.2.1 Computational Systematic Pharmacology Pharmacoanalytics on AD Clinical Data with Statistical Models

There might be a trending opinion that statistical models, such as linear regression models, are less powerful and currently out of date compared to all the advanced AI techniques, for example, machine learning and artificial neural networks. The purpose of listing statistical models under AI techniques is to remind and emphasize the concept that regardless of all the fancy names, the very basis and essence of AI techniques, models and algorithms are statistical models. The following examples show that the "simple" statistical model is still valuable in answering particular research questions.

The FDA has approved six drugs for the treatment of AD [26]; five of the drugs —donepezil, rivastigmine, galantamine, memantine, and memantine combined with donepezil, only temporarily alleviate Alzheimer's symptoms but are not able to alter the underlying lesion in the brain; the sixth drug, aducanumab, is claimed to reduce the beta-amyloid plaques in the brain only in people with mild cognitive impairment (MCI) or mild dementia due to AD. There is currently no cure for AD. Given the slow pace of drug development in AD, the researchers have put more effort into repurposing current drugs in combination therapy against AD.

AD's pathology and etiology are of great complexity and not fully understood. AD's onset is at the relatively late stay of the human life span (usually >65 years old) [18], and the patients could already have co-existing age-related diseases such as cardiovascular diseases, diabetes, hypertension, and others [27]. The patients could have been taking multiple medications. Comparing disease progression among patients with different prescriptions enables systematic pharmacology analysis in observational clinical data that will lead to drug repurposing opportunities. Additionally, the systematic pharmacology analysis study on different drug combinations and disease progression differences in AD patients will not only reveal the potential underlying mechanisms for AD onset and processing but also help to point out new directions of AD drug development [28].

32.2.1.1 Case Study 1

As illustrated in the case study below [29], the authors explored the synergic effect of antihypertensive drugs (aHTN) and cholinesterase inhibitors (ChEI) in AD patients.

Research Objective To investigate cholinesterase inhibitor (ChEI) and antihypertensive (aHTN) medications to explore the synergism effect in slowing cognitive decline in AD patients.

Data Source The data is obtained from the University of Pittsburgh Alzheimer's Disease Research Center from April 1983 to March 2015 [30]. Out of 4364 participants in the clinic, a total of 617 probable AD patients' cases were selected for the analysis. MMSE scores are used as a measure of their cognitive function. Please note that there are also similar datasets available for the public from the National Alzheimer's Coordinating Center (NACC, https://naccdata.org/).

Study Design

1. The effect of aHTN drugs on cognitive decline in patients with HTN and AD: The patients with hypertension were divided into nine groups according to the types of aHTN drugs they were taking, which could be a single drug or a combination of two to three drugs (this is referred as the set 1 patients in Fig. 32.2). The author would like to know if the MMSE score decreases slower over 4 years as the patients start taking aHTN medications. A mix-effect regression analysis with random intercept and trend was conducted to control for several covariates associated with cognitive declines, such as age, sex, years of education, and APOEε4 carrier status. The time after the first use of an aHTN drug was used as the primary predictor for the MMSE score that was power transformed.
2. The cognitive decline in the first 2 years with the synergistic effect of aHTN and ChEIs drugs in patients diagnosed with probable AD. A total of 419 patients who did not switch treatment (which means they remained ChEI

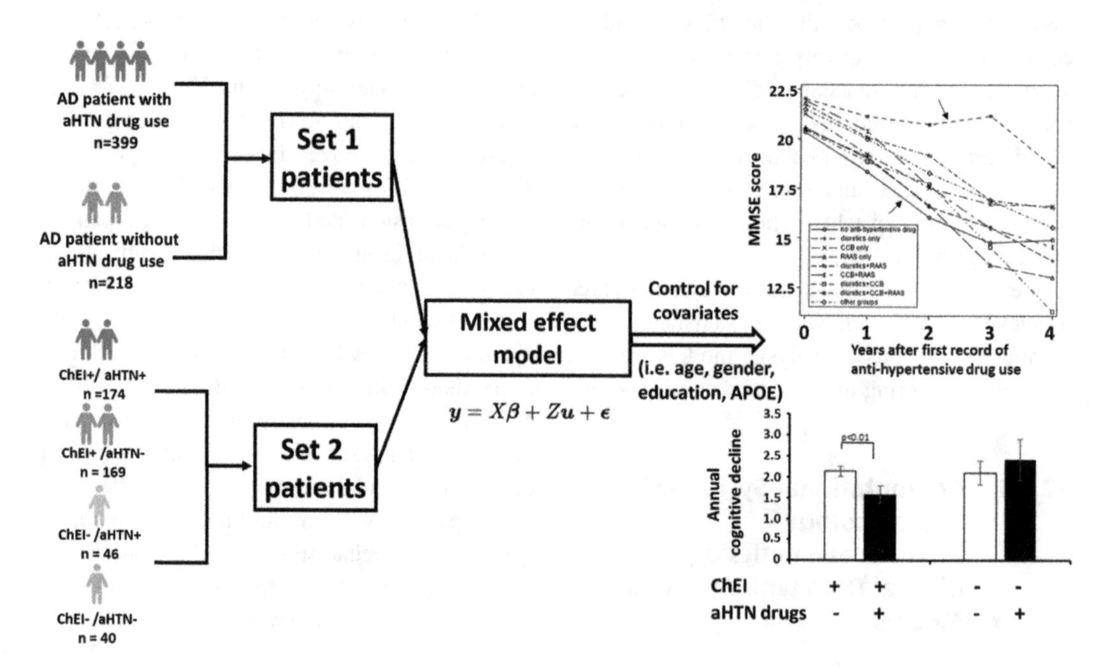

Fig. 32.2 Illustration of study design and outcomes of aHTN drugs reduce cognitive decline in AD patients, as shown in the upper left corner. Analysis with set 1 patients show diuretics + Calcium channel blocker (CCB) + renin-angiotensin-aldosterone system (RAAS) inhibitor (upper arrow) was associated with the slowest rate of cognitive decline among all groups, especially compared to the no aHTN group (lower arrow). The study result on set 2 patients (lower right corner) shows that aHTN drugs and ChEI synergistically slow the cognitive decline in AD patients (as illustrated by the right-side figure with a publisher permission [29])

users or nonusers throughout the entire time) were selected for this analysis from the original 617 patients. They were then divided into 4 groups (referred as the set 2 patients in Fig. 32.2). Similarly, a mixed-effect regression model was applied with controlling age, sex, years of education, and APOEε4 carrier status. All the data maintenance and management were done with SPSS, and the analysis was done with SPSS and SAS software.

Evaluation Metrics Both study designs use the MMSE score decline over different times to represent the real cognitive decline of the patients.

Outcomes

1. aHTN drugs on cognitive decline.
   As shown in the upper right corner in Fig. 32.2, when compared with the slope of cognitive decline in all groups (−2.15)—which refers to the average speed of decline over 4 years in all patients, the slope (−0.69) for the patient group which is taking the combination of 3 types of aHTN drugs (CCB + diuretics+ RAAS) was significantly ($p < 0.0001$) smaller in its absolute value, which means those patients have a slower cognitive decline
2. Synergistic effect of aHTN and ChEls drugs on AD patients
   The result was plotted as bar graph at the lower right corner in Fig. 32.2, it shows that, overall, ChEI users have a slower decline trend than non-ChEI users but without statistical significance ($p = 0.068$). However, within the ChEI users, the patients who take aHTN drugs have a significantly ($p < 0.01$) slower cognitive decline rate (−1.57) vs. those who do not take aHTN drugs (−2.09). Another interesting finding is that, for non-ChEI users, taking aHTN drugs is associated with a larger cognitive decline rate.**Summary**

The paper also applied computational methods to predict potential AD targets that might interact with aHTN drugs with a HTDocking program, an online-accessible package for high-throughput docking (http://www.cbligand.org/HTDocking/), which is another aspect of AI technique in drug discovery and mechanism study. However, this is beyond the scope of this section. Readers interested in this tool can refer to these publications [31–33].

With observational clinical data and a mix-effect regression model (essentially a multivariate linear model), the author concluded that aHTN drugs, by themselves or with ChEI drugs, are associated with a significant reduction of cognitive decline rate in AD patients. Those results indicate that, with an appropriate research hypothesis and dataset, the "old-fashion" statistical model is still valuable in clinical data analysis.

The overall advantage of the statistical models is that all the variables are explanatory, meaning we can attribute the effect on the outcome from inputs to a single real-world variable. For example, taking aHTN drugs will affect cognitive decline [34]. Moreover, a statistical model usually has fewer parameters, which is computationally friendly (it does not require days or weeks of computation with a supercomputer to finish the analysis). The major downfall of the statistical models is that the covariates that need to be controlled largely depend on previous knowledge [35]. In this case, the author controlled for age, sex, education years, and APOEε4 status, which have been reported previously to be associated with AD progression [36]. However, in clinical data, there are easily thousands of features in each patient's record; therefore, it is far from a complete list of all the covariates that might influence the outcome; which means, in this type of analysis, there will still be relatively large noise or variance intra- and inter- groups that might mask some important features in the analysis. In a case where the usefulness of certain covariates is still in debate, whether to pick up them or not is largely subject to the researchers, which can bring in inherent bias in the model.

This type of modeling may also suffer from insufficient samples in each subgroup [37]. For example, in this case, even starting with 4364 patients, only 400 to 600 samples meet the initial inclusive critique. In some subgroups, there might be 30 to 40 cases which will lower the power of detecting clinical significance since statistical significance is not shown.

The mixed-effect model is a linear regression model that cannot catch the non-linear relationship or interactions between the features and the outcomes [38]. Lastly, the model has less prediction power on individual patients since many features of each patient were ignored in the model. In other words, not all the information from the clinical record was fully used.

The mixed-effect model, together with other linear regression models, will have its best use in clinical studies, where the covariates associated with the outcome are well-defined, and each subgroup in the study is designed to have a relatively large sample size. In the real research scenario, starting the analysis with relatively straightforward models is worthwhile before moving into more sophisticated models.

32.2.2 Machine Learning and Artificial Neural Network (ANN) in AD Progression Prediction

The other applications of AI techniques in AD are disease diagnosis and clinical decision-making. Besides the new drug or existing drug repurposing in the treatment of AD, another critical need is to distinguish patients who will progress to severe dementia from MCI due to AD from patients who tend to be stable (sometimes referred to as sMCI). Fifteen percent of MCI patients will develop dementia after 2 years, and one-third of MCI patients will develop dementia due to Alzheimer's within 5 years [39]. At the same time, some MCI patients will revert to normal cognitive status or live on without additional cognitive level loss. Successfully predicting a patient's disease progression will facilitate applying early intervention to those patients with a higher risk of developing dementia to prevent or slow down the disease progression.

Machine learning (ML) is a field of study that allows computers to learn to complete tasks without being explicitly programmed. ML can be roughly divided into unsupervised, supervised, and reinforced learning, Here are a few recent reviews of conventional machine learning algorithms such as K-nearest neighbors (KNN), support vector machine (SVM), and Naïve Bayes, random forest/decision trees, etc. [40–42]. The essence of Machine learning (referring to supervised learning here) is no different from the statistical model we discussed in the last section: to fit a model to the existing data [43]. However, at the same time, they are much better at dealing with the non-linearity properties in the data and are more versatile in dealing with different data types, such as images [44].

The ANN is a subset of machine learning (Fig. 32.3a). It is inspired by the neuron system and mimics the way of signal transduction between biological neurons. The ANN usually consists of an input layer that takes the input features; an output layer that outputs the outcome from the network; and, an in-between, where the hidden layers are in between the two layers and the number of hidden layers can be as small as one or as large as 256 (Fig. 32.3b).

Deep learning (DL) is a branch of ANN. The development of deep learning models has made major advances in dealing with the problems that have troubled the AI community for years [45]. It is especially good at finding out the intricate data structures in high-dimensional data (clinical data/clinical images are usually very high-dimensional data) in scientific research, business and government applications, including audio and speech processing [46], visual/image data processing [47, 48], and natural language processing (NLP) [49], etc. The key feature of deep learning is that it does not need a feature selection/engineering process. They are obtained within the learning algorithm, and a few popular machine learning and deep learning packages are summarized in Table 32.1.

The deep learning models can be divided into supervised and unsupervised models (Table 32.2). A good review of various types of deep learning can be found in reference [50], and a good survey of DL applications used in health science can be found in this paper [51]. Our lab

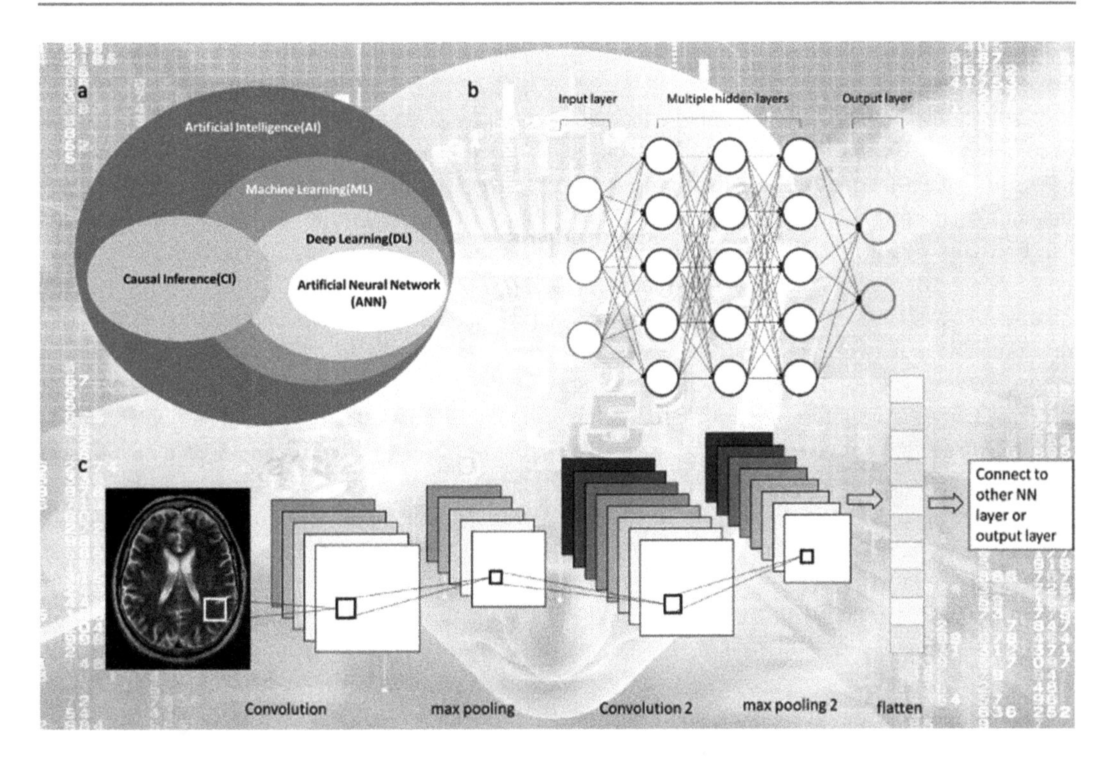

Fig. 32.3 Overview of AI technology. (**a**) a diagram showing the relationship between ANN, DL, ML, CI and AI; (**b**) the basic structure of an Artificial Neural Network; (**c**) a typical convolutional neural network architecture

Table 32.1 Popular Machine-/Deep-Learning (ML/DL) packages for the different platforms used in health science analytics

Category	Package name	Platform/interface	Link
ML	Scikit-learn	python	https://scikit-learn.org/stable/
ML	E1071	R	https://cran.r-project.org/web/packages/e1071/index.html
DL	Torch	Lua, LuaJIT, C	http://torch.ch/
DL	PyTorch	python	http://pytorch.org/
DL	Keras	python	https://keras.io/
DL	TensorFlow	Python	https://www.tensorflow.org/
DL	CNTK	Python, C	https://learn.microsoft.com/en-us/cognitive-toolkit/
DL	Caffe	C++, python, matlab	http://caffe.berkeleyvision.org/
DL	Theano	Python	https://github.com/Theano/Theano
DL	Deeplearning4jk	Java, Scala, Clojure	https://deeplearning4j.konduit.ai/
DL	MxNet	R/Python	http://mxnet.io/

Table 32.2 Popular deep learning (DL) models that are used in health science analytics

Model	Category
Deep (Dense) Neural Networks (DNN)	Supervised
Convolutional neural network (CNN)	Supervised
Recurrent neural networks (RNN)	Supervised
Long Short-Term Memory (LSTM)	Supervised
Deep belief networks (DBNs)	Unsupervised
Auto-Encoder (AE)	Unsupervised
Generative Adversarial Networks (GAN)	Unsupervised

has also published a review focused on deep learning application in drug discovery [14].

Among the supervised learning models, Convolutional Neural Network (CNN) is a specialized deep learning model that is used for imaging processing. A basic structure of CNN can be found in Fig. 32.3c. The central idea of how CNN works is down sampling- reducing the number of variables/features while still catching the connections between adjacent pixels. Its primary principle is to train and recognize specific patterns images—the information is stored in the "filters" used to convert the images in the convolutional layer—enough to distinguish different categories. A detailed review of the CNN model concept and architectures can be found in reference [52]. In biomedical research, it has been used for tumor detection with an histology image [53], neuro-imaging (including CT, MRI, and PET) [54–56], and cancer prediction with gene expression data (since gene expression can be presented as a heatmap, which can be viewed as images) [57].

32.2.2.1 Case Study 2

For this study, we will use an example [58] with CNN to process the brain image data from AD patients and make predictions on their disease progression.

Research Objective To build a CNN model to predict cognitive decline and if an MCI patient will eventually progress to AD.

Data Source Fluorodeoxyglucose (FDG, representing brain metabolism level) and florbetapir (referred to as AV-45, representing amyloid deposition) positron emission tomography (PET) image obtained from Alzheimer's Disease Neuroimaging Initiative (ADNI) [59] was used in this study. Those images were for 139 patients with AD, 171 with MCI patients (among which 79 were converted into AD and 92 were not-converted), and 182 normal subjects were selected. The critique for inclusion is that the normal control subjects and AD patients have baseline scans of FDG, AV-45 PET, and the MCI patients have the baseline scan and at least a 3-year follow-up clinical evaluation (which will track the cognitive decline of the patient).

Study Design The CNN was designed using MatConvNet (a MATLAB-based CNN package). The author rescaled and digitalized the 96 PET slices from both the FDG and AV-45 to get two $160 \times 160 \times 160$ 3D input volume for each patient with zero padding along the z-axis. Then a $7 \times 7 \times 7$ filter was used to convolute and combine the data from both PETs into a 64 pooled feature volume. Thus, the information of the FDG and the AV-45 PET from the same brain area of the same patient were combined in this step. The feature volume was then pooled, convoluted (128 filters), pooled, and finally convoluted into a 1D feature map that contains 512 features (this is a typical CNN architecture with two convolution layers). Furthermore, the feature map was fed into 2 output nodes with a full connection. It is worth mentioning that during the training of the model, only the baseline scan image of a normal control subject (referred to as NC) and AD was fed into the model. The MCI data was processed similarly to the AD and NC patients' data and fed to the trained model. Therefore, the CNN model was not exposed to any data from MCI patients. The model will score each patient and then predict if the patient will be an MCI converter or non-converter (Fig. 32.4).

Evaluation Metrics As a prediction model, sensitivity, specificity, accuracy, and Area under the curve (AUC) were used for model evaluation; the author also compared the performance of SVM and feature volume of interests-based method with data from either FDG or AV45 images. Both methods are considered conventional prediction methods.

Outcomes The author first showed that the AD classification (AD vs. normal) performance of the model is superior compared to the other two methods, with a sensitivity of 93.5%, a specificity of 97.8%, an accuracy of 96.0%, and an AUC of 0.98. These results indicate the CNN model did

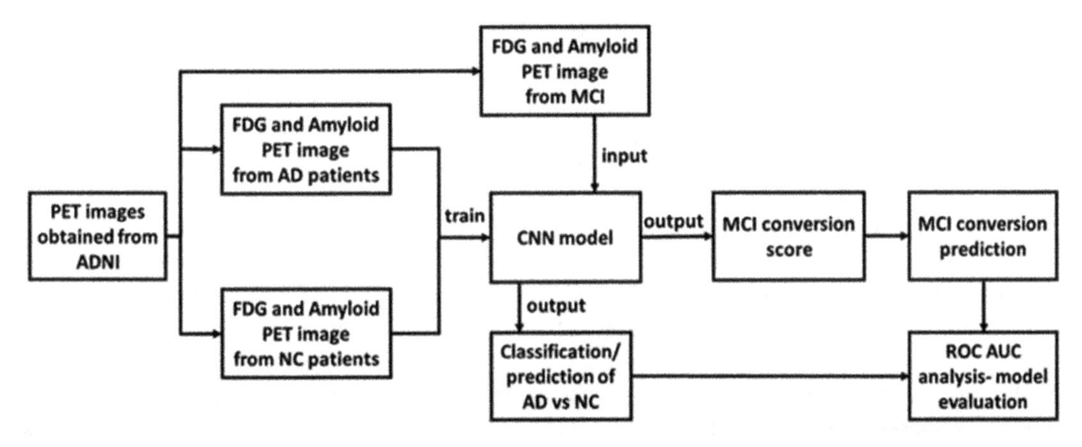

Fig. 32.4 A schematic diagram of the study design of building a CNN model using PET images from normal control subjects (NC), mild cognitive impairment (MCI) patients, and Alzheimer's disease (AD) patients for MCI conversion prediction

catch essential differences between PET images of normal and AD patients. It is a very accurate model.

For MCI conversion prediction, the CNN model achieved 81.0%, 87.0%, 84.2%, and 0.89 for sensitivity, specificity, accuracy, and AUC, respectively. Even though it is lower overall than the AD classification, given that MCI data was not included in the training dataset, this is also considered a relatively reliable and robust model performance.

Summary

The authors successfully established a CNN model with the information from the FDG and AV-45 PET scans in normal subjects and AD patients. It was able to predict MCI conversion in MCI patients with their baseline PET scans. This model will help select appropriate prodromal AD patients who benefit from early intervention [60].

Besides the excellent performance of the prediction model itself, the CNN model demonstrated its advantage of automatically discovering the appropriate features for the classification, which has to be carried out separately and manually for conventional machine learning methods such as SVM and VOI analysis. This advantage will largely avoid potential information loss and bias that would have been introduced during the feature engineering phase.

It is worth mentioning that the author used two convolution layers in the CNN architecture, which measure the "depth" of the model. Three or many more convolution layers (up to 256) can be seen in other applications, such as object recognition. More layers of convolution usually lead to the recognition of high-level patterns, but at the same time, more parameters within the model need to be learned which increase the model complexity. A simple CNN model can have 60 k parameters. However, a very deep CNN model (i.e., VGG-34) can easily have over 100 million parameters. The number of parameters is directly correlated with the needed computation power. Deep CNN could require weeks of computation before the model is trained [61].

The other downfall of deep CNN is that the features extracted in the end are low interpretability to human physicians—in other words—it is a "black box" [62]. This downfall is true for other deep learning models, especially in the case above. The feature combines FDG and AV-45 PET data. It has been transformed several times, making it very hard to continue any mechanism study or provide patients with MCI-specific treatment plans.

32.2.3 Causal Inference in AD Clinical Research

Although machine learning, especially deep learning, is powerful in finding the correlation between the features and making accurate predictions, this also reflects its weakness [63], especially in observative clinical datasets such as the data of AD we obtain from NACC or ADNI. For example, in exploring drug efficacy in AD patients, we found that the AD related drugs taken by the patients are positively correlated with the disease progression (lower MMSE scores). In other words, the more AD medication the patients take, the more likely they will have worse cognitive status. As the patients progress toward severe dementia due to AD, they will potentially be prescribed with more/stronger medication by the physicians-this is due to the fact that we don't have any drug that can cure AD. It makes sense that medication intake is positively correlated with the disease progression, but this does not give us any information on the drug efficacy or what type of medication we should recommend for a patient who is seeking help from a physician.

That is where causal inference (CI) comes into play. CI is a collection of algorithms based on machine learning, but it can identify confounding factors and point out potential causal relationships between variables and outcomes [64]. Figure 32.5 shows a simple diagram of causal relationships that can be studied by causal analysis. A few popular causal analysis packages are listed in Table 32.3.

The causal inference is trying to answer the following question: for a given patient, what will be the difference if one does action A versus not does action A, or whether the genetic mutation the patient is carrying will cause the onset of the disease [65].

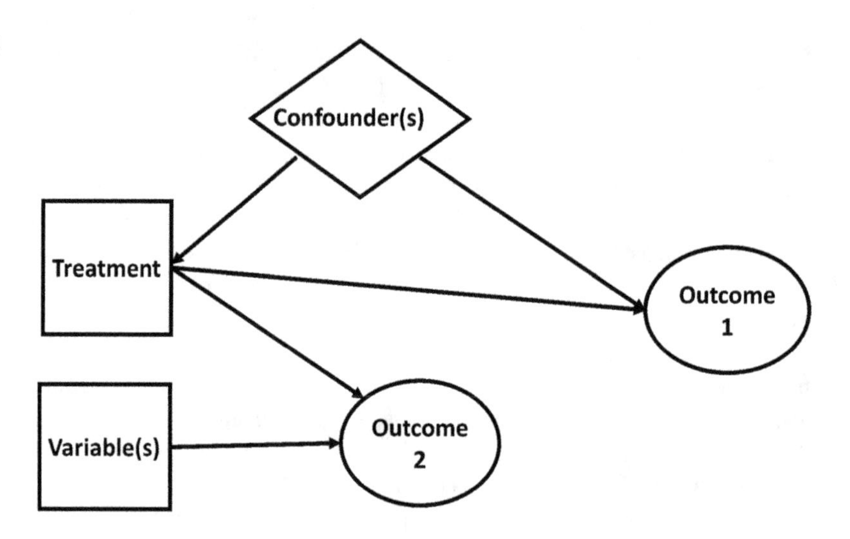

Fig. 32.5 In a simplified diagram of a causal relationship, the arrows indicate the causal relationship and its direction

Table 32.3 Popular causal analysis packages on different platforms used extensively in causal inference. The Web links are provided for detailed instructions

Package	Platform/interface	Link
Causalinference	Python	https://causalinferenceinpython.org/
Causallib	Python	https://github.com/IBM/causallib
Causalimpact	Python	https://github.com/jamalsenouci/causalimpact
DoWhy	Python	https://py-why.github.io/dowhy/
EconML	Python	https://www.microsoft.com/en-us/research/project/econml/
CausalImpact	R	http://google.github.io/CausalImpact/CausalImpact.html
Tetrad	Java application	https://www.ccd.pitt.edu/

The causal analysis has just begun drawing more attention in the AD research field [66], and APOE status is causally related to AD onset [67]. It is suggested that other risk factors, such as cardiovascular disease or diabetes, may also have a causal relationship with AD [59]. Determining which brain area or other genes have a causal relationship with AD or APOE is also very attractive research. If carried out successfully, the causal analysis will yield specific interventions that can be applied in clinical trials that may lead to new drug discoveries for AD patients.

32.2.3.1 Case Study 3

Here, we will dissect a research paper [68] that is trying to discover a causal relationship between the brain neuroimaging region and AD, and a causal relationship between the brain neuroimaging region and genes.

Study Objective To understand the genetic basis of brain structures and function and how it is related to AD and assess association and causal relationships among genetic variants, brain regions, and AD with DTI and genomic data from AD patients and normal subjects.

Data Source Diffusion Tensor Imaging (DTI) images ($91 \times 109 \times 91$) from 100 normal control subjects (NC) and 51 AD patients were obtained from ADNI for the following time points: baseline, 6 months, 12 months, and 24 months. DTI images from patients were processed with the linear and non-linear registration process (namely FMRIB's Linear Image Registration Tool and RNiftyReg), which mapped the image details to a standard brain while keeping the structure details from the original data. Some image augmentation techniques (Gaussian filter, translation, image flipping, random duplication of images) were utilized to overcome the small sample size limitation. Finally, data augmentation resulted in over 20 times more data than the original dataset.

Genetic data - specifically single-nucleotide polymorphism (SNP) - of the patients were obtained and preprocessed to remove any problematic data and individuals with too much missing data. A similar procedure was also done on the SNP level. Finally, 1,589,061 common SNPs in 36,480 genes genotyped in 151 individuals were included in this study.

Study Design First, a CNN model was set up with VGG-GAP architecture [69]. VGG-GAP is a very deep network with 16 convolution layers compared to the CNN with 2 convolution layers in the case study from the last section. The authors used this model for AD classification and prediction. They built up several models at each time point and used them to predict/classify AD at later time points; for example, one such CNN model used 151 patients' data from a 6-month time point to train the model and to predict the disease status with the 24-month patient images as testing data (Fig. 32.6).

The authors then performed a feature selection within the DTI images with a technique called prediction difference analysis; the basic principle behind this method is to try the same model with some piece of information from the DTI images removed and then compare the prediction difference with the original model. A larger model performance loss means high importance of that piece of data. In this case, each piece of the information unit is a $3 \times 3 \times 3$ patch. By doing this, we can obtain a heatmap of the regions contributing to the prediction or better distinguishing AD patients from NC, implying those regions may have functional importance in the disease etiology.

As a result, a total of twenty-three regions of interest (ROI) that substantially contribute to AD prediction were identified. The regions included the temporal lobe (the left temporal, medial, and right temporal lobes), ventricles and enlarged ventricle, occipital lobe, and prefrontal area. The ROI is then fed into the causal analysis. Using a Conditional Generative Adversarial Network (CGAN), the authors tested for the potential causal relationship between DTI image ROIs and AD disease at baseline, 6 months, 12 months, and 24 months [70].

In the CGAN model, for each assay, two variables were placed in the model for the causal

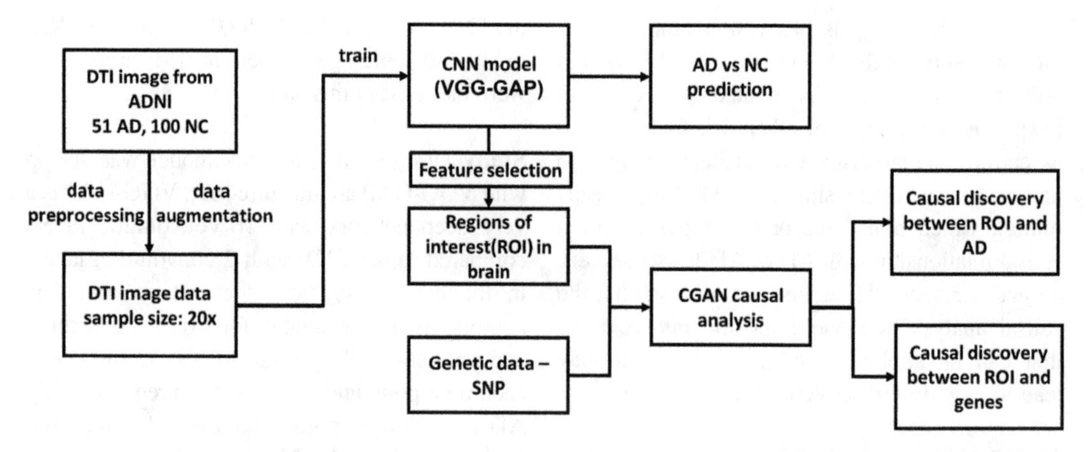

Fig. 32.6 A schematic diagram of the study design for using DTI images and genetic data for causal discovery with the CGAN model. *NC* normal cognition

relationship test. Usually, it is the disease status (Y) plus a brain region of interest (ROI) defined in the feature selection or SNP/gene from genetic data (X) after being converted into function principal component scores. The model first assumes the causal relationship could be in both directions (X causes Y and Y causes X) and then calculates a statistical value from both conditions and takes their difference. If there is no causal relationship, the new statistical value will be a normal distribution centered at 0. Otherwise, it is indicated to have a causal relationship.

Outcomes The CNN model trained in this case study yields an accuracy range from 0.76 to 0.91 without reporting sensitivity, specificity, and ROC-AUC.

In the causal analysis with DTI images, three different regions of interest show causation to AD: the part of the verticals and enlarged ventricle, the right temporal lobe, and the left temporal lobe. The causal discovery for genetic information identified 61 genes that had causal relationships with the brain imaging regions. Among them, CD33, COBL, and APP [71, 72] have also been reported in other literature multiple times, and a few other genes like FGF4, FRMD6, Dock9, H3F3B SCYL1, AKAP5, and PIGC all have been reported to at least be involved in regulating neuron or brain function. Those results might point to good opportunities to identify novel targets in AD.

Summary
This research used the DTI and genetic data from 151 patients (100 normal and 51 with AD). First, a very deep CNN model was built with DTI image information to predict/classify the AD status of the patient. Although the authors have done some data augmentation and balancing to enlarge the number of samples, the overall accuracy of the model was not the best due to the limited sample size. The deep feature selection provided information about the importance of brain regions that could be important for the prediction, thus indicating the functional importance of those regions in the etiology of AD. This study, to some degree, overcame the problem with CNN with the lack of interpretability. The causal analysis carried out with the CGAN model found some regions that have causal relationships with AD and identified a few genes that have causal relationships with neuroimaging. Some genes have already been reported to be related to AD causally, showing that the method is promising in discovering causal relationships within this limited dataset.

Causal models also come with their inherent limitations. For example, the CGAN model does not have a very intrinsic metric to evaluate the model performance [73]. Furthermore, in all the graphic analyses of causal models, there are a number of easily ignorable assumptions associated with each model. Another limitation of the causal analysis is that the putative causal relationship we got from the analysis is very hard and takes a long time to validate. There is no guarantee that any of them will work. The causal analysis also needs a large population/data to generate strong conclusions, which is often limited in the health science clinical data research scenarios.

32.3 Concluding Remarks

In this chapter, we reviewed and dissected three representative research AI techniques used in clinical data pharmacoanalytics studies for AD. We covered the linear statistical mixed-effect model, CNNs, an example of deep learning, and causal analysis with the CGAN model. We presented to the readers an idea of how clinical data can be mined and analytics studied with different models and also what are the advantages and drawbacks of each model.

We should avoid over-using machine learning in clinical data analytics. Overuse means unnecessary adoption of advanced ML or DL techniques when sophisticated ML models often offer only marginal accuracy gains. A number of ML/DL research papers have shown an AUC value greater than 0.8 or even 0.9, but it is not a guarantee of the robustness of the model, especially when the model is prone to overfitting due to a small sample size or a very large number of parameters in the models [74]. Therefore, it is suggested that, whenever appropriate, traditional statistical models should be conducted first and presented together with the ML/DL models to justify the use of the sophisticated models. The detailed protocol, including data preprocessing, model hyperparameters, datasets, and scripts, should be made available for better reproductivity.

As for the future aspect, the limited amount of data and data imbalance will still be a great challenge for clinical data pharmacoanalytics studies for a while due to the nature of heterogeneous clinical data sources and the hardship of collecting and managing those data [75]. Efforts have been made by clinicians and researchers to produce, store, and utilize large, multicenter international datasets with high-quality data [76], particularly in AD clinical data. NACC and ADNI have been good examples of large-scale database build-up. Algorithms and methods have also been developed to augment the dataset to achieve a larger sample size [77], which is critical for trying robust machine learning models, especially deep learning models.

Furthermore, understanding domain knowledge in the future will be of great importance [78]. The future direction of AI technology is going to be automation and modulation. With highly integrated model-building tools, one does not need a sophisticated understanding of the underlying detail of the models to build a workable model and conduct machine learning. This trend even emphasizes the importance of domain knowledge. One has to be an expert in a certain research field to raise valued research questions and hypotheses with the available data. Otherwise, it may easily end up in "garbage in, garbage out." It should be pointed out that the AI technique is still not a replacement for human critical thinking but more like an augmentation of human ability to deal with numerous amounts of data.

Finally, interpretability has long been an issue for ML and DL methods due to their feature transformation and non-linearity between features and outcomes. Currently, three major methods have been taken to improve the DL model. These methods are SHAP (SHapley Additive exPlanations) [79], LIME (Local Interpretable Model-agnostic Explanations) [80], and Anchors. These will remain a hot area of study for a while and will provide opportunities for new advancement in AI technology.

Acknowledgments The authors acknowledge the funding support to the Xie Laboratory and CDAR Center from the NIH (R56AG074951, R01DA052329 and P30PDA035778A).

Conflict of Interest All authors have no conflict of interest to declare.

References

1. Kumar Y, Koul A, Singla R, Ijaz MF (2022) Artificial intelligence in disease diagnosis: a systematic literature review, synthesizing framework and future research agenda. J Ambient Intell Humaniz Comput:1–28
2. Busnatu Ș, Niculescu AG, Bolocan A, Petrescu GED, Păduraru DN, Năstasă I, Lupușoru M, Geantă M, Andronic O, Grumezescu AM et al (2022) Clinical applications of artificial intelligence-an updated overview. J Clin Med 11(8):2265
3. Bian Y, Xie X-Q (2022) Artificial intelligent deep learning molecular generative modeling of scaffold-focused and cannabinoid CB2 target-specific small-molecule sublibraries. Cells 11(5):915
4. Hou T, Bian Y, McGuire T, Xie X-Q (2021) Integrated multi-class classification and prediction of GPCR allosteric modulators by machine learning intelligence. Biomolecules 11(6):870
5. Bian Y, Xie X-Q (2021) Generative chemistry: drug discovery with deep learning generative models. J Mol Model 27(3):71
6. Bian Y, Jing Y, Wang L, Ma S, Jun JJ, Xie X-Q (2019) Prediction of orthosteric and allosteric regulations on cannabinoid receptors using supervised machine learning classifiers. Mol Pharm 16(6):2605–2615
7. Bian Y, Wang J, Jun JJ, Xie X-Q (2019) Deep convolutional generative adversarial network (dcGAN) models for screening and design of small molecules targeting cannabinoid receptors. Mol Pharm 16(11):4451–4460
8. Xing C, Zhuang Y, Xu T, Feng Z, Xu E, Chen M, Wang L, Meng X, Xue Y, Wang JM, Liu H, McGuire TF, Zhao GP, Melcher K, Zhang C, Xu HE, Xie X-Q (2020) Cryo-EM structure of human cannabinoid receptor CB2-Gi signaling complex. Cell 180(4): 645–654
9. Liang T, Chen H, Yuan J, Jiang C, Hao Y, Wang Y, Feng Z, Xie X-Q (2021) IsAb: a computational protocol for antibody design. Brief Bioinform 22(5): bbab143
10. Chen M, Feng Z, Wang S, Lin W, Xie X-Q (2021) MCCS, a novel characterization method for protein-ligand complex. Brief Bioinform 22(4):bbaa239
11. Davenport T, Kalakota R (2019) The potential for artificial intelligence in healthcare. Fut Healthc J 6(2):94–98
12. Sancesario GM, Bernardini S (2018) Alzheimer's disease in the omics era. Clin Biochem 59:9–16
13. Yang Y, Yuan Y, Zhang G, Wang H, Chen Y-C, Liu Y, Tarolli CG, Crepeau D, Bukartyk J, Junna MR et al (2022) Artificial intelligence-enabled detection and assessment of Parkinson's disease using nocturnal breathing signals. Nat Med 28(10):2207–2215
14. Jing Y, Bian Y, Hu Z, Wang L, Xie X-Q (2018) Deep learning for drug design: an artificial intelligence paradigm for drug discovery in the big data era. AAPS J 20(3):58
15. Hu Z, Jing Y, Xue Y, Fan P, Wang L, Vanyukov M, Kirisci L, Wang J, Tarter RE, Xie X-Q (2020) Analysis of substance use and its outcomes by machine learning: II. Derivation and prediction of the trajectory of substance use severity. Drug Alcohol Depend 206: 107604
16. Chen M, Jing Y, Wang L, Feng Z, Xie X-Q (2019) DAKB-GPCRs: an integrated computational platform for drug abuse related GPCRs. J Chem Inf Model 59(4):1283–1289
17. Knopman DS, Amieva H, Petersen RC, Chételat G, Holtzman DM, Hyman BT, Nixon RA, Jones DT (2021) Alzheimer disease. Nat Rev Dis Primers 7(1):33
18. Guerreiro R, Bras J (2015) The age factor in Alzheimer's disease. Genome Med 7:106
19. Rajan KB, Weuve J, Barnes LL, McAninch EA, Wilson RS, Evans DA (2021) Population estimate of people with clinical Alzheimer's disease and mild cognitive impairment in the United States (2020–2060). Alzheimers Dement 17(12):1966–1975
20. Ienca M, Fabrice J, Elger B, Caon M, Scoccia Pappagallo A, Kressig RW, Wangmo T (2017) Intelligent assistive technology for Alzheimer's disease and other dementias: a systematic review. J Alzheimers Dis 56(4):1301–1340
21. Fabrizio C, Termine A, Caltagirone C, Sancesario G (2021) Artificial intelligence for Alzheimer's disease: promise or challenge? Diagnostics (Basel, Switzerland) 11(8):1473
22. Tahami Monfared AA, Byrnes MJ, White LA, Zhang Q (2022) Alzheimer's disease: epidemiology and clinical progression. Neurol Ther 11(2):553–569
23. Beekly DL, Ramos EM, Lee WW, Deitrich WD, Jacka ME, Wu J, Hubbard JL, Koepsell TD, Morris JC, Kukull WA (2007) The national Alzheimer's Coordinating Center (NACC) database: the uniform data set. Alzheimer Dis Assoc Disord 21(3):249–258
24. Fan C, Chen M, Wang X, Wang J, Huang B (2021) A review on data preprocessing techniques toward efficient and reliable knowledge discovery from building operational data. Front Energ Res 9: 652801
25. Semanik MG, Kleinschmidt PC, Wright A, Willett DL, Dean SM, Saleh SN, Co Z, Sampene E, Buchanan JR (2021) Impact of a problem-oriented view on clinical data retrieval. J Am Med Inform Assoc 28(5): 899–906
26. Cummings J, Lee G, Nahed P, Kambar MEZN, Zhong K, Fonseca J, Taghva K (2022) Alzheimer's disease drug development pipeline: 2022. Alzheimer's Dementia 8(1):e12295
27. Santiago JA, Potashkin JA (2021) The impact of disease comorbidities in Alzheimer's disease. Front Aging Neurosci 13:631770
28. van Bokhoven P, de Wilde A, Vermunt L, Leferink PS, Heetveld S, Cummings J, Scheltens P, Vijverberg EGB (2021) The Alzheimer's disease drug development landscape. Alzheimers Res Ther 13(1):186

29. Hu Z, Wang L, Ma S, Kirisci L, Feng Z, Xue Y, Klunk WE, Kamboh MI, Sweet RA, Becker J, Lv QZ, Lopez OL, Xie X-Q (2018) Synergism of antihypertensives and cholinesterase inhibitors in Alzheimer's disease. Alzheimers Dement (N Y) 4:542–555
30. Lopez OL, Becker JT, Wahed AS, Saxton J, Sweet RA, Wolk DA, Klunk W, Dekosky ST (2009) Long-term effects of the concomitant use of memantine with cholinesterase inhibition in Alzheimer disease. J Neurol Neurosurg Psychiatry 80(6):600–607
31. Liu H, Wang L, Lv M, Pei R, Li P, Pei Z, Wang Y, Su W, Xie X-Q (2014) AlzPlatform: an Alzheimer's disease domain-specific chemogenomics knowledgebase for polypharmacology and target identification research. J Chem Inf Model 54(4): 1050–1060
32. Zhang Y, Wang L, Feng Z, Cheng H, McGuire TF, Ding Y, Cheng T, Gao Y, Xie X-Q (2016) StemCellCKB: an integrated stem cell-specific chemogenomics knowledgebase for target identification and systems-pharmacology research. J Chem Inf Model 56(10):1995–2004
33. Xu X, Ma S, Feng Z, Hu G, Wang L, Xie X-Q (2016) Chemogenomics knowledgebase and systems pharmacology for hallucinogen target identification-Salvinorin A as a case study. J Mol Graph Model 70: 284–295
34. Cortese G (2020) How to use statistical models and methods for clinical prediction. Ann Translat Med 8(4):76
35. Wu W-T, Li Y-J, Feng A-Z, Li L, Huang T, Xu A-D, Lyu J (2021) Data mining in clinical big data: the frequently used databases, steps, and methodological models. Mil Med Res 8(1):44
36. Safieh M, Korczyn AD, Michaelson DM (2019) ApoE4: an emerging therapeutic target for Alzheimer's disease. BMC Med 17(1):64
37. Burke JF, Sussman JB, Kent DM, Hayward RA (2015) Three simple rules to ensure reasonably credible subgroup analyses. BMJ. Br Med J 351:h5651
38. Schober P, Vetter TR (2021) Linear regression in medical research. Anesth Analg 132(1):108–109
39. Petersen RC, Lopez O, Armstrong MJ, Getchius TSD, Ganguli M, Gloss D, Gronseth GS, Marson D, Pringsheim T, Day GS et al (2018) Practice guideline update summary: mild cognitive impairment: report of the Guideline Development, Dissemination, and Implementation Subcommittee of the American Academy of Neurology. Neurology 90(3):126–135
40. Singh A, Thakur N, Sharma A (2016) A review of supervised machine learning algorithms. In: 2016 3rd international conference on computing for sustainable global development (INDIACom): 2016. IEEE, pp 1310–1315
41. Mahesh B (2020) Machine learning algorithms-a review. Int J Sci Res (IJSR)[Internet] 9:381–386
42. Greener JG, Kandathil SM, Moffat L, Jones DT (2022) A guide to machine learning for biologists. Nat Rev Mol Cell Biol 23(1):40–55
43. Dhall D, Kaur R, Juneja M (2020) Machine learning: a review of the algorithms and its applications. In: Proceedings of ICRIC 2019. Springer, Cham, pp 47–63
44. Varoquaux G, Cheplygina V (2022) Machine learning for medical imaging: methodological failures and recommendations for the future. npj Digit Med 5(1):48
45. Pouyanfar S, Sadiq S, Yan Y, Tian H, Tao Y, Reyes MP, Shyu M-L, Chen S-C, Iyengar SS (2018) A survey on deep learning: algorithms, techniques, and applications. ACM Comput Surv (CSUR) 51(5): 1–36
46. Adeel A, Gogate M, Hussain A (2020) Contextual deep learning-based audio-visual switching for speech enhancement in real-world environments. Inform Fusion 59:163–170
47. Tian H, Chen S-C, Shyu M-L (2020) Evolutionary programming based deep learning feature selection and network construction for visual data classification. Inf Syst Front 22(5):1053–1066
48. Koppe G, Meyer-Lindenberg A, Durstewitz D (2021) Deep learning for small and big data in psychiatry. Neuropsychopharmacology 46(1):176–190
49. Young T, Hazarika D, Poria S, Cambria E (2018) Recent trends in deep learning based natural language processing. IEEE Comput Intell Mag 13(3):55–75
50. LeCun Y, Bengio Y, Hinton G (2015) Deep learning. Nature 521(7553):436–444
51. Esteva A, Robicquet A, Ramsundar B, Kuleshov V, DePristo M, Chou K, Cui C, Corrado G, Thrun S, Dean J (2019) A guide to deep learning in healthcare. Nat Med 25(1):24–29
52. Alzubaidi L, Zhang J, Humaidi AJ, Al-Dujaili A, Duan Y, Al-Shamma O, Santamaría J, Fadhel MA, Al-Amidie M, Farhan L (2021) Review of deep learning: concepts, CNN architectures, challenges, applications, future directions. J Big Data 8(1):53
53. Arevalo J, González FA, Ramos-Pollán R, Oliveira JL, Guevara Lopez MA (2016) Representation learning for mammography mass lesion classification with convolutional neural networks. Comput Methods Prog Biomed 127:248–257
54. Akkus Z, Galimzianova A, Hoogi A, Rubin DL, Erickson BJ (2017) Deep learning for brain MRI segmentation: state of the art and future directions. J Digit Imaging 30(4):449–459
55. Gao XW, Hui R, Tian Z (2017) Classification of CT brain images based on deep learning networks. Comput Methods Prog Biomed 138:49–56
56. Amini M, Pedram MM, Moradi A, Jamshidi M, Ouchani M (2022) GC-CNNnet: diagnosis of Alzheimer's disease with PET images using genetic and convolutional neural network. Comput Intell Neurosci 2022:7413081
57. Mostavi M, Chiu Y-C, Huang Y, Chen Y (2020) Convolutional neural network models for cancer type prediction based on gene expression. BMC Med Genet 13(5):44

58. Choi H, Jin KH (2018) Predicting cognitive decline with deep learning of brain metabolism and amyloid imaging. Behav Brain Res 344:103–109
59. Weiner MW, Veitch DP, Aisen PS, Beckett LA, Cairns NJ, Green RC, Harvey D, Jack CR, Jagust W, Liu E et al (2013) The Alzheimer's Disease Neuroimaging Initiative: a review of papers published since its inception. Alzheimers Dement 9(5):e111–e194
60. Dubois B, Feldman HH, Jacova C, Dekosky ST, Barberger-Gateau P, Cummings J, Delacourte A, Galasko D, Gauthier S, Jicha G et al (2007) Research criteria for the diagnosis of Alzheimer's disease: revising the NINCDS-ADRDA criteria. Lancet Neurol 6(8):734–746
61. Justus D, Brennan J, Bonner S, McGough AS (2018) Predicting the computational cost of deep learning models. In: 2018 IEEE international conference on big data (Big Data): 2018. IEEE, pp 3873–3882
62. O'Mahony N, Campbell S, Carvalho A, Harapanahalli S, Hernandez GV, Krpalkova L, Riordan D, Walsh J (2019) Deep learning vs. traditional computer vision. In: Science and information conference: 2019. Springer, pp 128–144
63. Voulodimos A, Doulamis N, Doulamis A, Protopapadakis E (2018) Deep learning for computer vision: a brief review. Comput Intell Neurosci 2018: 7068349
64. Pearl J (2010) An introduction to causal inference. The. Int J Biostat 6(2):7
65. Bellamy SL, Lin JY, Have TRT (2007) An introduction to causal modeling in clinical trials. Clin Trials 4(1):58–73
66. Pagoni P, Korologou-Linden RS, Howe LD, Davey Smith G, Ben-Shlomo Y, Stergiakouli E, Anderson EL (2022) Causal effects of circulating cytokine concentrations on risk of Alzheimer's disease and cognitive function. Brain Behav Immun 104:54–64
67. Liu CC, Liu CC, Kanekiyo T, Xu H, Bu G (2013) Apolipoprotein E and Alzheimer disease: risk, mechanisms and therapy. Nature reviews. Neurology 9(2):106–118
68. Liu Y, Li Z, Ge Q, Lin N, Xiong M (2019) Deep feature selection and causal analysis of Alzheimer's disease. Front Neurosci 13:1198
69. Simonyan K, Zisserman A (2014) Very deep convolutional networks for large-scale image recognition. arXiv preprint arXiv:1409.1556
70. Hong W, Wang Z, Yang M, Yuan J (2018) Conditional generative adversarial network for structured domain adaptation. In: Proceedings of the IEEE conference on computer vision and pattern recognition: 2018, pp 1335–1344
71. Bradshaw EM, Chibnik LB, Keenan BT, Ottoboni L, Raj T, Tang A, Rosenkrantz LL, Imboywa S, Lee M, Von Korff A et al (2013) CD33 Alzheimer's disease locus: altered monocyte function and amyloid biology. Nat Neurosci 16(7):848–850
72. Huang C-Y, Hsiao I-T, Lin K-J, Huang K-L, Fung H-C, Liu C-H, Chang T-Y, Weng Y-C, Hsu W-C, Yen T-C et al (2019) Amyloid PET pattern with dementia and amyloid angiopathy in Taiwan familial AD with D678H APP mutation. J Neurol Sci 398:107–116
73. Greenland S, Mansournia MA (2015) Limitations of individual causal models, causal graphs, and ignorability assumptions, as illustrated by random confounding and design unfaithfulness. Eur J Epidemiol 30(10):1101–1110
74. Volovici V, Syn NL, Ercole A, Zhao JJ, Liu N (2022) Steps to avoid overuse and misuse of machine learning in clinical research. Nat Med 28(10):1996–1999
75. Weissler EH, Naumann T, Andersson T, Ranganath R, Elemento O, Luo Y, Freitag DF, Benoit J, Hughes MC, Khan F et al (2021) The role of machine learning in clinical research: transforming the future of evidence generation. Trials 22(1):537
76. Bakouny Z, Patt DA (2021) Machine learning and real-world data: more than just buzzwords. JCO Clin Cancer Informat 5:811–813
77. Taylor L, Nitschke G (2018) Improving deep learning with generic data augmentation. In: 2018 IEEE symposium series on computational intelligence (SSCI): 2018. IEEE, pp 1542–1547
78. Dash T, Chitlangia S, Ahuja A, Srinivasan A (2022) A review of some techniques for inclusion of domain-knowledge into deep neural networks. Sci Rep 12(1): 1040
79. Lundberg SM, Lee S-I (2017) A unified approach to interpreting model predictions. Adv Neural Inf Proces Syst 30:4768–4777
80. Mishra S, Sturm BL, Dixon S (2017) Local interpretable model-agnostic explanations for music content analysis. In: ISMIR: 2017, pp 537–543

33 Bioinformatics: Theory and Applications

Anuj Gahlawat, Rajkumar. R, Tanmaykumar Varma, Pradnya Kamble, Aritra Banerjee, Hardeep Sandhu, and Prabha Garg

Abstract

From the mid-twentieth century, the digital age started changing every aspect of human life by utilizing information technology advancements. Informatics is a field of science that makes use of these signs of progress to transform data and information into required knowledge that can be used by humans efficiently. Bioinformatics, an interdisciplinary field of science, was developed mainly because of the parallel development of molecular biology, computer science and the recent emergence of big data. It analyses and interprets, a huge amount of biological data that has been generated by sequencing techniques and genome projects. The increased number of bioinformatics tools, biological big data and changes in computer architecture in recent times have raised the need for more expertise in the field. Thus, this chapter highlights the basic understanding of bioinformatics approaches including the databases, tools, and their diverse application in different fields such as biology and drug design and development. It also discusses diverse approaches to tackling biological complex networks and understanding disease conditions. The significance of bioinformatics has been growing due to continual research and making the availability of new and updated resources for the interpretation of biological data.

Keywords

Bioinformatics · Artificial intelligence · Machine learning · Big data · Molecular modelling · Omics

33.1 Introduction

Persistent efforts have been made around the world by researchers to decipher the functioning of the complex biological system. It is easy to understand the biology of unicellular organisms through the extraction of cell DNA, mRNA, proteins, and their metabolites [1]. It becomes tedious when extended to multicellular organisms where cells are differentiated after the cell division process. These differentiated cells are allocated with a specific function; but that function can be regulated at different levels (i.e., DNA, RNA, and protein) by various biological signals. Although each cell of an organism has identical genome composition, protein expression differs from cell to cell [1, 2]. The degree to

---

A. Gahlawat · Rajkumar. R · T. Varma · P. Kamble · A. Banerjee · H. Sandhu · P. Garg (✉)
Department of Pharmacoinformatics, National Institute of Pharmaceutical Education and Research (NIPER), S.A.S. Nagar, Punjab, India
e-mail: anuj_pip18@niper.ac.in; pip21_rajkumar@niper.ac.in; pip21_tanmaykumar@niper.ac.in; pip22_pradnya@niper.ac.in; pim20_aritra@niper.ac.in; hardeep_pip18@niper.ac.in; prabhagarg@niper.ac.in

G. Jagadeesh et al. (eds.), *The Quintessence of Basic and Clinical Research and Scientific Publishing*,
https://doi.org/10.1007/978-981-99-1284-1_33

which proteins are expressed inside the cell determines which function it will perform and how well it will accomplish it. The change in protein expression either up or down-regulation can cause major consequences/disease-like conditions for individuals such as the formation of malignant cells in cancer. Over time, biologists have used several experimental techniques to solve numerous mysteries and provided meaningful thoughts about how biological events occur inside the cells of animals, and plants. The knowledge gained from experimental procedures helps us to improve the yield and productivity of plant and animal useful products in the agriculture, chemical and pharmaceutical industries [3]. It also aids in the eradication of diseases such as poliovirus and smallpox by allowing one to learn how to manipulate their regulatory mechanisms to have the desired effect on the system [4].

The human brain appears to have a limited capacity to handle complicated biological networks present throughout the organism. It has necessitated the use of other technologies such as computational and statistical methods, to unwrap hidden information present within massive biological datasets. Due to the lack of computer literacy among biologists, the beginning of bioinformatics is hazy, making it difficult to pinpoint its precise beginning. The Fig. 33.1 depicts the highlight of potential accomplishments from various timelines, showing the evolution of bioinformatics through several eras, such as protein, DNA, computer, genomics, big data, and future perspectives. The term "Bioinformatics," has been in use since the late 1980s to refer computational analysis of genetic data and it can be broadly understood as an application of information technology in the biotic system [1, 5]. Initially, computational resources were utilized in the biology field for the analysis of proteins and extended to DNA. The growth of computer technology, software, and expertise enabled substantial applications in the field of molecular modelling, genome, and big data. As a result, researchers are developing novel strategies including whole-genome analysis, concept of personalized medicine, development of various diseases-based models, *etc.*

This chapter will give the reader a brief understanding of bioinformatics and different approaches including methodologies and techniques to tackle biological problems. Throughout the chapter, a diverse set of bioinformatics applications are highlighted in the field of biology, drug design and development.

33.2 Definition, the Distinction Between Related Disciplines and Their Goals

Bioinformatics is an interdisciplinary field of biology and computer technology that uses cutting-edge methods to organise and analyse the data that is beneficial for interpreting sophisticated biological systems. It generally refers to the design, development, and advancement of algorithms, and computational and statistical techniques for organizing and analysing experimental datasets and results [1, 5]. On the other hand, computational biology is a hypothesis-driven investigation carried out on experimental or simulated data to develop new algorithms and theoretical methods to address the unique challenges of biology. Computational biology requires advanced knowledge of biology, but bioinformatics also needs a deep understanding of programming and other skills like statistics [6].

Bioinformatics is used to develop predictive models on the biological data to exploit vast amounts of information such as functional and structural annotation of proteins/genes, evolutionary relationships among sequences, and defining genotype and expression level of different genes/proteins [1]. Nowadays, due to the advancement in omics techniques, a massive amount of data has been produced related to proteome, transcriptome, genome, metabolome, *etc.* To record, manage, and analyse these high throughput data specific to an organism/subject, bioinformatics is required [6]. Databases designed for omics data would be helpful to enhance the accuracy of available tools and assist in designing new tools by enabling curated training datasets for machine and deep learning approaches. The main aims of bioinformatics are as follows [1, 7]:

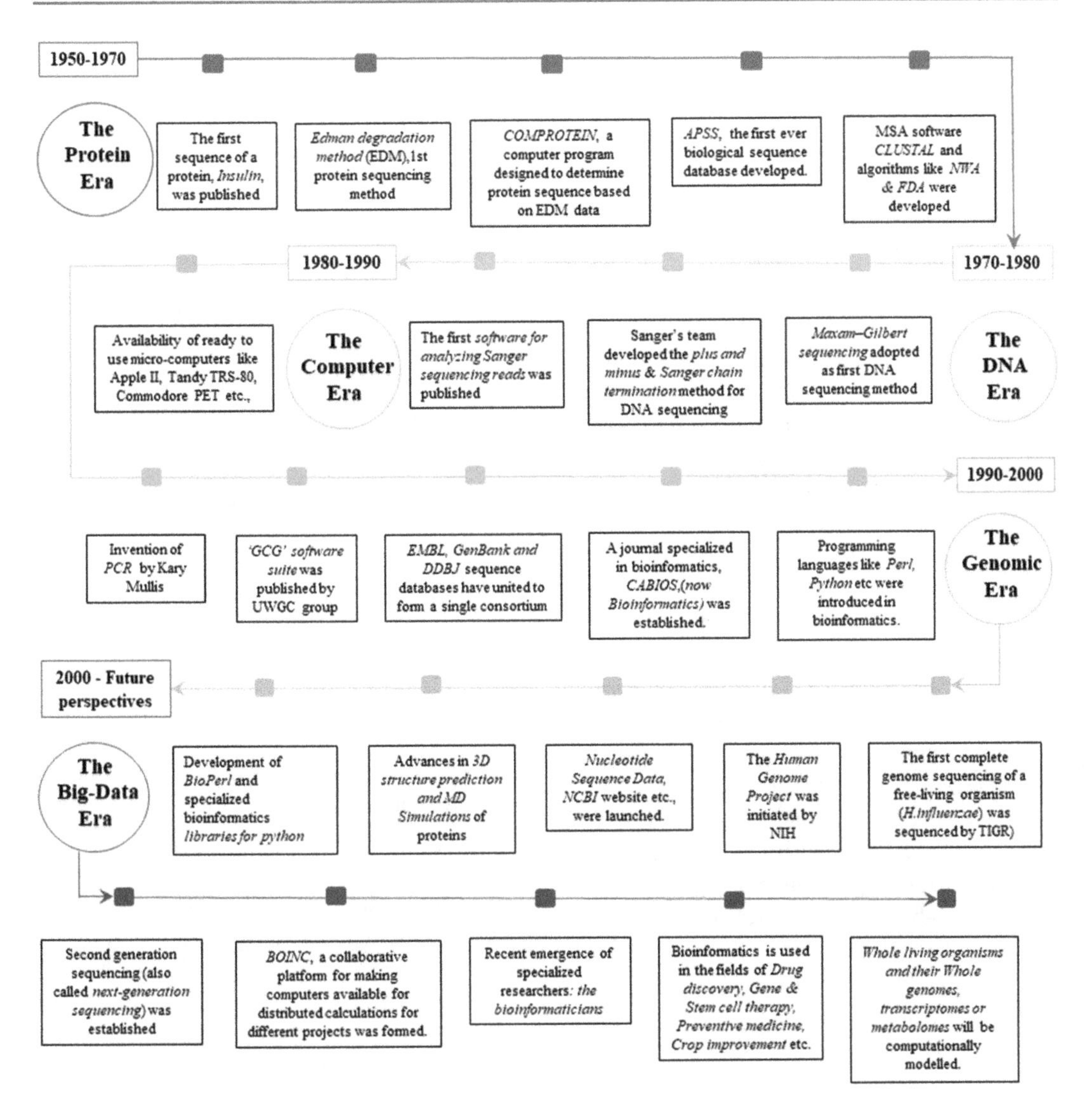

Fig. 33.1 Timeline plot showing the significant events in the field of bioinformatics. The figure uses abbreviations such as APSS (Atlas of Protein Sequence and Structure), MSA (Multiple Sequence Alignment), NWA (Needleman-Wunsch Algorithm), FDA (Feng-Doolittle Algorithm), PCR (Polymerase Chain Reaction), GCG (Genetics Computer Group software), EMBL (The European Molecular Biology Laboratory), DDBJ (DNA Data Bank of Japan), CABIOS (Computer Applications in the BIOSciences), PERL (Practical Extraction and Reporting Language), MD (Molecular Dynamics), NCBI (National Centre for Biotechnology Information), NIH (National Institutes of Health), TIGR (The Institute of Genomic Research)

- Data curation and collection are based on the biological significance, of the experimental/high throughput screening raw data.
- Documentation of curated data in different databases such as disease, organism, or geographical-specific databases.
- Databases must be updated with new experimental data to improve their usefulness.
- Development of new algorithms and tools to address biological problems.
- Development of predictive models using artificial intelligence techniques like machine learning, deep learning, *etc.*

- Computational analysis of curated data obtained from experimental methods or high throughput sequencing techniques such as genome, transcriptome, proteome, system biology, *etc.*

Bioinformatics has diverse applications in a variety of disciplines and industries including molecular biology, agriculture, genetics, biophysics, phylogenetic study, omics, and the pharmaceutical industry. In the agriculture sector, it can be used to understand how plant genome functions and then relevant genes are tweaked to boost nutrition value, increase product yield and provide prevention against biotic and abiotic stresses [3]. It is employed in molecular biology for structure prediction and function annotations of unknown proteins and genes. It is also helpful in establishing phylogenetic relationships among proteins/genes using their sequence, and structural and functional data. It gives information about up and down-regulated genes and their impact on biological functions, after genomic data analysis. It also tells us how biomolecules are communicating with one another in organisms [1, 2, 5]. In drug discovery and development, numerous modelling techniques such as molecular docking/dynamics, QSAR, pharmacophore, quantum methods, and machine/deep learning are utilised to identify and modulate the pharmacokinetic of lead/drug molecules and pharmacodynamic properties of target proteins/genes.

33.3 Applications of Bioinformatics

This section will enlighten the readers about various areas of bioinformatics, provide their brief ideology, and their applications in the biology and drug development field. The following subsections are included to explore the range of bioinformatics.

33.3.1 Biological Databases

Computer science provides a unique feature to organise, search, access, and analyse large-scale raw data into systematic tables using the Database Management System (DBMS). Different queries including insert, update, delete, join, *etc.* help in the time-dependent management of data in the databases. Biological databases are designed to keep a record of biological events in a well organised manner so that researchers can retrieve significant data as per their requirements. Data management aids in understanding hidden patterns by implementing machine learning and artificial intelligence on diverse datasets [8]. There are various types of databases that are described below along with their application.

33.3.1.1 Sequence Databases

These databases store sequence-related information about biological entities including DNA, RNA, and protein for different organisms/species. Examples include GenBank, European Nucleotide Archive (ENA), DNA Data Bank of Japan (DDBJ), GenPept, and UniProt. Secondary sequence databases are created by processing the sequence data using different alignment techniques such as multiple sequence alignment, Hidden Markov models, or fingerprints. It includes PROSITE, PRINTS, Pfam, and InterPro which store sequences with biological significant sites based on their conserved regions and divide them into families and domains. These databases have been utilized for phylogenetic/evolutionary studies, expression studies, mutation studies, functional annotation, and in the omics fields to uncover the mechanism of various disease conditions.

33.3.1.2 Structure Databases

The structural data of biological entities (DNA, RNA, or proteins) and their interactions with small or large molecules is stored in these databases. The RCSB PDB database is a well-known structural database, that provides structural coordinates, secondary structure content (α helix, β sheet, and loop), and conformation information about the entities. Secondary structural databases involve SCOP, and CATH that classify proteins into different classes and domains based on their common structural organization. These databases are used for functional annotation, mutation studies, structural analysis, and evolutionary studies. Structural data can be utilized to design novel molecules for a specific target using a variety of drug development techniques including molecular docking, molecular dynamics, pharmacophore, QSAR studies, and QM/MM methodologies [9].

33.3.1.3 Specialized Databases

These databases are designed to store specific biological data such as bibliographic, taxonomy, expression, similarity, interactions, pathways, enzyme assays, microRNA, small molecule, and vaccine design-specific information. Some examples of specialized databases are given in Table 33.1 [9].

33.3.2 Sequence and Phylogenetic Analysis

An enormous number of sequences are generated with the advent of high throughput sequencing techniques such as RNA sequencing, next generation sequencing (NGS), *etc*. Sequence and phylogenetic analyses have been performed to give biological significance to raw sequence data.

Sequence alignment and analyses involve the alignment of two or more sequences, with considering various phenomena like gaps, insertions, deletions, and translocations. Pairwise sequence alignment (PSA) and Multiple sequence alignment (MSA) are used to align two and multiple sequences respectively and an alignment score is assigned. It can be utilized for DNA/RNA or protein. In the nucleotide alignment, the identity/similarity score is determined on the bases of identical base pairs as only two nucleotide pairs (AT, and GC) exist in nature. However, the similarity score in protein alignment is predicted by taking into account the similar physicochemical characteristics of the amino acids [10]. To achieve the optimum protein sequence alignment various matrices such as Block Substitution Matrix (BLOSUM) and Point Accepted Mutation (PAM) were used. Sequences can be aligned in two ways a) global alignment, which uses the Needleman Wunsch algorithm to align complete sequences b) local alignment, which uses the Smith Waterman algorithm to locate the highest local match region in the sequences. The local alignment is suitable for aligning sequences of varying lengths. The tools used for sequence alignment involve BLAST, MUSCLE, T-Coffee, and Clustal Omega [11, 12].

Phylogenetic analysis is an evolutionary study of a group of organisms/species, to infer their evolution in diverse families. Multiple sequence

Table 33.1 List of some specialized databases with their data type and other descriptive information

Databases	Data type	Description
A Cluster of Orthologous Groups	Clustering	Pre-computed sequence similarity
PubMed, MEDLINE, BIOSIS	Bibliographic	Literature sources
ArrayExpress, BioStudies, Gene Expression Omnibus	Expression	Microarray and NGS data, experiment techniques
Taxonomy (NCBI, UniProtKB)	Taxonomy	Classification of organism/species
IntAct, DIP, IntEnz	Interaction	Interactions among nucleotides, proteins, and enzymes
Kyoto Encyclopedia of Gene and Genomes (KEGG)	Pathway	Published data about pathways, chemical reactions, metabolism, and connectivity between genes to their respective products (protein)
BRENDA, ENZYME	Enzyme	Enzyme classifications, functions, catalytic activity, disease information, enzyme assays
PubChem, ChemDB, DrugBank, Zinc	Small molecules	2D & 3D Coordinates, Physical & chemical properties, disease & bibliographic information
VIOLIN, AntiJen	Vaccine data	Quantitative binding data, kinetic, thermodynamic information, target prediction

alignment helps in the reconstruction of phylogenetic trees by considering gap penalties between the sequence alignments. Different algorithms such as neighbour join (NJ), minimum evolution (ME), Fitch-Marfoliash (FM), maximum likelihood, and maximum parsimony (MP) methods are used for phylogenetic trees [10, 11]. The sequence analysis methods have the following application:

- NGS itself makes use of sequence alignment to generate the entire length of sequences from short reads using either reference genome or de novo approaches. NGS utilized tools like Bowtie, GSnap, and Burrows Wheeler Aligner methods for short reads alignment.
- Genomics, proteomics, and transcriptomics utilized sequence analysis methods for personal and comparative studies, assigning functions to unknown sequences and structures.
- Appropriate sequence similarity between the sequences is used to predict 3D structure using homology modelling methods.
- It helps in identifying various regulatory elements such as binding site, promotor, enhancer, and silencer regions by analysis of conserved regions in the related sequences.
- Phylogenetic tree aids in understanding the origin of the evolution of different organisms, and geographical differences among species [11].

33.3.3 Gene Ontology (GO)

Ontology is the branch of philosophy that deals with how entities are categorized into pertinent classes and which of these entities are present at the most fundamental level. Gene Ontology gives us information on the biological domain in three aspects a) Cellular Component, b) Biological Process and c) Molecular Function [13].

Manual annotation of gene products based on scientific evidence is the most appropriate method. Since it demands domain expertise and more time, similarity search based computational predictions like BLAST/PSI-BLAST (Based on *e-value* or query coverage), *etc.*, are commonly employed. Recently, Artificial Intelligence (AI) techniques like Machine learning and Deep learning are used to overcome the challenges of mutations, and changes in structural and functional domains e.g., Blast2GO, DEEPred, *etc.* Enrichment analyses are often done using different statistical tests for GO to reduce noise in large datasets. For example, to characterize disease phenotypes and related gene products, GSEA (Gene Set Enrichment Analysis) is done using tools like g: Profiler and ClusterProfiler. GO is the most widely used annotation for analyzing high-throughput data, identification of new genes, establishing various disease-gene relationships, differential expression, distribution, and functionality of genes [13].

33.3.4 Secondary Structure Prediction

Secondary structure prediction methods are used, to determine which amino acid residues of the protein sequence are involved in the formation of different secondary structures such as helix, strand, coil, and turn. The Chou-Fasman technique uses heuristic approaches to estimate the statistical propensities of amino acids towards a certain secondary structure. There are other approaches like Garnier-Osguthorpe-Robson (GOR) and the Lim method which use a sliding window of nearby residues along with a variety of theoretical algorithms, including statistical data, graph theory, neural networks, logic-based machine learning approaches, and nearest neighbour techniques to predict secondary structures. Secondary structure prediction methods are also used to design 3D structure models when structural homologs are not available [14].

33.3.5 Protein 3D structure prediction

Proteins are composed of amino acids and are known for performing numerous biological functions. To perform functions, a protein must attain its quaternary structure. With the advancement in proteomics techniques, a huge amount of sequence data is available, but limited structural data is known due to biological complexity, and lack of experiment sensitivity. So, it is an extremely challenging task to predict the 3D structure of proteins based on their sequence information [7, 15]. Bioinformatics offers computational approaches such as homology or comparative modelling, threading or fold-recognition, and ab initio methods for structure prediction [16].

Homology Modelling It is based on the concept that homologue sequences (i.e. similar sequences) share structural similarity among themselves. It is useful when the query sequence has at least one template (a protein with a known sequence and structure) with desirable similarity. The sequences are aligned to themselves with the introduction of gaps between query and template sequences [16]. The aligned sequences help in identifying structural conserved regions inside the template structures and provide a high-quality homology model with biological-related functionality. The quality of the predicted structure depends on the degree of similarity between query and template sequences. The tools used in homology modelling involve Modeller, SWISS-Model, and PRIMO (Protein Interactive Modelling).

Fold recognition relies on the concept that non-homologue sequences (those with less than 25% pairwise sequence identity) can also share structure similarity. Threading is one of the methods utilized for fold recognition, it evaluates how well an unknown sequence fits into known fold structures of protein despite their low sequence similarity [15, 16]. Threading involves in following steps [17].

1. Multiple sequence alignment is obtained for query sequence (i.e. unknown structure) with SWISS-PROT protein sequences.
2. Alignments are subjected to a neural network to predict parameters such as secondary structure, and relative solvent accessibility. Later, a 1D structural profile is generated for the query sequence based on these parameters.
3. Dynamic programming is utilized to find an optimal match between the query structure profile and structure profile database of known folds.
4. The maximum matched fold is considered as a predicted fold and alignment of the query and predicted fold is used to model its 3D structure.

Ab initio methods are based on protein folding processes irrespective of their sequence similarity. According to the Anfinsen hypothesis, a protein sequence undergoes conformation changes to attain its native structure i.e., minimum free energy structure when kept in a biological condition. It is a unique, stable, and kinetically accessible structure [15]. These methods take protein sequences and start searching for native conformation using various energy-based functions, conformational search algorithms, and model selection methods to model a given protein sequence. **ROSETTA** produces a small fragment library for a given protein sequence using sequence conservation with structural databases. These fragments are assembled using Monte Carlo fragment assembly methods to generate low-resolution structural models. Then, to acquire the native structure of a given protein, atomic refinement of these low structure models is carried out using force fields like AMBER, GROMACS, and OPLS along with conformation search through the Monte Carlo method [18, 19]. **I-Tasser** uses threading methods to create structural assembly using profile-profile threading alignment. Like ROSETTA, the structural assembly is set up for atomic refinement to develop a model. Ab initio approaches are only useful for short protein sequences; otherwise, the search for the native structures will take a long time [16, 18, 19].

Structural information helps scientists to access protein function, protein-protein interactions, identification of novel therapeutics targets, enzyme kinetics, and the effect of losing structural identity in a cell, tissue, or organism [15, 16]. Such technical advancement may address various issues concerning biological functioning, tissue specialization, signalling routes, and disease causes. Furthermore, disease-related mutagenesis studies are another use of structure modelling in the aspect of identification of amino acids with relevant functions in a protein. It can be employed in molecular modelling of biological assemblies of protein complexes (e.g., whole viral 3D structure), and in protein-protein interaction investigations [20]. Homology modelling is also used to refine cryo-electron microscopy (EM) 3D structures. After examining the 3D molecule surface and density maps of cryo-EM, homology modelling is utilized to build an atomic 3D model.

33.3.6 Omics Techniques

The "Omics" suffix is used in biology to refer to a variety of molecular disciplines such as genomics, transcriptomics, proteomics, metabolomics, *etc*. It involves the characterisation, mapping and quantification of various messenger molecular entities including DNA, RNA, proteins, and metabolites, *etc*. Their names already give away the level at which they will operate like genomics, epigenomics, transcriptomics, proteomics, and metabolomics disciplines are involved in the investigation of genes, DNAs, RNAs, proteins, and metabolites respectively in the organisms [21].

One of the many ways that "omics" benefits the scientific community is by identifying a molecular entity as a biomarker that would be an indication of an underlying disease or illness condition. The identification of these biomarkers is done by comparing the compositions of different entities like DNA, RNA, protein, metabolites, *etc*. in the normal individual and the abnormal one. The higher variation in the entity between the normal and abnormal groups, easy access to the entity, outcome accuracy, and reproducibility, determine how useful a biomarker is for foretelling or monitoring ongoing disease conditions. Biomarkers are biological entities such as enzymes, receptors, mRNA, microRNA, noncoding RNA, antibodies, peptides, metabolites, *etc*. used to characterize a disease condition by measuring/quantifying their presence in both normal biological and pathological processes. It also acts as an indicator to measure pharmacological response to a therapeutic intervention. A collection of biomarkers such as gene expression, proteomics, and metabolic signatures can also be employed to increase their sensitivity against the undergoing changes or to predict the incidence of disease in the individual [21, 22]. Following are some of the omics technologies:

33.3.6.1 Genomics

Genomics involves the study of a whole gene pool present in an organism. Gene acts as a fundamental source of information which is then decoded into different regulatory macromolecules including DNA, RNA, and proteins inside each cell of the organism [21]. Recent advancements in high throughput methods such as NGS techniques have made it possible to sequence a large number of bases (up to billions) at an effective cost. The generated data is stored in raw and curated form in the different databases. Genomics helps in understanding how genes interact with each other and any abnormalities in the gene signalling can contribute to various disease conditions like heart disease, diabetes, cancer, *etc* [23]. Therefore, genomic research opens up a new avenue for the diagnosis, treatment, and cure of diseases. Genomics is categorised into the following fields based on its application:

Pharmacogenomics utilizes an individual genome sequence to assess drug effectiveness and safety. It can be used to reduce drug toxicity and to identify pathologic symptoms on the bases of individual genome sequences. It presents a novel idea of targeted therapy and personalised medicine for each individual.

Metagenomics deals with several species genomes that are known to exist together and show interaction among themselves in a certain environment. It is mainly implemented in the microbiology field where it is used to understand the effect of biotic (other microbes) and abiotic (pH, temp, *etc.*) environments on a particular microorganism.

Mitochondrial Genomics is used to establish genealogical/evolutionary relationships among multicellular organisms, as, during the fertilization process, mitochondrial DNA is passed on to offspring.

Plant Genomics is utilized in the agriculture field to identify links between desirable traits and gene signatures. It helps scientists to improve crop breeding, develop new hybrid variants, and reduce the environmental stress on plant species. It has been used to preserve endangered plant species, and improve the yield and quality of a variety of crops.

Genome analysis is also used in forensic science to find the culprit by matching DNA samples collected from the crime scene. This information can then be used as evidence in the legal system. There are other clinical applications of genomics such as gene discovery, diagnosis, and cure of rare monogenic disorders (cystic fibrosis, sickle cell anaemia, albinism, *etc.*) [24]. Numerous computational algorithms are utilized to predict coding regions of genes after the genome has been sequenced using high throughput methods. The prediction of coding regions aid in correlating genome data with transcriptomic/proteomics data to comprehend signalling processes under various stress and unstressed conditions [21]. Another well-known application of genomics involves the identification of single nucleotide polymorphism (SNP) markers and their analysis. SNPs have been aroused in an individual owing to the replacement, insertion, and deletion of a single nucleotide base. These are used to diagnose genetic abnormalities, track the inheritance of disease-associated genetic variants, and foretell individual responsibility toward drugs and toxins [24, 25].

33.3.6.2 Transcriptomics

Transcriptomics involves the study and analysis of an RNA pool that is being transcribed from the organism's genome inside a tissue or cell at a certain developmental stage. All forms of RNA such as messenger RNA (mRNA), transfer RNA (tRNA), ribosomal RNA (rRNA), double stranded RNA (dsRNA), and noncoding RNA (ncRNA) are kept under surveillance to observe biological changes. The transcriptome studies are time-dependent and tissue-specific phenomena i.e., different tissue cells have distinct gene expression at various cell stages due to the activation of certain transcription factors. mRNA transmits genetic information to proteins while other RNAs control gene expression, protein synthesis, and several biological activities in the cell. Transcriptome sequencing can be done at the cellular or tissue level using high throughput techniques like microarrays, RNA sequencing, next-generation sequencing, *etc.* It has the following major applications [26].

- Differential gene expression analysis uses for the characterization of different stages of cells or tissue. It also helps in identifying biomarkers that are differently expressed in healthy and disease states.
- Transcriptome study reveals gene function, and advances our knowledge of specific biological processes and molecular mechanisms.
- It can foretell the severity of a particular disease condition e.g. neurodegenerative disease, cancer, *etc.*
- It establishes a causative relationship between genetic variants and gene expression to address disease pathology.
- Cell-specific transcriptomes can be employed to illuminate cellular heterogeneity present among similar cells of a tissue e.g., the study of intratumor heterogeneity (ITH), and tumour microenvironment (TME) of cancerous cells.
- Its clinical application includes the prognosis, or diagnosis of complex diseases such as acute myeloid leukaemia, breast cancer, or cardiovascular diseases where the expression of multiple genes is affected.

Nowadays, transcriptomic profiling is utilized in clinical oncology as a cancer prognostic and diagnostic biomarker, and in predictive gene expression assays.

33.3.6.3 Proteomics

Proteins are made up of amino acids and play an essential role in the survival of an organism by regulating complex machinery. Like transcriptomics, proteomics is also a tissue-and time-specific phenomenon but in proteomes, complete protein expression data is analysed for a cell/tissue. Protein expression data can be deduced from the mRNA data, as it is well known to serve as a bridge between genes and proteins. Biological events like post-translational modifications, mutations, cleavage, complex formation, transcripts from variant mRNA and cellular location produce significantly different protein expression data from what is expected from mRNA [21]. It makes a solo study of proteins important to get a better understanding of the functioning of the biological system. High throughput methods such as mass spectroscopy, reverse phase microarray, *etc.*, are used for the quantification of proteins. Proteomics involves [21, 22]:

- Understanding of protein functions, complexes, and signalling at the cellular/tissue level.
- In the study of differential protein expression, that can be used to identify disease conditions, personalized medicine for individuals, biomarker identification, and heterogeneity between the cells.
- Identification of novel protein targets for drug design and development. It involves the isolation of antibodies signature, identification of antigens, and specific enzyme targets to halt grow of pathogens.
- Characterization of different disease conditions such as Alzheimer's, Parkinson's, various types of cancers, and liver cirrhosis, along with the make-up and functions of pathogens.
- Finding functions of unknown proteins, by finding their similarity with proteins whose functions are known.

33.3.6.4 Metabolomics

Metabolites are end products of cellular processes that occur in a living system. Metabolites are small molecules that are produced by biological processes and further show interaction with other biological entities. Metabolomics involves high throughput analysis of heterogeneous mixtures of body fluids or tissues for identification and quantification of each metabolite. Numerous analytical techniques including mass spectroscopy, electron ionization, Nuclear Magnetic Resonance (NMR), capillary electrophoresis, and high-performance liquid chromatography (HPLC) are utilized to obtain metabolomics data [21, 27]. It allows a global assessment of cellular processes by considering enzymes/receptor kinetics, regulation of enzymes, protein signalling, and change in metabolic reactions. Its concurrent research with other omics indicates change at the appropriate level (i.e. gene, RNA, protein), and how the function of a specific process gets affected at an individual level. Metabolomics studies can be cell, tissue, or organism specific [27–29].

- Toxicological studies can be conducted, by keeping the track of the rate of production of toxic and reactive metabolites.
- The presence of metabolites in excretory samples (urine, saliva, bile, feces and seminal fluid) makes them a reliable prognostic, diagnostic biomarker for various pathological states (cancer, diabetes, autoimmune, and coronary diseases).
- Metabolomics data such as target-based metabolite analysis, metabolite profiling, metabolite flux analysis, and metabolic fingerprinting help researchers understand the molecular mechanism behind the complex processes.

A metabolomic investigation offers the same advantages as proteomic and transcriptomic techniques like discovering novel drug targets, personalised therapeutics treatments, *etc.* All the omics techniques can also be used in diverse fields such as agriculture, microbiology, pharmaceuticals, and the food industry to provide a healthy life to an individual [23, 28]. Therefore, a collective implementation of these omics techniques i.e., multiomics can serve researchers by giving aggregate information about the network of unknown pathways since each molecular entity (genes, RNAs, and proteins) has a limited piece of information.

33.3.7 Molecular Modelling Approaches

The design and development of a new drug for a therapeutic intervention takes a lot of effort from scientists in various fields. It is a time-consuming and costly process. Identification of the underlying cause of biological dysfunction in the patients is the first step in the drug development process [30]. The causative mechanism can be individual-based or pathogen based, which further produces malfunctioning of normal biological processes. In both situations, the triggering points of aberrant biological processes are identified using omics approaches, evolutionary knowledge, and structural information. These triggering points include DNA, RNA, receptors, enzymes, ion channels, and transporters that are present at various levels and assist in managing usual body functions. After the target selection for a particular disease condition, an intervention is designed through some foreign chemical molecules. These foreign chemical molecules must have selectivity towards the target, without having interaction with other normal functioning biological processes, and must not be toxic or harmful to the environmental assemblies. Selective action of these chemical molecules on specific targets aid individual to regain normal body functions through setting up proper signalling, preventing structural changes, and modulating enzyme activity and gene expression. But in some cases, foreign chemicals are directly supplied to meet the deficiency of essential precursors or in the absence of biological pathways. Here, we discussed how exogenous chemical compounds can affect the biological system and observed changes in biological processes which are known as pharmacodynamics study.

Since these chemicals are exogenous substances, therefore, the body will respond toward these entities which is known as a pharmacokinetics study. It involves the absorption, distribution, metabolism, excretion, and toxicity study of compounds. Absorption is needed for compounds, to reach the destination/target site from the administered site through various membranes, and transporters. Bloodstream distributes compounds into different components like plasma proteins, tissues, *etc*. based on their lipophilicity and hydrophilicity, ionization, molecular size, and pH of the environment. During absorption and distribution, a compound goes through a variety of metabolic phase reactions to produce active, inactive, and toxic metabolites. These foreign chemicals and their metabolites need to be eliminated from the body after pharmacological action at a specific target. A drug should demonstrate desirable pharmacokinetics to produce noteworthy pharmacodynamic effects in an individual. At different stages of drug discovery and development, computational approaches are quite helpful and offer useful information. Here, we discuss some of the frequently used computational techniques in the drug development process.

33.3.8 Molecular Docking

The prime objective of molecular docking approaches is to determine the most effective way for small/large molecules to form complexes with a macromolecular partner (DNA, RNA, protein). It is performed by using a three-dimensional structure of molecular targets and small/large molecules. Molecular docking is based on two main phases a) Conformation search that involves the sampling of different conformations of a molecule (small/macromolecule that will be docked) to explore the conformation space defined by free energy landscape b) Scoring function used to approximate energy of each conformation in the docked complexes. Docking analyzes the conformation and orientation of molecules in the binding site of target macromolecules and provides potential poses with a score assigned by the scoring functions [31]. It is used in the drug design process. (1) to identify lead molecules and their optimization via docking-based virtual screening [32]; (2) to rank molecules according to their binding propensity with target based on docking score; (3) to understand the binding mechanisms such as conformation changes, binding poses, and potential interacting residues of various biological entities (drug-protein, protein-protein, DNA/RNA interactions); (4) to perform target fishing, where a single molecule is screened against several proteins to forecast cross targets and their associated adverse effects that can be seen among individuals; (5) to identify new therapeutic applications for already approved drugs by using the drug repurposing concept [33]; (6) to identify lead molecules that can interact with multiple potential targets of a disease condition, and modulate the disease via different mechanisms. It helps in preventing drug resistance among individuals.

33.3.8.1 Molecular Dynamics

Molecular dynamics is a computer approach for simulating the dynamic behavior of a molecular system as a function of time, with all entities in the simulation box treated as flexible. Molecular dynamics may be utilized to comprehend the variety of drug development problems. It is commonly used in biology to examine the conformational stability and flexibility of protein and ligand complexes. It can be implemented to understand dynamic processes over time, such as transport across the membrane, conformational changes, protein folding, or ligand binding. MD is not constrained to these though. The most practical use of molecular dynamics is to study the behavior of a system after applying controlled changes, such as adding or removing ligands, altering proteins via mutations or protonation states, pulling ligands out of binding pockets with biased force, applying mechanical force, using external potential, or altering the protein environment [34, 35]. Xiaoli et al. employed homology modelling, molecular docking, and molecular dynamics to explore the interaction between mutant T1 lipase and fatty acids such as caprylic, myristic, stearic, oleic, linoleic and linolenic acids. Their work demonstrated that mutant t1 lipase had an excellent binding affinity for linoleic acid and linolenic acid compared to other fatty acids. They showed that the binding affinity of fatty acids for mutant t1 lipase was rising as the number of carbon atoms and double bonds of the acyl chain of fatty increased [36].

33.3.8.2 Quantum Mechanics

The QM technique considers molecules to be collections of nuclei and electrons, with no notion of "chemical bonding." Understanding the atomic-level behavior of the system requires knowledge of quantum mechanics. QM techniques use the QM laws to solve the Schrödinger equation and estimate the wave function. The Schrödinger equation may be solved in terms of the movements of electrons, which in turn provide information about bonding as well as other explanatory variables including molecule structure and energy. However, approximations must be used since the Schrödinger equation can only be solved for systems with one electron (the hydrogen atom). The results obtained by QM methods are more accurate and reliable as compared to other methods [37].

In Drug Discovery, QM helps in obtaining precise geometries of small molecules, which can be further used in computational studies such as molecular docking and molecular dynamics. Structure-based drug design (SBDD) can benefit from using quantum mechanics with a few minor adjustments or approximations to reduce computing time and cost. Examples of such a modification include divide and conquer, QM/MM, and fragments-based approaches. These techniques have been used to calculate the binding free energies and the energies of protein-ligand interactions. Because QM takes into account the electrical changes that take place when a ligand binds to a receptor, it is incredibly helpful in the design of compounds. As a result, interactions including hydrogen bonds, halogen bonds, Cation-aryl and aryl-aryl interactions, which are significant in the biological system, are properly characterized. Particular molecular features, such as partial charges, bond orders, and molecular geometry analysis, may be calculated using ligand-based applications of QM techniques and utilized to parameterize force fields. Additionally, QM descriptors may be applied to the construction of QSAR models. Energy prediction of bioactive conformation and finding molecular similarities are the recent application of QM-based approaches.

33.3.8.3 Pharmacophore Modelling

The presence of complementary functional groups in each molecule is one of the mechanisms through which small or large molecules interact with one another. Pharmacophore is used in the drug development process, to represent molecules with an abstract description of molecular properties instead of individual functional groups and their original bonds. The abstract molecular properties include aromatic rings, hydrogen bond donors (HBD), hydrogen bond acceptors (HBA), negative and positive charge features, and hydrophobic features. Their combinations are used to describe crucial molecular interactions between the molecules by considering their steric and electronic aspects. In the pharmacophore, each feature is represented in three-dimensional space by a sphere having a tolerance radius. A variety of tools and software have been utilized such as Phase, Catalyst/Discovery Studio, MOE, and LigandScout for pharmacophore development.

Pharmacophores can be generated for ligands as well as for the binding site of enzyme targets. Virtual screening can be performed to identify new lead molecules for drug targets using ligand/structure-based pharmacophore. It has been implemented before molecular docking to filter out compounds that do not satisfy fundamental chemical and structural criteria for binding. Binding site-specific pharmacophores are utilized in the drug discovery field to identify potential lead molecules, similar binding sites, chances of cross-reactivity between different targets, and to identify new therapeutic targets. Pharmacophore modelling is also utilized to characterize ADMET (absorption, distribution, metabolism, excretion and tolerable toxicity) proprieties of lead molecules using the ADMET profile of well-known drugs/molecules. It can be used to predict potential interaction between the lead molecules and metabolizing enzymes [38].

33.3.8.4 Quantitative Structure Activity Relationship (QSAR)

Structure-activity analysis relationships (SAR) approach is founded on the notion that molecular structures are directly connected with biological activities, and hence, that molecular or structural alterations influence biological activities. QSAR is described as a method involving the building of computational or mathematical models by employing chemometric approaches to uncover substantial correlations between a set of structures and biological functions. A library of compounds with known biological activity is used to train statistical QSAR models that establish theoretical optimization of chemical functional groups to biological activity. Comparative molecular field analysis (CoMFA) and comparative molecular similarity indices analysis (CoMSIA) are the two most important approaches proposed for drug design. Like pharmacophores, the QSAR model also performed well in predicting physiochemical properties such as ADMET of compounds. It is used to modulate existing molecules to enhance their biological activity. It has been frequently employed in drug design to identify new novel lead compounds with high selectivity and specificity toward the drug target [39, 40].

33.3.9 Holistic Approaches in Bioinformatics

Except for the omics techniques, all techniques discussed above are reductionist approaches where a single entity or a few entities (gene, proteins, metabolites, *etc.*) are considered to characterize biological regulations. Biological systems have a complex network framework that enables the simultaneous transmission of signals at various entity levels. Therefore, holistic approaches like system biology are needed to characterize the perturbation of a single entity (gene, RNA, protein, drug, *etc.*) in a whole biological system.

33.3.9.1 System Biology

System biology helps in understanding all components of the biological system at the molecular level and predicts how an organism will respond to external alterations such as gene mutation or knockout. It mainly deals with heterogeneous data such as sequences, structures, function, subcellular localization, biological activity, *etc.*, that is generated by various types of experimental methods from several resources. System biology is utilized to build comprehensive systems via network biology including ecosystem, single or multiple organism systems, cell or tissue specific systems, a system of a group of biological entities, and their pathway and simulation analysis.

Systems biology offers insight into how evolution has shaped the phenotype by allowing researchers to examine how the genotype leads to the phenotype via modelling, simulations and network analysis. The development of methods to comprehensively investigate the amount of protein, RNA, and DNA on a gene-by-gene basis, as well as the post-translational modification and localization of proteins, is a part of systems biology. The emergence of high-throughput biology has compelled us to think about biological processes systematically and is more powerful when combined with conventional techniques [41, 42].

33.3.9.2 Network Biology

Network biology is an interdisciplinary science involving life sciences, mathematics, computational science, statistics, *etc.* Networks are one of the most common ways to visually represent various complex sets of objects and the relationship between them, derived basically from the graph theory. Biological networks are made of nodes (type of information) representing genes, proteins, metabolites, *etc.*, and edges (interaction patterns) representing genetic interaction, biochemical reaction, *etc.* They help to establish the relationship between different ecosystems, gene regulations, protein-protein interactions, metabolic pathways, and cellular signalling by utilizing the data obtained using modern high-throughput technologies, *etc.* Databases like KEGG (Kyoto Encyclopedia of genes and genomes), STRING, BioCyc, *etc.*, provide the necessary data for network development and tools like KEGGtranslator, ERGO, OmicsNet, *etc.*, are utilized for network analysis [43].

33.3.9.3 Pathway Modelling and Simulation Analysis (PMSA)

Pathway modelling is one of the finest ways to capture and convey our extant understanding of various biological processes. Here, a pathway model is described as a collection of biological interactions (such as those involving proteins and metabolites) pertinent to a specific context that have been carefully selected and arranged to represent a specific process. The models are generated by making an accurate description of the pathway, and utilizing the information about the elements and how they interact with each other. Once the initial topology of the model is defined, the mathematical model can be formulated using ordinary differential equations (ODEs), partial differential equations (PDEs), or stochastic differential equations, *etc.*, for simulations [44]. Cell Designer and Cytoscape are the widely used pathway modelling tools whereas General Pathway Simulator (GEPASI), Systems Biology Workbench, COmplex PAthway Simulator (COPASI), *etc.*, are used for simulating biochemical pathways and their kinetics.

All these holistic approaches have been applied in the field of drug discovery and development widely ranging from investigations of important gene/protein targets, protein/gene functions, signalling, identification of functional and non-functional associations between biomolecules, *etc.* Apart from being applied in studying phenotypes, designing metabolites, and flux-balance analysis, these methods are widely used in the drug discovery process to identify drug-drug/target interactions, identify new disease-causing pathways, predict drug toxicity effects at an early stage, host-pathogen interactions, drug repurposing, *etc* [33, 43, 45, 46].

33.3.10 Machine Learning, Data Mining, and Big Data Analytics

Machine learning (ML) is a branch of artificial intelligence (AI) that enables machines to learn from data and past experience while finding patterns to make predictions with minimal human intervention (*see* Chap. 31). Machine learning approaches allow computers to function independently without requiring explicit programming. It aids in the development of various tools that learn to anticipate and identify biological significant knowledge from databases. In bioinformatics, ML algorithms are widely utilized for prediction, classification, and feature selection tasks. ML techniques have some applied applications such as gene finding, genome annotations, protein structure prediction, gene expression analysis, complex interaction modelling, understanding biological systems, and drug discovery [47]. The machine learning concept is inexpensive and efficient in dealing with bioinformatics issues. There is a belief that machine learning will play a significant role in the future of bioinformatics.

Data Mining is a field of computer science and statistics, that utilises machine learning techniques and database systems to mine information from a very huge dataset and convert them into an understandable form that can be further used for its intended purpose. Data mining in drug discovery is useful for many aspects such as gene analysis, biomarker discovery, protein function analysis, finding new drug targets and drug optimization, *etc.* Some of the main data mining algorithms that are currently in use include C4.5, K-means, Support vector machine, *etc* [48]. The noise, biasness, and imperfectness of the current high throughput datasets make it hard to properly index, search and mine the required information. Linking the different databases containing information on the gene, RNA, proteins, expression, diseases, clinical outcomes, *etc.*, are helpful in drug discovery. It will be a boon for the researchers to save time and retrieve required information more accurately.

Big Data Analytics can be described in terms of four Vs (Volume, Velocity, Veracity, and Variety). a) First V stands for a huge Volume of data (petabyte/exabyte) that can be generated through various methods such as high throughput techniques data for cells, tissues or organisms. b) Second V stands for the Velocity at which data is being generated. c) Third V stands for the Veracity of data i.e., data correctness, as biasness in data can occur due to batch effect, different statistical models, and other experimental errors. d) Fourth V stands for the Variety of data since big data analytics deal with a diverse range of data such as sequences, structures, omics data, functional, biological activity data, 2D and 3D images, different interaction data (protein-protein/DNA/small molecules), pathways data, *etc.* It widely utilizes MapReduce, Fault-Tolerant graph, and streaming graph architectures in the bioinformatics field [49].

Like Machine learning (ML) and deep learning techniques, Big data analytics can be utilized to build predictive, classification models on the diverse nature of biological data. Additionally, it aids in predicting important features of data that will help in understanding ongoing biological changes. It can be implemented to analyse real-time data of organisms or species to even detect slight changes in the data. Big data offers numerous opportunities to address various issues in the drug discovery process. The unavailability of a universal format to represent big data, better visualization and analysis platforms, better architectures to handle heterogeneous data, data warehouses for dynamic storage and real-time retrieval, and lack of Hadoop or other cloud-based big data tools demonstrate a lack of support for the bioinformatics problems [48, 50]. These issues need to be addressed to use them for better research purposes.

33.4 Concluding Remarks

Bioinformatics has evolved as a key subject and a prominent research area in recent years, and is connected to various disciplines and techniques. The current chapter emphasises the strength of bioinformatics approaches along with their application in biological research and drug development. The availability of high-performance computational platforms has led researchers to develop several databases and tools to manage and analyse experimental and omics data. Furthermore, the chapter offers new career opportunities in the field of bioinformatics, to students and researchers to understand complex biological networks. In the upcoming decades, bioinformatics will be a new guide for molecular biologists and drug development researchers by providing a preliminary switch from the laboratory bench to futuristic computational methods.

Conflict of Interest The authors declare no conflict of interest.

References

1. Pathak RK, Singh DB, Singh R (2022) Introduction to basics of bioinformatics. In: Bioinformatics. Elsevier, pp 1–15
2. Payne JL, Wagner A (2019) The causes of evolvability and their evolution. Nat Rev Genet 20(1):24–38
3. Halewood M, Chiurugwi T, Sackville Hamilton R, Kurtz B, Marden E, Welch E et al (2018) Plant genetic resources for food and agriculture: opportunities and challenges emerging from the science and information technology revolution. New Phytol 217(4):1407–1419
4. Schatzmayr HG (2004) Poliovirus vaccine strains will continue to circulate long after wild strains have been eradicated. Bull World Health Organ 82(1):65
5. Gauthier J, Vincent AT, Charette SJ, Derome N (2019) A brief history of bioinformatics. Brief Bioinform 20(6):1981–1996
6. Domokos A (2008) Bioinformatics and computational biology. Bulletin of University of Agricultural Sciences and Veterinary Medicine Cluj-Napoca Horticulture, North America 6527:571–574
7. Bayat A (2002) Science, medicine, and the future: bioinformatics. BMJ 324(7344):1018–1022
8. Cannataro M, Guzzi PH, Tradigo G, Veltri P (2014) Biological databases. Springer, Springer handbook of bio–/neuro-informatics, pp 431–440
9. Sharma PK, Yadav IS (2022) Biological databases and their application. Elsevier, Bioinformatics, pp 17–31
10. Phillips A, Janies D, Wheeler W (2000) Multiple sequence alignment in phylogenetic analysis. Mol Phylogenet Evol 16(3):317–330
11. Saeed U, Usman Z (2019) Biological sequence analysis. Exon Publications, pp 55–69
12. Kuznetsov IB (2011) Protein sequence alignment with family-specific amino acid similarity matrices. BMC Res Notes 4(1):296
13. Ashburner M, Ball CA, Blake JA, Botstein D, Butler H, Cherry JM et al (2000) Gene ontology: tool for the unification of biology. The Gene Ontology Consortium. Nat Genet 25(1):25–29
14. Nishikawa K (1983) Assessment of secondary-structure prediction of proteins. Comparison of computerized Chou-Fasman method with others. Biochim Biophys Acta 748(2):285–299
15. Kuhlman B, Bradley P (2019) Advances in protein structure prediction and design. Nat Rev Mol Cell Biol 20(11):681–697
16. Kumar P, Halder S, Bansal M (2019) Biomolecular structures: prediction, identification and analyses. Elsevier, Encyclopedia of bioinformatics and computational biology, pp 504–534
17. Rost B, Schneider R, Sander C (1997) Protein fold recognition by prediction-based threading. J Mol Biol 270(3):471–480
18. Lee J, Freddolino PL, Zhang Y (2017) Ab initio protein structure prediction. In: From protein structure to function with bioinformatics. Springer, pp 3–35
19. Roterman-Konieczna I (2012) A short description of other selected ab initio methods for protein structure prediction. In: Protein folding in silico. Elsevier, pp 165–189
20. Bonarek P, Loch JI, Tworzydlo M, Cooper DR, Milto K, Wrobel P et al (2020) Structure-based design approach to rational site-directed mutagenesis of beta-lactoglobulin. J Struct Biol 210(2):107493
21. Misra BB, Langefeld CD, Olivier M, Cox LA (2018) Integrated omics: tools, advances, and future approaches. J Mol Endocrinol 62(1):R21–R45
22. Nimse SB, Sonawane MD, Song KS, Kim T (2016) Biomarker detection technologies and future directions. Analyst 141(3):740–755
23. Tan YC, Kumar AU, Wong YP, Ling APK (2022) Bioinformatics approaches and applications in plant biotechnology. J Genet Eng Biotechnol 20(1):106
24. Mooney SD, Krishnan VG, Evani US (2010) Bioinformatic tools for identifying disease gene and SNP candidates. Methods Mol Biol 628:307–319
25. Daly AK (2017) Pharmacogenetics: a general review on progress to date. Br Med Bull 124(1):65–79
26. Cable DM, Murray E, Shanmugam V, Zhang S, Zou LS, Diao M et al (2022) Cell type-specific inference of differential expression in spatial transcriptomics. Nat Methods
27. Wishart DS (2016) Emerging applications of metabolomics in drug discovery and precision medicine. Nat Rev Drug Discov 15(7):473–484
28. Yang Q, Zhang AH, Miao JH, Sun H, Han Y, Yan GL et al (2019) Metabolomics biotechnology, applications, and future trends: a systematic review. RSC Adv 9(64):37245–37257
29. Gomez-Casati DF, Zanor MI, Busi MV (2013) Metabolomics in plants and humans: applications in the prevention and diagnosis of diseases. Biomed Res Int 2013:792527
30. Lin X, Li X, Lin X (2020) a review on applications of computational methods in drug screening and design. Molecules 25(6):1375
31. Salmaso V, Moro S (2018) Bridging molecular docking to molecular dynamics in exploring ligand-protein recognition process: an overview. Front Pharmacol 9:923

32. Gahlawat A, Kumar N, Kumar R, Sandhu H, Singh IP, Singh S et al (2020) Structure-based virtual screening to discover potential lead molecules for the SARS-CoV-2 main protease. J Chem Inf Model 60(12):5781–5793
33. Kumar N, Gahlawat A, Kumar RN, Singh YP, Modi G, Garg P (2022) Drug repurposing for Alzheimer's disease: in silico and in vitro investigation of FDA-approved drugs as acetylcholinesterase inhibitors. J Biomol Struct Dyn 40(7):2878–2892
34. Hollingsworth SA, Dror RO (2018) Molecular Dynamics Simulation for All. Neuron 99(6):1129–1143
35. Kumar N, Garg P (2022) Probing the molecular basis of cofactor affinity and conformational dynamics of mycobacterium tuberculosis elongation factor Tu: an integrated approach employing steered molecular dynamics and umbrella sampling simulations. J Phys Chem B 126(7):1447–1461
36. Qin X, Zhong J, Wang Y (2021) A mutant T1 lipase homology modeling, and its molecular docking and molecular dynamics simulation with fatty acids. J Biotechnol 337:24–34
37. Jaladanki CK, Gahlawat A, Rathod G, Sandhu H, Jahan K, Bharatam PV (2020) Mechanistic studies on the drug metabolism and toxicity originating from cytochromes P450. Drug Metab Rev 52(3):366–394
38. Kaserer T, Beck KR, Akram M, Odermatt A, Schuster D (2015) Pharmacophore models and pharmacophore-based virtual screening: concepts and applications exemplified on hydroxysteroid dehydrogenases. Molecules 20(12):22799–22832
39. Hasan MR, Alsaiari AA, Fakhurji BZ, Molla MHR, Asseri AH, Sumon MAA et al (2022) Application of mathematical modeling and computational tools in the modern drug design and development process. Molecules 27(13):4169
40. Abdel-Ilah L, Veljović E, Gurbeta L, Badnjević A (2017) Applications of QSAR study in drug design. Int J Engineer Res Technol 6(06)
41. Kirschner MW (2005) The meaning of systems biology. Cell 121(4):503–504
42. Sharma M, Shaikh N, Yadav S, Singh S, Garg P (2017) A systematic reconstruction and constraint-based analysis of Leishmania donovani metabolic network: identification of potential antileishmanial drug targets. Mol BioSyst 13(5):955–969
43. Redhu N, Thakur Z (2022) Network biology and applications. Elsevier, Bioinformatics, pp 381–407
44. Hanspers K, Kutmon M, Coort SL, Digles D, Dupuis LJ, Ehrhart F et al (2021) Ten simple rules for creating reusable pathway models for computational analysis and visualization. PLoS Comput Biol 17(8):e1009226
45. Hillmer RA (2015) Systems biology for biologists. PLoS Pathog 11(5):e1004786
46. Tandon G, Yadav S, Kaur S (2022) Pathway modeling and simulation analysis. Elsevier, Bioinformatics, pp 409–423
47. Sandhu H, Kumar RN, Garg P (2022) Machine learning-based modeling to predict inhibitors of acetylcholinesterase. Mol Divers 26(1):331–340
48. Sarangi SK, Jaglan DV, Dash Y (2013) A review of clustering and classification techniques in data mining. Int J Engineer Bus Enterp Appl:140–145
49. Shukla R, Yadav AK, Sote WO, Junior MC, Singh TR (2022) Systems biology and big data analytics. Elsevier, Bioinformatics, pp 425–442
50. Dara S, Dhamercherla S, Jadav SS, Babu CM, Ahsan MJ (2022) Machine learning in drug discovery: a review. Artif Intell Rev 55(3):1947–1999

34 Divergent Approaches Toward Drug Discovery and Development

Summon Koul

Abstract

The drug discovery process aims to identify new chemical entities (NCEs) that are therapeutically beneficial for the management and/or cure of diseases. Such novel compounds have been identified using various strategies that include phenotypic drug discovery (PDD), target-based drug discovery (TDD) and serendipitous drug discovery (SDD). PDD is an empirical approach and has been a more successful strategy for discovering small molecule, first-in-class drugs. Phenotypic screening involves testing a molecule in cells, isolated tissues, or animals to evaluate the desired effect without exactly knowing the mechanism of action. TDD approach, on the other hand, has yielded more best in class drugs. It is a complex process that initiates the identification and validation of novel targets. To accelerate target analysis, high throughput screening (HTS) has played a pivotal role by cost-effectively screening large-scale compound libraries. The target identification is followed by synthesis, characterization, and screening of NCEs in assays relevant to the disease target to find therapeutic efficacy. Advanced techniques in molecular biology and genomics have made TDD a preferred approach to drug discovery in the pharmaceutical industry. SDD has also played an important role in the drug discovery process, particularly in the discovery of psychotropic drugs that have led to the management of several psychiatric disorders. This chapter highlights successful drug discoveries achieved so far by applying the three approaches mentioned above, citing a few examples of successful drugs for each strategy.

Keywords

Drugs · Receptors · Targets · Therapeutics · Diseases · Discovery and development

34.1 Introduction

A 'Drug' may be defined as a substance or product that is used or intended to be used to modify or explore physiological systems or alter the pathological conditions of the disease for the benefit of the recipient and is approved by regulatory health agencies around the world to treat diseases [1]. U.S. Food and Drug Administration (FDA) defines a drug as a substance recognized by an official pharmacopoeia intended for the diagnosis, cure, mitigation, treatment or prevention of

S. Koul (✉)
School of Consciousness, Dr. Vishwanath Karad MIT World Peace University, Pune, Maharashtra, India
e-mail: summon.koul@mitwpu.edu.in

G. Jagadeesh et al. (eds.), *The Quintessence of Basic and Clinical Research and Scientific Publishing*,
https://doi.org/10.1007/978-981-99-1284-1_34

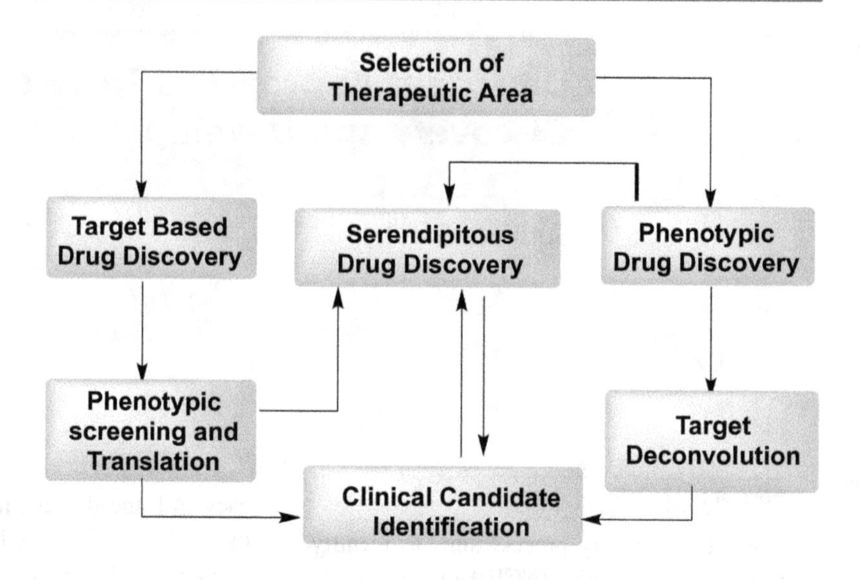

Fig. 34.1 Multiple approaches toward drug discovery

disease. A drug used as a medicine component affects the structure or any physiological function of the body [2]. In scientific literature, the term 'drug target' is most often used to describe the specific molecular targets, like proteins, nucleic acids, etc., that a drug interacts with to initiate a biological response [3]. Drugs bind to these specific targets and change the gene function of these targets, thus achieving disease treatment [4]. Examples of targets for drug action are receptors, ion channels, transporters and enzymes. Therefore, the target identification step is one of the most critical steps in the drug discovery and development process [5]. Identifying new drug targets and a rational understanding of the link between target biology and disease states plays a key role in the drug discovery process [6].

There are two main complementary approaches to drug discovery: phenotypic drug discovery (PDD) and target-based drug discovery (TDD). In addition, serendipitous drug discovery (SDD) has also played a significant role in identifying novel and efficacious drugs (Fig. 34.1).

This chapter provides a detailed explanation of the processes of these multiple approaches adapted to achieve discoveries of some of the successful drugs. Each approach is illustrated by citing examples of drugs with a brief description of their discovery.

34.2 Phenotypic Drug Discovery

This historical approach to drug discovery does not depend on the knowledge of specific drug target identification or a hypothesis about its role in disease. In PDD approach, therapeutic agents are initially selected by establishing their pharmacological actions in cells, tissues, or animals, followed by comprehending the mechanism of action. For example, bactericidal and bacteriostatic antibiotics have been identified without initial knowledge of the molecular targets involved in the disease [7]. After the active therapeutics are identified, disease-relevant phenotypes provide essential information about the translation between the observation and clinical outcome. The correlations between target agnostic screens in cellular signaling pathways or physiologically relevant disease models and clinical observations are also followed up in PDD approach [8]. Although this approach is primarily meant for small-molecule discovery, PDD has also played an important role in antibody drug discovery [9]. Recently, this strategy has been shown to have great potential to give insight into the complexity of diseases and their ability to deliver first-in-class drugs [10]. Some examples of game-changing therapeutic agents with novel mechanisms discovered by PDD approach are medicines for spinal muscular atrophy (SMA), cystic fibrosis (CF), and hepatitis C [11].

34.2.1 Risdiplam for Spinal Muscular Atrophy

Of note, SMA is a genetic disorder affecting the nervous system and is characterized by weakness of skeletal muscles. It is caused by the loss of motor neurons that control muscle movement. The discovery of risdiplam started with SMA screens that were carried out with simple cell based reporter gene assays. Since the mechanism of the disease was well understood, this helped researchers build a simple HTS campaign relevant to the disease model. Hence the molecular basis of SMA that warrants inappropriate exon splicing of SMN2 RNA, led the HTS design to identify small molecules that increase the inclusion of exon 7 during SMN2 pre-mRNA splicing, thus increasing levels of SMN2 protein [11, 12]. This drive led to the identification of a promising hit that was optimized to dial out *in-vitro* mutagenicity and phototoxicity associated with this series to give lead compound (RG7800). Finally, further optimization of key parameters led to the identification of risdiplam [13]. Risdiplam is the first oral medication sold under the brand name Evrysdi approved to treat SMA [14, 15].

34.2.2 Elexacaftor–Tezacaftor–Ivacaftor for Cystic Fibrosis

CF is a genetic disease that damages the lungs, digestive tract and other organs. It is a progressive, most lethal disorder that drastically decreases patients' average life expectancy. It is caused by mutations in the gene encoding the CF transmembrane conductance regulator (CFTR) protein that reduces CFTR channel function or interrupts intracellular folding of CFTR and insertion into the plasma membrane [16, 17]. High throughput screening strategy using target agnostic approach started from a library of 228,000 compounds, followed by optimization of various parameters. This led to the identification of compound classes that enhanced channel-gating properties of CFTR (potentiators) and improved the CFTR folding and plasma membrane insertion (correctors). FDA approved Ivacaftor, a CFTR potentiator, in 2012 that provided treatment for only 5% of CF patients. Subsequently, clinical trials using Ivacaftor in combination with CFTR corrector Tezacaftor led to the FDA approval of this dual combination in 2018. This combination is predicted to treat 46% of CF patients. Finally, the approval of the triple combination therapy Elexacaftor, Tezacaftor and Ivacaftor (two correctors, one potentiator) called Trikafta in 2019 is expected to treat 90% of CF patients [18, 19].

34.2.3 Daclatasvir for Hepatitis C

Hepatitis C is a liver inflammation caused by the hepatitis C virus (HCV) that can result in mild to severe hepatitis leading to lifelong illness including liver cirrhosis and cancer. In the early phase, the development of antiviral agents for the treatment of HCV was directed mainly toward inhibitors of the viral enzymes NS3 protease and the RNA-dependent RNA polymerase NS5B4 [20]. Later, a differentiating strategy was adopted to identify compounds that are functionally distinct from those acting on these enzymes. Toward this goal, a chemical genetics based mechanistically unbiased approach was carried out to identify hits for interfering with HCV replication [21]. This process led to the identification of a lead compound that was further optimized to achieve the first-in-class HCV NS5A replication complex inhibitor daclatasvir with a potent clinical effect. The discovery and development of daclatasvir provide one of the unique examples of the importance of phenotypic screening to identify lead molecules engaging targets with a novel and unique mechanism [22].

34.2.4 Metformin: First Line Medication for Type 2 Diabetes Mellitus

Metformin is a biguanide class of medication that is most frequently prescribed for type 2 diabetes mellitus (T2DM) [23]. The mechanism of this class of compounds is not completely understood. Hence metformin is one of the unique examples to highlight how a safe and efficacious drug can be used successfully even though the identity of the molecular target is not entirely understood. The discovery of metformin originates from a traditional herb "*Galega officinalis*" found in Europe that was shown in 1918 to lower blood glucose owing to the presence of guanidine. Several guanidine derivatives were synthesized and used to treat diabetes in 1920–1930. However, at this stage metformin though synthesized, was not tested for diabetes. The use of guanidine derivatives was later on discontinued due to side effects. Subsequently, in the 1940s, metformin was explored as an antimalarial agent. During clinical studies, metformin showed the potential to treat influenza and, at the same time, was found to lower blood glucose in some cases. This became the basis for using metformin to treat diabetes in 1957 for which the credit goes to French physician Jean Sterne. However, little attention was paid to metformin because of its lower potency than other biguanides like phenformin and buformin. Later, extensive studies and intensive scrutiny of various parameters provided a new rationale for adopting metformin as a treatment for Type 2 diabetes. Metformin continues to be the most prescribed glucose-lowering medicine worldwide to manage hyperglycemia in T2D patients [24].

Phenotypic approaches to drug discovery have had major success for neglected diseases, particularly in malaria and Human African trypanosomiasis (HAT) [25]. For example, the US FDA recently approved fexinidazole discovered through phenotypic screening as the first oral treatment for both HAT stages in children [26, 27].

34.3 Target Based Drug Discovery (TDD)

In the past three decades, due to advances in molecular biology, recombinant technology and genomics, TDD—in which a defined molecular target is assumed to have a key role in disease—has become the preferred approach to drug discovery in the pharmaceutical industry [10, 28]. Most of the first in class drugs approved by the US FDA have been successfully identified through TDD approach [7]. In TDD strategy, targets, usually "druggable" proteins, are validated with a particular therapeutic indication [11]. Target selection remains a crucial step in this approach because the selected target needs to be relevant to treat a specific disease. The approach applies not only to small molecule drug discovery but also to antibody drugs and other biologics. TDD approach also encompasses the discovery of gene therapy and nucleic acid-based therapeutics [7]. The most common drug targets of currently marketed drugs are described in the following sections.

34.3.1 Proteins

34.3.1.1 G Protein-Coupled Receptors as Drug Targets

G protein-coupled receptors (GPCRs), also known as 7-transmembrane receptors, are humans' largest family of receptors [29–31]. These receptors mediate most of the cellular responses to hormones and neurotransmitters. GPCRs are also responsible for vision, olfaction and taste. Based on their sequence homology and structural similarity, these receptors are divided into five families—rhodopsin (family A), secretin (family B), glutamate (family C), adhesion and frizzled/taste2. The rhodopsin family, which is the largest and most diverse, is characterized by conserved structural features across the members of the families [32]. Most of the GPCRs contain seven helices and three intracellular loops, some members of the rhodopsin family being an

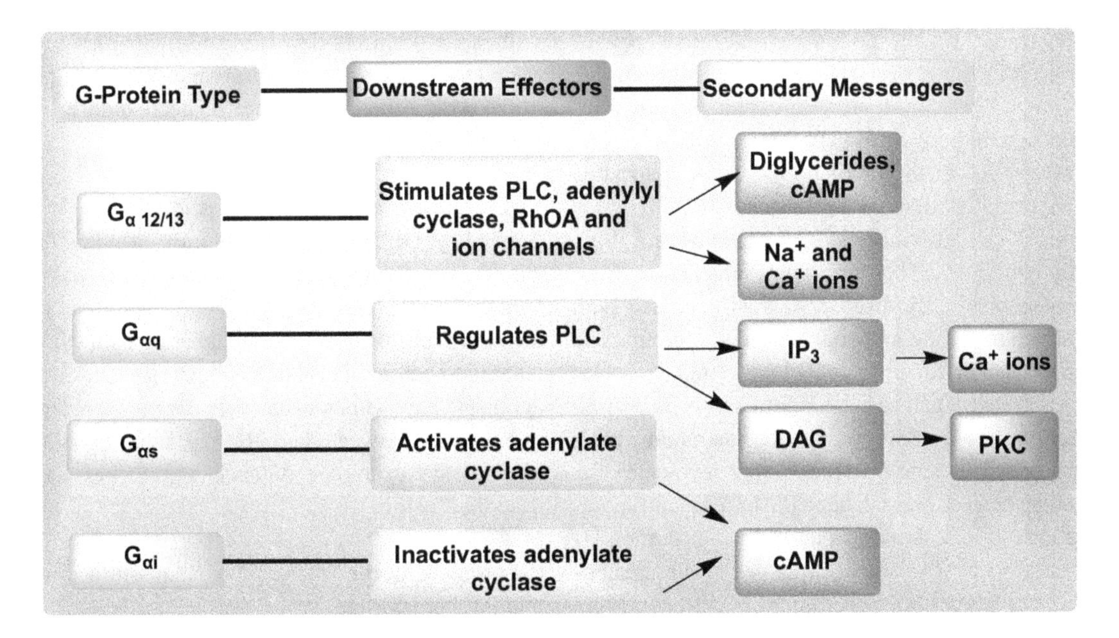

Fig. 34.2 G proteins and their secondary messenger signaling

exception that may have eight helices and four intracellular loops [33]. GPCRs and their respective G proteins are characterized by random diffusion within the cell membrane. These interact only upon receptor activation or form stable complexes in the absence of an agonist [34]. G proteins are classified into four subfamilies according to their α subunit: $G_{\alpha i}$, $G_{\alpha s}$, $G_{\alpha 12/13}$, and $G_{\alpha q}$. Subtypes $G_{\alpha i}$ and $G_{\alpha s}$ either inactivate or activate, respectively, adenylyl cyclase activity which converts adenosine triphosphate (ATP) into second messenger cyclic adenosine 3′, 5′ - monophosphate (cAMP) (Fig. 34.2). Calcium mobilizing hormones activate Gαq pathway. It stimulates phospho-inositol phospholipase C enzyme (PLC) that hydrolyses phosphatidylinositol biphosphate (PIP_2) into diacylglycerol (DAG) and inositol triphosphate (IP_3). IP_3 triggers the release of calcium ions from endoplasmic reticulum calcium channels into the cytosol, while DAG activates protein kinase C (PKC) [35]. $G_{\alpha 12/13}$ interacts with specific guanine nucleotide exchange factors (GEFs) and stimulates downstream effectors like PLC, adenylyl cyclase, ras homolog family member A (RhoA) and several ion channels. These, in turn, trigger the release of secondary messengers in the cells like diglycerides, cAMP, sodium and calcium ions [36]. Since the G proteins are expressed in most cell types and can induce diverse cellular signaling, they have an important role in disease pathogenesis. Thus GPCRs are considered by far the largest family of targets (approximately 35%) for approved drugs [37]. Additionally, GPCR targets are usually druggable and interact with numerous chemical entities. Since these receptors are expressed in the plasma membrane, molecular interactions in the extracellular space are highly facilitated.

Some examples of approved drugs that act on GPCR targets and are used to date are given in the following subsections.

Cetirizine: H_1 Receptor Antagonist for the Treatment of Allergy Symptoms

Histamine H_1 receptor belongs to the family of rhodopsin-like GPCR. The classic H_1 receptor antagonists are highly lipophilic, leading to the significant metabolism of these compounds [38]. These drugs possess a single strongly basic center and exist as lipophilic cations at physiological pH [39]. The major drawback of the first generation of H_1 receptor antagonists was an effect on the central nervous system due to high brain penetration and other off-target effects. Second-generation H_1 receptor antagonists, though highly lipophilic and basic, did not show the central nervous system effects [40, 41]. However, some of these compounds were associated with cardiotoxicity at higher doses [42, 43] which might be attributed to these molecules' high lipophilicity and basicity. Thus, novel H_1 receptor antagonists were discovered that contained both basic amino and an acidic carboxylic group. The first marketed drug of this generation was cetirizine, which exists predominantly as a zwitterion at physiological pH. Later, the FDA approved its single R-enantiomer levo-cetirizine as the newest antihistamine [44]. Cetirizine shows a clear differentiation from its first generation H_1 receptor antagonists, such as negligible metabolism and low CNS penetration.

Losartan: A Drug for Hypertension

GPCRs are activated by different ligands like hormones, ions, small molecules and vasoactive peptides [45]. Angiotensin peptides are examples of such kinds of ligands that bind to GPCRs called Angiotensin 1 (AT1) and Angiotensin 2 (AT2) receptors. The renin–angiotensin system (RAS), or renin–angiotensin–aldosterone system (RAAS), is an endocrine system that plays a key role in the regulation of blood pressure and fluid and electrolyte balance [46]. Overactivation of the RAS system results in high blood pressure. Several types of drugs interfere at different steps of this system to contain high blood pressure. These drugs can be angiotensin-converting enzyme (ACE) inhibitors, angiotensin II receptor blockers (ARBs), or renin inhibitors [47]. Losartan is an example of ARB that is used to treat hypertension [48]. Telmisartan is another ARB inhibitor with a longer action duration than losartan. Telmisartan (80 mg, QD) markedly reduces 24-h blood pressure relative to losartan (50 mg, QD). It has been proven to be especially beneficial in the last 6 h of the dosing interval [49]. ARB inhibitors have been proven to have additional therapeutic benefits beyond their blood pressure lowering effect. For example, telmisartan is the only ARB that has been shown to reduce cardiovascular morbidity in patients with atherothrombotic cardiovascular disease. Similarly, losartan and valsartan are beneficial for the second-line treatment of heart failure in patients with ACE inhibitor intolerance [50].

GPCR Targets and Implications in Cancer

Some key GPCRs have mutations linked with cancer initiation and progression. Therefore, some of these GPCRs have been exploited to develop drugs that can inhibit signaling pathways leading to cancer. This approach has used several agonists and antagonists that target specific interactions between GPCRs and their ligands [51]. Some examples of such approaches are described below.

Cabergoline for the Treatment of Neuroendocrine and Pituitary Tumors

Cabergoline is a dopamine receptor D1 agonist that has been used as a first-line medication to treat prolactin secreting benign tumors in pituitary glands. It suppresses tumor cell proliferation and induces cell death. Cabergoline also has a broader therapeutic efficacy for treating pancreatic neuroendocrine tumors [52]. Cabergoline is an ergot derivative that is potent and selective and has long-lasting inhibitory activity on abnormally high levels of prolactin in the blood secreted by the lactotroph cells in the anterior pituitary gland [53, 54].

Sonidegib for Locally Advanced and Metastatic Basal Cell Carcinoma (BCC)

Sonidegib is an orally administered hedgehog (HH) pathway inhibitor that works via smoothened receptor (SMO) antagonism. It is used to treat locally advanced basal-cell carcinoma in adults who have the recurrent disease following surgery or radiation therapy, or those who are not patients for surgery or radiation therapy [51, 55]. HH GLI signaling pathway plays a crucial role in cell proliferation and differentiation and gets activated in the pathogenesis of various types of tumors [56–58]. SMO receptor is the Class F GPCR, the main transducer of HH signaling. Vismodegib was the first HH pathway inhibitor approved by the FDA and EMA followed by sonidegib which was approved as a first-line

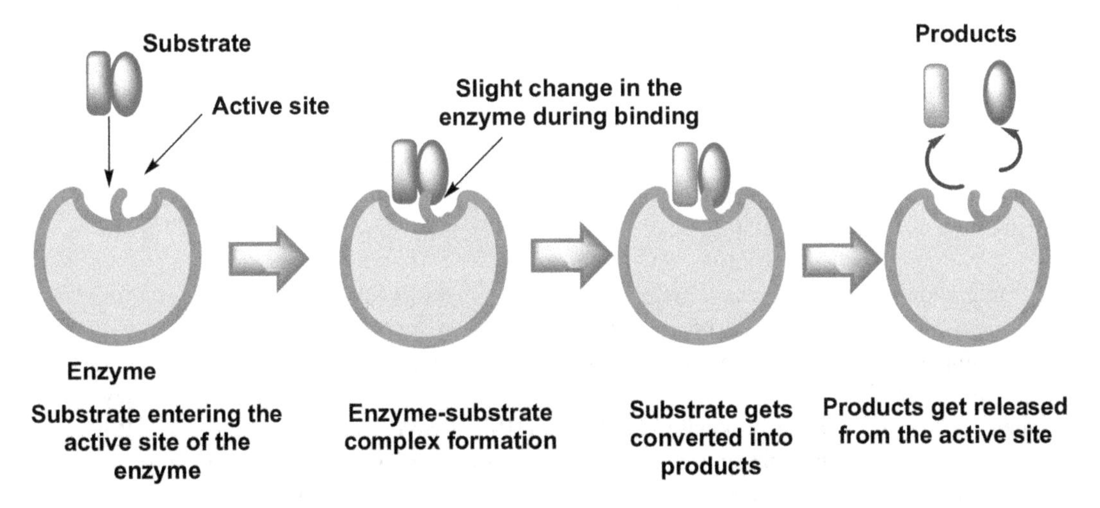

Fig. 34.3 Mechanism of enzyme-catalyzed reactions

treatment for adult patients with locally advanced BCC, [59]. Both sonidegib and vismodegib have similar efficacy and tolerability profiles, but a few of their pharmacokinetic parameters like the volume of distribution and half-life show several differences. The impact of these differences on clinical outcomes is not yet clear [60].

Zibotentan Under Clinical Trials for Prostate Cancer

Zibotentan is an oral specific endothelin (ET) A receptor antagonist from AstraZeneca that is under development for prostate cancer [51, 61]. The ETs and their receptors, referred to as ET axis, are implicated in several mechanisms that promote the growth and progression of various tumours [61, 62]. Several studies have shown that the ET axis is involved in the pathogenesis of prostate cancer. A significant correlation has been observed between increased levels of the ETA receptor expression and tumor grade and stage [63]. In Phase II clinical studies, Zibotentan monotherapy has shown a good tolerability profile and significant survival in patients with castration-resistant prostate cancer who were not having pain or mild symptoms of pain [64].

34.3.1.2 Enzymes as Therapeutic Targets

Enzymes are proteins that act as essential catalysts for several physiological processes like metabolism, cell growth, cell division, and cellular signaling pathways in our body [65]. Enzymes catalyze multiple-step biochemical reactions in the cells that would otherwise proceed at a slower rate and would not occur under ambient temperature and pressure compatible with life [66]. The first step in the enzymatic reaction is binding the substrate to the amino acid residues present in the active site of the enzyme through several interactions like hydrogen bonding, van der Waals' interactions, dipole-dipole interactions, etc. In the second step, the enzyme provides functional groups that can attack the substrate and carry out the enzyme-catalyzed reaction. After the reaction is complete, the products get released from the enzyme's active site [67] (Fig. 34.3). Drugs that can inhibit this catalytic reaction by blocking the enzyme's active site are known as enzyme inhibitors. Some of the drugs that directly bind to the active site where the substrate binds are known as competitive inhibitors while the drugs that bind to a site other than the active site of the enzyme are known as allosteric inhibitors. Some examples of enzyme inhibitors that have been approved as drugs by the FDA and are still in use are given below.

Imatinib for the Treatment of Cancer

Imatinib, a tyrosine kinase inhibitor, is an oral chemotherapy medication used to treat cancer especially chronic myeloid leukemia (CML) and acute lymphocytic leukemia (ALL), certain types of gastrointestinal stromal tumors (GIST), hypereosinophilic syndrome (HES), chronic eosinophilic leukemia (CEL), systemic mastocytosis, and myelodysplastic syndrome [68]. Tyrosine kinases are a family of enzymes that mediate important signal transduction processes, leading to cell proliferation, differentiation, migration, metabolism, and programmed cell death. Many studies have shown tyrosine kinases' important role in cancer pathophysiology [69]. The active sites of these enzymes have a binding site for ATP. Tyrosine kinases catalyze the transfer of the terminal phosphate from ATP to tyrosine residues on its substrates. Imatinib inhibits this process by binding to the ATP-binding site of the tyrosine kinase. This, in turn, inhibits the process of downstream signaling from the tyrosine kinase. This induces apoptosis and inhibits the further proliferation of cell lines. In CML patients, imatinib has improved the overall survival rate and it is the first effective oral therapy for GIST. This drug is an example of highly targeted cancer chemotherapy and is generally well tolerated [70].

Sitagliptin for T2DM

Sitagliptin is a dipeptidyl peptidase-4 (DPP-4) inhibitor used to manage T2DM. DPP-4 is a Type-II transmembrane glycoprotein expressed on cells throughout the body. When shed from the membrane, it circulates as a soluble protein in blood plasma and various body fluids [71]. DPP-4 plays a pivotal role in glucose metabolism, where it is responsible for the degradation of incretin hormones like glucagon-like peptide-1 (GLP-1). Due to the degradation of GLP-1, the level of this hormone goes down in the body, leading to several cascading effects like a decrease in glucose-dependent insulin secretion, impairment of β–cell function, deactivation of insulin biosynthesis and many more [72]. Thus, two approaches have been adopted to enhance endogenous GLP-1 *in-vivo*: injecting GLP-1 analogs like exenatide that stimulates the pancreas to secrete insulin or using DPP-4 inhibitors that enhance the level of incretin hormones by inhibiting DPP-4 responsible for degradation of GLP-1. Sitagliptin is a potent, selective, competitive and reversible inhibitor of DPP-4. The selectivity against DPP-8 and DPP-9, which share very similar homology with DPP-4 gives an advantage to sitagliptin being safe *in-vivo* [73]. Orally administered sitagliptin is an effective treatment option for managing patients with T2DM.

Atorvastatin for the Treatment of Hypercholesterolemia

Atorvastatin belongs to the family of statins that are 3-hydroxy-3-methyl-glutaryl-coenzyme A (HMG-CoA) reductase inhibitors. HMG-CoA reductase is anchored in the membrane of the endoplasmic reticulum. The key function of this enzyme is to convert HMG-CoA to mevalonic acid in the hepatocytes, which is the first and rate-limiting step in cholesterol biosynthesis. Atorvastatin competitively inhibits HMG-CoA reductase by binding to the enzyme's active site and inducing a conformational change in its structure. This results in reducing the formation of mevalonic acid, a precursor of cholesterol [74]. Thus, atorvastatin acts as a lipid lowering medication used in the primary and secondary prevention of coronary heart disease. In addition, statins in general may also inhibit rho-kinase and modulate vascular dysfunctions in atherosclerosis [75]. FDA has also approved atorvastatin to prevent cardiovascular events in patients with cardiac risk factors and for patients with abnormal lipid profiles [76].

34.3.1.3 Ion Channels as Drug Targets

Ion channels are transmembrane proteins that control the flow of ions across the cell membrane. These proteins are important drug targets that play a pivotal role in treating various pathophysiologies [77]. There are four types of ion channels (Fig. 34.4)—resting K^+ channels, voltage-gated, ligand-gated and signal-gated ion channels. K^+ channels can be considered a subtype of voltage-gated ion channels, and signal-gated ion channels can be categorized as ligand-gated ion channels. Resting K^+ channels are located in cell membranes that generate resting potential across the membrane. These channels are usually closed in the resting state

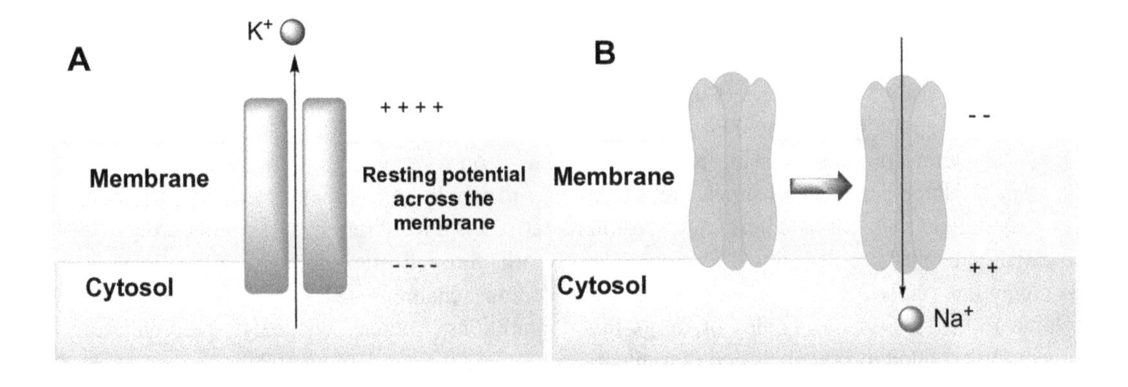

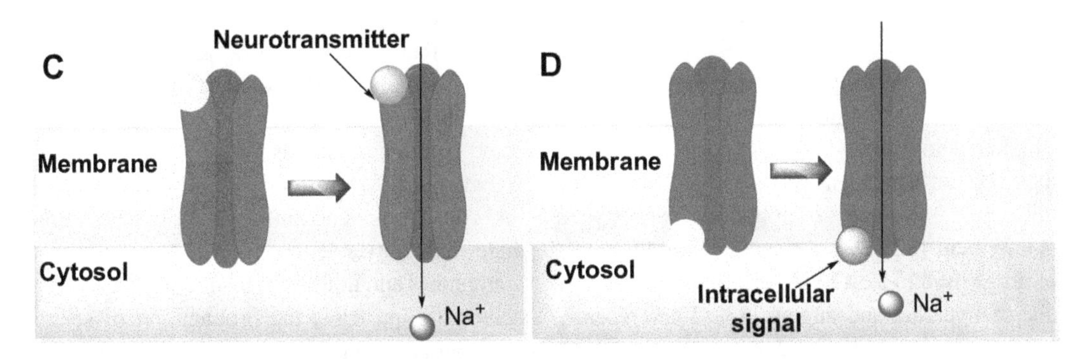

Fig. 34.4 Types of ion channels. (**a**) Resting K+ channels; always open. (**b**) Voltage-gated; opens or closes when there is a change in membrane potential. (**c**) Ligand-gated; opens or closes when an extracellular ligand like a neurotransmitter binds. (**d**) Signal-gated; opens or closes in response to an intracellular signal

and get opened after activation by stimulus [78]. Voltage-gated ion channels are more selective than ligand-gated ion channels for the permeation of ions. Voltage-gated ion channels allow passage of only one type of ion, while ligand-gated ion channels allow permeation of more than one type of ion. This flux of ions helps in the propagation of action potential along the membrane, which plays a crucial role in various physiological processes like neurotransmitter release, hormone secretion, muscle contraction, etc. Ligand-gated ion channels contain binding sites for specific extracellular ligands, such as neurotransmitters, and signal-gated ion channels respond to intracellular signals induced by an extracellular ligand binding to a separate receptor [79]. The widespread distribution of ion channels in the tissues and the physiological outcome of the opening and closing of each channel has made these desirable targets for drug discovery. After GPCRs, ion channels are the largest targets for existing drugs [30].

Voltage-Gated Ion Channels

Voltage-gated ion channels are transmembrane proteins that play a key role in the electrical signaling of cells. The membrane potential of a cell regulates the activity of these ion channels. The movement of ions occurs through open channels that lead to the development of an electrochemical gradient across cellular membranes. Voltage-gated ion channels are selectively permeable to ions such as Na^+, K^+, Ca^{2+} and Cl^-. Since Na^+ channels generate a longer action potential, these are important targets for local anesthetics. K^+ channels are the largest and most diverse voltage-gated ion channels and play the most important role in generating the resting membrane potential. Like Na^+ channels, Ca^{2+} channels have different activation and inactivation properties.

Hence subtle variations exist in electrical and chemical signaling processes mediated by these channels. Thus Ca^{2+} ion channels provide an important target for drugs that treat conditions such as hypertension, heart disease and anxiety disorders. Voltage-gated Cl^{-} channels are widely distributed in neurons which contribute to resting potential and control excitability. These channels also play an important role in regulating cell volume [80, 81]. Some examples of drugs that target voltage-gated ion channels and their mechanism of action are given below.

Verapamil

Verapamil is an L-type voltage-gated calcium channel blocker that mimics the cardiac effect of simple calcium ion withdrawal and thus reduces the calcium-dependent high-energy phosphate utilization, the force of contraction and oxygen requirement [82]. It is a phenylalkylamine class of drugs used to treat hypertension, supraventricular tachycardia and angina [83]. Very recently when oral verapamil was administered to patients with Type 1 diabetes, they were found to show improvement in endogenous beta cell function, showed fewer hypoglycemic incidences and had lower insulin requirements. Administration of verapamil helps in the preservation of beta cell function. Additionally, verapamil helps regulate the thioredoxin system and promotes anti-apoptotic, immunomodulatory and anti-oxidative gene expression in human islets. This justifies the beneficial effect of verapamil in patients with Type 1 diabetes [84].

Oxcarbazepine

Oxcarbazepine is an anticonvulsant drug approved in 2000 for the treatment of partial onset seizures. It inhibits abnormal electrical activity in the brain and thus helps reduce the incidence of seizures in epileptic patients [85–87]. Oxcarbazepine is known to exert its major pharmacological activity via its clinically relevant metabolite licarbazepine (monohydroxy derivative (MHD)]. This metabolite exists as a racemate in the blood even though it has both (*S*)-(+)- and the (*R*)-(−)-enantiomers. Since oxcarbazepine gets primarily metabolized to the hydroxyl derivative, it is less susceptible to involvement in drug-drug interactions, as has been a concern in the case of many anti-epileptic drugs [88]. Although the exact mechanism of how oxcarbazepine and its metabolite exert their effect on anti-epileptic activity is not entirely understood, it is believed to work through the blockade of voltage-gated sodium channels [85–87, 89]. Under normal conditions, sodium channels open and close, allowing the action potentials to propagate along the neurons. However, in epileptic conditions, excessive propagation of action potentials occurs, causing their repetitive firing and leading to seizure activity. It is believed that oxcarbazepine and MHD bind to the inactive state of the voltage-gated sodium channels inhibiting the seizure activity [90]. Since the receptor is not available for propagation of action potential for a long time, this helps stabilize hyperactive neuronal membranes and inhibits high-frequency repetitive neuron firing. Thus, the propagation of seizure activity in the central nervous system is controlled without disturbing normal neuronal transmission.

Ligand-Gated Ion Channels

Ligand-gated ion channels, also known as ionotropic receptors, are transmembrane ion channel proteins. When these receptors bind to a ligand such as neurotransmitters, the passage of ions through the transmembrane pore occurs transiently. These receptors are crucial for regulating electrochemical balance in the cells [91]. Ligand-gated ion channels are abundantly present in the central nervous system and regulate essential functions such as anxiety, seizures, and cognitive functions. These functions are closely associated with extracellular ligands like neurotransmitters - serotonin, acetylcholine, GABA, and glutamate. Contrary to these, the intracellular ligands are commonly present inside the cells and these act as secondary messengers like inositol triphosphate and adenosine triphosphate [92]. Below are some examples of drugs that work through inhibitory action by binding to the ligand-binding site.

Ondansetron and Granisetron

Ondansetron and granisetron are ligand-gated ion channel serotonin type 3 receptor ($5\text{-HT}_3\text{R}$) antagonists. This receptor is a cation-selective channel located in the peripheral and central nervous systems. These drugs show strong anti-emetic effects and are used to prevent chemotherapy and radiation therapy-induced nausea, as well as the impact of post-operative anesthetics. These are also used to treat irritable bowel syndrome [93–95]. These antagonists prevent the binding of serotonin to 5-HT_3 receptors on the vagal nerve terminals in the gastrointestinal system peripherally and the ventricular chemoreceptor trigger zone centrally, thus leading to strong anti-emetic effects. Ondansetron is a carbazole analog and is the first generation 5-HT_3 receptor antagonist. Granisetron is an indazole analog with higher receptor specificity and potency, which makes it a better alternative to ondansetron [96].

Perampanel

Perampanel is a non-competitive ligand-gated ion channel α-amino-3-hydroxy-5-methyl-4-isoxazolproponic acid glutamate receptor (AMPAR) antagonist that is the first anti-epileptic drug in the class of selective antagonists of this receptor [97]. Glutamate is a major excitatory neurotransmitter in the nervous system. It has been shown that the antagonism of AMPA receptor reduces overstimulation and promotes anti-convulsant effects in addition to inhibiting seizures and their spread. Also, AMPA receptor antagonists are believed to prevent neuronal death [98].

34.3.1.4 Nuclear Hormone Receptors as Drug Targets

Nuclear hormone receptors control a host of biological processes and are ideal targets for drug discovery. These receptors have the advantage of being regulated by small lipophilic molecules that can be easily replaced with a drug molecule of choice [99]. These receptors are part of a superfamily of ligand-dependent transcription factors responsible for regulating cell growth, development and metabolism. Nuclear hormone receptors have a characteristic feature of a conserved structural and functional organization across the superfamily. These members have two well-defined structural domains—a highly conserved DNA-binding domain (DBD) and a ligand-binding domain (LBD). DBD is located approximately along the center of polypeptides, while LBD resides along the C-terminal residues. The amino terminal domain varies considerably in sequence and size and in some cases, regulates transcriptional activation function (AF-1). LBD is responsible for mediating ligand binding, can form homo- or hetero-dimerization surfaces, and represses transcription in the absence of a ligand.

Additionally, when an agonist binds to LBD, it regulates ligand-dependent transcription activation. Since ligand binding and coactivator-derived motif interactions occur at this site, LBD has been the main focus of drug discovery [100, 101]. Steroid hormones like estrogen, progesterone, and nuclear hormones like all-trans retinoic acid, thyroid hormone, etc., activate nuclear receptors. Type I/III receptors like estrogen and progesterone receptors are bound to chaperone proteins like HSP90 in the cytoplasm. The chaperone gets released from the receptor after ligand binding. This leads to homo-dimerization of the receptor followed by its entry into the nucleus (Fig. 34.5). After entering the nucleus, the ligand-receptor complex gets associated with the coactivators that lead to binding followed by transcription of specific genes. In this way, steroid hormones regulate specific cellular processes. Nuclear hormone receptors, on the other hand, are located in the nucleus. The hormones diffuse across the plasma membrane and nuclear pores and bind to their respective receptors in the nucleus. They usually form heterodimers with retinoid X receptor (RXR) and in the absence of the ligand get associated with corepressors. This is followed by the binding of the ligand to LBD that leads to the dissociation of corepressors followed by the association of coactivators. This process leads to the activation of target genes followed by the regulation of specific cell processes [102, 103].

Some examples of drugs that target nuclear hormone receptors are given below.

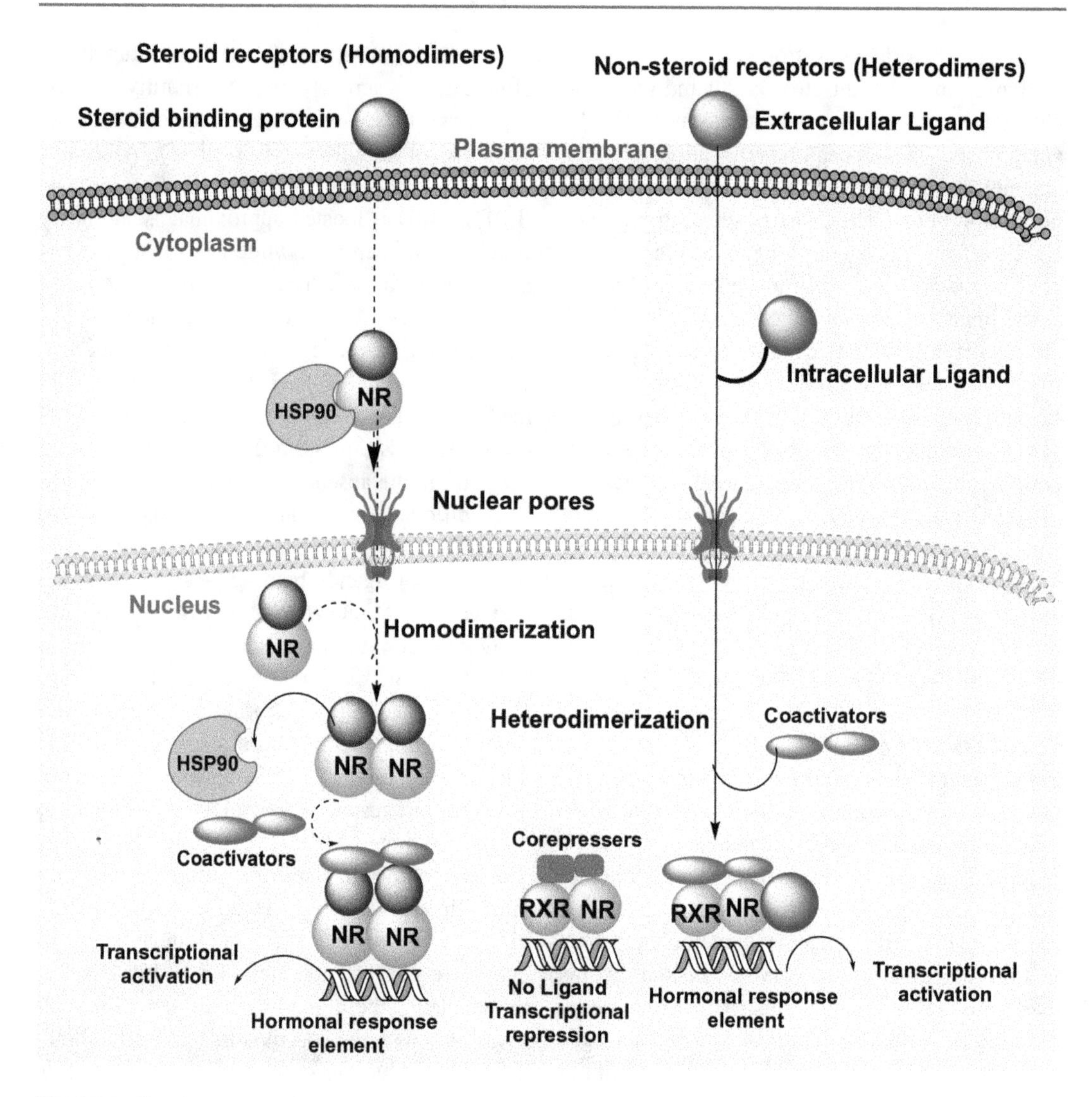

Fig. 34.5 Signaling pathway of nuclear receptors

Tamoxifen and Raloxifene

Tamoxifen and raloxifene are first and second generation selective estrogen receptor modulators (SERM), respectively, that have been approved for the treatment and prevention of postmenopausal osteoporosis. In this disease, progressive loss of bone mass and bone mineral density takes place because bone resorption occurs much faster than bone formation. Hence there is a high risk of bone fractures in such patients. Tamoxifen and raloxifene bind to estrogen receptors resulting in the estrogen agonistic effect on bone thus increasing bone mineral density and mass by decreasing bone resorption. Raloxifene antagonizes estrogen in the uterus and breast while tamoxifen has an estrogen-agonistic effect in the uterus. Additionally, tamoxifen is believed to be an effective treatment for metastatic breast cancer; tamoxifen and raloxifene are also known to reduce the risk of breast cancer [104, 105].

Bexarotene

Bexarotene is an oral synthetic anti-cancer agent used to treat cutaneous manifestations of T cell lymphoma (CTCL) at both early and advanced stages. It specifically binds to the retinoid X receptor (RXR) and does not show significant binding to the retinoic acid receptor (RAR), another major retinoic acid target. Bexarotene, after binding to RXR activates the receptor and its heterodimer, thus modulating a myriad of gene expression pathways. This eventually leads to the modulation of signaling pathways responsible for cell differentiation and apoptosis [106]. The most important mechanism of action of bexarotene for treating CTCL seems to be the apoptosis of neoplastic T cells [107].

34.3.2 Nucleic Acids

Nucleic acids—DNA and RNA are the storehouse of genetic information and take part in important biological activities like replication and transcription. The host's innate immune system plays a crucial role in defense against pathogens, and its activation contributes significantly to the subsequent activation of the adaptive immune system. Germline encoded pattern recognition receptors (PRRs) initiate the immune response by recognizing small molecular motifs conserved widely within pathogens, namely pathogen-associated molecular patterns (PAMPs) [108]. Under pathological conditions, mitochondrial DNA and cellular RNA release components in response to cell or tissue injury known as damage-associated molecular patterns. This process triggers the activation of PRRs like endosomal Toll-like receptors (TLRs), RIG-1 like receptors (RLRs), cytosolic DNA sensor proteins, etc. (Fig. 34.6). Thus activation of downstream signaling cascades initiates that induce innate

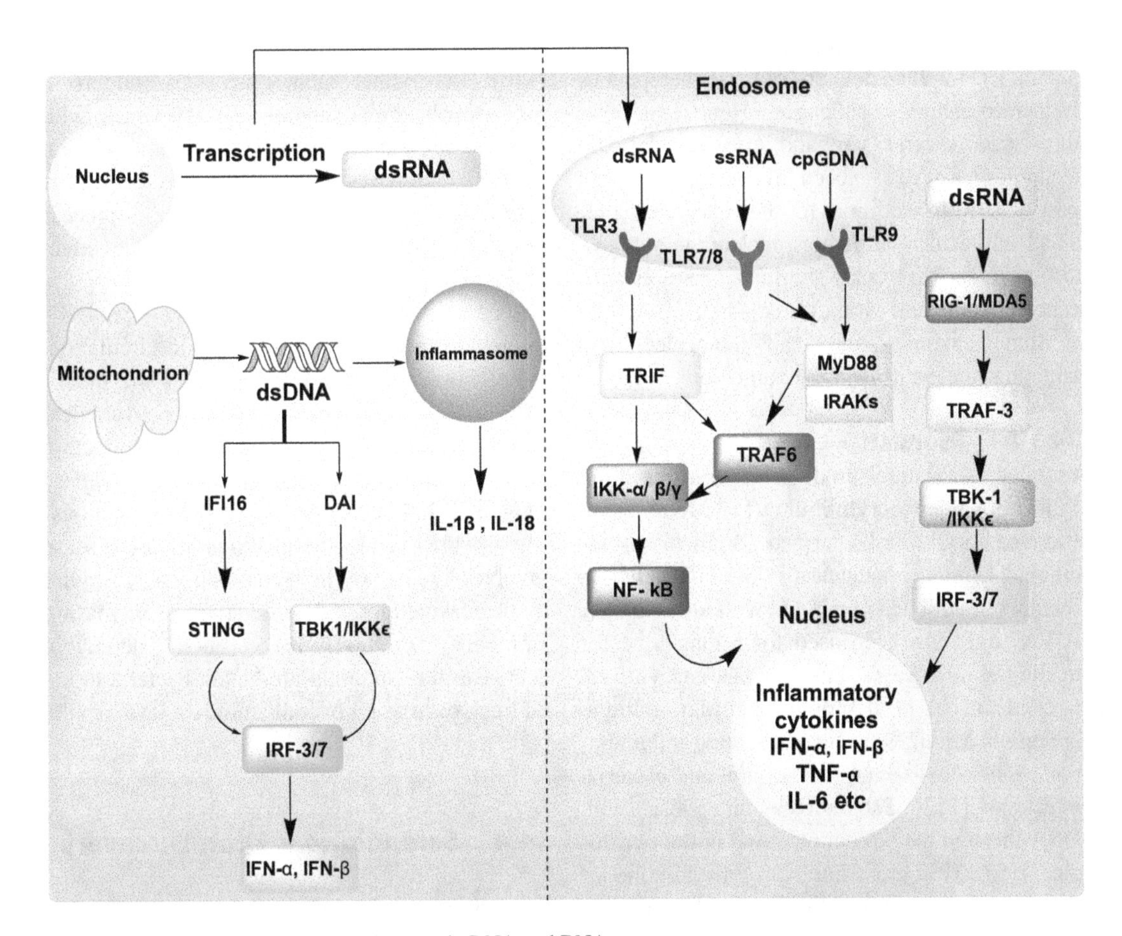

Fig. 34.6 Nucleic acid signaling pathways via DNA and RNA receptors

immune and inflammatory responses through the production of pro-inflammatory cytokines. Activation of TLRs leads to downstream cascading primarily via myeloid differentiation primary-response protein 88 (MyD88), tumor necrosis factor receptor-associated factor 6 (TRAF6), nuclear factor kappa B (NF-κB), and mitogen activated protein kinase (MAPK) pathways. On the other hand, dsRNA mostly leads to the activation of interferon (IFN) pathways that proceed through the phosphorylation of interferon regulatory factors (IRFs) [109].

Thus, nucleic acids are important to drug targets, especially for genetic diseases. The US FDA and European Medicines Agency have recently approved various nucleic-acid-based therapeutics. These therapeutics are known to target proteins and modulate gene expression leading to beneficial therapeutic effects. However, the degradation of nucleic acids by nucleases induces unfavorable physicochemical parameters, and the delivery of these therapeutics is, therefore, highly challenging. Hence to address this issue, several safe and effective delivery technologies have also been invented in the recent past [110]. In addition to these technologies, small molecules targeting nucleic acids have also been shown to play significant roles as anti-cancerous and anti-viral agents [111].

Some examples of small molecule drugs targeting nucleic acids are given below.

34.3.2.1 Psoralen

Psoralen is a natural product found in the dried fruits of Cullen corylifolium (L.) Medik (syn. Psoralea corylifolia L), termed "Buguzhi" in traditional Chinese medicine (TCM) [101]. It is a furanocoumarin and combined with ultraviolet-A (UVA) light has been used for a long time for treating skin diseases like psoriasis and vitiligo. Recently it has been shown to display multiple biological activities useful for treating inflammation, osteoporosis, cancer, and viral and bacterial infections [112]. Psoralen can intercalate with DNA through the 5,6 double bond in the pyrimidine ring. This cross-linking of pyrimidine of DNA duplex by either the furan or the pyrone moiety of the psoralen, followed by UVA activation at 365 nm, leads to DNA mutation. This photochemical reaction leads to the inhibition of DNA replication and stops further cell cycle. This characteristic property of psoralen helps to inactivate viruses and has thus been applied in the production of vaccines for treating viral infections [113, 114]. Other psoralen derivatives like 4-(Hydroxymethyl)-4,5,8-trimethylpsoralen (HMT) have also been explored in chemotherapeutics treatment.

34.3.2.2 Neomycin

Neomycin is an aminoglycoside antibiotic that is applied topically to treat superficial skin infections. It is also used to treat infections due to burns, wounds and ulcers. Neomycin demonstrates bactericidal activity against gram-negative aerobic bacteria and some strains of anaerobic bacteria that have not developed resistance. It is used orally to treat diarrhea and disinfect the bowels in hepatic encephalopathy [115]. Like other aminoglycosides, neomycin binds to ribosomal subunit 30S, interfering with bacterial protein synthesis. This process does not interfere with the initial steps required for peptide synthesis, but the bacterial translation process gets disrupted. This leads to the bactericidal effect of neomycin [116, 117].

The underlying liver disease sometimes leads to elevated ammonia levels in blood circulation. Ammonia at higher levels can cross the blood-brain barrier leading to neurological complications like hepatic encephalopathy. Also, higher ammonia levels in the brain increase glutamine and lactate levels resulting in neuronal edema [118]. Thus, the treatment of the disease involves a reduction in the ammonia levels either by decreasing ammonia production or increasing ammonia excretion. In addition, neomycin decreases the ammonia-producing bacteria in the GI tract, decreasing overall ammonia levels in the patient [119].

34.4 Serendipitous Drug Discovery

Serendipity has played a significant role in discovering many drugs used today [120–122]. It has been found that approximately 6% of all drugs in the market have been discovered by serendipity. These drugs have been discovered incidentally in the laboratory or clinical setting. In some cases, the therapeutics currently in use are the derivatives for serendipitous medicines. Serendipity has proved crucial, particularly in discovering drugs that act on the central nervous system. Also, anticancer drugs owe a share of approximately 35% to this phenomenon [123]. Many medications for infectious diseases result from accidental observation and an experiential quest for therapeutics active against pathogens followed by the clinical experience of the researchers [124]. The serendipitous discovery of drugs has laid a strong foundation for scientific advancement and benefit to society. Some examples of drugs discovered by serendipity are given below.

34.4.1 Plerixafor

Plerixafor is a highly specific G protein-coupled chemokine receptor CXCR4 antagonist approved for treating multiple myeloma. CXCR4 is specifically implicated in cancer metastasis and HIV-1 infection [125]. Plerixafor is a typical example of a serendipitous drug discovery that was not designed for the treatment of multiple myeloma. This drug was initially identified as an impurity while evaluating several commercially available cyclams (1,4,8,11-tetraazacyclotetradecane) for their potential inhibitory effects on the replication of human immunodeficiency virus (HIV). In this library, only one compound showed significant anti-HIV activity. After a thorough investigation, it was found that the remarkable activity was apparently due to an impurity. This impurity was purified and identified as the bicyclam with two cyclam rings tethered through a direct carbon–carbon linkage. However, this impurity could not be synthesized, so it was decided to synthesize a bicyclam derivative in which a propyl bridge linked two cyclam rings. This compound to a great surprise was found to be as potent as the one with a direct C-C bond. After executing SAR around the linker, replacing the propyl bridge with 1,4-phenylene-bis (methylene)] bridge led to a 100-fold increase in potency against replication of HIV-1 and HIV-2. After approximately three years, this compound was found to be a specific antagonist for the CXCR4 receptor. Finally, after a complete drug discovery and development program, this drug was approved by US FDA in 2008 for the treatment of multiple myeloma [126].

34.4.2 Lorazepam

Lorazepam is a benzodiazepine-based drug used to treat anxiety and sleep disorders, active seizures, alcohol withdrawal, and chemotherapy-induced nausea and vomiting [127]. The first benzodiazepine chlordiazepoxide was discovered by Leo Sternbach while working at Hoffmann-La Roche in 1955. This work was a part of the project being pursued at Roche for identifying tranquilizers. Since there was not much success for a very long time, the company decided to abandon the research activities on the project. Leo Sternbach was keen to pursue the project but, by the company decision, had to drop this pursuit. The synthesized compounds were left as such in the lab. After two years, when the management asked Leo Sternbach and his team to clean the lab space, one of his colleagues Earl Reeder observed very nice crystals of one of the compounds that were identified as chlordiazepoxide. When this compound was tested in animals, it showed astonishing anti-convulsant, muscle relaxant, and sedative effects. In 1960, this compound was marketed by Roche as Librium. Later further SAR studies were pursued to enhance the activity [128–131]. This led to the identification of Lorazepam with a relatively clean side effect profile approved by the FDA in 1977. By this time, researchers had successfully found out the mechanism of action of this class of drugs. They associated benzodiazepines with gamma-aminobutyric acid (GABA) receptors - GABA being a principal inhibitory neurotransmitter in the mammalian central nervous system [132].

34.4.3 Penicillin

During World War I, Alexander Fleming, serving in the Army Medical Corps as a captain, observed that the injured soldiers were dying of uncontrollable infections arising from the wounds. At that time, antiseptics were used to combat infections, but Fleming suspected the growth of anaerobic bacteria in deep wounds despite their use. In 1922, Fleming discovered an enzyme with weak antibacterial properties from his nasopharyngeal mucus infected with a cold and named it lysozyme. Even though this enzyme did not have strong antibacterial properties, this laid a foundation for Fleming's great discovery of penicillin. In 1928, Fleming initiated experiments involving staphylococcal bacteria. He had left an uncovered Petri dish near an open window contaminated with mold spores. His keen observation made him realize that the bacteria near the mold colonies were dying. He isolated this mold and identified it as a member of *Penicillium* genus. His further research led him to find that the isolated mold was effective against all Gram +ve bacteria responsible for major infections like pneumonia, meningitis, diphtheria, etc. Later on, Fleming found that it was not the mold that was killing the bacteria but some kind of juice produced by the mold, and he named this juice penicillin. Unfortunately, he could not isolate this juice in large quantities, but in 1940, Howard Florey and Ernst Chain were able to produce it for mass use during World War II [133].

34.4.4 Vinblastine

Vinblastine, an anticancer drug, was first serendipitously discovered while studying the antidiabetic effect of the extract of periwinkle plant. Researchers everywhere were curious to find an oral treatment for diabetes that could avoid insulin injections. Although the periwinkle plant was known to be a source of such oral medication, the oral administration of the plant extract in the rats did not show any effect on blood sugar or glucagon levels. To further improve their own results, Robert Noble, a professor at Western University, Canada, and his team injected the aqueous plant extract into the rats and found unexpected results. The autopsy results of the rats showed the presence of multiple abscesses. Further investigations revealed rapidly falling white blood count, granulocytopenia, and highly depressed bone marrow. The effect on white blood cell count increased the team's interest in continuing further studies. After rigorous efforts, the team was finally able to isolate, purify and crystallize the active ingredient from the plant extract and was named vinblastine because of the source being periwinkle (Vinca) and its effect on white blood cells. At that time, Eli Lilly had been engaged in a search for anti-cancer agents, including a wide range of plant extracts and was interested to see their effect on animal tumors. The researchers at Eli Lilly collaborated with Robert Noble, which led to the rapid advancement of vinblastine and its closely related analog vincristine. Finally, in the 1960s, the US FDA approved vinblastine and vincristine as chemotherapeutic agents for treating several types of cancer [134].

34.4.5 Warfarin

Warfarin is an anti-coagulant commonly used to prevent blood clots and stroke in people with abnormal heart rhythms, cardiovascular disease, or artificial heart valves [135]. It was discovered in the 1930s by Karl Paul Link—an American biochemist. Initially, a farmer from Wisconsin whose cattle were experiencing hemorrhagic disorder approached him with the problem. The affected cattle were consuming spoiled sweet clover hay. After investigation, Karl Paul and his team in 1939 identified dicumarol as the causative agent for hemorrhages. Wright and other researchers in the 1940s did the widespread therapeutic use of this compound. Finally, Link developed its analog warfarin as a rodenticide but was used as an anticoagulant in humans in the 1950s [136].

34.5 Concluding Remarks

Despite being an elaborate and time-consuming process, drug discovery has progressed well, leading to the discovery of several life-saving drugs. Several approaches like PDD, TDD, and SDD have been applied to arrive at such successful and challenging discoveries. However, more efforts are required to fully exploit the potential of drug discovery using robust computational chemistry and modern techniques. Drug discovery aims to identify safer and more efficacious drugs that can make global health care affordable and efficient. Unlike TDD, PDD does not rely on validating a specific target. This approach has provided several first-in-class drugs that have made us understand unanticipated mechanisms of action and improved patient quality of life. Although TDD has a high potential in terms of time and cost reduction compared to PDD, the latter has the advantage of unraveling the complex unknown biology that may provide solutions for unmet medical needs in the long run. Although many drugs have been discovered by a serendipitous approach that encourages researchers to change the focus to effective results, such discoveries may be less acceptable in the future due to unaffordable costs coupled with stringent guidelines for drug approvals.

Conflict of Interest The author has no conflict of interest to disclose.

References

1. Santos R, Ursu O, Gaulton A, Bento AP, Donadi RS, Bologa CG, Karlsson A, Al-Lazikani B, Hersey A, Oprea TI, Overington JP (2017) A comprehensive map of molecular drug targets. Nat Rev Drug Discov 16:19–34. https://doi.org/10.1038/nrd.2016.230
2. https://www.fda.gov/drugs/drug-approvals-and-databases/drugsfda-glossary-terms
3. Davis RL (2020) Mechanism of action and target identification: a matter of timing in drug discovery. iScience 23:101487. https://doi.org/10.1016/j.isci.2020.101487
4. Wang H, Wang J, Dong C, Lian Y, Liu D, Yan Z (2020) A novel approach for drug-target interactions prediction based on multimodal deep autoencoder. Front Pharmacol 10:1592. https://doi.org/10.3389/fphar.2019.01592
5. Horien C, Yuan P (2017) Drug development. Yale J Biol Med 90:1–3
6. Emmerich CH, Gamboa LM, Hofmann MCJ, Andresen MB, Arbach O, Schendel P, Gerlach B, Hempel K, Bespalov A, Dirnagl U, Parnham MJ (2021) Improving target assessment in biomedical research: the GOT-IT recommendations. Nat Rev Drug Discov 20:64–81. https://doi.org/10.1038/s41573-020-0087-3
7. Croston GE (2017) The utility of target-based discovery. Expert Opin Drug Discov 12:427–429. https://doi.org/10.1080/17460441.2017.1308351
8. Swinney DC (2020) Phenotypic drug discovery: history, evolution, future in phenotypic drug discovery, p 1–19. https://doi.org/10.1039/9781839160721-00001. eISBN: 978-1-83916-072-1
9. Gonzalez-Munoz AL, Minter RR, Rust SJ (2016) Phenotypic screening: the future of antibody discovery. Drug Discov Today 21:150–156. https://doi.org/10.1016/j.drudis.2015.09.014
10. Moffat JG, Vincent F, Lee JA, Eder J, Prunotto M (2017) Opportunities and challenges in phenotypic drug discovery: an industry perspective. Nat Rev Drug Discov 16:531–543. https://doi.org/10.1038/nrd.2017.111
11. Swinney DC, Lee JA (2020) Recent advances in phenotypic drug discovery. F1000Research 9:944. https://doi.org/10.12688/f1000research.25813.1
12. Ratni H, Scalco RS, Stephan AH (2021) Risdiplam, the first approved small molecule splicing modifier drug as a blueprint for future transformative medicines. ACS Med Chem Lett 12:874–877. https://doi.org/10.1021/acsmedchemlett.0c00659
13. Ratni H, Ebeling M, Baird J, Bendels S, Bylund J, Chen KS, Denk N, Feng Z, Green L, Guerard M, Jablonski P, Jacobsen B, Khwaja O, Kletzl H, Ko C-P, Kustermann S, Marquet A, Metzger F, Mueller B, Naryshkin NA, Paushkin SV, Pinard E, Poirier A, Reutlinger M, Weetall M, Zeller A, Zhao X, Mueller L (2018) Discovery of Risdiplam, a selective survival of motor neuron-2 (SMN2) gene splicing modifier for the treatment of spinal muscular atrophy (SMA). J Med Chem 61:6501–6517. https://doi.org/10.1021/acs.jmedchem.8b00741
14. O'Keefe L (2020) FDA approves oral treatment for spinal muscular atrophy (Press release). https://www.fda.gov/news-events/press-announcements/fda-approves-oral-treatment-spinal-muscular-atrophy
15. Evrysdi (risdiplam) for spinal muscular atrophy. SMA News Today (2020). https://smanewstoday.com/evrysdi-risdiplam/
16. Lee JA (2020) Phenotypic drug discovery: a personal perspective, in phenotypic drug discovery, p 009–020. https://doi.org/10.1039/9781839160721-FP009; eISBN: 978-1-83916-072-1
17. Lorenzo G, Maria F, Sante DG, Onofrio L, Arianna B, Valeria C, Carla C, Massimo C (2020)

The preclinical discovery and development of the combination of ivacaftor + tezacaftor used to treat cystic fibrosis. Expert Opin Drug Discov:1–19. https://doi.org/10.1080/17460441.2020.1750592
18. Collins FS (2019) Realizing the dream of molecularly targeted therapies for cystic fibrosis. N Engl J Med 381:1863–1865. https://doi.org/10.1056/NEJMe1911602
19. Middleton PG, Mall MA, Drevinek P, Lands LC, McKone EF, Polineni D, Ramsey BW, Taylor-Cousar JL, Tullis E, Vermeulen F, Marigowda G, McKee CM, Moskowitz SM, Nair N, Savage J, Simard C, Tian S, Waltz D, Xuan F, Rowe SM, Jain R (2019) Elexacaftor-tezacaftor-ivacaftor for cystic fibrosis with a single Phe508del allele. N Engl J Med 381:1809–1819. https://doi.org/10.1056/NEJMoa1908639
20. Francesco RD, Migliaccio G (2005) Challenges and successes in developing new therapies for hepatitis C. Nature 436:953–960. https://doi.org/10.1038/nature04080
21. Gao M, Nettles RE, Belema M, Snyder LB, Nguyen VN, Fridell RA, Serrano-Wu MH, Langley DR, Sun J-H, O'Boyle DR II, Lemm JA, Wang C, Knipe JO, Chien C, Colonno RJ, Grasela DM, Meanwell NA, Hamann LG (2010) Chemical genetics strategy identifies an HCV NS5A inhibitor with a potent clinical effect. Nature 465:96–100. https://doi.org/10.1038/nature08960
22. Belema M, Meanwell NA (2014) Discovery of Daclatasvir, a pan-genotypic hepatitis C virus NS5A replication complex inhibitor with potent clinical effect. J Med Chem 57:5057–5071. https://doi.org/10.1021/jm500335h
23. Nathan DM, Buse JB, Davidson MB, Ferrannini E, Holman RR, Sherwin R, Zinman B (2009) Medical management of hyperglycemia in type 2 diabetes: a consensus algorithm for the initiation and adjustment of therapy: a consensus statement of the American Diabetes Association and the European Association for the Study of Diabetes. Diabetes Care 32:193–203. https://doi.org/10.2337/dc08-9025
24. Bailey CJ (2017) Metformin: historical overview. Diabetologia 60:1566–1576. https://doi.org/10.1007/s00125-017-4318-z
25. Gilbert IH (2013) Drug discovery for neglected diseases: molecular target-based and phenotypic approaches. J Med Chem 56:7719–7726. https://doi.org/10.1021/jm400362b
26. Das AM, Chitnis N, Burri C, Paris DH, Patel S, Spencer SEF, Miaka EM, Castaño MS (2021) Modelling the impact of fexinidazole use on human African trypanosomiasis (HAT) transmission in the Democratic Republic of the Congo. PLoS Negl Trop Dis 15:e0009992. https://doi.org/10.1371/journal.pntd.0009992
27. https://www.sanofi.com/en/media-room/press-releases/2021/2021-07-19-05-30-00-2264542
28. https://www.technologynetworks.com/drug-discovery/articles/promises-and-challenges-of-target-based-drug-discovery-355809
29. Rang HP, Dale MM, Ritter JM, Flower RJ, Henderson G (2012) Chapter 2: how drugs act: general principles. Rang and Dale's pharmacology. Elsevier/Churchill Livingstone, Edinburgh; New York, pp 6–19. ISBN 978-0-7020-3471-8
30. Overington JP, Al-Lazikani B, Hopkins AL (2006) How many drug targets are there? Nat Rev Drug Discov 5:993–996. https://doi.org/10.1038/nrd2199
31. Landry Y, Gies JP (2008) Drugs and their molecular targets: an updated overview. Fundam Clin Pharmacol 22:1–18. https://doi.org/10.1111/j.1472-8206.2007.00548.x
32. Rosenbaum DM, Rasmussen SG, Kobilka BK (2009) The structure and function of G-protein-coupled receptors. Nature 459:356–363. https://doi.org/10.1038/nature08144
33. Oldham WM, Hamm HE (2008) Heterotrimeric G-protein activation by G-protein-Coupled receptors. Nat Rev Mol Cell Biol 9:60–71. https://doi.org/10.1038/nrm2299
34. Hein P, Bunemann M (2009) Coupling mode of receptors and G proteins. Naunyn Schmiedeberg's Arch Pharmacol 379:435–443. https://doi.org/10.1007/s00210-008-0383-7
35. Neves SR, Ram PT, Iyengar R (2002) G protein pathways. Science 296:1636–1639. https://doi.org/10.1126/science.1071550
36. Guo P, Tai Y, Wang M, Sun H, Zhang L, Wei W, Xiang YK, Wang Q (2022) Gα12 and Gα13: versatility in physiology and pathology. Front Cell Dev Biol 10:809425. https://doi.org/10.3389/fcell.2022.809425
37. Sriram K, Insel PA (2018) G Protein-Coupled Receptors as targets for approved drugs: how many targets and how many drugs? Mol Pharmacol 93:251–258. https://doi.org/10.1124/mol.117.111062
38. Sharma A, Hamelin BA (2003) Classic histamine H1 receptor antagonists: a critical review of their metabolic and pharmacokinetic fate from a birds eye view. Curr Drug Metab 4:105–129. https://doi.org/10.2174/1389200033489523
39. Pagliara A, Testa B, Carrupt P-A, Jolliet P, Morin C, Morin D, Urien S, Tillement J-P, Rihoux J-P (1998) Molecular properties and pharmacokinetic behavior of cetirizine, a zwitterionic H1-receptor antagonist. J Med Chem 41:853–863. https://doi.org/10.1021/jm9704311
40. Bousquet J, Campbell AM, Canonica CW (1996) H1-receptor antagonists: structure and classification. In: Simons FER (ed) Histamine and H1- receptor antagonists in allergic disease. Marcel Dekker, Inc, New York, pp 101–102
41. Estelle F, Simons R, Simons KJ (1991) Pharmacokinetic optimization of histamine H1-receptor antagonist therapy. Clin Pharmacokinet 21:372–393. https://doi.org/10.2165/00003088-199121050-00005

42. Woosley RL, Chen Y, Freiman JP, Gillis RA (1993) Mechanism of the cardiotoxic actions of terfenadine. J Am Med Assoc 269:1532–1536. https://doi.org/10.1001/jama.1993.03500120070028
43. Martinez LG, Gonzalez GEC (2018) Cardiotoxicity of H1-antihistamines. J Anal Pharm Res 7:197–201. https://doi.org/10.15406/japlr.2018.07.00226
44. Chen C (2008) Physicochemical, pharmacological and pharmacokinetic properties of the zwitterionic antihistamines cetirizine and levocetirizine. Curr Med Chem 15:2173–2191. https://doi.org/10.2174/092986708785747625
45. Rodrigues-Ferreira S, Nahmias C (2015) G-protein coupled receptors of the renin-angiotensin system: new targets against breast cancer? Front Pharmacol 6:1–7. https://doi.org/10.3389/fphar.2015.00024. http://www.frontiersin.org/Pharmacology/editorialboard
46. Fountain JH, Lappin SL (2019) Physiology, renin-angiotensin system. In: StatPearls [Internet]. https://www.ncbi.nlm.nih.gov/books/NBK470410/
47. Solomon SD, Anavekar N (2005) A brief overview of inhibition of the renin–angiotensin system: emphasis on blockade of the angiotensin II type-1 receptor. Medscape Cardiol:9. https://www.medscape.org/viewarticle/503909
48. https://www.accessdata.fda.gov/drugsatfda_docs/label/2018/020386s062lbl.pdf
49. Neutel J, Smith DH (2003) Evaluation of angiotensin II receptor blockers for 24-hour blood pressure control: meta-analysis of a clinical database. J Clin Hypertens 5:58–63. https://doi.org/10.1111/j.1524-6175.2003.01612.x
50. Dézsi CA (2016) The different therapeutic choices with ARBs. Which one to give? When? Why? Am J Cardiovasc Drugs 16:255–266. https://doi.org/10.1007/s40256-016-0165-4
51. Usman S, Khawer M, Rafique S, Naz Z, Saleem K (2020) The current status of anti-GPCR drugs against different cancers. J Pharm Anal 10:517e521. https://doi.org/10.1016/j.jpha.2020.01.001
52. Lin S, Zhang A, Zhang X, Wu ZB (2020) Treatment of pituitary and other tumours with Cabergoline: new mechanisms and potential broader applications. Neuroendocrinology 110:477–488. https://doi.org/10.1159/000504000
53. Colao A, Savastano S (2011) Medical treatment of prolactinomas. Nat Rev Endocrinol 7:267–278. https://doi.org/10.1038/nrendo.2011.37
54. Thapa S, Bhusal K (2020) Hyperprolactinemia. StatPearls [Internet]. https://www.ncbi.nlm.nih.gov/books/NBK537331/
55. https://dailymed.nlm.nih.gov/dailymed/drugInfo.cfm?setid=028312dc-d155-4fd5-8abd-6bb9f011d3cc
56. Ingham PW, Nakano Y, Seger C (2011) Mechanisms and functions of Hedgehog signalling across the metazoa. Nat Rev Genet 12:393–406. https://doi.org/10.1038/nrg2984
57. Choudhry Z, Rikani AA, Choudhry AM, Tariq S, Zakaria F, Asghar MW, Sarfraz MK, Haider K, Shafiq AA, Mobassarah NJ (2014) Sonic hedgehog signalling pathway: a complex network. Ann Neurosci 21:28–31. https://doi.org/10.5214/ans.0972.7531.210109
58. Carpenter RL, Lo HW (2012) Hedgehog pathway and GLI1 isoforms in human cancer. Discov Med 13:105–113
59. Brancaccio G, Pea F, Moscarella E, Argenziano G (2020) Sonidegib for the treatment of advanced basal cell carcinoma. Front Oncol 10:582866. https://doi.org/10.3389/fonc.2020.582866
60. Dummer R, Ascierto PA, Basset-Seguin N, Dréno B, Garbe C, Gutzmer R, Hauschild A, Krattinger R, Lear JT, Malvehy J, Schadendorf D, Grob JJ (2020) Sonidegib and vismodegib in the treatment of patients with locally advanced basal cell carcinoma: a joint expert opinion. J Eur Acad Dermatol Venereol 34:1944–1956. https://doi.org/10.1111/jdv.16230
61. Tomkinson H, Kemp J, Oliver S, Swaisland H, Taboada M, Morris T (2011) Pharmacokinetics and tolerability of zibotentan (ZD4054) in subjects with hepatic or renal impairment: two open-label comparative studies. BMC Clin Pharmacol 11:3. https://doi.org/10.1186/1472-6904-11-3
62. Nelson J, Bagnato A, Battistini B, Nisen P (2003) The endothelin axis: emerging role in cancer. Nat Rev Cancer 3:110–116. https://doi.org/10.1038/nrc990
63. Roh M, Abdulkadir SA (2010) Targeting the endothelin receptor in prostate cancer bone metastasis: back to the mouse? Cancer Biol Ther 9:615–617. https://doi.org/10.4161/cbt.9.8.11309
64. James ND, Caty A, Payne H, Borre M, Zonnenberg BA, Beuzeboc P, McIntosh S, Morris T, Phung D, Dawson NA (2010) Final safety and efficacy analysis of the specific endothelin A receptor antagonist zibotentan (ZD4054) in patients with metastatic castration-resistant prostate cancer and bone metastases who were pain free or mildly symptomatic for pain: a double-blind, placebo-controlled, randomised Phase II trial. BJU Int 106:966–973. https://doi.org/10.1111/j.1464-410X.2010.09638.x
65. Rufer AC (2021) Drug discovery for enzymes. Drug Discov 26:875–886. https://doi.org/10.1016/j.drudis.2021.01.006
66. Cooper GM (2000). The cell: a molecular approach, 2nd edn. The central role of enzymes as biological catalysts. https://www.ncbi.nlm.nih.gov/books/NBK9921/
67. Ma B, Kumar S, Tsai CJ, Hu Z, Nussinov R (2000) Transition-state ensemble in enzyme catalysis: possibility, reality, or necessity? J Theor Biol 203:383–397. https://doi.org/10.1006/jtbi.2000.1097
68. https://web.archive.org/web/20170116192526/https://www.drugs.com/monograph/imatinib-mesylate.html
69. Paul MK, Mukhopadhyay AK (2004) Tyrosine kinase-Role and significance in Cancer. Int J Med Sci 1:101–115. https://doi.org/10.7150/ijms.1.101
70. Joensuu H, Dimitrijevic S (2001) Tyrosine kinase inhibitor imatinib (STI571) as an anticancer agent

for solid tumours. Ann Med 33:451–455. https://doi.org/10.3109/07853890109002093
71. Deacon CF (2019) Physiology and pharmacology of DPP-4 in glucose homeostasis and the treatment of type 2 diabetes. Front Endocrinol 10:80. https://doi.org/10.3389/fendo.2019.00080
72. Barnett A (2006) DPP-4 inhibitors and their potential role in the management of type 2 diabetes. Int J Clin Pract 60:1454–1470. https://doi.org/10.1111/j.1742-1241.2006.01178.x
73. Gadsby R (2009) Efficacy and safety of Sitagliptin in the treatment of type 2 diabetes. Clin MedTher 1:53–62
74. Stancu C, Sima A (2001) Statins: mechanism of action and effects. J Cell Mol Med 5:378–387. https://doi.org/10.1111/j.1582-4934.2001.tb00172.x
75. https://www.spandidospublications.com/10.3892/etm.2020.9070
76. McIver LA, Siddique MS (2022) Atorvastatin. In: StatPearls [Internet]. https://www.ncbi.nlm.nih.gov/books/NBK430779/
77. Kaczorowski GJ, McManus OB, Priest BT, Garcia ML (2008) Ion channels as drug targets: the next GPCRs. J Gen Physiol 131:399–405. https://doi.org/10.1085/jgp.200709946
78. Kuang Q, Purhonen P, Hebert H (2015) Structure of potassium channels. Cell Mol Life Sci 72:3677–3693. https://doi.org/10.1007/s00018-015-1948-5
79. Alexander SPH, Mathie A, Peters JA, Veale EL, Striessnig J, Kelly E, Armstrong JF, Faccenda E, Harding SD, Pawson AJ, Southan C, Davies JA, Aldrich RW, Attali B, Baggetta AM, Becirovic E, Biel M, Bill RM, Catterall WA, Conner AC, Davies P, Delling M, Virgilio FD, Falzoni S, Fenske S, George C, Goldstein SAN, Grissmer S, Ha K, Hammelmann V, Hanukoglu I, Jarvis M, Jensen AA, Kaczmarek LK, Kellenberger S, Kennedy C, King B, Kitchen P, Lynch JW, Perez-Reyes E, Plant LD, Rash L, Ren D, Salman MM, Sivilotti LG, Smart TG, Snutch TP, Tian J, Trimmer JS, Eynde CVD, Vriens J, Wei AD, Winn BT, Wulff H, Xu H, Yue L, Zhang X, Zhu M (2021) The concise guide to pharmacology 2021/22: ion channels. Br J Pharmacol 178:S157–S245. https://doi.org/10.1111/bph.15539
80. Purves D, Augustine GJ, Fitzpatrick D (2001) Neuroscience, 2nd edn. Sinauer Associates; Voltage-Gated Ion Channels, Sunderland (MA). https://www.ncbi.nlm.nih.gov/books/NBK10883/
81. Ruiz MDL, Kraus RL (2015) Voltage-gated sodium channels: structure, function, pharmacology, and clinical indications. J Med Chem 58:7093–7118. https://doi.org/10.1021/jm501981g
82. Cox B (2014) Ion channel drug discovery: a historical perspective. Ion Channel Drug Discovery (RSC). https://doi.org/10.1039/9781849735087-00001
83. Fahie S, Cassagnol M (2022) Verapamil. In: StatPearls [Internet]. https://www.ncbi.nlm.nih.gov/books/NBK538495/
84. Xu G, Grimes TD, Grayson TB, Chen J, Thielen LA, Tse HM, Li P, Kanke M, Lin T-T, Schepmoes AA, Swensen AC, Petyuk VA, Ovalle F, Sethupathy P, Qian W-J, Shalev A (2022) Exploratory study reveals far reaching systemic and cellular effects of verapamil treatment in subjects with type 1 diabetes. Nat Commun 13:1159. https://doi.org/10.1038/s41467-022-28826-3
85. https://dailymed.nlm.nih.gov/dailymed/drugInfo.cfm?setid=4c5c86c8-ab7f-4fcf-bc1b-5a0b1fd0691b
86. Goldenberg MM (2010) Overview of drugs used for epilepsy and seizures: etiology, diagnosis, and treatment. P T 35:392–415
87. Rogawski MA, Löscher W, Rho JM (2016) Mechanisms of action of antiseizure drugs and the ketogenic diet. Cold Spring Harb Perspect Med 6: a022780. https://doi.org/10.1101/cshperspect.a022780
88. Schmidt D, Sachdeo R (2000) Oxcarbazepine for treatment of partial epilepsy: a review and recommendations for clinical use. Epilepsy Behav 1:396–405. https://doi.org/10.1006/ebeh.2000.0126
89. Czapinski P, Blaszczyk B, Czuczwar SJ (2005) Mechanisms of action of antiepileptic drugs. Curr Top Med Chem 5:3–14. https://doi.org/10.2174/1568026053386962
90. Abou-Khalil BW (2016) Antiepileptic drugs. Continuum (Minneap Minn) 22:132–156. https://doi.org/10.1212/CON.0000000000000289
91. Rao R, Shah S, Bhattacharya D, Toukam DK, Cáceres R, Pomeranz Krummel DA, Sengupta S (2022) Ligand-gated ion channels as targets for treatment and management of cancers. Front Physiol 8(13):839437. https://doi.org/10.3389/fphys.2022.839437
92. Jacob NT (2017) Drug targets: ligand and voltage gated ion channels. Int J Basic Clin Pharmacol 6: 235–245. https://doi.org/10.18203/2319-2003.ijbcp20170314
93. Machu TK (2011) Therapeutics of 5-HT3 receptor antagonists: current uses and future directions. Pharmacol Ther 130:338–347. https://doi.org/10.1016/j.pharmthera.2011.02.003
94. Thompson AJ, Lummis SC (2007) The 5-HT3 receptor as a therapeutic target. Expert Opin Ther Targets 11:527–540. https://doi.org/10.1517/14728222.11.4.527
95. Theriot J, Wermuth HR, Ashurst JV (2021) Antiemetic serotonin-5-HT3 receptor blockers. In: StatPearls [Internet]. https://www.ncbi.nlm.nih.gov/books/NBK513318/
96. Savant K, Khandeparker RV, Berwal V, Khandeparker PV, Jain H (2016) Comparison of ondansetron and granisetron for antiemetic prophylaxis in maxillofacial surgery patients receiving general anesthesia: a prospective, randomised, and double blind study. J Korean Assoc Oral Maxillofac Surg 42:84–89. https://doi.org/10.5125/jkaoms.2016.42.2.84

97. Chong DJ, Lerman AM (2016) Practice update: review of anticonvulsant therapy. Curr Neurol Neurosci Rep 16:39. https://doi.org/10.1007/s11910-016-0640-y
98. Franco V, Crema F, Iudice A, Zaccara G, Grillo E (2013) Novel treatment options for epilepsy: focus on perampanel. Pharmacol Res 70:35–40. https://doi.org/10.1016/j.phrs.2012.12.006
99. Sladek FM (2005) Nuclear receptors as drug targets: new developments in coregulators, orphan receptors and major therapeutic areas. Expert Opin Ther Targets 7:679–684. https://doi.org/10.1517/14728222.7.5.679
100. Li JJ (2006) Laughing gas, viagra and lipitor: the human stories behind the drugs we use. https://doi.org/10.1093/oso/9780195300994.001.0001
101. Schulman IG (2010) Nuclear receptors as drug targets for metabolic disease. Adv Drug Deliv Rev 30:1307–1315. https://doi.org/10.1016/j.addr.2010.07.002
102. Sever R, Glass CK (2013) Signaling by nuclear receptors. Cold Spring Harb Perspect Biol 5: a016709. https://doi.org/10.1101/cshperspect.a016709
103. https://www.cellsignal.com/pathways/nuclear-receptors
104. Quintanilla Rodriguez BS, Correa R (2022) Raloxifene. In: StatPearls [Internet]. https://www.ncbi.nlm.nih.gov/books/NBK544233/
105. Zhao L, Zhou S, Gustafsson J-A (2019) Nuclear receptors: recent drug discovery for cancer therapies. Endocr Rev 40:1207–1249. https://doi.org/10.1210/er.2018-00222
106. Rigas JR, Dragnev KH (2005) Emerging role of rexinoids in non-small cell lung cancer: focus on bexarotene. Oncologist 10:22–33. https://doi.org/10.1634/theoncologist.10-1-22
107. Pileri A, Delfino C, Grandi V, Pimpinelli N (2013) Role of bexarotene in the treatment of cutaneous T-cell lymphoma: the clinical and immunological sides. Immunotherapy 5:427–433. https://doi.org/10.2217/imt.13.15
108. Okude H, Ori D, Kawai T (2021) Signaling through nucleic acid sensors and their roles in inflammatory diseases. Front Immunol 11:625833. https://www.frontiersin.org/articles/10.3389/fimmu.2020.625833
109. Sun Q, Wang Q, Scott MJ, Billiar TR (2016) Immune activation in the liver by nucleic acids. J Clin Transl Hepatol 4:151–157. https://doi.org/10.14218/JCTH.2016.00003
110. Kulkarni JA, Witzigmann D, Thomson SB, Chen S, Leavitt BR, Cullis PR, Meel RVD (2021) The current landscape of nucleic acid therapeutics. Nat Nanotechnol 16:630–643. https://doi.org/10.1038/s41565-021-00898-0
111. Wang M, Yu Y, Liang C, Lu A, Zhang G (2016) Recent advances in developing small molecules targeting nucleic acid. Int J Mol Sci 17:779. https://doi.org/10.3390/ijms17060779
112. Ren Y, Song X, Tan L, Guo C, Wang M, Liu H, Cao Z, Li Y, Peng C (2020) A review of the pharmacological properties of psoralen. Front Pharmacol:11, 571535. https://doi.org/10.3389/fphar.2020.571535
113. Hearst JE, Thiry L (1977) The photoinactivation of an RNA animal virus, vesicular stomatitis virus, with the aid of newly synthesized psoralen derivatives. Nucleic Acids Res 4:1339–1348. https://doi.org/10.1093/nar/4.5.1339
114. Hanson CV, Riggs JL, Lennette EH (1978) Photochemical inactivation of DNA and RNA viruses by psoralen derivatives. J Gen Virol 40:345–358. https://doi.org/10.1099/0022-1317-40-2-345
115. https://www.pediatriconcall.com/drugs/neomycin/800
116. Jana S, Deb JK (2006) Molecular understanding of aminoglycoside action and resistance. Appl Microbiol Biotechnol 70:140–150. https://doi.org/10.1007/s00253-005-0279-0
117. Mingeot-Leclercq MP, Glupczynski Y, Tulkens PM (1999) Aminoglycosides: activity and resistance. Antimicrob Agents Chemother 43:727–737. https://doi.org/10.1128/AAC.43.4.727
118. Wijdicks EF (2016) Hepatic encephalopathy. N Engl J Med 375:1660–1670. https://doi.org/10.1056/NEJMra1600561
119. Patidar KR, Bajaj JS (2013) Antibiotics for the treatment of hepatic encephalopathy. Metab Brain Dis 28: 307–312. https://doi.org/10.1007/s11011-013-9383-5
120. Huang P, Chandra V, Rastinejad F (2010) Structural overview of the nuclear receptor superfamily: insights into physiology and therapeutics. Annu Rev Physiol 72:247–272. https://doi.org/10.1146/annurev-physiol-021909-135917
121. Meyers MA (2007) Happy accidents: serendipity in modern medical breakthroughs. Korean J Radiol 8: 263. https://doi.org/10.3348/kjr.2007.8.4.263
122. Sneader W (2005) Drug discovery: a history. Wiley. ISBN: 978-0-470-01552-0
123. Hargrave-Thomas E, Yu B, Reynisson J (2012) Serendipity in anticancer drug discovery. World J Clin Oncol 3:1–6. https://doi.org/10.5306/wjco.v3.i1.1
124. Campbell WC (2005) Serendipity and new drugs for infectious disease. ILAR J 46:352–356. https://doi.org/10.1093/ilar.46.4.352
125. Wu B, Chien EY, Mol CD, Fenalti G, Liu W, Katritch V, Abagyan R, Brooun A, Wells P, Bi FC, Hamel DJ, Kuhn P, Handel TM, Cherezov V, Stevens RC (2010) Structures of the CXCR4 chemokine GPCR with small-molecule and cyclic peptide antagonists. Science 330:1066–1071. https://doi.org/10.1126/science.1194396
126. Clercq ED (2009) The AMD3100 story: the path to the discovery of a stem cell mobilizer (Mozobil). Biochem Pharmacol 77:1655–1664. https://doi.org/10.1016/j.bcp.2008.12.014

127. Ghiasi N, Bhansali RK, Marwaha R (2022) Lorazepam. In: StatPearls [Internet]. https://www.ncbi.nlm.nih.gov/books/NBK532890/
128. Baenninger A, Costa e Silva JA, Hindmarch I, Moeller H, Rickels K (2004) Good chemistry: the life and legacy of valium inventor Leo Sternbach. McGraw-Hill, New York
129. Sternbach LH (1972) The discovery of librium. Agents Actions 2:193–196. https://doi.org/10.1007/BF01965860
130. Sternbach LH (1979) The benzodiazepine story. J Med Chem 22:1–7. https://doi.org/10.1021/jm00187a001
131. Miller NS, Gold MS (1990) Benzodiazepines: reconsidered. Adv Alcohol Subst Abuse 8:67–84. https://doi.org/10.1300/J251v08n03_06
132. Wick JY (2013) The history of benzodiazepines. Consult Pharm 28:538–548. https://doi.org/10.4140/TCP.n.2013.538
133. Tan SY, Tatsumura Y (2015) Alexander Fleming (1881-1955): discoverer of penicillin. Singap Med J 56:366–367. https://doi.org/10.11622/smedj.2015105
134. Duffin J (2000) Poisoning the spindle: serendipity and discovery of the anti-tumor properties of the Vinca alkaloids. Can Bull Med Hist 17:155–192. https://doi.org/10.3138/cbmh.17.1.155
135. https://www.drugs.com/monograph/warfarin-sodium.html
136. Mueller RL, Scheidt S (1994) History of drugs for thrombotic disease. Discovery, development, and directions for the future. Circulation 89:432–449. https://doi.org/10.1161/01.CIR.89.1.432

35 Role of Nonclinical Programs in Drug Development

Anup K. Srivastava and Geeta Negi

Abstract

Developing a molecule into a new drug for safe and effective human use is a complex, time and resource-consuming process requiring collaboration of multidisciplinary experts. The nonclinical assessment of new drugs is an integral part of the drug development process providing a stepwise characterization of a pharmaceutical to support different phases of clinical development. The objectives of the nonclinical program are synchronized with the clinical development of drug candidates, ranging from providing a basis for selecting doses for first-in-human (FIH) trials to informing clinical monitoring of potential adverse effects. The overall aim is to translate data from in vitro and/or in vivo systems into an understanding of the risk to the intended patient population. A typical nonclinical program encompasses studies to evaluate pharmacology, pharmacokinetics, general toxicology, genetic toxicology, reproductive and developmental toxicology, carcinogenicity, and special toxicities if any. However, nonclinical programs to support the development and approval of drugs vary with the therapeutic indication, patient population as well as molecular size of the drug (small molecules vs biotherapeutics). ICH guidance provides a strategic framework for a minimum standard for the nonclinical program in support of clinical development across the various regulatory bodies. This chapter highlights the pertinent points to consider for the conduct of nonclinical studies in accordance with current guidance recommendations.

Keywords

Drug development · ICH guidance · Nonclinical safety assessment · Toxicology · Small molecules · Biotherapeutics

Disclaimer: This article reflects the views of the authors and should not be construed to represent US FDA's views or policies.

A. K. Srivastava
Division of Pharmacology-Toxicology for Immunology and Inflammation, Office of New Drugs, Center for Drug Evaluation and Research, US Food and Drug Administration, Silver Spring, MD, USA
e-mail: Anup.Srivastava@fda.hhs.gov

G. Negi (✉)
Division of Pharmacology-Toxicology for Cardiology, Hematology, Endocrinology and Nephrology, Office of New Drugs, Center for Drug Evaluation and Research, US Food and Drug Administration, Silver Spring, MD, USA
e-mail: Geeta.Negi@fda.hhs.gov

35.1 Introduction

Drug development is an arduous, expensive, and challenging process with intricacies at each step. It involves concerted efforts from experts in all disciplines to bring a drug candidate from

G. Jagadeesh et al. (eds.), *The Quintessence of Basic and Clinical Research and Scientific Publishing*,
https://doi.org/10.1007/978-981-99-1284-1_35

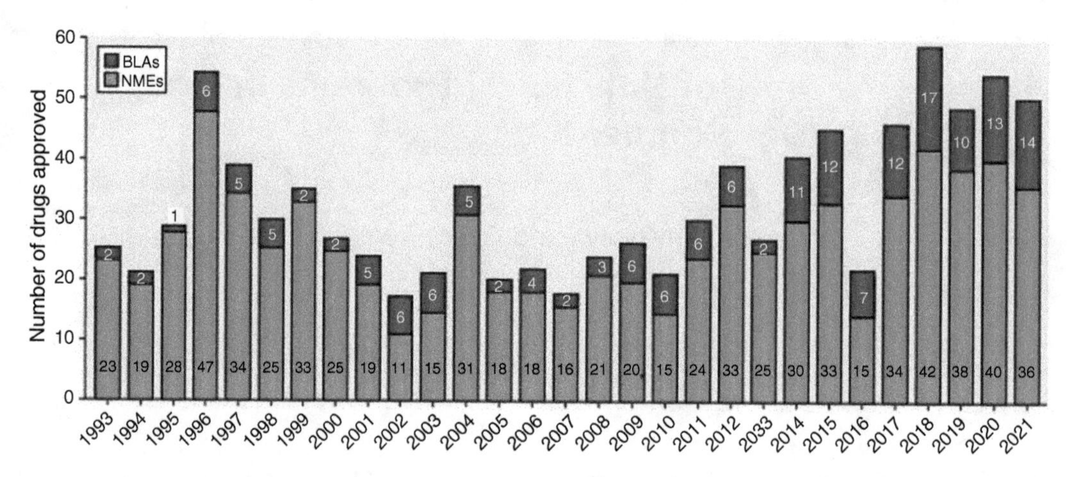

Fig. 35.1 List of new drugs approved by FDA's Center for Drug Evaluation and Research (CDER) since (1993) [4]

laboratory to market. Safety issues are considered one of the foremost causes of attrition in all stages of drug development with the majority of safety-related attrition occurring before the first-in-human (FIH) studies [1]. The cost of bringing a new drug to market from 2010 to 2022 varied from around $161 million to $4.54 billion with an average cost of $2 billion [2, 3]. This process encompasses drug discovery, nonclinical studies and clinical trials, regulatory filings, and post-marketing surveillance. The Center for Drug Evaluation and Research (CDER) of the US Food and Drug Administration (FDA) approved 50 novel therapeutics in 2021 as compared to 53 approvals in 2020 (Fig. 35.1) [4]. The 5-year average approval by CDER is around 51 drugs per year. Of all the 50 therapeutics approved by CDER in 2021, 36 are small molecules and 14 are biologics/biotherapeutics. Based on therapeutic areas, oncology drugs accounted for maximum new approvals with 15 (30%) followed by neurology drugs being a distant second with 5 (10%) new approvals, and infectious diseases and cardiovascular diseases were tied for the third spot with 4 (8%) approvals each (Fig. 35.2) [4].

Nonclinical and clinical development of drug candidates are closely intertwined from the early first-in-human (FIH) studies through the application for marketing authorization. The main objective of the nonclinical program is to characterize the safety of drug candidates to support specific phases of clinical drug development as shown in Table 35.1 (for small molecules) and Table 35.2 (for biotherapeutics). Generally, these studies (1) identify potential target organs of toxicity and potential reversibility following cessation of dosing; (2) identify minimum recommended start dose (MRSD) and maximum recommended human dose (MRHD) for first-in-human (FIH) clinical studies and subsequent dose-escalation schemes; (3) identify safety parameters for clinical monitoring; and (4) determine toxicity endpoints not amenable to an evaluation in clinical studies such as genetic toxicity, reproductive and developmental toxicity, and carcinogenicity which are included in the product label. The objective of this chapter is to provide an overview of the pivotal nonclinical programs in the drug development process. However, given the scope and complexity of the topic, this chapter only highlights selected aspects of the typical nonclinical safety testing paradigm to support the registration of new pharmaceuticals. Several specific topics and related ICH and FDA guidance are discussed in greater detail elsewhere in this volume as indicated by the specific chapter number at appropriate places in the present chapter.

35.2 Guidances for Conducting Nonclinical Studies

Nonclinical safety characterization of drug candidates is generally required by regulatory

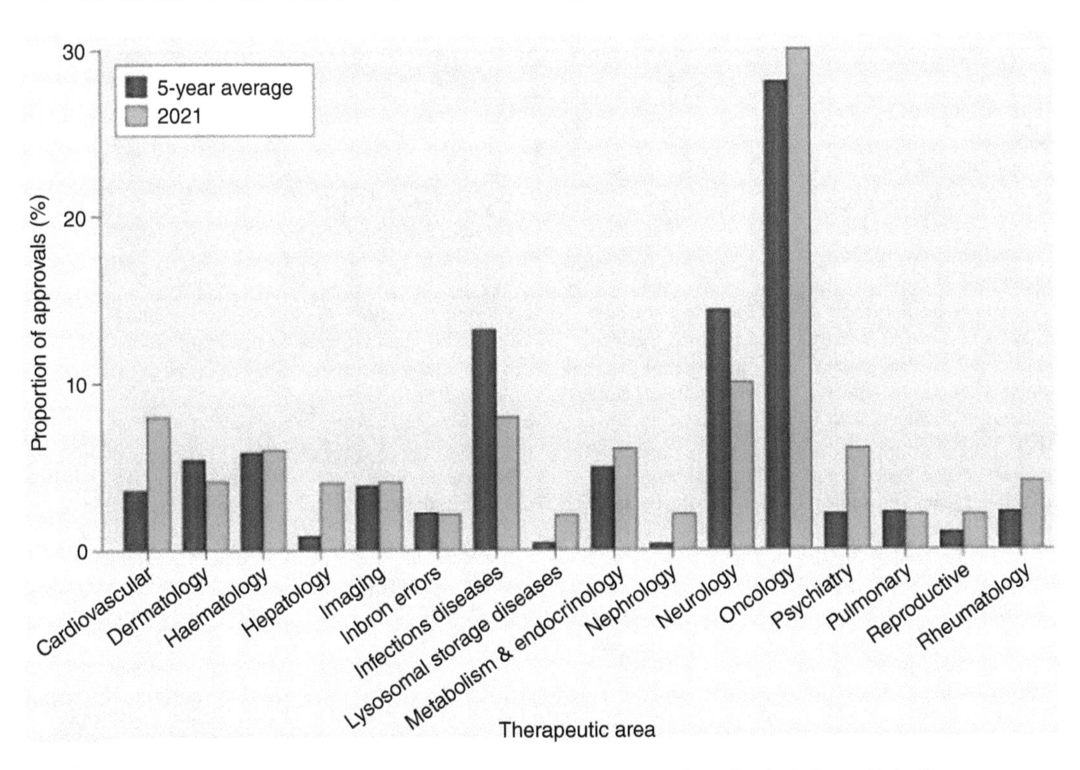

Fig. 35.2 CDER new drug approvals based on therapeutic areas. Indications that include multiple disease areas are classified under one category [4]

agencies, but the specific guidelines for the conduct of these studies vary widely by region. There was a global effort in the 1990s to standardize nonclinical testing and this led to the formation of a compendium of harmonized guidance produced by the International Conference on Harmonization (ICH). The ICH guidelines have been categorized in four groups with a specific topic code assigned to each group: Quality (Q), Efficacy (E), Safety (S) and Multidisciplinary (M) guidelines [5]. The ICH guidance "M3 (R2) Nonclinical Safety Studies for the Conduct of Human Clinical Trials and Marketing Authorization for Pharmaceuticals (2010)" is the master guidance that recommends a basic framework of nonclinical studies to support each phase of clinical development [6]. Similarly, guidance ICH "S6 (R1) Preclinical Safety Evaluation of Biotechnology-derived Pharmaceuticals" recommends the appropriate nonclinical safety studies for the safety evaluation of biotechnology-derived pharmaceuticals (biotherapeutics). However, the timing of the nonclinical studies relative to the clinical development of biotherapeutics is provided by M3 (R2).

For anticancer pharmaceuticals intended to treat patients with advanced cancer, the ICH guidance "S9 Nonclinical Evaluation for Anticancer Pharmaceuticals" describes the type and timing of nonclinical studies in relation to the development of anticancer drugs in patients with advanced cancer and references other guidance as and when required. The ICH S9 applies to small molecules and biotherapeutics. Of note, ICH M3 (R2) or ICH S6 guidance, as applicable, remains the primary guidance for guiding the development of therapies intended for nonterminal non-cancer indications, cancer prevention, treatment of symptoms or side effects of

Table 35.1 General nonclinical program for development of small molecules

Pre-phase 1	Pre-phase 2	Pre-phase 3	Pre-marketing
Pharmacology (ICH S7A)			
Safety pharmacology (ICH S7A, S7B)			
PK: Basic ADME studies such as in vitro and in vivo metabolism, and protein binding (ICH S3A)		PK: Complete ADME studies (ICH S3A)	
Single-dose or short-term (2-week or 4-week) toxicity studies in 2 species (rodent and non-rodent) to support duration of a clinical trial (ICH M3(R2))	Repeat-dose toxicity studies in 2 species to support dose/duration of Phase 2 clinical studies (ICH M3(R2))	Chronic toxicity studies: 6-month in rodent and 9-month in non-rodent (ICH S4A, ICH M3(R2))	Toxicity specific mechanistic study
In vitro genetic toxicity: gene mutation to support single-dose FIH trial and chromosomal damage in mammalian system to support repeat-dose clinical trials (ICH S2A, S2B)	Complete battery of genetic toxicity studies (ICH S2B, ICH M3(R2))	Male and female fertility study (ICH S5A) Embryofetal development study (ICH S5A)	Carcinogenicity study is dependent on duration of clinical dosing, patient population and other biological activities of the product (ICH S1A)
Local tolerance (part of acute toxicity studies (ICH M3 (R2)))			Pre- and post-natal development and offspring fertility study (ICH S5A)
Phototoxicity (initial assessment of phototoxic potential if required)		Complete phototoxic assessment (if required) (ICH M3 R2)	
Immunotoxicity (part of standard toxicity studies) (ICH S8, ICH M3 (R2))		Complete immunotoxicity assessment (if required) (ICH S8)	

chemotherapies, studies in healthy volunteers, vaccines, radiopharmaceutical or cellular or gene therapy [7].

European Medicine Agency (EMA) "Guideline on strategies to identify and mitigate risks for first-in-human and early clinical trials with investigational medicinal products" was adopted by CHMP (Committee for Medicinal Products for Human Use) on July 20, 2017 [8], and aligns with ICH M3(R2), S6(R1) and S9 guidance. This guidance refers to ICH M3(R2), S6(R1) and S9 for the design of safety pharmacology, PK/ADME, and toxicity studies. In addition, ICH experts developed a series of Questions and Answers to facilitate the implementation of the M3(R2) and S9 Guidance [6, 9].

35.3 General Considerations for the Conduct of Nonclinical Studies to Support Clinical Studies

35.3.1 Pharmacology

These studies consist of primary pharmacodynamic (PD), secondary pharmacodynamic, and safety pharmacology studies with "lead drug candidate". Generally, primary PD studies demonstrate the mechanism of action, the binding affinity of a drug candidate to desired therapeutic targets in different species, and the in vivo efficacy of a drug candidate in various animal disease

Table 35.2 General nonclinical program for development of biotherapeutics

Pre-phase 1	Pre-phase 2	Pre-phase 3	Pre-marketing
Pharmacology studies to support rationale and selection of relevant species (ICH S6)			Toxicity specific mechanistic study
In vitro studies to determine receptor occupancy, receptor affinity and/or pharmacological effects in humans and other species		Male and female fertility study parameters evaluated in chronic studies Embryofetal development study (ICH S5A)[a]	Enhanced pre-and post-natal development study (ePPND) (ICH S5A)[a]
Tissue cross-reactivity (TCR) studies			Carcinogenicity—weight-of-evidence[b]
Safety pharmacology (ICH S7A)			
Single-dose or 2-week study, repeated-dose studies to support duration of a clinical trial (ICH S6/ICH M3(R2))	Repeat-dose toxicity studies in a relevant species to support Phase 2 study.	Chronic toxicity studies (up to 6-month) in relevant species. (ICH S6)	
Absorption and distribution (ICH S6)			
Immunogenicity (part of repeat-dose studies) (ICH S6)			
Local tolerance with intended clinical formulation (part of general toxicity studies) (ICH S6)			

[a]If drug candidate is pharmacologically active in both rodents and rabbits, then embryofetal development (EFD) toxicity studies should be conducted in both species, unless embryofetal lethality or teratogenicity has been identified in one species. If drug candidate is pharmacologically active only in non-human primates (NHPs), then an enhanced PPND (ePPND) study with dosing from gestation day (GD) 20 to birth could be considered, rather than separate EFD and/or PPND studies. Generally, ePPND study could be conducted in parallel to Phase 3 clinical study and its report could be submitted at the time of marketing application, provided there are adequate measures to prevent pregnancy in women of childbearing potential (WOCBP)

[b]Generally, weight-of-evidence (WoE) approach is considered for biotherapeutics carcinogenicity assessment. If WoE does not suggest carcinogenic potential, then no additional nonclinical testing is recommended. However, if WoE suggests carcinogenic potential, then sponsor could propose additional nonclinical studies to mitigate the concern, or the label should clearly highlight this concern

models. PD activity is evaluated using in vitro and/or in vivo assays. These in vitro assays are designed to determine receptor occupancy, receptor affinity, and/or pharmacological effects in humans and other species, thus, assisting in the selection of relevant animal species for in vivo toxicity studies. In vivo studies elucidate the mechanism of action, assess pharmacological activity, and justify the proposed use of biotherapeutic in clinical studies [10].

Secondary PD studies investigate the mode of action and/or effects of drug candidates that are not related to their desired therapeutic target and are considered as "off-target" effects usually associated with small molecules. In contrast, as biotherapeutics are selected based on their high specificity to the therapeutic target, secondary PD studies evaluating off-target effects are not routinely conducted [11, 12].

Safety pharmacology studies identify undesirable PD effects of drug candidates on specific physiological functions that may be clinically relevant. The core battery of safety pharmacology studies consists of the cardiovascular, central nervous system and respiratory system assessment and these are generally standalone studies for small molecules. However, for biotherapeutics, standalone safety pharmacology studies are not required because these are more target specific and therefore, safety pharmacology parameters could be incorporated into the general toxicity studies [13]. Supplemental safety pharmacology studies are recommended when the core battery safety pharmacology studies are not able to address all the concerns associated with the drug candidate and these studies are conducted on a case-by-case basis.

35.3.2 Pharmacokinetics (PK) and Toxicokinetics (TK)

For small molecules, PK studies are conducted both in vitro and in vivo in multiple species using radiolabeled and unlabeled drugs. These studies characterize the ADME profile of the drug candidate (for details, see Chap. 13). TK is the generation of PK data in toxicity studies and is used in the interpretation of toxicity findings and their relevance to clinical safety in humans. In addition, in vitro metabolic and plasma protein-binding data in animals and humans should be evaluated prior to Phase 1 studies [14]. Further, extensive PK characterization in test species and in vitro data relevant to potential drug interactions should be available prior to Phase 3 clinical studies. PK data can be used to compare human and animal metabolites and can determine if additional testing is warranted. Additional nonclinical characterization of a human metabolite is only warranted when the particular metabolite(s) is observed at exposures greater than 10% of total drug-related exposure at steady-state and at disproportionately higher levels in humans than in any of the animal test species. Metabolite characterization studies should be conducted to support Phase 3 clinical studies [15].

In contrast, limited ADME characterization is required for biotherapeutics because these are not metabolized by cytochrome P450 enzyme system, unlike small molecules. Instead, these are catabolized into small peptides and amino acids that may be either excreted or reused for protein synthesis. PK/TK profile of biotherapeutics is characterized in a relevant animal species via the route of administration relevant to the anticipated clinical studies [16].

35.3.3 Toxicity Studies

Toxicity studies form an integral part of the nonclinical drug development program for both small molecules and biotherapeutics. It includes acute to chronic toxicity studies, reproductive toxicity studies, genotoxicity studies, carcinogenicity studies, and special toxicity studies such as immunotoxicity, phototoxicity, local tolerance, and juvenile toxicity studies (for details, *see* Chaps. 5–8). The conduct of definitive nonclinical studies which are utilized for establishing the safety characteristics, must be in compliance with good laboratory practice (GLP) standards as per 21 CFR Part 58 [17]. Generally, acute (single-dose) and repeat-dose toxicity studies are conducted in one and/or two relevant species of duration equal to or greater than the proposed duration of clinical studies (Table 35.3, adopted from ICH M3(R2) [6]). In addition, the route of administration and dosing regimen in the toxicity studies should mimic the clinical study design and should characterize the TK profile of the drug candidate, when feasible. Recovery period is included in study designs of selected repeat-dose toxicity studies to determine the potential reversal or worsening of pharmacological/toxicological effects and/or potential delayed toxic effects. For biotherapeutics that induce prolonged

Table 35.3 Recommended duration of repeat-dose toxicity studies to support marketing

Duration of clinical study	Rodent	Nonrodent
Up to 2 weeks	1 month	1 month
>2-weeks to 1 month	3 months	3 months
>1 month to 3 months	6 months	6 months
>3 months	6 months	9 months

Adopted from ICH M3R(2) [5]

pharmacological/toxicological effects, the duration of the recovery period should take into consideration the duration of the drug candidate's pharmacological and/or toxicological effects. Generally, for FIH studies safety margins are calculated based on body surface area or body weight (*see* Chap. 13).

Small Molecules Repeat-dose toxicity studies with small molecules characterizes the safety profile of drug candidates over the entire dose range and duration of the study from no observed adverse effect level (NOAEL) up to the maximum tolerated dose (MTD). In cases where MTD cannot be established, a limit dose of 1000 mg/kg/day is considered an acceptable high dose in all species, provided that a mean exposure margin of tenfold clinical exposure is achieved [6]. However, if 1000 mg/kg/day does not yield a tenfold exposure over clinical exposure and clinical dose is over 1 g/day, then the high dose in the toxicity studies should be based on tenfold exposure margin or 2000 mg/kg/day or maximum feasible dose (MFD), whichever is lower. Generally, single-dose and/or repeat dose toxicity studies are conducted in relevant rodent and non-rodent species to support Phase 1 clinical studies. Acute toxicity studies determine the potential target organs of toxicity and potential for acute toxicity in humans, estimate safe acute doses for humans and estimate doses for repeat-dose toxicity studies, and account for species differences in toxicity. Of note, lethality should not be an intended endpoint in acute toxicity studies [6]. However, acute toxicity data could be obtained from dose-escalation studies or short-duration dose-range-finding studies that determine MTD in the relevant test species [18] (also *see*, Chap. 5).

Repeat dose toxicity studies characterize the drug's safety profile following a treatment duration consistent with the duration of the proposed clinical study and the NOAEL identified in these studies is used to calculate safety margins for proposed clinical doses. Short term studies also aid in the selection of appropriate doses for chronic toxicity studies. As per ICH S4A, chronic toxicity studies employ a dosing period of 6-months in rodents and 9-months in non-rodents (e.g., dogs, monkeys) to determine potential target organs of toxicity following chronic dosing and reversibility of toxicities following cessation of dosing, determine NOAEL for calculating safety margins for clinical studies exceeding 3-month treatment duration, and determine potential clinical risk to patients in long-term clinical studies (Table 35.3) [19]. Chronic toxicity studies are typically conducted prior to the initiation of Phase 3 clinical studies and after efficacy data is obtained from Phase 2 proof-of-concept clinical studies. Longer duration toxicity studies often result in the emergence of new safety signals that could be clinically monitored in Phase 3 studies involving a significantly large number of diseased patients relative to earlier clinical studies (see Chap. 20).

Biotherapeutics Selection of pharmacologically relevant animal species plays a critical role in toxicity evaluation of biotherapeutics as it helps in predicting acceptable risks and expected toxicities in humans. A biotherapeutic is considered pharmacologically relevant in a particular animal species when the expression of the target antigen or epitope or orthologous drug receptor and tissue cross-reactivity profile of a biotherapeutic in that animal species is similar to humans [20–22].

Single-dose toxicity studies provide information regarding dose-related systemic and/or local toxicity and help in the selection of doses for repeat-dose toxicity studies. These studies could potentially be used to evaluate safety pharmacology parameters of biotherapeutics. Generally, repeat-dose toxicity studies in at least one pharmacologically relevant species are recommended to support the clinical studies. If the biotherapeutic is pharmacologically active in 2 relevant species (one rodent and one non-rodent), then at least short-term (up to 1 month) general toxicity studies are conducted in both species. If the toxicity findings from short-term studies are "similar" or these findings are attributed to the well-understood mechanism of action of the biotherapeutic, then longer-term toxicity studies (chronic studies up to 6 months treatment duration) in one species are usually considered sufficient. Rodent species should be considered for chronic studies unless there is a scientific justification for using non-rodents. If a drug candidate is pharmacologically active in only one species, then the general toxicity studies should be conducted in single relevant species [23]. In cases where there are no pharmacologically relevant species, the use of relevant transgenic animals engineered to express the human target antigen, or the use of homologous proteins could be considered. The success of using a transgenic animal model expressing human target antigen for safety evaluation relies on the extent to which the antigen-antibody interaction has PD effects similar to those expected in humans. Although the use of homologous proteins may yield useful information, there could be significant differences in homologous protein and clinical product in terms of their production process, range of impurities/contaminants, PK, and exact pharmacological mechanisms, and this approach could substantially add to the cost of product development. In instances where transgenic animal models or homologous proteins could not be possibly used for safety evaluation, a limited repeat-dose toxicity study of ≤14 days duration in a single species could be conducted to evaluate functional endpoints (e.g., cardiovascular and respiratory). Interestingly, surrogate antibodies have been previously used for the evaluation of developmental and reproductive toxicity (DART) studies' endpoints and licensure of monoclonal antibodies such as infliximab and efalizumab [20, 24].

The toxicity of biotherapeutics is most often related to their targeted mechanism of action and is manifested as "exaggerated or excessive pharmacology", also referred to as "on-target toxicity". High-dose selection for biotherapeutics involves PK-PD approaches that identify (1) a dose that provides maximum intended pharmacologic effect in preclinical species and (2) a dose that provides approximately tenfold exposure multiples over the maximum systemic exposure to be achieved in clinical studies. The higher of these two doses is selected for the high-dose group in toxicity studies unless there is an adequate justification for using a lower dose (e.g., maximum feasible dose). Toxicity studies are designed to identify a NOAEL which is used to calculate safety margins for the proposed clinical doses. Toxicity studies with biotherapeutics may result in the formation of anti-drug antibodies (ADA) which may affect the PK/PD parameters, incidence and/or severity of adverse effects, complement activation, or appearance of new toxicities related to immune complex formation and deposition. Although the formation of ADAs in animals is not predictive of potential immunogenicity in humans, the impact of the ADAs on the validity of the study results should be evaluated.

35.3.4 Genotoxicity Studies

Nonclinical drug development involves a comprehensive assessment of genotoxic potential of small molecule drug candidates. Genotoxicity studies consists of a standard battery of in vitro and in vivo assays because no single assay is capable of detecting all relevant genetic damage induced by drug candidates. ICH S2R1 recommends two options for the standard test battery [25]. The first option includes (1) a test to detect gene mutation in bacteria (Ames assay); (2) an in vitro cytogenetic test to detect chromosomal damage (by either in vitro metaphase chromosome aberration test or in vitro micronucleus test), or an in vitro mouse lymphoma Tk gene mutation assay; and (3) an in vivo micronucleus assay or chromosomal aberrations in metaphase cells to detect chromosomal damage. The second option includes (1) a test to detect gene mutation in bacteria (Ames assay); (2) An in vivo assessment of genotoxicity in two different tissues, usually an assay for micronuclei in rodent hematopoietic cells and a second in vivo assay. The second in vivo assay is typically a DNA strand breakage assay in the liver, or it could potentially use other tissues with adequate justification. A gene mutation assay in bacteria (Ames test) should be conducted prior to FIH Phase 1 clinical study and a negative Ames test is considered sufficient to support all single-dose clinical studies. To support repeat-dose clinical studies, genotoxicity assays capable of detecting chromosomal damage in mammalian systems should be completed. Overall, the standard battery of genotoxicity studies should be completed prior to Phase 2 clinical studies. Negative results in the standard battery provide adequate evidence of safety regarding the genotoxic potential of the tested drug candidates. However, drug candidates that test positive in the standard test battery, depending on their therapeutic use, may require more extensive testing and are considered potential human carcinogens and/or mutagens.

In contrast, genotoxicity studies are not required for biotherapeutics as these are not expected to pass through the cellular and nuclear membranes of intact cells due to their large size and interact directly with DNA or other chromosomal material (For more information, *see* Chap. 7).

35.3.5 Reproductive and Developmental Toxicity

Reproductive toxicity testing of pharmaceuticals in animals is critical to the progression of clinical development. Testing of pharmaceuticals for potential human reproductive or developmental risk presents a complex phase of drug development because of the inherent complexity of the reproductive lifecycle. Following the thalidomide tragedy in 1960s, several safety testing procedures have been introduced for evaluating the potential reproductive toxicities of drugs. The first set of regulations was issued by US FDA in 1966 describing a series of animal studies entailing all stages of reproduction and development including general fertility and embryofetal development and extending to peri- and post-natal development [26]. Subsequently with the establishment of ICH, a harmonized guidance for reproductive safety assessment was finalized in 1993 and has been implemented by several regulatory agencies besides FDA, EMA and Japan. The guidance has gone through several rounds of iterations since then to align it with the rapidly emerging new science. The current version of ICH S5(R3) Detection of developmental and reproductive toxicity for human pharmaceuticals (Feb 2020), also provides the flexibility of utilizing in vitro, ex vivo and non-mammalian in vivo assays under limited circumstances [27].

The complexity of the reproductive cycle makes it very difficult to design a single nonclinical study to accurately depict the effects on each phase in a reproductive cycle. Therefore, to obtain a panoramic assessment of all phases of reproduction, the complexity of the reproductive cycle is divided into individual components which can then be studied individually or together. ICH S5 (R3) guidance describes a testing strategy to ensure drug exposure through critical periods of development and assess immediate as well as latent adverse effects. The following studies have been recommended in the guidance (1) Fertility and initial embryofetal development study in one species, usually in rats (2) Embryofetal development studies in two species, usually rats or mice and rabbits and (3) Pre-and post-natal development study in one species, usually in rats (also, see Chap. 6) [27]. For biotherapeutics, the need for developmental and reproductive toxicity studies is guided by product specific attributes, clinical indication and intended patient population while the study design is guided by factors such as species specificity, immunogenicity, biological activity, and long half-lives [16].

The timing of reproductive assessment as described in ICH M3(R2) guidance should be consistent with the study population being exposed. Generally, men and women of non-childbearing potential can be enrolled in

Table 35.4 Nonclinical studies supportive of WOCBP enrollment in different regions

Region	Clinical phase	Supporting nonclinical studies
United States	Phase 1 and 2	WOCBP can be enrolled without conducting embryo-fetal development studies with the provisions of use of highly effective methods of contraception to prevent pregnancy
	Phase 3	Female fertility study and embryo-fetal development studies
European Union	Phase 1 and 2	Embryo-fetal development studies
	Phase 3	In addition to embryo-fetal development studies, assessment of female fertility
Japan	Phase 1–3	Female fertility and embryo-fetal development studies

Adopted from ICH M3R(2) [5]

phase 1 and phase 2 studies without the need for a dedicated animal fertility study as long as there are no concerns identified in the reproductive organs in the repeat dose studies. For the inclusion of women of childbearing potential (WOCBP) in early trials, the nonclinical requirements differ among different regions (Table 35.4). US FDA does not require an embryofetal development study if there are appropriate provisions in place to prevent pregnancy in the clinical study. In the EU and Japan, however, WOCBP may be enrolled in clinical studies in the absence of embryofetal developmental studies only for a short duration (e.g., 2 weeks) or if such data is available from a preliminary study, then WOCBP can be studied in a trial of up to 3-months duration. Prior to initiation of a large-scale clinical trial (for example, phase 3), fertility study as well as embryofetal development studies are required to be completed to support the enrollment of men and women of childbearing potential in the trial. The timing of pre- and post-natal development study is consistent across ICH regions and is required at the time of submission of marketing application [28].

These studies are conducted with the primary purpose of identifying risk of reproductive and developmental effects by integrating the information from pharmacology, pharmacokinetics and toxicology studies as well as data from clinical studies. Our understanding of the relevance of reproductive toxicities for human risk assessment continues to evolve but significant progress is still needed to improve the predictivity of animal findings. The use of several alternative methods for assessing potential reproductive toxicities in drug discovery and development has advanced significantly in the recent past and has started the ball rolling for the qualification and integration of alternative testing approaches for the detection of reproductive toxicities. Robust validation data will expedite this shift towards a new paradigm by minimizing variance, bias, and the potential for false-positive and false-negative results [29, 30].

35.3.6 Carcinogenicity

Rodent carcinogenicity studies are a crucial part of the clinical development of new chemical entities mainly due to the inherent limitations of clinical trials in assessing human risk during the course of drug development. Animal testing is one of the principal mechanisms for identifying drugs with a carcinogenic potential prior to exposure in a broader population. Prior to the ICH harmonization process in 1991, different regulatory bodies had region-specific requirements for carcinogenicity testing of pharmaceuticals. The first ICH Harmonized Guideline for the standardization of regulatory expectations for carcinogenicity studies was finalized in 1995 [31]. Currently, there are three ICH Guidelines, namely S1A the Need for Long-term Rodent Carcinogenicity Studies of Pharmaceuticals (1995), S1B(R1) Testing for Carcinogenicity of Pharmaceuticals (2022), and S1C(R2) Dose Selection for Carcinogenicity Studies of Pharmaceuticals (2008) providing recommendations on appropriate strategies for testing the carcinogenic potential of pharmaceuticals intended for human use [31–33].

Carcinogenicity potential assessment generally takes place later in the drug development cycle for small molecule drugs for chronic human administration. Per ICH S1, for the drugs whose expected human-use profile involves administration for at least six months or drugs intended to be used intermittently but repeatedly in the treatment of a chronic or recurrent disease such as depression, anxiety or allergy, these studies are generally conducted in parallel with phase 3 and are submitted with the marketing application. In cases where drugs are intended for life threatening or severely debilitating diseases, these studies are often conducted post-approval to accelerate the market availability of drugs for such serious conditions. There are instances where such studies may not be required at all, such as for drugs intended for the treatment of advanced systemic disease where life expectancy is short. Early assessment of carcinogenicity potential in advance of phase 3 clinical trials may be necessary if a cause of concern for potential carcinogenic effects has been identified based on the drug class or based on the signal for genotoxicity or short-term toxicity studies. One such example is peroxisome proliferation–activated receptors (PPARs) agonists. PPAR activation has been associated with a high prevalence of neoplasms necessitating rodent carcinogenicity data to be submitted to the US-FDA before initiating chronic clinical studies.

A standard battery of carcinogenicity testing consists of two-year rat and mouse studies. Alternatively, a 2-year study may be conducted only in rats and complemented with a 6-month transgenic mouse assay (these are described in detail in Chap. 8). Since the carcinogenicity studies with new agents are conducted to meet regulatory requirements for approval, compliance with applicable GLP regulations is an essential prerequisite. The Carcinogenicity Assessment Committee (CAC) was created within US FDA in 1991 to ensure consistency in standards among the FDA's review divisions by reviewing the carcinogenicity protocols as well as the final study outcomes [34]. The animal carcinogenicity study protocols are submitted and reviewed under a Special Protocol Assessment (SPA) to assess whether they are adequate to meet scientific and regulatory standards. ICH S1C guidance addresses the dose selection criteria for such studies establishing a rational basis for the high dose selection [32]. Historically maximum tolerated dose (MTD) has been used as a basis of dose selection to maximize the sensitivity of the study. Besides MTD, doses for rodent carcinogenicity studies can be selected based on pharmacokinetic endpoints (25-fold and 50-fold rodent: human AUC exposure ratio for 2-year rodent and 6-month rasH2-Tg mouse study respectively), saturation of absorption; pharmacodynamic endpoints (e.g., inhibition of hemostasis); maximum feasible dose (e.g., due to limited solubility); and limit dose.

Unlike small molecules, standard carcinogenicity testing is considered inappropriate for biotherapeutics. An assessment of carcinogenicity potential should be conducted on a case-by-case basis, with regard to pharmacologic activity, intended patient population and treatment duration. If any potential for risk is identified from the existing nonclinical and clinical data, in vitro and in vivo studies in relevant experimental systems and/or assessment of cellular proliferation in chronic toxicity studies may be considered. ICH S6(R1) guidance also describes a weight-of-evidence (WoE) based approach for evaluating the carcinogenic potential for biotherapeutics by integrating data from target biology/pharmacology, toxicology including immunotoxicity or relevant data from literature [16]. If the integrated evidence suggests conducting a rodent study is unlikely to add further value to the overall carcinogenicity assessment, then the rodent carcinogenicity studies may not be warranted.

The carcinogenicity studies are not only amongst the most challenging studies during the drug development cycle but are also the longest and most expensive studies requiring 600 to 800 animals per study. The debate around the value of rodent carcinogenicity studies has gained traction in recent years due to questionable human relevance of these studies [35, 36]. Although these rodent studies have provided a sensitive method for detecting a human carcinogen, poor human specificity has limited their predictivity for human risk. Both false negative (e.g cigarette smoke) and false positive (e.g., cetirizine, loratadine, doxylamine) cases observed in the standard cancer hazard testing paradigm have severely complicated the successful extrapolation of human carcinogenic hazards from rodent bioassays. Lately, the three Rs (reduce/refine/replace) principle has encouraged several changes within regulatory toxicology testing, especially for carcinogenicity testing. WoE based integrative approach for evaluating the carcinogenic potential for biotherapeutics was extended to all pharmaceuticals with the recent addendum in ICH S1B(R1) [33]. Paradigm shifting alternative methods are being pursued vigorously to move the focus away from rodent carcinogenicity studies, but as of now, despite several limitations and costs, there is no suitable substitute for rodent carcinogenicity studies which remain a gold standard to inform clinical risks [37].

35.3.7 Juvenile Toxicity

Until a few decades ago, the drug development process lacked safety/efficacy assessment in pediatric population and several drugs with therapeutic benefits to children are not labeled for use in this population. The lack of pediatric information on drug labels has encouraged the practice of off-label use in the field of pediatrics which continues to raise several concerns. Scaling down an adult dose for pediatric dosing, as cutting the pills in half, may result in changes in effectiveness, under/overdosing and unique adverse events not identified in adults, such as effects on organ systems that are still undergoing development in children (i.e., skeletal growth and CNS). In 1975 only 22% of the approved drugs contained information on pediatric use on the drug labels which increased merely to 46% in 2009 [38].

Several scientific, legislative, and regulatory initiatives in recent years have promoted the practices to generate progressive data to inform the use of therapeutics in pediatric populations. In the United States, the two most comprehensive pieces of legislation which are considered pivotal for encouraging pediatric safety assessments for pharmaceuticals are the Best Pharmaceuticals for Children Act (BPCA) and the Pediatric Research Equity Act (PREA). BPCA was enacted in 2002 to provide a financial incentive to the pharmaceutical industry to conduct pediatric studies voluntarily. BPCA offers 6 months of additional marketing exclusivity when pediatric safety studies are performed upon FDA request [39]. PREA enacted in 2003 requires that the pharmaceutical companies submitting a new drug application (NDA) or a biologics licensing application (BLA) for a new active ingredient, indication, route of administration, dosing regimen or dosage form to assess safety and effectiveness in pediatric patients. Both were reauthorized in 2007 for five years with the FDA Amendments Act (FDAAA) and were subsequently reauthorized permanently in 2012 with the FDA Safety and Innovation Act (FDASIA) 2012. European Union implemented an EU Pediatric Regulation (Reg 1901/2006/EU 1901/2006), Medicinal Products for Pediatric Use in 2007 with similar goals of providing a regulatory framework for the development of pharmaceutical products and their safe and effective use in the pediatric population. Although the European Union and United States legislation differs in the timing of the pediatric development plan, assessment and evaluation process, exemptions, and marketing exclusivity process, the two regulatory agencies work in a collaborative environment to facilitate the efficient development of pharmaceuticals for pediatric use [40].

The adoption of pediatric regulations has certainly increased the number of clinical trials in pediatric populations making nonclinical juvenile animal studies an essential part of the pediatric clinical development plan. US-FDA was the first to publish guidance on Nonclinical Safety Evaluation of Pediatric Drug Products in 2006 providing a general consideration for conducting juvenile animal studies. A similar guidance was published by the EMA in 2008, Guideline on the Need for Nonclinical Testing in Juvenile Animals of Pharmaceuticals for Pediatric Indications. The ICH S11 Guidance "Nonclinical Safety Testing in

Support of Development of Pediatric Medicines" endorsed by ICH Steering Committee in 2014 and finalized in 2020, harmonized the recommendations across all the ICH regions [41]. A foremost objective of juvenile animal toxicity studies is to address safety concerns relevant to the intended pediatric population that are not adequately addressed in general toxicity or reproductive toxicity studies and are impractical to be addressed in a pediatric clinical trial. Therefore, the need for a nonclinical juvenile study should be based upon a thorough evaluation of the clinical context together with data available from nonclinical and clinical studies. An integrated assessment of the available data also informs the design and timing of juvenile animal studies.

Juvenile studies are conceptually too complex to have a standardized study design, particularly considering the factors involving differences among intended patient age, diverse endpoints of interest or target organs of toxicities, and technical limitations. Thus, instead of implementing a standardized study protocol, several variations in study design can be considered on a case-by-case basis. These studies should be designed based on a sound scientific rationale and should adequately assess the stages of growth and development of organ systems of concern. Usually, juvenile studies are conducted in one relevant species, preferably rodents. However, when scientifically justified, other species may be considered but a thorough understanding of organ system development compared to humans is critical for designing a useful study [42]. In some cases, when the drug is intended to treat a chronic disease in children only or to be used in children first without any prior data in adults, chronic studies may be conducted in juvenile animals, in one or two species, with dosing initiated at the appropriate developmental age.

The general principles of juvenile toxicity studies for biotherapeutics are similar to small molecules, but unique characteristics of biotherapeutics present some specific challenges owing to their high species specificity and immunogenic potential in nonclinical species. The selection of species for these studies is often limited by target specificity, restricting the selection to only non-human primates (NHPs) as the most suitable species. However, if relevant, rats are the preferred species for juvenile studies with biotherapeutics. Also, immunogenic response to human biotherapeutics in animals may further complicate the study design as well as the interpretation of data. In specific circumstances, for example, when there are no relevant species, alternative approaches may be more appropriate such as the use of a surrogate molecule or genetically engineered mice [43].

35.4 Concluding Remarks

A carefully designed nonclinical strategy provides a scientific foundation to guide the successful "bench to bedside" transition of drug candidates. Preclinical assurance of safety prior to exposing humans to new compounds in clinical trials derives from in vitro, ex vivo and in vivo nonclinical studies which provide a basic understanding of the pharmacologic and toxic effects of the new pharmaceuticals. Later in the development when the stakes are even higher, it is crucial that the nonclinical program remains ahead of the next clinical progression to higher and/or longer exposures in order to identify and characterize potential toxicities and their reversibility, identify the target organs of toxicity, safety margins at clinical doses and identify possible prodromal markers for clinical monitorability of the identified adverse responses. Several regulatory region-specific and harmonized guidances have been published to facilitate and accelerate the drug development process. The ICH guidance on different domains for nonclinical testing provides a framework for defining the essential elements to support and move forward at various stages of development. Although these guidances provide a basic framework for the types, timing, and extent of data essential for regulatory decisions, many development programs require a tailored approach.

The drug development process has evolved and expanded rapidly in recent decades resulting in clinically significant improvements in several therapeutic areas. Pivotal events, unfortunately involving some tragic incidents in history, have facilitated a roadmap to circumvent the pitfalls of the past and opened the door to an era that offers the prospect of improved assessment of human risk and improved regulatory decision-making. Technological advancements, enactment of laws and evolution of guidances and their implementation have also set a high standard for nonclinical safety assessment of new pharmaceuticals. The area continues to evolve with several alternative assays being developed and accepted by the regulatory bodies to develop safe drugs more quickly as well as reduce, and perhaps in the future, replace the use of animals.

Conflict of Interest The authors declare that there are no conflicts of interest.

References

1. Hughes JP, Rees S, Kalindjian SB et al (2011) Principles of early drug discovery. Br J Pharmacol 162(6):1239–1249
2. Schlander M, Hernandez-Villafuerte K, Cheng CY et al (2021) How much does it cost to research and develop a new drug? A systematic review and assessment. PharmacoEconomics 39(11):1243–1269
3. Wouters OJ, McKee M, Luyten J (2020) Estimated research and development investment needed to bring a new medicine to market, 2009-2018. JAMA 323(9): 844–853
4. Mullard A (2021) FDA approvals. Available from: https://www.nature.com/articles/d41573-022-00001-9. Accessed 16 Aug 2022
5. ICH Guidelines. Available from: https://www.ich.org/page/ich-guidelines. Accessed 30 Nov 2022
6. ICH M3(R2) (2009) Guidance on nonclinical safety studies for the conduct of human clinical trials and marketing authorization for pharmaceuticals. Available from: https://database.ich.org/sites/default/files/M3_R2__Guideline.pdf. Accessed 16 Aug 2022
7. S9 Implementation Working Group (2018) ICH S9 guideline: nonclinical evaluation for anticancer pharmaceuticals questions and answers. Available from: https://database.ich.org/sites/default/files/S9_Q%26As_Q%26As.pdf. Accessed 16 Aug 2022
8. EMEA/CHMP/SWP/28367/07 Rev 1 (2017) Guideline on strategies to identify and mitigate risks for first-in-human and early clinical trials with investigational medicinal products. Available from: https://www.ema.europa.eu/en/documents/scientific-guideline/guideline-strategies-identify-mitigate-risks-first-human-early-clinical-trials-investigational_en.pdf. Accessed 10 Nov 2022
9. ICH S9 (2009) Nonclinical evaluation for anticancer pharmaceuticals. Available from: https://database.ich.org/sites/default/files/S9_Guideline.pdf. Accessed 16 Aug 2022
10. Andrade EL, Bento AF, Cavalli J et al (2016) Non-clinical studies required for new drug development - Part I: early in silico and in vitro studies, new target discovery and validation, proof of principles and robustness of animal studies. Braz J Med Biol Res 49(11):e5644
11. Papoian T, Chiu H-J, Elayan I et al (2015) Secondary pharmacology data to assess potential off-target activity of new drugs: a regulatory perspective. Nat Rev Drug Discov 14(4):294–294
12. Jenkinson S, Schmidt F, Rosenbrier Ribeiro L et al (2020) A practical guide to secondary pharmacology in drug discovery. J Pharmacol Toxicol Methods 105: 106869
13. ICH S7A (2000) Safety pharmacology studies for human pharmaceuticals. Available from: https://database.ich.org/sites/default/files/S7A_Guideline.pdf. Accessed 25 July 2022
14. ICH S3A (1994) Note for guidance on toxicokinetics: the assessment of systemic exposure in toxicity studies. Available from: https://database.ich.org/sites/default/files/S3A_Guideline.pdf. Accessed 25 July 2022
15. US FDA (2020) Safety testing of drug metabolites guidance for industry. Available from: https://www.fda.gov/media/72279/download. Accessed 25 July 2022
16. ICH S6(R1) (1997) Preclinical safety evaluation of biotechnology-derived pharmaceuticals. Available from: https://database.ich.org/sites/default/files/S6_R1_Guideline_0.pdf. Accessed 25 July 2022
17. 21 CFR Part 58 good laboratory practice for nonclinical laboratory studies. Available from: https://www.accessdata.fda.gov/scripts/cdrh/cfdocs/cfcfr/CFRSearch.cfm?CFRPart=58. Accessed 25 July 2022
18. Robinson S, Delongeas JL, Donald E et al (2008) A European pharmaceutical company initiative challenging the regulatory requirement for acute toxicity studies in pharmaceutical drug development. Regul Toxicol Pharmacol 50(3):345–352
19. ICH S4 (1998) Duration of chronic toxicity testing in animal. Available from: https://database.ich.org/sites/default/files/S4_Guideline.pdf. Accessed 16 Aug 2022
20. Chapman K, Pullen N, Graham M et al (2007) Preclinical safety testing of monoclonal antibodies: the significance of species relevance. Nat Rev Drug Discov 6(2):120–126
21. Hall WC, Price-Schiavi SA, Wicks J et al (2008) Tissue cross-reactivity studies for monoclonal antibodies: predictive value and use for selection of relevant animal species for toxicity testing. In:

Preclinical safety evaluation of biopharmaceuticals, pp 207–240
22. Subramanyam M, Rinaldi N, Mertsching E, et al Selection of relevant species. In: Pharmaceutical sciences encyclopedia, pp 1–25
23. Prior H, Clarke DO, Jones D et al (2022) Exploring the definition of “similar toxicities”: case studies illustrating industry and regulatory interpretation of ICH S6(R1) for long-term toxicity studies in one or two species. Int J Toxicol 41(3):171–181
24. Clarke J, Leach W, Pippig S et al (2004) Evaluation of a surrogate antibody for preclinical safety testing of an anti-CD11a monoclonal antibody. Regul Toxicol Pharmacol 40(3):219–226
25. ICH S2(R1) (2011) Guidance on genotoxicity testing and data interpretation for pharmaceuticals intended for human use. Available from: https://database.ich.org/sites/default/files/S2%28R1%29%20Guideline.pdf. Accessed 18 Sept 2022
26. Raheja KL, Jordan A, Fourcroy JL (1988) Food and Drug Administration guidelines for reproductive toxicity testing. Reprod Toxicol 2(3–4):291–293
27. ICH S5(R3) (2020) Detection of reproductive and developmental toxicity for human pharmaceuticals. Available from: https://database.ich.org/sites/default/files/S5-R3_Step4_Guideline_2020_0218_1.pdf. Accessed 18 Sept 2022
28. Dorato MA, Buckley LA (2006) Toxicology in the drug discovery and development process. Curr Protoc Pharmacol Chapter 10: p Unit10.3
29. Avila AM, Bebenek I, Bonzo JA et al (2020) An FDA/CDER perspective on nonclinical testing strategies: classical toxicology approaches and new approach methodologies (NAMs). Regul Toxicol Pharmacol 114:104662
30. Nikolaidis E (2022) Chapter 6 - Relevance of animal testing and sensitivity of end points in reproductive and developmental toxicity. In: Gupta RC (ed) Reproductive and Developmental Toxicology, 3rd edn. Academic, pp 93–106
31. ICH S1A (1995) Guideline on the need for carcinogenicity studies of pharmaceuticals. [Cited 2022]. Available from: https://database.ich.org/sites/default/files/S1A%20Guideline.pdf. Accessed 18 Sept 2022
32. ICH S1C(R2) (2008) Dose selection for carcinogenicity studies of pharmaceuticals. Available from: https://database.ich.org/sites/default/files/S1C%28R2%29%20Guideline.pdf. Accessed 18 Sept 2022
33. ICH S1B(R1) (2022) Testing for carcinogenicity of pharmaceuticals. Available from: https://database.ich.org/sites/default/files/S1B-R1_FinalGuideline_2022_0719.pdf. Accessed 18 Sept 2022
34. US FDA (2002) Carcinogenicity study protocol submissions. Available from: https://www.fda.gov/regulatory-information/search-fda-guidance-documents/carcinogenicity-study-protocol-submissions. Accessed 18 Sept 2022
35. Knight A, Bailey J, Balcombe J (2006) Animal carcinogenicity studies: 2. Obstacles to extrapolation of data to humans. Altern Lab Anim 34(1):29–38
36. Alden CL, Lynn A, Bourdeau A et al (2011) A critical review of the effectiveness of rodent pharmaceutical carcinogenesis testing in predicting for human risk. Vet Pathol 48(3):772–784
37. Bourcier T, McGovern T, Stavitskaya L et al (2015) Improving prediction of carcinogenicity to reduce, refine, and replace the use of experimental animals. J Am Assoc Lab Anim Sci 54(2):163–169
38. Balan S, Hassali MAA, Mak VSL (2018) Two decades of off-label prescribing in children: a literature review. World J Pediatr 14(6):528–540
39. Breslow LH (2003) The Best Pharmaceuticals for Children Act of 2002: the rise of the voluntary incentive structure and congressional refusal to require pediatric testing. Harvard J Legis 40(1):133–193
40. Penkov D, Tomasi P, Eichler I et al (2017) Pediatric medicine development: an overview and comparison of regulatory processes in the European Union and United States. Ther Innov Regul Sci 51(3):360–371
41. ICH S11 (2020) Nonclinical safety testing in support of development of paediatric pharmaceuticals. Available from: https://database.ich.org/sites/default/files/S11_Step4_FinalGuideline_2020_0310.pdf. Accessed 30 Oct 2022
42. Faqi AS (2017) Chapter 11 - Juvenile toxicity testing to support clinical trials in pediatric population. In: Faqi AS (ed) A comprehensive guide to toxicology in nonclinical drug development, 2nd edn. Academic, Boston, pp 263–272
43. Morford LL, Bowman CJ, Blanset DL et al (2011) Preclinical safety evaluations supporting pediatric drug development with biopharmaceuticals: strategy, challenges, current practices. Birth Defects Res B Dev Reprod Toxicol 92(4):359–380

Drug Repurposing: Strategies and Study Design for Bringing Back Old Drugs to the Mainline 36

Alejandro Schcolnik-Cabrera

Abstract

Every year, the success rate of the development of new medications covering different fields of human health gets more limited. This is particularly challenging for conditions imposing a substantial impact on the life expectancy of patients and the costs related to healthcare for subjects and public health. A novel approach countering this limitation is drug repurposing, which is a term that concerns the ability of any given compound already on the market to be used in a different condition other than the originally intended. Drug repurposing has marked a paradigm in research and development for the pharmaceutical industry, and currently, several branches concentrate on the research of how existing drugs could lead to alternative pathways by altering other targets. Thanks to the expansion of -omics data coming from pharmacology research, massive information is gathered in specialized databases which could then be used in the selection process to find molecule candidates for drug repurposing. However, a proper selection of compounds is mandatory beforehand. In this chapter, I cover the generalities of drug repurposing and enlist specific databases that are advantageous for repurposing medications, after which, I provide readers with a general protocol for the evaluation of compounds. Finally, I present some examples of successful repurposed drugs and their clinical implications.

Keywords

Drug repurposing · Drug databases · Study design · Cancer · Diabetes · Tuberculosis

Abbreviations

ADA	Adenosine deaminase
AI	Artificial intelligence
AIDS	Acquired immunodeficiency syndrome
BioGRID	Biological General Repository for Interaction Datasets
CLUE	CMap and LINCS Unified Environment
CMap	Connectivity Map project
CRISPR	Clustered Regularly Interspaced Short Palindromic Repeats
DL	Deep learning
DR	Drug repurposing
FDA	Food and Drug Administration
GDA	Gene-Disease Association

A. Schcolnik-Cabrera (✉)
Département de Biochimie et Médicine Moléculaire, Université de Montréal, Montréal, QC, Canada

Division of Immunology-Oncology, Hôpital Maisonneuve-Rosemont Research Centre, Montréal, QC, Canada
e-mail: adrian.alejandro.schcolnik.cabrera@umontreal.ca

G. Jagadeesh et al. (eds.), *The Quintessence of Basic and Clinical Research and Scientific Publishing*,
https://doi.org/10.1007/978-981-99-1284-1_36

ICs	Inhibitory concentrations
IDH1	Isocitrate dehydrogenase 1
KEGG	Kyoto Encyclopedia of Genes and Genomes
ML	Machine learning
MT-DTI	Molecule transformer-drug target interaction
ODs	Orphan diseases
ORCS	Open Repository for CRISPR Screens
R & D	Research and development
ReDO	Repurposing Drugs in Oncology
SIDER	Side Effect Resource
siRNA	Small interfering RNA
TCA	Tricarboxylic acid cycle
TTD	Therapeutic Target Database
VDA	Variant-Disease Association

36.1 Introduction

The development rate of novel drug candidates for treating conditions of critical medical interest such as cancer, chronic metabolic diseases, and infectious diseases face an important shortage and the cost-effective results are poor [1–4]. Indeed, even though every year several papers are published regarding the *in vitro* and *in vivo* evaluation of alternative medications, the pathway to actually test, use and distribute drugs in human diseases is extensive [5]. Besides, due to the rise in the longevity of the population, which implies an increasing trend of both prevalence and incidence of such diseases year by year, as well as the fact that some subjects do not respond appropriately to compounds routinely in use [6, 7], it is required to implement a different therapeutic approach to secure the treatment of patients. In this setting, drug repurposing has gained attention as an alternative for covering the limited success of compounds across all medical fields.

Drug repurposing (DR) is a process in which any clinically available compound, whose target and mechanism of action is known, is used for a different indication other than the initially intended purpose, hence the term "repurposing" [8]. By following a DR approach, pharmaceutical companies and the health care system ensure the reduction in research and development (R & D) costs of compounds, and because the pharmacokinetic, pharmacodynamic, and toxicology profiles are well known, less time is required to make the product available for patients [9]. The traditional drug discovery process is laborious, involving around 10 to 20 years of research and ~US$2.7 billion in investments, and up to 90% of the candidate drugs fail to enter the market because of limited efficacies and undesired secondary effects [5, 10]. Therefore, most experimental drugs fail when they are tested on human beings, particularly in clinical phases II and III [11]. However, DR reduces R & D by up to 5 to 7 years, and the cost could be less than 10% of that required for novelty drugs, which altogether makes repurposable compounds more attractive and a safer investment for R & D [12].

The rigorous R & D process in drug development is highly time-consuming, and the final implementation of any novel compound involves a high risk for investors in the pharmaceutical industry. Knowing this possibility, the implementation of DR protocols provides users with a better management option for each compound. Thus, the key objective of this chapter is to provide readers with the basic information about biomedical research involving the DR strategy. In particular, I will show how to design a protocol covering pre-clinical and clinical phases, starting with searching strategies using publicly available drug databases. In order to do so, an example of a theoretical DR candidate will be described.

36.2 What Types of Drug Repurposing Approaches Exist?

Historically, successful DR options were first discovered through serendipitous approaches in opportunistic situations, when researchers found that certain compounds have failed for their primary intended purpose, but instead had a strong potential for evaluation in an alternative disease [13, 14]. This DR tactic is called 'drug rescue', and its best example is the phosphodiesterase-5 blocker sildenafil citrate. Sildenafil was originally tested for hypertension and angina pectoris [15, 16], but Pfizer Inc. used the retrospective data collected from volunteers in phase I clinical trials and identified its potential for being used in erectile dysfunction [13]. Another example is tamoxifen, which is a routine therapy for estrogen receptor-positive breast cancer patients. A subgroup analysis in this patient population who also suffered from cardiovascular events showed efficacy in reducing the risk of an acute coronary syndrome [17]. Currently, research focusing on the systematic mining of data for drug rescue from retrospective studies in resources including CancerQuest, NCBI Gene, and UniProt, have found secondary uses for other medications. For instance, the adenosine deaminase (ADA) inhibitor dipyridamole, which is prescribed for limiting the formation of blood clots, after a direct inference and with the use of logical interactions of the drug indications, was found to be a potential anticancer agent for acute lymphoblastic leukemia, due to the overexpression of ADA in this pathology [16].

Another resource to find DR candidates is polypharmacology, which is defined as the potential of any given compound to have multiple targets. This term must not be confused with polypharmacy, which is defined as the prescription of multiple drugs for the benefit of the patient. Polypharmacology has unveiled formerly unexpected off-target effects of existing drugs, modifying selected pathways and gene families of treated cells [18, 19]. The fact that drugs are not static molecules, but instead have a highly dynamic activity, means that a myriad of pathways could be affected by them [20]. Polypharmacology implies that one compound is capable of modulating simultaneous cellular pathways, and thus, has more than one possible use [19] such as aspirin, which acts as antipyretic and analgesic but could be added to the treatment regimen against rheumatoid arthritis and colon cancer [21]. Indeed, the deep evaluation and understanding of the mechanisms of action of medications allow researchers to find interesting and unexpected off-target effects, even for already commercially available drugs. Unlike the traditional model in drug discovery, which follows the "one drug-one gene-one disease" paradigm to focus on one condition while avoiding off-target outcomes [19], the multi-target binding allows researchers to expand the impact of the drug [22]. It is a fact that most compounds exhibit a wide spectrum of target activity, either within the expected mechanism, or as a side effect, and this reality could be commercially exploited by pharmaceutical companies to maximize the potential uses of the drugs under research. However, most of the recent attention in the polypharmacology area of DR has relied on kinase inhibitors, particularly those from the serine/threonine and phosphoinositide classes, due to their wide effects on malignant cells [21]. In complex diseases like cancer and Alzheimer's disease, polypharmacology could also be seen as the concomitant treatment with multiple and versatile drugs acting on separated targets from the networks that together regulate the disease [21, 22].

Taken together, DR is useful for preserving the work, investment, and generated knowledge related to drug candidates, which might be abandoned otherwise. In the end, through DR it is possible to provide more patients with additional and more accessible therapy options for their conditions and diseases.

36.3 Benefits and Limitations of Repurposing

Over the recent years, there has been an increasing trend for launching fewer new molecules for treating diseases, following a steep decline of ~50% for each clinical phase [11]. Additionally, average costs related to the R & D of novel compounds, especially small molecules and targeted therapies in oncology, are expected to be on continuous expansion [23]. Instead, DR has been gaining a more relevant niche and nowadays its presence is not negligible. The number of both DR and DR database publications has been increasing over time in a significant way, especially during the last 10 years [24]. Now, there have been established multiple startup enterprises focusing exclusively on DR, and big biotech and pharma companies have created specific branches for researching DR [11]. This has led the pharmaceutical industry to invest up to 10–50% of its total expenditures for R & D solely on DR [25].

The execution of DR offers multiple advantages over the usual *de novo* drug development, such as decreasing the failure rate in clinical evaluation and surpassing safety analysis in preclinical phases, since this information is already known [13]. It is estimated that around 30% of the compounds entering clinical trials ultimately can have a new therapeutic use, implying that even previously unknown DR candidates will finish being tested in an alternative condition [14]. Also, the costs related to DR are more affordable. As a comparison, the reformulation of an existing compound could be as 5 times more expensive as the relaunching of a repurposed drug [18]. Besides, DR has shown an attractive use in orphan diseases (ODs), defined as conditions present in <1 in 2000 to 10,000 subjects, where the lack of interest in R & D investment is common due to the low profits that are expected [12]. The US Food and Drug Administration (FDA) even stimulates pharmaceutical research for compounds intended for the >7000 types of ODs with the application of special regulations beneficial for the industry, and with the implementation of the Rare Disease Repurposing Database [26]. Due to the recent COVID-19 pandemic leading to the high demand for drug research, the Coronavirus Treatment Acceleration Program from the FDA has facilitated access to alternative antimicrobial therapies also based on DR, including remdesivir and nitazoxanide [27], implying that the general population is directly benefited from DR under situations of global emergency.

However, DR is not a perfect approach as it does not fit equally well for all the compounds. As expected, not all diseases disturb tissue homeostasis in the same way, thus confounding the drug-target interaction for each compound and therefore complicating data interpretation if reliance is based on a single molecular activity profile [18]. The blind confidence in DR without further validation is a mistake that must be avoided. Based on the fact that the bioactivity of the drug-target interaction is already known, we could infer and predict further modulations of additional targets, but in reality, the mechanism of action of any given drug is highly dependent on specific contexts and tissues, which under real scenarios demands proper corroboration [13]. Another aspect that is frequently unexploited by researchers is the intellectual property of the drug intended to be repurposed, which therefore means that the selected compounds for DR must be free of patents beforehand [28]. Whether the final formulation of the repurposed medication is patented is another story; but the pharmaceutical industry could be supported by gaining new patents over the novel formulation/doses. Plus, certain resources for DR, such as -omics databases, contain profiles mostly from selected cell lines and they tend to come from limited labs and studies [13], complicating a generalized application of the compound and demanding an initial critical analysis of the therapy options for repurposing.

36.4 Computational Methods to Screen Drugs for Repurposing

The rationale behind DR relies on the complete understanding of the mechanisms of action of the drug, which includes the targets, interactions, and side effects in the subject. The surge of massive data coming from cellular and *in vivo* models has prompted the generation of user-friendly databases collecting the information related to each drug, whose integration employs systematized algorithms easing the screening process for the DR researcher [29]. This is highly valuable because it is estimated that every year over two million research papers are published, limiting our capacity to critically evaluate new data from each compound. However, although data from -omics studies are more abundant every day, the information generated from the patients' physiopathology, and more importantly from clinical studies, can still be scarce for some novel molecules.

Many procedures can be followed when searching for candidates for repurposing, but the most common ones require starting a computational screening of databases. As explained above, multiple specialized databases allow a quick and precise pre-selection of drugs depending on the interest of the researcher. Hence, the screening process using computational methods in the form of databases has the most relevant role, and they allow us to unveil unknown capabilities of compounds [24]. Computational methods are classically categorized as either drug-based or disease-based according to the inference methodology [30], but more recently network-based databases integrate the data more accurately. Additionally, with the evolution of machine learning and deep learning unbiased screening algorithms, the analysis and prediction of drug-drug interactions and even of drug indications has systematically proliferated, deserving their own category as well [29].

In the next paragraphs, I shall cover the most relevant classifications of databases for DR that are currently in use, according to the computational approach employed, and a brief description of machine learning and artificial intelligence will be covered as well (*see* Chap. 31). However, for a deeper and more detailed description of the existing databases, we invite the reader to refer to Tanoli et al. and Masoudi-Sobhanzadeh et al. [13, 24].

36.4.1 Drug-Based Databases

These repositories are based on three main characteristics. First, on chemical similarity, where shared chemical features of two different drugs are expected to yield similar biochemical activity and effectiveness. Second, on molecular activity similarity, where the mechanism of action of the drug in the biological system, supported by -omics experiments such as RNAseq or gene expression microarrays, creates a signature which is then compared against other drugs to evaluate their relationship. Third, on the molecular docking of the compounds, where the goal is to identify previously unidentified associations between the compounds and their possible targets, based on *in silico* modeling and simulations of the physical interactions between them using their 3D structures. If the target is associated with the physiopathology of a certain disease, then the compound could be tested as a candidate for DR [18, 30].

An example of a molecular activity similarity-based approach for DR is the Connectivity Map project (CMap), which collects experimental transcriptomics data from multiple cell types and contains >1,500,000 gene expression profiles from ~5000 small molecule drugs and ~3000 genetic reagents [30, 31]. The creators have housed the data in the cloud-based user-friendly infrastructure named CMap and LINCS Unified Environment (CLUE) (https://clue.io), which depicts a catalog of signatures and off-target profiles of each compound [31]. When similar, such signatures could represent novel biological connections that can be further analyzed by the researcher to build relationships between drugs [30, 31]. Importantly, in CMap when the user finds a shared connection by the compound of interest across several cell types, it could imply that the drug aims to share core cellular events, such as ribosomal function, while cell type-related signature patterns suggest a specialized feature of the medication [31]. On the other hand, DrugBank (https://www.drugbank.ca/) is a comprehensive database that offers, among others, basic chemical information about >500,000 compounds, their gene and protein targets, their 3D structure, and even the pathways associated with each drug [24]. This robust resource was originally released in 2006 for predicting both drug metabolism and interaction. With the addition of data from pharmacogenomics, pharmacometabolomics, pharmacotranscriptomics, and pharmacoproteomics, it now allows users to evaluate *in silico* drug design and molecular docking [32]. Plus, DrugBank has its own DR tool that facilitates even more compound screening.

36.4.2 Disease-Based Databases

These types of databases contain mostly gene-disease associations, but sometimes they can provide clinical data and expected side effects of the therapies applied [15].

An example of a disease database is DisGeNET (https://www.disgenet.org), which is a discovery and versatile platform deeply covering a collection of genes and their variants related to human pathologies, including symptoms, comorbidities, and diseases [33]. DisGeNET, which originally was implemented as a plugin in Cytoscape in 2010, integrates data automatically mined from scientific resources such as research papers and databases including Orphanet, Uniprot, and ClinGen, to provide users with Gene-Disease Association (GDA) and Variant-Disease Association (VDA) information [34]. In DisGeNET, several metrics can be used to prioritize genotype-phenotype relationships, and both the scripts and even an R package are available for further evaluation [35]. The most recent release of the database as of September 2022 (V7.0) contains 1,134,942 GDAs, including 21,671 genes and 30,170 diseases, and 369,554 VDAs, involving 194,515 variants and 14,155 diseases.

Regarding the clinical information, ClinicalTrials (https://clinicaltrials.gov) is a web-based repository supported by the National Library of Medicine and the National Institutes of Health in the US, that allows almost every type of user, such as researchers, health care workers, patients and the public in general, access to information collected from both private and public clinical studies across diseases of human interest, including adverse effects of the interventions. Launched in 2000, ClinicalTrials informs about the registry and the results from initiated, ongoing, and completed clinical studies, either cancelled or finished at term, but it relies on researchers and sponsors for submitting the information to the website [36]. ClinicalTrials currently lists 422,060 studies from the US and from 221 countries, of which 77% are interventional, particularly employing drugs or other biologics.

As its name implies, the Side Effect Resource (SIDER), which was developed in 2010, is a database comprising information about side effects extracted from drug labels (http://sideeffects.embl.de). By summarizing the presence of adverse reactions gathered from placebo-controlled clinical trials and both phase III and IV trials, SIDER is suitable for text-mining data, and predicting and elucidating the relationship between drug targets and their side effects [37]. The most recent version (SIDER 4.1) covers by September 2022, 5868 side effects from 1430 medications generating 139,756 drug-side effects pairs, being only 39.9% of the pairs with frequency information. Based on the range of frequencies of drug-side effect pairs, the resource lists 24,562, 16,765, and 11,784 as frequent, infrequent, and rare, respectively.

36.4.3 Network-Based Databases

Any biological interaction is network-shaped, which makes it coherent trying to evaluate the DR capacity of compounds taking into consideration not only the drug itself but the disease and the possible targets altogether [18]. By understanding the biological pathways that are implicated in the cellular response of the compound and even of combined treatments, it is feasible to comprehend the whole mechanism of action and the therapeutic effects of each medication [13, 38]. These types of databases reduce even more the times and costs of the drug R & D processes, by executing complex network-based prioritization methods using multiple types of interactions and data [18], and with the implementation of models of the effects of diseases and therapies under multiple biological networks, their potential could be exploited even more [38]. An additional useful approach in R is to use the SAveRUNNER network-based algorithm for DR, which predicts drug-disease associations according to a similarity measure [39].

As a freely accessible, curated biomedical repository, the Biological General Repository for Interaction Datasets (BioGRID) integrates gene and protein function and interaction networks [40]. BioGRID was launched in 2002 and it contained data related to both gene and protein roles in the development and physiopathology of diseases beyond human origin, acting as an integrated cross-species database [40]. In 2019 there was an update allowing BioGRID housing chemical interaction data from the DrugBank database, such as chemical-protein interactions for human drug targets [41]. The current version (4.4.213) includes the data of 2,537,592 protein and genetic interactions from 80,848 publications, plus 29,417 chemical interactions and 1,128,339 post-translational modifications (last checked in September 2022) (https://thebiogrid.org). Additionally, the Open Repository for Clustered Regularly Interspaced Short Palindromic Repeats (CRISPR) Screens (ORCS) database, an extension of BioGRID, contains data from both human and mouse cell lines emerging from CRISPR studies [41].

DrugNet (http://genome.ugr.es:9000/drugnet) is a network-based drug-disease database that was established using the ProphNet generic web-based prioritization method which finds associations of interconnected drugs, diseases, and proteins using as starting material for the integration of heterogeneous data and complex networks [18, 24, 38]. DrugNet simultaneously employs data from drugs, targets, and diseases, and generates networks that prioritize drug-disease and disease-drug analysis for discovering novel drug uses [18, 24]. Thus, DrugNet has shown its usefulness for predicting drug-protein interactions for polypharmacology purposes [18]. DrugNet facilitates the searching job by allowing the recovery of results in the form of datasheets for further analysis, and every single entry is linked to secondary databases such as DrugBank and Disease Ontology.

PATHOME-Drug (http://statgen.snu.ac.kr/software/pathome/) is a recently launched subpathway-based polypharmacology DR database, which makes use of drug-related transcriptomes to identify subparts of signaling cascades from the phenotype that get modified upon treatment and incorporates current data from approved compounds to enrich the subparts [42]. PATHOME-Drug integrates data from comprehensive resources such as DrugBank and PharmGKB (see Chap. 19) and allows users to have access to source codes and data as well.

Finally, the Kyoto Encyclopedia of Genes and Genomes (KEGG) (http://www.genome.jp/kegg/) is a highly comprehensive biological pathway database that contains information from the target pathways of several compounds, as well as their interactions in the cell, integrating chemical, genomic and systematic functional data [13]. By using KEGG it is possible to model signaling pathways, where nodes are the representation of a given protein or gene, and how they interact with each other is presented through edges, which are weighted according to the stresses they receive, including activation and inhibitory signals [43]. The KEGG database holds 552 pathways from 970,216 references, where there could be found 18,966 metabolites and other chemical substances, as well as 1289 disease-related network elements (last checked in September 2022). KEGG also holds a drug-target interaction database termed the KEGG Drug Database, which indicates the specific node from the pathway that is being affected by the drug of interest. The KEGG Drug Database covers as of September 2022, 11,965 drugs from which 6454 have their targets annotated, and 1983 of them are linked to FDA drug labels (https://www.genome.jp/kegg/drug/).

36.4.4 Machine Learning and Artificial Intelligence Survey for Repurposing

The striking growth of publicly available datasets related to signaling in diseases, therapeutics, side effects, and possible targets, has demanded the development of state-of-the-art methods for accelerating DR in an unbiased, systematic, and efficient way [44]. By implementing in the databases machine learning (ML), deep learning (DL), and artificial intelligence (AI) algorithms, such as those relying on matrix completion and factorization, as well as on network propagation, it is possible to systematically identify DR candidates from the information related to their off-target effects from big data repositories [13]. Thus, ML, DL, and AI can be seen as robust tools to maximize the recovery and integration of data for the DR researcher (also see Chap. 31).

The realization that data coming from biological networks can be vectorized implies that ML unsupervised and supervised algorithms are useful to elucidate hidden relationships between drugs and targets and to do so, we can make use of graph-embedding tools such as network propagation to reduce the dimensional representation of graphs while preserving its properties and connections [38]. Indeed, after finding the information from databases like KEGG, DisGeNET, or DrugBank, the algorithms could then represent it in the form of informative feature vectors through different representation approaches like sequences, graphs, and text. Further DL processing using recurrent, and graph deep neural networks provides the summarized data to researchers for proper validation [45]. Some approaches currently in use include the molecule transformer-drug target interaction (MT-DTI) model that is based on pre-trained drug-target interactions, which illuminates the affinity of a determined drug and the possible targets [44]; and Mol2vec, which is an unsupervised ML language model that identifies molecular substructures as words and drugs as sentences for further training and uses in combination with the protein-vector ProtVec algorithm for interaction prediction [45, 46].

36.5 Protocol to Evaluate Repurposable Molecules

A regular DR protocol consists of three basic steps which should be followed with any candidate before moving into the late stages of development, approval, and commercialization, being **(a) hypothesis**, where the identification of the candidate compound which will cover the desired indication or the idealization of a novel use for a specific compound, is done; **(b) validation**, in which all the pre-clinical analyses including cell-based pharmacogenomic approaches are performed to evaluate the mechanism of action, the doses, and the overall feasibility of the molecule; and **(c) efficacy confirmation**, by testing the drug in phase II clinical trials if previous data already exists related to safety from the original indication of the DR candidate [47]. A summary of a general protocol is shown in Fig. 36.1.

36.5.1 Hypothesis

This is the most important step. Based on the presumptive information of any drug of interest, or on the disease that we want to target, we should carefully select from an integrative database the possible options for DR. It is suggested to corroborate the results using more than one database and search beforehand for possible side-effects that are already reported for the DR candidates, which will aid us to pre-select medications for further evaluation. The more information available, the better.

36.5.2 Validation

The first step is to corroborate whether the drug does what it is predicted to do. If our hypothesis covers a particular type of target and disease, like the mutated form of isocitrate dehydrogenase I (IDH1) in human glioma (Fig. 36.2a, b), the most logical approach is to start testing the DR candidates in human cell lines of this same condition. Following this example and based on the DeepMap portal (https://depmap.org/portal/interactive/), the Becker cell line could be useful to this end because of its high *IDH1* expression (Fig. 36.2c). Further information about the cell line to be employed should be analyzed in resources such as Expasy (https://web.expasy.org/cellosaurus/CVCL_1093). Hence, we could validate the drugs using Becker cells following the next steps.

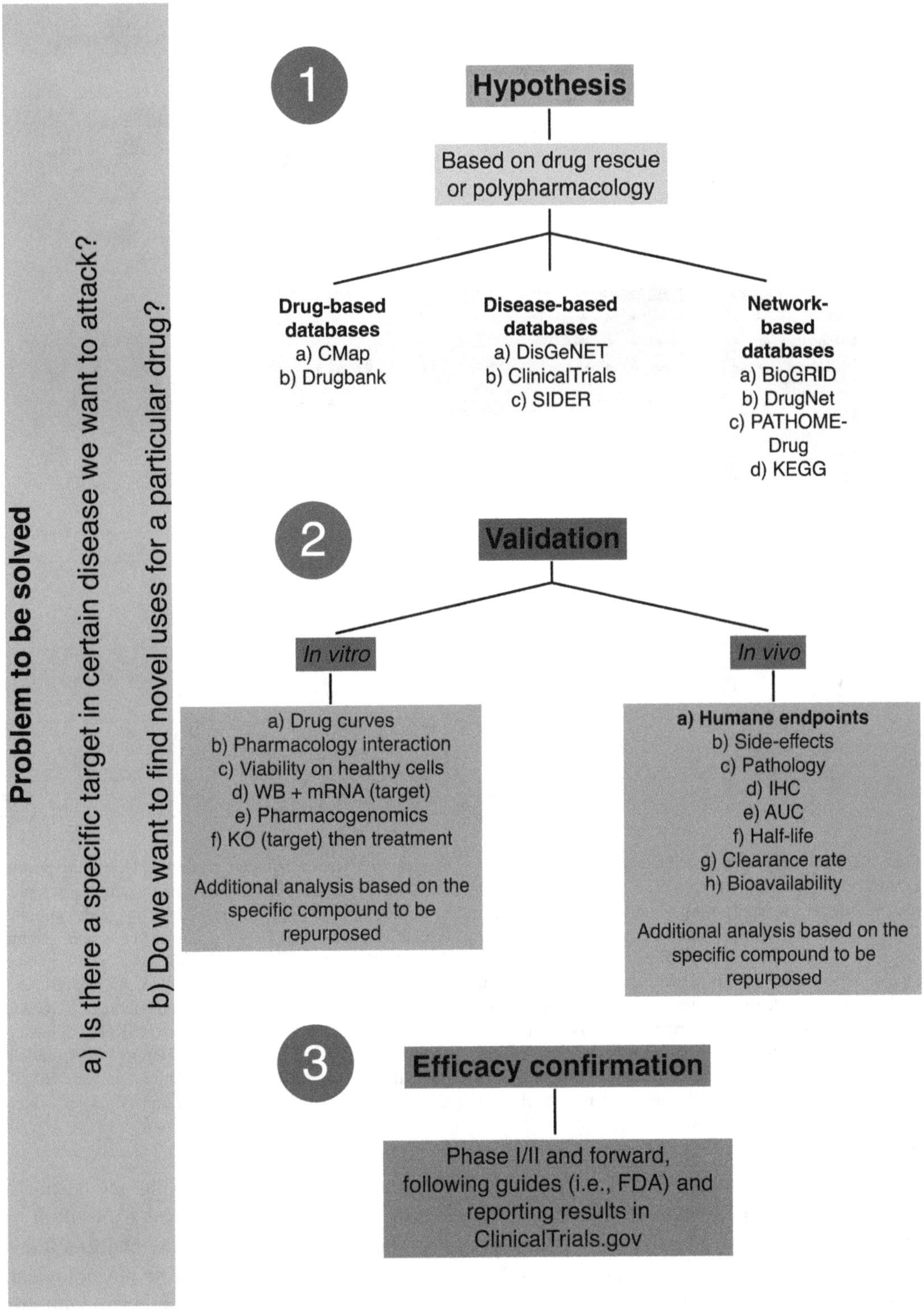

Problem to be solved

a) Is there a specific target in certain disease we want to attack?

b) Do we want to find novel uses for a particular drug?

1 **Hypothesis**

Based on drug rescue or polypharmacology

Drug-based databases
a) CMap
b) Drugbank

Disease-based databases
a) DisGeNET
b) ClinicalTrials
c) SIDER

Network-based databases
a) BioGRID
b) DrugNet
c) PATHOME-Drug
d) KEGG

2 **Validation**

In vitro

a) Drug curves
b) Pharmacology interaction
c) Viability on healthy cells
d) WB + mRNA (target)
e) Pharmacogenomics
f) KO (target) then treatment

Additional analysis based on the specific compound to be repurposed

In vivo

a) Humane endpoints
b) Side-effects
c) Pathology
d) IHC
e) AUC
f) Half-life
g) Clearance rate
h) Bioavailability

Additional analysis based on the specific compound to be repurposed

3 **Efficacy confirmation**

Phase I/II and forward, following guides (i.e., FDA) and reporting results in ClinicalTrials.gov

Fig. 36.1 Protocol for identifying and evaluating the suitability of a drug for repurposing. The regular procedure is straightforward, starting with a hypothesis pre-selecting a candidate in databases based on either drug rescue or polypharmacology; following with the experimental validation of the candidate both *in vitro* and *in vivo*; and finally confirming the efficacy in real human subjects suffering the condition that we want to target

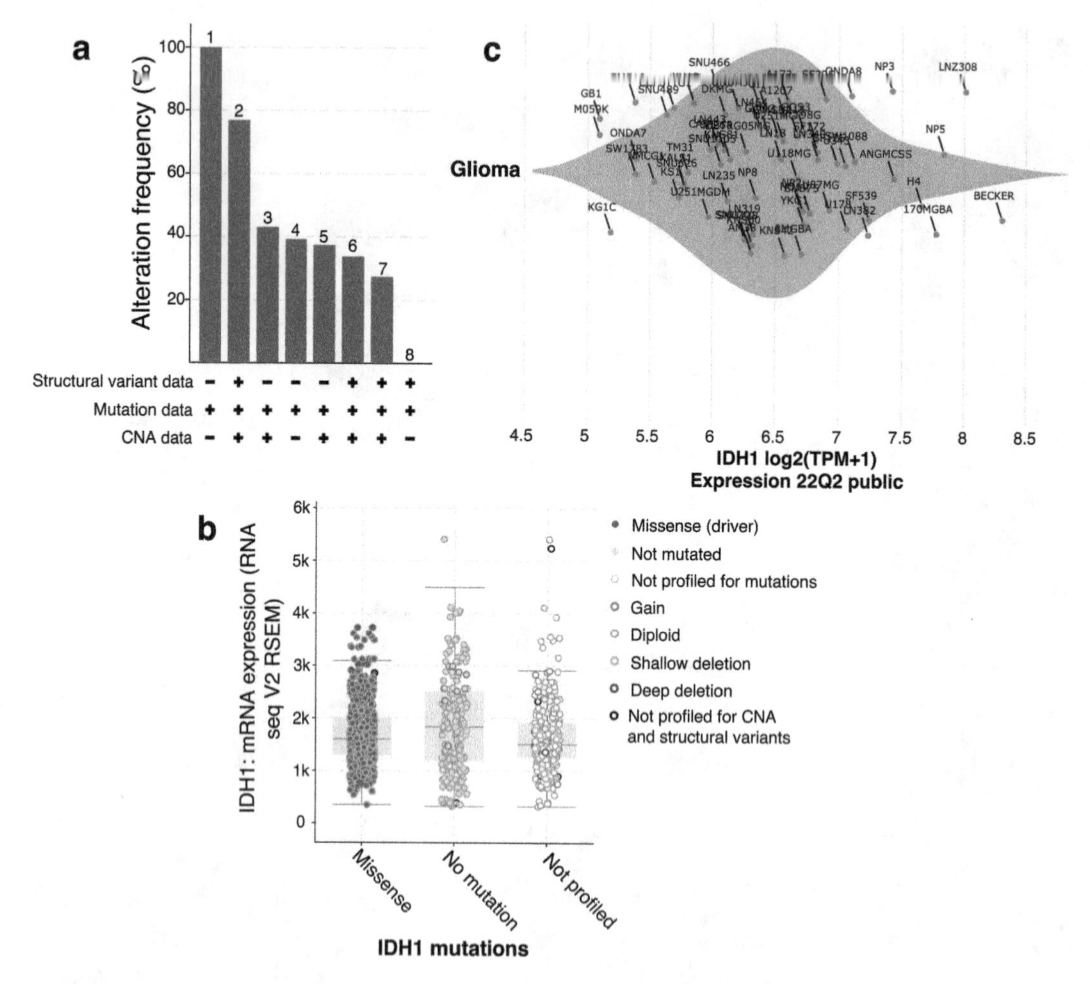

Fig. 36.2 Selection of a cell line for drug repurposing evaluation in oncology. (**a**) An initial analysis identifies a potential target for repurposing, which in this case is the IDH1 enzyme in glioma subjects. The alteration frequency for IDH1 is shown here in (**a**), while the type of alteration/mutation is shown in (**b**). Data shows that the main alterations are missense mutations and overexpression of the enzyme. (**c**) The mean differential expression of IDH1 shows that the human cell line with the highest expression is Becker. (**a**, **b**) were concluded from cBioPortal, covering 3862 samples from 3459 patients across 8 clinical studies. (**c**) was concluded from the DeepMap portal, covering 70 human glioma cell lines. In a, numbers represent the next tumor subtypes: 1: Low-grade gliomas (UCSF, Science 2014); 2: brain lower grade glioma (TCGA, PanCancer Atlas); 3: brain lower grade glioma (TCGA, Firehose Legacy); 4: diffuse glioma (GLASS Consortium, Nature 2019); 5: merged cohort of LGG and GBM (TCGA, Cell 2016); 6: glioma (MSKCC, Clin Cancer Res 2019); 7: glioma (MSK, Nature 2019); and 8: pilocytic astrocytoma (ICGC, Nature Genetics 2013). *CNA* copy number alterations, *TPM* transcripts per million

36.5.2.1 Drug Curves

A basic evaluation involves drug curves in a logarithmic scale to identify the inhibitory concentrations $(IC)_{10-50}$, or the drug doses that reduce cell numbers by 10–50%, respectively, as compared to the control which should be the vehicle of the drug. As a matter of comparison, concomitant drug curves using the standard treatment should be performed as well. It is important to remember that the concentrations that we will test later should be physiological, meaning that when a patient is treated with this compound, it will be possible to find it in circulation at the same range we are evaluating it. Here, if the DR candidates exert their antineoplastic activity in glioma cells in a range within what is expectable in subjects, and also if that same range has demonstrated a lack of toxicity, then we could conclude that the drugs are feasible for repurposing, and we could move forward.

36.5.2.2 Pharmacology Interaction

An important part of DR is not only finding novel compounds but also testing whether current treatments could be used in combination with repurposable drugs to improve their effectiveness. Therefore, the next step should be the assessment of drug interaction between our candidates and the standard treatment at the different ICs found in the previous step, and based on the final cell numbers and using isobologram analyses, we could interpret the pharmacology interaction as an antagonist, additive or synergistic. We should keep the best doses based on a possible synergistic interaction for further studies.

36.5.2.3 Safety in Healthy Cells

Even though we must support the pre-selection of DR candidates based on current data showing safety in patients, it is always recommended to corroborate its suitability in healthy cells using our compound under the same doses we are planning to test it in the disease we are interested in. Since we are showing here an example with human cancer cells, an appropriate approach should be to simultaneously test the final doses in human fibroblast cells and even in healthy leukocytes collected from donors. Including healthy neurons in this case in specific could be an interesting idea. We expect healthy cells a higher survival rate than neoplastic ones.

36.5.2.4 Effects on the Desired Targets

If we suspect possible polypharmacology effects by our DR candidates, we could expect that such medication can attack more than only the desired enzyme we originally aimed to. Although neither mRNA nor protein expression of the targets is a logical way to show quantitatively the effects of the drug, a more accurate approach is to demonstrate it at both levels. Plus, by doing so, we could also decipher whether any possible differences seen in these biomolecules are generated by transcriptional or translational regulations, which could lead us to additional analysis to improve our knowledge about the drug. If the results demonstrate that our targets are even more affected when assayed in a combinatory approach between the candidate(s) and the standard treatment evaluated at synergistic doses, then we could encourage the simultaneous use of both drugs.

36.5.2.5 Global-Scale Outcomes

If possible, massive analysis of treated cells such as RNAseq and RNA microarrays must be done to illuminate alternative pathways that are altered due to the drug and the specific dose we are testing. Then, the subsequent profile against what is already known for such current treatment should be compared, to corroborate consistent data and delineate new targets that explain our results.

36.5.2.6 Validation of the Target

In theory, if what we find indicates that the DR candidates aim specifically for the target we want, then by attacking the same target beforehand it is expected a lack of response of cells upon treatment with our candidate(s). If the target is not a housekeeping gene/protein, then knocking it out should not kill the cells directly but would instead protect the cells from the treatment. We could first evaluate the effects on cell numbers after, for example, the use of a small interfering RNA (siRNA) or CRISPR against *IDH1*. Next, after demonstrating the feasibility of establishing a stable cell line with the downregulation of this target, we should add the same doses of the DR candidate(s). Alternatively, we may deplete the media of treated cells from the substrate of the target, which in this case could be citrate as IDH1 isomerizes citrate in the Krebs cycle, and then could add labeled citrate to compare the saturation kinetics of the substrate between control and treated cells.

36.5.2.7 *In Vivo* Experiments

A vital step when repurposing a compound is its evaluation with *in vivo* models. In this scenario, we must make use in an ethical way of mice inoculated with the same cell line we are using, and to analyze variability seen in real subjects and the real structure of the tumors, it would be desirable to evaluate mice with patient-derived xenotransplants as well [48].

Since we are aiming at DR in oncology in this example, the most elementary exploration is the progression of tumor volume over time, when comparing non-treated against treated mice with our candidate, with the standard treatment, and also with both drugs together. It is fundamental, however, to take into consideration humane endpoints to identify signs of distress and suffering in the model of study, including but not limited to alterations in their social interaction, weight loss, lack of appetite, abnormal respiration, and piloerection upon the treatment we are adding. When possible, circulating markers indicative of suffering and inflammation, such as interferon-γ, C-reactive protein, tumor necrosis factor-α, and interleukin-6, must be quantified as well [49]. If the clinical examination of mice suggests any of these signs through the timeline of the study, their sacrifice is mandatory, and therefore an alternative approach should be taken, such as different inoculation routes, shorter treatment times, or the use of another vehicle(s). If these tactics are not enough, then the compounds cannot be further studied. Otherwise, we could continue with the pathological analysis of both tumor slices and healthy surroundings, and to do so the aid of a pathologist working under blindness conditions must be warranted to maximize a critical analysis of the characterization of the tumor and to grade it correctly. Immunohistochemistry aiming at the target of our DR candidate, IDH1 in our example, and also of prognostic markers such as Ki-67 for cell division, are highly recommended. The pharmacological characterization of the compounds of interest, performing analyses such as the area under the curve, half-life, clearance rate, and bioavailability, must be evaluated as well when used either alone or in combination with the standard treatment. At this point, if we had >1 DR candidate, we could select the most promissory one based on our data for further evaluation.

Importantly, supplementary experimentation based on the particular type of candidate should be performed to enrich our understanding of its mechanisms of action and its efficacy under the conditions we are evaluating. In the case we are presenting here since we are planning to test potentially antineoplastic medications, we could expect to assess cell cycle and apoptosis, to identify whether the reduction in cell numbers seen at the doses we are proposing is associated directly with cell death or with a cytostatic effect. If we suspect a possible secondary pathway explaining the phenotype, we are seeing in our treated cells either with the DR candidate alone or when used in combination with the standard therapy, such as a potential reduction in the cell migration capacity, the stemness, or in the angiogenic potential of cells, we must then perform the corresponding assays. For this case specifically, when we pretend to target a metabolic enzyme, some useful experiments that could be implemented are the analysis of TCA intermediates by GC/MS and LC/MS, and the effects on the energetic phenotype from glycolysis and oxidative phosphorylation. For additional metabolic experiments in cancer mouse models treated with DR compounds, the user is invited to read [50].

36.5.3 Efficacy Confirmation

Finally, and only after validating our DR candidate at the doses and conditions we are testing without relevant secondary effects using *in vivo* models, we could proceed to test its efficacy with real patients. The bottleneck in this step, and the most important part that we should focus on, is the recruitment and caring of patients. According to the FDA, for phase II clinical studies, considering that the medication we want to repurpose already has substantial information from phase I clinical research, we should recruit up to several hundred people suffering from the condition we want to address. In this sense, for our example, we must test the DR candidate in hundreds of patients with glioma, and the study usually lasts from several months up to 2 years. If our compound surpasses this step, meaning that the efficacy was checked, we can move into phase 3 recruiting up to 3000 volunteers with the disease, lasting from 1 to 4 years. Finally, after monitoring the efficacy and possible adverse reactions, our compound enters phase 4 with several thousand volunteers and we can request the final approval by the right governmental regulatory instance, like the FDA, for its massive implementation (also see Chaps. 20–22).

We must ensure transparency in our data. A correct way to do so is to share the information from our study in ClinicalTrials, which as discussed above, compiles data from clinical studies. Importantly, Clinicaltrials.gov supports users with the US and international guides and policies about regulations of clinical trials, as well as how to register the generated data on the website.

36.6 Examples of Drug Repurposing in Clinical Conditions of Critical Interest

In the next subsections, I will briefly cover some DR efforts for cancer, diabetes, and tuberculosis, three clinical disorders of critical interest in human medicine. A further description of additional compounds that have been repurposed for these conditions is shown in Table 36.1.

36.6.1 Cancer

Cancer is one, if not the most, targetable condition in DR research. Only in 2022 and in the US, ~2 million new cancer patients are expected to be diagnosed, with estimations of >600,000 deaths directly related to any type of cancer [70]. In the world, however, data from GLOBOCAN 2020 suggested 19.3 million new cancer cases and ~10 million cancer deaths, and alarming, cancer remains the first-second leading cause of death according to the World Health Organization, with a non-stopping increasing trend worldwide [71]. Due in part to the high rates of resistance over time using the conventional therapies and to the low success rate of around 6.7% for novel compounds in phase I trials [72], cancer is an excellent model for testing old medications and providing them a novel use [73–75]. This has led proposals like the Repurposing Drugs in Oncology (ReDO) Project database (https://www.anticancerfund.org/en/redo-db) to bring back to the light previously non-anticancer drugs to be evaluated against malignant cells, based on their widespread clinical use, their good toxicology profile, their known mechanism of action that has proved to inhibit a particular pathway, their evidence about efficacy at physiological dosing, and finally, their overall strong evidence *in vitro*, *in vivo* and with human data [72, 76]. The ReDO Project initially suggested mebendazole, nitroglycerin, cimetidine, clarithromycin, diclofenac, and itraconazole as potential candidates for DR in oncology, and as of today it enlists 366 drugs against 1144 unique molecular targets, 85.5% of which are off-patent, 92.1% show *in vitro* evidence, and 65.8% have been tested in clinical trials [72]. Diabetes and cancer are highly interconnected, as they are both metabolic diseases and, not surprisingly, the risk factors and the metabolic pathways implicated in their physiopathology are also related between both [51]. This has allowed antidiabetic drugs to be tested in cancer, including thiazolidinediones and importantly, biguanides such as metformin and phenformin [77].

36.6.2 Diabetes

Around 537 million adults suffer from diabetes in the world, of which 75% live in low and middle-income countries, and the numbers are expected to continue following a cumulative trend [78]. Concomitantly, a limited number of novel options have been launched against diabetes, and people living under unfavorable conditions have limited access to therapies that could otherwise restrict the progression of associated conditions

Table 36.1 Examples of drugs repurposed for treating cancer, diabetes, and tuberculosis

Condition	Repurposed drug	Original indication	Original mechanism (s) of action	Mechanism(s) of action in the repurposed condition	Reference
Cancer	Metformin	Diabetes	Improvement of circulating glucose uptake by targeting AMPK in muscle cells and thus increasing their expression of GLUT4; inhibition of liver gluconeogenesis through the interaction with PGC-1α	Risk reduction of cancer in patients with type 2 diabetes through targeting the upstream regulator of AMPK, the LKB1 protein kinase. It is also a cytostatic agent; promotes DNA damage; decreases progenitor capacity of CSCs; has high activity against hormone receptor-positive breast cancer luminal cells; inhibits mTOR signaling and protein synthesis, and blocks the mitochondrial complex I	[51–53]
	Disulfiram	Alcoholism	Irreversible blockade of the aldehyde dehydrogenase, limiting the oxidation of alcohol and increasing 5–10 times circulating levels of acetaldehyde. This produces an unpleasant sensitivity towards alcohol	When used in combination with copper, it forms the anti-cancer metabolite copper diethyldithiocarbamate complex which immobilizes NPL4 and blocks both the p97-dependent ubiquitin-proteasome system and NF-κB, to inhibit stemness, metastasis, angiogenesis, and drug resistance. It has a broad spectrum against malignancies	[54, 55]
	Atovaquone	Antimicrobial against *Plasmodium falciparum*, *Toxoplasma gondii,* and *Pneumocystis jirovecii*	Inhibition of the CoQ10-dependence of mitochondrial complex III	Inhibition of the CoQ10-dependence of mitochondrial complex III shifting metabolism from OXPHOS into glycolysis without promoting EMT; high activity against CSCs; promotion of oxidative stress; reduction in hypoxia-regulated genes like *ALDH1C*; promotion of stress response-related genes like *ATF3*	[56, 57]

(continued)

Table 36.1 (continued)

Condition	Repurposed drug	Original indication	Original mechanism (s) of action	Mechanism(s) of action in the repurposed condition	Reference
Diabetes	Bifeprunox	Schizophrenia, Parkinson's disease and psychosis	Atypical antipsychotics act as a partial dopamine D2 agonist, decreasing the overstimulated dopamine receptors and stimulating underactive ones. It also acts as a 5-HT1A agonist to counter extrapyramidal symptoms	5-HT1A agonist when insulin secretion is impaired	[58]
	Niclosamide ethanolamine	Anthelmintic against *Taenia solium*, *Taenia saginata*, *Fasciolopsis buski*, *Diphyllobothrium latum*, and *Schistosoma* spp.	Uncoupling mitochondria of parasites prevent the synthesis of ATP in adult worms	Mitochondrial uncoupling agent; promoter of lipid metabolism; increase in whole-body energy expenditure; amelioration of diabetes-related muscle wasting blockade of autophagy-related proteins including FoxO3a, LC3B-II, and p-ULK1	[59, 60]
	Pentoxifylline	Atherosclerosis-, diabetes-, inflammatory- and functional-related circulatory diseases, including cerebrovascular insufficiency, intermittent claudication and peripheral arteriopathy	It is an anti-inflammatory and anti-oxidant compound acting as a PDE inhibitor that increases cAMP levels, lowering the blood viscosity by promoting erythrocyte flexibility; blocking neutrophil activation; limiting plasma fibrinogen, and ablating erythrocyte/platelet aggregation	Improvement of insulin resistance and reduction of proteinuria in diabetic nephropathy	[61, 62]
Tuberculosis	Bortezomib	Resistant multiple myeloma and mantle cell lymphoma	It is a boronic acid derivative acting as a reversible 26S proteasome inhibitor, blocking the aberrant proteasome-mediated proteolysis in malignant cells which induces	*M. tuberculosis* is the only bacterial pathogen with a proteasome compartment. Bortezomib has bactericidal activity and blocks the mycobacterial caseinolytic proteases P1 and P2. To avoid	[63, 64]

(continued)

Table 36.1 (continued)

Condition	Repurposed drug	Original indication	Original mechanism (s) of action	Mechanism(s) of action in the repurposed condition	Reference
			apoptosis and cell cycle arrest at the G_2-M phase	toxicities, the dipeptidyl-boronate derivative of bortezomib is 100-fold less active against human proteasomes and still is effective against *M. tuberculosis*	
	Simvastatin	Preventive agent against myocardial infarction, coronary death, hypercholesterolemia, and hyperlipidemias types IIa, IIb, III, and IV	A competitive HMG-CoA reductase inhibitor belonging to the statin class of drugs, which reduces the endogenous synthesis of cholesterol, LDL, and VLDL in the liver, modulating lipid levels	It improves the *in vivo* activity of isoniazid, rifampicin, and pyrazinamide, which are first-line anti-tuberculosis drug regimens, by increasing bacillary killing and limiting the number of lung CFUs. This effect is linked to the promotion of autophagy and phagosome maturation in macrophages, as well as to a reduction in membrane cholesterol and mevalonate	[65, 66]
	Ebselen	Ménière disease, and both types 2 and 1 diabetes	It is a molecule mimicking the glutathione peroxidase activity thus reducing ROS. It inhibits lipid peroxidation and promotes both the oxidation of glutathione thiol and the reduction of hydrogen peroxide into water. Among others, it also targets NOS, glutamate dehydrogenase, and lactate dehydrogenase	It blocks the antigen 85 complex, which synthesizes elements of the outer and inner leaflets of the mycobacterial outer membrane. It also reduces the secretion of the membranolytic agent ESAT-6 through the virulence determinant ESX-1, which is required for granuloma formation and therefore for mycobacterial survival	[67–69]

AMPK AMP-activated protein kinase, *GLUT4* glucose transporter 4, *PGC-1α* peroxisome proliferator-activated receptor gamma co-activator 1α, *LKB1* liver kinase B1, *NPL4* nuclear protein localization 4, *NF-κB* nuclear factor κ-light-chain-enhancer of activated B cells, *CSCs* cancer stem cells, *OXPHOS* oxidative phosphorylation, *EMT* epithelial-to-mesenchymal transition, *5-HT1A* serotonin-1A receptor, *PDE* cyclic-3′,5′-phosphodiesterase, *cAMP* cyclic adenosine monophosphate, *HMG-CoA* hydroxymethylglutaryl-coenzyme A, *LDL* low-density lipoprotein, *VLDL* very low-density lipoprotein, *ROS* reactive oxygen species, *NOS* nitric oxide synthase, *CFUs* colony-forming units such as blindness, amputations, cardiovascular disease, and end-stage kidney disease [3]. Plus, the chronic use of conventional treatments such as sulfonylureas has important adverse effects, including weight gain, pancreatic β-cell overload, and hypoglycemia [79]. Thus, approaches are trying to bring DR molecules to diabetes patients, such as data mining and the Therapeutic Target Database (TTD) which have given us interesting candidates [58]. Medications like salicylate reduce the low-grade constant state of inflammation seen in diabetes, improving insulin sensitivity; the antiallergic drug amlexanox has demonstrated activity against upstream kinases of the NF-κB pathway, promoting energy expenditure and limiting steatosis; the second-generation bile acid sequestrant agent colesevelam, which minimizes serum lipid concentration while stabilizing glycemic control [79]; and the lipid control and cardiovascular protective drug icosapent ethyl, which diminishes triglycerides levels, additionally prevents insulin resistance and glucose intolerance related to the high-fat diet, lowering the risk for developing type 2 diabetes [80]. With the analysis of data from genome-wide association studies, metabolomics, and proteomics, Zhang et al. reported 992 potential anti-diabetic protein targets in humans, and after selecting 35 targetable proteins with drug projects information with the use of TTD, they found in CMap phenoxybenzamine and idazoxan against the α-2A adrenergic receptor in type 2 diabetes [58].

36.6.3 Tuberculosis

Tuberculosis is a worldwide-spread infectious disease generated by the bacilli *Mycobacterium tuberculosis*. It is still highly prevalent in subjects living under adverse conditions, such as those in underdeveloped countries with a high population density, and more recently in immunosuppressed patients affected by the acquired immunodeficiency syndrome (AIDS). The first-line anti-tuberculosis compounds, named rifampicin, pyrazinamide, isoniazid, and ethambutol, are generally successful to eradicate the bacterium, but when multidrug- and extensively drug-resistant strains surge, limited therapy options are available [81]. Tuberculosis has not gained as much attention as other diseases such as cancer for the identification of DR candidates. Nevertheless, there are some databases for repurposing in infectious diseases, such as ReFRAME (https://reframedb.org), which is an open access catalog established mainly through data mining by researchers at the California Institute for Biomedical Research in La Jolla [82]. In ReFRAME, at least 66 chemicals are suggested to have anti-tuberculosis activity. Now it is known that the antibacterial drugs vancomycin and linezolid, and even the anti-inflammatory compound celecoxib, have anti-tuberculosis properties [83]. The fluoroquinolone moxifloxacin, on the other hand, has a high bactericidal activity and when the classical anti-tuberculosis agent ethambutol was substituted by moxifloxacin in a phase IIb clinical trial, there were greater chances of culture-negative sputum collected from Africa and North America patients [84]. Interestingly, even though the oxazolidinone antibiotic linezolid has proven to be beneficial in multidrug-resistant and extensively resistant tuberculosis strains, the presence of hematologic and neurotoxicity side effects have discouraged its use alone [85]. However, when combined with pretomanid and bedaquiline, 90% of treated subjects reported a favorable outcome at 6 months of use [86], stimulating the research for combinatory regimens using DR agents in tuberculosis. Additional non-antibiotic medications, including the cyclooxygenase-2-inhibitors etoricoxib and ibuprofen, and more recently the cytochrome bc1 complex blocker telacebec [81, 87], are promissory anti-tuberculosis compounds as well.

36.7 Concluding Remarks

There is high interest by both the pharma industry and clinicians to evaluate the potential DR capacities of old compounds, a tendency that continues rising over time due to all the proven benefits of DR. Given the fact that every day more data are being gathered from pharmacology -omics studies in human cells and subjects, the systematic and unbiased analysis covering them is a task that has demanded the proliferation of highly specialized databases which now ease the pre-selection of potential DR candidates. However, current databases for DR can still be improved. There are some limitations including the constant requirement for updates; the addition of plug-ins that facilitate the direct search for DR candidates; the fact that some databases lack an adequate internal classification grouping the compounds they present to facilitate their access; and the programming interface, which sometimes could make it hard for researchers to perform an adequate data mining for finding a suitable drug for their specific interests [24]. Hence, any researcher interested in DR must perform an adequate identification of compounds according to the target/disease they aim for, or in an inverse way, to analyze the potential DR capacity of the medication they work with. The databases and the protocol shown here are basic approaches for this end. Even though I did not cover them in this chapter, further pathway enrichment of the biological effects related to the use of the compound of interest can be done using tools such as REACTOME, the Drug Gene Interaction Database, and even with gene ontology terms [39].

Acknowledgments Alejandro Schcolnik-Cabrera would like to thank the Fonds de Recherche Santé du Québec (FRQS) for the granted postdoctoral fellowship (307595).

Conflicts of Interest None.

References

1. Cherla A, Renwick M, Jha A, Mossialos E (2020) Cost-effectiveness of cancer drugs: comparative analysis of the United States and England. EClinicalMedicine 29-30:100625
2. Hilal T, Gonzalez-Velez M, Prasad V (2020) Limitations in clinical trials leading to anticancer drug approvals by the US Food and Drug Administration. JAMA Intern Med 180(8):1108–1115
3. Taylor SI (2020) The high cost of diabetes drugs: disparate impact on the most vulnerable patients. Diabetes Care 43(10):2330–2332
4. Miller BR, Nguyen H, Hu CJ, Lin C, Nguyen QT (2014) New and emerging drugs and targets for type 2 diabetes: reviewing the evidence. Am Health Drug Benef 7(8):452–463
5. Hodos RA, Kidd BA, Shameer K, Readhead BP, Dudley JT (2016) In silico methods for drug repurposing and pharmacology. Wiley Interdiscip Rev Syst Biol Med 8(3):186–210
6. Global Burden of Disease Cancer Collaboration, Kocarnik JM, Compton K, Dean FE, Fu W, Gaw BL, Harvey JD, Henrikson HJ, Lu D, Pennini A et al (2022) Cancer incidence, mortality, years of life lost, years lived with disability, and disability-adjusted life years for 29 cancer groups from 2010 to 2019: a systematic analysis for the global burden of disease study 2019. JAMA Oncol 8(3):420–444
7. Spann SJ, Ottinger MA (2018) Longevity, metabolic disease, and community health. Prog Mol Biol Transl Sci 155:1–9
8. Malik JA, Ahmed S, Jan B, Bender O, Al Hagbani T, Alqarni A, Anwar S (2022) Drugs repurposed: an advanced step towards the treatment of breast cancer and associated challenges. Biomed Pharmacother 145: 112375
9. Pushpakom S, Iorio F, Eyers PA, Escott KJ, Hopper S, Wells A, Doig A, Guilliams T, Latimer J, McNamee C et al (2019) Drug repurposing: progress, challenges and recommendations. Nat Rev Drug Discov 18(1): 41–58
10. Pulley JM, Rhoads JP, Jerome RN, Challa AP, Erreger KB, Joly MM, Lavieri RR, Perry KE, Zaleski NM, Shirey-Rice JK et al (2020) Using what we already have: uncovering new drug repurposing strategies in existing omics data. Annu Rev Pharmacol Toxicol 60:333–352
11. Novac N (2013) Challenges and opportunities of drug repositioning. Trends Pharmacol Sci 34(5):267–272
12. Juarez-Lopez D, Schcolnik-Cabrera A (2021) Drug repurposing: considerations to surpass while re-directing old compounds for new treatments. Arch Med Res 52(3):243–251
13. Tanoli Z, Vaha-Koskela M, Aittokallio T (2021) Artificial intelligence, machine learning, and drug repurposing in cancer. Expert Opin Drug Discov 16(9):977–989
14. Cavalla D, Singal C (2012) Retrospective clinical analysis for drug rescue: for new indications or stratified patient groups. Drug Discov Today 17(3–4):104–109
15. Tanoli Z, Seemab U, Scherer A, Wennerberg K, Tang J, Vaha-Koskela M (2021) Exploration of databases and methods supporting drug repurposing: a comprehensive survey. Brief Bioinform 22(2): 1656–1678
16. Tari LB, Patel JH (2014) Systematic drug repurposing through text mining. Methods Mol Biol 1159:253–267
17. Choi SH, Kim KE, Park Y, Ju YW, Jung JG, Lee ES, Lee HB, Han W, Noh DY, Yoon HJ et al (2020) Effects of tamoxifen and aromatase inhibitors on the risk of acute coronary syndrome in elderly breast cancer patients: an analysis of nationwide data. Breast 54: 25–30

18. Martinez V, Navarro C, Cano C, Fajardo W, Blanco A (2015) DrugNet: network-based drug-disease prioritization by integrating heterogeneous data. Artif Intell Med 63(1):41–49
19. Rivero-Garcia I, Castresana-Aguirre M, Guglielmo L, Guala D, Sonnhammer ELL (2021) Drug repurposing improves disease targeting 11-fold and can be augmented by network module targeting, applied to COVID-19. Sci Rep 11(1):20687
20. Hernandez-Lemus E, Martinez-Garcia M (2020) Pathway-based drug-repurposing schemes in cancer: the role of translational bioinformatics. Front Oncol 10:605680
21. Reddy AS, Zhang S (2013) Polypharmacology: drug discovery for the future. Expert Rev Clin Pharmacol 6(1):41–47
22. Chaudhari R, Fong LW, Tan Z, Huang B, Zhang S (2020) An up-to-date overview of computational polypharmacology in modern drug discovery. Expert Opin Drug Discov 15(9):1025–1044
23. Schlander M, Hernandez-Villafuerte K, Cheng CY, Mestre-Ferrandiz J, Baumann M (2021) How much does it cost to research and develop a new drug? A systematic review and assessment. PharmacoEconomics 39(11):1243–1269
24. Masoudi-Sobhanzadeh Y, Omidi Y, Amanlou M, Masoudi-Nejad A (2020) Drug databases and their contributions to drug repurposing. Genomics 112(2): 1087–1095
25. Zhu Y, Che C, Jin B, Zhang N, Su C, Wang F (2020) Knowledge-driven drug repurposing using a comprehensive drug knowledge graph. Health Informatics J 26(4):2737–2750
26. Sardana D, Zhu C, Zhang M, Gudivada RC, Yang L, Jegga AG (2011) Drug repositioning for orphan diseases. Brief Bioinform 12(4):346–356
27. Nainwal LMS (2022) FDA Coronavirus Treatment Acceleration Program: approved drugs and those in clinical trials. Coronav Drug Discov:249–264
28. Begley CG, Ashton M, Baell J, Bettess M, Brown MP, Carter B, Charman WN, Davis C, Fisher S, Frazer I et al (2021) Drug repurposing: misconceptions, challenges, and opportunities for academic researchers. Sci Transl Med 13(612):eabd5524
29. Schcolnik-Cabrera A, Juarez-Lopez D, Duenas-Gonzalez A (2021) Perspectives on drug repurposing. Curr Med Chem 28(11):2085–2099
30. Dudley JT, Deshpande T, Butte AJ (2011) Exploiting drug-disease relationships for computational drug repositioning. Brief Bioinform 12(4):303–311
31. Subramanian A, Narayan R, Corsello SM, Peck DD, Natoli TE, Lu X, Gould J, Davis JF, Tubelli AA, Asiedu JK et al (2017) A next generation connectivity map: L1000 platform and the first 1,000,000 profiles. Cell 171(6):1437–1452.e17
32. Wishart DS, Knox C, Guo AC, Cheng D, Shrivastava S, Tzur D, Gautam B, Hassanali M (2008) DrugBank: a knowledgebase for drugs, drug actions and drug targets. Nucleic Acids Res 36(Database issue):D901–D906
33. Pinero J, Sauch J, Sanz F, Furlong LI (2021) The DisGeNET cytoscape app: exploring and visualizing disease genomics data. Comput Struct Biotechnol J 19: 2960–2967
34. Pinero J, Ramirez-Anguita JM, Sauch-Pitarch J, Ronzano F, Centeno E, Sanz F, Furlong LI (2020) The DisGeNET knowledge platform for disease genomics: 2019 update. Nucleic Acids Res 48(D1):D845–D855
35. Pinero J, Bravo A, Queralt-Rosinach N, Gutierrez-Sacristan A, Deu-Pons J, Centeno E, Garcia-Garcia J, Sanz F, Furlong LI (2017) DisGeNET: a comprehensive platform integrating information on human disease-associated genes and variants. Nucleic Acids Res 45(D1):D833–D839
36. Tse T, Fain KM, Zarin DA (2018) How to avoid common problems when using ClinicalTrials.gov in research: 10 issues to consider. BMJ 361:k1452
37. Kuhn M, Letunic I, Jensen LJ, Bork P (2016) The SIDER database of drugs and side effects. Nucleic Acids Res 44(D1):D1075–D1079
38. Rintala TJ, Ghosh A, Fortino V (2022) Network approaches for modeling the effect of drugs and diseases. Brief Bioinform 23(4)
39. MotieGhader H, Tabrizi-Nezhadi P, Deldar Abad Paskeh M, Baradaran B, Mokhtarzadeh A, Hashemi M, Lanjanian H, Jazayeri SM, Maleki M, Khodadadi E et al (2022) Drug repositioning in non-small cell lung cancer (NSCLC) using gene co-expression and drug-gene interaction networks analysis. Sci Rep 12(1):9417
40. Stark C, Breitkreutz BJ, Reguly T, Boucher L, Breitkreutz A, Tyers M (2006) BioGRID: a general repository for interaction datasets. Nucleic Acids Res 34(Database issue):D535–D539
41. Oughtred R, Stark C, Breitkreutz BJ, Rust J, Boucher L, Chang C, Kolas N, O'Donnell L, Leung G, McAdam R et al (2019) The BioGRID interaction database: 2019 update. Nucleic Acids Res 47(D1):D529–D541
42. Nam S, Lee S, Park S, Lee J, Park A, Kim YH, Park T (2021) PATHOME-Drug: a subpathway-based polypharmacology drug-repositioning method. Bioinformatics
43. Peyvandipour A, Saberian N, Shafi A, Donato M, Draghici S (2018) A novel computational approach for drug repurposing using systems biology. Bioinformatics 34(16):2817–2825
44. Yang F, Zhang Q, Ji X, Zhang Y, Li W, Peng S, Xue F (2022) Machine Learning Applications in Drug Repurposing. Interdiscip Sci 14(1):15–21
45. Pan XQ, Lin X, Cao D, Zeng X, Yu PS, He L, Nussinov R, Cheng F (2022) Deep learning for drug repurposing: methods, databases, and applications. WIREs Comput Mol Sci 12(4):e1597
46. Jaeger S, Fulle S, Turk S (2018) Mol2vec: unsupervised machine learning approach with chemical intuition. J Chem Inf Model 58(1):27–35
47. Cavalla D (2019) Using human experience to identify drug repurposing opportunities: theory and practice. Br J Clin Pharmacol 85(4):680–689

48. Day CP, Merlino G, Van Dyke T (2015) Preclinical mouse cancer models: a maze of opportunities and challenges. Cell 163(1):39–53
49. Seemann S, Zohles F, Lupp A (2017) Comprehensive comparison of three different animal models for systemic inflammation. J Biomed Sci 24(1):60
50. Schcolnik-Cabrera A, Chavez-Blanco A, Dominguez-Gomez G, Juarez M, Vargas-Castillo A, Ponce-Toledo RI, Lai D, Hua S, Tovar AR, Torres N et al (2021) Pharmacological inhibition of tumor anabolism and host catabolism as a cancer therapy. Sci Rep 11(1): 5222
51. Olatunde A, Nigam M, Singh RK, Panwar AS, Lasisi A, Alhumaydhi FA, Jyoti Kumar V, Mishra AP, Sharifi-Rad J (2021) Cancer and diabetes: the interlinking metabolic pathways and repurposing actions of antidiabetic drugs. Cancer Cell Int 21(1):499
52. Evans JM, Donnelly LA, Emslie-Smith AM, Alessi DR, Morris AD (2005) Metformin and reduced risk of cancer in diabetic patients. BMJ 330(7503): 1304–1305
53. Dowling RJ, Goodwin PJ, Stambolic V (2011) Understanding the benefit of metformin use in cancer treatment. BMC Med 9:33
54. Kannappan V, Ali M, Small B, Rajendran G, Elzhenni S, Taj H, Wang W, Dou QP (2021) Recent advances in repurposing disulfiram and disulfiram derivatives as copper-dependent anticancer agents. Front Mol Biosci 8:741316
55. Skrott Z, Mistrik M, Andersen KK, Friis S, Majera D, Gursky J, Ozdian T, Bartkova J, Turi Z, Moudry P et al (2017) Alcohol-abuse drug disulfiram targets cancer via p97 segregase adaptor NPL4. Nature 552(7684): 194–199
56. Fiorillo M, Lamb R, Tanowitz HB, Mutti L, Krstic-Demonacos M, Cappello AR, Martinez-Outschoorn UE, Sotgia F, Lisanti MP (2016) Repurposing atovaquone: targeting mitochondrial complex III and OXPHOS to eradicate cancer stem cells. Oncotarget 7(23):34084–34099
57. Skwarski M, McGowan DR, Belcher E, Di Chiara F, Stavroulias D, McCole M, Derham JL, Chu KY, Teoh E, Chauhan J et al (2021) Mitochondrial inhibitor atovaquone increases tumor oxygenation and inhibits hypoxic gene expression in patients with non-small cell lung cancer. Clin Cancer Res 27(9):2459–2469
58. Zhang M, Luo H, Xi Z, Rogaeva E (2015) Drug repositioning for diabetes based on 'omics' data mining. PLoS One 10(5):e0126082
59. Cai Y, Zhan H, Weng W, Wang Y, Han P, Yu X, Shao M, Sun H (2021) Niclosamide ethanolamine ameliorates diabetes-related muscle wasting by inhibiting autophagy. Skelet Muscle 11(1):15
60. Tao H, Zhang Y, Zeng X, Shulman GI, Jin S (2014) Niclosamide ethanolamine-induced mild mitochondrial uncoupling improves diabetic symptoms in mice. Nat Med 20(11):1263–1269
61. Leehey DJ, Carlson K, Reda DJ, Craig I, Clise C, Conner TA, Agarwal R, Kaufman JS, Anderson RJ, Lammie D et al (2021) Pentoxifylline in diabetic kidney disease (VA PTXRx): protocol for a pragmatic randomised controlled trial. BMJ Open 11(8):e053019
62. Han SJ, Kim HJ, Kim DJ, Sheen SS, Chung CH, Ahn CW, Kim SH, Cho YW, Park SW, Kim SK et al (2015) Effects of pentoxifylline on proteinuria and glucose control in patients with type 2 diabetes: a prospective randomized double-blind multicenter study. Diabetol Metab Syndr 7:64
63. Bibo-Verdugo B, Jiang Z, Caffrey CR, O'Donoghue AJ (2017) Targeting proteasomes in infectious organisms to combat disease. FEBS J 284(10): 1503–1517
64. Moreira W, Santhanakrishnan S, Ngan GJY, Low CB, Sangthongpitag K, Poulsen A, Dymock BW, Dick T (2017) Towards selective mycobacterial ClpP1P2 inhibitors with reduced activity against the human proteasome. Antimicrob Agents Chemother 61(5): e02307-16
65. Skerry C, Pinn ML, Bruiners N, Pine R, Gennaro ML, Karakousis PC (2014) Simvastatin increases the in vivo activity of the first-line tuberculosis regimen. J Antimicrob Chemother 69(9):2453–2457
66. Parihar SP, Guler R, Khutlang R, Lang DM, Hurdayal R, Mhlanga MM, Suzuki H, Marais AD, Brombacher F (2014) Statin therapy reduces the mycobacterium tuberculosis burden in human macrophages and in mice by enhancing autophagy and phagosome maturation. J Infect Dis 209(5):754–763
67. Azad GK, Tomar RS (2014) Ebselen, a promising antioxidant drug: mechanisms of action and targets of biological pathways. Mol Biol Rep 41(8):4865–4879
68. Favrot L, Grzegorzewicz AE, Lajiness DH, Marvin RK, Boucau J, Isailovic D, Jackson M, Ronning DR (2013) Mechanism of inhibition of Mycobacterium tuberculosis antigen 85 by ebselen. Nat Commun 4: 2748
69. Osman MM, Shanahan JK, Chu F, Takaki KK, Pinckert ML, Pagan AJ, Brosch R, Conrad WH, Ramakrishnan L (2022) The C terminus of the mycobacterium ESX-1 secretion system substrate ESAT-6 is required for phagosomal membrane damage and virulence. Proc Natl Acad Sci U S A 119(11): e2122161119
70. Siegel RL, Miller KD, Fuchs HE, Jemal A (2022) Cancer statistics, 2022. CA Cancer J Clin 72(1):7–33
71. Sung H, Ferlay J, Siegel RL, Laversanne M, Soerjomataram I, Jemal A, Bray F (2021) Global Cancer Statistics 2020: GLOBOCAN estimates of incidence and mortality worldwide for 36 cancers in 185 countries. CA Cancer J Clin 71(3):209–249
72. Pantziarka P, Bouche G, Meheus L, Sukhatme V, Sukhatme VP, Vikas P (2014) The repurposing drugs in oncology (ReDO) project. Ecancermedicalscience 8:442
73. Schcolnik-Cabrera A, Dominguez-Gomez G, Chavez-Blanco A, Ramirez-Yautentzi M, Morales-Barcenas R, Chavez-Diaz J, Taja-Chayeb L, Dueaas-Gonzalez A (2019) A combination of inhibitors of glycolysis, glutaminolysis and de novo fatty acid synthesis decrease the expression of chemokines in human colon cancer cells. Oncol Lett 18(6):6909–6916

74. Juarez M, Schcolnik-Cabrera A, Dominguez-Gomez-G, Chavez-Blanco A, Diaz-Chavez J, Duenas-Gonzalez A (2020) Antitumor effects of ivermectin at clinically feasible concentrations support its clinical development as a repositioned cancer drug. Cancer Chemother Pharmacol 85(6):1153–1163
75. Singhal S, Mehta J (2001) Thalidomide in cancer: potential uses and limitations. BioDrugs 15(3): 163–172
76. Pantziarka P, Verbaanderd C, Sukhatme V, Rica Capistrano I, Crispino S, Gyawali B, Rooman I, Van Nuffel AM, Meheus L, Sukhatme VP et al (2018) ReDO_DB: the repurposing drugs in oncology database. Ecancermedicalscience 12:886
77. Jiang W, Finniss S, Cazacu S, Xiang C, Brodie Z, Mikkelsen T, Poisson L, Shackelford DB, Brodie C (2016) Repurposing phenformin for the targeting of glioma stem cells and the treatment of glioblastoma. Oncotarget 7(35):56456–56470
78. Ogurtsova K, Guariguata L, Barengo NC, Ruiz PL, Sacre JW, Karuranga S, Sun H, Boyko EJ, Magliano DJ (2022) IDF diabetes Atlas: global estimates of undiagnosed diabetes in adults for 2021. Diabetes Res Clin Pract 183:109118
79. Turner N, Zeng XY, Osborne B, Rogers S, Ye JM (2016) Repurposing drugs to target the diabetes epidemic. Trends Pharmacol Sci 37(5):379–389
80. Khankari NK, Keaton JM, Walker VM, Lee KM, Shuey MM, Clarke SL, Heberer KR, Miller DR, Reaven PD, Lynch JA et al (2022) Using Mendelian randomisation to identify opportunities for type 2 diabetes prevention by repurposing medications used for lipid management. EBioMedicine 80:104038
81. Adeniji AA, Knoll KE, Loots DT (2020) Potential anti-TB investigational compounds and drugs with repurposing potential in TB therapy: a conspectus. Appl Microbiol Biotechnol 104(13):5633–5662
82. Janes J, Young ME, Chen E, Rogers NH, Burgstaller-Muehlbacher S, Hughes LD, Love MS, Hull MV, Kuhen KL, Woods AK et al (2018) The ReFRAME library as a comprehensive drug repurposing library and its application to the treatment of cryptosporidiosis. Proc Natl Acad Sci U S A 115(42):10750–10755
83. An Q, Li C, Chen Y, Deng Y, Yang T, Luo Y (2020) Repurposed drug candidates for antituberculosis therapy. Eur J Med Chem 192:112175
84. Gillespie SH (2016) The role of moxifloxacin in tuberculosis therapy. Eur Respir Rev 25(139):19–28
85. Sharma D, Dhuriya YK, Deo N, Bisht D (2017) Repurposing and revival of the drugs: a new approach to combat the drug resistant tuberculosis. Front Microbiol 8:2452
86. Thwaites G, Nguyen NV (2022) Linezolid for drug-resistant tuberculosis. N Engl J Med 387(9):842–843
87. de Jager VR, Dawson R, van Niekerk C, Hutchings J, Kim J, Vanker N, van der Merwe L, Choi J, Nam K, Diacon AH (2020) Telacebec (Q203), a new antituberculosis agent. N Engl J Med 382(13): 1280–1281

Part VII

Literature Search/Researching Ideas

37 Empowering Efficient Literature Searching: An Overview of Biomedical Search Engines and Databases

Pitchai Balakumar, Joanne Berger, Gwendolyn Halford, and Gowraganahalli Jagadeesh

Abstract

Literature searching may be defined as a comprehensive exploration of existing literature for the purpose of enriching knowledge and finding scholarly information on a topic of interest. A systematic, well-defined and thorough search of a range of literature on a specific topic is an important part of a research process. The key purpose of a meticulous literature search is to formulate a research question and problem by critically analysing the existing literature with a primary target of filling the gaps in the literature and opening a path for further research. Here, we discuss potential ways to construct a commendable biomedical literature search using a variety of databases and search engines.

Keywords

PubMed/Medline · Scopus · Embase · Web of Science · PubChem · Biomedical databases

37.1 Introduction

Literature search, an important part of the never-ending active learning process for any research, involves not only a researcher's gathering of information on a specific topic of interest; but also provides insight on gaps in the biomedical literature, enabling the formulation of a potential research problem. A meticulous literature search helps the researcher interpret ideas, perceive shortcomings of existing

Disclaimer: *This article reflects the views of the authors (JB, GH, GJ) and should not be construed to represent the US FDA's views or policies.*

P. Balakumar (✉)
Professor & Director, Research Training and Publications, The Office of Research and Development, Periyar Maniammai Institute of Science & Technology (Deemed to be University), Vallam, Tamil Nadu, India
e-mail: directorcr@pmu.edu; pbalakumar2022@gmail.com

J. Berger · G. Halford
FDA Library, Office of Data, Analytics and Research (ODAR), Office of Digital Transformation (ODT), US Food and Drug Administration, Rockville, MD, USA
e-mail: joanne.berger@fda.hhs.gov

G. Jagadeesh
Retired Senior Expert Pharmacologist at the Office of Cardiology, Hematology, Endocrinology, and Nephrology, Center for Drug Evaluation and Research, US Food and Drug Administration, Silver Spring, MD, USA

Distinguished Visiting Professor at the College of Pharmaceutical Sciences, Dayananda Sagar University, Bengaluru, Karnataka, India

Visiting Professor at the College of Pharmacy, Adichunchanagiri University, BG Nagar, Karnataka, India

Visiting Professor at the College of Pharmaceutical Sciences, Manipal Academy of Higher Education (Deemed-to-be University), Manipal, Karnataka, India

Senior Consultant & Advisor, Auxochromofours Solutions Private Limited, Silver Spring, MD, USA
e-mail: GJagadeesh2000@gmail.com

G. Jagadeesh et al. (eds.), *The Quintessence of Basic and Clinical Research and Scientific Publishing*, https://doi.org/10.1007/978-981-99-1284-1_37

research and explore new research opportunities, while a systematic and in-depth literature review helps in proposing a novel hypothesis and meaningful research.

Identifying a potential research problem is a major concern for any active researcher. Once formulated, a research question needs to be addressed by specific experiments based on the existing literature. As soon as a researcher identifies the research problem, seeking more of the existing literature and analysing it deeply may further strengthen the research approach and support the hypothesis [1]. Whether for obtaining pertinent information on research topics, or for grant applications, theses, drugs, chemical substances, treatments, mechanistic insights, signalling, disorders, pathobiologies, clinical studies, or other purposes, literature searching is an established component of academic activities and research, demanding adequate resources. The critical phases of effective research planning include efficient literature search and review. Information on a particular area of research can grow daily. While the amount of information available in modern literature is extremely high, a researcher has to employ appropriate search engines and/or databases to obtain the precise information they actually seek [2–7].

Beginning a literature search with the support of modern tools can help a researcher identify potential gaps in the topic, while a web-based search is the first step to obtain pertinent information in the literature that can educate the novice and budding researchers on a research topic and greatly support an established researcher identifying unmet research problems. An efficient researcher should know how to pursue appropriate types of biomedical search engines and databases to mine the wealth of information in the medical, paramedical, biological, and biomedical health sciences. In this chapter, we discuss the use of several most commonly employed biomedical search engines and databases for efficient literature searching. In addition, we briefly describe numerous supporting databases and search engines frequently used by biomedical scientists for finding pertinent information.

37.2 Literature Search Using PubMed/Medline

PubMed has been available online to the public since 1996. It has been developed and maintained by the National Center for Biotechnology Information (NCBI) at the U.S. National Library of Medicine (NLM) within the National Institutes of Health (NIH). PubMed is a freely available database primarily supporting the search and retrieval of literature pertaining to medical, biomedical and life sciences, with the aim of providing up-to-date information. Citations in PubMed mainly stem from the biomedicine and health fields, as well as closely related disciplines such as life sciences, behavioural sciences, chemical sciences, and bioengineering. The PubMed database currently has more than 34 million citations and abstracts of biomedical literature. Its home page is shown in Fig. 37.1 (https://pubmed.ncbi.nlm.nih.gov/).

Although PubMed does not normally include full-text articles, it links to the full text when available from published sources (open access) or through PubMed Central (PMC) [8]. PubMed facilitates searching across NLM literature resources such as Medline [9], PMC [10], and Bookshelf [11].

Medline, the primary component of PubMed, is the online counterpart to the MEDical Literature Analysis and Retrieval System (MEDLARS) that originated in 1964. Medline is the premier bibliographic database of NLM that contains more than 29 million references to journal articles in life sciences with a concentration on biomedicine. A characteristic feature of Medline is that the records are indexed with Medical Subject Headings (MeSH) of NLM [9]. Medline includes literature published from 1966 to present, and selected coverage of literature prior to that period. Oldmedline subset in PubMed represents journal article citations from two print indexes such as Cumulated Index Medicus (CIM) and Current List of Medical Literature (CLML). The Oldmedline contains citations from 1960 through 1965 CIM print indexes and the 1946 through 1959 CLML print indexes. Of note, Oldmedline records that are included in the Medline database could be searched via PubMed [12].

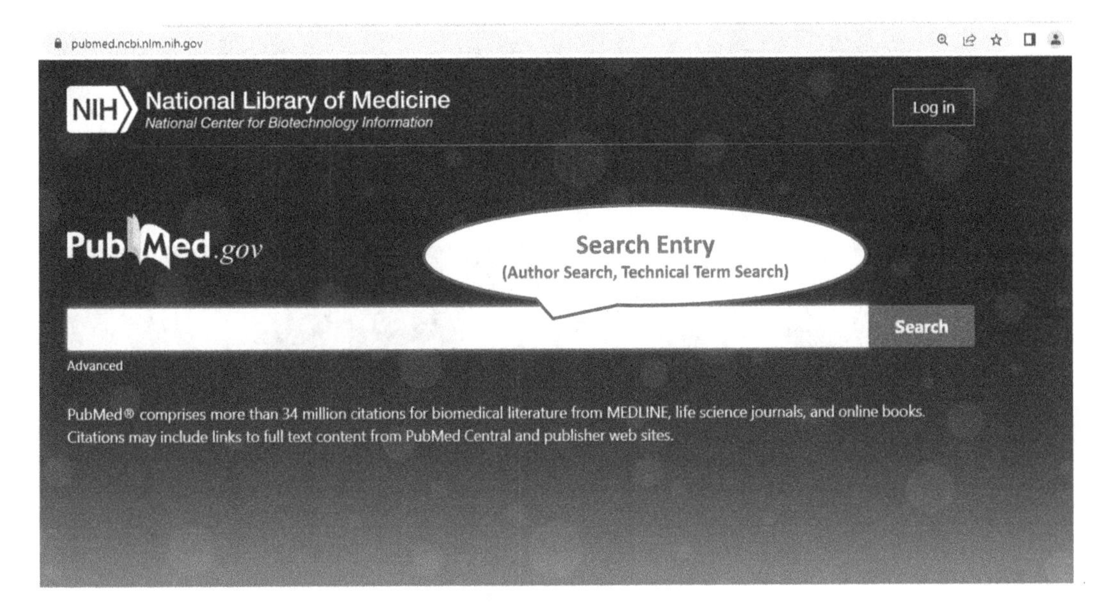

Fig. 37.1 The home page of PubMed for biomedical literature searching

While Medline is known as the largest component of PubMed consisting primarily of citations from journals selected for Medline and articles indexed with MeSH, citations for PMC articles make up the second largest component of PubMed. PMC is a free full-text archive of biomedical and life sciences journal literature at the U.S. NIH's NLM. It has been available to the public online since 2000, and has been developed and maintained by the NCBI at NLM [10]. Since its inception in 2000, PMC until time contains more than seven million full-text records spanning centuries of biomedical research and life science research of late 1700s to present [10]. The search page through PMC is available at: https://www.ncbi.nlm.nih.gov/pmc/ (Fig. 37.2).

The third and final component of PubMed is citations for books and some individual chapters available on Bookshelf, which is a full text archive of books, reports, databases, and other documents related to biomedical sciences, and health and life sciences [8]. The search page for Bookshelf [11] is shown in Fig. 37.3.

NLM is actively committed to developing PubMed which remains a trusted and highly accessible source of biomedical literature to

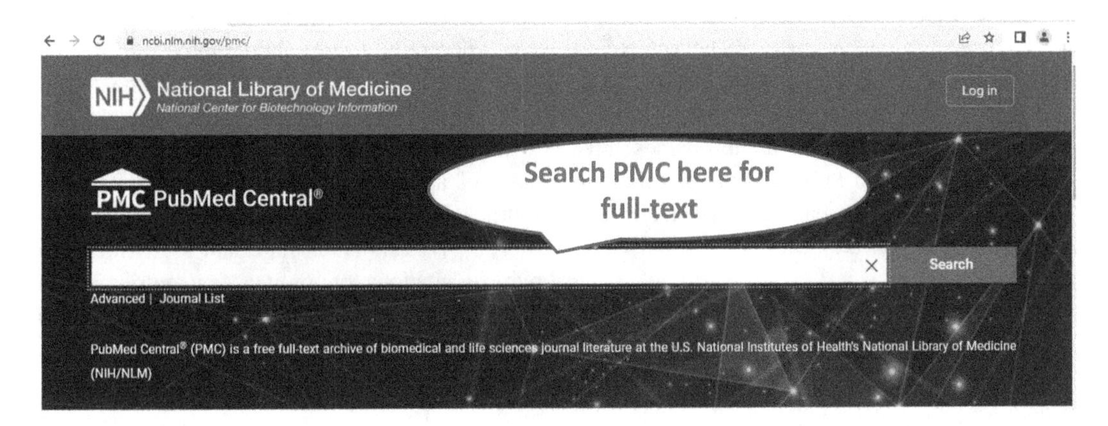

Fig. 37.2 The home page of PMC used for retrieving free full-text articles from biomedical and life science journals

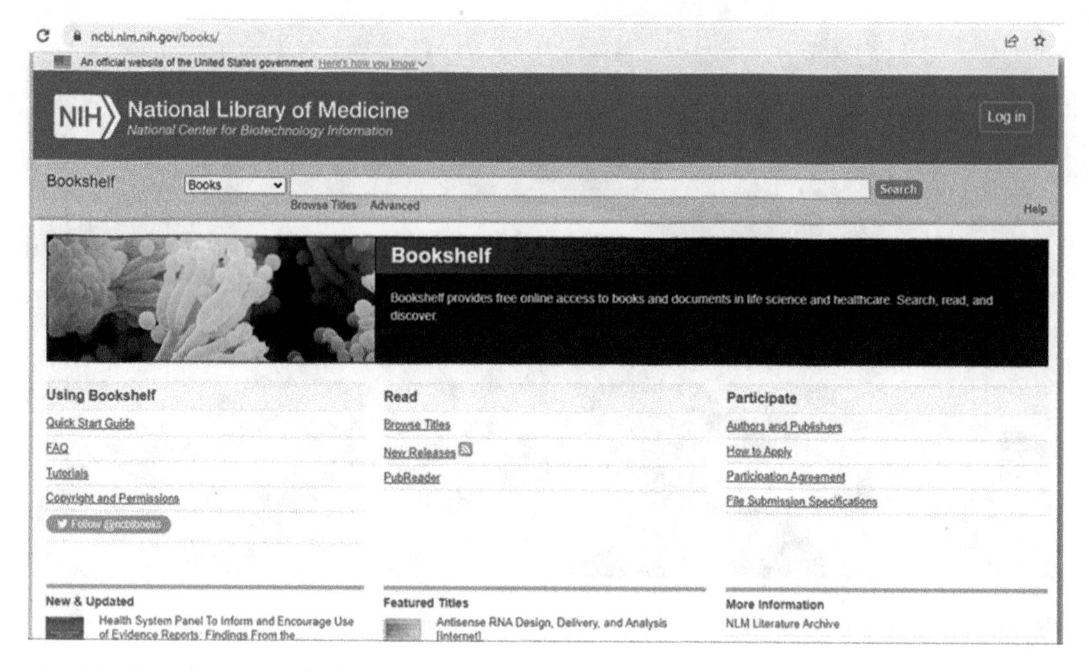

Fig. 37.3 The home page of NLM's Bookshelf

date. Below, we discuss some of its key features that assist in the efficient search and retrieval of desired literature.

37.2.1 Search by Authors

Using the Author search box, enter the author's last name and initials without punctuation before clicking the 'Search' button. Alternatively, enter the author's name, (e.g., Night) along with a search tag [au]; for instance, Night[au], which will retrieve papers authored by individuals with the last name Night. Names can be entered as either the last name along with initials, (e.g., Jagadeesh G), or the full name (Pitchai Balakumar). Note that full author names have not been included on PubMed citations prior to 2002; hence, full name searches retrieve citations only from 2002 forward [13].

37.2.2 Search by Journals

Search for journal names by entering either a full journal title, (e.g., *Pharmacological Research*), or a journal title abbreviation, (e.g., *Pharmacol Res*). Use the journal's ISSN number for best results, (e.g., for *Pharmacological Research*, enter 1043-6618).

37.2.3 PubMed Single Citation Matcher

The single citation matcher is available on the PubMed homepage. It can be used to find PubMed citations, and it has a fill-in-the-blank form for the purpose of quickly searching a citation when only a few bibliographic information about an article is available such as journal name, authors' first or last name, etc. For instance, a total of three citations were retrieved for the entry of author first name (Pitchai Balakumar), Journal (Pharmacological Research), Year (2021) (Fig. 37.4). This is an important tool which can retrieve citations on only a few entries of its bibliographic details [14]. While searching, one may input information for or leave blank any additional field (Fig. 37.4).

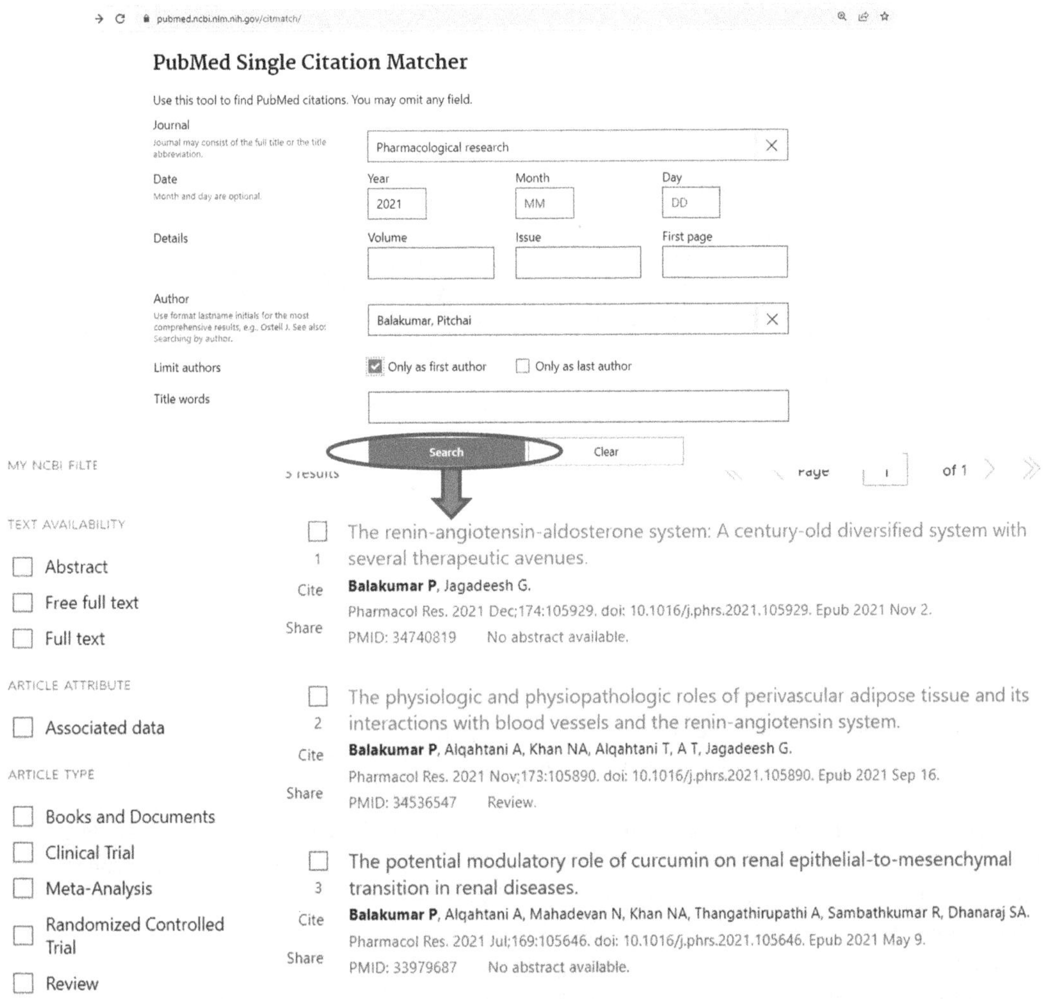

Fig. 37.4 The use of PubMed's single citation matcher in retrieving citations with the inputs of only a few bibliographic details

37.2.4 Search PubMed Using the MeSH Database

MeSH stands for Medical Subject Headings, a hierarchically-organized vocabulary maintained by NLM, used both for indexing and for searching for biomedical science and health-related information [15]. One can use the MeSH database to identify Medical Subject Headings (MeSH) which help find literature indexed with the MeSH term. The MeSH database can be effectively used to find MeSH terms, including subheadings, and pharmacological actions, and then a PubMed search can be built. A researcher might continue an efficient search by including additional terms to the PubMed Search Builder. The PubMed Search Builder on the right side of the screen can be used to add the selected MeSH term to the box, and then search can be done. The MeSH homepage search is shown in Fig. 37.5 (https://www.ncbi.nlm.nih.gov/mesh/).

37.2.5 Search PubMed Using Truncation

The searcher may use "wildcard" symbols, such as an asterisk (*) at the end of a keyword. This is

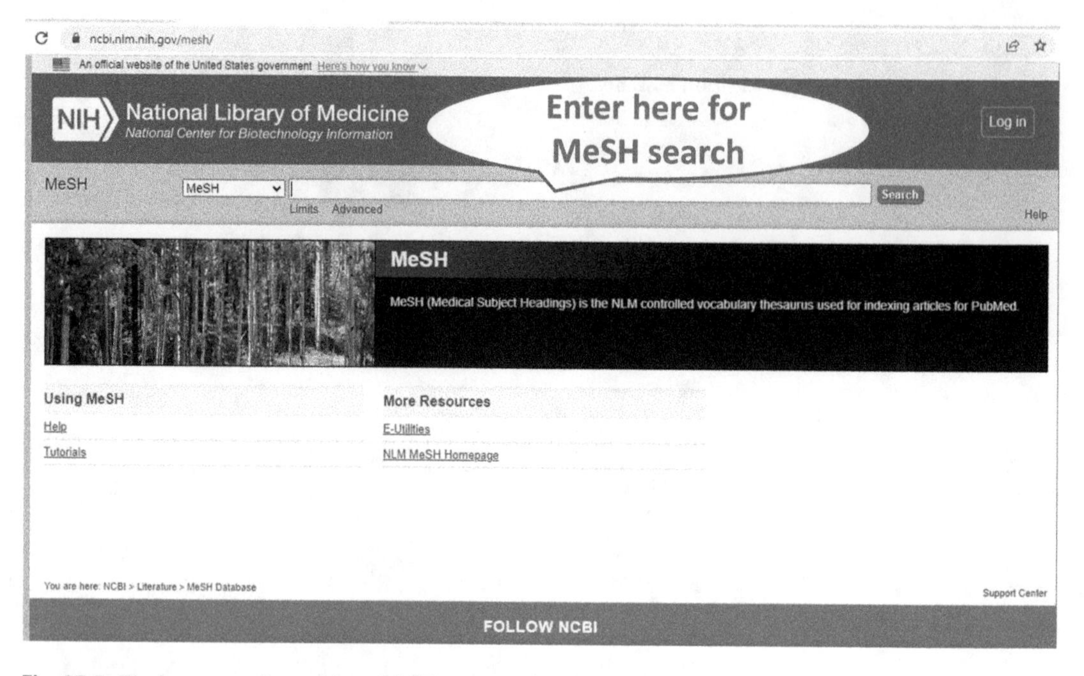

Fig. 37.5 The home page for entering a MeSH term search

known as "truncation." The wildcard symbol is a word truncation device which instructs the search engine to gather variations of a particular search term. In PubMed, the researcher may wish to use the asterisk at the root of a word to find multiple endings. For example: an entry of the search term, epid* would retrieve entries for all documents containing the words: epidemic, epidemics, epidemiology, epidemiological, and other terms such as epidermal and epidermidis, among others. Hence more focused search terms should be included while using truncation. For instance: an entry of hypertens* rather than hyper* would retrieve more focused results that include the documents having the words: "hypertensive" and "hypertension."

The truncation search may be done in different formats such as (1) enclosing the phrase in double quotes: "primary hypertens*"; (2) employing a search tag: primary hypertens*[tiab], where tiab means 'limit to **ti**tle or **ab**stract [Title/Abstract]; (3) using a hyphen: primary-hypertens* (Fig. 37.6). It should be noted that a minimum four characters must be provided in the truncated term, and it should be the last word in the phrase. Also important is that truncation can turn off the automatic term mapping and the process that includes the MeSH term in the MeSH hierarchy. For instance, angina* will not map to the MeSH term "ischemic heart disease" or include any of the more specific terms, e.g., acute coronary syndrome, myocardial infarction or heart attack [16].

37.2.6 The PubMed Search History, and Advanced Search by Combining Search Terms with Boolean Operators

The search history can be located at the "Advanced Search" option within PubMed. The searcher may wish to go to the "Advanced Search" page to combine searches. Boolean operators are commonly used to combine or exclude search terms, while a comprehensive search of PubMed can be done using both controlled vocabulary (MeSH) and any keyword terms. In PubMed, a searcher may use the Boolean operators such as AND, OR, and NOT. Of note, Boolean operators are used in upper case (AND, OR, NOT). For instance, for the search terms: primary hypertension and secondary

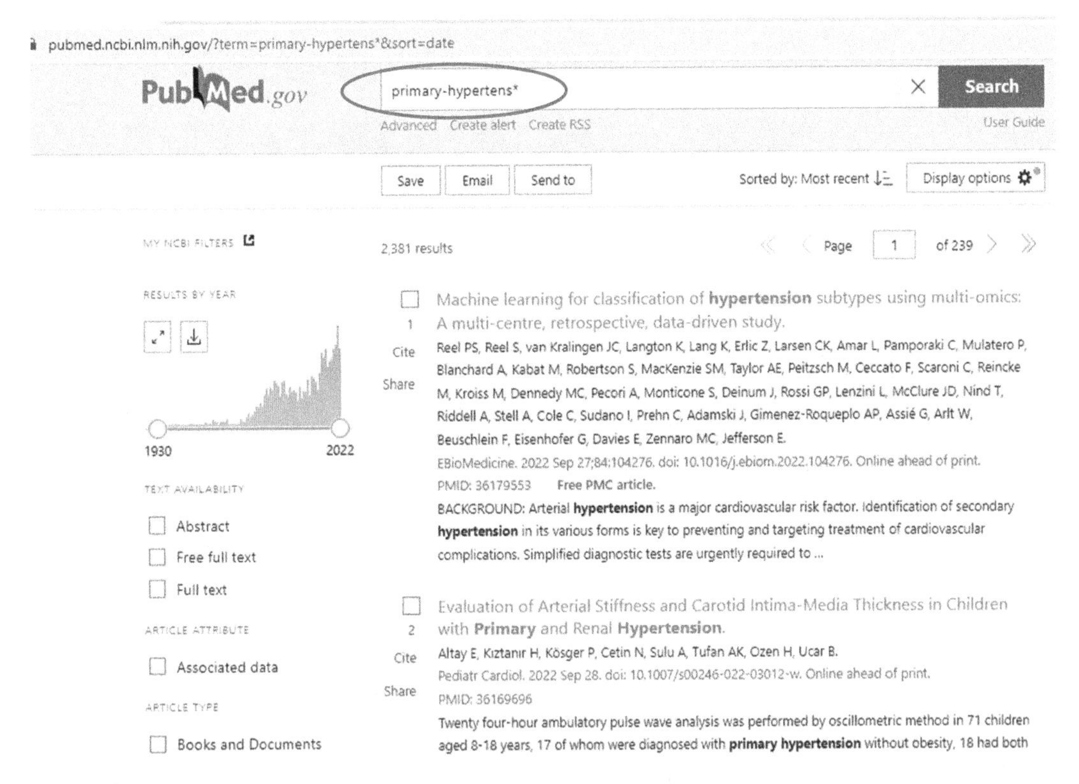

Fig. 37.6 PubMed search using truncation

hypertension (Fig. 37.7), (1) AND retrieves results that include both the search terms (primary AND secondary hypertension); (2) OR retrieves results that include at least one of the search terms (primary OR secondary hypertension); (3) NOT excludes the retrieval of terms from the search (primary NOT secondary hypertension) [17].

37.2.7 PubMed Search Using Filters

A searcher may wish to use various filters (Figs. 37.8 and 37.9) to narrow the search results by text availability (Abstract, Free full text, Full text); article attribute (Associated data); article type (Books and Documents, Clinical Trial, Meta-Analysis, Randomized Controlled Trial, Review, Systematic Review); publication date (1 year, 5 years, 10 years, Custom Range); species (Humans, Other Animals); several languages; sex (Female, Male); journal category (Medline); and age (Newborn: birth–1 month, Infant: birth–23 months, Infant: 1–23 months, Preschool Child: 2–5 years, Child: 6–12 years, Adolescent: 13–18 years, Adult: 19+ years, Young Adult: 19–24 years, Adult: 19–44 years, Middle Aged + Aged: 45+ years, Middle Aged: 45–64 years, Aged: 65+ years, 80 and over: 80+ years).

To effectively apply filters, run a PubMed search for an item, followed by clicking the appropriate filters along the sidebar (Fig. 37.9). This results in a check mark, indicating the activation of the chosen filter(s). Subsequent search options will be filtered until the selected filters are turned off, or the browser data is cleared [18].

37.2.8 Selecting and Modulating the Display Format of Search Results

After searching for an item, results are displayed by default in the summary format. One can

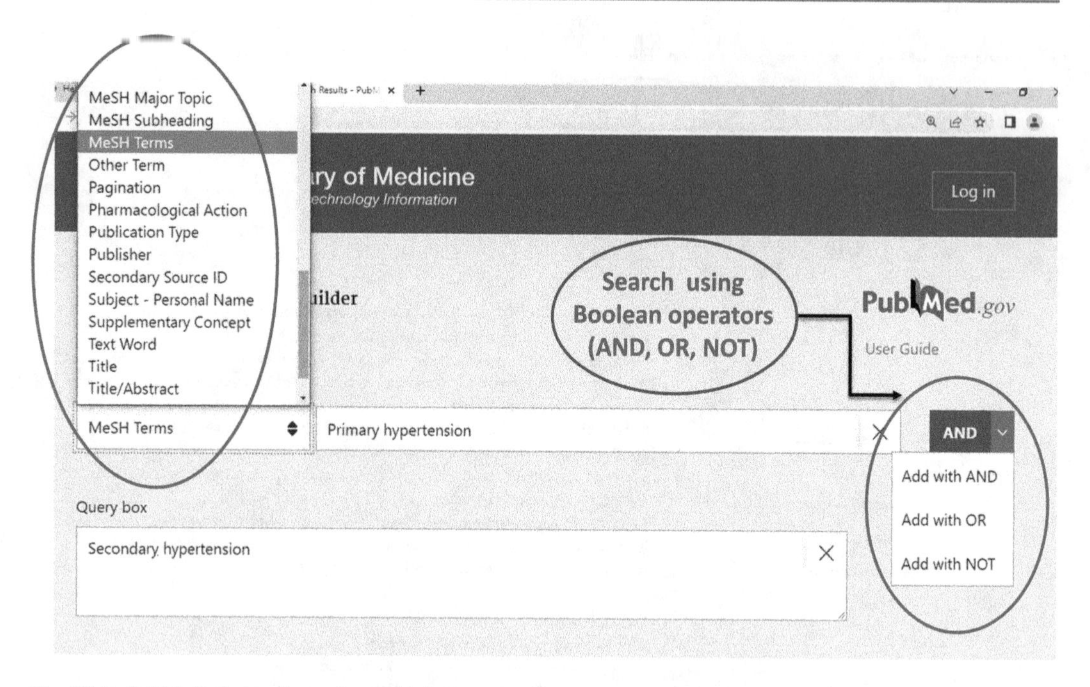

Fig. 37.7 PubMed advanced search, combining search terms with Boolean operators

change the display of results using "Display options" just below the "Search" button (Fig. 37.10). Four major display options are available: (1) Format (Summary, Abstract, PubMed, PMID); (2) Sort by (Best match, Most recent, Publication date, First author, Journal); (3) Per page (10, 20, 50, 100, 200); and (4) Abstract snippets (Show or Hide) [16]. Notably, the summary format includes snippets, by default, from the citation abstract that can be turned off by deselecting "Abstract snippets." Selecting one or more options in the display format will allow the searcher to obtain the desired format.

37.2.9 Search Using PubMed Clinical Queries

The PubMed Clinical Queries tool uses predefined filters to refine PubMed searches on clinical or disease-specific areas. The key filter category includes (1) Clinical studies and (2) COVID-19 (Figs. 37.11 and 37.12). The "Clinical studies" filter category has further filters such as (a) Therapy, (b) Clinical Prediction Guides, (c) Diagnosis, (d) Etiology, and (e) Prognosis (Fig. 37.11). The 'Clinical Studies' along with these filters can be further searched under the 'Scope' of Broad or Narrow (Fig. 37.11). On the other hand, the 'COVID-19' filter category possesses further filtering options such as (a) Treatment, (b) General, (c) Mechanism, (d) Transmission, (e) Diagnosis, (f) Prevention, (g) Case Report, (h) Forecasting, and (i) Long COVID (Fig. 37.12) [19].

37.3 Literature Search Using Embase

Embase (https://www.elsevier.com/solutions/embase-biomedical-research) is a comprehensive biomedical literature database consisting of published peer-reviewed literature, in-press publications, and conference abstracts. It contains over 41 million indexed records, 8100 journals, and over three million conference abstracts.

Embase includes three databases: Embase (1974–present), Embase Classic (1947–73), and MEDLINE (1966–present). Embase Classic and MEDLINE also include some older articles dating back to 1907. Searching the combined three is

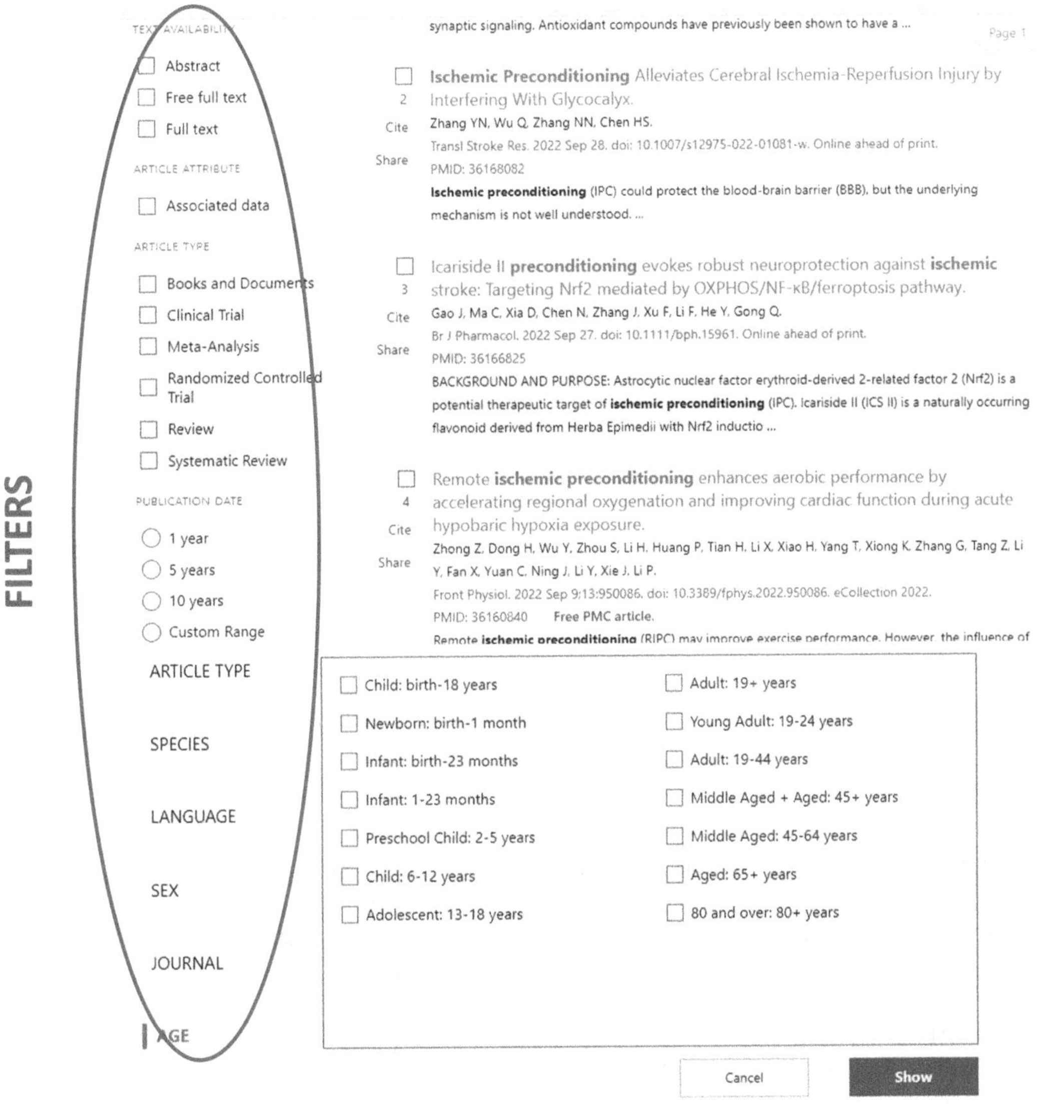

Fig. 37.8 PubMed search using various filters

the default, but each can be searched individually. Newer content includes Preprints, launched in 2021, with early versions of manuscripts shared on the MedRXiv and BioRXiv public preprint servers; and PubMed-not-MEDLINE articles, launched in 2018, with new records that are part of the NLM's PubMed collection but are out of MEDLINE's scope.

Embase contains features designed to help users structure a search query for precision. The Emtree thesaurus enables the use of natural language to find terms that describe Embase records. There are also tailored search forms for drug, device, disease, and citation information, and three newer search forms, PICO, PV Wizard, and Medical Device. These forms provide the following functionality:

- Drug, Device, and Disease search forms: Designed for advanced searching to enhance precision, using options not available on the quick and advanced search forms, which are unique subheadings for drugs, devices, and

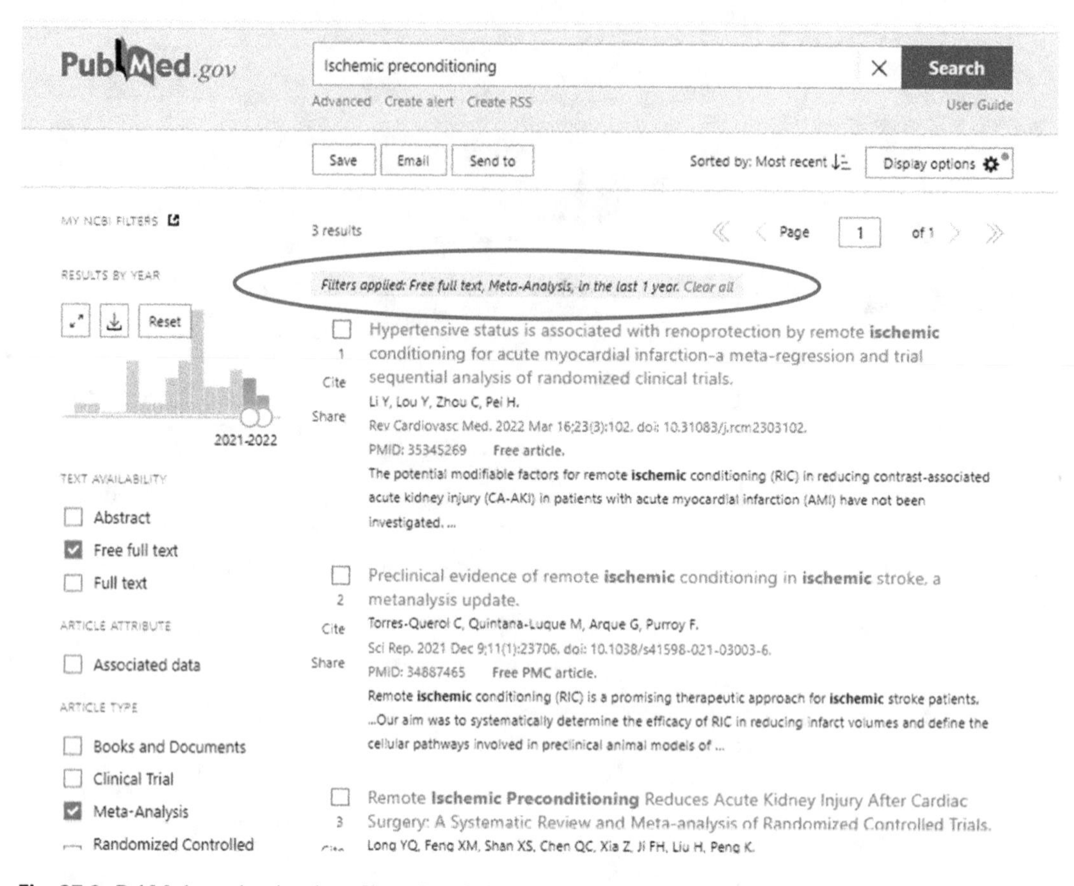

Fig. 37.9 PubMed search using three filters, Free full text, meta-analysis and 1 year

diseases, unique search fields for drugs and devices, and various drug routes of administration.

- PICO: Enables a searcher to build a search using the PICO framework, i.e., patient, intervention, comparison/control, and outcome. This form can also be used to reduce time spent building a search query by using tools that automatically add all or selected Emtree terms, e.g., synonyms, to the search query, and shows results at each step of building a search.
- PV Wizard: Guides a searcher in constructing a comprehensive pharmacovigilance search query. It includes five key elements: drug name, alternative drug names, adverse drug reactions, special conditions, and limits, such as human subjects. Embedded in the PV Wizard form is the European Medicines Agency's (EMA) Medical Literature Monitoring search service, which monitors 300 chemical and 100 herbal active substance groups to identify suspected adverse reactions to medicines authorized in the European Union.
- Medical Device: Enables a searcher to conduct searches on medical device adverse effects and find information about a manufacturer's other products.
- Citation Information: Enables a searcher to search within a specific area of an article, e.g., author's name.

Other Embase features include Triple indexing, a three-part subject phrase search consisting of an Emtree drug, device, or disease term, a subheading, and a term linked to the subheading (e.g., 'aspirin'/'adverse drug reaction'/'hypertension'), to provide more precise retrieval of relevant records. Index Miner displays a list of all indexed terms that appeared in the search results. These terms can be searched within the result set for

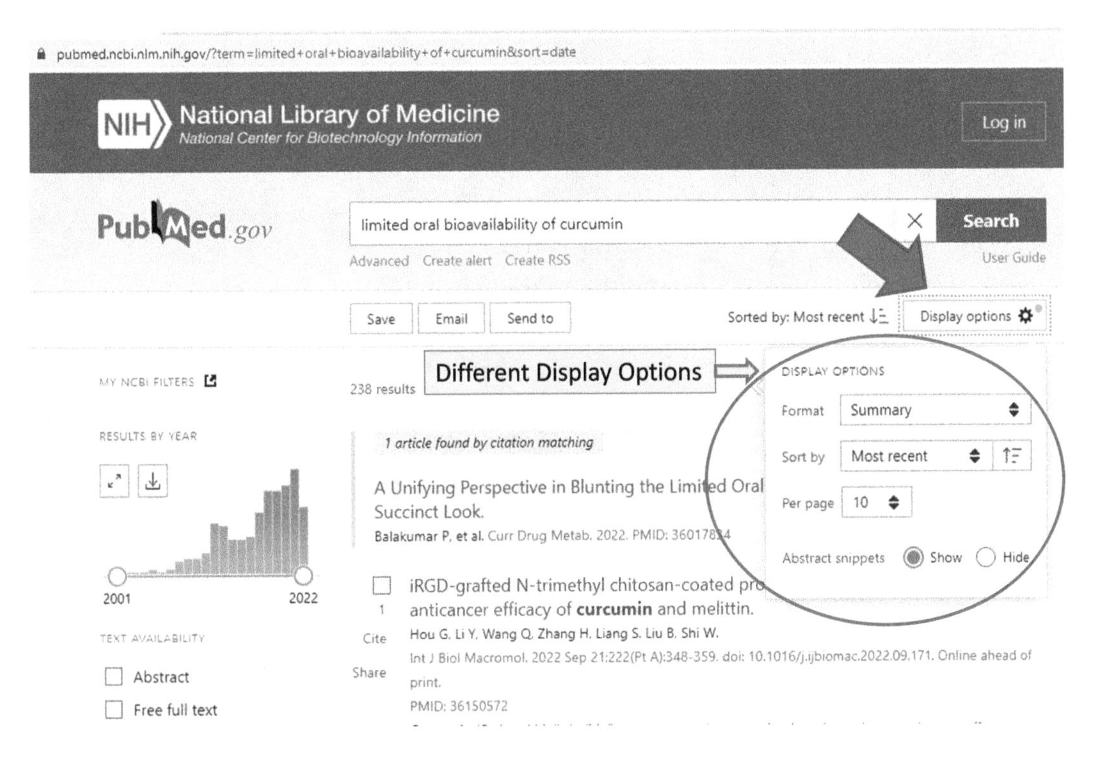

Fig. 37.10 PubMed search with different display options

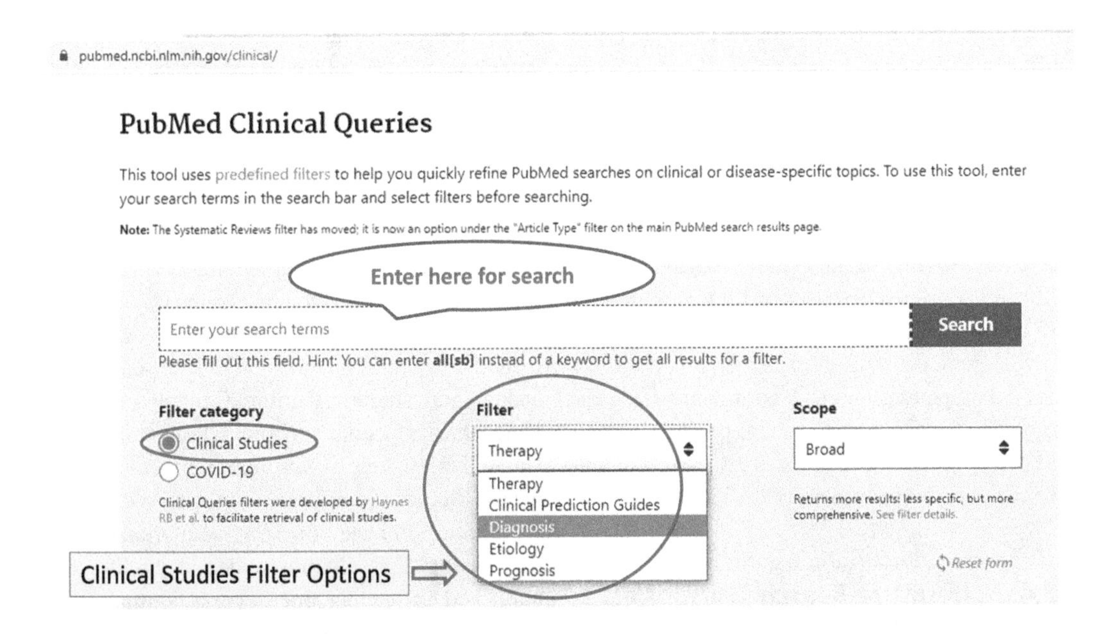

Fig. 37.11 PubMed Clinical Queries tool with the filter category Clinical Studies

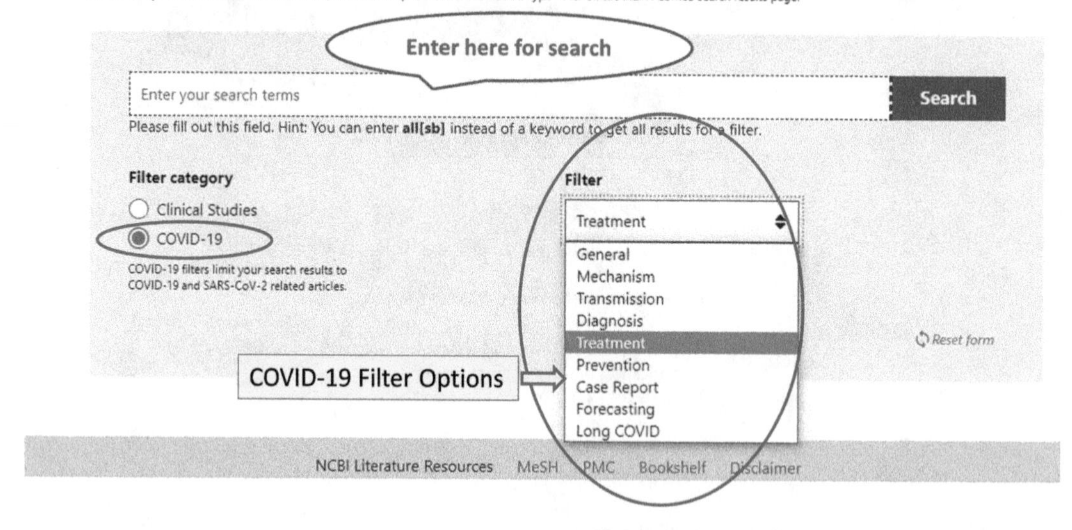

Fig. 37.12 PubMed Clinical Queries tool with the filter category COVID-19

added precision. Table 37.1 shows different types of Embase searches.

The **first and second searches** include preferred and natural language (keyword) terms for searching primary hypertension and treatment. Essential hypertension and drug therapy are preferred Emtree terms. The **third and fourth searches** use only Emtree preferred terms. The **fifth and six searches** use an Emtree synonym and a natural language term or a keyword. Primary hypertension is an Emtree synonym for essential hypertension, and treatment is a keyword. The three search types are performed with and without limits. The limits are review, English language and 5-year coverage period.

37.4 Literature Search Using Web of Science

Web of Science (WoS), a subscription database suite from Clarivate (https://clarivate.com/webofsciencegroup) covers more than 34,200 journals internationally, in 254 subject categories, comprising 1.89 billion cited references. With articles going back to 1900, WoS provides access to citations in the sciences, social sciences, arts, and humanities.

The WoS suite consists of several databases: the Core Collection, including scholarly journals, books, and proceedings; Biological Abstracts, with literature in the life sciences; BIOSIS Index, with literature on preclinical and experimental research, methods and instrumentation, and animal studies; Current Contents Connect, with tables of contents from scholarly journals; Data Citation Index, with research data sets and data studies; Derwent Innovations Index, containing information on and citations to patents; SciELO Citation index, with literature published in leading open access journals from Latin America, Portugal, Spain, and South Africa; KCI-Korean Journal Database; and Medline. In addition, WoS has a tab specifically for searching for results by researcher name, with tools to help exclude researchers with the same names. It also

Table 37.1 Embase search example: treatment of primary hypertension, mapping option: "searching as broadly as possible"

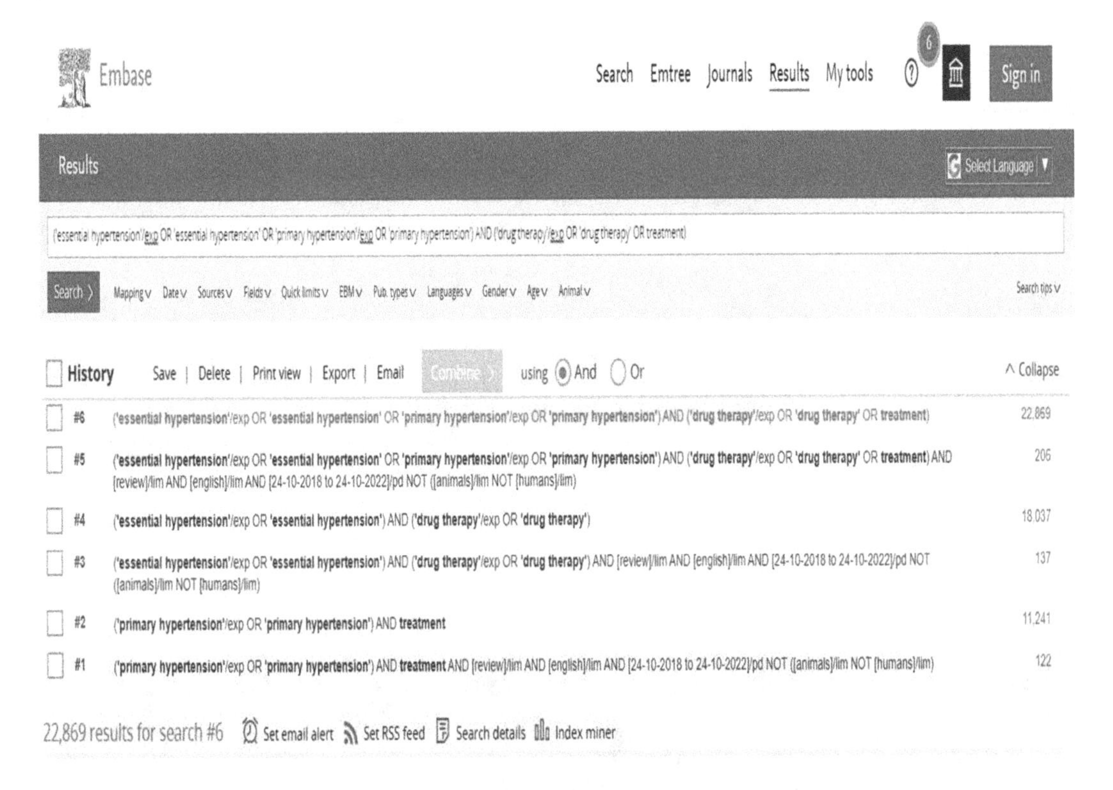

has special templates for Advanced, Cited Reference, and (Chemical) Structure searching.

In the search results, the "Analyze Results" feature can be used to generate a bar chart or tree map with data from the results (see Figs. 37.13 and 37.14, https://www.webofscience.com/wos/woscc/basic-search). Other research impact and citation tools from Clarivate, such as Journal Citation Reports, InCites, Essential Science Indicators and EndNote Online, can be used from the WoS interface.

37.5 Alternate Medline/PubMed Search Using BibliMed

BibliMed (http://www.biblimed.com/appli/index.php?lang=en) is an alternate search engine that could be used for literature search. BibliMed is a smart Medline interface and an intuitive PubMed alternative. It can be employed for literature searching using Keyword, Substance, Journal, Date, Author, Issue, Volume and Title (Fig. 37.15).

Typing a single word or phrase in the BibliMed search box enables the searcher to select the best auto-suggested MeSH term. This will further allow the searcher to choose options from subheadings for more focused literature search. For instance, selecting the MeSH term Hypertension opens 43 subheadings as shown in Fig. 37.16.

Moreover, as shown in Fig. 37.17, BibliMed also offers the option to open an article directly from PubMed, as well as to view h5-index and SCImago Journal Rank (SJR) metrics (*See* Chap. 40 for explanations of these metrics). Various filters include: (1) Synthesis (Systematic

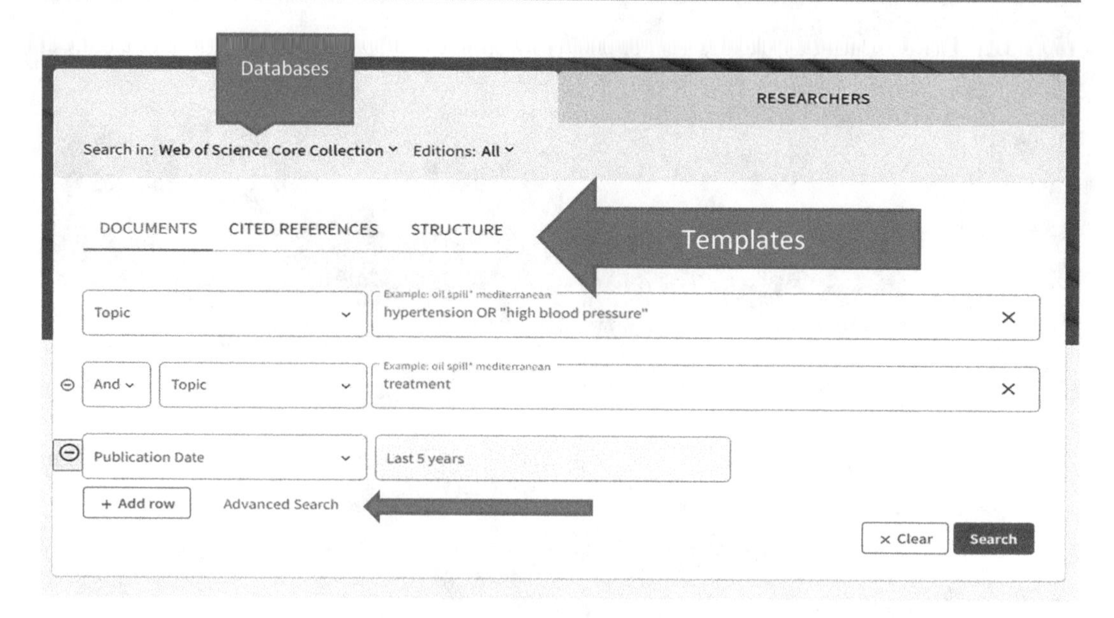

Fig. 37.13 Web of Science landing page: basic search

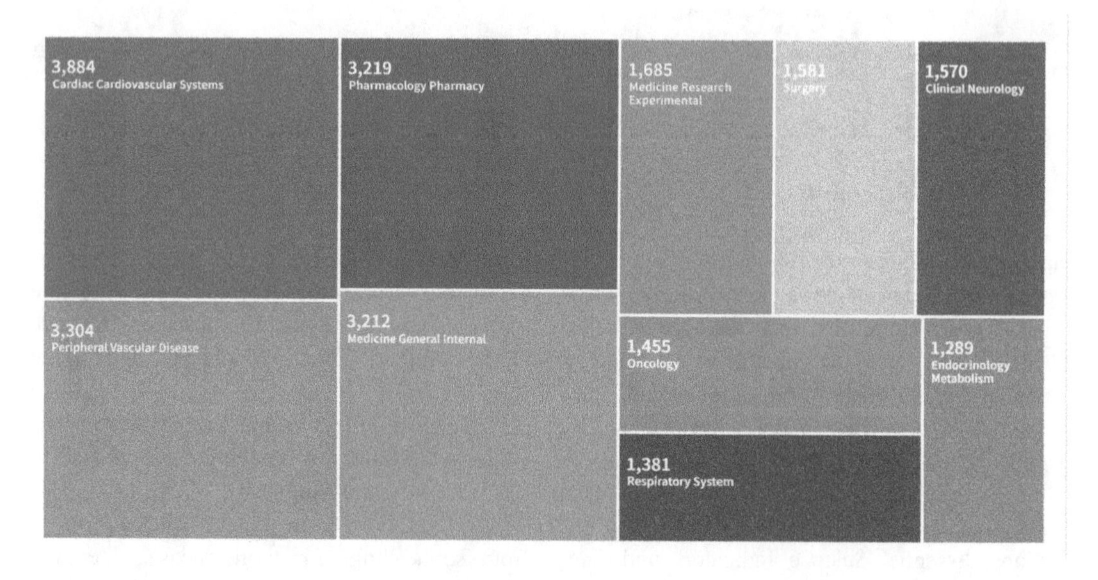

Fig. 37.14 Web of Science: tree map with Top 10 categories of search results (from the search in the above figure)

Review, Meta-Analysis, Cochrane Meta-Analysis, Guidelines, Consensus Conference); (2) Clinical Studies (Clinical Trial, Randomized Controlled Trial, Cross-Over Study); (3) Epidemiological Studies (All, Cross-Sectional Study, Cohort-Prospective or not, Case-Control Study); (4) Multi Selection (Case Reports, Editorial, Letter, Meta-Analysis, Multicenter Study, Randomized Controlled Trial, Review, Systematic Review).

37.6 Scopus Literature Search

Scopus (https://www.scopus.com/home.uri) is considered one of the largest abstracts and citation databases. Scopus is a product of Elsevier,

Fig. 37.15 The homepage for BibliMed search

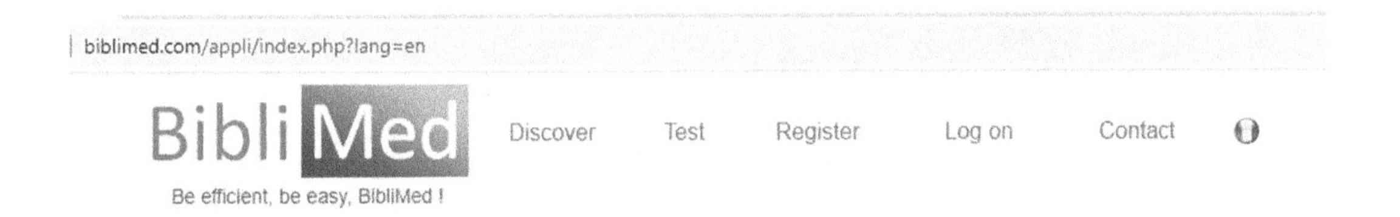

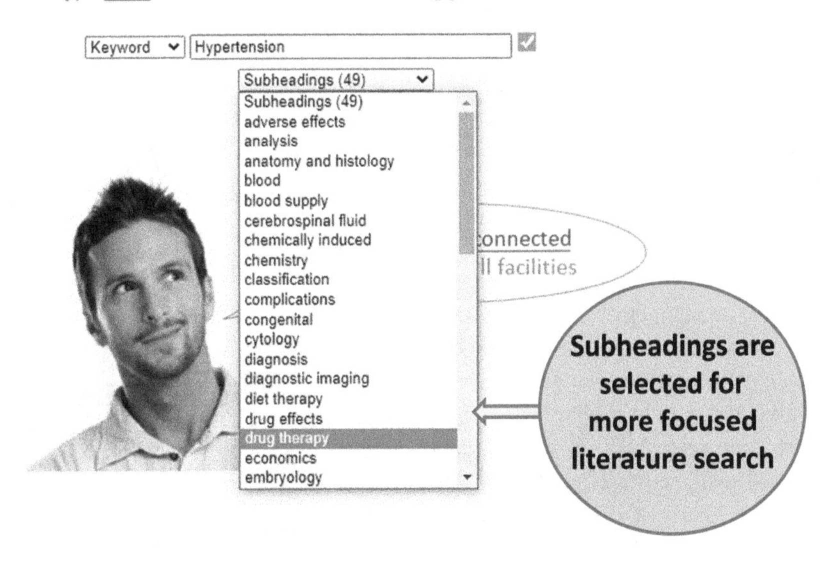

Fig. 37.16 BibliMed and its key feature of accessing subheadings options for more focused literature searching

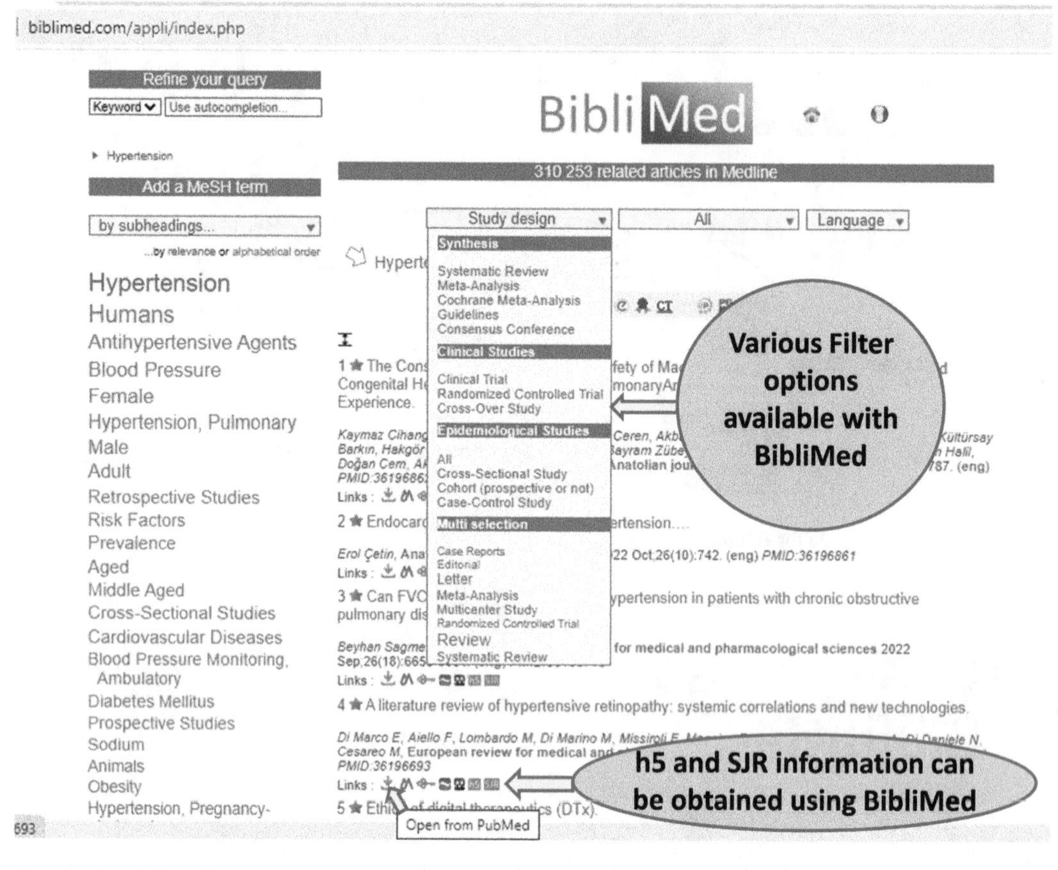

Fig. 37.17 Various filter options and other key features of BibliMed

and it has been launched in 2004. Scopus is considered the world's largest abstract and citation database of peer-reviewed research literature with over 22,000 titles from more than 5000 international publishers [20]. In addition to literature search, Scopus offers citation searching and h-index, which are its important features.

Scopus combines a comprehensive abstract and citation database with enriched data and scholarly literature across many disciplines. In addition, it can help researchers identify subject experts, and gain access to reliable data, analytical tools, and metrics [21]. The key features of Scopus are listed in Box 37.1. Scopus provides literature in the fields of medicine, science, technology, social sciences and the arts and humanities, from which one can search for specific information on topics, authors, journals or books [22].

Box 37.1 Key Features of Scopus [22]

1. *Researchers and authors*: Scopus helps researchers and authors meet demands to stay productive and upsurge their research output.
2. *Editors and reviewers*: Scopus provides insights into authors' areas of expertise, helping to ease the review process. Scopus provides authors citations and h-index data.
3. *Educators and students*: Scopus permits educators and students to access the previous and current scientific articles, enhancing their learning within and beyond the classroom.

(continued)

Box 37.1 (continued)

4. *Librarians*: Scopus enables librarians to help and support in accomplishing the literature need of researchers, students, and faculty. Scopus greatly supports accessing reliable data, metrics, and analytical tools.
5. *Research administrators*: Scopus provides crucial insights on research impact and output, helping with strategic decisions.

Major search options in Scopus include searching for a document, an author, an affiliation, and advanced search. The document search can be done within article title, abstract, keywords, source title, chemical name, Chemical Abstracts Service (CAS) number, ISSN, DOI, references, conference, and others. The affiliation search provides information on the documents published by a college, institution or university. The author search can be done by entering last name (mandatory) and first name (optional). In addition to retrieving documents, the author search provides information on 'hindex' for an author and 'citations' for the author's published documents. The citation details can be exported in Excel sheet. Moreover, various Scopus metrics for a document include (1) percentile score, (2) Field-Weighted citation impact, (3) Views count for the current and previous years, and (4) PlumX metrics such as number of readers, abstract views, etc. [23]. The home page for an author search is shown in Fig. 37.18 [20].

CiteScore is the key metric provided by Scopus and is calculated annually. CiteScore 2021 calculation methodology has been updated. CiteScore 2021 counts the citations received in 2018–2021 for regular articles, reviews, book chapters, conference papers, and data papers published in 2018–2021, and divides the citations by the number of publications published in 2018–2021 [24]. In addition to literature search, Scopus provides information on journal source titles, highest percentile, citations for previous 4 years (for example, 2018–2021), documents published by a journal for previous 4 years (2018–2021), % cited, Source Normalized Impact per Paper (SNIP), and SJR [24].

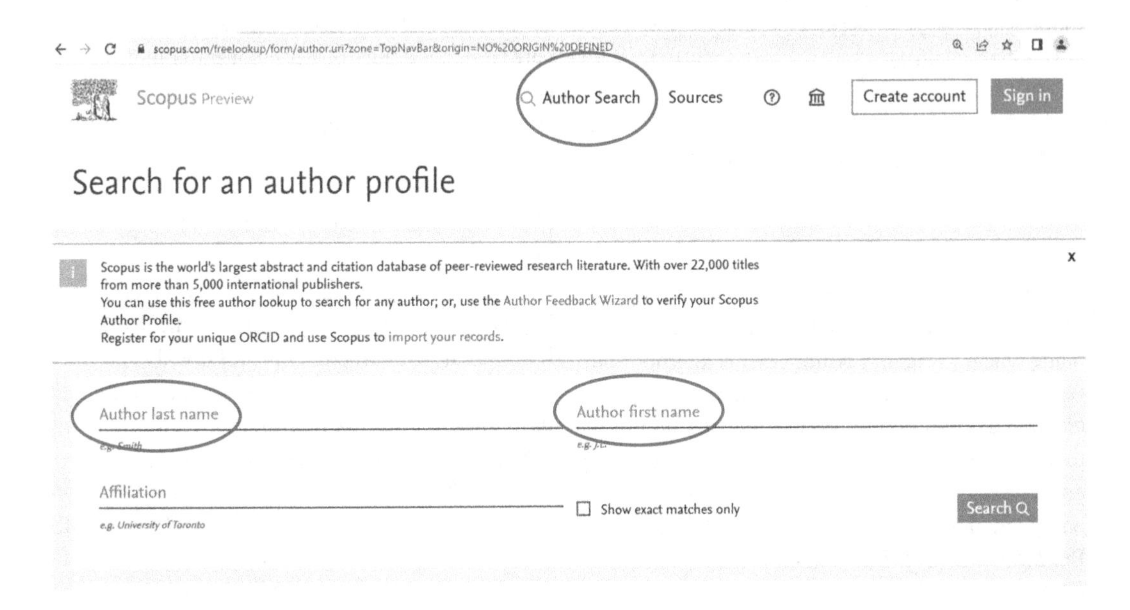

Fig. 37.18 The home page for an author search in Scopus

37.7 Literature Search Using Google Scholar

Google Scholar (https://scholar.google.com/) is a freely accessible web search engine that provides a simple way to search broadly for desired scholarly literature. Google Scholar searches across many disciplines and sources for articles, abstracts, theses, books, and more, from academic publishers, universities, professional societies, and online repositories, among others [25]. In addition, Google Scholar metrics enable authors to quickly view overall citations, citations within the last 5-years, and metrics such as the h-index, and i10-index, and also journal h5-index and h5-median [26].

37.8 Literature Search Using JSTOR

Part of the nonprofit organization ITHAKA, JSTOR (https://about.jstor.org) contains over 12 million journal articles, 100,000 books and millions of images and primary sources in 75 disciplines. It includes both free and subscription content; the free content includes open access journals, books, images, and media, as well as more than 39,000 research reports from over 140 policy institutes.

37.9 Literature Search Using SciFinder®

SciFinder® is produced by CAS. It is the most comprehensive database for chemical literature, searchable by author, topic, and substances by name or CAS Registry Number. It is an important research tool for chemistry, biochemistry, nanotechnology, physics, materials science, environmental science, chemical engineering, and other engineering and science disciplines. SciFinder® is complementary to other databases such as WoS, PubMed, etc. [27]. SciFinder® provides access to CAS REGISTRY℠ with substance information such as organic and inorganic molecules, DNA, RNA, proteins, polymers and Markush structures. SciFinder® affords convenient access to (1) organic and inorganic substances, (2) DNA and protein sequences, (3) experimental properties and properties predicted by state-of-the-art technology, (4) ^{13}C and ^{1}H NMR spectra and mass spectra, (5) bioactivity and target indicator (protein, enzyme, glycoprotein, among others) information, (6) regulatory information, (7) substances found in patents, journals and reputable web sources around the world, etc. [28]. CAS SciFinder-n is the latest scientific information solution from CAS, a division of the American Chemical Society [29].

37.10 Literature Search Using CINAHL Ultimate, MEDLINE Ultimate and BIOSIS Previews

CINAHL Ultimate (EBSCO product) is the largest collection of leading nursing and allied health literature resources, providing full text for many of the most-used scientific journals in the CINAHL index. CINAHL Ultimate covers more than 50 nursing specialties and includes quick lessons, research instruments, evidence-based care sheets, and other features. Subjects covered in the EBSCO database include: (1) ambulatory care nursing, (2) cardiovascular nursing, (3) critical care nursing, (4) emergency medical technicians, (5) gerontologic nursing, (6) nuclear medicine technicians, (7) occupational therapy, (8) oncologic nursing, (9) pediatric nursing, (10) physical therapy, and (11) psychiatric nursing [30].

MEDLINE Ultimate (EBSCO product) is the largest collection of leading biomedical full-text journals. It offers medical professionals and relevant researchers access to evidence-based and peer-reviewed full-text contents from more of the uppermost biomedical journals [31].

BIOSIS Previews® is produced by the WoS Group. Available from EBSCO, BIOSIS Previews® is considered an expansive index to life sciences and biomedical research from journals, books, meetings, and patents, while the database covers experimental and preclinical research, methods, instrumentation, animal studies, and more [32].

37.11 Literature Search Using The Cochrane Library

Owned by Cochrane and published by Wiley, the Cochrane Library is a collection of databases having different types of high-quality, independent evidence to assist in healthcare decision-making (https://www.cochranelibrary.com). The databases of the Cochrane Library include: (1) the Cochrane Database of Systematic Reviews (CDSR), (2) the Cochrane Central Register of Controlled Trials (CENTRAL), and (3) Cochrane Clinical Answers (CCAs) [33]. In addition, it has a federated search feature that incorporates results from external databases (Fig. 37.19).

(1) The CDSR is the database for systematic reviews in health care, while it includes Cochrane Reviews (systematic reviews), protocols for Cochrane Reviews, editorials and supplements. The CDSR is built throughout the month, and the new and updated reviews and protocols are continuously published when ready [34]. (2) The CENTRAL database is a source of reports of randomized and quasi-randomized controlled trials. The CENTRAL records often include a summary of the article, and do not contain the full text of the article. It is published monthly [34]. (3) The CCAs provide a clinically-focused entry point to rigorous research from Cochrane Reviews. They are designed to inform point-of-care decision-making, while each CCA comprises a clinical question, a short answer, and data for the outcomes from the Cochrane Review deemed most relevant to practising healthcare professionals [34].

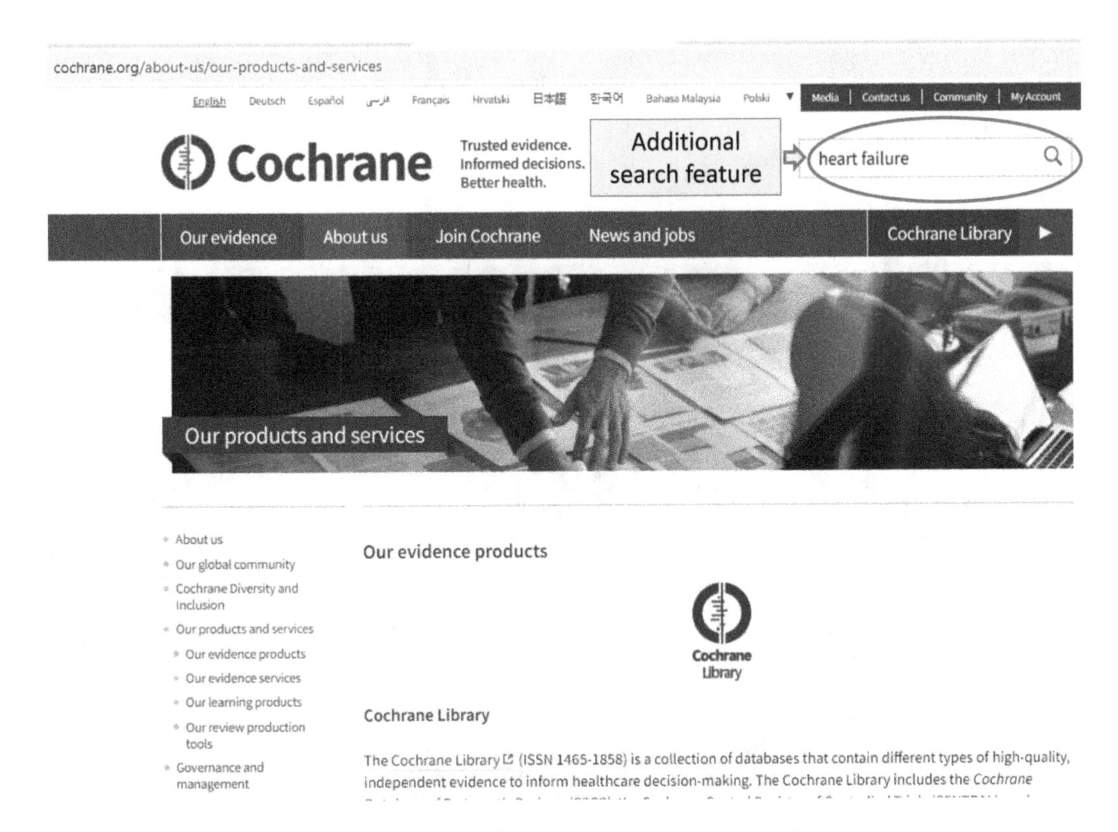

Fig. 37.19 The home page for Cochrane Library that includes an additional search feature

37.12 Literature Search Using International Pharmaceutical Abstracts

Produced by the WoS Group, 'International Pharmaceutical Abstracts (IPA)' is a comprehensive database providing indexing and abstracts for pharmaceutical, and medical journals published globally. This database is the product of EBSCO [35]. It covers the entire spectrum of pharmaceutical information, with subject coverage including: (1) biopharmaceuticals and pharmacokinetics; (2) legal, political and ethical issues; (3) new drug delivery systems; and (4) pharmacist liability. While IPA covers the entire spectrum of drug therapy and pharmaceutical information, inclusion of the study design, number of patients, dosage, dosage schedule, and dosage forms are features of IPA's clinical studies abstracts [35].

37.13 Other Databases: PubChem

Launched in 2004, PubChem, an open chemistry database from the National Institutes of Health (NIH), has become a key resource for chemical information for students, scientists, and the general public. While PubChem mostly contains small molecules, it also contains information on larger molecules such as nucleotides, lipids, carbohydrates, peptides, and chemically-modified macromolecules. This database collects information on chemical structures, chemical and physical properties, biological activities, health, safety and toxicity information, patents, and more [36]. PubChem is considered the world's largest collection of freely accessible chemical information, while searches on chemicals can be performed by name, structure, molecular formula, and other identifiers (https://pubchem.ncbi.nlm.nih.gov/). The retrieving details include: (1) structures, (2) names and identifiers, (3) chemical and physical properties, (4) spectral information, (5) chemical vendors, (6) drug and medication information, (7) pharmacology and biochemistry, (8) safety and hazards, (9) toxicity, (10) literature, (11) patents, and (12) taxonomy, among others. The homepage for PubChem is shown in Fig. 37.20.

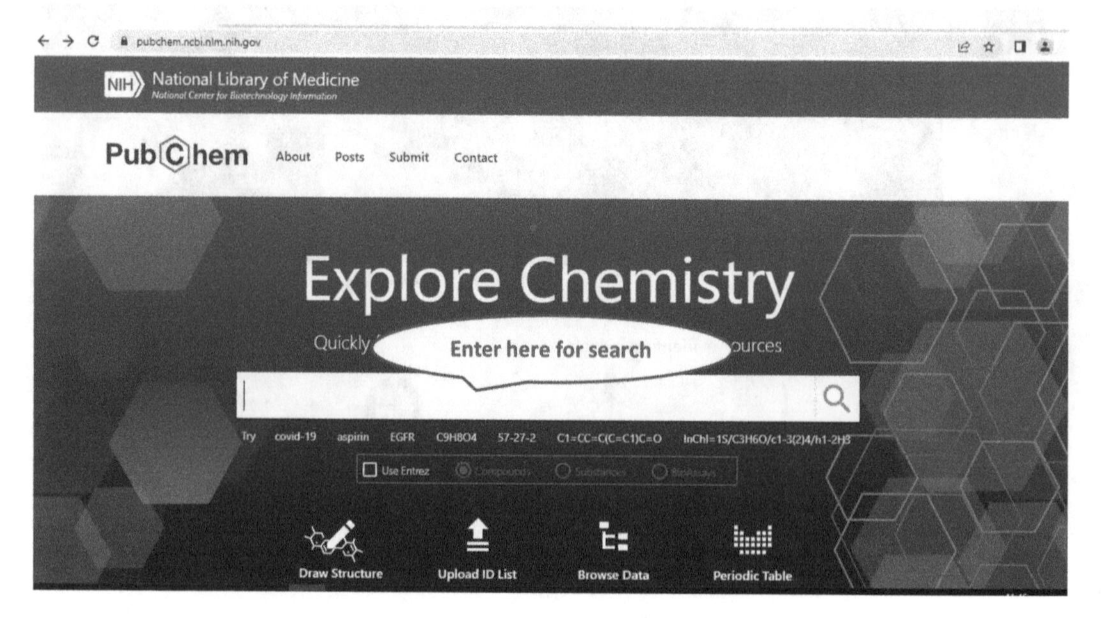

Fig. 37.20 The home page for searching PubChem

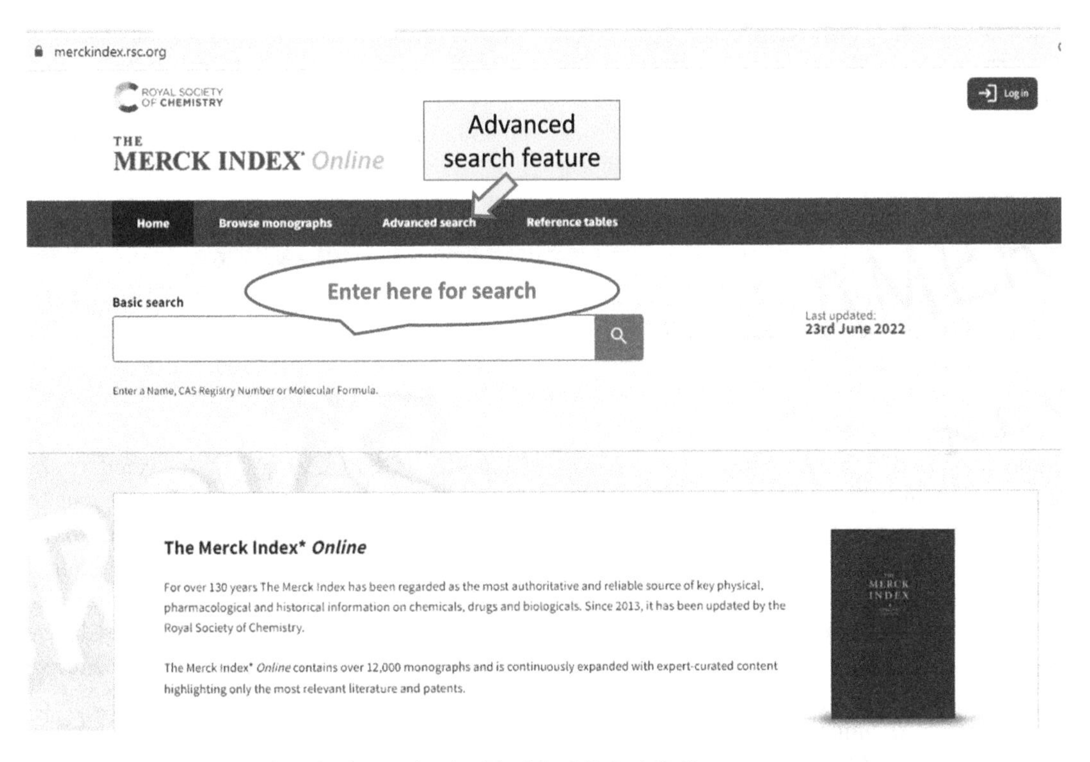

Fig. 37.21 The homepage for a basic search using The Merck Index* *Online*

37.14 Other Databases: The Merck Index

The Merck Index has been the source of information on chemicals, drugs and biologicals for over 120 years. It is now available online from the Royal Society of Chemistry [37]. The Merck Index* *Online* offers the same information as the print edition in an easily searchable full-text database. This resource contains over 11,500 monographs [37]. Considered Chemistry's Constant Companion™, the coverage of the Merck Index* *Online* includes: (1) human and veterinary drugs, (2) biotech drugs and monoclonal antibodies, (3) substances used for medical imaging, (4) biologicals and natural products, (5) plant and herbal medicines, (6) nutraceuticals and cosmeceuticals, (7) laboratory reagents and catalysts, (8) dyes, colour and indicators, (9) environmentally significant substances, (10) food additives and nutritional supplements, (11) flavours and fragrances, and (12) agricultural chemicals, pesticides and herbicides, and industrial and specialty chemicals [37]. The homepage for performing a basic search using The Merck Index* *Online* is shown in Fig. 37.21.

37.15 Other Databases: PharmaPendium

PharmaPendium is a comprehensive source of preclinical and clinical pharmaceutical data and information from regulatory documents [38]. It provides comparative regulatory-based evidence in a single database, apprising researchers of critical drug development activities. Subscribers can get access to searchable FDA and EMA regulatory documents, extracted pharmacokinetic, safety, efficacy, and metabolizing enzyme data, and more [39]. PharmaPendium helps searchers to extract data from FDA drug approval documents back to 1938 and EMA drug approval documents back to 1995, FDA Advisory Committee meeting reports, *Meyler's Side Effects of Drugs*, and journal articles [38]. In addition,

PharmaPendium helps organize and connect to: (1) pharmacokinetic data, (2) metabolizing enzymes and transporters data, (3) drug safety data, (4) FAERS (FDA Adverse Event Reporting System) data, (5) chemistry search, (6) clinical trial data, (7) pharmacology data, and (8) toxicity, and drug-drug interaction data [38].

37.16 Other Databases: AHFS Clinical Drug Information

The American Hospital Formulary Service® Drug Information (AHFS DI) from the American Society of Health-System Pharmacists® (ASHP) (www.ashp.org) provides evidence-based information for safe and effective drug therapy. The AHFS Drug Information has been the widely used drug information and safety resource for over 60 years [40]. With AHFS CDI (Clinical Drug Information), the information is now digital and incorporates information on current drug shortages, and real-time updates on drugs and safety. Using AHFS CDI, a searcher is able to (1) access drug shortages information, (2) get quick access to comprehensive and evidence-based drug information, (3) perform a full-text search, (4) access thousands of monographs, (5) utilize linking to over a dozen related resources and databases, (6) get off-label uses, (7) view incorporated FDA safety data, and more [40].

37.17 Other Databases: Facts and Comparisons®

Facts and Comparisons is a drug referential resource, bringing together evidence-based content and drug comparative tools and tables [41]. Using the Facts and Comparisons Product Availability tool, a searcher is able to find important drug information for patients on drug access, with additional information such as dosing, labeler, and Orange Book AB ratings from the US FDA. Moreover, the easy-to-use drug comparisons tool allows a seeker to search at a time up to four drugs and subsequently select various patient-relevant information such as adverse drug effects, considerations for use during pregnancy conditions, drug-drug interactions, contraindications and precautions, and others [41]. Overall, in Drug Facts and Comparisons®, drugs are searchable by generic or brand name and are divided into related therapeutic or pharmacologic groups for easy comparison [41].

37.18 Other Databases: Clinical Pharmacology

Clinical Pharmacology (powered by ClinicalKey®) delivers information that can be used to make medication decisions based on safety and efficacy [42]. Using Elsevier's Clinical Pharmacology, a searcher can retrieve information tailored for the continuum of care. Clinical Pharmacology provides knowledge to help recommend effective drug therapy [42].

37.19 Other Databases: PharmGKB

PharmGKB is the NIH-funded resource database that provides information on how human genetic variation could affect responses to medications. It is a wide-ranging resource that curates knowledge with regard to the impact of genetic variation on drug response for researchers and clinicians (https://www.pharmgkb.org/). It helps with searching on pharmacogenomics, the study of the relationship on how our body responds to medications because of genetic variations. PharmGKB gathers, curates and disseminates information on clinically actionable gene-drug associations and genotype-phenotype relationships.

37.20 Other Databases: Knovel

Knovel (https://app.knovel.com/kn), from Elsevier, contains engineering and technical reference books, databases, tools and calculators, and a template for searching material properties. It covers the areas of biochemistry, biology, and biotechnology; chemistry and chemical engineering; engineering management and leadership; food science; general engineering and product administration; pharmaceuticals, cosmetics, and toiletries; and safety and industrial hygiene.

37.21 Searching Through "Food Science and Technology Abstracts"

Developed by IFIS (originally International Food Information Service) and indexed in WoS, Ovid, and EBSCOHost, Food Science and Technology Abstracts (FSTA) (https://www.ifis.org/fsta) contains 1,700,000 abstracts from journals, trade publications, conference proceedings, and patents relating to food science, food technology, nutrition, and related areas.

37.22 Searching Through "Toxicology Literature" from NLM (Formerly TOXNET)

Most of the NLM's content formerly in TOXNET has been integrated into other NLM products (see https://www.nlm.nih.gov/toxnet/index.html). Repositioned subject areas include comprehensive toxicology, toxicology literature, breastfeeding and drugs, chemicals, developmental toxicology, drug-induced liver injury, chemical releases and mapping, comparative toxicogenomics, household product safety, occupational exposure to chemicals, dietary supplements, and risk assessment.

37.23 FDA Resources for Drug Information

37.23.1 Drugs@FDA

This online database provides information about FDA-approved brand name and generic prescription and over-the-counter human drugs and biological therapeutic products [https://www.fda.gov/drugs]. It includes information on most of the drug products approved since 1939. Moreover, the majority of patient information, labels, approval letters, reviews, and other information are also available, with this resource, for drug products approved since 1998 [43].

37.23.2 Orange Book

The publication *Approved Drug Products with Therapeutic Equivalence Evaluations* is commonly known as the 'Orange Book', which identifies drug products that are approved on the basis of safety and effectiveness by the FDA. The Orange Book also contains therapeutic equivalence evaluations for the approved multisource prescription drug products [43] [https://www.accessdata.fda.gov/scripts/cder/ob/index.cfm].

37.23.3 Purple Book Database

The 'Purple Book' is available both as lists (lists of Licensed Biological Products with Reference Product Exclusivity and Biosimilarity or Interchangeability Evaluations) and a searchable online database. The FDA has transitioned the Purple Book to a searchable online database which encompasses information on biological products, including biosimilar and interchangeable biological products, licensed (approved) by the FDA. Currently, the searchable database comprises information on all FDA-licensed biological products regulated by the Center for Drug Evaluation and Research (CDER), including licensed biosimilar and interchangeable products, and their reference products, and FDA-licensed allergenic, cellular and gene therapy, hematologic, and vaccine products regulated by the Center for Biologics Evaluation and Research (CBER) [44] [https://purplebooksearch.fda.gov/].

37.23.4 Inactive Ingredient Database

FDA's Inactive Ingredient Database, updated quarterly, provides information on inactive ingredients in FDA-approved drug products, which can be used by industry in drug product development [45]. It can be searched by entering any portion of a name (at least three characters). Results are displayed alphabetically, sorted first by ingredient, next by route of administration and dosage form. Other fields include CAS number, Unique Ingredient Identifier, potency amount, and maximum daily exposure. Contents of the database are available as a download provided as delimited text and Excel files [46].

37.24 Product Information Using DailyMed

The NLM provides DailyMed database to the public. It contains labelling for the products such as (1) FDA-approved products (prescription drug and biological products for human use, drug products, and biological products), (2) non-prescription (over-the-counter) drug and biological products for human use, (3) certain medical devices for human use, (4) medical gases for human and animal use, (5) prescription and non-prescription drugs for animal use, and (6) additional products regulated, but not approved, by the FDA such as certain medical devices, cosmetics, dietary supplements, medical foods and unapproved prescription and non-prescription products [47].

37.25 DrugBank Online: The Database for Drug and Drug Target Information

Launched in 2006, DrugBank Online (https://go.drugbank.com/about) is a free-to-access resource and a comprehensive database encompassing information on drugs and drug targets. Being a combined bioinformatics and a cheminformatics resource, DrugBank Online amalgamates detailed information on drug data (i.e. chemical, pharmacological and pharmaceutical) with comprehensive drug target (i.e. sequence, structure, and pathway). It is used by the students, academicians, drug industry, medicinal chemists, pharmacists, physicians, and the general public [48].

37.26 Micromedex®

Micromedex®, from Merative, is a suite of databases focusing on medication, disease, and toxicology management (https://www.micromedexsolutions.com/micromedex2/100.1.3.738/WebHelp/MICROMEDEX_2.htm?navitem=headerHelp#Home_Page/Home_Page.htm). It includes resources related to drug interactions, IV compatibility, drug identification, drug comparison, and pediatrics; as well as access to the RED BOOK, which includes product and pricing information on prescription and over-the-counter medications, nutraceuticals, bulk chemicals, medical devices, and medical supplies. Additional tools and resources include: (1) an extensive list of medical calculators, (2) Black Box warnings, (3) comparative tables, (4) a “do not confuse” drug list, (5) drug classes, (6) drug consults (evidence-based articles covering a wide range of drug therapy topics, providing specific guidance), and (7) drug risk evaluation and mitigation strategy (REMS) program information.

37.27 Concluding Remarks

To facilitate the proficiency in finding a target information in the ocean of published literature, a searcher needs to conduct a smart and efficient literature search employing right search engines and databases. In this chapter, we have discussed several search engines and biomedical databases. A researcher should be familiar with the types of these needed to retrieve the desired information. Yet, while selecting the appropriate database(s) is an important aspect of literature searching, it is equally essential to systematically map out a search strategy to achieve the necessary proficiency. Ideally, results of a focused biomedical literature search provide essential information and current knowledge on a relevant topic within the world of published research. To add to the complexity of literature searching, the quantity of information being published is rapidly expanding daily, making the search process—not to mention locating the necessary information—increasingly challenging.

While this chapter is by no means complete, since databases and search engines—both free and fee-based—are being developed continually by a variety of organizations, we hope we have added tools and techniques to the beginning researcher's toolbox, and potentially provided a refresher for more experienced researchers.

Conflict of Interest No conflict of interest is declared.

References

1. Grewal A, Kataria H, Dhawan I (2016) Literature search for research planning and identification of research problem. Indian J Anaesth. 60(9):635–639
2. Balakumar P, Marcus SJ, Jagadeesh G (2012) Navigating your way through online resources for biomedical research. RGUHS J Pharm Sci 2(3):5–27
3. Jagadeesh G, Murthy S, Gupta YK, Prakash A (eds) (2010) Biomedical research—from ideation to publication, 1st edn. Wolters Kluwer Health-Lippincott Williams and Wilkins
4. Balakumar P, Murthy S, Jagadeesh G (2007) The basic concepts of scientific research and communication (A report on preconference workshop held in conjunction with the 40th annual conference of the Indian Pharmacological Society-2007). Indian J Pharmacol 39:303–306
5. Balakumar P, Jagadeesh G (2012) The basic concepts of scientific research and scientific communication (A report on the pre-conference workshop held in conjunction with the 63rd annual conference of the Indian Pharmaceutical Congress Association-2011). J Pharmacol Pharmacother 3:178–182
6. Balakumar P, Inamdar MN, Jagadeesh G (2013) The critical steps for successful research: The research proposal and scientific writing: (A report on the pre-conference workshop held in conjunction with the 64th annual conference of the Indian Pharmaceutical Congress-2012). J Pharmacol Pharmacother 4: 130–138
7. Balakumar P, Srikumar BN, Ramesh B, Jagadeesh G (2022) The critical phases of effective research planning, scientific writing, and communication. Phcog Mag 18:1–3
8. The PubMed Search. https://pubmed.ncbi.nlm.nih.gov/about/. Accessed 24 Sep 2022
9. The PubMed Search. https://www.nlm.nih.gov/medline/medline_overview.html. Accessed 24 Sep 2022
10. The PubMed Search. https://www.ncbi.nlm.nih.gov/pmc/about/intro/. Accessed 24 Sep 2022
11. The PubMed Search. https://www.ncbi.nlm.nih.gov/books/. Accessed 24 Sep 2022
12. The PubMed Search. https://www.nlm.nih.gov/databases/databases_oldmedline.html. Accessed 24 Sep 2022
13. The PubMed Search. https://pubmed.ncbi.nlm.nih.gov/help/#author-search. Accessed 25 Sep 2022
14. The PubMed Search. https://pubmed.ncbi.nlm.nih.gov/citmatch/. Accessed 25 Sep 2022
15. The PubMed Search. https://www.nlm.nih.gov/mesh/meshhome.html. Accessed 26 Sep 2022
16. The PubMed Search. https://pubmed.ncbi.nlm.nih.gov/help/. Accessed 27 Sep 2022
17. The PubMed Search. https://pubmed.ncbi.nlm.nih.gov/help/#combining-with-boolean-operators. Accessed 28 Sep 2022
18. The PubMed Search. https://pubmed.ncbi.nlm.nih.gov/help/#help-filters. Accessed 28 Sep 2022
19. The PubMed Search. https://pubmed.ncbi.nlm.nih.gov/clinical/. Accessed 4 Oct 2022
20. The Scopus Search. https://www.scopus.com/freelookup/form/author.uri?zone=TopNavBar&origin=NO%20ORIGIN%20DEFINED. Accessed 8 Oct 2022
21. The Scopus Search. https://www.elsevier.com/solutions/scopus?dgcid=RN_AGCM_Sourced_300005030. Accessed 7 Oct 2022
22. The Scopus Search. https://www.elsevier.com/solutions/scopus/academic-institutions. Accessed 7 Oct 2022
23. The Scopus Search. https://www.scopus.com/search/form.uri?zone=TopNavBar&origin=searchbasic&display=basic#basic. Accessed 8 Oct 2022
24. The Scopus Search. https://www.scopus.com/sources. Accessed 8 Oct 2022
25. Google Scholar Search. https://scholar.google.com/intl/en/scholar/about.html. Accessed 9 Oct 2022
26. Google Scholar Search. https://scholar.google.com/citations?view_op=metrics_intro&hl=en. Accessed 9 Oct 2022
27. https://ucsd.libguides.com/scifinder/about. Accessed 10 Oct 2022
28. https://www.cas.org/products/scifinder/content-details#substance. Accessed 10 Oct 2022
29. https://blogs.cranfield.ac.uk/library/introducing-scifinder-n-and-an-invitation-to-an-online-demo/. Accessed 10 Oct 2022
30. https://www.ebsco.com/products/research-databases/cinahl-ultimate. Accessed 11 Oct 2022
31. https://www.ebsco.com/products/research-databases/medline-ultimate. Accessed 11 Oct 2022
32. https://www.ebsco.com/products/research-databases/biosis-previews. Accessed 11 Oct 2022

33. The Cochrane Library. https://www.cochrane.org/about-us/our-products-and-services. Accessed 12 Oct 2022
34. The Cochrane Library. https://www.cochranelibrary.com/about/about-cochrane-library. Accessed 12 Oct 2022
35. International Pharmaceutical Abstracts. https://www.ebsco.com/products/research-databases/international-pharmaceutical-abstracts. Accessed 14 Oct 2022
36. PubChem Search. https://pubchemdocs.ncbi.nlm.nih.gov/about. Accessed 15 Oct 2022
37. The Merck Index Search. https://www.rsc.org/Merck-Index/info/rsc-database-introduction. Accessed 15 Oct 2022
38. PharmaPendium. https://www.elsevier.com/solutions/pharmapendium-clinical-data. Accessed 16 Oct 2022
39. PharmaPendium. https://www.elsevier.com/__data/assets/pdf_file/0005/1234814/PLS_DDD_PP_FS_PharmaPendium_base_Fact_Sheet_WEB_rebrand.pdf. Accessed 16 Oct 2022
40. AFHS Clinical Drug Information. https://www.ahfsdruginformation.com/ahfs-clinical-drug-information/. Accessed 16 Oct 2022
41. Drug Facts and Comparisons®. https://www.wolterskluwer.com/en/solutions/lexicomp/facts-and-comparisons. Accessed 17 Oct 2022
42. Clinical Pharmacology powered by ClinicalKey®. https://www.elsevier.com/solutions/clinical-pharmacology. Accessed 23 Oct 2022
43. The FDA Resources. https://www.fda.gov/drugs/drug-approvals-and-databases/resources-information-approved-drugs. Accessed 29 Oct 2022
44. The Purple Book. https://www.fda.gov/drugs/therapeutic-biologics-applications-bla/purple-book-lists-licensed-biological-products-reference-product-exclusivity-and-biosimilarity-or. Accessed 29 Oct 2022
45. https://www.accessdata.fda.gov/scripts/cder/iig/index.cfm. Accessed 29 Oct 2022
46. https://www.fda.gov/drugs/drug-approvals-and-databases/inactive-ingredients-approved-drug-products-search-frequently-asked-questions#purpose. Accessed 29 Oct 2022
47. The DailyMed. https://dailymed.nlm.nih.gov/dailymed/about-dailymed.cfm. Accessed 30 Oct 2022
48. DrugBank Online. https://go.drugbank.com/about. Accessed 30 Oct 2022

Literature Reviews: An Overview of Systematic, Integrated, and Scoping Reviews

38

John R. Turner

Abstract

Literature reviews are a main part of the research process. Literature Reviews can be stand-alone research projects, or they can be part of a larger research study. In both cases, literature reviews must follow specific guidelines so they can meet the rigorous requirements for being classified as a scientific contribution. More importantly, these reviews must be transparent so that they can be replicated or reproduced if desired. The rigorous requirements set out by the National Science Foundation (NSF) and the Preferred Reporting Items for Systematic Reviews and Meta-Analyses (PRISMA) aim to support researchers in conducting literature reviews as well as address the replication crisis that has challenged scientific disciplines over the past decade. The current chapter identifies some of the requirements along with highlighting different types of reviews and recommendations for conducting a rigorous review.

Keywords

Literature review · Integrated review · Systematic review · Scoping review · Cooper's taxonomy · Synthesis

J. R. Turner (✉)
University of North Texas, Denton, TX, USA
e-mail: john.turner@unt.edu

38.1 Introduction

The field of science has been critically reviewed in the last decade due to problems with reproducing and replicating published research studies. Some estimate this irreproducibility to range from 75% to 90% [1], depending on the discipline and type of study. This replication crisis has affected multiple disciplines (social, behavioral, medical) due to their inability of reproducing or replicating published studies [2]. This replication crisis has produced a new emphasis on promoting more robust research practices [3] within the social and behavioral sciences. Many researchers and administrators are concerned that this replication crisis may contribute to, rather than lessen, misinformation or distortions from reality [4].

To address this reproduction crisis, the National Science Foundation (NSF) implemented a subcommittee to identify what actions they could make to further promote robust research practices. Some recommendations made by the NSF committee include: requiring researchers to provide enough transparent information so that reviewers could reproduce the results; conducting research to evaluate different approaches of replication; emphasizing reporting congruent with robust research practices such as details relating to "conceptualization, operationalization, experimental control over other potential independent variables, statistical power, execution, analysis, interpretation" [5]. Other researchers have called

G. Jagadeesh et al. (eds.), *The Quintessence of Basic and Clinical Research and Scientific Publishing*,
https://doi.org/10.1007/978-981-99-1284-1_38

for an awareness of the current system (replication crisis) as it relates to research in general and preclinical research specifically [1].

The first main point relating to transparency is mostly relevant to all reviews of the literature. This point is highlighted in the NSF committee report by stating that science should: "Make reproduction possible, efficient, and informative, researchers should sufficiently document the details of the procedures used to collect data, to convert observations into analyzable data, and to analyze data" [5].

Transparency applies to literature review studies as it relates to aiding researchers in either reproducing or replicating a published study. If the data (literature), data retrieving mechanisms, search terms, context, and problem are not clearly identified it becomes more difficult to reproduce or replicate a study. To improve the robustness of literature reviews, and to help improve the reporting of literature reviews, the Preferred Reporting Items for Systematic Reviews and Meta-Analyses (PRISMA) was formed. Researchers published The PRISMA 2020 statement (revision from 2009) that provides guidelines for researchers when conducting systematic reviews or meta-analyses. The guidelines provided by PRISMA address these transparency issues and help to reduce the replication crisis in academia. Their guidelines are designed to "help systematic reviewers transparently report why the review was done, what the authors did, and what they found" [6]. The PRISMA website (https://www.prisma-statement.org) is a great resource for researchers looking to conduct any type of literature review.

This chapter aids researchers in conducting literature reviews by identifying what a literature review is, what the purpose of a literature review is, and what the benefits of a literature review are compared to other methods. The characteristics of a literature review are then presented along with identifying three main types of literature reviews (systematic, integrated, scoping) along with checklists for each to help aid researchers in their planning stages. Because one essential component of a literature review involves synthesizing the data (literature), we discuss different techniques of synthesis along with providing a few simple steps to help researchers get started with the synthesis process. We conclude with general recommendations for researchers to aid in writing clearly and concise literature reviews.

38.2 Literature Reviews

A literature review is probably the most common academic writing activity that is performed by scholars and graduate students. Imel [7] identified a literature review as being either part of a larger study or as a research effort on its own. As a part of a larger study, Imel [7] identified the literature is "the foundation for the study." It has been suggested that the literature review for a larger research project as setting the "context of the study, clearly demarcates what is and what is not within the scope of the investigation, and justifies those decisions" [8]. As a stand-alone research method, literature reviews can identify future research, point out gaps or discrepancies in the literature, highlight unresolved issues, or provide new perspectives [7]. Bryman [9] also highlighted that literature reviews can be used as "a means of showing why your research questions are important."

38.3 What Is a Literature Review

A literature review in its most fundamental structure provides an account of what has already been published in the peer-reviewed literature [10]. The purpose of a literature review is to "convey to the reader what knowledge and ideas have been established on a topic, and what their strengths and weaknesses are" [10]. The primary purpose of a literature review is NOT to portray a list of what has been written, a literature review should:

- be organized around and related directly to the thesis or research question you are developing,
- synthesize results into a summary of what is and is not known,

- identify areas of controversy in the literature, and
- formulate questions that need further research [10].

Additionally, literature reviews should also address the following concerns:

- Is the topic selected of interest to your audience, the industry and/or the publication you are targeting?
- Does the literature make a significant, value-added contribution to new thinking in the field [11]?
- Does the literature add value to the field? What value?
- Does the literature identify a discrepancy in the literature, identify a gap in the literature, identify conflicting/contradictory views in the existing literature, or identify a change in common trends?

38.3.1 Purpose for Exploring Literature

The purpose of a literature review is to generate novel ideas and to present new knowledge. Literature reviews are designed to explore the literature and identify:

- What is already known about this area?
- What concepts and theories are relevant to this area?
- What research methods and research strategies have been employed in studying this area?
- Are there any significant controversies?
- Are there any inconsistencies in findings relating to this area?
- Are there any unanswered research questions in this area [9]?

38.3.2 Benefits of a Literature Review

The main benefits that can be gained from conducting a literature review include the following:

- You need to know what is already known in connection with your research area because you do not want to be accused of reinventing the wheel.
- You can learn from other researchers' mistakes and avoid making the same ones.
- You can learn about different theoretical and methodological approaches to your research area.
- It may help you develop an analytic framework.
- It may lead you to consider the inclusion of variables in your research that your might not otherwise have thought about.
- It may suggest further research questions for you.
- It will help with the interpretation of your findings.
- It gives you some pegs on which to hang your findings.
- It is expected [9]!

38.4 Taxonomy of Literature Reviews

Cooper [12] presented a taxonomy of six characteristics distinguishing between literature reviews. This taxonomy highlights the different options available to the researcher when planning a literature review. These characteristics include focus, goal, perspective, coverage, organization, and audience (see also [13]).

38.4.1 Focus

The focus of the literature review includes four potential categories (research findings, research methods, theories, practices or applications). Is the review focused on research findings, research methods, theories, or practices and applications? These four categories represent the main focus of any literature review.

38.4.2 Goal

The goal of the literature review includes three main categories (integration, criticism, and identification of central issues). Goals represent what the authors hope to accomplish [14]. Under the integration category, Cooper [14] provided three sub-categories (generalization, conflict resolution, and linguistic bridge building). Does the author plan to form general statements from multiple perspectives, resolve a conflicting issue, or bridge the gap between theories or disciplines? Alternative goals could be for authors to critically analyze the literature or to identify new critical issues around a topic or phenomenon.

38.4.3 Perspective

What perspective does the author(s) intend on taking for their literature review? Cooper [14] provided two main perspectives: neutral representation and espousal of position. Do the authors intend on remaining neutral and only reporting on the data or do they intend on taking a stand or position on the issue? The authors perspective should be made clear at the beginning of the review process.

38.4.4 Coverage

What coverage of the literature is being proposed? Coverage includes "the extent to which reviewers find and include relevant works in their paper" [14]. Is the goal to provide an exhaustive review of the literature (e.g., dissertation), an exhaustive with selective citation (e.g., purposeful review), a representative coverage of the literature, or a central or pivotal review of the literature (e.g., highly cited or seminal literature)?

38.4.5 Organization

How do the authors plan to organize the literature review and present it in a comprehensive and coherent manner? Cooper [14] presented three categories for the review's organization: historical, conceptual, and methodological. The organization of a literature review will, in part, be dictated on the coverage and focus of the review as well as whether it is organized historically (chronologically), by concept (theory and construct), or by methods used. The organization may also be influenced by a discipline's preference or a particular format dictated by a journal.

38.4.6 Audience

The audience relates to who the primary target is for the final synthesis of the review. Cooper [14] recognized four categories for audience characteristics: specialized scholars, general scholars, practitioners or policymakers, and the general public. While there may be one primary audience selected, it is important to understand that there will be secondary audiences that may also be interested in comprehensive reviews. Authors should try to meet the needs of multiple potential audiences when writing their literature review.

38.4.7 Summary of Cooper's Taxonomy

Authors need to identify each of the six characteristics in their introduction section of the literature review. Figure 38.1 provides an overview of these characteristics. These six characteristics provided by Cooper [14] should also help authors structure and organize their review during the planning stages. As stated by Cooper [14]:

> It is important that reviewers [authors] explicitly state what the foci, goals, perspective, coverage, organization, and audiences are for their work. This statement should appear early in the review so that the reader can construct an appropriate frame of reference for evaluating the effort.

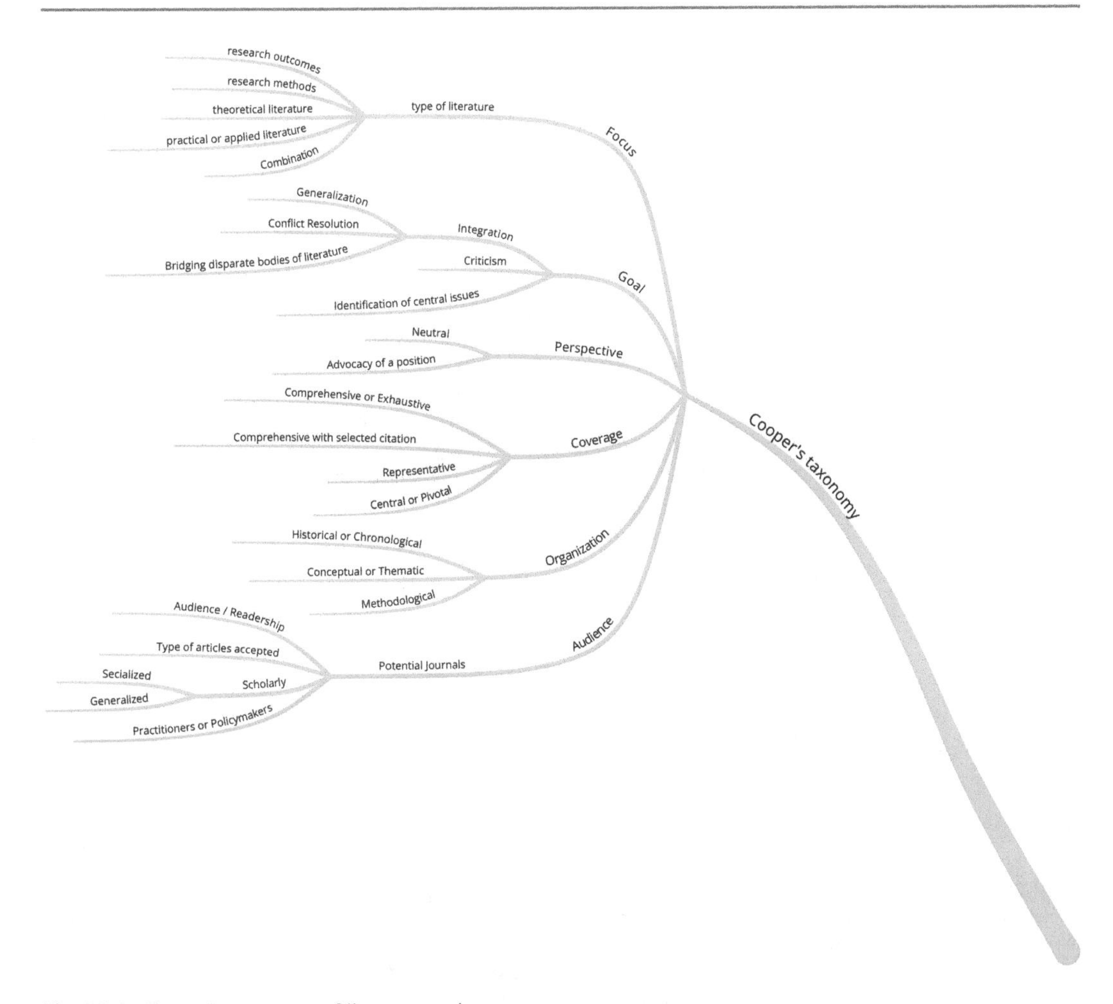

Fig. 38.1 Cooper's taxonomy of literature reviews

38.5 Types of Literature Reviews

38.5.1 Systematic Reviews

A systematic review is defined as "a review that uses explicit, systematic methods to collate and synthesize findings of studies that address a clearly formulated question" [6]. Systematic reviews are best when used to address specific research questions rather than for exploratory purposes. Systematic reviews provide depth whereas scoping reviews (see below) provide the breadth of coverage.

A checklist to help guide authors when conducting a systematic review is provided by PRISMA-S at https://osf.io/y765x. Authors are encouraged to download and print this checklist to use as a guide when planning a scoping review (Table 38.1).

38.5.2 Integrated Literature Reviews

Integrated literature reviews provide a review, critique, and synthesis of a prespecified body of literature that is integrated into a new framework or provides a new perspective on a topic [11]. While systematic reviews are best for addressing specific research questions, integrated literature reviews are best for integrating different bodies of literature (e.g., interdisciplinary, theory and practice) into a coherent framework or theory.

Table 38.1 Evaluating systematic literature reviews

Section	Checklist item
Information sources and methods	
Database name	Name each individual database searched, stating the platform for each
Multi-database searching	If databases were searched simultaneously on a single platform, state the name of the platform, listing all of the databases searched
Study registries	List any study registries searched
Online resources and browsing	Describe any online or print source purposefully searched or browsed (e.g., tables of contents, print conference proceedings, websites), and how this was done
Citation searching	Indicate whether cited references or citing references were examined, and describe any methods used for locating cited/citing references (e.g., browsing reference lists, using a citation index, setting up email alerts for references citing included studies)
Contacts	Indicate whether additional studies or data were sought by contacting authors, experts, manufacturers, or others
Other methods	Describe any additional information sources or search methods used
Search strategies	
Full-search strategies	Include the search strategies for each database and information source, copied and pasted exactly as run
Limits and restrictions	Specify that no limits were used, or describe any limits or restrictions applied to a search (e.g., date of time period, language, study design) and provide justification for their use.
Search filters	Indicate whether published search filters were used (as originally designed or modified), and if so, cite the filter(s) used
Prior work	Indicate whether search strategies from other literature reviews were adapted or reused for a substantive part of all of the search, citing the previous review(s)
Updates	Report the methods used to update the search(es) (e.g., rerunning searches, email alerts)
Dates of searches	For each search strategy, provide the date when the last search occurred
Peer review	
Peer review	Describe any search peer review process
Managing records	
Total records	Document the total number of records identified from each database and other information sources
Deduplication	Describe the processes and any software used to deduplicate records from multiple database searches and other information sources

PRISMA-S An Extension to the PRISMA Statement for Reporting Literature Searches in Systematic Reviews

Some items for authors to consider when conducting an integrative literature review are provided in Table 38.2.

38.5.3 Scoping Reviews

Scoping reviews are more exploratory and useful for providing a broad overview of new and emergent lines of research. Scoping reviews are used "to identify knowledge gaps, set research agendas, and identify implications for decision-making" [16].

A checklist to help guide authors when conducting a scoping review is provided as an open source document by PRISMA-ScR at https://www.equator-network.org/wp-content/uploads/2018/09/PRISMA-ScR-Fillable-Checklist-1.pdf. Authors are encouraged to download and print this checklist to use as a guide when planning a scoping review (Table 38.3).

38.6 Synthesizing the Literature

A synthesis contrasts existing knowledge with new ideas while providing a new representation (e.g., framework, model, theory) of a concept, phenomenon, or construct. The synthesis stage is what sets a literature review apart from a simple review or summary of the literature.

Table 38.2 Checklist for literature review [8, 15]

Category	Checklist item
Planning	What will the integrative literature review address?
	What is the focus of the review?
	What is the goal of the review?
	What perspective will be taken by the authors?
	What coverage will the data come from?
	How will the review be organized?
	Who is the primary and secondary audience for this review?
Methodology	What literature was reviewed?
	How was the literature searched?
	What databases were used and why were these databases?
	What search terms were used and why (associated with problem or phenomenon)?
	What inclusion criteria were used?
	What exclusion criteria were used?
	How were query results reduced if too many articles were returned on the initial search?
	Is a figure or table provided showing the search process and identifying the total number of articles included in the review?
	What content and information will be obtained from the literature (typically there should be a spreadsheet on what information should be collected)?
Writing	Report on what has been done.
	Report on what has not been done.
	Critique literature and identify deficits or gaps.
	Did the authors provide a concept map of the main constructs for the review?
	What synthesis techniques were used?
	How was the data synthesized?
	What new framework or model is being presented?
Conclusion	List future research studies that you have identified.
	What contribution does this study make to the discipline?
	What are the strengths and weaknesses of the review?
	How might this review inform practice?

A distinction between analysis and synthesis is in order here. Analysis is the process of breaking something down into its constituent components or parts.

Analysis = taking an idea, argument or piece of research apart to see how it was constructed. Doing the same with multiple ideas, arguments and pieces of research. [15]

In contrast, synthesis involves making connections from the components or parts identified in the analysis stage.

Synthesis = re-assembling the pieces of an idea (s), argument(s) or piece(s) of research in a different way to support new claims, interpretations and proposals for research. [15]

When synthesizing the literature, broad coverage is required to represent several perspectives across a wide range of disciplines. In addition, comprehensive knowledge of the topic is required for the authors to "dig beneath the surface of an argument and see the origins of a piece of research or theory" [15].

Synthesis is the backbone of a literature review, regardless of the type. Synthesis is where literature reviews provide new knowledge and value to one's discipline and practice. While there are several techniques and recommended steps for synthesizing the literature, a few general techniques are provided below along with recommended steps for conducting integrative and systematic reviews. These techniques and

Table 38.3 Checklist for scoping reviews [16]

Section	PRISMAScR checklist item
Title	
Title	Identify the report as a scoping review
Abstract	
Structured summary	Provide a structured summary that includes (as applicable): the background, objectives, eligibility criteria, sources of evidence, charting methods, results, and conclusions that relate to the review questions and objectives
Introduction	
Rationale	Describe the rationale for the review in the context of what is already known. Explain why the review questions/objectives lend themselves to a scoping review approach
Objectives	Provide an explicit statement of the questions and objectives being addressed with reference to their key elements (e.g., population or participants, concepts, and context) or other relevant key elements used to conceptualize the review questions and/or objectives
Methods	
Protocol and registration	Indicate whether a review protocol exists; state if and where it can be accessed (e.g., a Web address); and if available, provide registration information, including the registration number
Eligibility criteria	Specify characteristics of the sources of evidence used as eligibility criteria (e.g., years considered, language, and publication status), and provide a rationale
Information sources	Describe all information sources in the search (e.g., databases with dates of coverage and contact with authors to identify additional sources), as well as the date the most recent search was executed
Search	Present the full electronic search strategy for at least one database, including any limits used, such that it could be repeated
Selection of sources of evidence	State the process for selecting sources of evidence (i.e., screening and eligibility) included in the scoping review
Data charting process	Describe the methods of charting data from the included sources of evidence (e.g., calibrated forms or forms that have been tested by the team before their use, and whether data charting was done independently or in duplicate) and any processes for obtaining and confirming data from investigators
Data items	List and define all variables for which data were sought and any assumptions and simplifications made
Critical appraisal of individual sources of evidence	If done, provide a rationale for conducting a critical appraisal of included sources of evidence; describe the methods used and how this information was used in any data synthesis (if appropriate)
Synthesis of results	Describe the methods of handling and summarizing the data that were charted
Results	
Selection of sources of evidence	Give numbers of sources of evidence screened, assessed for eligibility, and included in the review, with reasons for exclusions at each stage, ideally using a flow diagram
Characteristics of sources of evidence	For each source of evidence, present characteristics for which data were charted and provide the citations
Critical appraisal within sources of evidence	If done, present data on critical appraisal of included sources of evidence (see 'Critical appraisal of individual sources of evidence' above)
Results of individual sources of evidence	For each included source of evidence, present the relevant data that were charted that relate to the review questions and objectives
Synthesis of results	Summarize and/or present the charting results as they relate to the review questions and objectives
Discussion	
Summary of evidence	Summarize the main results (including an overview of concepts, themes, and types of evidence available), link to the review questions and objectives, and consider the relevance to key groups

(continued)

Table 38.3 (continued)

Section	PRISMAScR checklist item
Limitations	Discuss the limitations of the scoping review process.
Conclusions	Provide a general interpretation of the results with respect to the review questions and objectives, as well as potential implications and/or next steps
Funding	
Funding	Describe sources of funding for the included sources of evidence, as well as sources of funding for the scoping review. Describe the role of the funders of the scoping review

steps are listed to help guide researchers as they begin this synthesis journey.

developing metatheory across theoretical domains through future research [11, 17].

38.6.1 General Synthesis Techniques

Torraco [11, 17] presented five common techniques for synthesizing the literature. These include providing a research agenda, a taxonomy, alternative models or conceptual frameworks, meta-analysis, and a metatheory. Each of these are described below:

- A *research agenda* that flows logically from the critical analysis of the literature. The research agenda should pose provocative questions (or propositions) that give direction for future research.
- A *taxonomy or other conceptual classification of constructs* is often developed as a means to classify previous research. They, in turn, lay the foundation for new theorizing (see [18]).
- *Alternative models or conceptual frameworks*–new ways of thinking about the topic addressed by the integrative review. Alternative models or conceptions proposed by the author should be derived directly from the critical analysis and synthesis provided.
- *Meta-analysis*–Summary of a collection of comparable research studies generated through statistical analysis of analytic results from individual studies to arrive at a combined average effect size for the purpose of integrating research findings.
- *Metatheory*–The integration and synthesis of a literature review can provide the basis for developing metatheory across theoretical domains through future research [11, 17].

38.6.2 Steps for Systematic Reviews

Additional synthesizing steps that relate to formulating the problem are provided by Cooper [13]. Because these steps begin with formulating the problem, they are best suited for systematic reviews. The steps for research synthesis include:

- Formulating the problem
- Searching the literature
- Gathering information from studies
- Evaluating the quality of studies
- Analyzing and integrating the outcomes of studies
- Interpreting the evidence
- Presenting the results [13].

38.6.3 Steps for Integrated Reviews

To help synthesize information for contrasting different theoretical perspectives (integrated literature review), Leedy, Ormrod [19] identified the following guidelines:

- Compare and contrast varying theoretical perspectives on the topic.
- Show how approaches to the topic have changed over time.
- Describe general trends in research findings.
- Identify discrepant or contradictory findings, and suggest possible explanations for such discrepancies.

- Identify general themes that run throughout the literature.

38.7 Concluding Remarks

For literature reviews, the literature is the data that is being analyzed. What data and how the data is to be examined is necessary to report clearly to meet the requirements of being transparent. Aside from being transparent in reporting what was done, identifying the steps taken to collect the data reduces the author's bias in the process. For example, if a researcher is not transparent in where the data came from, it could be assumed that the literature reviewed was already familiar to the author, reporting only known literature. This is a form of author bias that is too common in published works.

A second major concern in reporting what was done is the issue of randomness in the scientific process. As a research study, literature reviews do include random data. This random data comes from the literature reviewed. By selecting relevant databases and search terms to retrieve the literature, and by reporting the steps taken to retrieve this literature, the study's data meets the needs of being random (not pre-selected by the authors). When the data retrieval process is not reported clearly, or when authors select the data for their study with no methodological process guiding this decision, the data is selective rather than random. This lack of randomness results in a lack of coverage of the literature, bypassing the benefits of having a broad coverage of the literature with multiple perspectives to contrast and synthesize.

Literature reviews are an exercise in prose where writing clearly becomes essential. Writing clearly represent clear thinking [20]. To present a synthesis of many research studies one must be clear and concise, organized and methodological, requiring the reader to be able to effortlessly follow the thinking of the author.

In identifying common issues that reviewers have with problematic submissions, Ragins [20] highlighted three main issues: foggy writing, undefined concepts and jargon, and lack of a story. Foggy writing occurs when language is used inaccurately (e.g., large words used incorrectly). One point of consideration is that the authors are writing to the general public (e.g., students, patients, practitioners) rather than to their colleagues. Trying to write to your colleagues produces language that is hard to understand by those outside of your area of expertise, making it nearly impossible for most readers to follow. Writing complexly and abstractly only detracts from the research findings and implications. Foggy writing is best described in the following:

> Foggy writing may be due to writers' insecurities, their misperceptions about writing, or their lack of clarity about what they want to say, why they want to say it, and who their reader is. [20]

To avoid foggy writing, it is recommended for authors to take extra time to think through their ideas as they are writing them and to have colleagues read for clarity. Making complicated or complex issues understandable is an art all writers struggle to accomplish. Extra care should be taken to present all ideas, concepts, and theories so that people outside of your discipline can follow along and remain engaged.

Use of undefined concepts and jargon results in the readers being confused. The author assumes the readers are in-their-head and have the same train-of-thought [20]. In general, it is recommended not to use jargon. It becomes necessary for researchers to evaluate what they have written for clarity. Authors can step away from their article for a few days before evaluating or they can read their written words aloud to a peer not working on the same study. This form of evaluating one's work becomes an important one that is practiced by even the best writers today:

> There comes a point when you must judge what you've written and how well you wrote it. I don't believe a story or a novel should be allowed outside of your study or writing room unless you feel confident that it's reasonably reader-friendly. [21]

Authors often get too caught up in the details that they miss the main point. This lack of story loses the reader's attention and interest. Each story, including research studies, must have a beginning, middle, and end. This point is highlighted in the following:

> Papers should offer a clear, direct, and compelling story that first hooks the reader and then carries the reader on a straightforward journey from the beginning to the very end of the manuscript. [20]

Authors should provide a road map or visuals that aid readers in following the story. Other techniques include the usage or creation of metaphors. Visuals are useful aids that can convey meaning and emotions when used correctly. Types of visuals could include concept maps, chronological timelines, network maps, and figures [22] that help relay the information to the readers. Metaphors and analogies are other techniques that authors can use to support reader's understanding. Metaphors and analogies help to relay the meaning when the discussion is more abstract. Metaphors and analogies project "meaning onto a second thing in a manner that does not require elaboration" [23].

Literature and systematic reviews provide a synthesis of several research findings that span long periods of time. The synthesis from these works often provides new theories and models and novel solutions to today's problems. As our world becomes more entangled and globally interconnected, our problems become more complex. As this complexity increases, our capability of addressing complexity diminishes. This point is highlighted by Simon [24] in his concept of bounded rationality: "The meaning of rationality in situations where the complexity of the environment is immensely greater than the computational powers of the adaptive system" [24]. To counter this complexity, Ashby derived what is now called "Ashby's Law" which simply states that "only variety can destroy variety" [25]. Synthesizing or integrating new models from several research studies is one method of providing this increase in variety that Ashby was calling for. Science is now being challenged, more so than in the past, to develop new and novel theories and models for today's complex problems. This point is highlighted by Reen [26]: "Then and now, challenging problems require new forms of knowledge integration," and this integration can come from rigorous integrated and systematic reviews.

Conflict of Interest No conflicts of interest.

References

1. Begley GC, Ioannidis JPA (2015) Reproducibility in science: improving the standard for basic and preclinical research. Circ Res 116:116–126. https://doi.org/10.1161/CIRCRESAHA.114.303819
2. Romero F (2019) Philosophy of science and the replicability crisis. Philos Compass 14:e12633. https://doi.org/10.1111/phc3.12633
3. Van Bavel JJ, Mende-Siedlecki P, Brady WJ, Reinero DA (2016) Contextual sensitivity in scientific reproducibility. Proc Natl Acad Sci U S A 113:6454–6459. https://doi.org/10.1073/pnas.1521897113
4. Ioannidis JPA (2012) Why science is not necessarily self-correcting. Perspect Psychol Sci 7:645–654. https://doi.org/10.1177/1745691612464056
5. Bollen K, Cacioppo JT, Kaplan RM, Krosnick JA, Olds JL (2015) Social, behavioral, and economic sciences perspectives on robust and reliable science. https://www.nsf.gov/sbe/AC_Materials/SBE_Robust_and_Reliable_Research_Report.pdf. Accessed 15 Sept 2022
6. Page MJ, McKenzie JE, Bossuyt PM, Boutron I, Hoffmann TC, Mulrow CD et al (2021) The PRISMA 2020 statement: an updated guideline for reporting systematic reviews. BMJ 372. https://doi.org/10.1136/bmj.n71
7. Imel S (2011) Writing a literature review. In: Tonette RS, Hatcher T (eds) The handbook of scholarly writing and publishing. Jossey-Bass, San Francisco, CA, pp 145–160
8. Boote DN, Beile P (2005) Scholars before researchers: on the centrality of the dissertation literature review in research preparation. Educ Res 34:3–15. https://doi.org/10.3102/0013189X034006003
9. Bryman A (2008) Social research methods. Oxford University Press, New York, NY
10. Taylor D (n.d.) The literature review: a few tips on conducting it. https://advice.writing.utoronto.ca/types-of-writing/literature-review/. Accessed 1 Oct 2022
11. Torraco RJ (2005) Writing integrative literature reviews: guidelines and examples. Hum Resour Dev Rev 4(3):356–367. https://doi.org/10.1177/1534484305278283
12. Cooper HM (1988) Organizing knowledge syntheses: a taxonomy of literature reviews. Knowl Soc 1:104–126. https://doi.org/10.1007/BF03177550

13. Cooper H (2010) Research synthesis and meta-analysis. 4th ed. Applied social research methods series, vol 2. Sage, Los Angelas, CA
14. Cooper H (2003) Psychological bulletin: editorial. Psychol Bull 129:3–9. https://doi.org/10.1037/0033-2909.129.1.3
15. Hart C (2018) Doing a literature review: releasing the research imagination, 2nd edn. Sage, Los Angelas, CA
16. Tricco AC, Lillie E, Zarin W, O'Brien K, Colquhoun H, Kastner M et al (2016) A scoping review on the conduct and reporting of scoping reviews. BMC Med Res Methodol 16:15. https://doi.org/10.1186/s12874-016-0116-4
17. Torraco RJ (2016) Writing integrative literature reviews: using the past and present to explore the future. Hum Resour Dev Rev 15:404–428. https://doi.org/10.1177/1534484316671606
18. Doty DH, Glick WH (1994) Typologies as a unique form of theory building: toward improved understanding and modeling. Acad Manag Rev 19:230–251. https://doi.org/10.5465/AMR.1994.9410210748
19. Leedy PD, Ormrod JE (2005) Practical research: planning and design, 8th edn. Pearson, Upper Saddle River, NJ
20. Ragins BR (2012) Reflections on the craft of clear writing. Acad Manag Rev 37:493–501. https://doi.org/10.5465/amr.2012.0165
21. King S (2000) On writing: a memoir of the craft. Scribner, New York, NY
22. Torraco RJ (2016) Research methods for theory building in applied disciplines: a comparative analysis. Adv Dev Hum Resour 4:355–376. https://doi.org/10.1177/1523422302043008
23. Pendleton-Jullian AM, Brown JS (2018) Design unbound: designing for emergence in a white water world. MIT Press, Cambridge, MA
24. Simon HA (2019) The sciences of the artificial [reissue of 3rd ed.]. MIT Press, Cambridge, MA
25. Klir GJ (2009) W. Ross Ashby: a pioneer of systems science. Int J Gen Syst 38:175–188. https://doi.org/10.1080/03081070802601434
26. Reen J (2020) The evolution of knowledge: rethinking science for the anthropocene. Princeton University Press, Princeton, NJ

How to Create a Bibliography

39

Rohan Reddy, Samuel Sorkhi, Saager Chawla, and Mahadevan Raj Rajasekaran

Abstract

This chapter describes the fundamental principles and practices of referencing sources in scientific writing and publishing. Understanding plagiarism and improper referencing of the source material is paramount to producing original work that contains an authentic voice. Citing references helps authors to avoid plagiarism, give credit to the original author, and allow potential readers to refer to the legitimate sources and learn more information. Furthermore, quality references serve as an invaluable resource that can enlighten future research in a field. This chapter outlines fundamental aspects of referencing as well as how these sources are formatted as per recommended citation styles. Appropriate referencing is an important tool that can be utilized to develop the credibility of the author and the arguments presented. Additionally, online software can be useful in helping the author organize their sources and promote proper collaboration in scientific writing.

R. Reddy · S. Sorkhi · S. Chawla
Department of Urology, University of California at San Diego, San Diego, CA, USA
e-mail: roreddy@stanford.edu; Srs03427@creighton.edu; stchawla@ucsd.edu

M. R. Rajasekaran (✉)
Department of Urology, University of California and VA San Diego Healthcare System, San Diego, CA, USA
e-mail: mrajasekaran@health.ucsd.edu; mrajasekaran@ucsd.edu

Keywords

Referencing · Citations · Harvard and Vancouver styles · Plagiarism · Referencing software · Scientific writing

39.1 Introduction

Bibliographies are essential to scientific research, as they provide a comprehensive list of the sources that have been used in the research and writing process. Including a bibliography is important for several reasons. Citations in works submitted for publication are closely scrutinized by reviewers and publishers for the following reasons:

- Bibliographies help acknowledge the contributions of other researchers and scholars whose work has informed or influenced the research being presented. Comprising a crucial aspect of academic integrity, they ensure that credit is given where it is due and that the original sources of information and ideas are properly cited [1].
- Bibliographies are an invaluable resource, as they allow readers to explore the inspiration behind the paper and learn more about the topic. They help increase the credibility and rigor of the research, as they demonstrate that the research is well-informed and based on a broad range of sources [2].

G. Jagadeesh et al. (eds.), *The Quintessence of Basic and Clinical Research and Scientific Publishing*,
https://doi.org/10.1007/978-981-99-1284-1_39

- Bibliographies also serve as a useful reference tool for other researchers who may be interested in exploring similar topics or building upon the research presented in the paper. By providing a clear and comprehensive list of sources, bibliographies help to facilitate further research and contribute to the advancement of knowledge in a particular field.
- Overall, bibliographies are an essential component of scientific research papers, as they help to acknowledge the contributions of other researchers, provide a valuable resource for readers, and facilitate further research and the advancement of knowledge [3].

An important distinction to be made is that a bibliography is a list of sources that have been used in the research and preparation of a work, such as a book, web article, or essay, while referencing is the act of citing those sources within the body of the work. References should include a sufficient amount of data points, curated in a specific style, allowing a reader to identify the source material. It is critical to be transparent and clear when citing the source of a specific argument. Bibliographies can act as a guide for readers as they try to understand an investigator's train of thought. Furthermore, science is a discipline that builds upon itself as time passes. Giving credit to previous publications in the field is a sign of respect and holds the author accountable for interpreting the findings correctly. This chapter describes the principles and process of creating a bibliography and referencing list.

39.2 Choosing Your Sources

Understanding where to find sources of credible information and their purpose in a research paper is a crucial first step in creating a bibliography. When a research question or topic of interest has been established (see Chap. 1), many researchers will opt to conduct a preliminary search through common search engines (see Chap. 37) such as Google Scholar. Creating parameters for search results can further improve the reliability of the information [4]. After you have gathered potential sources, a critical assessment of the source's purpose in your paper is needed. Is the source providing context for an argument? Is the resource from a reliable source? Consider the author's background, the methods by which the data was gathered, and the references the author has cited [4].

These resources can be categorized into primary and secondary sources. Primary sources are original research materials that provide firsthand information on a topic, while secondary sources are materials that interpret or analyze primary sources. Primary sources provide raw data and information that form the basis of research. Primary sources are typically considered more reliable and credible than secondary sources because they are based on direct observations and experiments. Examples of primary sources in scientific writing include research articles, conference proceedings, and technical reports (Table 39.1). These types of sources typically include original data and findings, as well as the methods used to collect and analyze the data.

Secondary sources provide interpretation or analysis of primary sources. Examples of secondary sources in scientific writing include review articles, textbooks, and encyclopedias. These types of sources are useful for providing context and background information on a topic, and for synthesizing and summarizing the findings of primary sources as shown below in Table 39.1 [4].

Table 39.1 Comparing and contrasting primary and secondary sources of bibliography

Primary source	Secondary source
Original sources that are reporting results for the first time • Official documents • Numerical Data • Works of fiction	Interpretation/discussion of primary source results • Textbooks • Research Papers • Indexes

39.3 Fundamental Principles of Citation

It can be overwhelming for an individual to understand what referencing style to choose. While there are pros and cons to each style, it is important to understand the basic elements that are necessary when referencing another's work. Four core elements should be extracted from each source: title, date, author, and source (publishing company or the DOI). A Digital Object Identifier, DOI, is a group of numbers and letters that uniquely identify a publication. These are normally associated with journal publications [5]. For example, in a PubMed indexed article, the DOI can be found below the Journal Title. When citing a website, locate the "about" page or scroll to the bottom to extract these elements [5] (See Chap. 37).

During the process of researching and writing the paper, keep track of all the sources being used. Organize the sources alphabetically by the author's last name. If a source does not have an author, use the title of the source instead. Format the references according to the specific citation style requested by the publisher. This may involve utilizing a specific font and font size, indenting certain lines, and using italics or quotation marks for specific types of sources. Create a heading for the reference list labeled "Works Cited". Based on the referencing style chosen, format the reference list accordingly. It is important to accurately cite all sources in the reference list to give credit to the original authors and to help the readers locate these sources in an efficient manner [6].

39.4 How to Know Which Information to Reference?

In scientific writing, it is important to properly cite the sources that you have used in your research. This includes not only direct quotations, but also any ideas, data, or other information that you have obtained from these sources. To know what information to reference in scientific writing, it is essential to carefully review and evaluate the sources that you have used. This includes reading the sources thoroughly and taking detailed notes on the ideas and information that you find most relevant and important to your research. As you write your paper, be sure to carefully consider the purpose and context of each piece of information that you include. If you are unsure whether or not to include a particular piece of information in your paper, it is usually a good idea to err on the side of caution and provide a citation.

Any information that is not considered common knowledge should be associated with a proper citation. This can be tricky as "common knowledge" is a subjective idea. If you are paraphrasing or describing an established theory, definition, or model, from another paper, the work should be included in the reference list at the end of your paper. This includes figures or statistics that are provided as evidence in your writing, even if they are simply a source of inspiration for your work [3].

Normally it is not required to cite a reflection on personal experiences or narratives. However, if further evidence is provided to bolster the perspective, proper acknowledgment of the author's work needs to be given. Furthermore, a paper may reference the commentary of another publication in an effort to provide emphasis to its own argument. If you are directly taking from another source, quotation marks followed by an in-text citation need to be placed around the statement [7]. In the Harvard System, the in-text citation is placed before the final punctuation mark [8]. Additionally, when paraphrasing an important argument or finding from a source a citation is needed.

39.5 Referencing Styles

There are many referencing styles, and the specific style will depend on the institutional standards that are set in place. Despite the wide

Table 39.2 Table showing various popular referencing styles

Referencing style	Disciplines commonly used
APA (American Psychological Association)	Psychology, education, social sciences
MLA (Modern Language Association)	Literature and humanities
Chicago/Turabian	History, Business, Fine Arts
Harvard	Business, Economics, and Natural Sciences
IEEE (Institute of Electrical and Electronics Engineering)	Computer Science and Engineering
Vancouver	Medicine and Health Sciences

variety, the basic principles that constitute a reference still stand. Each referencing style will contain similar data points that help the reader identify the publication. Additionally, the author needs to stick to one referencing style throughout the paper. Switching between styles can confuse the readers and make it difficult to identify source material.

When writing an academic research article, citing referenced work will occur at two different places. First, citations will be found within the text you have written. These references within the text, also known as in-text citations, will contain brief pieces of information about the source. Secondly, included source material will be referenced at the end of the paper in what is known as a reference list. This list will include much more information about the source material. In-text citations can help guide a reader to the specific citation that is listed in the reference list.

There are two main styles of citing references as shown in the Table below (Table 39.2): the Vancouver System and the Harvard System. The Vancouver Style is widely used in the biomedical field utilizing in-text citations and a numerical reference list. The order of the references in your bibliography will depend on the order in which they appear in your text. An in-text citation using the Vancouver Style should include a number located within round brackets or a superscript number with no brackets. Each of these numbers should align with your reference list, allowing the reader to locate the full reference [9].

The Harvard Style In-text citations will contain the author's last name and the year of publication. If a book is being cited, it is important to include a specific page number after the year of publication. Furthermore, if the source material contains multiple authors, we can save space in our in-text citation by referring to the first author's name followed by "et al.". The reference list (found at the end of the paper) should be listed in alphabetical order, which varies from the Vancouver system which lists references based on the order each reference is found within the text [10, 11].

39.6 Quality Referencing

Creating an accurate references page for any scientific publication is a time-consuming process. It can help to start the process of gathering and organizing your references at the beginning of your project.

To find quality evidence, it is vital to stay current with new literature in the field [12]. Previous work is usually the inspiration for making changes in the experimental design or reporting new findings. The references that are listed will determine the validity of the author's argument. Thus, it is highly recommended that the author read through and understand the basic ideas presented in each paper. Reading the entire paper reduces the chance of misrepresenting data [12]. References are listed as contextual evidence for the arguments the author is making. A majority of the references for scientific papers will include journal publications that can be found through keyword searches on Google Scholar or PubMed. Make sure to take note of the names of the journals as they can provide insight into the reputability of the author's work. Findings published in the New England Journal of Medicine or Nature are subjected to multiple rounds of peer review and editing, which adds to the validity of the work. In scientific research, this evidence should be peer-reviewed and published by a reputable source [4].

Furthermore, scientists must maintain an objective stance when writing. A reference list that is heavily self-cited or lacks balance in providing evidence for all sides can affect the accuracy of the author's interpretations and findings. Situating your work within the proper context of a field can be achieved by reading the current literature that is being published. Reference lists may grow or shrink as the researcher becomes increasingly familiar with previous work that influences their experimental design or findings. Finally, keeping a concise and organized bibliography helps the researcher better understand the limitations of their study and the future direction of experiments. The foundation of scientific writing is to introduce new perspectives into an existing field. Thus, the references that are cited function as supporting pillars of existing knowledge, reinforcing the author's findings.

39.7 Online Software

Technology can be a useful tool for mitigating common mistakes when creating a bibliography, such as incorrect formatting. With the advent of the internet, many online resources can assist in constructing a bibliography. Popular software includes Endnote, Mendeley, and Zotero [13]. Specifically, Zotero, a free referencing tool, can "collect, organize, cite and share their research sources" [14]. Available for both Windows and Mac, Zotero can be downloaded by going to https://Zotero.org/download/. This free-to-use software easily integrates into Microsoft Word and promotes collaboration as it can store and cite a large number of articles and text. Creating a citation can be done by clicking on the "green plus sign" and choosing the type of publication from the drop-down list (Fig. 39.1). The publication's info can then be entered into the corresponding fields (Fig. 39.2). Additionally, if an article has an associated DOI it can be added to the reference library by clicking on "Add Item (s) by Identifier." Zotero promotes a collaborative environment as multiple individuals can also upload and cite articles within the same library. When creating large databases of publications, Zotero offers a web capture feature that can directly extract bibliographic elements from a website or journal. These elements can then be curated into the multitude of referencing styles that are included in the program.

On the other hand, Endnote is a paid reference management software program that is available for Windows and Mac. It includes a range of features for organizing and citing sources in research papers and other documents. Some of

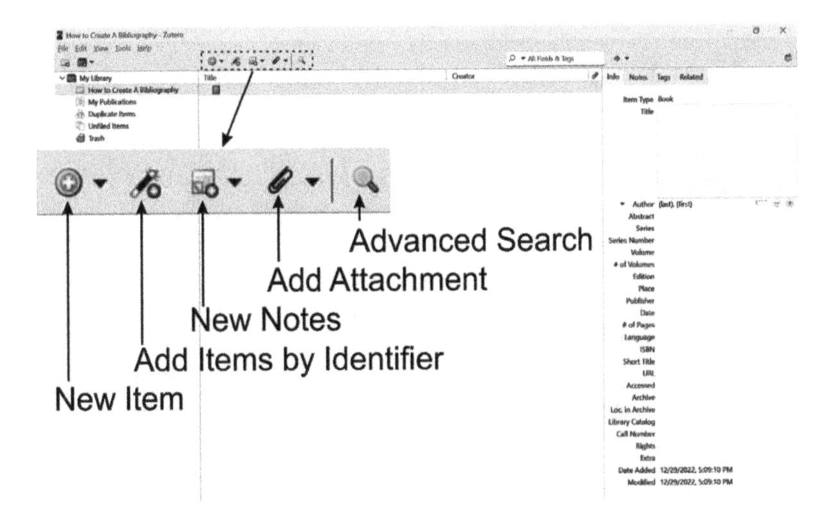

Fig. 39.1 Schematic showing annotated Zotero homepage for creating referencing libraries

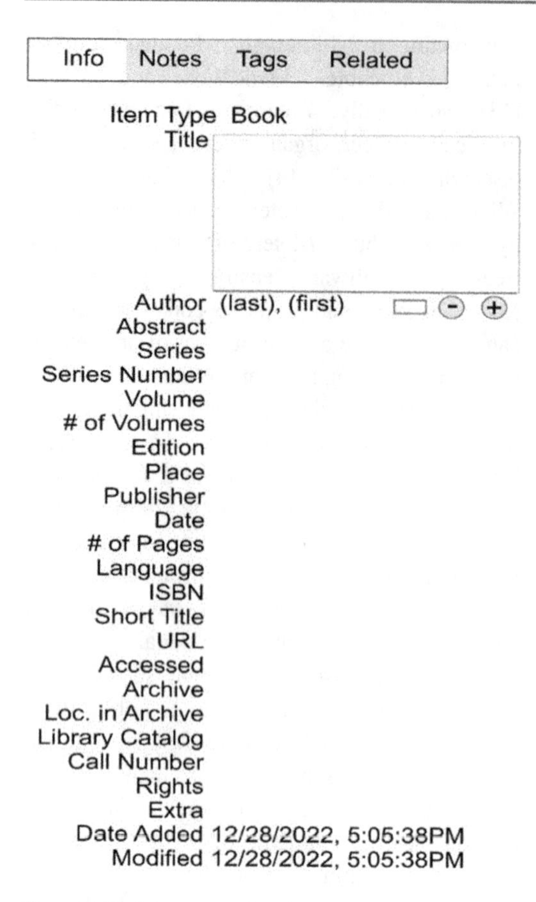

Fig. 39.2 Schematic showing Zotero's referencing section

the main features of Endnote include a personal library for storing and organizing sources [15]. This allows users to create a personal library of sources, which can be organized using tags, groups, and keywords. Additionally, Endnote supports a wide range of citation styles, including Vancouver and Harvard Systems. Similar to Zotero, Endnote allows users to share their library with other Endnote users. They take this feature one step further by having a web-based version that allows users to access their library from any device with an internet connection [16].

The assistance of technology in creating a bibliography does not take away from the intellectual process of curating and gathering resources for a publication. Rather, programs such as Zotero can help streamline the process of gathering and citing large numbers of references. However, its quick workflow also has the potential to influence the author to overcite previous work, which can dilute the credibility of the evidence and increase the possibility of leaving out relevant work.

39.8 Do's and Don'ts of Referencing

Here are some general do's and don'ts for referencing journal publications:

Do's:

- Do include all the relevant information for each journal article you reference, such as the author's name, the title of the article, the title of the journal, the volume and issue number, and the publication date.
- Do use the appropriate citation style for your field or discipline. Different fields have different citation styles, so make sure you are using the correct one.
- Do format your references correctly according to the citation style you are using. This may involve using a specific font and font size, indenting certain lines, and using italics or quotation marks for certain types of sources.
- Do double-check your reference list to make sure it is accurate and complete.Don'ts:
- Don't omit any necessary information from your references. Make sure to include all the relevant information for each journal article you reference.
- Don't use a journal article as a source if you have not read it yourself. It is important to read and understand the sources you are using in your research.
- Don't plagiarize by copying and pasting information from a journal article into your work without properly citing it. Plagiarism is the act of using someone else's ideas or words without giving them credit (see below).
- Don't use a journal article as a source if it is not relevant to your research. Make sure to only include sources that are relevant and contribute to your paper or project.

39.9 Plagiarism

While appearing deceivingly simple, plagiarism plagues an academic world that has strived to maintain its traditional values of respect, ethics, and morality (see Chap. 59). Many academicians are beginning to reside in a state of limbo where they are starting to question the purpose and practicality of adhering to the laws of plagiarism. However, the current practice still abides by such laws, so it is paramount that budding academicians learn and understand the essential framework behind plagiarism and the proper steps necessary to produce original work.

After a thorough review, there is a clear need for a singular definition/framework of plagiarism. Currently, each institution defines plagiarism in its own way, which may offer some benefit to that specific institution given the type of work they produce. Nonetheless, the lack of a universally accepted definition leaves room for individual interpretation/misinterpretation [2]. Upon review of various definitions, two concepts surfaced as common denominators in each interpretation: (1) the intent of the author producing the work and (2) the concept that plagiarism is based on inappropriately utilizing someone else's product rather than their idea.

Thus, plagiarism is defined as: The practice of utilizing someone else's work and attempting to take ownership of the idea by maintaining its originality. As Angélil-Carter [17] succinctly denotes, "the true plagiarist writes to conceal the sources". While all acts of plagiarism are looked down upon and usually met with some form of justice/remediation, the aforementioned intent of the author must be taken into consideration when judging each act.

In 1995, Howard [18] set out to delineate the different forms of plagiarism that are commonly encountered: patchwork writing, non-attribution, and outright cheating. In each form, respectively, there is an increase in the level of malintent and consciousness of the author. The first form, patchwork writing, was later termed "omission paraphrasing" by Barrett and Malcolm [19] and was used to describe the process of selectively modifying excerpts taken from a single source in an attempt to adapt the author's work into what is needed for an assignment. Some academics such as Introna et al. [20] are warming to the idea that patchwork writing still requires a certain degree of hard work and understanding of the material to be able to write about the topic, which is particularly impressive (and thus understandable) for those writing in a second language. Rather than approaching each potential plagiaristic event with an inflexible, black and white lens, academicians must consider the novice researcher that are navigating a relatively novel world filled with hidden pitfalls. While this is the least damning form of plagiarism, it exposes writers to criticism and ultimately results in a loss of credibility, so it's wise to err on the side of caution and avoid this practice. The second form, non-attribution, was described by Howard [18] as situations where students were aware of the need for citations but succumbed to pressures such as academic workload or the need to plagiarize to pass a course. Finally, cheating is plagiarism with full awareness and intent.

39.9.1 How Can You Avoid Plagiarism?

Two simple methods can be utilized to avoid plagiarism, and choosing either is perfectly sound:

1. When in doubt, cite the author's work and remember to include the citations in a bibliography or list of references
   OR
2. Use quotation marks to differentiate between your words and the source author's words. Again, list your in-text citations in the bibliography or list of references [21].

39.10 How to Cite Internet-Based References

The internet has had a significant impact on the way that sources are cited in scientific papers.

One of the major changes has been the proliferation of electronic sources, such as web-based articles and online databases, which have largely replaced print sources as the primary means of accessing scientific information [22]. As a result, it has become increasingly common to include URLs and other electronic identifiers in citations, in addition to traditional bibliographic information such as the author, title, and publication year [23].

The internet has also made it easier for researchers to access and share information, leading to an increase in the use of online collaboration tools and platforms for publishing and disseminating research. This has led to the development of new citation styles and guidelines for citing electronic sources, such as the American Psychological Association's (APA) guidelines for citing web-based material [24].

Citing websites in published papers is often done with the assumption that they will remain accessible for an extended period of time. However, this is not always the case, as many of these websites tend to become unavailable within a few years of the paper's publication. The impact of this issue is not fully understood [25].

The rise of the open access movement and the proliferation of online communication and archiving platforms, such as blogs, Research Gate, and LinkedIn, has led to an increase in the use and functionality of references. In the digital age, new forms of scholarly sources have become available, adding complexity to the process of formatting and citing. For example, it is important to include correct URL links and access dates for web-based materials to enable readers to easily access these sources in the future [26, 27].

39.11 Concluding Remarks

Citing references is an invaluable practice that is imperative to academic and intellectual discourse. Proper citations demonstrate to the reader that you have explored current literature regarding the topic, which is crucial to scientific research given that novelty is an essential criterion for publication. They also help to acknowledge the contributions of other researchers and scholars whose work has informed or influenced the research being presented, provide a valuable resource for readers, and facilitate further research and the advancement of knowledge. It is important to carefully consider the sources that are included in a bibliography and to use a specific referencing style to ensure that the sources are accurately and appropriately cited. To produce a high-quality bibliography, choose reliable and credible sources, categorize them as primary or secondary sources, and understand the different types of sources that are available. By following these principles, researchers can create a bibliography that enhances the credibility and rigor of their research paper.

Conflict of Interest None declared.

References

1. AWELU (2022) The functions of references. Lund University. https://www.awelu.lu.se/referencing/the-functions-of-references/. Accessed 28 Dec 2022
2. Masic I (2013) The importance of proper citation of references in biomedical articles. Acta Inform Med 21(3):148–155
3. Neville C (2012) Referencing: principles, practice and problems. RGUHS J Pharm Sci 2(2):1–8
4. Horkoff T (2015) Writing for success 1st Canadian edition: BCcampus. Chapter 7. Sources: choosing the right ones. https://opentextbc.ca/writingforsuccess/chapter/chapter-7-sources-choosing-the-right-ones/. Accessed 28 Dec 2022
5. A guide to database and catalog searching: bibliographic elements. Northwestern State University. https://libguides.nsula.edu/howtosearch/bibelements. Updated 5 Aug 2021; Accessed 28 Dec 2022
6. Research process: step 7: citing and keeping track of sources. University of Rio Grande. https://libguides.rio.edu/c.php?g=620382&p=4320145. Updated 12 Dec 2022; Accessed 28 Dec 2022
7. Pears R, Shields G (2022) Cite them right. Bloomsbury Publishing
8. Harvard system (1999) Bournemouth University. http://ibse.hk/Harvard_System.pdf. Accessed 28 Dec 2022
9. Gitanjali B (2004) Reference styles and common problems with referencing: Medknow. https://www.jpgmonline.com/documents/author/24/11_Gitanjali_3.pdf. Accessed 28 Dec 2022
10. Williams K, Spiro J, Swarbrick N (2008) How to reference Harvard referencing for Westminster

Institute students. Westminster Institute of Education Oxford Brookes University
11. Harvard: reference list and bibliography: University of Birmingham. 2022. https://intranet.birmingham.ac.uk/as/libraryservices/library/referencing/icite/harvard/referencelist.aspx. Accessed 28 Dec 2022
12. Limited SN (2017) Responsible referencing. Nat Methods 14(3):209
13. Hensley MK (2011) Citation management software: features and futures. Ref User Serv Q 50(3):204–208
14. Ramesh A, Rajasekaran MR (2014) Zotero: open-source bibliography management software. RGUHS J Pharm Sci 4:3–6
15. Shapland M (2000) Evaluation of reference management software on NT (comparing Papyrus with Procite, Reference Manager, Endnote, Citation, GetARef, Biblioscape, Library Master, Bibliographica, Scribe, Refs). University of Bristol
16. EndNote 20. Clarivate. 2022. https://endnote.com/product-details. Accessed 28 Dec 2022
17. Angélil-Carter S (2014) Stolen language? Plagiarism in writing. Routledge
18. Howard RM (1995) Plagiarisms, authorships, and the academic death penalty. Coll Engl 57(7):788–806
19. Barrett R, Malcolm J (2006) Embedding plagiarism education in the assessment process. Int J Educ Integr 2(1):38–45
20. Introna L, Hayes N, Blair L, Wood E (2003) Cultural attitudes towards plagiarism. University of Lancaster, Lancaster, pp 1–57
21. Neville C (2016) The complete guide to referencing and avoiding plagiarism. McGraw-Hill Education (UK), London, p 43
22. Campion EW, Anderson KR, Drazen JM (2001) Internet-only publication. N Engl J Med 345(5):365
23. Li X, Crane N (1996) Electronic styles: a handbook for citing electronic information. Information Today, Inc.
24. APA (2005) Concise rules of APA style. APA, Washington, DC
25. Snyder PJ, Peterson A (2002) The referencing of internet web sites in medical and scientific publications. Brain Cogn 50(2):335–337
26. Barroga EF (2014) Reference accuracy: authors', reviewers', editors', and publishers' contributions. J Korean Med Sci 29(12):1587–1589
27. Standardization IOf (1997) Information and documentation—bibliographic references—part 2: electronic documents or parts thereof, ISO 690-2: 1997. ISO

Use of Bibliometrics to Quantify and Evaluate Scientific Output

40

Joanne Berger

Abstract

Bibliometrics, the quantification of such scientific research outputs as journals, articles, social media mentions, and more, has taken on increased importance as a means of evaluating researcher productivity, collaboration, and impact. This chapter provides an overview of bibliometrics in scientific research, including a brief history of the discipline, real-world applications, use in studies, and descriptions of standard and alternative metrics. Also discussed are several efforts of interested stakeholders to create metrics filling specific needs and more broadly, a climate for more responsible use of all bibliometrics. It concludes with bibliometrics' contributions to positive developments in scientific research, including the combination of metrics with qualitative forms of assessment, the tracking of research in the COVID-19 pandemic, and the generation of more information on the researchers themselves.

Disclaimer: *The views expressed herein are those of the author and should not be construed to represent the U.S. FDA's views or policies.*

J. Berger (✉)
FDA Library, Office of Data, Analytics and Research (ODAR), Office of Digital Transformation (ODT), US Food and Drug Administration, Rockville, MD, USA
e-mail: joanne.berger@fda.hhs.gov

Keywords

Bibliometrics · Scientific research · Publications · Citation analysis · Alternative metrics · Research impact

40.1 Introduction

Scientific evaluation incorporating traditional peer review is costly, time-consuming, and open to bias, leading some evaluators to increase their reliance on data to help streamline the process and attempt to make it more equitable. The discipline of bibliometrics, a "science of science," is based on the premise that a researcher's publications and the data they generate—such as the number of times they have been cited—can be used to answer questions about the research they represent [1]. Bibliometric data, arrived at through various calculations, or metrics, have been used to measure researcher productivity, influence, collaboration networks, and other factors.

Even with this expanded use, bibliometrics were intended to complement, not replace, qualitative measures of research evaluation. They do not measure the quality of research, the impact of research on the public, nor do they account for an author's motivation in citing a particular article. Additionally, they only relate to a researcher's publications, not activities such as teaching and mentoring, peer review, and committee

G. Jagadeesh et al. (eds.), *The Quintessence of Basic and Clinical Research and Scientific Publishing*,
https://doi.org/10.1007/978-981-99-1284-1_40

participation [1]. Yet, some have argued that metrics have resulted in an overemphasis on quantity over quality, that they have been used to create policies and incentive systems that favor some researcher demographics over others, and even that they have been misused or extrapolated to be indicators of quality.

Other factors add to the complexity of using publications as proxies for research. For example, not all scientific fields contribute equally to the corpus of published scientific literature. Researchers in fields that publish more frequently have more opportunities to be cited. Many metrics incorporate normalization to make the data usable across fields and other variables [2]; however, normalization does little or nothing to account for the role of demographics, such as author age and gender, not to mention geographical differences.

Metrics have limits, necessitating that researchers use combinations of them to make up for deficiencies of individual metrics. One important limitation common to several metrics involves time span; citation activity within a field may fluctuate wildly from year to year. Since many calculations cover only a 5-year window, these fluctuations can strongly skew the data up or down. Adding metrics that show an author's collaborations, or recent web postings (see Sect. 40.4.3), can help in showing a fuller picture of their research activity. Further, the emphasis on this window excludes so-called "sleeping beauty" papers that become widely cited many years after publication when the scientific community recognizes them as innovative or seminal [2].

Finally, bibliometrics have exerted a strong influence on researchers' subject and publication choices—and spurred some unethical behaviors—in the name of boosting citations, (see Sect. 40.7 of this chapter).

In the executive summary of *The Metric Tide* report, James Wilsdon, University of Sussex, UK and Chair of the review group, stated, "Metrics hold real power: they are constitutive of values, identities and livelihoods" [3]. Acknowledging this power, several groups have created guidelines and manifestos for more responsible use of bibliometrics. This chapter aims to discuss commonly employed bibliometrics and how their use in scientific research has evolved over time.

40.2 Early Roots

Researchers have long sought to measure aspects of scientific research. As early as 1873, French-Swiss botanist Alphonse de Candolle attempted to find environmental factors in scientific societies, such as religion, that contributed to the scientific strength of nations. In 1926, mathematician, chemist, and statistician Alfred Lotka, known for his work in population dynamics, developed a mathematical equation for studying chemistry researchers' productivity. It is among the earliest evidence that scientific fields are characterized by skewed distributions [4].

Belgian information scientist Paul Otlet in 1934 is credited with the development of "bibliomètrie," which he defined as measurements related to the publication and reading of books and documents [5]. Also that year, chemist turned librarian Samuel Bradford of the London (UK) Science Museum, discovered that literature in a subject area was concentrated in a core set of journals, a phenomenon known as Bradford's Law of Scattering. It was applied to estimate the number of journals one needs to read to attain a specific level of literature coverage on a subject [4].

The 1950s brought early harbingers of technological tools used for bibliometrics today. Early in the decade, American psychologists Robert Daniel and Chauncey Louttit created a science map by hand based on journal-to-journal citations for the psychology literature [4]. In 1955, American linguist and computer scientist Eugene Garfield began indexing journal literature using punch cards. His company was renamed Institute for Scientific Information (ISI) in 1960, when it began publishing what is now known as the Science Citation Index™ (SCI). It was the forerunner of the current Web of Science database suite from Clarivate and tracked the number of times articles were cited in the scientific literature.

In 1955, Garfield mentioned the utility of an "impact factor" to "evaluate the significance of a particular work and its impact on the literature and thinking of the period" [6]. He and colleague Irving H. Sher developed it initially to help select source journals for the index. "To do this, we simply re-sorted the author citation index into the journal citation index. From this simple exercise, we learned that initially a core group of large and highly-cited journals needed to be covered in the new Science Citation Index," Garfield stated. Yet, Garfield and Sher had not begun to use it for its current purpose until several years later, when they sampled the 1969 SCI to publish the first journal ranking using the impact factor [7].

Development in bibliometrics in the ensuing years has been reflected in its nomenclature. The term "bibliometrics" in English was coined in 1969 by British librarian Alan Pritchard, to replace the term, "statistical bibliography" [8]. He applied it to the counting of books, articles, journals, citations, and any statistically measurable element of publications, regardless of discipline [9]. Bibliometrics has since been subdivided into names that describe its application to research areas, (e.g., "scientometrics," "informetrics"), and since the advent of the Internet, "webometrics," "netometrics," and "cybermetrics," which sometimes have been used interchangeably [9]. Later, incorporating views on social media, internet metrics related to research have become known as "altmetrics."

40.3 Applications

40.3.1 Real-World Applications

Today, bibliometrics are applied to many activities related to scientific research. More information on some of these will appear in later sections of this chapter.

Performance Evaluation Citations and measures of visibility on the internet can shed light on researcher productivity and influence, enabling evaluators to make inferences about their impact.

Hiring Job candidates may highlight publications in high impact factor journals on their CVs [10]. Evaluators may use metrics to help them quickly glean information from long publication lists.

Funding Institutions use analyses of published literature in justifications for funding, as well as awards of grants, fellowships, and honors. They have also been used to measure the impacts of such funding. For example, the Obama Administration implemented STAR METRICS (Science and Technology in America's Reinvestment Measuring the Effects of Research on Innovation, Competitiveness, and Science) which aimed to track the effects of federal research grants and contracts on growth in such areas as employment, publications, and economic activity [11].

Budgeting Library staff have long used bibliometric rankings to help decide which journal subscriptions to purchase, renew, or discontinue [12].

Benchmarking Publishing and citing practices vary among scientific fields, so institutions have sought to generate comparisons among publications, authors, groups, departments, institutes, and universities in the same field over time to determine levels of influence in advancing scientific research [13, 14].

Selection of Peer Reviewers, Collaborators, and Journals Bibliometric data can aid researchers in the selection of potential peer reviewers and collaborators by pointing to highly-cited authors and the authors they cite, as well as authors with exposure in major journals. Newer researchers who have not yet been cited can also use metrics to help decide where to publish when less familiar with journals in a field [12].

Research Trend Analysis and Policy Setting Bibliometrics can also be used to identify trending topics within large publication sets, as well as relationships between topic areas [1].

40.3.2 Bibliometric Studies

Bibliometric studies are designed to answer specific questions. They can be considered to fall under four main themes, productivity, collaboration, research topics, and citation impact. Often, these themes are combined within a study [1].

Productivity Productivity analyses can involve counts of publications, or types of publications, often published within a particular time period [1].

Collaboration These studies discuss the networks of scientific research, or which scientists are working together to produce that research. Co-authorship often serves as a proxy for collaboration. Collaborations can be measured at the author, institution, or country levels [1].

Topic Topic studies might be used to understand which topics are gaining ground among researchers in a field, how that emphasis changes over time, and relationships between topic areas. They often employ either citation- or text mining-based approaches. Citation analysis approaches use cited references as a measure of similarity; while text mining approaches use the words in article titles, abstracts, or the full text to categorize large sets of articles by topic [1].

Citation Impact Which articles, or article sets, have had the most influence on advancing science? The number of articles citing a particular article is one indicator of its impact. You could also count the times the articles in a set—for example, authors at a particular research institution, published in a particular journal, or from a certain country—have been cited [1].

An example of a combined productivity and topic analysis was conducted by Panpan Wang and Deqiao Tian of Beijing Institute of Biotechnology. They searched published literature in the Web of Science database or submitted to four preprint platforms, for global trends in COVID-19 research. They found that the U.S. had published the most reports, followed by China. U.S. reports in Web of Science mostly were in the categories of non-pharmaceutical interventions, treatment, and vaccines; while the bulk of China's reports were on clinical features and complications, virology and immunology, epidemiology, and detection and diagnosis. Overall, they found countries with the most publications were concentrated in Asia, Europe, and North America, and called for greater worldwide research cooperation [15].

Russell et al. of Indiana University School of Medicine, Indianapolis, IN, sought to study changes in several bibliographic variables in four major musculoskeletal science journals over 30 years. Increases in several variables (number of authors, institutions, and countries) pointed to a trend toward greater collaboration. In addition to making comparisons related to specific journals, the authors focused on gender, noting an increase over time in female authors, female first authors, and corresponding authors. They concluded that this could be the result of more women attaining doctoral degrees in science, medicine, and engineering [16].

The next section describes the individual metrics and key points related to their use in research evaluation.

40.4 Brief Descriptions of Metrics

It is important to note that there are variations of all metrics described in this section. In addition, the names for the same metric can vary among information sources

40.4.1 Basic Metrics

Raw Citation Count The number of citations an article or set of articles has received [17].

Publication Count Total number of publications for an individual researcher or group of researchers, for example, a department, college, or institution [18].

Citation Percentile As stated in the Metrics Toolkit, "the position of a paper or group of papers with respect to other papers in a given discipline, country, and/or time period, based on the number of citations they have received. Expressed as a percentile or awarded an honorific "Highly Cited" honorific based upon percentile rankings" [18].

Normalized Citation Scores According to Ioannidis et al., "...normalization can be seen as a process of benchmarking that is needed to enhance comparability across diverse scientists, fields, papers, time periods, and so forth" [2]. Many metrics have several normalized versions, still more have been proposed.

40.4.2 Citation-Based Metrics

Impact Factor The journal impact factor aims to demonstrate which journals have the highest impacts on their respective fields, as measured by the number of times individual articles have been cited by other articles. It is calculated by taking the number of times that articles published in a journal over a 2-year period were cited by indexed journals in the following year and dividing that by the total number of citable items in that journal for the same 2-year period, illustrated in Eq. 40.1:

$$\frac{\textit{Number of citations received in 2022 by items published in the journal in 2020} - \textit{2021}}{\textit{Number of citable items published in the journal in 2020} - \textit{2021}\ [21]} \quad (40.1)$$

Given the broad adoption of the impact factor, it is important to note that success in publishing in a journal with a higher impact factor does not imply that a journal article is more worthy of citation. Many factors can affect whether a particular article is published, including the journal's editorial policy, its scope, frequency of publication, and even language. For example, journals with greater publication frequency have greater visibility, which often means more citation opportunities.

Key Points

- Assists in the selection of journals; both in publication and subscription decisions.
- Encourages authors to submit to a small subset of journals, overwhelming those journals' selection committees while reducing submissions to other journals in the field [20].
- Articles written in well-established fields are likely to be cited more frequently than are articles in emerging fields. The search for journals with high impact factors can lead scientists to restrict their research to areas that will encourage high citation counts [20].
- The number can be driven up by one highly-cited article [21].

Eigenfactor® Score The Eigenfactor® Score was developed by researchers Jevin West and Carl Bergstrom at the University of Washington, Seattle in 2007. In the U.S., it measures a journal's importance to the scientific community in terms of the number of times articles from the previous 5 years have been cited in the current Journal Citation Reports™ (see Sect. 40.5.2 of this chapter) year, shown in Eq. 40.2:

$$\begin{aligned} \textit{EF for Year } x = \; & \textit{Proportion of weighted citations from articles published in Year } x \\ & \textit{to articles published in Years } x-1, x-2, x-3, x-4 \textit{ and } x-5 \\ & \textit{minus self-citations.} \end{aligned} \quad (40.2)$$

It is weighted towards citations from highly-cited articles. In this way, the Eigenfactor® Score considers the entire citation network of journals and articles, rewarding journals that are cited by other influential journals. It also reduces the impact of journal self-citation by removing references from articles that cite another article in the same journal [22].

Key Points
Helps in finding the journals which are driving the conversations in a particular field.

- Removes self-citations.
- Like the impact factor, strongly influenced by the journal's subject area [23].

h-index As a response to the criticisms of other metrics being not useful for measuring the impacts of individual authors, the h-index is calculated by examining the work of one author and deriving their citation impact. An author's h-index is defined as the number of papers with a citation number greater than h. For example, a researcher with an h-index of 10 has 10 publications that have each been cited a minimum of 10 times [24, 25].

$$\textit{Author's h-index} = \textit{h papers cited} \geq \textit{h times} \tag{40.3}$$

Among its variations are an hm-index, accounting for multi-authored manuscripts.

Key Points

- Aims to give a more accurate picture of an author's impact on a research field.
- Simple to calculate.
- The h-index can increase but never decrease. For example, even if an author stops publishing, their h-index continues to be impacted by new citations. This can give more senior authors an unfair advantage over early-career researchers.

CiteScore CiteScore is a family of metrics counting citations to articles, reviews, conference papers, book chapters and data papers (peer-reviewed document types) in a 4-year period, divided by the total number of publications in Scopus within that same period. It includes the SciImago Journal Rank and Source-Normalized Impact Per Paper described below in this section [26, 27].

$$\frac{\textit{Citations to CiteScore publication types in Scopus database over 4 years}}{\textit{Documents in Scopus database over 4 years}} \tag{40.4}$$

Key Points

- The citation window in 2021 was expanded from 1 to 4 years to enable more robust scores for older publications.
- Publications from the most recent year are included in the calculation so these receive impact scores earlier.
- This metric is only used in the Scopus database.

Scimago Journal & Country Rank The Scimago Journal & Country Rank (SJR), from Scimago Lab, is used to score the prestige of scientific journals, based on both the number of times journal articles have been cited, as well as on the prestige of the citing journals, which is estimated by Google's PageRank algorithm. The SJR is based on journals indexed in Elsevier's Scopus database, and includes the total number of documents of a journal in the calculation's denominator, in contrast to the journal impact factor, which includes document categories deemed "citable" [28, 29].

Key Points

- Inclusion of all article types, exclusion of self-citations, and the use of a computer algorithm make it less easy for editors to use various

tactics to boost the metric [28]. See Sect. 40.7 of this chapter.
- Use is limited to articles published after 1996.
- May underestimate the prestige of journals with large numbers of non-citable articles relative to journals without them.

Source Normalized Impact per Paper Also using indexed journals in the Scopus database, the Source Normalized Impact per Paper (SNIP) measures the citation potential of journal articles by considering citation patterns within a subject field. If a typical article in one field has dozens of citations, for example, it will be weighted differently from another in a field where the typical article is cited fewer than 10 times. This weighting enables the comparison of articles in different subject areas [30]. Newer subject areas with fewer publications are on a more even playing field with more established ones, because the impact is always normalized to the state of the field at publication time.

Key Points

- Can be used to compare publications in different subject areas or fields.
- Useful for comparing newer fields with smaller collections of publications.
- Normalization of the citation counts leads to a more complicated calculation of citation impact.

Article Influence Score The Article Influence Score is used to represent the average influence of the journal's articles over 5 years. It is calculated by dividing a journal's Eigenfactor® Score by the total number of articles published in that journal in the 5-year period [22]. Journals that score greater than a mean score of 1.00 can be said to have an above-average influence in their area.

$$\frac{\textit{Journal Eigenfactor}^{\circledR}\ \textit{score}}{\textit{Number of journal's published articles over 5 years}} \quad (40.5)$$

Key Points

- Can also be used to measure cost effectiveness in situations where the author pays to publish their article, as in the Gold open access model. Article Influence Scores can be plotted against the cost of publishing and offer authors the chance to see which open access journals with fees are the most influential.
- Like the impact factor, strongly influenced by the journal's subject area [23].

i10- and h5 Indexes These metrics are calculated from Google Scholar, and are, respectively, the number of publications with at least 10 citations, and the number of papers with the same or greater number of citations over a 5-year period. The i10-index is used in Google Scholar Citations, while the h5 can be found on Google Scholar's Metrics page [31, 32] (Table 40.1).

40.4.3 Alternative Metrics

Altmetrics Since its inception in 2010, the field of altmetrics has endeavored to measure research impact beyond citations. It has expanded to include data from news sources, patents, and social media platforms. Some of these include microblogs (e.g., Twitter, Weibo), blogs and blog aggregators, online reference managers (e.g., Mendeley), scholarly tools (e.g., Dimensions, F1000Prime, Publons, Web of Science), a wide variety of public repositories (e.g., arXiv, GitHub, Figshare, Vimeo, YouTube), wikis (e.g., Wikipedia), as well as social networking platforms (e.g., Facebook, LinkedIn, Reddit) [33]. Some authors, such as Wouters et al., encourage disaggregating the number of various measures under one term [33].

Examples of altmetrics providers include Altmetric.com, PlumX, and Crossref Event Data. The oldest provider, Altmetric.com, uses weighted counts for each data source, which contribute to an Altmetric Attention Score (AAS)

Table 40.1 Comparison of several major citation-based metrics

Metric	Measures	Originating database	Coverage period	Weighting	Excludes self-citations
Article Influence Score	Journals	Multiple dbs	5 years	Yes	Yes
CiteScore	Journals	Scopus	4 years/ variations	Yes	No
Eigenfactor® Score	Journals	WOS/JCR	5 years	Yes	Yes
h-index	Authors	WOS	Career	Variations[a]	Variations
h5 index	Journal	Google Scholar	5 years	No	No
i10-index	Authors	Google Scholar[b]	5 years	No	No
Impact Factor	Journal	WOS	2 years/ variations	Variations	Variations
SJR	Journal	Scopus	3 years	Yes	No
SNIP	Articles (citation potential)	Scopus	3 years	Yes	No

dbs databases, *WOS* Web of Science, *JCR* Journal Citation Reports™, *SJR* Scimago Journal & Country Rank, *SNIP* Source Normalized Impact Per Paper

[a] Existing and proposed variations of this metric have different capabilities in these areas

[b] Metric is used in Google Scholar Citations

[34]. Several studies have compared the AAS with citation counts. Argyro Fassoulaki at Aretaieio University Hospital, Athens, Greece and coauthors found that a strength of the AAS is the impact of a particular article in real time, whereas citations are more reliable for highlighting its scientific impact over time [34].

Key Points

- Researchers receive some feedback about their work, often within a week of publication, much more quickly than via citations [35].
- They may help—though not completely—to make up for a shortage of global and publisher-independent (grey literature) usage data [35].
- The impacts that altmetrics measure, who is using them, and how the research is being used often are not clear.
- Data may be incomplete or inaccurate, e.g., missing links or metadata erroneously entered. As a result, altmetrics may count views from people who have never actually read the research, or outputs that are ambiguously connected to the research [35].

Online Access Counts Any metric that relies on citation counts will necessarily be delayed because of the lag time between publication of the original article and the authorship and publication of a citing article. To reduce this delay and pull "real time" statistics, some authors prefer to count the times their articles have been downloaded. This allows them to see an article's immediate impact [36].

Key Points

- Online publication ahead of print (epubs), and open access have boosted the counts, increasing the chances of an article's being read and cited.
- Research has shown a strong, positive correlation between download counts and citation counts and impact factors [36].
- Little or no information is given on how downloaded articles are being used.

Usage Some journal publishers have sought to differentiate download counts from access counts and citation counts, often employing the word "usage" in this count. The rationale behind this article-level metric is that most of these downloads are from readers making use of the content in practice, as opposed to researchers citing the content in their papers.

Key Points

- Usage counts, rather than citations, may better reflect interest in articles from more practical fields, as well as those of general interest [37].
- It may be impossible to know whether an article was used in ways other than as a download for later reading.

40.5 Databases and Analytical Tools

The products in this section do not comprise an exhaustive list and exclude those devoted solely to data visualization.

The databases described in the first subsection are data sources and generate metrics based on their own data. Consequently, scores will differ depending on the content of the database being used [38]. Some of the database providers also offer tools for more refined data analyses, often at additional cost. Some of these appear in Sect. 40.5.2.

40.5.1 Databases

Web of Science Web of Science, from Clarivate (https://clarivate.com/webofsciencegroup), is a suite of databases providing access to citations in the sciences, social sciences, arts, and humanities (see Chap. 37 of this book for more information). The database itself contains several tools for bibliometric analysis. Users can create citation reports on individual authors, who can be disambiguated from other authors with the same names. The report shows the number of published items, number of times cited for each, and the author's h-index, and can be displayed in either list or graphic forms. Web of Science also contains a template for cited reference searching, offering best practices for using the parameters to conduct a tailored search. Clarivate's additional bibliometric tools are described in this chapter's next section.

Scopus Scopus, launched by Elsevier in 2004 (https://www.scopus.com/home.uri), also enables users to compare journal influence and impact. In addition, it features the ability to compare countries in particular areas of publishing, and rank journals in a particular country for influence within that country. Like Web of Science, Scopus offers an author search option. It uses Elsevier-developed journal, author, and article-level metrics, described earlier in this chapter.

Because Scopus and Web of Science have differences in journal coverage, authors have been encouraged to run searches in both systems, if possible, to ensure that they are missing nothing vital in their counts [12].

Google Scholar Google Scholar, launched by Google in 2004 (https://scholar.google.com/), is a free search engine/database. In addition to much of the content in Web of Science and Scopus, Scholar contains publications beyond the scope of those databases; the set of documents its results draw from is not clear [39]. Searches on author names may be less reliable than in the aforementioned databases. Scholar also contains a Metrics page, where users can view top-cited publications. Results can be sorted by subject categories and subcategories.

Dimensions Dimensions is a database from Digital Science & Research Solutions Inc., London, UK, launched in 2018 to enable users to explore connections among a wide variety of sources (https://www.dimensions.ai/dimensions-data/). It links publications and citations with such outputs as grants, patents, clinical trials, datasets, and policy papers, and is available in free and subscription versions. The product and its data are available free to the scientometric community for use in developing future indicators [40].

Overton Built and maintained by Open Policy Ltd in London, UK, Overton is a searchable index of policy documents, guidelines, think tank publications and working papers from more than 182 countries (https://www.overton.io/).

References, people's names, and key concepts are extracted from these and linked to related news, academic research, think tank and other policy content so subscribers can find where their work is being cited or mentioned, and where it is potentially influencing policy [41].

40.5.2 Analytical Tools

Journal Citation Reports™/InCites™/Essential Science Indicators™ Clarivate's Web of Science is the home of Journal Citation Reports™ (JCR), a tool for comparing journal metrics (https://jcr.clarivate.com/jcr/home). Journals can be ranked by the number of articles they have published and the number of citations to those journal articles in a given year, as well as by the journals' impact factors.

In 2012, Thomson Reuters added a benchmarking tool, InCites™, to Web of Science (https://clarivate.com/webofsciencegroup/solutions/incites/). It enables institutions to compare their citation counts and influence in a research field with those of similar institutions. Individual authors can also use this information to compare their research output with that of colleagues within their institution and with other researchers in their fields. InCites™ employs author name disambiguation, along with persistent identifiers, to ensure other similarly named authors' works are not erroneously included and that the individual author's own name variations are not missed in the count. **Essential Science Indicators™**, another Web of Science component, focuses on field-based metrics and emerging areas of research, and displays a map view by "Top," "Hot," and "Highly Cited" papers (https://clarivate.com/webofsciencegroup/solutions/essential-science-indicators/).

SciVal Like InCites, SciVal, from Elsevier, enables users to compare the research performance of institutions and researchers from 234 countries (https://www.scival.com). Users can also find collaborators, demonstrate research impact, and discover new areas of research [42].

Publish or Perish This is a free software program from Harzing.com that retrieves and analyzes citations to demonstrate citation impact (https://harzing.com/resources/publish-or-perish). The program is the brainchild of Anne-Wil Harzing, a professor of International Management at Middlesex University, London, UK [43].

iCite, from the National Institutes of Health's Office of Portfolio Analysis, is "a tool for accessing a dashboard of bibliometrics for papers associated with a portfolio" (https://icite.od.nih.gov/). It was launched in response to analysis needs related to the COVID-19 pandemic. Users can type in a PubMed search query or upload PubMed IDs of articles they want to analyze. The output is a report table with article-level information [44].

VOSviewer software, developed by CWTS B.V. in Leiden, The Netherlands, is used to construct and visualize bibliometric networks. It also has text mining capabilities for constructing and visualizing co-occurrences of key terms in a body of work (https://www.vosviewer.com/) [45].

The **Science of Science (Sci2) Tool** is a combined data analytics and visualization tool developed by Indiana University, the National Science Foundation, and the James S. McDonnell Foundation (https://sci2.cns.iu.edu/user/index.php). Its website describes it as follows: "it supports the temporal, geospatial, topical, and network analysis and visualization of scholarly datasets at the micro (individual), meso (local), and macro (global) levels [46].

scite/Other Machine Learning Tools To address the problem of negative citations being counted the same as supportive ones, several tools have used machine learning to categorize citations based on context. The output of one such citation index, scite, (https://www.scite.ai/) based in Brooklyn, NY, is a display of the textual context from the cited work, and a classification from the product's deep learning model as to whether the citation presents supporting or contrasting evidence, or simply mentions, a referenced work. It also provides information about corrections and retractions, if applicable [46].

Other contextual tools include the web-based search service **Colil** (Comments on Literature in Literature, Database Center for Life Science, Chiba, Japan, http://colil.dbcls.jp/browse/papers/ [47] and **SciRide Finder** (SciRide.org, Boston, MA), both of which enable users to see citation context from articles in PubMed Central [48].

40.6 Supporters and Detractors

The standard metrics are widely recognized. Yet, since the information they provide is limited, bibliometrics supporters and detractors differ in their opinions on how and to what extent they should be used in evaluations.

Supporters see them as a mitigating factor against overt and covert politics and biases that creep into peer reviews. They argue that the use of metrics can be reassuring to researchers concerned about being judged by a system in which connections are valued more highly than achievements. In other words, the presence of this information could help keep peer reviewers accountable for using sources beyond their own experiences and networks, while complementing and possibly strengthening the credibility of the peer review process [49].

Evidence exists that the metrics agree with peer reviewers' judgements, and that they are more transparent and reproducible. Bibliometrics and peer review together present a more complete picture of a researcher's impact, as each counterbalances the strengths and weaknesses of the other [1].

Detractors say that bibliometrics promote biases in favor of more prolific authors, more mature fields of study, and journals with higher impact factors; and against new authors, nascent fields and research in new areas that involve some risk of rejection from publishers or lack of interest from fellow scientists. New authors typically require time to gain attention, as do journals which publish potentially groundbreaking research in less-populated fields, all of which mean further delayed gratification in terms of citation counts [12]. Alternative metrics have made some headway in remedying the time lag issues.

On top of this, first authors receive more attention in both citations and social media, and evidence of gender bias in citation exists [50, 51]. And, importantly, there is the overall feeling that these biases have changed scientists' publishing behavior: opting for paper quantity versus quality, and me-too research over innovation [20]. Even Garfield remarked that the tendency for authors to seek out better known, higher impact journals perpetuates a "vicious circle" that will last as long as researchers perceive that publication in these journals will give their work the greatest exposure to other scientists [7].

In 2021, Utrecht University in the Netherlands put impact factor supporters and detractors at odds when it announced a policy of no longer using the metric in hiring or promotion decisions. Instead, candidates would be evaluated based on such factors as level of public engagement, teamwork, and dedication to open science. This decision prompted a letter signed by over 170 researchers in the Netherlands arguing that the move would result in more arbitrary and difficult hiring and promotion. However, 383 additional scientists wrote a letter applauding the move. Reasons included the possible reduction of researcher stress, fraud, and the impact factor's overall influence on publication and personnel decisions [52].

Ironically, the strong viewpoints on both sides of the debate seem to indicate that bibliometrics are here to stay and will continue to evolve. According to James Griesemer of the University of California, Davis, "The war on metrics seems to be over: metrics will not go away, even if which metrics are today's favorites will face a constant churn. We can describe the central tension in various ways, but they boil down to the idea that expert judgment should play a central role in evaluating research content and that quantitative data should play a role in measuring research productivity. Judgment cannot be automated, yet productivity can in some ways be measured" [53]. Similarly evolving are the means of manipulating the metrics to tell the desired stories, as well as efforts to use metrics in more responsible ways. The next two sections will address these areas, respectively.

40.7 Gaming the System

There are many ways authors have tried to "game the system" to boost their citation counts. On the most basic level, an author or group of authors might publish papers in different journals based on the same work, thus repurposing one study into several opportunities to be cited [54].

The impact factor is susceptible to manipulation by both editors and authors. For the former, publishing reviews, which can generate higher than average citation numbers, and the use of articles as vehicles for citing other articles in the same journal, e.g., editorials, have been known to drive up impact factors [55].

Self-citations are a point of contention in the bibliometrics space. They can be used honestly, for example, to differentiate an author's previous work from their current work. That said, besides the obvious reasons of increasing citation counts, authors self-cite in order to provide evidence to peer reviewers that they have strong backgrounds in their fields. Similarly, researchers have formed groups of friends or colleagues who arrange to cite each other. The process is known as "clubbing," and the groups as "citation rings" or "citation cartels." There is even an agency offering citations to researchers who pay its fee, creating a network of citations in the manner of a Ponzi scheme. While there are cases of papers being retracted upon discovery of the aforementioned groups, authors may risk engaging with them because peer reviewers rarely check all citations [56].

Predatory journals have also given rise to fake metrics companies intended to facilitate authors choosing to publish in these journals (*see* Chap. 59). Along with fraudulent document and object identifiers, they have become "the heralds of an entire parallel scholarly apparatus, a crooked ancillary economy," according to Finn Brunton of New York University [57].

Gaming has taken hold in the retractions sphere as well. The average time for a publication to be retracted is 3 years, a year longer than the time for citations to be counted toward a journal's impact factor. This incentivizes journal editors to slow the process to keep an article in the impact factor calculation. A slower process also reduces the likelihood that a retraction might affect an author's career or an institution's funding decisions [58].

Sometimes retracted papers continue to be cited as if the retraction never happened. Andrew Wakefield's discredited 1998 *Lancet* paper linking autism and vaccines was second on the list of the blog *Retraction Watch's* top 10 cited retracted papers in 2016. Ironically, the blog's editor, Ivan Oransky of New York University's Carter Journalism Institute, stated that journals with the highest impact factors tended to have the highest rate of retraction [58].

All that said, gaming the system may be more than just an unintended consequence. According to Griesemer, "Conversions of metrics into standards not only invite 'gaming the system,' they also practically construct 'gaming' as the new form of practice..." [53].

40.8 Responsible Use of Metrics

Acknowledging metrics as part of the scientific research ecosystem, several collaborations of stakeholders have produced manifestos and other documents aimed at using them responsibly for research evaluation. Concerned scientists' organizations at the December 2012 meeting of the American Society for Cell Biology launched the San Francisco Declaration on Research Assessment (DORA). Its focus was on ending the use of the impact factor in assessing the works of individual scientists. DORA states that the impact factor must not be used "as a surrogate measure of the quality of individual research articles, to assess an individual scientist's contributions, or in hiring, promotion, or funding decisions" [20].

A collaboration of UK researchers, in their report, *The Metric Tide*, proposed "responsible metrics" to describe appropriate uses of research metrics. They described them in terms of the following qualities: robustness, humility, transparency, diversity, and reflexivity. The latter quality refers to updating the indicators in response to their systemic and potential effects. The report offered 20 specific recommendations for action by UK stakeholders and spawned a blog, *Responsible Metrics* [3].

More recently, the Leiden Manifesto, named after the location of the conference from which it emerged, set forth 10 principles for best practices in research assessment. These are:

1. Quantitative evaluation should support qualitative, expert assessment.
2. Measure performance against the research missions of the institution, group, or researcher.
3. Protect excellence in locally relevant research.
4. Keep data collection and analytical processes open, transparent, and simple.
5. Allow those evaluated to verify data and analysis.
6. Account for variation by field in publication and citation practices.
7. Base assessment of individual researchers on a qualitative judgement of their portfolio.
8. Avoid misplaced concreteness and false precision.
9. Recognize the systemic effects of assessment and indicators.
10. Scrutinize indicators regularly and update them [10].

The Initiative for Open Citations, or I40C, a collaboration between publishers, researchers, and others, aims to make citation data more widely available and machine-readable via "a global public web of linked scholarly citation data to enhance the discoverability of published content, both subscription access and open access," for the benefit of researchers without access to costly subscriptions. Future plans include the ability to build services over this open data and the creation of a citation graph that can be used to explore connections between fields and follow the evolution of ideas and scholarly disciplines [59].

An Altmetrics Manifesto was published in 2010, calling for new approaches to using data and indicators from the web in research assessment and other applications. Annual altmetrics conferences bring together users and providers. The EU-funded ACUMEN project created assessment guidelines for researchers and evaluators [33], and in the U.S., the National Information Standards Organization's (NISO) Alternative Assessment Metrics Project continues to create structure, standards, nomenclature, and best practices around altmetrics research [4, 60].

40.9 Differing Perceptions, Alternative Ideas

Much has been reported about the ways standard metrics can be improved or supplemented by other measures. Julia Lane of the National Science Foundation in the U.S. has advocated for metrics to be made more scientific, that is, "grounded in theory, built with high-quality data, and developed by a community with strong incentives to use them." She cited the Lattes Platform, Brazil's national research and science database (https://lattes.cnpq.br/), as an example of good practice in assessing researcher credentials. To build it, federal agencies joined with researchers, created incentives to use the database, and established a system to ensure that researchers with similar names are credited correctly [61].

Bruno Frey and Margit Osterloh of the University of Zurich, Switzerland, stated that offering incentives matched to scientists' main performance drivers such as curiosity, autonomy, and peer recognition might motivate researchers to do quality work rather than just try to manipulate the metrics. These incentives could include honorary doctorates, professorships and fellowships, membership in scientific academies, and recommendations for awards from reputable award-giving institutions [62].

Other suggestions include measures that address evaluators' specific goals, for example, if they want to reduce the burden on the peer review system, they could encourage candidates for promotion to submit for evaluation their five best papers rather than their full body of work. In addition to being less demanding on peer reviewers, this method could in turn motivate researchers to write higher-quality papers, according to West, a developer of the Eigenfactor® Score [62].

Another type of effort aimed at correcting the problems of established bibliometrics involved developing indicators for quality attributes. In this vein, a group of researchers in India proposed the **Original Research Publication Index** (ORPI), an indicator of the overall value of an individual's research output, to be used in conjunction with some form of peer review. Its formula was designed to assess a researcher's originality over time by measuring original articles versus total publication output and overall citations, with all journals being considered equal. More weight was placed on such factors as first authorship and original articles published in low-impact journals [63].

More recently, investigators in Amsterdam and Sweden proposed an independence indicator, assessing both whether a researcher has developed collaboration networks, and their level of thematic independence [64].

40.10 Concluding Remarks

The history of bibliometrics shows it is a discipline in a near-constant state of evolution. It was a daunting subject to write about, since the sheer volume and publishing frequency of bibliometrics-related papers meant that some important ones would be missed. Yet, I also found searching for bibliometrics literature to be a "meta" experience, as I used bibliometrics tools to find highly cited papers on bibliometrics and bibliometric studies.

Adding to the complexity of the discipline are its many applications, including hiring, promotion, funding, budgeting, resource selection, policy setting, finding research trends, and collaborations. Over time, however, its use in its use in performance evaluation continues to be controversial. Concerns about misuse of quantitative publication measures for quality assessment remain, as do issues of unintended consequences, (e.g., gaming), calls for transparency about the specifics of their use, and the weight they are assigned.

While calls for responsible use of bibliometrics, exemplified by DORA, *The Metric Tide*, the Leiden Manifesto, and others are useful aids for organizations in developing best practices; the onus of selecting the best evaluative metrics is in on the individual organizations. Institutional stakeholders must decide for themselves which information is most important to gather, and which metrics are most relevant to that information. Conversely, this may mean excluding some metrics from certain types of evaluative criteria, e.g., the impact factor's exclusion from hiring and promotion criteria at Utrecht University [52].

Since all metrics have limits, the use of multiple metrics in combination, or the consideration of metrics together with qualitative forms of assessment, (e.g., non-publishing-related research activities, engagement with the organization) are likely to best serve both the organizations and the candidates being evaluated. Tools and databases used, researcher demographics, fields of study, novelty, and qualitative evidence of impact also are important considerations. As Diana Hicks of the Georgia Institute of Technology, Atlanta, and coauthors stated in an article on the Leiden Manifesto, "The best decisions are taken by combining robust statistics with sensitivity to the aim and nature of the research that is evaluated" [10].

Aside from evaluation, the past few years have shown promising developments bibliometrics research. Bibliometric studies related to the COVID-19 pandemic have pointed to trends in research topics, countries where the research is being conducted, research gaps, and other information useful for future research [15, 65, 66]. Similarly, bibliometric studies aimed at increased understanding of the researchers themselves have focused on demographic inequalities, as well as emerging, expanding, and declining interest in specific topics and fields [16, 67]. Collaboration networks shed light on the globalization of research.

Another promising development involves partnerships between organizations in the scholarly publishing space to prevent or solve potential problems. For example, Clarivate and Retraction Watch joined forces to address concerns about possible research misconduct, (e.g., plagiarism, image manipulation, fake peer review), among authors on Clarivate's preliminary list of 2022 Highly Cited Researchers (https://clarivate.com/highly-cited-researchers/). Assisted by Retraction Watch and its retractions database, Clarivate analysts searched for evidence of misconduct in all publications by researchers on the list. Researchers found to have committed misconduct in formal proceedings conducted by a research institution, government agency, or publisher would be excluded. David Pendlebury, head of research analysis at Clarivate, stated that the company expanded its qualitative analysis in this way "to ensure the Highly Cited Researchers list reflects genuine, community-wide research influence. . .part of a wider responsibility across the whole research community to better police itself and uphold research integrity" [68].

All of this—knowledge about the diversity of researchers, greater understanding of global contributions to research, and efforts to tell a more complete story by combining bibliometrics and qualitative analysis—help make research more robust and trustworthy.

Acknowledgements Many thanks to Ya-Ling Lu, Center for Tobacco Products, U.S. FDA, for valuable feedback on a draft of this manuscript; and Prof. Pitchai Balakumar, The Office of Research and Development, Periyar Maniammai Institute of Science & Technology, India; and Dr. Gowraganahalli Jagadeesh, Center for Drug Evaluation and Research; for assistance with and helpful comments on this manuscript.

Conflict of Interest No conflict of interest has been declared.

References

1. Belter CW (2018) Providing meaningful information: part B—bibliometric analysis. In: DeRosa A (ed) A practical guide for informationists: supporting research and clinical practice. Elsevier, pp 33–47
2. Ioannidis JP, Boyack K, Wouters PF (2016) Citation metrics: a primer on how (not) to normalize. PLoS Biol 14(9):e1002542
3. UK Research Innovation (2015) The metric tide: review of metrics in research assessment
4. van Raan A (2019) Measuring science: basic principles and application of advanced bibliometrics. In: Springer handbook of science and technology indicators. Springer, pp 237–280
5. Rousseau R (2014) Forgotten founder of bibliometrics. Nature (London) 510(7504):218
6. Garfield E (2006) Citation indexes for science. A new dimension in documentation through association of ideas†. Int J Epidemiol 35(5):1123–1127
7. Garfield E (2003) The meaning of the impact factor. Int J Clin Health Psychol 3(2):363–369
8. Ball R (2018) Introduction and history. In: Ball R (ed) An introduction to bibliometrics: new developments and trends. Elsevier, pp 7–14
9. De Bellis N (2009) Bibliometrics and citation analysis: from the science citation index to cybermetrics. Scarecrow Press, Lanham, MD
10. Hicks D, Wouters P, Waltman L, de Rijcke S, Rafols I (2015) Bibliometrics: The Leiden Manifesto for research metrics. Nature (London) 520(7548):429–431
11. Macilwain C (2010) What science is really worth: spending on science is one of the best ways to generate jobs and economic growth, say research advocates. But as Colin Macilwain reports, the evidence behind such claims is patchy. Nature 465(7299):682–685
12. Berger J, Baker C (2013) Bibliometrics: an overview. Rajiv Gandhi Univ Health Sci J Pharm Sci 4(3):81–92
13. Reedijk J, Moed HF (2008) Is the impact of journal impact factors decreasing? J Doc 64(2):183–192
14. Thijs B, Glänzela W (2009) A structural analysis of benchmarks on different bibliometrical indicators for European research institutes based on their research profile. Scientometrics. 79(2):377–388
15. Wang P, Tian D (2021) Bibliometric analysis of global scientific research on COVID-19. J Biosaf Biosecurity 3(1):4–9
16. Russell AF, Loder RT, Gudeman AS, Bolaji P, Virtanen P, Whipple EC et al (2019) A bibliometric study of authorship and collaboration trends over the past 30 years in four major musculoskeletal science journals. Calcif Tissue Int 104(3):239–250
17. Guide to Metrics: University of Liverpool. https://www.liverpool.ac.uk/open-research/responsible-metrics/guide-to-metrics/
18. Metrics Toolkit. https://www.metrics-toolkit.org/metrics/
19. Larivière V, Sugimoto CR (2019) The journal impact factor: a brief history, critique, and discussion of adverse effects. In: Glänzel W, Moed HF, Schmoch U, Thelwall M (eds) Springer handbook of science and technology indicators. Springer, pp 3–24
20. Alberts B (2013) Impact factor distortions. American Association for the Advancement of Science, p 787
21. Dimitrov JD, Kaveri SV, Bayry J (2010) Metrics: journal's impact factor skewed by a single paper. Nature. 466(7303):179
22. About the Eigenfactor Project: EIGENFACTOR.org. http://www.eigenfactor.org/about.php

23. Finch A (2012) 10—Citation, bibliometrics and quality: assessing impact and usage. In: Campbell R, Pentz E, Borthwick I (eds) Academic and Professional Publishing. Chandos Publishing, pp 243–267
24. Bar-Ilan J (2008) Which h-index?—a comparison of WoS, Scopus and Google Scholar. Scientometrics 74 (2):257–271
25. Hirsch JE (2005) An index to quantify an individual's scientific research output. Proc Natl Acad Sci U S A 102(46):16569–16572
26. CiteScore: Elsevier. https://service.elsevier.com/app/answers/detail/a_id/30562/supporthub/scopus/session/
27. Elsevier. Scopus LibGuide: Metrics: Elsevier. https://elsevier.libguides.com/Scopus/metrics
28. Falagas ME, Kouranos VD, Arencibia-Jorge R, Karageorgopoulos DE (2008) Comparison of SCImago journal rank indicator with journal impact factor. FASEB J 22(8):2623–2628
29. Scimago Journal & Country Rank. https://www.scimagojr.com/
30. Leydesdorff L, Opthof T (2010) Scopus's source normalized impact per paper (SNIP) versus a journal impact factor based on fractional counting of citations. J Am Soc Inf Sci Technol 61(11):2365–2369
31. Teixeira da Silva JA (2021) The i100-index, i1000-index and i10,000-index: expansion and fortification of the Google Scholar h-index for finer-scale citation descriptions and researcher classification. Scientometrics 126(4):3667–3672
32. Google Scholar Metrics. https://scholar.google.com/intl/en/scholar/metrics.html
33. Wouters P, Zahedi Z, Costas R (2019) Social media metrics for new research evaluation. Springer handbook of science and technology indicators. Springer, pp 687–713
34. Fassoulaki A, Vassi A, Kardasis A, Chantziara V (2020) Altmetrics versus traditional bibliometrics: short-time lag and short-time life? Eur J Anaesthesiol 37(10):944–946
35. Bornmann L (2015) Alternative metrics in scientometrics: a meta-analysis of research into three altmetrics. Scientometrics 103(3):1123–1144
36. Meho LI (2007) The rise and rise of citation analysis. Phys World 20(1):32–36
37. Fenner M (2013) What can article-level metrics do for you? PLoS Biol 11(10):e1001687
38. Van Noorden R (2010) Metrics: a profusion of measures. Nature 465(7300):864–866
39. Leydesdorff L, Wouters P, Bornmann L (2016) Professional and citizen bibliometrics: complementarities and ambivalences in the development and use of indicators: a state-of-the-art report. Scientometrics 109(3):2129–2150
40. Dimensions: Digital Science & Research Solutions Inc. https://www.dimensions.ai/dimensions-data/
41. Overton: Open Policy Ltd. https://www.overton.io/
42. SciVal: Elsevier (2022). https://www.scival.com/landing
43. Harzing AW (2022) Publish or perish: Harzing.com. https://harzing.com/resources/publish-or-perish
44. iCite: Office of Portfolio Analysis, National Institutes of Health. https://icite.od.nih.gov/
45. VOSviewer: CWTS B.V. https://www.vosviewer.com/
46. Nicholson JM, Mordaunt M, Lopez P, Uppala A, Rosati D, Rodrigues NP et al (2021) scite: a smart citation index that displays the context of citations and classifies their intent using deep learning. Quant Sci Stud 2(3):882–898
47. Fujiwara T, Yamamoto Y (2015) Colil: a database and search service for citation contexts in the life sciences domain. J Biomed Semant 6(1):38
48. Volanakis A, Krawczyk K (2018) SciRide Finder: a citation-based paradigm in biomedical literature search. Sci Rep 8(1):6193
49. Weingart P (2005) Impact of bibliometrics upon the science system: inadvertent consequences? Scientometrics 62(1):117–131
50. Ross MB, Glennon BM, Murciano-Goroff R, Berkes EG, Weinberg BA, Lane JI (2022) Women are credited less in science than men. Nature 608 (7921):135–145
51. Wulf K (2022) The scholarly kitchen. Society for Scholarly Publishing. https://scholarlykitchen.sspnet.org/2022/08/17/still-ambiguous-at-best-revisiting-if-we-dont-know-what-citationsmean-what-does-it-mean-when-we-count-them/?informz=1&nbd=b5b819f2-03a8-4861-b0c3-805df08cf68b&nbd_source=informz. [Cited 2 Aug 2022]
52. Singh Chawla D (2021) Scientists at odds on Utrecht University reforms to hiring and promotion criteria. Nat Index. https://www.nature.com/nature-index/news-blog/scientistsargue-over-use-of-impact-factors-for-evaluating-research#:~:text=Researchers%20based%20in%20the%20Netherlands,arbitrariness%20in%20promotions%20and%20hiring. [Cited 26 Apr 2023]
53. Griesemer J (2020) Taking Goodhart's law meta: gaming, meta-gaming, and hacking academic performance metrics. In: Biagioli M, Lippman A (eds) Gaming the metrics: misconduct and manipulation in academic research. The MIT Press, pp 77–87
54. Abbott A, Cyranoski D, Jones N, Maher B, Schiermeier Q, Van Noorden R (2010) Do metrics matter? Many researchers believe that quantitative metrics determine who gets hired and who gets promoted at their institutions. With an exclusive poll and interviews, Nature probes to what extent metrics are really used that way. Nature 465(7300):860–863
55. Reedijk J (2012) Citations and ethics. Angew Chem Int Ed 51(4):828–830
56. Bik E (2022) Science Integrity Digest 2022. [cited August 2, 2022]. https://scienceintegritydigest.com/2022/03/23/citation-statistics-and-citation-rings/#more-3043
57. Brunton F (2020) Making people and influencing friends: citation networks and the appearance of significance. In: Biagioli M, Lippman A (eds) Gaming the metrics: misconduct and manipulation in academic research. The MIT Press, pp 243–250

58. Oransky I (2020) Retraction watch: what we've learned and how metrics play a role. In: Biagioli M, Lippman A (eds) Gaming the metrics: misconduct and manipulation in academic research. The MIT Press, pp 141–148
59. I4OC. Initiative for open citations. https://i4oc.org/
60. NISO (2022) NISO alternative assessment metrics (altmetrics) initiative. National Information Standards Organization, Baltimore, MD. https://www.niso.org/standards-committees/altmetrics
61. Lane J (2010) Let's make science metrics more scientific. Nature 464(7288):488–489
62. Braun T, Bergstrom CT, Frey BS, Osterloh M, West JD, Pendlebury D et al (2010) How to improve the use of metrics. Nature 465(17):870–872
63. Saxena A, Thawani V, Chakrabarty M, Gharpure K (2013) Scientific evaluation of the scholarly publications. J Pharmacol Pharmacother 4(2):125–129
64. van den Besselaar P, Sandström U (2019) Measuring researcher independence using bibliometric data: a proposal for a new performance indicator. PLoS One 14(3):e0202712-e
65. Ioannidis JPA, Bendavid E, Salholz-Hillel M, Boyack KW, Baas J (2022) Massive covidization of research citations and the citation elite. Proc Natl Acad Sci U S A 119(28):e2204074119
66. Yu Y, Li Y, Zhang Z, Gu Z, Zhong H, Zha Q et al (2020) A bibliometric analysis using VOSviewer of publications on COVID-19. Ann Transl Med. 8 (13):816
67. Chatterjee P, Werner RM (2021) Gender disparity in citations in high-impact journal articles. JAMA Network Open 4(7):e2114509-e
68. Clarivate names world's influential researchers with highly cited researchers 2022 list [press release]. London, UK

Part VIII

Academic Writing and Publishing

How to Write a Scientific Paper

41

Michael J. Curtis

Abstract

A scientific paper is a report of research, prepared from the investigator's experimental findings, and intended to contribute to knowledge. The publication process is predicated by the content (the data) and the requirements of the publication (which is normally a journal that exists in paper form and/or online) whom you wish to publish your work. The requirement of the investigator is to map their data to the structure required by the journal. It is essential, therefore, to read and understand the journal's Instructions to Authors. This means the subject matter must map to the journal's scope, and the manuscript structure must map to the journal's needs. If the investigator can do this, the manuscript may then be sent by the journal for peer review. This means there is no single formula for creating a publishable item; *publishability depends on the content, presentation and the requirements of the publication*. As a journal editor, I frequently reject items without peer review because they are not in scope ('off topic') or do not follow key construction requirements (e.g., Results and Discussion presented as a single section rather than as separate sections). The author must decide whether their research is appropriate for their chosen journal. In many cases there may not be sufficient data, or data of an appropriate type to justify a publication in all but the least discerning of journals. The investigator may learn lessons from this. Here I provide guidance on navigating the process of writing a scientific paper.

Keywords

Communication · Experimental planning · Instructions to authors · Peer review · Scientific research · Transparency

41.1 Introduction

Writing a scientific paper requires thorough understanding of the process of getting a scientific paper published. The publisher curates a repository for your data, but there is an important context that drives the process. In research, careers are built on the visibility and perceived value of your papers. The paper itself therefore must contain elements that your peers (other investigators) would regard as valuable. Valuable content requires effective planning and execution of experiments. It is possible to publish *any* data set *somewhere,* but the value of doing so is entirely dependent on *where* the data set is

M. J. Curtis (✉)
Cardiovascular Division, Faculty of Life Sciences and Medicine, School of Cardiovascular Medicine & Sciences, King's College London, Rayne Institute, St Thomas' Hospital, London, UK
e-mail: michael.curtis@kcl.ac.uk

G. Jagadeesh et al. (eds.), *The Quintessence of Basic and Clinical Research and Scientific Publishing*,
https://doi.org/10.1007/978-981-99-1284-1_41

published. Not all repositories of data are equal for reasons explained later.

To publish in a repository that will attract readers, the outcomes are expected to be novel and interesting. They may even be expected to be important. The rewards earned by the investigator are obtained by publishing the findings. The recognition of findings that are valuable hinges upon the visibility of the published findings and the provenance of the repository (journal) in which the findings are published. The value of a repository can be understood by answering the question 'is the journal well-regarded?'. In many countries the employer, usually a university, measures the value of the employee based on the number and impact of their research publications. The impact is measured using the journal impact factor (JIF). It is worth taking a moment to consider this because whether we regard a JIF as meaningful or not, it plays a pivotal role in our employability. There are several different JIFs but the one with greatest currency is the one reported by Web Of Science (https://www.webofscience.com/wos/). The JIF is a number that represents the total number of papers a journal publishes in a 2-year window divided into the total number of times these papers are cited in a 2-year window. Thus, a journal with a JIF of 5 is a journal that publishes papers that are cited, on average, 5 times a year by other published papers in their bibliography (list of papers cited in the text). Experienced investigators therefore select journals in which to publish their work largely based on JIF.

Journal publishers sell their journals to universities, pharmaceutical companies, and individuals (the reader groups) as part of their commercial activity. Most journals are now available online, and readers can pay for access to individual papers if they wish, avoiding a journal subscription. Other journals allow free downloading to individuals as part of an Open Access mandate (that involves institutions or individuals paying a fee for their work to be published). The reader groups will not have the means nor inclination to subscribe to every journal that has a paywall. The choice of which journals to buy is therefore made based on perceived quality of the journal. The measure of quality that is used to make this decision is the JIF.

This rubric of dependency on JIF as the arbiter of quality has created a process that makes use of peer review of work to ensure that a level of quality is met before a manuscript is accepted for publication. It has also created a hierarchy of journals that investigators and employers rank, based on the JIF. Consequently, a journal with a high JIF will receive many manuscripts and will operate consistently stringent peer review. This operates as follows. Submitted manuscripts will be sent to leading experts who will evaluate the quality of the manuscript. They will almost certainly ask for changes, clarifications and perhaps further experiments before the work can be published. Journals with high JIFs receive many more manuscripts than they can publish. They therefore reject many. Some journals reject >90% of manuscripts received. The higher the JIF, the more demanding the peer review. This is because it is believed, not unreasonably, that the more stringent the peer review the more likely it is that papers that are accepted for publication are those of the highest quality. All journal editors aim to increase and maintain a high JIF for the perceived 'health' of the journal. The publisher can consequently sell the content to a larger market, and charge more for access.

The investigator therefore must select a target journal with the highest JIF into which they consider their manuscript stands a chance of acceptance. This is a challenging calculation for an investigator. The journals with the highest JIFs, such as Science and Nature (and many others) look for a certain magical ingredient they call *impactfulness* when conducting peer review. Before the editor sends the manuscript to experts for evaluation, they will consider whether, if the findings are true, they will have impact in the field and beyond. Many investigators are poor judges of the impactfulness of their findings and aim too high when using JIF to select a target journal to send their manuscript. It is best, if you are unsure, to seek guidance on target journal selection from a senior colleague.

The key objective of this chapter is to explain to an investigator engaged in research how to plan and execute the successful preparation of a publishable manuscript.

41.2 Before Writing a Paper the Work Must Be Completed

Most of the effort that goes into writing a publishable paper is the generation of good quality data. For high JIF journals (and by this, I mean a JIF of 3 or more, although some institutions rate a JIF as 'high' if the journal is ranked in the top quartile of journals in the field, so-called Q1) the expectation will be that the manuscript outlines a clear hypothesis that is justified by important unanswered questions. This information will form the Introduction of the paper. Therefore, I will give some guidance first on how to conduct publishable research. Before that it is important to note that not all journals are equal, and there are many vanity/predatory journals that take a fee for publishing any item submitted, however poor. Publishing in such journals will provide the author with nothing of value because the JIF will be small or even non-existent—Web Of Science will not even list a JIF for many publications.

Conducting publishable research requires tacit expertise. Quality data requires that the experiments undertaken are fit to address the hypothesis. This means that chosen techniques must be appropriate, and the experimental models validated. Some general issues relevant to appropriate model selection have been addressed, using studies on cardiac arrhythmias as the exemplar discipline [1] and will not be repeated here. After selecting the approach, it is critical to design and analyse the experiments appropriately. This is an emerging area of importance and there are several key guidance articles published for pharmacology that can be applied to other scientific disciplines (e.g., [2, 3]). Finally, the findings should be discussed fairly so it is important to have a clear view of what the findings mean. I will provide more detailed guidance on this below. One's ability to navigate through identifying a question and how it may be answered, evolve experimental planning and execution, and undertake balanced analysis and interpretation are the key skills that represent the tacit expertise of the effective investigator.

If the investigator wishes to aim as high as possible (in terms of JIF) when selecting a journal, there is an increased risk the manuscript will be rejected by peer review. The investigator will then have to aim a little less high and select an alternative journal. This has implications about the preparation of the manuscript. In my lab we create a master draft and then edit it to fit the requirements of our first (then second and so on) target journal. The strategy is to write a full master document that contains everything we wish to include, and then the submitted versions will be edited down with removal of extraneous material to ensure the journal's word limits are not exceeded. Planning is key and a master draft is recommended.

Finally, in this introduction, I include these notes on good practice. First it is unethical to send a manuscript to more than one journal at once. This is partly based on the "Ingelfinger rule", an editorial edict made by Franz Ingelfinger, editor of New England Journal of Medicine in 1969, to not consider publication of work published elsewhere. Likewise, it is unethical to leave out awkward findings that don't map to your narrative. I will return to the latter, later.

41.3 General Strategic Advice

The intention of the investigator is to generate a paper that will be read and appreciated; therefore it will help to put yourself in the shoes of the intended reader of your paper. The first people who read your work will be the experts who will undertake the peer review of the submitted manuscript. How much general information would you, an expert, need to see when attempting to understand a research paper in your area? As well as the inclusion of sufficient detail, the readability of the manuscript will be a critical part of this assessment. It is wise to use as few abbreviations as possible. Likewise, if you have numerous study groups, avoid coding them. The reader does not want to have to go back to the Methods, repeatedly, to remind themselves which group is which. Keeping the paper uncluttered with long sentences and needless distractions such as digression into irrelevant areas is key, and the manuscript should contain only the content that is necessary. Hereon I will address how to assemble each section of the manuscript.

41.4 How to Write an Abstract

When I was a young investigator, I would leave the abstract till last when writing a paper, hoping that a narrative would emerge once the main part of the paper was written. Now I recommend starting the writing with the abstract. One assumes that the lead writer has an overview of the work, so it should be possible to set out some simple sentences summarizing the reason for the work, what was done, what was found and what it may mean. It is necessary to keep the length to that prescribed by the journal by editing the abstract after it has been made coherent. Even if you select a journal after you have created a good draft of the manuscript, ensure the abstract fits the journal Instructions. Thus, print the instructions and refer to them often. Many journals, especially high JIF journals will reject a manuscript without explanation ('desk reject') because it goes against Instructions. This is justified by an editor, not unreasonably, who will deem an investigator who cannot follow simple instructions to be the author of poor-quality work. Many journals require a structured abstract, which means that you may be asked to break the abstract into sections, typically 'Aims and objectives; Methods; Results; Conclusions (or similar variants)'. This can help the author focus on structuring the abstract to communicate the key points. The identity of the subheadings in the abstract may depend on the type of work undertaken. However, the general purpose of the abstract is always to summarize the reason for the work, what was done, what was found and what it may mean.

41.5 How to Write an Introduction Section

Many journals have a word limit which may apply to each section (Introduction, Methods, etc.,) or the manuscript as a whole. Make sure that you do not exceed the limits, or a desk reject may result. With an Introduction you are explaining why you did the research and how you planned your general approach. It is a good tactic to not include detailed background information here. The readership will be your peers: investigators interested in your area of research. Thus, when I am writing about new findings for a new antiarrhythmic drug, I would be unwise to start by explaining that the heart is a four-chambered organ that lives in the thorax. Instead, I would note that prevention of cardiac arrhythmias (I would specify the type I study—ventricular) represent an important unmet clinical need. I would mention a few 'headline' failed clinical trials (*without* any details such as patient numbers and doses). Then I would outline my proposed solution to the problem and my general approach (selection of animal model, etc.). A common mistake investigators make is to write a long unstructured review of the general area, at too simple a level, and pad out the text with needless details such as 'in a randomized, double blind trial of 600 patients, 290 in the treatment arm and 310 in the controls, 35 mg of drug X was administered and it was found that... ..'. Not even in a review article is this level of granularity of value. Ask yourself what information you would need, as a reader familiar with the general area, to be able to understand what problems are being addressed, and how. Always put yourself in the shoes of the reader when writing a manuscript, especially at the pre-submission proof stage.

41.6 How to Write a Methods Section

Here the structure will normally be dictated by the journal so ensure you follow the guidance. The better journals now require detailed information on a range of specific (e.g., animal ethics and approvals) and generic issues (experimental design and analysis). British Journal of Pharmacology has numerous requirements on key issues that can only be accommodated at the experimental planning stage [2]. This includes adherence to a randomized design, adequate group sizes and blinded analysis. If your research did not incorporate these design aspects your paper is automatically excluded from this journal. Increasingly these requirements are likely to become the norm across research so, as noted earlier, acquiring tacit expertise is pivotal if you wish to write a paper. You need to have data worthy of publication to justify attempting to publish the work. If your data are of good provenance, then it is useful to you and the reader if you create multiple subsections as repositories for the protocols. These can be combined later if you find it would help the flow of the manuscript to do so.

41.7 How to Write a Results Section

There are several key issues concerning the presentation of results that will help or hinder acceptance by the referees of a journal, some of which are nuanced. For example, some journals encourage methodological detail in figure legends whereas other journals expect all the methods to be presented in the Methods section only. As always, ensure you follow instructions.

At all times it is essential to make the presentation accessible to the reader (the first of whom will be the referees). It is best to order the Results section to map to the Methods section, with matching subheadings. Do not repeat methodological detail in Results (unless the journal requires this—i.e., in figure legends), and if you feel the need to do so this may mean that you failed to do so in the Methods section, and are compensating inappropriately. Authors can become lost when writing their results. It will help to repeatedly re-read the Methods and Results sections to ensure that a reader would be able to reproduce the experiment and its analysis. This is one of the issues to be addressed by a pre-submission proofreading process that you will need to undertake.

To facilitate accessibility, figures and tables should be simple. Tables should occupy no more than one page, and not be made so wide they need to be printed on their side. Numerical data should map to the rules of mathematics, especially with respect to precision. For example, we measure heart rate in beats/minute, so it is false precision to give a mean heart rate with one (or more) decimal places, just because your software can calculate the values to whatever number of decimal places you choose. Figures should be drawn when they show findings more clearly than presentation in text or tables would do. Many journals allow composite figures which comprise, for example, some western blot images and some graphs. It is critical to prepare the presentation according to the journal's requirements, and the most informative way to learn about this is to browse papers in your target journal to see what others do. Histogram graphs should have an appropriate Y axis that begins at zero, not a random larger value that makes the histogram look as if there are big differences between groups when there may be little or none.

It may be that during the execution of the study you included some experiments that were not well-considered. Perhaps you measured a variable that transpires to be of little interest. There is no point in including data in a manuscript if they do not map to your narrative or supply important background information about the quality of the study. In my laboratory we always report the size of an ischaemic region in hearts subjected to regional ischaemia, not because we expect a drug to change the values but to confirm that the size is similar between experimental groups, and sufficient to generate the cardiac arrhythmia, ventricular fibrillation, a lethal arrhythmia that is our primary variable of interest (e.g., [4]). However, we do not report numbers of ventricular premature beats, minor arrhythmias that are normally unaffected by drugs with strong effects on ventricular fibrillation [5]. There will be countless equivalent issues for all investigators working in any field, and being skilled at judging what is important and what is not is part of the constellation of tacit expertise that is acquired through experience. A manuscript is likely to be rejected by a journal if the data presented are primarily a list of findings or a catalogue of outcomes without clear relevance to a narrative.

Despite the caveats about inclusion of superfluous data, it is important to not leave out data that do not fit the narrative. One of the most likely reasons for the abundance of false positive findings in the literature is the exclusion from a manuscript of other findings from the study that would not be expected if the hypothesis were true [6]. In fact, it may be the case that the data set overall does not support the hypothesis and may even support a conclusion that the hypothesis is false, were all the findings included. It is essential that the investigator accept this and report the data accordingly. It is not as well-recognized as it should be that falsifying a hypothesis (showing it to be false) is the most powerful tool in experimental science. Sadly, the reward system of publication and its fondness for positive findings that are more likely to be cited, adding to a journal's JIF, acts to dissuade many researchers from attempting to publish a 'negative' finding. It can be challenging, and the less well-educated of our colleagues undertaking peer review may reject a 'negative' finding on the grounds that it isn't 'interesting'. A paper from my lab was rejected by three journals for this reason and was eventually accepted by a fourth journal [7] when an editor overruled his referees who were demanding the impossible (because our findings appeared to differ from those of a published study from another laboratory). It transpired, with the test of the passage of time, that our findings were correct, the hypothesis was false, and a class of drug that had been hyped based on the earlier 'positive' outcome never became an antiarrhythmic drug. We learned from this that when one has faith in one's data derived from studies that are blinded and randomized and conducted with care, and undertaken without regard for the outcome, then one has a duty to see the work published. It takes time and training to acquire the confidence to pursue this agenda. In the meantime, when you are learning your trade as a research scientist it is essential to not manipulate your data, by excluding (or changing) awkward data sets to generate a more pleasing outcome.

Therefore, when preparing a Results section, it is essential to understand your data, how it was generated, why and what it is telling you. On that basis you will be able to present your findings appropriately.

41.8 How to Write a Discussion

The Discussion is more than a summary of findings. It requires careful planning so as to put the findings into context. It is helpful to start by creating some subheadings to help guide your topic selection. Some journals like authors to include an executive summary, and preparing this can be very helpful to the author to help shape the focus of the narrative. It is best to not restate your objectives and methods using the same language as used in the Introduction and Methods. There is no formula for preparing a discussion which is why it can be the most challenging part of the manuscript to complete.

My best advice, therefore, is to include more than is necessary in the first draft of the Discussion, then be prepared to delete needless text as you reread the manuscript. Be ruthless and ensure that sentences and paragraphs fit into the subsections with the text mapped to the subsections' headings. Include a subsection on limitations of the study. Here, briefly note the issues that cannot be resolved by your study, owing to its design. Much of my work makes use of rats, and referees would regularly ask me to defend the selection of this species. I learned to pre-empt this by including a subsection in Methods on species and model selection. In the Discussion I would include, in the limitations subsection, some comment about the next step in our project being confirmation of key findings in a second species. In time I learned to decide when to include a second species in the study itself, from the start. Thus, peer review and anticipating its process, can help your practice as a research scientist. Overall, one must determine what is sufficient and necessary to generate a publishable manuscript, and the issues one chooses to cover, or feel obliged to cover in the Discussion, are critical in this regard.

At a more generic level, a Discussion should avoid the following egregious errors. First, you are not writing a review of the literature. Of course, you must refer to relevant published findings, but you should use them and your data to address an idea or hypothesis, fitting their consideration into your narrative. Do not use the fact that other investigators have made similar findings to you as an argument in favour of the correctness and value of your work, as this is false equivalence. Ensure that you note what is novel about your findings, but avoid self-aggrandizement. Finally, there is an inherent tendency to want to state how one's findings may be important, but you must avoid hyping your data. Many referees read a manuscript abstract and then jump to the final part of the paper to see if the work is likely to have 'impact' (interest to the expert and, hopefully, the general reader). Therefore, the Discussion should neither fizzle out into some comment about a nuanced aspect of the study, nor should it make boldly optimistic or unjustifiable claims. It should end with a clear and defensible message.

41.9 Before Submitting Your Manuscript

It would be wise to ask a colleague to proofread your manuscript before you submit it for consideration. Ideally, ask several colleagues to do this, and invite them to be brutal. After that, proofread the manuscript yourself, and take ownership of it because it is your responsibility to do so. It is surprising how many manuscripts are triage rejected because the authors have made typographical errors or formatting errors in the first few pages of the manuscript, including low quality pixelated figures, or figures with tiny unreadable axis labels, or figures that don't fit on the page. Often an author of a rejected manuscript will send the manuscript to a second journal without reformatting to the new journal's requirements. I have personally seen a cover letter to the editor that states that 'we wish to submit our manuscript to (journal name)' and the journal name is the name of another (presumably the previously selected) journal. This is unprofessional and indicates that the content of the paper may have been generated without due care and attention. If you want to publish your findings, have pride in them. And good luck.

41.10 Conclusion

To write a paper one must have findings worth publishing. Learning how to write a paper should be part of the process of learning how to be a scientist, which involves understanding how to generate and test a hypothesis, plan experiments and execute them, and understand and communicate the meaning of the results. This requires training. The ability to write clearly in the chosen language used by the intended publication is essential. Finally, one must be honest with one's data, and not committed to publish at any cost.

Acknowledgements None.

Conflict of Interest Statement None.

References

1. Curtis MJ, Hancox JC, Farkas A, Wainwright CL, Stables CL, Saint DA, Clements-Jewery H, Lambiase PD, Billman GE, Janse MJ, Pugsley MK, Ng GN, Roden DM, Camm AJ, Walker MJA (2013) The Lambeth Conventions (II): guidelines for the study of animal and human ventricular and supraventricular arrhythmias. Pharmacol Ther 139:213–248
2. Curtis MJ, Bond RA, Spina D, Ahluwalia A, Alexander SPA, Giembycz MA, Gilchrist A, Hoyer D, Insel P, Izzo AA, Lawrence AJ, MacEwan DJ, Moon LDF, Wonnacott S, Weston AH, McGrath JC (2015) Experimental design and analysis and their reporting: new guidance for publication in BJP. Br J Pharmacol 172: 2671–2674
3. Curtis MJ, Alexander S, Cirino G, Docherty JR, George CH, Giembycz MA, Hoyer D, Insel PA, Izzo AA, Ji Y, MacEwan DA, Sobey CG, Stanford SC, Teixeira MM, Wonnacott S, Ahluwalia A (2018) Experimental design and analysis and their reporting II: updated and simplified guidance for authors and peer reviewers. Br J Pharmacol 175:987–993
4. Wilder CDE, Pavlaki N, Dursun T, Gyimah P, Caldwell-Dunn E, Ranieri A, Lewis HR, Curtis MJ

(2018) Facilitation of ischaemia-induced ventricular fibrillation by catecholamines is mediated by β1 and β2 agonism in the rat heart in vitro. Br J Pharmacol 175:1669–1690
5. Curtis MJ, Walker MJA (1988) Quantification of arrhythmias using scoring systems: an examination of seven scores in an in vivo model of regional myocardial ischaemia. Cardiovasc Res 22:656–665
6. Williams M, Mullane K, Curtis MJ (2018) Addressing reproducibility: peer review, impact factors, checklists, guidelines, and reproducibility initiatives. In: Williams M, Mullane K, Curtis MJ (eds) Research in the biomedical sciences. Elsevier, New York, pp 197–306
7. Rees SA, Curtis MJ (1995) A pharmacological analysis in rat of the role of the ATP-sensitive potassium channel as a target for antifibrillatory intervention in acute myocardial ischaemia. J Cardiovasc Pharmacol 26: 319–327

42 Preparing and Structuring a Manuscript for Publication

Diego A. Forero

Abstract

The writing of scientific manuscripts is a common activity for researchers around the world. In this context, preparing and structuring a manuscript adequately takes time, training and practice. In this chapter, I highlight multiple recommendations from the recent scientific literature and from my expertise in the writing and review of a large number of articles. The common sections of original articles are briefly reviewed, in addition to key suggestions for structuring manuscripts. I also discuss interesting proposals from recent papers oriented to increase the quality of the manuscripts and improving research productivity, aiming to guarantee research integrity and study reproducibility. Novel authors would benefit from training activities, feedback from colleagues and the constant practice of writing.

Keywords

Writing · Authorship · Scientific articles · Research articles · International journals · Research integrity

D. A. Forero (✉)
School of Health and Sport Sciences, Fundación Universitaria del Área Andina, Bogotá, Colombia
e-mail: dforero41@areandina.edu.co

42.1 Introduction

Publishing in international scientific journals is the main platform for sharing academic ideas and results from research projects, taking into account the high importance of the review process by expert peers [1]. On the other hand, the writing of articles is a challenge for many scientists, particularly for students and early career researchers [2]. Achieving the adequate structure of a manuscript takes time and increases the probability of it being accepted by an indexed scientific journal [3]. Additionally, the standards of scientific manuscripts have been evolving and there are novel aspects, such as preprints and open data, that need to be considered [3]. In this chapter, I will cover several key topics related to preparing and structuring a manuscript for publication. In addition, I will highlight multiple recommendations from the recent scientific literature [3–5] and my expertise in the writing and review of a large number of articles.

42.2 Manuscript Organization

In this section, I will briefly review the traditional key elements of an original manuscript, such as title, abstract, introduction, methods, results and discussion. Other chapters in this book will provide further details about some of those specific sections (see Chaps. 43–47, among others).

G. Jagadeesh et al. (eds.), *The Quintessence of Basic and Clinical Research and Scientific Publishing*,
https://doi.org/10.1007/978-981-99-1284-1_42

42.2.1 Title

The title of a manuscript will be indexed in the main bibliographic databases and will be included in the reference lists of other articles. There are some main types of titles: the declarative, which includes the main result of the study, or the descriptive (or neutral), which does not describe the main findings, in addition to the interrogative or compound styles. It is suggested that the title should be concise, informative and accurate [6].

42.2.2 Abstract

The abstract is another element of a manuscript that will be indexed in bibliographic databases. Commonly, journals have limits on word counts for abstracts and journals in the clinical sciences frequently ask for structured abstracts (in which this section is subdivided into Introduction, Methods, Results and Discussion). It is recommended that the abstract should be informative and consistent with the main methods and results of the manuscript [7].

42.2.3 Introduction

It has been proposed that the Introduction section should have the structure of a funnel: starting with the background information, highlighting the knowledge gap, putting forward the hypothesis or main question and describing the proposed approach or solution [8]. In the Introduction section, it is important to use transition phrases, which helps the author to have a clearer story [8].

42.2.4 Materials and Methods

It is important to include in this section key and sufficient information about the methodology used in the study. In the health sciences, it is fundamental to provide details about the individuals analyzed and the approaches used for their recruitment and evaluation. It is recommended to include citations to papers describing development or validation of methods used [9]. Statistical analysis is also a key subsection, providing information about tests and software used [10].

42.2.5 Results

It is advised that the presentation of results should be concordant with the approaches reported in the Methods section. In addition to the text, tables and figures [11] are essential ways for the presentation of results in scientific articles. This section should be supported in the statistical analyses of the data obtained and must not contain descriptions of methods or a discussion of findings [12].

42.2.6 Discussion

The Discussion section might be seen as an inverted funnel: starting with a discussion of the main results of the study and moving to potential explanations and implications for those findings, in addition to commenting on the contributions to the field. This section should be fair and balanced. Usually, limitations of the study and recommendations for future works are also included in this section [13].

42.2.7 Conclusions

In some journals, it is possible to include a Conclusions section and in other cases it might be included in the Discussion section. It should describe clearly the main findings of the study, in order to provide a take-home message to the readers [14]. Similar to the Discussion section, it should be written in a fair and balanced way.

42.2.8 References

This section should contain references to the citations (articles, books, among others) included in the previous segments of the manuscript. The authors should check that all citations in the text have the corresponding references. It is recommended to include relevant citations and to read the publications being cited [15]. Although there are hundreds of citation types, the Vancouver style is quite common in journals in the health sciences.

42.2.9 Conflict of Interest

Potential conflicts of interest involve both financial and non-financial aspects. Common examples of financial conflicts of interest are being employed or being supported by companies potentially benefiting from the sale of products related to the article submitted [16]. An important number of journals in the health sciences ask explicitly for the declaration of potential conflicts of interest.

42.3 General Recommendations

One of the initial suggestions for structuring a manuscript is to identify the standards of the specific field: the length and complexity, among other aspects, of the different sections vary depending on the scientific areas [1]. In addition to the specific field, the type of manuscript (original or review, among other categories) is a major aspect to have into account [5]. Complementing the information provided in other chapters of this book, there are several articles with guidelines for the different types of manuscripts [14, 17, 18]. Another key consideration is to have information about the requirements, or guidelines for authors, of the candidate journal (or journals). Usually, this information is found on the journal´s website and downloading related papers from the same journal might provide further information [19]. There are several online tools (created by major publishers, among others) that facilitate the selection of the candidate journals (Table 42.1), according to multiple criteria and taking the title and abstract as major inputs. Presubmission inquiries to the editors are another option of interest for some manuscripts [20]. A cover letter is commonly required by journals for the submission of manuscripts and it should contain information about the potential significance of the work and the possible relevance for the readers [21]. An initial part of the editorial evaluation by the journal is to run an automatic analysis of text similarity, to identify manuscripts with potential plagiarism [21].

It is important to highlight that having adequate collaborations for the development of the research activities and the writing of the manuscripts is a key element, in providing the variety of scientific expertise and academic profiles needed [22]. Perhaps one of the most important elements to define in a manuscript is related to the authorship [23, 24]. The International Committee of Medical Journal Editors (ICMJE, www.icmje.org) defines four main criteria, broadly used around the world: Substantial contributions, drafting the work, final approval and agreement to be accountable. Adherence to these main criteria would guarantee an adequate process of recognizing authorship [25]. In case some people do not meet all four ICMJE criteria they might be listed in the Acknowledgements section. Moshontz et al. [26] provided several important suggestions for writing a manuscript in large collaborations (such as those involving dozens of authors).

It is possible that developing an adequate literature review [27] (see Chap. 38) might be an important challenge for some authors when writing a scientific manuscript. In this context, it is important to use adequately the main available bibliographic databases, such as PubMed and Scopus (among others) [28–30], in addition to reading major and recent reviews on the specific topic. In this context, umbrella reviews, in addition to systematic reviews and meta-analyses,

Table 42.1 Key digital resources to facilitate the writing of articles in the health and biological sciences

Online resource	Website	Use
JANE	jane.biosemantics.org	To identify journals and authors with similar articles
Journal Suggester (Taylor and Francis)	authorservices.taylorandfrancis.com/journal-suggester	To identify candidate journals
Journal Finder (Elsevier)	journalfinder.elsevier.com	To identify candidate journals
Journal Suggester (Springer Nature)	journalsuggester.springer.com	To identify candidate journals
Journal Finder (Wiley)	journalfinder.wiley.com	To identify candidate journals
Scimago Journal Rank	scimagojr.com	To identify ranking of journals
NLM Catalog	ncbi.nlm.nih.gov/nlmcatalog	To identify indexing of journals
Journal Citation Reports[a]	jcr.clarivate.com	To identify IF of journals
Directory of Open Access Journals	doaj.org	To identify OA journals
ORCID	orcid.org	IDs for researchers
Publons	publons.com	Information about peer reviewers
Open Science Framework	osf.io	Data repository
GitHub	github.com	Code repository
figshare	figshare.com	Data repository
Google Scholar	scholar.google.com	Search of citations
Mendeley	mendeley.com	Management of references
Zotero	zotero.org	Management of references
EndNote[a]	endnote.com	Management of references
bioRxiv	biorxiv.org	Preprints repository
medRxiv	medrxiv.org	Preprints repository
ICMJE recommendations	icmje.org/recommendations/browse/	International criteria for writing manuscripts
MeSH on Demand	meshb.nlm.nih.gov/MeSHonDemand	Selection of MeSH
Equator Network	equator-network.org	Guidelines for reporting
Altmetric	altmetric.com	It provides alternative metrics
PubPeer	pubpeer.com	To comment on published articles
Retraction Watch	retractionwatch.com	It provides information about retracted articles

[a] Commercially available

provide an interesting opportunity for accessing evidence synthesis of high value [31]. In terms of quality, adherence to commonly used reporting guidelines is of high relevance [3, 32]. Examples of these guidelines are the Preferred Reporting Items for Systematic Reviews and Meta-Analyses (PRISMA) [33], Consolidated Standards of Reporting Trials (CONSORT) [34], STrengthening the Reporting of OBservational studies in Epidemiology (STROBE) [35], Transparent Reporting of a multivariable prediction model for Individual Prognosis Or Diagnosis (TRIPOD) [36], COnsolidated criteria for REporting Qualitative research (COREQ) and Animal Research: Reporting of In Vivo Experiments (ARRIVE) [37], among others (please visit www.equator-network.org for a comprehensive list).

Preprints have revolutionized the way of scientific communication [38], as it has been observed in the recent pandemic, allowing the immediate sharing of results around the globe (before they are peer-reviewed by the journals). Ettinger et al. [39] have developed recently a

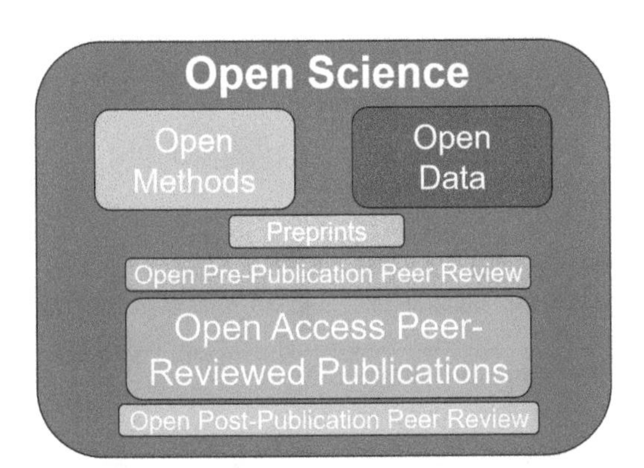

Fig. 42.1 An overview of the different dimensions and components of Open Science. Figure previously published under an open access license (Attribution 4.0 International (CC BY 4.0) https://creativecommons.org/licenses/by/4.0/ . Forero DA, Lopez-Leon S, Perry G. A brief guide to the science and art of writing manuscripts in biomedicine. J Transl Med. 2020 Nov 10;18(1):425. https://doi.org/10.1186/s12967-020-02596-2)

useful guideline for preprinting. Commonly used platforms for sharing preprints are: OSF Preprints (osf.io/preprints), bioRxiv (www.biorxiv.org), medRxiv (www.medrxiv.org) and PsyArXiv (psyarxiv.com), among others. In addition, Registered Reports, in which a protocol is reviewed before the study is carried out, are becoming more common in multiple research areas and Henderson et al. [40] developed ten rules for its writing.

In terms of Open Science approaches (Fig. 42.1), there are several commonly used licenses for Open Access of articles [41], which facilitate broad dissemination [42]: such as the Attribution 4.0 International (CC BY 4.0; creativecommons.org/licenses/by/4.0/), the Attribution-NonCommercial 4.0 International (CC BY-NC 4.0; creativecommons.org/licenses/by-nc/4.0) and the Attribution-NoDerivatives 4.0 International (CC BY-ND 4.0; creativecommons.org/licenses/by-nd/4.0). CC BY 4.0 allows for sharing and adapting the material, giving appropriate credit, CC BY-NC 4.0 does not allow its use for commercial purposes and CC BY-ND 4.0 does not allow sharing of the transformed material [41].

Sharing open data is increasingly required by international journals and publishers, in addition to being associated with a higher scientific impact (receiving more citations) [43]. In the context of Open Data (Fig. 42.1), Wilson et al. [44] wrote several valuable recommendations for sharing data, such as following the Findable, Accessible, Interoperable and Reusable (FAIR) guiding principles [45], providing adequate metadata and documentation and following the specific standards for the different scientific fields. They also highlighted the importance of employing the CC0 License (creativecommons.org/share-your-work/public-domain/cc0) for sharing data [44]. In terms of general open data repositories, there are multiple options, such as OSF (www.osf.io), Zenodo (zenodo.org) and Figshare (www.figshare.com), among others [46]. An additional advantage of depositing open data in major repositories is that the main bulk of the supporting data is easily and continuously available, allowing the authors to focus on major figures and tables for the body of the manuscript (avoiding overcrowded or too complex manuscripts).

One of the most important elements in scientific communication, including the writing of manuscripts, is aiming to guarantee research integrity [47], which involves multiple steps to avoid scientific malpractices such as plagiarism, fabrication, hiding conflicts of interest, ghost authorship or guest authorship [48–52], in addition to facilitating an adequate sharing of results, avoiding spin (overstating inappropriately findings) [53] and promoting the use of an anti-bias and inclusive language [54, 55] (see Chap. 59 for more details). The use of software, such as EndNote (endnote.com), Mendeley (mendeley.com) or Zotero (zotero.org), is quite helpful for saving time for authors, particularly when changing citation styles (see Chap. 39).

42.4 Specific Recommendations from the Scientific Literature

In this section, I will revisit major elements from key published articles about structuring manuscripts and improving writing productivity. Mensh et al. [5] have proposed ten simple rules

Rule	Sign it is violated
1: Focus on one big idea	Readers cannot give 1-sentence summary.
2: Write for naive humans	Readers do not "get" the paper.
3: Use context, content, conclusion structure	Readers ask why something matters or what it means.
4: Optimize logical flow	Readers stumble on a small section of the text.
5: Abstract: Compact summary of paper	Readers cannot give the "elevator pitch" of your work after reading it.
6: Introduction: Why the paper matters	Readers show little interest in the paper.
7: Results: Why the conclusion is justified	Readers do not agree with your conclusion.
8: Discussion: Preempt criticism, give future impact	Readers are left with unanswered criticisms and/or questions on their mind.
9: Allocate time wisely	Readers struggle to understand your central contribution despite your having worked hard.
10: Iterate the story	The paper's contribution is rejected by test readers, editors, or reviewers.

Fig. 42.2 A summary of the ten simple rules for structuring papers, proposed by Mensh and Kording [5]. Figure previously published under an open access license (Attribution 4.0 International (CC BY 4.0) https://creativecommons.org/licenses/by/4.0/. Mensh B, Kording K. Ten simple rules for structuring papers. PLoS Comput Biol. 2017 Sep 28;13(9):e1005619. https://doi.org/10.1371/journal.pcbi.1005619)

for structuring papers (Fig. 42.2): having a clear title, writing for humans, using the context-content-conclusion scheme, employing a logical flow, having an abstract with complete information, highlighting the importance of the paper in the introduction section, presenting the results supported by the figures (and tables), discussing the relevance to the field, dedicating time to major issues (such as title, abstract and figures), and getting feedback. Additionally, the authors highlighted the possible consequences when these recommendations are not considered, such as the readers having problems understanding the paper or the manuscript being rejected by the editors and reviewers [5].

Zhang [19] developed ten simple rules for writing research articles (Box 42.1): designing a project with a paper in mind, producing fewer but more significant publications, selecting an adequate journal, using a logical flow, being concise, making it complete, having a polished writing, revising the draft, having feedback from colleagues and taking into consideration the comments of the reviewer. Meo [14] provided several suggestions for the writing of scientific papers, such as having an unambiguous title, containing an abstract with a summary of the main points of the work, explaining the rationale of the study, including details of the methods used, reporting both positive and negative results, recommending further research and acknowledging people who helped with the work.

Box 42.1 Ten Simple Rules for Writing Research Papers, as Proposed by Zhang [19]

"Rule 1: Make It a Driving Force
Rule 2: Less Is More
Rule 3: Pick the Right Audience
Rule 4: Be Logical
Rule 5: Be Thorough and Make It Complete
Rule 6: Be Concise
Rule 7: Be Artistic
Rule 8: Be Your Own Judge
Rule 9: Test the Water in Your Own Backyard
Rule 10: Build a Virtual Team of Collaborators"

Cook [56] shared twelve tips for a better chance of having a manuscript published: planning it early, defining authorship in advance, having control of the writing, adhering to reporting standards, using reference management software, polishing the writing, selecting an adequate journal, following the instructions of the journal, resubmitting the non-accepted manuscript, recognizing and answering the questions of the reviewers and having input from colleagues. Bourne [1] developed ten simple rules for getting published: reading many papers, being objective about your work, benefiting from good reviewers, improving your English, learning to live with rejection, having a good organization, starting to write at the beginning of the project, being a reviewer, selecting the journal and aiming for high quality.

Barroga et al. [4] wrote guidelines for creating logical flow in the writing of scientific manuscripts. They recommend starting with a draft outlining the manuscript, to develop a well-written article, through the use of concise words, clear sentences and adequate transitions. Oshiro et al. [2] characterized the perceived barriers, by a sample of authors, for the preparation of manuscripts and they found that several of them (such as following the format of the journals) were not affected by the expertise of the scientists.

Busse et al. [57] provided several recommendations for addressing common pitfalls in the writing of manuscripts, such as what to do when the introduction is too vague or when the citations are inadequate for supporting the claims made. They also highlighted the importance of describing in detail the methods used and justifying their selection, in addition to facilitating a checklist for manuscript quality (Fig. 42.3). Aga et al. [58] wrote an article with suggestions for novice authors and they also highlighted common mistakes in the different sections of the manuscript. They also emphasized the importance of the use of reporting guidelines and described key elements of the submission and editorial process [58]. Huston et al. [59] created a guide for publishing in the health sciences and they underscored several key issues, such as: applying collaborative writing, giving appropriate attributions, developing a compelling story, and refining the manuscript. Forero et al. [3] developed a brief guide to writing articles in biomedicine and they stressed the usefulness of digital resources (Table 42.1) for the development of different activities related to manuscript organization and submission, such as Journal/Author Name Estimator (jane.biosemantics.org), Scimago Journal Rank (scimagojr.com), Open Researcher and Contributor ID (orcid.org) and MeSH on Demand (meshb.nlm.nih.gov/MeSHonDemand), among others. Ecarnot et al. [60], Iskander et al. [61] and Vitse et al. [21] put forward additional valuable suggestions for the writing and organization of scientific manuscripts, such as major elements needed to promote study reproducibility, further criteria for selecting a journal or complementary strategies to respond to reviewers.

Peterson et al. [62] developed ten simple rules for increasing writing productivity (Box 42.2): defining a time for writing, having an adequate working environment, writing a draft first and editing it later, developing writing habits, being responsible, looking for feedback, thinking about your manuscripts during other times, practicing a lot, being positive and reevaluating your writing practices. In addition, Forero and Moore [63] proposed nine key elements for a higher productivity in research activities (Box 42.3): avoiding unnecessary meetings, having clear priorities, focusing on science, preventing to depending too much on collaborations, developing adequate research collaborations, avoiding to aim too low or too high, enjoying the writing of manuscripts and grant applications, preserving the importance of being the principal investigator and creating a positive working environment.

- Did you write to a specific type of reader (readers of target journal)?
- Did you follow the introduction guide (upside-down triangle)?
- Is the research aim specific and informative?
- Have you included a thorough description of the dependent variables, the independent variables, and every covariate or descriptive variable you included in models and present in the results?
- Have you justified each of your methods, including your sampling approach, a justification of why you included each variable, your measurement approach and statistical analysis?
- Have you described your results in words that convey the direction of associations?
- Are table titles and figure legends complete?
- Does the text of the results section summarize key findings from tables and figures rather than repeating them exactly?
- Did you avoid repeating detailed results in the discussion section and avoided presenting new results in the discussion?
- Have you included limitations and implications, defending your approach where appropriate?
- Are the future steps you present specific?
- Are the research aims, methods, results and discussion are consistent in addressing the same (and all) research aims throughout?
- Is your title informative and interesting?
- Have you cited appropriately throughout the manuscript?

Fig. 42.3 Checklist for manuscript quality, developed by Busse and August [57]. Figure previously published under an open access license (Attribution 4.0 International (CC BY 4.0) https://creativecommons.org/licenses/by/4.0/. Busse C, August E. How to Write and Publish a Research Paper for a Peer-Reviewed Journal. J Cancer Educ. 2021 Oct;36(5):909–913. https://doi.org/10.1007/s13187-020-01751-z)

Box 42.2 Ten Simple Rules for Improving Writing Productivity in Scientists, as Developed by Peterson et al. [62]

"Rule 1: Define your writing time
Rule 2: Create a working environment that really works
Rule 3: Write first, edit later
Rule 4: Use triggers to develop a productive writing habit
Rule 5: Be accountable
Rule 6: Seek feedback and ask for what you want
Rule 7: Think about what you're writing outside of your scheduled writing time
Rule 8: Practice, practice, practice
Rule 9: Manage your self-talk about writing
Rule 10: Reevaluate your writing practice often"

Box 42.3 Nine Considerations for Higher Efficiency and Productivity in Research Activities, as Proposed by Forero and Moore [63]

"1. Unnecessary meetings disrupt scientific productivity.
2. Lack of clear priorities affects science.
3. Having more than one job is good for money but negative for productivity.
4. Depending too much on collaborations is undesirable.
5. Lack of collaborations is negative for research.
6. Aiming too high or too low is counterproductive.
7. Never become tired of writing grants or papers.
8. Do not delegate being the Principal Investigator.
9. Try to create a positive working environment."

Taking into account that English is the current international language of science, there are some published articles with valuable recommendations for non-Anglophone authors [64–67], such as writing simple sentences to increase readability and adequate use of punctuation marks (such as the comma and the semicolon).

Of particular interest for trainees, Sharma and Ogle [68] recently provided twelve tips for students wanting to write and publish (Box 42.4): having a clear motivation, being realistic, reading a lot, exploring research opportunities, talking to professors, broadening your horizons, knowing the submission process, focusing on details, recognizing that the submission is not the end of the process, identifying that it takes time, considering alternative options to presenting research and being started with the writing. Finally, few authors have written articles with suggestions for authors to respond to comments from reviewers and editors [69–71].

Box 42.4 Twelve Tips for Students Who Wish to Write and Publish, as Proposed by Sharma and Ogle [68]

"Tip 1: Find your why
Tip 2: Play to your strengths and be realistic
Tip 3: Be well read
Tip 4: Revisit missed opportunities
Tip 5: Talk to the doctors around you
Tip 6: Broaden your horizons
Tip 7: Get to grips with the submission process early
Tip 8: Pay attention to the details
Tip 9: Remember that submission is not the end
Tip 10: The process cannot be rushed
Tip 11: Consider the alternative paths to presenting research
Tip 12: Start writing."

42.5 Concluding Remarks

In this chapter, I highlighted several suggestions available in the scientific literature for better structuring of manuscripts and for enhancing writing productivity [4, 5, 62]. Accomplishing an adequate structure of a manuscript involves multiple rounds of refinement, incorporating international standards and following the recommendations for the authors from the target journal, among other aspects [3, 5]. Other chapters from this book will provide readers with further information about writing specific sections of manuscripts (see Chaps. 43–47, among others).

As it has been previously suggested, prospective authors are specially invited to practice writing and enjoy the pleasure of communicating their scientific ideas and results [63]. In addition, taking advanced workshops about writing and asking for feedback from experienced colleagues help to develop those valuable skills [1, 3, 72].

Acknowledgments DAF has been previously supported by research grants from MinCiencias and DNI-Areandina.

References

1. Bourne PE (2005) Ten simple rules for getting published. PLoS Comput Biol 1(5):e57
2. Oshiro J, Caubet SL, Viola KE, Huber JM (2020) Going beyond "not enough time": barriers to preparing manuscripts for academic medical journals. Teach Learn Med 32(1):71–81
3. Forero DA, Lopez-Leon S, Perry G (2020) A brief guide to the science and art of writing manuscripts in biomedicine. J Transl Med 18(1):425
4. Barroga E, Matanguihan GJ (2021) Creating logical flow when writing scientific articles. J Korean Med Sci 36(40):e275
5. Mensh B, Kording K (2017) Ten simple rules for structuring papers. PLoS Comput Biol 13(9): e1005619
6. Bahadoran Z, Mirmiran P, Kashfi K, Ghasemi A (2019) The principles of biomedical scientific writing: title. Int J Endocrinol Metab 17(4):e98326
7. Tullu MS (2019) Writing the title and abstract for a research paper: being concise, precise, and meticulous is the key. Saudi J Anaesth 13(Suppl 1):S12–SS7
8. Annesley TM (2010) "It was a cold and rainy night": set the scene with a good introduction. Clin Chem 56(5):708–713
9. Annesley TM (2010) Who, what, when, where, how, and why: the ingredients in the recipe for a successful Methods section. Clin Chem 56(6):897–901
10. Dwivedi AK (2022) How to write statistical analysis section in medical research. J Investig Med 70(8): 1759–1770

11. Rougier NP, Droettboom M, Bourne PE (2014) Ten simple rules for better figures. PLoS Comput Biol 10(9):e1003833
12. Annesley TM (2010) Show your cards: the results section and the poker game. Clin Chem 56(7): 1066–1070
13. Annesley TM (2010) The discussion section: your closing argument. Clin Chem 56(11):1671–1674
14. Meo SA (2018) Anatomy and physiology of a scientific paper. Saudi J Biol Sci 25(7):1278–1283
15. Penders B (2018) Ten simple rules for responsible referencing. PLoS Comput Biol 14(4):e1006036
16. Dunn AG, Coiera E, Mandl KD, Bourgeois FT (2016) Conflict of interest disclosure in biomedical research: a review of current practices, biases, and the role of public registries in improving transparency. Res Integr Peer Rev 1:1
17. Forero DA, Lopez-Leon S, Gonzalez-Giraldo Y, Bagos PG (2019) Ten simple rules for carrying out and writing meta-analyses. PLoS Comput Biol 15(5): e1006922
18. Gregory AT, Denniss AR (2018) An introduction to writing narrative and systematic reviews—tasks, tips and traps for aspiring authors. Heart Lung Circ 27(7): 893–898
19. Zhang W (2014) Ten simple rules for writing research papers. PLoS Comput Biol 10(1):e1003453
20. Lengauer T, Nussinov R (2015) How to write a presubmission inquiry. PLoS Comput Biol 11(2): e1004098
21. Vitse CL, Poland GA (2017) Writing a scientific paper—a brief guide for new investigators. Vaccine 35(5):722–728
22. Zeng A, Fan Y, Di Z, Wang Y, Havlin S (2021) Fresh teams are associated with original and multidisciplinary research. Nat Hum Behav 5(10):1314–1322
23. Kornhaber RA, McLean LM, Baber RJ (2015) Ongoing ethical issues concerning authorship in biomedical journals: an integrative review. Int J Nanomedicine 10: 4837–4846
24. Patience GS, Galli F, Patience PA, Boffito DC (2019) Intellectual contributions meriting authorship: survey results from the top cited authors across all science categories. PLoS One 14(1):e0198117
25. Zimba O, Gasparyan AY (2020) Scientific authorship: a primer for researchers. Reumatologia 58(6):345–349
26. Moshontz H, Ebersole CR, Weston SJ, Klein RAJS, Compass PP (2021) A guide for many authors: writing manuscripts in large collaborations. Soc Personal Psychol Compass 15(4):e12590
27. Shaffril HAM, Samsuddin SF, Samah AA (2021) The ABC of systematic literature review: the basic methodological guidance for beginners. Qual Quant 55(4): 1319–1346
28. Gusenbauer M, Haddaway NR (2020) Which academic search systems are suitable for systematic reviews or meta-analyses? Evaluating retrieval qualities of Google Scholar, PubMed, and 26 other resources. Res Synth Methods 11(2):181–217
29. Gusenbauer MJS (2019) Google Scholar to overshadow them all? Comparing the sizes of 12 academic search engines and bibliographic databases. Scientometrics 118(1):177–214
30. Smalheiser NR, Fragnito DP, Tirk EE (2021) Anne O'Tate: value-added PubMed search engine for analysis and text mining. PLoS One 16(3):e0248335
31. Fusar-Poli P, Radua J (2018) Ten simple rules for conducting umbrella reviews. Evid Based Ment Health 21(3):95–100
32. Ioannidis JP, Greenland S, Hlatky MA, Khoury MJ, Macleod MR, Moher D et al (2014) Increasing value and reducing waste in research design, conduct, and analysis. Lancet 383(9912):166–175
33. Page MJ, Moher D, Bossuyt PM, Boutron I, Hoffmann TC, Mulrow CD et al (2021) PRISMA 2020 explanation and elaboration: updated guidance and exemplars for reporting systematic reviews. BMJ 372:n160
34. Schulz KF, Altman DG, Moher D, Group C (2010) CONSORT 2010 statement: updated guidelines for reporting parallel group randomised trials. BMC Med 8:18
35. von Elm E, Altman DG, Egger M, Pocock SJ, Gotzsche PC, Vandenbroucke JP et al (2007) The Strengthening the Reporting of Observational Studies in Epidemiology (STROBE) statement: guidelines for reporting observational studies. PLoS Med 4(10):e296
36. Collins GS, Reitsma JB, Altman DG, Moons KG (2015) Transparent reporting of a multivariable prediction model for individual prognosis or diagnosis (TRIPOD): the TRIPOD statement. BMC Med 13:1
37. Percie du Sert N, Ahluwalia A, Alam S, Avey MT, Baker M, Browne WJ et al (2020) Reporting animal research: explanation and elaboration for the ARRIVE guidelines 2.0. PLoS Biol 18(7):e3000411
38. Abdill RJ, Blekhman R (2019) Tracking the popularity and outcomes of all bioRxiv preprints. Elife 8:e45133
39. Ettinger CL, Sadanandappa MK, Gorgulu K, Coghlan KL, Hallenbeck KK, Puebla I (2022) A guide to preprinting for early-career researchers. Biol Open 11(7):bio059310
40. Henderson EL, Chambers CD (2022) Ten simple rules for writing a Registered Report. PLoS Comput Biol 18(10):e1010571
41. Carroll MW (2013) Creative Commons and the openness of open access. N Engl J Med 368(9):789–791
42. Langham-Putrow A, Bakker C, Riegelman A (2021) Is the open access citation advantage real? A systematic review of the citation of open access and subscription-based articles. PLoS One 16(6):e0253129
43. Colavizza G, Hrynaszkiewicz I, Staden I, Whitaker K, McGillivray B (2020) The citation advantage of linking publications to research data. PLoS One 15(4):e0230416
44. Wilson SL, Way GP, Bittremieux W, Armache JP, Haendel MA, Hoffman MM (2021) Sharing biological data: why, when, and how. FEBS Lett 595(7):847–863
45. Wilkinson MD, Dumontier M, Aalbersberg IJ, Appleton G, Axton M, Baak A et al (2016) The FAIR guiding principles for scientific data management and stewardship. Sci Data 3:160018

46. Forero DA, Curioso WH, Patrinos GP (2021) The importance of adherence to international standards for depositing open data in public repositories. BMC Res Notes 14(1):405
47. Shaw DM, Erren TC (2015) Ten simple rules for protecting research integrity. PLoS Comput Biol 11(10):e1004388
48. Bradshaw MS, Payne SH (2021) Detecting fabrication in large-scale molecular omics data. PLoS One 16(11): e0260395
49. Fang FC, Steen RG, Casadevall A (2012) Misconduct accounts for the majority of retracted scientific publications. Proc Natl Acad Sci U S A 109(42): 17028–17033
50. Perez-Neri I, Pineda C, Sandoval H (2022) Threats to scholarly research integrity arising from paper mills: a rapid scoping review. Clin Rheumatol 41(7): 2241–2248
51. Gaudino M, Robinson NB, Audisio K, Rahouma M, Benedetto U, Kurlansky P et al (2021) Trends and characteristics of retracted articles in the biomedical literature, 1971 to 2020. JAMA Intern Med 181(8): 1118–1121
52. Morreim EH, Winer JC (2023) Guest authorship as research misconduct: definitions and possible solutions. BMJ Evid Based Med 28(1):1–4
53. Chiu K, Grundy Q, Bero L (2017) 'Spin' in published biomedical literature: a methodological systematic review. PLoS Biol 15(9):e2002173
54. Huang GC, Truglio J, Potter J, White A, Hunt S (2022) Anti-bias and inclusive language in scholarly writing: a primer for authors. Acad Med 97(12):1870
55. Flanagin A, Frey T, Christiansen SL, Committee AMAMoS (2021) Updated guidance on the reporting of race and ethnicity in medical and science journals. JAMA 326(7):621–627
56. Cook DA (2016) Twelve tips for getting your manuscript published. Med Teach 38(1):41–50
57. Busse C, August E (2021) How to write and publish a research paper for a peer-reviewed journal. J Cancer Educ 36(5):909–913
58. Aga SS, Nissar S (2022) Essential guide to manuscript writing for academic dummies: an editor's perspective. Biochem Res Int 2022:1492058
59. Huston P, Choi B (2017) A guide to publishing scientific research in the health sciences. Can Commun Dis Rep 43(9):169–175
60. Ecarnot F, Seronde M-F, Chopard R, Schiele F, Meneveau NJEGM (2015) Writing a scientific article: a step-by-step guide for beginners. Eur Geriatr Med 6(6):573–579
61. Iskander JK, Wolicki SB, Leeb RT, Siegel PZ (2018) Successful scientific writing and publishing: a step-by-step approach. Prev Chronic Dis 15:E79
62. Peterson TC, Kleppner SR, Botham CM (2018) Ten simple rules for scientists: improving your writing productivity. PLoS Comput Biol 14(10):e1006379
63. Forero DA, Moore JH (2016) Considerations for higher efficiency and productivity in research activities. BioData Min 9:35
64. Yakhontova T (2021) What nonnative authors should know when writing research articles in English. J Korean Med Sci 36(35):e237
65. Yakhontova T (2020) Conventions of English research discourse and the writing of non-Anglophone authors. J Korean Med Sci 35(40):e331
66. Yakhontova T (2020) Punctuation mistakes in the english writing of non-Anglophone researchers. J Korean Med Sci 35(37):e299
67. Yakhontova T (2020) English writing of non-Anglophone researchers. J Korean Med Sci 35(26):e216
68. Sharma RK, Ogle HL (2022) Twelve tips for students who wish to write and publish. Med Teach 44(4): 360–365
69. Annesley TM (2011) Top 10 tips for responding to reviewer and editor comments. Clin Chem 57(4): 551–554
70. Guyatt GH, Brian Haynes R (2006) Preparing reports for publication and responding to reviewers' comments. J Clin Epidemiol 59(9):900–906
71. Kotz D, Cals JW (2014) Effective writing and publishing scientific papers, part XII: responding to reviewers. J Clin Epidemiol 67(3):243
72. Busse CE, Anderson EW, Endale T, Smith YR, Kaniecki M, Shannon C et al (2022) Strengthening research capacity: a systematic review of manuscript writing and publishing interventions for researchers in low-income and middle-income countries. BMJ Glob Health 7(2):e008059

43 Writing a Scientific Article

Asdrubal Falavigna, Vincenzo Falavigna, and Maria Eduarda Viapiana

Abstract

Medical practice follows evidence-based medicine, which has the "3 Es": Patient's Expectation, Physician's Experience and Evidence from the Literature. Ethical and reproducible evidence is critical to treating patients better. Physicians need to balance their experience with the best evidence from the literature to plan for a better patient diagnosis and treatment. It becomes increasingly necessary and important to publish since evidence should be based on firm methodological foundations and sustained by statistical proof, which confers force and power on the concepts. The merit belongs not to the person who discovers something but to the one who describes it and convinces the world that it is valid. In this chapter, we outline the internal consistency, grammar, writing style, structure and individual components according to the order in which a scientific paper is formulated. All the crucial steps described have grounding support for the knowledge of writing a scientific paper. Ultimately, to write well, one must practice writing, reading other articles and receiving feedback from a multidisciplinary team. The person who learns how to write a scientific paper automatically qualifies to take a critical look by reading many published scientific articles daily.

A. Falavigna (✉) · V. Falavigna · M. E. Viapiana
Department of Neurosurgery, Caxias do Sul University, Caxias do Sul, Brazil
e-mail: afalavig@ucs.br; vffalavigna@ucs.br; melviapiana@ucs.br

Keywords

Scientific writing style · Scientific article · Grammar · Article structure · Successful scientific publications

43.1 Introduction

Writing a scientific article is the last stage in a research study. Researchers are always encouraged to write and publish their results, despite knowing that publication is a consequence of an overall scientific and investigative attitude. A scientific publication, besides the willingness or not to publish, requires a need for global efforts to increase the conditions for research training and maximize human resources [1]. To contribute significantly to the global share of publications, education in research is essential, relatively inexpensive and highly effective [2, 3].

Publication of a scientific study is a critical moment when the work comes alive and the research methodology and results are shared with a scientific community. The scientific report will be effective when it allows study reproducibility [3]. Reproducibility can only be achieved if the paper is consistent in grammar and structure,

G. Jagadeesh et al. (eds.), *The Quintessence of Basic and Clinical Research and Scientific Publishing*,
https://doi.org/10.1007/978-981-99-1284-1_43

has clear and precise information, and has a simple, concise, and objective vocabulary [4].

The journal reviewers can detect bias, unsatisfactory design, inappropriate statistics, and ethical problems in the study that may threaten the research and lead to rejection of the paper [5]. It is never too soon or too late to learn how to write a scientific article. The recommendations to increase the chances of the article being accepted by the journal can be summarized in four points: clarity, accuracy, communicability and consistency. The structure of the writing presented in this chapter allows one to weigh all the strengths of the article. Although the form of articles may differ somewhat in different journals, most have a similar basic structure to help achieve a logical presentation of research [6]. The objective of this paper is to describe a simple, concise and precise way to write a scientific article.

43.2 Internal Consistency

The internal consistency depends on the organization of the article and the methodological congruence between the different sections to answer the research question (see Chap. 1). The introduction section reviews the literature and justifies the research question, while the methodology is designed based on the hypotheses and objectives (discussed in Chap. 1). The results show the answer to the question and the discussion explains the difference and similarities between the paper's results with what has already been published in the literature and the author's opinions. There should be a concordance between the objectives, title, and methodology established and the object of study to give a precise answer.

43.3 Grammar and Writing Style

Good grammatical language is required to understand the paper and follow the author's logic. Recommendations to make your article more attractive are described in Table 43.1. In summary, the language must be simple, clear, concise and objective. And for this to be effective, some points should be followed, such as (1) prefer the use of the active voice instead of the passive; (2) write short sentences because excessively elaborate and long sentences tend to confuse the readers; (3) a new paragraph must be started when changing the subject; (4) write the text in a logical sequence following the original question of the article.

43.4 Structure of a Scientific Paper

The article must be structured appropriately to allow reproducibility and complete understanding by the readers. The paper is divided into specific sections to provide unique information. The section names may vary according to different journals, but the order and content are organized to provide a logical presentation of the study. The sequences in which the paper is read and written

Table 43.1 Basic grammar, style, and structural characteristics of an article

Recommendations to make your article more attractive
• Choose a good title
• Structured and consistent summary or abstract
• Clear, simple, objective and concise
• Start a new paragraph when you change the subject or introduce a new idea
• Organize paragraphs to create a logical sequence of ideas based on the research question and the results
• Straightforward and organized material and method
• Self-explanatory results
• Short paragraphs
• Attractive discussion
• Keep the article clearly focused

Table 43.2 Suggested structure and order of reading and writing scientific papers

Order of reading	Order of writing
• Title, abstract/summary, keywords	• Introduction
• Authorship and acknowledgement	• Methods
• Disclosure and conflicts of interest	• Results
• Introduction	• Discussion
• Methods	• Conclusion
• Results	• References
• Discussion	• Title, abstract/summary, keywords
• Conclusion	• Authorship and acknowledgement
• References	• Disclosure and conflicts of interest

are somewhat different (Table 43.2). The writing sequence follows the steps of the ongoing research and personal preference.

43.5 Order of Writing a Scientific Paper

Before you start writing, select the target journal for your manuscript. Journals have different guidelines and presentation styles. By doing that, you will save a lot of time and energy in revising the format of a manuscript later.

The article is written according to the ongoing research and authors' preferences (Table 43.2): Introduction, Methods, Discussion, Results, References, Conclusion, Abstract and Keywords, Title, Authors, Disclosure, Conflict of Interest, and Acknowledgements.

43.5.1 Introduction

The introduction justifies the study. The introduction attracts the scientific community and informs clearly about the current knowledge gaps and the opportunities to run further research to validate the research question. The author should clarify the reasons why he/she thought it worthwhile to elaborate on the work that will be developed. The hypothesis is described in this section. The hypothesis represents the objective of the study to reject the null hypothesis ($x = y$) in favor of the alternative hypothesis (x is different from y).

The length of the introduction is usually related to the complexity of the study. Generally, it is around half a page and no more than one and half page. An introduction is often written in the present tense because it reports the most recent information on the subject. The authors should select a reference that is current and relevant to justify the gap in knowledge and the importance of the research. The authors must remember that the journals always cap the number of references to 30 or 40 at maximum.

The authors should describe the proposed research question by using the FINER criteria: Feasible, Interesting, Novel, Ethical, and Relevant [7]. They must explain closely the research question because it drives all the structure of the study: type of study, methodology applied, population studied, sample size calculation, time available, equipment, funding, instruments or questionnaire to measure the primary and secondary outcome or endpoint and implementing the work.

The introduction follows the three paragraphs rules to report the background, the justification of the study and the objectives expected (Table 43.3). In the first paragraph, the authors review the general background of the topic using the most relevant and updated literature. The second paragraph emphasizes the unknown aspects of the subject, controversies in the literature and description of the clinical question formulated. Finally, the third paragraph should explain the logic of their hypothesis, develop straightforward research questions and briefly introduce the methodology. If more than one research question is presented, primary and secondary objectives should be described separately.

Table 43.3 Three paragraph rules of the introduction and the questions to be answered

First introduction paragraph (WHAT)—Background
• What is the topic of the study? • What are the characteristics and causes of the chosen topic?
Second introduction paragraph (WHY)—Justification for the study
• Why is it crucial to perform this study? • What are the objectives behind developing the study? • Why will the results benefit the science? • Is there a clear, focused, and answerable study question? • Is the study question innovative or relevant?
Third introduction paragraph (HOW)—Objectives and hypothesis
• What is the study question? • Does it matter? • How will the research question be answered? • What will be the methodology or strategy used?

43.5.2 Methods

Considering that the hypothesis was presented in the introduction, the methods section is essential for the manuscript's success because it explains the study design. The validity and credibility of the results refer mainly to the methods applied to the work. Many authors underestimate how difficult it is to do so correctly, and consequently, flaws in the methodology are one of the leading causes of paper rejections.

The methodology describes and explains the steps-by-steps of how the research was done. When well-designed, it has the potential to convince readers of the validity and reproducibility of the study. Reproducibility is synonymous with the research team's reliability, trustworthiness, and competence because it allows us to find similar results if the researchers follow the methodology under a similar clinical or surgical condition. The authors should describe the interventions used, the statistical tests used, as well as the results obtained so that other researchers can examine their results for methodological rigor and allow reproducibility.

The methods section must be written in the past tense, as the actions have already been taken and follow a chronological sequence. It is the only paper section in which the authors cannot try to save space by using excessively succinct language. There is no apparent limit because the sequence of steps needs to be well described to allow reproducibility. At the same time, it must be objective, transparent, chronological and concise. The well-established procedures should have bibliographic citations rather than writing everything down in detail.

The methods section must be well-organized and related to the general and specific objectives of the project [8, 9]. The authors usually replicate similar subsections used in methodology to describe the results and discussion sections to make it more transparent and easier to read. For example, in a clinical study, we consider dividing the methods section into the subsections of study design, population analysis, intervention and control group, outcome and statistical analysis (Table 43.4).

43.5.3 Results

The purpose of the results section is to report what was observed during the study through a transparent and objective description of the findings. It should not be a simple accumulation of tables or charts but confirm or rule out the hypothesis presented at the beginning of the research. The process should be quantitative, and a study's results should be presented logically.

Results are always presented in the past tense. The findings are easier to read when they follow the same order as the subsections presented in the methods section and have the text organized with titles and subheadings.

Table 43.4 Subsections of the methodology that need to be explored to describe the clinical study

Subsections of the methodology	Topics to be described
Research design	• Study design, e.g. prospective or retrospective • Justification of the choice of methods and techniques • Duration and follow-up period • Single or multicenter study and environment (trauma center or rural hospital) • Ethics committee approval number, informed consent, trial registration numbers
Study population	• How patients were recruited and the size of the sample • Inclusion and exclusion criteria • Classification of the population sample and the randomization, if applicable
Intervention and control group	• Recommended to use the product's generic name • Product brand names, their manufacturers and locations
Measurement of outcomes	• Primary and secondary endpoints, measurement of outcome and the time points of measurements • Describe the data collection methods and measuring instruments • References should be given for the measuring techniques, including their validation • Clinical charts, opinion surveys, or other measurement instruments
Statistical analysis	• Specify the statistical method used for sample calculation and the statistical tests • Indicate the computer software used to analyze the results

The authors must supply a visual or graphic summary of the study flow accompanied by a text that briefly describes their content and by a number that is used to refer to them. Self-explanatory tables, graphs, figures, and images can help simplify data presentation. For example, a study flowchart shows patient enrollment, allocation, follow-up and analysis, allowing readers to get an overview of how the study was conducted. It is important to identify the format of the summary tables and the journal figures chosen for publication, in other words: colors, lines, titles and variables. The summary tables or figures should be numbered consecutively. The title should be concise, elaborating the texts below them with notes and abbreviations. It is advisable not to have more than six summary tables and figures (for instance, 4 and 2 or 3 and 3).

Relevant results that prove or reject the study hypothesis should be communicated to the readers in clear, objective language and using numbers instead of adjectives (e.g., very, rarely, often, generally). Information from tables or figures should not simply be repeated in the text. Instead, emphasize the main findings that will later be aligned with the hypothesis and the objectives in the discussion section. There is less misinterpretation if the author provides absolute numbers and lets the reader interpret for themself whether the phenomenon is rare, very rare, or infrequent.

Authors should avoid interpreting the findings or drawing conclusions in the results section. The former belongs to the discussion section and the latter to the conclusion section. Deviations from the study plan should also be explained.

43.5.4 Discussion

The discussion section is a key and complex part of a scientific article because it shows the researcher's understanding of the topic, where the author should provide a clear explanation for the results obtained. The main objective is to place the research findings in the context investigated, explain their meaning and emphasize their importance compared to the literature to justify or reject the hypothesis. Therefore, the discussions should be constructed hierarchically, starting with the most essential and primary result, then addressing the results of the secondary objective in the following paragraphs.

The background must be written in the present tense, and the description of the research findings must be in the simple past. Whenever possible, the content should be put into the same divisions or subheadings used in the methods and results sections.

After all the results have been obtained, the findings are contrasted with the available evidence and compared with those from previously published studies. At this point, the discussion will transform the hypothesis into a conclusion because, based on the results, the author establishes his position regarding the question formulated at the beginning of the work and defends his opinions.

The first paragraph reports the most important findings of the primary outcome and their impact on the investigated topic. The present results, particularly when seen in the light of previously reported results, should be interpreted. Differences in the general information or prior knowledge accompanied by the bibliographies should be emphasized, and the negative and positive effects should be compared. This helps the authors to establish their position regarding the questions formulated in the introduction and to defend their thesis. It also allows the authors to transform their hypothesis into a conclusion later.

The following paragraphs interpret and analyze the results of secondary outcomes, always accompanied by the respective bibliographies. It is essential to emphasize the scientific relevance of the results and the study's negative and positive points. Then, following the same structure, present and discuss the secondary outcomes when applicable.

The second-last paragraph highlights the study's strengths and weaknesses. The study's strengths should be addressed to convince the readers of the validity and reliability of the conclusions. The strengths should be firmly emphasized, describing the new aspects discovered and the gaps in the literature that were answered by the study. It is equally important to declare the study's limitations and weaknesses. If you do not critique your study, the reviewers will likely do it. Explain with transparency why such a flaw exists and the decisions taken by the authors to overcome those limitations.

The last paragraph may comment on future implications and the subsequent research study proposed by the research group based on the present investigation.

Authors should avoid presenting additional information not touched upon in the work, independent of its relevance. Likewise, discussing aspects not shown in the present study, irrespective of its significance, is also inappropriate. Not only do such debates lack substance and lead to no conclusion, but they also generally draw criticism from reviewers and editors.

43.5.5 Conclusion

The conclusion should answer the research questions formulated in the introduction and highlight the impact the results will have on the scientific environment. It should be written in a brief, simple and understandable way.

A reliable and convincing conclusion can only come from a well-written, well-designed and well-conducted work. On the other hand, the manipulation of data and extrapolation of the results will reflect an unreal conclusion, with a loss of reliability. Neither self-aggrandizement nor apologies for the work's imperfections are appropriate in this section. Using tables, graphs, citations, figures, and the inclusion of new data is not recommended. The last sentence of the conclusion can highlight the need for further studies to understand more on the relevant topic of research.

43.5.6 References

References provide the basis for everything in the manuscript being conveyed to the reader. In your bibliographic review, concentrate on the studies that are most often repeated and that you think are relevant to the specific circumstances of your work. In this sense, look for references that will touch on the same points that are similar to those of your research so that you can compare, discuss, and support your theoretical justification. Authors should always know precisely from where the data were extracted when using references. Reference management software can help to catalog, organize, and format the references according to the publication rules of different scientific journals. Constantly inserting references as the paragraphs are written saves time and helps to avoid situations where some references are left out or mixed up

43.5.7 Title, Abstract/Summary and Keywords

When an article is published, people usually find it through the keywords. Later, they will only read the title and summary and, based on the content, decide whether to read the article. Thus, the title, abstract and keywords are important parts of an article.

43.5.7.1 Title

It is likely that some of the readers will not read the entire article, but there is no doubt that most will read the title. A title is a highly simplified and condensed description of a study that must call the reader's attention to take an interest in the work and read the full article. Elements of the title include the main topic, study design (e.g., randomized trial, cohort study, prospective versus retrospective study), number of patients, primary outcome, and follow-up time. A successful title should attract the target audience, if they are doing a database search. The words selected for the title are essential for the initial evaluation of the paper and to be found by the reader interested in the subject.

Avoid having words and phrases which do not contain useful information. These include constructions such as: "about," "presentation of a new case of," "considerations about," "contributions to the knowledge on," "study of," "study about," "influence of," "interest of," "investigations about," "our personal experience with," "new contributions to," "observations on," and "about the nature of."

Appealing titles are usually precise, specific, and relatively short. The type of study should be included in the title, especially if they are a randomized trial and systematic review, with and without meta-analysis, and the expressions "randomized clinical trial" and "systematic review and meta-analysis". Before the end of the work and having the results, the title should be modified for publication to include the changes achieved by the intervention (Fig. 43.1).

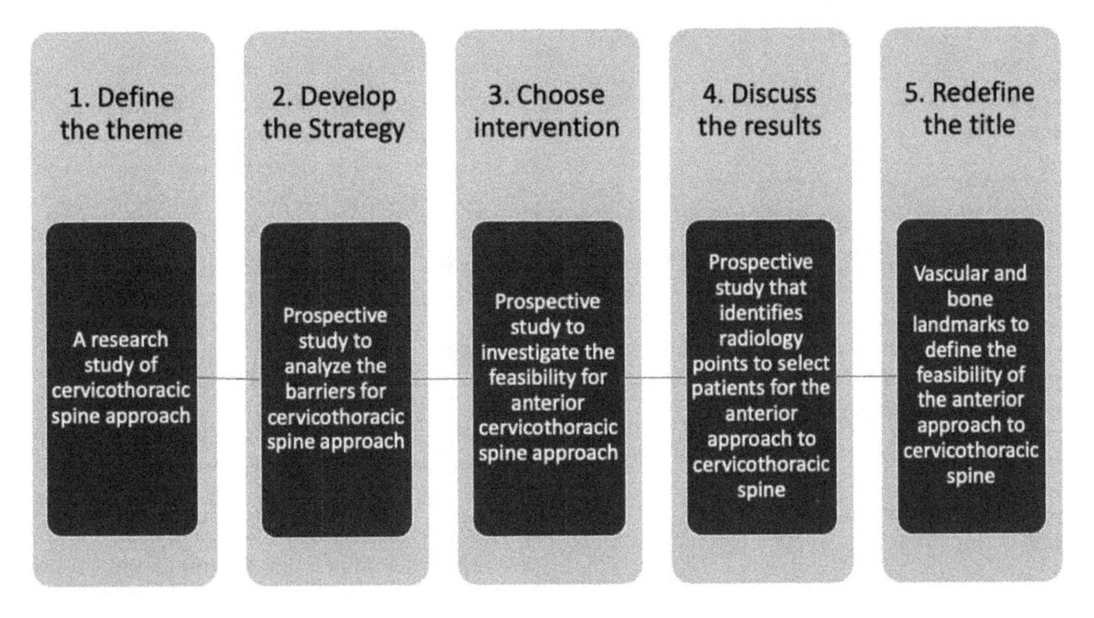

Fig. 43.1 Title evolution during the scientific research

Table 43.5 Tips for writing an abstract

Tips for writing an abstract
• Obey the limits proposed for most summaries: 250 and 500 words
• Excessively long summaries are no longer appropriate for their function
• Include synonyms of the words of the title to make it easier to find the work using search engines
• Make sure that the essential points of the work are presented in summary and ensure consistency
• Avoid describing data that are not in the text
• Do not use abbreviations unless they are justified
• Do not cite references in summary

43.5.7.2 Abstract or Summary

The abstract or summary is a miniature copy of the work that invites you to continue reading it after the title has called the reader's attention to the paper. Most of the journals allow reading the title and its summary free of charge as a way of picking up possible readers. It is a stand-alone part of the manuscript where the author describes the basic procedures, the main findings and the conclusion. It must contain information from several sections of the article; therefore, it is recommended that it be the last to be written. The abstract's structure can vary according to each journal; most limit it to 200–300 words, as extended abstracts lose their function. A tip to make it easier to find the work using the search engines is to include synonyms of the title words and references should not be used in the abstract (Table 43.5). In addition, abbreviations should not be used unless justified and no bibliographic citations.

The abstract structure is the following as suggested:

- Introduction: The clinical question's background and the work's primary purpose must be clearly written.
- Methods: describe the study design used to achieve the objective(s).
- Results: shows the main findings and their analyses.
- Conclusion: present the result related to the answer to the primary research question.

43.5.7.3 Keywords

The list of keywords provides an opportunity to add words besides those already used in the title, which will be utilized to index the work. It facilitates the location and alerts readers to your paper through search engines, increasing the article's visibility. Selecting keywords that are Medical Subject Headings (MeSH) is recommended. Some keywords are already used in the title to facilitate readers' seeing your paper.

43.5.8 Authorship and Acknowledgement

The research team should define the authors and who will be recognized before starting the study. The distribution and responsibility of each research team member were usually known when the study was planned. The authors can adjust during the research but generally do not change the authorship. It is important to have this agreement written and signed by everyone involved.

Authorship is serious accountability because all the authors have a similar responsibility for everything that was done and written in the paper. All authors must have agreed and approved the final version of the manuscript. The authors have made scientific and intellectual contributions to hypothesis formulation, study design, statistical analysis, interpretation, and discussion.

Many people may have helped with the research and prepared the manuscript for publication but do not qualify as authors based on the ICMJE guidelines. Nevertheless, their contributions should be acknowledged in this section [10]. Examples of such contributions are people who supplied special equipment or substantial technical help, provided significant help writing the manuscript, and critically read and commented on the manuscript but did not participate in the article's planning, implementation and drafting stages.

43.5.9 Disclosure and Conflict of Interest

The presence of a conflict of interests or any financial support should be declared when sending the article for review to the journal and later publication. There is a conscious and unconscious influence in the analysis of statistical results. Some journals may require that these be declared in either the methods section or a separate conflict of interest section.

43.6 Concluding Remarks

The main recommendations needed for writing an article adequately can be highlighted in four points: be clear, accurate, consistent, and logical. The investigative question needs good reasons to research it and an appropriate methodology to answer it. Therefore, the results section focuses on the findings of the research and the discussion should highlight it and compare it with the literature information and the author's experience. In the end, the article looks like a chain linked by the research question, each part of the paper being expected to perform a particular function. The best way to learn to write is through writing itself. Aside from that, paying attention while reading other articles and asking for feedback from experienced writers also helps.

References

1. Falavigna A, Botelho RV, Teles AR, da Silva PG, Martins D, Guyot JP et al (2014) Twelve years of scientific production on medline by Latin American spine surgeons. PloS One 9:e87945. https://doi.org/10.1371/journal.pone.0087945
2. Falavigna A, Khoshhal KI (2019) Research education: is it an option or necessity? J Musculoskelet Surg Res 3:239–240. https://doi.org/10.4103/jmsr.jmsr_34_19
3. Falavigna A, Martins Filho DE, Avila JM, Guyot JP, Gonzáles AS, Riew DK (2015) Strategy to increase research in Latin America: project on education in research by AOSpine Latin America. Eur J Orthop Surg Traumatol 25:S13–S20. https://doi.org/10.1007/s00590-015-1648-8
4. Falavigna A, De Faoite D, Blauth M, Kates SL (2017) Basic steps to writing a paper: practice makes perfect. Bangkok Med J 13:114–119. https://doi.org/10.31524/bkkmedj.2017.02.019
5. Falavigna A, Blauth M, Kates SL (2017) Critical review of a scientific manuscript: a practical guide for reviewers. J Neurosurg 128:1–10. https://doi.org/10.3171/2017.5.JNS17809
6. International Committee of Medical Journal Editors (1997) Uniform requirement for manuscript submitted to biomedical journals. N Engl J Med 336:309–315. https://doi.org/10.1056/NEJM199701233360422
7. Cummings SR, Browner WS, Hulley SB (2001) Conceiving the research question. In: Hulley SB, Cummings SR, Browner WS (eds) Designing clinical research: an epide- miologic approach. Lippincott Williams & Wilkins, Philadelphia, PA, pp 17–23
8. Bernstein J, McGuire K, Freedman KB (2003) Statistical sampling and hypothesis testing in orthopaedic research. Clin Orthop Relat Res 413:55–62. https://doi.org/10.1097/01.blo.0000079769.06654.8c
9. Bhandari M, Tornetta P (2004) Issues in the hierarchy of study design, hypothesis testing, and presentation of results. Tech Orthop 19:57–65. https://doi.org/10.1097/00013611-200406000-00003
10. Erlen J, Sminoff L, Sereika S, Sutton L (1997) Multiple authorship: issues and recommendations. J Prof Nurs 13:262–270. https://doi.org/10.1016/s8755-7223(97)80097-x

How to Present Results in a Research Paper

44

Aparna Mukherjee, Gunjan Kumar, and Rakesh Lodha

Abstract

The results section is the core of a research manuscript where the study data and analyses are presented in an organized, uncluttered manner such that the reader can easily understand and interpret the findings. This section is completely factual; there is no place for opinions or explanations from the authors. The results should correspond to the objectives of the study in an orderly manner. Self-explanatory tables and figures add value to this section and make data presentation more convenient and appealing. The results presented in this section should have a link with both the preceding methods section and the following discussion section. A well-written, articulate results section lends clarity and credibility to the research paper and the study as a whole. This chapter provides an overview and important pointers to effective drafting of the results section in a research manuscript and also in theses.

Keywords

Tables · Figures · Statistical analyses · Qualitative study · Report · Data

44.1 Introduction: Importance of Results in a Research Paper

The "Results" section is arguably the most important section in a research manuscript as the findings of a study, obtained diligently and painstakingly, are presented in this section. A well-written results section reflects a well-conducted study. This chapter provides helpful pointers for writing an effective, organized results section. Often, we are not cognizant of the fact that the reader has not been as involved in the study through its lifecycle as the investigator, and hence, it is important to present the study findings in a manner that is clear and logical for everybody to follow. A haphazard, directionless, and over-elaborative description of the results can be the Achilles heel of many good studies. Coherently presented data, on the other hand, can go a long way in inspiring the readers' confidence in the credibility and robustness of the study. It should be a step-by-step approach, taking the readers

A. Mukherjee · G. Kumar
Clinical Studies, Trials and Projection Unit, Indian Council of Medical Research, New Delhi, India
e-mail: aparna.sinha.deb@gmail.com; gunjan2587@gmail.com

R. Lodha (✉)
Department of Pediatrics, All India Institute of Medical Sciences, New Delhi, India
e-mail: rlodha1661@gmail.com

G. Jagadeesh et al. (eds.), *The Quintessence of Basic and Clinical Research and Scientific Publishing*,
https://doi.org/10.1007/978-981-99-1284-1_44

along the plan of analysis that was described in the methods section. After reading the results, the readers should be able to draw their own conclusions about the study.

This chapter will take the reader through the steps of writing an impressive results section of a manuscript.

44.2 What to or Not to Include in the Results Section?

Results are the core of any research paper. It is a section to narrate the observations based on the data you have collected. One would have collected an exhaustive quantity of data in a study, however, it is not necessary to present all the data at once. Finding the right balance is important—too much information might obscure the pertinent findings, whereas too little hampers the reliability and understandability of the study. One needs to present an overall description of the evidence that has been generated with the help of text, figures, and graphs in a logical order. The instructions to authors given by the journal of choice should be thoroughly read and adhered to.

Box 44.1 What does not belong to the results section?

- Results that do not answer any of the research questions
- The methods that you might have left out of materials and methods sections, such as details of statistical tests, definitions, plan of the study, etc
- Interpretation of the results
- Data that are not collected in your study
- Comparison with the results of other studies
- Opinions
- Explanations for the findings
- Limitations and pitfalls of the current research

44.3 General Guidelines for Writing the Results

44.3.1 Style of Writing

The style of writing should be crisp and fluent; unnecessary use of adjectives and adverbs is discouraged. The language mentioned in the instructions to authors (English UK/English American) should be adhered to. The sentences should not be unnecessarily long and should be kept simple. It is advisable to use past tense when describing the results as the events being reported have already occurred [1]. However, present tense is used when referring to table/figure in the manuscript. The current convention is to use the term 'study participants' instead of patients. Write with clarity and brevity, as shown in the following example:

(a) It is clearly seen that the average height of children in group A is markedly higher than that of group B.
(b) The mean (SD) height of children in group A [120 (5.8) cm] was higher as compared to that in group B [110 (3.2) cm], $p = 0.04$.

Sentence 'b' will be more acceptable to the scientific and medical community of readers.

44.3.2 Organization of the Results

The description and organization of findings may vary according to the study type and the specific journal's instructions. However, the broad outlines are similar. Qualitative studies, meta-analyses and systematic reviews have a different approach as compared to quantitative studies and will be discussed later in this chapter.

44.3.2.1 The Starting Point

The duration/period of the study should be mentioned. The details of study participants, including the numbers of potential participants screened, numbers found eligible, numbers excluded with reasons for exclusion, numbers of

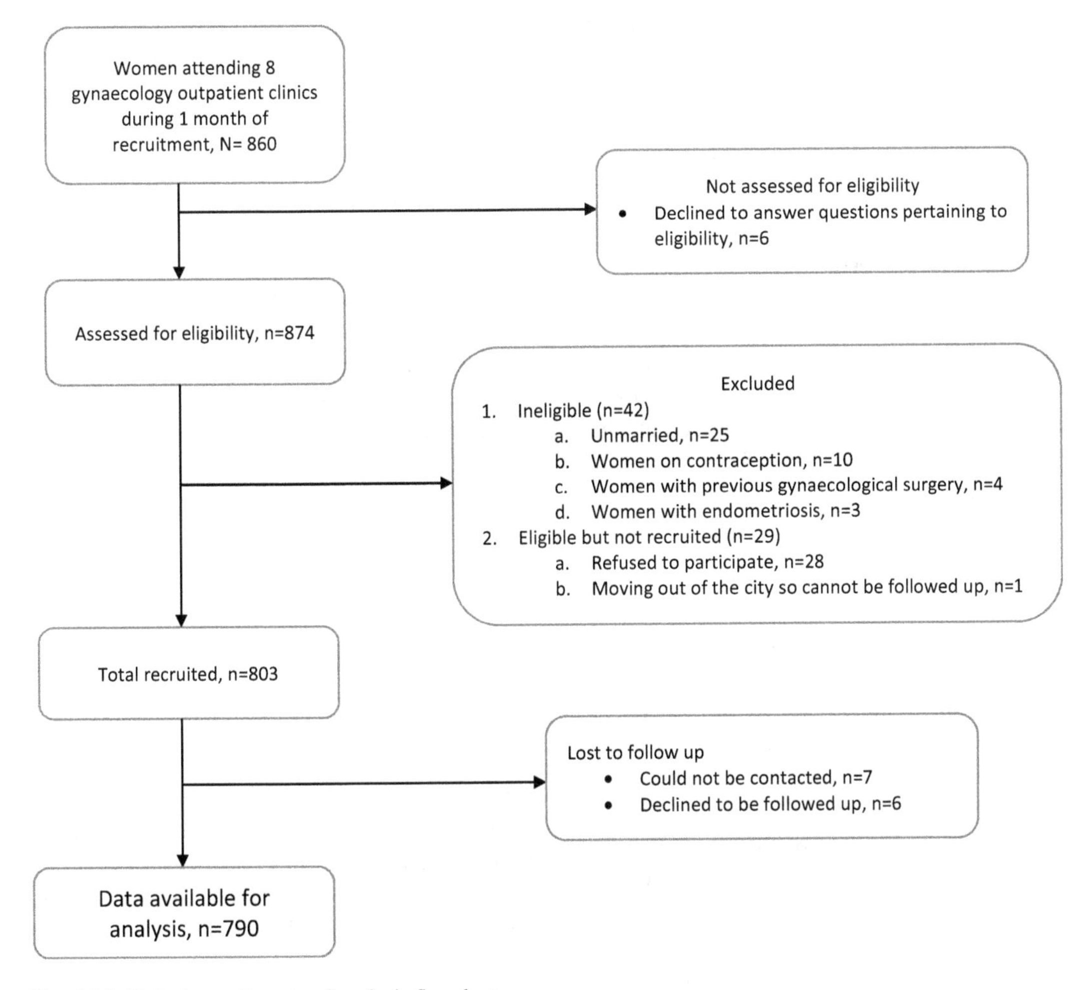

Fig. 44.1 Patient recruitment and analysis flowchart

cases and controls or the numbers randomized, numbers lost to follow-up or trial deviates, and such are usually enumerated in the opening paragraph of this section. The use of a flow diagram to show these parameters may be useful. An example of a basic flow diagram for an observational study is given in Fig. 44.1. An example of describing enrolment in the text is as below:

"Three hundred [175 (58.3%) girls] children with tuberculosis, aged 6 months to 15 years of age, were enrolled after screening 800 children attending the pediatric outpatient department of our hospital."

Substantial duplication of information in text, figures and tables should be avoided. If a study flow chart is used, only the key points could be highlighted in the text while referring to the figure.

Thereafter, a baseline clinico-demographic description of study participants can be presented in a table or text. In case of a randomized controlled trial, presenting the p- values to compare the baseline characteristics of the intervention and control arms is considered superfluous. If the randomization has been done correctly, differences, if any, between the two arms are only due to chance, irrespective of the p value.

Indicate the number of participants with missing data for each variable of interest. In the case of non-interventional studies, such as cohort studies, numbers or summary measures regarding exposures and potential confounding factors should be mentioned here. In case-control studies, the baseline characteristics of cases and controls are to be compared, preferably in a table.

In intervention studies, it is desirable to report adherence to the study interventions before presenting the outcomes.

44.3.2.2 Reporting Guidelines

The guidelines for reporting studies differ according to the study design. Details of these guidelines can be accessed at: http://www.equator-network.org/. Many journals require a checklist to be submitted as per these guidelines. Following are the guidelines for the most common study designs:

- ***Observational Studies:*** The STROBE (Strengthening the Reporting of Observational studies in Epidemiology) guidelines are applicable to both case control and cohort studies [2]. It is advisable to include a flow diagram indicating the study outline. It is essential to report the number or the summary measure of exposure categories and outcome events.
- ***Randomized Controlled Trials (RCT):*** CONSORT (Consolidated Standards of Reporting Trials) statement for reporting should be followed for a RCT. A study flow diagram is essential [3]. Mention the study periods, including the dates coinciding with start and end of study recruitment and follow-up. If the trial is terminated prematurely, reasons for doing so should be clearly specified. Remember to document and summarize all adverse events, expected as well as unexpected, in both the intervention and control groups.
- ***Diagnostic studies:*** The STARD (Standards for Reporting of Diagnostic Accuracy Studies) guidelines are to be followed while reporting diagnostic studies [4]. A flow diagram can be included to depict the enrollment of participants. Provide a '2 X 2' table for the results obtained from the diagnostic test under study and the reference or gold standard. The calculated sensitivity, specificity, positive predictive value, negative predictive value, and likelihood ratios are to be reported as applicable. It is expected that one would report any adverse events encountered while performing the tests, both the one under study and the reference.
- ***Meta analysis and systematic review:*** PRISMA (Preferred Reporting Items for Systematic Reviews and Meta-Analyses) is an evidence-based minimum set of items for reporting in systematic reviews and meta-analyses. PRISMA primarily focuses on the reporting of reviews evaluating the effects of interventions, but can also be used as a basis for reporting systematic reviews with objectives other than evaluating interventions (e.g., evaluating etiology, prevalence, diagnosis or prognosis) [5, 6].
- ***Qualitative research:*** COREQ (Consolidated criteria for reporting qualitative research) is a 32-point consolidated checklist primarily for reporting interviews and focused group discussions [7].

44.3.2.3 Sequence of Presentation

A proper sequence for the presentation of the observations of the study improves the comprehensibility of the data. Often, the authors are inclined to exclusively present the findings where the statistical tests return significant p values. At times, all the data collected are presented without any predetermined order; the main outcomes are lost somewhere in this maze of information. It becomes difficult for the readers as well as the reviewers to follow such a chaotic arrangement.

Addressing the Objectives

It is a cardinal rule to address the primary objective of the study first. The primary question asked in the study needs to be answered clearly, using the statistical plan as described in the methodology section. The secondary outcomes or endpoints, if any, can follow (see Chap. 1).

Maintain congruence between the sequence followed in the methodology and results section. It is advisable to go from simpler to more complicated results. Subgroup analysis should come in last. Refrain from reporting any finding in the results section without first describing the corresponding methodology and statistical analysis plan in the preceding methodology section. Also, refrain from excluding the observations from any test or investigation which has been already mentioned in the methodology section. In case *post hoc* analyses are performed, the same should be stated explicitly.

To increase clarity and readability, the Results section can be divided into appropriate subheadings; however, the same should be done as per the journal's format and instructions.

44.3.3 Important Points to Remember

While concentrating on the more interesting results, we frequently overlook certain simple but essential attributes of this section. Some of the points to remember while drafting the results section are:

(a) Decimal points
The computer programs used for statistical analysis usually return results involving decimal points which may look like 3.4562789. The instrument or assay used to measure the parameters provided as inputs for the programs may not be sensitive enough to be able to give such outputs. It makes sense to round off to the decimal point which reflects the sensitivity of your measuring instrument or assay. For example, if the birth weight in a study is measured in grams, then the mean value for the weight of the groups of studied neonates should have only one decimal place rather than 2 or more decimal places. In most instances, one or two places after the decimal point are sufficient. However, it is prudent to consult the targeted journal's requirements and be consistent with the decided format throughout the report.

(b) p value
The significance of the statistical tests applied is usually presented as p values; the actual p value should be written instead of just stating <0.05. Never state a p value as 0.000 even if the statistical program does return such a value. To convey appropriate meaning, such values should be stated as <0.0001 (also, see Chap. 29).

(c) Confidence Intervals
The advantage of confidence intervals over p-values in reporting hypothesis testing is that the result is stated directly at the level of data measurement. Confidence intervals provide information about statistical significance, as well as the direction and strength of the effect. It is a good practice to report confidence intervals wherever possible.

(d) Choose your words carefully
The word 'significant' when used in the results section carries a different connotation than its literary meaning. Here the significance has to be statistically proven i.e., $p < 0.05$ or appropriate confidence interval. Hence, one should be careful while declaring any result as significant. At the same time, it is essential to include the statistically non-significant data for the stated objectives. Refrain from using indecisive words such as 'about' or 'approximately' while describing the results.

(e) Confounders
In scenarios where confounders are expected to alter the results, it is helpful to provide the unadjusted as well as the confounder-adjusted estimates and their precision (e.g., 95% confidence interval). Add a note listing the confounders which were adjusted for. The reasons for deciding on the confounders and the process applied to adjust for them should be described in the methods section.

(f) Negative results
Very often the researchers as well as the reviewers are biased toward the positive findings of the study. However, keep in mind that it is equally important to publish the negative results obtained from any study as they may help to prove or disprove proposed hypotheses. Omitting negative results is considered unethical.

(g) Text–table dichotomy
It is important to ensure that text and tables are complementary to each other and not merely repetitive. Describing all parameters that are depicted in the table is not required. Only some salient features and concise description in the text is sufficient to inform the reader as to what is described in the tables; for example: "Three hundred [175 (58.3%) girls] children with tuberculosis, aged 6 months to 15 years of age, were enrolled. The mean (SD) age of the children was 110 (15.3) months. Table ... shows the baseline characteristics of the enrolled children."

(h) Numbers in a sentence
It is conventional not to start a sentence with digits, rather the numerical value is spelled out in words. Values less than ten are usually spelled out.

44.3.4 Data Presentation Tools

44.3.4.1 Tables

Tables are extremely helpful in communicating a large amount of complicated data in a structured fashion. But that does not mean the tables themselves are to be inordinately complicated. Tables are most informative when they are simple and organized. Avoid redundancy by not repeating the same information in both text and tables. The tables should be numbered in the same chronological order as they are referenced in the text. Do not insert any table without the corresponding reference in the text. Depending on the journal, tables may be presented sequentially at the end of the manuscript after the 'References' or may be submitted as a separate file [8].

The number of tables:

The maximum number of tables that can be included depends on the requirement of the journal. Usually, 3 to 4 tables are permitted for original, full-length manuscripts; a lesser number is allowed for short communications. The remaining data may be submitted as supplementary material in case of online publications.

Requisites of a good table:

Each table should be self-explanatory and complete in itself. One should be able to comprehend the data presented in a table without consulting the text. Apart from the rows and columns containing data, a table contains a heading or legend, row and column headers and footnotes. A precise but informative heading follows the table number. Do not use abbreviations in the heading. Each row and column header should be able to explain what the row/column contains. Abbreviations mentioned in the table and any other explanatory notes should be expanded in the footnote. Do not forget to mention what summary statistic is being presented such as N (%), mean (SD) or median (IQR) and the units of measurement either in the appropriate column/row header or in the footnote (Table 44.1). Mention the denominator applied for each variable of interest, especially if they are not uniform. Do not just give percentages if the denominator is less than 100, rather give the actual observed values. Do not combine disparate variables in the same table. It is better to split a long and unwieldy table into two for clarity.

44.3.4.2 Graphs and Charts

Visual display of data by means of figures immediately makes the data prominent and more appealing to the reader and helps in longer retention of the information. Not all data are amenable to being presented as figures. One must judge carefully as to which data can be best presented in the form of figures. It is usually the important information which is presented in this format—messages which you want to catch the attention of the readers. Figures can be in colour or grey tone. Almost all journals charge extra for coloured figures. The number of figures allowed depends on the targeted journal.

Figures include charts like simple line diagrams, scatter plots, radiographs, images, photographs, maps, etc. As with tables, a figure should also be self-explanatory with an informative but precise heading. Other components of a figure include legends, data labels, axis titles, etc. [8]. It is noteworthy that text, tables and figures

Table 44.1 Factors associated with mortality among hospitalized COVID-19 patients

Characteristic	*Mortality n(%)*	*Odds Ratio*[a] *(95% CI)*	*P value*[a]
Age	529 (7.4)	(Reference)	–
• 18–39 years (n = 7169)	1455 (13.5)	2 (1.8, 2.2)	<0.001
• 40–59 years (n = 10,760)	1973 (21.2)	3.4 (3.0. 3.7)	<0.001
• ≥60 years (n = 9322)			
Gender	2613 (15.2)	1.2 (1.1, 1.2)	<0.001
• Male (n = 17,240)	1344 (13.4)	(Reference)	–
• Female (n = 10,008)			
Vaccinated with Anti-SARS-CoV-2 vaccine	1306 (21.9)	(Reference)	–
• Unvaccinated (n = 5964)	85 (15.5)	0.7 (0.5, 0.8)	<0.001
• Vaccinated with one dose (n = 550)	29 (9.5)	0.4 (0.3, 0.6)	<0.001
• Vaccinated with two doses (n = 305)			
WHO ordinal scale[b]4 or above at admission	3101 (26.3)	10.3 (9.2, 11.4)	<0.001
• Yes (n = 13,860)	433 (3.4)	(Reference)	–
• No (n = 13,049)			
BMI	387 (9.8)	(Reference)	–
• Underweight and normal weight (n = 3944)	766 (10.5)	1.1 (0.9, 1.2)	0.26
• Overweight and obese (n = 7307)			

SARS Severe Acute Respiratory Syndrome, CoV-2 Coronavirus 2, WHO World Health Organization, BMI Body Mass Index. Adapted with permission from *Mukherjee A, Kumar G, Turuk A, et al; NCRC Study team. "Vaccination saves lives: a real-time study of patients with chronic diseases and severe COVID-19 infection". QJM. 2022 Sep 2:hcac202*

[a]*Odds ratio and p value calculated via bivariate logistic regression*

[b]*https://www.who.int/docs/default-source/documents/emergencies/minimalcoreoutcomemeasure.pdf*

serve different purposes in presenting information, however, repetition of data should be avoided. All figures should be cited in the text and numbered in the order of citation/appearance in the manuscript. All figures and pictures should be submitted as separate files in the form of an image file (.jpg, .ppt, .gif, .tif or .bmp) with a minimum resolution of 300 dpi to ensure good print quality; the authors should refer to the target journal's instructions to choose the correct file type and the desired resolution.

Data charts are highly effective tools to summarize numerical data for better visual presentation. Choosing the type of graph appropriate for your data is critical. This depends on various factors: the type of data (continuous or categorical), the number of groups or variables involved, and the intent of creating the graph. For example, when one wants to demonstrate the composition or break up of a dataset or groups within the dataset, one can use pie charts or stacked bar charts. Barcharts or line plots are helpful in comparing data. Line plots can be utilized to illustrate time trends. Distribution of data may be demonstrated by histograms, and scatter plots. The relationship or correlation between two variables can be demonstrated by scatter plots. Commonly used data charts are [9] (Fig. 44.2a–h):

(a) Pie chart: Pie chart shows classes or groups of data in proportion to the whole data set, particularly to depict large data sets, e.g., epidemiological surveys (Fig. 44.2a). These are best used when the number of classes/groups is 3 to 7. One should avoid using the pie chart where there are only two groups, e.g., gender.
(b) Bar charts: Bar charts may be horizontal or vertical. The height or length of the bars represents the measurement. By this method the same variable can be compared across groups or time points (Fig. 44.2b). Stacked bars can also be made to make intra-group comparisons like in males and females or to show the composition of each group. It is better not to compare two variables in the same chart if the values of one variable overshadow or dwarf the other. Also, it is prudent to avoid clubbing too many variables or categories in the same chart-this makes the chart unreadable. If there are more than 5 groups to be compared, it is better to use horizontal bar charts.

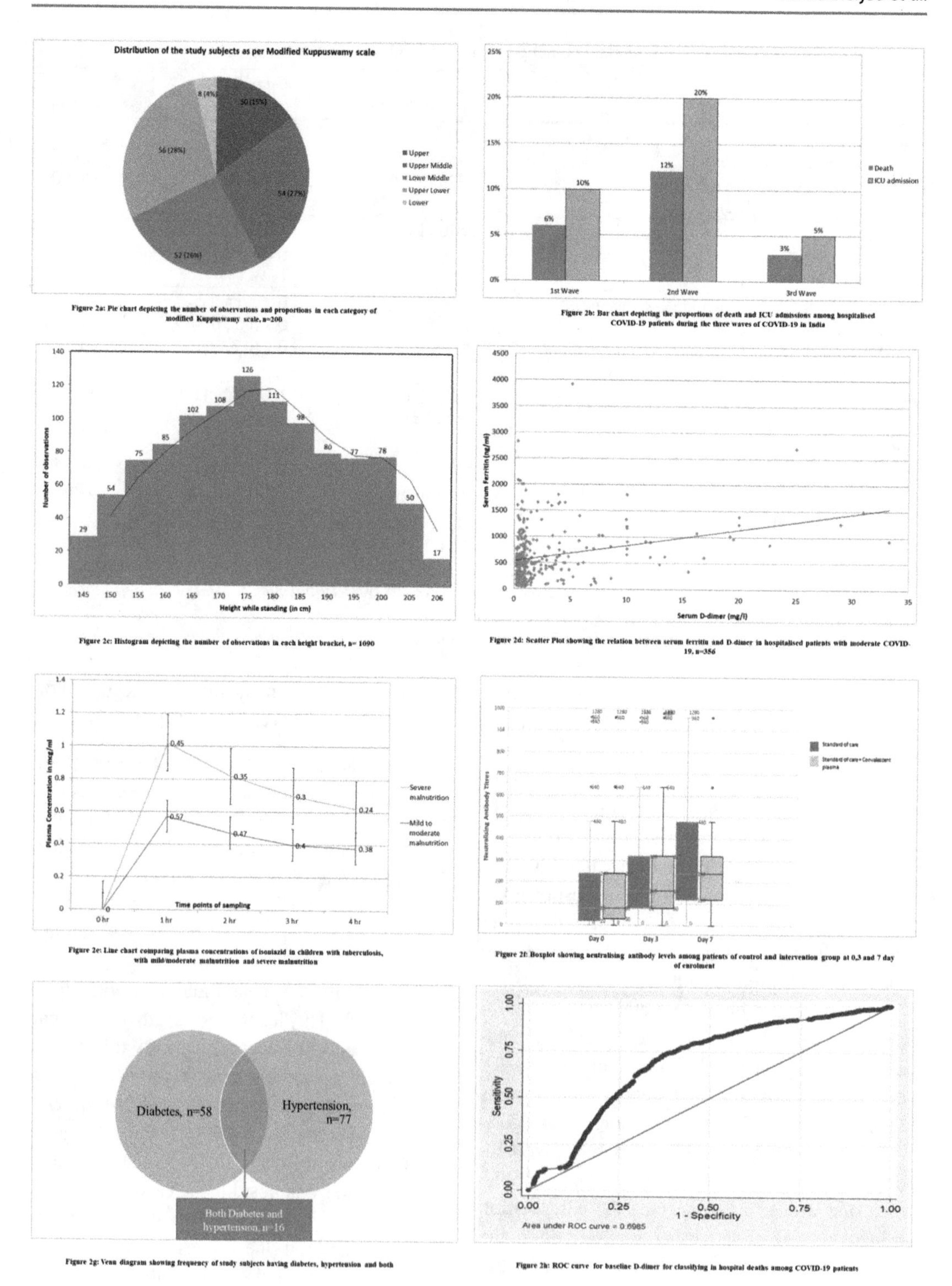

Fig. 44.2 (**a**–**h**) Commonly used charts and figures in research papers

(c) Histogram: A histogram is used to show the distribution of numerical data; the entire range of a continuous variable is divided consecutively into non-overlapping groups known as class intervals. The height of the vertical rectangles for each class interval represents the frequency or density of the variable, depending on whether the class intervals are of equal or unequal width (Fig. 44.2c).
(d) Scatter plots: Scatter plots can be used to present measurements on two variables that are related; the values of the variables on the y-axis are dependent on the values of the variable plotted along the x-axis (Fig. 44.2d).
(e) Line plots: Line plots are similar in some ways to the scatter plots, with the condition that the values of the x variable have their own sequence (Fig. 44.2e). Line plots can be used to depict time trends.
(f) Box plots: Box plots are used to depict the numerical data in the form of median and interquartile range (which forms the box); sometimes whiskers are added which may denote the maximum and minimum value. Outliers can also be depicted in the box plot (Fig. 44.2f).
(g) Venn diagram: A Venn diagram is a type of chart that shows how different data sets relate to or overlap each other through intersecting portions of circles (Fig. 44.2g).
(h) ROC curves: Receiver Operating Characteristic curve is a plot of the true positive rate (sensitivity) against the false positive rate (1-specificity) for the different possible cut-offs of a diagnostic test. The area under the curve is a measure of the accuracy of the test under question (Fig. 44.2h).

Box 44.2 Common errors in tables and figures

Tables

- Excessive repetition of information in text and tables
- Tables not self-explanatory and complete
- Disparate characteristics and comparisons clubbed together in one table
- Tables not cited in the text
- The table number and the chronological order of their reference in text do not match
- Inaccurate arithmetic—numbers do not add up; discrepancies in data

Figures

- Graph not plotted to scale
- Data not properly labelled
- Omission of proper legends
- Not cited in the text
- Not numbered as per the reference in text

Images

Different types of photographic images are used in a research paper; they include clinical photographs of patients; imaging investigations such as radiographs, ultrasonography, CT scan/MRI scan, radionuclide studies; intra-operative findings; surgical specimen; pathology images-cytopathology, histopathology, special stains, immunohistochemistry, etc.; laboratory investigations such as PCR results, gel/blot images; tracing of investigations such as ECG, EEG, EMG, etc. Publishers would like the images to be of excellent quality, undistorted, i.e., maintaining the original proportions, and with adequate labelling.

Obtaining informed, written consent is imperative for publishing clinical photographs. Maintaining patient confidentiality is of paramount importance; the identity of the person should be protected by obscuring any identification mark and covering the eyes. As with other types of figures, images should also have a descriptive legend. At times, it is more informative and meaningful to combine many images in a single figure, e.g., CT scan/MRI scan images; each of the images should be identified separately. The use of arrows or other markers may be helpful to highlight important findings.

Maps

Maps may be used in some manuscripts to add a crucial demographic context to a research paper, e.g., multi-centre research projects or National level surveys where the country-wise results or participation can be easily illustrated. The use of Geographical Information Systems also helps researchers present the inequities in the communities by mapping and presenting this data. A map if added to the study should be well labelled and the footnotes should be clearly indicative of the markers being used on the maps (Figs. 44.3 and 44.4).

Box 44.3 Tips to make good quality tables and figures

- Caption or legend for the figure or image appears below the graphic, and above the table, though the author instructions should be thoroughly read and adhered to
- Include the word "Figure" or "Table" along with associated number in the caption, the same number being given as reference in text
- Make the caption clear and comprehensive even if it is lengthy
- In case the table has been recreated from someone else's published work, it should be mentioned along with the citation
- Provide a list of all abbreviations and symbols at the end of the table/graph/chart
- Do not distort the image
- Include a copyright statement below any image used

44.4 Writing the Results in a Meta-Analysis and Systematic Review

A systematic review aims to answer a defined research question by gathering the already available evidence from studies that adhere to pre-defined eligibility criteria. A meta-analysis further summarises the results of these studies by applying various statistical tests. Unlike the other study designs, where data is collected by the researchers, the systematic review and the meta-analysis collates data already collected by multiple researchers in order to generate evidence that is usually stronger than that generated from individual studies.

The results in a systematic review or meta-analysis are started with a description of the various studies included in the review as opposed to a description of the participants in a conventional quantitative study followed by pooled baseline characteristics of the units of study (e.g., study participants). It also provides some details of the excluded studies in addition to the reasons.

The most important part of the results in a meta-analysis is the visual display of information using a forest plot, which provides a wealth of information. A forest plot takes all the relevant studies asking a common question and displays them on a single axis, along with a pooled result from all the studies included.

An example of a forest plot is given in Fig. 44.4 [10].

44.5 Writing the Results of a Qualitative Study

Qualitative research is gaining recognition and is now being increasingly used in healthcare research, specifically in social and cultural contexts. Unlike quantitative research which analyses the data as trends, frequencies, p values and odds via various statistical tests, qualitative research tends to collect data through the experiences, views, and dialogues of the participants. A few examples of qualitative data include audio recordings and transcripts from interviews or focussed group discussions (FGD), interview questionnaires with open-ended responses, video recordings, images, emails, minutes, case notes, etc. [11].

The results in a qualitative research manuscript are usually presented as a central theme, or categories extracted from the data collected.

North
- Postgraduate Institute of Medical Education & Research, Chandigarh
- Medanta Institute of Education and Research, Gurugram, Haryana
- Christian Medical College, Ludhiana, Punjab
- Pandit Bhagwat Dayal Sharma Post Graduate Institute of Medical Sciences, Rohtak, Haryana

West
- All India Institute of Medical Sciences, Jodhpur, Rajasthan
- Rajasthan University of Medical Sciences, University of Medical Sciences, Jaipur, Rajasthan
- Mahatma Gandhi Medical College and Hospital, Jaipur, Rajasthan
- Sardar Patel Medical College, Bikaner, Rajasthan
- Dr. D. Y. Patil Medical College, Hospital & Research Centre, Pune, Maharashtra
- Smt. NHL Muncipal Medical College, Ahmedabad, Gujarat
- CIMS Hospital, Ahmedabad, Gujarat
- Sumandeep Vidyapeeth and Institution, Deemed to be University & Dhiraj Hospital, Vadodara, Gujarat
- GMERS Medical College and Hospital, Himmatnagar, Gujarat

South
- National Institute of Mental Health & neurosciences, Bengaluru.
- Bowring and Lady Curzon Medical college and Research Institute, Bengaluru, Karnataka
- Gulbarga Institute of Medical Sciences, Kalburgi, Karnataka
- St. John's Medical College & Hospital, Bengaluru
- Gandhi Medical College, Secunderabad, Hyderabad

North- East
- North-Eastern Indira Gandhi Regional Institute of Health and Medical Sciences, Shillong
- Naga Hospital Authority, Kohima, Nagaland.

Central
- All India Institute of Medical Sciences, Bhopal, Madhya Pradesh
- Gandhi Medical College, Bhopal, Madhya Pradesh
- King George's Medical University, Lucknow, Uttar Pradesh
- Banaras Hindu University, Varanasi, Uttar Pradesh
- J.N. Medical College, Aligarh Muslim University, Aligarh, Uttar Pradesh
- Government Institute of Medical Sciences, NOIDA, Uttar Pradesh

East
- All India Institute of Medical Sciences, Bhubaneswar, Odisha
- Hitech Medical college, Bhubaneswar, Odisha
- IMS & SUM Hospital, Bhubaneshwar, Odisha
- Institute of Post-Graduate Medical Education and Research, Kolkata, West Bengal
- The Medical College, Kolkata, West Bengal
- Infectious Diseases & Beliaghata General Hospital, West Bengal
- College of Medicine and Sagore Dutta Hospital, Kamarhati, West Bengal
- Tata Medical Centre, Kolkata
- All India Institute of Medical Sciences, ESI hospital, Raipur, Chattisgarh
- ESI Hospital, Raipur, Chattisgarh
- Shaheed Nirmal Mahto Medical College, Dhanbad
- Government Medical college, Jagdalpur

Fig. 44.3 Map illustrating the institutes participating in the National Clinical Registry for COVID-19, India. Adapted with permission from *Mukherjee A, Kumar G, Turuk A, et al; NCRC Study team. "Vaccination saves lives: a real-time study of patients with chronic diseases and severe COVID-19 infection". QJM. 2022 Sep 2:hcac202*

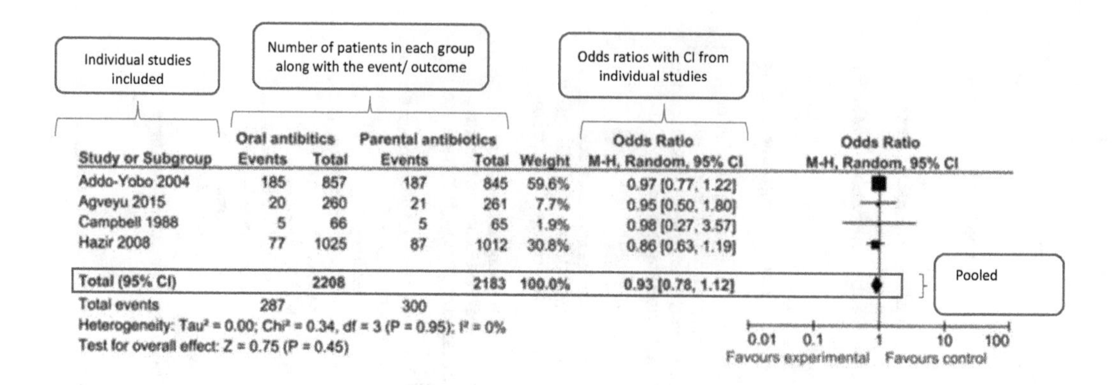

Fig. 44.4 Figure adapted from "Lodha R, Randev S, Kabra SK. Oral Antibiotics for Community acquired Pneumonia with Chest indrawing in Children Aged Below Five Years: A Systematic Review. Indian Pediatr. 2016 Jun 8;53(6):489-95. doi: 10.1007/s13312-016-0878-0. PMID: 27376603". (Reproduced with permission)

They are further elaborated as sub-themes, codes and dialogues. The themes should be chosen in a way that they provide an answer to the research question [12]. The main focus lies in reporting an identifiable pattern in the data rather than the number of participants in the categories, which is usually in the form of quotes from the participants rather than tables or any statistical analysis. Choosing quotes to be added to the manuscript could be tricky, and the selected quote must be long enough to be convincing [13] and clear to make sense to the reader. The quotes can be used judiciously to provide strong evidence for the results described.

The following extracts are from interviews conducted among the family members of a deceased organ donor in order to understand the barriers towards organ transplantation [14], shown as examples of how data from a personal interview can be presented:

The interview discusses perceptions of the family of the deceased that lead to the denial of the family to donate organs.

- "She (mother of the deceased) was not an educated lady and did not have any knowledge on organ donation and hence decided not to donate her son's organs". (interviewee-Deceased's family member)—Poor knowledge concerning organ donation
- "They (family members) were afraid that if we opt to donate the organ of her (mother of the deceased) deceased son, the healthcare team would stop working hard to retrieve her son back to life and concentrate only on organ donation" (interviewee- Deceased's family member)—Mistrust on the healthcare
- "When we had some doubts, we were replied in an improper Tamil (local regional language) which we were not able to understand easily" (interviewee- Deceased's family member)—Poor communication

Following are a few quotes as an example to present data from a FGD on public concerns for sharing and governance of personal health information [15]:

- "Cause you don't know what the next institution's gonna do or who they're gonna give it [the data] to"- an interviewee from FGD
- "I would be worried at some point that people would be identified. If it's going so many places, so many people are involved, so many people seeing that data, that would be a little worrisome to me"- interviewee from FGD

A conclusive summary of results could be provided after the actual themes, sub-themes and codes are presented to highlight the evidence that has been generated out of the data collected. It should be remembered that, unlike quantitative research which aims to remove any subjective influences from the researchers, subjectivity is a core characteristic of qualitative research, where the researchers are directly involved in creating evidence out of the information collected. This also tells why there will always be some element of discussion in the results section, unlike quantitative research.

44.5.1 Data Display in Qualitative Research

Similar to quantitative research, data display tools including flow charts, tables, graphs, charts, images and maps are utilised to display the data in qualitative research. Various additional data display tools uniquely utilised for qualitative research are given below:

(a) ***Boxed display***
Boxed display refers to the text provided in a box. This is usually used in circumstances where a certain narrative that is considered most important is boxed in order to highlight it. Text is provided in a simple text box and the name of the interviewee is given along with the actual narrative in codes.

(b) ***Decision tree***
The decision tree is a graphical description of possible answers to a particular research question. It helps in the evaluation and comparison of various outcomes/solutions. An example of a decision tree is given in Fig. 44.5a.

(c) ***Matrix***
Matrix is created based on the interpretation of transcripts generated from personal interviews or focused group discussions. It is made by creating a set of categories that are made as per the responses provided in the interview. This gives a visual dimension to the patterns being observed, without giving a quantitative angle to the qualitative data [16].
A hypothetical example of a tally matrix is shown in Fig. 44.5b. The matrix tally clearly shows the differences in priorities of men and women.

(d) ***Word art***
Word art visually elaborates the responses retrieved from the interviews and focused group discussions. The more commonly provided responses are shown in a bigger size and the less common responses are in a smaller size. This visually highlights the important responses without actually quantifying them. An example of word art is provided in Fig. 44.5c.

44.6 Value of Review/Revision—By Self and Others

Finally, devote time to check your data thoroughly. Mistakes such as incongruities within the result section or erroneous calculations cast doubts on the credibility and authenticity of the study in the minds of reviewers as well as readers. It is wise to get your results (or the whole paper, if possible) reviewed by a peer or colleague before submission to any journal. Many a time, a neutral perspective of an uninvolved party uncovers shortcomings that are overlooked by the authors themselves.

44.7 Concluding Remarks

The results section is the core of a research paper where the study data and analyses are presented in an organized, uncluttered manner such that the reader can easily understand and interpret the findings. It is often embellished with self-explanatory tables and figures which assist in presenting data in addition to the text. This section is completely factual; there is no place for opinions or interpretations from the authors. The results presented should have a link with both the preceding methods section and the following discussion section. A well-written, articulate results

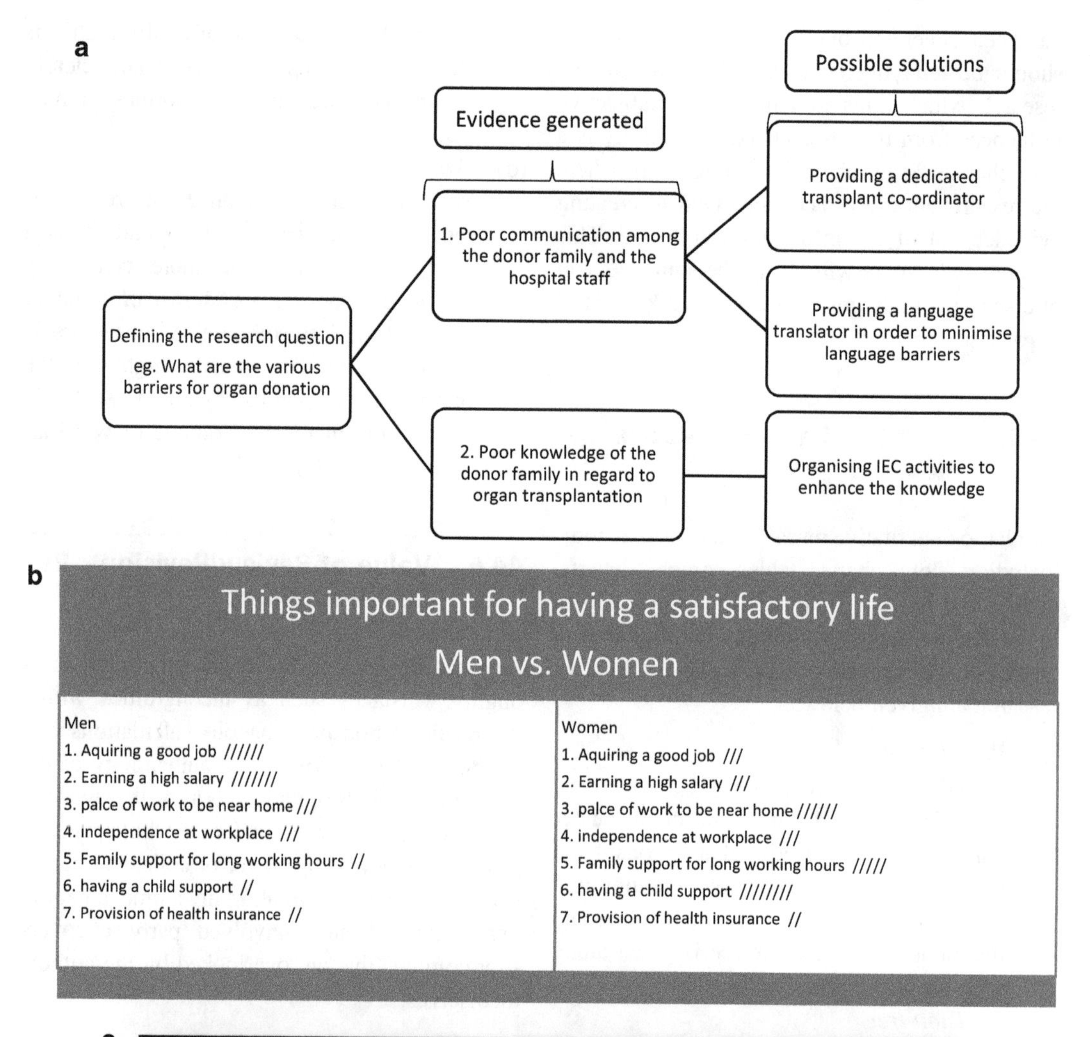

Fig. 44.5 (a) A decision tree to evaluate the barriers to organ donation. (**b**) A tally matrix showing the responses from men and women to understand the important things to have a satisfactory life. (**c**) Word art elaborating the symptoms of patients who died at home after discharge from the hospital, as ascertained by verbal autopsy interviews

section lends clarity and credibility to the research paper and the study as a whole.

Acknowledgments The book chapter is derived in part from our article "Mukherjee A, Lodha R. Writing the Results. Indian Pediatr. 2016 May 8;53(5):409-15." We thank the Editor-in-Chief of the journal "Indian Pediatrics" for the permission for the same.

Conflict of Interest None.

References

1. Kallestinova ED (2011) How to write your first research paper. Yale J Biol Med 84(3):181–190
2. STROBE. STROBE. [cited 2022 Nov 10]. https://www.strobe-statement.org/
3. Consort—Welcome to the CONSORT Website. http://www.consort-statement.org/. Accessed 10 Nov 2022
4. Korevaar DA, Cohen JF, Reitsma JB, Bruns DE, Gatsonis CA, Glasziou PP et al (2016) Updating standards for reporting diagnostic accuracy: the development of STARD 2015. Res Integr Peer Rev 1(1):7
5. Page MJ, McKenzie JE, Bossuyt PM, Boutron I, Hoffmann TC, Mulrow CD et al (2021) The PRISMA 2020 statement: an updated guideline for reporting systematic reviews. BMJ 372:n71
6. Page MJ, Moher D, Bossuyt PM, Boutron I, Hoffmann TC, Mulrow CD et al (2021) PRISMA 2020 explanation and elaboration: updated guidance and exemplars for reporting systematic reviews. BMJ 372:n160
7. Consolidated criteria for reporting qualitative research (COREQ): a 32-item checklist for interviews and focus groups I EQUATOR Network. https://www.equator-network.org/reporting-guidelines/coreq/. Accessed 10 Nov 2022
8. Aggarwal R, Sahni P (2015) The results section. In: Aggarwal R, Sahni P (eds) Reporting and publishing research in the biomedical sciences, 1st edn. National Medical Journal of India, Delhi, pp 24–44
9. Mukherjee A, Lodha R (2016) Writing the results. Indian Pediatr 53(5):409–415
10. Lodha R, Randev S, Kabra SK (2016) Oral antibiotics for community acquired pneumonia with chest indrawing in children aged below five years: a systematic review. Indian Pediatr 53(6):489–495
11. Anderson C (2010) Presenting and evaluating qualitative research. Am J Pharm Educ 74(8):141
12. Roberts C, Kumar K, Finn G (2020) Navigating the qualitative manuscript writing process: some tips for authors and reviewers. BMC Med Educ 20:439
13. Bigby C (2015) Preparing manuscripts that report qualitative research: avoiding common pitfalls and illegitimate questions. Aust Soc Work 68(3):384–391
14. Vincent BP, Kumar G, Parameswaran S, Kar SS (2019) Barriers and suggestions towards deceased organ donation in a government tertiary care teaching hospital: qualitative study using socio-ecological model framework. Indian J Transplant 13(3):194
15. McCormick JB, Hopkins MA (2021) Exploring public concerns for sharing and governance of personal health information: a focus group study. JAMIA Open 4(4):ooab098
16. Groenland -emeritus professor E. Employing the matrix method as a tool for the analysis of qualitative research data in the business domain. Rochester, NY; 2014. https://papers.ssrn.com/abstract=2495330. Accessed 10 Nov 2022

Communicating Results of Quantitative Research

45

Jane E. Miller

Abstract

In this chapter, I show how to apply expository writing techniques and principles for writing about numbers to communicate effectively about the results of quantitative research. Using examples from the biomedical literature, I demonstrate how to write a clear narrative with numbers as evidence, introducing the question, describing individual facts and patterns, and maintaining a focus on the topic and context at hand. The chapter starts with basic principles for writing about numbers including specifying the context and several dimensions of units. It then discusses how to choose and design complementary tools (prose, tables, charts, and maps) to communicate results, with guidance about how to make exhibits self-contained and how to organize numbers in those exhibits to match the associated narrative description. Next, the chapter introduces principles for comparing two or more numbers, including specifying direction, magnitude, and statistical significance, and how to summarize complex patterns. Those principles are demonstrated for presenting results of both bivariate and multivariate analyses, with examples of how to coordinate tables or charts with prose. The chapter ends by emphasizing the importance of conveying both the substantive and statistical significance of numeric findings.

Keywords

Communication · Effect Size · Multivariate Regression · Statistical Significance · Substantive Importance · Tables

45.1 Introduction

45.1.1 Overview of Communicating Quantitative Research Results

Writing about statistical analyses is a common task for biomedical researchers. Results of such analyses routinely inform decisions of medical practitioners and researchers, included in materials such as research papers, grant proposals, infographics and fact sheets about medications or treatment options, and in conference presentations. Despite this widespread need, few people are formally trained to write about quantitative information. Statisticians learn to calculate statistics and interpret the findings, but rarely are taught to describe them in ways that readers with less quantitative training can grasp. Communications specialists learn to write for varied audiences, but rarely are taught to deal with statistical analyses. Biomedical researchers

J. E. Miller (✉)
Bloustein School of Planning and Public Policy, Rutgers, The State University of New Jersey, New Brunswick, NJ, USA
e-mail: jem@rutgers.edu

G. Jagadeesh et al. (eds.), *The Quintessence of Basic and Clinical Research and Scientific Publishing*,
https://doi.org/10.1007/978-981-99-1284-1_45

bring expertise on their topic, but may have very little experience with statistics or communication.

In this chapter, I show how to apply expository writing techniques and principles for writing about numbers to communicate effectively about the results of quantitative studies. Using examples of topics from the biomedical literature, I demonstrate how to write a clear narrative with numbers as evidence, introducing the question, describing individual facts and patterns, and maintaining a focus on the topic and context at hand. I illustrate these principles with examples of "poor," "better," and "best" descriptions – samples of ineffective writing annotated to point out weaknesses, followed by concrete examples and explanations of improved presentation.

45.1.2 Audience Considerations

Results of quantitative biomedical research are of interest to a spectrum of audiences, including:

- physicians, pharmacists and other health practitioners, health policymakers, and other "applied audiences" who may have little statistical training but want to understand and apply results of statistical analyses about issues that matter to them;
- readers of biomedical journals, who often vary substantially in their familiarity with statistics;
- reviewers for a grant proposal or research paper, some of whom are experts on your topic but not the methods, others of whom are experts in statistical methods but not your topic;
- the general public, many of whom have no training in biomedical topics or statistical analysis but want to know what the results mean for their own health and health care.

Clearly, these audiences require very different approaches to communicating about quantitative research. Statistically trained readers will expect that you explain not only the methods and findings but also how those methods fit your question. Avoid common pitfalls such as overreliance on jargon and equations, labeling variables with acronyms from computerized databases, or writing about statistics in generalities without helping readers see how those statistics fit your specific topic and data. See Evergreen [1] and Miller [2], Chap. 13 for guidelines on how to adapt your writing to suit non-technical audiences, including formats such as fact sheets and infographics.

45.2 Fundamentals for Writing About Numbers

There are several fundamental principles for communicating quantitative information, including setting the context, and specifying several aspects of units.

45.2.1 Set the Context

Context is essential for all types of writing. Few stories are told without conveying "who, what, when, and where," or "the W's" as they are known in journalism. Without them your audience cannot interpret your numbers and will probably assume that your data describe everyone in the current time and place (e.g., the entire population of India in 2022). To set the context for the statistics you report, begin with a topic sentence that introduces the concepts and the W's, each of which requires only a few words or a short phrase that can be easily incorporated into the sentence with the numbers.

Poor: "37.7% of mothers bypassed their nearest facility for childbirth."

From this statement, readers cannot tell the date or place to whom the statistic pertains. Also, avoid starting a sentence with a numeral.

Better: "In 2014, 37.7% of all mothers bypassed their nearest facility for childbirth."

This version conveys the topic, subgroup, and date but not the location.

Best: "In 2014, just over one-third (37.7%) of all mothers in three selected districts within the state of Gujarat (Dahod, Sabarkantha, and Surendranagar) bypassed their nearest facility for childbirth" [3].

This statement conveys when, where, and what in a short, simple, sentence.

When writing a description of numeric patterns that spans several paragraphs, periodically reference the W's again. For longer descriptions, this will occur naturally as the comparisons you make vary from one paragraph to the next. In a detailed analysis of childbirth in facilities, you might compare different types of facilities at the same time and place, the same topic in other places or other times, and the use of facilities for childbirth compared to other types of health care. Discuss each of these points in separate sentences or paragraphs, with topic sentences to introduce the purpose and context of the comparison. Then incorporate the pertinent W's either into that sentence or the sentence reporting and comparing the numbers.

45.2.2 Specify Units and Categories

For quantitative information, it is essential to convey the units in which variables are measured and the categories in which they are classified, which is related to the level of measurement of each variable.

45.2.2.1 Level of Measurement

"Level of measurement" refers to the mathematical precision with which a variable is measured and thus which types of comparisons and calculations can be done with their values as well as how they are communicated in tables, charts, maps, and sentences (Table 45.1) [4]. Variables can be classified into one of four levels of measurement: nominal, ordinal, interval, and ratio, listed in order from least to most mathematically precise. Variables can be either "quantitative" (numerical) or "qualitative" (non-numerical). Qualitative variables capture attributes that differ in *nature* but not extent, so those attributes cannot be quantified.

Table 45.1 Definitions of levels of measurement, and types of mathematical operations that can be performed with their values[a]

		Level of measurement			
		Least precise --→ most precise			
		Nominal	Ordinal	Interval	Ratio
Definition		Named categories, *no* inherent # order	Ordered categories	Continuous, both – and + values possible	Continuous, values <0 *not* possible
Type of comparison	Math				
Same or different	= or ≠	YES	YES	YES	YES
Greater or less than	< or >	NO	YES	YES	YES
Subtraction	–	NO	NO	YES	YES
Division	÷	NO	NO	NO[b]	YES
		Qualitative	**Quantitative**		
		Categorical[c]		**Continuous**	

[a]All four types of comparisons can be conducted on measures of *frequency of occurrence* (counts or percentage distribution) of values of variables at any level of measurement. See Miller [4], Chap. 10

[b]Division is a sensible approach to comparing values of interval variables that assume either only positive *or* only negative values, resulting in positive ratios

[c]Numbers used as codes (abbreviations) for categories of nominal or ordinal variables cannot be used in calculations; see Miller [20], Chap. 4

Example: Blood type (e.g., Type O, Type B) is a *qualitative* variable because it captures a characteristic that differs in nature (*quality* or type) but not in quantity.

Blood type is classified based on the presence or absence of two antigens. We cannot say someone who has Type A blood has "less" of a blood type than someone who is Type AB.

Example: Weight in kilograms is a *quantitative* variable. We can figure out who is heavier and how much heavier by comparing the values of body weight for different cases.

Knowing the numeric value for a case, along with the units in which it was measured quantifies the value of "weight" for each case.

Categorical Variables

Categorical variables are those whose values represent separate categories or classifications, including those that describe some kind of state (status), condition, or situation, and those that group values into ordered ranges. There are two types of categorical variables: nominal and ordinal. "Nominal" variables capture some quality that can be *described* but *not quantified*, so they are qualitative variables. The characteristic being described is classified into named categories that do *not* have a natural order in terms of the extent or degree of the characteristic they are measuring.

Example: Eye color is identified as "brown," "blue," and other named colors.

There is no inherent order to those categories in terms of "eye color-ness" (the extent of having an eye color), therefor eye color is a nominal variable.

"Ordinal" variables measure attributes that can be classified into categories that have an inherent rank order in terms of the extent or degree of the characteristic they are measuring, but for which the distance between categories cannot be quantified. Thus, ordinal variables are quantitative categorical variables.

Example: Blood pressure is often classified into "normal," (systolic blood pressure (SBP) <120 mmHg *AND* diastolic blood pressure (DBP) <80 mmHg); "pre-hypertensive" (SBP 120 to 139 *OR* DBP 80 to 90); "Stage 1 hypertension" (SBP 140 to 159 *OR* DBP 90 to 99); and "Stage 2 hypertension" (SBP ≥160 *OR* DBP ≥100) [5].

Although there is a clear rank order to the blood pressure classifications, the width of the different ranges (in mmHg units) varies, so those categories are ***not*** *"one unit apart", making the variable ordinal, not continuous.*

Continuous Variables

Continuous variables are quantitative variables with values spaced equally from one another along a spectrum, with the concept measured in the units of that variable. They include the interval and ratio levels of measurement. Whereas interval variables can take on either positive or negative values as well as zero, for ratio variables, values below zero are not possible.

Example: Some biochemical reagents and biological material must be stored at temperatures below 0° Celsius to prevent the degradation of proteins, nucleic acids, whereas antibodies, enzymes, and cell culture media are stored at refrigerator temperatures (+4 °C) [6].

Temperature in degrees Celsius is an interval variable, with values both above and below zero possible. Although those values can be compared using subtraction, division ***doesn't*** *make sense (e.g., a ratio of −10 °C to +10 °C.)*

Example: Results of many physical examinations and lab tests such as blood pressure, pulse, or serum cholesterol can only take on positive values.

Variables measuring those phenomena are at the ratio level of measurement.

45.2.3 Units

There are several aspects of units that must be conveyed when communicating research results: the units of observation and the system and scale in which the variable was measured. Make a note of the units for any numeric facts you collect from other sources so you can use and interpret them correctly. Then incorporate units of observation and measurement into sentences with the pertinent numbers.

45.2.3.1 Unit of Observation

The first aspect of units to identify is the "unit of observation," also known as the "unit of analysis" or "level of aggregation." Common examples include individuals, families, institutions such as hospitals or schools, or geographic units such as districts or nations. Values for different levels of aggregation *cannot* be directly compared with one another, so ensure that any numbers you compare are for equivalent units of observation.

Example: In 2016, 73% of healthcare facilities in India had basic water services, meaning water was available from an improved source on the premises [7]. At about the same date, only 25% of households in India had drinking water on the premises and about 16% of rural households had piped water access [8].

*If access to clean water is measured as the percentage of facilities with such services, a **facility** is the unit of analysis. If it captures the percentage of households with water access, a **household** is the unit of analysis.*

Unit of observation is the only aspect of units that pertain to nominal variables.

45.2.3.2 System of Measurement

There are different systems of measurement for many things we quantify. Most of the world uses the metric system (e.g. weight in kilograms, distance in meters and kilometers), but the U.S., Britain, and some other countries use the Imperial system of measurement (e.g., weight in pounds, distance in feet and miles).

Example: "In 2015-16, per capita public spending on health care ranged from ₹400 to ₹1800 across districts within the state of Odisha, India (data from Chatterjee and Smith [9])."

This sentence conveys units of measurement (rupees), units of observation (per capita means "for each person"), and the W's to which the information pertains.

45.2.3.3 Scale of Measurement

Having identified which system of measurement is being used, specify the scale of measurement. "Scale" (or "order of magnitude") refers to multiples of units, with each order of magnitude *ten times* as large as the next lower order of magnitude. This aspect of units is especially important when labeling tables and charts because very large or very small values are often rounded to other scales to reduce the number of digits and decimal places.

Example: For an 82 kg adult male, the lethal dose of acetaminophen is 27.2 g, that for fentanyl is 2.5 mg, and for carfentanil, 41 μg [10].

*The lethal dose for acetaminophen is measured in **grams** (g.), which is an order of magnitude higher than that for fentanyl (**milligrams**; mg.) which is almost an order of magnitude higher than that for carfentanil (**micrograms**; μg). Clearly conveying the relevant scale for each of those drugs is vital for correctly communicating their respective fatal dosages.*

Example: Population for districts within a country might be reported in millions of persons, whereas the global population is typically reported in billions.

For numeric values to be compared, both the system of measurement and scale of measurement must be consistent with one another. E.g., if one study measured body weight in kilograms and another study measured weight in pounds and ounces, the data must be converted into the same system of measurement to compare the different values with one another.

45.3 Tools for Presenting Numbers

45.3.1 Choosing Tools

An important early consideration when writing about quantitative research is which tool or combination of tools you will use to convey each component of your results. The main tools—prose, charts or visualizations, tables, and maps—have different, albeit overlapping, advantages and disadvantages. Prose (sentences and paragraphs) is the most effective for describing patterns and explaining how numeric results answer the research question. However, prose is a very inefficient way to organize a lot of numbers (such as mean values for each of the 20 locations and several outcomes) because the numbers get lost among the words needed to report the units and specify the context for each number.

In contrast, tables and charts are excellent ways to organize many numbers. Tables use a grid to present numbers in a predictable way, guided by labels and notes within the table, whereas charts use axis labels, legends, and notes to accomplish those tasks. Tables are preferred to charts for presenting *precise* numeric values, such as if readers will be using your paper as a source of data for comparing with other studies, locations, or dates, or conducting a different type of computation. A chart is very effective at portraying the *general shape and size* of a pattern, making it easy to observe whether trends are rising or falling, or which values exceed others and by how much; however, it is difficult to see precise values from a chart [2].

Example: Table 45.2 and Fig. 45.1 present data on three indicators of socioeconomic disadvantage for three districts of Gujarat, India. Both the table and chart compare the percentage of the population that is rural [11], is below the poverty line [11], or is illiterate [12].

It is much easier to see that the rank order of the three districts is ***different*** *for each of those indicators from the chart than from the table. For instance, Dahod had the highest percentage of each of the three outcomes, Surendranagar was second highest in terms of illiteracy and poverty, whereas Sabarkantha was second highest in percentage rural.*

However, if a reader wanted to conduct calculations within those data or in conjunction with data on other districts or time periods, they would need the precise values shown in the table. For instance, in 2015, the poverty rate in Dahod was 2.18 times as high as that in Sabarkantha.

Maps are by far the most effective way to convey spatial patterns, whether numbers, rates or other information for geographic units, location of particular features such as medical facilities, or distance from such features.

Example: From Fig. 45.2, it is evident that high rates of poverty are concentrated in the southwestern districts of Odisha, India, whereas the lowest rates are in the eastern-central districts. *The map conveys the clusters of adjacent high- and low-poverty areas far more clearly than either a table or chart could.*

Table 45.2 Characteristics of the study districts Gujarat, India

Characteristics	Dahod	Sabarkantha	Surendranagar (%)
% rural[a]	91	85	72
% < poverty line[b]	72	33	47
% illiterate[a]	41	24	28

Adapted from Salazar et al. [3], Table 1, with permission
[a]Sample Registrar of India. Districts of Gujarat Socio Economic Survey 2002-03 [11]
[b]Commissionerate of Rural Development, Gujarat. 2015 [12]

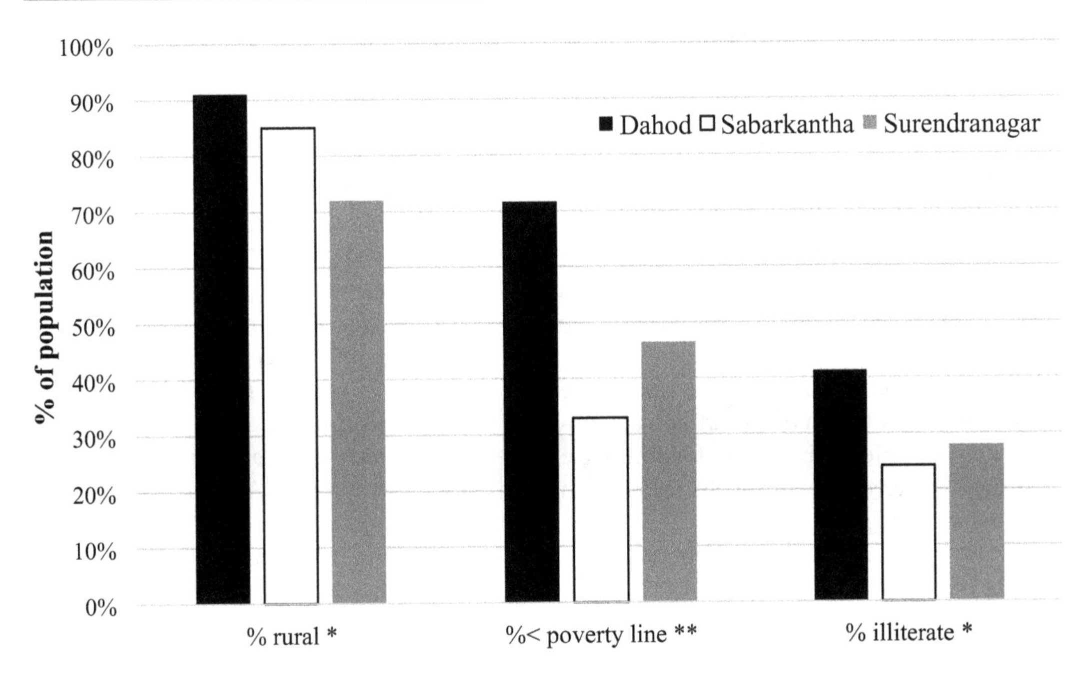

Fig. 45.1 Socioeconomic characteristics of study districts, Gujarat, India. **Notes**: (a) Sample Registrar of India. Districts of Gujarat Socio Economic Survey 2002-03 [11]. (b) Commissionerate of Rural Development, Gujarat. 2015 [12]. (Data from Salazar et al. [3], Table 1, with permission)

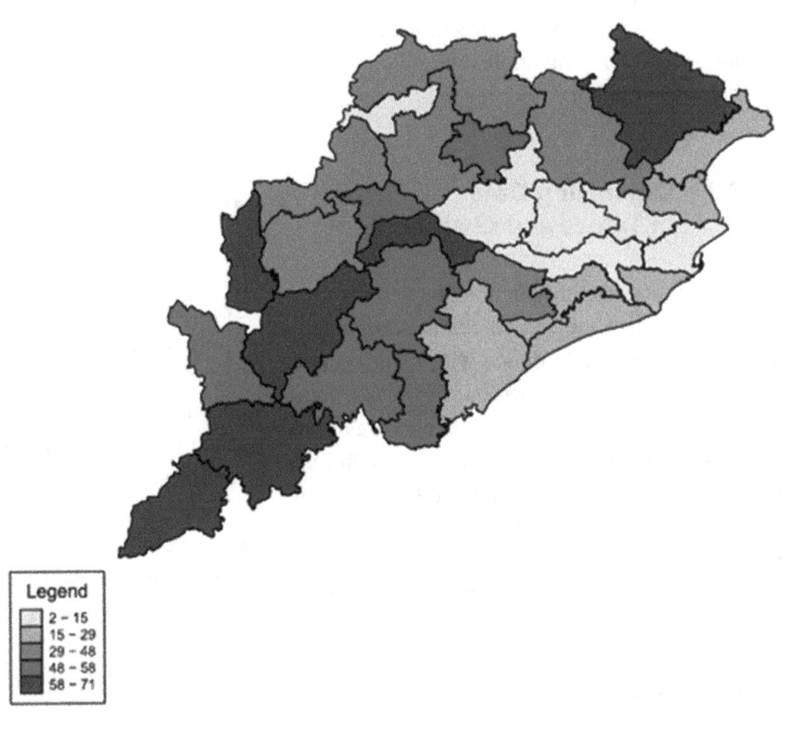

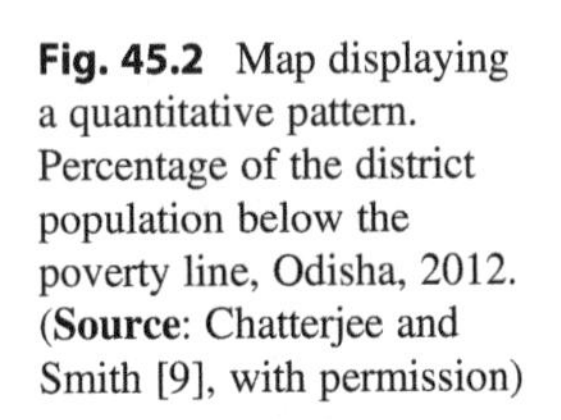

Fig. 45.2 Map displaying a quantitative pattern. Percentage of the district population below the poverty line, Odisha, 2012. (**Source**: Chatterjee and Smith [9], with permission)

45.3.2 Creating Self-Contained Exhibits

Each exhibit – whether a table, chart, map, or diagram – should be designed so that readers can interpret it without reference to the text.

- Write a title to reflect the specific contents of that specific table, chart, or map. Use the W's as a checklist to ensure that you identify the topic, context, and methods for that exhibit.
- In the row and column labels of a table or the axis labels and legend of a chart or map, provide a short, informative phrase to identify the concept captured by each variable, along with its units or categories.
- Use footnotes to define terms, abbreviations, and symbols, and to identify the data source.

Table 45.3 is an example of a well-constructed table of multivariate regression results, annotated with footnotes to point out how the layout and labeling make it easy to interpret every number in the table using only the information therein.

See Miller [13], Chaps. 5 and 15 for additional guidelines and examples of structure and labeling of tables to present univariate, bivariate, and multivariate information.

45.3.3 Organizing Data in Tables and Charts

Another important, but often neglected, aspect of creating effective tables and charts is organizing them in ways that coordinate with the associated prose. Most research papers aim to make points about either substantive (theoretically organized) patterns or empirical patterns, or some combination of those criteria, thus the exhibits that contain the associated quantitative information should be organized to be consistent with those objectives. Unless the purpose of a document is to make it easy for readers to look up information for their own calculations, rarely should either alphabetical order or order of items from a questionnaire or other source document be used to organize data [14].

Example: As of early August 2022, Europe had the highest number of reported COVID-19 cases, followed by the Americas, Western Pacific nations, South-East Asia, Eastern Mediterranean, and Africa [15].

To coordinate with that description, Fig. 45.3a portrays the WHO regions in descending empirical order of case counts on the vertical axis of the chart. Listing them instead in alphabetical order (Fig. 45.3b) would require the reader to visually jump around the chart to follow the rank order listed in the description.

Example: The informational inserts that accompany medications list possible adverse effects grouped first into life-threatening symptoms and less urgent ones, then in descending order of frequency *within* those groups.

The first thing consumers need to know is whether a symptom they are experiencing requires immediate medical attention or can wait (a theoretical criterion). Within each of those categories, organizing most- to least-frequent helps users find the symptom more quickly (empirical order).

45.4 Comparing Two or More Numbers

Comparing two or more numbers is an extremely common task when writing about quantitative findings, whether contrasting a couple of numeric facts in an introduction or conclusion section, or describing the relationship between two or more variables in the results section. This section presents four principles to help you write more effectively about such contrasts: (1) reporting and interpreting numbers, (2) specifying the direction and magnitude of an association, (3) conveying results of inferential statistical tests, and (4) summarizing patterns involving many numbers.

Table 45.3 Coefficients from OLS regression of child's birth weight in grams on father's and grandmother's co-residence with mother, KwaZulu-Natal, South Africa, 2000–2003[a]

	Unadjusted		Adjusted[b,c]	
Variable[d]	Coeff.	Std. err.[e]	Coeff.	Std. err.
Father's survival and residence (not co-residing with mother[f])[g]				
Co-residing with mother	138.2	(23.1)**	58.6	(25.5)*[h]
Deceased	14.5	(45.9)	23.8	(45.3)
Maternal grandmother's survival and residence (not co-residing with mother)				
Co-residing with mother	−88.1	(17.8)**	−8.9	(20.1)
Deceased	−57.5	(24.4)*	−45.8	(24.2) +
Bio-demographic characteristics[i]				
Male[j]			98.6	(16.3)**
First-born child			−99.8	(22.1)**
Mother's age at child's birth (20–24 years)				
<20			2.2	(23.1)
25–29			72.3	(24.0)**
30–34			68.7	(29.9)*
35+			100.1	(33.4)**
Social and economic characteristics				
Mother's education (high school)				
No education			7.6	(41.4)
Primary school			−24.6	(22.8)
Higher education			-25.1	(46.8)
Mother's travel (spends most nights in the homestead)				
Mother spends time away regularly	−35.0	(26.8)		
Household wealth and shocks				
Wealth quintile (1st = poorest)[k]				
2nd			58.1	(26.4)*
3rd			90.1	(36.1)*
4th			69.3	(28.4)*
5th = wealthiest			83.2	(32.0)**
Experienced economic shock			−45.7	(24.2) +
Experienced health shock			12.39	(23.7)
Constant			3148.18	(49.12)**
R^2	0.01		0.05	

Adapted from Cunningham et al. [17], Table 2, with permission

Annotations about table structure and labelling

[a] The table title mentions (a) the type of model (OLS), (b) the concept measured by the dependent variable and its units, and (c) who, when, and where were studied

[b] Includes adjustments for missing values, isigodi (traditional administrative unit) and year-of-birth. **Notes**: N = 3993. Omitted category in parentheses. +p < 0.10; *p < 0.05; **p < 0.01. Data from Africa Centre Demographic Information System, KwaZulu-Natal, South Africa: children born 2000–2003

[c] A footnote specifies what was controlled for in the adjusted model

[d] Each variable is labeled with a brief phrase that clearly identifies the concept being measured

[e] The type of inferential statistical information is labeled

[f] Each category of the variables is labeled to clearly convey the nature of that classification

[g] A footnote explains that the omitted category of each is shown in parentheses

[h] The meaning of the symbols for levels of statistical significance are identified in a footnote

[i] Subheadings within the column of variables identify thematic groupings of variables in the analysis

[j] Gender (and other 2-category variables) are labeled to make it clear which category is being modeled

[k] The rank order of wealth quintiles is specified as from poorest to wealthiest

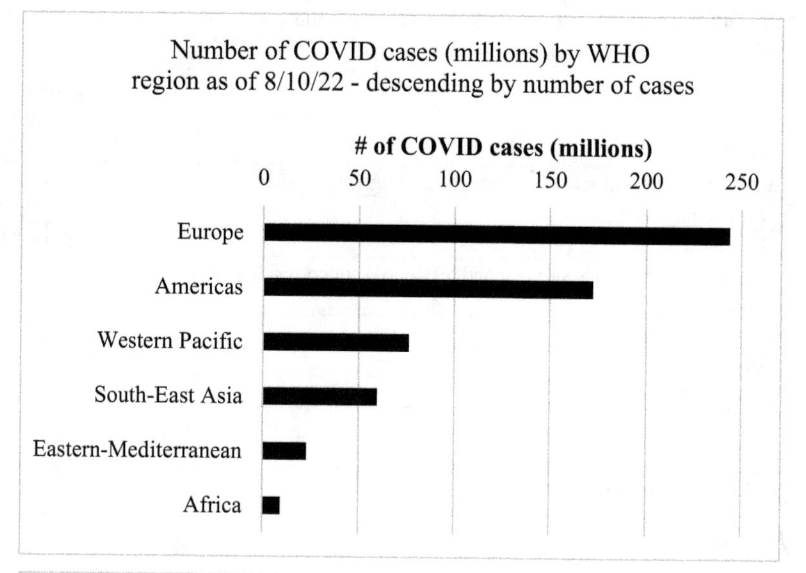

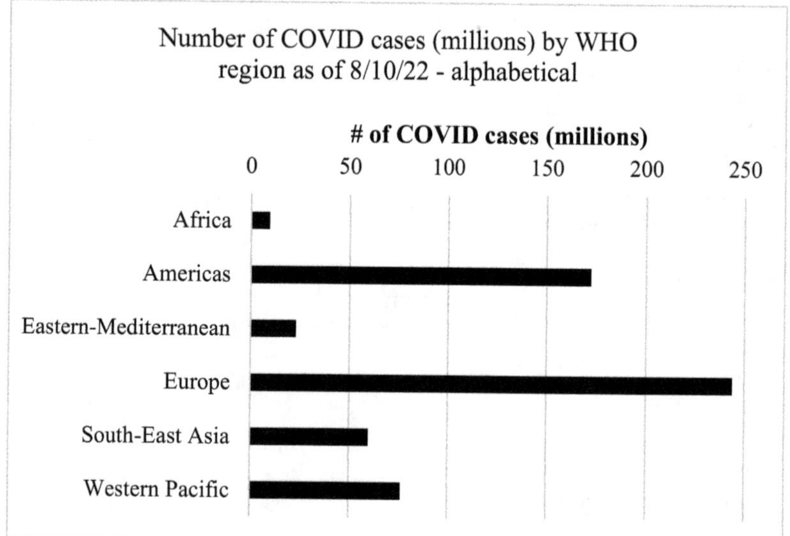

Fig. 45.3 Alternative ways of organizing data on the number of cases of COVID by region. (**a**) descending by number of cases; (**b**) alphabetical. (**Notes**: Data from WHO [15], with permission)

45.4.1 Report and Interpret

As you present evidence to describe a pattern or test a hypothesis, both report and interpret the numbers. Reporting the numbers is an important first step: By including the numbers in the text or exhibit, you give your readers the data with which to perform additional calculations or to compare your numbers with that for other times, places, or groups.

However, if you stop there, you leave it to your readers to figure out how those data answer the research question. An isolated number that has not been introduced or explained leaves its explication entirely to your audience. Those who are unfamiliar with your topic are unlikely to know which comparisons to make or to have the information for those comparisons immediately at hand. Consequently, after reporting the raw data, interpret it. To help readers grasp the meaning of the numbers you report, provide the relevant data and explain the results of the comparisons. Consider an introduction to a report on poverty in India.

Poor: "In 2018-19, gross domestic product (GDP) in Odisha, India was about 6.9 billion U.S. dollars (data from [9])."

From this sentence, it is difficult to assess whether GDP was high or low, stable or changing quickly. To most people, $6.9 billion sounds like a lot of money, but a key question is "compared to what?"

Better: "In 2018-19, per capita gross domestic product (GDP) in Odisha, India was about 1,450 U.S. dollars."

This simple translation of total GDP into a per capita figure takes a large number that is difficult for many people to fathom and converts it into one they can relate to. However, to readers in America (in whose monetary units GDP is reported), $1,450 sounds very low. If readers knew the poverty threshold in India, they could do a benchmark calculation, but you will make the point more directly if you compute it for them.

Best: [To follow the preceding sentence.] "Between 2005 and 2012 (the latest figure available at the time), the percentage of the Odisha population in poverty fell from 59% to 33%, the fastest among all Indian states [9].

This description conveys the implications of that per capita GDP figure by showing how it maps into the poverty rate, describes the trend over time, and puts it in geographic context.

Although it is important to interpret quantitative information, it is also essential to report the original data upon which they are based. If you only present a ratio or percentage change, for example, you will have provided an incomplete picture. Suppose a report stated that the poverty rate in India declined by 45% over the past 10 years, but did not report the poverty rate for either year. A 45% decrease is consistent with many possible combinations: from 18% in the earlier year to 10% ten years later, or from 90% to 50%, for example. The first pair suggests a very low incidence of poverty, the last pair an extremely high rate. Unless the rates themselves are mentioned, readers can't determine whether India has nearly eradicated poverty or continues to face a substantial poverty problem; nor could they compare India's current poverty rate with those from other times or places.

45.4.2 Specify Direction and Magnitude

To describe an association between two variables, explain both its shape and size rather than simply stating whether they are correlated. In other words, *which* value is higher and how much higher? For instance, in which district is poverty highest (direction)? *How much* higher than in adjacent districts (magnitude)? Is poverty rising, falling, or level across time (direction)? Was the decline in poverty steep or shallow (magnitude)?

45.4.2.1 Direction

The wording for conveying the direction of association depends on the level of measurement of the variables. For nominal independent variables such as blood type or region, describe the direction of association by specifying which category has the highest or lowest value of the dependent variable, and how other categories compare.

Poor: "Type of residence was negatively associated with vaccination status in Ethiopia in 2006."

Type of residence and vaccination status are both nominal variables. Thus the shape of their association ***cannot*** *be described using wording that implies an inherent order to their categories ("negatively associated").*

Better: "In Ethiopia in 2006, children living in urban areas were more likely to have completed all four vaccinations (Table 45.4)."

This version names the group of interest, date, and country, but does not specify the comparison group. Since Table 45.4 also reports vaccination statistics for India, it would be easy to mistakenly conclude that the comparison was between urban areas in Ethiopia and urban areas in India rather

Table 45.4 Prevalence of complete vaccination, by parental literacy, residence, and presence of community health workers, Ethiopia and India, 2006

	% completed all 4 vaccinations	
	Ethiopia	India
Maternal literacy		
Illiterate	85.9	95.0
Literate	95.2	95.2
Paternal literacy		
Illiterate	83.7	94.6
Literate	93.3	95.5
Type of residence		
Rural	87.9	88.4
Urban	91.7	97.3
Community health workers		
No	85.8	93.5
Yes	91.3	96.6

The four types of vaccinations included BCG, MCV, DPT3, and polio
Adapted from Lee et al. [16], Table 1, with permission

than across residential locations within Ethiopia.

Best: "In Ethiopia in 2006, children living in urban areas were more likely than those in rural areas to have completed all four vaccinations (Table 45.4) [16]."

This version conveys direction by identifying which type of residential location had the higher vaccination rate, and naming the category against which it is being compared.

For quantitative (ordinal, interval, or ratio level) variables such as age group or year for which the values have an inherent numeric order, phrasing for direction of association can take advantage of that order. Patterns involving quantitative variables can be described in terms of positive (direct) or negative (inverse) associations.

Poor: "Mother's age and birth weight were correlated."

This sentence conveys the concepts and that their values move together, but does ***not*** *specify either the direction or magnitude of that pattern.*

Better: "As the mother's age at the time she gave birth increased, average birth weight also increased (Table 45.3) [17]."

Better, version 2: "Mother's age at birth was positively associated with her child's birth weight."

Both of these "better" versions work! Mother's age at birth is a quantitative variable, so we don't have to name and compare specific categories to convey the direction of association between age and birth weight.

45.4.2.2 Magnitude

An association can be large – a given change in one variable is associated with a big change in the other variable – or small – a given change in one variable is associated with a small change in the other. For bivariate associations, describing the size of an association is quite straightforward. The size of a difference between two values can be calculated in any of several ways, including subtraction, division, or percentage difference or change. In the biomedical literature, ratios and percentage differences are widely used. As you describe the sizes of differences between values, mention the pertinent units, whether the original units of measurement (as in subtraction) or multiples (as for ratios). See Miller [2, 13] for suggestions on how to choose among and write about the results of these different types of quantitative comparisons.

Poor: "In both Ethiopia and India, child vaccination rates varied by maternal literacy (Table 45.4)."

This sentence doesn't convey which literacy level had a higher vaccination rate – in other words, it fails to express either the direction or magnitude of the association.

Better: "In both Ethiopia and India, children of literate mothers were more likely than those whose mothers were illiterate to have completed their vaccinations (Table 45.4)."

Although this version conveys the direction of the associations, the size (magnitude) of the association is still unclear.

Best: "In Ethiopia, children of literate mothers were about 11% more likely than those whose mothers were illiterate to have completed their vaccinations. Although having a literate mother was also associated with higher child vaccination in India, the advantage was less than 1% (Table 45.4) [16]."

This version explains that although the ***direction*** *of the association between literacy status and vaccination was consistent in the two countries, the* ***magnitude*** *was quite different: In India, the vaccination difference by literacy status was trivially small, whereas the corresponding disparity in Ethiopia was large enough to merit attention in programs seeking to improve vaccination rates. In situations like these, failing to convey the size of the difference can yield misleading conclusions about the importance of that pattern for informing program design.*

45.4.3 Statistical Significance

Statistical significance is a formal way of assessing whether observed associations are likely to be explained by chance alone, based on the way the study sample was selected from the population [18]. The *p*-value measures the probability of *incorrectly* concluding that the true population parameter is equal to the null value based on the sample statistic. Inferential statistical results can be reported as standard errors of the estimates, test statistics, confidence intervals (CI), *p*-values, or symbols for levels of statistical significance (Miller [13], Chap. 11). The conventional significance level (α) is 0.05, which corresponds to a 95% confidence interval.

Example: In Ethiopia, having a literate mother is associated with statistically significantly higher odds of complete vaccination (adjusted odds ratio (AOR) = 4.84; 95% CI: 1.75 to 13.36; $p < .01$; Table 45.5), but having a literate father is *not* (AOR = 1.28; 95% CI: 0.79 to 2.08).

In Table 45.5, inferential statistical information is reported both as the confidence interval around each point estimate, and with symbols to denote which results are statistically significant at the $p < 0.01$ *and 0.05 levels. For odds ratios, the null value ("no difference between groups") = 1.0. Since the lower and upper 95% confidence limits for the AOR for literate mothers are both above 1.0, that result* ***is*** *statistically significant at* $p < .0.05$*. However, the 95% CI for the AOR for literate fathers includes values below and above 1.0, so the observed AOR is* ***not*** *statistically significantly different from 1.0 (because the CI encompasses the "no difference in odds" value). See below for more on odds ratios.*

Note that I mention statistical significance *after* direction and magnitude because many authors overlook those two essential aspects of statistical results while emphasizing solely statistical significance; see Substantive Significance below, and Chaps. 29 and 30 in this book.

Table 45.5 Adjusted odds ratios (and 95% confidence intervals) from multilevel logistic models of the association between maternal literacy and child completion of four vaccinations in Ethiopia, children aged 4–5 years old, 2006–7

	Adjusted odds ratios (95% confidence interval)
Fixed effects	
Maternal literacy (illiterate)	
Literate	4.84 (1.75 to 13.36)**
Paternal literacy (illiterate)	
Literate	1.28 (0.79 to 2.08)
Presence of community health care workers (none)	
Yes	1.80 (0.72 to 4.46)
Interaction: Mother literate × Presence of community health care workers	0.29 (0.09 to 0.96)*
Random effect	
Community variation (Std. error)	0.331 (0.222)
Intraclass correlation	9.1

Notes: The four types of vaccinations included BCG, MCV, DPT3, and polio
Reference categories are shown in parentheses
*: $p < 0.05$. **: $p < 0.01$
All models were additionally adjusted for maternal age, gender of the child, sibling status, wealth level, type of residence, and region
Adapted from Lee et al. [16], Table 2, with permission

45.4.4 Summarize Patterns

Often answering a research question requires describing a pattern involving a relationship among two or three variables, such as comparison of several dimensions of socioeconomic status across multiple geographic areas, or time trends in each of several diseases. Typically, the data for such patterns are reported in tables and charts, which provide an efficient way to organize lots of numbers. However, if you only report those numbers in an exhibit, readers have to figure out for themselves what that evidence says. An important step in communicating your results is to summarize the patterns shown in your exhibits, and relate them back to the main research question.

When summarizing the pattern portrayed in a table or chart, writers often make one of two opposite mistakes: (1) they report every number from that exhibit in their description, or (2) they describe a few arbitrarily chosen numbers to contrast without considering whether those numbers represent an underlying general pattern. Neither of those approaches adds much to the information presented in the exhibit, and both can confuse or mislead readers. As you write, describe the big picture, rather than reiterating all of the little details.

To guide you through the steps of writing an effective description of a pattern involving three or more numbers, I devised the "generalization, example, exceptions" approach; abbreviated "GEE." The aim is to identify and describe the general shape of a pattern, illustrate that pattern using a representative numeric example, and then explain and illustrate any exceptions.

45.4.4.1 Generalization

For a generalization, come up with a description that characterizes a relationship among most, if not all, of the numbers. If the pattern fits most of the data points, it is a generalization. The few numbers or associations that don't fit are exceptions (see below).

Generalization: "In each of the three study districts in 2002, the percentage rural was higher than the percentage poor, followed by the percentage illiterate (Fig. 45.1)."

This sentence introduces a geographic comparison, names the concepts (three dimensions of socioeconomic disadvantage), states the rank order of those *phenomena, and specifies the figure in which the pattern is portrayed. No numbers yet – that comes in the second step. That pattern characterizes all three districts, so it is a generalization. To frame all three dimensions in terms of **disadvantage**, I reported the percentage **illiterate** instead of % literate (as did the original article by Salazar et al.* [3]*). I also re-ordered the outcomes in descending order of prevalence to support an empirical description.*

45.4.4.2 Example

Having described a generalizable pattern, illustrate it with numbers from the associated exhibit. This step ties your generalization to the specific numbers upon which it is based. It links together the narrative and exhibit. By reporting a few illustrative numbers, you show your audience where in the exhibit those numbers come from as well as explaining the comparison involved. Readers can then assess whether the pattern applies to other times, groups, or places using other data from the exhibit. To illustrate the above generalization:

Example: "For example, residents of Dahod were more than twice as likely to live in rural areas (91%) as to be illiterate (41%)".

This sentence reports the direction and magnitude of one relationship (example) that fits the generalization introduced above.

45.4.4.3 Exceptions

Sometimes your generalizations will capture all the relevant variation in your data. However, where biomedical and social phenomena are concerned, often there will be important exceptions to the generalization. Substantively small discrepancies can usually be ignored, but if some parts of a pattern depart substantially from your generalization, they should be described.

To portray an exception, explain its overall shape and how it *differs* from the generalization you have described and illustrated in your preceding sentences. Is it higher or lower? By how much? If a trend, is it moving toward or away from the pattern you are contrasting it against? In other words, describe both the direction and magnitude of the difference between the generalization and the exception. Finally, provide numeric examples from the table or chart to illustrate the exception. Use phrases such as "however" or "on the other hand" to differentiate an exception from a generalization; "conversely" or "on the contrary" can be used to point out when one pattern is in the opposite direction of another.

Exception [To follow the above generalization and example]: "However, the rank order of the three districts varied depending on the socioeconomic outcome: Although Dahod had the highest rates of each of the three outcomes, Surendranagar ranked second highest in terms of illiteracy and poverty, but third (behind Sabarkantha) in terms of percentage rural.

*By using the words "However" and "although," this description helps readers to see the different rank orders of the three districts for the different measures of socioeconomic disadvantage. This is an exception in **direction**.*

Example to quantify the exception: "Dahod's disadvantage was much larger for poverty and illiteracy than for percentage rural: Residents of Dahod were more than twice as likely as those in Sabarkantha to be poor or illiterate, but only about 10% more likely to live in rural areas.

*The first phrase conveys that although the direction of association is the **same** for the three indicators of disadvantage, the sizes of those disparities are quite **different**. The second phrase illustrates that point by describing the results of calculations that quantify the size of the inter-district gap in each of those indicators.*

Exceptions can also be in terms of statistical significance, as when all but one association is statistically significant (not shown). For a step-by-step guide to arriving at a GEE, see Miller [2], Appendix A.

45.5 Reporting Results of Multivariate Regressions

The same general principles for reporting bivariate associations apply equally to describing the results of multivariate statistical models. In the next few paragraphs, I provide a quick overview of how to interpret coefficients from ordinary least squares (OLS) and logistic regression. For an excellent intuitive description of multivariate regression, see Allison [19]. For more in-depth guidelines about how to write about such models, see Miller [13], Chaps. 9 and 15, or Miller and Rodgers [20].

Research journals will expect detailed coefficients and statistical significance information from multivariate models to be reported in a table. Format your tables following the guidelines earlier in this chapter for effective titles, row and column labels, interior cells, and footnotes, along with these special considerations for multivariate tables, which are demonstrated in Table 45.3:

- In the title, name the type of statistical model, the concept captured by the dependent variable along with its units and/or category name, and a phrase summarizing the concepts measured by the independent variables.
- Label each column of effect estimates to convey whether it contains standardized or unstandardized OLS coefficients, log-odds or odds ratios, etc.
- Label columns of statistical test results to identify which type of information is presented—test statistic (name the specific type!), standard error, *p*-value, confidence interval, etc.

45.5.1 Ordinary Least Squares Regression

Ordinary least squares regression ("OLS," also known as "linear regression") is used to estimate the association of one or more independent variables with a continuous dependent variable such as birth weight in grams or blood pressure in mm Hg. Coefficients (also known as "effect estimates," "parameter estimates", or "β's") on continuous and categorical independent variables are interpreted differently from one another.

For a continuous independent variable, the unstandardized coefficient[1] from an OLS regression is an estimate of the slope of the relationship between the independent and dependent variables: the marginal effect of a one-unit increase in that independent variable on the dependent variable, holding constant all other variables in the model. In tables and exhibits that report OLS coefficients, specify the units in which the dependent and independent variables are measured. Unstandardized OLS coefficients are in the same units as the dependent variable: "For each additional child in the household, the family experienced on average 0.05 fewer criminal victimizations ($p < 0.05$) [21]."

With categorical variables such as age group or district, one category is selected as the reference (or "omitted") category for the regression model, and is the basis of comparison for the other categories of that variable.[2] The OLS coefficient measures the absolute difference in the dependent variable for the independent variable category of interest compared to the reference category, taking into account the other variables in the model. In the text, specify the reference group as you interpret the coefficients, and mention the units of the dependent variable.

Example: "Infants born to mothers aged 35 years or older weighed on average 100 g more at birth than did children born to mothers aged 20–24 years ($p < 0.001$; Table 45.3) [17]."

The sentence conveys the dependent variable (birth weight) and its units (grams), the independent variable (age group) and the categories of that variable that

[1] A standardized coefficient estimates the effect of a one-standard-deviation increase in the independent variable on the dependent variable, where that effect is measured in standard deviation units of the dependent variable [13].

[2] Dummy variables (also known as "binary," "dichotomous," or "indicator" variables) are defined for each of the other categories, each coded 1 if the characteristic applies to that case, and 0 otherwise. A dummy variable is not defined for the reference group (hence the name "omitted category"), resulting in (n – 1) dummies for an n-category variable [13].

are being compared, as well as the direction, magnitude, and statistical significance of the pattern.

To reduce the chances of misinterpreting coefficients on categorical variables, label the dummy variables in your data set, text, and exhibits to reflect the values they represent, e.g., "Male," not "Gender."

45.5.2 Logistic Regression

Logistic regression (also known as a "logit model") is used to estimate the effects of several variables on a categorical dependent variable such as bypassing a local facility for childbirth or completing a series of vaccinations [22]. The coefficient for a continuous independent variable in a logit model measures the marginal effect of a one-unit increase in that variable on the log-odds of the modeled category of the dependent variable—controlling for the other variables in the model. However, many readers find it easier to assess effect size from logit models using the odds ratio (abbreviated OR) rather than the log-odds because the units of an OR are interpretable as simple multiples of odds, or can be easily transformed into percentage differences (Miller [13] Chap. 9). Consequently, the odds ratios for each independent variable are often presented instead of the log-odds,[3] as in Table 45.5.

An adjusted odds ratio (AOR) on a continuous independent variable is an estimate of the change in relative odds of the outcome for a one-unit increase in the independent variable, taking into account (adjusted for) the other variables in the model.

Example: "For each additional year of mother's education, her odds of bypassing a local facility for childbirth increased by 5% ($p < 0.05$; not shown [3])."

*The sentence names the concept captured by the independent variable (educational attainment) and its units (years), and the category of the dependent variable being modeled (those who **did** bypass a facility). Mother's education is a continuous independent variable, measured in years. The adjusted odds ratio is 1.05 which translates into a 5% increase in odds for each additional year of schooling.*

For a categorical independent variable, the odds ratio compares the odds of the modeled category of that variable to the odds of that outcome for those in the reference category.

Example: "Women who experienced complications during pregnancy had nearly 1.9 times the odds of bypassing a local facility for childbirth, compared to those who did not experience pregnancy complications (NS; not shown [3])."

Experiencing pregnancy complications is a categorical (yes/no) variable, so the interpretation of the adjusted odds ratio conveys both the group of interest ("yes complications" group, in the numerator of the AOR) and the reference category ("no complications" group, in the denominator of the AOR). The AOR is greater than 1.0, reflecting higher odds of the outcome in the numerator group than in the denominator group.

Note that the example sentence avoids referring to "numerator" and "denominator" which I explain here as important behind-the-scenes info that the readers do not need to read about!)

Example: "Women residing in Dahod had only about two-thirds the odds of bypassing a local facility for childbirth as did residents of Sabarkantha (AOR = 0.65), but the difference was not statistically significant (not shown [3])."

*An AOR < 1.0 corresponds to **lower** odds in the group of interest (numerator) than the reference group (denominator). Concepts (independent variable and dependent variable), categories, direction, magnitude, and statistical significance, all in one short, simple sentence.*

[3] The odds ratio is calculated by exponentiating the logit coefficient: odds ratio = $e^{\beta} = e^{\text{log-odds}}$.

45.5.3 Net Effect of an Interaction

Statistical interactions, also known as "effect modifications," occur when the effect of one independent variable on the outcome depends on the value of a second independent variable. In a multivariate regression, interactions are specified using separate terms for the "main effect" of each independent variable and other term(s) for their interaction (Miller [13], Chap. 16). However, the coefficients on those terms cannot be interpreted in isolation from one another: To determine the overall effect of the interaction requires combining the coefficients (for OLS regression) or the odds ratios (from logistic regression) for the pertinent main effects and interaction terms for each **combination** of the two IVs. Perform those computations behind the scenes, presenting and interpreting the overall shape of the pattern in a chart and accompanying text rather than expecting them to do the calculations while reading your paper.

Example: Lee et al. investigated whether the association between maternal literacy and completing the series of four vaccinations was *different* for women who lived in a district with community health workers (CHW) than for those whose area lacked such workers. They found such a pattern in Ethiopia, but not in India. The AORs for the main effect of maternal literacy is 4.84, the main effect of community health workers is 1.80, and the interaction term is 0.29 (Table 45.5) [16].

For women in the reference category for ***either*** *of the independent variables involved in the interaction (literate but live in an area with no CHW,* ***or*** *illiterate but live in an area with a CHW), the odds ratio is simply that on the other variable: 4.84, and 1.80, respectively, when each is compared with illiterate residents of areas without CHW. However, for women who are* ***both*** *literate* ***and*** *live in an area with a CHW (not in either reference category), calculating the odds ratio involves multiplying the two main effects ORs with the interaction OR = 4.84 × 1.80 × 0.29 = 2.53.*

When we calculate the interaction by mathematically combining those AORs and presenting them graphically (Fig. 45.4), the shape of the overall pattern becomes apparent.

Figure 45.4 includes a bar for each of the four possible combinations of literacy and the presence of a community health worker. The two categories of literacy (literate, illiterate) are labeled on the x-axis, while the two categories of the presence of CHW (yes/no) are identified in the legend. Women who are illiterate ***and*** *living in a community without CHW are the reference category, so the odds ratio for that group is by definition 1.0. The relative heights of the bars convey the predicted OR for each group. No need for readers to do the math and then compare three odds ratios in their heads!*

The "generalization, example, exception" technique is an extremely effective way to summarize the shape of an interaction pattern.

Example: "As predicted, and children of literate women were more likely than those born to illiterate women to have received all four vaccinations, regardless of whether or not they lived in areas that had community health workers (CHW; Fig. 45.4). However, the relationship between the presence of CHW and vaccination completion varied by maternal literacy. Although among illiterate women, those in areas without CHW were about 1.8 times as likely as those with CHW to have complete vaccination, among literate women, those in areas without CHW were only about half as likely as those with CHW to have complete vaccination (OR = 2.53 and 4.84, respectively, when each was compared with the reference category)."

The first sentence conveys a generalization about the relationship between one independent variable (maternal literacy) and the dependent variable (completion of the vaccination series), and refers to the figure in which the pattern is portrayed. By using the word "however," the second sentence introduces the exception: that the

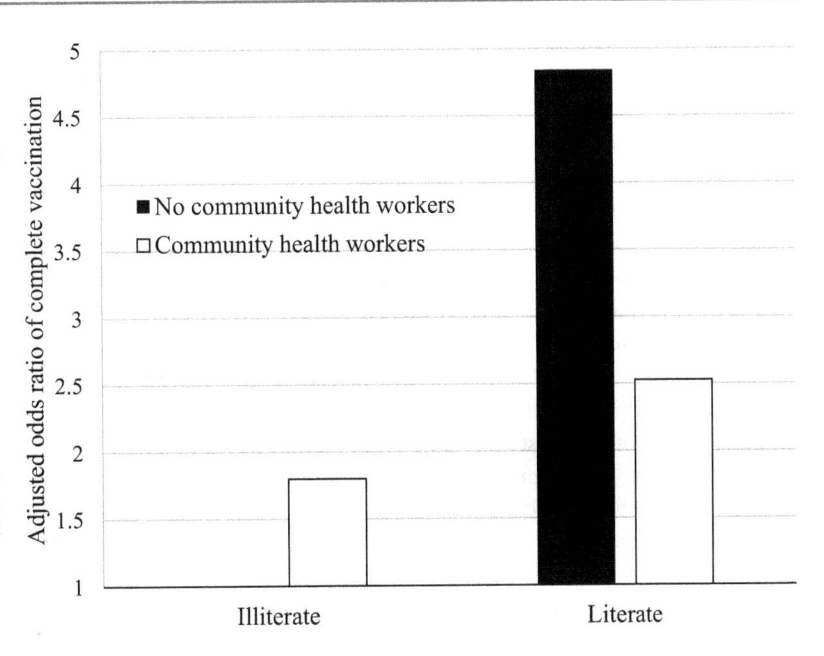

Fig. 45.4 Net effect of interaction between maternal literacy and presence of community health care workers on odds of child completion of four vaccinations, Ethiopia, 2006-7. (**Notes**: Based on estimates from Lee et al. [16], Table 2, Model 2.2, with permission). The reference category is illiterate women living in areas without community health care workers. The model also adjusted for maternal age, gender of the child, sibling status, wealth level, type of residence, and region. The four types of vaccinations included BCG, MCV, DPT3, and polio

relationship between the second independent variable (presence of CHW) and the dependent variable ***differs*** *according to the value of the first independent variable. The third sentence illustrates that point by reporting the direction and magnitude of the association between CHW and vaccination status for each of the two literacy levels.*

45.6 Substantive Significance

Many researchers and others who are communicating the results of quantitative research frequently are imprecise in their use of the term "significant" or "significance," failing to specify whether they mean *statistical* significance or *substantive* (practical) significance. Both are valuable aspects of the importance of a numeric result, and each provides a different perspective on what that finding means and how it can be applied to answer a real-world question [4]. As a consequence, some top peer-reviewed journals have strengthened requirements for reporting and interpreting effect size as well as statistical significance [23].

45.6.1 What Questions Can Statistical Significance Answer?

Statistical significance answers a very specific, technical question: "How likely it would be to obtain an estimate at least as large as that based on the *sample* if the true value of that statistic was 0 in the *population* from which that sample was drawn?" The *p*-value measures the probability of incorrectly concluding that the true population value is 0 based on the sample statistic. If the *p*-value associated with an estimate is less than the specified significance level (α), we say that the result is "statistically significant."

45.6.2 Substantive Significance

Although statistical significance is a *necessary* component of reporting statistical results that are based on a *sample* rather than the full population, it *cannot* address questions about the *substantive* significance of those findings, which pertain to the effect size and direction. Put differently statistical significance does *not* tell us whether the size of the effect is of practical significance: for instance big enough to represent a clinically meaningful difference. This is such a contentious issue that an entire supplement to the journal *The American Statistician* was devoted to the theme "Statistical Inference in the 21st Century: A World Beyond $p < 0.05$ [24]."

A finding that is *not* statistically significant can be highly salient if that result pertains to the main variable you are investigating –in other words, very substantively significant. If the *lack* of statistical significance runs contrary to theory or previously reported results, report and interpret the direction and size of the association, and the lack of statistical significance. In such situations, the lack of a difference between groups answers a central question in the analysis, so highlight it in the discussion section of the paper.

Example: Although maternal literacy was a statistically significant predictor of vaccination completion in Ethiopia (Table 45.5), it was not a statistically significant predictor in India (OR = 1.84, 95% CI: 0.96 to 3.51; Lee et al., Table 45.3, not shown) [16].

The statistically significant finding for Ethiopia was consistent with theory, but the result for India did ***not*** *meet conventional criteria for statistical significance. However, the India result was moderately* ***large*** *and in the* ***predicted direction,*** *so it should be compared with predictions and with the result for Ethiopia in both the results and discussion sections of the paper.*

Standard deviations and other empirical measures of distribution provide useful criteria for assessing whether an observed change in a biomedical outcome is clinically meaningful. The same size effect is considered more practically important if it is equivalent to a substantial share of a standard deviation of the deviation than if it corresponds to only a small fraction of a standard deviation, Miller [4], Chap. 15.

45.6.3 Presenting both Practical Importance and Statistical Significance

To provide a more complete perspective on the importance of a numeric result, present information about *both* the practical importance of your findings *and* whether they were statistically significant. In the results section, report and interpret effect size, direction, and statistical significance. In the discussion section, summarize the direction and size of those results in ways that convey their importance in broader biomedical context and application, as well as the extent to which the association can be interpreted as cause-and-effect, and the generalizability of the findings [25]. For results that are intended to inform the design of an intervention (such as treatment with a drug), the discussion section should also address the extent to which the independent variable can be modified or used to target a subgroup of interest. For additional guidelines on communicating statistical and substantive significance of quantitative research findings, see Miller [4] Chap. 15, and Chaps. 29 and 30 in this book.

45.7 Concluding Remarks

Writing an effective paper, grant proposal, or oral presentation about a quantitative research study involves integrating good expository writing techniques with principles for presenting numeric evidence. In the methods section, provide detail about the study design, data collection, preparation, variables, and analytic methods, which affect the interpretation and application of your results; see Miller [13] Chap. 13. In the results section, systematically review the statistical evidence, organizing it into separate paragraphs or sub-sections for each major component of your research question. Use topic and transition sentences to keep readers oriented to the objectives and logical flow of the analysis. Good introductory and transition sentences are especially important when presenting a series of exhibits, each of which addresses one part of your analysis. For instance, the first paragraphs of the results might present key univariate statistics such as mean values of the dependent variables and main independent variables. Then a separate paragraph should introduce and describe bivariate relationships between those variables before transitioning to a section conveying the purpose of the multivariate analysis and interpreting those results. Throughout, maintain an orientation to the specific concepts you are studying. By using these principles for writing about quantitative research, you will be able to more clearly communicate the purpose and meaning of your results to audiences interested in biomedical research.

References

1. Evergreen SDH (2020) Effective data visualization: the right chart for the right data, 2nd edn. Sage Publications, Thousand Oaks, CA
2. Miller JE (2015) The Chicago guide to writing about numbers, 2nd edn. University of Chicago Press, Chicago
3. Salazar M, Vora K, Costa AD (2016) Bypassing health facilities for childbirth: a multilevel study in three districts of Gujarat, India. Glob Health Action 9: 32178. https://doi.org/10.3402/gha.v9.32178
4. Miller JE (2021) Making sense of numbers: quantitative reasoning for social research. Sage Publications, Thousand Oaks, CA
5. National High Blood Pressure Education Program (2004) The seventh report of the Joint National Committee on prevention, detection, evaluation, and treatment of high blood pressure. National Heart, Lung, and Blood Institute (US), Bethesda, MD. Table 3, Classification of blood pressure for adults https://www.ncbi.nlm.nih.gov/books/NBK9633/table/A32/. Accessed 24 Oct 2022
6. Vaniotis G (2020) Laboratory cold storage temperature guide. https://blog.labtag.com/laboratory-cold-storage-temperature-guide/. Accessed Sept 2022
7. World Health Organization (2019) Wash in health care facilities, global baseline report 2019. https://apps.who.int/iris/bitstream/handle/10665/311620/9789241515504-eng.pdf. Accessed Sept 2022
8. Bhowmick S, Ghosh N, Saha R (2020) Tracking India's progress in clean water and sanitation: a sub-national analysis. ORF Occasional Paper No. 250, Observer Research Foundation. https://www.orfonline.org/research/tracking-indias-progress-in-clean-water-and-sanitation-a-sub-national-analysis-67139/#_edn54
9. Chatterjee U, Smith O (2021) Going granular: equity of health financing at the district and facility level in India. Health Syst Reform 7(2):e1924934. https://doi.org/10.1080/23288604.2021.1924934
10. Baxter C, Sharp G (2022) Understanding the magnitude of emerging threats: grams, milligrams, micrograms, and nanograms. Fed Resrces. https://www.federalresources.com/understanding-the-magnitude-of-emerging-threats-grams-milligrams-micrograms-and-nanograms/. Accessed Sept 2022
11. Sample Registrar of India. Districts of Gujarat (2011). www.census2011.co.in/census/state/districtlist/gujarat.html. Socio Economic Survey 2002–03. Add-on lists 2008–09
12. Commissionerate of Rural Development, Gujarat. www.ses2002.guj.nic.in/. Accessed by Salazar et al., 15 Jan 2015
13. Miller JE (2013) The Chicago Guide to Writing about Multivariate Analysis, 2nd edn. University of Chicago Press, Chicago
14. Miller JE (2007) Organizing data in tables and charts: different criteria for different tasks. Teach Stats 29(3): 98–101. https://doi.org/10.1111/j.1467-9639.2007.00275.x
15. World Health Organization (2022) WHO Coronavirus (COVID-19) Dashboard. https://covid19.who.int/. Accessed 10 Aug 2022
16. Lee HY, Oh J, Heo J, Abraha A, Perkins JM, Lee JK, Tran TGH, Subramanian SV (2019) Association between maternal literacy and child vaccination in Ethiopia and southeastern India and the moderating role of health workers: a multilevel regression analysis of the Young Lives study. Glob Health Action 12(1): 1581467. https://doi.org/10.1080/16549716.2019.1581467
17. Cunningham SA, Elo IT, Herbst K, Hosegood V (2010) Prenatal development in rural South Africa: relationship between birth weight and access to fathers and grandparents. Popul Stud 64(3):229–246. https://doi.org/10.1080/00324728.2010.510201
18. Utts J, Heckard R (2014) Mind on statistics, 5th edn. Cengage, Brooks Cole, Independence, KY
19. Allison PD (1999) Multiple regression: a primer. Sage Publications, Thousand Oaks, CA
20. Miller JE, Rodgers YV (2008) Economic importance and statistical significance: guidelines for communicating empirical research. Fem Econ 14(2):117–149. https://doi.org/10.1080/13545700701881096
21. Fauth RC, Roth JL, Brooks-Gunn J (2007) Does neighborhood context alter the link between youth's

after-school time activities and developmental outcomes? A multilevel analysis. Dev Psychol 43(3): 760–777
22. Powers D, Xie Y (2000) Statistical methods for categorical data analysis. Academic Press, San Diego, CA
23. Amrhein V, Greenland S, McShane B (2019) Retire statistical significance. Nature 567:305–307. https://www.nature.com/articles/d41586-019-00857-9
24. Multiple authors. The American Statistician. 73(Supplement 1):2019 "Statistical Inference in the 21st century: a world beyond $p < 0.05$." https://www.tandfonline.com/toc/utas20/73/sup1
25. Miller JE (2023) Beyond statistical significance: a holistic view of what makes a research finding 'important'. Numeracy 16(1):Article 6. https://doi.org/10.5038/1936-4660.16.1.1428

How to Efficiently Write a Persuasive Discussion Section

46

Pitchai Balakumar, Ali Alqahtani, Kumaran Shanmugam, P. K. Srividhya, and Karupiah Sundram

Abstract

The discussion section, a systematic critical appraisal of results, is a key part of a research paper, wherein the authors define, critically examine, describe and interpret their findings, explaining the significance of results and relating them to the research question/s. Considered one of the most critical components of the IMRaD (Introduction, Methods, Results, and Discussion) system, the discussion section deals with all about what the results of the research mean. While formulating a discussion section of the manuscript, the researcher needs to consider the broader context of the study's research findings. In particular, a proficient discussion section should relate closely to the results and obtained data, reviewing the strengths and weaknesses of the research study with a balanced approach. The discussion section should also evaluate the applicability of results and their outcomes, relating them to existing literature and future directions in a holistic manner. Specifically, the discussion section should address the questions posed in the introduction, explaining how the results obtained in the study support the answers. The challenges and limitations of a study are also crucial parts of a good discussion. Finally, the discussion section closes with concluding remarks on a research study and its future directions. The discussion section is the author's opportunity to help readers understand their research outcomes. In light of this context, this chapter presents a systematic approach to writing an effective discussion section.

Keywords

IMRaD · Research question · Persuasive discussion · Results interpretation · Strength and limitations · Future directions

46.1 Introduction

Clarifying the meaning of obtained results to the reader is the primary objective of the discussion section of a scientific paper [1]. The discussion section allows the reader to grasp the direct

P. Balakumar (✉)
Professor & Director, Research Training and Publications, The Office of Research and Development, Periyar Maniammai Institute of Science & Technology (Deemed to be University), Vallam, Tamil Nadu, India
e-mail: directorcr@pmu.edu; pbalakumar2022@gmail.com

A. Alqahtani
Department of Pharmacology, College of Pharmacy, King Khalid University, Abha, Kingdom of Saudi Arabia

K. Shanmugam · P. K. Srividhya
Periyar Maniammai Institute of Science & Technology (Deemed to be University), Vallam, Tamil Nadu, India

K. Sundram
Faculty of Pharmacy, AIMST University, Bedong, Malaysia

G. Jagadeesh et al. (eds.), *The Quintessence of Basic and Clinical Research and Scientific Publishing*,
https://doi.org/10.1007/978-981-99-1284-1_46

relevance and validity of the study and its applicability in a particular field of research [2]. Discussion section has direct links with introduction and results sections, such as attempting and responding to questions raised in introduction, and interpreting study results [3].

The writing in the discussion section needs to be precise and unambiguous. Nevertheless, writing a research paper's discussion section is difficult, necessitating adequate training and a balanced approach to analysing and interpreting the results and concluding the research outcomes [4]. A researcher should be familiar with writing a scientific paper (see Chaps. 41–43), which is indeed considered a major task of an academic teacher and a scientist. Certainly, a researcher needs to describe the study outcomes meticulously. This is done by writing a persuasive discussion section using simple, clear, meaningful, and effective language throughout the text [5]. A good discussion section majorly includes the study outcomes, significance of the study, importance of results that are appraised, key results which are consistent or controversial relative to existing literature, strength and weakness of the study, major limitations, suggestions for additional research, concluding remarks, future perspectives and further recommendations [6]. The presentation of results and discussion sections should not have similarities. The results section (see Chap. 44) primarily consists of text, tables and figures, presenting the data in detail (see Chap. 45) without interpreting the results [7], whereas the discussion section is formulated with the interpretation of results and implication of findings. It is perpetually recommended to practice a pre-submission review process with expert colleagues to obtain feedback on the discussion and the entire manuscript draft before it is submitted to an indexed journal for consideration and publication (see Chap. 47).

The nascent and postgraduate research students often face difficulties in making logical arguments in the discussion section while constructing the research paper [8], necessitating systematic and holistic guidance in writing a persuasive discussion. Intensive training is needed in basic research methodologies and scientific writing concepts at all levels of researchers and budding academic teachers (see Chaps. 1 and 4) [9–13]. The primary aim of this chapter is to describe the main aspects of efficiently writing the discussion section of a research paper. In addition, the do's and don'ts of the discussion section are depicted in this chapter.

46.2 Key Components of a Discussion Section: An Overview

The discussion section is integral to a research paper [14]. In general, no defined format is available for writing the discussion section of a research paper. Nevertheless, the discussion section is recommended to be written using three major fragments, which include (a) an introductory/first paragraph, (b) an intermediate paragraph having the major contents in a few sub-paragraphs, describing the study outcomes and interpretation of results, among others, and (c) finally a paragraph on conclusions and future perspectives of the study [5, 6]. In discussion, the results obtained in a research study can be described in order either from most important findings focusing on key issues to least important or in a systematically logical direction. Overall, a persuasive discussion section is expected to critically examine the results, describe how the data impacts the field of study, relate up-to-date literature, and consider suggesting future directions [14–17]. The structure of the discussion section may be formulated as shown in Box 46.1.

Box 46.1 Key Points To Be Considered While Structuring the Discussion Section

1. Start the discussion by stating the major findings, and answering to research questions
2. Highlight whether the null hypothesis is accepted or rejected
3. Discuss the results relevant to the previous body of literature by citing most specific references
4. The previous work may be discussed in support or counter to the results obtained in the current study
5. Express the uniqueness of the study coherently
6. A brief discussion on the principle and advantage of methods employed in the study may be considered
7. Discuss unexpected results and also minor findings with possible explanations
8. Non-specific, loose and sweeping statements should be avoided
9. Discuss the strength and major pitfalls of the study and its outcomes
10. Discuss the relevance of the obtained results to the current practice of the field
11. Finally, end the section by highlighting the study's key findings and stating the concluding remarks
12. The need for additional studies and future directions can be highlighted at the end
13. Generally, use the present tense and active voice while writing the discussion section
14. All together, be specific in writing the discussion section, and strictly avoid technical jargon and unfamiliarised acronyms
15. Do not write too short or too lengthy a discussion section. Editors and reviewers are more likely to favourably consider a manuscript with a well-written discussion section

46.3 An Introductory Paragraph of the Discussion Section

The beginning paragraph earns the most prominent position in the discussion section of a research paper. Essentially the lead/introductory paragraph of the discussion section summarizes the key findings signifying the primary outcomes, briefly addressing the research questions and concisely supporting the answers with results. Mainly the discussion section can start by presenting the imperative outcomes of the study with a brief statement on why the study is of much importance [6, 18].

Answering the research questions stated in the introduction section is considered the foremost task of the discussion section, whereas pertinent answers can be presented in the first few lines after restating the research question. Therefore, the beginning of a paragraph in the discussion section must present the most interesting findings. In this context, it is essential to note that the introductory paragraph should not be presented with the complete results of the study.

46.4 The Intermediate Paragraphs: Core Segments of the Discussion Section

The "divide-and-conquer paradigm" can be employed in the intermediate paragraphs of the discussion section to describe an optimal solution for a problem and to solve the given problem posed in the introduction section of the manuscript [5]. The intermediate part of the discussion section has main and subparagraphs with the most focused information relating the answers to the research questions. From paragraph two onwards (after the introductory paragraph), it is necessary to describe how the answers are supported by the results obtained in the study. The authors should briefly explain how the research findings validate or contradict the previously published reports.

Authors need to answer the questions of the study using the same words/key terms employed in the introduction section. It is essential to express the relevant results after stating the answers by elucidating how the results support the answers. Moreover, the authors need to present the importance of the results and influences of the study in the field by explaining how the answers fit within the existing literature and providing a piece of new knowledge in the field of research [6]. While interpreting the results, it is advisable that authors consider moving beyond the data; nevertheless, far speculative statements not supported by the obtained results should be principally avoided. Relating the findings obtained in a research study with the existing literature often makes the discussion section more interesting and pertinent. In this framework, citing highly relevant and most recently published papers available in the literature will demonstrate the author's expertise in the field, convincing the readers that the discussion is up-to-date.

Only those findings which are presented in the result section should be described in the discussion segment in relevance to the existing literature. It is important to note that the results as such should not be repeated in the discussion. The contradicting or challenging results, if any, should be briefly discussed with proper justification. Describing in the discussion the findings which are not presented in the results section and not stated in the methods section would confuse the readers, and hence, such practice should be avoided. At the same time, appraising the minor results of the study should not be avoided in the discussion section. Likewise, describing the study's unexpected findings in a lucid style makes a good discussion point.

Throughout the discussion, highlighting the study's strengths based on the results forms a good discussion. At the same time, a discussion section should include balanced information about the study's limitations in addition to highlighting its strength. Every research study possesses its own limitations and weaknesses. Such limits should be lucidly spelled out at the end of the intermediate paragraphs of the discussion section (before the concluding remarks). Because limitations are valuable to understand the findings, translating the importance of potential errors, formulating new research questions, and opening a way for further research [6]. The authors need to describe the direct impact of limitations on the study findings and brief the attempts made to minimize those limitations and errors [6]. Finally, a research paper with a potentially result-driven discussion will likely receive appreciation from the reviewers and the readers.

46.5 The End Paragraph: Concluding Remarks of the Discussion Section

The final segment of the discussion section is the conclusion, which states in a few focussed sentences the critical outcomes of a research study and the author's final opinion/s based on the study results. The prime importance of the study findings is indicated in this segment, answering the research questions stated in the introduction section. It is essential to state clearly and concisely the one or two most important aspects of the study outcomes. Indeed, conclusion statements should have high clarity and brevity of the key outcomes of the study directly supported by the data. More importantly, the conclusion must evidently state whether the study's key findings support the hypothesis described in the introduction. At the end of the conclusion, it is appropriate to express the necessity for further studies in the form of future directions in a realistic and meaningful way [5, 19]. Box 46.2 shows the key points to formulate concluding remarks.

Box 46.2 Key Points to Formulate the Concluding Remarks

1. In this study, we addressed the current progress and challenges in.......................... ....
2. In summary, the present study reveals/shows/demonstrates ....
3. These findings provide new insights into/Our results bring a new light on..................... ....
4. Based on these findings, we confirm/suggest/propose/recommend that....................
5. Nevertheless, further studies are required to validate/examine/conclude.................. .... .
6. Though additional studies are needed, the present study contributes to a better understanding of.......................... ....

Taken together, the conclusion section highlights all the important assessments and outcomes of the research study and their meaning relative to the current knowledge. The Do's and Don'ts of the discussion section are provided in Box 46.3.

Box 46.3 The Do's and Don'ts of the Discussion Section

(A) The Do's
1. The initial paragraph of the discussion section should instruct the reader on the importance of the study, its key outcomes, and how such developments fill the knowledge gaps in the relevant field of research. Important findings of the study should be stated at the end of the first paragraph
2. The intermediate paragraphs must consider focusing on the higher order cognitive domain of Bloom's Taxonomy, critically analysing and interpreting the result outcomes and stating what the study findings mean and why such findings are of great importance
3. A healthy discussion is expected to relate the study findings with the already published literature. In this connection, a few pertinent recent references can be cited
4. Authors should consider citing references of previously published studies with results contradictory to their current study outcomes. This would yield a balanced approach to updating the knowledge in that area of research. Moreover, such kind of practice would receive favourable consideration of the paper by reviewers and editors
5. The limitations of the study (*e.g.*, less sample size, no additional parameters or interventions or variables, *etc.*) should be pitched in the discussion section before being pointed out by the editors/reviewers
6. Authors should specify the strengths and limitations of the study and the significant effect they might have on the observed results
7. At the end of the discussion, while providing the concluding remarks, it is considered essential to provide recommendations for additional studies. Such practice would open a new way for further research to expand the current knowledge in the field
8. Authors should write the discussion section in a sequence so that the gatekeepers and readers can easily understand the flow of the outcomes
9. Authors should meticulously check for spelling, punctuation, syntax, and grammatical errors
(B) The Don'ts
1. Do not repeat results in the discussion as they are already in the results section
2. Discussing something that is not presented in the results section should be avoided
3. While writing the study limitations, the authors should not offer suggestions that could have been effortlessly addressed in their current study by performing simple experiments. (The limitations to be mentioned by the authors should always be beyond their level of research protocol/performance/expertise)
4. Overstating the study strength and understating the limitations should be generally avoided
5. Expressing very general statements and providing too many overviews from several published papers while presenting the current findings should be avoided
6. Authors should not provide superfluous and redundant details in the discussion aiming to increase the length of the discussion section
7. Authors should not cite references from unreliable and non-peer reviewed sources
8. Authors should not write the discussion section without reading the journal author's instructions thoroughly (the suggested word limits must be contained)

46.6 Concluding Remarks

The discussion section debates and puts forth arguments relevant to the research questions stated in the introduction section. Authors need to highlight the significance of the results and specify whether the study findings support or reject the null hypothesis. Successfully passing the gatekeepers (editors and peer reviewers) is considered a tough job, demanding a strong, viable, well-written and persuasive discussion section in the research article. Moreover, the authors need to point out how the results obtained in the study can be directly related to existing literature and applied to future studies. Comparing and linking the results with articles published in predatory or hijacked/cloned journals should be firmly avoided. The implications and applicability of the results should be presented realistically. It is also imperative to state in the discussion section the major strength and limitations of the study protocols and outcomes. In the discussion section, it is essential to remind the gatekeepers why the paper is to be favourably considered for publication in the target journal. Finally, the concluding remarks should provide a venue to set a valid stage for future research directions.

Acknowledgements The authors wish to thank Dr. Gowraganahalli Jagadeesh of the US Food and Drug Administration (Division of Pharmacology & Toxicology, Office of Cardiology, Hematology, Endocrinology, and Nephrology, Center for Drug Evaluation and Research), Silver Spring, MD 20993, USA, for his expertise suggestion and critical reading of this manuscript.

Conflict of interest No conflict of interest is declared.

References

1. Hess DR (2004) How to write an effective discussion. Respir Care 49(10):1238–1241
2. Jenicek M (2006) How to read, understand, and write 'Discussion' sections in medical articles. An exercise in critical thinking. Med Sci Monit Int Med J Exp Clin Res 12(6):SR28–SR36
3. Gray JA (2019) Discussion and conclusion. AME Med J 4:26
4. Walsh K (2016) Discussing discursive discussions. Med Educ 50:1269–1270
5. Şanlı Ö, Erdem S, Tefik T (2013) How to write a discussion section? Turk J Urol 39(1):20–24
6. Ghasemi A, Bahadoran Z, Mirmiran P, Hosseinpanah F, Shiva N, Zadeh-Vakili A (2019) The principles of biomedical scientific writing: discussion. Int J Endocrinol Metab 17(3):e95415
7. Bahadoran Z, Mirmiran P, Zadeh-Vakili A, Hosseinpanah F, Ghasemi A (2019) The principles of biomedical scientific writing: Results. Int J Endocrinol Metab 17(2):e92113
8. Bitchener J, Basturkmen H (2006) Perceptions of the difficulties of postgraduate L2 thesis students writing the discussion section. J English Acad Purposes 5(1): 4–18
9. Jagadeesh G, Murthy S, Gupta YK, Prakash A (eds) (2010) Biomedical research - from ideation to publication, 1st edn. Wolters Kluwer Health-Lippincott Williams and Wilkins, Philadelphia
10. Balakumar P, Murthy S, Jagadeesh G (2007) The basic concepts of scientific research and communication (A Report on preconference workshop held in conjunction with the 40th annual conference of the Indian Pharmacological Society-2007). Indian J Pharmacol 39:303–306
11. Balakumar P, Jagadeesh G (2012) The basic concepts of scientific research and scientific communication (A report on the pre-conference workshop held in conjunction with the 63rd annual conference of the Indian Pharmaceutical Congress Association-2011). J Pharmacol Pharmacother 3:178–182
12. Balakumar P, Inamdar MN, Jagadeesh G (2013) The critical steps for successful research: The research proposal and scientific writing: (A report on the pre-conference workshop held in conjunction with the 64th annual conference of the Indian Pharmaceutical Congress-2012). J Pharmacol Pharmacother 4: 130–138
13. Balakumar P, Srikumar BN, Ramesh B, Jagadeesh G (2022) The critical phases of effective research planning, scientific writing, and communication. Pharmacogn Mag 18:1–3
14. Al-Shujairi YBJ, Tan H, Abdullah AN, Nimehchisalem V, Imm LG (2019) Moving in the right direction in the discussion section of research articles. J Language Commun 6(2):23–38
15. https://filipodia.com/2017/02/23/the-dos-and-donts-of-the-discussion-section/. Accessed 01 Dec 2022
16. Höfler M, Venz J, Trautmann S, Miller R (2018) Writing a discussion section: how to integrate substantive and statistical expertise. BMC Med Res Methodol 18:34
17. Clarke M, Chalmers I (1998) Discussion sections in reports of controlled trials published in general medical journals: Islands in search of continents? JAMA 280(3):280–282
18. Barroga E, Matanguihan GJ (2021) Creating logical flow when writing scientific articles. J Korean Med Sci 36(40):e275
19. Ng KH, Peh WC (2009) Effective medical writing. Pointers to getting your article published: writing the discussion. Singap Med J 50(5):458

Submitting a Manuscript for Publication, the Peer Review Process, and Getting Past the Gatekeepers

47

Pitchai Balakumar and Gowraganahalli Jagadeesh

Abstract

The publication serves the vital purpose of letting the work or ideas known worldwide to researchers and other academia. One's findings acquire meaning only when these are used further for the betterment of society and the advancement of knowledge. Research helps to disseminate the knowledge to the relevant places. The ultimate goal of scientific research is publication. Considered an integral part of scientific publishing, peer review is the systematic procedure employed in the shortest possible timeframe to meticulously assess the quality of a submitted manuscript before it is considered for publication in an indexed journal. Peer reviewers volunteer their time to help improve the quality of the submitted manuscripts. The expert reviewers in the relevant area of research critically evaluate the manuscript for its originality, validity, consistency, and significance to help editors decide whether or not a manuscript can be considered for publication in their journal. Only after the manuscript meets the submission criteria and scope of the journal, the editors will invite potential peer reviewers close to the field of research to review the manuscript and receive recommendations on the overall scientific integrity of the manuscript for further consideration before a manuscript is accepted, revised, or rejected. Different types of peer review adopted by biomedical journals to confirm the validity of the manuscript include single-blind, double-blind, and open peer review systems. In this chapter, we delineate the principles of manuscript preparation, journal submission, and peer review systems and their importance in scientific publishing.

P. Balakumar (✉)
Professor & Director, Research Training and Publications, The Office of Research and Development, Periyar Maniammai Institute of Science & Technology (Deemed to be University), Vallam, Tamil Nadu, India
e-mail: directorcr@pmu.edu; pbalakumar2022@gmail.com

G. Jagadeesh
Retired Senior Expert Pharmacologist at the Office of Cardiology, Hematology, Endocrinology, and Nephrology, Center for Drug Evaluation and Research, US Food and Drug Administration, Silver Spring, MD, USA

Distinguished Visiting Professor at the College of Pharmaceutical Sciences, Dayananda Sagar University, Bengaluru, Karnataka, India

Visiting Professor at the College of Pharmacy, Adichunchanagiri University, BG Nagar, Karnataka, India

Visiting Professor at the College of Pharmaceutical Sciences, Manipal Academy of Higher Education (Deemed-to-be University), Manipal, Karnataka, India

Senior Consultant & Advisor, Auxochromofours Solutions Private Limited, Silver Spring, MD, USA
e-mail: GJagadeesh2000@gmail.com

G. Jagadeesh et al. (eds.), *The Quintessence of Basic and Clinical Research and Scientific Publishing*,
https://doi.org/10.1007/978-981-99-1284-1_47

Keywords

Manuscript preparation · Cover letter · Manuscript submission · Peer review · Types of peer review · Editor's decision

47.1 Introduction

There was a belief that teaching students were the most critical aspect of an academic job where teachers have devoted all their energy to teaching students (excellence in teaching). Nowadays, the bar is raised for entry into the teaching profession. Actively involved in research and eventually publishing the scientific findings are part of the job descriptions at all academic career levels, starting from instructor/lecturer to professor and above. The publication is the gateway to the successful selling of scientific findings. A substantial body of published works helps advance one's academic career as they are considered for an academic appointment and a promotion. It creates requirements for a job and aids in recognition. Publishing helps establish an individual as an expert in their field of science. Research funding requests are evaluated based on peer-reviewed publications. Prior research experience and preliminary findings supported by publication(s) open the door for grants, while research funding is essential to facilitate high-quality research. Without publication, there would be no grant. Teachers, researchers, and science administrators need to understand why publication matters a lot.

Though timely submitting the thesis and completing the research project are considered critical, what really matters is how one tells the story of the project/clinical investigation in a clear, succinct, simple language, weaving the previous work done in the field, answering the research question, and addressing the hypothesis set forth at the beginning of the study. In addition, the research findings should be wrapped in the form of an excellent scientific paper. Writing an article is one of the components of all research projects, while research outcomes are measured through a quality publication. Since the number of submissions and the number of researchers/clinicians vying for space are more than the availability of journals, one can stand out from the competition with a high quality write-up.

The peer review system assesses the quality of a submitted manuscript that is considered an integral part of scientific publishing. The expert reviewers carefully assess the manuscript for its originality, validity, and error-free write-up before it is considered for publication in an indexed journal of repute. With the help of reviewers' comments, the editors decide whether or not to consider the manuscript for publication. In this chapter, we describe the journal selection process based on the nature of the study and the type and area in which the manuscript (MS) is prepared. The steps involved in getting the MS to the editors' desk are sequentially explained. In addition, the peer review process involving the author, editor, and reviewer toward a successful publication of the paper is described in the chapter.

47.2 The Selection of a Journal

There are three main problems in the publication of a paper: (1) Typographical errors and grammar problems- grammar and spelling issues markedly interfere with the clarity of scientific contents, while sloppy writing is by far the biggest problem with submissions; (2) Structural issues of a paper, inconsistencies, and issues with the clarity and tone, whereas the subject matter is not well stated by the author/s. There may be a lack of authority as a result of insufficient understanding of the existing literature; and (3) Choosing the right kind of journal for submission within the scope of the journal. All of these may serve as a bottleneck in delaying the publication of a scientific paper.

During the writing process or before you begin to write, consider selecting a suitable journal for submission. This is based on the type of publication (original research article, short/rapid communication, review, state-of-the-review, mini-review, letter to the editor, clinical case, methods, technical or laboratory notes, book review, and others). For publishing an original research article, consider the novelty of new findings (incremental or additional, conceptual or theoretical advances) and the significance of the study (practical applications for a specific field or across many fields). The novelty may be described as: (a) the findings are reported for the first time, (b) although the findings have been reported earlier, controversy exists, (c) the current work extends the previous findings, and (d) the largest study of the research question [1]. The author must address the impact/significance of the work and the target audience (general or subject specialty focused). Equally important factors to be considered while selecting a journal include, but are not limited to, Web of Science-Science Citation Index Expanded (SCIE)/Social Sciences Citation Index (SSCI)-Clarivate Analytics journal impact factor, SCImago journal and country rank, abstracting/indexing service databases (PubMed/Medline, Scopus, Embase), the professional society journal, subscription/open access, page charge, publishing frequency, review time frame, acceptance/rejection rate, and target audience. Do not publish in predatory, hijacked/cloned, and non-indexed journals. Additional factors to consider in selecting the right kind of journal are listed in Box 47.1.

Box 47.1 Factors to be considered while selecting a journal: Where to publish?

1. Language- English only
2. Traditional journals or Open access journals (OAJ) (https://www.doaj.org/)
   – Visibility, cost, prestige, and speed
   – In OAJ, everyone can read your paper- this increases visibility and attracts more citations; however, an open access fee is applicable
3. Publishing frequency- preferably monthly or semi-monthly
4. Time frame- Time of submission to acceptance (review time) and publication. It varies from month to year
5. Does the journal offer fast-tracking for publication? (Published ahead of printing, accepting rapid communication). Take advantage of it
6. Journal with well-indexing and is readily accessible across the globe
7. Acceptance rate: Prestige journals have a low acceptance rate
8. Has the journal published articles similar (area of research) to yours? Browse a few issues to understand the nature of articles published in the target journal
9. Highest impact factor for that particular field (Web of Science Q1 rated journals)
10. Before you write the MS and before the submission of the MS, read "Information for Authors" of the target journal to determine the journal's restrictions. A good fit is essential
11. Understand the scope of the journal, and its audience—clinical, experimental, theoretical, new or modified techniques and methods
12. Journal format is different for each journal-*Review* the length of the MS allowed (word limit), Abstract size and style (structured/non-structured), Number of Figures and Tables allowed, Harvard/Vancouver Citation formatting (in text and reference), Number of references allowed, Type of abbreviations allowed, and Structure of the paper—IMRaD style? If not, write Supplementary Methods, including Supplementary Figures, and Data

Furthermore, several easy-to-use online resources help you decide which journal would be most pertinent to your research. These journal websites (there are many more) allow you to paste the Abstract or Title of your paper and match the right fit.

The JANE (Journal/Author Name Estimator): JANE compares the author's document to Medline documents to find the best match. http://jane.biosemantics.org/

Elsevier: http://journalfinder.Elsevier.com/

Springer Nature: https://journalsuggester.springer.com/

Edanz: https://www.edanz.com/journal-selector

BMJ: https://authors.bmj.com/before-you-submit/how-to-choose-a-journal/

47.3 Before Submission of a Manuscript

A well-written MS has a clear, valuable, and exciting message to potential readers. It is systematically presented and logically constructed. These criteria sway the mindset of editors and reviewers to consider reading the MS and assess its value for publication. A few journals encourage an initial approach before considering a final MS for a peer review. Authors might contact the editor or journal office by sending an outline of the MS consisting of the abstract, introduction, and the general area or outline of the work. The reason for doing this is some of the high-impact prestigious journals receive too many MS, the journal may have a similar paper in the pipeline, and the journal is unlikely to publish a type of MS. Furthermore, the editor may be willing to consider a shorter version or correspondence. Early declining to publish saves the time and effort of both editors and authors.

47.3.1 Similarity of Words

Before submitting a MS to an indexed journal, thoroughly check its quality more than twice. Check the MS for the similarity of words (or plagiarism). Once the writing is done, run plagiarism checking software to ensure that your document is mostly (score > 95% unique) original. Rewrite sentences to make sure that it is exceptional. A plethora of plagiarism detectors are available online. It takes work to select the best among them. A few online plagiarism tools/software (free or paid) are as follows:

1. Dupli Checker: https://www.duplichecker.com/
2. Smallseotools: https://smallseotools.com/plagiarism-checker/
3. Ithenticate: it is commercial software being used by most publishers: http://ithenticate.com/
4. Plagscan: This is a commercial, but a free trial is also available. http://www.plagscan.com/
5. Turnitin: https://www.turnitin.com
7. The Viper: http://www.scanmyessay.com/viper/Release/ViperSetup.exe
8. http://plagiarismdetector.net/

47.3.2 The Cover Letter

A well-written MS is sold through an effective and meaningful cover letter, which should create a positive first impression with journal editors. It acts as a guide for selling the author's work to the editor. The cover letter must explain the clinical, investigational, or experimental relevance of the original research work and provide background, rationale, and research outcomes to justify why the journal should consider the MS for peer review and eventual publication.

Most journals receive more papers than they can publish. The editor may not necessarily be an expert in your field of super-specialization. It creates the first impression for journal editors if the letter is addressed personally. The letter should include the MS title/publication type and highlight the study's important findings. It should be followed by a brief statement on the work's novelty/significance/relevance. To expedite the publication process, the cover letter is expected to include some "must-have" statements such as original unpublished, not submitted to other journals, authors agreeing on the MS, key outcomes of the study, no conflict of interest, authorship contributions, and source of funding.

47.3.3 Submission Checklist

Before submitting a manuscript, thoroughly check its quality, evaluate its contents critically, and ask yourself—could anything be done further better? Then, follow through with the journal's author guidelines in preparing the MS to prevent rejection of the submission before even being considered for peer review. Finally, only submit the same MS to one journal at a time. The critical checklist is listed in Box 47.2.

Box 47.2 Checklist before submitting the MS to an indexed journal

Make sure that:
1. You have followed the Instructions for Authors as per the target journal
2. The MS has gone through spell and grammar checks
3. The MS has been checked for plagiarism (score should be >95% unique)
4. You have completed the online registration for the submission process for the target journal
5. All authors have consented to the publication of the study. Signed consent forms/Author declaration forms are ready (forms are downloaded from the journal website)
6. The copyright transfer form is signed

(continued)

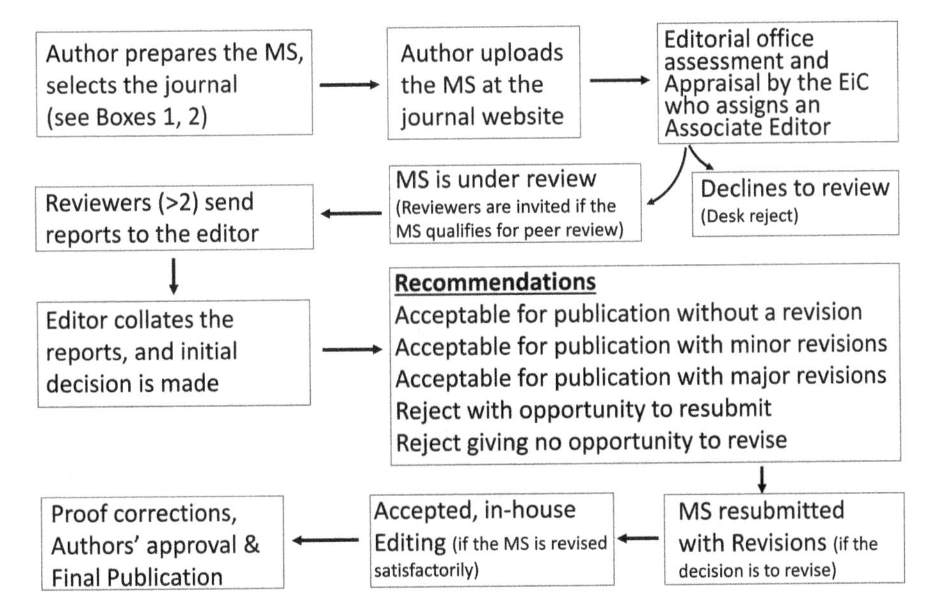

Fig. 47.1 The publication process and the peer review (EiC: editor-in-chief)

Box 47.2 (continued)

7. The required files are in the correct file format, figures are of appropriate resolution or size
8. Separate files are generated for Figures (individual files), Abstract, Key Points, or Highlights
9. Finally, Graphical Abstract and a well-written persuasive cover letter are ready
10. The main MS file has: The title page, Contributing Authors with their institute affiliation and Email addresses, designated Corresponding author, Abstract, Keywords, List of abbreviations, all the key sections of the MS (IMRaD format), Acknowledgment, Funding source, Conflict of interest, Authors' contribution statement, References, Table(s), and Figure legend (no figures)
11. Once all files are ready, log into your account at the journal website
12. Follow online instructions and upload files in the order listed. The process will take nearly 30 min to an hour
13. Once completed, the system collects all the files and rearranges and builds up a PDF file of the submission. It prompts you to check your email to approve the file. Then, download the PDF file and check for the content and accuracy
14. If everything is good, 'approve' the submission
15. The editorial manager informs all co-authors of the submission

A part of the publishing task is over. Wait for the Editorial decision on the MS

47.4 The Publication Process and the Peer Review System

The publication process and the peer review system are depicted in Fig. 47.1.

47.4.1 The Key Role of the Editor and the Editorial Office

All papers are checked for completeness at the editorial office before being submitted to the plagiarism detection system and word similarities. If the paper is considered unsuitable for the journal (not within its scope, not unique, very poorly written, and does not advance the field of research), it will be rejected outright before sending it for peer review. This is known as 'desk reject'. Preventing 'desk reject' is easy, and it often happens because the authors are formatting the MS and cutting the corners in a hurry to submit. There could be multiple reasons, such as formatting style, poorly written, faulty research topic, inappropriate study design, insufficient data, discussion, and illogic conclusion [2].

Once the MS enters the peer review, you have some chance to have it accepted by a journal, suppose the MS passes the initial screening procedure. In that case, the editorial office first sends the abstract of the MS to a panel of reviewers (2 to 5) seeking their willingness to review the MS. Once they agree to review, the reviewers can access the entire MS and comment on the suitability and accuracy of the MS for publication. Authors can have the current submission status by checking the journal's Editorial manager frequently.

Peer review is a process of subjecting authors' research to the experts' evaluation. The process separates the wheat from the chaff, the good from the bad. Passing the gatekeepers is very tough as they scrutinize the findings and do not just let anybody publish whatever they want. In this way, the journals keep up the standard by shortlisting papers of much higher quality to publish. Peer review remains the foundation of publishing and an essential element of the quality publication process. Critical assessment of a MS is vital to peer review and the publication process. The articles that have not gone through peer review are not cited and the findings are not taken seriously by fellow scientists. Furthermore, peer-reviewed publications build a credible body of knowledge in the field and allow everyone in the field to trust the journal's publishing practice.

Both the editor and the publication managers coordinate in selecting reviewers (two or more) based on the following:

- Journal editorial board
- Experts in the specific area (from an internal database created by the journal, reference section of the MS, biomedical databases, or editor's personal knowledge)
- Experts from different geographic regions
- Experts identified using a database like Scopus based on their h-index and other metrics.

Authors can qualify or disqualify some referees, but the final decision in selecting reviewers for the MS rests with the editor. Journals use different types of peer review. Broadly, one of the three variations is being practiced. Single-blind, where the reviewer knows who the author is, but the author does not know who the reviewers are. This is the most common form of peer review system followed, and the proponents argue that reviewer anonymity allows for providing objective feedback to the authors. The double-blind procedure is less practiced where the authors and reviewers do not know their identities each other. This is time-consuming in unmasking the authors details and assigning a code number to each MS. However, this procedure can remove the author and regional bias (Mathew effect) [3] that existed with the single-blind.

On the contrary, the reviewers can figure out the authors of the MS based on the text, references, preprint, or conference presentation. Reviewers are reported to award higher marks when the MS is from a famous author or lab [3]. In the case of the 'open' (or non-blind) method, again less practiced, the authors and reviewers' identity is disclosed.

During the peer review system, there may be a possibility of intentionally delaying the publication of the MS and/ or using the contents of the unpublished material for personal gain. While sending out the MS to reviewers, the journal mostly instructs the reviewers to treat the MS as privileged, and confidential and not to disclose to anyone during the review and once the review is done. Scientists worldwide spend millions of hours peer-reviewing manuscripts for scholarly journals every year. This is time-consuming and labor-intensive work. Most of the journals recognize the contributions of their reviewers by publishing their names in a special section of an issue of the journal.

What does the peer reviewer look at the MS? The reviewers look for originality, relevance/ significance, study design, methodology, presentation of results, the strength of discussion in relation to previous findings, possible conflicting findings, study limitations, the strength of the conclusions, and the overall quality of the MS, including the language quality and write-up. The other reasons why the MS gets rejected associate with the wrong statistical analyses, leading to a wrong conclusion and a weak discussion because of insufficient understanding of the subject and outdated literature [4]. A few journals ask authors to include in the article a brief statement on novelty and significance, stating 'what is new' and 'what is relevant'. Peer reviewers may or may not detect fraudulent data, plagiarism, or image alteration. Detection of data manipulation at a later stage, even after the publication, may have consequences. It might result in the retraction of a published paper.

The Editor receives a minimum of two reports from reviewers before reaching a decision for a submitted MS. A confidential message (not for the author) to the editor suggesting a recommendation with reasons on the MS, and a detailed review of the paper with constructive critique to the authors are submitted by peer reviewers. The editor will collate reviewers' responses that could vary from rejection to minor or major revision of the MS and rarely to its direct acceptance on the first decision. All reviewers may not be in agreement in recommending the MS for publication, and in such a case the editor has to address the divergent opinions of the reviewers and arbitrate. Based on reviewers' recommendations, the editor sends a letter, usually a variation of the five decisions, from rejection to acceptance, with or without revision (Fig. 47.1).

If the MS is accepted, it is sent for production. On the other hand, if the MS is rejected or considered for a revision (minor or major), the editor who handles the MS should communicate with the authors with constructive comments received from the reviewers to help the author improve the MS. At the same time, reviewers should also be communicated with an email on the outcome of their reviews.

47.4.2 The Authors' Role After the Peer Review

Authors generally feel frustrated if their MS is not accepted or asked to undertake major revisions with additional experiments. Most of the time, the papers fall into this category. A small number of MS comparatively are accepted with minor revisions. Some of the prestige journals have very low acceptance rates. Novice researchers face extreme difficulties publishing their research work in a reputed journal. Publishing in quality, high impact journals is the need of the hour for budding scientists as they need to build a track record and expertise in their field [5]. Although peer review enhances the quality of a paper, the blog authors argue that not all revisions improve the paper. Sometimes there could be contradicting reviewers' views for the same section of the MS [5]. In such divergent opinions, the editor guides on addressing the query, or the author might ask the editor to weigh in and adjudicate. Honest, constructive reviews and motivating feedback help authors. On the contrary, there are instances of criticism, and rude and inappropriate comments by reviewers [6, 7]. Reviewers should practice a well-mannered and constructive review as they are also authors. They should be judicious in asking authors to do additional experiments [8].

Rejection and revision are norms in the publication of scientific articles. Even some Nobel Prize-winning scientists' papers were rejected [9]. Before attempting to revise, authors should check the journal's revision guidelines. If revisions are required and asked to respond to reviewers' comments, authors should meticulously respond to individual comments with appropriate answers keeping in view to get the paper accepted and eventually published. Use 'track changes' or a different color to clarify where changes were made to the revised text in the MS. If necessary, a separate page listing responses to each reviewers' comments be included. It is a norm to thank the reviewer if there is a compliment for the work. Also, it is common to offer a rebuttal of reviewer comments that the author disagrees with the reviewer. In such a case, the author should show politeness by stating, "we do not agree with the reviewer's views or we respectfully disagree with the reviewer's opinion" and then state the valid reason for the disagreement.

The authors should return the revised MS and response letter within the requested time period. The Editor-in-Chief or the handling Associate Editor reviews the reviewers' comments along with the resubmitted MS and the authors' line-by-line responses to ensure that all comments have been appropriately addressed. The editor may send the revised manuscript and the author's responses to reviewers' comments to the original reviewers for a second reading. If revisions are acceptable, the editor might issue an 'acceptable' letter and the uploaded files will be transferred to the publisher's production department for publication. Upon receipt of the acceptance letter for publication, the author should relax and wait for proofs to arrive.

The production department revises language and style. Next, the paper is typeset and proofread by professional proofreaders who identify grammatical/typographical/syntax errors and highlight any inconsistencies. Typeset proofs are sent to authors to clarify the queries raised during the production stages. Once the authors finally approve the galley proof, the publisher publishes the article with final bibliographic details, ending the publication journey of a MS.

47.5 Concluding Remarks

Writing a publishable and citable paper is an arduous job that needs meticulous planning, hard work, and persistence. Finally, if you are passionate about research, you must pursue and persist in research and publications. Each paper is a stepping stone for additional and continuity of research. Demonstrate your dexterity at the bench work, write well and enjoy the benefit of publication and the citation that follows with it. The editors always want to publish good quality papers which advance science and knowledge. The editors often judge between a considerable number of high-quality articles. The submitted MS is more likely accepted if it is meticulously prepared, describes the research that advances the field using clear and concise language, and strictly follows ethical standards.

Conflict of interest None declared.

References

1. Prakash A (2010) Working with journals: cover letter to the journal, peer review and revision process. In: Jagadeesh G, Murthy S, Gupta YK, Prakash A (eds) Biomedical research, from ideation to publication. Wolters Kluwer, New Delhi, pp 491–499
2. Bordage G (2001) Reasons reviewers reject and accept manuscripts: the strengths and weaknesses in medical education reports. Acad Med 76:889–896
3. Brainard J (2022) Reviewers award higher marks when a paper's author is famous. Science 377(6612):1251. https://doi.org/10.1126/science.ade8714
4. Gewin V (2018) The write stuff. Nature 555:129–130
5. Merga MK, Mason S, Morris JE. Tips for negotiating the peer-reviewed journal publication process as an early-career researcher. http://blogs.lse.ac.uk/impactofsocialsciences/2018/11/07/tips-for-negotiating-the-peer-reviewed-journal-publication-process-as-an-early-career-researcher/. Accessed 25 Nov 2022
6. Editorial. (2020) Communication is key to constructive peer review. Nature 582:314
7. Clements JC (2020) Don't be harsh in peer review. Nature 585:472
8. Drubin DG (2011) Any jackass can trash a manuscript, but it takes good scholarship to create one (how MBoC promotes civil and constructive peer review). Mol Biol Cell 22:525–527
9. Kotsis SV, Chung KC (2014) Manuscript rejection: how to submit a revision and tips on being a good peer reviewer. Plast Reconstr Surg 133:958–964

Writing a Postgraduate or Doctoral Thesis: A Step-by-Step Approach

48

Usha Y. Nayak, Praveen Hoogar, Srinivas Mutalik, and N. Udupa

Abstract

A key characteristic looked after by postgraduate or doctoral students is how they communicate and defend their knowledge. Many candidates believe that there is insufficient instruction on constructing strong arguments. The thesis writing procedure must be meticulously followed to achieve outstanding results. It should be well organized, simple to read, and provide detailed explanations of the core research concepts. Each section in a thesis should be carefully written to make sure that it transitions logically from one to the next in a smooth way and is free of any unclear, cluttered, or redundant elements that make it difficult for the reader to understand what is being tried to convey. In this regard, students must acquire the information and skills to successfully create a strong and effective thesis. A step-by-step description of the thesis/dissertation writing process is provided in this chapter.

Keywords

Thesis · Dissertation · Postgraduate · Doctoral · Hypothesis · SMART objectives

48.1 Introduction

The foundation of the entire postgraduate or doctoral research program is disciplinary knowledge. At most universities, one of the main requirements is that the research introduces or expands a novelty that contributes to the advancement of the subject [1]. Even though the writing is a clear component of higher-level coursework and is frequently acknowledged as a source of significant concern for students, it is commonly undervalued. Gaining proficiency in academic writing is necessary for earning a Master or Doctor of Philosophy (Ph.D.) degree and improving the employability of Master or doctoral graduates, who must demonstrate this competency during their professional career. Universities, corporations, Non-government Organizations (NGOs), and government agencies demand efficient writing skills from research job candidates, making this talent increasingly crucial in the global job market. While research degree

U. Y. Nayak · S. Mutalik
Department of Pharmaceutics, Manipal College of Pharmaceutical Sciences, Manipal Academy of Higher Education, Manipal, Karnataka, India
e-mail: usha.nayak@manipal.edu; ss.mutalik@manipal.edu

P. Hoogar
Centre for Bio Cultural Studies, Directorate of Research, Manipal Academy of Higher Education, Manipal, Karnataka, India
e-mail: praveen.hoogar@manipal.edu

N. Udupa (✉)
Shri Dharmasthala Manjunatheshwara University, Dharwad, Karnataka, India
e-mail: udupa1553@gmail.com

G. Jagadeesh et al. (eds.), *The Quintessence of Basic and Clinical Research and Scientific Publishing*,
https://doi.org/10.1007/978-981-99-1284-1_48

programs frequently emphasize subject-specific information, academic writing development is usually underestimated [2]. University's success is increasingly being measured by quality publications, which are also considered one of the requirements for 'academic promotion and funding competitions. Lack of journal publications reduces prospects for knowledge expansion. As a result, producing a strong thesis for doctorate candidates is essential that is good in writing, with structural format, original research, with proper analyses of data and interpretation.

The thesis should be an independent work written entirely by the research candidate, supported by the supervisor or guide in all aspects. It outlines a specific issue that the candidate has tackled, sometimes as part of a larger team and with the support and guidance of academic mentors. It motivates and describes the problem, identifies a clear gap for a potential novel academic contribution through critical analysis, evaluates existing solutions, and lays out a hypothesis, a suggested cause for the issue, or a proposed solution. The work done to choose the theory is also sufficiently explained and justified by the thesis [3]. An idea of the thesis must meet all Master or Ph.D. requirements, adhere to disciplinary standards and expectations, and exhibit advanced writing skills. It must demonstrate critical thinking and uphold a high level of formal literacy, having both accuracy and persistence. Writing for a doctoral degree thesis puts candidates through emotional endurance tests, prompts identity changes, and reassigns them to modern social and scholarly networks [1]. These factors make writing a thesis challenging and call for academic, interpersonal, and emotional guidance. It is also essential that the ability to explain ideas and analyse the data is more critical for a Ph. D. than the amount of writing a student can produce [4]. Through the various writing experiences obtained while pursuing the degree, the candidate will acquire knowledge on how to articulate their ideas, how to arrange their work so that readers of all types may connect with it, and how to develop the best layout so that it is evident to fit all the data together. They also learn what is proper, expected, and helpful in writing contexts. The list of writing abilities that can be applied outside of the academic environment is extensive and is a take-home learning gained through the Ph.D. degree. Time management is another significant point worth highlighting. The need for a routine and the willpower to follow it seems to be the first requirement for time management. The effectiveness of patterns and habits varies markedly from person to person. Without a schedule to finish the writing, imperatives like promises and deadlines may not be enough [5]. Thus, it becomes essential to set a timeline.

The students who seem to be writing frequently for journals or newsletters admit with disappointment that they remain engaged in the same chapter they worked on the week before. Many become overconfident about how soon they can complete specific aspects of their thesis. It appears that there is a connection between writing habits and time management, as well as an awareness of the scope of each writing assignment. Therefore, success in doctorate writing involves figuring out what works for each individual or adjusting to what one's life permits [1]. In this chapter, we will explain how to write a Ph.D. thesis, so that postgraduate or doctoral students can be appropriately guided.

48.2 Steps in Constructing and Structuring a Thesis

Ph.D. students are not likely to have produced numerous theses at the time of writing up; typically, they would have written one in a final undergraduate year and potentially another during post-graduation. However, the work is substantially longer in duration and covers a broader range of topics, necessitating greater attention to detail than any predoctoral studies/work. Most Universities demand that the entire dissertation or its component sections be "publishable" quality. This serves as the yardstick for choosing the information to be included in the thesis. Candidates should consider concepts of quality, credible and novel ideas for thesis writing and should consider the questions like; Are the stated experiments accurate? Is the data acceptable to the scientific community? Would the entire framework hold up to peer review [3]? The most effective method for creating a strong thesis is likely to be thinking through writing. However, frequently the student cannot adequately explain the significance of the study's findings until the study is completed and the results are analysed and interpreted.

A thesis should be as specific and unambiguous as possible; it should not be just a list of questions and answers. Thus, it is always preferable to develop a thesis writing structure before beginning to write. When comparing the thesis with research papers, it is suggested that the main distinction between the two is the amount of meta-discourse used in the thesis. Since a thesis is a longer document than a research article, it is essential to include sections that inform the reader of what is to come and that make connections to other sections related to the topic the author is covering at that time [6]. A typical thesis comprises different chapters, Introduction, Literature Review, Material and Methods, Results, Discussion, Summary and Conclusion, each of which will be covered in more detail in the following sections. It is crucial to understand that a doctoral thesis is not constrained to any one chapter or part. Institutional standards and guidelines, supervisor and researcher preferences all play a role in determining how many chapters a thesis should include. Ideally, the thesis is organised primarily like a scientific research report, with discrete chapters for the introduction, methodology, results, and discussion; this is known as the IMRaD (Introduction—Method—Results—and—Discussion) model. The standard thesis often begins with an introductory chapter, a literature review, and a series of chapters that follow the IMRaD structure and concludes with a general summation chapter. While it is unsurprising that the IMRaD structure has managed to capture the interest of researchers because it is a frequently used format, many researchers have recently emphasised the attention towards alternative writing styles, particularly to research that uses a qualitative approach [6].

The main accomplishment of the candidate is to strike a balance between their points of view and the standards of their discipline. The goal of this book chapter is to assist Ph.D. students in having a visualisation of what and how the thesis should look like, as well as to provide ways of defining common frameworks. The use of frameworks like IMRaD indicates that the written work reflects authentic research contextualised inside the discussion and that the epistemology is based on legitimate methodologies. Students can use IMRaD as a checklist to identify the sections of their thesis that require development and to make sure that the anticipated aims and objectives are clearly visible. Although the initial structure plans are frequently not final, the basic framework of IMRaD remains the same. Students may need to rearrange their strategy as they gradually get a better understanding of their subject. Still, they must be careful to avoid concluding that doctoral writing is a compromised negotiation [1]. Editing the thesis and thoroughly checking for grammatical and typographical errors come as the final steps. In the following sections, a detailed step-by-step explanation is given on writing the different chapters in a doctoral thesis.

48.2.1 Thesis Introduction Chapter

The introduction is the action of introducing something or initiating by presenting the essential details before in-depth discussion, writing, and delving into it. It is an elementary act, part, or portion which would be in use of every spare of our routine life and also in scientific research, such as thesis and paper writing of various types of Graduate, Postgraduate, M.Phil, and Ph.D. It is a fundamental and unseparated part of any kind of thesis. An introduction starts with a statement grabbing the interest and curiosity of the readers. It introduces the central problem and highlights the importance of the investigation. It specifies the research question along with the approach that will be taken to address it. Citing verified evidence emphasises how serious the research issue is [7]. In the introductory part, "Create a Research Space" (CaRS) strategy proposed by Swales is frequently used to explain the presence of an introductory section. It focuses on three characteristics or general strategies that should be presented in the introduction: (1) why it matters, (2) identifying a research topic's gap, and (3) outlining the uniqueness it will bring. These techniques are excellent for seizing readers' attention and highlighting the significance of the research [1].

This is a highly important chapter in the thesis equally as methodology, results and discussion, etc., mainly because of its position placed in the thesis. This has been always named Introduction chapter presented at the beginning. Since it comes in as Chap. 1, author must present the research topic in a diligent manner which should create interest in the reader. The author should provide clarity on the topic and the elements associated with it by defining it, describing the elements and their interaction, relationship, and differences. The systematic organization of background or summary of the existing research in the respective field must be presented. This chapter is also the place where the author has to provide a detailed account of the specific research problem, problem statement, hypothesis, objectives, scientific rationale, and clarify his or her position on the topic and the approach. Finally, the overview of the thesis and related aspects would be accommodated here. For the sake of easy understanding and the widely accepted structure, Chap. 1 of the thesis could be taken as a funnel or inverse triangle. It begins with a broad and general subject related to the research topic or title of the thesis, then narrows down to research questions, hypothesis, objectives, variables and the problems that are going to be solved in the research (see Chap. 1). It can also be said as the travel from divergent to convergent or details to micro details.

In simple terms, the author must explain what exactly a reader is going to read about the research topic, its importance, relevance, how it is going to contribute to society's welfare, and what are the policy implications.

The general structure of the introduction chapter is shown below (Box 48.1).

Box 48.1 The general structure of the introduction chapter is as follows, but not limited to:

- Introduction to the Problem
- Statement of the Problem
- Review of Literature
- Research Gap
- Purpose of the Study
- Research Questions
- Research Hypothesis
- Objectives of the Study
- The Significance/Rationale of the Study
- Definition of the Keywords (respective topic)
- Limitations and Delimitations

48.2.1.1 Introduction to the Problem

The author must establish his/her research territory by defining, explaining and describing the topic chosen for the study or research. It must provide a wider and clearer understanding of the topic to the reader and must be presented from a broader to the specific of the research. Broader may be in terms of geography, population, etc., to specific means to the specified population of geography. This is not the summary of the thesis or a brief version of each chapter. It is a place for introducing the research and its topic and the thesis. The statements, quotes, or definitions from other authors can be placed in the section, which will provide the stronghold, understanding, importance, and relevance of the study topic.

48.2.1.2 Statement of the Problem

The statement of the problem is the central or focal point of the whole research or the thesis because everything further in the thesis goes around this and tries to address issues of multiple concerns and features. A statement of the problem is highly specific to the study usually written in six to eight paragraphs with a strong aim for each one. This section can be identified and written with Topic, Gap, evidence, Deficiencies, and Audiences:

- Topic: need to state the specific problem both from theoretical and practical points of view.
- Gap: Clearly mention that of not solving or not attempting this specific problem in previous research works.
- The evidence: must be written that the researchers indicate the problem prevails by referring to the earlier studies.
- Deficiencies: must be demonstrated how you as an author solved the problem and how the identified gap was filled.
- Audience: whom your study is directed to, and where it would be useful.

Note:

- There is no need of providing resources or references for every single statement made here except for the evidence section.
- The mention of a research gap in two places; the introduction and statement of the problem must not be confused. In the introduction section, it is a border gap but in the statement of the problem, it is a specific one where the author needs to clearly define and provide details.

48.2.1.3 Review of Literature

The process of reviewing literature will help the author to understand the various aspects of the research topic including what, where, how and among whom the studies were conducted. This will give an idea of methodological applications, study implications and an overview of the respective research area. At the same time, this process will provide opportunities to identify the research density (overworked) and the gaps, which further leads the researcher to consider those aspects of the research.

48.2.1.4 Research Gap (Niche)

The author must clearly state the research gaps identified from the literature review and cite pertinent references. Here one must include methodological gaps, research topic gaps, theoretical gaps, etc., which in turn help in constructing a strong background to conduct research.

48.2.1.5 Purpose of the Study

Indicate what you want to learn, what you try to find, and what you want to reveal from this research.

48.2.1.6 Research Questions

One of the key components of a thesis is the research question. It concentrates on the study, governs the approach, directs all of the phases of research and analysis, and then provides the solution to the problem. The research question needs to elaborate on and reflects a greater comprehension of the subject. As a researcher, demonstrate how the study will bridge a gap in the knowledge base. A Ph.D. thesis should concentrate on issues such as: (1) What, Where, and (2) Why is the subject important? (3) What is the issue and how can it be resolved? (4) If the problem is not resolved, will it continue to exist? (5) Who is negatively impacted by the problem? (6) Does this problem support or refute previously held beliefs? An effort should be made in the thesis (in the discussion section) to answer all such questions [7].

In a research question, you specify precisely what you want to learn from your work. Your research paper, dissertation, or thesis should be guided by a well-chosen research question. All research questions ought to be focused on a single problem or issue that can be researched using primary and secondary sources and can be answered within the time frame and practical constraints. Specific enough to answer in-depth and complex enough to develop the answer over the course of a paper or thesis. Relevant to your field of study and society as a whole.

48.2.1.7 Research Hypothesis

An unconfirmed theory is called a hypothesis. It makes it possible to estimate or forecast an outcome or relationship under particular circumstances. To evaluate its validity and dependability, it must undergo thorough testing after which, if proven, it turns into a scientific theory. A hypothesis does not have to be a component of a study; it merely aids the researcher in seeing the issue more clearly. The hypothesis is a declaration and an educated guess that the student is not sure of and should be validated to see if it confirms the claim being made in the research or not. Therefore, the candidate should emphasise the proposed hypothesis and how it relates to the thesis topic [7].

The author must respond to the research questions posed in the preceding part of this section. For example, if one of the questions is "Is there any statistically significant relationship between parents' educational status and children's nutritional status in India?" then the hypothesis would be "There is no statistically significant relationship between children's nutritional status in India and their parents' educational status." Use only null hypotheses. It is a type of statistical hypothesis that makes the case that a given set of observations does not have any statistical significance.

48.2.1.8 Objectives

The objectives force us to be precise about methods and to define key terms of the research protocol/proposal. The objectives will specify what scientific questions the study is designed to answer. It is developed logically from the topic, research questions, and hypothesis. The statement of objectives is essential for selecting the factors to be investigated, the response variables to be measured, the data needed to describe the effects of the factors, and the kind of statistical analysis required. Thus, when the study is completed, the results will be compared to the objectives and research questions. Unless the objectives are defined, the project cannot be initiated. The thesis should compare the results to the objectives specified in the beginning.

The objectives should cover the entire breadth of the project. Take into consideration-contributing factors, and variables (drug treatment). Arrange them in a logical sequence. Group the objectives: they are sometimes organized into hierarchies: primary, secondary, and exploratory. Or General and specific. For additional details, see Chaps. 1 and 4.

48.2.1.9 Significance/Rationale of the Study

The author is expected to provide details in this section on how his or her research has a large impact on other studies, the area of study as a whole, and some other key individuals or the specific population. You can achieve this by asking yourself how and why this study would be essential. In the conclusion chapter of the theses, researchers frequently discuss any gaps in the literature that they discovered while conducting their research. They can serve as evidence of the importance of your research.

48.2.1.10 Definition of the Keywords

The author has to define the discipline specific keywords theoretically and operationally. The keywords are nothing but the different variables of the thesis. The theoretical definitions come from earlier publications in the respective area of research, where the author has to quote and cite definitions. This is also a process of placing the study in the research area. Then the task is to define the selected keywords, and what they mean in the present study. This can be called an operational or empirical definition. The whole study will be abiding by the operational definition, its meaning, and scope.

48.2.1.11 Limitation and Delimitation

Although they appear to be synonymous, limitation and delimitation are not. Particularly in research, their meanings, scope, and effects differ. They are essential components of any study that spans time. Some members of the scientific community have the incorrect idea that they would like to conceal it or not document it, which is an unacceptable practice. They are not as bad as people think they are. They are essential to record not only to demonstrate the study's shortcomings, but also to demonstrate the research's clear understanding, methodology, focus, and scope. The study's deficiencies are primarily the result of external factors or factors beyond the researcher's control. They could be the researcher's theoretical and practical constraints. Delimitation is a factor that is entirely within the researcher's control and is used to define or focus on a specific research problem or incident. It mostly talks about the research question and the scope of the research aim.

The author is obligated to document all factors beyond their control, such as the duration of the study, access to funding, equipment, participants, and so on under the limitation section and the consideration of the limitations with a clear comprehension of the study's requirements as outlined in the delimitation section.

48.2.2 Literature Review

All literature cannot be covered in the introduction chapter. Usually, a separate chapter with a detailed literature search on the problem statement, and existing literature related to the proposed work and methodology will be included. The students become more familiar with the ideas and works of others through the literature review process. It is, therefore, imperative to read relevant literature from reliable sources as much as possible. It is a segment that is present in a Ph.D. programme from the start to the finish, and thus it becomes vital to compile literature and information as and when it is gathered from the standard databases. Furthermore, it is crucial to understand that every claim will have a counterargument. Therefore, preparing for it will aid in effectively shaping the thesis. The next point that ought to cross the student's mind is HOW am I going to approach my study so that I can find a solution? It is important to consider WHY you want to conduct the study. What is the theoretical foundation of the research problem's investigation? To conduct the study scientifically, students must establish a research or experimental design [8]. If the student wants to use any figures and tables in the literature section of their thesis, permission from the publisher is required.

48.2.3 Aims and Objectives

The literature review is followed by defining the problem statement and arriving at the objective (s) of the study. The possible solutions for the unanswered questions are to be discussed to derive a hypothesis.

One of the most crucial components of the thesis is how the research aim and objectives are developed. This is so because the purpose and goals of the research dictate its depth and scope [9]. The goals would then be focused on how to accomplish the proposed study's goals [10]. Aims are declarations of intention. They are typically expressed in broad terms. They outline the results you anticipate getting from the endeavour. Contrarily, objectives should be clear statements that outline measurable consequences, such as the procedures that will be followed to bring about the intended result.

Aims and objectives usually describe the main focus of any research project. Basically, objectives are the action that the researcher will undertake to achieve the specified aspect of the research or project. In other words; the outcome researcher wants to achieve by conducting research is called the research objective. Objectives are the guiding force for any research or project. There may be multiple objectives in a research or project. The author must write clearly and specifically what the study answer or give the solution for a problem as an aim of the study and the specific mention how those answers or solutions were achieved as objectives. It means breaking up the most important ideas or aspects of your research aim into several smaller parts that together represent the whole.

Objectives may be designed based on the SMART approach; Specific, Measurable, Achievable, Realistic, Time constrained [11].

- Specific: Does the action you plan to take have any ambiguity, or is it focused and clear?
- Measurable: How will you track your progress and determine when the action has been completed?

- Achievable: do you have the help, assets and offices expected to do the activity?
- Relevant: Is the action crucial to achieving your research objective?
- Time-bound: Are you able to finish the action in the time you have available while still working on other research projects?

It is also a best practice of using verbs at the beginning of the objective writing that help convey your intent, in addition to following SMART. For example;

- In the case of Understanding and organising information verbs can be used Review, Identify, Explore, Discover, Discuss, Summarise, and Describe.
- In the case of solving problems using information verbs can be used Interpret, Apply, Demonstrate, Establish, Determine, Estimate, Calculate and Relate.
- Reaching conclusion from evidence verbs can be used Analyse, Compare, Inspect, Examine, Verify, Select, Test, Arrange.

48.2.4 Materials and Methods

The methodology is the most crucial part of the research design. Knowing the difference between methodology and method is important for researchers. While method relates to the systematic organisation and measurement of your research, methodology describes the theoretical interpretation of the research. The approaches used for various studies vary depending on the subject [8].

The methods used for data collection and analysis must be discussed and explained in your research methodology in the past tense. The methodology chapter or section is an essential component of your thesis, dissertation, or research paper. It explains what you did and how you did it, allowing readers to assess the reliability and validity of your research and the topic of your dissertation. This provides the opportunity for sharing how you directed your examination and why you picked the tools, methods, and techniques you picked. It is also where you should demonstrate that your research was rigorously carried out and that it can be replicated.

In experimental research, chemicals, equipment, biosamples, and other materials are used as materials and methods. You describe the study in detail in this section. If it's a plant or animal, include its scientific name. For a better understanding, you can look at previous theses or publications that are closely related to your research. The following are some examples: Specify the location from where the chemicals/samples were purchased, obtained, or harvested. Include information about the age, sex, group size, treatment and control groups. For reproducibility, this information will provide precise recommendations. Chemicals require information such as brand and manufacturing location, food or chemical/analytical grade, the molarity of chemicals concentration and enzyme activity.

48.2.5 Results

It is always beneficial, to begin with, a brief abstract-style introduction outlining the goals of the chapter. A brief description would enable the reader in recalling the research topic, questions, and hypothesis so they could comprehend the findings more fully while reading this chapter. Write a summary paragraph that is comparable to the introduction at the end of the Results chapter [8]. Theoretically, everything must have been planned out from the beginning, but in practice, things may not have turned out as intended, and a student may need to embellish their work with a narrative [3].

The output of the experiments is to be disseminated to the reader with good clarity. Three to four years of efforts of the candidates can be recognized by the way of representing the results obtained in the experiments. Compiling the results after completing the entire duration of Ph.D., will be a tedious task and it may not be ideal. Candidate should be capable of analysing the experimental results as and when performed and data collected. At the same time, the data is to be processed and interpreted and documented in written format. This will assist and show the way for the next experiments. The flow of writing the results may vary with the disciplines. However, the results should follow the methods mentioned in the previous section, ideally with the same titles and subtitles. This will help the reader to follow the stepwise process and link to the previous experiments. A systematic way of representing data is the beauty of writing skills.

The data presented should be clear, precise and concise. Usually while writing the results, a question may arise about whether or not to include negative results. The positive results are those which are in agreement with the hypothesis. If the results are not as per the hypothesis, then it may require further study. After reperforming the experiment, if the results are not good or relevant, they can be considered negative results. It is always preferred to include the negative results. This may be of marginal interest, however, it may help future scientists, so that such experiments need not be repeated by others. Hence always mention the limitations of the experiments in the Results and Discussion sections.

The results that can not be adequately presented in text form, can be represented either in table or graphics (figure) form. The tables are used to represent numerical data, and figures are used for comparing the meaningful relationships between the groups or to study the trend. Figures are easy to understand and convey the message quickly. It is always better to avoid duplicating the results in tables and figures. Sometimes, due to unavoidable reasons, tables and figures may be required, especially when qualitative and quantitative/numerical comparison is involved in the results (for example, see Chap. 45). While representing the values in a table, restrict it to two decimal points and maintain uniformity throughout. The tables and figures must be simple and self-explanatory. Figures may be a line graph, scattered plot, bar chart, histogram or pie chart (see Chaps. 29 and 44). Special emphasis is to be given to the table title, subtitles, units, figure axis titles and magnification. Also, the abbreviations used in tables and figures are to be defined. User friendly colour codes are to be used in the figures. More importantly, cite the tables and figures at appropriate text locations.

The statistical comparison of all the results (see Chaps. 29 and 30) is a must to include in a thesis. The number of repeated trials, statistical terminologies (mean, standard deviation, standard error of the mean, measure of variability, p value, F-value), symbols, colour codes if any, are to be given as footnotes in the table and the figure legends.

48.2.6 Discussion

Writing the discussion in a thesis is a very crucial and difficult part. The discussion section is the most significant part of the thesis. The thesis's central argument and the culmination of results can be found here. Here the student will discuss why the issue raised is important, what people have done to address it, where opportunities exist, and what novelty was chosen about contributing to this area. Here candidates will discuss their ideas, criticisms, comprehension, experimental strategy, data analysis and summary, visualisations, insights, and conclusions.

The discussion of the results includes the explanation for the practical experimentations or interpretations of the results, such as their usefulness and relevance to the current situation. The results are linked to the hypothesis. The discussion should address whether the results are fully or partially related to the hypothesis. Do not repeat the results, instead highlight the significant aspects. The discussion part should give information on the findings of the experiments and the reason for those results with supporting literature if any. Why the results are important and how they may affect the outcome of the research, is there any added value for the existing knowledge to be discussed? The scientific basis/evidence from the experiment is to be provided while discussing every aspect of the results. If there is no supporting literature, it should be clarified that the results obtained are unique and/or different from the existing ones.

It is very important to maintain research integrity while discussing the thoughts on the results. If the results are not according to the existing literature, in spite of repeated experiments, the researcher should have the courage to write the rebuttals to the literature in the thesis. Sometimes the published literature might not have been done under proper experimental conditions or authors might have missed the step(s) in the methodology section. Rebuttals will help to correct such errors. Occasionally, some of the candidates may tend to do falsification of data if the results are not according to the literature. Such kind of misconduct is an unethical practice and may ruin the future of the candidate (see Chaps. 58 and 59). Instead, write the limitations of the experiments and the possible reasons for obtaining different results [10]. Publishing the research work in the journals before thesis submission will also help in writing a good discussion for the thesis. The reviewers comments will help to improve the quality of the representing the results.

Statistical differentiation of data and discussion will enhance the confidence level of the thesis presentation. Further, the inclusion of scope for future studies or suggestions to improve the results will also encourage to take the study forward and corrections to be made in the method/procedure followed [12]. Future perspectives can also be a part of the conclusion of the study.

48.2.7 Summary and Conclusions

Writing the introduction and the conclusion sections is equally difficult. One significant distinction, though, is that in the conclusion chapter, questions raised in the introduction are addressed. Even if the "Conclusions" requires a paragraph that may describe the entire argument, it is crucial to keep in mind that it is not a synopsis of the Introduction. It could be helpful to quickly recapitulate the study questions and hypotheses in the conclusion chapter to connect them to the discussion of the results. In order for the scientific community to accept and acknowledge the thesis' contribution, the Conclusion part, like the other chapters, should be written in a scientific style. Instead of discussing the limitations, it is preferable to emphasise that the thesis has a valid and measurable result [8].

The conclusion section should also summarize the entire research including the problem statement, methodology, results and discussion in a concise manner. The conclusion should indicate whether objectives are fulfilled and the hypothesis is confirmed or refuted. Only the important research findings are mentioned without repeating the sentences. How the results are encouraging for the clinical settings is to be discussed. The conclusion should also include the future directions of the study. As readers usually prefer to see the abstract and conclusion, the outcome of the study must be conveyed in such a way that readers agree with the researcher's views.

48.2.8 Referencing Section in the Thesis

Studies in science are motivated by earlier work. Therefore, no one can claim that their research was carried out independently, without referencing the work of other scholars. The information sources that were used by researchers must be listed under the References chapter. They are the key components of research as they link various information sources and help the readers spot knowledge gaps and deepen their understanding of a particular subject. They also help in verifying the originality of the arguments presented in the thesis. As a result, every researcher needs to cite sources to support their findings [13].

Any research in the modern era begins with the earlier work, or there will be some reading material for their hypothesis, and they are called references for the work [14]. The references can be anything that is referred to obtain the data for the thesis work. The primary source of references are journals or periodicals and sometimes books. Books are said to be a lesser priority as the change in knowledge is higher in journals and magazines than in textbooks. However, in recent days, the references can be a diary or a lab notebook of a scientist, photographs, figures, diagrams, or even videos. Every statement which brings forward the point of discussion should preferably have a reference. In the research communications, except the abstract, results (there could be an exception) and the conclusion will have references as they originated from the thesis work. Depending on the design of the thesis outline, generally, the references will be at the end. However, some universities accept the references chapter-wise. The chapter-wise referencing will make the references repeat and end up in extra pages in the thesis. Traditionally, the references come under the side heading of the bibliography. However, in recent days the most acceptable side-heading is 'References'. Whatever the outline format of a thesis, the critical aspect is citing those references in the right place, and keeping the references at the end of the thesis is challenging. Hence, the method of citing plays an essential role in being learned and interpreted before starting the thesis. There are several standards referencing styles to cite the works. However, it is ideal to follow the referencing style suggested by the respective Universities where the Ph.D. work is being done.

Citing references in research communication plays an important role. It validates the idea of the argument in the hypothesis, enables the readers or future researchers to follow the original work, provides credit to the original author or scientist, or inventor, avoids plagiarism to a certain extent, keep-up the academic honesty and the research integrity, and it proves the extent of reading and interpretation that has happened over time [15].

There are multiple ways of citing references in the text and the bibliography. For details, readers should read Chap. 39. There are many referencing styles, such as the Modern Language Association (MLA), the American Psychological Association (APA), the Chicago Manual of Style (CMS), the Vancouver system, the Harvard system, the American Medical Association (AMA) Style, the American Chemical Society (ACS) Style, the American Institute of Physics (AIP) Style, the American Political Science Association (APSA) Style, the Institute of Electrical and Electronics Engineers (IEEE) Style, the Turabian style, the Modern Humanities Research Association (MHRA) style, Oxford system, etc. [16]. The most common one is Vancouver and Harvard referencing method in biomedical sciences. However, American Psychological Association (APA) format is also followed in many instances. Both Harvard and Vancouver reference styles are parenthetical referencing in which the references in the text are inserted using parenthesis or brackets. The Harvard system is also called 'Author–Date system', the text citation uses using sir name followed by the year, and in the bibliography, references are sorted alphabetically and then chronologically. The Vancouver system uses numbers for citation in the text, and the references are sorted numerically in the bibliography [15].

To simplify the citing to reduce the ambiguities, a large group of publishers and societies came forward and introduced Digital Object Identifier (DoI) with the ISO standard (ISO 26324) by forming International DOI Foundation (IDF) (https://www.doi.org/) [17, 18]. The DoI connects digitally to the article of origin. Hence, DoI will play the most important role in the future referencing system. Before we understand the important components to be cited in the reference, we need to understand the fundamental need for references in the thesis or any research communication. To know the basic objectives of referencing one needs to know the different components to be cited in the references, following are a few important ones.

- Journal article: Authors name, the title of the article, year, volume, issue, page number/article number (for online journals), DoI (for digital archives).

- Textbook/book: Editors, book title, publisher, place and year of publication. DoI or ISBN (International Standard Book Number).
- Book-Chapter: Author names of the book chapter, followed by writing 'in' Title of the book, publisher, place and year of publication. DoI or ISBN (International Standard Book Number), page range.
- Websites/blogs/social media: Author name, year, page title. Available at: URL (Accessed: Day Month Year).
- Images/videos/ podcasts: Author/presenter name, year, title (video/podcast). Day/month/ year. URL with accessed day month year.

Overall, managing references is the most crucial Herculean task. However, nowadays, computer tools make it easier. Hence, the researcher should have proper knowledge of these tools, and it is also essential to have the skills to manage them. Looking at the complexity of citing and managing references, scientists in informatics have developed electronic tools for managing references in the thesis and journal publications. These tools are highly versatile and convenient in modern-day science communication. Many such tools include EndNote, Mendeley, RefWorks, Zotero, and ReadCube [19–21]. Most Universities across the world used EndNote and Mendeley. EndNote is a product by Clarivate Analytics (formerly Thomson Reuters), currently a subscription-based product. However, EndNote is one of the best reference tools with high compatibility with almost all journals and publishers [22]. Currently, Mendeley is not just helping to manage references, but it also helps as social media for the researcher to share their findings or research updates.

48.3 Concluding remarks

Writing a thesis or dissertation is a scientific art, that involves the depiction of the research skills of a scholar. After performing detailed experiments for 1 to 3 years, it is extremely difficult to summarize the results in a crisp manner. It is obvious that the candidate might be struggling for the selection of required data to be included in the thesis. For a decision to be made, the candidate has to check the results against the objectives and hypotheses. Special attention must be paid to ethics as well. Most ethical issues have unique codes of conduct, such as how to handle human/ animal or tissue samples or carry out clinical trials. If these regulations apply, they must be followed precisely within the thesis. Even without evidence of deceit, plagiarism can result in the revocation of a Ph.D. award years after it has been granted. Hence care must be taken while writing the thesis and including the reported literature. Special attention is to be given during the final checking of the thesis. The candidate should ask someone else to read the thesis who is not familiar with the current research problem, for readability and grammar check. Overall, the meticulous planning and execution of the research work may help in writing a good thesis.

Conflict of interest No conflict of interest exists.

References

1. Carter S, Guerin C, Aitchison C (2020) Doctoral writing: practices, processes and pleasures. Springer, Singapore. https://doi.org/10.1007/978-981-15-1808-9
2. Odena O, Burgess H (2017) How doctoral students and graduates describe facilitating experiences and strategies for their thesis writing learning process: a qualitative approach. Stud High Educ 42:572–590. https://doi.org/10.1080/03075079.2015.1063598
3. Stefan R (2022) How to write a good PhD thesis and survive the viva, pp 1–33. http://people.kmi.open.ac.uk/stefan/thesis-writing.pdf
4. Barrett D, Rodriguez A, Smith J (2021) Producing a successful PhD thesis. Evid Based Nurs 24:1–2. https://doi.org/10.1136/ebnurs-2020-103376
5. Murray R, Newton M (2009) Writing retreat as structured intervention: margin or mainstream? High Educ Res Dev 28:541–553. https://doi.org/10.1080/07294360903154126
6. Thompson P (2012) Thesis and dissertation writing. In: Paltridge B, Starfield S (eds) The handbook of english for specific purposes. John Wiley & Sons, Ltd, Hoboken, NJ, pp 283–299. https://doi.org/10.1002/9781118339855.ch15
7. Faryadi Q (2018) PhD thesis writing process: a systematic approach—how to write your introduction.

Creat Educ 09:2534–2545. https://doi.org/10.4236/ce.2018.915192
8. Faryadi Q (2019) PhD thesis writing process: a systematic approach—how to write your methodology, results and conclusion. Creat Educ 10:766–783. https://doi.org/10.4236/ce.2019.104057
9. Fisher CM, Colin M, Buglear J (2010) Researching and writing a dissertation: an essential guide for business students, 3rd edn. Financial Times/Prentice Hall, Harlow, pp 133–164
10. Ahmad HR (2016) How to write a doctoral thesis. Pak J Med Sci 32:270–273. https://doi.org/10.12669/pjms.322.10181
11. Gosling P, Noordam LD (2011) Mastering your PhD, 2nd edn. Springer, Berlin, Heidelberg, pp 12–13. https://doi.org/10.1007/978-3-642-15847-6
12. Cunningham SJ (2004) How to write a thesis. J Orthod 31:144–148. https://doi.org/10.1179/146531204225020445
13. Azadeh F, Vaez R (2013) The accuracy of references in PhD theses: a case study. Health Info Libr J 30:232–240. https://doi.org/10.1111/hir.12026
14. Williams RB (2011) Citation systems in the biosciences: a history, classification and descriptive terminology. J Doc 67:995–1014. https://doi.org/10.1108/00220411111183564
15. Bahadoran Z, Mirmiran P, Kashfi K, Ghasemi A (2020) The principles of biomedical scientific writing: citation. Int J Endocrinol Metab 18:e102622. https://doi.org/10.5812/ijem.102622
16. Yaseen NY, Salman HD (2013) Writing scientific thesis/dissertation in biology field: knowledge in reference style writing. Iraqi J Cancer Med Genet 6:5–12
17. Gorraiz J, Melero-Fuentes D, Gumpenberger C, Valderrama-Zurián J-C (2016) Availability of digital object identifiers (DOIs) in web of science and scopus. J Informet 10:98–109. https://doi.org/10.1016/j.joi.2015.11.008
18. Khedmatgozar HR, Alipour-Hafezi M, Hanafizadeh P (2015) Digital identifier systems: comparative evaluation. Iran J Inf Process Manag 30:529–552
19. Kaur S, Dhindsa KS (2017) Comparative study of citation and reference management tools: mendeley, zotero and read cube. In: Sheikh R, Mishra DKJS (eds) Proceeding of 2016 International conference on ICT in business industry & government (ICTBIG). Institute of Electrical and Electronics Engineers, Piscataway, NJ. https://doi.org/10.1109/ICTBIG.2016.7892715
20. Kratochvíl J (2017) Comparison of the accuracy of bibliographical references generated for medical citation styles by endnote, mendeley, refworks and zotero. J Acad Librariansh 43:57–66. https://doi.org/10.1016/j.acalib.2016.09.001
21. Zhang Y (2012) Comparison of select reference management tools. Med Ref Serv Q 31:45–60. https://doi.org/10.1080/02763869.2012.641841
22. Hupe M (2019) EndNote X9. J Electron Resour Med Libr 16:117–119. https://doi.org/10.1080/15424065.2019.1691963

Part IX

Communication Skills

Poster Presentation at Scientific Meetings

49

K. Gokulakrishnan and B. N. Srikumar

Abstract

The poster presentation is an integral part of the scientific journey. It plays an important role in developing the research career of the researcher, especially novice researchers and budding scientists. The process of poster presentation begins with the abstract submission to the conference secretariat. Once the abstract is accepted for presentation, the preparation of the poster begins. Before making the poster, it is essential to draft, review and revise the content in IMRaD format without much emphasis on the discussion. Then design the poster using presentation or any other designing software, review and revise the content in the layout before printing it. When designing the poster's layout, organize illustrations and text using a grid plan, placing most significant findings at eye level and using muted background colours. It is better to keep the content less, with good illustrations to attract the viewers' attention. Once the content and design are completed, it is important to edit carefully and review meticulously. Before the presentation at the conference, rehearse the presentation to improve the flow of thoughts. Utilize the opportunity to develop the science and network with peers. Poster presentations, when done well, can significantly contribute to scientific and career advancement. This article aims to provide a comprehensive approach to preparing a poster and making a successful presentation.

Keywords

Communication · Poster presentation · Conference · Scientific meeting · Career advancement · Networking

K. Gokulakrishnan (✉)
Department of Neurochemistry, National Institute of Mental Health and Neuro Sciences, Bengaluru, Karnataka, India
e-mail: gokul@nimhans.ac.in

B. N. Srikumar
Department of Neurophysiology, National Institute of Mental Health and Neuro Sciences, Bengaluru, Karnataka, India
e-mail: kumarasri@nimhans.ac.in

49.1 Introduction

Poster presentation in scientific meetings is a way of effectively communicating the research findings in a nutshell, taking advantage of the poster as a visual aid. Notably, the poster presentation offers a forum for close interaction. But it becomes difficult if the author must describe and explain the poster's content to a succession of viewers. Therefore an effective poster is self-contained and self-explanatory so that the viewers can proceed on their own while promoting discussion where required [1]. The poster presentations offer several advantages. The first and foremost advantage is that the format permits

G. Jagadeesh et al. (eds.), *The Quintessence of Basic and Clinical Research and Scientific Publishing*,
https://doi.org/10.1007/978-981-99-1284-1_49

research presentations that might still be in progress. It enables the presenter to confidently present the research findings, interact with peers and experts in the field, and provide an opportunity for one-on-one interaction. This will help the presenter to come up with ideas to address the potential gaps and expand the research work. It also triggers networking, exchange of ideas and creates opportunities to collaborate with peers. Unlike oral presentations, the time spent on a poster is determined by the viewer rather than the author [1]. The advantages of the poster presentation are given in Box 49.1. While there is a notion that poster presentation is done only by the novice in the field, this may not be true, particularly in larger society meetings and conferences, where even established scientists use posters to communicate. This article provides a comprehensive approach to preparing a poster and making a successful presentation.

Box 49.1 Advantages of a poster presentation

- Work in progress can also be presented
- Builds confidence to present the research findings
- Promotes interaction with peers and experts in the field
- Provides an opportunity for a one-on-one interaction
- Ideas to address the potential gaps and expand the research work
- Triggers networking, exchange of ideas, and creates opportunities to collaborate with the peers

49.2 Planning for a Poster Presentation

49.2.1 Abstract Submission

It is often not necessary to complete the proposed research and submit an abstract for poster presentation at a conference or scientific meeting. Therefore, a poster presentation can be made even with preliminary findings, which can give further directionality to the proposed work. Identifying a scientific meeting or conference that would be appropriate to present the work and capitalize the scope is essential. It is worthwhile investing time in preparing a good and relevant abstract so that it will be selected for presentation. Make the title of the poster interesting and concise so that it will grasp the attention of the attendees [2]. Carefully go through the guidelines given by the organizers for the submission of abstracts. Depending on the conference guidelines, the abstract may need to be structured. Irrespective of the structuring requirement, the abstract should contain the following: It should begin with the background or rationale of the work, followed by aims/objectives, methodology in brief, results, and conclusion. Pay attention if there is a word limit for the abstract and restrict the abstract within the limit without compromising scientific content. Before submitting to the conference secretariat, obtaining the inputs and suggestions of the mentors and expert colleagues is often helpful. Then, submit the abstract according to the Conference guidelines.

49.3 Preparing the Poster

Before starting the preparation for the poster, it is critical to go through the guidelines given by the conference secretariat. This would include the dimension of the poster and the poster material. Sometimes, they might offer detailed guidelines for poster preparation and presentation. It is always better to start the poster preparation well ahead of allowing sufficient time for each of the following:

- Drafting content
- Reviewing draft content
- Revising content
- Designing the poster
- Reviewing and revising content in a layout
- Obtaining final approval
- Printing

With the advent of technology, making a poster can be quite easy with presentation software programs such as Microsoft PowerPoint. Butz et al., have provided a detailed description of using PowerPoint to make a poster [3]. This will also facilitate the ease of drafting and reviewing the content. The presentation software can easily convert the final content to a large poster size. The availability of large size printers and page-layout software permits the economical production of effective and attractive posters on a single sheet that can be easily carried to conferences in a poster tube or as a folded one [1]. Use line borders to separate areas. Avoid reflective, plastic-coated paper. Nowadays, a miniature poster, which can be handed out to the viewers, may be a good idea to attract collaborations.

49.3.1 The Layout of the Poster

Consider organizing illustrations and text using a grid plan. Arrange materials in columns rather than rows–this format is more straightforward for viewers to read. Place the most significant findings at eye level immediately below the title bar; place supporting data and/or text in the lower panels. Use muted background colors—shades of gray are also effective [1].

49.3.2 Content of the Poster

One of the most common mistakes is too much content on the poster, which makes it appear cluttered. Always remember that a poster is not a manuscript; hence content should be selected judiciously [2]. Begin with the essential elements that correspond to those of the abstract and include the following:

Header with title, author(s), institution(s), indicate the presenting author, and provide the presenting and corresponding author's contact information. Similar to a research article, follow the IMRaD format (Introduction, Methods, Results and Discussion/Summary) [4]; however, a detailed discussion may not be required. Include acknowledgements and references as appropriate. An abstract is not required on the poster. There are also innovative ways of poster presentation, including the QR code to access the poster or additional details virtually [5, 6]. There can be no single way to design an effective poster. We present an example in Fig. 49.1, which is meant to serve as a general guide when preparing your poster.

An effective poster distributes illustrations and text equally and should not be a page-by-page printout of a journal article or a slide presentation [1]. Minimize the text because the attendees may not have the time to read; instead, use figures or flow charts [2]. Box 49.2 provides the do's and don't's of poster presentation.

Box 49.2 Dos and Don'ts of a poster presentation

Dos	Don'ts
• Go through the guidelines	• Do not use too much text
• Determine the audience, focus of the event	• Do not clutter the poster with a lot of content
• Plan ahead, start with a skeleton and draft the content, use optimal range of font size	• Do not use small font size or small figures
• Arrange materials in columns	• Do not forget to remove the poster after the presentation
• Place significant findings at eye level	
• Use captivating visual elements	
• Use colours judiciously	
• Revise meticulously	
• Rehearse the presentation	
• Paste the poster on time and be available during the poster session	

49.3.3 Attract the Attention of the Viewers

The viewer's attention span depends on the presentation and content [7]. Therefore, it is important to capture the attention of the viewers—which leads us to the popular 10–10

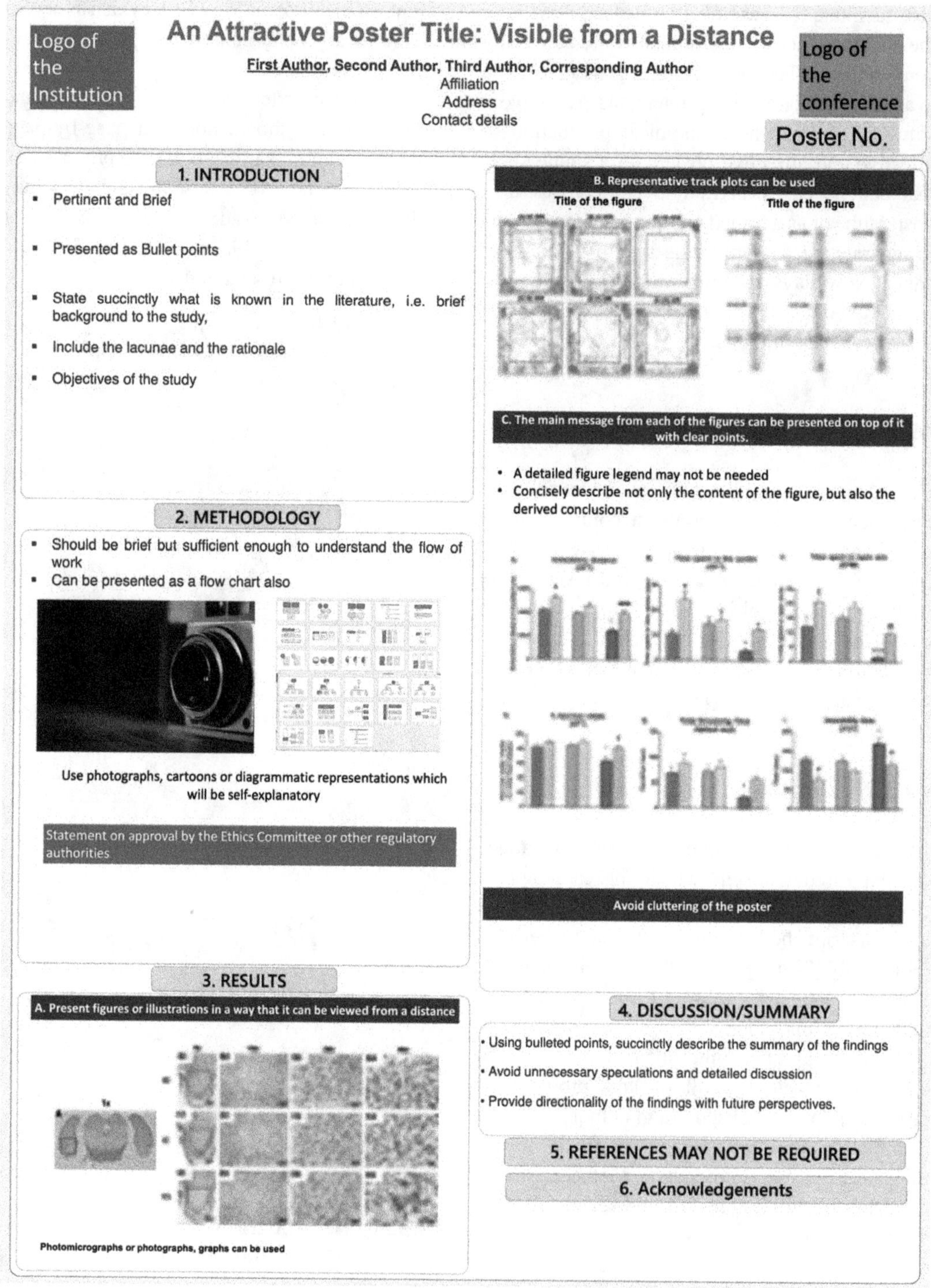

Logo of the Institution

An Attractive Poster Title: Visible from a Distance

First Author, Second Author, Third Author, Corresponding Author
Affiliation
Address
Contact details

Logo of the conference

Poster No.

1. INTRODUCTION

- Pertinent and Brief
- Presented as Bullet points
- State succinctly what is known in the literature, i.e. brief background to the study,
- Include the lacunae and the rationale
- Objectives of the study

2. METHODOLOGY

- Should be brief but sufficient enough to understand the flow of work
- Can be presented as a flow chart also

Use photographs, cartoons or diagrammatic representations which will be self-explanatory

Statement on approval by the Ethics Committee or other regulatory authorities

3. RESULTS

A. Present figures or illustrations in a way that it can be viewed from a distance

Photomicrographs or photographs, graphs can be used

B. Representative track plots can be used

Title of the figure

Title of the figure

C. The main message from each of the figures can be presented on top of it with clear points.

- A detailed figure legend may not be needed
- Concisely describe not only the content of the figure, but also the derived conclusions

Avoid cluttering of the poster

4. DISCUSSION/SUMMARY

- Using bulleted points, succinctly describe the summary of the findings
- Avoid unnecessary speculations and detailed discussion
- Provide directionality of the findings with future perspectives.

5. REFERENCES MAY NOT BE REQUIRED

6. Acknowledgements

Fig. 49.1 Sample poster

rule—attendees spend only 10 s scanning posters as they stroll by from a distance of 10 ft. [2, 8]. Accordingly, use captivating visual elements (graphs, photographs, illustrations, and even cartoons) to attract attendees to your poster, and sparingly use tables. Each section and illustration should have meaningful titles without needing text such as figure legends [1, 2]. Colours are effective, however, excess use will make the poster gaudy; it is better to use contrasting colours for backgrounds and foregrounds (use dark colours on white or pale backgrounds and light colours on dark backgrounds) [1]. Finally, determine the event's audience and focus and prepare the poster accordingly.

49.3.4 The Title

The title is critical to attract attention of the attendees. For example, the title can be a decisive question, the scope of the study, or highlighting a new finding. While it is possible that some scientists may arrive at specific posters depending on their interests, others may see the title and decide to visit the poster on the spot. Therefore, the title should be catchy and lettered in a font (at least 1-in. high) that is visible from a distance, short and understandable to a broader audience [1].

49.3.5 Introduction

It is good to keep the introduction pertinent, brief and presented as bullet points. State succinctly what is known in the literature, i.e., a brief background to the study, move on to the lacunae and the rationale, and end with the objectives.

49.3.6 Materials and Methods

Unlike a manuscript, the materials and methods section should be brief but sufficient enough to understand the flow of the work and techniques involved. It could also be presented as a flow chart. The use of photographs or a diagrammatic representation of the methodology which will be self-explanatory may also be an option. Alternatively, the methodology and study design could be merged into the pertinent results. It is important to include a statement on the Ethics Committee or other regulatory approvals for the study (if appropriate).

49.3.7 Results

Present figures or illustrations in a way that can be viewed from a distance and avoid too many figures, which would make the poster crowded. The main message from each of the figures can be presented on top of it with clear points. A detailed figure legend may not be necessary but instead, that space can be used to integrate text that would normally appear in the body (Results and Discussion) of a manuscript. Concisely describe not only the content of the figure but also the derived conclusions [9, 10].

49.3.8 Summary/Conclusions

Using bulleted points, succinctly describe the summary of the findings in this section. Avoid unnecessary speculations and detailed discussion. Provide directionality of the findings with future perspectives.

49.3.9 Edit Ruthlessly and Review Meticulously

First, ensure that the main message is clear and the content is self-explanatory. Avoid long sentences and use phrases or bulleted points in active voice and plain language, wherever possible. Avoid jargon, but some technical language may be appropriate to attract the aficionado. Invite colleagues to review the layout format and explain the main message.

It is advisable to review the poster in a A4/letter size sheet printout before printing the final full-size poster. The focus for crispness and completeness of text, includes logical line breaks, appropriate coloration, avoid pixelation and artifacts. Wait for a few days and then proofread again, and request someone unfamiliar with the poster to review it [2].

49.4 Presentation of the Poster at the Conference

Judge the competence level of the audience and customize the presentation. It is important to rehearse the presentation, which will improve the flow of thoughts and focus on the point of discussion. Be cognizant of the presentation guidelines (if any) that are provided by the organizers. The posters should either be fastened on the board or pinned to the board based on the guidelines provided by the Conference Secretariat and the presenter should be available near the poster at the given time and removed once the presentation is completed.

49.5 Concluding Remarks

A poster presentation is an opportunity to showcase the research work even if it is still in progress. It provides a platform for effective interaction between the presenter and the experts in the field, thereby creating an opportunity to address the gaps and expand the research work. Often other participants share their own bench-level experiences with the presenter validating the findings, helping in troubleshooting difficult experiments, providing new ideas for future research projects and promoting networking with contemporaries [10]. Poster presentations when listed on the curriculum vitae serve as indicators of research experience and performance [8, 10]. Several scientific societies offer attractive incentives to promote young scientists in the form of 'best poster awards' or 'travel awards', which could strengthen the presenters' curriculum vitae. Sometimes, this platform may be used by even journal editors to seek manuscripts or by academia and industry leaders to offer career opportunities. Therefore, the poster presentation is a very important part of the scientific journey and can be highly beneficial if put to good use.

Conflict of Interest None.

References

1. Society for Neuroscience, Poster Presentations (2022). https://www.sfn.org/meetings/neuroscience-2022/call-for-abstracts/presenter-resources/poster-presentations. Accessed 2 Dec 2022
2. Hamilton CW (2008) At a glance: a stepwise approach to successful poster presentations. Chest 134:457–459. https://doi.org/10.1378/chest.08-1078
3. Butz AM, Kohr L, Jones D (2004) Developing a successful poster presentation. J Pediatr Health Care 18:45–48. https://doi.org/10.1016/j.pedhc.2003.08.006
4. Alexandrov AV, Hennerici MG (2007) Writing good abstracts. Cerebrovasc Dis 23:256–259. https://doi.org/10.1159/000098324
5. Gray AL, Curtis CW, Young MR, Bryson KK (2022) Innovative poster designs: a shift toward visual representation of data. Am J Health Syst Pharm 79:625–628. https://doi.org/10.1093/ajhp/zxac002
6. Faridi E, Ghaderian A, Honarasa F, Shafie A (2021) Next generation of chemistry and biochemistry conference posters: animation, augmented reality, visitor statistics, and visitors' attention. Biochem Mol Biol Educ 49:619–624. https://doi.org/10.1002/bmb.21520
7. Bradbury NA (2016) Attention span during lectures: 8 seconds, 10 minutes, or more? Adv Physiol Educ 40:509–513. https://doi.org/10.1152/advan.00109.2016
8. Erren TC, Bourne PE (2007) Ten simple rules for a good poster presentation. PLoS Comput Biol 3:e102. https://doi.org/10.1371/journal.pcbi.0030102
9. Boullata JI, Mancuso CE (2007) A "how-to" guide in preparing abstracts and poster presentations. Nutr Clin Pract 22:641–646. https://doi.org/10.1177/0115426507022006641
10. Ecoff L, Stichler JF (2015) Disseminating project outcomes in a scholarly poster. HERD 8:131–138. https://doi.org/10.1177/1937586715583463

50 Strategies for the Preparation and Delivery of Oral Presentation

B. N. Srikumar and K. Gokulakrishnan

Abstract

Communicating the research findings represents the culmination of the scientific endeavour. An oral presentation is one of the several forms of science communication and needs to be tailor-made depending upon the audience and ambience. While this might appear to be a daunting task to a novice, it is a fact that for most scientists, an oral presentation is an art that develops with practice. This chapter intends to provide pointers for developing the content required for a scientific presentation and strategies for its successful delivery. Although several kinds of presentations may appear different, the basic requirements and the challenges are mostly common. In this chapter, we discuss the strategies to unify the requisites for different presentations, preparing the content in general and specifically for a scientific presentation at a conference, and tips to deliver a good talk. Similar to other forms of scientific communication, the rewards of a successful oral presentation are multi-fold. First and foremost, it serves to disseminate the research findings, foster discussion through scrutiny and consequent hypothesis generation for further scientific advancement. It also provides an opportunity to network and can facilitate career growth. Therefore, oral presentation represents an important aspect of scientific communication at multiple levels and is successful when the preparation and delivery are well-planned and executed effectively.

Keywords

Communication · Platform presentation · Oral presentation · Conference · Scientific meeting · Content delivery

50.1 Introduction

Communicating science is fundamental to the research domain and happens at every stage of the research endeavour. This forms the basis for training research students through seminars, journal club presentations, and work progress presentations at the in-house level. An oral presentation involves the exchange of information with the audience. There are several forms of oral presentation—could be a one-way (didactic), or a two-way (Socratic or Dialectic) presentation [1]. The different contexts where presentations are required to be made include: student

B. N. Srikumar (✉)
Department of Neurophysiology, National Institute of Mental Health and Neuro Sciences, Bengaluru, India
e-mail: kumarasri@nimhans.ac.in

K. Gokulakrishnan
Department of Neurochemistry, National Institute of Mental Health and Neuro Sciences, Bengaluru, India
e-mail: gokul@nimhans.ac.in

G. Jagadeesh et al. (eds.), *The Quintessence of Basic and Clinical Research and Scientific Publishing*,
https://doi.org/10.1007/978-981-99-1284-1_50

presentations as part of the academic program: seminars, journal club reviews; presentations at conferences or scientific meetings; lecture presentations, defending presentations such as thesis defense, presentations to funding agencies or review committees; presentations as part of job interviews; and presentations for a general audience to engage the public. The presenter should find out the context and audience before preparing for the presentation. This chapter intends to cover the basic aspects of preparing the content and making a presentation.

50.2 Mine and Collate the Content for a Presentation

It is important to have an idea about the purpose of the presentation, which will determine the nature and content. Before commencing the preparation, it is good to mine the content through proper literature search and collate it, tailoring it for the purpose of the presentation and the intended audience.

50.2.1 Unifying the Requisites of Different Presentations

Although the several kinds of presentations may appear different, they possess the same requisites and challenges for successful delivery [2]. Start with a skeleton of the content, then collate the information required for the first draft of the content. It is good to review the content before getting an opinion from peers or mentors. Based on the feedback, revise the content. In addition, it is advisable to make a mock presentation, which might help in further revising the content based on the flow and then finalizing it.

50.2.2 Preparing the Content (In General)

It is important to know much more than what is there on the slides. Therefore, invest enough time to study in detail the relevant literature or material before starting to make the slides. Before embarking on the preparation, it is good to know beforehand the time allotted for the presentation, the ambience and the audience. The content should reflect the time given for the presentation and should be appropriate to the context and audience. It has two components—content and style [1]. The content will be predominantly in three sections—introductory slides, main content, and concluding remarks. The presentation should take the listener from the known to the unknown [3]. The take-home message should be very clear at the end of the presentation. The actual content and its arrangement to an extent depend on the purpose of the presentation. For example, if the presentation is meant to educate (and in situations, where these talks usually extend beyond 45–50 min), you could use content/activities on the slides that will shake the audience and bring back their attention. However, this approach may not be appropriate for a shorter talk to a more informed audience or when you are (or your science is) being assessed or examined. Remember that content on the slide is only meant to be a mnemonic and a good presenter always talks more than what is there on the slide. The data in the presentation should be concise and the presentation itself should be tailor-made to the audience [4]. Once a draft presentation is ready, it is good to practice the delivery of the presentation and shape it further depending on the flow, clarity, and time. The key points to prepare for an optimal presentation are summarized in Boxes 50.1 and 50.2 provides the rule of thumb for the content.

It is good to capture one idea per slide [2, 5]. Use meaningful images where appropriate, and avoid using too much text. It is always a good practice to give due credit to the source of the images and obtain copyright permission, where applicable. Images should be clear, non-pixelated, and legible even when viewed from a distance. The text can be presented as bullet points in readable font sizes and should never be paragraphs. Whole sentences should be avoided unless required in the context. Do not put content if you are not going to talk about it and be thorough with the content that is on the slide. Give full reference on the slide wherever possible since, unlike published material, there will be no bibliography or reference list at the end of the document. Harvard format of referencing may be acceptable if the full reference cannot be given for the want of space.

Box 50.1 Key points to prepare the presentation

- Understand the ambience and audience beforehand
- The presentation should take the audience from the known to the unknown and should have a take-home message
- The presentation should address a question and explain its relevance
- Contain crucial information
- The content should be based on the purpose of the presentation—Is it to educate, inform or persuade?
- It should have concise data
- It should be tailor-made for the target audience
- Rehearse the presentation and time it—reshape the content of required
- Prepare to be able to present at any time and even without the supporting content
- Engage the audience

Box 50.2 Rule of thumb for the content

Know much more than the content on the slides
- One idea per slide
- Meaningful images where relevant and possible
- Animations where appropriate (Don't overdo it)
- Placeholders or prompts for conversation points
- Should not be verbatim
- Should not contain whole sentences
- Legible font sizes
- Images should be clear and legible from a distance with good visibility

50.2.3 Planning for a Presentation at a Conference: *Abstract Submission*

The requirements of an abstract for an oral presentation are similar to poster presentations (see the previous chapter for more details).

50.2.4 Preparing Content for a Conference Presentation

Presentation at a conference is a professional way of sharing scientific information. It could be sharing of an observation, detailed findings from a research project or introduction of a novel hypothesis. It could also be a comprehensive sharing of the current knowledge or state of the art and providing future directions. These presentations make conferences memorable for both presenters and the audience. The attention of the audience can be difficult to maintain, particularly with growing scientific data and technology-based distractions. Therefore a good conference presentation continuously engages the audience with a simple delivery style with a clear central message, supported by a seamless flow of ideas, and should be like a good story that they will remember. While a novice presenter might be tempted to dump the data in the presentation to avoid criticism, this might be counter-productive since a scientific presentation is ultimately a form of communication, with the audience expecting a central message that arises from a logical interpretation of the data [6, 7]. A good presentation always promotes healthy discussions, exchange of scientific thoughts, fosters networking, and transforms the current thinking of both the presenter and audience. Presentations can fall into several categories such as excellent, good, or modest, with the latter being the most common. This could be because the content and subject

A.

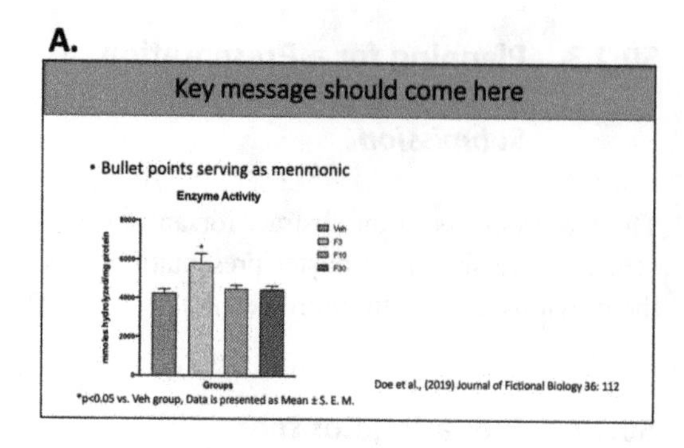

B.

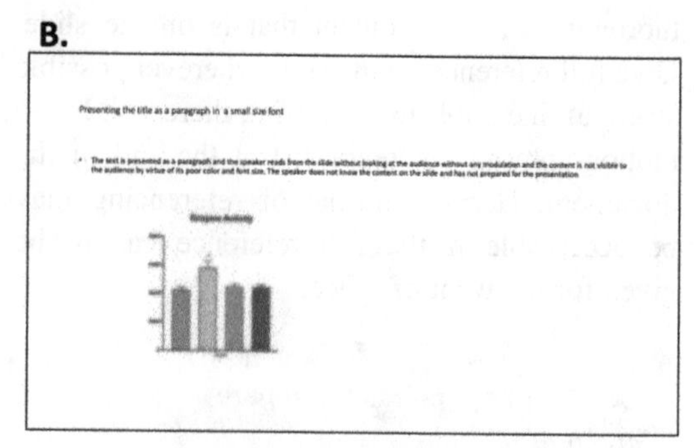

Fig. 50.1 Illustration of an effective (**a**) and an ineffective (**b**) slide (a fictional figure generated for an illustrative purpose)

itself were average, the delivery was not effective, or both. Hence, the young presenters should put their efforts to make their science clear and visible, which might facilitate their career growth and scientific network [2, 8]. For the conference presentations that contain data from original research, it is advised to follow the accepted abstracts and the IMRaD format: Introduction, Methods, Results and Discussion (Conclusions). Box 50.3 summarizes the content required for a conference presentation and Fig. 50.1 illustrates an effective and ineffective slide in a presentation.

50.3 Delivering a Successful Presentation

गीती शीघ्री शिरःकम्पी तथालिखितपाठकः।.
अनर्थजञोऽल्पकण्ठश्च षडेते पाठकाधमाः॥.
Gītī śīghrī śiraḥkampī yathālikhitapāṭhakaḥ,
anarthajño'lpakaṇṭhaśca ṣaḍete pāṭhakādhamāḥ.

As described in the above Sanskrit quote—the presenter should focus on the intonation, pace, avoid unnecessary movements and reading from the slide, know (more than) what is there on the slide and talk with a well-modulated voice, which can be heard even if the microphone fails. In totality, the presenter should capture the audience's attention and focus on key messages.

Before the start of your session, acquaint yourself with the podium and learn how to operate the audio-visuals (collar/podium/hand-held microphone, computer, slide changer, laser pointer, or mouse). Get to the podium as soon as your turn comes or as soon as you are introduced by the moderator. You could briefly thank the organizers for the opportunity before commencing the talk.

To capture and retain the attention of the audience, present slides at a rate of one slide per minute. Do not talk too much (that is much more than what is there on the slide) and at the same time do not hide behind the content on the slide (use the content to convey the key message and convince the audience). Avoid reading the entire content on the slides, instead talk around the content. Try to use the pointer only when necessary and do not constantly point at the text that is being read.

To retain the audience's attention, modulate the voice, always maintain eye contact with the audience and pause when needed, allowing time to settle your nerves [9, 10]. For a conference presentation, go with the flow that is described in Box 50.3. For an effective message delivery, identify a few people in different sections of the audience and speak as if you are personally talking to one of them at a time and alternate between them. Do not—bury your head into the podium, look at only the screen, monitor, or notes

Box 50.3 Content required for an oral presentation in a conference

Slide	Content	Remarks
Title slide	The full title of the presentation Presenting author's details (if there is more than one author) Affiliation Email address of the presenting author	All authors can be listed if space permits, else they can be listed in the acknowledgements slide
Conflict of interest OR disclaimer	It is good to disclose the conflicts of interest (if any) or any disclaimer that no product or service is being endorsed or advertised	
Introduction	Background of the study using bullet points Bring out the lacunae in the background and state the objectives of the study	
Methods	Should be brief but sufficient enough to understand the flow of the work and techniques involved Use flowcharts, diagrams or other visual aids Avoid too much text Statement on ethical/regulatory compliance	Methods pertaining to the Results can be presented together on the same slide if required
Results	The flow of results should mirror the objectives The title of the slide should reflect the findings presented on the slide Use graphs, pictures or photomicrographs of good resolution with large visible fonts	Use tables sparsely, where required
Discussion	Detailed discussion is not required Study limitations if any may be included	
Summary (conclusions)	State succinctly the findings from the study	Future directions can be included
Acknowledgments	Include study funders, participants and people who are not listed as co-authors	

[9–11]. You should have practiced your talk enough to remember it or you should be familiar with the flow and the content on the slides. Do not assume that 'it will come to me'—so practice! practice!! practice!!!

One of the important aspects of a scientific presentation is handling the discussion effectively. While most of the audience will be nice to the presenter, but one should be prepared to handle a hostile audience as well. Listen carefully, and if the question or comment is not clear, ask again (maybe even to rephrase it). The presenter should maintain calm and composure, and answer the questions with the highest degree of scientific rigour and humbly accept any ignorance and criticism [12]. The snippets for an effective presentation are given in Box 50.4.

Box 50.4 Snippets for an effective presentation

- Capture audience attention
- Stay focused on the key messages
- Familiarize yourself with the podium and the audiovisuals
- Warm up the audience and focus on your opening
- Confident delivery
- Creative and clever
- Think on your feet
- Do not read from the slides
- Be prepared for a confident discussion
- Have a backup plan: keep the presentation in a pen drive or on the cloud

50.4 Concluding Remarks

Although making an oral presentation may appear to be an intimidating task, there is no other way to advance science and be recognized in the field than to persevere and succeed from ideation to presentation. An oral presentation is an art that is learned with time. 'Practice makes a man perfect'—therefore it is important to practice the preparation of the content and delivery of the presentation to be an effective presenter. The presenter should make the groundwork before embarking on the content preparation, which itself should be customized to the context, audience, and purpose of the presentation. The content delivery should engage the audience and convey the key messages. While the presentation in itself is rewarding, several scientific societies offer attractive awards to encourage young and aspiring researchers. Training to successfully deliver a scientific talk goes a long way in developing the network and career growth of the presenter.

Conflict of Interest None.

References

1. Nundy S, Kakar A, Bhutta ZA (2022) How to give an oral presentation? In: Nundy S et al (eds) How to practice academic medicine and publish from developing countries? Springer, Singapore, pp 357–366. https://doi.org/10.1007/978-981-16-5248-6_38
2. Alexandrov AV, Hennerici MG (2013) How to prepare and deliver a scientific presentation. Teaching course presentation at the 21st European Stroke Conference, Lisboa, May 2012. Cerebrovasc Dis 35:202–208. https://doi.org/10.1159/000346077
3. Rubio Pérez I (2022) How to prepare an oral presentation for a scientific congress. Cir Esp (Engl Ed) 101: 5077. https://doi.org/10.1016/j.ciresp.2022.01.015
4. Lortie CJ (2017) Ten simple rules for short and swift presentations. PLoS Comput Biol 13:1–6. https://doi.org/10.1371/journal.pcbi.1005373
5. Kwok R (2013) Communication: two minutes to impress. Nature 494:138–138. https://doi.org/10.1038/nj7435-137a
6. Papanas N, Maltezos E, Lazarides MK (2011) Delivering a powerful oral presentation: all the world's a stage. Int Angiol 30:185–191. https://doi.org/10.13140/RG.2.1.5129.4169
7. Yang J (2010) Mastering the big talk-preparing an oral presentation. Gastrointest Endosc 71:1275–1276. https://doi.org/10.1016/j.gie.2010.04.002
8. Leira EC (2019) Tips for a successful scientific presentation. Stroke 50:E228–E230. https://doi.org/10.1161/STROKEAHA.119.025337
9. Hartigan L, Mone F, Higgins M (2014) How to prepare and deliver an effective oral presentation. BMJ 47:g2039. https://doi.org/10.1136/bmj.g2039
10. Horiuchi S, Nasser JS, Chung KC (2022) The art of a scientific presentation: tips from steve jobs. Plast Reconstr Surg 149:533–540. https://doi.org/10.1097/PRS.0000000000008849
11. Rubenson D (2020) Why your scientific presentation should not be adapted from a journal article. Nature. https://doi.org/10.1038/d41586-020-03300-6
12. Waljee JF, Larson BP, Chang KWC, Ono S, Holland AL, Haase SC, Chung KC (2012) Developing the art of scientific presentation. J Hand Surg Am 37:2580–2588.e2. https://doi.org/10.1016/j.jhsa.2012.09.018

Part X

Research Support (Grantsmanship)

Grant Process and Peer Review: US National Institutes of Health System

51

Shamala Srinivas and Ranga V. Srinivas

Abstract

The National Institutes of Health (NIH), a part of the Department of Health and Human Services of the United States is one of the largest public funders of grants appropriated by the United States Congress for every fiscal year. The mission of NIH is to support science in pursuit of knowledge about the biology and behavior of living systems and to apply that knowledge to extend healthy life and diminish illness and disability. NIH grants are awarded to non-profit and for-profit organizations, universities, hospitals, research foundations, governments, and their agencies, and occasionally to individuals both within and outside the United States, to support scientists at every stage of their careers. Before writing a grant application, several resources available to investigators include Program Officials in each institute and center (ICs) as guides, the principal liaison between investigators/ scientists seeking funding, and the NIH. Funding decisions are dependent on several factors including the peer review evaluation for the scientific and technical merit of the project proposed. This chapter is an introduction to the fundamentals of the NIH grant process.

Keywords

NIH · Grant process · Funding opportunity announcement · Application submission · Peer review · Review criteria

51.1 Introduction

National Institutes of Health (NIH) a part of the United States Department of Health and Human Services (DHHS), the United States (US) medical agency is the largest public funder in the world, investing more than $41 billion a year to enhance life, and reduce illness and disability. The mission of the NIH is to support science in pursuit of knowledge about the biology and behavior of living systems and to apply that knowledge to extend healthy life and reduce illness and disability. The NIH is made up of 27 institutes and centers (ICs) of which 24 of them have the authority to fund. Each IC has its own specific research agenda, often focusing on particular diseases or body systems. A grant is used whenever the NIH IC anticipates no/ substantial programmatic involvement with the recipient during

S. Srinivas (✉)
Scientific Review and Policy, Division of Extramural Activities, National Cancer Institute Shady Grove, National Institutes of Health, Rockville, MD, USA
e-mail: hiitssks@gmail.com

R. V. Srinivas
Review Branch, National Institute on Alcohol Abuse and Alcoholism, National Institutes of Health, Bethesda, MD, USA
e-mail: hiitssks@gmail.com

G. Jagadeesh et al. (eds.), *The Quintessence of Basic and Clinical Research and Scientific Publishing*,
https://doi.org/10.1007/978-981-99-1284-1_51

the performance of the financially assisted activity. Most governmental funding agencies in the United States are mission driven. Hence, if funding is sought from a specific agency, it is important to ensure that research is within the mission of the agency.

The budget is appropriated by the United States Congress for every fiscal year. Budget appropriation for NIH occurs in phases. In the formulation phase, NIH examines scientific and budget priorities to develop a budget within NIH and DHHS guidelines. In the presentation phase, NIH representatives including the Director of the NIH and some of the IC Directors defend the budget at the congressional hearings; and then the US Congress appropriates the funds for the NIH. In the execution phase, after the grant applications undergo two levels of review, initial peer review followed by the second level of Council review, funds are obligated. The majority of competing awards are selected by paylines. A payline is a percentile or impact score up to which applications will likely be funded with minimal review by the IC. Typically, grant funding is based on several factors including peer review, filling gaps in the IC research portfolio, novel or promising scientific approach, and commitment to New/Early-Stage Investigators.

Types of funding supported by the NIH include- (a) Grants: Federal financial assistance including money, property, or both to an eligible entity to perform approved scientific activities with little or no government involvement; (b) Cooperative Agreement: A support mechanism where the NCI and extramural scientists/ clinicians work together during the performance of the research; (c) Research and Development Contracts: Used to obtain or procure cancer research services and other resources needed by the federal government; and (d) Other Transactions: Unique type of legal instrument (s) other than contracts, grants or cooperative agreements that generally are not subject to Federal laws and regulations that apply specifically to procurement contracts or grants. Grants do not have substantial program involvement during the performance of the financially assisted activities and Cooperative Agreements, a grant mechanism (in the U Series, see Types of Competing Grants), used when there will be substantial NIH scientific or programmatic involvement. Substantial involvement means that, after the award, IC scientific or program staff will assist, guide, coordinate, or participate in project activities.

NIH grants are awarded to non-profit and for-profit organizations, universities, hospitals, research foundations, governments, and their agencies, and occasionally to individuals. Foreign institutions and international organizations are eligible to receive mainly research project grants. Other programs, such as the Small Business Innovation Research (SBIR) Grants, Small Business Technology Transfer (STTR) Grants, and minority program grants are established for certain categories of applicants. Foreign institutions are not eligible to apply for some NIH training grants, center grants, small business grants, or construction grants.

NIH offers funding programs to support scientists at every stage of their careers starting from graduate school to senior scientists. Research Project Grants (example: R01, R21, R03) are the "bread and butter" of the research community. The Small Business Innovation Research (SBIR)-R41/42 and Small Business Technology Transfer (STTR)-R43/44 programs are highly competitive programs that encourage domestic small businesses to engage in Federal Research/Research and Development (R/R&D) with the potential for commercialization. Research congressionally mandated set aside funding. Program Project (P01)/Center Grants (P20, P30, P50 Series) supports a broadly based, multidisciplinary, often long-term research program which has a specific major objective or a basic theme. The purpose of this chapter is to introduce an investigator seeking grant support to the resources and process of the NIH grants.[1]

51.2 Funding Opportunity Announcement

All grant applications to be submitted have to be in response to a published Funding Opportunity Announcement (FOA) which can be a program announcement (PA), program announcement with special receipt or review (PAR), and request for application (RFA). All FOAs have the same format taken directly from regulation—2 CFR Part 200 spells out the information that must be included in a FOA. Before writing a grant application, it is imperative that the proposed research is responsive to a FOA published by that agency. Each type of grant program has its own set of eligibility requirements. Before submitting an application any new related notice within the FOA has to be verified by the investigator (s) submitting a grant application so that any changes in the FOA are followed. The information in the notice augments/supersedes the information in the FOA. In the case of NIH, there are several FOAs published throughout the year. A PA is a generic announcement in which most of the ICs participate, is usually open for three years, and does not have a specific research topic—meaning it is 'investigator-initiated'. These may be submitted throughout the year. In NIH, the most common FOAs for the investigator-initiated research is the research project grants (RPG) that include: (a) Research project Grant (R01), (b) NIH Exploratory/Developmental Research Grant (R21), and (c) NIH Small Research Grant Program (R03).

Every NIH FOA contains the following sections:

Part 1. Overview Information
Part 2. Full Text of Announcement
Section I. Funding Opportunity Description
Section II. Award Information
Section III. Eligibility Information
Section IV. Application and Submission Information
Section V. Application Review Information
Section VI. Award Administration Information
Section VII. Agency Contacts
Section VIII. Other Information

The FOA informs an applicant of the facts about the preparation and submission of a grant application in terms of (a) the purpose and description, (b) application due date (s), scientific review due date, Council date, expiration date of the FOA, (c) how to submit the application electronically, (d) budget information, (e) project period, (f) eligibility of applicants, (g) required registrations to submit an application, (h) page limits, (i) application submission information (j) review criteria, and (f) award administration information. This is the best possible guide for preparing grant applications and should be read carefully and followed precisely.

The authorized organization representative (AOR)/signing official (SO) holds the authority to legally bind the institution and assume responsibility for adhering to all federal grant administration requirements. They maintain all the institutional profile data. Grant applications are funded in a competitive system based on scientific merit. To apply for a grant, both the investigator and the institution or small business must be registered in advance with multiple entities listed below.

[1] List of Abbreviations:

AOR	Authorized organization representative
CSR	Center for Scientific Review
CFR	Code of Federal Regulation
DRR	Division of Receipt and Referral
eRA	Electronic Research Administration
FOA	Funding Opportunity Announcement
IC	Institutes and Center
NIH	National Institutes of Health
SAM	System for Award Management
SBIR	Small Business Innovation Research
SO	Signing official
STTR	Small Business Technology Transfer
UEI	Unique entity identifier

Organization

1. **Unique Entity Identifier (UEI) formerly DUNS**—Data Universal Numbering System; must be completed for SAM registration (next step)
2. **SAM**—System for Award Management; consolidates federal procurement systems and the Catalog of Federal Domestic Assistance (CFDA); must be completed before Grants.gov registration—must be renewed annually
3. **Grants.gov**—must register to submit applications—can work on Grants.gov and Commons registrations at the same time
4. **eRA** (electronic Research Administration) **Commons**—must register in eRA to apply for grants at any NIH IC (also several other HHS agencies)

51.2.1 Application Submission Systems

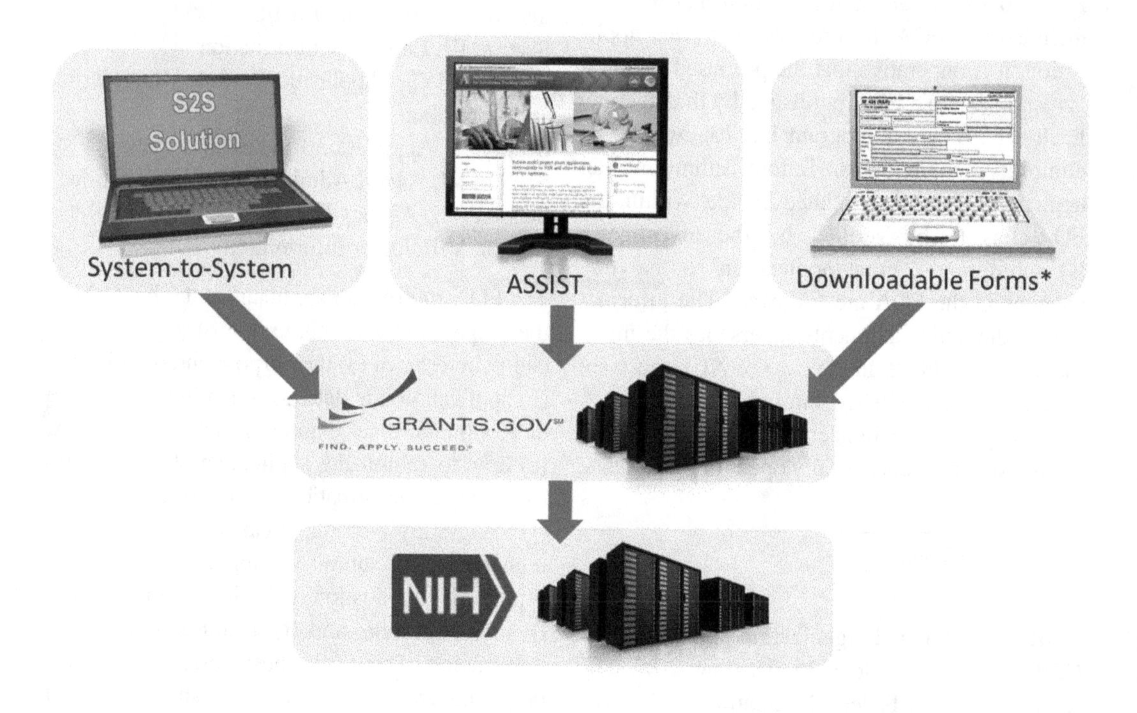

AOR at Grants.gov—called Signing Official (SO) in eRA; must enter AOR Grants.gov credentials in ASSIST to submit.

ASSIST—Application Submission System & Interface for Submission Tracking; custom NIH developed system. Pre-submission validation, pre-population of eRA Commons profiles, and the pre-submission preview of the application in NIH format.

System to System—there are commercial service providers (some at grantee institutions) that offer application prep and submission services. Many of the same advantages to ASSIST but are also integrated with the institution's internal systems.

Downloadable forms—very manual process; only get 'generic' Grants.gov validations but do not include most NIH validations, no preview phase.

51.3 Grant Process

A research project grant application should be submitted when an investigator is capable of running an independent research project, as an independent, tenured/non-tenured investigator who has lab space with access to appropriate equipment.

51.3.1 Types of Grants

Grants are available for various types of research including person based (e. g.: fellowships, career development, training, etc.) and project based. The following list represents frequently used project-based research grant programs by most of the ICs at the NIH.

Activity code	Description
R01	**NIH Research Project Grant Program (R01)** • Used to support a discrete, specified, circumscribed research project • Generally awarded for 3–5 years • Advance permission is required for $500 K or more (direct costs) in any year • No specific dollar limit unless specified in FOA • Renewable
R03	**NIH Small Grant Program (R03)** • Provides limited funding for a short period of time to support a variety of types of projects, including pilot or feasibility studies, collection of preliminary data, secondary analysis of existing data, small, self-contained research projects, development of new research technology, etc. • Limited to two years of funding • Direct costs generally up to $50,000 per year • Not renewable
R21	**NIH Exploratory/Developmental Research Grant Award (R21)** • Encourages new, exploratory and developmental research projects by providing support for the early stages of project development. Sometimes used for pilot and feasibility studies • The combined budget for direct costs for the two-year project period usually may not exceed $275,000 • No preliminary data is generally required • Not renewable
K99/R00	**NIH Pathway to Independence (PI) Award (K99/R00)** • Provides up to five years of support consisting of two phasesProvides 1–2 years of mentored support for highly promising, postdoctoral research scientists • Up to 3 years of independent support contingent on securing an independent research position • Award recipients will be expected to compete successfully for independent R01 support from the NIH during the career transition award period • Eligible Principal Investigators include outstanding postdoctoral candidates who have terminal clinical or research doctorates who have no more than 4 years of postdoctoral research training • PI does not have to be a U.S. citizen • Foreign institutions are not eligible to apply

A project period is a total time for which support of a project has been programmatically approved. The project period usually consists of a series of individual budget periods for one year each (i.e., 4/1/2021–3/31/2026). The budget period is the funded increment of time the grant recipient is authorized to spend the funds awarded (i.e., 4/1/2021–3/31/2022). The recipient is required to submit an annual progress report at the end of each budget year to request funding for a subsequent budget year within the competitive segment. Once awarded, the continuation of each successive budget period is subject to satisfactory performance, availability of funds, and program priorities.

The NIH also supports Training—(T series) providing institutional research training opportunities to trainees at the undergraduate, graduate, and postdoctoral levels; Fellowships—(F series) provide individual training opportunities to supported fellows at the undergraduate, graduate, postdoctoral and senior career levels; Career Development Awards (K series)—provides individual and institutional career development opportunities to supported candidates at the postdoctoral, early career, and mid-career levels, and Research Education (R00, R25)—provides research education opportunities including courses, research experiences, mentoring, curriculum development and outreach.

51.3.2 Preparation of Grant Application

Before writing a grant application, several resources available to investigators include Program Officials in each ICs as guides, the principal liaison between investigators/scientists seeking funding and the NIH. The most important aspect to grant writing is a clear idea of what, why, how, who, and where? Good grant applications should be based on a good concept. Consider why the proposed project is important and if successful, what will the project accomplish. How novel is the technology, techniques, or approaches to be used (also see chepter "The Roadmap to Research: Fundamentals of a Multifaceted Research Process")? Deliberate the expertise and experience of the personnel needed to complete the goals of the project and adequacy of where the project will be conducted. Sample grant applications are available at https://www.niaid.nih.gov/grants-contracts/sample-applications.

Once an application is received by the NIH, it is assigned a Grant Number that will be its identifier throughout the life of the grant.

1	R01/R21/ R03	CA	240148	1	A1, S1
Type	Mech	IC	Serial #	Support Year	Suffix

Suffix does not always occur in a grant number. The A1 suffix indicates resubmitted application (as an amended application) and the S suffix indicates a supplement to an existing grant.

51.4 Referral of Grant Applications

All grant applications submitted to the NIH first go to the Division of Receipt and Referral (DRR) within the Center for Scientific Review (CSR). Applications compliant with NIH policies are assigned to an NIH Institute or Center and a scientific review group for evaluation of scientific and technical merit. After receiving the application, (1) The DRR checks the application for completeness, (2) Determines the area of research and decides which specific NIH Institute or Center to assign it to for possible funding, (3) Assigns an application identification number (4) Assigns application to a specific study section, also known as a Scientific Review Group (SRG) or to an IC for review.

Transfer of an application from one IC to another based on the mission of an IC is possible. Referral for peer review occurs within ICs.

51.5 Peer Review of Grant Applications

Grant applications undergo rigorous two-stage review. The first stage of the review is carried out primarily by non-federal government scientists while the second stage is performed by Advisory Councils or Boards. Grant applications are either reviewed by a chartered committee or a Special Emphasis Panel.

The core values of peer review drive the NIH to seek the highest level of ethical standards, and form the foundation for the laws, regulations, and policies that govern the NIH peer review process. The NIH dual peer review system is mandated by statute in accordance with section 492 of the Public Health Service Act and federal regulations [1] governing "Scientific Peer Review of Research Grant Applications and Research and Development Contract Projects." NIH policy is intended to promote a process whereby grant applications submitted to the NIH are evaluated based on a process that strives to be fair, equitable, timely, and free of bias.

The first level of review is carried out by a Scientific Review Group (SRG; also referred to as study sections) composed primarily of non-federal scientists who have expertise in relevant scientific disciplines and current research areas. The second level of review is performed by IC National Advisory Councils or Boards. Councils are composed of both scientific and public representatives chosen for their expertise, interest, or activity in matters related to health and disease. Only applications that are recommended for approval by both the SRG and the Advisory Council may be considered for funding. Final funding decisions are made by the IC Directors.

Applicants can use eRA Commons to (1) Retrieve assignment information, (2) Find contact information for the assigned program and scientific review officers, (3) Find review meeting and council meeting dates, (4) Withdraw submitted applications from consideration, see Withdraw Your Application.

Initial peer review meetings are administered by either the Center for Scientific Review (CSR) [2] or one of the NIH ICs [3]-as specified in the funding opportunity announcement. A list of CSR study sections [4], their membership rosters, and the topics reviewed by these study sections can be found on the CSR website. Applicants may use the CSR Assisted Referral Tool (ART) CSR Assisted Referral Tool (ART) [5] to identify CSR study sections that might be appropriate for the review of your application. The CSR coordinates the reviews for most R01s, fellowships, and small business applications, as well as some PAs, PARs, & RFA's. Institutes and Centers coordinate the review of applications that have Institute-specific features such as program project grants, training grants, career development awards, and responses to RFAs.

Each FOA specifies all of the review criteria and considerations that will be used in the evaluation of applications submitted for that FOA. The RFAs and certain PARs may include additional review criteria and considerations. Other types of funding opportunities (e.g., for construction or fellowship applications) may use different review criteria and considerations (See the Review Criteria at a Glance [6]). Unless the FOA specifies otherwise, standard NIH review procedures will be followed, including the NIH scoring system described in NOT-OD-09-024 [7].

All peer review meetings are announced as a notice in the Federal Register [8]. The meetings are closed to the public, although some meetings may have an open session; the Federal Register provides the details of each meeting.

51.5.1 Peer Review Roles and Meeting Overview

Each SRG is led by a Scientific Review Officer (SRO). The SRO is an NIH extramural staff scientist (usually with an advanced degree such as a Ph.D. or MD) and the designated federal official responsible for ensuring that each application receives an objective and fair initial peer review and that all applicable laws, regulations, and policies are followed. The SROs analyze the content of each application and check for completeness. They document and manage conflicts of interest; recruit qualified reviewers based on scientific and technical qualifications and other considerations, including authority in their scientific field, dedication to high quality, fair, and objective reviews, ability to work collegially in a group setting, experience in research grant review, a balanced representation based on gender, and geographical location within the USA, assign applications to reviewers for critique preparation and assignment of individual criterion scores. The SROs also attend and oversee administrative and regulatory aspects of peer review meetings and prepare summary statements for all applications reviewed.

A Scientific Review Group (SRG) is made up of a chair, reviewers, and other NIH staff. The Chair serves as moderator of the discussion of the scientific and technical merit of the applications under review and also serves as a peer reviewer for the meeting.

Reviewers including the chair declare (1) Conflicts of Interest (Managing Conflict of Interest in NIH Peer Review of Grants and Contracts [9]) with specific applications following NIH guidance. Reviewers (2) receive access to the grant applications approximately six weeks prior to the peer review meeting. (3) Ensure they maintain the confidentiality of peer review information (See Integrity and Confidentiality in NIH Peer Review [10]). (4) Prepare a written critique (as directed by the Scientific Review Officer) for each application assigned, based on review criteria and judgment of merit. (5) Assign a numerical score to each scored review criterion (see Review Criteria at a Glance [6]). (6) Make recommendations concerning the scientific and technical merit of applications under review, in the form of final written comments and numerical scores. (7) Make recommendations concerning protections for human subjects; inclusion of women, minorities, and children in clinical research; the welfare of vertebrate animals; and other areas as applicable for the application (see guidance for reviewers on Human Subjects Protection [11] and Inclusion [12], Human Embryonic Stem Cells [13], and Vertebrate Animals [14]. (8) Make recommendations concerning the appropriateness of budget requests (see Budget Information for Reviewers [15]).

Other NIH Staff is federal officials who have need-to-know or pertinent related responsibilities and are permitted to attend closed review meetings. NIH /Center staff or other federal staff members wishing to attend an SRG meeting must have advance approval from the responsible SRO. These individuals may provide programmatic or grants management input at the SRO's discretion.

Applicants must maintain the integrity of the peer review process by not contacting reviewers to influence the outcome of the review; not sending information directly to a reviewer, and not accessing information related to the review. There are consequences to any of these actions (See Integrity and Confidentiality in NIH Peer Review [10]).

Applications submitted in support of the NIH mission are evaluated for scientific and technical merit through the NIH peer review system based on the review criteria described in the FOA.

Overall Impact: Reviewers will provide an overall impact score to reflect their assessment of the likelihood for the project to exert a sustained, powerful influence on the research field(s) involved, in consideration of the following review criteria, and additional review criteria (as applicable for the project proposed).

Scored Review Criteria: (1) Significance (2) Investigator(s) (3) Innovation (4) Approach, and (5) Environment.

Additional Review Criteria: As applicable for the project proposed, reviewers will evaluate the following additional items while determining scientific and technical merit and providing an overall impact score but will not give separate scores for these items.

- Study Timeline (specific to applications involving clinical trials)
- Protections for Human Subjects
- Inclusion of Women, Minorities, and Children
- Vertebrate Animals
- Biohazards
- Resubmission
- Renewal
- Revision

Additional Review Considerations: As applicable to the project proposed, reviewers will consider each of the following items, but will not give scores for these items and should not consider them in providing an overall impact score.

- Applications from Foreign Organizations
- Select Agent
- Resource Sharing Plans

- Authentication of Key Biological and/or Chemical Resources
- Budget and Period Support

Scoring: The NIH utilizes a 9-point rating scale (1 = exceptional; 9 = poor) for all applications; the same scale is used for overall impact scores and criterion scores (Scoring Guidance [16]).

Before the SRG meeting, each reviewer assigned to an application gives a separate score for each of (at least) five review criteria (i.e., Significance, Investigator(s), Innovation, Approach, and Environment for research grants and cooperative agreements; see Review Criteria at a Glance [6]). For all applications, the individual scores of the assigned reviewers and discussant(s) for these criteria are reported to the applicant.

In addition, each reviewer assigned to an application gives a preliminary overall impact score for that application. In many review meetings, the preliminary scores are used to determine which applications will be discussed in full at the meeting. For each application that is discussed at the meeting, a final impact score is given by each eligible committee member (without conflicts of interest) including the assigned reviewers. Each member's score reflects his/her evaluation of the overall impact that the project is likely to have on the research field(s) involved.

The final overall impact score for each discussed application is determined by calculating the mean score from all the eligible members' final impact scores and multiplying the average by 10; the final overall impact score is reported on the summary statement. Thus, the final overall impact scores range from 10 (high impact) to 90 (low impact). Numerical impact scores are not reported for applications that are not discussed (ND), which may be reported as ++ on the face page of the summary statement and typically rank in the bottom half of the applications.

Applicants just receiving their scores, or summary statements should consult the Next Steps [17] page for detailed guidance. Applicants seeking advice beyond that available online may want to contact the NIH Program Official listed at the top of the summary statement.

An application may be designated Not Recommended for Further Consideration (NRFC) by the SRG if it lacks significant and substantial merit; presents serious ethical problems in the protection of human subjects from research risks; or presents serious ethical problems in the use of vertebrate animals, biohazards, and/or select agents. Applications designated as NRFC do not proceed to the second level of peer review (National Advisory Council/ Board) because they cannot be funded.

Summary Statement: Applications that are not discussed at the meeting will be given the designation "ND" (which may be reported as ++ on the face page of the summary statement) as an overall impact score, but the applicant, as well as NIH staff, will see the written comments and scores from the assigned reviewers and discussants for each of the scored review criteria as feedback on their summary statement.

51.5.1.1 Understanding the Percentile

- A percentile is the approximate percentage of applications that received a better overall impact score from the study section during the past year (see blog on Paylines, Percentiles and Success Rates [18]).
- For applications reviewed in ad hoc study sections, a different base may be used to calculate percentiles.
- All percentiles are reported as whole numbers.
- Only a subset of all applications receives percentiles. The types of applications that are percentiled vary across different NIH Institutes and Centers.
- The summary statement will identify the base that was used to determine the percentile.

51.5.1.2 Appeals

NIH established a peer review appeal system (see NOT-OD-11-064 [19]) to provide investigators and applicant organizations the opportunity to seek reconsideration of the initial review results if, after consideration of the summary statement, they believe the review process was flawed for reasons of either bias of a reviewer, conflict of interest, absence of appropriate expertise, or factual errors by one or more reviewers that could have substantially altered the review outcome. This policy does not apply to appeals of the technical evaluation of R&D contract projects through the NIH peer review process, appeals of NIH funding decisions, or appeals of decisions concerning extensions of the MERIT award.

51.5.2 Second Level of Review: Advisory Council or Board

The Advisory Council/Board of the potential awarding Institute/Center performs the second level of review. Advisory Councils/Boards are composed of scientists from the extramural research community and public representatives (NIH Federal Advisory Committee Information [20]). Members are chosen by the respective IC and are approved by the Department of Health and Human Services. For certain committees, members are appointed by the President of the United States.

51.6 Post-Review

51.6.1 Not Funded-What Are the Next Steps?

The NIH receives thousands of applications for each application receipt round and competition for funding can be fierce. If the original application is not funded, applicants may resubmit the application, making changes that address reviewer concerns, or they may submit a new application. Once an applicant receives a summary statement, they are directed to information on Next Steps [17], and they may contact the NIH program official assigned to their application for guidance.

Fundable Score-What are the next steps?

Some of the ICs publish paylines [21] as part of their funding strategies [22] to guide applicants on their likelihood of receiving funding. Application scores can only be compared against the payline for the fiscal year when the application will be considered for funding, which is not necessarily the year when it was submitted. There may be a delay of several months to determine paylines at the beginning of fiscal years. If the application is assigned to an IC that does not announce a payline, the program official listed at the top of the summary statement may be able to guide the likelihood of funding. After the Advisory Council meeting, if an application results in an award, the applicant will be working closely with the program officer of the funding Institute or Center on scientific and programmatic matters and a Grants Management Officer on budgetary or administrative issues. The Grants Management Specialist will contact the applicant to collect the information needed to prepare the award.

51.7 Grant Recommendation Process

- NIH program staff members examine applications and consider the overall impact scores given during the peer review process, percentile rankings (if applicable), and the summary statements in light of the Institute/Center's priorities.
- Program staff provides a grant-funding plan to the Advisory Board/Council. Council members have access to applications and summary statements pending funding for that IC in that council round.
- Board/Council members conduct a Special Council Review of grant applications from investigators who currently receive $1 million or more in direct costs of NIH funding to support Research Project Grants (https://grants.nih.gov/grants/guide/notice-files/NOT-OD-22-049.html [23]). This additional review is to determine if additional funds should be provided to already well-supported investigators and do not represent a cap on NIH funding.
- The Advisory Board/Council also considers the Institute/Center's goals and needs and

advises the Institute/Center director concerning funding decisions.
- The Institute/Center director makes final funding decisions based on staff and Advisory Council/Board advice.

How to Volunteer to Be a Reviewer: For those interested in volunteering on NIH review panels, please see: Becoming a Peer Reviewer [24]. https://grants.nih.gov/grants/peer/becoming_peer_reviewer.htm. For more details about Peer Review, visit Peer Review Policies & Practices—https://grants.nih.gov/policy/peer/index.htm. [25].

51.8 Concluding Remarks

The chapter describes the structure of NIH and how to approach grant funding. The NIH grant process is dependent on the integrity of both applicants and peer reviewers. Maintaining the security and confidentiality of the process is essential for safeguarding the exchange of scientific opinions and evaluations without fear of reprisal; protecting trade secrets or another proprietary, sensitive and/or confidential information described in the grant applications; providing reliable input to the agency about research projects to support; and safeguarding the NIH research enterprise against the misappropriation of research and development to the detriment of national or economic security. In addition, maintaining integrity in the peer review process is important for maintaining public trust in science.

Conflict of Interest None.

References

1. Code of Federal Regulation (2007) https://www.govinfo.gov/content/pkg/CFR-2007-title42-vol1/pdf/CFR-2007-title42-vol1.pdf. Accessed 3 Oct 2022
2. About CSR (2022) https://public.csr.nih.gov/AboutCSR. Accessed 9 Sept 2022
3. National Institutes of Health (2022) http://www.nih.gov/icd. Accessed 7 Oct 2022
4. Study Sections (2022) https://public.csr.nih.gov/StudySections/Pages/default.aspx. Accessed 10 May 2022
5. Assisted Referral Tool (ART) (2022) https://art.csr.nih.gov/ART/selection.jsp. Accessed 19 Jun 2022
6. Criteria at a Glance (2021) https://grants.nih.gov/grants/peer/guidelines_general/Review_Criteria_at_a_glance.pdf. Accessed Aug 2022
7. Enhancing Peer Review (2009) https://grants.nih.gov/grants/guide/notice-files/not-od-09-024.html. Accessed 11 Jul 2022
8. Federal Register Notice. https://www.govinfo.gov/app/frtoc/today. Accessed 19 Jun 2022
9. Managing Conflict of interest in NIH Peer Review of Grants and Contracts (2019) https://grants.nih.gov/policy/peer/peer-coi.htm. Accessed 2 May 2022
10. Integrity and Confidentiality in NIH Peer Review. https://grants.nih.gov/policy/research_integrity/confidentiality_peer_review.htm. Accessed 5 Jul 2022
11. Guidelines for the Review of Human Subjects Section (2013) https://grants.nih.gov/grants/peer/guidelines_general/Review_Human_Subjects_20130508.pdf. Accessed 10 May 2022
12. Guidelines for the Review of Inclusion on the Basis of Sex/Gender, Race, Ethnicity, and Age in Clinical Research (2019) https://grants.nih.gov/grants/peer/guidelines_general/Review_Human_Subjects_Inclusion.pdf. Accessed 1 Aug 2022
13. Applications Proposing Use of Human Embryonic Stem Cells (2016) https://grants.nih.gov/grants/peer/guidelines_general/human_embryonic_stem_cells.pdf. Accessed 4 May 2022
14. Work Sheet for Applications Involving Animals (2017) https://grants.nih.gov/grants/olaw/VASchecklist.pdf. Accessed 25 Oct 2022
15. Budget and Period of Support Information (2012) https://grants.nih.gov/grants/peer/guidelines_general/budget_information.pdf. Accessed 12 Aug 2022
16. Scoring Guidance. https://grants.nih.gov/grants/policy/review/rev_prep/scoring.htm. Accessed 19 May 2022
17. Applicant Guidance: Next Steps and Frequently Asked Questions. https://grants.nih.gov/faqs#/next-steps.htm. Accessed 19 Jul 2022
18. Paylines, Percentiles and Success Rates (2011) https://nexus.od.nih.gov/all/2011/02/15/paylines-percentiles-success-rates/. Accessed 12 Jun 2022
19. Appeals of NIH Peer Review (2011) https://grants.nih.gov/grants/guide/notice-files/NOT-OD-11-064.html. Accessed 1 Sept 2022
20. NIH Federal Advisory Committee Information. https://ofacp.od.nih.gov/. Accessed 10 Sept 2022
21. Paylines. https://grants.nih.gov/grants/glossary.htm#Payline. Accessed 4 Jul 2022
22. NIH funding strategies. https://grants.nih.gov/policy/nih-funding-strategies.htm#strategies. Accessed 10 Sept 2022
23. NIH's Policy on Special Council Review of Research Applications (2022) https://grants.nih.gov/grants/

guide/notice-files/NOT-OD-22-049.html. Accessed 8 Oct 2022
24. Becoming a Peer Reviewer (2015) https://grants.nih.gov/grants/peer/becoming_peer_reviewer.htm. Accessed 1 Oct 2022
25. Peer Review Policies and Practices (2020) https://grants.nih.gov/policy/peer/index.htm. Accessed 25 Oct 2022

Rigor and Specifics in Writing Research Proposals

52

Cynthia E. Carr

Abstract

Writing grant proposals requires careful attention to the audience, which is usually comprised of reviewers. Grant reviewers are different from the usual academic journal reader on at least two dimensions: (1) they can award funding on the basis of their reading; and (2) they may come from a variety of different backgrounds and different levels of expertise in your field. The writing of the research proposal therefore should take this into account through the selection of detail, language used, and appropriate rhetoric. In this chapter, I will examine the use of logic, detail, and linguistic relevance to convert team background, knowledge and intention into a winning proposal tailored to the expected level of understanding of the reviewer.

Keywords

Grant writing · Audience · Rhetoric · Writing strategies · Word choice · Reviewers

52.1 Introduction

A key aspect of creating an excellent grant proposal is the proper tailoring of the message to your audience. An elegant, thoughtful, engaging set of arguments for a project may appear simple; however, the production of such text is anything but easy. Simplicity and engagement in the text are the results of a fairly complex and involved process of study, analysis and skilled writing often resulting in numerous versions (even numerous submissions) before the final, perfect proposal emerges.

What is presented of rigor, or the precision of the method, and specifics, meaning how much detail is to be provided, is a function of how well the material is expressed by the Principal Investigator (PI) or the research team combined with the expected level of understanding of the audience perhaps modified by how much the audience cares about the topic. More and denser information may be provided to an audience that understands the jargon or cares enough to try to understand, and this too can be influenced by a skilled writer. The confluence of expression, audience knowledge and audience concern can bring one to the sweet spot of the argument, where the reviewer understands the project and its importance and remembers the project despite reading several others much like it.

One might paraphrase what has become something of a saying about scientific explanations and note that to accomplish these three elements, one needs to understand the project well enough to explain it to Granny.[1]

C. E. Carr (✉)
Research Support Services, School of Medicine, The University of California, Irvine, CA, USA
e-mail: cecarr@hs.uci.edu

[1] The heart of this paraphrase is often attributed to Einstein, however this it may need to be updated in the

G. Jagadeesh et al. (eds.), *The Quintessence of Basic and Clinical Research and Scientific Publishing*,
https://doi.org/10.1007/978-981-99-1284-1_52

It is worth exploring this saying further. Grant proposals are analyzed and recommended by reviewers, of course, not grandma, however, anyone who has taught has learned how much one needs to know in order to engage students in material. Similarly, explaining your project in an engaging way to someone with no background requires mastery. Why? Because when providing information to a mind with no background, one must select details and string them together carefully. The person only knows a particular topic based on what you explain. Unpack a process[2] that the audience finds illogical and they are left scratching their heads in speculation. Overwhelm them with too many details and they have gone back to their smartphones. Even if a perfectly logical process is presented with a small number of well-chosen details, the audience will still be lost if the project appears irrelevant.

When explaining your project to Granny you will need to embrace all the processes, details, and minutiae of it, and then let them all go. From all that knowledge, you will pick out a few important strands that represent the pith of the project that conveys the most salient information to help the reviewer understand what must be done, why it must be done, and how you will do it. Then, ornament that pith with the judicious use of details that convey how you, the PI, have a deep understanding of the processes involved with the project such that you can supervise, manage, and carry it out.

There are three elements, with many related and contextual substeps in between, to buttress appropriate rigor and specifics in a winning grant proposal:

1. Logic
2. Judicious Selection of Detail
3. Relevance

This article will primarily review these three items to present guidance on how to provide the rigor and specifics that will convince the reviewer that your project is the one to fund.

next generation given the growing numbers of women holding PhDs.

[2] When I refer to processes in this piece, I will always be talking about how you will perform your investigations, the action of your work.

52.2 Master the Project

It bears repeating, over and over, that one must know as much as possible about the proposal material. As might be imagined, this starts with knowledge of the methods and all about the materials, as well as the actual activities necessary to make the project work, but this is not all. Information is also required about how colleagues and contributors will play their parts, and how all of this will transpire under the specific conditions to be found at your institution. It is even important to know the many arguments against the project very well in order to counter them as needed.

Of course, this means that you have read deeply of the relevant publications available on your topics and methods. You know the related disciplinary controversies, what you think about them, and how to be clever in the way you (do or do not) provide your opinion. You know the top scholars in your areas of interest, especially those who must be cited, and who can be safely ignored. But that is only the start.

Of course, you have experience with the methods, right? You have training in that specific type of microscope, know how to prep the samples for the sequencer, have listed the relevant reagents in the budget, and are familiar with how to handle and manage the mutant mice. There may be colleagues of complementary training working with you, however, what is being highlighted is the importance of the PI having knowledge of all processes involved in the project and mastery over what will happen in their own lab. If an advanced person in your discipline should be able to run a specific type of machine, then be sure you also know how to run it, or if you are an early-stage investigator, arrange for a mentor who can train you, perhaps as part of the project. In this case, you will want a letter from this person to include with the proposal.

Of course, you also know how your institution deals with the challenges that might be presented by your project, and you know when to mention them, even if fleetingly. Including relevant costs in the budget and budget justification related to specific technical and institutional processes will help demonstrate the administrative acumen that undergirds grants administration for the PI. For example, a line to pay for the disposal of the minor radioactive waste created by the project, or how you will actually acquire the subject eyeballs (and whether IRB is needed or not) is just a little more evidence for the reviewer that you are the kind of capable person who can manage the large sum of money you are requesting to lead the team and carry out the very specific and complex project you are preparing.

Now on to our three basic steps for deploying the appropriate level of rigor and specifics.

52.3 Logic

I know that you have moved through some very challenging years of training to even be at the point of contemplating submitting a medical research grant proposal. I also know that the training you have undergone has caused you to incorporate very specific logical systems of thought into your personal cognitive systems. Further, you are surrounded every day by teachers, mentors, and colleagues who share in these systems. I do not doubt that you are a logical person—that is not the grant writing problem.

The grant writing problem here is whether your logical thoughts are being expressed clearly in your proposal in such a way that people who do not share in your conceptual systems and training—and lack the context to do so—are still able to glean sufficient understanding of your work to influence them into giving you the award. In my previous book, *The Nuts and Bolts of Grant Writing* [1], I spend an entire chapter discussing the importance of logic models (See that book for more detail on this topic). Considering that the book was directed toward faculty, this might be somewhat surprising as the readers were expected to have advanced degrees, and therefore to be logical people. The problem however is that disciplinary thinking, even the thinking within specific research groups, may or may not be obvious to everyone else, and especially, may not be obvious to the reviewers from various sponsors. Things that may appear obvious to you, because you work in this field daily, may not be so obvious to others, even others who work in a related field. Further, if you are working on a real breakthrough your logic may be entirely new, and therefore even more important to explain carefully.

For this reason, I recommend that anyone considering a grant proposal, and particularly when one is new to the project at hand,[3] sketch out the entire project as a series of steps. Seeing the process as an illustration can help you see the project differently. It can help you see where steps may have been missed, or where additional resources are needed. Is there a process that you assume everyone knows about which is part of the arcane knowledge of your subdiscipline? Are there steps that might appear so obvious that they are not included? On a more mundane level, is there proper refrigeration at your satellite site to accommodate the medicines required? What materials will be needed to physically secure the confidential data? Do prospective patients have access to travel money to get to the medical center?

Show the plan to your mentors and helpful colleagues, ask for feedback and then incorporate the feedback into your project. It can also be a good idea to show the plan to colleagues in other subfields because they may better represent the reviewers you will get and may be able to point out areas that need more description.

Be sure that the grant proposal reflects all the key steps in some way. A key step may be a big obvious point, like who will build the next generation mammography machine you might be designing, but it could also be a very small but critical process that might be hard to put in place, like access to a specific cell line, or having the correct collaborator on board who is expert at a specific technique. It may also be key to explain why something that appears obvious to others is not necessary to your project. This is particularly true if you are contemplating a new technique where it may be important to present additional

[3] This would include both novices to grant writing and experienced PIs who are contemplating an entirely new project.

information about why older techniques are no longer needed. Remember that the reviewers may have been using the old techniques for many years, and may need a thorough explanation.

Additionally, use this logic model to have compassion for your reviewers. Your grant proposal may be read after a long day at work, or just before heading out on the first day off for weeks, and despite training, education, acumen and good intentions, your reviewer has a human brain that may be tired. It behooves you to make the logic of your project obvious not to yourself, but to your reader, who is reading from an unknown vantage point and knowledge base. Know and express all the important steps, write them clearly, and be sure that the sequence makes sense. Carefully highlight key points. Show your work to others to be sure you nailed it.

This logic model will form the strong tree trunk and main branches of your proposal. It will provide the structure of the argument and always be available to compare with the developing proposal to be sure all relevant processes are represented appropriately and that they appear in the correct logical order.

Everyone wants to know what the logic model should look like. At first, it should look like whatever works for you to roughly conceptualize it. Drawings on a piece of copy paper with notes arranged on it is a most likely first effort. In the beginning, the most important thing is to capture all your thoughts. Scribbles, cross-outs, and multiple pages are part of the process. Later, when you want to show it to someone else, a simple graphic program like PowerPoint, and even non-graphic programs like Word and Excel can be used to effectively represent logic models.

52.4 Judicious Selection of Detail

In the previous section, when I advised that you write out all the steps of your project, did I also mean that every single step should appear in your proposal? Not necessarily.

While detail is extremely important, it can also be overwhelming. I once received a resume from a job applicant who included 5 pages of detail about all the things they had done over the last 10 years. There was no summing up of items or classification, just lists of sponsors, projects, and tasks accomplished in great detail. Now, not only was this a very dry read, but it also assumed that I, as an interviewer, would be doing a great deal of work—I was supposed to read all these minutiae (a chore) and create my own analysis of what all of these items meant in terms of a possible staff member. Of course, I did make an analytical judgment: I concluded that this person had little idea how to do the job because they did not know how to categorize, organize, or explain their own experience. This person certainly did not come off as an analytical thinker. The fact that they could not create a narrative of themselves for me, the representative of a potential employer, meant that they could not conceptualize the job they had been doing and therefore were not likely to understand the job on offer. This was obvious to all of us on the committee and that person did not get an interview.

Similarly, once you have your logic model, think carefully about how much detail to include. All those steps need to be organized, conceptualized, and converted into a persuasive narrative explanation. What this will look like is two or three (likely no more than four) narrative threads that represent processes needed to be undertaken to carry the study out—in US National Institutes of Health (NIH) proposals, these are often called Specific Aims (see Chap. 51). These form the core of the proposal and should be explained very well but economically. Consider which steps need to be highlighted, and which can be conceptually consolidated for rhetorical efficiency. This will depend on not only your discipline, mentors, colleagues, and theoretical position, but also, your personal sense of style. There will be many correct answers to this puzzle, even across your career, so think carefully and experiment with different modes of expression if this seems needed.

Once you have worked out the conceptualization of the project processes, other details can be added, or selected for deletion. There are three main functions of additional detail at this point:

1. to provide information that buttresses your argument
2. to provide information that demonstrates mastery over the topic, and
3. details that ground the project in the real world.

All detail provided should buttress the work except where the risk or limitations involved in the project are significant and must be explained,[4] so I will concentrate on items 2 and 3.

52.4.1 Details That Demonstrate Mastery

Particularly if you have few publications and no completed grant projects, it is useful to provide a few telling details that only could be related by someone who has participated in the kind of processes described by the proposal. This lets the reviewer know that you have done this before, perhaps as part of your advisor's lab. It may not be necessary to write about that lab experience (you will probably do this elsewhere in the proposal), it is only necessary to relate the information that allows an expert reader to understand that you have experience with this process. It is also not necessary to provide such telling detail for each and every technique or procedure described, perhaps only for the most difficult, or most important of them.

Such details show mastery, they do not tell about it. Similar to advice typically given to journalists, one does not tell the reader about how experienced one is in the lab, one demonstrates it by presenting information that could not be known any other way. For example, if I were writing about how to drive a car I might naturally provide a great deal of information about how to use the accelerator, the brakes, and the steering. I might write about speed limits, one-way streets, traffic rules, and where one might drive a car. If I also added something like the following line, you would know I had actually driven a car:

> The complementary relationship between acceleration and braking is never more obvious than in the process of taking a curve where one slows the car down from the onset to the apex and then uses the accelerator to gradually speed through the exit. The pleasing way the car responds to this handling makes the process far simpler to execute than the novice may think.

I am discussing how to drive a car, however, by providing observation about how a car responds to the process of driving on a curve I let the reader know that not only have I driven a car, but I have also performed a little more advanced driving and reflected on this enough to make cogent observations. In the above quote, I have shown something about competency in driving, rather than having just told about it.

52.4.2 Details That Ground the Project in the Real World

Details can also be used to break up the monotony of the scientific descriptive language by adding interest to the topic. As a rule, these details appropriately appeal to the reader's senses or emotions. Scientific language, by its very nature, is not used much outside specific academic or industry circles and so often has limited sensory and emotional reach. While you do not want to turn your proposal into tabloid journalism, a few grounded details can create dynamism in the text.

For example, when I wrote "dynamism" above it was no less correct than the following: "A few grounded details can bring color, depth, and appropriate feeling into your proposal." When I wrote dynamism, however, many readers would have to make a mental translation to derive the meaning of "something lively." Reading "color, depth, and feeling" however, the average reader knows immediately what was being expressed—the images can burst into the mind. In fact, I felt I had to add the adjectival modifier "appropriate," a Latinate English word, because we are, after all, discussing scientific expression. Academic

[4] Even in this case, the point of the discussion of risk and limitation is to thoroughly explain how the factors in question will be thoughtfully mitigated through the project plan.

writing, and particularly medical terminology in English, is largely based on Latin and Greek, of course. Linguistically this is like writing with brakes in English. To add in a few more immediate Anglo-Saxon words ("thread," "simple," or "false" or example) is to add just a little acceleration to the sentence.

As another example, when writing up my dissertation on plastics facilities, I made a point of locating one such factory that was very close to my university, giving the street location and what was made there. The work dealt with many chemical names, so in this specific example, I translated the product plastic polyethylene terephthalate (PET) into the product: plastic water bottles. Not only are these easily understandable mostly Anglo-Saxon words, but they describe a very common item and one that carries a substance required for life: "water." Two of the three words have immediate emotional connotations in English, and the whole set together has a completely different modern connotation: trash. By locating the facility in space for my readers and making the chemical into a familiar product, the work became much more interesting to them and demonstrated to the ones who were paying attention that I understood my subject matter. In other words, this was a paragraph where I was explaining some of the main concepts in language my granny would understand.

I promise that even the most expert scientific reviewer will appreciate details that make the project more concrete and that connect the work to the social or sensory world. The art of the endeavor is when you, with your vast knowledge of a research topic and highly developed discretion can effectively select the correct details to make the project more vital and alive for your reviewers (once or twice during the proposal perhaps), without sacrificing one jot of intellectual caliber or finesse.

52.4.3 An Example of Effective Use of Details

It is ironic to note that proposal writers often spend a great deal of space discussing their projects in relatively vague terms and never seem to get to the actual work that will take place. One can nail the relevance and endlessly discuss the relationship of the project to this or that process or theoretical school and never seem to describe the actual concrete actions that must be accomplished in order to complete the investigation. Even though this is the heart of the study and comprises what will be funded, one oddly must often search the application to find these critical details.

Of course, a logic model militates against this common problem because it gathers together all the steps necessary to perform the project. These steps are then converted into the narrative that explains exactly what will be done and why. Below is an example from the NIH of such a paragraph taken from a proposal that showed high scientific transparency and rigor.[5]

> Aim 3: Male and female mice will be randomly allocated to experimental groups at age 3 months. At this age the accumulation of CUG repeat RNA, sequestration of MBNL1, splicing defects, and myotonia are fully developed. The compound will be administered at 3 doses (25%, 50%, and 100% of the MTD) for 4 weeks, compared to vehicle-treated controls. IP administration will be used unless biodistribution studies indicate a clear preference for the IV route. A group size of n = 10 (5 males, 5 females) will provide 90% power to detect a 22% reduction of the CUG repeat RNA in quadriceps muscle by qRT-PCR (ANOVA, α set at 0.05). The treatment assignment will be blinded to investigators who participate in drug administration and endpoint analyses. This laboratory has previous experience with randomized allocation and blinded analysis using this mouse model [refs].

[5] This paragraph is reproduced as given by NIH. "[ref]" refers to areas where references were originally given, however these are not relevant here as the construction of this paragraph is being examined, not its claims. See this and other scientific rigor examples at "Resources for Preparing Your Application: Scientific Rigor Examples" National Institutes of Health https://grants.nih.gov/policy/reproducibility/resources.htm#scientific

Their results showed good reproducibility when replicated by investigators in the pharmaceutical industry [ref].

Let's break this down.

Male and female mice will be randomly allocated to experimental groups at age 3 months.

It might seem obvious to say that the subjects will be randomly assigned on the basis of sex, however by providing this information the PI assures the reviewer of their knowledge of this area. Note that how the random process is coordinated is a level of detail unnecessary to this proposal, so it is encapsulated in the word "randomly" and does not comprise an entire sentence.

At this age the accumulation of CUG repeat RNA, sequestration of MBNL1, splicing defects, and myotonia are fully developed.

The PI lets the reviewer know why the mice are assigned at 3 months. These topics will have already been unpacked and discussed in the proposal, and the PI is connecting the dots for the reviewer. By reminding the reviewer of the previous material while explaining the actual investigational process the various aspects of the proposal are woven together and the reviewer understands how the theory of the project is going to be tested. This is the heart of the proposal.

The compound will be administered at 3 doses (25%, 50%, and 100% of the MTD) for 4 weeks, compared to vehicle-treated controls.

Here, the PI clearly lets the reviewer know the basic research plan and dosage. Again, use of the controls does not have to be unpacked here—it forms a 4-word clause. It might be a little more appropriate to mention the name of the compound here, just to remind the reader.

IP administration will be used unless biodistribution studies indicate a clear preference for the IV route.

The PI offers some information on risk management and a possible way the experiment could change slightly. If the IV route were controversial, or a poorer choice, a sentence defending the decision might be included here.

A group size of n = 10 (5 males, 5 females) will provide 90% power to detect a 22% reduction of the CUG repeat RNA in quadriceps muscle by qRT-PCR (ANOVA, α set at 0.05).

An extremely important justification for the number of animals to be used is offered, the relationship between the number of animals and the statistical requirements for adequate power. This should be a standard part of the justification of any study.

The treatment assignment will be blinded to investigators who participate in drug administration and endpoint analyses.

The team will be working toward the most rigorous methods of the experimental method, and one that takes some organization and effort to deploy.

This laboratory has previous experience with randomized allocation and blinded analysis using this mouse model [refs]. Their results showed good reproducibility when replicated by investigators in the pharmaceutical industry [ref].

Blinding a subgroup of investigators is sufficiently difficult that additional details are provided to demonstrate that this team is up to the task. Not only have they done it before, but their results showed the kind of reproducibility expected from the blinding, meaning they did the operation very well.

One could imagine that a more junior team that had not performed this sort of blinding before might provide information about a member of the team who had participated in a similar process and would be coordinating it for the team, or of a mentor brought on to advise on this process.

52.5 Relevance

I have included relevance here because all your hard work on appropriate rigor and detail will be for naught if your reviewer does not bother to follow your argument. Of course, they will read your proposal, but if they do not see the point it will just be another proposal read, and believe me, once your proposal is put down, there will be several to be reviewed next. The key is to make your project relevant to the point that it is remembered even when ten or fifteen other proposals have been reviewed after yours.

There are many ways to adjust your logic and judicious detail to make very obvious to your reviewer what is important in the proposal, so experiment with the writing. One thing is certain: if you leave it to the reviewer to figure out what is important, there is no guarantee they will do so. A useful rule of thumb is that nothing in your proposal is as obvious to others as it is obvious to you, so explain what is important and why. Begin the proposal with a robust justification of the work, and once you have your logical sequences set out and your relevant details provided, be sure to remind the reviewer occasionally about how and why the project is significant, and repeat a few other key points as needed.

A wise mentor once advised me that it is not enough to just set your study out, one must metaphorically take the hand of the reader and gently lead them to each item, reminding them along the way of why you are making the journey. I have been doing this in this chapter—I keep repeating the names of my steps (logic, detail, relevance) as well as the point of the chapter: developing rigor and specifics. I present these topics and then I make them touchstones by circling back to how my suggestions may support the development of rigor and specifics in your writing.

In this vein, I would make one more suggestion, which is to allow the slightest bit of emotion into your proposal. I suggest that this be done through word and example choice. First, if you are writing in English, remember that the heart of the language is Anglo-Saxon, not Latin or Greek. Of course, as a physician, you use a great deal of Latin and Greek, but these words are less likely to emotionally register with the reader. In an academic grant proposal one is not trying to create waves of emotion, however subtle word choices can cause a message to become slightly more meaningful.

Specifically, in the few paragraphs describing the relevance of the project, the use of the Anglo-Saxon can remind the reviewer of the human dimension of the work. Mentions of heart in cardiological relevance discussions; kidney alongside renal physiology; use of eye alongside retina and the macula: these are examples of grounding language and bring subtle emotional nuances, even for doctors. In this way, you humanize your work and reify its impact and importance to others.

For example, rather than just citing large numbers of cases that need assistance, like, "5 million Americans experience age-related macular degeneration yearly" one might rephrase to something like, "Every year 3 million American elders, including grandmothers and grandfathers, will never see their children or grandchildren again due to age-related macular degeneration." It is true that by specifying the blind as a subset of those suffering from macular degeneration the number of the afflicted is reduced, however a more specifically human aspect of the disease is introduced at the same time. Although it seems obvious that when elderly people become blind they won't see their grandchildren, this is a logical detail that tends to get lost within medical discussions. It may suit the study to highlight it, and therefore provide increased emotional resonance for what you are trying to accomplish. The revision did not use that many more words, and it is not necessary to discuss the matter further, but the stakes of the project have been given emotional valence by introducing socio-emotional language through Anglo-Saxon words that relate the medical issues to the reality of the patients—the people you serve.

Finally, it is important to note that rigor is not necessarily a function of big words—it is a function of precision in language. For example, in the last sentence of the former paragraph, I debated whether to use "Anglo-Saxon verbiage" or "Anglo-Saxon words." In the US "verbiage" is a fancy way of indicating "words" however it can also mean verbal excess. In a discussion of the emotional cadences of expression, "words" appeared more precise and fitting, even though it is a very elementary word in English.

It is useful to be aware of such distinctions because the non-medical reviewer, or even the medical reviewer from a different specialty, ethically cannot fund a proposal they cannot understand. So, the PI must calibrate the proposal language to the likely scientific understanding of the potential reviewer and adjust vocabulary as necessary. It is worth emphasizing that adjusting the vocabulary to accommodate the lay or non-specialist reader is not dumbing it down, so keep the logical sequences intact. It would be folly, for example, to treat a reviewer from a computer or social media foundation as unintelligent just because they don't have a medical degree. Individual reviewers will expect a logical argument from you in lay language, and if you can make your excellent project come alive, so much the better for your funding chances.

52.6 Concluding Remarks (A Few More Tips)

Do your research and find out whether you can predict who might be reviewing the proposal, and the level of knowledge of the reviewers. For example, a proposal going to a scientific sponsor, like the NIH, which will try to recruit reviewers as specifically knowledgeable about proposed projects as possible, might need to be written at a level comparable to a high-calibre scientific journal. If the sponsor is a foundation with no medical research personnel on staff the logic and details will need to be written with less technical rhetoric, even lay language.

If you can get access to a funded proposal, so much the better. Ask your advisor or mentors for the opportunity to examine their successful proposals, and help them write their proposals if you can. Large sponsors may make example proposals available online (the NIH provides examples), or it might be worth asking the sponsor for an example of a funded proposal. Alternatively, one could examine lists of awards and write to those PIs, requesting the opportunity to read their successful proposals. Colleagues of your advisors and mentors might be more willing to share, as might be your university colleagues. Another option might be to form a collegial review group and provide reviews for each other. Tracking who gets funded from such activities will provide a wealth of data for your proposal writing.

Finally, while all the advice in the world may be offered, the only perfectly written grant proposal is the one that has been funded. Just as a great deal of preparation is needed to write a proposal well, don't allow perfect to be the enemy of good as they say. At some point stop writing and submit the proposal. Even if you go unfunded you will get excellent feedback. And if the proposal is not even reviewed? Well, that is data, too! Go back and work more, organize better. Write, let others read, make revisions, and submit again. I have worked with hundreds of PIs and I can say that the ones who get funded the most are also often the ones who are rejected the most—because they tend to submit many proposals. Like any other skill, one's proposal writing becomes better with practice.

I wish you luck in all your endeavors, and especially, in your grant writing.

Conflict of Interest None.

Reference

1. Carr CE (2015) The Nuts & Bolts of Grant Writing. SAGE, Thousand Oaks, CA

53 Writing Research Grants Proposals: From an Indian Perspective

Wajeeda Bano and K. Byrappa

Abstract

Higher education covers both teaching and research. It contributes directly or indirectly to the establishment of the country as a power among nations of the world. In the nation building process, science and technology play a key role in every sector of human life. It is still widely accepted that research is a luxury and that a poor country cannot afford it. Funding for research is shrinking worldwide. In India, the success rate of research proposals has come down to a meager 7%. Therefore, writing a research proposal with a higher success rate needs a lot of planning and background studies. The participation of industries or other laboratories of national and international repute are important. The transdisciplinary approach is also another major parameter for successful project proposal writing. This chapter covers various issues such as how to deal with a research problem, research methodology, and its societal or academic impact. In addition, we discuss preparation of research project proposal.

Keywords

Research proposals · Funding · Research and development · Funding agencies · Indian perspective · Fellowships

53.1 Introduction

Research is to investigate, experiment, test explore, fact-finding, examine, probe, unravelling the mysteries of nature, and understanding the phenomenon or a problem systematically. It adds to our knowledge, addresses gaps in our knowledge and provides a voice to the individuals to discuss debate and initiate policy making. Here it is appropriate to mention a quote from Sam Veda: *The light of knowledge is fully capable of destroying the darkness of ignorance.* This also helps us in overcoming all the difficulties and in achieving success in all our endeavors. Higher Education and Research would contribute directly or indirectly to the establishment of the country as a power among nations of the world. In the Nation Building Process, Science and Technology play a key role in every sector of human life. It is the USA, which holds up to 72% of the technology in the world, in other words, it is a Superpower in a real sense. Who holds the keys to Science and Technology? It's the one who has a strong bearing on research. It is still widely accepted that research is a luxury and that a poor country cannot afford it. Research covers

K. Byrappa has died before the publication of this book.

W. Bano
Department of Studies and Research in Economics, Mangalore University, Mangalagangothri, Konaje, Karnataka, India
e-mail: wajeedabano@gmail.com

G. Jagadeesh et al. (eds.), *The Quintessence of Basic and Clinical Research and Scientific Publishing*,
https://doi.org/10.1007/978-981-99-1284-1_53

various issues such as how to deal with a research problem, and its societal or academic impact. It contributes to nation building and what parameters are used to assess the quality of research, which is very important in the current context since research funding is shrinking everywhere. A researcher poses questions and tries to answer them through experiments, collects data (both primary and secondary) and answers questions. Any good research topic seeking grants should have a proper goal setting and should be specific, measurable, attainable, and realistic and time framed. According to Joseph Erlanger (1944), a Nobel Prize winner described that an individual can make a difference, but as a team can make a miracle. This applies so well in the current context that good research can be achieved if there is collaboration or team work. Many times, scientific discoveries are just accidental, and a simple curiosity has led to several discoveries. According to an American Physicist Joseph Henry (1840) the seeds of great discoveries are constantly floating around us, but they only take root in minds well prepared to receive them. There are several examples like C.V. Raman, Thomas Edison, Michael Faraday, Frederick Banting, John J.R. MacLeod, Alexander Fleming, etc. who converted their curiosities into great scientific discoveries. This chapter discusses the availability of research funds and various sources, the target groups, thrust areas, and strategies to be adopted in conceiving excellent research ideas and its presentation to obtain research grants irrespective of the discipline from an Indian perspective.

53.2 Research and Development

Research and Development (R and D) activities can be defined as any systematic and creative work undertaken in order to increase the stock of knowledge and use knowledge to devise new applications which enhance productivity growth [1]. R and D activity has three distinguished features: basic research, applied research and experimental development. Basic science is sometimes called "pure" or "fundamental" science. Through basic science, researchers try to answer fundamental questions about how life works. Examples like how cells talk to each other. What controls gene activity? How do proteins fold so they can work properly? How do diseases develop? Similarly, one can pose a question like why do so many Americans win the Nobel Prizes? According to experts, the answer is that in the USA, it is a strong historic investment in basic science, academic freedom for researchers and patience to see results [2].

The present article has been divided into two parts. Part 1 discusses various funding agencies and Part 2 describes good research grant proposal writing, both from an Indian perspective.

As a measure of critical rigor, quality education and overall capacity, research in higher education institutions (HEIs) plays a vital role in a country's R and D ecosystem. However, India's R and D expenditure to GDP is 0.69 %, the USA spends 2.8%, Korea spends 4.2%, Israel spends about 4.3%, China spends nearly 2.1% and Japan spends about 3.2%. Accordingly, India's spending is the lowest among emerging economies. At the core of this issue of low spending for India is the unavailability of comparable data in relation to research spending. Currently, there is no publicly available consolidated data on the institution-wise R and D expenditure and research grants made to specific HEIs by any of the national funding agencies. Moreover, the HEIs themselves neither publish consolidated data on grants received from the funding agencies nor the details of the R and D expenditure made by them. Hence, there is a visible paucity of information about the scenario of R and D funding in the higher education sector of the country. Globally, funding models for university research are often classified based on the idea of the degree to which they are supported by internal or external funding [3, 4]. Governments typically have two main modes of direct investment in university-based Rand D: Institutional and Project-based. Institutional investment can help ensure stable long-run investment in research, while project-based investment can promote competition within the research system and strategic target areas. Institutional funding generally provides institutions with more scope to shape their own research agenda, while project funding provides governments with more scope to steer research toward certain fields or issues. This is attributed to the fact that in most cases, it is a lump sum amount (a common block grant) for both education and research [5]. On the other hand, project funding may allow governments to target the best research groups or support structural change [6].

Research funding is defined as a grant obtained for conducting scientific research generally through a competitive process. Conducting research requires applying for grants and securing research funding. A research grant is a sum of money awarded to support a particular activity and is not to be paid back except in some exceptional cases.

53.3 Part 1: Research Funding

This part discusses the research funding from an Indian perspective (also described in the next chapter) and the strategies to be adopted in the search for appropriate funding and in turn the writing of a suitable research proposal.

In India, there are three major funding sources for research: Central Government, State Governments, and Industries. In the Central Government funding, scientific research is carried out by two groups: the first group is R and D performing bodies inter alia including the Department of Atomic Energy (DAE), Department of Space (DOS), Defence Research and Development Organization (DRDO), Council of Scientific and Industrial Research (CSIR) and Indian Council of Agricultural Research (ICAR), and in the second group of R and D funding falls the Department of Science and Technology (DST), Department of Biotechnology (DBT), Ministry of Earth Sciences (MES), etc. (R and D Statistics 2019-20 Report). Although the primary role of R and D performing group is to undertake R and D activities, they also sponsor some amount of extramural research in the areas of their interest. On the other hand, the R and D funding group is primarily engaged in its major role of promoting scientific research in extramural or sponsored mode. Research carried out by the Public Sector, Private Sector and Non-Governmental Organization is supported mainly with their own resources. The academic sector performs R and D through both intramural as well as extramural sources [7].

53.3.1 Types of Research Funding

In India, every funding body has a set of goals and also thrust areas identified to suit its activities. Decisions are made on the applicant's ability to fit the proposed research activities to the interests of the funding body. Following are the types of projects called by different agencies and provided with funding in India:

- Minor Research Projects and Major Research Projects (it applies to all the HEIs in India, the minor research projects are given to young faculty as seed money to initiate their research irrespective of the specialization or subject).
- Funding for Collaborative Research Projects (it is the most popular method of getting major research funding in the current context, but the important aspect is the identification of the right collaborator or collaborating institutes).
- Institutional research projects and institutional performance based funding for infrastructure development funded by the Department of Science and Technology, Government of India (Fund for Improvement of Science and Technology Infrastructure in Universities and Higher Educational Institutions (FIST) Programs; Promotion of University Research and Scientific Excellence (PURSE).
- Travel Grants, International Collaborations and exchange programmes by various funding agencies like the University Grants Commission (UGC); Department of Science and Technology (DST); Council of Scientific and Industrial Research (CSIR); All India Council of Technical Education (AICTE) and so on.
- Book Writing proposals
- Funding to organize Conferences / Seminars / Workshops/ Training and Skill programs, etc. (several National and State Government

agencies provide funding to propagate science and technology).
- Short term Research Studentship.
- National Task Force Projects (usually awarded to major interdisciplinary nationally coordinated research proposals involving multi-institutions).
- Funding for R and D system – Industry Cooperation, Technology Development Programs and International Technology Transfer Program (these research projects proposals are becoming very popular in the recent years with a higher success rate owing to their direct impact on science and technology and society).
- Consultancy Promotion Program (HEIs are encouraging in a big way the consultancy programs with industries and corporates).

Student Oriented Funding (2 to 5 years)
- Post-Doctoral Fellowship and DST INSPIRE Fellowship.
- Junior Research Fellowship, Project Fellowship, Project Assistantship, Senior Research Fellowship, Research Assistantship, Research Associateship, etc., awarded either directly by various R and D funding agencies or through various research projects.
- Women Scientists, Fast Track Young Scientist and Kothari Fellowship, Ramanujan Fellowship, J.C. Bose National Fellowships and Swarnajayanti Fellowship, etc.National Funding Agencies in India

Following are the important funding agencies of the Government of India offering grants for research proposals both under major and minor research projects:
- All India Council for Technical Education (AICTE)
- Council of Scientific and Industrial Research (CSIR)
- Defense Research and Development Organization (DRDO)
- Department of Atomic Energy (DAE)
- Department of Ayurveda, Yoga and Naturopathy, Unani, Siddha and Homeopathy (AYUSH)
- Department of Biotechnology (DBT)
- Department of Coal (DOC)
- Department of Ocean Development (DOD)
- Department of Science and Technology (DST)
- Department of Scientific and Industrial Research (DSIR)
- Indian Council of Medical Research (ICMR)
- India Meteorological Department (IMD)
- Indian Space Research Organization (ISRO)
- Ministry of Comm. and I. T. (MOCIT) Dept. of I.T.
- Ministry of Environment, Forests and Climate Change
- Ministry of Food Processing Industries (MFPI)
- Ministry of Non-Conventional Energy Sources (MNES)
- Ministry of Power, Central Power Research Institute (CPRI)
- Ministry of Social Justice and Empowerment (MOSJE)
- Ministry of Water Resources (MOWR)
- Petroleum Conservation Research Association (PCRA)
- University Grants Commission (UGC)
- Drugs and Pharmaceutical Research, Dept. of Science and Technology
- Ministry of Electronics and Information Technology (MEIT)
- Ministry of Statistics and Program Implementation (MSPI)
- Department for Promotion of Industry and Internal Trade (DPIIT)
- Ministry of Skill Development and Entrepreneurship (MSDE)
- Ministry of Chemicals and Fertilizers (MCF)
- Ministry of Micro, Small and Medium Enterprises (MSME)
- National Center for Biological Research, TIFR
- Ministry of Food Processing Industries (MFPI)
- Ministry of Tribal Affairs (MTA)

53.3.2 Role of Some Selected Funding Agencies in India

India's Ministry of Science and Technology is the dominant agency that is responsible for funding Science, Technology, Engineering and Mathematics (STEM) R and D in India. DST, DBT, SERB and CSIR are under the Ministry of Science and Technology, while the Indian Council of Medical Research (ICMR) is an autonomous body under the Ministry of Health and Family Welfare. Recognizing the need to promote Research and Development in emerging areas of Science and Technology in the Indian Innovation Ecosystem, the Department of Science and Technology has also set up three Technology Missions or Divisions to promote research activities in specific thematic areas such as Technology Mission Program on Clean Water and Clean Energy; Nano Science and Technology Mission; and National Supercomputing Mission (https://dst.gov.in/document/budget/2022-23).

Important Schemes are:

1. Swarnajayanti Fellowship
2. INSPIRE Faculty Fellowship
3. Mission on Nano Science and Technology
4. Women Scientists Scheme

The first three initiatives provide selected scientists with a fellowship amount every month. Awardees of the INSPIRE Faculty Fellowship receive Rs. 1,25,000 per month as salary. Swarnajayanti Fellowship is made in addition to the regular salary drawn from the employing institute. Applications for Swarnajayanti Fellowship and INSPIRE Faculty Fellowship are invited annually. All three fellowships are applicable to Indian citizens only. The quality of candidates' research proposals is a crucial factor that is common for all three schemes. Past academic performance of candidates and publications (at least 3) with high Impact Factor are two determinants, Candidates under INSPIRE Faculty Fellowship are selected by a three tier selection process by - the Expert Committee by Indian National Science Academy (INSA), Apex Level Committee (INSA) and finally, the INSPIRE Faculty Award Council (DST). The three funding initiatives differ in terms of thematic concentrations. The INSPIRE Faculty Fellowship funds scientists' research projects across all broad areas of Science. On the other hand, Swarnajayanti Fellowships are awarded to scientists for vital research projects in the realm of Science and Technology. The Nano Mission projects specifically focus on the application of Nano Science and Technology towards a greater understanding of basic research and Science and Technology development.

All three initiatives have funded researchers at nodal IITs and IISERs and IISc, Bangalore. However, out of the three schemes, All India Institute for Medical Sciences (AIIMS), New Delhi researchers have a major share to have engaged with INSPIRE Faculty Fellowship. Furthermore, some researchers from several sector-specific HEIs have also received the INSPIRE Fellowship, such as - Indian Agricultural Research Institute, New Delhi and Punjab Agricultural University, Ludhiana, Punjab. Similarly, Swarnajayanti Fellowship has also been awarded to researchers at one sector-specific institute like the Tata Institute of Fundamental Research, Mumbai.

Women form an important section of the workforce, more particularly in the science and technology domain. However, a large number of well-qualified women get left out of science and technology activities due to various circumstances which are usually typical to the gender. The challenges faced by them are several but most often the "break in career" arises out of motherhood and family responsibilities. To address such issues, the Department of Science and Technology (DST) launched "Women Scientists Scheme (WOS)" during 2002-03. This initiative primarily aimed at providing opportunities to women scientists and technologists between the age group of 27 to 57 years who had a break in their career but desired to return to the mainstream. The following three categories of fellowships with research grants are available for Indian citizens:

- Women Scientist Scheme-A(WOS-A): Research in Basic/Applied Science.
- Women Scientist Scheme-B (WOS-B): science and technology interventions for societal benefit.
- Women Scientist Scheme-C (WOS-C): Internship in Intellectual Property Rights (IPRs) for self-employment.

Other Schemes

Apart from these schemes, the Department of Science and Technology (DST) has also established the Science, Technology and Innovation Policy Fellowships Program (DST-STI- PFP) to encourage technology studies, Science and Technology Innovation-related data and Monitoring and Evaluation in the field of Science and Technology Policy. It offers three varying categories of awards, namely STI-Senior Fellow, STI Post-doctoral Fellow and STI-Young Policy Professional to promote the engagement of researchers with India's Science and Technology ecosystem across different age groups. In addition, the DST also runs several schemes, such as the Fund for Improvement of Sand T Infrastructure in Higher Educational Institutions (FIST) Program in order to contribute to the establishment of necessary research and development infrastructures in Higher Education Institutes and laboratories in India.

53.3.2.1 Department of Biotechnology (DBT)

The DBT functions under the Ministry of Science and Technology, and it equally supports the growth of basic research in India in the field of biological sciences. DBT has divided its Schemes and Programs for research and development into four thematic areas:

1. Medical Biotechnology.
2. Agriculture, Animal and Allied Sciences.
3. Knowledge Generation and Research, New Tools and Technologies.
4. Energy, Environment and Bio-Resource Based Applications.

Important Schemes

1. Ramalingaswami Re-Entry Fellowship
2. S Ramachandran - National Bioscience Award for Career Development
3. Har Gobind Khorana - Innovative Young Biotechnologist Award (IYBA)

Out of these three, the first two schemes, i.e. Ramalingaswami Re-entry Fellowship and S. Ramachandran Award, are open for scientists up to the age of 45 years. The age restriction for IYBA is lower, i.e. 35 years with an age relaxation of 5 years for reserved and backward class people, Women and Physically Handicapped candidates. An excellent academic career and significant background in research are some of the common selection criteria for the three schemes. To date, several researchers at IITs, IISERs and AIIMS, New Delhi have been funded. Funds from these three schemes have also been utilized at other institutions such as the Indian Institute of Science, Bangalore and the Institute of Nano Science and Technology, Mohali. DBT also runs various schemes and awards to encourage high quality research in the Indian Science and Technology ecosystem. In particular, the M.K. Bhan—Young Researcher Fellowship Program provides independent research grants to young scientists for post-doctoral research work in the country, a scheme under the 'Building Capacities' initiative.

53.3.2.2 Science and Engineering Research Board (SERB)

It funds relevant and quality research projects to fulfill the overarching aim of ensuring social and economic progress via scientific research. SERB has established the following three categories of research funding: 'Awards and Fellowships,' 'Schemes and Programs' and 'Partnership Programs'. The following are the important schemes and fellowships:

1. Core Research Grant (CRG)
2. Ramanujan Fellowship
3. JC Bose Fellowship

Out of the above three schemes, the Core Research Grant and Ramanujan Fellowship are awarded annually to exceptional scientists while selections under the J C Bose Fellowship are made twice a year. Applicants need to be nominated by the heads of institutions or J C Bose Fellows Indian scientists are eligible to apply for the Ramanujan Fellowship only if they are below the age of 40 years and working abroad at the time of nomination. Furthermore, they will not remain eligible for the fellowship in case of their selection for a regular position in a university or institute. In the case of CRG, the Program Advisory Committee (PAC) selects the researchers, while for J.C. Bose Fellowship; Search-cum-Selection Committee is the decision-making body. Both J.C. Bose and Ramanujan Fellowship schemes fund projects belonging to all broad areas of science. SERB awards the Core Research Grant via eight established thematic divisions - Chemical Sciences, Earth and Atmospheric Sciences, Engineering Sciences, Mathematical Sciences, Life Sciences, Physical Sciences, Exponential Technology and Quantitative Social Sciences. Funds from Ramanujan Fellowship have been utilised for research activities at several sector-specific HEIs such as the Indian Institute of Nano Science, Indian Institute of Geomagnetism, Indian Statistical Institute and Indian Institute of Science, Bangalore. Apart from the Core Research Grant, Ramanujan and J.C. Bose Fellowships, SERB also runs various other schemes and awards. Some of these programs specially focus on promoting multi-institutional research on pressing issues, such as the Intensification of Research in High Priority Area (IRHPA).

Awards, Fellowships and Grants

SERB has also established specific awards, fellowships and grants to boost the engagement of researchers belonging to certain groups. For instance, the Women Excellence Award and POWER (Promoting Opportunities for Women in Exploratory Research) are two such programs that aim to mitigate the gender disparity in Science and Technology research funding. Furthermore, SERB also provides financial support to facilitate research by extending grants for international travel support during research presentations and organizing seminars.

53.3.2.3 Council of Scientific and Industrial Research (CSIR)

CSIR funds R and D activities to promote innovation and progress in diverse science and technology fields such as Oceanography, Geophysics, Chemical Technology and Biotechnology. Its main objective is to assist projects with the potential for scientific and societal impact that leads to economic development. It lays special emphasis on converting fundamental research into value-added technologies that boost collaboration among stakeholders from HEIs and within the industry. CSIR has established 8 Theme Directorates to target particular projects and problems in specific sectors of STEM research as under:

- Aerospace, Electronics, Instrumentation and Strategic Sector.
- Civil Infrastructure and Engineering.
- Mining, Minerals, Metals and Materials.
- Energy (Conventional and Non-Conventional) and Energy Devices.
- Chemicals and Petrochemicals.
- Ecology, Environment, Earth and Ocean Sciences and Water.
- Agri, Nutrition and Biotech.
- Healthcare.

With such provisions and policies, CSIR plays an influential role in science and technology human resource development via its fellowships and grants for research projects and specific schemes that encourage research enthusiasts and young scientists to pursue doctoral and post-doctoral research. Important schemes of CSIR are :

(a) Shanti Swarup Bhatnagar Prize (SSB)
(b) Emeritus Scientist Scheme (ES)

In particular, CSIR's Shanti Swarup Bhatnagar Prize recognizes and awards exceptional scientists and their contributions to the field. Researchers selected for the SSB Prize are awarded Indian Rupees 500,000. Further, they receive INR 15,000 per month honorarium up to the age of 65 years. Under the ES scheme, selected scientists are offered a monthly allowance of INR 20,000 for the duration of the award.

It also provides funds to the selected researchers for the purchase of equipment, housing and medical allowances and contingency expenditures. Further, the grant also partially covers expenditure for foreign travel - up to INR. 30,000 or 50% of the airfare (whichever is less). Researchers up to 45 years of age are eligible to apply for the SSB prize. Under the Emeritus Scientist Scheme, researchers are eligible to receive funding for the grant duration until they attain the age of 65 years.

53.3.2.4 Indian Council of Medical Research (ICMR)

The ICMR is the top body in India for the conception, coordination and promotion of biomedical research in the country. It strives to establish programs that encourage innovation which can be implemented in the public health system in the long term. These initiatives are shaped around the following seven thrust areas: Communicable Diseases, Tribal Health, Reproductive and Child Health, Nutrition, Non-Communicable Diseases, Basic Medical Sciences and Traditional Medicine. In particular, ICMR awards several Senior Research Fellowships and Junior Research Fellowships/Research Associates based on UGCNET score to promote medical research.

(a) Ad-hoc Project Funding
(b) Emeritus Scientist Scheme
(c) Cohort Studies

All three initiatives differ in terms of the amount allotted for research. The Emeritus Scientist Scheme provides a Honorarium of INR 100,000 per month. Under Ad-Hoc Project funding, researchers get a maximum of INR 15000000 for the entire duration of the grant. Contrarily, researchers involved in Cohort Studies receive INR 20,000,000 per year. Researchers involved in the Ad-Hoc Project Funding and Cohort Studies provide for travel and contingency grants as well, further covering equipment and overhead charges Applications under the ICMR- Emeritus Scientist Scheme can be made on a rolling basis. The same is considered in two batches: applications sent from January to June are evaluated in July, while those sent between July and December are evaluated in January.

The ICMR-Emeritus Scientist Scheme is available for researchers who have retired or about to retire from a permanent position from the University. As such, the minimum age criterion for eligibility is 60 years under this scheme. On the contrary, for Ad-Hoc project funding, scientists must hold regular employment in a Medical College or Research Institute or Laboratory or a registered Semi-Governmental Institute. All three initiatives fund research in the fields of biomedical research. Mostly scientists who belong to AIIMS, New Delhi and IISc, Bangalore have received funding under this scheme.

Other Schemes

Apart from the schemes evaluated above, ICMR has also established different categories of projects to promote research in specific thematic areas. For instance, Task Force Studies are nationally coordinated projects wherein scientists across the country collaborate on time-bound and specific objective oriented projects of national interest. Similarly, the National Registry maintains a systematic framework to collect clinical and non-clinical data on disease control. The Indian Council of Medical Research also aims to encourage research practices among undergraduate students by awarding Short Term Studentships each year. In addition, ICMR also invites proposals for the establishment of Centers of Advanced Research (CAR) to promote research on a particular subfield of medicine. The scheme entails the development of Research and Development facilities, centered on an eminent scientist.

53.3.2.5 Indian Council of Agricultural Research (ICAR)

The ICAR sponsors and supports short-term result-oriented Extramural Research Projects intending to fill critical gaps in the scientific field or in the resolution of problems limiting production and value addition in agriculture, animal husbandry and fisheries. These projects can be submitted by ICAR Institutes, State Governments, Agricultural and other Universities, Public, quasi-public and private institutions, capable of undertaking research in the above areas (https://icar.org.in). There will be two modes for accepting project proposals. The first mode will be by inviting proposals on critical gaps identified by the Subject Matter Divisions (SMDs). The second mode will be voluntary proposals in relevant areas to be submitted by the investigators. Priority Areas for research will be identified for five years by each SMD and widely circulated. The projects will be invited not only from the National Agricultural Research System (NARS), but also from other Institutions or Universities. The scheme shall also cover collaborative projects with foreign institutions. The proposal for short term projects on critical gaps will be invited, scrutinized and processed by the SMDs.

53.3.2.6 India Council for Technical Education (AICTE)

The AICTE was established in the year 1945 as an Apex Advisory body at the national level to promote proper planning and coordinated development of technical education systems throughout the country (https://aicte-india.org/). The Council has been performing its regulatory, planning and promotional functions through its Bureaus, namely: Administration; Finance; Planning and Coordination; Under Graduate Studies; Post Graduate Education and Research; Faculty Development; Quality Assurance; and Research and Institutional Development Bureaus; and through its Regional Offices located in various parts of the country.

1. Research and Institutional Development Schemes:
   (a) Modernization and Removal of Obsolescence Scheme
   (b) (MODROBS) Research Promotion Schemes (RPS)
2. Industry-Institute Interaction Schemes
   (a) Industry Institute Partnership Cell (IIPC)
   (b) Entrepreneurship Development Cells (EDC)

National Facilities in Engineering and Technology with Industrial Collaboration (NAFETIC) under Research and Institutional Development Schemes, MODROBS started to equip technical institutions with modern infrastructural facilities in laboratory(s)/workshop(s)/computing facilities to enhance functional efficiency for teaching, training and research purposes. whereas RPS aims to create a research ambience by promoting research in technical disciplines and innovations in established and emerging technologies; and to generate Masters and Doctoral degree candidates of IIPC, EDC and NAFETIC are to develop, train and collaborate with institutions. The Council invites fresh proposals annually from AICTE approved technical institutions: University Departments, Government Institutions, Grant-in-aid Institutions and accredited institutions in the private sector for financial assistance for schemes operated by the RID Bureau. For five year old institutions in Jammu and Kashmir State and North-Eastern States, the accreditation criterion is not mandatory. Research proposals in Engineering and Technology, Architecture, Town Planning, Management, Pharmacy, Hotel Management and Catering Technology, Applied Arts and Crafts are invited.

53.3.2.7 Department of Ayurveda, Yoga and Naturopathy, Unani, Siddha and Homoepathy (AYUSH)

The Department of AYUSH has introduced a Scheme for extra-mural research in addition to the intra-mural research undertaken by four Research Councils for Ayurveda and Siddha, Unani, Homoeopathy, Yoga and Naturopathy set up by the Ministry of Health and Family Welfare three decades ago (https://www.ayush.gov.in/). The off-take and output from this scheme have so far been limited and have not been able to meet the standards for scientific enquiry and outcome effectively. The Department has taken up a series of programs/interventions wherein evidence-based support for the efficacy claims is needed. Safety, quality control and consistency of products are also very much required.

The important schemes are:

1. Extra-mural Research (EMR) project Scheme of AYUSH Systems of medicine and accreditation of Organizations for Research and Development in the fields of AYUSH.
2. Golden Triangle Partnership (GTP) Scheme for validation of traditional ayurvedic drugs and development of new drugs.

53.3.2.8 Funding in Social Science Research in India

Funding from the government of India: Social science research is immensely funded by the government of India and its other agencies like the Indian Council for Social Science Research (ICSSR) and UGC. Central Universities of India are funded by the Central Government through the UGC which in turn recovers its grants from the Ministry of Human Resource Development (MHRD).

The UGC strives to promote teaching and research in emerging areas in Humanities, Social Sciences, Languages, Literature, Pure sciences, Engineering and Technology, Pharmacy, Medical, Agricultural Sciences, etc. The emphasis would be supporting such areas that cut across disciplines and subjects such as health, gerontology, environment, biotechnology, stress management, WTO and its impact on the economy, history of science, Asian philosophy and many other areas as would be identified by subject experts.

UGC will provide support to permanent or regular, working or retired teachers in the Universities and Colleges (Under Section 2 (f) and 12 B of UGC Act, 1956). Colleges and Universities. The quantum of assistance for a research project is provided under:

I. Major Research Project (for Sciences including Engineering and Technology, Medical, Pharmacy Agriculture.—INR 1,200,000. And Humanities, Social Science, Languages, Literature, Arts, Law and allied disciplines—INR 1,000,000).
II. Minor Research Project up to INR 500,000.

There are a variety of important research grants and schemes offered by the UGC in India. Notable ones are:

For Faculty:

1. Dr. D.S. Kothari Research Grant for newly recruited faculty members
2. Research Grant for in service Faculty members:
3. Fellowship for Superannuated faculty members

For Students:

1. Savitribai Joyti Rao Phule for Single Girl Child
2. Dr.S. Radhakrishnan Post-Doctoral Fellowship

53.3.2.9 Indian Council of Social Science Research (ICSSR)

ICSSR provides grants for projects, fellowships, international collaboration, publications, capacity building, surveys, etc. to promote research in social sciences in India. The promotion of social science research is one of the major objectives of the ICSSR. The research grant is for direct financial support to research projects undertaken by Indian social scientists. ICSSR provides funding to Indian scholars to conduct cutting edge research in various fields of social sciences that have theoretical, conceptual, and methodological and policy implications (https://icssr.org/). The Research Projects may belong to any of the social science disciplines or may be multi-disciplinary. The ICSSR awards two types of research projects on the basis of the scope, duration of the study and budget:

(a) Major Project—Duration from 12 months to 24 months with a budget of INR 500,000 to 1,500,000.
(b) Minor Project—Duration from 6 months to 12 months with a budget up to INR 500,000. ICSSR Research Institutes or Institutes of national importance as defined by the MHRD/UGC recognized Indian universities/deemed to be universities under

12B, etc. are eligible to apply. However, other registered organizations with established research and academic standing may collaborate with any of the above-mentioned institutions for the implementation of the study. The professional social scientists who are regularly employed or retired as faculty in a UGC recognized Indian university/deemed to be university/ colleges, senior government and defense officers (not less than 25 years of regular service) and persons with proven Social Science expertise possessing a Ph.D. degree or equivalent research work in any social science disciplines and demonstrable research experience through publications of books/ research papers/reports can also apply, The applications will be invited through an advertisement on ICSSR website and if required in print media. Applications are scrutinized by the ICSSR Secretariat/and or by a Screening Committee in respect of eligibility. Eligible applications are then examined by the Expert Committee(s). Shortlisting of the program is done for interaction or presentation at ICSSR (in person or through technology). The expert committee (s) make(s) recommendation for the award of studies and also suggest a budget for the recommended studies.

(c) Important schemes.

International Collaboration: These programs of the ICSSR provides an opportunity for both Indian and foreign scholars in the field of social sciences to interact and research. It has been envisaged to promote academic links among social scientists in India and abroad. The council is one of the implementing agencies of the social science component of the Cultural Exchange Agreements (CEPs) and Educational Exchange Programs (EEPs) signed between the Government of India and the respective governments of other countries.

Cultural Exchange Program

ICSSR has ongoing Cultural Exchange Programs with Thailand, France and China. Under these CEPs, the ICSSR has institutional collaborations with the National Research Council of Thailand (NRCT), Bangkok, Foundation Maison des Sciences de L'Homme (FMSH), Paris and the Chinese Academy of Social Sciences (CASS), Beijing.

ICSSR has established bilateral collaborations with the Vietnam Academy of Social Science (VASS), Vietnam; the National Science Foundation (NSF), Sri Lanka; the Economic and Social Research Council (ESRC), UK; The Netherlands Scientific Organization (NWO), The Netherlands; German Research Foundation (DFG), Germany; UNIL-ALH (University of Lausanne – Associated Leading House), Switzerland; Japan Society for Promotion of Science (JSPS), Japan; and National Institute for the Humanities and Social Sciences (NIHSS), South Africa, Swedish Research Council for Health, Working Life and Welfare (FORTE), Sweden, Academy of Sciences and Arts of Bosnia and Herzegovina (ANUBiH), Sarajevo, Bosnia and Herzegovina. Within the framework of bilateral collaboration, the concerned organizations undertake activities such as the exchange of scholars, joint seminars, joint research projects, exchange of books and periodicals, etc.

The ICSSR has also engaged itself in multilateral collaboration like Bonn Group under Indo-European Research Networking Program in Social Sciences with participating agencies like ICSSR, India; ANR, France; DFG, Germany; ESRC, UK; and NWO, Netherlands. The objective of this networking is to promote and strengthen of the social sciences within and between the five countries by providing top up funding to allow joint research activities for internationally excellent research in relevant areas. It has been a part of the EU-India Platform for the Social Sciences and Humanities (EqUIP), which brought together research funding and support organizations in Europe and India to develop a stronger strategic partnership. Through this partnership, the EqUIP has launched its first research funding call on "Sustainability, Equity, Well-being and Cultural Connections". Besides, the ICSSR has been participating in the activities of international social science organizations, such as, the International Federation of Social Science Organizations (IFSSO), International Social Science Council (ISSC), Science Council of Asia (SCA), UNESCO, etc.

Financial Assistance for Participation in International Seminars/Conferences and Data Collection Abroad

ICSSR also provides financial assistance to social scientists who intend to visit abroad for data collection or consulting archival material in connection with their research work. This scheme is aimed at providing financial assistance only for cases where data collection abroad is fully justified in the approved original proposal of the research study under which this assistance is sought.

Financial Assistance for Organizing International Seminars/Conferences/Workshops in India

ICSSR has its endeavor to encourage dialogue and discussion on important issues of social science relevance, promotes and provides financial assistance to organize national and international conferences or seminars in India. The ICSSR's Seminar Grant Scheme aims at facilitating research in different disciplines of social sciences and interdisciplinary areas. The seminars provide opportunities to national and international social science researchers, academicians and scholars to exchange views and opinions, address and debate research questions on themes of contemporary and policy relevance and generate academic research output on important social issues.

Training and Capacity Building: the ICSSR Training and Capacity Programme (TCB) provides grants to the social science faculties for organizing research methodology and capacity building programme for young researchers and junior faculties in various social science disciplines, Publication Grant/Subsidy for Publication of Doctoral Thesis/Research Report/Papers Presented in Seminars/Conferences, Etc.

The ICSSR's Seminar Grant Scheme aims at facilitating research in different disciplines of social sciences and interdisciplinary areas. The seminars provide opportunities for national and international social science researchers, academicians, and scholars to exchange views and opinions, address and debate research questions on themes of contemporary and policy relevance and generate academic research output on important social issues

Training and Capacity Building: the ICSSR Training and Capacity Programme (TCB) provides grants to the social science faculties for organizing research methodology and capacity building programme for young researchers and junior faculties in various social science disciplines, Publication Grant/Subsidy for Publication of Doctoral Thesis/Research Report/Papers Presented in Seminars/Conferences, Etc.

53.3.2.10 Funding from Other Agencies (Non-Government Funding)

There are various non-state funders of social science research in India. If we compare the share of funding from government of India and other non-state funders, then the this share is very small, but still they are providing assistance for developing social science research in India, which is highly commendable. These agencies include International multilateral agencies (the World Bank, ADB), bilateral agencies (DFID, CIDA, USAID, NORAD), the UN as well as domestic and foreign foundations. These agencies have been funding studies on poverty, employment, education and health. Some other prominent non-state funders of social science research are the Ratan Tata Trust, the Bill and Melinda Gates Foundation, Ford Foundation, ICICI Centre for Child Health and Nutrition Foundation etc.

53.3.3 Overseas Research Funding

Almost all institutions and universities in India rely on government funding, but some institutions go for collaboration between government and non- government funding for the development of their research intellects. This proves to be very beneficial to the society and human resources of our economy to update their knowledge and skills. For instance, Delhi University and Jawaharlal Nehru University's School of Social Sciences are collecting project specific grants from various international donor agencies. Students pursuing Ph.D. degree can now apply for funding from a Dutch organization—Sephis—which funds scholars pursuing their doctoral research in developing countries (also see Chap. 54).

53.4 Part 2: Preparation of Research Project Proposal

The two most important components of any research project are ideas (*see* Chap. 2) and execution. The successful execution of the research project depends not only on the effort of the researchers, but also on the available infrastructure to conduct the research. It is an art that can be learned only by practicing. The most important requirement is having an interest in the particular subject, thorough knowledge of the subject, and finding out the gap in the existing knowledge (*see* Chap. 1). The second requirement is to know whether your research can be completed with internal resources or requires external funding. The next step is finding out the funding agencies which provide funds as described in Part 1 (Sect. 53.3), for your research topic, preparing research grant and submitting the research grant proposal on time (see Chap. 54).

A grant proposal or application is an important document or set of documents that are to be submitted to an organization with the explicit intent of securing funding for a research project proposed. Before beginning the writing of the proposal, one needs to know what kind of research he/she will be doing and why? A researcher may have a topic or experiment in mind, but taking the time to define what is the ultimate purpose of the proposal is very essential to convince others to fund that project.

Many local, national, and international funding bodies can provide grants necessary for research. However, the priorities for different funding agencies on the type of research may vary and this needs to be kept in mind while planning a grant proposal. Apart from this, different funding agencies have different timelines for proposal submission and limitations on funds. Therefore, one has to identify one's needs and focus. Believe that someone wants to give money to the researchers based on the research topic and the investigator's potential. Always consider the long-term goals of the host institution. Start with the end in mind (top-down approach). Identify the strength of the researcher and your own strength (the CV). Identify the host institution's strength or profile. Create a comprehensive plan - not just a proposal of what the investigator wishes to achieve. There are several basic steps in writing a good research proposal for grants (*see* Chap. 52). These steps are briefly described below for the benefit of the reader:

In the First Step, identify the needs of the researcher(s). Answering the following questions may help in the identification of the needs:

- Are you undertaking preliminary or pilot research or Post-doctoral research?
- Do you want a fellowship in residence at an institution that will offer some programmatic support or other resources to enhance your project? Do you want funds for a large research project that will last for several years and involves multiple investigators or researchers?

The Second Step is to think about the focus of the research / project. What is the topic? Identify a Topic – its novelty and why you want it? Its impact on academics, industry and society. Then identify a suitable funding agency to fit your topic. Target funding source that has an interest in your organisation and program. Whether your proposal receives funding will depend on whether your purpose and goals closely match the priorities of granting agencies. Even if you have the most appealing research proposal in the world, if you don't send it to the right institutions or the appropriate funding agency, then you're unlikely to receive funding. There are many sources of information about granting agencies and grant programs. Most universities and many schools within universities have Offices of Research, whose primary purpose is to support faculty and students in grant-seeking endeavors [8]. In this step, seek answers to the following questions:

- What are the current trends?
- What are the current limitations, gaps, challenges, and restrictions?
- What subject areas were awarded funding recently?
- What are the future plans of the funder?
- Who are the decision-makers/assessors?
- What review processes do they follow? What groups are working in India and abroad in your area of research?
- What thrust areas the funding agency has identified?

Remember one thing you are not begging money from the funder. You have the knowledge, ideas, skill and expertise. They have the money to utilize your ideas, and expertise.

In the Third Step, the investigators have to work on the background of the topic / Project title proposed. The most important aspect of research is the idea. After having the idea in mind, it is important to refine the idea by going through the literature and finding out what has already been done on the subject and what are the gaps in the research. FINER (stands for feasibility, interesting, novel, ethical, and relevant) framework should be used while framing research questions (also *see* Chap. 1). It's national status, international status, and other groups engaged in similar research both in India and abroad are to be discussed. How different is the proposal being prepared from the others already under investigation? Still why do you want it? How will you convince the funding body about the present proposal? The quality of any research project improves by having a subject expert onboard and it also makes acceptance of grants easier. The availability of the facility for the conduct of research in the department and institution should be ascertained before planning the project. Inter-departmental and inter-institutional collaborations are the key to performing good research [9].

The Fourth Step deals with planning out the objectives (preferably not more than five). Study design should include details about the type of study, methodology, sampling, blinding, inclusion and exclusion criteria, outcome measurements, and statistical analysis. It is better to give an abstract of data available in the literature, what you want to follow, and how innovative it is. Why you prefer this particular methodology—justify. If possible, give the schematic chart or a flow chart or flow diagrams illustrating the methodology.

The Fifth Step has the two most important parts of any research proposal, i.e. methodology and budget proposal. In budget planning (do not be over-ambitious, instead be more realistic). Under this, the first step is to find out what is the monetary limit for the grant proposal and what are the fund requirements for the proposed project. If these do not match, even a good project may be rejected based on budgetary limitations. The budgetary layout should be prepared with prudence and only the amount necessary for the conduct of research need to be asked. The administrative cost to conduct the research project should also be included in the proposal. The administrative cost varies depending on the type of research project. Research fund can generally be used for the following requirement but are not limited to these; it is helpful to know the subheads under which budgetary planning is done. Recurring grants involving Research and Technical Staff, Contingency, Consumables, Travel, Books/Journals, Seminars or Conferences or Workshops Service Charges, Publications charges, and Miscellaneous. The non-recurring grants cover equipment, laboratory infrastructure development (Caution: do not duplicate any equipment).

Most Important in the Fifth Step is that every item mentioned under recurring and non- recurring grants needs a proper justification. Under non-recurring identify the usage hours of equipment to be procured or infrastructure to be built.

Availability of the same for others (both within and outside the host institution). The investigators have to convince the expert committee that no such sophisticated equipment (if any) is available in a radius of about 100 or 150 kms. Do not show that the equipment proposed is just to equip the investigator's laboratory or institution. Time frame to complete the project since in some institutions or individual cases, the first two years are spent only to recruit the research staff and procure equipment, only limited time is available for research, and such things should not repeat. Once the research funding is granted, the fund allotted has to be expended as planned under budgetary planning at the earliest with transparency, integrity, fairness, and competition, which are the cornerstones of public procurement and should be remembered while spending grant money. Year-wise planning and progress expected are to be projected graphically in a standard and convincing manner. Submission of the annual report, and audited expenditure statement are to be carried out on time.

The Sixth Step deals with the final stage of the project, and once the research project is completed, the completion report has to be sent to the funding agency without any delay with all the accomplishments well documented, copies of the papers published, patents filed, and audited statement of expenditure. Most funding agencies also require periodic progress reports and the project should ideally progress as per the Gantt chart. The completion report has two parts. The first part includes a scientific report which is like writing a research paper and should include all subheads (review of literature, materials and methods, results, conclusion including implications of research). The second part is an expenditure report including how much money was spent. Was it according to budgetary layout or there was any deviation? If so, what reasons for the deviation. Any unutilized fund has to be returned to the funding agency. Ideally, the allotted fund should be post-audited by a professional (chartered accountant) and an audit report along with original bills of expenditure should be preserved for future use in case of any discrepancy. This is an essential part of any funded project that prevents the researcher from getting embroiled in any accusations of impropriety.

Sharing of scientific findings helps Human Resource Development, Collaboration and Exchange of ideas, Knowledge sharing, Patent possibilities, R and D possibilities – industries interest, impact on science and technology, society and thus help in scientific advancement is the ultimate goal of any research project. Publications of findings are part of any research grant and many funding agencies have certain restrictions on publications and presentation of the project completed out of research funds. It is imperative that during the presentation and publication, the researcher has to mention the source of funding. When the funding is obtained, then it would become the responsibility of the researcher to follow the code and conditions set by the concerned institution while carrying out the research. In fact, the research often involves a great deal of cooperation and coordination among many different people in different disciplines and institutions, ethical standards promote the values that are essential to collaborative work, such as trust, accountability, mutual respect, and fairness. For example, many ethical norms in research, such as guidelines for authorship, copyright and patenting policies data sharing policies, and confidentiality rules in peer review, are designed to protect intellectual property interests while encouraging collaboration. Therefore, it is very important that researcher must be aware of research ethics, code and responsibility when the research project proposal for funding is prepared.

53.5 Concluding Remarks

In a country like India with a fast-growing economy, research and development activities have become an integral part of higher education. Research provides knowledge, addresses the gaps in the knowledge, and also facilitates nation building and self-sufficiency in technology. However, research needs a collective approach or teamwork on an inter-disciplinary topic of national and international importance, and the researchers have to showcase their expertise in obtaining grants to carry out research on the topic of their choice, which has direct relevance to technology and the national economy. Therefore, any research project needs proper planning, and selection of an appropriate funding agency to facilitate the research. The article has been prepared from an Indian perspective and the researchers can get first-hand information on the criteria used to select the research topic, framing the objectives, planning the budget, and designing the methodology and submitting to an appropriate funding agency. Similarly, knowledge of the available funding agencies for various activities related to research is described in this chapter keeping in mind the researchers, faculty, students and entrepreneurs from an Indian perspective.

Conflict of Interest The authors declare that there is no conflict of interest.

References

1. UNESCO (2015) Manual 1984 and Frascati Manual, OECD, UNESCO
2. S.P. Alliance. https://www.sciencephilanthropyalliance.org/s2sfiles. Accessed 9 Jan 2023
3. Geuna A, Martin BR (2003) University Research Evaluation and Funding: An International Comparison, Minerva. https://doi.org/10.1023/B:MINE.0000005155.70870.bd
4. Martin BR, Irvine J, Isard PA (1990) Trends in UK government spending on academic and related research: a comparison with F R Germany, France, Japan, the Netherlands and USA. Sci Public Policy 17(1):3–13. https://doi.org/10.1093/spp/17.1.3
5. Jongbloed BL (2015) The funding of research in higher education: mixed models and mixed results. The Palgrave International Handbook of Higher Education Policy and Governance, Education, All India Council for Technical. (n.d.). https://aicte-india.org/INR500000. Accessed 12 Jan 2023
6. Frans Vught HB (2015) Governance Models and Policy Instruments. The Palgrave International Handbook of Higher Education Policy and Governance
7. Mehta BA (2022) Research Funding for STEM Higher Education Institutions: an analysis of India vs International Models
8. Hill TW. https://writingcenter.unc.edu/tips-and-tools/grant- proposals-or-give-me-the-money/. Accessed 12 Jan 2023
9. Neema S (2021) Research funding—why, when, and how? Indian Dermatol Online J 12:134–138

54 Funding Opportunities (Resources) in Biomedical Sciences: Indian Perspective

Monojit Debnath and Ganesan Venkatasubramanian

Abstract

India contributes substantially to global research and development activities in the domain of biomedical sciences. Indian scientists have demonstrated remarkable innovations in health research over the past few years and especially during the COVID-19 pandemic. There are multiple ministries and departments of the Government of India to cater to the needs of the scientists, however, the funding resources are to a greater extent limited considering the size of the population and are very competitive. Further strengthening the infrastructure and enhancing the resources for cutting-edge health research remains one of the top priorities of the Government of India. The government agencies offer different types of funding opportunities, this includes start-up research grants for young scientists, individual-centric research grants and awards to established scientists, special grants, and awards to women researchers, etc. The grants are provided for basic, as well as translational research. Besides, there are attractive funding opportunities for 'out-of-box' research ideas or 'high risk-high reward' type of research questions. However, the priorities, theme of research, timelines, application process, etc., of the funding agencies are different. The researchers often find it difficult to keep track of the funding resources, announcements, funding process, criteria, eligibility, and so on. A printed or electronic document having comprehensive details about the funding opportunities, funding mechanisms, guidance on the preparation of grant applications, etc., will be a useful resource for Indian researchers. This chapter is aimed at providing detailed information on the funding resources, offered by government agencies for biomedical research.

Keywords

Biomedical research · Government agencies · Grant · Awards · Fellowship · Research · Development

54.1 Introduction

India has emerged as an important player in the domain of basic and translational research in recent years. India makes a phenomenal contribution to international biomedical research through capacity building, i.e., training a large number of doctoral students and publishing many scientific articles. India has more than 50,000 institutions

M. Debnath (✉)
Department of Human Genetics, National Institute of Mental Health and Neuro Sciences, Bengaluru, India
e-mail: monozeet@gmail.com

G. Venkatasubramanian
Department of Psychiatry, National Institute of Mental Health and Neuro Sciences, Bengaluru, India

G. Jagadeesh et al. (eds.), *The Quintessence of Basic and Clinical Research and Scientific Publishing*,
https://doi.org/10.1007/978-981-99-1284-1_54

and universities offering higher education, and this sector remains one of the biggest in the world. The Government of India (GOI) recently started reforms by initiating National Research Foundation (NRF) to catalyse, facilitate, coordinate, seed, grow and mentor research in India [1].

India has multiple ministries and governmental funding agencies to support and oversee research and development (R&D) activities of various disciplines. India has 216.2 researchers per one million people, which is alarmingly less than other developed and developing nations. The number of scientists and clinicians working in the field of biomedical sciences is relatively fewer than in other fraternities. India's total health research budget in the past few years has ranged from 0.01% to 0.02% of the Gross Domestic Product (GDP), which is also significantly less considering the size of the Indian population and the challenges. The National Health Policy of India-2017 has emphatically highlighted the need to strengthen the country's health research [2]. The GOI has renewed interest in boosting research infrastructure and funding opportunities for R&D activities in the domain of biomedical sciences. Several new schemes, both fellowships and research grants, have been launched over the past few years to engage more scientists and clinicians in biomedical research.

Multiple agencies provide funding in the form of (1) scholarships to support young researchers, (2) research grants/fellowships/associateships to established and independent researchers, (3) grants for developing skills and knowledge, and (4) travel grants for attending international academic and scientific events. Different funding agencies have specific interests/focus, roadmaps, application processes, thrust areas and timelines. A print and/or electronic document providing comprehensive information about funding opportunities in India for biomedical scientists is needed. Herein this chapter, an attempt has been made to prepare a single consolidated document on the funding opportunities for biomedical scientists in India which will serve as a quick reference document.

54.2 Funding Opportunities: Types and Prerequisites

There are several funding opportunities to cater to the needs of researchers at various stages of their careers. There are fellowship schemes for early career researchers with a recent Ph.D. degree or Master's in Medicine/MPH or equivalent degree, without regular positions to pursue postdoctoral research and establish themselves as independent researchers. Some funding agencies provide financial assistance to young scientists with regular positions to initiate research activities as independent researchers. There are opportunities for young and established researchers to obtain extramural funding from various government agencies. Besides these, established researchers have opportunities to apply for academic research enhancement awards or fellowship programmes. Detailed information about the funding schemes, eligibility criteria, and other requirements are highlighted in the following sections.

54.2.1 Research Grants for the Established Researchers

The GOI formed the Ministry of Science and Technology in 1971 to formulate the rules and regulations, and laws relating to Science and Technology (S&T) in India. This Ministry created different departments which promote basic research in science, including biomedical science and provide financial assistance to persons engaged in such research; these include the Department of Biotechnology (DBT), Department of Science and Technology (DST), Science and Engineering Research Board (SERB), and Council for Scientific and Industrial Research (CSIR). Besides the Ministry of S&T, the GOI also has the Ministry of Health and Family Welfare (MOHFW) which was formed in 1947 to frame health policies in India. A Department of Health Research (DHR) was created in 2007 under the MOHFW. DHR aims to bring modern health technologies to people through research and innovations related to diagnosis, treatment methods and vaccines for prevention. The Indian Council of Medical Research (ICMR), the apex body for formulating, coordinating and promoting biomedical research, is also under the administrative supervision of DHR. In addition to these ministries, the GOI launched the Ministry of AYUSH in 2014, which is responsible for the education, research and propagation of traditional medicine systems in India. The Defence Research and Development Organization (DRDO) of the Ministry of Defence, GOI, provides financial assistance to promote cutting-edge biomedical research relevant to defence services through Life Science Research Board (LSRB). In addition, India Alliance (DBT/Wellcome Trust India Alliance), funded by the DBT, GOI and the Wellcome Trust, United Kingdom, serves as an independent, dynamic public charity, and funds research in health and biomedical sciences in India. These four ministries of the GOI as well as the India Alliance provide funding for biomedical research.

54.2.1.1 Funding Support by the Ministry of Science and Technology

Department of Biotechnology

Biotechnology Industry Research Assistance Council (BIRAC)

BIRAC is a non-profit Public Sector Enterprise set up by the DBT to strengthen and empower emerging Biotech enterprises to undertake strategic research and innovation, addressing nationally relevant product development needs. The primary objective of BIRAC is to bridge the gaps in industry-academia innovation research and facilitate novel, high-quality affordable product development through cutting-edge technologies. Over the past few years, BIRAC has initiated numerous schemes under different theme areas. For conducting translational research, and to encourage/support academia to develop technologies/products of societal/national importance and their subsequent validation by an industrial partner, BIRAC has initiated Promoting Academic Research Conversion to Enterprise (PACE) programme. The PACE scheme has two components, such as Academic Innovation Research (AIR) and Contract Research Scheme (CRS) [3]. AIR supports projects with well-established proof-of-concept principles leading to prototype development or a product/technology of national relevance or commercial potential. The applicants must have completed at least one extramurally funded project in the same area as the proposed project. The applicant must have authored one publication as the first or lead author or filed patents in the same research area. The duration of AIR is 24 months and the total budget cap is INR 50,00,000/−. The CRS aims to validate a process or prototype (developed by the academia) by the industrial partner/LLP. Academia (Public or private Institute, University, NGO or Research Foundation) having a well-established support system for research shall be the primary applicant. There is no time limit for the duration and no ceiling for funding under CRS projects.

Department of Science and Technology

DST provides financial support towards (1) Scientific and Engineering Research, (2) Technology Development, (3) International S&T Cooperation, and (4) Women Scientists schemes.

Under Scientific and Engineering Research, the DST supports theme-based research in biomedical science. These are "Science and Technology of Yoga and Meditation (SATYAM)" and "Cognitive Science Research Initiative (CSRI)".

Cognitive Science Research Initiative

The scheme aims to facilitate a platform for the scientific community to work for better solutions for challenges about cognitive disorders. The main objectives include (I) understanding the nature or origins of mental disorders, of physiological, social and neurochemical origins, (II) designing better learning tools and educational paradigms, (III) designing better software technologies and artificial intelligence devices, and (IV) streamlining social policy formulation and analysis. Under this scheme, DST funds individual R&D projects, multi-centric Mega projects, and Postdoctoral Fellowships. The project is tenable for a maximum period of three years. However, postdoctoral fellowships are given to young scientists below 40 years of age for a period of 2 years [4].

Science and Technology of Yoga and Meditation

This scheme aims to foster research on the impact of yoga and meditation on physical and mental health as well as cognition in healthy people and patients with disorders. Two basic themes are considered under the SATYAM scheme: (I) investigations on the effect of yoga and meditation on physical and mental health and well-being, and (II) investigations on the effect of Yoga and Meditation on the body, brain, and mind in terms of the basic processes and mechanisms [5]. Scientists/academicians having regular positions and research backgrounds in 'Yoga and Meditation' are eligible to apply. The project is tenable for a maximum period of 3 years.

Science and Engineering Research Board

SERB has several programmes as well as schemes and some of them support biomedical research.

Core Research Grant (CRG)

CRG encourages emerging and eminent scientists to an individual-centric competitive mode of funding. It is either given to an individual researcher or a group of researchers working in a recognized institution. The applicants must hold regular academic/research positions in recognized R&D institutions in India. Faculty recruited through the UGC-faculty recharge programme, INSPIRE faculty, as well as Ramanujan and Ramalingaswami Fellows are eligible to apply, provided they have at least 3.5 years of tenure remaining at the time of submission of application. Investigators should have Ph.D. degree in Science/Engineering or MD/MS/MDS/MVSc. degree. This scheme is flexible concerning the age, grant amount and duration [6].

Scientific and Useful Profound Research Advancement (SUPRA)

SUPRA is designed for high quality research proposals with new hypotheses or ones that can challenge existing knowledge and provide 'out-of-the-box solutions'. The investigators should hold regular academic/research positions in recognized academic institutions or national laboratories in India with at least 5 years of service remaining. Faculty recruited through the UGC-faculty recharge programme are also eligible to apply. The investigators should have a Ph. D. degree in Science/Engineering or MD/MS/MDS/MVSc. degree. This scheme provides support to individual researchers or a group of researchers working in recognized institutions in India. The funding is provided for a period of 3 years and extendable up to 5 years subject to performance evaluation. In the case of projects with a total budget exceeding INR 80,00,000/, 50% of the non-recurring and consumables cost should be shared by the host institution [7].

High Risk-High Reward

This scheme aims at supporting proposals that are conceptually new and risky, challenge existing hypotheses, and provide 'out-of-the-box' thinking on important problems and if successful, expected to have a paradigm-shifting influence on S&T. This may be in terms of formulating new hypotheses or scientific breakthroughs which aid in emergence of new technologies. The applicant must hold a regular academic/research position in a recognized institution. This scheme is tenable for a period of 3 years, however, in exceptional cases, the duration can be extended up to 5 years and there is no prescribed budget limit [8].

Promoting Opportunities for Women in Exploratory Research (POWER) Grant

POWER aims to encourage emerging and eminent women researchers to undertake R&D activities in frontier areas of science and engineering [9]. The investigators should be Indian citizens, holding regular academic/research positions in recognized academic institutions or national laboratories in India. Faculty recruited through the UGC-faculty recharge programme, INSPIRE faculty, Ramanujan and Ramalingaswami Fellows are eligible to apply, provided they have at least 3.5 years of tenure remaining at the time of submission of application. The amount of funding is up to INR 60,00,000/– for 3 years.

SERB-POWER Translation Grant

This scheme is envisaged to encourage women researchers to translate innovative ideas, discoveries and inventions. This is awarded to women investigators who have clearly demonstrated the translational potential of their research (patents, etc.) and have established contact with industrial partners for fast-track graduation to a prototype stage and beyond. Women PIs who hold or have applied for patents/IP from SERB supported project, a regular academic/research position in a recognized academic institution or national laboratory with at least 4 years of service and Indian citizens are eligible to apply for this grant. Male Co-PI is allowed and the PI, as well as Co-PIs, should hold a Ph.D. degree in Science/Engineering or MD/MS/MDS/MVSc. The funding is provided for 2 years with a budget of up to INR 30,00,000/– [10].

Council of Scientific and Industrial Research

CSIR provides financial assistance to promote research in the fields of S&T and medicine. Research grants are provided to Professors/Experts in regular employment in Universities, IITs, post-graduate institutions, etc. Priority is given to multi-disciplinary projects involving inter-organizational cooperation and to schemes having relevance to the research programmes of CSIR laboratories. Applications for research grants can be submitted at any time during the year and the duration is 3 years or less as proposed by the investigator [11].

54.2.1.2 Funding Support by the Ministry of Health and Family Welfare

Both DHR and ICMR provide financial assistance to promote research in the field of medicine, public health and allied areas.

Grant-In-Aid (GIA) Scheme of DHR

This scheme provides funding for research studies to identify the existing knowledge gap and to translate the existing health leads into deliverable products. The priority areas include communicable and non-communicable diseases, reproductive and child health, biomedical imaging and processing, innovation in health technologies, disease modelling, etc.

Researchers in regular employment in universities, medical colleges, postgraduate institutions, and recognized R&D organizations are eligible. Three different categories of grants are available [12].

1. Research studies with emphasis on public health: This scheme funds projects on non-communicable diseases focussing on disease burden, risk factors, diagnosis and treatment, etc. of major diseases. The maximum duration of this category is 3 years. The total budget for individual PI-driven projects is INR 50,00,000/– to INR 3,00,00,000/–, however, for multi-centric /network projects the total budget is INR 50,00,000/– to INR 10,00,00,000/–.
2. Translational Research Projects: This scheme is meant for translating already identified leads into products and processes in the area of human healthcare (from bench to bedside), through coordination among the agencies involved in basic, clinical and operational research for use in the public health system. The duration of this kind of project ranges from 1 to 4 years and the budget ranges from INR 4,00,00,000/– to INR 10,00,00,000/–.
3. Inter-sectoral coordination including funding of joint projects: This scheme aims to promote joint/collaborative research projects with other agencies involved in bio-medical/health research in the country for optimum use of resources and transfer of knowledge. The duration is 2–3 years per project and the budget ranges from INR 50,00,000/– to INR 10,00,00,000/–.

Ad-Hoc Project Scheme of ICMR

ICMR provides financial support through ad-hoc projects to promote research in the field of biomedical sciences to researchers having regular employment in medical colleges, research institutes, Universities, and recognized R&D organizations. ICMR calls for research proposals in a theme-based manner and the themes include: basic medical sciences, epidemiology and communicable diseases, biomedical informatics, reproductive biology and maternal and child health, cancer biology, neuroscience, sexually transmitted infections, non-communicable diseases, innovation and translation research, etc. [13].

54.2.1.3 Funding Support by Ministry of Ayurveda, Yoga and Naturopathy, Unani, Siddha and Homeopathy (AYUSH)

AYUSH offers a wide range of holistic treatments covering preventive, promotive, curative, rehabilitative, and rejuvenatory needs. The "AYURGYAN Scheme" of AYUSH ministry supports education, research and innovation. This scheme aims to develop opportunities for scientific scrutiny of AYUSH systems for the benefit of users, researchers, practitioners, common people at large, and so on. There are multiple priority areas of research under AYUSH with a focus on integrated health research, experimental research, epidemiological research, clinical research, translational research, etc. Medical, scientific and R&D Institutions, Universities, etc. with adequate infrastructure and technical expertise can apply for this scheme. The PI should hold a regular position and have a minimum five-year research experience in the concerned field. This scheme generally considers only one PI and not more than two co-investigators. The applications are received in four quarters, such as the first week of March, June, September and December [14].

AYUSH also offers a Centre of Excellence (ACE) programme [15]. Organizations with exceptional contributions at the national and international levels are considered for the ACE programme as recognition for their substantial contribution in the field of AYUSH services like research and drug development. One of the main objectives of this scheme is to create a network of organizations that will accelerate research in AYUSH systems. This scheme is generally tenable for 3 years.

54.2.1.4 Funding Support by the Ministry of Defence

The Life Science Research Board (LSRB) of DRDO was initiated to support and strengthen the research base in the area of life sciences to meet the national needs of defence services. The grant-in-aid scheme of LSRB supports various areas including Life support and biomedical devices (LSBD), Soldier Health and Drug Development (SH&DD) and Physiology of Extreme Environment and Behavioral Sciences (PEE&BS) [16]. This grant is offered to IITs, Universities, Colleges, eminent scientists working in reputed R&D organizations, etc.

54.2.1.5 Funding support by India Alliance

Team Science Grant

This scheme is meant for a team of researchers who bring together complementary skills, knowledge, and resources to address an important health challenge for India. It funds high-risk, high-reward research work [17]. The project should be multi-institutional and interdisciplinary in its approach. A minimum of 3 investigators should participate, including one as PI who will manage and lead the team. The applying team should bring different expertise or disciplines to address the research problem. No more than 2 investigators can be from one institution, and a minimum of two institutions should be included. The PI and co-PIs should have PhD/MD/MBBS-MS/MPH or equivalent degrees, while the PI should have at least 5 years of experience and Co-PIs should have at least 3 years of experience in running an independent research group. All the investigators should have obtained competitive grants/fellowships and tenure at their respective host institutions for the entire grant duration. There are no age or nationality restrictions for the applicants. This project is tenable for 5 years and the total budget is up to INR 10,00,00,000/−.

Clinical and Public Health Research Centres (CRCs)

CRC aims to improve clinical and public health research ecosystems in India to bridge gaps in human resources, supervision, mentorship, equipment and administration. This scheme is envisioned as research-oriented centres established with a focus on major biomedical research problems and preferably involving multiple institutions [18]. The CRCs are meant to promote clinical/public health research and develop physician-scientists. Eligible institutions should provide evidence of access to an in-house or external Clinical Study Design Unit (CSDU) and relevant training programmes. The proposal should have components of human resource training and mentorship. Training of MD and Ph.D. students should be a critical part of the proposed project. The total duration of the project is 5 years.

54.2.2 Academic Research Enhancement Award/Fellowships for the Established Researchers

54.2.2.1 Funding Support by the Ministry of Science and Technology

Department of Biotechnology

Har Gobind Khorana-Innovative Young Biotechnologist Award

This is a career-oriented award to identify and nurture outstanding young scientists with innovative ideas who are desirous of pursuing research in the frontier areas of Biotechnology. This award is given to candidates whose age is below 35 years, but an age relaxation of 5 years is given to SC/ST/OBC, women and differently abled candidates [19] The applicant must have a Ph.D. in any branch of Life Sciences or Master's degree in Medicine or Dentistry (MD/MS/MDS). The applicant should be an Indian citizen with excellent academic credentials and track record, high-impact peer-reviewed publications or should have developed technologies, Indian and/or international patents. The proposed research proposal for this award should be of very high scientific quality with innovativeness. This award carries a fellowship of INR 1,00,000/– per annum for candidates who have a permanent faculty position and Rs. INR 75,000/– per month for those candidates who are not holding permanent positions. Besides this, an amount of INR 10,00,000/– is given for purchasing equipment and INR 10,00,000/– per year for consumables.

Tata Innovation Fellowship

This is a highly competitive scheme of DBT to recognize and reward scientists with an outstanding track record in Biological Sciences/Biotechnology for finding innovative solutions to major problems in health care and other allied areas of life sciences [20]. This fellowship is awarded to Indian citizens, below the age of 55 years. The applicant should possess a Ph.D. degree in Life Sciences or a Master's/equivalent degree in medical sciences, engineering or biotechnology. The applicant should have outstanding contributions and publications in the specified area and a deep commitment to finding innovative solutions to major problems in health care and other areas of Life Sciences. The applicant must have a regular/permanent position in a University/Institute/Organization and should be engaged in R&D activities and should have spent at least five years in India before applying for the fellowship. This scheme offers an amount of INR 25,000/– as a fellowship per month, in addition to a regular salary from the host institute and a contingency grant of INR 6,00,000/– per annum. This fellowship is for three years and extendable by another 2 years based on a fresh appraisal.

Janaki Ammal National Women Bioscientist Award

This award (young category) is given to Indian women scientists below 45 years of age who have made significant contributions to different branches of science, including biomedical sciences. Nominations for the applicants forwarded through the Head/Executive authority of the Institutes/Scientific Organizations are considered for selection. The work (of the last 5 years) for which nomination is made must have been carried out in Indian institutes and acknowledged in the publications. This carries a cash prize of INR 1,00,000/– and a Research Grant of INR 25,00,000/–per annum for a period of 5 years [21].

Department of Science and Technology

Swarnajayanti Award

This scheme is offered by DST to a selected number of young scientists who are under 40 years of age with a proven track record and to enable them to pursue basic research in frontier areas of S&T. The project should have an innovative research idea with the potential for making an impact on R&D in the discipline. This scheme has a fellowship of INR 25,000/– per month in addition to the salary drawn from the host institute and the applicants are given additional monetary support for research work. This fellowship is given for a period of five years. The applicant should possess Ph.D. in Science/Engineering or a Master's degree in Medicine [22].

Science and Engineering Research Board

Early Career Research Award (ECRA)

This award aims to provide quick research support to young researchers who are in their early careers pursuing exciting and innovative research in frontier areas of science. The applicant should hold a Ph.D. degree in Science/Engineering or an M.D./M.S. degree in any area of Medicine and must hold a regular academic/research position in a recognized academic institution or national laboratory. The upper age limit is fixed at 37 years, however, for SC/ST/OBC/Physically challenged and women candidates, the upper age limit is 40 years. This is a one-time award and carries a research grant of up to INR 50,00,000/– for a period of 3 years [23].

SERB-POWER Fellowship

This fellowship supports outstanding women researchers and innovators working in Indian academic institutions and R&D laboratories, holding Ph.D. degrees in any branch of science and engineering. The nominee should be an Indian citizen between 35–55 years of age with an excellent record of research performance as an independent investigator. A fellowship of INR 15,000/– per month in addition to regular income and a research grant of INR 10,00,000/– per annum are given for a period of three years [9].

Council of Scientific and Industrial Research

Young Scientist Award

CSIR launched this scheme in 1987 to promote excellence in various fields of S&T. Each award consists of a citation, a cash prize of INR 50,000/– and a plaque. Awardees are entitled to a special honorarium of INR 7500/– per month till the age of 45 years and also a research grant of INR 25,00,000/– over a period of five years. Any scientist, engaged in research in any of the CSIR institutes, who is not more than 35 years of age, as reckoned on 26th September of the preceding year, is eligible for the award. The nominee should be a regular scientific staff of the CSIR system and should have joined the CSIR laboratory on or before the 26th of September of the previous year [24].

India Alliance

Intermediate Fellowship

This fellowship is offered to researchers holding permanent positions but who are either newly established or in the process of establishing themselves as independent investigators. There are two categories of this fellowship: Basic Biomedical Research Fellowship and Clinical and Public Health Research Fellowship [25]. Researchers with 4 to 15 years of post-Ph.D. research experience are eligible to apply under the biomedical research category for five years with a total budget cap of INR 3.6 crores and a salary up to INR 13.8 lakhs per annum. Clinicians and allied health professionals with 0–15 years of research experience post-qualifying degree are eligible to apply for the Clinical and Public Health Research Fellowship for five years with a total budget cap of INR 3.8 crores and salary of up to INR 16.8 lakhs per annum.

Senior Fellowship
This fellowship is offered to researchers who have established themselves as independent investigators and now would like to lead a larger programme of research or expand their research programme in India. There are two categories of this fellowship: Basic Biomedical Research Fellowship and Clinical and Public Health Research Fellowship [25]. Researchers having 4 to 15 years of post-Ph.D. research experience are eligible to apply under the biomedical research category for 5 years with a total budget cap of INR 4.5 crores and a salary of up to INR 16.1 lakhs per annum. Clinicians and allied health professionals with 0 to 15 years of research experience post-qualifying degree are eligible to apply for the Clinical and Public Health Research Fellowship for five years with a total budget cap of INR 4.7 crores and salary of up to INR 19.5 lakhs per annum.

54.2.3 Fellowships for the Postdoctoral Students/Early Career Researchers

54.2.3.1 Funding Support by the Ministry of Science and Technology

Department of Biotechnology

Research Associateship (RA)
DBT-RA is aimed to train and nurture young researchers and generate a critical mass of trained manpower in modern areas of biology and biotechnology. This fellowship is given for a period of 2 years (up to 4 years in exceptional cases) to researchers who have a Ph.D. degree in Science/Engineering or MD/MS degree in Medicine and a good academic record. The upper age limit is 40 years for male candidates and 45 years in the case of women candidates [26].

Biotechnology Career Advancement and Re-orientation Programme (BioCARe)
This is to promote Women Scientists in science and in an attempt to enhance their participation in Biotechnology research. This programme is mainly for the career development of unemployed female scientists, including those who are not in regular positions or had a career break up to the age of 45 years for whom it is the first extramural research funding [27].

Ramalingaswami Re-entry Fellowship
This aims to attract high-quality Indian brains working abroad to pursue their research interests in Life Sciences, Biotechnology and other related areas in India by providing them an attractive avenue to pursue their R&D interests in Indian Institutions [28]. This is a senior fellowship programme, and awardees are to be considered equivalent to Assistant Professor/Scientist-D. The applicant should possess a Ph.D./MD/ equivalent degree with an outstanding track record as reflected in publications or other recognitions and with at least 3 years of post-doctoral research experience, of which the last two years should be from an overseas laboratory. Only Indian nationals working overseas are eligible to apply. Researchers up to 45 years of age on the closing date of application are eligible to apply. Fellows are entitled to take up teaching/research assignments and supervise doctoral/MS students. The fellows are given a consolidated monthly remuneration of INR 1,00,000/– per month and INR 18,500/– per month as HRA. In addition, an amount of INR 10,00,000/– for the first and second year; INR 7,50,000/– for third and fourth year, and INR 5,00,000/– for the fifth year are given as a research/contingency grant.

Department of Science and Technology

Women Scientists Scheme
This scheme is to encourage women in the S&T domain, preferably those who have had a career break and not having regular employment, to explore the possibility of re-entry into the profession. There are two categories under this scheme: (1) Women Scientist Scheme-A (WOS-A), given to those who have Ph.D. in Basic or Applied Sciences/MD or equivalent degree, and (2) Scientist Scheme-B (WOS-B) for applicants who have either M.Phil./M.Tech./M.Pharm./MVSc.

or M.Sc. in Basic or Applied Sciences/MBBS/ B.Tech. or an equivalent degree. A grant amount of INR 30,00,000/–, including a stipend of INR 55,000/– is sanctioned for WOS-A. However, for the WOS-B scheme, INR 25,00,000/–, including a monthly salary of INR 40,000/– is given to the candidates for the first category, while the candidates selected for the latter category are given INR 20,00,000/–, including a monthly stipend of INR 31,000/– [29].

INSPIRE Faculty Fellowship

This scheme aims to provide attractive opportunities to young achievers/post-doctoral fellows to carry out independent research and develop independent scientific profiles. The applicants should have a Ph.D. degree in Science/Engineering/Medicine and should possess a minimum of 60% (or equivalent CGPA) marks throughout their academic life (Class 12 onwards). The upper age limit is 32 years, however, for SC/ST and women candidates it is 37 years and for persons with benchmark disabilities, the upper age limit is 42 years. The applicants should have publications in highly reputed journals, demonstrating outstanding research potential. The INSPIRE Faculty fellow is entitled to a consolidated monthly salary of INR 1,25,000/– and a research grant of INR 7,00,000/– per annum for 5 years. This scheme is for a maximum period of 5 years [30].

Science and Engineering Research Board

National Postdoctoral Fellowship (N-PDF)

N-PDF is offered to Indian citizens who are either holders of Ph.D./MD/M.S. or are about to submit a Ph.D./MD/M.S. thesis. The maximum age limit is 35 years, while it is 40 years for SC/ST/OBC/ Physically handicapped and women candidates. The N-PDF is tenable only for two years and an amount of INR 55,000/– per month is given as a fellowship. In addition, the candidate will receives INR 2,00,000/– per annum as a research grant [31].

Ramanujan Fellowship

This is meant for brilliant Indian scientists, below 40 years of age and who want to return to India from abroad and take up scientific research positions in India [32]. The applicant should have a Ph.D. degree in Science/Engineering or MD in Medicine and have adequate professional experience. The applicant should be working abroad at the time of nomination. This fellowship is tenable for 5 years and has a consolidated monthly salary of INR 1,35,000/– and a research grant of INR 7,00,000/– per annum.

Council of Scientific and Industrial Research

CSIR-Nehru Science Postdoctoral Fellowship

This fellowship is provided to pursue postdoctoral research in Science, Engineering, Agriculture and Medicine to students who have obtained Ph.D. within the last 3 years or are about to submit a Ph.D. thesis. The maximum age limit is 32 years, while it is 37 years for SC/ST/OBC/ Physically handicapped and women candidates. This fellowship is tenable for two years and extendable for another year. The awardees are given INR 65,000/– per month as a fellowship and a contingency grant of INR 3,00,000/– per annum. This fellowship is given to Indian nationals, Overseas Citizens of India (OCI) and persons of Indian Origin (PIO) (80%) and the remaining 20% to foreign nationals [33].

54.2.3.2 Funding Support by the Ministry of Health and Family Welfare

Department of Health Research

Young Scientist Scheme

This is offered for a period of three years and there are two categories. Category A is given to applicants who are less than 35 years and hold MD/MS/MDS or Ph.D. in biomedical sciences. Awardees with medical degrees will receive a stipend of INR 70,000/– per month and non-medical awardees will receive a stipend of INR 60,000/– per month and INR 10,00,000/– as a contingency grant. For category B, the applicants should be 30 years of age and hold

MBBS/BDS/MVSc. or M.Sc. in biomedical sciences and have at least 5 years of research experience in the bio-medical field. Awardees with medical degrees will receive a monthly stipend of INR 55,000/– and non-medical applicants will receive INR 50,000/– and a contingency grant of INR 10,00,000/– per year [34].

Indian Council of Medical Research

Research Associates

This is awarded to young scientists to carry out research in the field of Biomedical Sciences at Medical Colleges, Research Institutes, Universities and recognized R&D organizations. Applicants should have a Ph.D./MD/MS/MDS/ equivalent degree or 3 years of research, teaching, design or development experience after MVSc./ M.Tech./M.Pharm./ME with at least one research paper in SCI journal. Based on qualification and experience, one of the 3 pay levels is given: RA-I (INR 47,000/–), RA-II (INR 49,000/–) and RA-III (INR 54,000/–). The upper age limit is 40 years for male and 42 years for female candidates [35].

54.2.3.3 Funding Support by India Alliance

Early Career Fellowship

This fellowship is for postdoctoral scientists who wish to undertake high-quality research and establish themselves as independent researchers in India. There are two categories of early career fellowships; one is called Basic Biomedical Research Fellowships, while the other is known as Clinical and Public Health Research Fellowships. This fellowship is given for five years. The amount of fellowship is up to INR 10.3 lakhs per annum and the total budget cap is INR 1.7 crores under the Basic Biomedical Research fellowship and is given to the candidates who are either in the final year of Ph.D. studies or have no more than 4 years of postdoctoral research experience. The candidates under the Clinical and Public Health Research category are paid an amount of up to INR 13.4 lakhs as salary and the total budget cap is INR 1.9 crore. Candidates with MD/MS/MPH or equivalent clinical/public health degree with 0–15 years of research experience post-qualifying degree are eligible to apply for the Clinical and Public Health Research Fellowship. These fellowships do not have age or nationality restrictions [36].

Clinical Research Training Programme

This fellowship is offered to develop physician scientists as a way forward for improving clinical and public health research ecosystems in India. This program funds mentored research training fellowships for medical graduates (MBBS) and Post-graduates (MD/MS/equivalent) to provide training opportunities in diverse areas of research relevant to India's health priorities. It encourages the trainee to receive M.Phil. and/or Ph.D. degree or other certification at the exit point. The value of the fellowship is INR 50,00,000/– for a 3- to 4-year period [37].

54.2.3.4 Funding Support by University Grant Commission (UGC)

Dr. D.S. Kothari Postdoctoral Fellowship

This fellowship is provided by the University Grant Commission (UGC) for pursuing postdoctoral studies in sciences [38]. This fellowship is awarded to unemployed candidates with Ph.D. The upper age limit for applicants is 35 years and the relaxation in age for SC/ST/OBC/ Women candidates is as per GOI norms. This fellowship is tenable for three years and INR 43,800/– for first year, INR 45,000/– for second year and INR 46,000/– for the third year are given as stipend, along with a contingency grant of INR 1,00,000 per annum.

54.2.4 Grants for Skill Development/Knowledge Enhancement

In addition to providing financial assistance such as research grants, fellowships, awards, etc., the Indian funding agencies also fund Universities, Academic Institutions, Colleges, etc. to organize scientific/academic events such as workshops, symposia, seminars, conferences, etc., for developing skill and knowledge.

Financial assistance is also provided by the funding agencies like DBT, SERB, CSIR and ICMR to young (Ph.D. students, Research Associates, Resident doctors, etc.) as well as established scientists as 'travel grants" for presenting their research papers in international scientific events such as conferences, workshops, short-term schools, courses, training programmes. This scheme provides up to full reimbursement of the actual airfare from the airport (nearest to the place of work) to the venue of the event and back.

DBT started the Unique Methods of Management of Inherited Disorders (UMMID) initiative [39] for providing training to clinicians working in Government hospitals on Biochemical Genetics, Cytogenetics, Molecular Genetics, Clinical Genetics and Comprehensive critical care. SERB initiated the 'VRITIKA' scheme [40] to provide opportunities to promising PG students from universities/Colleges to get exposure and hands-on research skills from institutes of national importance such as IITs, IISc., NITs, CSIR, ICMR, etc.

54.3 Roadmap and the Priority Areas of the Indian Funding Agencies

Until recently, there were different task forces under various funding agencies to evaluate, support and monitor the research projects of various niche areas, which included, for example, Human Genomics and Genome Analysis, Cancer Biology, Infectious Disease Biology, Neuroscience, and so on. However, most of the funding agencies have removed such thrust areas of research and task forces. Funding agencies like DBT and ICMR call for applications for theme-based topics, based on the relevance and priority of the GOI's 'Swasth Bharath' mission. The themes of grant application calls vary across these funding agencies and new themes are proposed every year.

The roadmap of DBT in recent years is to emphasize the promotion of excellence and innovation for facilitating basic, early and late-stage translation research, establishing national and international partnerships, capacity building and public-private partnerships. Keeping this mission in mind, the DBT allotted huge funds for research programmes like the Accelerator Program for Discovery in Brain Disorders using Stem cells (ADBS), Genome India: Cataloguing the Genetic Variation in Indians, Human Microbiome Initiative of Select endogamous populations of India, etc. The mission of the ICMR is to increase focus on research on the health problems of the vulnerable, disadvantaged, and marginalized sections of society, to harness and encourage the use of modern biology tools in addressing health concerns of the country and encourage innovations and translation related to diagnostics, treatment, methods/vaccines for prevention. The DHR also encourages clinical and translational research that benefits the community.

Unlike DBT, DHR and ICMR, funding agencies like India Alliance and SERB, especially the CRG scheme do not have very stringent requirements. These two funding agencies do not ask for theme-based research proposal submission and are flexible as well as open to various kinds of research ideas from basic to translational and para-clinical to clinical.

54.4 Application Procedure and Familiarity with All Competitive Grants

54.4.1 Mode of Application and Format

The application procedure and the format for preparing the grant proposals vary to some extent across the funding agencies. All the funding agencies have moved to the online mode of project submission. The first step of the project submission across funding agencies is the creation of a login ID and completion of the registration process by providing information related to the current position and affiliation as well as academic, training and employment history. The application for projects/fellowships should be submitted through online portals of funding agencies, such as India Alliance (https://grants.indiaalliance.org/Login.aspx), DST (https://onlinedst.gov.in/Login.aspx), SERB (https://www.serbonline.in/SERB/HomePage), ICMR (https://epms.icmr.org.in/), DHR https://dhrschemes.icmr.org.in/userLogin), and DBT (https://dbtepromis.nic.in/Login.aspx). The formats of grant applications for these funding agencies are available on their respective websites.

Most of the funding agencies such as SERB, DST, DBT and LSRB providing financial assistance for extra-mural projects ask for a full application. However, all the theme-based research grants, supported by DBT, DHR and ICMR mandate the submission of a Letter of intent (LOI) or concept note as a first step. Based on the merit of the research questions put forwarded in the LOI or concept note, full proposals are invited. India Alliance-DBT Wellcome Trust always follows a two-step application process, i.e., a preliminary followed by a full application.

54.4.2 Programme Announcements

Except for the India Alliance and CRG scheme of SERB, most of the funding agencies do not have a fixed timeline for announcing project submission. The India Alliance launches its application request in July for Early Career Fellowship (ECF) while for Senior and Intermediate Fellowship (SIF) in June (first round) and February (second round) for Basic Biomedical Research. For Clinical and Public Health Research, the applications for ECF and SIF are launched in January of each calendar year. The SERB has streamlined the application process for its CRG scheme over the past few years and the call for application for the CRG programme is generally made in March/April of each calendar year. The announcements for the theme-based calls for applications are made at various time points by the funding agencies.

The announcements for the grant applications are generally published on the website of the respective funding agencies. Currently, there does not exist a single portal for searching programme announcements and requests for applications. Besides, there is a lack of email alert service on announcements of grant applications by the funding agencies. The researchers need to browse the websites of the funding agencies for announcements around the year.

54.4.3 Preparation of a Grant Application

A successful grant application and funding depend on multiple parameters, which include but are not restricted to (1) the area of research (basic, translational, clinical, pre-clinical, etc.), (2) the type of research (exploratory studies, small research grant, basic research grant for definitive studies, academic research enhancement award, clinical trial planning, etc.), (3) research ideas (fundamental hypothesis-driven versus out-of-box), (4) choice of the funding agency, and (5) potential and credentials of the applicants.

54.5 Thrust Areas and Types of Research

Unlike other nations, for example, the USA, where the type of research grants has been categorised as R01 for a basic grant for definite studies for 5 years, R03 for a small research grant, R15 for academic research enhancement award, R21 for exploratory/developmental grant (see Chap. 51), etc., there is lack of such categorization owing to the existence of multiple funding agencies in India. Funding agencies like India-Alliance, DBT and SERB accept research proposals in the areas of basic, translational, pre-clinical and clinical studies. However, studies having a major focus on clinical and translational research are apt for DHR and ICMR. In recent times, DBT, DHR and ICMR have identified research areas pertinent to vaccine development as a priority programme. The Coalition for Epidemic Preparedness Innovations (CEPI) is a global alliance between public, private, philanthropic, and civil society organizations to accelerate the development of vaccines against emerging infections and enable equitable access to these vaccines for affected populations during outbreaks. The DBT, in partnership with CEPI, has initiated the Ind-CEPI mission and has been issuing calls for proposals for projects to develop specific vaccine candidates or research that can directly support vaccine development [41].

Young scientists can apply for small research grants either for exploratory or innovative studies to Har Gobind Khorana-Innovative Young Biotechnologist Award of DBT, ECF scheme of India-Alliance, or the ECRA scheme of SERB. Basic research grants for definitive studies ranging from 3–5 years can be submitted mainly to India-Alliance, SERB, DBT and DHR. Mid-level and senior scientists have multiple opportunities to apply for fellowships or awards that are considered academic research enhancement awards to funding agencies like India-Alliance (Intermediate and Senior Fellowships), Swarnajayanti Fellowships of DST, Tata Innovation Fellowships of DBT, SERB-POWER Fellowships, etc. ICMR and DHR also participate in implementing and monitoring clinical trials.

54.6 Research Ideas, Selection of Funding Agencies and the Applicant's Potential

Research ideas or the proposed research questions constitute one of the major determinants of successful grant applications. The research questions should be unique and novel and should be based on strong hypotheses (*see* Chap. 1). Research proposals for replication studies do not get favourable responses from the Indian funding agencies. Research proposals having 'out-of-the-box' ideas are most appropriate for the SUPRA scheme of SERB. Overall, a research project that has unexplored research questions, a strong hypothesis, sound methodologies, and highly anticipated deliverables are generally supported by Indian funding agencies.

Most of the funding agencies have restructured their funding mechanisms in recent times. A few funding agencies like India-Alliance, CSIR and SERB offer opportunities to individual researchers to frame their research areas, hypothesis and questions on a broad range of topics. However, funding agencies like LSRB, DST, DBT, DHR and ICMR decide their priority areas of research and ask the researchers to submit proposals on their prescribed themes. For example, the DST has been funding two theme-based programmes, such as CSRI and SATYAM on yearly basis for the past several years. However, DBT, DHR and ICMR do not have such fixed theme-based research schemes, rather these funding agencies call for proposals on a varied range of topics based on relevance and priority. Therefore, researchers who wish to test their specific hypothesis and research questions can target India-Alliance and SERB, while, those who fulfil the requirements of theme-based research of various funding agencies can prepare their research questions to fit into the proposed domain of the theme-based research of funding agencies like DBT, DHR, and ICMR. Researchers interested in working on a specific research question, interdisciplinary in nature and aiming to address important health challenges across communities of the different geographical regions of India may look for multi-centric or networking projects. Multi-centric or networking projects, can be submitted to India-Alliance under "Team Science Grants" and this is open to a broad range of ideas. Alternatively, DBT and ICMR also call for multi-centric or networking projects but such calls are generally theme based. Research proposals that aim to develop a proof-of-concept for a process/product can be submitted to the BIRAC-PACE scheme of DBT.

The skills, expertise, and track record of the investigators are also taken into consideration while granting the research projects. The India-Alliance pays special attention to the academic/research track record as well as the publication potential of applicants even for the early career fellowship. Investigators with a good track record in obtaining research funds/fellowships on innovative research proposals, consistently working on a given topic and with publications in high-impact journals have an edge over other applicants in the successful grant application. Investigators having preliminary data on the topics of their proposed research or the proposed research questions, supported by data from a pilot study also adds value to the submitted application.

54.7 Resource Materials for Preparing a Grant Proposal

A funding agency needs to make publicly available the number and kind of research projects it has sanctioned and also make a repository having full details of the sanctioned projects. This offers multiple benefits to the researchers, such as, providing a template to novice researchers, serving as a basis to refine and improve the research hypothesis and questions, avoiding duplication of the research works, identifying a potential area of research for further testing/validation, etc. Unlike the NIH (*see* Chap. 51), there is a lack of an extensive database in India to track down funded grants from various funding agencies. Some of the funding agencies have stored a list of funded projects; however, this is not adequate. For example, SERB has created a separate portal, i.e., SERB-Project Information System & Management (SERB-PRISM) to provide a whole range of information like funding details, status, research summary, and project output details such as publications and patents of the projects sanctioned by SERB from 2011 onwards (https://prism.serbonline.in/).

All the funding agencies receive a large number of research proposals and only less than 10% of projects are generally being considered for funding. A project rejected by a funding agency should not be submitted without any modification to another funding agency as this may not get a favourable review from the second one. The proposed hypothesis and the research questions of the rejected project should be analysed critically and the comments of the grant reviewer should be addressed adequately before submitting the same project to another funding agency. The projects rejected by one funding agency are seldom funded by other funding agencies.

54.8 Concluding Remarks

The funding agencies of the GOI offer multiple opportunities to Indian researchers, both young and established, to conduct basic as well as translational research by providing grant-in-aid and fellowships/awards, to organize academic events/workshops, etc., to develop international collaborations through bilateral-exchange programmes, to attend and/or to present their research findings in international scientific events by providing travel grants. The diverse funding mechanisms cater to the needs of a large number of researchers in India. However, a novice researcher often finds it difficult to identify an appropriate funding agency to submit his/her first proposal due to lack of (1) a centralized repository of funded projects, (2) structured timelines, (3) notifications/alerts through NIC email services, (4) funding agency-specific standard template, as well as frequent changes in announcements, especially for theme-based calls. An e-platform of GOI having all the important details on the funding announcements, timelines, mechanisms, templates, etc., will serve as an important resource for all researchers.

Conflict of Interest None.

References

1. https://www.psa.gov.in/nrf. Accessed 13 Dec 2022
2. https://www.nhp.gov.in/nhpfiles/national_health_policy_2017.pdf. Accessed 15 Dec 2022
3. https://birac.nic.in/desc_new.php?id=286. Accessed 8 Dec 2022
4. https://dst.gov.in/cognitive-science-research-initiative-csri. Accessed 6 Dec 2022
5. https://dst.gov.in/science-and-technology-yoga-and-meditation. Accessed 6 Dec 2022
6. https://www.serbonline.in/SERB/emr?HomePage=New. Accessed 6 Dec 2022

7. https://www.serbonline.in/SERB/Supra?HomePage=New. Accessed 10 Dec 2022
8. https://serbonline.in/SERB/HRR. Accessed 12 Dec 2022
9. https://www.serbonline.in/SERB/serbPowerInstructions. Accessed 8 Dec 2022
10. https://www.serbonline.in/SERB/Spt. Accessed 8 Dec 2022
11. https://csirhrdg.res.in/Home/Index/1/Default/2762/62. Accessed 11 Dec 2022
12. https://dhrschemes.icmr.org.in/gia. Accessed 6 Dec 2022
13. https://main.icmr.nic.in/content/information. Accessed 6 Dec 2022
14. https://main.ayush.gov.in/schemes-2/central-sector-scheme/ayurgyan-scheme/. Accessed 12 Dec 2022
15. https://main.ayush.gov.in/schemes-2/ayush-centre-of-excellence-ace-programme/. Accessed 12 Dec 2022
16. https://www.drdo.gov.in/life-sciences-research-board/eligibility. Accessed 13 Dec 2022
17. https://www.indiaalliance.org/team-science-grants. Accessed 8 Dec 2022
18. https://www.indiaalliance.org/clinical-public-health-research-centres. Accessed 8 Dec 2022
19. https://dbtindia.gov.in/schemes-programmes/building-capacities/awards/innovative-young-bio-technologist-award-iyba. Accessed 6 Dec 2022
20. https://dbtindia.gov.in/schemes-programmes/building-capacities/awards/tata-innovation-fellowships. Accessed 12 Dec 2022
21. https://dbtindia.gov.in/scientific-decision-units/hrd-awards-fellowships/hrd-2-biocare. Accessed 6 Dec 2022
22. https://dst.gov.in/scientific-programmes/scientific-engineering-research/human-resource-development-and-nurturing-young-talent-swarnajayanti-fellowships-scheme. Accessed 9 Dec 2022
23. https://serbonline.in/SERB/ecr. Accessed 6 Dec 2022
24. https://csirhrdg.res.in/Home/Index/1/Default/2687/67. Accessed 14 Dec 2022
25. https://indiaalliance.org/fellowshiptype/basic-biomedical-research-fellowships. Accessed 8 Dec 2022
26. https://ra.dbtindia.gov.in/#. Accessed 3 Dec 2022
27. https://dbtindia.gov.in/scientific-decision-units/hrd-awards-fellowships/hrd-2-biocare. Accessed 2 Dec 2022
28. https://dbtindia.gov.in/schemes-programmes/building-capacities/building-critical-mass-science-leaders/ramalingaswami-re. Accessed 6 Dec 2022
29. https://dst.gov.in/scientific-programmes/scientific-engineering-research/women-scientists-programs. Accessed 8 Dec 2022
30. https://online-inspire.gov.in/Account/INSPIREProgramme#. Accessed 14 Dec 2022
31. https://serbonline.in/SERB/npdf. Accessed 6 Dec 2022
32. https://www.serbonline.in/SERB/Ramanujan_fellowship?HomePage=New. Accessed 10 Dec 2022
33. https://csirhrdg.res.in/Home/Index/1/Default/2688/55. Accessed 13 Dec 2022
34. https://dhrschemes.icmr.org.in/staticweb/pdf/DHR/Downloads/HRD/HRD-FINALGuideline.pdf. Accessed 14 Dec 2022
35. https://epms.icmr.org.in/fellowshippage. Accessed 9 Dec 2022
36. https://indiaalliance.org/fellowshiptype/basic-biomedical-research-fellowships. Accessed 7 Dec 2022
37. https://www.indiaalliance.org/clinical-public-health-research-training-programme. Accessed 7 Dec 2022
38. https://www.ugc.ac.in/pdfnews/1181594_revised-DSKPDF.pdf. Accessed 5 Dec 2022
39. https://dbtindia.gov.in/scientific-decision-units/computational-biology/ummid-initiative. Accessed 13 Dec 2022
40. https://acceleratevigyan.gov.in/programs/abhyas/vritika. Accessed 13 Dec 2022
41. https://cepi.net/get_involved/cfps/. Accessed 10 Dec 2022

Part XI

Career Development

55 Mentorship in Biomedical Sciences

Gemma Cox, Lauren E. Smith, and Savithiri Ratnapalan

Abstract

This chapter aims to provide an overview of mentorship within the context of biomedical research. After a brief introduction about mentoring and the rationale for mentorship in biomedical sciences, the chapter describes different types and models of mentoring relationships. The process of mentorship in biomedical sciences is detailed with suggestions for best practice tips for mentees and mentors. The challenges to successful mentorship and suggestions for managing conflict are highlighted along with a description of when and how to end unproductive mentoring relationships. The chapter concludes with a description of mentoring networks and suggestions for augmenting mentorship in biomedical research in the future.

Keywords

Mentorship · Peer mentorship · Biomedical research · Career development · Mentoring network · Virtual mentoring

55.1 Introduction

Beginning a graduate program in biomedical sciences can be isolating and confusing at the outset. The arduous process of conducting a research study coupled with the complexity of methods, highly competitive environments, and pressure to produce high impact results can leave both graduate students and junior faculty alike feeling overwhelmed. Navigating new systems and networks blindly can be a discouraging process [1]. Effective mentorship plays a critical role in establishing career success within such a challenging environment. It is an imperative mechanism by which career satisfaction, retention, research productivity, and professional success are achieved and is thus deemed a core duty of higher academic institutions [2]. This chapter discusses the value of mentorship within biomedical research and provides important insights for both mentees and prospective mentors to be able to cultivate positive mentoring relationships.

G. Cox
Hospital for Sick Children, University of Toronto, Toronto, ON, Canada
e-mail: gemma.cox@sickkids.ca

L. E. Smith
Health Services Research, Institute of Health Policy, Management and Evaluation, University of Toronto, Toronto, ON, Canada
e-mail: laurenellen.smith@mail.utoronto.ca

S. Ratnapalan (✉)
Department of Paediatrics and Dalla Lana School of Public Health, University of Toronto, Toronto, ON, Canada

Divisions of Emergency Medicine, Clinical Pharmacology and Toxicology, Hospital for Sick Children, Toronto, ON, Canada
e-mail: savithiri.ratnapalan@sickkids.ca

G. Jagadeesh et al. (eds.), *The Quintessence of Basic and Clinical Research and Scientific Publishing*,
https://doi.org/10.1007/978-981-99-1284-1_55

55.2 Understanding Mentorship

The landscape of biomedical graduate education and research in the twenty-first century is rapidly evolving; the quantity of knowledge has exponentially grown, the resources and methods available for research continue to fuel change, and the boundaries between disciplines have blurred. At the same time, funding allocation has receded and there is a shift towards large scale multidisciplinary research work [3]. In this climate, mentorship is increasingly recognized as a major strategic priority for academic institutions to produce quality research outputs and maintain competitiveness [4].

The meaning of mentorship, and how it differs from other relationships between faculty and students such as teaching and academic training have to be explored. Broadly defined, professional mentorship can be understood as a collaborative and educational relationship that aims to facilitate the acquisition of essential competencies and skills over time. Mentors generally serve two broad functions for their mentees; one related to career development, by establishing the mentor as a coach to provide advice and enhance professional performance and development. The second is a psychosocial function, where the mentor acts as a role model and support system for the mentee [5]. A mentor's role is to teach, sponsor, encourage, counsel, challenge, support and befriend an unskilled or inexperienced person to achieve their academic, personal, and professional goals [3]. See Fig. 55.1: Formal and Informal Roles of Mentorship. Mentoring relationships may be formally assigned by the department or program, sought out by the prospective mentee, or informally evolve due to exposure, circumstance and mutual compatibility between a novice and a senior faculty member [5].

There are multiple opportunities for faculty to incorporate aspects of mentorship into their educational roles. Perhaps the most common and visible is that of the academic supervisor. They are also referred to as academic advisors in some universities. The role of the academic supervisor is to oversee and guide a student to complete their specific academic program. These individuals are formally assigned based on their areas of interest and expertise, and the role is task-oriented and time limited. The supervisors aim to "guide a student through the requirements of their academic program, establish expectations, provide evaluative assessments, navigate policies and procedures, promote academic integrity and sound research design, and generally assist their students in meeting and completing various milestones and tasks that are part of the academic program in a timely manner" [6].

Although faculty may choose to limit their responsibilities to academic supervision, many effectively combine the role of mentor and supervisor which has been shown to positively impact the overall graduate student experience [7]. However, graduate students and especially doctoral students should avoid haste and not request their academic supervisors to be their thesis supervisors and research mentor before exploring their own research agenda and identifying the resources they need.

55.3 The Role of the Mentor

Mentorship is distinguished as a more expansive role compared to an academic supervisor role in encompassing personal, career, and professional skill development. Doctoral supervision and mentoring go hand in hand where mentors who supervise academic proposals and dissertations not only focus on research and technical skill development but also the personal and professional development and psychosocial needs of the mentee [8]. Good mentorship extends beyond project supervision and bestowment of sound experimental design. A strong mentor serves to role model the ethical conduct of research with integrity, in order to make a positive contribution to both science and the community [9]. As noted in one faculty guideline, "First and foremost, mentors socialize students into the culture of the discipline, clarifying and reinforcing—

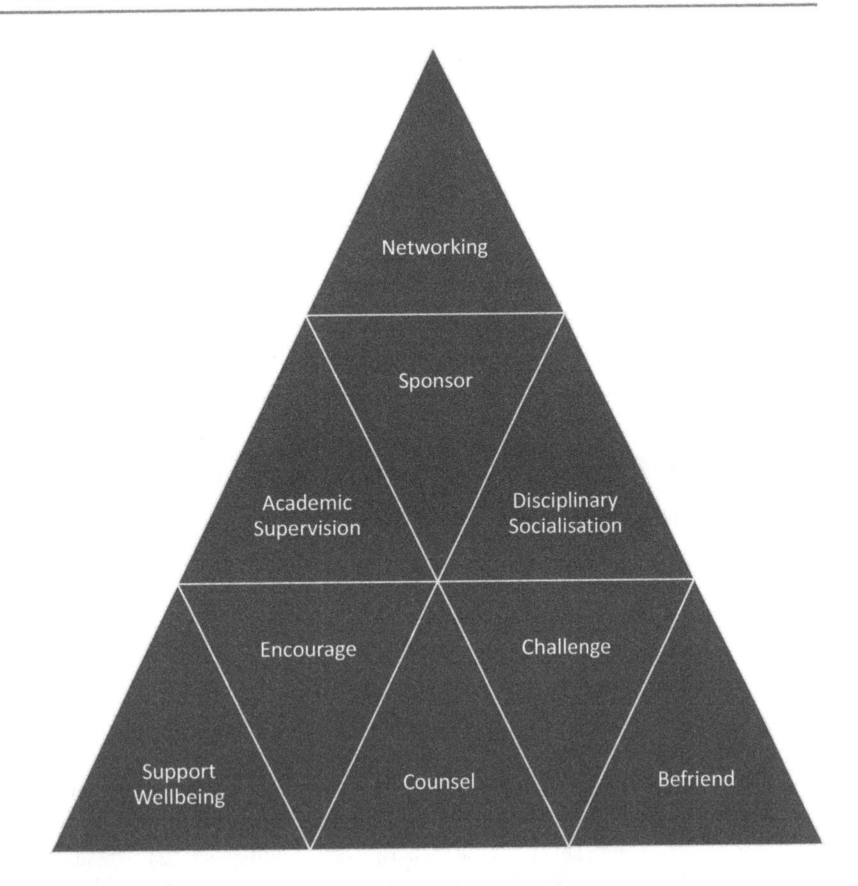

Fig. 55.1 Formal and informal roles of mentorship

principally by example—what is expected of a professional scholar" [3]. As a result, it often cultivates a more holistic and personal relationship between mentor and mentee.

Throughout the relationship, the mentor's role may evolve, but generally accepted responsibilities include continuing to offer support and resources to master the relevant theoretical knowledge, methodological or technical skills required for their research and to consistently offer guidance and support with respect to long term personal and professional development [10, 11]. Additional responsibilities include time management, conflict resolution, career planning, networking and introduction to the research or institutional culture of a specific discipline [11]. Although mentoring involves advising, teaching, coaching, sponsoring and advocating for mentees each of these can be different roles and should not be used interchangeably. For example, facilitating a small group discussion should not be called group mentoring [12].

Coaching usually addresses skills, performance or behavior in some aspect of an individual's work or life [13]. Coaching relationships have clear objectives which have to be achieved within a limited time. Any individual student, junior or senior faculty may identify a need for coaching at various times in their life and seek a coach to improve their performance. The need for coaching in a particular domain may be identified by individuals, peers, mentors, mentees or the institution. A coach is an expert in coaching a particular domain and coaching can be a voluntary or a paid endeavor where the coaches are paid for their services.

55.4 The Rationale for Mentorship Within Biomedical Sciences

Biomedical research work routinely involves setbacks, and a need for iteration to troubleshoot problems. It is possible that some will lead to failed experiments. Novice scientists will need to develop the skill to learn from their failures to shape and strengthen their future work [14]. Mentors can help novice scientists to overcome failed experiments, clarify misunderstood concepts, and reframe failure as a learning opportunity. The positive impact of effective mentorship is well established within biomedical sciences [15]. A growing body of literature consistently documents major benefits to the mentee, the mentor, the institution, and the greater scientific community [5].

55.4.1 Student Mentees

For the mentee, effective research mentorship is a strong indicator of career success, individual motivation and productivity, and professional satisfaction [16] and facilitates the development of self-esteem, competence, and decisiveness about one's career path [17, 18]. It supports advancement in research activities, conference presentations, grant writing, and research methodology [3]. These benefits are seen across all stages of career development. Through the foundational lens of mentorship, undergraduate and graduate students come to understand both the purpose and practice of research [19, 20]. While fundamental technical skills and scientific theory can be taught didactically, the quintessential functions of a researcher as a professional are too nuanced for the classroom, thus mandating apprenticeship and role modeling [4]. At the doctoral level, mentorship facilitates professional recognition in the form of publications, presentations, and networking and is valued as one of the most important attributes of high-quality education [21].

55.4.2 Faculty Mentees

Mentorship is especially important for the early careers of junior scientists. The transition from trainee to faculty "marks a point at which the acquisition of independence begins to supersede the importance of conducting elementary science or developing technical capabilities" [2]. For junior faculty, mentorship facilitates academic advancement using the mentor's own expertise, networking, sponsorship, and institutional knowledge. With guidance, the mentee develops professional competency, defined broadly as "career planning, communication skills, research and scholarship skills, managerial and leadership skills, negotiating and networking skills, and navigating the institutional culture" [2]. Such skills are not explicitly taught and cannot be learned easily through self-study. Further, strong mentorship fosters the development of professional leaders by imparting guidance through business and administration, financial acumen, managerial skills, team building, and interpersonal communication—critical qualities needed for success in today's academic environment [4].

55.4.3 Mentors

The rewards of mentorship from the mentor's perspective are no less significant especially with successful research mentoring. In addition to increased productivity, mentors stand to gain a sense of fulfillment, acquisition of new knowledge, and refinement of leadership and communication skills [16]. Mentorship offers an opportunity to engage with the next generation of scientists and researchers resulting in bi-directional knowledge translation and allows senior faculty to remain abreast of new techniques and avenues for further study. In the rapidly evolving culture of academia today, the expectation for all faculty to remain lifelong learners renders mentorship an important avenue to achieving this goal [4]. In addition, the process of mentorship can result in the discovery of engaged and like-minded trainees interested in pursuing the mentor's area of work, who may ultimately collaborate on further research or manuscript writing. Generating successful scholars reflects well on the professional status of the mentor and thus helps further their own career.

Perhaps most importantly, mentors often find personal satisfaction and reward in the knowledge that they have helped shape the next generation of researchers thus contributing personally to the future of their field [22]. Informal mentoring from many shadow mentors helps the research process in biomedical sciences. These informal or shadow mentors have to be cautious of becoming ghost advisors to several mentees. Three recipients of 'The annual Nature Awards for Mentoring in Science' discuss some issues faced by female faculty where they become ghost advisors [23]. The authors illustrate this point by the following narration: "*Joe Famous Scientist is the mentor of record but the mentee has to call on another individual to provide guidance and hands on mentoring. Often, the ghost advisor, who does the work but does not receive any formal credit, is female. Such work is time-consuming, but more importantly, there are two nefarious outcomes: the work of the ghost advisor advances the career of Joe Famous Scientist, as the mentee's work is now polished and ready for primetime, while simultaneously detracting from the time the ghost advisor has to dedicate to his or her own work, thus potentially undercutting the career and advancement of the ghost advisor who is putting in the time and effort. The ghost advisor is torn with wanting to help the mentee, but in helping the student, the ghost advisor perpetuates the system*" [23].

55.4.4 Institutions

Mentorship can be seen as an evidence-based investment in the organization's future. Mentorship enables tangible short-term benefits, such as increasing scientific output, attracting grant funding and facilitating publication in high-impact journals [24]. Mentorship is extremely important for student retention [25] and results in increased research productivity, faster academic promotion, and greater faculty retention and career satisfaction [4]. By maintaining collective engagement within the discipline of academia, mentorship assures collective benefits for the field of biomedical science as a whole, which benefits the wider society [4].

55.4.5 The Role of Mentorship in Enhancing Diversity in Biomedical Sciences

Diversity is required to build diverse research teams and increase research productivity, However unequal representation of individuals from Black or African American, or Hispanic backgrounds, and indigenous populations still exists in many Science, Technology, Engineering and Mathematics (STEM) fields particularly in the United States [26]. Evidence has shown that mentoring is critically important for the successful outcome of underrepresented individuals and women in STEM fields and strong mentorship can increase recruitment, retention, and persistence in science careers [26]. Gender differences in mentorship are known to perpetuate male-favored advantages in STEM disciplines [26], thus obstructing the advancement of diversity. Enhancing diversity requires a supportive web of invested mentors and peers who are committed to advancing the careers of individuals from underrepresented groups [22]. This is especially true in biomedical research where there exists significant underrepresentation at the faculty and Principle Investigator levels [26].

55.5 Types of Mentoring Relationships

55.5.1 Formal Mentorship

Formal mentoring relationships are intentionally created as part of a defined program within a larger organization. Such programs typically include specific objectives, formalized participant selection and matching processes, a predetermined timeline for the relationship, and guidelines for meeting frequency [5]. Mentors are given predetermined responsibilities according to the specific institutional arrangements and policies designed to meet their mentees' development needs for academic progress and ultimate success [27]. Examples of these responsibilities include stewarding knowledge, imparting skills, instilling values, and orientating mentees to the cultural and social aspects of their profession [28]. Formalized mentorship is beneficial in that it operates within a predetermined infrastructure to achieve specific organizational goals, and is easily amenable to systematic review and study. Additionally, formal programs can alleviate barriers of entry for underrepresented groups to improve diversity and inclusivity; in this way, leveling access to experienced mentors across the playing field regardless of one's racial, cultural, or gender based affiliation. At least theoretically, formal mentorship programs offer a level of 'quality assurance' by defining expected levels of engagement for mentors and mentees in advance.

One of the downsides to formal mentorship programs is that the connection between mentor and mentee is not organic and thus likely to suffer incompatibilities. This may result in prolonged pairings between mentors and mentees who are unable to reciprocally benefit from each other. The structured nature and connection with the academic organization may also limit a mentee's willingness to be vulnerable or open, for fear of retribution. Finally, the inherent structure of formal mentorship programs may limit flexibility in tailoring the mentor's role to the mentee's individual needs [29].

55.5.2 Informal Mentorship/Shadow Mentorship

By comparison, an informal mentorship arises spontaneously, is mentee driven as a result of overlapping values, interests, and career goals. Inherently flexible, these relationships allow the mentee to define their own needs and tailor the role of their mentor accordingly. The major downside to informal mentorship is that it relies heavily on the mentee to find a mentor, and is subject to disparity regarding access to quality mentors. Informal mentorship pairings typically occur because the mentor and mentees readily identify with one another, share similar backgrounds, or function within similar demographic settings. While this may result in more organic connection, it can perpetuate the already well-established disparities in regards to diversity within the biomedical field [30].

Mentoring relationships outside of traditional mentoring roles where a faculty mentors a student to help them succeed have been called shadow mentorship [30]. This is often an unseen and unaccounted part of faculty activity especially for females and under- represented minority (URM) faculty. Shadow mentors help mentees on multiple topics such as navigating career transitions, time management, resolving personal conflict, self-advocating, applying for a new position, steps to improve career readiness, overcoming imposter syndrome, building professional and social networks in graduate school and beyond, and seeking outside help for financial support or health issues [30]. Shadow mentoring is critical for student retention. However, shadow mentoring lessen mentors' 'traditional' productivity' as shadow mentors are often not part of the research team and remain unacknowledged in research publications. Furthermore, as a friendship based relationship, shadow mentorship may inadvertently detract from the overall research direction if their guidance contradicts that of a formally assigned mentor. Davis- Reyes et al. suggest that trainees can invite shadow mentors to serve as formal mentors, such as on thesis committees, where appropriate and that "faculty, departments, and institutions must recognize their URM colleagues not only for their diversity, equity, and inclusion expertise, but also for their scientific expertise, and they should recommend URM colleagues accordingly" [30] Such considerations should not prevent informal or shadow mentorship which may contribute to the overall mentorship needs of the mentees. However, mentees and institutions should not be rely on shadow mentors to fill the role of an effective research mentor.

55.5.3 Group Mentorship

Four primary types of group mentoring has been described in the literature as, one-to-many mentoring, many-to-one mentoring, and many-to-many mentoring and peer group mentoring [12]. One mentor mentoring many students is common in research laboratories and research groups. This method is also suitable when there is a limited availability of senior mentors and when a faculty mentor has several mentees at the same time [28]. Many to many mentoring allows for rotating between mentors and learning different approaches from different mentors [31]. Peer group mentoring occurs when a group of students mentor each other without assigning the mentor role to one or two people in the group [32] or it can involve junior faculty members forming a peer group to help career advancement [12]. Advantages of group mentoring include flexibility, inclusiveness, shared knowledge, interdependence, broader vision of the organization, widened external networks and team spirit in addition to personal growth, and friendships [11]. Successful groups build the capacity for exchanging ideas and information, as well as conflict management and team work which are of immense value to novice researchers. However, one to one faculty mentorship is often needed in addition to group mentoring for completion of research projects at a doctoral level.

55.5.4 Online Mentoring and Virtual Mentoring

Virtual mentoring became the norm during the recent Covid 19 pandemic where online platforms were used for communication between mentors and mentees. This provides flexibility and is more geographically friendly. There are multiple virtual platforms for online mentoring and many universities provide them for their faulty. Online or virtual mentorship follows the similar rules and expectation of the mentorship process of face to face mentoring. Particular attention should be pain to maintaining confidentiality and boundaries [33]. There are some potential risks including miscommunication, slower development of the mentoring relationship, trust and confidence [33]. One way to mitigate this would be to shift to working as doctoral dyads online which has shown to be very beneficial [34]. On- line mentoring will increase touch points for individual attention and supplement face to face mentoring which may seem rushed if there are multiple mentees and the mentor usually does one to many group mentoring [34].

55.5.5 Near-Peer and Peer-Peer Mentorship

Peer mentorship occurs between individuals who are roughly equal in age, status and power, with the goal of providing psychosocial support [35]. One of the increasingly popular forms of mentorship, near-peer mentorship can be defined as a process where an older or more able peer tutors a younger peer who is close to his or her social, professional, and/or age level [36]. The proximity in age, academic or social standing, or stage of career offers a unique support system and has been shown to promote both career advancement and academic success [36, 37]. It is particularly valuable for the more junior mentees experiencing challenges with major transitions; for example, the academic expectations of graduate school after completing ones undergraduate degree, or acclimatizing to newfound responsibility and independence as a junior faculty [26].

Major advantages of the incorporating a peer mentorship approach into formal mentorship programs include peer mentor availability, as there is generally a higher number of available peers compared to faculty. Common perspective between peer mentors and mentees may also result in increased likelihood to follow given advice and ultimately improved efficacy [35]. Thus peer and near peer mentorship remains an important player in the overall mentorship network. Peer support and mentoring has also shown positive effects amongst doctoral student but did not prevent attrition [38].

55.5.6 Reverse Mentoring

Reverse mentoring can occur between two peers or within the traditional mentor mentee pair of faculty and student. Students and junior faculty may be able to mentor the senior mentor in some aspects such as a new technique or new technology or the use of artificial intelligence in research. Where both parties act in the capacity of mentor and mentee two-way learning is enables and the gap between generations are bridged [39].

55.6 Models of Mentorship

Mentorship operates under different models, serving different purposes at different points in time [40]. Expectations of mentoring styles are changing and in some cases shifting away from the classic model [28]. Although the appropriate mentorship model will depend on the personalities, time availability, resources and workplace arrangements of both the mentor and mentee [28], the apprenticeship models will primarily guide the discussion as it is most common in the biomedical sciences discipline.

The apprenticeship or classic model is the most well-known in biomedical sciences where there is a hierarchy of professional positions, and the student is mentored by a more experienced professional. This model is less personal than other models and it is within the professional relationship that mentees learn from mentors. In general, mentors are advised to resist cloning and support student mentees to grow and progress. The cloning model is based on role modeling by the mentor, providing the insider scoop, sponsoring, providing gradual socialization, and grooming the mentee into the senior role. This relationship may be between faculty and student or junior and senior faculty and is very important for succession planning [41]. The nurturing model creates a safe, open environment in which mentees can discuss personal issues, learn, and try things for themselves, with their mentors acting as resources and facilitators. Although the research mentor can also provide some nurturing, it is often difficult for a single faculty to be an academic supervisor, research mentor and provide support for personal issues. The friendship model occurs when mentors and mentees are close to or at the same professional level; rather than being involved in a hierarchical relationship, they are often peers.

55.7 The Research Mentorship Process

While mentorship style varies according to field, institution, and personal inclination, the success of all mentorship relationships hinges on three common fundamentals, mentorship is relational and developmental, includes phases and transitions and is expected to yield career and psychosocial outcomes. Mentoring relationship developments have been described by Kram as overlapping phases of initiation, cultivation, separation, and redefinition [18] and as five phases consisting of, rapport building, direction setting, progress making, winding down, and moving on as described by Clutterbuck [42]. Adequate and appropriate preparation by both mentee and mentor prior to commencing a mentoring relationship will be useful to identify potential incompabilities ahead of time. Regular reflexion and progress tracking throughout the relationship will help identify early signs of unproductive mentoring relationships that may have to be modified or terminated.

55.7.1 Preparation and Reflection

We would like to take a step back to examine the preparation and reflection that has to occur prior to initiation of a mentoring relationship. The mentee is responsible for taking initiative through completing appropriate courses and self-reflection ahead of time. Mentees should also find out about their mentors current research, specific expertise, the number of graduate students currently being mentored and their research areas. It is important that mentees accept ownership of their career development. This includes establishing their own goals and timetables prior to attempting to self-identify their unique needs from the mentor. Resources that are available within the university that one needs to succeed in a graduate program should be identified and used. For example, the writing center will be a useful resource for a mentee who needs coaching in scholarly writing or the librarians for learning to perform literature searches.

Once a faculty member accepts to be the research mentor for a graduate student, it becomes a formalized relationship with responsibilities and expectations of outcomes regardless of whether the mentor was assigned by the institution or approached by the mentee. As such, mentors should take this responsibility seriously and prepare for it. Prior to connecting with their new mentee, it is helpful for a mentor to reflect on their desired role in the project and their own mentoring style. Development of an individualized approach should aim to identify personal goals, motivation, and engagement norms for the relationship, as well as assessment tools (if any) that will be used to evaluate both the mentee's learning and the effectiveness of the mentorship pairing [26]. This allows the mentor to intentionally define and tailor the kind of mentor they wish to be for that particular mentee, rather than relying purely on prior mentoring experiences. It is also important that mentors carefully examine their other commitments and their suitability to mentor the student prior to accepting a new research mentee.

55.7.2 Rapport Building

At the outset of the relationship, it is essential for both mentors and mentees to transparently outline their expectations. These initial conversations serve to shape the path forward and establish a foundation for an effective relationship. In addition to clarifying their project and personal goals, a mentee should consider the potential mentor's past experience, availability, resources, level of engagement, and professional network, to help align their expectations accordingly. Aligning and clarifying expectations fosters mutual trust and respect, ensures progress through creation of established timelines, and helps to avoid the challenges associated with miscommunication [26]. It can be helpful to utilize mentorship contracts as a means of catalyzing important discussions around goals and expectations. Box 55.1 provides useful topics to cover in mentorship contracts and Box 55.2 provides best practice suggestions for mentees and mentors. It is important that at the beginning of the relationship both mentors and mentees acknowledge that the mentoring relationship might end early due to personal incompatibility, mentor-mentee, misalignment of goals, or other reasons.

Box 55.1 Topics for Research Mentorship Contracts

1. Establishing goals and expected outcomes
2. Clarifying expectations of both parties and negotiating inputs, outputs and timelines
3. Discussion of preferred communication methods and expected turnaround time for responses.
4. Arranging regular meeting and or communications
5. Discussion about confidentiality

Box 55.2 Best Practice Tips for Mentees and Mentors

<u>Ten Tips for Mentees</u>

1. Schedule short frequent meetings well in advance
2. Create an agenda for each meeting, and share agenda and documents well in advance.
3. Follow-up each meeting with a document summarizing the discussion and action points
4. Learn to listen and minimize talking
5. Ask permission and record the meeting conversation or take notes
6. Always follow up, close the loop
7. Provide updates and reminders about deadlines
8. Be honest and open

(continued)

Box 55.2 (continued)

9. Have a growth mind set and learn to take criticisms
10. Learn to say no and do not over commit
11. Under promise and over deliver

Ten tips for Mentors

1. Learn about your mentee's experience, expertise, needs, goals and objectives
2. Convey your expectations of the mentee such as timelines, arranging meetings, etc.
3. Explore what the mentee is expecting from you
4. Carefully assess and convey what you can offer realistically
5. Discuss professional boundaries and confidentiality
6. Provide resources
7. Provide honest feedback and action plans
8. Encourage and teach how to deal with setbacks and failures
9. Be accessible and encouraging
10. Facilitate and prioritize personal and professional growth and psychosocial wellbeing

55.7.3 Direction Setting: Identifying Goals

During the first few meetings, the mentor and mentee should identify individual and mutual goals, the nature of support to be provided, a mutually agreed upon direction for the research project, and the essential skills required to achieve completion. It can be helpful to use a goal framework such as SMART (Specific, Measurable, Achievable, Relevant, Time-bound) to ask targeted questions and clarify the necessary resources, effort, and time commitment needed. These goals and expectations must be discussed at the outset to ensure the suitability of a potential mentor-mentee pairing [43]. In the context of biomedical research, the above efforts serve to clarify the nature of a mentee's research project and may uncover opportunities for collaboration and authorship, while avoiding potential overlap or competition with other lab members [26].

55.7.4 Progress Making

It is expected that the mentor will be available for regular meeting times for discussion and assessment of progress over the duration of the relationship, as well as remain reasonably accessible for unanticipated meetings. It would be the mentee's responsibility to book these meetings with the mentor's administrative staff. During meetings, the mentor should create a supportive and open environment to facilitate provision of appropriate guidance, assessment of progress, and constructive feedback. This includes acknowledgment of successful achievements. Mentors should fairly and honestly acknowledge the mentee's contribution within publications, teaching materials, presentations, or posters. Provision of positive feedback and honest feedback when appropriate is an important component of maintaining mutual respect and trust. One of the most important ways in which a mentor can support their mentee is by ensuring a psychologically safe and supportive environment. Mentors should "Encourage students to try new techniques, expand their skills, and discuss their ideas, even those they fear might seem naive or unworkable.... [as well as] let their students know that mistakes are productive because we learn from our failures" [3]. Fostering a growth mindset is an essential way to nurture self-sufficiency. At the same time, the mentor should show enthusiasm, optimism, and genuine interest in the mentee's work, as this can be an important source of inspiration. Mentees should be reminded of not only the personal value of their work, but its contribution to the professional community and the general public as a whole [3].

Both the mentor and mentee are required to keep track of progress, and when setbacks are encountered, aim to understand them. Should the mentee fail to meet expectations, the mentor is expected to transparently communicate, seek to understand, and offer guidance about how to get back on track. This is also the time to explore setbacks and teach the mentee to troubleshoot problems as they arise in their research project. Mentors should also be able to explore the reasons as to why mentees fall behind in their work. If there is a consistent pattern of disengagement, it may be prudent to provide resources such as student services and also arrange for further assistance as needed. It is always better to have difficult conversations early to avoid academic failures.

If mentorship is to be understood as a socialization process, the importance of developing collaborative social and professional networks cannot be overstated [26]. The role of the mentor extends beyond academic supervision and project oversight. To offer guidance through potential career paths, the mentor must also provide networking opportunities for new researchers to collaborate with [like-minded] professionals and navigate the funding process [9]. For example, if the mentor intends to connect with various experts in their field, they may invite the mentee to join the meeting to establish an introduction. In addition to making introductions, mentors may enhance networking opportunities by involving their mentees in editing, journal activities, conference presentations, and grant writing [3]. Such activities not only offer the chance to establish professional connections, but also allow for the development of communication and networking skills. As the relationship evolves, the mentor's role may focus increasingly on professional development by socializing mentees into the culture of their specific discipline, department, or field of research [3].

Finally, facilitating development of the mentee's professional network plays an important role in removing barriers and enhancing diversity within biomedical research. Such networking has been widely shown to improve the professional success of underrepresented individuals as it addresses implicit bias, macro and micro aggressions, and social isolation [26].

55.7.5 Winding Down and Moving On

The mentor should have discussions about the next steps in the mentee's career as an independent researcher or scientist when the research project is completed and the thesis is ready for submission. At this stage, the mentee should be well socialized in the field and able to navigate the social network well. The mentee has to be mentored on moving on further in their career which may take them on an entirely new path. Many doctoral students seek post-doctoral training positions upon completion of their academic program [44]. During the post-doctoral training, the mentee starts another apprentice model to practice and reinforce the learnings and experiences from the doctoral mentor prior to becoming an independent researcher. Although it is possible to continue with the same mentor, students are encouraged to seek post- doctoral positions elsewhere to expand their network. There has to be a separation and redefinition of the relationship regardless of whether the mentee finds another mentor for their post -doctoral position or continues with the same mentor. Mentoring relationships, models and types change and modify as novice scientist progress through different stages of their career to become independent researchers and mentors for others.

55.8 Mentorship as a Shared Responsibility

Research mentorship is a shared responsibility between the mentor, the mentee and the institution. Most research mentors have established areas of research interests and may expect the mentee to focus on their area of research when they agree to be the research mentor for a graduate student. Given mentors often hold multiple roles including esteemed professionals, researchers and academics, effective mentors must be able to balance their workload, research, and teaching while working closely with their mentees [45]. As a mentor, it is therefore important to be realistic about the time requirements when deciding who to take on as mentees [23]. Effective mentors seek opportunities for a mentee to succeed and encourage their mentees to set high, yet realistic standards for themselves [28]. In some perspectives, successful mentoring involves challenging mentees and not making their life easier [23]. In this approach, mentees are given the chance to fail and to learn from failures. Humiliating a mentee in public or private is not acceptable and their successes must be adequately celebrated [23]. Importantly the mentor must serve as a positive role model demonstrating successful behaviours not just preaching them [45].

Mentees need to be in 'the driver's seat', be active participants in the mentoring process and learn to manage to be successful [46]. Effective mentees are good communicators, thoughtful, motivated and quick learners [23]. Ideally, their goals and talents are well aligned [23]. The mentee must be sensitive to the time constraints of the mentor, and be proactive in establishing and working towards their own goals [28].

Successful mentorship necessitates institutional and program level commitment and engagement [3]. It is the duty of these higher organizational structures to create a culture that supports the efforts of individual mentors and mentees by ensuring that mentorship is valued by both students and faculty alike. At this level, responsibilities include providing appropriate infrastructure, evidence-based resources, and training for mentors; affirming mentorship as a core component of education by developing formalized mentoring plans; arranging opportunities for engagement with alumni who can speak to a variety of career possibilities; and formally evaluating existing mentorship programs based on site specific goals and metrics. Establishing a mentorship committee to oversee this process may prove useful in this regard. Mentorship programs and relationships may be assessed using outcomes like research publications, grants, individual mentor evaluations, awards, and academic retention.

The purpose of mentoring is to create a reflective environment in which the mentee can address issues of career, personal growth, and relationships [42]. Key to a successful mentoring relationship is for both parties to recognize the impact of biases, prejudices, privilege, and power on the relationship and acquire skills to manage them respectively [9]. In the end, the mentor–mentee relationship depends on the two people involved [23] and both parties must be open and adaptive. A gracious and humble attitude is required from both to help mentees learn from failures and enable mentees to admit to mistakes without the fear of negative consequences from their mentor [23]. Similarly, the mentee should be able to challenge the views of their mentors [23] to have productive conversations.

55.9 Navigating Mentorship Challenges

Different institutions, disciplines, professions and people may have different challenges to a successful mentorship. In this section, we will discuss a few challenges pertaining to mentoring in the biomedical field. One ubiquitous challenge is the limited availability of mentors. Sometimes mentors have significant workloads and are not available to provide adequate support to the mentee [28]. Assigning mentors has been used as a way to circumvent this issue. However, when mentors are assigned, rather than selected by the mentee, the mentorship relationship may be less likely to be based on mutual respect and shared interest, and may not produce the expected outcomes [28]. This is particularly important when a mentee is assigned a relatively new mentor who values being a mentor and is eager to dedicate time to this training but the mentee prefers another well-established mentor. Many students continue to join the research teams of highly accomplished scientists with several research projects and trainees due to the prestige of the science and the mentor. Many thrive in that environment but some struggle. The problem does not get resolved as many of these highly accomplished scientists, do not have the time or the willingness to go through mandated mentorship programs and spend time and effort with their struggling mentees [23].

Further challenges may arise due to a lack of clarity and understanding of the roles of the mentor and mentee, and the boundaries of the relationship [28]. This could be a result of poor communication, lack of commitment, personality differences, competition within the research team, or a lack of experience of the mentor [47]. There could be other serious ethical issues such as the betrayal of confidence or confidentiality, conflict of interest, or imbalance of power issues that may also arise leading to a lack of access to opportunities, especially by underrepresented groups [28]. If any of these are identified, early action to resolve the issue or seek an alternative mentor is recommended.

It is important to remember that "mentoring is not coaching, induction, or training, a passive undertaking, therapy, a one-way street, a cure-all, a bandage that metaphorically binds a wound, or a one-time intervention to fix a problem" [11].

55.9.1 Managing Conflicts and Challenges

As a way of combatting potential challenges, many organisations have formalized mentorship programs [28]. Training, awareness raising of potential ethical issues, evaluations, and guidelines for behaviour can also help mentoring relationships be successful [28]. Many of the potential challenges can be prevented and managed with good planning and communication. All universities provide multiple resources for graduate students. As mentioned earlier graduate students should preplan and prepare for a mentoring relationship by learning about the potential mentor. There is no mandated training for new mentees and most learn by trial and error. Peer or near peer mentoring becomes very useful in this stage to help socialize with their research environment and provide information on other informal or shadow mentors who may help with conflict management. There are other resources to access such as student services and ombudsperson or program director if there are conflicts and major challenges for graduate students once a mentoring relationship has been formalized with a research mentor. Mentees should reflect on and examine their needs to identify resources that can help them in the research endeavor [10]. As mentioned earlier, it would be very useful for students to invite some of their informal mentors to be part of their thesis committee to have continued access to these mentors.

55.9.2 Ending the Mentoring Relationship

Research mentoring relationships may have to be terminated prematurely in some cases because the mentor or mentee or both believe that their relationship is not effective. Mentee dissatisfaction with the mentorship they receive is common and has been as high as 60% in a clinician investigator training program [48]. Mentee dissatisfaction may be due to unrealistic expectations of coaching and hand holding from the mentee or because of a lack of commitment from the mentor. A lack of transparent pathways for problem escalation in mentorship relationships makes it difficult for both mentors and mentees to terminate mentoring relationships [49].

Certain steps can be followed when a mentee considers breaking up the relationship [50]. The mentee should carefully evaluate if they are getting what they need from the mentoring relationship, what is lacking or where things need to improve. They should also arrange a meeting to discuss their concerns with the mentor to alleviate the issues [50]. However, if a mentee decides that the relationship is unproductive, then they should gracefully break up without dragging it out or staying in an unproductive relationship [50].

If a mentee wishes to end the relationship the mentor should be given ample warning and a few week notice which allows both to wrap up any collaborative work and achieve some sort of closure. Being straightforward helps the termination be less complicated and hurtful [51]. A neutral third party such as a program director, student advisor, or an ombudsperson should be consulted and be involved in the termination process if a mentee wishes to terminate the relationship due to unacceptable behavior such as breach of confidentiality or romantic issues. If a mentor terminates the relationship, the mentee should accept that another mentor may be more suited for the role and move on.

The following examples illustrate the point when a mentoring relationship may be terminated mutually. Some mentees may find the research work that they are currently involved in with their mentors uninteresting or unstimulating and may want to explore a different research focus. In these situations, the mentee should reflect on what his or her research project will be and write a summary of their project and explore who works in that area and may approach some informal mentors or peers for advice. When a mentee approaches the research mentor with a draft protocol to examine the feasibility of the project and ask for advice on how to proceed, the mentor has a clear idea of the problem and may suggest an alternate mentor. Leaving on good terms is always good as the mentor is very likely to be willing to be involved in the mentee's future career even if the research mentoring relationship is terminated. Doctoral students are expected to formulate their research questions, try to identify the most appropriate methodology and discuss the feasibility of completing the research project with their mentors. The mentors are expected to review the methodology and guide mentees to complete their research projects and publish them. Mentors must also provide open and honest discussions about feasibility and their suitability to be the research mentor or the sole supervisor for the project, especially if the most appropriate methodology is not one of their areas of expertise.

A study examining reasons for attrition amongst doctoral students found the difference between those who completed the program and those who did not show attrition was likely if the mentees did not feel that they were moving forward, were experiencing too much distress, or their research project did not make sense to them [38]. This study also states that although support from peers had a positive role, the peer support will not have any impact on dissertation progress and lead to doctoral completion [38].

55.9.3 Future of Mentorship as a Team-Based Endeavor

Traditionally, academic mentorship has been conceptualized as a unidimensional and formalized relationship between one mentor and their assigned mentee. However, if we are to understand mentorship in part as a multifaceted socialization process, it follows that one mentor alone cannot fulfill all of the roles and responsibilities required. To be truly effective, mentoring must be individualized. This requires relationships to be multi-dimensional, with the goal of addressing numerous disparate objectives and providing a holistic support system [10, 26]. Through daily interaction and functioning, mentees will often naturally encounter many different kinds of mentors in addition to their formally assigned one, each of whom provides their own unique interpersonal dynamic, expertise, and pedagogy. For example, the formal academic supervisor might mentor a graduate student through the key elements of research design and the publication process, while another senior colleague might hold the familiarity with the employment landscape du jour, and offer important career development strategies and networking opportunities. The mentee might also naturally feel drawn to a peer or slightly senior level colleague, who can facilitate acclimatization and orientation to a new environment or career transition. Thus it is important to encourage the development of Mentoring Networks, which consist of multiple mentors from different backgrounds and contexts, to ensure holistic and comprehensive support to each mentee [26].

The second year of a doctoral program in biomedical sciences is especially challenging for students due to its complex, competitive and dynamic nature and 9% of students officially leave the program within the first two years [52]. It is estimated that 25 to 40% of students drop out of doctoral programs in the sciences [52]. Lack of mentorship has been implicated in the high attrition rates among doctoral students [25]. Laboratory work in biomedical sciences demands long hours and is almost always punctuated with setbacks and failure. Research mentors in biomedical sciences have to ensure mentees' personal and professional wellness to nurture future scientists who would not only complete the program but also be interested and equipped with the skills for a career in science.

A study from a biomedical campus found that 92% of doctoral students identified (29%) or strongly considered (63%) non research careers for their post degree employment [53]. The authors question the rationale for doctoral programs to continue to prepare students for a traditional academic career path despite the inadequate supply of research-focused faculty positions and advocate for a broader doctoral curriculum that prepares trainees for a wide range of science-related career paths. One could argue that doctoral students from biomedical sciences should have access to e-mentors from science-related fields. As such, it is important creative mentoring strategies such as distance mentoring and virtual or online mentoring can be used to combat some of the issues with limited time for mentoring. These online modalities have been the main modes of mentoring during the recent pandemic and have shown significant promise in supporting doctoral students [34, 54]. Our mentorship models may need to be modified and some long held beliefs about research and research mentoring have to be reassessed as our students' needs evolve and change over time.

55.10 Concluding Remarks

Traditional mentorship models are in need of modification, especially in regards to diversity and inclusivity. Mutual reflexivity is an important predictor of mentoring success. This process of reflection can also alert the mentee and mentor to warning signs and predictive factors which may indicate the chosen mentoring relationship is not productive and may have to be terminated early. Alternatives models such as informal or shadow mentorship, group, online, peer and near-peer mentorship, and network mentorship provide significant support to mentees in biomedical sciences and should be actively cultivated. Combining multiple forms of mentorship is advisable and has been associated with successful outcomes.

Conflict of Interest None to declare.

References

1. Cypress BS (2020) Fostering effective mentoring relationships in qualitative research. Dimens Crit Care Nurs 39(6):305–311
2. Spence JP, Buddenbaum JL, Bice PJ, Welch JL, Carroll AE (2018) Independent investigator incubator (I(3)): a comprehensive mentorship program to jumpstart productive research careers for junior faculty. BMC Med Educ 18(1):186
3. Rackham Graduate School UoM (2020) How to mentor graduate students: a guide for faculty. https://rackham.umich.edu/downloads/how-to-mentor-graduate-students.pdf. Accessed 10 Nov 2022
4. Choi AMK, Moon JE, Steinecke A, Prescott JE (2019) Developing a culture of mentorship to strengthen Academic Medical Centers. Acad Med 94(5):630–633
5. American Psychological Association Taskforce (2012) Introduction to mentoring: a guide for mentors and mentees. American Psychological Association, Washington, DC
6. Toronto Uo (2016) Supervision guidelines for faculty adapted from School of Graduate Studies, Graduate Supervision Guidelines: Student Edition
7. Barnes BJ (2009) The nature of exemplary doctoral advisors' expectations and the ways they may influence doctoral persistence. J Coll Stud Retent Rese Theor Pract 11(3):323–343
8. Schunk DH, Mullen CA (2013) Toward a conceptual model of mentoring research: integration with self-regulated learning. Educ Psychol Rev 25(3):361–389
9. Rockey SJ (2014) Mentorship matters for the biomedical workforce. Nat Med 20(6):575
10. Montgomery BL (2017) Mapping a mentoring roadmap and developing a supportive network for strategic career advancement. SAGE Open 7(2): 2158244017710288
11. Mullen CA, Klimaitis CC (2021) Defining mentoring: a literature review of issues, types, and applications. Ann N Y Acad Sci 1483(1):19–35
12. Huizing RL (2012) Mentoring together: a literature review of group mentoring. Mentor Tutor 20(1):27–55

13. Koopman R, Englis PD, Ehgrenhard ML, Groen A (2021) The chronological development of coaching and mentoring: side by side disciplines. Int J Evid Based Coach Mentor 19(1):137
14. Simpson A, Maltese A (2017) "Failure is a major component of learning anything": the role of failure in the development of STEM professionals. J Sci Educ Technol 26(2):223–237
15. Brinton TJ, Kurihara CQ, Camarillo DB, Pietzsch JB, Gorodsky J, Zenios SA et al (2013) Outcomes from a postgraduate biomedical technology innovation training program: the first 12 years of stanford biodesign. Ann Biomed Eng 41(9):1803–1810
16. Pfund C, Byars-Winston A, Branchaw J, Hurtado S, Eagan K (2016) Defining attributes and metrics of effective research mentoring relationships. AIDS Behav 20(Suppl 2):238–248
17. Day R, Allen TD (2004) The relationship between career motivation and self-efficacy with protégé career success. J Vocat Behav 64(1):72–91
18. Kram KE (1983) Phases of the mentor relationship. Acad Manage J 26(4):608–625
19. Lovitts BE (2002) Leaving the ivory tower: the causes and consequences of departure from doctoral study. Rowman & Littlefield Publishers, pp 131–166
20. Sands RG, Parson LA, Duane J (1991) Faculty mentoring faculty in a public university. J High Educ 62(2):174–193
21. Golde C, Bueschel A, Jones L, Walker GE (2006) Apprenticeship and intellectual community: lessons from the Carnegie Initiative on the Doctorate. In: Conference proceedings of the National Conference on Doctoral Education and the Faculty of the Future Cornell University. The Carnegie Foundation for the Advancement of Teaching, Ithaca, NY, pp 1–28
22. Sorkness CA, Pfund C, Ofili EO, Okuyemi KS, Vishwanatha JK, Zavala ME et al (2017) A new approach to mentoring for research careers: the National Research Mentoring Network. BMC Proc 22:83–88. https://doi.org/10.1186/s12919-017-0083-8
23. Clynes M, Corbett A, Overbaugh J (2019) Why we need good mentoring. Nat Rev Cancer 19(9):489–493
24. Freel SA, Smith PC, Burns EN, Downer JB, Brown AJ, Dewhirst MW (2017) Multidisciplinary mentoring programs to enhance junior faculty research grant success. Acad Med 92(10):1410–1415
25. Maddox S (2017) Did not finish: doctoral attrition in higher education and student affairs. University of Northern Colorado
26. Diggs-Andrews KA, Mayer DCG, Riggs B (2021) Introduction to effective mentorship for early-career research scientists. BMC Proc 15(Suppl 2):7
27. Stoeger H, Balestrini DP, Ziegler A (2021) Key issues in professionalizing mentoring practices. Ann N Y Acad Sci 1483(1):5–18
28. Burgess A, van Diggele C, Mellis C (2018) Mentorship in the health professions: a review. Clin Teach 15(3):197–202
29. Clutterbuck D (2005) How formal should your mentoring programme be? Clutterbuck Associates, Burnham, pp 1–3
30. Davis-Reyes B, Starbird C, Fernandez AI, McCall T, Hinton AO Jr, Termini CM (2022) Shadow mentoring: a cost-benefit review for reform. Trends Cancer 8(8): 620–622
31. Levine RB, Hebert RS, Wright SM (2003) The three-headed mentor: rethinking the classical construct. Med Educ 37(5):486
32. Hadjioannou X, Shelton NR, Fu D, Dhanarattigannon J (2007) The road to a doctoral degree: co-travelers through a perilous passage. Coll Stud J 41(1)
33. McReynolds MR, Termini CM, Hinton AO Jr, Taylor BL, Vue Z, Huang SC et al (2020) The art of virtual mentoring in the twenty-first century for STEM majors and beyond. Nat Biotechnol 38(12):1477–1482
34. Mullen CA (2020) Online doctoral mentoring in a pandemic: help or hindrance to academic progress on dissertations? Int J Mentor Coach Educ 10:137–159
35. Collier P (2017) Why peer mentoring is an effective approach for promoting college student success. Metrop Univ 28(3):9–19
36. Singh S (2010) Near-peer role modeling: the fledgling scholars education paradigm. Anat Sci Educ 3(1): 50–51
37. Anderson MK, Tenenbaum LS, Ramadorai SB, Yourick DL (2015) Near-peer mentor model: synergy within mentoring. Mentor Tutor 23(2):116–132
38. Devos C, Boudrenghien G, Van der Linden N, Azzi A, Frenay M, Galand B et al (2017) Doctoral students' experiences leading to completion or attrition: a matter of sense, progress and distress. Eur J Psychol Educ 32(1):61–77
39. Chen Y-C (2013) Effect of reverse mentoring on traditional mentoring functions. Leadersh Manag Eng 13(3):199–208
40. Ratnapalan S (2010) Mentoring in medicine. Can Fam Physician 56(2):198
41. Bourke L, Waite C, Wright J (2014) Mentoring as a retention strategy to sustain the rural and remote health workforce. Aust J Rural Health 22(1):2–7
42. Megginson D (2006) Mentoring in action: a practical guide. In: Development and learning in organizations
43. Zachary LJ, Fischler LA (2009) The mentee's guide: making mentoring work for you. Wiley, Hoboken, NJ, pp 1–15
44. Camacho S, Rhoads R (2016) Panel: collective bargaining issues concerning post-doctorates-breaking the silence: the unionization of post-doctorate workers at the University of California. J Collect Bargain Acad 11:72
45. Klausmeier RL Jr (1994) Responsibilities and strategies of successful mentors. Clearing House 68(1):27–29
46. Straus SE, Chatur F, Taylor M (2009) Issues in the Mentor–Mentee relationship in academic medicine: a qualitative study. Acad Med 84(1):135–139

47. Straus SE, Johnson MO, Marquez C, Feldman MD (2013) Characteristics of successful and failed mentoring relationships: a qualitative study across two academic health centers. Acad Med 88(1):82–89
48. Ng E, Wang X, Keow J, Yoon J-Y (2015) Fostering mentorship for clinician-investigator trainees: overview and recommendations. Clin Invest Med 38:E1–E10
49. Collins K, Oliver SW (2017) Mentoring: what matters most? Clin Teach 14(4):298–300
50. O'Hara C (2014) How to break up with your Mentor. Harv Bus Rev:29
51. Morrow KV, Styles MB (1995) Building relationships with youth in program settings: a study of Big Brothers/Big Sisters. Public/Private Ventures, Philadelphia, pp 115–119
52. Maher MA, Wofford AM, Roksa J, Feldon DF (2020) Exploring early exits: doctoral attrition in the biomedical sciences. J Coll Stud Retent Res Theor Pract 22(2): 205–226
53. Fuhrmann CN, Halme D, O'sullivan P, Lindstaedt B (2011) Improving graduate education to support a branching career pipeline: recommendations based on a survey of doctoral students in the basic biomedical sciences. CBE Life Sci Educ 10(3):239–249
54. Termini CM, McReynolds MR, Rutaganira FU, Roby RS, Hinton AO Jr, Carter CS et al (2021) Mentoring during uncertain times. Trends Biochem Sci 46(5): 345–348

56 Commercializing the Technology: Transitioning from the Academic Lab to the Market

A. Sankaranarayanan

Abstract

Technology commercialization is one of the major factors contributing to the economy of a country. The various stages of this process are disclosure of the invention, evaluation of the technology, commercialization in partnership with the industry or creation of a startup company. Protection of intellectual property and licensing arrangements are central to this commercialization process. Currently, research and development in universities and public-funded institutions are not focused on innovations relevant to the industry. The changes that are necessary to improve this at the academic institutions, their organisational setup, the quality of their faculty and their research focus and the policies and initiatives at the national level are discussed in this chapter.

Keywords

Academic innovation · Technology commercialization · Stages of technology commercialization · Critical factors in technology commercialization · Indian National Policies and Initiatives

56.1 Introduction

The economy of any country heavily depends on the products and services offered in its markets. The importance of innovative science and technology in this regard is increasingly appreciated in the present knowledge economy. The technology transfers ultimately contribute to the economy in the form of introduction of new products/ services to the market, generation of jobs and national productivity. In developed economies, research conducted in academic institutions play a major role in contributing to industrial growth and progress and wealth. While academic institutions are normally concerned with teaching and research, their potential in contributing to industrial growth and economic development in countries like India is still evolving. The effective transfer of academic innovations to the industry is a complex process and depends on the organisational setup in academic institutions and the supportive national policies. The processes of technology transfer, the critical factors involved and the evolving Indian national policies and initiatives are presented in this chapter.

56.2 Academic Institutions: Changing Trends

Universities and academic institutions have their core functions in education and research. It is

A. Sankaranarayanan (✉)
Vivo Bio Tech Ltd., Hyderabad, Telangana, India
e-mail: sankar@vivobio.com

G. Jagadeesh et al. (eds.), *The Quintessence of Basic and Clinical Research and Scientific Publishing*,
https://doi.org/10.1007/978-981-99-1284-1_56

generally accepted that "academic freedom" allows these institutions to carry out research of their choice. The generation of new knowledge in such institutions mostly ends as research publications or might lead to a Ph.D. However, such a traditional model may not be sustainable in the coming days. The future would be for academic institutions that would carry out innovative research programs and generate technology that would contribute to the economy. This model would also sustain and provide resources to run and maintain these institutions. The current trend in developed economies is such a paradigm where university research contributes to new products and services and contributes GDP and job creation. However, all academic research need not have a commercial motive! There would always be projects that address "basic research" without any commercial motive.

In the domain of pharmaceuticals, research might mean the discovery of a new innovative drug, a new formulation or even a process to improve the product qualities like stability, etc. (see Chap. 1). The development would involve the creation of a new product or a process out of this research with a defined benefit over what is currently available in the market.

56.3 Building an Innovative Pipeline

How to strengthen academic research to have products and services at the core of its objectives? And how to encourage publicly funded institutions to focus on industrial applications? Technology transfers and commercialization are partnership driven and hence it is important to involve corporations and industry leaders in academic research programs right in the beginning. Palmintera et al. [1] made relevant recommendations to academicians on how to build an innovation pipeline with an industry bias. Primarily, academic institutions should build excellent research strengths with an emphasis on niche areas that are of interest to local industries. They should make efforts to get industry members involved in their research centres. Inviting corporations and entrepreneurs to build their research capacity and programs in academic institutions would enrich the academic research programs. Academic institutions should promote "entrepreneurial culture" and be aware of the benefits of promoting such activities. Academic research and its orientation with industrial bias could be promoted by various forms of corporate relationships. Industry sponsored research, sponsoring faculty/student internships, inviting industry to participate in peer review panels for research grants, creation of endowed chairs or scholarships with industry contribution, providing adjunct professorships to academically qualified persons, etc., are the ways to develop such relationships.

56.4 The Stages of Technology Commercialization

Technology commercialization is the process of transforming an innovative new product or a service all the way from the conceptualisation stage to a marketable product.

Let us consider an academic research program for the development of a new drug for treating Covid-19. Since there was an acute need for such a drug during the pandemic, it was reasoned that a strategy to bring a drug quickly to the market would have commercial success. Few investigational new drugs were already available for treating Severe Acute Respiratory Syndrome (SARS) caused by a related virus, SARS CoV-1. Since these drugs have been shown to have good safety profiles, it was easier to develop one of them as a therapy for Covid-19. Due to the genetic similarities between the two viruses, it was expected that these investigational drugs would have a high probability of activity against SARS CoV-2; one has to generate data on their anti-viral activity and choose the best among them for further development. After successfully completing initial proof-of-concept studies in the lab, efforts would be taken to identify an industry partner for carrying out further developments, applying for market registration, etc. The success of the project mainly depends on the novelty of the idea, the timely introduction of the new product, and the commercial potential that would be attractive to an industry partner.

The stages of commercialization of technology [2] are listed below:

1. Disclosure of the Innovation
2. Evaluation of the Technology
3. Technology Implementation
4. Licensing

56.4.1 Disclosure of the Invention

The innovation for commercialization would first be shared with the Institutional Technology Licensing Office (TLO) or similar offices of the Institution. Hence, the first step in commercializing technology is to document the innovation for review. Clear disclosure is essential to assess the market opportunity and to identify industry partners. Hence, the researcher needs to have a good system of documenting his/her observations and maintaining laboratory records. This is critical particularly when one wants to market the product. Proper documentation is also essential when one applies for patents as this data forms the basis for the patent application and it establishes priority.

The full disclosure also helps in devising strategies for intellectual property protection to take full commercial advantage of the product. If the innovation has been already published or presented elsewhere, that should be disclosed. If there had been any sponsor for the research or was supported by a grant, and if any contract/agreement has already been executed with the sponsor, that should be made explicit in the disclosures. It would be essential to show with examples, how this new invention or technology has applications. Complete disclosure will enable the industry partner to develop a marketing strategy to commercialise this technology. That would also facilitate applying for legal protection of the invention.

The Technology Licensing Office (TLO) shall have a "Technology Disclosure Form" for this purpose so that no essential information is missed out.

56.4.2 Evaluation of the Technology

The technology commercialization involves evaluation and implementation phases. The experts in the TLO and the research team collaborate in these evaluations. The activities and milestones achieved in these two phases would vary depending on the nature of the innovation.

Firstly the technology is evaluated for innovative elements of the technology and how this translates into competitive advantages and market opportunities. If it is encouraging, the next step would be to draw strategies to establish a competitive market advantage. The technology is also assessed for possible barriers to market entry and the presence of any technology risks. At this stage, it is also necessary to assess the market size to justify the required investments to be made to market the product. The patentability is assessed followed by necessary patent applications. If prospective industry partners are also involved, a "Non-disclosure Agreement" shall be executed. This is an essential step to protect intellectual property rights (see Chap. 57).

56.4.3 Technology Implementation

On successful completion of the assessment to confirm the market potential, industry funding sources are explored. Several industry partners may be approached for the same; and the TLO plays an important role in this activity. The investors need to be convinced of the viability of the commercialization of the product. The industry funding sought shall be based on milestone achievements while developing a marketable product. A business model is drawn taking into account all the processes involved and the projection of costs to bring the product to market. These would include a thorough analysis and estimate of the cost of the product development, clinical evaluation and registration of the product for marketing, etc. At this stage, based on the technology, appropriate intellectual property protection in the form of patent filing shall be initiated. In the commercialization process, the last step is the transfer of intellectual property rights to a commercial partner. This ultimately takes the technology from the laboratory to the market. Appropriate licensing is central to this transfer and is granted to an existing company or a startup created for this purpose.

While academic institutions and their faculty may be proficient in educational and research activities, they may not have the wherewithal to manage intellectual property, fiscal evaluations, negotiate and manage the licensing process and agreements with industry. All these aspects are managed by the commercial expertise available with the TLO and related offices.

56.4.4 Licensing

Licensing agreements would need discussions and negotiations with the industry partners regarding contract terms and conditions, the fee that would be charged, a memorandum of understanding, etc. The kind of licensing would depend on the nature of the technology. The exclusive license shall be drawn when the technology is licensed to an external agency or company. With the non-exclusive license, the university has the option of transferring its technology to more than one company or different geographic regions or from time to time, as and when needed. Each project/technology is unique and the licensing arrangement is drawn to give maximum benefit to both parties concerned. The licensing officer might also like to explore if there are other companies with an interest in the technology to evaluate the full commercial value of the license. One of the routes the commercialisation may take shall be as a Startup Company. It could be either by individual efforts of the inventor and his associates or with the support of a business incubator of the institution. When forming a startup company is the most suitable way for the technology transfer, the original research team itself becomes its founder. To attract investors, a startup must be of interest to its customer - the customers in this case for a startup is the public and private markets and the company's potential acquirers. The evolving startup company shall convince prospective investors at appropriate times and bring them in at various stages of the growth of the Startup. The product life cycle and the investment needed at various stages by the startups have been elaborated by Jordan [3].

56.5 Attributes of a Successful University in Technology Commercialization

Several factors have been identified to be of critical importance in the process of commercialization of technology in academic institutions. Massachusetts Institute of Technology (MIT), is one of the most successful institutions in the US in carrying out innovative research and commercialization of technology including the generation of several spinoff organizations. In a critical analysis of MIT, O'Shea et al. [4] identified and highlighted several features that had contributed to its success. Some of these are:

- Emphasis on interdisciplinary research
- Networking between government, industry and academia to facilitate entrepreneurship
- Organisational structures like Technology Licensing Office to guide and support industrial licensing and "deal" making
- Policies and practices to encourage startups
- The positive attitude of academics toward commercializing technology

It was recognized that attracting research funding from the industry plays an important role in the generation and conduct of industry-related research with a high chance for commercialization. Even basic research conducted with industry collaboration has high relevance to the needs of the industry. Another critical factor in innovation and technology transfer was the quality of human resources. Faculty with a keen eye for identifying and pursuing areas of research with potential for commercialisation and the ability to attract motivated students create an environment for research and innovation. The organisational structure also permits the recruitment and retention of highly accomplished and motivated faculty. For example, the MIT faculty were leaders in their own chosen areas of research and some of them even pioneered unexplored fields of research with high potential for spinoffs.

Even the most accomplished and innovative faculty would need organisational structure and support to commercialise the technology developed by them. Typically, an office (TLO) dealing with matters related to patents, copyright, and licensing provides this support. The TLO works closely with the scientists/faculty and evaluates the inventions from a commercial perspective and proactively looks for licensing opportunities. Such offices play a critical role in interacting with venture capitalists to explore interests in the innovations and technologies ongoing in the University.

Institutional policies that are supportive of commercialization are also relevant. The policies are designed to handle conflicts of interests that are likely to arise. For example, the Faculty may not hold equity in a company that sponsors research at the Institution. Similarly, a faculty may not have any consulting activities without the knowledge of the University. It also makes it mandatory that discussions with out-side agencies are always preceded by Non-disclosure Agreements (NDA). In addition to TLO, Institutional entrepreneurial support is also provided for a spinoff in the form of grants and advice for promising innovations.

The mission of the University, in addition to promoting education in science and technology, should have an emphasis on addressing real-world problems. Encouraging entrepreneurship should be one of its thrust areas. The academic culture in addition to aiming for a high level of academic achievement should also have the mindset for interaction with the industry. Only such culture and attitudes help develop innovative research with an industry focus.

56.6 Indian Academic Innovation and Commercialization

Patenting and technology commercialization activities in Indian universities are getting more emphasis in recent times. However, the achievements at the national level are far from satisfactory, particularly considering the various initiatives and policies introduced from time to time. The Indian Patent Act which was introduced as early as 1970 supported the concept of "reverse engineering". This had a positive impact, particularly on the chemicals and pharmaceutical industry. Later in 1988 the Technology Information Forecasting and Assessment Council (TIFAC) under the Department of Science and Technology was established to encourage the development of R & D infrastructure and technology commercialization. Beyond the IPR issues, the Ministry of Science and Technology took steps to promote entrepreneur generation. For this purpose, the National Science and Technology Entrepreneurship Development Board (NSTEDB) was created in 2009. Under the aegis of the NSTEDB, Technology Business Incubators (TBI) were created at Indian Institute of Technology- Delhi, Kanpur, Mumbai and Chennai and at Banaras Hindu University and Delhi University [5].

The New Millennium India Technology Leadership Initiative (NMITLI) is one of the largest public-private partnership efforts in the R & D domain in the country, financed by the government of India. Its objective is to synergise the best competencies of publicly funded R & D institutions, academia and private industry. It has so far evolved about 60 largely networked projects in biotechnology and pharmaceuticals. Finally, to support the technical institutions of the country and foster research innovation and entrepreneurship through training in various emerging areas, the All India Council for Technical Education Training and Learning academy was started in 2018. As of today, there are eleven such academies have been established [6].

In spite of all these initiatives, successful industry-academy collaborations have been few and far between. However, there had been a few shining examples. In the mid-50s the Government of India asked the Central Drug Research Institute (CDRI) at Lucknow to develop a safe contraceptive for its National Family Planning Program. Steroidal contraceptives available at that time were associated with a variety of side-effects such as nausea, cramps, headaches and breakthrough bleeding, etc. Hence the CDRI scientists wanted to design a contraceptive without these side effects that would be more acceptable to women. Their strategy was to design a molecule that would selectively bind to oestrogen receptors of reproductive tissues. Their innovation worked and the result was a successful contraceptive, Centchroman® (INN: Ormeloxifene). The CDRI licensed centchroman to a private company HLL in 1991 for development and marketing. It was launched as "Saheli" and under the National Program, the contraceptive is available free of cost in all primary healthcare centres [7]. This is really a remarkable story.

In spite of all these developments and advances, Indian universities and academic institutions have a long way to go in innovation and technology transfer. Based on a study of patent activities and revenue generated in 40 Indian Universities/Institutes from 2013 to 2018, Ravi and Janodia [5] concluded that the practice of IP generation and Technology Transfer is underdeveloped in the country. They suggested the following to remedy this: (a) Universities and Institutes should leverage their expertise in specific domains and pursue interdisciplinary research to attain high value through knowledge commercialization, (b) focus on commercially viable research programs and (c) identify mechanisms to reach out to and collaborate with the industry through exhibitions, conferences and research partnerships.

56.7 Concluding Remarks

Technology commercialization from publicly funded academic institutions is an important component contributing to the economy. The stages of technology commercialization and the factors influencing them have been presented. The quality of innovative research programs and commercialization of technology though gaining momentum in the country needs to progress to a much larger extent. The government by its innovative policies and their efficient implementation should bring in necessary changes in the days to come.

Conflict of Interest The author declares no conflict of interest.

References

1. Palmintera D, Joy J, Xiang EX (2007) Technology transfer and commercialization partnerships. Innovation Associates Inc. www.InnovationAssociates.us. Accessed 5 Jan 2023
2. Michigan Technological University/Innovation and Commercialization. https://www.mtu.edu/research/innovation/. Accessed 1 Jan 2023
3. Jordan JF. First online date: 2018-06-29. Innovation, commercialization and the successful startup (Working Paper). https://kilthub.cmu.edu/articles/journal_contribution/Innovation_Commercialization_and_the_Successful_Startup/6471581. Accessed 1 Jan 2023
4. O'shea RP, Allen TJ, Morse KP et al (2007) Delineating the anatomy of an entrepreneurial university: the Massachusetts Institute of Technology experience. R & D Manag 37:1–16
5. Ravi R, Janodia MD (2022) Factors affecting technology transfer and commercialization of university research in India: a cross-sectional study. J Knowl Econ 13:787–803
6. Srivastava P, Sunita C (2012) Technology commercialization: Indian perspective. J Technol Manag Innov 7: 121–131
7. Kumar N, Pant G, Kulkarni SR (2018) The journey of the world's first non-steroidal contraceptive from Academic venture to National Family Program. Curr Sci 115:1638–1640

57 Patent Law Fundamentals for Biomedical Scientists

Srikumaran Melethil

Abstract

The major objective of this 2-part chapter is to introduce, demystify and advance the knowledge of patent law among biomedical scientists. Part I presents a discussion on the significance of understanding patent law concepts among scientists, followed by an introductory excursion into fundamental patent law concepts and terminology in the United States, such as intellectual property, novelty, anticipation, utility, non-obviousness, enablement, person of ordinary skill in the art, patent infringement, specification, and claims. Part II is geared more toward pharmaceutical scientists and discusses patent litigation between brand names and generic companies to illustrate the application of these concepts in patent prosecution and litigation. It is hoped that this chapter will serve as a starting point for biomedical scientists to foray into the area of patent law.

Keywords

Patent law · Biomedical scientists · Intellectual property · Non-obviousness · Commercialization of research · Novelty

57.1 Introduction

Patent law, at its heart, is mostly (if not, all) about commercialization of an inventor's intellectual property (IP). Simply put, it is about seeking financial gain from the creativity (inventions) of an inventor. Such gain also provides peer-recognition incentives to develop new products for the market, which is beneficial for the overall well-being of society. Development and marketing of pharmaceuticals, automobiles, computers, and communication and entertainment devices such as mobile phones and video games comprise a small sample of such useful products. Patent law attempts to provide a paradigm to strike a balance between the economic interests of the individual (inventor) and society at large.

The major objective of this introductory chapter on United States patent law is to introduce, demystify [1] and advance the knowledge of patent law among biomedical scientists, especially those involved in the discovery and development of pharmaceuticals. Therefore, the primary focus is to explain fundamental patent law concepts and

Disclaimer: Information and opinions presented in this chapter are strictly for educational purposes. It should not be construed as legal advice. Readers should consult a patent practitioner, such as a patent attorney or agent for specific advice on their patent needs.

S. Melethil (✉)
University of Missouri – Kansas City (Schools of Pharmacy and Medicine), Kansas City, MO, USA
e-mail: melethils@umkc.edu

G. Jagadeesh et al. (eds.), *The Quintessence of Basic and Clinical Research and Scientific Publishing*,
https://doi.org/10.1007/978-981-99-1284-1_57

procedures, and their applications using pharmaceutical patent litigation cases. To promote understanding of the subject matter, a comparison is made to the daily activities of research scientists. For example, the procedure to obtain a patent resembles the steps a scientist undertakes to develop a research proposal to a funding agency or to publish a paper in a peer reviewed journal.

An obvious question may relate to the presence and significance of this chapter on law, an apparent stranger in a book, otherwise devoted to science. A thorough understanding of patent law concepts is important to all biomedical scientists for the following three reasons: First, they enable them to better understand and appreciate the commercial potential of their research. Traditionally, scientists, especially in academia, have focused on the originality, quality and impact of their work, key requirements to obtain financial support for their research and ensuing publications in leading journals in their respective fields. Academic recognition also means job security (i.e., tenure, mobility) and financial success. But this culture of science is changing. During the last two to three decades, the commercialization of scientific efforts in academic institutions has become important, as seen by the establishment of technology transfer ("tech transfer") offices in a growing number of universities in the United States (US). The primary function of these offices is to assist in the commercial success of the ongoing research at their respective institutions (e.g., obtaining patents, establishing start-up companies). Such success has recently become part of the tenure dossier of faculty. Budgetary shortfalls from traditional sources are the single most important reason for this change in the academic culture [2]. Second, such knowledge enables scientists to communicate more effectively and work with patent attorneys and agents who traditionally develop strategies to prepare patent applications to obtain a patent from the United States Patent and Trademark Office (USPTO) for their inventions. Such interactions are inevitable, because, in this author's opinion, patent law involves more science than any other area of legal practice. Due to this, a science background is required to become a patent attorney or patent agent. Later in this chapter, a case is presented where a company lost its patent on a multi-billion dollar ("blockbuster") drug. In this author's opinion, this loss was likely due to a lack effective communication between pharmaceutical scientists and patent attorneys at that company (see The Prilosec Case, Part II). Third, a scientist with an understanding of patent law can have additional career advancements and upward mobility in today's marketplace, especially at start-ups, where a scientist often has interdisciplinary responsibilities.

At the outset, a key point must be made: There is considerable tension at the interface of law and science for both scientists and lawyers alike. This cultural clash between the disparate disciplines of law and science can be best stated in the following quote from a publication of the National Academy of Sciences [3]: "Because there is a general lack of understanding of each culture, these interactions often lead to a cognitive friction that is both disturbing and costly to society." Therefore, any effort to introduce patent law to scientists is a formidable task. To make this introductory chapter on patent law more understandable to a community of scientists, analogies to the field of science are often made to better explain patent law concepts and jargon to the readership.

Patent law basics are covered in Part I of this chapter. Discussion of selected cases is included in Part II to illustrate how the patent law concepts presented in Part I are applied in legal disputes (patent litigation). The following references published by the author provide additional information to better understand this chapter [4, 5].

57.2 Part I: Patent Law Fundamentals

57.2.1 What Is Intellectual Property (IP)?

The world intellectual property organization (WIPO) defines IP as follows [6]: It "refers to creations of the mind, such as inventions; literary and artistic works; designs; and symbols, names and images *used in commerce* (emphasis added)."

WIPO goes on to explain the social, legal, and commercial issues related to IP [6]:

> [IP] is *protected in law* by, for example, patents, copyright and trademarks, which enable people to earn recognition or financial benefit from what they invent or create. By striking the right balance between the interests of innovators and the broader public interest, the IP system aims to foster an environment in which creativity and innovation can flourish (emphasis added).

For a quick read on patents and other types of IP, such as copyrights and trademarks, the reader is referred to a publication authored by Miller and Davis [7].

57.2.2 Intellectual Property Explained

A simple example is presented to illustrate the legal meaning of IP. Inventor A constructs a mouse trap. Is that IP? Yes, from the ordinary meaning of the phrase "IP", because it is a product of inventor A's mind (intellect), and s/he owns it. But is it IP according to patent law? For example, can Inventor A make and sell his/her mouse trap on the open market ("inventions ... used in commerce" from WIPO's definition)? It depends [8] on answers to certain key questions: (1) Are there other mouse traps sold on the market? (2) If the answer is Yes, are any of them "patent protected" (more on patent rights later)? and (3) Is A's mouse trap *legally* different (with respect to design, material and technology used, etc.) from all other patented mouse traps on the market? For now, if the answer to question 3 is "yes", then Inventor A can market the mousetrap, preferably with patent protection to prevent others from copying and selling A's mousetrap on the open market. A leading entrepreneur however does not believe in protecting his inventions via patents [9].

57.2.3 Inventive Steps

In patent law, an invention is created by a two-step process, namely conception and reduction to practice.

57.2.3.1 Conception Defined

Conception has been defined as the *complete performance* of the *mental part* of the *inventive act* and it is "the formation in the mind of the inventor of a definite and permanent idea of the *complete* and *operative* invention as it is thereafter to be applied in practice ... [10] (emphasis added).

This is a critical step and *must* be done by the inventor(s).

57.2.3.2 Reduction to Practice Defined

Reduction to practice may be an actual reduction [i.e., making the invention] or a *constructive* reduction which occurs when a patent application on the claimed invention [patent application] is filed [11] (emphasis added).

Unlike conception, reduction to practice may be performed by anybody working under the direct and close supervision of the inventor.

57.2.3.3 Conception and Reduction to Practice Explained

To better understand these two steps, consider the previous example: Inventor A conceives an idea for a mouse trap (along with all the necessary details and drawings). Next, Inventor A has two options: S/he can make it himself/herself or hire somebody else (Contractor B) to build the mouse trap (reduction to practice). In the latter option, inventor A must provide Contractor B with all the *critical* details of making the mousetrap. Contractor B is considered an extension of Inventor A's hands and does not have any intellectual input. Note that Contractor B is *not considered* an inventor.

57.2.4 Constitutional and Policy Bases for Patents Defined [12]

One of the powers (responsibilities) granted to the Congress by the US Constitution (in pertinent part) is "To promote the *Progress of Science* and *useful Arts*" by securing for *limited Times* to ... *Inventors* the *exclusive Right* to their respective ... *Discoveries*" (emphasis added).

57.2.4.1 Constitutional and Policy Bases Explained

The US Constitution [13] provides the legal basis to establish a patent system to promote scientific development and the commercialization of such development for economic success in the US. Based on the authority delegated by Congress, the USPTO has developed a reward system that gives inventors exclusive rights to their inventions for a limited time with the provision that they publicly disclose their inventions. "It is hoped that such disclosure will promote innovations to the patented invention." This is sometimes called the patent bargain: limited-time monopoly in exchange for full disclosure.

57.2.5 Rights of a Patentee (Patent Owner) Defined [14]

A patent, in simple terms, is a *property right* granted by the US government to a patent holder for a limited time.

Ownership of a patent gives the patent holder the right *to exclude* others from:

1. making,
2. using,
3. offering for sale, selling, or
4. importing into the United States
   the invention claimed in the patent.

57.2.5.1 Patentee Rights Explained

In simple terms, others are legally prohibited from the four activities stated above. These rights are "negative "in nature in that they are exclusionary. As noted with the mouse trap example, obtaining a patent on his/her mouse trap does not automatically give Inventor A the right to make and sell it on the open market.

Like all other property owned by inventor A (such as land, house, automobiles, bank accounts, etc.), s/he can sell, bequeath, transfer (assign) or license (allow another party to make and market the mouse trap for a fee) the mousetrap, during the life of the patent. Currently, this limited time (patent term) is 20 years from the effective filing date of the patent application [15]. Given the time needed for patent prosecution (the process used by the USPTO to evaluate a patent application and issue a patent, the effective patent life is less than 20 years. For example, patent prosecution on average takes 3.4 years for a drug and 4.4 years for a biological [16].

57.2.6 Patenting of Inventions (Patent Eligibility)

An invention must meet the following two major criteria to obtain an utility patent. First, it must belong to one of the four legal ("statutory") categories, namely, process, machine, manufacture, or composition of matter (see next section for more details). Second, the invention must be directed at the patent-eligible subject matter. For example, "abstract ideas, laws of nature, and natural phenomena (including products of nature)", referred to as judicial exemptions, are patent ineligible, "because they are the basic tools of scientific and technological work" and "granting them patent rights may impede innovation rather than promote it [17]". Inventions directed at nuclear weapons are also patent ineligible by law [18]: "No patent shall ... be granted for any invention or discovery which is useful *solely* in the utilization of special nuclear material or atomic energy in an atomic weapon (emphasis added)."

The facts of the ultimate patenting of the genetically engineered oil eating bacteria would provide insights into certain legal steps involved in obtaining a patent [19]. The USPTO at first rejected the patent application for this oil eating bacteria stating that it did not fit *into* any of the four statutory categories. After the applicant legally challenged this rejection, the case worked its way to the US Supreme Court, the final arbiter of all legal disputes including patent matters. In a close (5 - 4) decision, it overruled the USPTO and a lower court, with that now famous and often quoted line in patent law circles, “Anything under the sun that is made by man is patentable [20].” More details of this breakthrough case in the biotechnology area and the inventor can be found in this reference [21].

57.2.7 Classification of Utility Patents

Classification [22]: Utility patents are classified into the following categories: “Whoever invents or discovers *any new and useful process*, *machine*, *manufacture*, or *composition of matter*, or *any new and useful improvement thereof*, may obtain a patent therefor, . . . (emphasis added).”

57.2.7.1 Patent Classification Explained

Accordingly, the mouse trap being a machine, is eligible for a utitlity patent.

57.2.8 Types of Patents

Patents are broadly classified based on the subject matter of the invention, namely, utility, plants, and designs.

57.2.8.1 Patent Classification and Types Explained

Utility patents are the focus of this chapter and include the four (process, machine, manufacture and composition of matter) statutory categories mentioned in Sect. 57.2.7. Examples of each of these categories are a machine (e.g., the hypothetical mouse trap discussed), a manufacture (e.g., oil eating bacteria), a composition of matter (e.g., a new drug) and a process (e.g., a method to treat pain, discussed in Sect. 57.2.11.1).

57.2.9 Patent Eligibility Requirements

57.2.9.1 Overview of Patent Eligibility Requirements

In addition to being patent eligible (discussed in Sects. 57.2.6 and 57.2.9), the invention must also be new (referred to as the “novelty” requirement), be useful (referred to as the “utility requirement”) and be non-obvious (“the non-obviousness” requirement. The Specification section of a patent application must also include sufficient details to enable (referred to as the “enablement” requirement) a person having ordinary skill in the art, a PHOSITA, (explained in Sects. 57.2.9.4 and 57.2.9.5) to make and use the invention (“practice the invention”). In the mouse trap example, Inventor A must provide enough details for a PHOSITA to make the mouse trap to meet the enablement requirement

Meeting the non-obviousness requirements is the biggest obstacle to obtaining a patent. As stated, these commonly used words, such as non-obviousness and enablement have distinct legal definitions (legal constructs) and might therefore be confusing to individuals new to patent law. Each of these requirements are discussed next in detail.

57.2.9.2 Novelty/Anticipation Defined

This requirement states (in pertinent part):

> A person shall be entitled to a patent *unless* the claimed invention was patented, described in *a* printed publication, or public use, on sale, or otherwise . . . available to the public before the effective filing date the claimed invention; . . . [23].

57.2.9.3 Novelty/Anticipation Explained

Note that all the exceptions stated recognize the concept of priority, i.e., being the first to invent, which is crucial in obtaining a patent. The USPTO has further clarified the meaning of novelty and anticipation:

A claimed invention [i.e., an application for a patent] may be rejected [by the USPTO] . . . when the invention is *anticipated* (or is "not novel") over *a disclosure* that is available as prior art (emphasis added). An invention (claim) to be rejected on the basis that it was anticipated, requires that ". . . *a* disclosure must *teach* [describe] every element required by the claim . . . [24]. The words, *a disclosure*, in the previous quote (reference 24) refers to a single reference, as was explained in the following court case: "A claim is anticipated only if each and every element as set forth in the claim is either found, either expressly or inherently described in a single prior art reference [25].

57.2.9.4 Non-obviousness Defined

From a conceptual perspective, non-obviousness may be distinguished from novelty (anticipation) in that this requirement takes a broader consideration of prior art (i.e., more than a single prior art disclosure, such as pertinent publications, patents, and disclosures can be combined to reject an invention (claim [26])):

1. A *patent for a claimed invention may not be obtained*,
2. notwithstanding that the claim is not identically disclosed as set forth in the [novelty section],
3. if the differences between the claimed invention and the prior art are such that the claimed invention as a whole,
4. would have been obvious before the effective filing date of the claimed invention
5. to *a person having ordinary skill in the art [PHOSITA]*,
6. to which said claimed invention pertains.

57.2.9.5 Prior Art and PHOSITA Explained

A central question then is: How are novelty and non-obviousness determined in patent law? Two steps are involved in this process. The first step is to conduct a "prior art search" to identify *all* existing information relevant to a given invention. In fact, a patent examiner conducts such a search to evaluate the patentability of the invention described in a patent application. This is like a literature search conducted by scientists prior to starting a research project to determine critical issues relating to the proposed project, such as its originality, significance, experimental design, materials, and methods. In the second step, the determination of non-obviousness and novelty is done from the perspectives of a "person having ordinary skill in the art" (often referred to as a PHOSITA) after having evaluated the information gathered from the prior art search. The USPTO defines a PHOSITA as follows [27]:

The person of ordinary skill in the art is *a hypothetical person* who is presumed to have known the relevant art at the *time of the invention*. Factors that may be considered in determining the level of ordinary skill in the art may include: (a) "type of problems encountered in the art;" (b) "prior art solutions to those problems;" (c) "the rapidity with which innovations are made;" (d) "sophistication of the technology; and" (e) "educational level of active workers in the field. . . . In many cases, a PHOSITA will be able to fit the teachings [information] of multiple patents [and/or publications] together like pieces of a puzzle (emphasis added)."

Understandably, a PHOSITA is someone knowledgeable of the subject matter, like a reviewer of a scientific manuscript, who comments on its publication merits. In the mousetrap example, another mousetrap maker or those with formal training (such as a degree or apprenticeship) in mousetrap making would be a PHOSITA. In the biotechnology area, a PHOSITA is often someone with a Ph.D. degree in a related field, like molecular biology or biochemistry.

57.2.10 Enablement Defined

Enablement [28]: The patent application (specification section) should provide sufficient information for a PHOSITA to be able to make and use ("practice") the invention described in the patent application: "The specification shall contain a *written description* of the invention, and of the manner and process of making and using it, in such full, clear, concise, and exact terms as to *enable* any person skilled in the art to which it pertains, or with which it is most nearly connected, to make and use the same and shall set forth the *best mode* contemplated by the inventor of carrying out his invention. [29].

57.2.10.1 Comments on Enablement

The evolving nature of patent law in the US can be noted from the fact that on November 3, 2022, the US Supreme Court agreed to hear arguments to "fine tune" the proper standard to be used for enablement. (https://www.natlawreview.com/article/supreme-court-to-consider-enablement-requirement visited November 26, 2022).

57.2.11 Overview and Significance of Patent Claims

Claim(s) in a patent must clearly define the invention and are therefore an important, if not the most important, part of a patent application [30]. Further, the "specification shall conclude with one or more claims particularly pointing out and distinctly claiming the subject matter which the inventor . . . regards as the invention [31]. Claims define the boundaries of an invention (claim), formally referred to as its "metes and bounds". This is analogous to a real estate property (like a home) where the owner erects a fence to clearly mark its boundaries. Patent infringement occurs when somebody encroaches on the "metes and bounds" of an invention (patent claim). To continue the analogy, infringement of a patent is conceptually like trespassing on somebody's property.

57.2.11.1 Explanation of Claims

Selected claims from a patent relating to a pharmaceutical product to treat pain are shown next [32]. Consider each claim as a separate invention; note that the claims are related.

What is claimed is [numbers refer to individual claims]:

[Claim] 1. A *method* of effectively treating pain in humans *comprising* orally administering to a human on a once a-day basis an oral sustained release dosage form containing an *opioid analgesic or salt thereof* which upon administration provides a time to reach maximum plasma concentration (*Tmax*) of said opioid in *about 2 to about 10 h* and a maximum plasma concentration (C_{max}) which is *more than twice* the plasma level of said opioid at *about 24 h* after administration of the dosage form, and which dosage form provides effective treatment of pain for about 24 h or more after administration to the patient (emphasis added).

Claim 1 is called an *independent* claim because it is not dependent on any other claim. It is also a "broad" claim in that it does not state the specific opioid and includes all opioids used to treat pain. The patent title describes the invention as a whole: a method to treat pain over a period of 24 h in humans by administering a sustained release dosage form (could be a tablet, capsule, etc.) containing an opioid (specific opioid not mentioned) once a day. The claims also list other *metes and bounds* of the invention:

1. the time of maximum concentration (T_{max}) occurs between 2 and 10 h after administration, and
2. the maximum plasma concentration (C_{max}) observed following oral administration is more than twofold the concentration observed 24 h after dosing.

A competitor who markets a product that overlaps the details in Claim 1 above should expect an infringement lawsuit from the owner of this patent.

[Claim] 2. The method of Claim 1, wherein T_{max} occurs in about 2 to about 8 h after oral administration of said dosage form."

Claim 2 is a dependent claim because it depends on Claim 1. It claims everything in Claim 1, except that the T_{max} time range is narrower, occurring between 2–8 h, versus the 2–10 h in claim 1. It is a "defensive" strategy in claim drafting. If Claim 1 is rejected by the examiner for some reason (e.g., not supported by the information presented in the Specification), the narrower range is more likely to be allowed by the patent examiner, and the inventor obtains a patent on the narrower range T_{max} range. Note that these ranges ("boundaries") are based on a clinical understanding of treating pain with opioids on the part of the inventor.

[Claim] 3. The method of Claim 1 wherein T_{max} occurs in about 6–8 h after oral administration of said dosage form.

[Claim] 3 further shortens the T_{max} interval.

1. The method of Claim 1 wherein that said opioid analgesic is morphine sulfate.

As explained, Claim 1 includes (claims) all opioid analgesics. However, claim 4 is narrower than Claim 1 in that it is directed at (claims) morphine sulfate, and has a greater chance be being allowed by the patent. (Remember, when one files a patent examiner application, the applicant is not sure which claims will be allowed by the patent examiner, like when an author submits a manuscript to a peer-reviewed journal). Therefore, the patent applicant makes all claims vital to the invention supported by the Specification.

57.2.12 Process for Obtaining a Patent (Patent Prosecution)

The first step usually is to file *a non-provisional* patent application with the USPTO disclosing the invention in great detail, as required by the agency. Drafting this application requires an in-depth knowledge of both, the scientific aspects of the invention, and the legal format and submission requirements. Understandably, it is a joint and challenging task for the inventor(s) and patent professionals, such as patent attorneys or agents. A detailed discussion of patent prosecution is beyond the scope of this introductory chapter on patent law. Instead, a short and modified outline of this process focused on scientific details is discussed below to generally introduce the reader to the application process; procedural formalities and legal requirements, such as the filing of specific forms and inventor oath or declaration, have been omitted to avoid confusion for a starting reader. Detailed instructions can be found in standard texts, such as that by Sheldon [33].

The non-provisional application should contain Specification and Drawings (if needed) sections [34]. The Specification section (abbreviated here to avoid confusion) should have the following technical and procedural details in the order shown below [35]:

(1) Title of the invention… (2) Cross-reference to related applications, [This can have implications for the effective filing date of the patent application] …, (7) Background of the invention, (8) Brief summary of the invention, (9) Brief description of the several views of the drawing, [if drawings are included], (10) Detailed description of the inventions, (11) A claim or claims…, and (13) Sequence Listing [for nucleotides and/or amino acid sequences, filed separately from the specifications].

Biomedical scientists would immediately recognize the similarity between the Specifications and Drawings sections of a patent application and a scientific manuscript prepared for publication in a peer-reviewed journal. As in a manuscript, the specification section includes experimental details, such as materials and methods used, results obtained, and conclusions relating to the invention. The Specification section ends with the claims.

57.2.12.1 Patent Application Review

The application is reviewed by USPTO experts (i.e., patent examiners) in the subject matter of the invention. This part is analogous to a peer review of a scientific manuscript or grant application. During the patent prosecution process, the inventor (or his agent, usually the patent attorney or agent) has limited opportunities, including a face to face to interview with the examiner, to respond to ("rebut") a patent examiner's in-writing ("office actions") criticisms ("rejections" and/or "objections") relating to the patent application. The difference between a rejection and an objection is explained below:

"The refusal to grant claims because the subject matter is unpatentable is called a rejection; The term "rejected" is used by a patent examiner when the substance of the patent claims sought are deemed unallowable under U.S.C. 101, 102, 103 and/or 112. If the form of the claim (as distinguished from its substance is improper), an "objection" is made. An example of a matter of form as to which an objection is made is dependency of a claim on a previously rejected claim [36].

All rejections and objections by the examiner must be resolved prior to the granting ("allowance") of a patent. Final rejections by the USPTO can be challenged in the courts, as noted with the oil eating bacteria case [20].

57.2.13 Provisional Patent Application (PPA)

Briefly, a PPA is primarily for placeholder purposes, i.e., to establish a priority date for the invention [37]. It is like a non-provisional patent application, but a major difference is that a PPA application does not need to state any claims. It is important, however, that the Specification should support (provide data) the claims to be listed in a future non-provisional application, which must be filed within 12 months from the filing date of the PPA to benefit from its filing date priority. It automatically expires 12 months from the filing date, is not examined and cannot be extended.

57.3 Part II: Drug Patent Litigation (Applications of Patent Law Fundamentals)

57.3.1 Overview of Patent Litigation

In this part, litigation ("drug war" [38]) between generic and brand name (innovator) companies, where the patent owner (brand name company) lost, is used to provide a glimpse of how patent law concepts described in Part I are applied in the court. These cases mostly involve blockbuster drugs (annual sales of $1 billion or more) because they also generate considerable business and public interest. Brief summaries of each case are presented to explain the pertinent patent law principle involved in each of these cases. Legal citations are provided to allow the reader to pursue an in-depth reading of these cases, though it is not required to understand the information presented in this chapter. The legal basis for these cases is the Hatch-Waxman Act which is briefly described next.

57.3.2 The Hatch-Waxman Act [39]

This Act was passed in 1984 to primarily provide a pathway to market generic versions of patented drugs in the US. It has two major objectives aimed at balancing the interests of the stakeholders, namely, the brand name and generic drug companies, and the consumer (patient): (1) To encourage innovation by providing better patent protection to brand name drug companies, i.e., the development of new drugs, and (2) to foster competition in the pharmaceutical industry by providing a legal and regulatory pathway to market (hopefully, less expensive) generic versions of brand name drugs.

57.3.2.1 Hatch-Waxman Procedural Details

Under this Act, a new drug application (NDA) submitted to the FDA by an innovator (brand name) drug company for approval of its new drug must also include patent numbers and expiration dates of all patents that claims, either the drug (active ingredient and/or composition or formulation) or the method of use (i.e., indication). The FDA is *required* to list these patents in the FDA's "Orange" book, a commonly used abbreviation for its lengthy formal title, Approved Drug Products with Therapeutic Equivalence Evaluations. The Orange book can be easily found online at the FDA website (www. FDA. gov). This information serves as a public notification of any patent protection afforded a drug in the US.

When an amended new drug application (ANDA) is submitted to the FDA by a generic company for approval of a generic version of a patented drug, it must certify to one of the following:

1. the drug has not been patented,
2. patent on the drug has expired,
3. the generic version of the drug will not be marketed prior to the expiration of the patent(s) on the drug, or
4. the generic version of the drug will not infringe the patent(s) covering the drug or the patent(s) are invalid (i.e., patents listed in the Orange Book) This is commonly called a Paragraph IV certification based on the Roman numeral nomenclature used in the Hatch-Waxman Act.

The generic company must also notify the patent holder about its ANDA and explain why the *generic version will not infringe the patent (s) listed in the Orange Book* or why *these patent (s) is/are invalid.* This sets the stage for litigation.

FDA's role in patent matters with respect NDA and ANDA applications is ministerial, i.e., the agency must list the patents included in an NDA application. In addition, FDA approvals of NDA and ANDA applications are independent of patent issues. Patent issues between brand name and generic companies are settled in Federal courts.

57.3.3 Patent Litigation: Specific Cases

57.3.3.1 The Prilosec Case [40]: A Generic Company "designed around" a Dosage Form to Avoid Patent Infringement

Legal Background: For Kremers Urban Development Co. (KUDCo) to infringe Claim 1 of the '505 patent, its product must have all the components ("elements') cited in Claim 1 of the '505 patent listed below.

Case Details: Omeprazole is the active ingredient of the proprietary drug Prilosec® marketed by Astra Aktiebolag (Astra) and patent protected by US Patent No. 4,786,505 (the '505 patent) and US Patent No. 4853, 230 (the '230 patent). At the time of litigation (decided in 2002), it had an annual worldwide sale of $6 billion; US sales accounted for $4 billion. KUDCo., a small generic company, submitted its ANDA application to the FDA for approval of generic omeprazole with a Paragraph IV certification that its product would not infringe Astra's patents. Astra disagreed and filed a patent infringement lawsuit.

Claim 1 of the '505 patent became the deciding issue for Astra's infringement allegation, which reads as follows (in pertinent part):

An oral pharmaceutical preparation *comprising*, a core region *comprising* an effective amount of a material selected from the group consisting of omeprazole plus an *alkaline reacting compound*, an alkaline omeprazole salt plus an alkaline reacting compound and an omeprazole salt alone; (emphasis added).

As a procedural matter in general, Paragraph IV certification by a generic company and the counter arguments by the patent holder are decided in court by an evaluation of the pertinent patent claims.

The KUDCo. microtablet has three components: a core, a sub-coat and an enteric coat. It was found that the sub-coat and the enteric coat of this microtablet did not differ from that claimed in the '505 patent. But it does not have an alkaline reacting compound in its core like Astra's tablet. Therefore, the two tablets are different. Thus, the court ruled that KUDCo did not infringe Astra's patents and it could legally market generic versions of omeprazole, a big win for this small generic company at that time.

For those readers who are drug formulators, omeprazole is acid labile, and the addition of the alkaline reacting compound was likely to protect it during the tablet manufacturing process. KUDCo using newer technology designed around Prilosec®. The '505 patent was issued in 1988, more than a decade before this litigation. Though speculative, if the pharmaceutical scientists and patent attorneys at Astra had worked closely, they might have better seen this major weakness of the '505 patent and might have reformulated their tablet without the alkaline reacting agent. Readers may recall from Part I that excluding others is the major right of a patentee.

For completeness's sake, three other generic companies, namely Andrx Pharmaceuticals, Cheminor Drugs and Genpharm, Inc. also tried to market their generic versions of Prilosec®. They failed to do so because their tablets were found to infringe Astra's patents.

57.3.3.2 The Prozac® Case (Double Patenting Invalidity) [41]

Legal Background: In lay terms, the law of double patenting prohibits issuing two patents for the same invention. Even common sense would support this prohibition because it would unfairly extend the life of a patent on a given invention. Legally, this is stated as:

> "[T]he extension of exclusive rights [patent protection] through claims in a later patent that are *not patently distinct* from claims in an earlier patent" (41). The *italicized* segment relates to the criteria for novelty and non-obviousness requirements discussed earlier.

Case Details: Fluoxetine is the active ingredient of Prozac®, Eli Lilly's proprietary blockbuster drug. It is used to treat depression and anxiety. Barr Laboratories submitted an ANDA in December 1995 for generic fluoxetine with Paragraph IV certification challenging the validity of Lilly's patents. In response, Lilly brought legal action alleging that Barr's ANDA application infringed its patents. Eli Lilly had two patents to protect Prozac®; US Patent No. 4,626,549 (the '549 patent, which issued on December 12, 1986) and the US Patent No 4, 590, 213 (the'213 patent, which issued on May 20, 1986). The court compared the following two critical claims listed below to determine if they were "patentably distinct":

> "A *method of blocking the uptake of serotonin by brain neurons in animals* comprising the administering to said animal of fluoxetine [claim 7, the '549 patent, the later patent'']," and

> "A *method for treating anxiety in a human subject in need of such treatment* which comprises the administration to such human an effective amount of fluoxetine or norfluoxetine or pharmaceutically acceptable salts thereof [claim 1, the '213 patent, the earlier patent"]

These two claims are the same because the mechanism of action of fluoxetine is by blocking serotonin uptake in the brain [41] They are, therefore, not patentably distinct. Barr Laboratories won the case, giving it legal authority to market its generic version of Prozac®.

57.3.3.3 The Prometheus Case (Patentable Subject Matter, a "101" Issue) [42]

Legal Background: "[L]aws of nature, natural phenomena, and abstract ideas are not patentable" (e.g., $E = mC^2$) (42).

Case Details: Mayo Clinic (herein after Mayo) used diagnostic tests sold by Prometheus Laboratories (hereinafter Prometheus) based on the latter's two patents: U.S. No. 6,355,623 (the '623 patent), and 6,680,302 (the '302 patent) Mayo stated in 2004 that it planned to market its own version of a similar diagnostic test. Prometheus filed an infringement suit against Mayo.

Claim 1 of the '623 patent (vital to the case) states:

> A *method of optimizing therapeutic efficacy* for treatment of an immune-mediated gastrointestinal disorder, comprising: (a) administering a drug providing 6-thioguanine (6-TG) to a subject having said immune-mediated gastrointestinal disorder; and (b) determining the level of 6-thioguanine in said subject having said immune-mediated gastrointestinal disorder, wherein the level of 6-thioguanine less than about 230 pmol per 8 × 10 red blood cells indicates a need to increase the amount of said drug subsequently administered to said subject and wherein the level of 6-thioguanine greater than about 400 pmol

per 8 × 10 red blood cells indicates a need to decrease the amount of said drug subsequently administered to said subject (emphasis added).

In pertinent part, the trial court concluded that claim 1 of the '623 patent, which identifies the therapeutic range for 6-TG, covered natural laws, and thus, the subject matter of the '623 patent was not patentable (invalid), and ruled in favor of Mayo. However, the appeals court reversed the trial on the issue of patentability and ruled that the '623 patent is valid, but Mayo's method infringed the patented method claim of Prometheus. On appeal by Mayo, the Supreme Court agreed to listen to arguments on the patentability of Claim 1 of the '623 patent. This Court, concurred with the lower (trial) court, and ruled the Prometheus patent invalid, stating [42]:

Anyone who wants to make use of these laws must first administer a thiopurine drug and measure the resulting metabolite concentrations, and so the combination amounts to nothing significantly more than an instruction to doctors to apply the applicable laws when treating their patients.

In other words, the stated invention (claim) is a fundamental law and, therefore, cannot be patented (a "101" issue). This decision shook up the diagnostic industry, with one author saying, "The new patent eligibility analysis provided in *Mayo* has narrowed the breadth of patent eligibility for diagnostic methods] [43].

57.3.3.4 The Bayer Case (Obvious to Try) [44]

Legal Background: When an invention is a result of solving a problem using methods that would have been obvious to a PHOSITA based on information in the prior art ("teachings"), then that invention fails to overcome the non-obviousness barrier and becomes patent ineligible. The Supreme Court had laid down the following standard for such situations in the KSR case [45]:

"When there is a design need or *market pressure* to solve a problem, and there are a finite number of identified, predictable solutions, a person of ordinary skill has a good reason to pursue the known options within his or her technical grasp. If this leads to the anticipated success, it is the product not of innovation but of ordinary skill and common sense. In that instance, the fact that a combination was obvious to try might show that it was obvious under §103 (emphasis added)."

Often patent decisions made in cases involving one technical/scientific discipline (subject matter) are applied to cases involving a different subject matter. Patent(s) in the KSR case dealt with automobiles (specifically, accelerator pedals) and this standard was applied to the Bayer drug case.

Case History: Bayer markets a patent protected drug Yasmin® [46], a female contraception drug. Barr, planning to market a generic version of this drug, filed its ANDA application along with a Paragraph IV certification that the claim central to the patent directed at Yasmin® is invalid. Bayer files suit alleging infringement. The outcome of the case depended on the validity of the Claim 1 (the invention in the '531 patent), which reads:

"A pharmaceutical composition comprising from about 2 mg to 4 mg of *micronized* drospirenone particles, about 0.01 mg to about 0.05 of 17α-ethynlestradiol, and one or more pharmaceutically acceptable carriers, the composition being in an oral dosage form exposed to the gastric environment upon dissolution and the composition being effective for oral contraception in a human female [marketed as Yasmin®] (emphasis added.)"

Bayer had solved a particular drug formulation problem relating to the drug drospirenone by using methods suggested in the prior art. Specifically, since the drug is poorly water soluble, Bayer scientists used micronized particles of the drug particles (see Claim 1 of the '531 patent) to improve the dissolution properties of Yasmin® (and hence its bioavailability or gastrointestinal absorption). The patent examiner, as might be expected, rejected this claim as obvious during patent prosecution based on prior art. In response to the rejection, Bayer countered by citing another prior art publication that did not support ("teaches away" from) micronization, because the increased surface area resulting from micronization, could promote greater destruction. The examiner then allowed the claim and allowed the patent to issue. After a detailed examination of the science involved, the court sided with Barr, invalidating the '531 patent. The legal victory allowed Barr to market its generic version of Yasmin®.

57.4 Concluding Remarks

Legal complexities and nuances have been omitted in this chapter to promote an understanding of patent law fundamentals among the anticipated readership consisting of biomedical scientists. It is hoped that such understanding would encourage them to include the patentability of their research as one of the indices of innovative research. It is also hoped that the information provided here will promote more effective communication between biomedical scientists and patent attorneys and agents.

References

1. Everyday words such as comprising, consisting, novelty, anticipation, and obviousness are legal constructs in patent law
2. Personal Observations
3. Anon (2001) A convergence of science and law. National Academy Press, Washington, D.C.
4. Patent Law Basics For Scientists (So, you want to make money from your inventions?). https://youtu.be/LbdmqjjVU_0
5. Melethil S (2005) Patent issues in drug development: perspectives of a pharmaceutical scientist attorney. AAPS J 7(3):E723–E727
6. What is intellectual property? https://www.wipo.int/about-ip/en/
7. Miller AR, Davis MH (1990) Intellectual property: patents, trademarks and copyright, 2nd edn. West Publishing Co., St. Paul, MN
8. Legal humor: all answers to legal questions should begin with the phrase, *it depends*
9. Elon Musk, CEO of SpaceX does not believe in obtaining patents related to materials used to build the company's spaceships. https://www.cnbc.com/2022/09/21/why-elon-musk-says-patents-are-for-the-weak.html. Accessed 26 Nov 2022
10. Conception, Manual of Patent Examination Procedure (MPEP) 2138.04
11. Reduction to practice MPEP 2138.05
12. United States Constitution, Article I, §8, cl 8
13. The United States Constitution was ratified in 1788 and has been in force since 1789
14. 35 United States Code (U.S.C). 261 Ownership; assignment
15. The effective filing date may be earlier than the actual filing date if priority to a previous application is invoked
16. Burger, J, Dunn, JD, Johnson, MM, Karst, KR, Shear, WC. How drug life-cycle management patent strategies may impact formulary management. https://www.ajmc.com/view/a636-article. Accessed 12 Nov 2022
17. Patent subject matter eligibility MPEP 2106
18. Atomic Energy Act of 1954. Section 151(a) (42 U.S.C. 2181(a)
19. Microorganisms having multiple compatible degradative energy-generation plasmids and preparation thereof, US Patent No. 4,259,444. Inventors: Ananda M. Chakrabarthy
20. Diamond v. Chakrabarty, 447 U.S. 303 (1980)
21. Davey N, Rade RR (2021) Chakravarti, Anand Mohan 'Al' Chabrabarty 1938-2020. Nat Biotechnol 39:18
22. 35 U.S.C. § 101 – Inventions patentable
23. 35 U.S.C. § 102 - Conditions of patentability; novelty (paragraph (a)(1))
24. Anticipation – Application of 35 U.S.C. 102, MPEP 2131
25. Verdegaal Bros. v. Union Oil Co. of California 814 F. at 628, 631
26. 35 U.S.C. § 103 - Conditions of patentability; non-obviousness subject matter
27. Level of Ordinary Skill in the Art, MPEP 2141.03
28. 35 U.S.C 112 (a) Specification
29. This requirement is seldom a cause for rejection
30. "To coin a phrase, the name of the game is the claim." Giles Rich, then Chief Judge of the Federal Circuit (CAFC), The Extent of the Protection and Interpretation of Claims, American Perspectives, 21 Int'l Rev. Indus. Prop. & Copyright L., 497, 499 (1990)
31. 35 U.S.C 112 (b) [A] patent application specification shall conclude with one of more claims particularly pointing out and distinctly claiming the subject matter which the inventor . . . regards as the invention
32. Method for treating pain by administering 24-hour oral opioid formulations. US Patent No. 5, 672, 360, issued 9/30/1997. Inventors: Richard S. Sackler, Robert F Kalco and Paul Goldenhelm; Assignee: Purdue Pharma L.P
33. Sheldon JG (2022) How to write a patent application, 3rd edn. Patent Law Institute, New York
34. 37 Code of Federal Regulations (C.F.R.) 1.77 (a) Arrangement of application elements
35. 37 C.F.R.) 1.77(b) Arrangement of application elements

36. Quinn G. Understanding the patent process: rejection vs. objections. https://ipwatchdog.com/2016/04/02/patent-process-rejections-vs-objections/id=67761/. Accessed 3 Dec 2023
37. Provisional application for patent. https://www.uspto.gov/patents/basics/types-patent-applications/provisional-application-patent. Accessed 18 Nov 2022
38. McLean B. A Bitter Pill Prozac made Eli Lilly. Then along came a feisty generic maker called Barr Labs. Their battle gives new meaning to the term 'drug war'. https://money.cnn.com/magazines/fortune/fortune_archive/2001/08/13/308077/index.htm. Accessed 26 Nov 2022
39. Drug Price Competition and Patent Term Restoration Act of 1984, Public Law 98–417
40. Astra Aktiebolag v. Kremers Urban Development Co. 222 F. Supp2d 423 (S.D.N.Y. 2002) (defendants that lost are not included in the citation)
41. Eli Lilly & Co. v. Barr Laboratories, Inc., 222 F.3d 973 (2000) (US Court of Appeals, Federal Circuit)
42. Mayo Collaborative Servs. v. Prometheus Labs., Inc. 566 U.S. 66 (2012) (US Supreme Court case)
43. Dorn BR (2013) *Mayo v. Prometheus*: a year later. ACS Med Chem Lett 4(7):572–573
44. Bayer Pharma v. Barr Laboratories, (575 F3d. 1341) (2009) For a less legalistic and brief summary of the case, visit https://www.courtlistener.com/opinion/208257/bayer-schering-pharma-ag-v-barr-laboratories/. Accessed 5 Dec 2022
45. KSR Int'l Co. v. Teleflex Inc. 550 U.S. 398, 421 (2007) (US Supreme Court case)
46. Pharmaceutical composition for use as a contraceptive. US Patent No. 6,787, 531 (the '531 patent)

Part XII

Research Integrity/Ethics

Research Integrity: Responsible Conduct of Research 58

Amruthesh C. Shivachar

Abstract

The legitimacy of research in any field is boosted by three crucial elements: research ethics, responsible behavior and integrity. These are the very aspects that make up the three faces of a prism lens through which a researcher investigates analyze/solve any research problems. None of these three faces is separable, each one of them is dependent on one another. In general, it is historically assumed in human behavior that "good ethical" thinking will lead to the "responsible conduct or behavior" that will enhance the "integrity" of a person's character. This notion is intricately applicable to any research endeavor: good research ethics will reflect in the responsible conduct of research (RCR), which in turn establishes research integrity (RI). Therefore, the purpose of this chapter is to review the influence of research ethics and the RCR on RI. The author also thinks this overview will provide a basic knowledge on the global efforts for the development and establishment of universal codes or unified *norms* for research ethics, and RCR for RI worldwide. By emphasizing the efforts on RCR in different regions and countries, the chapter will guide a beginning researcher such as a graduate student, and an established researcher from deviating from any ethical boundaries.

Keywords

Bioethics · Code of conduct · History of research ethics · Global ethics · Responsible conduct of research · Research misconduct

58.1 Introduction

Historically, morality, ethics and integrity are regarded as the three characteristics of an individual in the society. While morality and ethics are often used intermittently, the definitions of morality and ethics differ significantly [1]. Morals constitute the personal beliefs of an individual while ethics refer to a set of principles or standards or rules for the professional conduct. Morals are often based on culture and religion while ethics are based on logic and reason. Nonetheless, morals and ethics are used to distinguish "right from wrong" conduct of an individual, and a group or organization, respectively.

There are well-established ethical rules for research involving animals and human subjects, children, embryonic stem cells, and for the protection of data and privacy matters [2]. However,

A. C. Shivachar (✉)
Department of Pharmaceutical Sciences, College of Pharmacy and Health Sciences, Texas Southern University, Houston, TX, USA
e-mail: Amruthesh.shivachar@tsu.edu

G. Jagadeesh et al. (eds.), *The Quintessence of Basic and Clinical Research and Scientific Publishing*,
https://doi.org/10.1007/978-981-99-1284-1_58

research integrity (RI) mainly deals with *reliability, honesty, respect* and *accountability* in research [3]. The *reliability* of the research is dependent on the use of sound methodology, experimental design, data analysis and the resources [4]. The *honesty* deals with reviewing and reporting research data in an unbiased and transparent way. The *respect* involves acknowledging the abilities, qualities, and achievements of colleagues, motivation and hopes of subjects and the needs of society. Respect also involves appreciation of cultures and the environment encompassing such cultures. Last but not the least is *accountability,* which includes every level in research from formulating the idea, conducting research, data analyses, supervision, mentoring and publishing [5]. Both ethical rules and RI are designed to provide researchers with a set of guidelines for RCR. However, for some research communities, research ethics are considered one aspect of integrity and for others, RI is an ethical part of the research [6]. This is because the definitions of research ethics and RI are overlapping and often confusing. So, it is imperative to understand what constitutes research ethics and research integrity. As we go into details, the definitions for research ethics and research integrity are dependent on each other, and both constitute the fundamental *norms* for the RCR. However, the perceptions on accomplishing RI vary among different cultures and ethnicity in the entire world. These variations need to be normalized, requiring the development for International guidelines for RI across the globe. The key objective (s) of this chapter will be to highlight the regional and cultural variations on RCR across the globe. For the sake of brevity and convenience we will review the perceptions in Europe, Asia and the United States of America as representative countries/regions and compare them with the ongoing global efforts.

58.2 Perceptions on RI in Europe

Historically, European policymakers have attempted to unify and harmonize ethical standards across Europe. These standards have been highlighted in the European Code of Conduct for RI. The European Federation of Academies of Sciences and Humanities constituted "All Europe Academies (ALLEA), which includes more than 50 Academies from countries from all over the Europe [7]. ALLEA consists of multilateral stakeholders' organizations that are permanent member Academies such as European Science Foundation (ESF), and many others (please see ALLEA updated report 2017 for a complete list of permanent member Academies) [7–9].

The member Academies of ALLEA consist of research performing organizations that operate as learned societies and think tanks. These organizations are self-governing communities of natural, and social sciences and the humanities. The overall purpose of ALLEA is to provide resource of intellectual excellence and expertise to improve the conditions of science and scholarship across Europe.

In a survey conducted by the ESF they found that there are a wide range of RI standards among European countries that need to be harmonized [7]. These survey results are the basis for ALLEA to release its first report on RI Standards in 2011 as "The European Code of Conduct for Research Integrity" [8]. In this document ALLEA lists "Reliability, Honesty, Respect and Accountability" as four principles of Research Integrity and considered this document to be a "living document" and to be reviewed every 3–5 years and revised, if necessary. Indeed in 2017 ALLEA released a revised version of the "European Code of Conduct" after participating in a stakeholder meeting of its permanent member organizations held in Brussels in November 2016 [10]. In the revised version, in addition to the four standard principles of RI, ALLEA added several aspects such as Research Environment, Training, Supervision and Mentoring, Research Procedures, Safeguards, Data Practices and Management, Collaborative Working, Publication and Dissemination and Reviewing, Evaluating and Editing. These guidelines appear to be parallel to those developed in 2016 by the International Council for harmonization (ICH) of technical requirements for pharmaceuticals for human use (ICH-E6) across the globe [11].

In the revised report in 2017 ALLEA focused on the definition of fabrication, falsification and plagiarism (the so-called FFP categorization) as the three traditional research misconducts and several other aspects of research misconduct as well [10]. ALLEA kept the original definitions of Fabrication for making up results, *Falsification* for manipulating research data or materials, and *Plagiarism* for taking someone else's ideas without giving due credit to the original creator. In its revised report, ALLEA added several "unacceptable research practices" into the list including, manipulating authorship, republishing, or withholding research results, falsely accusing other researcher of misconduct and exaggerating the importance of findings (for a complete list please see the revised report 2017) [10]. In this report ALLEA's inclusion of "establishing or supporting journals ('predatory journals') that undermine the quality control of research" as one of the new unacceptable practices is worth mentioning, given the escalation of a number of predatory journals in the recent years from few European and Asian countries. The revised report also includes a section on guidelines to deal with research misconduct with integrity and fairness as well [10].

Although ALLEA and other European agencies have tried to come up with a harmonized Codes for ethical conduct of research that can be used across Europe, it is that almost every European country follows code of conduct for RCR and RI [12]. However, by developing their own standards, these countries may be taking a risk with collaborating institutions abroad [13, 14]. As Bendiscioli and Garfinkel in their EMBO report put it "there are general trends of common principles of integrity and definitions of misconduct, yet differences remain [13]". Overall there are different approaches taken by different countries in Europe for implementing RCR and research misconduct: few countries emphasizing on broad values of RI while others focus on negative behaviors [15–17].

58.3 Perceptions on RI in Asia

In Asia, it is entirely a different story. Unlike in Europe, my search did not yield any published documentation for the establishment of a universal and harmonized Codes for ethical conduct of research across Asia. The lack of such a harmonized documentation may be due to various barriers, including language, culture and societal beliefs. Nonetheless, it appears that each Asian country may have its own code of conduct developed within the individual Institutions or organizations. There is a vast divergence in ethical standards established by each country in Europe. In the interest of space, I will mainly focus on the development of Codes for Research integrity in two major Asian countries, China and India, where the research has been progressing in a very rapid rate in the past decade.

58.3.1 Perception on RI in China

China is one of the oldest civilizations not only in Asia but also in the world with a history of scientific and technological inventions. However, the inventions were hindered by the cultural revolution in the 1960–1970s. Chinese science progression resumed slowly after the cultural revolution. Today, China is placed second in scientific publications and citations, and leads the world mainly in nanotechnology and several other scientific discoveries [18]. This progression in science and technology also lead to several social and ethical issues in China. To address these challenges, Chinese Society of Medical Ethics (CSME) has been established in late 1980s. Since then China has established various medical ethics committees at hospitals and universities to review research with human subjects. In the late 1990s China established the Chinese Association for Science and Technology (CAST) to develop code of conducts to combat the growing issues of plagiarism and other types of misconducts among research community. In the meantime, Chinese government also started its own initiatives to address the widespread scientific misconduct that lead to the establishment of National "Natural Science Foundation of China" (NSFC), which is under the control of State Council of China [18]. Additionally, many different government and non-government organizations, including the Ministry of Science and Technology (MOST), and the Ministry of Education (MOE), the Chinese Academy of Engineering (CAE), and the Chinese Academy of Sciences (CAS) were established to address the ongoing ethical issues in the country from 2002 to 2006 [19, 20]. In 2007, MOST adopted misconduct rules for investigating research misconduct as well as a definition of misconduct in government-funded research [21–23]. According to MOST, the definition of research misconduct was beyond FFP and included submitting false resumes and violating rules pertaining to research with human or animal subjects.

58.3.2 Perceptions on RI in India

In India on the other hand, the perception on research ethics is complex. The various sorts of misconduct, discriminatory practices and colonial attitudes in hierarchical institutions can be chronologically categorized as: pre-independence era, post-independence era, crisis of 1970s and the present crisis on predatory publications [24].

Under British rule during the pre-independence era, there has been an importation of European social, cultural and authoritarian principles to Indian scientific communities. Along with these the Indian scientific communities also experienced an inter-regional contentions and politicization. There was an influx of scientists from Europe to India to pursue their research using the local resources including native Indian scientist workers. The European scientists often failed to acknowledge the contribution of Indian scientists in their scientific discoveries. Eventually when the British government established its colonial rule in 1858, also got the control over Indian education and scientific endeavor. This created a system of distrust between the management and Indian scientific community causing facilitated misconduct both at the management and scientific community. Despite these rifts a great deal of contribution by the Indian scientific community to global scientific discovery occurred between 1900–1940s.

In 1947 when India was granted independence, Indian scientific community was suffering from social, gender, religious and regional divisions [25, 26]. These events greatly influenced Indian scientific community in post-independence era, while government priority was focused on mainly nation-building. By late 1960s, there was a growing issue on research misconduct and discrimination among scientists in the hierarchal Indian institutions. Although Indian science in the 1970s was deeply troubled in political turmoil, the country's political leadership had accepted science as a central feature of their vision. In 1980s increase in unscientific behavior and practices raised concerns among the Indian scientific community and lead to the establishment of the "Society for Scientific Values" (SSV) [27]. The SSV was charged to investigate cases of scientific misconduct brought to its notice by whistle-blowers and to publicly punish those found guilty. However, SSV has no formal regulatory power and its investigations have often been neglected by the scientific community. Nonetheless, the SSV has played a key role involving scientific misconduct, publication and science management misconduct.

Besides SSV, there are other agencies in India such as Center for Scientific and Industrial Research (CSIR), Indian Council for Medical Research (ICMR), and University Grants Commission (UGC) [28–30]. These agencies have put out their own guidelines for publication of scientific/technical/biomedical data and results, making them available in the public domain and in the administration of scientific establishments at all levels. Beyond academic and publication guidelines, these agencies also emphasized on honesty and scientific validity of the work being published. These agencies also provided guidelines to pursue new ideas to credit original inventors, mutual respect, conflict of interest, education and mentorship, and social responsibility [30, 31].

Despite the establishments of all these guidelines by various agencies, the current crisis in India seem to be "pop-up" of "predatory journals [24]." It appears that more than two-thirds of such publications coming from India are mainly from institutions that lack financial resources. Furthermore, the pressure from the management to speed up the production of research output is playing a role in this behavior among scientific community. It is alarming that the publication in predatory journals is spreading to scientific communities in many other developing countries.

To overcome these short-comings, India must establish strict policies and regulations and ensure prompt implementations to avoid further damage to RCR and RI. It is therefore imperative to introduce ethical policies in the educational system very early in any field of research (for details on this subject, see Chap. 59).

58.4 Perceptions on RI in America

Unlike Asian and European countries, the research ethics in the United States of America are being consistently developed from 1945 to present. For convenience, this section is divided into two subsections: Sect. 58.4.1 will focus on human subject research and Sect. 58.4.2 will focus on laboratory research practices or ethics in bench work.

58.4.1 Human Subject Research

The Nuremberg Code was developed in 1948 after an American Tribunal discovered that German officials conducted medical experiments on thousands of concentration camp prisoners without their consent, causing numerous prisoners' death and many were permanently disabled. According to the Nuremberg Code "the voluntary consent of the human subject is absolutely essential," and the "benefits of research must overweigh the risks."

Another milestone in US history occurred when the US Public Health Services (PHS) conducted a study on Syphilis in Tuskegee from 1932 to 1972. In this study, a group of low-income African-American males, 57% of whom were infected with syphilis, were monitored for 40 years. Free medical examinations were given to them; however, subjects were not told about their disease. The participants were denied treatment even though a proven cure (penicillin) became available in the 1950s and the study continued until 1972. Many subjects died of syphilis during the study and few others permanently disabled. The study was forced to stop in 1973 by the US Public Health Services. Because of the bad publicity from the Tuskegee Syphilis Study, the US Congress passed the National Research Act in 1974 and created the National Commission for the Protection of Human Subjects of Biomedical and Behavioral Research. The commission was charged to identify the basic ethical principles for the conduct of biomedical and behavioral research involving human subjects.

Although Nuremberg Code established the basic rules to help protect human subjects involved in research, but these rules were found to be inadequate to cover complex situations. To avoid the limitations of Nuremberg codes, the Belmont Report was introduced in 1976 by the National Commission for the Protection of Human Subjects of Biomedical and Behavioral Research [32]. The Belmont report describes basic ethical principles for conducting research that involve human subjects. It also sets forth guidelines to assure these principles are followed throughout the research process. The Belmont Report consists of three ethical principles: (a) Respect for persons; (b) Beneficence and (c) Justice. The first ethical principle in the Belmont Report talks about *respect for persons* and consists of two parts: the first is the recognition that people are autonomous and entitled to their own opinions and choices, unless detrimental to others. The second part is the recognition that due to various reasons, not all people are capable of self-determination and instead require protection. Thus, the Report promotes the idea that in most cases, individual subjects enter research voluntarily and with adequate information. The Report's second ethical principle is *beneficence*, which means that people are treated in an ethical manner not only by respecting their decisions and protecting them from harm. This principle identifies two general and complementary rules: (a) Do not harm; and (b) Maximize possible benefits and minimize possible harms. It is imperative to achieve both in any study involving human subjects.

According to this principle it is the obligation of scientific investigators and members of their institutions to think about both maximizing benefits and reducing risks in their research. Also, it is the obligation of the society to recognize longer term benefits and risks that may result from the improvement of knowledge and development of novel medical, psychotherapeutic, and social procedures, including cancer treatments.

The last part of the Belmont Report is *justice,* which talks about who should receive the benefits of research and who should bear its burdens [32]. The Belmont Report considers the need to scrutinize people's economic status, race, ethnicity etc., as factors before selecting for any study. Thus, climinating thc possibility of bccoming research subjects due to their socioeconomic position rather than their connection to the problem. The *justice* part of Belmont report demands therapeutic devices and procedures developed from public funds must not provide advantages only to those who can afford them.

Belmont report is the basis for the Department of Health and Human Services (DHHS) and the Food and Drug Administration (FDA) to issue regulations on the use of human subjects in research in 1981. The DHHS issued the Code of Federal Regulations (CFR) Title 45 (public welfare), and Part 46 (protection of human subjects). The FDA also issued a similar Code of Federal Regulations with slight modification to fit into its mission {(CFR Title 21 (food and drugs), Parts 50 (protection of human subjects), and 56 (Institutional Review Boards)}. In 1991, the DHHS regulations (45 CFR Part 46, Subpart A) were formally adopted by many other departments and agencies that conduct or fund research involving human subjects. This Federal Policy for the Protection of Human Subjects is often termed as "Common Rule" or the "Federal Policy".

The main elements of this federal policy include: (a) requirements for assuring compliance by research institutions; (b) requirements for researchers obtaining and conducting informed consent; (c) requirements for Institutional review board (IRB) membership, function, operations, review of research and record keeping; and (d) additional protections for certain vulnerable research subjects including pregnant women, prisoners, and children.

Additionally, certain federally and privately sponsored research is subject to the regulations of the FDA at 21 CFR Parts 50 and 56. FDA regulations provide protection on human subjects when a drug, device, biologic, food additive, color additive, electronic product, or other test article subject to FDA regulation is involved. FDA regulations and the provisions of the "Common Rule" are largely overlapping, although some significant differences exist, despite both being aimed to provide protections to human subjects (also, see Chap. 23).

58.4.2 Laboratory Research Practices

Bad research practices became a public issue in the early 1980s in the United States. Research misconduct cases in several institutions prompted Congress to pass the Health Research Extension Act in 1985. This Act added Sect. 493 to the PHS Act, requiring applicant institutions to establish an administrative process to review reports of scientific fraud. In 1989, the PHS office created the Office of Scientific Integrity (OSI) in the office of the Director of National Institutes of Health (NIH), Office of Scientific Review (OSIR), and Office of the Assistant Secretary of Health (OASH) to deal with research misconduct. In 1992 all these three offices were consolidated into the Office of Research Integrity (ORI) which is headed by the Assistant Secretary of Health [33–37]. The ORI underwent several structural reforms and became the sole organization within NIH to prevent research misconduct and promote RI through oversight, education, and review of institutional findings. In the early 2000s, the ORI began its research on RI to develop a research community focused on the RCR. In 2004, ORI published *"The Introduction to RCR"* and introduced RCR program in graduate and, post-doctoral training and then expanded to other scientific community within the NIH [35–37]. The ORI then collaborated with major research-intensive Universities to expand the RCR awareness among the next generation scientific community.

While the NIH was using the above mentioned "top-down" approach via ORI, other entities, such as National Science Foundation (NSF) and National Academy of Science in the research community simultaneously made efforts to fight research misconduct and to foster RI within scientific community. Among them the National Academy of Science, Engineering and the Institute of Medicine created the Committee on Science, Engineering, and Public Policy (COSEPUP) [38–40]. The panel completed its report and released it as *Responsible Science: Ensuring the Integrity of the Research Process. The panel recommended* steps for reinforcing responsible research practices (NASNAE-IOM, 1992). The report distinguishes three categories of behaviors that can compromise the integrity of the research process. *Responsible Science* formally places the primary responsibility for strengthening the RCR on individual researchers and research institutions. *The panel* also believes that the integrity of research depends on creating and maintaining a system and environment for research in which institutional educational programs, and incentive structures support responsible conduct of an individual researcher. The report defined misconduct as "fabrication, falsification, or plagiarism in proposing, performing, or reporting research. The committee added another category "Questionable research practices" to include "actions that violate traditional values of the research enterprise" and considered counterproductive to the research process. These definitions of COSEPUP were adopted by a unified Federal policy in 2000.

Several years later, the COSEPUP was commissioned to develop new standards while reaffirming the recommendation from Responsible Research Report. In its second report published in 2017 as a book entitled *"Fostering Integrity in Research,"* the panel made eleven recommendations with a special effort on better refinements and harmonization of the definitions of research misconduct and its use [41]. The major change was that the panel replaced "questionable Research practices" with "detrimental Research practice," to include abusive actions taken by research institutions and journals in addition to the actions of individual researchers.

This report is divided into three parts and consists of nine chapters (see the report for details [41]). The first part focuses on integrity in research. The second part covers research misconduct and detrimental research practices. The third part considers how the research enterprise can better foster integrity in research.

Another latest development in the US that is worth-mentioning is the creation of an "Open Researcher and Contributor Identification Initiative (ORCID). ORCID is a nonproprietary alphanumeric, 16-digits code to uniquely identify authors and contributors of scholarly communication [42]. ORCID's website and services provide resources to look up authors and their bibliographic output. ORCID was first announced in 2009 as a collaborative effort by publishers of scholarly research "to resolve the author name ambiguity problem in scholarly communications." ORCID, Inc. is an independent nonprofit organization founded in the United States of America, with an international board of directors. ORCID was created as an initiative to fortify the validity and integrity of academic publishing through author name disambiguation [42, 43]. The aim of ORCID was to provide a unique code for humans to address the issues involving multiple persons with the same name. Furthermore, ORCID can solve problems of name change (such as with marriage), name ordering conventions, name suffixes, and middle initials, to name a few. It is hoping that implementation of ORCID system worldwide will solve this issue globally.

Overall, in the last three decades a consistent effort has been made by the US government,

NIH, NSF, and other entities that are integral components of the US "research enterprise" to foster integrity in research [44–46]. The creation of unified and harmonized policies and regulations on research ethics helped to foster research integrity in biomedical, behavioral, and medical research. Additionally, efforts were also made to disseminate these policies in research institutions and Academic institutions across all fifty-one states in the US [47]. These guidelines not only protected the research subjects but also provided awareness in the minds of researchers in academic institutions on RCR and research misconduct.

58.5 Perceptions on Fostering Global RI

Although RCR to foster RI is a global issue, the current policies and regulations developed are specific for a region or a country as reviewed in previous sections in this chapter. There is a need for developing a unified and harmonized policy and regulation(s) on research ethics, and RCR for fostering integrity in global research [48]. The creation of such a global RI will increase transparency in Research collaborations across national boundaries to the advancement of science worldwide [48]. International collaborations are now arguably the new *norm* as expected by many funding agencies. Additionally, a generation of new researchers who grew up with the power of the internet and big data is entering the research workforce. As a result, the way we influence or foster research cultures towards RCR and RI is also changing. We should also develop software systems with Artificial intelligence (AI), validated algorithms, and automated agents to assist the detection and prevention of research misconduct [49]. The development of such a software program may help reduce the rate of occurrence of research misconduct. These changes present special challenges for RCR because they may involve substantial differences in regulatory and legal systems, organizational and funding structures, research cultures, and approaches to training [50, 51]. It is therefore important that researchers be aware of such differences, as well as issues related to integrity that might arise in International research collaborations.

In a step towards establishing international awareness on fostering RI, the ORI from the US and ESF from Europe joined together to organize the first World Conference on Research Integrity (WCRI) which was held in Portugal in 2007. Subsequently, more conferences were held on a regular basis in 2010, 2013, 2015, 2017, 2019, and 2022 each one had a different theme and was held in various places in the world. Each meeting was attended by an increasing number of delegates and countries, suggesting a growing interest on fostering global RI. The last WCRI meeting was held in Cape Town, South Africa in 2022, one year later due to the COVID-19 Pandemic issues in 2021. The 2022 conference was on '*Fostering Research Integrity in an Unequal World*' and attended by RI stakeholders across all disciplinary fields from the basic and applied natural and biomedical sciences to the humanities and social sciences. Among other attendees, there were RI stakeholders including researchers, institutional leaders, national and international policymakers, funders and journal editors as well. The conference also included subthemes, such as RI as a driver of research excellence and public trust, Ethical best practice in authorship, publication and the use of research metrics and other topics.

Since this conference took place in an African country for the first time, some additional subthemes particularly relevant to many African and Low and Middle Income Countries (LMICs) were also included. Among them, Colonial legacies and research integrity: moving forward by building equity into research, Counteracting plagiarism in multicultural and multilingual contexts and Institutionalizing RCR education and training, curriculum development and implementation of these in low resource settings are noteworthy.

58.6 Concluding Remarks

It is tremendously intimidating for researchers to realize that there are currently so many guidelines and regulations to follow in order to conduct research properly and maintain RI. Although following the local and institutional guidelines are good enough to conduct research *in situ*, additional guidelines need to be followed in collaborative research, especially if the collaborating institution is in another country and bounded by its own *norms* and guidelines for RCR. The traditional *norms* that were once considered the gold standards for establishing collaborations between institutions are not enough anymore. There should be further discussions necessary on the policies and guidelines of each Institution or country on the RCR and developing a unified working policy is needed before starting any interinstitutional collaborations within the country. These steps need to be further escalated when collaboration involves more countries with different ethical policies for fostering RI and research misconduct. In this regard, the implementation of World Conferences on RI is the right step towards developing a unified global policy for RCR to foster integrity in global research. Another step is the creation and implementation of ORCID system as an initiative to strengthen the validity and integrity of academic publishing. The latter becomes one of the big issues when publishing a collaborative research work performed by a team of international investigators with different naming systems. A valid ORCID account for each researcher will solve this issue. Although obtaining an ORCID account is free and yet optional for any researcher, it is becoming a mandatory requirement for submission of manuscripts for ORCID journals such as Springer Nature and others. At present, there are inconsistencies across the world in implementing these ORCIDs despite the number of current account holders having increased to several millions, if not billions, since 2012. A third step is the involvement of World Health Organization (WHO), which recently released its Code of conducts for global RI in 2017 [52]. Such a move from WHO as an intergovernmental organization (IGO) is encouraging to motivate international researchers to conduct research in collaboration with the international community using the WHO developed codes for responsible research conduct. Taken together these recent developments have the potential to set a common ground for global collaborative research where everyone will be protected by universal codes of RI and research misconduct.

Conflict of Interest The author states no conflict of interest related to this topic.

References

1. Difference between morals and ethics [Internet]. https://keydifferences.com/difference-between-morals-and-ethics.html. Accessed 20 Sept 2022
2. Research ethics and integrity [Internet] [University of Edinburgh]. https://www.ed.ac.uk/files/atoms/files/research_ethics_and_integrity_awareness.pdf. Accessed 10 Nov 2022
3. Guidelines and polices for the conduct of research in the Intramural research program at NIH [Internet]. NIH Office of the Director 7th Edition Nov. 2021
4. Enhancing reproducibility through rigor and transparency [Internet]. https://grants.nih.gov/policy/reproducibility/index.html. Accessed 20 Sept 2022
5. Fostering integrity in research [Internet]. https://www.nap.edu/catalog/21896/fostering-integrity-inresearch. Accessed 10 Nov 2022
6. NIH policies and procedures for promoting scientific integrity [Internet]. https://www.nih.gov/sites/default/files/about-nih/nih-director/testimonies/nih-policies-procedures-promotingscientific-integrity-2012.pdf. Accessed 10 Nov 2022
7. European Science Foundation/ALLEA European code of conduct for research integrity [Internet]. http://www.esf.org/activities/mo-fora/completed-mo-fora/research-integrity.html, https://doi.org/10.1016/S0140-6736(13)60759-X. Accessed 10 Nov 2022
8. European Science Foundation and All European Academies (2011) The European code of conduct for research integrity [Internet]. http://ec.europa.eu/research/participants/data/ref/h2020/other/hi/h2020-ethics_code-of-conduct_en.pdf. Accessed 10 Nov 2022
9. Giorgini V, Mecca JT, Gibson C, Medeiros K, Mumford MD, Connelly S, Devenport LD (2015) Researcher perceptions of ethical guidelines and codes of conduct. Account Res 22:123–138. https://doi.org/10.1080/08989621.2014.95560
10. ALLEA (All European Academies) (2017) The European code of conduct for research integrity (Revised Edition) [Internet]. https://www.allea.org/wp-content/uploads/2017/05/ALLEA-European-Code-of-Conduct-for-Research-Integrity-2017.pdf. Accessed 15 Jan 2023
11. International Council on Harmonization (ICH-E6) [Internet]. https://www.ema.europa.eu/en/documents/

scientific-guideline/ich-e6-r1-guideline-good-clinical-practice_en.pdf. Accessed 30 Jan 2023
12. Bosch X (2010) Safeguarding good scientific practice in Europe. EMBO Rep 11:252–257. https://doi.org/10.1038/embor.2010.32
13. Bendiscioli S, Garfinkel MS. Governance of research integrity: options for a coordinated approach in Europe Science Policy Program EMBO Report [Internet]. EMBO Science Policy Program. www.embo.org/science-policy@embo.org. Accessed 20 Sept 2022
14. Bonn NA, Godecharle S, Dierickx K (2017) European universities' guidance on research integrity and misconduct. J Empir Res Hum Res Ethics 12(1):33–44
15. Desmond H, Dierickx K (2021) Research Integrity codes of conduct in Europe: understanding the divergence. Bioethics 35(5):414–428. https://doi.org/10.1111/bioe.12851
16. Rabesandratana T (6603) France introduces research integrity oath. Science 377:251. https://doi.org/10.1126/science.add9092
17. European Commission, Directorate-General for Research and innovation (2019) Mutual learning exercise (MLE) on research integrity – final report. Publications Office of the European Union, Luxembourg. https://doi.org/10.2777/72096. http://europa.eu
18. Resnik D, Zeng W (2010) Research integrity in China: problems and prospects. Dev World Bioeth 10(3): 164–171. https://doi.org/10.1111/j.1471-8847.2009.00263x
19. Jordan S, Gray PW (2013) Research integrity in Greater China: surveying regulations, perceptions and knowledge of Research Integrity from a Hong Kong perspective. Dev World Bioeth 13(3):125–137. https://doi.org/10.1111/j.1471-8847.2012.00337.x
20. Yang W (2013) Research integrity in China. Science 342:1019. https://doi.org/10.1126/science.1247700
21. Qiu J (2015) Safeguarding research integrity in China. Natl Sci Rev 2:122–125. https://doi.org/10.1093/nsr/nwv002
22. National Natural Science Foundation of China. Strengthening view research integrity [Internet]. http://www.nsfc.gov.cn/Portal0/InfoModule_396/28832.htm. Accessed 10 Nov 2022
23. Chou C, Lee IJ. Fudano J (2023) The present situation of and challenges in research ethics and integrity promotion: experiences in East Asia. Account Res 1–24. https://doi.org/10.1080/08989621.2022.2155144
24. Shahare M, Roberts LL (2020) Historicizing the crisis of scientific misconduct in Indian science. Hist Sci 58(4):485–506. https://doi.org/10.1177/007327532093090
25. Krishna VV (1991) The emergence of the Indian scientific community. Sociol Bull 40:89–107
26. Raj K (2002) Circulation and the emergence of modern mapping: Great Britain and early colonial India, 1764–1820. In: Markovits C, Pouchepadass J, Subrahmanyam S (eds) Society and circulation: mobile people and itinerant cultures in South Asia 1750–1950. Permanent Black, New Delhi, pp 23–54
27. Jayaraman KS (1987) Healthy scientific environment promoted by Society in India. Nature 326:535
28. CSIR Guidelines for Ethics in Research and in Governance [Internet]. https://www.csir.res.in/notification/csir-guidelines-ethics-research-and-governance. Accessed 15 Jan 2023
29. UGC Academic Integrity and Research Quality [Internet]. https://www.ugc.ac.in/e-book/Academic%20and%20Research%20Book_WEB.pdf
30. UGC Academic Integrity and Research Quality Bahadur Shah Zafar Marg New Delhi – 110002. Dec 2021 [Internet]. https://www.ugc.ac.in/ebook/Academic%20and%20Research%20Book_WEB.pdf. Accessed 15 Jan 2023
31. Chaddah P, Lakhotia SC (2018) A policy statement on "Dissemination and Evaluation of Research Output in India" by the Indian National Science Academy, New Delhi. Proc Indian Nat Sci Acad 84:319–329. https://doi.org/10.16943/ptinsa/2018/49415
32. BELMONT REPORT: Ethical Principles and Guidelines for the Protection of Human Subjects of Research (/ohrp/regulations-and-policy/Belmont-report/index.html) [Internet]. /ohrp/sites/default/files/the-belmont-report-508c_FINAL.pdf. Accessed Jan 2023
33. Chris B, Pascal CB (1999) The history and future of the office of research integrity: scientific misconduct and beyond. Sci Eng Ethics 5:183–198
34. ORI (Department of Health and Human Services, Office of Research Integrity) (1994) ORI policy on plagiarism. Office of Research Integrity Newsletter 3(1)
35. Goodwin E (2010) ORI (Department of Health and Human Services, Office of Research Integrity).Case summary: August 24
36. ORI (2014) Office of Research Integrity Newsletter 22(4)
37. Potti A. ORI case summary 2015 [Internet]. https://ori.hhs.gov/content/case-summarypotti-anil. Accessed 20 Nov 2022
38. NSF (National Science Foundation) (2002) Research misconduct policy. Federal Register 67: 11937. Mar 18
39. NSF (2009) NSF's implementation of section 7009 of the America COMPETES act, 2009. Fed Regist 74(160):42126–42128
40. OSTP. Scientific integrity [Internet]. http://www.whitehouse.gov/administration/eop/ostp/library/scientificintegrity. Accessed 15 Jan 2023
41. Fostering Integrity in Research: Committee on Responsible Science Committee on Science, Engineering, Medicine, and Public Policy and Global Affairs. A Consensus Study Report of The National Academies of Science, Engineering and Medicine [Internet]. www.nationalacademies.org

42. Open Research Contributor ID Wikipedia [Internet]. https://en.wikipedia.org/wiki/ORCID. Accessed 23 Jan 2023
43. Teixeira JA (2020) ORCID-issues and concerns about its use for academic purposes and research integrity. Ann Libr Inf Stud 67:246–250
44. NIH (National Institutes of Health) (2009) Update on the Requirement for Instruction in the Responsible Conduct of Research. http://grants.nih.gov/grants/guide/notice-files/not-od-10-019.html. Accessed 20 Nov 2022
45. HHMI (Howard Hughes Medical Institute) - Research Policies: Scientific Misconduct (2007) [Internet]. www.hhmi.org/sites/default/files/About/Policies/sc_200.pdf. Accessed 15 Jan 2023
46. National Academies of Sciences, Engineering, and Medicine (2016) Optimizing the Nation's investment in academic research: a new regulatory framework for the 21st century. The National Academies Press, Washington, DC
47. Steneck NH (1994) Research universities and scientific misconduct—history, policies, and the future. J High Educ 65(3):310–330
48. GRC (Global Research Council) (2013) Statement of Principles on Research Integrity [Internet] [Accessed Jan 2023]. http://grc.s2nmedia.com/statement-principles-research-integrity. Accessed Aug 15 2022
49. Fawcett T, Haimowitz I, Provost F, Stolfo S (1998) AI approaches to fraud detection and risk management. AI Mag 19(2) (© AAAI)
50. Resnik DB (2013) Plagiarism among collaborators. Account Res 20(1):1–4
51. Resnik DB, Master Z (2013) Policies and initiatives aimed at addressing research misconduct in high-income countries. PLoS Med 10:e1001406. https://doi.org/10.1371/journal.pmed.1001406
52. Global code of conduct for research in resource-poor settings [Internet]. www.globalcodeofconduct.org/. Accessed 20 Jan 2023

Publication Integrity, Authorship, and Misconduct

59

Subhash C. Lakhotia

Abstract

Research is an outcome of innate human inquisitiveness and self-driven efforts to know the unknown. The new knowledge and technology generated through research are essential for societal development and progress. To achieve sustainable development, research and communication of its output must be rooted in integrity and appropriate ethical conduct on part of all the stakeholders. Research communication involves authors, peer-reviewers, editors, publishers, and readers. This chapter considers the various ethical norms developed by the research community itself for all those involved in research communication and the reasons for the misconduct commonly encountered in the dissemination of research output. To minimize misconduct, it is necessary that research communication does not remain a 'for-profit' commercial activity, and the assessment of researchers and their institutions is based, instead of on where the research output is published, on what is published. Good quality and useful research that promotes societal welfare needs passion rather than fashion or compulsion.

Keywords

Authorship · Peer-review · Impact factor · Predatory journals · Preprints · Publisher ethics

59.1 Introduction

The innate curiosity of humans and the self-driven efforts to know the unknown are at the root of research. In its current form, research involves systematic and creative investigations in any domain of knowledge, which can be related to philosophical or materialistic issues, or anything in this universe that can be perceived by one or more of our sensory systems. The new knowledge generated through research has owner/s, who discover something novel, and recipients who learn about the discovery through communication made by the owner/s. Currently, research publications are the major mode of communication between the owner/s and recipients. Such communications, while providing personal satisfaction/pleasure to the discoverer, also serve the important social responsibility of spreading new knowledge, based on which the next levels of questions about the unknown are pursued.

The dawn of organized research correlates with the establishment of universities and other academic institutions, which have witnessed global growth in the past few centuries. Organized research has generally been supported

S. C. Lakhotia (✉)
Cytogenetics Laboratory, Department of Zoology, Banaras Hindu University, Varanasi, Uttar Pradesh, India
e-mail: lakhotia@bhu.ac.in

G. Jagadeesh et al. (eds.), *The Quintessence of Basic and Clinical Research and Scientific Publishing*,
https://doi.org/10.1007/978-981-99-1284-1_59

by society in one form or the other. The need for societal support for financing research by governmental agencies, philanthropic individuals and organizations has substantially increased because of the enhanced cost of research, especially in science, technology, health, and agriculture domains. Consequently, the classical self-satisfying curiosity-driven research efforts are now often 'market-driven', regulated formally and informally by state and/or commercial institutions [1]. This has provided a formal framework and financial support needed to undertake challenging research and has thus catalyzed unprecedented rapid technological advances. However, a flip side of the vastly expanded research efforts is the visible adverse effects on research integrity and ethical conduct, especially since the resources available for conducting research have not kept pace with the increase in the number of researchers and the cost of undertaking research. The resulting unhealthy competition, combined with the lure of quick material gains, has in many cases led to erosion in ethical values since human nature also follows the laws of thermodynamics and therefore, tends to indulge in selfish activities. These enhance the 'Social Entropy'.

In earlier times the curiosity-driven research output by individuals was published as monographs/books or as transactions/proceedings of meetings of academic societies. With the increasing volume of research output, well-organized research journals have become the major conduit for disseminating research output. Originally the research journals were managed and published by academic institutions, but during the past several decades, the share of commercial publishing houses in research publications has increased exponentially. The digital revolution, beginning in the late 1990s, metamorphosed the research publication ecosystem. While the digital mode has made the task of publication easier and swifter, it has also ushered in more challenging ethical issues [2–8]. The increasing interdependence between science, industry, and business, and the consequent conflict between the somewhat divergent 'values' in science and business [9], too have contributed to unwarranted practices in research and communication of its output.

To maintain research integrity, essential for sustainable development, the research community evolved its own set of conduct rules to safeguard against unwarranted behavior of researchers since this has a cascading negative impact on the profession and society [1–7, 10–17].

This chapter examines issues about integrity and ethics relating to the dissemination of information about new knowledge obtained by owner/s of research to the recipients through different modes of research publications. Issues of integrity and ethics in the conduct of research have been discussed elsewhere in this book (see Chap. 58).

59.2 Types of Research Publications

The modes of research publications have evolved over the centuries. Contemporarily, the non-commercial original research output is generally available as articles in serially published research journals. There are several categories of research publications. Different research journals publish some but not the other categories (see Chaps. 41 and 47). Research articles can communicate the results of original research or may review the present state of the available information in a given domain. The original research articles can be full-length papers or brief communication (see Chaps. 42 and 43). Similarly, the review articles can be a full or a mini-review, or state-of-the-art review. Several journals also publish articles in categories like Letter to Editor, Perspective/Commentary/Opinion, Conference/Meeting Report, Book Review, Personal Notes (e.g., obituary), hypothesis article, methods, *etc.* The Letter to Editor category generally includes a comment by a reader on an article published in that journal; however, in some journals (e.g., Nature, Current Science) a Letter to Editor is also used to rapidly communicate new and significant research findings. In some cases, focused short reviews by subject experts are also published as Letter to the Editor. Perspective/Commentary/Opinion articles can be of historical nature or a general discussion on an issue of contemporary interest. Conference/Meeting reports cover some well-known major scholarly meetings and describe what the different speakers presented about their research. Book reviews are generally written following an invitation from the editor to a person who has profound knowledge in the area covered in the book.

All standard journals have well-defined editorial policies regarding the types of articles that are published and the style of presentation in the given journal. Therefore, a prospective author must be familiar with the 'guidelines for authors' provided by the journal and prepare the manuscript accordingly. Most established journals now require the manuscripts to be submitted only in online soft copy format.

With reference to the state of the manuscript when it becomes available to readers, a classification of the published works into the following versions has been suggested [18]: (a) Preprint, a research paper prior to peer review and publication in a journal; (b) Postprint or peer reviewed accepted manuscript (Author Accepted Manuscript or AAM), prior to any type-setting or copy-editing by the publisher; (c) Version of Record (VOR), the final published version of a scholarly research paper following formatting and any other additions by the publisher; and (d) e-Print: a research paper posted on a public server in any of the above three versions.

The most common current mode of scholarly communication involves peer review before publication in a journal or book. Accordingly, the editor shares the submitted manuscript with a few experts (peers) in the field to seek their informed opinion, and based on the final recommendation emerging from the peer review, the manuscript gets accepted or rejected. Publication of preprints before the peer review is becoming increasingly popular in recent years as an alternative mode. Unfortunately, during the past two decades, a third mode of publication in 'predatory' or bogus journals has also mushroomed. This is highly undesirable because it severely vitiates the scholarly publication ecosystem.

59.3 Trustworthiness of Research Requires Integrity, Transparency Objectivity, and Following of Ethical Norms in the Conduct of Research As Well as Communication of its Output

Ethics define social norms for desirable and undesirable behavior. It should be noted that ethics and law or ethical and legal behavioral requirements are not synonymous or inter-changeable since in some cases what may be ethical may not be legal and vice-versa [9]. Behavioral integrity, one of the most essential components for the sustenance of human social organization and order, lies at the root of ethics and ensures the trustworthiness of the individual. An inherent, although often unstated, assumption underlying the huge number of published research articles is that the information is trustworthy [19]. Trustworthiness demands integrity of the researcher's actions in the conduct of research, and responsible and transparent communication of its output in open mode [14, 19, 20]. Integrity is essential to fulfilling a researcher's obligation to society to honor the trust placed by others and to be accountable for the public money used. Integrity also ensures the researcher's accountability to oneself to not violate scientific values and ethics and to act in ways that serve humanity.

Research and academic integrity have been variously defined and attributed [14, 15, 21]. In simple terms, research integrity generally implies that the conduct of research and dissemination of its output employs holistic and transparent practices based on appropriate ethical values. Promotion of research integrity and the consequent trustworthiness of the research output requires that (a) all contributions to scholarly activity are recognized without any bias for research topics and fields, (b) wholesome training is imparted to researchers in the relevant methods, and (c) appropriate experimental design and statistics are used. The other determinants of the trustworthiness of published research are: (a) rigorous oversight to ensure scientific integrity, (b) a transparent peer review system, and (c) an objective assessment system that relies on the quality of the research output rather than the currently common practice of using quantitative bibliometric parameters.

Inappropriate research integrity leads not only to widely known misconducts like falsification and/or fabrication of data, plagiarism but also to other less commonly discussed detrimental research practices like inappropriate authorship, inadequate data management, inadequate/improper citations, or incomplete sharing of research findings. While individual researcher/s are usually targeted for research misconduct, research supervisors, funding agencies, the administrative system at academic institutions, peer reviewers, journal editors, and publishers also have significant roles in promoting integrity and transparency or the lack of these [14, 19, 20, 22–24]. Some of these are considered below.

59.4 Authorship

Unlike in the past when single author monographs were the major mode of research communication, contemporary research output is often multi-authored because of the need for diverse expertise. While the multi-authorship in a research paper promotes the much desired interdisciplinary approach, unfortunately, it also provides greater scope for certain unethical practices that are often not focused.

In order to ensure ethical practices, the Committee on Publication Ethics (COPE, https://publicationethics.org/) has mandated that to be an author in a research publication, one must contribute to one or more of the following tasks associated with the research: (a) conception and/or design of the work, (b) acquisition, analysis, and/or interpretation of data generated/collected during the work, and (c) drafting/editing the work or revising it critically and thus contributing intellectually to the content. Every researcher should be familiar with the COPE guidelines to avoid unethical authorship practices.

In current practice, one or more authors is/are identified as first author/s, one or more authors as corresponding author/s, and others, if any, as co-authors. Those who carried out the bulk of the primary work that forms the basis of the article are defined as the first author/s. The corresponding author is primarily responsible for communication with the journal during manuscript submission, peer review, and publication process, and also ensures the journal's administrative requirements like providing authorship details, ethics committee approval, documentation related to clinical trials, collecting information on conflict of interest issues, etc. The corresponding author continues to be the primary contact for any issue related to the published paper. The order of authorship should be a joint decision of co-authors and accordingly, the name (s) of the corresponding author(s) may come anywhere in the authors' list. The corresponding author must ensure that all other authors have seen the final manuscript and approved it for submission to the identified journal. To avoid inappropriate authorship (see below), the contributions of each author in a multi-author article should be clearly defined. Authorship in research communication should be decided in good time to avoid disputes at a later stage. Adding name/s as author/s of a manuscript during revision is permitted if suitably justified but removing one or more author names after submission is generally not permitted. As a part of their collective responsibility, all authors should be aware of the contributions of each co-author.

Non-adherence to the COPE guidelines has resulted in a variety of unethical authorships, some of which are noted here. *Guest* or *Honorary* or *Gift authorship*, a not uncommon practice, is provided to someone who made no useful contribution to the study but is included to improve chances of the work getting accepted for publication, or on a feeble/indirect affiliation with the study (e. g., head of the department/institution). This unethical practice also involves the inclusion of names of colleagues as co-authors, on the understanding that they will also reciprocate, with the sole objective of boosting each other's publication lists. A related unethical practice concerns the publication of research work embodied in a student's thesis for the award of an academic degree. In many cases, the research papers emanating from the work included entirely in one thesis are published with additional names, besides those of the examinee and supervisor/co-supervisor, as co-authors. The thesis document itself, however, does not record the specific contributions of others who are named as co-authors. If the other authors had indeed contributed to the published work, their specific contributions must be clearly defined in the thesis. In the absence of a such clear statement in the thesis, it becomes an unethical practice that only helps others to increase the number of research publications in their profiles without any specific contribution.

Ghost authorship includes unethical instances when the person/s, who actually participated in the research, data analysis, and/or writing of a manuscript, is/are excluded as author/s in a manuscript or while filing a patent application (see Chap. 57). Accordingly, the thesis supervisor must not publish the work embodied in the thesis without the concerned student's name, although under certain conditions the student may be specifically permitted by the supervisor to publish the work without the supervisor/co-supervisor being co-author/s.

A more serious unethical practice is *Surrogate authorship* when manuscripts are written or got written by someone else without any original data of one's own; such plagiarized or 'fabricated' papers are typically published in 'predatory'/bogus journals.

A different class of *Ghost authorship*, which is not unethical, is when a person contributing to writing a paper, as a professional writer (on a payment or honorary basis), or as a helping colleague, is not on the author list. Such help should, however, be acknowledged.

Since authorship has to be transparent and requires accountability, publishing research articles anonymously or under different names is improper.

59.5 Ethical Conventions of Publications

Authors need to follow ethical practices both during the submission of a research article for publication and after its publication. A manuscript can be submitted to only one journal at a given time. Submission to another journal can be made only when either the submitted article is rejected by the editor or withdrawn by the authors. The same set of data/results cannot be published multiple times, except when necessary to include in subsequent work or review, in which case appropriate reference to the original source is essential. 'Salami' or segmented publications, usually characterized by similarity of hypothesis, methodology, and results but not similarity of text, are unethical since they merely enhance the number of published articles without any intellectual advancement [25].

Since all advances in knowledge build upon pre-existing knowledge, due attribution is required to acknowledge the authority, based on which some statement is made or the findings of the study are interpreted. This is done through citations in the text at appropriate locations and by providing the relevant details in the reference list. The journal's defined format for citations in text and the listing of references must be strictly followed. Citations of earlier published works in an original research article or a review article need great care since it is implied that the authors have read the work and find it appropriate to cite. Great care is needed to avoid mis-citations, especially when using a software to manage citations and the reference list (see Chap. 39). The convenience of finding citations online while writing the text can lead to the unethical practice of citing a work without reading. It is generally preferred that the original work be cited instead of a broad review so that due attribution is provided to the original research. A cap on the number of references in an article by the journal policy demands greater care and vigilance in selecting a mix of older and recent original and review articles of relevance. References to unethical publications and irrelevant self-citations must be

Table 59.1 Types of plagiarism and their general features [13, 28–30, 33]

Broad category	General features
Direct or verbatim or complete plagiarism ('stealing')	Full copying of whole or parts of a text and/or data of a publication and passing on the same as one's own novel findings
Mosaic or patchwork plagiarism	Copying and pasting some sentences from another author's publication without using quotation marks; citing the original source without using the quotation marks for the verbatim copied text can also be counted as plagiarism
Source plagiarism	Misleading citation of the source of the published information *Secondary*: copying from a secondary source but attributing to a primary source cited within the secondary, leading to the secondary source missing the due attribution while generating a false impression of the depth of the literature survey *Invalid source*: citing an incorrect or even a non-existent source to increase the number of references cited and to unjustifiably generate an impression of an in-depth study
Translational plagiarism	Publishing a translated version of whole or substantive parts of an earlier publication of one's own or of others without attribution to the original source
Paraphrasing	Modifying the copied idea/statement from a published work by rephrasing the sentence/s and using synonyms, without attribution to the source
Replication plagiarism	Submission of the same manuscript to more than one journal and getting them published
Idea plagiarism	Using another person's ideas (published or unpublished) for one's own study/interpretation without giving appropriate credit to the source
Self-plagiarism (duplicate, dual, overlapping, repetitive or redundant or text recycling)	Re-publishing substantive parts of one's own earlier publications without attribution to the earlier publications or reuse of data and/or text from one's own published work without appropriate attribution, with intention of misleading others about one's 'productivity'
Fragmented or piecemeal or/and 'salami' publication	A similarity of hypothesis, methodology, and results, but without text-plagiarism, in multiple publications to merely enhance the number of published articles

avoided. A correct reference list is essential for reliable citation meta-analyses. The reference list also reflects the quality of the work and its presentation [26, 27].

Serious types of misconduct in research communications include the widely discussed issues of plagiarism (text, images, ideas, etc.), data and/or image manipulations, or suppression of some data that are not in conformity with the preferred hypothesis [13, 27–32].

Although the specific definition of plagiarism varies in different contexts, the use of language, ideas, results, explanations, or text of another author without due attribution to the original source, and thus giving an impression of these being one's own, is generally considered as an act of plagiarism. Plagiarism is a violation of the intellectual property rights of others and is, therefore, a serious breach of academic integrity. Some acts of unintentional plagiarism happen either due to ignorance and/or to less than the required level of care while preparing the manuscript. Intentional plagiarism is an act of design with the sole purpose of taking credit, which is actually due to some other author/s, and, therefore, is serious misconduct and is often also termed as 'stealing'. Table 59.1 summarizes some general features of different acts of plagiarism.

Researchers must honour the intellectual property of others. Any creative output of a person's mind is that person's intellectual property, and, therefore, needs appropriate acknowledgment, typically in the form of citation in research publications. Research misconducts like text, data, or idea plagiarism essentially imply using someone else's intellectual property without appropriate attribution and thus violate the intellectual property rights of others. While industrial patents have a limited lifespan, intellectual property rights remain eternally with the person whose creative activity generated it in the first place. Rules and conventions for citation and reuse of intellectual property have been codified as 'Creative Commons Licenses' (https://creativecommons.org/licenses/), which researchers and authors should be familiar with while publishing the research output or making other communication.

When deliberate misconduct like inappropriate authorship, plagiarism, data/image manipulation or suppression of data, *etc.* on the part of one or more authors is detected and established, the involved authors attract punitive actions by the journal, funding agencies, and/or their employers. The punitive actions can range from the expression of concern or retraction of the paper by the journal, stoppage of research grants by funding agencies, to disciplinary action by the host institution. However, more damaging is the stigma that such persons suffer.

Instances of misconduct in research and communication of its output appear more common in bio-medical sciences because of the inherent diversity of biological systems which makes the interpretations more subjective than objective. Additionally, the digitalization of data and images and the various software available for image processing make it relatively easy and attractive to unethically tweak the original data. The enormous volume and diversity of information in big data 'omics' sciences also provide scope for unethical conduct in reporting or interpreting. Because of the alarmingly increased instances of such misconduct, it is not uncommon for authors to be asked to provide the original sets of raw data and images for verification.

Inadvertent errors may persist in published articles. These may be pointed out by readers or may be discovered by authors themselves. It is primarily the corresponding author's responsibility to publish erratum/corrigendum in the same journal where the original article was published.

Another less commonly understood ethical convention relates to sharing of new material/s described in a published paper. Once published, the authors are generally obliged to share the described new physical or chemical, or biological material with desiring fellow researchers so that they can use the same to further advance the knowledge. Not sharing the material because of fear of competition or otherwise is unethical. The journals where the concerned article is published can publish an expression of concern at this unethical practice or may even retract the paper.

59.6 Integrity in Peer Review and Editorial Handling of Submitted Manuscripts

Realizing the need for quality control of published material in scientific journals, some editors started the peer review system in the 1800s but it became more common around the 1960s [9]. It is now an integral part of research publication so that some experts or peers in the given research area are invited by the editor to scrutinize the submitted manuscript's usefulness, quality, the rigour of experimental design and methods used, interpretations and conclusions arrived at, and its overall worthiness for publication [34–37]. A general impression is that the more stringent the peer review system, the better the standing of the research journal in the field because the critical comments and suggestions help authors in improving the quality and reliability of the research output. Good peer review is expected to filter out manuscripts with poor/unreliable data and/or interpretations. Several modes of peer review systems are currently in practice [38, 39]. The most common form is the single-blind system where either only the author or only the reviewer knows the identity of the reviewer or author/s, respectively; the more common form of the single-blind review is where the author does not know the reviewers' identity. In the double-blind system, neither author nor reviewers know each other's identity. A more recent review system is the open format in which the reviewer's comments and author responses can be available during the review process itself or when the article is published; if agreed upon, the reviewer's identity can also be disclosed [34, 37, 39]. For a detailed discussion on this subject refer to Chap. 47.

A major advantage of peer review, for which it is accorded so much importance, is the experts' feedback to authors. In recent years, the open peer review system has gained greater popularity since, in this mode, the reviewers' comments tend to be less harsh and more constructive [8, 37]. A wider practice of an open review system will also curb the growth of predatory and sub-standard journals [23, 24]. More recently a post-publication peer review, which involves review after the publication of the article as an un-peer reviewed preprint, is being widely discussed as a better alternative to the conventional pre-publication peer review [35, 38–40]; also see https://www.leidenmadtrics.nl/articles/the-growth-of-open-peer-review; http://f1000research.com/.

The conventional peer review system's efficiency and/or utility have been questioned because of some inherent and some unwarranted unethical factors [8, 35–37]. A major concern is a delay in publication beyond a reasonable time. Other factors that seriously affect the basic purpose of the review process is reviewers' bias, which can stem from one or more of the following: (a) competitive conflict when a reviewer works on a similar research question, (b) geographical/racial/gender/language bias, or (c) new interpretation in the manuscript being at variance with the reviewer's views. These biases lead to unjustified rejections and/or to idea-plagiarism or even to unjustified acceptance. Peer-reviewer's approach to the assigned task also affects its utility and usefulness. Indifferent reviews are of little use to the editor or author while unduly harsh, sometimes even derogatory, reviews can be very damaging, especially to young researchers [37]. In experimental studies, it is a common experience that reviewers ask for additional studies to buttress the observations and claims made in the study. While this can improve the quality of the final published paper, asking for an 'endless' series of additional studies or for confirming the results using 'the latest' tools/methods can be unjustifiable and may delay the publication or lead to its rejection on less than reasonable grounds. It is also not uncommon that the multiple reviewers examining the same manuscript come out with varying or even clearly opposing reasons and recommendations, which add to the frustration of the author/s. Failure of the pre-publication peer review system is glaringly noticeable in the increased incidences of misconduct of one or the other kind on part of authors and, therefore, retractions of published articles by editors. Retractions appear to be more frequent in cases of journals that have a high reputation in the field. Given such limitations of the pre-publication peer review system, preprint archives and the post-publication peer review process are gaining favor.

To be effective, the peer review system must follow norms of ethical conduct. The reviewers are expected to have good knowledge (theoretical, methodological, and practical) in the given domain and should be able to apply their understanding in a critical but constructive manner while assessing a manuscript [34, 37]. Peer reviewers must not indulge in 'idea-plagiarism' to unauthorized exploitation of the confidential unpublished information in the manuscript being reviewed [33, 41]. The peer reviewers should respond immediately to the editor's invitation to either accept or decline the assignment, and if accepted, the comments should be provided to the editor in time to complete the peer review process swiftly. A reviewer should provide a detailed report in a critical but constructive manner so that authors can benefit and improve the quality and wholesomeness of their work. If the reviewer disagrees with a new interpretation being put forth by the author/s, the reviewer's personal bias should not take the upper hand. The review process should not become a battleground. Essentially, the reviewer should act like a friend, philosopher, and guide [42].

Reviewing a manuscript is an art. A simple way to learn this art is for the reviewers to remember that they are also or have been in the author's shoes. Formal training in the peer review process is helpful, especially for younger researchers.

While senior researchers can generally provide a more holistic review because of their experience and knowledge, sometimes being out of touch with current developments or their own bias can become negative factors. Mid-career researchers are expected to be more conversant with contemporary literature and open to new ideas. Editors, therefore, should exercise due caution in identifying the potential reviewers for a given manuscript. When authors are required to suggest potential reviewers for their submitted manuscript, as some journals do, the suggested names should be beyond any conflict of interest and the suggested reviewers should be knowledgeable in the given field.

The journal editor's task is more arduous because of the great responsibility of ensuring adherence to ethics at all steps of processing of the submitted manuscript. The first step involves checking for any apparent misconduct, especially text or image plagiarism/manipulation, which is usually done through appropriate software. If there are multiple editors in the team, the chief editor assigns the manuscript to one of the editors having the closest expertise in the subject matter of the manuscript. The next step in the editorial process involves identifying the potential reviewers, the most important, but sometimes painful task. The identified reviewers should not only be experts in the given domain but should not have any 'conflict of interest' with the authors or their affiliation or with the subject matter of the manuscript. This task sometimes becomes painful because one or more of the identified reviewers may not respond or decline or after agreeing, do not send the report in time or the provided report is indifferent. Finally, based on the reviewers' reports, the editor has to take a call for rejection, revision (major or minor), or acceptance of the submission for publication. The decision to convey rejection or to ask for a revision of a submitted manuscript involves the editor's greater intellectual involvement with the manuscript as well as the reviewers' reports since the decision needs to be conveyed to authors with adequate justification. The editor should not act only as a post office between the reviewers and authors but should act as an adjudicator. The latter role is often missing in contemporary editorial practices, which can leave the authors in a state of frustration. Neither the editor nor reviewers should reject a submission just because the authors' interpretation is at variance with the contemporarily popular views in the given field. In my personal experience, there have been occasions when our manuscript was rejected by a reviewer/editor simply because our interpretations of the observed results were contrary to the prevailing views, although in subsequent periods our interpretations indeed turned out to be not incorrect. Editor and reviewer should understand that responsibility for the novel views/interpretations advanced in the manuscript lies with the authors; editor and reviewer need only to satisfy themselves that the interpretation advanced by authors is logical and buttressed with appropriate experimental support and/or reasoning, without any obvious misconduct. If subsequent research does not support the authors' proposed views, editors and reviewers are not held responsible. Editors and reviewers also need to be conscious of the fact that knowledge is never static; today's solidly supported/believed theory may be turned over in the future as we get a deeper understanding through continuing prodding. A glaring example is the about turn in molecular biology during the past two decades which led to the strongly held view that 'selfish' or 'junk' DNA in our genomes is useless being replaced with the current belief that such DNA sequences are actually critical for the biological complexities of diverse organisms [43]. Just as authors need to be unbiased about their hypotheses, reviewers and editors also need to be more liberal with authors' novel interpretations.

59.7 Publishers' Ethics and Predatory Journals

Originally, research journals were maintained and published by academic institutions or learned societies as a societal responsibility [9, 44]. With the increasing numbers of researchers and the consequent increase in demand for the publication of research results, commercial publishers entered the arena, and over the years have taken near complete control of research journals and other publications. The business economy moved research publications from being a 'not-for-profit' societal responsibility to a 'for profit' mode with levy of a variety of charges (article processing charges (APC), open access charges (OAC), *etc.*) that are paid by authors or authors' institutions/funders to publish their research output. The shrewd sense of business on part of the commercial publishers also hoodwinked authors to mistakenly believe that sharing reprints (hard or soft copy) of their published work with fellow academics may not be permitted by the copyright transferred to the publisher in lieu of their publishing the article; this has led to a collection of hefty open access charges by commercial publishers from authors or readers [45]. One of the seriously damaging unethical outcomes of commercial profit-making interests in research publications has been the birth and uncontrolled proliferation of sub-standard and predatory journals. In the race for increasing profit margins, journals and publishers have contributed to the misplaced belief that the journal impact factor is the most important bibliometric marker to indicate the quality of the research output. Numerous studies have indicated limitations and adverse consequences of using the impact factor as the major indicator of the quality of research by an individual or an institution [23, 24, 46–51]. Along with the predatory journals, many spurious sites that provide an 'impact factor' have also mushroomed [52] because of the undue importance given to the impact factor as a bibliometric index of quality (see Chap. 40).

The Declaration on Research Assessment (DORA, https://sfdora.org/read/) was issued in 2012 during the annual meeting of the American Society of Cell Biology in San Francisco (USA) and since then has been widely signed across the globe. This declaration suggested a set of guidelines for publishers which included reducing the emphasis on impact factor as a promotional tool, developing individual article-level metrics, promoting responsible authorship, removing or reducing the constraints on the number of references in research articles, and, where appropriate, mandate the citation of primary literature rather than reviews to give due credit to the group(s) who first reported a finding. Unfortunately, most of these well-meaning ethical guidelines have not been followed by journal publishers since most journals continue to highlight their yearly impact factors to boost their commercial interests and thus continue to vitiate the system.

J. Beall coined the term '**predatory journals**' around 2008 for journals (*the Predator*) which have little or no peer review system and are primarily focused on making money from the gullible authors (*the Prey*). He collated a list of such journals and their publishers (see https://www.immunofrontiers.com/list-of-predatory-journals-and-trusted-resources-2022) to alert authors and the scholarly publishing community against such grossly unethical practices [53, 54]. The serious damages inflicted on the scholarly publishing ecosystem by such blatantly unethical practices by authors as well as publishers/editors and the underlying causes have been widely discussed, and many corrective steps have been suggested [23, 24, 55–65]. As discussed in the recent IAP report [23], it is not possible to have 'black' and 'white' lists of journals since there is a continuum of journals ranging from typical 'predatory' to bad to poor to 'good' journals. A seriously damaging unethical practice adopted by some unscrupulous publishers is to 'hijack' or 'clone' an established journal's website so that unsuspecting authors submit their manuscripts to such sub-standard/bogus publishers [23]. To mislead authors, many bogus journals are deliberately named to closely resemble those of established journals. Therefore, authors need to carefully ascertain the nature and quality of the journal while submitting their manuscripts for publication.

As long as the assessment policies for academic jobs, promotions, and awards/rewards continue to rely primarily on quantitative parameters like the number of published articles, the impact factor of the journal, and the citation index, it will be very difficult to curb the scourge of predatory and money-making publication ecosystem.

59.8 Preprints

Preprints are research output manuscripts shared by researchers, privately or publicly, without peer review for rapid dissemination to and/or seeking feedback from fellow workers. Perhaps the first informal communication system (in some ways analogous to the more recent preprint system) was the publication of the annual Drosophila Information Service (DIS) in March 1934, for sharing information about research material and reagents available in different laboratories across the world and new unpublished research findings; this practice has been continued since then without interruption. The usefulness of free sharing of new information amongst researchers was clearly stated in the Preface in the first issue of DIS: "An appreciable share of credit for the fine accomplishments in Drosophila genetics is due to the broadmindedness of the original Drosophila workers who established the policy of a free exchange of material and information among all actively interested in Drosophila research. This policy has proved to be a great stimulus for the use of Drosophila material in genetic research and is directly responsible for many important contributions. In over twenty years of its use, no conspicuous abuse has been experienced" (https://www.ou.edu/journals/dis/). As a researcher in *Drosophila* genetics, I am not aware of any instance of misuse resulting from the free sharing of new findings and information about new resources during the nearly 90 years of DIS publication.

Information Exchange Groups initiated by the National Institutes of Health (USA) in 1961 for circulating biological preprints were shut down in 1967 because established journals, including Nature, Science, and others, felt threatened and refused to publish material already shared as preprints [66]. The arXiv preprints repository is popular in Physics, Mathematics, Computer Science, etc. since 1991 [67]. The Social Sciences Research Network, and Research Papers in Economics were launched in 1994 and 1997, respectively, for the informal sharing of preprints and other research publications amongst the member community. The Academia.edu and ResearchGate network platforms, which allow sharing of research output at any stage, started in 2008 [18]. During the past decade, many preprint servers have been launched, with their popularity increasing exponentially across the globe [18, 66–68]. Interestingly, unlike the hostility shown by established journals to the Information Exchange Groups and other similar efforts in the 1960s, most publishers now permit the submission of articles posted at preprint servers for publication in their journals and many funding agencies also consider preprints for assessment [66]; also see https://asapbio.org/funder-policies and Chaps. 51–54 in Grantsmanship of this book.

The currently followed practices at preprint servers offer many advantages to authors and readers. These include rapid dissemination (often within a day) of information free of cost to authors as well as readers for eternity, claim for priority, facilities for searching and citing (through DOI numbers), possibility to revise the earlier version, post-publication onsite peer-review, and promotion of the spirit of sharing.

Since instances of misconduct leading to the retraction of papers published even in the so-called 'high-impact and prestigious' journals are not uncommon, the veracity and usefulness of the pre-publication peer review system are also being increasingly debated. It may be noted that since preprints cannot be withdrawn after publication, the chances of misconduct in preprints published on authenticated preprint servers are less. In fact, while the Covid-19 pandemic saw a huge jump in preprints at bioRxiv and medRxiv preprint servers, no major public health hazard resulted because of misleading publication, unlike the several cases of serious health hazards due to misconduct in conventional refereed journal publications [67]. The various advantages provided by preprints have, therefore, strengthened arguments for publication as preprints that undergo post-publication peer review. Consequently, the various preprint servers, Pub-Peer (https://pubpeer.com/), Retraction Watch (https://retractionwatch.com/), etc. have also become popular.

Given the increasing popularity and recognition of preprint publications, unscrupulous publishers have created 'predatory' preprint servers [69] that charge money and may soon disappear in thin air. As in the case of predatory journals, authors must carefully examine and assess the veracity of sites claiming to be preprint servers.

59.9 Concluding Remarks

Research is a creative activity carried out primarily for self-satisfaction. Communication of the research output not only adds to the self-satisfaction of the researcher but is also an altruistic act necessary for the welfare and progress of society. A variety of factors promote unethical practices in this otherwise honest and greatly self-satisfying intellectual activity. Besides the general erosion of ethical and moral values in the contemporary materialistic society, several other factors, having their origins within the research and academic ecosystem, also promote misconduct in research and its communication. The pressure to publish to qualify for a job, promotion, or recognition, the undue emphasis on the impact factor, and the lure of monetary gains or of gaining 'prestige' are some of the major factors that incentivize unethical conduct in research communication [70, 71]. The absence of well-defined and stringent national policies in most countries and wide variations among those existing in different countries also make it difficult to curb academic misconduct [72]. The really important consideration in promoting good research should be what is published rather than where it is published [47, 51]. Good quality and useful research that promotes societal welfare needs passion rather than fashion or compulsion.

Acknowledgments I sincerely thank Prof. Rajiva Raman (Banaras Hindu University, Varanasi, India), Prof. Pradeep Burma (Delhi University, New Delhi, India), Dr. Richa Arya (Banaras Hindu University, Varanasi, India), Dr. Arun Tripathi (Central University of Tibetan Studies, Varanasi, India) and Dr. Animesh Banerjee (Institute of Molecular and Cell Biology, Singapore) for their useful comments and suggestions on the manuscript. I acknowledge the support of the Science & Engineering Research Board (Govt. of India, New Delhi) to me as SERB Distinguished Fellow.

References

1. Lakhotia SC (2021) Philosophy and ethics of research in science. In: Handbook on academic and research integrity. University Grants Commission, New Delhi, pp 8–17
2. Lakhotia SC, Chaddah P (2019) Ethics of research. In: Muralidhar K, Ghosh A, Singhvi AK (eds) Ethics in science education, research and governance. Indian National Science Academy, New Delhi, pp 35–43
3. Armond ACV, Gordijn B, Lewis J, Hosseini M, Bodnár JK, Holm S, Kakuk P (2021) A scoping review of the literature featuring research ethics and research integrity cases. BMC Med Ethics 22(1):50. https://doi.org/10.1186/s12910-021-00620-8
4. Reijers W, Wright D, Brey P, Weber K, Rodrigues R, O'Sullivan D, Gordijn B (2018) Methods for Practising ethics in research and innovation: a literature review, critical analysis and recommendations. Sci Eng Ethics 24(5):1437–1481. https://doi.org/10.1007/s11948-017-9961-8
5. Nazemian S, Balash F, Balash R (2017) Psychological factors underlying unethical research. The Eurasia Proceedings of Educational & Social Sciences. pp 211–215
6. Edwards MA, Roy S (2017) Academic research in the 21st century: maintaining scientific integrity in a climate of perverse incentives and hypercompetition. Environ Eng Sci 34:51–61. https://doi.org/10.1089/ees.2016.0223
7. Belle N, Cantarelli P (2017) What causes unethical behavior? A meta-analysis to set an agenda for public administration research. Public Adm Rev 77(3): 327–339. https://doi.org/10.1111/puar.12714
8. Banerjee S, Mande SC (2022) Is the peer review system in scientific publishing broken? Curr Sci 122(8): 877–878
9. Resnik DB (1998) The ethics of science: an introduction. Routledge, New York
10. Wager E, Kleinert S (2012) Cooperation between research institutions and journals on research integrity cases: guidance from the committee on publication ethics (COPE). Maturitas 72(2):165–169. https://doi.org/10.1016/j.maturitas.2012.03.011
11. Sovacool BK (2008) Exploring scientific misconduct: isolated individuals, impure institutions, or an inevitable idiom of modern science? J Bioethical Inquiry 5(4):271–282. https://doi.org/10.1007/s11673-008-9113-6
12. Lakhotia SC, Chandrasekaran S (2019) Ethics of publication. In: Muralidhar K, Ghosh A, Singhvi AK (eds) ETHICS in science education, research and governance. Indian National Science Academy, New Delhi, pp 65–84
13. Lin W-YC (2020) Self-plagiarism in academic journal articles: from the perspectives of international editors-in-chief in editorial and COPE case. Scientometrics 123(1):299–319. https://doi.org/10.1007/s11192-020-03373-0

14. Kretser A, Murphy D, Bertuzzi S, Abraham T, Allison DB, Boor KJ, Dwyer J, Grantham A, Harris LJ, Hollander R, Jacobs-Young C, Rovito S, Vafiadis D, Woteki C, Wyndham J, Yada R (2019) Scientific integrity principles and best practices: recommendations from a scientific integrity consortium. Sci Eng Ethics 25(2):327–355. https://doi.org/10.1007/s11948-019-00094-3
15. Macfarlane B, Zhang J, Pun A (2014) Academic integrity: a review of the literature. Stud High Educ 39(2): 339–358. https://doi.org/10.1080/03075079.2012.709495
16. World Medical Association (2013) World medical association declaration of Helsinki: ethical principles for medical research involving human subjects. JAMA 310(20):2191–2194. https://doi.org/10.1001/jama.2013.281053
17. Moreno JD, Schmidt U, Joffe S (2017) The Nuremberg code 70 years later. JAMA 318(9): 795–796. https://doi.org/10.1001/jama.2017.10265
18. Knowledge Exchange Preprints Advisory Group, Tennant J (2018) The evolving preprint landscape: introductory report for the Knowledge Exchange working group on preprints
19. Moher D, Bouter L, Kleinert S, Glasziou P, Sham MH, Barbour V, Coriat A-M, Foeger N, Dirnagl U (2020) The Hong Kong principles for assessing researchers: fostering research integrity. PLoS Biol 18(7): e3000737. https://doi.org/10.1371/journal.pbio.3000737
20. Roje R, Reyes Elizondo A, Kaltenbrunner W, Buljan I, Marušić A (2022) Factors influencing the promotion and implementation of research integrity in research performing and research funding organizations: a scoping review. Account Res:1–39. https://doi.org/10.1080/08989621.2022.2073819
21. Huberts LWJC (2018) Integrity: what it is and why it is important. Public Integrity 20:S18–S32. https://doi.org/10.1080/10999922.2018.1477404
22. Boughton SL, Kowalczuk MK, Meerpohl JJ, Wager E, Moylan EC (2018) Research integrity and peer review—past highlights and future directions. Res Integr Peer Rev 3(1):3. https://doi.org/10.1186/s41073-018-0047-1
23. The InterAcademy Partnership (2022) Combatting predatory academic journals and conferences
24. Lakhotia SC (2022) Combating predatory journals and conferences. Curr Sci 122(10):1121
25. Supak SV (2013) Salami publication: definitions and examples. Biochem Med (Zagreb) 23(3):237–241. https://doi.org/10.11613/bm.2013.030
26. Gasparyan AY, Yessirkepov M, Voronov AA, Gerasimov AN, Kostyukova EI, Kitas GD (2015) Preserving the integrity of citations and references by all stakeholders of science communication. J Korean Med Sci 30(11):1545–1552. https://doi.org/10.3346/jkms.2015.30.11.1545
27. Berenbaum MR (2021) On zombies, struldbrugs, and other horrors of the scientific literature. Proc Natl Acad Sci U S A 118(32):e2111924118. https://doi.org/10.1073/pnas.2111924118
28. Sharma H, Verma S (2020) Insight into modern-day plagiarism: the science of pseudo research. Tzu Chi Med J 32(3):240–244. https://doi.org/10.4103/tcmj.tcmj_210_19
29. Chambers L, Michener C, Falcone T (2019) Plagiarism and data falsification are the most common reasons for retracted publications in obstetrics and gynaecology. BJOG Int J Obstet Gynaecol 126(9): 1134–1140. https://doi.org/10.1111/1471-0528.15689
30. Eaton SE, Crossman K (2018) Self-plagiarism research literature in the social sciences: a scoping review. Interchange 49(3):285–311. https://doi.org/10.1007/s10780-018-9333-6
31. Boutron I, Ravaud P (2018) Misrepresentation and distortion of research in biomedical literature. Proc Natl Acad Sci U S A 115(11):2613–2619. https://doi.org/10.1073/pnas.1710755115
32. Laraway S, Snycerski S, Pradhan S, Huitema BE (2019) An overview of scientific reproducibility: consideration of relevant issues for behavior science/analysis. Perspect Behav Sci 42(1):33–57. https://doi.org/10.1007/s40614-019-00193-3
33. Chaddah P (2021) Ethics in research publications: fabrication, falsification, and plagiarism in science. In: Academic integrity and research quality. University Grants Commission, New Delhi, pp 18–33
34. Ali PA, Watson R (2016) Peer review and the publication process. Nurs Open 3(4):193–202. https://doi.org/10.1002/nop2.51
35. Kelly J, Sadeghieh T, Adeli K (2014) Peer review in scientific publications: benefits, critiques, & a survival guide. EJIFCC 25(3):227–243
36. Onken J, Chang L, Kanwal F (2021) Unconscious bias in peer review. Clin Gastroenterol Hepatol 19(3): 419–420. https://doi.org/10.1016/j.cgh.2020.12.001
37. Le Sueur H, Dagliati A, Buchan I, Whetton AD, Martin GP, Dornan T, Geifman N (2020) Pride and prejudice – what can we learn from peer review? Med Teach 42(9):1012–1018. https://doi.org/10.1080/0142159X.2020.1774527
38. Harms PD, Credé M (2020) Bringing the review process into the 21st century: post-publication peer review. Ind Organ Psychol 13(1):51–53. https://doi.org/10.1017/iop.2020.13
39. Peh WCG (2022) Peer review: concepts, variants and controversies. Singap Med J 63(2):55–60. https://doi.org/10.11622/smedj.2021139
40. O'Sullivan L, Ma L, Doran P (2021) An overview of post-publication peer review. Scholarly Assessment Reports 3(1):6. https://doi.org/10.29024/sar.26
41. Chaddah P (2011) E-print archives ensure credit for original ideas. SciDevnet Oct 17. http://www.scidev.net/global/communication/opinion/e-print-archives-ensure-credit-for-original-ideas.html
42. Lakhotia SC (2013) Peer review: then and now. Curr Sci 105(6):745–746

43. Lakhotia SC (2018) Central dogma, selfish DNA and noncoding RNAs: a historical perspective. Proc Indian Natl Sci Acad 84(2):315–427. https://doi.org/10.16943/ptinsa/2018/49347
44. Lakhotia SC (2014) Societal responsibilities and research publications. Proc Indian Natl Sci Acad 80(5):913. https://doi.org/10.16943/ptinsa/2014/v80i5/47963
45. Lakhotia SC (2022) Do we need to spend substantial amounts on 'open access'? Confluence. http://confluence.ias.ac.in/do-we-need-to-spend-substantial-amounts-on-open-access/
46. Lawrence PA (2003) The politics of publication. Nature 422(6929):259–261. https://doi.org/10.1038/422259a
47. Berenbaum MR (2019) Impact factor impacts on early-career scientist careers. Proc Natl Acad Sci 116(34): 16659–16662. https://doi.org/10.1073/pnas.1911911116
48. Garfield E (1998) From citation indexes to informatics: is the tail now wagging the dog? Libri 48:67–80. https://doi.org/10.1515/libr.1998.48.2.67
49. Kumar A (2018) Is "impact" the "factor" that matters. . .? (part I). J Indian Soc Periodontol 22:95–96
50. Lakhotia SC (2014) Research, communication and impact. Proc Indian Natl Sci Acad 80(1):1–3. https://doi.org/10.16943/ptinsa/2014/v80i1/55078
51. Chaddah P, Lakhotia SC (2018) A policy statement on "dissemination and evaluation of research output in India" by the Indian National Science Academy (New Delhi). Proc Indian Natl Sci Acad 84(2): 319–329. https://doi.org/10.16943/ptinsa/2018/49415
52. Gutierrez FRS, Beall J, Forero DA (2015) Spurious alternative impact factors: the scale of the problem from an academic perspective. BioEssays 37(5): 474–476. https://doi.org/10.1002/bies.201500011
53. Beall J (2013) Predatory publishing is just one of the consequences of gold open access. Learned Publishing 26(2):79–84. https://doi.org/10.1087/20130203
54. Beall J (2012) Predatory publishers are corrupting open access. Nature 489(7415):179
55. Dadkhah M, Bianciardi G (2016) Ranking predatory journals: solve the problem instead of removing it! Adv Pharm Bull 6(1):1–4. https://doi.org/10.15171/apb.2016.001
56. Jain N, Singh M (2019) The evolving ecosystem of predatory journals: a case study in Indian perspective. In: Jatowt A, Maeda A, Syn SY (eds) Digital libraries at the crossroads of digital information for the future. Springer, Cham, pp 78–92
57. Alexandru-Ionut P, PetriÅŸor A-I (2016) Evolving strategies of the predatory journals. Malays J Libr Inf Sci 21(1):1–17. https://doi.org/10.22452/mjlis.vol21no1.1
58. Leena G, Jeevan VKJ (2022) Disrupting predatory journals. Curr Sci 122:396–401. https://doi.org/10.18520/cs/v122/i4/396-401
59. Duc N, Hiep D, Thong P, Zunic L, Zildzic M, Donev D, Jankovic S, Hozo I, Masic I (2020) Predatory open access journals are indexed in reputable databases: a revisiting issue or an unsolved problem. Med Arch 74(4):318–322. https://doi.org/10.5455/medarh.2020.74.318-322
60. Grudniewicz A, Moher D, Cobey KD, Bryson GL, Cukier S, Allen K, Ardern C, Balcom L, Barros T, Berger M (2019) Predatory journals: no definition, no defence. Nature 576:210–212
61. Panda S (2020) Predatory journals. Indian J Dermatol Venereol Leprol 86(2):109–114. https://doi.org/10.4103/ijdvl.IJDVL_22_20
62. Richtig G, Berger M, Lange-Asschenfeldt B, Aberer W, Richtig E (2018) Problems and challenges of predatory journals. J Eur Acad Dermatol Venereol 32(9):1441–1449. https://doi.org/10.1111/jdv.15039
63. Patwardhan B (2019) Why India is striking back against predatory journals. Nature 571:7
64. Priyadarshini S (2017) India tops submissions in predatory journals. Nature India Published online 6 September 2017. https://doi.org/10.1038/nindia.2017.1115
65. Lakhotia SC (2017) The fraud of open access publishing. Proc Indian Natl Sci Acad 83(1):33–36. https://doi.org/10.16943/ptinsa/2017/48942
66. Cobb M (2017) The prehistory of biology preprints: a forgotten experiment from the 1960s. PLoS Biol 15(1):e2003995. https://doi.org/10.1371/journal.pbio.2003995
67. Ginsparg P (2021) Lessons from arXiv's 30 years of information sharing. Nat Rev Phys 3(9):602–603. https://doi.org/10.1038/s42254-021-00360-z
68. Eisen MB, Akhmanova A, Behrens TE, Harper DM, Weigel D, Zaidi M (2020) Implementing a "publish, then review" model of publishing. eLife 9:e64910. https://doi.org/10.7554/eLife.64910
69. Moore A (2020) Predatory preprint servers join predatory journals in the paper mill industry. . . . BioEssays 42(11):2000259. https://doi.org/10.1002/bies.202000259
70. Lakhotia SC (2017) Mis-conceived and mis-implemented academic assessment rules underlie the scourge of predatory journals and conferences. Proc Indian Natl Sci Acad 83(3):513–515. https://doi.org/10.16943/ptinsa/2017/49141
71. Paul H (2018) The scientific self: reclaiming its place in the history of research ethics. Sci Eng Ethics 24(5): 1379–1392. https://doi.org/10.1007/s11948-017-9945-8
72. Resnik DB, Rasmussen LM, Kissling GE (2015) An international study of research misconduct policies. Account Res 22(5):249–266

Index

G. Jagadeesh et al. (eds.), *The Quintessence of Basic and Clinical Research and Scientific Publishing*,
https://doi.org/10.1007/978-981-99-1284-1

C

D

H

J

K

O

P

S

T

U

V

W

Printed in the USA
CPSIA information can be obtained
at www.ICGtesting.com
LVHW081519111123
763678LV00003B/3